TREATISE ON INORGANIC CHEMISTRY

SOLE DISTRIBUTORS FOR THE U.S.A. AND CANADA:

D. VAN NOSTRAND COMPANY, INC.

120 Alexander Street, Princeton, N.J.
257 Fourth Avenue, New York 10, N.Y.
25 Hollinger Road, Toronto 16, Canada

FOR THE BRITISH COMMONWEALTH EXCEPT CANADA:

CLEAVER-HUME PRESS, LTD.

31, Wright's Lane, Kensington, London, W.8.

TREATISE ON
INORGANIC CHEMISTRY

by

H. REMY

PROFESSOR OF INORGANIC CHEMISTRY,
UNIVERSITY OF HAMBURG (GERMANY)

Translated by

J. S. ANDERSON, F.R.S.

PROFESSOR OF INORGANIC CHEMISTRY,
UNIVERSITY OF MELBOURNE (AUSTRALIA),
FORMERLY DEPUTY CHIEF OFFICER, ATOMIC ENERGY RESEARCH
ESTABLISHMENT, HARWELL (ENGLAND)

Edited by

J. KLEINBERG

PROFESSOR OF INORGANIC CHEMISTRY,
UNIVERSITY OF KANSAS, LAWRENCE,
KANSAS (U.S.A.)

VOLUME II
SUB-GROUPS OF THE PERIODIC TABLE
AND GENERAL TOPICS

ELSEVIER PUBLISHING COMPANY
AMSTERDAM   LONDON   NEW YORK   PRINCETON
1956

PRINTED IN THE NETHERLANDS BY N.V. DRUKKERIJ G. J. THIEME, NIJMEGEN

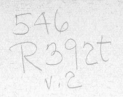
PREFACE

Since its first German edition in 1931, this book has been repeatedly revised and supplemented in accordance with the progress of chemistry; but its original plan has proved entirely satisfactory as a framework for new knowledge. In particular the systematic treatment of inorganic substances on the basis of the Periodic System has so far proved its worth as to have found its way into many other textbooks. Equally successful, when dealing with particular classes in each group, has been found the method of giving first a general introduction and then, before the detailed discussion of each substance, a survey of such topics of wider significance as may present themselves in its connection.

It is likely that future textbooks will give increasing prominence to the behaviour of groups of substances, as opposed to the discussion of individual ones; and in successive editions of this book, such chapters have become more numerous and fully developed. Each starts from some special property which for its further explanation needs the general theory, or else points the way to it. Thus is taken into account the line of thought of the chemist, progressing from particulars to generalities; and such knowledge then throws further light on subsequent problems.

The present translation is based essentially on the 7th and 8th German editions, but with further revision, improvement and supplementation before translation; notably in the sections on the chemical bonds, radioactivity, nuclear chemistry and the transuranic elements. I am very grateful to Professor J. S. Anderson, F.R.S., for the translation, carried out with great keenness and insight into the author's mind, and for suggesting many valuable improvements and additions to meet the needs of the wider circle of readers to whom the book is now addressed. And to the Editor, Professor Kleinberg, for his great interest my thanks are equally due.

Hamburg, October 1955   H. REMY

PREFACE TO VOLUME II

Advances in research made in recent years have a close bearing on many of the topics dealt with in this Volume. This has been borne in mind in preparing the new edition for translation. It is no longer permissible, for example, to mention compounds of the so-called 'anomalous oxidation states' of metals such as chromium and manganese only incidentally, as was formerly common. They must be incorporated in the systematic discussion of compounds, since it has been shown that many of them, far from having 'anomalous' character, are quite stable and easily obtainable under certain conditions. Above all, however, the treatment of radioactivity, isotopy and nuclear chemistry has been considerably extended. Whereas formerly only *one* chapter was devoted to these subjects, they have now been expanded into *four* chapters. In conclusion with this, a special chapter has been devoted to the transuranic elements. I should like to mention that the translator has incorporated several valuable improvements and additions, particularly in the sections on the platinum metals and in the chapter on the lanthanides.

Because of the volume of new work being published, it has not always been easy to make a selection of the most important results without going beyond the stated limit of the book's aims. I hope I have succeeded in make this selection in such a way that the two-fold aim set before this book will be achieved—to help to afford students a basic knowledge of modern inorganic chemistry and a comprehensive insight into its foundations and methods, and, in addition, the hope that it will be of use to supplement and deepen the knowledge of professional chemists and as a reference book.

Hamburg, April 1956   H. REMY

TRANSLATOR'S FOREWORD

During the past ten or fifteen years there has been a marked resurgence of interest in inorganic chemistry in Great Britain and the United States. The need for a detailed understanding of the chemical relationships between the elements in many pure and applied research problems has become apparent, and increased attention has been given to the chemistry of the elements in university courses. In consequence, many have felt the need for a comprehensive text in the English language setting out the subject in its present state of development. Such a work needs to cover the whole of the factual material and to bring it into proper relationship with the relevant thermodynamic, kinetic and structural data.

Amongst German texts, successive editions of Remy's *Lehrbuch der anorganischen Chemie* have gone far to meet this need, and it is to be hoped that this translation will prove valuable to a wide circle of readers. The translation has been based upon Professor Remy's latest revision of this standard work, and has been brought up to date as far as is practicable in a rapidly changing field of activity. With the author's permission, a few passages have been modified by the translator. Thermodynamic conventions have been changed where necessary, to follow American usage.

The author is indebted to Mr. J. D. M. McConnell B.Sc. for undertaking the task of proof correction, and for his care in the final preparation of the text.

Melbourne, September 1955   J. S. ANDERSON

CONTENTS

Chapter 5. Sixth Sub-Group of the Periodic System: Chromium, Molybdenum, Tungsten (Wolfram), and Uranium

Chapter 6. Seventh Sub-Group of the Periodic System: Manganese Group

Chapter 7. Eighth Sub-Group of the Periodic System: Metals of the Iron Group and Platinum Metals

A. The Metals of the Iron Series

Chapter 13. Artificial Radioactivity and Nuclear Chemistry

Chapter 14. The Transuranic Elements

ABBREVIATED CONTENTS OF VOLUME I

SOME IMPORTANT PHYSICAL CONSTANTS

Absolute temperature of the melting point of ice, $T_{0°C} = 273.16°K$.

Atmosphere (normal pressure), $1 \text{ atm.} = 1.013250 \cdot 10^6 \text{ dyne cm}^{-2}$.

Avogadro's number, $N_A = 6.0238 \cdot 10^{23}$.

Boltzmann's constant, $k = 1.38026 \cdot 10^{-16}$.

Elementary quantum of electricity, $e = 4.8022 \cdot 10^{-10} \text{ e.s.u.} = 1.60186 \cdot 10^{-20}$ e.m.u. $= 1.60186 \cdot 10^{-19}$ coulomb.

  Specific charge on the electron $= \dfrac{e}{m} = 1.7591 \cdot 10^8 \text{ coulomb g}^{-1}$.

Units of energy

$1 \text{ erg} = 10^{-7} \text{ joules}$ (watt-seconds) $= 2.777778 \cdot 10^{-14} \text{ kWh} = 0.239006 \cdot 10^{-7} \text{ cal.}$

$1 \text{ liter-atmosphere} = 1.013278 \cdot 10^9 \text{ erg} = 101.3278 \text{ joules} = 2.81466 \cdot 10^{-5} \text{ kWh}$ $= 24.2180 \text{ cal.}$

$1 \text{ cm}^3\text{atm.} = \dfrac{1 \text{ l-atm.}}{1000.028} = 0.101325 \text{ joules} = 2.81458 \cdot 10^{-8} \text{ kWh} = 2.42177 \cdot 10^{-2} \text{ cal.}$

1 cal (thermochemical gram-calorie) $= 4.1840 \text{ joule} = 1.16222 \text{ kWh} = 0.041292$ l-atm. $= 41.293 \text{ cm}^3\text{atm.}$

$1 \text{ cal}_{15°} (15° \text{ calorie}) = 4.1855 \text{ joules} = 1.00036 \text{ cal.}$

1 ev (electron-volt) $= 1.6020 \cdot 10^{-19} \text{ joules} = 3.829 \cdot 10^{-20} \text{ cal.}$
  1 ev per molecule is equivalent to 23.064 kcal per mol.

Energy of a light quantum (photon) of wave length λ cm, $\dfrac{hc}{\lambda} = \dfrac{1}{\lambda} \cdot 1.98574 \cdot 10^{-16}$ erg $= \dfrac{1}{\lambda} \cdot 1.23954 \cdot 10^{-4} \text{ ev.}$

  This corresponds to $\dfrac{1}{\lambda} \cdot 11.9617$ joules per mol or $\dfrac{1}{\lambda} \cdot 2.85892$ cal per mol.

Faraday (electrical charge per gram equivalent), $1 \mathfrak{F} = 96493 \text{ coulombs}$. This represents 26.804 ampere-hours.

Gas constant, $R = 0.082054 \text{ l-atm.} = 8.3144 \text{ joules} = 1.9872 \text{ cal.}$

Velocity of light, $c = 2.9979 \cdot 10^{10} \text{ cm sec}^{-1}$.

Liter (volume of 1 kg of air free water at its temperature of maximum density), $1 \text{ l} = 1000.028 \text{ cm}^3$.

Molar volume of ideal gas (at $0°$, 760 mm pressure) $= 22.414 \text{ l} = 22414.5 \text{ cm}^3$.

Acceleration of gravity (normal value, at sea level and $45°$ latitude), $g_0 = 980.665$ cm sec^{-2}.

Quantum of action (Planck's constant), $h = 6.6238 \cdot 10^{-27} \text{ erg sec.}$

Smythe factor $\left(= \dfrac{\text{physical atomic weight}}{\text{chemical atomic weight}} \right) = 1.000279$.

INTRODUCTION

In Volume I, the elements of the *Main Groups* of the Periodic System were taken as comprising all the elements (inclusive of hydrogen) with atomic numbers either 1 or 2 units larger, or between 5 and 1 units smaller, than the atomic number of an inert gas. All the other elements are relegated to *Sub-groups* of the Periodic System, except for the 14 elements immediately following lanthanum, which make up the special series of *lanthanide* elements, and also the elements following uranium, which again constitute a special series, the *transuranic elements*, or *uranides*.

Thus the *Sub-groups of the Periodic Table* (cf. Table II, in the appendix to Vol. I) include the elements with atomic numbers 21 to 30, 39 to 48, 57, 72 to 80 and 89 to 92. They make up 2 half-periods of the system, each of 10 elements, and also 10 elements in a long sub-period of 24 elements in all; in the latter case, 14 of the elements belong to the lanthanide series. There are also 4 elements of a further sub-period, beginning with actinium ($Z = 89$). From the standpoint of atomic structure, it is possible that out of the elements 89 to 92, only actinium is properly regarded as a Sub-group element. It is possible, but not proven, that the next element (thorium) begins another series which, like the lanthanide series, is associated with the filling of f-levels (the *actinide* series). As has been stated in Vol. I, however, the first three elements (thorium, protactinium, and uranium) are usually considered to be Sub-group elements for practical reasons (i.e., because of their chemical similarities), whereas the elements following uranium are assigned to a special group.

There is a more or less close relationship between the Sub-group elements and the corresponding Main Group elements (i.e., those belonging to the same family of the Periodic System). The resemblance is closest at the place where the bifurcation into Main Groups and Sub-groups commences—i.e., in the third family. With increase in atomic number and in group number, the resemblance diminishes from Group III to Group VII until, in Group VIII, it has completely disappeared. With continued increase of atomic number, we reach Sub-group I, which is written as the next row in the short-period form of the Table. Certain similarities between Main Group and Sub-group reappear at this stage, and these become stronger as one passes to Group II, where the Sub-groups merge again into the Main Groups. In order to fit the Sub-group elements into the same 8 families as the Main Group elements, 3 very closely similar elements in every sub-period of 10 must be included within a single group. These 'triads' are Fe, Co, Ni–Ru, Rh, Pd and Os, Ir, Pt, respectively.

In every case, the elements of the Sub-groups bear an especially close resemblance to the *second* element of the corresponding Main Group. This has already been referred to several times in Vol. I, in considering the Main Groups. This similarity also diminishes from Group III to Group VII, disappears in Group VIII, and reappears in Group I to become still more marked in going from Group I to Group II.

There is a particularly close resemblance between the elements of each Sub-group among themselves. In the middle of the Periodic System (Groups III, IV and V), this similarity is hardly less close than that exhibited between the heavier elements of the Main Groups. In one particular case, in fact (zirconium and hafnium), it is greater than that exhibited by any pair of homologues in the Main Groups. In general, however, elements belonging to the same Sub-group do not show such close agreement in the matter of valence states as do the Main Group elements. Many of the elements of the Sub-groups are able to change their valence state readily. This ready variability of valence reaches its maximum in the VIIIth Sub-group.

In addition to their readily variable valence, the elements of the Sub-groups possess several other characteristics. Thus most of them are *paramagnetic* in a number (sometimes the majority) of their compounds. They can also form *colored elementary electrolytic ions*—a property not displayed by any Main Group elements.

Similarities between adjacent (consecutive) elements are far more evident in the Sub-groups than in the Main Groups of the Periodic System. In fact, such resemblances are frequently more important in the Sub-groups than the similarities between homologues (i.e., between elements standing one above the other in the Periodic System). Thus in its properties, and in the types and chemical behavior of the compounds it forms, *iron* resembles its neighbors, *manganese* and *cobalt*, more closely than it does its homologues, *ruthenium* and *osmium*. The same is true of *cobalt* and of *nickel*. These last elements, together with iron, belong to Sub-group VIII of the Periodic System. Resemblances in the horizontal direction are most strongly marked within and immediately in the neighborhood of Group VIII.

It formerly seemed that the occurrence of the Sub-groups constituted a break in the regular character of the Periodic System. It can now be seen that it is a necessary consequence of the general laws governing the disposition of electrons within the atom. As is shown by spectroscopic observations, the Sub-groups in the Periodic System arise from the fact that, as the nuclear charge of the atoms increases and successive electrons are added, it is not a valid generalization that (to use a graphic form of expression), successive electrons are added 'outside already filled shells'. At certain stages in the sequence of the elements, the electrons enter orbits 'inside' already filled shells. A more precise form of expression is that, at certain stages in the sequence, although levels with the principal quantum number n are already occupied, further increase in the nuclear charge is associated with the entry of electrons into orbits of lower principal quantum number, $n-1$ or $n-2$. It is possible to deduce from spectroscopic evidence at what element this first occurs*. It is also possible from the theory of atomic structure to state why this should take place.

Attention has already been drawn (in discussing Main Group II of the Periodic Table) to a characteristic difference between the spectra of singly ionized calcium and of neutral potassium, which are fully analogous in other respects. The spectral terms show that in

---

* This conclusion had already been reached by Ladenburg (1920), from the paramagnetism and color of the ions of the Sub-group elements, even before the proof that 'intermediate' electron shells were formed had been established by the analysis of spectroscopic data.

Ca^+, the $3d$-orbit (or the $3d$-level) does not represent a higher energy content of the atom than the $4p$-level, as in K, but a lower energy state. It may be seen from a comparison of the energy levels in the potassium atom and the singly ionized calcium atom (in Fig. 59, p. 252 of Vol. I) that the $3d$-energy level would have to drop only a little lower relatively, or the $4s$-level be relatively raised by a little, for the $3d$-level to represent a state of lower energy than the $4s$-level. It may be seen that in going from the arc spectrum of potassium to the spark spectrum of calcium there is both a considerable rise in the energy of the $4s$-level (the value for Ca^+ is divided by 4, to allow for the unscreened nuclear charge), and a decrease in the energy of the $3d$-state. Therefore it must be anticipated that in the next element, *scandium*, the energy of the $4s$ state in the spectrum of the doubly ionized atom, Sc^{++}, will have overtaken that of the $3d$ state. This expectation has been fully borne out by the term analysis of the scandium spectrum. The ground state of the 19th electron in scandium is no longer a $4s$-level, as it is in potassium and calcium, but is a $3d$-level. The 20th electron is the first one to enter a $4s$-level, in its normal state, as can be deduced from the spectrum of singly ionized scandium. The 21st electron in scandium also occupies a $4s$-level. We thus arrive at the following electron configuration for scandium in its ground state:

$$1s^2 \quad 2s^2 2p^6 \quad 3s^2 3p^6 3d \quad 4s^2$$

With further increase in the nuclear charge, a stage is soon reached in which not only is the 19th electron bound in a $3d$-state, but the same is true first of some, and ultimately of all the electrons subsequently added, up to the maximum number of electrons that can be accommodated in $3d$-states (i.e., 10). Whereas in scandium the 20th and 21st electrons are bound in $4s$-states, it has been inferred from the spectrum of *titanium*, the next element, that both the 19th and the 20th electrons are bound in $3d$-levels, while the 21st and 22nd electrons in titanium are in $4s$-levels. For *vanadium*, the spectrum shows that the 19th, 20th and 21st electrons are bound in $3d$-levels, with the 22nd and 23rd electrons in the $4s$-level.* The same is true of the elements which follow next in the sequence. All the elements up to nickel contain, at the most, 2 electrons in levels with principal quantum number 4. The other 'outer' electrons are bound in $3d$-levels. (See Table II, Vol. I, appendix.)

The fact that one electron in scandium is bound in a d-level, instead of in a p-level as with the elements of the IIIrd Main Group, has only a very slight influence on the chemical properties of scandium itself. The determining factor for most chemical properties is the relative ease with which electrons are split off. This is particularly true of the valence properties exhibited in electrovalent compounds. The fact that calcium has a maximum electrovalence of $+2$—i.e., that it can give up only the two electrons which are bound in $4s$ states in the ground state of the calcium atom, and not the electrons occupying $3p$ states—is conditioned by the much tighter binding of the latter, as compared with the former. Scandium, in its ground state, has one electron in a $3d$-level; but since this is bound only a little more tightly than the electrons of the $4s$-level, the chemical character is not much influenced. Analogous considerations apply to the elements that follow. However, it is probable that the greater lability of the valence states of the Sub-group elements, as compared with the Main Group elements, is to be associated with the different electron configurations in their atoms.

One important consequence of the assignment of electrons to levels of lower principal quantum number ('inner shells') is that in the element following 8 places after argon (i.e., iron, $Z = 26$), the 'outermost shell' has not yet attained the configuration of the inert gases, represented by the symbol $ns^2\, np^6$. Since both in scandium itself and in the elements which follow scandium, some of the electrons are always bound in $3d$-levels, there are at first not enough electrons available for a new 'inert gas configuration' to be built up (which, in this case, would contain two $4s$ and six $4p$ electrons). *Only when it is impossible to accommodate any more electrons in $3d$-levels can the levels with principal quantum number 4 be filled up*; and a new 'inert gas shell' be built up. The fact that krypton follows 18 places after argon shows that $18 - (2 + 6) = 10$ electrons in all must be assigned to the $3d$-level. Exactly the same applies in the next long period. The occupation of an 'inner shell' begins again with *yttrium*, as is shown by the spectral terms. Thus from yttrium onwards, electrons enter the $4d$-levels, instead of the $5s$- and $5p$-levels. Again, 10 electrons in all enter $4d$-levels, so that xenon, the

---

* This is for the *neutral* vanadium atom. In the singly ionized vanadium atom, the 22nd electron is also bound in a $3d$-level.

next inert gas, follows 18 (= 10 + 2 + 6) places after krypton. The building up of inner electron shells starts again in the next period (8th series) with lanthanum. In this case, however, there are *two* groups of energy levels with principal quantum number less than 6, available for occupation. These are the 4*f*-level and the 5*d*-level. The next inert gas (radon) is therefore encountered only when all these 'inner shells' have been occupied and the 6*p*-levels thereafter filled. As has already been stated in Vol. I, the elements in which the inner shells are being progressively filled are known generally as *transition elements*.

In the sixth period of the Table, between lanthanum and the row of Sub-group elements beginning with hafnium, there is interposed yet another sub-period, consisting of elements of special characteristics—namely those that closely resemble lanthanum (the *lanthanide* series). This fact shows that the two electron shells concerned are built up in turn. Since the elements immediately following lanthanum all have the same valence as lanthanum (except in so far as a few of these elements can be quadrivalent or bivalent, as well as trivalent), it may be inferred that the 'innermost' of the two shells is filled up first—i.e., the 4*f*-level is the first to be filled. This conclusion is confirmed by the term analysis of the spectra. Electrons in the 4*f*-level are so much more firmly bound than those in the 5*d* and 6*s* levels that they do not, in general affect the valence properties of the elements. Thus the valence remains generally constant while the 4*f*-level is being filled up. The *number* of elements in the lanthanide series can also be deduced. By analogy with its lighter homologues, radon, the next inert gas, should have two electrons in 6*s*- and six electrons in 6*p*-levels. Radon is separated by 32 places from the preceding inert gas, krypton, and since by analogy with the filling of the 3*d* and 4*d*-levels, 10 electrons must be required to fill the 5*d*-levels, to give a set of elements resembling those of the earlier Sub-groups. It follows from the theory of atomic structure that 14 (= 32—[10 + 2 + 6]) elements should be contained in the lanthanide series. In agreement with this, it has been found that lanthanum-like properties extend just to the element of atomic number 71, standing 14 places after lanthanum. The next element, hafnium (atomic number 72), no longer has the properties of a lanthanide element, but—like lanthanum itself —has the properties of a Sub-group element. With hafnium, the series of 10 elements of 'Sub-group character' thus continues. In the third long period, this series thus comprises lanthanum ($Z = 57$), hafnium ($Z = 72$), and the elements following hafnium, as far as mercury ($Z = 80$).

Finally, it may be expected that the filling of an inner shell ought to begin again at some stage in the series following radon. In radium ($Z = 88$), the most loosely bound electrons in the ground state of the atom belong to the 7*s* level, but with actinium ($Z = 89$), the assignment of electrons to *d*-levels (6*d*-levels) begins again. Actinium therefore has the character of a Sub-group element, and is the first member of a sequence of elements with Sub-group properties in the fourth long Period. However, in elements following actinium the filling of the 5*f*-levels commences—possibly even with the next element, thorium ($Z = 90$). Hence, actinium is followed either immediately or later by the sub-period of *actinide elements*, corresponding to the filling up of the 5*f*-levels, in just the same way as lanthanum is succeeded by the lanthanides, because of the filling of 4*f*-levels.

Table II (Vol. I, Appendix) summarizes the distribution of the electrons over the various energy levels in the ground state of the atoms. The electronic configurations given in the table are mostly based on spectroscopic data (including X-ray spectra).

In a few cases, the electronic configurations cannot be regarded as proved. It is possible that for $_{40}$Zr the configuration of the outer electrons should be $4d^2 5s^2$ (analogous to $_{22}$Ti), instead of the assignment $4d^3 5s$ given in the Table. Seaborg considers that the following are the most probable electronic configurations (beyond the xenon shell) for the elements 58 to 63:

$_{58}$Ce	$_{59}$Pr	$_{60}$Nd	$_{61}$Pm	$_{62}$Sm	$_{63}$Eu
$4f^2 6s^2$	$4f^3 6s^2$	$4f^4 6s^2$	$4f^5 6s^2$	$4f^6 6s^2$	$4f^7 6s^2$

In addition to the configurations given in Table II, the following electron assignments (beyond the preceding inert gas configurations) also have to be considered as possible for elements 43, 69 and 70, and 90 to 95:

$_{43}$Tc $\quad$ $_{69}$Tm $\quad$ $_{70}$Yb $\quad$ $_{90}$Th $\quad$ $_{91}$Pa $\quad$ $_{92}$U $\quad$ $_{93}$Np $\quad$ $_{94}$Pu $\quad$ $_{95}$Am

$4d^6 5s \quad 4f^{12}5d6s^2 \quad 4f^{13}5d6s^2 \quad 6d^2 7s^2 \quad \begin{cases} 5f6d^2 7s^2 \\ 6d^3 7s^2 \end{cases} \quad 6d^4 7s^2 \quad \begin{cases} 5f^5 7s^2 \\ 6d^5 7s^2 \end{cases} \quad 5f^6 7s^2 \quad 5f^7 7s^2$

[See Dawson, *Nucleonics*, 10 (1952) 39]

The electronic configurations as tabulated in Vol. I, Appendix, only take account of the subdivision of the energy levels by the one subsidiary quantum number l—i.e., by the azimuthal or angular quantum number. On the basis of the spectroscopic data it is possible to subdivide the energy levels further, in terms of the two other subsidiary quantum numbers m and s (cf. Vol. I, p. 120). This further analysis is important in that the possible values of the four quantum numbers, together with the Pauli principle (Vol. I), determine the possible number of electrons in the various levels, in the ground state of the atom, for every value of the nuclear charge. The whole of the Periodic System, in its broad essentials, can therefore be based theoretically on a *single principle*. It was shown in Vol. I, Chapter 4, that the maximum number of electrons in the electron shells of the inert gases,— and therefore the length of the so-called 'short periods'—could be deduced from the Pauli principle. The occurrence of energy levels with the angular momentum quantum number $l = 2$ similarly results in the formation of the 'long periods' of 18 elements. For $l = 2$, m may take on the values $-2, -1, 0, +1, +2$, and since each of these corresponds to two possible values of s, there are thus 10 possible combinations of quantum numbers to be added to the 8 already provided by the quantum numbers $l = 0$ and $l = 1$. Similarly, for $l = 3$, the possible values of m are $-3, -2, -1, 0, +1, +2$ and $+3$. For each of these, s may be $+\frac{1}{2}$ or $-\frac{1}{2}$, giving 14 combinations of quantum numbers which become available through the existence of energy states with the angular momentum quantum number $l = 3$. It is this set of 14 quantum number combinations which gives rise to the existence of the lanthanide series, and therefore results in the length of the period increasing from 18 to 32. The existence and predicted extent of the actinide series is explained analogously.

It remains to be considered why the filling of 'inner shells' should take place only at certain definite places in the sequence of the elements. This can be understood in terms of the same reasoning as was used in Vol. I, Chap. 4 to justify the statement that the energy of an electronic orbit outside the $1s$-level in the heavier atoms varies with the angular momentum quantum number l in a different manner from that which would be expected simply by making allowance for the relativistic variation of electronic mass with velocity. For clarity, the reasoning will be developed in terms of the Bohr-Sommerfeld atom model. It follows from this that in potassium, for example, the 19th electron is very considerably more strongly bound in a $4s$-level than in a $4f$-level. The latter, in the Bohr-Sommerfeld theory, corresponds to a 4_4 orbit, i.e., a circular orbit in which, in the case of potassium, the effective nuclear charge acting on the electron is completely screened except for about one unit of charge. The $4s$-level, however, corresponds to a Bohr-Sommerfeld 4_1 orbit—i.e., an orbit of high ellipticity. The electron in this orbit must be more tightly bound than in the 4_4 orbit, since the strongly elliptic 4_1 orbit penetrates deeply into the inner, screening orbits, so that for part of the time the electron closely approaches the nucleus of the atom (with its charge of 19 units, in the case of potassium). For the same reason the 19th electron in potassium is more tightly bound in the 4_1 orbit than in a 3_3 orbit. However, the binding force on an electron in a 3_3 orbit increases considerably in going from potassium to calcium. Assuming that the nuclear charge acting on an electron in a 3_3 orbit of potassium is fully screened, except for a single unit of charge, the effective nuclear charge would be about 2 in calcium and about 3 in scandium. It follows from eq. (11)

of Vol. I, Chap. 3, that the binding energy of an electron in a 3_3 orbit increases with the square of these numbers. The binding energy of an electron in a 4_1 orbit increases much more slowly, since it is largely determined by the *total* nuclear charge, which rises only from 19 to 21 in passing from potassium to scandium, and so increases by a small fraction. In consequence, the binding energy of an electron in the circular 3_3 orbit will eventually become greater than that of an electron in the highly elliptic 4_1 orbit.

From the foregoing, the number of electrons required to fill an 'inner shell' determines the number of elements in the 'sub-periods'. The positions at which the Sub-groups branch off from the Main Groups also coincide with the stages at which the filling of 'inner shells' begins. However, the *termination* of the sub-periods of elements with Sub-group properties does *not* coincide with the stage at which the filling of the 'inner shells' is complete. The 'inner shells' are already complete in copper, silver, and gold, as is shown by the spectra of these elements (cf. Chap. 8). However, both these elements and those immediately following them (zinc, cadmium, and mercury) must unquestionably be assigned to the Sub-groups on the evidence of their chemical behavior. Thus the concepts of '*Sub-group elements*' and '*transition elements*' are not identical.

Since the branching between Sub-groups and Main Groups begins with Group III, our discussion of the Sub-group elements will also start with Sub-group III. The Sub-groups will then be dealt with in order, as far as Sub-group VIII. Further increase in atomic number leads to the elements of Sub-groups I and II, which belong to the next series of the Periodic Table. The Sub-groups are thereby completed, and the next elements in the sequence belong to the Main Groups once more (IIIrd Main Group).

In this connection, it should be remembered that the expression 'third Sub-group', 'fourth Sub-group', etc. is an abbreviation for 'Sub-group of the IIIrd (or IVth) family or group of the Periodic System'. At first sight it might appear more logical to give the name of 'first Sub-group' to the one in which elements with the Sub-group characteristics first appear in the Periodic Table—i.e., the one known as Sub-group III. The latter is the accepted and better terminology, however, since it brings out the fact that the elements in that Sub-group are part of the IIIrd family in the Periodic System, and are in many respects very similar to the main group elements.

The next elements to be discussed are the 14 lanthanum-like elements ($Z = 58$ to $Z = 71$) which are interposed between lanthanum and hafnium in the sixth series of the Periodic Table—i.e., immediately following the point at which the Sub-group branches off in this series. This group of 14 elements displays a particularly close relationship; their oxides all have the character of '*rare earths*'. They cannot be assigned to the Sub-groups of the Periodic System, but form a horizontally related group on their own. This *lanthanide series* is discussed in Chap. 10.

Of the *actinide series*, only the *transuranic elements* will be dealt with in a special chapter. The first three members (thorium, protactinium and uranium) are discussed within the framework of the IVth to VIth Sub-groups, for the reasons already given.

As has already been stated, the elements in which 'inner shells' are being elaborated (i.e., in which *d* and *f* levels are being filled up) are known collectively as *transition elements*. This name is therefore to be applied to the elements from $_{21}$Sc to $_{28}$Ni, from $_{39}$Y to $_{46}$Pd, from $_{57}$La to $_{78}$Pt, and to all the elements of atomic number 89 and over that are at present known—including those formed artificially in nuclear reactions: i.e., to $_{89}$Ac and the following actinides. There are 8 elements in each of the first two transition series (Sc—Ni and Y—Pd), and 22 elements in the third transition series which includes the 14 lanthanide elements also. The last transition series is incomplete, because of the increasing instability of the elements of very high atomic number. In each period, the transition elements begin with the element at which the branching of Sub-groups from Main Groups of the Periodic System commences*. Except for the last, incompletely known series, the transition series end with an element standing two places before the point where Sub-groups and Main Groups reunite again (at Zn, Cd, or Hg, respectively).

In those Sub-groups of the Periodic System which are composed of transition elements, it is a general rule that the chemical resemblance between the second and third elements in each column is much closer than that between the first and second elements. Where (as in Sub-groups IV to VI) there are 4 elements in any Sub-group, the resemblance between second and third elements is closer than that between third and fourth.

This is partly due to the so-called 'lanthanide contraction' (p. 482), as a result of which there is very little difference in atomic radius and ionic radius between the 2nd and 3rd element in each Sub-group. This can be seen more clearly from Fig. 1 than from the usual plot of the periodicity of atomic radius (cf. Fig. 3, Vol. I). In Fig. 1, the atomic radii of Sub-group elements belonging to the same column of the Periodic Table are plotted one above another. To make the data still more closely comparable, the atomic radii of those elements which (unlike the majority) do not crystallize in structures with coordination number 12 are corrected to coordination number 12 by means of the f_{CN} factors given in Table 42, p. 215, Vol. I. The numerical data of Tables 5, 8, 12 and so on show that the same relationship is to be found for ionic radii as for atomic radii. It may be seen from Fig. 1 that where there is an especially far-reaching similarity in chemical behavior and in the properties of the compounds of a homologous pair of transition elements, the pair have almost identical atomic radii. The figure shows, however, that very close chemical similarity must be conditioned by some other factors besides the analogous configurations of the outer electron shells and the small difference in atomic or ionic radii. The highest degree of similarity in behavior and in the properties of compounds is presented by the pair zirconium-hafnium. The values of atomic and ionic radii (Tables 8 and 12) for this pair are not so close, however, as for the pair niobium-tantalum, although the latter pair undoubtedly displays less chemical

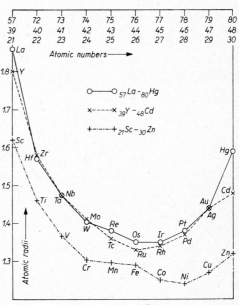

Fig. 1. Atomic radii of the elements of the Sub-groups in Ångstrom units. (Radii for coordination number 12.)

* It is probable that in the *metallic* state valence electrons are first found in *d*-levels not in Group III, but already in certain elements of Main Group II (namely, Ca, Sr, Ba and Ra). These elements are therefore often included among the transition elements when the formation and properties of intermetallic compounds are under consideration.

similarity (cf. p. 88). In the elements which follow the transition elements, these relationships may be reversed. Thus in the Sub-group Cu-Ag-Au, the similarity between the 2nd
and 3rd element is hardly closer than that between the 1st and 2nd (apart from the properties of mixed crystal formation between the metals). In the Sub-group Zn-Cd-Hg, the
1st and 2nd elements have a distinctly closer resemblance than the 2nd and 3rd. If allowance is made for the differences in polarizing power of the ions, the discrepancy between the
pairs Zr-Hf and Nb-Ta disappears, as is shown later (p. 77). This has the effect of displacing
the dotted curve of Fig. 1 downwards by a small amount. When this shift is carried out,
the resulting differences in radius between the points of the curves in Fig. 1 very satisfactorily represent the relationships within the individual columns of the transition elements.
In order to remove the discrepancies presented by the Sub-groups which follow the transition elements, it would be necessary to shift the dotted curve by a much larger amount. It
is very questionable whether differences in polarizing properties would justify such a large
displacement.

All the elements of the Sub-groups are purely metallic in character. Since an
understanding of many of their properties has come only from some knowledge of
the nature of the metallic state, a discussion of the individual Sub-groups will be
preceded by a survey of the general properties of metals.

CHAPTER 1

METALS AND INTERMETALLIC PHASES

1. General

Metals [1] differ from non-metals chiefly in respect of the following characteristic properties. The metals display a peculiar luster ('metallic luster') which is the consequence of their high reflectivity for visible light. Their optical transparency is very low, so that they are opaque even in thin layers. They are mostly ductile, and can therefore be worked by rolling, pressing, hammering, etc. Above all, the metals are distinguised by their high thermal conductivity and good electrical conductivity. The latter increases with falling temperature.

These differences between the metals and non-metals apply to the solid and liquid states. In the gaseous state they vanish. Thus mercury vapor is colorless, transparent, and a non-conductor. The fact that most metals are monatomic in the gaseous state cannot be taken as a characteristic property. For example, the inert gases—typical non-metals—are monatomic, whereas diatomic molecules occur to a considerable extent in the vapors of the alkali metals.

It is not possible to draw a sharp distinction between metals and non-metals. Elements which obviously are intermediate in properties between metals and non-metals are called *semi-metals*. The typical semi-metals are distinguished from the metals, among other properties, in that their electrical conductivity increases in going from the solid to the liquid state, whereas that of metals decreases on melting. Semi-metals also differ from typical metals in that they lack ductility. Elements which are typically metallic in all respects except for their brittleness are called *brittle metals*.

The occurrence of non-metallic properties among the elements is clearly dependent on their place in the Periodic System. Non-metals (including hydrogen) occur only in the Main Groups of the Periodic Table, and the region occupied by non-metals is separated from that of the metals by the diagonal belt running through boron, silicon, arsenic, tellurium, and astatine, in the form of Periodic System as set out in Table II, Vol. I, Appendix. Of the elements on this diagonal, boron and silicon are non-metals, arsenic, tellurium (and possibly astatine) are semi-metals. All the Main Group elements to the right of this diagonal are *non-metals*, all the elements to the left are *metals*. Among these, the ones adjacent to the lower part of the diagonal (germanium, antimony, bismuth, and probably polonium) have the character of brittle metals. All the elements of the Sub-groups, as well as all the lanthanide and transuranic elements, are metals. Thus the majority (about four fifths) of the elements are metallic in character.

Chemical elements with metallic character are termed *pure metals*, as distinct from *alloys**
Alloys (cf. Vol. I, Chap. 13) can contain minor amounts of non-metals as constituents, as
well as metals and semi-metals. Certain chemical compounds, composed of metals and non-
metals, may display a typical metallic luster and high electrical conductivity, while being
distinctly brittle—e.g., many carbides, nitrides and borides. These are not considered to
be metals or brittle metals, but will be referred to subsequently as '*quasi-metallic*' compounds.

Many elements can exist in both a metallic and a non-metallic modification. Thus a
metallic form of phosphorus (black phosphorus) is known, as well as the non-metallic
allotropes. The cubic modification of tin (grey tin) has the properties of a non-metal, in
contradistinction to the ordinary tetragonal tin. It has, indeed, long been a habit in chemistry to use the terms 'metal' and 'non-metal' to designate two *classes* of matter. Strictly
speaking, however, these names characterize a *state*, not a class of matter.

(a) Preparation of Metals

The following are the principal processes used for the isolation of metals from
their compounds [1–3].

(*i*) *Reduction by Chemical Means.* Hydrogen is generally used as reducing agent in
the laboratory, whereas carbon (usually in the form of coke) is employed on the
technical scale. The usual starting materials are the *oxides* of the metals. Sulfides
may be converted into oxides by heating them in a stream of air ('roasting'). The
lower the heats of formation of oxides, the more readily are they reduced.

A rough indication of the reducibility of any element is afforded by its position in the
electrochemical series. Oxides of metals which are more weakly electropositive than zinc
can be reduced either by hydrogen or by carbon, but hydrogen has no practical importance
as a reducing agent for the more strongly electropositive metals. Some of them can, indeed,
be liberated from their oxides by hydrogen, but only at very high temperatures. Zinc oxide
can easily be reduced by carbon; indeed, even the alkali metals can be obtained from their
oxides by reduction with carbon. Most of the other strongly electropositive metals react
with carbon, to form carbides; the reduction of the oxides is thereby vitiated. It is often
possible to facilitate reduction by carbon, and restrict the formation of carbides, if alloys are
formed instead of the pure metals. Use is made of this, for example, in the production of
iron-manganese alloys (ferromanganese, p. 211).

In cases where reduction by carbon is unsuccessful because of carbide formation,
aluminum is often used as a reducing agent for the production of technically pure
metals (aluminothermic reduction). Magnesium and calcium are also employed,
but less frequently.

Calcium, magnesium, and sodium are mostly used for the reduction of *metallic halides*.
The halides of the heavy metals can usually be readily reduced in a current of hydrogen.

When metals are prepared in the laboratory by chemical reduction processes, it is usual
to obtain them first in powder form, or in a spongy state, and it is often a matter of difficulty
to melt them down to a compact mass. Melting is facilitated by the addition of fusible substances (e.g., borax or alkali chlorides) which dissolve away the impurities that hinder agglomeration. Since they protect the heated metal from access of air, they also hinder the fresh
formation of oxide or nitride on the surface of the metallic grains. Very pure metal powders
can readily be transformed into the compact state by heating them. It is not necessary that
they should be heated to the melting point, since the metallic grains sinter together as they
recrystallize, often far below their melting point. This is the basis of the *powder metallurgical
process* for the production of moulded articles from metal powders (also known as 'metal
ceramics'[18–20]).

* The term 'metal' is often used in the same sense as 'pure metal'—e.g., in the expression
'metals and their alloys'.

(ii) Electrolysis of Solutions or Melts. The electrolytic deposition of metals from aqueous solution [*14–16*] is used industrially largely for the preparation of pure metals.

There is usually no difficulty involved in the electro-deposition from aqueous solutions of those metals which do not decompose water. With metals that cannot be deposited from acid solutions, it is usually possible to prevent the formation of oxide on the cathode (which interferes with the deposition) by adding substances that form complex compounds (e.g., oxalates or cyanides). Such additions are also frequently used in the electrodeposition of the more noble metals, since they favor the deposition of the metal in a compact form.

Electrodeposition of metals from solution is very widely used for the protective coatings on objects made of readily corroded metals ('galvanostegy'—e.g., nickel and chromium plating on steel). Electrodeposition is also used to form surfaces of closely defined relief, as determined by the profile of the object ('matrix') on which electrodeposition is effected ('galvanoplastics').

When electrodeposition from aqueous solution is not possible, because of the strongly electropositive character of the metal, and when it is difficult to achieve reduction by chemical means, recourse is often had in the laboratory to electrolysis of non-aqueous solutions— e.g., of solutions in pyridine.

The principal method employed for the industrial preparation of the strongly electropositive metals, consists of electrolytic deposition from melts [*17*]. Aluminum, magnesium, sodium, and calcium, in particular, are prepared on a large scale by this means.

The melts used for this purpose contain the oxides or halides of the metals to be isolated; these are mixed with other compounds which lower the melting point and enhance the conductivity without themselves being decomposed electrolytically under the given conditions.

The potentials needed for electrodeposition from melts bear no simple relation to the deposition potentials which are valid in aqueous solutions (Vol. I, p. 156 ff.), since the hydration energies of the ions enter into the latter. Just as both hydration energies and ionization energies are involved in deposition from aqueous solution, so the forces between the ion to be discharged and the other components of the melt are important, as well as the ionization energy, in electrodeposition from melts. The decomposition voltage of a compound in a melt may therefore be substantially modified by the addition of other compounds. The variation of decomposition potential with temperature, which is often widely different for different ions, must also be borne in mind when considering the processes governing the electrolysis of molten salts. For example, below 600° the decomposition potential of NaCl is greater than that of $CaCl_2$, but above 600° it is smaller. This makes it possible to prepare metallic sodium by the electrolysis of a fused mixture of sodium chloride and calcium chloride (Vol. I, p. 165).

(iii) Thermal Decomposition of Compounds. It is possible to prepare many metals in a particularly pure state by the thermal decomposition of suitable compounds.

The oldest large scale industrial process using thermal decomposition for the preparation of a metal is the production of nickel by decomposition of nickel carbonyl (p. 308). Iron carbonyl is used similarly for the preparation of especially pure iron ('carbonyl iron', p. 253). In both cases, the advantage offered by the method lies in the great volatility of the carbonyls, which makes it possible to effect a particularly complete separation from all impurities.

Thermal decomposition is used in other cases when unusually great difficulty is involved in preparing the pure metal by chemical reduction. Thus titanium and zirconium are obtained in a state of poor ductility by reduction of their chlorides (Kroll process), or as powders by reduction of their oxides. It is difficult to melt the powders into a compact form, owing to the impurities present, and if this is effected the metal is still not completely pure. These metals can, however, be obtained directly in the compact state, and absolutely pure, by the thermal decomposition of their halides (preferably the iodides), in the filament growth process of Van Arkel and De Boer (p. 65).

Von Bolton's method for the preparation of such metals as vanadium, niobium, and tantalum in a pure state is also based on thermal decomposition. This involves heating the oxide, or metal containing the oxide as impurity, by means of a strong electric current (cf. pp. 105, 111). This method can be used if the oxides of the metals are electrical conductors at high temperatures, provided that the metals do not vaporize too rapidly at the dissociation temperature of their oxides.

The study of the industrial preparation of the metals from their ores is known as *extraction metallurgy* [4–17]. This deals not only with the methods involved in the actual preparation, but also with the processes of concentrating the metalliferous ores in the crude ores ('beneficiation'), and with the processes of purification ('refining') to which the crude metals must be submitted, since they are generally first obtained in an impure state, and with the determination of their purity ('assaying').

By far the most important impurity in metals is oxygen, since a small oxide content can drastically alter the physical properties of a metal (e.g., the electrical conductivity and the ductility), if the oxide is present in the metal in solid solution. The same is often true of nitrogen. It is frequently possible to determine the oxide and nitride content by volatilizing the metal as chloride, by heating it in a stream of chlorine, or by dissolving the metal in an acid which does not attack the oxide or nitride. Other impurities, and especially other metals, are often detected and determined spectroscopically, as well as by chemical methods.

The testing of metals for their mechanical and technical properties is a matter of great practical importance [21–25]. From the results obtained, it is often possible to infer the impurities that are present, since these may influence the mechanical properties in a high degree [2].

(b) Theory of the Metallic State [26–30]

To explain the high electrical conductivity of metals, it has long been assumed that, within a metal, there are almost freely mobile electrons, which transport the current. This assumption is supported by a great number of experimental observations—e.g., the observation that no material transport is associated with the conduction of electricity in metals, as is the case with electrolytic conduction*, and also that electrons may be emitted from metals under the influence of ultraviolet light (photoelectric effect) or by heating to high temperatures (Richardson or thermionic effect).

If there are freely mobile electrons present in a metal, they ought to behave like the molecules of a gas within an enclosing vessel. On the basis of this 'electron gas' concept (Riecke, 1898), Drude (1902) succeeded in deriving the Wiedemann-Franz law theoretically, and in explaining qualitatively the thermoelectric effect.

The Wiedemann-Franz law (1883) is of fundamental importance in the theory of the metallic state, since it demonstrates a connection between two of the most characteristic properties of metals—their high electrical conductivity and good thermal conductivity. It states that the electrical and thermal conductivities are proportional to one another: more precisely, the ratio $\dfrac{\lambda}{\varkappa \cdot T}$ is the same for all metals (where λ is the thermal con-

---

* It is possible in certain cases for some electrolytic conductivity to be superimposed on the electronic conductivity.

ductivity, and $\varkappa$ the electrical conductivity). The numerical value of this quotient obtained from the Drude theory is about 30% too small, as was shown by Lorentz (1905). It is only on the basis of the quantum mechanical theory that it has been possible to obtain the numerically correct Wiedemann-Franz ratio from the theory of metals.

The 'electron gas' theory of metals led to another conclusion that was in clear contradiction with the facts. The heat content of a monatomic gas (and therefore of an electron gas within a metal) should be $\frac{3}{2}RT$ according to classical theory. Its atomic heat or molecular heat at constant volume is thus $\frac{3}{2}R = 2.98$ cal. For metals, the atomic heat at constant volume is given by the theory of heat as $\frac{6}{2}R = 5.96$ cal., without making any allowance for the heat content of the electron gas, since each metallic atom must be assigned 6 degrees of freedom (3 for the kinetic and 3 for the potential energy of atomic vibration). As experience shows (Dulong and Petit rule), the atomic heats of the metallic elements are close to 6 cal.*, whereas they ought to be about 9 cal. if the quantity of heat required according to classical theory is assigned to the electron gas. This discrepancy was removed when the heat content of the electron gas was calculated according to quantum theory, in conjunction with an extension of the Pauli principle.

Fermi (1926) made the assumption that the Pauli principle held good not only for the electrons in an atom or molecule, but quite generally for any closed system of electrons—hence for the electrons in a metal.** It follows from this that such a gas is 'degenerate' at sufficiently low temperatures; it has a different—and indeed higher—energy content than it would have according to classical theory. Whereas, on the basis of latter, a gas would have zero energy at absolute zero, *only one particle* can have zero energy, even at absolute zero, if the Pauli principle is extended to gases. All other particles must have higher energies, different from one another. As a result of this, within the range of degeneracy, the increment of energy necessary to raise the temperature by some definite amount is less than in the 'normal' range. The greater the number of particles per unit volume, and the smaller the mass of the individual particles, the wider is the range of temperature over which a 'Fermi gas' is in the degenerate state. On account of the extraordinarily minute mass of its particles, an electron gas is still degenerate even at the highest attainable temperatures. Its specific heat is, consequently, practically zero (more precisely, $\frac{1}{100}$ R per g atom of metal at ordinary temperature). The fact that the atomic heats of the metals are practically no greater than may be calculated from the thermal vibrations of their atoms is thus accounted for.

Sommerfeld (1927) showed that Fermi's theory of the distribution of energy among the individual massive particles ('Fermi statistics') eliminated the discrepancies arising from the electron theory of metals in its original form. The new theory of the metallic state was elaborated and extended by the theoretical work of others (Nordheim, Bloch, Peierls, Borelius, Wilson, Brillouin), and it is now possible to deduce from the theory many of the properties of metals and the phenomena associated with them—e.g., the thermoelectric, Volta (contact potential) and thermionic effects, and the thermal and electrical conductivities, on a quantitative basis. The theory has also interpreted the magnetic properties of the metals

* That the atomic heats are rather greater than 6 cal., according to the Dulong-Petit rule, is due to the fact that the rule relates to the atomic heats measured at *constant pressure*. On account of the work expended in expansion, these are distinctly greater than the atomic heats at *constant volume*—about 3–10% at ordinary temperature, depending on the coefficient of expansion and the compressibility of the metals.

** It was later shown by Dirac that Fermi's generalization could be directly deduced from wave mechanics.

(cf. Vol. I p. 306), and is important for an understanding of some of the special features of the chemical behavior of metallic substances, since it makes it clear that the binding of valence electrons in a compact metal is quite different from that in an isolated atom.

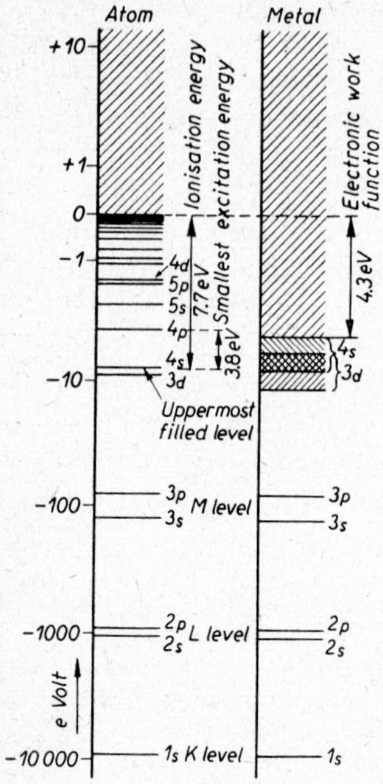

Whereas the electrons in a free atom are distributed over a few, discrete energy levels, in a compact metal such a statement is true only of those electrons which are so deeply screened within the interior of atoms that they are not influenced by neighboring atoms. By contrast, every one of the uppermost energy levels occupied by electrons in the normal state of the free atom is broadened into an *energy band*, which is composed of a multiplicity of closely spaced levels.

Fig. 2. Energy levels of copper.

This is conditioned by the fact that, in a metal, the outer electrons belong jointly, in a sense, to all the atoms. According to the Pauli principle, no two of these electrons can be assigned to identically the same energy level. Fig. 2 illustrates the situation for the case of copper. At the left are set out the discrete levels of a free copper atom, taking the energy of the singly ionized atom as zero. The level indicated as 4s is the uppermost occupied level in the ground state of the copper atom, and is tenanted by only one electron in the case of copper (cf. p. 361). The levels lying above this represent energy states of excited atoms—i.e., energy states to which electrons may be raised by the absorption of energy. The distance from the 4s level to the uppermost of these measures the ionization energy of the atom (cf. Vol. I, p. 111). Adjacent to this uppermost level is a continuous band of energies, as follows from the fact that an electron that is ionized off may bear with it any amount of unquantized translational energy, in addition to the energy of ionization. As may be seen from the right hand side of Fig. 2, the effect of bringing the copper atoms together into a regular lattice structure is that the 4s level is extended into an energy band made up from a multiplicity of energy states lying very close together. The same is true of the 3d levels, whereas the deeper-lying levels are essentially unaffected. The levels lying above the 4s state, which are unoccupied in the ground state of the free atom, are likewise broadened into bands. Indeed, in the case of copper these overlap one another in such a way that a practically unrestricted range of energy values is accessible to the electrons, immediately above the band of 4s electrons.

Among other consequences of the broadening of the energy levels containing the valence electrons into an energy band, is that the minimum work which must be expended to remove an electron from the solid metal is substantially less than the ionization energy of the free atom. This so-called work-function can be determined from photoelectric or thermionic measurements. For copper, it amounts to 4.3 e.v., whereas the ionization energy of the copper atom is 7.7 e.v. The upper boundary of the energy bands containing the 4s electrons in metallic copper thus lies about 3.4 e.v. higher than the corresponding level in the copper atom.

In the normal state, at 0° K, the electrons occupy only those energy bands that correspond with the energy levels occupied by electrons in the ground state of the free atom. If all the levels in such an energy band are occupied by electrons, these electrons can transport neither electricity nor heat as long as they remain in that band. Even though they may be freely mobile (in the sense of a corpuscular theory) they can, nevertheless, not be accelerated by an electric field, for this would imply augmenting their energy. Unless the increment of energy were large enough to raise the electrons into another band, any such increase of energy would raise an electron to an energy level already occupied by another electron; this, according to the Pauli principle, is impossible.

If only half of the energy levels of a band are occupied by electrons, the empty levels are available to receive electrons with enhanced energy. Practically all the electrons belonging to a band can then be accelerated by an applied electric field, and thereby contribute to the conduction of current. If more than half the energy levels are occupied, the number of electrons which contribute to the transport of current is *less* than when the band is just half full. If less than half the levels are occupied, it is also similarly less. The electrical conductivity of a metal is thus not dependent on the total number of mobile electrons present in unit volume, but upon that number for which unoccupied energy levels are still available. This is known as the *effective electron number* of the metal.

The theory represents the relation between the specific electrical conductivity $\varkappa$ and the effective electron number per cm³ of the metal, n_{eff}, by the following formula:

$$\varkappa = \frac{e^2}{2mv} \cdot n_{eff} \cdot l \tag{1}$$

Here, e, m, v and l denote the charge, mass, velocity and mean free path of the electrons. In wave mechanics the mean free path of the corpuscular theory has the meaning of a quantity which measures the ability of the metal to set up stationary electric waves in the free space between the atoms. This ability is adversely affected by randomly distributed foreign atoms built into the metal (cf. p. 11 *et seq.*) and by lattice imperfections (cf. p. 18), as well as by the vibration of the atoms about their mean positions in the crystal lattice. Since these vibrations increase with rise of temperature, the metallic conductivity diminishes with rise of temperature. The diminution of conductivity brought about by foreign atoms and other lattice imperfections is, on the other hand, independent of the temperature. The more strongly the lattice of the metal is distorted, the more is its normal temperature-dependent resistivity overlaid by an additional resistivity which is independent of temperature.

It was established as long ago as 1864, by Matthiessen, that impurities raise the specific resistance of a metal by an amount which is independent of temperature (even when the impurities are minor amounts of metals with conductivities greater than that of the host metal). As has since been shown, Matthiessen's rule is true only for those impurities which are incorporated through the formation of mixed crystals; that is, that are built into the crystal lattice of the metal, usually in random distribution (cf. p. 11 *et seq.*). By contrast, the specific conductivity of an inhomogeneous alloy—that is a mere mixture of different crystallites—is derived additively from the specific conductivity of its components.

The *specific thermal conductivity* λ of a metal, on the same theory, is given by

$$\lambda = \frac{\pi^2}{6} \cdot \frac{k^2 T}{mv} \cdot n_{eff} \cdot l \tag{2}$$

Here T denotes the absolute temperature and k the so-called Boltzmann constant—i.e., the quotient of the gas constant R and Avogadro's number $\mathcal{N}_A$. Dividing eqn. (2) by eqn. (1), one obtains:

$$\frac{\lambda}{\varkappa T} = \frac{\pi^2}{3} \left(\frac{k}{e}\right)^2 \tag{3}$$

i.e., the Wiedemann-Franz law results. When the appropriate numerical values* are substituted, the quantity on the right hand side of eqn. (3) is found to be $2.4443 \cdot 10^8$. At ordinary temperature ($T = 273.15 + 18°$) it follows that $\lambda/\varkappa = 291.15 \cdot 2.4443 \cdot 10^8 = 7.12 \cdot 10^{10}$. The average of observations on a considerable number of high conductivity metals yields $\lambda/\varkappa = 7.11 \cdot 10^{10}$. Equation (3) thus also yields the numerically correct value for the proportionality factor of the Wiedemann-Franz law.

If one multiplies the specific electrical conductivity of the individual elements by their atomic volumes, and plots the resulting 'atomic conductivities' as a function of the atomic number, a periodic curve is obtained, resembling the atomic volume curve (Vol. I, p. 15, Fig. 2). The metals copper, silver, and gold lie on pronounced peaks, however, as do the alkali metals. It is possible to explain the favored conductivity of the metals of both Main Group I and Sub-group I of the Periodic System, revealed in this way according to the electron theory of metals, as follows. As free atoms, the metals of Main Group I and Sub-group I contain one electron in their outermost shell, and in each case this is an s-electron (with subsidiary quantum number $l = 0$). According to the Pauli principle, however, a shell with the wave mechanical subsidiary quantum number $l = 0$ can contain two electrons (cf. Vol. I, p. 120). In the atoms of the elements considered, it is therefore only half filled in the atomic state, while all the shells lying below it are fully occupied. Wave mechanical calculation of the potential distribution in the lattices of these metals indicates that the energy band in the solid metal, corresponding to this shell, is also only about half filled. For these metals, the number of electrons per cm³, n_{eff}, available for conduction of the current is, therefore, practically equal to the number of valence electrons n, whereas with other metals n_{eff} is invariably smaller than n.

(c) Semi-Conductors and Non-Conductors

As has already been explained, those substances in which all the energy bands of the solid state are fully occupied, are quite unable to conduct the electric current. If, however, a little above the uppermost energy band occupied by the valence electrons in the normal state there is another energy band, which is empty in the normal state, it is possible for electrons to be raised to the empty band by accession of thermal energy. Although the electron gas is degenerate, its heat capacity is not actually zero, but is only very small in comparison with the heat capacity of a normal gas. As soon as electrons have been transferred to the higher-lying energy band, both these and also the electrons of the band which they have left (which is now no longer fully occupied) can take over the transport of current. The higher the temperature to which such a substance is raised, the more electrons enter the upper energy band, and the greater accordingly, is its conductivity. A substance of this sort is called a *semi-conductor* or, if an element is concerned, a *semi-metal*. The characteristic of such a substance is that its conductivity increases with rising temperature, instead of falling, as for the true metals.

As the gap between the lowest unoccupied and the highest occupied energy level widens,

$$* \qquad k = \frac{R}{N_A} = \frac{1.986 \cdot 4.186 \cdot 10^7}{6.025 \cdot 10^{23}} = 1.380 \cdot 10^{-16};$$

$$e = \frac{4.800 \cdot 10^{-10}}{2.99776 \cdot 10^{10}} = 1.601 \cdot 10^{-20};$$

$$\frac{\pi^2}{3} \left(\frac{k}{e}\right)^2 = \frac{9.8696}{3} \left(\frac{1.380}{1.601} \cdot 10^4\right)^2 = 2.4443 \cdot 10^8.$$

the number of electrons which are elevated to the former by any given increase in temperature diminishes greatly. With a large spacing of the energy bands, no significant conductivity is observed at ordinary or at moderately elevated temperatures. The substance is then described as a *non-conductor* (or insulator) or, referring to an elementary substance, a *non-metal*. A non-conductor is thereby distinguished from a semi-conductor only in *quantitative* respects; no sharp distinction is to be drawn between them (Wilson [27]). It follows, further, that it is possible for as many freely mobile electrons to be present in a solid non-conductor as in a well-conducting metal. This does not exclude the possibility that there are also solid non-conductors or non-metals of which the lack of conductivity is conditioned by the absence of mobile electrons from the crystal lattice, or by their vanishingly small number.

The greater is the distance between the uppermost occupied and the lowest unoccupied energy level in the free atom, so much the wider, in general, will the interval be between the corresponding energy bands. The former, as is known, is measured by the smallest excitation potential of the atom (cf. Vol. I, p. 113 *et seq.*). Borelius (1939) has pointed out that all elements with minimum excitation potentials appreciably over 6 e.v. are non-conductors. Those with minimum excitation potentials in the neighborhood of 6 e.v. are semi-conductors, and those with excitation potentials significantly less are metals.

In addition to the increase of conductivity with temperature, it is characteristic of semi-conductors that the conductivity is very markedly increased by the presence of traces of impurity. In this respect also semi-conductors display the reverse effect to that of true metals. This is explained on the assumption that the foreign atoms constitute additional, discrete energy levels, which lie between the energy bands. This makes it easier for electrons to undergo transitions into a higher band.

(d) Superconductivity [31]

The electrical conductivity of many metals increases suddenly to an enormous degree when they are cooled to very low temperatures; these metals then oppose no further measurable resistance whatever to the flow of the current. This phenomenon is known as *superconductivity*. An electric current, once induced, continues to flow in a closed circuit formed by a superconducting metal, since no electromotive force is necessary to maintain it. Superconductivity was discovered in 1911 by Kamerlingh Onnes, with mercury, for which the specific resistance fell sharply at 4.2°K to an immeasurably small value. Later the phenomenon was observed for Ga,In, Tl, Sn, Pb, Ti, Th, Nb, Ta, and Mo. Alloys of these metals with others which are not themselves superconductors may likewise be capable of superconductivity. Compounds having metallic conductivity (nitrides, carbides, silicides and borides of many transition elements) may also become superconducting at low temperatures.

The origin of superconductivity has not yet been explained. There is much evidence that in the superconducting state the transport of current is not brought about by those electrons which are normally freely mobile in the metal, but that other electrons, which do not usually participate in conduction, also play a part.

The further study of superconductivity seems likely to contribute to a deeper insight into the nature of the metallic state. In particular the explanation of superconductivity is of considerable significance for a full understanding of the magnetic properties of the metals, since it has been shown that there is a close relationship between the latter and the phenomenon of superconductivity. The thermal conductivity is uninfluenced by the onset of electrical superconductivity.

The electrical conductivity of an ideal metallic single crystal should, theoretically, become infinite at 0°K, i.e., the specific resistance should become zero. However, superimposed on the temperature-dependent resistance of real metals there is usually an additional resistance due to the lattice imperfections usually present, and this (in accordance with Matthiessen's rule) does not disappear at absolute zero (residual resistance). In the superconducting state not only the temperature-dependent resistance, but also the residual resistance has disappeared. A fundamental distinction also exists between superconductivity and the ordinary conductivity of the ideal single crystal, in that the latter approaches continuously and asymptotically to infinity, as a limiting value; this is a consequence of the theory, and is confirmed by experiment when the residual resistance is deducted from the observed resistivities. Superconductivity, however, sets in discontinuously in such a manner that in cooling a few hundredths of a degree the resistance of the metal falls from a quite measurable to an immeasurably small value.

(e) Electrolytic Conduction of Current in Metals [38, 39]

Since the valence electrons can move almost freely between the ions of the metallic lattice, they are responsible essentially, but not quite exclusively, for the transport of current. A certain fraction of the current—although it is extraordinarily small—can be due to the motion of the metallic ions. With alloys, accordingly, at very high current densities, it is often possible to observe a shift in the concentration of the alloy components—i.e., a process corresponding to electrolysis—as a consequence of the passage of the current. The amount of metal transported in the direction of the positive current (i.e., enriched at the cathode), in gram-atoms per Faraday, is termed the transport number of the corresponding metal in the alloy. The transport number has a negative sign if it is defined by reference to that metal which remains behind, as compared with the other, and so is enriched at the anode. According to measurements by Jost (1935–36) and Seith (1934) this amounts, e.g., for copper in a Cu-Au alloy, to $7.4 \cdot 10^{-11}$ (at 750°), for Pd in a Pd-Au alloy $1.6 \cdot 10^{-11}$ (at 900°), and for Au in a Pt-Au alloy of low gold content $1.3 \cdot 10^{-10}$ (at 180°). Considerably larger effects have been observed with liquid alloys (Kremann, 1923 and later, Schwarz, 1931 and later). This is because the mobility of the ions in liquids or in molten metals is substantially greater than in solid metals (as can be directly shown from measurements of diffusion rates), whereas the mobility of the electrons diminishes with decreasing order of atomic arrangement, and especially on the transition from the solid to the liquid state.

It was formerly widely assumed that the relative electro-positivities determined which component of an alloy would be enriched at the cathode and which at the anode. However, many experimental results are difficult to reconcile with this assumption; thus sodium and potassium, in the electrolysis of dilute amalgams, are enriched at the anode, although they are much more electropositive than mercury. Wagner (1932–33) considers that the *charge* of the alloy components (electrochemical valence) appears to be the decisive factor for migration in solid metals. Schwarz has advanced the theory that in liquid alloys the metal ion with the greatest charge density (ratio of charge to volume) should invariably migrate to the cathode; this appears to be in harmony with facts.

2. Mixed Crystals and Intermetallic Phases

It is a consequence of the particular nature of the metallic state that the regularities governing the formation of compounds by metals with each other should differ from those involved in the formation of compounds between non-metals, or between non-metals and metals [32–37]. As has already been stated (see Vol. I, Chap. 13), the compositions of intermetallic compounds frequently do not conform to simple stoichiometric proportions. They may contravene the law of Constant Proportions, their composition often being variable within rather wide limits ('non-Daltonide compounds'). There is, however, no sharp demarcation between the intermetallic compounds and other types of compounds. Examples of variability in composition are also known among the compounds of metals with non-metals. Many compounds of metals with non-metals, or even of non-metals with each other, are closely related in composition, structure, and to some extent in properties, with the typical intermetallic compounds. On the other hand, there are intermetallic compounds that correspond entirely in their composition to salts and salt-like compounds—i.e., to normal valence compounds—and represent a transition to this type. In general, however, this is not true of the compounds of the metals among themselves. Among alloys involving the metals of the Sub-groups, in particular, types of substances are frequently found which are so differ- ent from the typical chemical compounds that it is necessary to give particular consideration to the problem whether it is desirable or correct to regard them as chemical compounds at all (see p. 19 *et seq.*). The name '*intermetallic phases*' has come into use as a more general designation for those substances appearing in intermetallic systems, which differ structurally from the components out of which they are formed (and incidentally from mixed crystals of these components). Many of these intermetallic phases are closely related to mixed crystals, and were formerly regarded as such. Before characterizing the intermetallic phases more closely, it is therefore necessary to consider the most important properties of me- tallic mixed crystals.

3. Mixed Crystals

(a) General Aspects

Mixed crystals are distinguished by the fact that the atoms or ions of some element are incorporated into the lattice of another substance, without any change of structure; this may occur in any proportion within the limits within which mixed crystal formation takes place.

For example, all the silver atoms in a silver crystal could be successively replaced by gold atoms, without alteration of the crystal structure, until eventually, by exchanging the last silver atom for a gold atom, one obtained pure gold, which has the same crystal structure as silver. This is an example of *unrestricted miscibility*. Only a proportion of the silver atoms in the crystal lattice of silver may be replaced by copper atoms (up to about 0.2 atom per cent at the ordinary temperature); conversely, in the crystal lattice of copper, only a limited pro- portion of copper atoms are replaceable by silver (up to 0.03 atom per cent). In this case there is *restricted miscibility*; at room temperature there exists a miscibility gap stretching from 0.2 to 99.97 atoms per cent Cu*.

Incorporation of the foreign atoms in the crystal lattice can occur either in such a way that these replace individual atoms of the basic lattice, or that the foreign atoms are embedded in 'holes' in the basic lattice; the resulting structures are known as *substitutional* or *interstitial* mixed crystals, respectively (Fig. 3). Inter- stitial mixed crystals are also referred to as solid solutions**.

Substitutional mixed crystals are by far the most common. The formation of *interstitial mixed crystals* has been chiefly observed in the incorporation of non-metal atoms of small radius, such as C, N, H, in the metallic lattice. In both types the lattice constant is changed by the

* The figures given refer to the equilibrium state. With increase of temperature, the mutual miscibility of Ag and Cu increases considerably (Fig. 5, p. 16). On rapid cooling, unmixing may not occur, or may take place incompletely. One then obtains supersaturated mixed crystals, which contain appreciably more foreign atoms incorporated in their basic lattice than would be present in the equilibrium state. Such metastable supersaturated mixed crystals very frequently occur in alloys (cf. here p. 27). Because diffusion rates are very small, it is often quite impossible to remove the supersaturation at ordinary or at slightly elevated temperatures.

** As long as the structural difference between substitutional and interstitial mixed crystals remained unknown, the term 'solid solution' was employed in the same sense as 'mixed crystal'. Today it is customary to distinguish solid solutions (interstitial mixed crys- tals) from mixed crystals in the narrower sense (substitutional mixed crystals).

incorporation of the foreign atoms. The change takes place continuously, and is approximately proportional to the number of foreign atoms incorporated and, when substitution takes place, also to the difference in radius between foreign atom and host (Vegard 1921)

Foreign atoms are ordinarily incorporated in the basic lattice in completely random distribution (cf. Fig. 3b and c). In such cases, the specific electrical conductivity diminishes with increasing content of foreign atoms.

This can be explained on the basis of wave mechanics in that the randomly distributed foreign atoms disturb the formation of stationary waves by the conductivity electrons of the

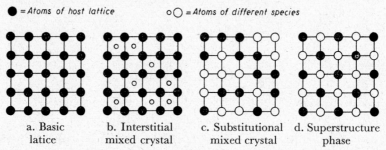

● = Atoms of host lattice   ○○ = Atoms of different species

| a. Basic | b. Interstitial | c. Substitutional | d. Superstructure |
| lattice | mixed crystal | mixed crystal | phase |

Fig. 3. Formation of mixed crystals by building atoms of a different species into a crystal lattice (schematic).

metallic lattice; i.e., the quantity l in eq. (1) p. 7, is decreased (Nordheim 1928). The ductility is changed in a similar way to the electrical conductivity by incorporation of foreign atoms. Alloys of two metals which are unrestrictedly miscible in the solid state therefore display a minimum both of conductivity and of ductility at a composition of 50 atom per cent.

Mixed crystals differ characteristically from inhomogeneous mixtures by their greatly diminished electrical conductivity. This may be seen from the example of the Cu-Ag alloys, given in Fig. 4. Within the mixed crystal range* an addition of copper to silver, or an

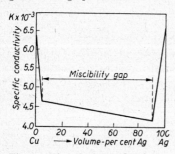

Fig. 4. Electrical conductivity of Cu-Ag alloys.

addition of silver to copper brings about a strong lowering of specific conductivity. In the region of the miscibility gap, however, the conductivity increases smoothly from that of the silver mixed crystals saturated with copper** to that of the copper mixed crystals saturated with silver, according to the law that the specific conductivity of an inhomogeneous crystal aggregate is made up additively from the conductivities of the components (in the case considered, Cu-Ag mixed crystals and Ag–Cu mixed crystals; cf. p. 7). Conductivity measurements can often be used to determine the range of miscibility of alloy components simply and accurately.

The ability to form mixed crystals is characteristic of the metals. Unrestricted or very extensive miscibility in the solid state is particularly common among metals of the Subgroups of the Periodic System. Unbroken mixed crystal series are only occasionally met with among binary alloys of which both components belong to the Main Groups of the Periodic System (cf. the tables on pp. 574, 575 and 576 of Vol. I).

Grimm's rule, which is valid for mixed crystals of salt-like compounds, states that chemi-

* The miscibility ranges of Fig. 4 are those for 700°, since the conductivities plotted were measured (at 0°) on alloys quenched from 700°.
** Supersaturated at the temperature of measurement (cf. previous footnote).

cally analogous materials can only form mixed crystals (at ordinary temperature) if the ions mutually replacing one another do not differ too widely in radius (not more than 5%, according to experiment). It is not generally valid for metallic crystals. For example, copper, which has an atomic radius about 10% less than that of Au, forms an unbroken series of mixed crystals with gold, whereas it is only slightly miscible with silver, which has almost exactly the same atomic radius as gold.

(b) Superstructure Phases

In mixed crystals with a composition corresponding to, or close to, a simple stoichiometric ratio, it is often found that an ordered distribution of both kinds of atoms among the lattice points sets in at some definite temperature, (cf. Fig. 3d), whereas at higher temperatures there is a random distribution. Since the ordered distribution of the different sorts of atoms among the lattice points results in additional interference lines on the X-ray diagram (cf. p. 23), it is said that a *superstructure phase* is formed in such cases.

Formation of the superstructure phase is attended with a sudden and fairly large increase in the electrical conductivity. This may be explained by the removal of that disturbance of the electron waves which was produced by the randomly incorporated foreign atoms. The formation of superstructure phases is also accompanied by a small increase in the density (Grube 1931). The ductility is also greater for an ordered atomic arrangement than for a disordered.

The superstructure phases frequently have a lower crystallographic symmetry than the corresponding mixed crystals with random atomic distribution. For example, with crystals of the composition CuAu, a compression of one of the three crystallographic axes accompanies the transition from the state of the random to that of the ordered atomic distribution. The cubic face-centered lattice of the mixed crystals changes into a tetragonal face-centered phase ($c/a = 0.93$) when the superstructure phase is formed. Transformation of a mixed crystal into a superstructure phase often leads to the formation of a crystal lattice which is typical of salt-like compounds. Thus the superstructure phase CuPd forms a crystal lattice of cesium iodide type (cf. Vol. I, p. 211). This is formed from the cubic face-centered lattice of the corresponding mixed crystal by the c axis of the original lattice contracting until it is only 0.7 of the a axis, as the ordered atomic arrangement is formed.

The superstructure phases take an intermediate position between mixed crystals and chemical compounds. They are often considered as a special case of mixed crystal formation. Since, however, discontinuous changes in properties accompany the transition from random to ordered atomic distribution, and since the appearance of the superstructure phases is fundamentally bound up with a simple stoichiometric ratio of the components, at least for the ideal case (see below), superstructure phases are also frequently regarded as chemical compounds. They may then conveniently be distinguished from the chemical compounds proper (p. 20) as *superstructure compounds*.

Examples of superstructure phases can be found, especially, in Tables 30 and 40 (pp. 248, 366). In these tables superstructure phases are indicated by enclosing their formulas in square brackets.

Superstructure phases are found not only among alloys corresponding exactly in composition to the simple stoichiometric ratio of the completely ordered state, but also for alloys with compositions lying close to that of the ideal superstructure phase. The properties characteristic of the superstructure phase still persist (though in weaker degree) if a proportion of the atoms of one component in the ordered structure is replaced by atoms of the other component. With increasing incorporation of excess atoms of the other component in the lattice (which occurs in random manner) the typical properties of the superstructure phase nevertheless soon disappear.

(c) Resistance Limits of Mixed Crystals

Binary mixed crystals of which one component, in contrast to the other, is practically unattacked by chemical reagents, display so called resistance limits, as found by Tammann in 1919; that is, chemical attack takes place only when the proportion of the component susceptible to chemical attack exceeds a definite amount. This amount depends upon the nature of the chemical attack, but very often comes to a whole-number multiple of $\frac{1}{8}$ atom-% ('$n/_8$-molar rule'). Thus a Cu-Au alloy is attacked by ammonium polysulfide only when it contains less than $\frac{3}{8}$ atom-% Au. Only those Cu-Au alloys react with silver nitrate solution, depositing silver, of which the gold content lies below $\frac{1}{8}$ atom-%. Resistance limits can only be observed if the alloy is composed of uniform mixed crystals, and is completely homogenized by sufficiently prolonged heating ('annealing'); furthermore, the reactions should be investigated at temperatures such that interchange of places by atoms within the crystal lattice is practically non-existent.

The resistance limits also show up in the galvanic solution potentials of the mixed crystals. The solution potentials of alloys composed of a *mixture* of different substances are independent of the relative proportions of these substances; but with *homogeneous* alloys—i.e., for those in which uniform mixed crystals are present—it is observed that the solution potential depends upon the composition. This often happens in such a way that the solution potential jumps, at a definite ratio of concentrations of the components, from that of the baser component to that of the nobler. Thus Mn-Cu mixed crystals with less than 0.5 atom-% Cu display the solution potential of pure manganese, but above 0.5 atom-% Cu, that of pure copper. Cu-Au alloys have the same solution potential as pure gold, if they contain up to 0.75 atom-% Cu; for higher copper contents, the potential falls linearly to that of pure copper. Ag-Au alloys show the solution potential of pure silver, right up to high gold content; if, however, they are dipped in nitric acid before making the potential measurements, they possess the solution potential of pure gold, even up to high silver content. The potential of the baser metal is set up, as follows from this, when its atoms are present in the *surface* of the mixed crystal. A silver-gold alloy can give the surface potential of either gold or silver, at will, according as silver atoms are dissolved out of the surface by means of nitric acid, or whether they are allowed to diffuse into the surface again by annealing. In the case of the Cu-Au alloys, immersion in the (air-containing) solutions for measurement of the solution potential suffices to dissolve the baser atoms out of the surface. Corresponding relations apply to the case of the Mn-Cu alloys. The occurrence of potential jumps can be explained in that the atoms lying below the surface are without action only as long as the lattice points of the surface are still sufficiently densely occupied after the baser atoms have been removed. The occurrence of resistance limits towards chemical reactions can be explained correspondingly.

Tammann [40] assumed that the requisite conditions for the protective action of the nobler atoms are only realized when these are present in the crystal lattice in *ordered* arrangement. With the Cu-Au mixed crystals, and in other instances, X-ray investigation has confirmed this assumption by the observation of superstructure lines. In a number of mixed crystals, however, which likewise display resistance limits (e.g., Ag-Au mixed crystals with 0.25 and 0.5 atom-% Au), no superstructure lines could be detected by X-rays. The question of the connection of the resistance limits with the ordered atomic distribution thus needs further clarification. Homogeneous substances in which not only is there no ordered distribution of different sorts of atoms over the lattice points, but no sort of regular spatial order at all—i.e., glasses—display *no* resistance limits.

4. Intermetallic Phases

(a) General Aspects

In alloys, crystalline species separated by phase boundaries from the crystals of the alloy components are called intermetallic phases*. The intermetallic phases

* A more general name is *intermediate phases*; this can also be applied to systems with non-metals.

therefore differ discontinuously in structure from the alloy components and their mixed crystals. They can, accordingly, only appear in regions of concentration in which, at the temperature concerned, the constituents are immiscible in the solid phase, or in which mixed crystals are not stable. In the phase diagrams of alloys the intermetallic phases are generally separated from the pure metals or from their mixed crystals (and from each other, if several intermetallic phases exist) by broader or narrower ranges of inhomogeneity—i.e., by ranges of concentration in which the solidified alloy consists of a mixture of crystallites of different composition and structure.

Examples of intermetallic phases of this kind are provided by the intermetallic compounds occurring in the phase diagrams of Vol. I, Chap. 13. The systems cited there involve intermetallic phases which separate directly from the melt. Such phases can also be formed however, by transformations which take place subsequently, involving the crystals first deposited. The primary crystals, still in contact with the melt, may either interact with it on further cooling, to form crystals of different composition and structure ('peritectic transformation'), or else, after solidification is complete, a transformation sets in, with alteration of composition and structure. It is not uncommonly found that a phase, deposited from the melt, undergoes a structural transformation on further cooling, without any alteration in composition. This process corresponds to the allotropic transformations of ordinary compounds. It leads on to the case in which a mixed crystal phase, on cooling, undergoes a structural transformation without change in composition, whereby an intermetallic phase appears in a region of concentration in which, at higher temperatures, only mixed crystals occur.

In the phase diagrams of binary alloys, of which both components belong to the Main Groups of the Periodic System, intermetallic phases are usually found to have quite narrowly limited ranges of homogeneity, separated from the homogeneity range of adjacent phases by a broad region of immiscibility. With alloys formed by metals of the Sub-groups, intermetallic phases with broad homogeneity ranges are common; the regions of inhomogeneity between them are often only narrow, and in certain cases are hardly noticeable*.

In Fig. 5 are set out the phase ranges of some solidified binary alloys at certain temperatures. The range of homogeneity of any one phase (indicated in the diagram by Greek letters) may either widen or diminish in breadth when the temperature is lowered. The ranges of homogeneity of mixed crystal phases of the pure components generally grow narrower as the temperature is lowered. The regions of immiscibility are indicated by shading in Fig. 5. The mode of presentation used in Fig. 5, which represents a cross section of the phase diagram of the alloy system in question, at some one temperature, makes the phase relations very clear, especially in the more complex cases. Thus the significantly different degree of miscibility of the components in the systems Cu-Ag and W-Pt is prominently displayed by the different breadths of the miscibility gap.

The concentration range of any one phase region at some definite temperature will be denoted subsequently as a range of existence. If the various ranges of existence in an alloy system at any one temperature are plotted out, we obtain a cross section through the phase diagram at this temperature. If the phase cross sections set out in Fig. 5 are regarded as a general scheme, the phase cross sections of any desired binary alloys can be derived from

---

* By the 'homogeneity range' of a phase one understands that range of concentration and temperature within which the homogeneous phase is stable. With regard to the inhomogeneous regions of a phase diagram, in addition to the concentration ranges in which the solidified alloy consists of a mixture of crystals of different structure and composition, must be considered the regions included by the liquidus and solidus curves, in which melt and crystals coexist.

them by making appropriate changes in the breadth of the phase ranges. Thus, if the range of the α-phase of Fig. 5a is contracted practically to nothing, the resulting alloy type is like that of Sn-Pb for which a phase diagram is given in Fig. 98 Vol. I (p. 564). In the case

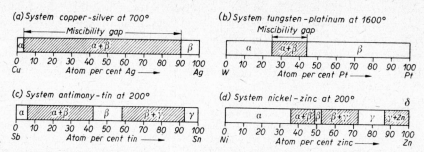

Fig. 5. Ranges of existence of phases in solid binary alloys. (a) and (b) Alloys with limited miscibility of components in the solid state, but without intermetallic phases. (c) Alloy system with one intermetallic phase (β). (d) Alloy system with two intermetallic phases, (β) and (γ).

represented in Fig. 5d, the δ phase (Zn) has zero homogeneity range: zinc cannot take up nickel in appreciable amounts to form mixed crystals, as may be seen in that the inhomogeneous region γ + δ (i.e., γ + Zn) reaches right up to 100% Zn. In the same way, where compounds of constant composition occur—that is intermetallic phases of infinitely narrow homogeneity range—two different inhomogeneity regions abut directly on each other. Their point of contact then gives the chemical composition of the compound. If the homogeneity ranges of the intermediate phases are contracted to zero, Fig. 5c represents the phase ranges of solidified alloys of the type of the Mg-Pb alloys, and Fig. 5d that of the type of Mg-Cu alloys (cf. Vol. I, Figs. 100, 101, p. 567). This form of diagrammatic representation may be extended to systems with as many intermetallic phases as required.

(b) Intermetallic Phases in Copper Alloys

As an example of alloys in which a considerable number of intermetallic phases appear —some, indeed, with considerable ranges of homogeneity—, Fig. 6 (p. 17) represents the phase diagram cross sections of some technically very important alloy systems of copper. Those phases of the different alloys which correspond to one another structurally—i.e., in which the arrangement of the atoms in the crystal lattice are similar—are indicated in the diagram by Greek letters without any subscript*. It is evident that in these systems, more or less radical changes take place when the temperature is lowered**. Not only are the phase boundaries displaced by temperature, but with decreasing temperature some of the phases stable at higher temperatures disappear, and new ones appear. To follow the course of these changes in detail reference must be made to the complete phase diagram. That of the system Cu-Zn is represented in Fig. 45, on p. 372.

There is a close relationship between the alloys represented in Fig. 6, as was shown by Westgren and Phragmén. Several of the phases appearing in them correspond to each other

* Where the same Greek letters have been used for structurally-nonequivalent phases, these are distinguished in Fig. 6 by Arabic subscript numerals; those phases peculiar to the systems Cu-Zn are marked with the subscript $_1$, those appearing only in the Cu-Sn systems with $_2$, and those only in the system Cu-Al with the subscript $_3$.

** This is true provided that cooling is sufficiently slow, so that the transformations giving rise to the phases stable at low temperature have time to occur. It is of considerable technical importance that one can frequently hinder the transformations by rapid cooling ('quenching') so that the phases stable at high temperatures persist at lower temperatures, in a metastable state. This is also of importance for investigating the properties of the 'high temperature' phases: this can often be performed at ordinary temperature, with the use of quenched samples.

structurally as the diagram shows*. In all three cases, the copper mixed crystals (α-phase, face-centered cubic) are first formed. With a higher content of the second component, an intermetallic phase with a body-centered cubic structure is formed, in which both kinds of atoms are distributed at random over the lattice points (β-phase). Whereas this phase disappears completely in the Cu-Al and Cu-Sn systems at lower temperatures, the only change it undergoes in the system Cu-Zn is that, below 470°, an ordered distribution of the atoms sets in (β'-phase). With a still higher content of the second component, a phase (the γ-phase) appears in all three systems, having a structure which resembles that (discussed further below) of α-manganese. The δ-phases in these three systems do not correspond structurally; their structures are not yet fully explained. The ε-phase appearing in the systems Cu-Zn and Cu-Sn has a crystal lattice with hexagonal close packing. The ε-phase occurring at high temperatures in the system Cu-Al (marked in Fig. 6 as ε_3) is structurally

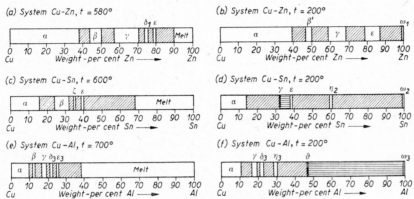

Fig. 6. Range of existence of phases in copper alloys. Phases denoted by the same Greek letters (without subscripts) correspond structurally to each other. α, copper mixed crystals (face-centered cubic). β, γ, and ε are phases of the Hume-Rothery type. β, (body-centered cubic)—$\overline{ZnCu}$, $\overline{SnCu_5}$ and $\overline{AlCu_3}$ respectively. β', superstructure form of β. γ, (cubic, very large cell, cf. p. 24)—Zn_8Cu_5, Sn_8Cu_{31} and Al_4Cu_9, respectively. ε, (hexagonal close packed)—$\overline{Zn_3Cu}$ or $SnCu_3$.

different from these, being cubic. The intermetallic phases occurring in the Cu-Sn and Cu-Al systems, with a higher content still of the second component (η_2 or η_3 and ϑ respectively) have quite narrow ranges of homogeneity, and can be represented by the formulas SnCu (nickel arsenide structure**), AlCu (rhombic, related structurally to the γ-phase), and Al_2Cu (with a tetragonal structure). In all three systems the second component is able to take up only a little copper into its crystal lattice (ω-mixed crystal phase).

The intermetallic phases η_2 or η_3 and ϑ respectively, occurring with high contents of the second component, may obviously be regarded as chemical compounds without further qualification, because of their stoichiometrically simple and but slightly variable composition. The intermetallic phases appearing with a predominant copper content, having in some cases quite extensive ranges of homogeneity can, however, also be regarded as chemical compounds. The marked variations in composition which they display can be related to defects of order of atoms in the crystal lattices of these compounds. As will be shown below, the idea that these phases are intermetallic compounds is not only convenient for the chemist, in that it makes it possible to denote them by chemical formulas, but it also brings to light certain regularities which govern the appearance of these phases.

* The phases correspond in so far as the spatial distribution of the lattice points is concerned. The distribution of the different kinds of atoms over the lattice points may differ.
** The structure deviates from the nickel arsenide type in that additional copper atoms are incorporated, in a regular fashion, in the NiAs lattice. The structure and the composition (45 atom-% Sn) therefore correspond better with the formula Sn_5Cu_6.

(c) Lattice Imperfections and Defects of Order

Crystal lattices in which a part of the atoms occupy 'wrong' positions are said to be *defective*. Lattice imperfections in the narrower sense are said to be present when more or less extensive regions of a lattice are irregularly distorted, by the displacement of atoms from their proper positions. Such lattice imperfections can arise because perfect ordering of the atoms has been achieved only to an incomplete extent during the formation of the crystal, or because the order has been subsequently disturbed by external agencies.—e.g., deformation by mechanical working.

If atoms occupy 'wrong' sites in a crystal lattice which is not irregularly distorted, or if a proportion of lattice sites (distributed at random) is not occupied by atoms, these are said to be lattice defects. In a lattice which is not distorted, atoms are said to be located in 'wrong' positions if they are either distributed at random over interstitial positions or, in a lattice composed of two sorts of atoms, if they take up, at random, sites which ought, for reasons of symmetry, to belong to the other kind of atom. In Fig. 7 (next page) the three types of lattice defect are illustrated schematically .Lattice defects of the first two types can occur in the crystal lattices of elements, as well as of compounds. Lattice defects of the third kind are possible only in the crystal lattices of compounds.

Lattice sites, distributed at random*, which are unoccupied by atoms, are termed 'lattice holes' or 'vacant sites'. They should not be confused with 'interlattice' or 'interstitial' positions—i.e., the free interstitial spaces in an assembly of spherical atoms in contact with each other, to which a crystal approximates.

Lattice imperfections in the narrower sense are also frequently found in non-metallic substances. With metals, indeed, on account of their good powers of crystallization, such imperfections rarely occur intrinsically in any considerable concentration. They are usually produced secondarily, as a result of deformation by working. As against this, it is typical of metals that defects of order may be present to a considerable degree. This is very closely connected with the capacity of the metals for forming mixed crystals. In doing so, atoms of one metal may be exchanged for those of another, or atoms may be incorporated in interstitial positions. However, lattice defects are in no sense restricted to metallic lattices only. In fact, it has been shown that they are quite usual in ionic crystals also, though their concentration in these is usually very minute**.

Lattice defects have no effect on the *chemical composition* of typical salt-like compounds, for the reason that the electrostatic forces between the ions require that the charges of opposite sign shall exactly compensate each other. If, therefore, in the crystal lattice of a salt AB, a certain proportion of the ions A takes up positions belonging structurally to the ions B, the requirement that the charges shall balance implies that a corresponding proportion of the ions B must occupy the places of A-ions. A corresponding compensation operates for the inclusion of ions in interstitial positions, and for the occurrence of vacant sites. Intermetallic compounds are not affected by any requirement of compensation of charges, except in so

* If one removes atoms from a crystal lattice in *regular fashion*, a new structural type is produced. Thus by leaving the cube centers and the mid-point of the edges of every unit cell unoccupied by atoms, (i.e., if one considers the chloride ions omitted from Fig. 44, p. 209 of Vol. I), the rock salt structure becomes a face-centered cubic structure.

** The number of lattice defects increases with rise of temperature.

far as they approximate the salts in constitution. In the crystal lattice of an intermetallic compound AB, the atoms B can therefore partially occupy the lattice sites of A-atoms, or be built into interstitial positions, or vacant sites may be left in the sub-lattice of A-atoms without the occurrence of anything corresponding in the sub-lattice of B-atoms. Each of these

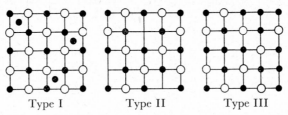

Type I   Type II   Type III

Fig. 7. Defects in the crystal lattice of a compound (schematic). Type I. Component present in stoichiometric excess is incorporated in interstitial positions. Type II. Vacant sites in sublattice of component present in deficient amount. Type III. Lattice sites belonging to one constituent are occupied by atoms of the other component, which is present in stoichiometric excess (substitutional type).

processes results in an excess of B-atoms in the compound. If the proportion of component A in the compound is progressively increased, the atoms of component B may, for example, be displaced from their position in the A sub-lattice, and finally A atoms may enter the B-sub-lattice; a compound now results which contains an excess of component A. It does not even follow that the lattice type corresponding to the stoichiometric composition will be distinguished by a particular freedom from lattice defects. The stoichiometric composition may arise simply because the different sorts of lattice defect—e.g., vacancies in the B-lattice, and incorporation of B-atoms in interstitial positions—happen to compensate in their effects.

(d) Intermetallic Phases as Chemical Compounds

No sharp demarcation can be drawn between intermetallic compounds with a narrow range of homogeneity and ordinary compounds. Intermetallic phases of this kind have therefore always been regarded as chemical compounds*. It has now been shown that many of these intermetallic phases, regarded as typical compounds, possess a homogeneity range which, although very narrow, is still of perceptible extent. This fact, in conjunction with analogous observations made upon compounds with the non-metals, has made it necessary to drop constancy of composition as the criterion of a chemical compound.

There are special reasons (which have been discussed in Vol. I, Chap. 9) why the typical 'salt-like' and 'adamantine' compounds (Vol. I, p. 293 ff.) display a strictly stoichiometric composition. When typical intermetallic compounds display practically stoichiometric compositions, their behavior is generally determined not so much by the nature of the valence forces, as by the atomic radii of the components and by the crystal structure. If, by reason of the different atomic radii of its structural units, the crystal lattice permits no appreciable defects of order, a substance of practically constant composition necessarily results. If lattice defects may be tolerated in substantial amount, the composition may vary correspondingly. From the chemical standpoint there is no real difference between the two cases. There is therefore no objection to the extension of the concept of 'chemical compound' to include intermetallic phases with homogeneity ranges of any magnitude whatever, as long as they may be referred to some structure with a definite ratio of the components. Such an idealized structure, obtained by eliminating all defects of order, then furnishes the *chemical formula* of the compound in question**.

* The first systematic experiments on the ability of the metals to combine with one another were described by Tammann (1906 and later).

** Variability of composition is indicated in formulas by a stroke placed over the formula (cf. Vol. I, p. 586).

If, starting from an intermetallic phase with an ordered arrangement of the atoms, the atoms of different sorts are permitted to change places to an increasing extent, the number of lattice defects increases continuously until a completely random distribution of atoms among the lattice sites is eventually reached. Except at high temperatures, this is probably rarely fully achieved in crystallized substances, since an increasing tendency towards the ordering of the atoms in the crystal lattice generally becomes apparent with decrease of temperature. No demarcation can be drawn, accordingly, between intermetallic phases with (practically) completely random atomic distribution and those with considerable numbers of lattice defects. It therefore appears desirable to extend the concept of the chemical compound to include such substances also.

It is, admittedly, not possible from the crystal structure to assign an idealized formula to an intermetallic phase with a completely random distribution of the atoms. As a rule, however, other criteria can usually be invoked to assign an ideal formula. For example, there are intermetallic phases, with a random distribution of atoms, which have a very narrow homogeneity range—i.e., practically constant composition. In this case, the formula is given by the composition (Example: $AlAg_3$, which displays constancy of composition, but contains both sorts of atoms distributed quite at random over the lattice points).

When intermetallic phases possess rather wide ranges of homogeneity, although their idealized formulas correspond to chemical compounds of definite composition, they are often spoken of as mixed crystals of the corresponding compound with one or other component. This view (of mixed crystals) is not logically valid, however, in cases where variability of composition is due to the occurrence of vacant sites in the lattice of one component only. It may indeed happen that a compound is stable only if vacant sites are present in the sub-lattice of one component, so that one component is necessarily present in excess, as compared with the idealized formula. The composition of such a compound may be practically constant or may be markedly variable but it necessarily departs from that given by the idealized structure. The stability of a crystal lattice may depend not only on vacant sites, but on other lattice defects also. Hence it is often quite impossible to establish formulas of intermetallic compounds on purely analytical grounds, even when they are of nearly constant composition; it can often only be done in conjunction with structure determination.

It is not desirable to restrict the concept of chemical compounds to those intermetallic phases which are of fixed composition, for the following reason. As methods of investigation have been continually refined, it has been established in a constantly increasing number of instances that intermetallic phases which were formerly regarded as constant in composition, have a perceptibly extended range of existence. Considering the marked capacity of the metals for mutual replacement in their crystal lattices, it seems doubtful whether intermetallic phases of *strictly* constant composition exist at all, except possibly at very low temperatures. It appears, moreover, that an intermetallic phase with a narrow range of existence —i.e., a compound in the old sense—frequently corresponds in some related system to a structurally similar phase with a broad homogeneity range, which resembles the former sufficiently closely in properties for the two phases to be regarded as completely analogous.

Thus, the ε-phase in the system Sn-Cu, which (because of its narrow homogeneity range) has always been regarded as a compound $SnCu_3$, corresponds to an ε-phase with a broad existence range in the system Zn-Cu. Both phases agree not only in their crystal structures (hexagonal closest packing), but are also related in other properties; moreover, as will be shown later, there is a close constitutional relationship between them.

Intermetallic phases can be brought into the concept of the chemical compound if the latter is defined as follows: *a chemical compound is a composite, homogeneous substance, the properties of which cannot be transformed continuously into those of its constituents by changes in the composition.*

The ordered arrangement of the atoms within the crystal lattice is no more an essential characteristic of chemical compounds than is constancy of composition. There are chemical compounds which at higher temperatures have a disordered, and at lower temperatures an ordered distribution of atoms in the crystal lattice, without the structure undergoing any change in other respects. For instance β' brass is the superstructure phase, stable at ordinary temperature, of the β-brass, ZnCu, which is stable at higher temperatures.

If a homogeneous substance, made up of the constituents A, B, and C can be transformed continuously, by altering the composition, into the constituent C, but not into A or B, then it is to be regarded as a mixed crystal of the element C with a compound of A and B. For example, the compound Cd_3Ag forms an unbroken series of mixed crystals with Mg, as has been shown by Laves (1936).

(e) The Hume-Rothery Rule

The occurrence of the phases marked as β, γ and ε in Fig. 6 is not peculiar to the alloy systems cited there. Phases structurally equivalent t) these occur, rather, in a large number of alloys of the metals Cu, Ag and Au. The ability to form intermetallic phases with the structure of β-, γ- and ε-brass can be regarded as a characteristic property of the metals of Sub-group I of the Periodic System. In the intermetallic phases of the type of β-, γ- and ε-brass, the metals of Sub-group I (and other metals which correspond to them in function —see below) are termed 'metals of type I'; the metals which form these phases with them are termed 'metals of type II'. Be, Mg, Al, Ge, Sn, Sb, Zn, Cd and Hg can function as metals of type II. The formation of these phases is governed by a rule proposed by Hume-Rothery in 1926. *The ratio of the total number of valence electrons to the total number of atoms determines the structure of the phase which is formed. With a ratio of 3 : 2, the body-centered β -phase appears, with 21 : 13 the cubic γ-phase, and with 7 : 4 the hexagonal ε-phase.*

Electrons considered here as valence electrons are those which cannot be assigned to a complete shell—i.e., with Cu, Ag, and Au, one electron each (cf. p. 361 *et seq.*), with Zn, Cd and Hg, two electrons each (cf. p. 425), and with the metals of the Main Groups as many electrons as are ionized off exerting the maximum electropositive valence.

Hume-Rothery's rule holds in nearly all instances where the compounds have constant composition, or where, with variable composition, the different atoms are regularly ordered in the crystal lattice, so that it is possible to assign idealized formulas to the compounds without difficulty*. In those cases where variable composition and random distribution of

TABLE I

HUME-ROTHERY COMPOUNDS

Ratio of *valence electrons : atoms*					
21 : 14 or 3 : 2		21 : 13		21 : 12 or 7 : 4	
β Phases		γ-Phases		ε-Phases	
BeCu	ZnCu	Al_4Cu_9	Zn_8Cu_5	Be_3Cu	Zn_3Cu
$\overline{AlCu_3}$		Sn_8Cu_{31}	Cd_8Cu_5	$GeCu_3$	
$\overline{SnCu_5}$				$SnCu_3$	
				$\overline{Sb_3Cu_{13}}$	
MgAg	$\overline{ZnAg}$		Zn_8Ag_5	Al_3Ag_5	Zn_3Ag
	$\overline{CdAg}$	—	$\overline{Cd_8Ag_5}$	$SnAg_3$	Cd_3Ag
			Hg_8Ag_5	Sb_3Ag_{13}	
MgAu	ZnAu		Zn_8Au_5	Al_3Au_5	Zn_3Au
	CdAu		Cd_8Au_5		Cd_3Au

* The γ-phase occurring in the system Cu-Hg, with the composition CuHg, is an exception. There are, moreover, a few compounds which have, indeed, the composition required by the rule, but differ in structure—e.g., $AlAg_3$, $AlAu_3$ and $SiCu_5$ form structures of the β-manganese type (p. 24).

atoms over the lattice points occur simultaneously, the compositions demanded by the rule lie within the measured homogeneity range, or so close to it that the formula based on the rule can be regarded as the ideal formula of the compound. Table 1 gives a summary of a series of compounds which obey the Hume-Rothery rule.

As Westgren and Ekmann have shown (1931), the Hume-Rothery rule can be extended to the metals of the VIIIth Sub-group and manganese; the number of their valence electrons can be considered to be 0 in compounds with metals of type II. Thus AlMn, AlFe, AlCo, and AlNi form structures of the type of the β-phase, and the compounds Fe_5Zn_{21}, Co_5Zn_{21}, Ni_5Zn_{21}, Rh_5Zn_{21}, Pd_5Zn_{21}, Pt_5Zn_{21}, those of the γ-phase type.

(f) Other Rules for the Formation of Intermetallic Compounds

The strongly electropositive metals (alkali and alkaline earth metals) also frequently form compounds with the metals of the Sub-groups of the Periodic System, which correspond in composition and structure to β-brass (Zintl, 1933). The Hume-Rothery rule does not, however, hold for these compounds which, in every case, have an ordered arrangement of the atoms; the concentration of valence electrons (= ratio of total number of valence electrons to total number of atoms) in these compounds can vary considerably.

At high temperatures the metals of the first kind (Mn, Fe, Co, Ni, platinum metals—except Ru and Os—, Cu, Ag, Au) form only mixed crystals with one another. On cooling, these often form superstructure phases (Dehlinger).

The alloys which the metals of the second kind (especially Zn, Cd, Hg, Mg, Al, and Sn) form with each other occupy, according to Dehlinger (1935), an intermediate place between the alloys formed by metals of the 1st kind with each other, and the alloys of the strongly electro-positive metals with the metals of the 1st and 2nd kinds. The metals of the Sub-group II of the Periodic System form no compounds with each other, and are extensively miscible in the solid state only when the difference in atomic radius is small (Cd-Hg). Mg, on the other hand, forms compounds with them, having an ordered atomic distribution and narrow homogeneity range, if the difference in atomic radius is large (Mg-Zn) or if the chemical similarity is only slight (Mg-Hg). Otherwise if the difference in atomic radius is small, extended ranges of mixed crystals are formed, and superstructure phases appear on cooling (cf. Table 45, p. 426–7). Al and Sn behave like magnesium towards the metals of Sub-group II, except that they have a yet more marked tendency to form compounds, whereas the ability to form mixed crystals is weaker. As compared with the metals of Sub-group II, Mg, Al, and Sn possess large atomic radii and high polarizability. According to Dehlinger, both mutual miscibility and the formation of compounds by these metals of the second kind are determined by the atomic radius and the polarizability of the atoms. In alloys formed by metals of the first kind with one another, on the other hand, atomic radii are never critically important.

5. Structures of the Metals and of Intermetallic Compounds

The pure metals mostly have very simple crystal structures. They crystallize preferably in structures with closest packing of atoms, or in structures resembling those of closest packing. Of the 54 metals of the Main and Sub-groups of the Periodic System*, 14 (at ordinary temperature) form structures of the magnesium type (hexagonal closest packing); face-centered cubic structures (cubic closest packing) and body-centered cubic structures are each formed by 13, whereas 2 (Sb, Bi) crystallize in the antimony type, which can be regarded as a deformed cubic structure. Of the pure metals with known structures**, only 6 (U, Mn, Hg, Ga, In and β-Sn) form unique structures which are not observed for other elements,

* For crystal structures of the lanthanides see p. 487.
** The structures of Ra, Sc, Ac and Pa are still unknown.

though some of them are found among intermetallic compounds. The low temperature modification of tin (a-Sn), which does not display purely metallic character, and also germanium (for which the same is true) have the diamond structure.

Intermetallic compounds often have the same structures as are observed for pure metals, except that where the atoms are distributed in ordered fashion over the lattice points, superstructure forms of these lattices result. Crystallographically, these are to be considered as special lattice types, but if the differences between the different sorts of atoms are neglected, they are transformed into such structures as are found for the pure metals.

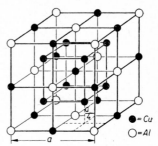

Fig. 8. Unit cell of AlCu$_3$ structure (metastable superstructure form of the β-phase of the Al-Cu system; obtained by quenching to 300° the disordered form of the β-phase, which is stable above 570°).

Thus β-brass, as already stated, forms a body-centered cubic structure—that is, a crystal lattice that differs structurally both from that of copper (face-centered cubic) and that of zinc (hexagonal close packed), but is observed for numerous other pure metals. In this, Cu and Zn atoms are distributed at random over the lattice points, so that the crystallographic symmetry of the crystal lattice is the same as if it were built up from only one sort of atom. If, however, the two sorts of atoms become distributed in an ordered manner over the lattice positions (as is the case in the β'-brass phase which forms on annealing), the body-centered cubic structure is transformed into a crystal lattice of cesium iodide type*. The compound AlCu$_3$, which also belongs to the β-brass type, forms the structure shown in Fig. 8 when the atoms are regularly ordered. It may be seen that this also becomes a body-centered cubic structure if the difference between the atoms is ignored, or if they are distributed at random.

The superstructure form of the body-centered cubic structure, shown in Fig. 8, has also been found in the compounds LaMg$_3$, CeMg$_3$, PrMg$_3$, in which the atoms are always distributed in regular order over the lattice positions (Rossi, 1934). These compounds do not belong to the β-brass group, however, since the valence electron concentration is not the determining factor in their formation.

A structure which was first discovered for the compound NaTl, by Zintl (1932), can also be considered as a superstructure of the body-centered cubic lattice (cf. Fig. 9). A series of alkali compounds (LiZn, LiCd, LiAl, LiGa, LiIn, NaIn, NaTl) crystallizes according to this type. Formation of these compounds is also not determined by the valence electron concentration, so that they do not belong to the β-brass group. In the crystal lattice of the NaTl type the atoms are always regularly distributed over the lattice positions, this occurring in such a way that each sort of atom, considered by itself, occupies the positions of the diamond structure.

Structures such as those last mentioned, which can be derived from another structure through the ordered occupation of a proportion of the lattice points by another kind of atom, are called 'superstructure types', because in their X-ray diffraction patterns, additional lines (superstructure lines) are superimposed on the lines of the basic structure. The superstructure lines arise from the fact that the lattice points occupied by atoms with different numbers of electrons reflect or scatter X-rays with different intensity. This leads to the appearance of additional interferences, for the same reason that the inclusion of ions with the same number of electrons in a crystal lattice, instead of atoms with different electron numbers, leads to the disappearance of X-ray interference lines, as was shown in Vol. I, p. 213.

The structure of CaSn$_3$, illustrated in Fig. 108, (p. 577, Vol. I), and of a series of other

---

* It can readily be seen from Fig. 48, p. 211 (Vol. I) that the CsI structure passes into a body-centered cubic structure if the distinction between the atoms disappears.

compounds of analogous composition, is derived from the cubic face-centered lattice as basic type. The same lattice appears in the superstructure phases of the mixed crystals of metals crystallizing in the face-centered cubic structure (Examples, [AuCu$_3$], [PdCu$_3$], [PtCu$_3$]).

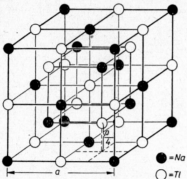

The structures of γ-brass, and of the alloy phases corresponding to it, are very similar to that of the modification of manganese stable at the ordinary temperature (a-manganese structure). The a-manganese structure is closely related to the body-centered cubic lattice, but has its atoms less closely packed, and less symmetrically arranged than in the latter. The unit cell of the a-manganese structure is made up of a cube containing no less than 58 atoms ('giant cell'). According to Laves (1934), the compound Mg$_{17}$Al$_{12}$ has the same structure. The unit cell of γ-brass, and of the other γ-phases cited on p. 21, is not quite identical, but is quite similar, and contains 52 atoms. In the majority of these γ-phases, the atoms are distributed over the lattice points in an ordered manner. Difference in symmetry between compounds of different composition can arise from this cause; however, if the differences between the kinds of atoms are overlooked, the structures are identical.

● = Na
○ = Tl

Fig. 9. Unit cell of NaTl ($a = 7.742$ Å). The following phases are isostructural with NaTl:—

NaIn ($a = 7.30$ Å)
LiIn ($a = 6.79$ Å)
LiGa ($a = 6.19$ Å)
LiAl ($a = 6.36$ Å)
LiCd ($a = 6.69$ Å)
LiZn ($a = 6.21$ Å)

A few compounds which, on the basis of the Hume-Rothery rule, ought to form structures of the type of β-brass (ZnCu), actually crystallize in the β-manganese type. This possesses a structure related to the face-centered cubic structure, containing 20 atoms per unit cell. The compounds crystallizing in the β-manganese type (AlAu$_3$, SiCu$_5$, Zn$_3$Co) have narrow ranges of homogeneity and a random distribution of the atoms.

Cubic giant cells with 112 atoms are formed by the compounds NaZn$_{13}$, KZn$_{13}$, KCd$_{13}$, RbCd$_{13}$ (Ketelaar 1937, Zintl 1938). The high coordination number [$_{24}$] of cadmium or zinc atoms around the alkali metals in these structures is noteworthy.

A structural type which represents a combination of hexagonal close packing with cubic close packing, as shown in Fig. 10, was discovered by Laves (1939) for the compound Ni$_3$Ti. Fig. 10 shows not a single unit cell of the cubic closest packing (a face-centered cube), but a section from the crystal lattice which brings out the relationship to the hexagonal closest packing. The close packed structures differ in that the centers of the atoms in cubic close packing lie exactly above those of every third layer, and not of alternate layers as in hexagonal close packing. This difference can be symbolized by the schemes ACACAC and ABCABC, respectively; the corresponding scheme for the Ni$_3$Ti structure is then ABACABAC.

Intermetallic compounds of the general formula AB$_2$ very frequently crystallize in crystal lattices which are characterized by the following structural plan: every A atom is equidistant, or nearly equidistant, from 4 other A atoms, and each B atom from six other B atoms. Each A atom is surrounded by 12 B atoms at equal, or nearly equal, distances, and each B atom by 6 A atoms. Crystalline species built up on this model are termed *Laves-phases*, since they have been principally investigated by Laves (1934 and onwards). Three types of Laves phases are known, which may be represented by the compounds MgZn$_2$, MgCu$_2$, and MgNi$_2$ (cf. Table 2). The MgZn$_2$ structure is derived from the crystal lattice of zinc (hexagonal closest packing) in such a way that groups, each of 7 Zn atoms forming two tetrahedra with a common vertex, are removed, and the center point of each of these tetrahedra is replaced by a magnesium atom. The Mg atoms then build up among themselves a lattice of mercury or wurtzite type. The MgCu$_2$ structure is derived in a similar way from the crystal lattice of copper (cubic close packing). It is obtained from this by removing half the tetrahedra of copper atoms, (of which one can consider the copper lattice to be built up), and placing Mg atoms at the centers of the missing tetrahedra. The Mg atoms, by themselves, then form a lattice of zinc blende or diamond type. The MgNi$_2$

structure can be regarded as a combination of the $MgZn_2$ structure with the $MgCu_2$ structure. In the Laves structures, both A atoms and B atoms taken by themselves form a

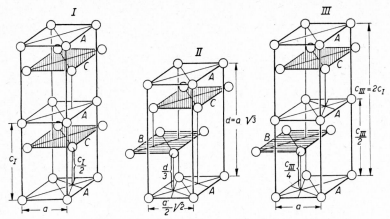

Fig. 10. The three simplest forms of closest packing.

I. Hexagonal closest packing (magnesium type).
II. Cubic closest packing (face-centered cubic structure, argon or copper type).
III. Ni_3Ti type.

I represents a doubled unit cell of the hexagonal close packed structure, with cell dimensions a and c (cf. Vol. I, p. 249, Fig. 57). Ideally, $c = \dfrac{2\sqrt{6}}{3} a = 1.633\, a$.

The lattice planes of the face-centered cubic structure shown in II are perpendicular to the body diagonal d of the face-centered unit cell of side a (cf. Vol. I, p. 210, Fig. 46).

The structure shown in III can be regarded as a mixture of hexagonal and cubic packing. If the three possible positions of the lattice planes are denoted by A, B, C, the sequence in hexagonal closest packing is ACACAC (identical with ABABAB, as may be seen by shifting the coordinate axis by $c/2$). The sequence in cubic closest packing is ABCABC, and in the hexagonal structure of Ni_3Ti it is ABACABAC. For atoms of a given size, the unit cell of III has its c axis twice as long as in the hexagonal close packed structure I.

continuous packing of spheres; however, if the atoms are regarded as spherical there is no contact between an A atom and a B atom at any point. This demonstrates that the affinity between atoms of the same kind plays at least as important a role in the formation of the Laves-compounds as does the affinity between A and B atoms. The ratio of the atomic radii

TABLE 2

LAVES-PHASES

$MgZn_2$-Type		$MgCu_2$-Type		$MgNi_2$-Type
$MgZn_2$	$MnBe_2$	$NaAu_2$	$AgBe_2$	$MgNi_2$
$CaMg_2$	$ReBe_2$	KBi_2	$MgZnNi$	$TiCo_2$
$TiMn_2$	$FeBe_2$	$MgCu_2$	$MgZn_{1.3}Co_{0.7}$	$MgAlCu$
$TiFe_2$	$MgAlCu$	$CaAl_2$	$MgZn_{1.2}Ag_{0.8}$	$MgSi_{0.5}Cu_{1.5}$
VBe_2	$MgAl_{1.1}Ag_{0.9}$	$PbAu_2$	$MgSi_{0.2}Ni_{1.8}$	$MgZn_{1.6}Ag_{0.4}$
$CrBe_2$	$MgSi_{0.5}Cu_{1.5}$	$BiAu_2$	$(Fe_{0.5}Be_{0.5})Be_2$	$Mg(Zn,Cu)_2$
$MoBe_2$		$TiBe_2$	$(Pd_{0.5}Be_{0.5})Be_2$	
WBe_2		$TiCo_2$	$(Cu_{0.9}Be_{0.1})Be_2$	
WFe_2		ZrW_2	$(Au_{0.5}Be_{0.5})Be_2$	

is also a significant factor. According to Dehlinger and Schulze (1939), a Laves phase will be formed between an A atom with very strong tendency for metallic binding and a B atom with an incomplete penultimate electron shell* and with a radius about 25% smaller than that of the A atom.

6. Structures of Quasi-Metallic Compounds of Non-Metals

Compounds of the non-metals B, C, N, and H with the transition metals generally have a quasi-metallic character (high conductivity, in some cases even superconductivity). If the ratio of the atomic radii of non-metal** and metal is small enough (< 0.59), regular structures are formed, as shown by Hägg (1931), which are derived either from one of the three types (face-centered cubic, body-centered cubic, hexagonal close packed) most commonly found for the pure metals or, less often, from a simple hexagonal lattice with the axial ratio $c/a \sim 1 : 1$ (see Fig. 11) which is not observed for the pure metals: the non-metal atoms are inserted into interstices of the lattice. The compounds so resulting correspond in almost every case to one of the following formulas: M_4X; M_2X; MX; MX_2. The compounds M_4X and M_2X tend to have wide, the compounds MX and MX_2 narrow ranges of homogeneity. The non-metal atoms X are usually incorporated as *pairs* in the crystal lattice of the compounds MX_2.

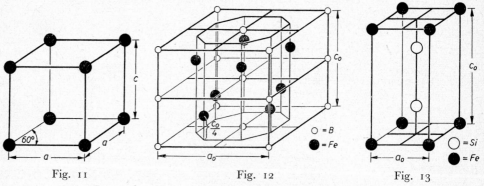

Fig. 11   Fig. 12   Fig. 13

Fig. 11. Simple hexagonal cell.

Example: Tungsten with 50 atom-% carbon forms a simple hexagonal lattice ($a = 2.90$, $c = 2.83$ Å), in which the C atoms are inserted—probably occupying the lattice points with

$$\text{coordinates } \frac{a}{3}, \frac{2a}{3}, \frac{c}{2}.$$

Fig. 12. Unit cell of the compound Fe_2B ($a_0 = 5.10$, $c_0 = 4.24$ Å). Isostructural with Fe_2B are Co_2B ($a_0 = 5.01$, $c_0 = 4.21$ Å), Ni_2B ($a_0 = 4.98$, $c_0 = 4.24$ Å) and Al_2Cu ($a_0 = 6.05$, $c_0 = 4.88$ Å).

Fig. 13. Unit cell of the compound $\overline{FeSi_2}$ ($a_0 = 2.68$, $c_0 = 5.12$ Å). The composition of the compound may vary between 68.8 and 72.1 atom-% Si. It therefore always contains an excess of Si atoms, which are incorporated in the crystal lattice at random, in place of Fe atoms.

---

* By an electron shell is meant here a group of electrons with the same principal quantum number n, and the same azimuthal quantum number l (or Bohr subsidiary quantum number k). It must be borne in mind that outer shells that are complete in the *atomic* state, are frequently resolved in the *metallic* state. Thus the ns shell of Be and Mg is split up into an ns and a np level, each with one electron (cf. here Vol. I, p. 239 *et seq.*).

** The non-metals cited have the following atomic radii in the compounds with the metals: B 0.97 Å, C 0.77 Å, N 0.71 Å, H 0.45–0.46 Å.

If the atomic radius ratio X : M > 0.59, only small amounts of the non-metal can be inserted into the interstices of the lattice. With higher contents of non-metal, quite varied structural types result, in some cases with complicated structures. In Figs. 12 and 13 are shown two compounds of this kind (Fe_2B and $\overline{FeSi_2}$). Al_2Cu, Co_2B and Ni_2B have structures like Fe_2B. The compound $\overline{FeSi_2}$ has a homogeneity range from 68.8 to 72.1 atom-% Si, and thus always has a higher Si-content than corresponds to the ideal formula $FeSi_2$, deduced from the structure. This is explained on the assumption that the compound is only stable when a proportion of the Fe atoms in the crystal lattice, randomly distributed, is replaced by Si atoms. The varying composition is explained by the greater or lesser extent of replacement of Fe atoms by Si atoms.

7. Improvement of Alloys

The mechanical properties of metals and alloys can be improved by mechanical working (hammering, rolling, wire-drawing, etc.), as has already been discussed (Vol. I, p. 570 *et seq.*). In addition to toughening by plastic deformation in the cold (*cold working*) it is very important that a considerable increase in hardness and strength can be brought about with certain alloys by heat treatment. This involves quenching from high temperatures, followed by an annealing at low temperatures, or by standing at ordinary temperature, and is known as precipitation hardening or age hardening. This hardening process, which was discovered by Wilm in 1911 for an aluminum-copper alloy, is distinguished from the long known hardening of steel (cf. p. 268), in that the considerable increase in hardness and strength is not obtained immediately on quenching, but only after standing or on moderate heating (annealing) of the quenched alloy*.

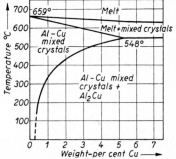

Fig. 14. Limiting composition of aluminum crystals containing copper, as function of temperature.

The oldest and best known representative of the age-hardening alloys is Duralumin (cf. Vol. I, p. 348). It has been found, however, that there are many age-hardening alloys. Alloys can always be age-hardened if they consist of mixed crystals which are nearly saturated at high temperatures, and which can be brought into the condition of supersaturated mixed crystals by quenching. The age-hardening depends on the fact that decomposition of these mixed crystals sets in on aging, with the result that a second crystalline phase separates out in a highly disperse state**. It is assumed that the increase in hardness, and the other properties connected with this, are produced by the internal stresses which arise within the crystal lattice of the mixed crystal, through the formation of the new crystal phase in numerous, extremely minute, individual regions.

* Related to precipitation hardening is the process known in the steel industry as tempering. See p. 268 for further information on this point.

** It seems that the hardening is caused not so much by the actual decomposition of the mixed crystals, as rather by the state which immediately precedes this. In this state, at many points in the lattice of the mixed crystals, the atoms go over to the position proper to the second crystal phase,—at first within small domains. Once decomposition has proceeded so far that the second kind of crystal can be detected by X-rays, the hardness decreases again.

The case in which miscibility in the solid state diminishes with fall of temperature is very common. This decrease of miscibility comes particularly strongly in evidence in the system Al–Cu, as Fig. 14 shows. Aluminum alloyed with copper, accordingly displays a particularly marked increase in hardness. The same is true of copper containing beryllium.

The requisite conditions for age-hardening properties are also fulfilled if an alloy of some particular composition consists of an intermetallic phase, stable only at high temperatures, which decomposes into two other phases on cooling, provided that this can be obtained in the metastable state at ordinary temperature by quenching. In this case also the decomposition can be carried out in such a way that the decomposition products appear in a highly disperse state, analogous to the age-hardening of a mixed crystal phase.

The tensile strength of alloys is augmented by age-hardening, just as is the hardness. The tensile strength of finely crystalline annealed pure aluminum is about 8 kg/mm². It can be raised to about 25 kg/mm² by cold working. Ordinary copper-containing aluminum (not age-hardened mixed crystals) has strengths up to 25 kg/mm², and after cold working up to 40 kg/mm². By age-hardening, its strength can be increased to over 50 kg/mm², and after cold working to 60–70 kg/mm². The tensile strength of an age-hardened beryllium-copper alloy with 2.5% Be is no less than 130–150 kg/mm², and thus reaches a figure which is reached or exceeded only by the best types of steel.

References

1 A. E. VAN ARKEL (Editor), *Reine Metalle; Herstellung, Eigenschaften, Verwendung*, Berlin 1939, 574 pp.

2 C. J. SMITHELLS, *Impurities in Metals*, London 1930, 190 pp.

3 H. FUNK, *Darstellung der Metalle im Laboratorium*, Stuttgart 1938, 183 pp.

4 F. SAUERWALD, *Physikalische Chemie der metallurgischen Reaktionen, ein Leitfaden der theoretischen Hüttenkunde*, Berlin 1930, 142 pp.

5 L. QUEVRON and L. OUDINÉ, *Cours de Métallurgie*, Paris 1938, 320 pp.

6 A. JACQUET and D. TOMBECK, *Eléments de Métallurgie*, Paris 1934, 263 pp.

7 C. G. JOHNSON, R. S. DEAN, M. S. MET and J. L. GREGGS, *Metallurgy*, Chicago 1938, 149 pp.

8 E. L. RHEAD, *Metallurgy*, London, 1935, 382 pp.

9 E. GREGORY, *Metallurgy*, London 1932, 284 pp.

10 D. M. LIDDELL and G. E. DOAN, *The Principles of Metallurgy*, New York 1932, 626 pp.

11 L. LOSANA, *Lezioni di Metallurgia*, Turin 1941, 470 pp.

12 M. REY, *Equilibres chimiques et Métallurgie*, Paris 1940, 259 pp.

13 D. M. LIDDELL (Editor), *Handbook of Nonferrous Metallurgy*, 2 vols., New York and London 1945.

14 R. MÜLLER, *Allgemeine und technische Elektrometallurgie*, Vienna 1932, 580 pp.

15 J. BILLITER, *Elektrometallurgie wässeriger Lösungen*, 2nd. Ed., Halle 1923, 244 pp.

16 V. ENGELHARDT (Editor), *Die technische Elektrolyse wässeriger Lösungen* (Vol. I and II of *Handbuch der technischen Elektrochemie*), Vol. I, Leipzig 1931, 613 + 331 + 448 pp.; Vol. II, Leipzig 1933, 451 + 328 pp.

17 *Idem*, Vol. III, *Die technische Elektrolyse im Schmelzfluss*, Leipzig 1934, 565 pp.

18 F. SKAUPY, *Metallkeramik; die Herstellung von Metallkörpern aus Metallpulvern; Sintermetallkunde und Metallpulverkunde*, 3rd Ed., Berlin 1943, 250 pp.

19 R. KIEFFER and W. HOTOP, *Pulvermetallurgie und Sinterwerkstoffe*, 2nd Ed., Vienna 1948, 412 pp.

20 F. BENESOWSKY (Editor), *Pulvermetallurgie*, Vienna 1953, 320 pp.

21 W. GUERTLER, *Metall-technisches Taschenbuch*, Leipzig 1939, 370 pp.

22 R. MITSCHE and M. NIESSNER, *Angewandte Metallographie*, Leipzig 1939, 229 pp.

23 R. GLOCKER, *Materialprüfung mit Röntgenstrahlen*, 2nd Ed., Berlin 1936, 386 pp.

24 E. SCHIEBOLD, *Spannungsmessung an Werkstücken*, Leipzig 1938, 216 pp.

25 H. O'NEILL, *The Hardness of Metals and Its Measurement*, London 1934, 292 pp.

26 H. FRÖHLICH, *Elektronentheorie der Metalle*, Berlin 1936, 386 pp.

27 A. H. WILSON, *The Theory of Metals*, Cambridge 1936, 272 pp.

28 W. Hume-Rothery, *The Metallic State; Electrical Properties and Theory*, London 1931, 372 pp.

29 L. Brillouin, *Les Statistiques Quantiques et leurs Applications*, Paris 1930, 404 pp.; translation into German by E. Rabinowitsch, *Die Quantenstatistik und ihre Anwendung auf die Elektronentheorie der Metalle*, Berlin 1931, 530 pp.

30 L. H. Darken and W. H. Gurry, *Physical Chemistry of Metals*, New York 1953, 535 pp.

31 K. Steiner and P. Grassmann, *Supraleitung*, Braunschweig 1937, 139 pp.

32 G. Masing, *Grundlagen der Metallkunde in anschaulicher Darstellung*, 2nd Ed., Berlin 1941, 138 pp.

33 F. Sauerwald, *Lehrbuch der Metallkunde des Eisens und der Nichteisenmetalle*, Berlin 1929, 452 pp.

34 U. Dehlinger, *Chemische Physik der Metalle und Legierungen*, Leipzig 1939, 174 pp.

35 G. Masing (Editor), *Handbuch der Metallphysik*, Leipzig 1935–41, Vol. I, 520 pp., Vol. II, 737 pp., Vol. III, 228+538 pp. (incompl.).

36 W. Hume-Rothery, *The Structure of Metals and Alloys*, London 1936, 120 pp.

37 N. F. Mott and H. Jones, *The Theory of the Properties of Metals and Alloys*, Oxford 1936, 326 pp.

38 K. E. Schwarz, *Elektrolytische Wanderung in flüssigen und festen Metallen*, Leipzig 1940, 95 pp.

39 R. Kremann, *Elektrolyse geschmolzener Legierungen*, Stuttgart 1926, 62 pp.

40 G. Tammann, *Die chemischen und galvanischen Eigenschaften von Mischkristallreihen und ihre Atomverteilung*, Leipzig 1919, 239 pp.

CHAPTER 2

THIRD SUB-GROUP OF THE PERIODIC SYSTEM;
SCANDIUM, YTTRIUM, LANTHANUM, AND ACTINIUM

Atomic numbers	Elements	Symbols	Atomic weights	Densities	Melting points	Specific heats	Valence states
21	Scandium	Sc	44.96	3.1	ca. 1400°		III
39	Yttrium	Y	88.92	4.34	ca. 1500°		III
57	Lanthanum	La	138.92	6.18	826°	0.045	III
89	Actinium	Ac	227				III

1. Introduction

a) General

The third Sub-group of the Periodic System contains three rare elements, *scandium, yttrium,* and *lanthanum,* together with a radioactive element, *actinium,* which occurs in Nature in extraordinarily minute quantities.

Except that their valency is higher by one unit, the elements of Sub-group III bear a close resemblance in properties to the elements of the IInd Main Group, that immediately precede them (calcium, strontium, barium and radium respectively). They resemble these elements, indeed, more closely than they do the elements of the third Main Group.

The atomic cores remaining after ionization of the three valence electrons for the IIIrd Sub-group elements are like those formed from the Main Group II elements, except that they bear a charge one unit greater, and that the higher nuclear charges bring about a stronger contraction of the electron shells. The ions Sc^{3+}, Y^{3+}, La^{3+}, and Ac^{3+}, like Ca^{2+}, Sr^{2+}, Ba^{2+} and Ra^{2+} have the 'inert gas' configuration, whereas the ions Ga^{3+}, In^{3+} and Tl^{3+} differ from the inert gas atoms, and therefore also from Al^{3+}, in that they contain electrons in d-levels. The filling up of the d-levels is accompanied by a contraction of the atoms, with the result that there is a smaller difference in radius between the Al^{3+} ion and the Ga^{3+} or In^{3+} ions than between the Al^{3+} ion and the ions of scandium and its homologues. In the *neutral atoms*, the configuration of the outer electron shells is the same for gallium, indium, and thallium as it is for boron and aluminum (s^2p), whereas scandium and its homologues have the configuration ds^2. These various influences act in opposite directions, with the result that—depending on the properties invoked in making the comparison—either the aluminum homologues in the Main Group, or those in the Sub-group, may seem to bear the closer relation to aluminum itself.

Like boron and aluminum, the elements of the third Sub-group are always trivalent in their normal compounds. They have a very strong tendency to form salt-like compounds. The electropositive character increases continuously from aluminum to lanthanum and actinium.

According to Neumann's measurements, the decomposition potential of lanthanum chloride at 800° is about 0.31 volts greater than that of aluminum chloride. The temperature coefficient is relatively high for lanthanum chloride, so that at room temperature the difference should be still greater.

Like all the elements of the Sub-groups, the elements of Sub-group III are purely metallic in character. They all form solid, very involatile and very insoluble oxides. These oxides are purely basic in character. In this respect they differ to a considerable extent from the typically amphoteric oxide of aluminum. Their basic properties are far more strongly developed than are those of the homologues of aluminum in the Main Group: gallium oxide and indium oxide, like aluminum oxide, are amphoteric. There is a distinct increase in basic strength from scandium oxide to lanthanum oxide; the latter is only slightly inferior to calcium oxide in its basic character.

The infusible metal oxides were formerly known generally as 'earths', by analogy with aluminum oxide (German, Tonerde; clay earth). This name has been retained for the oxides of scandium, yttrium, and lanthanum, which are especially typical of the 'earths'. The metals from which these typical earths are derived are conveniently known as the 'earth metals'.

Immediately following lanthanum in the Periodic System is a long series of elements which all display a far-reaching similarity in chemical character to lanthanum itself. These are known as the *lanthanides* (this name was proposed by Goldschmidt), and as has already been mentioned, they constitute a special series of elements within the Periodic System; this group will be discussed in Chap. 10. The oxides of scandium, yttrium, and lanthanum, together with those of the lanthanides, are all grouped together under the name of the *rare earths*, because they are found in Nature in substantial amounts only in very few places. A characteristic of the rare earths is that a large number of the group invariably occur associated with each other in Nature. Scandium is the only one which also occurs in minerals free from the other rare earths, or at least with a low content of the other members. Scandium definitely occupies a unique position with respect to the other rare earth elements; this arises from its closer resemblence to aluminum.

The electropositive trivalence of the metals of the IIIrd Sub-group can be explained in the same way as that of aluminum, namely by their position with respect to the inert gases. As neutral atoms, they all contain 3 electrons more than the preceding inert gas. Although one of these is bound differently from that in aluminum (as a d electron instead of a p electron), it is nevertheless relatively easily removed by ionization, like the two s electrons. As is shown in Table 3, the energy which must be expended in the case of scandium to ionize off a fourth electron is considerably greater than the sum of the energies of ionization of the three most loosely bound electrons.

The strongly electropositive character of the IIIrd Sub-group metals is shown, for example, in their high affinity for oxygen, and quite generally by the high heats of formation of their salt-like compounds (cf. Table 4). Heats of formation, in so

TABLE 3

IONIZATION ENERGIES FOR THE ELEMENTS OF THE IIIRD SUB-GROUP

		Scandium	Yttrium	Lanthanum
I	$M \rightarrow M^+ + e$	154	150	129 kcal/g atom
II	$M^+ \rightarrow M^{2+} + e$	295	283.7	262.5 ,,
III	$M^{2+} \rightarrow M^{3+} + e$	571.8	470	440.1 ,,
Sum I + II + III		1021	904	832 kcal/g atom
	$M^{3+} \rightarrow M^{4+} + e$	1697		kcal/g atom

far as they are at present known, are greater than those of the corresponding compounds of aluminum—in some cases considerably greater. The strongly electropositive character is also evident, in the case of lanthanum, by its behavior towards hydrogen, which combines with this metal, as with the alkaline earth metals, to form a solid *salt-like hydride*. Lanthanum hydride differs from the alkaline earth hydrides, however, in that its hydrogen content varies with the temperature and pressure, so that the formula LaH_3 represents only the limiting composition reached when the uptake of hydrogen is a maximum. Lanthanum hydride thus forms a transition towards the quasi-metallic hydrides, having the properties of solid solutions, which are encountered in the subsequent Sub-groups of the Periodic System. (There is as yet no detailed knowledge of the hydrides of Sc, Y, and Ac.)

TABLE 4

HEATS OF FORMATION OF COMPOUNDS OF ELEMENTS OF THE IIIRD SUB-GROUP
(in kcal per g-equivalent)

Element	Oxide	Fluoride	Chloride	Bromide	Iodide	Sulfide	Nitride
Scandium	—	—	73.6	63.5	—	—	25.0
Yttrium	—	—	78.3	—	47.7	—	23.8
Lanthanum	89.8	—	87.9	—	55.8	49	24.2

The carbides of the elements of Sub-group III (which are obtained by heating the oxides with carbon in an electric furnace) do not correspond in composition to aluminum carbide, but to the carbides of the alkaline earth metals: they are *acetylides*, of the composition MC_2, in which the metals are present in the electropositive *bivalent* state*.

Scandium and its homologues have only a very slight tendency to form *covalent* compounds, and especially compounds with organic radicals. It has been reported that alkyl compounds of these metals can be obtained, in the form of addition compounds with ether, by the reaction of the anhydrous chlorides with Grignard reagents (Pletz, 1938). The acetylacetonates—e.g., $Sc(C_5H_7O_2)_3$ (m.p. 187°, sublimes at 200° in a vacuum) can probably also be considered as covalent compounds. This poor capacity for forming typical organometallic compounds (alkyl and aryl compounds) shows itself not only in the elements

* Direct experimental proof of this has been obtained only for the compounds YC_2 and LaC_2. It is not known whether scandium forms an acetylide or a carbide of some other type.

of Sub-group III, but right through the other transition elements up to and including Sub-group VIII. Chromium is the only transition element for which such compounds can be readily obtained. (See p. 345 for the corresponding compounds of platinum.) In contrast with this behavior, alkyl and aryl compounds are known for all the elements of Sub-groups I and II. Organometallic compounds of the Sub-group II metals are formed very easily indeed.

(b) Crystal Structure of the Metals

In the metallic state, the elements of Sub-group III crystallize with the hexagonal closest packed structure (as do the metals of Sub-group IV). Sc and La can also crystallize in cubic closest packing (face-centered cubic structure). The face-centered cubic form of lanthanum is the stable modification at high temperatures. Table 5 shows the lattice constants, densities calculated therefrom, and apparent atomic radii of the elements. The apparent ionic radii are shown also. Densities as determined pycnometrically are given in the chapter-heading Table, p. 30.

TABLE 5

CELL DIMENSIONS, DENSITIES, ATOMIC RADII AND IONIC RADII OF ELEMENTS
OF SUB-GROUP III

	α-Sc	β-Sc	Y	α-La	β-La	Ac
Cell dimensions, Å	$a = 3.30$ $c = 5.24$	$a = 4.53$	$a = 3.63$ $c = 5.75$	$a = 3.74$ $c = 6.06$	$a = 5.30$	—
Density (X-ray)	3.02	3.20	4.47	6.19	6.19	—
Atomic radius r (C.N. = 12)	1.62 Å	1.60 Å	1.80 Å	1.86 Å	1.87 Å	
Radius of M^{3+} ion		0.83 Å		1.06 Å	1.04 Å	1.1 Å

(c) Alloys

Lanthanum is the only member of Sub-group III for which the behavior towards other metals has as yet been extensively studied. It has a great capacity for forming alloys. The alloy systems almost invariably involve not only mixed crystals, but also compounds, which in many cases are numerous. It may be assumed that scandium, yttrium, and actinium will resemble lanthanum in this respect. The alloys of lanthanum, and the intermetallic compounds formed in the various systems, will be found summarized in Table 53, p. 489.

(d) Radioactivity

Actinium occupies a special position among the elements of Sub-group III—not on account of its chemical properties, but because it is a strongly radioactive element, is therefore quite short lived, and is very difficult to obtain in weighable quantities. For this reason, the elements scandium, yttrium, and lanthanum will first be discussed in the following pages. Actinium will then be considered by itself.

2. Scandium (Sc); Yttrium (Y); Lanthanum (La)

(a) Occurrence

Scandium, yttrium and *lanthanum*, together with a whole series of other elements of very similar chemical behaviour, occur in a few minerals which are found in large amounts in only a few places—chiefly in the Scandinavian peninsula and also (in the European region) in the Urals; outside Europe there are important occurrences in the United States, in South America (chiefly in Brazil), in India, and in a few districts in Australia. Scandium, yttrium, and lanthanum, together with the closely similar elements, make up the series of *Rare Earth elements*. The oxides of these elements, the *rare earths*, are subdivided into two groups. One of these is named the *yttria earth sub-group*, after the mineral *ytterbite* or *gadolinite*, first discovered at Ytterby in Sweden, in which the oxides of this sub-group predominate. The other is known as the *cerite earth sub-group*, after the mineral *cerite*, which contains principally these oxides. Scandium and yttrium oxides belong to the yttria earth sub-group, whereas lanthanum oxide belongs to the cerite earths.

In addition to lanthanum oxide, the cerite earths comprise the oxides of cerium, praseodymium, neodymium and samarium. To the group of yttria earths belong not only the oxides of scandium and yttrium but also the ytterbium earths, the erbium earths and the terbium earths. These in turn are mixtures of oxides which are remarkably alike chemically.

Scandium generally occurs in quite minor amounts in the typical minerals of the rare earths. Only *one* mineral rich in scandium has been discovered. This is the Norwegian *thortveitite*, $Sc_2Si_2O_7$, which has also been found in Madagascar. According to Urbain (1922) this contains only quite small amounts of the other rare earths.

Yttrium occurs as the principal constituent not only in *gadolinite*, $Be_2Y_2FeSi_2O_{10}$, but also especially in *xenotime* (*ytter spar*), YPO_4, and in *euxenite*, a niobate and tantalate of yttrium. In all these minerals yttrium is partially replaced by other rare earth elements, especially those of the yttria, earth group. *Lanthanum* is found especially in *cerite*, *orthite*, and *monazite*, though in subordinate amount as compared with cerium, the principal constituent of all these.

(b) History

In 1794, Johann Gadolin discovered a new 'earth' in a mineral, *ytterbite*, later re-named *gadolinite*, which had been found at Ytterby, in Sweden, in 1788. For this oxide, Ekeberg introduced the name of *yttria*. Soon afterwards (1803) Klaproth and, independently, Berzelius isolated from another Swedish mineral yet another new earth, which was named *ceria*. The mineral accordingly acquired the name of *cerite*. Mosander, one of the pupils of Berzelius, subsequently succeeded in separating both ceria and yttria into simpler oxides. The former he resolved into *lanthanum*, *cerium* and *didymium* oxides and from the latter he separated *terbia* and *erbia*, retaining the name *yttria* for the main constituent. Yttrium, the element giving rise to yttria, was first obtained in the free state (although impure) in 1828, by Wöhler, by reduction of the chloride with metallic sodium. *Lanthanum,*—so named by Mosander (after the Greek λανθάνειν, the hidden one) because it had been difficult to discover, for lack of specific reactions,—was isolated by the discoverer, by reducing the chloride with potassium. It was first obtained in large amounts in 1875, by Hillebrand and Norton, from the electrolysis of the molten chloride.

The preparation of large amounts of lanthanum and other rare earth elements by Hillebrand and Norton made it possible to determine their specific heats. These, on the basis of the Dulong and Petit law, decisively characterized the atomic weights and valence of the elements concerned. Because of similarities between their compounds and those of the alkaline earths and magnesium, it had been considered that the rare earth elements were bivalent, so that their oxides were formulated as MO. The atomic weights deduced from the specific heats could, however, be reconciled only with a formulation of the oxides as M_2O_3. The investigations of Cleve, on the isomorphism between their salts and those of the trivalent elements, especially bismuth, carried out at about the same time, led to the same conclusion; that the elements of the rare earths are trivalent.

The existence of an element with the properties of scandium had already been predicted in 1871 by Mendeléeff, from the Periodic Law. He called the expected, but as yet unknown homologue of boron and aluminum *eka-boron*. Eight years later (1879), Nilson separated a new oxide from Swedish gadolinite and euxenite, and gave the name *scandium* to the element contained in it. This proved to be identical with the eka-boron predicted by Mendeléeff.

The agreement between the properties predicted for ekaboron and those found for scandium may be seen from the following comparison.

Ekaboron	*Scandium*
Atomic weight about 44.	Atomic weight found 44.96.
Density greater than 3.	Density 3.1.
Oxide Ek_2O_3.	Oxide Sc_2O_3.
Oxide, carbonate, and phosphate insoluble in water and alkalis.	Oxide, carbonate and phosphate insoluble in both water and alkalis.
Sulfate sparingly soluble in water. It should form double sulfates not isomorphous with the alums.	The sulfate is readily soluble in water, but not very soluble in sulfuric acid. It forms double sulfates not isomorphous with the alums.

(c) Preparation and Properties

Little is as yet known of the properties of scandium and yttrium in the elementary state, because of the difficulty of preparing the metals pure. Lanthanum has been prepared in large amounts, and was first investigated by Muthmann (1902).

Muthmann used the electrolysis of the molten chloride to prepare lanthanum and other metals of the rare earths. During the process of fusion, he joined the carbon electrodes by a thin carbon rod, which was raised to a bright white heat by the passage of current. When the salt had been melted thereby, the rod was knocked away, and the current passing through the melt then sufficed to maintain it in the molten state. This process was later improved, notably by Trombe (1932), especially by the addition of suitable substances (KCl and CaF_2) which lower the melting point of the electrolyte. According to Weibke (1939) a current yield of 50% can be achieved by operating at 1000° with a sufficiently high current density.

According to B. S. Hopkins (1933), small amounts of lanthanum can be obtained by electrodeposition from an ethanolic solution of $LaCl_3$ at a mercury electrode, and subsequent thermal decomposition of the amalgam obtained. Another method (Klemm, 1937) is to decompose the chloride by liquid metallic potassium in an argon atmosphere.

The so-called 'lanthanum Mischmetall' is more readily obtained than pure lanthanum. The preparation of this starts with the residues from the rare earths worked up for the gas mantle industry, omitting the difficult separation of lanthanum from its congeners. For many technical purposes, 'mischmetall' is more suitable than lathanum itself, since it is, in general, more reactive than the latter.

According to Muthmann, lanthanum is a ductile metal, white on freshly polished surfaces, but at once showing tarnish colors when exposed to air. In dry air, it becomes covered with a steel-blue coating which protects it from further oxidation. In moist air, it is gradually transformed into the white hydroxide. The density of lanthanum is 6.15, and its melting point is 826° (Kremers, 1923). Its hardness is rather greater than that of tin. The heat of combustion per gram atom is greater than that of aluminum. Lanthanum combines energetically with molten aluminum, forming a beautifully crystalline compound $LaAl_4$, which is completely stable in air. It also alloys very readily with platinum. In a current of nitrogen or ammonia, the metal is converted to the nitride LaN, a black friable mass, readily decomposed by water to form ammonia. Lanthanum nitride is formed together with the oxide when the metal is burned in air, especially if the combustion is carried out at a relatively low temperature.

Lanthanum also combines energetically with hydrogen. According to Muthmann, lanthanum glows with a reddish yellow light when it is heated to 240° in hydrogen, and forms a black product having a composition corresponding roughly to LaH_3.

Sieverts (1923) found that lanthanum mischmetall (i.e., the mixture of lanthanum with the other metals of the rare earths, including about 10% Ce) reacts with hydrogen even at ordinary temperature, breaking up into grey-black lamellae. The product of reaction is not spontaneously inflammable in air. The heat of formation was found to be 384.6 cal per gram of lanthanum mischmetall (Sieverts, 1928). Taking the atomic weight of pure lanthanum as the average atomic weight of mischmetall, this corresponds to about 17.8 kcal per g-equivalent of 'lanthanum'. The heat of formation of the 'lanthanum hydride' is thus similar, in order of magnitude, to the heats of formation of the alkali and alkaline earth hydrides. The volume occupied by hydrogen in lanthanum hydride also agrees with that in those hydrides, as was pointed out by Biltz (1928). In these respects, therefore, lanthanum hydride is related directly to the alkali and alkaline earth hydrides. The composition of the product obtained by saturating lanthanum with hydrogen at room temperature is close to that required by the formula LaH_3. Sieverts found, however, that the hydrogen content of the preparations was dependent upon the pressure, and that for a given hydrogen content the pressure increased with rising temperature. In contrast with Muthmann, therefore, he considers that the hydride represents a *solid solution* phase. Lanthanum hydride undoubtedly represents an interesting transition between the salt-like hydrides formed by the alkali and alkaline earths metals and the typical metallic hydrides with all the characteristics of solid solutions, such as are formed in the Sub-groups of the Periodic System.

Yttrium was described by Popp (1864), who obtained it by reducing the chloride with sodium, as a grey-black metallic powder, which did not become oxidized, in dry air, but was slowly attacked by water. Kremers, who prepared it (1925) by electrolysis of a mixture of YCl_3 and NaCl in a graphite crucible clad with molybdenum sheet, states that it ignites at 470° when it is heated in air. It does not burn with so white a light as do magnesium and aluminum, the light emitted having a reddish tinge. It is readily soluble in dilute acids, but is not attacked by alkaline solvents.

Metallic scandium was first prepared in the pure state by Fischer (1937), by the electrolysis of a fused mixture of KCl, LiCl, and $ScCl_3$, using molten zinc as the cathode. The zinc was then subsequently removed by vacuum distillation. Scandium is light grey, with a metallic luster; it has a density of 3.1, melts at about 1400°, and is perceptibly volatile at that temperature.

3. Compounds of Scandium, Yttrium, and Lanthanum

The compounds of scandium, yttrium, and lanthanum correspond to the general type $M^{III}X_3$, so that they are analogous in composition to the compounds of aluminum. In their chemical behavior, however, they are in many respects much closer to the compounds of the alkaline earths. Similarity to these latter increases progressively from scandium to lanthanum compounds. This is apparent, for example, with the oxides and hydroxides which, unlike aluminum oxide, are insoluble in caustic alkalis and have a clearly marked basic character in spite of their insolubility in water. The oxides M_2O_3, white powders, combine with water to form the hydroxides $M(OH)_3$. Lanthanum oxide does this with such vigor that it may be slaked like quick lime. These metals also resemble the alkaline earth metals in the stability of their carbonates in contact with water at ordinary temperature; when heated, the carbonates are readily decomposed. The sulfates $M_2(SO_4)_3$ also begin to decompose when they are ignited. Their dissociation pressures at $900°$ are:

$$Sc_2(SO_4)_3 \text{ 11 mm}, \quad Y_2(SO_4)_3 \text{ 3 mm}, \quad La_2(SO_4)_3 \text{ 2 mm Hg}.$$

In the case of lanthanum sulfate it is difficult to achieve complete conversion into the oxide by ignition. The nitrates $M(NO_3)_3$ are easily decomposed by heating.

The chlorides, nitrates, and acetates of scandium, yttrium, and lanthanum are readily soluble in water. The fluorides, carbonates, oxalates, and phosphates, as well as the oxides and hydroxides, are insoluble. The sulfates display a marked decrease of solubility from the very soluble scandium sulfate to the sparingly soluble lanthanum sulfate. The sulfides of scandium, yttrium, and lanthanum, like those of the alkaline earth metals, can be obtained only by dry methods, and are decomposed when they are boiled in water. Many of the salts mentioned have a tendency to form double or complex salts with alkali salts. Other sorts of addition compounds are also formed—e.g., with ammonia. These elements also have a marked affinity for water in their salts, as is apparent in that not merely the very soluble salts, but also the sparingly soluble simple salts, crystallize for the most part in heavily hydrated form.

The aqueous solutions of the salts contain the colorless ions Sc^{+++}, Y^{+++} and La^{+++}. Aqueous solutions of scandium sulfate form an exception to this rule; these contain predominantly complex ions, possibly $[Sc(SO_4)_3]^{3-}$.

Scandium, yttrium, and lanthanum combine with certain non-metals to form compounds which do not conform in composition to the ordinary valence compound type, and are related rather to the intermetallic compounds. Examples of such compounds are YB_6 and LaB_6; $ScSi_2$, YSi_2, and $LaSi_2$; LaS_2 and $LaSe_2$ (which exist as well as La_2S_3 and La_2Se_3; the only known sulfides and selenides of Sc and Y are of the type M_2R_3).

Of the silicides, $LaSi_2$ is isotypic with $ThSi_2$. YSi_2 (and also $SmSi_2$) crystallizes in a related structural type. The diffraction patterns appear to indicate, however, that the crystal structure of $ScSi_2$ is quite different (Brauer, 1952).

(a) Oxides

The oxides Sc_2O_3, Y_2O_3, and La_2O_3 are obtained as snow white, loose powders by ignition of the hydroxides, carbonates, or oxalates. Whereas scandium oxide obtained by ignition can be dissolved only with difficulty in cold dilute acids, La_2O_3 dissolves readily in acids even when it has been strongly ignited. The densities of the oxides, from X-ray measurements of cell dimensions, are: Sc_2O_3 3.09; Y_2O_3 5.01; La_2O_3 6.48. Pycnometric determinations give rather different values, but this may be due chiefly to impurity of the preparations investigated.

The oxides Sc_2O_3 and Y_2O_3 crystallize in the cubic system. Their structure (Sc_2O_3 or Mn_2O_3 type) can be derived from that of CaF_2 by taking one quarter of the anions out of the crystal lattice, and displacing the remaining ions a little from their places, La_2O_3 forms the hexagonal crystal lattice, $a = 3.945$ Å, $c = 6.151$ Å, shown in Fig. 15.

Scandium oxide (and other scandium compounds) can best be prepared pure by a *solvent extraction* method (Fischer, *Z. anorg. Chem.*, 249 (1942), 146). A hydrochloric acid solution of the crude oxide is treated with NH_4SCN, and extracted with ether. The etheral solution, which contains scandium in the form of the thiocyanate, is

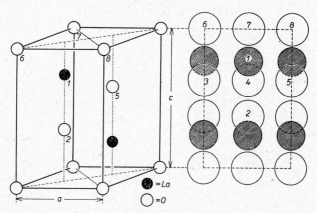

Fig. 15. Unit cell of lanthanum oxide, La_2O_3, with projection of the atomic positions on the a–c plane.

$$a = 3.945 \text{ Å}, \qquad c = 6.151 \text{ Å}, \qquad La \leftrightarrow O = 2.43 \text{ Å}.$$

evaporated, and the thiocyanate is decomposed by nitric acid. Scandium can then be precipitated in the form of hydroxide from the dilute nitrate solution. This procedure can be made applicable also to scandium preparations which contain iron.

(b) Hydroxides

The hydroxides $Sc(OH)_3$, $Y(OH)_3$ $La(OH)_3$, are obtained as slimy white precipitates when solutions of the corresponding salts are precipitated with alkali hydroxide or ammonia. They are insoluble in excess of the precipitant, and dry in air to masses with the appearance of porcelain, and compositions corresponding to the formulas given. Lanthanum hydroxide is a fairly strong base; it absorbs CO_2 avidly from the air, and liberates NH_3 from ammonium salts. Yttrium hydroxide possesses the same properties but to a lesser degree, and the basic character of scandium hydroxide is much more weakly developed. The hydroxides have been prepared in the crystalline state by Fricke (1947–48). When they are heated, they first form the metahydroxides $MO(OH)$; complete loss of water, forming the oxide, sets in only at higher temperatures.

(c) Peroxy Compounds

If lanthanum salts are precipitated with alkali in the presence of hydrogen peroxide, a compound is obtained which may be regarded as a basic lanthanum salt of hydrogen peroxide. Neglecting its water content, its composition corresponds to the formula

La(OH)$_2 \cdot$OOH (lanthanum dihydroxy hydrogen peroxide). Hydrogen peroxide is liberated from this by dilute sulfuric acid, or by carbonic acid. Yttrium forms an analogous compound.

(d) Halides

The fluorides of Sc, Y, and La are insoluble in water. The other halides are readily soluble in water and in alcohol.

(*i*) *Fluorides*. The *fluorides* are thrown down as precipitates when hydrofluoric acid is added to solutions of the corresponding salts. *Scandium fluoride* has a strong tendency to form complex fluorides (*fluoroscandates*), some of which are soluble, so that scandium is not precipitated from its solutions by neutral fluorides, such as NH$_4$F.

Most of the fluoroscandates belong to the type M$^{I}_3$[ScF$_6$], but compounds of the types M$^{I}_2$[ScF$_5$] and M^{I}[ScF$_4$] have also been described. The soluble *ammonium hexafluoroscandate*, (NH$_4$)$_3$[ScF$_6$], is deposited in octahedral crystals when a neutral solution of a scandium salt, treated with ammonium fluoride, is evaporated. Sc(OH)$_3$ is not precipitated from solutions of this complex salt by NH$_3 \cdot$H$_2$O even on boiling, but is precipitated by NaOH or KOH. The complex is also decomposed by acids. The K salt is less soluble than the NH$_4$ salt, and the Na salt still more sparingly soluble. *Scandium fluoride*, ScF$_3$, ($d = 2.5$) crystallizes with a rhombohedral structure, *yttrium fluoride*, YF$_3$ ($d = 4.01$) cubic, and *lanthanum fluoride*, LaF$_3$, ($d = 4.49$) hexagonal. YF$_3$ is remarkable in that it forms mixed crystals with CaF$_2$. These have the fluorite structure, in which the supernumerary F$^-$ ions are distributed at random over the interstitial positions. The double compound NaYF$_4$, is dimorphous. The high-temperature form (β-NaYF$_4$, $d^{25} = 3.87$) also crystallizes with the fluorite structure ($a = 5.45$ Å). In this case, the positions occupied by the metal ions are filled at random by Na$^+$ and Y^{3+} ions. a-NaYF$_4$ ($d^{25} = 3.87$) has a more complicated crystal structure. *Yttrium oxyfluoride*, YOF, is also dimorphous. β-YOF, which is metastable at ordinary temperature ($d^{25} = 5.18$) forms a fluorite structure in which the O$^=$ and F$^-$ ions are statistically distributed over the anionic lattice positions ($a = 5.363$ Å). a-YOF crystallizes in a structure which is related to this, but is tetragonally distorted (Hund, 1950–51).

(*ii*) *Chlorides*. The *chlorides* can be obtained anhydrous by heating the oxides, mixed with carbon, in a stream of chlorine. Lanthanum chloride is more readily obtained by heating anhydrous lanthanum oxide with an excess of ammonium chloride (Hopkins and Yntema); the volatilization of excess ammonium chloride serves to dehydrate the lanthanum chloride without hydrolysis occurring. The chlorides form white, deliquescent masses, with the densities ScCl$_3$ 2.39, YCl$_3$ 2.67, LaCl$_3$ 3.818, and melting points ScCl$_3$ 960°, YCl$_3$ 721°, LaCl$_3$ 862°. They crystallize as hydrates from aqueous solution—ScCl$_3$ and YCl$_3$ usually with 6H$_2$O, LaCl$_3$ with 7H$_2$O. The extent of hydrolysis in 0.1-N solution at 14° is: for ScCl$_3$ 0.9%, YCl$_3$ 0.01%, LaCl$_3$ 0.003%. Elimination of HCl also takes place readily if the chlorides are dehydrated by heating them in air. This occurs especially easily with scandium chloride, which is thereby converted into an insoluble oxychloride; this is resistant towards acids and alkalis.

The chlorides have only a small tendency to form double chlorides with the alkali metal chlorides. Meyer could obtain such a double chloride, in small white crystals, only with cesium chloride—Cs$_3$LaCl$_6 \cdot$ 4H$_2$O. The corresponding scandium compound could not be obtained pure because of its excessive solubility. A pyridinium double chloride of lanthanum, [C$_5$H$_5$NH]$_3$LaCl$_6 \cdot$ 2C$_2$H$_5$OH was obtained from alcoholic solution.

Anhydrous LaCl$_3$ crystallizes with a hexagonal structure, as also does LaBr$_3$. LaI$_3$ is orthorhombic. Melting points of the bromides and iodides are as follows:—

ScBr$_3$	YBr$_3$	LaBr$_3$	ScI$_3$	YI$_3$	LaI$_3$
960°	904°	763°	945°	1000°	761°

ScBr$_3$ has a density of 3.91.

(e) Nitrates

Scandium nitrate crystallizes when a solution of $Sc(OH)_3$ in dilute nitric acid is evaporated down on the water bath and cooled. It forms colorless, prismatic, very deliquescent crystals, of the composition $Sc(NO_3)_3 \cdot 4H_2O$, and can be dehydrated by heating it on a water bath. On heating somewhat more strongly, N_2O_5 is eliminated in stages, and the basic nitrates $Sc(OH)(NO_3)_2 \cdot H_2O$ and $ScO(NO_3)$ are formed. The pure oxide is obtained by gentle ignition. Scandium nitrate is readily soluble in water and alcohol. No double nitrates are known.

Yttrium and *lanthanum nitrates* crystallize as the hexahydrates, in large deliquescent prisms. They are also very soluble in water and alcohol. When they are heated, basic salts are first obtained, and then the oxides. Lanthanum nitrate forms well crystallized double salts of the type $M^I_2[La(NO_3)_5]$ with magnesium nitrate, and with alkali nitrates. The ammonium double nitrate, $(NH_4)_2[La(NO_3)_5] \cdot 4H_2O$ is very suitable for purifying lanthanum by fractional crystallization.

(f) Sulfates

When a solution of scandium hydroxide or carbonate in dilute sulfuric acid is evaporated to a syrupy consistency, *scandium sulfate* separates out on standing for some hours in the cold, as colorless crystals of the hexahydrate, $Sc_2(SO_4)_3 \cdot 6H_2O$. This compound is insoluble in absolute alcohol, but very soluble in water, (80 g of $Sc_2(SO_4)_3$ in 100 g of water at 12°, according to Crookes, 39.9 g in 100 g of water at 25° according to Wirth), and quite soluble in aqueous alcohol. It is not very soluble, however, in concentrated sulfuric acid. 1 molecule of water is lost on exposure to air, leaving $Sc_2(SO_4)_3 \cdot 5H_2O$, which is stable in contact with the solution at 25°. The anhydrous sulfate can be obtained by heating to 250°. Loss of SO_3, with the formation of a basic salt $Sc_2O(SO_4)_2$, sets in at a dull red heat. The pure oxide is obtained by ignition at a full yellow heat.

With the alkali sulfates, scandium sulfate forms complex salts—$M^I[Sc(SO_4)_2]$, $M^I_4[Sc_2(SO_4)_5]$ and $M^I_3[Sc(SO_4)_3]$. Solutions of the neutral sulfate also contain the scandium for the most part in the form of the complex ions corresponding to these salts. This follows from the abnormal analytical reactions of the sulfate solutions, from which scandium is only incompletely or slowly precipitated by the precipitants which are otherwise characteristic—e.g., sodium thiosulfate or oxalic acid. The comparatively low conductivity of the solutions is also indicative of auto-complex formation. Transport experiments have, in fact, shown that in solutions of its sulfate scandium migrates to a considerable extent towards the anode (Meyer). The complex acid $H_3[Sc(SO_4)_3]$ separates out from the solution of scandium sulfate in concentrated sulfuric acid.

Yttrium sulfate crystallizes from solutions of the oxide in sulfuric acid in colorless monoclinic crystals of the octahydrate, $Y_2(SO_4)_3 \cdot 8H_2O$, which is only moderately soluble. The solubility at 16° is 7.47 g and at 95°, 1.99 g of $Y_2(SO_4)_3$ in 100 g of water. It can easily be dehydrated to the anhydrous salt, $Y_2(SO_4)_3$, a white powder of density 2.52. The complex acid $H_3[Y(SO_4)_3]$ crystallizes from concentrated sulfuric acid solutions. Complex salts— e.g., $(NH_4)_4[Y_2(SO_4)_5]$, $K_8[Y_2(SO_4)_7]$, $K_6[Y_4(SO_4)_9]$—separate from mixed solutions of yttrium- and alkali sulfates.

Lanthanum sulfate generally forms an enneahydrate, $La_2(SO_4)_3 \cdot 9H_2O$, which is even less soluble than the hydrated sulfate of yttrium. Its solubility is: at 0° 3.0 g, at 14° 2.6 g, and at 100° 0.7 g of $La_2(SO_4)_3$ in 100 g of water. The anhydrous salt, a white hygroscopic powder of density 3.6, dissolves copiously in water, however, but as soon as the solution is warmed a little, the sparingly soluble hydrate separates out. Other hydrates of lanthanum sulfate have been obtained, in addition to the enneahydrate, but are stable only under certain conditions. Lanthanum sulfate forms double salts with the alkali sulfates, most of them being very sparingly soluble, especially when the alkali sulfates are present in excess. These are mostly of the types $M^I[La(SO_4)_2]$ and $M^I_3[La(SO_4)_3]$. The acid $H_3[La(SO_4)_3]$, corresponding to the latter type, has also been obtained.

(g) **Carbonates**

The neutral carbonates of Sc, Y, and La are obtained by precipitation from the appropriate solutions with alkali carbonate in the cold. If precipitation is carried out in hot solution, precipitates contaminated with basic salt are obtained. The neutral carbonates are hydrated. *Scandium carbonate*, $Sc_2(CO_3)_3 \cdot 12H_2O$, is relatively easily soluble in an excess of alkali carbonate solution. *Yttrium carbonate*, $Y_2(CO_3)_3 \cdot 3H_2O$, is also soluble in excess of alkali carbonate, and double salts of the type $M^I[Y(CO_3)_2]$ (+ crystal water) can be obtained from the solutions. *Lanthanum carbonate*, $La_2(CO)_3 \cdot 8H_2O$, occurs native as the rhombic *lanthanite*. This carbonate is also soluble in concentrated alkali carbonate, and on diluting the solution crystalline double carbonates—e.g., $K[La(CO_3)_2] \cdot 6H_2O$—separate. The carbonates lose CO_2 when heated, being converted first into basic carbonates, and then, on ignition, into the oxides. Loss of CO_2 from scandium carbonate occurs so readily that it cannot be thermally dehydrated without loss of CO_2.

(h) **Acetates**

The acetates of Sc, Y, La, obtained by the action of acetic acid on the several hydroxides or carbonates, are all quite soluble. Ammonia precipitates basic acetates from the solutions, and if the precipitation is carried out cautiously these may be obtained in colloidal form as gels, or also as sols. Basic lanthanum acetate, prepared in this way, has the property of turning blue with iodine. This is due to the formation of an adsorption product, analogous to the iodine-starch complex.*

(i) **Oxalates**

The oxalates of Sc, Y, and La are precipitated by adding oxalic acid to neutral or weakly acid solutions of the corresponding salts. The oxalates are formed as cheesy precipitates which generally become crystalline, with the compositions $Sc_2(C_2O_4)_3 \cdot 5H_2O$, $Y_2(C_2O_4)_3 \cdot 9H_2O$, $La_2(C_2O_4)_3 \cdot 10H_2O$. Their solubility in water decreases from scandium to lanthanum oxalate, and amounts to 0.20 mg of anhydrous salt in 100 g of water at 25°, in the case of the latter. The solubility is much greater in dilute strong acids—e.g., 100 ml of 1-N sulfuric acid at 25° dissolves 115 mg of scandium oxalate, 173 mg of yttrium oxalate, or 398 mg of lanthanum oxalate. The solubility in concentrated alkali oxalate solutions is also significant. Scandium oxalate is the most soluble in these, and double salts of the type $M^I_3[Sc(C_2O_4)_3] \cdot 5H_2O$ separate on cooling. Yttrium oxalate is less soluble, but also forms complexes—e.g., $K_3[Y(C_2O_4)_3] \cdot 9H_2O$. No oxalato-complexes of lanthanum are known; it therefore dissolves but little in hot ammonium oxalate solution, and separates unchanged on cooling. The differing solubility of the oxalates in a hot saturated solution of ammonium oxalate provides an important means of separating the rare earths from one another.

4. **Analytical (Scandium, Yttrium, Lanthanum)**

Scandium, yttrium, and lanthanum can be precipitated from strongly acid solutions by means of oxalic acid. They share this property with the other elements belonging to the rare earth group, and with thorium, but differ thereby from aluminum and from the alkaline earth metals which they otherwise resemble.

Scandium differs from its heavier homologues (and from all the other rare earths) in being precipitated by *thiosulfate*, a reaction due to the formation of an insoluble basic salt by

* It is not merely the colloidal character of the basic lanthanum acetate or starch that determines the formation of the blue adsorption product. This seems to be associated with some special atomic grouping present in these and other substances which give blue colorations with iodine. Thus other colloidal lanthanum salts do not give the coloration—even lanthanum salts of other organic acids (with the exception of the propionate). The elements praseodymium and neodymium, neighbors of lanthanum in the Periodic System, behave like lanthanum in this respect.

the hydrolysis of scandium thiosulfate. It can also be separated from the other rare earths, which form insoluble fluorides, by the solubility of its fluoride in ammonium fluoride solution. The separation of scandium from other rare earths by extracting it from ammonium thiocyanate solution by means of ether (Fischer, 1942, see above) affords a better separation.

Yttrium and lanthanum can also be separated from one another without excessive difficulty by use of differences in solubility of their salts. They (including scandium) can be identified, and tested for purity, by determining the density of the oxides or, more exactly, by determining the equivalent weight. The blue coloration given by iodine with the surface-active basic acetate affords a fairly characteristic reaction for the detection of lanthanum.*

In practice, however, the problem is invariably not merely to separate these three elements from each other, but from the whole group of other rare earths, and often from thorium as well. The methods required for this will be discussed later in this volume.

5. Actinium (Ac)

Actinium occurs in very minute amounts in uranium ores, their actinium content being only about 0.15 mg per ton, or $^1/_{300}$ of the radium content. Actinium was discovered by Debierne in pitchblende residues. It is obtained from these by working them up for the rare earths, and especially for lanthanum, with which actinium associates itself and from which it has only recently been possible to separate it by ion exchange methods, or by selective solvent extraction. Prior to 1950, the most concentrated actinium preparations contained about 1 to 2% of Ac_2O_3 in La_2O_3 (Perey, 1946). It is very similar to lanthanum in its reactions, but is even more strongly electropositive.

Actinium is *radioactive*. It disintegrates, first emitting a β-ray of very low energy, to form radioactinium (RdAc), which in turn disintegrates by emission of an α-particle, forming *actinium* X, an isotope of radium. Actinium X (^{223}Ra) disintegrates in the same manner as radium, first forming a gaseous disintegration product, (*actinium emanation*, AcEm, *actinon*), from which solid radioactive products (AcA, AcB, etc.) are deposited. Owing to the softness of the β-rays from actinium itself, its radioactivity is usually measured by allowing these α-particle emitting products to grow. With actinium C, as with radium C, a branching of the disintegration series takes place, but the two branches ultimately lead to a common final decay product, AcD. Actinium is not really the first member of this decay series, but is in its turn generated by the radioactive decay of an isotope of uranium, actinouranium, ^{235}U. *Uranium Y* and *protactinium* occur thereby as intermediate products and actinium is formed by the α-particle decay of the latter. Since protactinium has the atomic weight 231, it follows from the theory of radioactive disintegration that the atomic weight of actinium—which it has not yet been possible to determine directly—should hypothetically be 227.

Pure actinium preparations can now be more conveniently prepared by nuclear transmutation from radium, in the atomic pile, than from natural sources. Hagemann (*J. Am. Chem. Soc.* 72 (1950), 768) has described the preparation of 1.3 mg of pure actinium in this way, from 1 g of radium in the form of $RaBr_2$. The nuclear reaction involved may be written as

$$^{226}Ra + neutron \rightarrow {}^{227}Ra \rightarrow {}^{227}Ac + \beta\text{-particle.}$$

* As already mentioned, some other substances also give a blue coloration with iodine—viz., basic praseodymium acetate and various organic substances. Basic neodymium acetate gives a violet color with iodine.

The actinium was quantitatively separated from radium by formation of its inner complex salt with thenoyltrifluoroacetone ($C_4H_4S \cdot CO \cdot CH_2 \cdot CO \cdot CF_3$, 'T.T.A.'), which is soluble in organic solvents.

The position of actinium in the Periodic System follows not only from its close analogy to lanthanum in general chemical behavior, but especially from its observed trivalence. This was established by von Hevesy, by determining the diffusion constant of the ion. According to Nernst, there is a simple relationship connecting the diffusion constant of a salt, the electrolytic mobility of its ions, and their valence. Since the mobilities for elementary ions (except for H^+ and Li^+) vary only within rather narrow limits, it is possible from the diffusion coefficient to draw some conclusions about the valence. Von Hevesy was able to determine the valence of a considerable number of radioelements by this means.

6. Compounds of Actinium

Compounds of actinium prepared in the pure state (on a microgram scale) include the following. (These are all isotypic with the corresponding compounds of lanthanum.)

Halides. AcF_3 (hexagonal, $a = 4.27$, $c = 7.53$ Å; for LaF_3 $a = 4.14$, $c = 7.326$ Å) is isotypic with LaF_3, the rare earth fluorides, UF_3, NpF_3, and PuF_3. $AcCl_3$ (hexagonal, $a = 7.62$, $c = 4.55$ Å; for $LaCl_3$ $a = 7.468$, $c = 4.366$ Å), isotypic with UCl_3, $NpCl_3$, $PuCl_3$, and the rare earth metal chlorides. $AcOF$ (cubic, CaF_2-type structure, $a = 5.93$ Å; for $LaOF$ $a = 5.76$ Å). $AcOCl$ (tetragonal, $PbClF$ structure, $a = 4.27$, $c = 7.07$ Å; for $LaOCl$ $a = 4.113$, $c = 6.871$ Å).

Oxide. Ac_2O_3 (hexagonal, $a = 4.07$Å, $c = 6.29$ Å) is isotypic with La_2O_3.

Sulfide. Ac_2S_3 (cubic, Ce_2S_3 structure, $a = 8.97$ Å; for La_2S_3 $a = 8.706$ Å). (See Fried, Hagemann and Zachariasen, *J. Am. Chem. Soc.*, 72 (1950), 771.)

CHAPTER 3

FOURTH SUB-GROUP OF THE PERIODIC SYSTEM: TITANIUM, ZIRCONIUM, HAFNIUM, AND THORIUM

Atomic numbers	Elements	Symbols	Atomic weights	Densi- ties	Melting points	Boiling points	Specific heats	Valence states
22	Titanium	Ti	47.90	4.49	1725°	3260°	0.113	II, III, IV
40	Zirconium	Zr	91.22	6.52	2100°	about 3600°	0.068	II, III, IV
72	Hafnium	Hf	178.50	13.31	2300°	about 5200°	0.0341	IV
90	Thorium	Th	232.05	11.71	1800°	about 4200°	0.034?	III, IV

1. Introduction

(a) General [1]

The elements of Group IVB of the Periodic System—titanium, zirconium, hafnium, and thorium—closely resemble the Sub-group III elements of the same series—scandium, yttrium, lanthanum, and actinium—except that they are commonly quadrivalent.

The elements of the IVth Sub-group are also closely allied in many respects to the elements of the IVth Main Group. They differ from them in that (like all elements of the Sub-groups) they never function as the electronegative constituents of compounds. (An exception to this statement is found with rhenium, which may exist as a uninegative ion.)

Except for titanium, which is fairly easily reducible to compounds derived from lower valence states, the elements of the IVth Sub-group are almost exclusively tetrapositive in their compounds. In Sub-group IVB there is no tendency for the elements of higher atomic weight to behave as dipositive, as is found in the IVth Main Group.

The elements of Sub-group IV all form solid, very non-volatile and very insoluble oxides, and, unlike the corresponding compounds of the Main Group, can readily react with hydrogen peroxide, with the formation of peroxides or other peroxy compounds notable for their relatively great stability.

Compounds of the lower valence states of titanium are the first encountered in the series of the elements in which *colored elementary electrolytic ions* are present. The ability to form colored elementary electrolytic ions persists through the subsequent Sub-groups, and disappears just before the Sub-groups merge with the Main

Groups again, i.e., in the second Sub-group. *None* of the elements of the Main Groups of the Periodic System can form colored elementary electrolytic ions.

The elements of Sub-group IV differ sharply from those of the IVth Main Group in their behavior towards hydrogen, but fairly closely resemble the elements of Sub-group III (scandium, yttrium, lanthanum), in that they are able to absorb considerable quantities of hydrogen. The amounts absorbed depend upon temperature and pressure.

Titanium and zirconium at ordinary temperature can absorb up to 2, and thorium more than 3 gram atoms of hydrogen per gram atom of metal. With increase of temperature, the ability to combine with hydrogen diminishes. Above 1000°, the amount of hydrogen absorbed is approximately proportional to the square root of the hydrogen pressure. This indicates that hydrogen is dissolved in the metal under these conditions in the atomic form. With decreasing temperature, the relation between hydrogen uptake and pressure changes. This is connected with the fact revealed by X-ray investigations (Hägg, 1931), that hydrogen atoms are merely incorporated in the almost unaltered metallic lattice as long as the amount absorbed is small, whereas in the uptake of larger quantities of hydrogen, phases are formed with a different crystal structure, having the nature of non-Daltonide compounds. Titanium, for example, takes up 33 atom-% of hydrogen in solid solution. With higher hydrogen contents a new phase appears, with a homogeneity range between 50 and 66.7 atom-% H. It has a face-centered cubic structure, within which the H atoms are probably arranged like the F^-ions in the fluorite structure (cf. Vol. I). The lattice points at the disposal of the H atoms need not all be occupied, but at least half of them must be. The composition of the compound can vary, accordingly, between the limits TiH_2 and TiH. The apparent radius of the H atoms in the structures of the metal hydrides of the IVth Sub-group is 0.45 Å.

The structures of ThH_2 and of ZrH_2 have been determined by the method of *neutron diffraction*, which is the only diffraction technique which enables the positions of hydrogen atoms to be determined in the presence of heavy atoms. These compounds form body-centered tetragonal structures (for ThH_2 $a = 4.10$ Å, $c = 5.03$ Å; for ZrH_2 $a = 3.52$ Å, $c = 4.449$ Å), which are related to the cubic fluorite structure. Every metal atom is surrounded by 8 H atoms and each H atom by a flattened tetrahedron of metal atoms (Th—H = 2.41 Å), and it is considered that there are also metallic bonds between each Th (or Zr) atom and the nearest metal atoms around it.

The question formerly posed, whether metallic hydrides of this type were 'solid solutions' or 'compounds' is thus explained in that the quantity of hydrogen taken up determines whether only a solid solution or a compound is formed.

Among the oxides of the metals of Group IVB, titanium dioxide is more acidic than basic; zirconium and hafnium dioxides are more basic than acidic, whereas thorium dioxide is purely basic in character. The basic character of the oxides (or of the hydroxides corresponding to them) thus increases with increasing atomic weight within the group, in conformity with the general rule.

The *heats of formation* of the dioxides of the Group IVB are larger than those of the dioxides of the IVth Main Group—in some cases quite considerably so. Table 6 shows the heats of formation of some simple compounds of the metals of the Group IVB.

TABLE 6

HEATS OF FORMATION OF COMPOUNDS OF GROUP IVB METALS
(kcal per g-equivalent of metal)

[TiO$_2$]	Rutile	55.7	TiCl$_4$ liq.	46.5	[TiN]	26.8	[TiC]	28?
[ZrO$_2$]	monocl.	64.5	—	—	[ZrN]	27.4	[ZrC]	11.2
[HfO$_2$]	monocl.	67.9	—	—	—	—	—	—
[ThO$_2$]	cub.	73.2	—	—	[Th$_3$N$_4$]	26.0	[ThC$_2$]	11.4

The following compounds of the Group IVB metals with non-metals crystallize with the NaCl structure. TiC $(a = 4.315 \text{ Å})$, TiN $(a = 4.225 \text{ Å})$, TiO $(a = 4.235 \text{ Å})$, ZrC $(a = 4.687 \text{ Å})$, ZrN $(a = 4.63 \text{ Å})$, HfC $(a = 4.458 \text{ Å})$ and also HfN. ThO$_2$ crystallizes in the *fluorspar* type $(a = 5.57 \text{ Å})$. ZrO$_2$ and HfO$_2$ also crystallize with this structure when they contain certain impurities (see p. 69 *et seq.*). Structures of the *brucite* (or CdI$_2$) type (layer lattices) have been established for TiS$_2$ $(a = 3.40 \text{ Å}, c = 5.69 \text{ Å})$, TiSe$_2$ $(a = 3.53 \text{ Å}, c = 6.00 \text{ Å})$, TiTe$_2$ $(a = 3.79 \text{ Å}, c = 6.45 \text{ Å})$, ZrS$_2$ $(a = 3.68 \text{ Å}, c = 5.85 \text{ Å})$, and ZrSe$_2$ $(a = 3.79 \text{ Å}, c = 6.18 \text{ Å})$. The compounds TiSe and TiTe crystallize with the NiAs structure $(a = 3.56, c = 6.22 \text{ Å}, \text{ and } a = 3.83, c = 6.39 \text{ Å}, \text{ respectively})$. This structure can undergo continuous transition into the brucite type, through the removal of metal ions in a statistically random manner throughout the crystal lattice, from the middle sheet of metal ions which differentiates the NiAs type structure (Fig. 43) from the brucite type structure (Fig. 61, Vol. I). The halides of the IVth Sub-group elements mostly possess molecular crystal lattices.

In many respect the relationship between titanium, especially in its trivalent state, and aluminum, standing in the preceding series of Group III, is closer than that between scandium and aluminum. Thus the Ti^{3+} ion has the same ability to form alums as has the Al^{3+} ion. The chemistry of chromium and of iron also display analogies with the chemistry of titanium. In the quadrivalent state, titanium occupies a position roughly intermediate between aluminum and silicon. It differs from both, however, in having yet more strongly marked tendency to form double- and complex compounds. This tendency is altogether stronger among the elements of the Sub-groups than in those of the Main Groups.

From the standpoint of atomic structure, the fact that the normal and also the maximum valence of titanium, zirconium and thorium is four is a consequence of their position relative to the preceding inert gases argon, krypton, and radon. These elements attain the electron-number of these inert gases by losing four electrons. Since the inert gas structures represent particularly stable electron configurations, the loss of more than four electrons (resulting in a valence higher than four) is not possible. These elements cannot assume a negative charge, according to the Kossel-Lewis theory, because the nearest following inert gases have have electronic configurations which could not be built up by accepting a small number of electrons.

Hafnium, unlike the Group IVB metals just considered, is separated by more than four places from xenon, the inert gas preceding it. However, even before hafnium was discovered, it had been deduced from the Bohr theory of atomic structure that its outer shell must be built up in the same way as that of titanium, zirconium, and thorium, so that hafnium must also normally be a quadrivalent element.

With titanium and zirconium, the term analysis of the spectra furnishes us with evidence as to the structure of the atoms, based immediately on experimental data. It is thereby possible to state how the 19th electron in titanium and the 37th electron in zirconium are bound, in the normal state of the atoms of these elements. The lowest term found in the spectrum ascribed to Ti^{3+} is not, as in the case of potassium and of singly ionized calcium, an s term (cf. Vol. I, Fig. 59, p. 252), but a d term. This implies (cf. Vol. I, p. 117) that in the state of lowest energy the electron circulating round the Ti^{4+} core (i.e. the 19th electron) occupies an orbit with the subsidiary quantum number $l = 2$—in this case a 3d orbit. We also know from the term analysis that the 20th electron of the titanium atom, in the normal state, occupies a 3d orbit. The 21st, and also the 22nd electrons, on the other hand, are bound in 4s orbits. With zirconium the 37th, 38th and 39th electrons are bound in 4d orbits; only one, the 40th electron, here occupies a 5s orbit. As can be seen from Table 7, the energies to be expended for ionization of the valence electrons are smaller for the elements of the Group IVB than for those of the IVth Main Group. The elements of the Sub-group consequently are more strongly electropositive than the elements of the Main Group.

TABLE 7

IONIZATION ENERGIES FOR GROUP IV ELEMENTS

(in kcal per g atom)

	IVth Main Group					IVth Sub-Group			
	C	Si	Ge	Sn	Pb	Ti	Zr	Hf	Th
$M \rightarrow M^{1+} + e$	258.1	186.2	186.4	168.2	170.1	156.9	159.5	—	—
$M^{1+} \rightarrow M^{2+} + e$	559.1	374.7	366	334.6	344.8	313	322.0	341	—
$M^{2+} \rightarrow M^{3+} + e$	1097.9	768.1	786	702.7	735.4	636	553.1	—	—
$M^{3+} \rightarrow M^{4+} + e$	1478.9	1035.2	1049	933	968	992.4	779	—	678

Total:

	C	Si	Ge	Sn	Pb	Ti	Zr	Hf	Th
$M \rightarrow M^{4+} + 4e$	3394.0	2364.2	2387	2138	2218	2098	1814	—	—

(b) Crystal Structure of the Metals

In the elementary state, and at ordinary temperature, titanium, zirconium, and hafnium form crystal lattices of the magnesium type (hexagonal closest packing). The edge lengths for the unit cells are, for Ti $a = 2.95$, $c = 4.69$ Å; for Zr $a = 3.23$, $c = 5.14$ Å; for Hf $a = 3.32$, $c = 5.46$ Å. Thorium, like lead, crystallizes in the face-centered cubic structure (cubic closest packing), $a = 5.12$ Å.

Above 885°, titanium changes into a body-centered cubic modification ($a = 3.3$ Å). Zirconium undergoes the same transformation at 862° ($a = 3.61$ Å at 865°).

The apparent atomic and ionic radii of the elements of the IVth Sub-group are collected in Table 8; the ionic radii apply to the quadrivalent ions, M^{4+}.

TABLE 8

APPARENT ATOMIC AND IONIC RADII OF GROUP IVB ELEMENTS

Element	Titanium	Zirconium	Hafnium	Thorium
Atomic radius, Å	1.49	1.58	1.57	1.82
Ionic radius, Å	0.64	0.87	0.84	1.10

(c) Alloys

In the liquid state, the metals of the IVth Sub-group are apparently good solvents for other metals, and therefore form alloys. The preparation and the metallographic investigation of these alloys is rendered difficult, however, by the high melting points of the metals and by their great reactivity—e.g., with nitrogen and carbon. Our knowledge of alloys of the metals of Group IVB (cf. Table 9), is, accordingly, still very fragmentary. In many cases it is merely known that alloys are formed, without more precise knowledge of their character. Even in the cases in which statements are made in Table 9 about miscibility, on the basis of metallographic investigations, the equilibrium diagrams are known in some cases only very incompletely. The present state of our knowledge gives the impression that the metals of Sub-group IV have little tendency to form mixed crystals but are very disposed to form chemical compounds with other metals. It is notable that the number of compounds which are formed with any given metal is invariably small.

Apparently no compounds are formed with the alkali and alkaline earth metals, so these can be used to liberate the Sub-group IV metals from their halides or oxides. No mixed crystals are formed in this process. Relationships with the remaining metals, in so far as studies of the relevant systems are extant, are set out in Table 9.

In addition to the compounds listed in Table 9, reference may be made to the compounds Ti_4Pb, Ti_4Bi and Ti_3Pt. β-Ti (i.e., the high temperature modification) forms a complete range of mixed crystals with V, Nb, Ta, Mo, and W. α-Ti, however, can incorporate these metals in its crystal lattice only to a very limited extent. It has also been found that there is complete miscibility with Zr and Hf only above the transition temperature of titanium.

A few non-metals are also included in Table 9, as well as the metals; namely those which form compounds of quasi-metallic type with the metals of Sub-group IV. The *nitrides* of the type MN also display pronouncedly metallic properties. Like the borides of the type MB and the carbides of the type MC of this group, they possess metallic conductivity, and indeed these nitrides and borides are considerably better conductors than the pure metals, as the data of Table 10 show. In Table 10 are included the corresponding compounds of elements of Group VB which likewise possess metallic conductivity.

TABLE 9

MISCIBILITY AND COMPOUND FORMATION OF GROUP IVB METALS
WITH OTHER METALS AND SOME NON-METALS

						Main Groups				
	Be	B	Al	C	Si	Sn	P	Sb	S
Ti	Be_2Ti	$s > o$ TiB TiB_2	s $o?$ Al_3Ti 1355° AlTi	$s > o$ $Ti_2C?$ TiC 3450°	$s > o$ Ti_2Si $Ti_2Si_3?$ $TiSi_2$ 1760°	Alloy	Ti_2P TiP	Ti_4Sb TiSb $TiSb_2$	Ti_2S TiS Ti_2S_3 TiS_2 TiS_3
Zr	—	ZrB ZrB_2	$s > o$ Compnds	ZrC $ZrC_2?$	$ZrSi_2$	—	ZrP ZrP_2		ZrS_2
Th	—	ThB_4 ThB_6	s $o?$ Al_3Th* 880°	ThC 2625° ThC_2 2655°	$ThSi_2$	—	ThP Th_3P_4		ThS Th_2S_3 ThS_2 Th_3S_7

(Continued)   TABLE 9

	Cr	Mn	Fe	Co	Ni	Cu	Ag	Au	Zn	Hg
Ti	*s ∞* Cr_2Ti []? Cr_3Ti_2[]?	— Mn_2Ti $MnTi_2$ — Mo,W	*s > o* Fe_3Ti 1397° Fe_2Ti 1530° FeTi $FeTi_2$	— Co_2Ti CoTi $CoTi_2$	*s > o* Ni_3Ti 1378° NiTi $NiTi_2$ ~1060°	*s > o* Cu_3Ti $\overline{CuTi}$ $\overline{Cu_3Ti_2}$? $\overline{CuTi_3}$	*liq ~ o* *s o*	*liq ~ o* *s > o* Au_6Ti Au_2Ti $AuTi_3$	— — Zn_3Ti ZnTi	— — Alloy
Zr	— $ZrCr_2$	— Mo_3Zr W_2Zr	*s > o* Fe_3Zr_2 1640°	—	*s > o* Ni_4Zr Ni_3Zr	*s > o* Cu_3Zr	— Compnds ?	*s > o* Alloy	—	— Alloy
Th	—	— —	— —	— —	*s > o* Ni_9Th[] 1350° Ni_5Th 1530° Ni_5Th_2? NiTh 1200° $NiTh_2$[] 1050°	—	*s ~ o* Ag_3Th Ag_5Th_3	—	—	*liq > o* —

MEANING OF SYMBOLS

Italicized data refer to miscibility, symbols in ordinary type show the compounds formed.

Miscibility	∞	complete
	< ∞	restricted miscibility
	> o	very slight miscibility
	o	no detectable miscibility.

Where data are given for the solid state only, miscibility in the liquid state is generally complete.

Compound formation	o	no compounds formed.
	—	phase diagram not known or incompletely known.

Melting points or temperatures of decomposition are given immediately below the formulas to which they refer.

Compounds marked * melt incongruently. Compounds marked [] are formed by a reaction taking place in the solid state or (in less common instances) from mixed crystals formed primarily, or by structural transformation taking place without change of composition (cf. pp. 15, 339). Superstructure phases are indicated by enclosing their formulas in square brackets. Temperatures given for such compounds are temperatures of transition from the ordered to the disordered atomic distribution, or the temperatures at which the characteristic X-ray diffraction pattern of the superstructure phase vanishes.

TABLE 10

SPECIFIC ELECTRICAL CONDUCTIVITIES OF THE METALS OF GROUPS IVB AND VB
AND OF THEIR COMPOUNDS SHOWING METALLIC CONDUCTION

(ohm^{-1} cm^{-1} at room temperature)

	Titanium	Zirconium	Hafnium	Vanadium	Tantalum
Metal	$1.11 \cdot 10^4$	$2.24 \cdot 10^4$	$3.07 \cdot 10^4$	$0.6 \cdot 10^4$	$6.7 \cdot 10^4$
Nitride, MN	$4.6 \cdot 10^4$	$7.4 \cdot 10^4$	—	$1.16 \cdot 10^4$	$0.74 \cdot 10^4$
Boride, MB	$6.6 \cdot 10^4$	$10.9 \cdot 10^4$	$10.0 \cdot 10^4$	$6.2 \cdot 10^4$	—
Carbide, MC	$0.52 \cdot 10^4$	$1.58 \cdot 10^4$	$0.92 \cdot 10^4$	$0.64 \cdot 10^4$?	$1.0 \cdot 10^4$

2. Titanium (Ti)

(a) Occurrence

Titanium exists in Nature chiefly as the dioxide, TiO_2, which occurs in different modifications—usually as *rutile*, less commonly as *anatase* and *brookite*. Naturally occurring titanium dioxide is generally more or less contaminated with iron oxide; iron ores likewise often contain considerable quantities of titanium dioxide. A compound of the oxides of titanium and iron also occurs, however, as a definite mineral—titaniferous ironstone or *ilmenite* $FeTiO_3$. *Perowskite*, a calcium titanate $CaTiO_3$, and *titanite* or *sphene*, a calcium titanoxysilicate $CaTiO[SiO_4]$ may also be mentioned. Titanium dioxide also very frequently occurs in combination with the rare earths. It is very widespread in small amounts, so that almost every soil contains detectable amounts of titanium (over $\frac{1}{2}\%$ on the average).

(b) History

Titanium was first discovered in the form of the dioxide. Gregor, in England, found in a Cornish iron sand, menaccanite, the oxide of a new element, in 1789. Kirwan first called this 'menachine'. Klaproth, in 1795, quite independently discovered that *rutile* was the oxide of a new metal, which he named *titanium* after the minor planet then newly discovered. He soon recognized that the metal discovered by Gregor in menaccanite was identical with that found by him. In 1822 Wollaston found the metallic-looking titanium carbo-nitride in blast furnace slag, and mistook this for the pure metal. This erroneous view was generally accepted, and persisted for a long time, even though shortly afterwards (in 1825) Berzelius succeeded in preparing the true elementary titanium, (though not in a pure state), by reduction of potassium fluorotitanate with sodium. Even Berzelius did not doubt that the blast-furnace titanium of Wollaston was pure, crystalline titanium; he called his own product 'amorphous titanium'. Wöhler, however, in 1849, proved by their combustion in a stream of chlorine that the blast-furnace crystals were those of a compound of titanium with carbon and nitrogen, with the composition Ti_5CN_4.

(c) Preparation

On account of the great affinity of titanium, not only for the reducing agents usual in metallurgy, such as carbon and aluminum, but also for relatively inert gases such as nitrogen, it is impossible to prepare the metal in the pure state by the older methods. It has been achieved only in recent times by the thermal decomposition of thorium tetraiodide by the 'filament' method described under zirconium (p. 65-6). This process is still in use for the preparation of titanium of a specially high degree of purity.

Almost pure titanium was obtained for the first time by Hunter (1910), by heating a mixture of titanium tetrachloride and metallic sodium in a steel bomb:

$$TiCl_4 + 4Na = 4NaCl + Ti.$$

According to Billy (1921), the reaction of titanium tetrachloride with sodium hydride is more suitable:

$$TiCl_4 + 4NaH = Ti + 4NaCl + 2H_2;$$

the metal is freed from absorbed hydrogen by heating in vacuum to 800°. A process for preparing titanium by the reduction of titanium dioxide by metallic calcium has been described by Kroll (1937). The metal obtained in this way can be worked when hot, but displays cold-brittleness on account of its content of titanium monoxide, TiO.

In the United States, pure titanium is now prepared industrially chiefly by the Kroll process. This involves the reduction of titanium tetrachloride with molten magnesium at about 850° in steel reaction vessels, using helium or argon as a protective atmosphere. The metallic sponge which is first obtained is melted down under argon in an electric arc. It is possible to produce compact blocks 6 m long and 14 cm in diameter by this means.

A modification of the Kroll process is that developed by Madde and Eastwood, of the Batelle Memorial Institute (1950). The reduction furnace is combined with the remelting furnace in this form of the process, so as to obtain compact titanium in a single operation. Molten magnesium is first pumped into an electrically heated vessel, from which it is transferred under a pressure of argon into the steel reduction chamber. TiCl$_4$ vapor is forced into the reduction chamber from a second storage vessel. The mixture of liquid MgCl$_2$ with suspended solid Ti, formed by the reaction

$$2Mg_{liq} + TiCl_{4gas} = Ti_{solid} + 2MgCl_{2liq}$$

flows continuously into the arc furnace. The MgCl$_2$ and that portion of the Mg which has not reacted are there vaporized, and condense outside the furnace. The Ti fuses to a compact block, and as this grows by accretion of metal it is withdrawn through the base of the furnace. It serves also as one electrode for the arc, which is struck between the titanium and a tungsten rod.

Pure titanium has not until recently been required for technological purposes. An alloy of titanium with iron, ferro-titanium, melting below 1400°, with a titanium content of 10–25%, has hitherto generally been produced. Ferrotitanium can readily be obtained by reduction of rutile with carbon in the presence of iron or, if the carbon content of the alloy so produced is considerable, by reduction with aluminum. A slight aluminum content of ferrotitanium is advantageous for most technical purposes.

(d) Properties

Titanium, in the compact state, is a metal resembling steel in appearance, usually hard and brittle in the cold, and only workable at a red heat. When quite pure, however, it possesses considerable ductility even when cold. Pure titanium also takes an excellent polish, and retains its brilliant luster even longer than does chromium. In the form of powder it is grey to black. Titanium melts at about 1725°. It distils in the electric arc of the Moissan furnace, and the boiling point has been calculated as 3260° from the variation of the vapor pressure with temperature. The density of titanium is 4.49. For crystal structure see p. 47.

The atomic heat C_p of metallic titanium increases from 1.278 at —219.7°, to 5.880 at 0° (Kelley, 1944), and to 6.507 at 200°. Between 200° and 800° it can be represented, according to Jaeger and Rosenbohm (1936), by the expression $C_p = 4.1964 + 0.018208 t - 0.39114 \cdot 10^{-4} t^2 + 0.29223 \cdot 10^{-7} t^3$. The atomic heat of the cubic modification, above the transition temperature, is 7.525, and is independent of the temperature. The molar heat of titanium in the vapor state, according to Gilles and Wheatley (1951) has a distinct maximum ($C_p = 6.578$) at 135° K and a minimum ($C_p = 5.096$) at 990° K. At the boiling point of titanium, it is 8.034, at 4000° K 8.667, and at 5000° K 9.708. This anomalous temperature dependence of the molar heat is a consequence of the ease with which the Ti atom indergoes excitation. Similar behavior is shown by iron. The thermal conductivity of titanium is about 0.036 cal/cm·sec·degree, and the electrical resistivity (for very pure titanium) $\varrho_0 = 0.475 \cdot 10^{-4}$. The specific magnetic susceptibility $\chi_{20} = +3.18 \cdot 10^{-6}$.

At low temperatures, titanium is fairly stable in air. Above a red heat, however, it burns in a stream of oxygen, with considerable evolution of heat, to the dioxide (see p. 45). It unites with nitrogen at a higher temperature (above 800°), forming a nitride. Nitride, as well as oxide, is accordingly formed on heating titanium in air. Titanium is most readily attacked by the halogens—by chlorine, for example, a little above 300°; by fluorine already at 150°. At higher temperatures, titanium also reacts readily with other non-metals and some of the compounds thus formed are remarkable for the stability towards chemical reagents. The carbide TiC and the titanium carbonitride Ti_5CN_4, as well as the nitride TiN, are particularly worthy of note. The silicides Ti_2Si and $TiSi_2$, obtainable in the form of iron-grey crystals may also be mentioned.

Hydrogen is vigorously absorbed by powdered titanium. According to Sieverts (1929), 1 g of titanium can take up 407 cc of hydrogen at room temperature (= 1.74 g atoms of hydrogen per g atom of Ti; cf. p. 45); at 1000° only 66 cc. Titanium expands markedly (up to a maximum of 15.5%) on the absorption of hydrogen, and assumes a lighter color although it retains its metallic luster. According to Sieverts (1931), the heat of formation of titanium hydride amounts to 18.0 kcal per g atom of hydrogen, and so exceeds the heat of formation of sodium hydride (cf. Vol. I, p. 155). Titanium that has been heated in a stream of hydrogen catches fire in the air, and the absorbed hydrogen burns with a luminous flame to water.

Oxygen can also be taken up in titanium, in solid solution, in considerable amounts (up to 30 atom-%). The crystal lattice thereby undergoes only a slight expansion, chiefly in the direction of the c axis. If greater amounts of oxygen are absorbed, oxides are formed (cf. Table on p. 54).

Titanium alloys avidly with iron at high temperatures, forming the compound Fe_3Ti. The compound of aluminum with titanium has an analogous composition, Al_3Ti (lustrous, silver white, quadratic leaflets); it is obtainable by the alumino-thermic method. Alloys of titanium with other metals have as yet received relatively little study (cf. Table 9, p. 48–9).

Titanium dissolves in acids less readily than does iron. Thus it dissolves in dilute hydrochloric acid only on heating, forming the violet trichloride. It is oxidized by hot nitric acid to insoluble 'b-titanic acid', corresponding in this respect exactly to tin. Hydrofluoric acid is the best solvent for titanium and for titanium compounds that are attacked only with difficulty.

(e) Applications [2–8]

Titanium finds application chiefly in the form of its alloy with iron, ferrotitanium, as an additive to steel. Titanium steels have notably good strength and elasticity; a content of less than 0.1% of titanium is generally sufficient to achieve these properties. The action of titanium depends partly on its ability to prevent the separation of oxygen and nitrogen on cooling, which leads to the formation of blow holes. Its use, in the form of its aluminum alloy, has accordingly been recommended as an additive to copper and its alloys also. Its action here is analogous to that of phosphorus in the phosphor bronzes. The notable ability of titanium to combine with nitrogen and oxygen is also used to remove the last traces of gases from evacuated vessels ('gettering' of lamps, valves, etc.)

Since it became possible to prepare pure titanium, with good cold-working properties, on a large scale, the technical applications of the metal have become increasingly important. It furnishes alloys which almost equal stainless steel in strength and ductility, but have still greater corrosion resistance and are about 40% lighter than steel. The production of titanium now exceeds 10 thousand tons per year. There is no doubt that it will become a most important material in the construction of aircraft, ships, automobiles, and marine machinery.

Porous titanium and titanium carbide have been recommended as catalysts for the preparation of nitrogen-hydrogen and nitrogen-oxygen compounds. They are said to have high activity and long life. Titanium carbide, like molybdenum carbide, has recently achieved considerable importance through its use for the production of the so-called 'hard metals', or *cutting alloys*. These are produced by sintering titanium, molybdenum, or tungsten carbide with metals like cobalt or nickel. These last provide a tough matrix in which are embedded the carbides named, which are remarkable for their hardness. The cutting alloys are far more effecient than the high speed steels (p. 268). Because of their outstanding resistance to abrasion they can even be used for many purposes for which it was formerly necessary to use diamonds.

Titanium salts and especially the alkali double fluorides and oxalates find application as mordants in the dyeing of textiles and leather. A solution of iron(II) and titanium(III) chloride, readily obtained by dissolving ferrotitanium in hydrochloric acid, has been recommended as a non-injurious bleaching agent for silk and wool.

Finely divided titanium dioxide, prepared or worked in a certain way, is manufactured as a mineral pigment (titanium white [9]); it has better covering power than zinc white, and is more permanent than white lead. In combination with iron oxide and other oxides titanium dioxide serves for the production of colored glazes on porcelain and earthenware (Bunzlau earthenware).

Titanium(III) chloride solutions have proved to be very useful in various ways as reducing agents in volumetric analysis. Their use does, however, necessitate working in an inert atmosphere, on account of the tendency of titanium(III) compounds to undergo auto-oxidation. Titanium(III) chloride solutions have also been found very suitable in many cases for preparative purposes.

3. Compounds of Titanium

Titanium is usually electropositive and quadrivalent in its compounds, but may also function as trivalent and, in a few compounds as bivalent; the compounds of bivalent titanium are prepared only with difficulty, however, and are rather unstable in aqueous solution.

The existence of titanium(I) compounds has not been established, although certain observations point to their formation (e.g., the observation that there is a maximum volatility in the Ti-S system at a composition corresponding roughly to the formula Ti_2S).

Quadrivalent titanium has a strong tendency to form anions (acido-ions), such as $[TiO_3]^=$ (titanate ion), $[TiF_6]^=$ (fluorotitanate ion), $[TiCl_6]^=$ (chloro-titanate ion), and $[Ti(SO_4)_3]^=$ (sulfatotitanate ion). Acido salts are also derived from trivalent titanium. These are, however, not strongly complexed; they cannot be recrystallized from pure water without undergoing decomposition.

The soluble titanium(IV) compounds have a strong tendency to undergo hydrolytic decomposition. Incomplete hydrolysis may give rise to compounds of the type $TiOX_2$. Compounds containing the radical $[TiO]^{II}$ are called *titanyl* compounds.

The oxidation potential Ti^{3+}/Ti^{4+} (referred to the normal hydrogen electrode at $0°$ C) is $+0.04$ volt. If a platinum foil is dipped in a solution containing equal concentrations of Ti^{3+} and Ti^{4+} ions, and combined with a normal hydrogen electrode to form a galvanic cell, the current therefore flows through the wire joining the two electrodes from the normal hydrogen electrode to the platinum foil dipping in the titanium salt solution. Hydrogen is evolved at the former, while Ti^{3+} ions are oxidized to Ti^{4+} ions at the latter. Titanium(III) ions have so strong a reducing action, therefore, that in some circumstances they can evolve hydrogen from acid solutions. For the couple Ti^{2+}/Ti^{3+} the oxidation potential $E^0 = +0.37$ volt, for the couple Ti^{2+}/Ti^{4+}, $E^0 = +0.20$ volt. The normal potential of Ti in contact with a Ti(II) salt solution is $+1.75$ volt. For the process $Ti + 6F^- = [TiF_6]^= + 4e$, $E^0 = +1.24$ volt; for $Ti + 2H_2O = TiO_2 + 4H^+ + 4e$, $E^0 = +0.95$ volt; and for $Ti^{+++} + H_2O = [TiO]^{++} + 2H^+ + e$, $E^0 = +0.1$ volt.

SUMMARY OF THE SIMPLEST COMPOUNDS OF TITANIUM

Fluorides	Chlorides	Bromides	Iodides
—	$TiCl_2$, black	$TiBr_2$, black	TiI_2, black-brown
TiF_3, violet	$TiCl_3$, violet	$TiBr_3$, black	TiI_3, dark violet
TiF_4, white	$TiCl_4$, colorless, liq.	$TiBr_4$ amber	TiI_4, red brown

Oxides	Sulfides	Phosphides	Carbides
TiO, gold	$TiS_{<1}$, grey	Ti_2P?	Ti_2C?
Ti_2O_3, violet	TiS, dark brown	TiP	TiC, black
Ti_4O_7?, dark blue	Ti_2S_3, green black		
TiO_2, white	TiS_2, brassy	**Nitride**	
	TiS_3, graphite-like	TiN, gold	

(a) Titanium(II) Compounds

Compounds of bivalent titanium can be prepared by energetic reduction of titanium(IV) or titanium(III) compounds. Thus titanium(II) chloride, $TiCl_2$, is produced from titanium tetrachloride by means of sodium amalgam, as a black powder which is slowly decomposed by water with the evolution of hydrogen. Aqueous solutions which contain titanium(II) chloride (together with titanium(III) chloride) can be prepared by dissolving TiO in (old dilute hydrochloric acid. Ti^{++} ions are fairly rapidly oxidized to Ti^{+++} ions by water at ordinary temperature; the solutions are much more stable at lower temperatures. In the pure state, $TiCl_2$ ($d = 3.13$) is best obtained, according to Ruff (1923), by thermal decomposition of $TiCl_3$, or by heating $TiCl_4$ with Ti shavings (Klemm, 1942). $TiBr_2$ ($d = 4.31$) and TiI_2 ($d = 4.99$) can most conveniently be prepared by direct union of their components. By treating $TiCl_2$ with liquid ammonia, Schumb (1933) obtained the double compound $TiCl_2 \cdot 4NH_3$, as a pearl grey powder which, like anhydrous $TiCl_2$, reacted with water, evolving hydrogen.

$\overline{TiO}$ ($d = 4.93$, m.p. $1750°$) can be obtained by heating a mixture of TiO_2 and Ti. It has the NaCl type crystal structure, but a certain proportion of the lattice sites (distributed at random) is vacant. At the one end of the homogeneity range, these vacant sites represent missing Ti atoms, and at the other extreme O atoms (Ehrlich, 1939). When TiO is dissolved in dilute hydrochloric acid, partial oxidation occurs:

$$Ti^{++} + H^+ = Ti^{+++} + \tfrac{1}{2}H_2.$$

(b) Titanium(III) Compounds

The titanium(III) compounds are readily obtainable, by reducing soluble titanium(IV) compounds with zinc and acid, or electrolytically. The solutions contain violet colored titanium(III) ions; these have a strong tendency to revert to titanium(IV) ions. Titanium(III) salt solutions are considerably stronger reducing agents than stannous salt solutions (cf. their oxidation potentials).

In the presence of *acceptors*, the oxidation of titanium(III) compounds by means of molecular oxygen consumes more oxygen than is necessary to increase their charge by one unit; they can thus act as 'auto-oxidants' (cf. auto-oxidation, Vol. I, Chap. 16). The ability of trivalent titanium to bring about indirect oxidation by means of atmospheric oxygen is so great that even water can play the part of an acceptor in the presence of titanium(III) compounds in alkaline solution, and is thereby converted to hydrogen peroxide. If the hydrogen peroxide is removed by combination with $Ca(OH)_2$ immediately as it is formed, it is found that for each atom of Ti^{III} oxidized, $\tfrac{1}{2}$ molecule ($= 2$ equivalent) of oxygen is taken up, and $\tfrac{1}{2}$ molecule of H_2O_2 is formed:

$$Ti(OH)_3 + \tfrac{1}{2}O_2 + H_2O = Ti(OH)_4 + \tfrac{1}{2}H_2O_2.$$

In the presence of other substances which can function as acceptors, titanium(III) compounds also consume 2 equivalents instead of 1 equivalent per atom of Ti^{III} in their oxidation by molecular oxygen. In the oxidation with chromic acid or permanganate as much as 3 equivalents instead of 1 equivalent of oxygen may be taken up.

(i) *Titanium(III) halides*: *Acidotitanates(III)*. Anhydrous titanium(III) chloride is obtained as a violet powder by passing the vapor of titanium tetrachloride, mixed with much hydrogen, through a red hot tube—best in a 'hot and cold tube' arrangement, according to the method of St. Claire-Deville. When it is heated in hydrogen to about $700°$ it decomposes into $TiCl_2$ and $TiCl_4$. Titanium(III) chloride is obtained in solution by reducing a solution of titanium(IV) salt in hydrochloric acid by means of zinc, or by dissolving metallic titanium in hydrochloric acid.

A *violet hexahydrate*, $TiCl_3 \cdot 6H_2O$, crystallizes from the solution. If the concentrated solution is covered with a layer of dry ether, and saturated with hydrogen chloride gas, a *green hexahydrate*—i.e., a salt having the same composition as the violet hydrate—is obtained from the green ethereal solution. The violet salt has a similar constitution to the violet chromic chloride(cf. Chap. 5)—i.e., $[Ti(H_2O)_6]Cl_3$; the green hydrate probably has a constitution like that of one of the green chromium(III) chlorides. It thus provides an example of *hydrate isomerism* (for further discussion see under chromium).

Added to a dilute gold salt solution, a titanium(III) chloride solution gives an intense violet color; this depends on the adsorption of colloidal gold on titanium dioxide hydrate. The formation of the adsorption product is exactly analogous to that of the 'Purple of

Cassius' by means of stannous chloride. As little as 0.0001 mg of gold in 2 cc of water can be detected by means of the reaction with titanium(III) chloride.

On treatment with liquid ammonia, anhydrous $TiCl_3$ adds on 6 molecules of NH_3, forming a colorless double compound which readily loses $4NH_3$. The remaining two molecules of ammonia are held in a tightly bound state.

Pentachloro-complex salts of the type $M^I_2[TiCl_5 \cdot H_2O]$ are derived from the green titanium(III) chloride. The fluoro salts $M^I_2TiF_5$ and $M^I_3TiF_6$ are very readily obtained. K_2TiF_5 decomposes when it is sublimed in a high vacuum. According to Ehrlich (1952–53) it is probable that a tetrafluorotitanate is thereby formed, along with the hexafluorotitanate:

$$2K_2TiF_5 = KTiF_4 + K_3TiF_6.$$

Pure TiF_3 has a magnetic moment of 1.75 Bohr magnetons. It follows that the Ti^{III} atom in this compound has one unpaired electron, and that it is therefore present in the undissociated compound in the form of a Ti^{3+} ion.

Hexacido salts of titanium(III) also exist even in cases where their parent compounds TiX_3 are not known. This is so, for example, with the hexathiocyanato titanates(III)— $M^I_3[Ti(SCN)_6] \cdot 6H_2O$.

$TiBr_3 \cdot 6H_2O$ forms reddish violet crystals (m.p. 115°) which are readily soluble in alcohol and acetone, as well as in water.

Anhydrous $TiBr_3$ exists in 2 modifications. It decomposes reversibly at 400°, according to the equation $2TiBr_3 \rightleftharpoons TiBr_2 + TiBr_4$. TiF_3 may be obtained as a dark blue powder by treating Ti or $TiCl_3$ with HF (Ehrlich, 1951). It is very unreactive chemically, and begins to sublime at 900° in a vacuum.

(ii) Titanium(III) sulfate and Double Sulfates. In the electrolytic reduction of a sulfuric acid solution of titanium(IV) sulfate, an inky black-violet colored solution is first obtained*. The solution becomes lighter with continued reduction, and when practically all the titanium is converted to the trivalent state, it has a transparent pure violet color. An *acid titanium(III) sulfate*, with the composition $3Ti_2(SO_4)_3 \cdot H_2SO_4 \cdot 25H_2O$ can be isolated from the solution, in the form of a silky, glistening violet, crystalline powder. This may be converted to anhydrous neutral titanium(III) sulfate, $Ti_2(SO_4)_3$, by fuming down with sulfuric acid; a deep blue colored compound, of unknown composition, is formed as an intermediate. *Neutral titanium(III) sulfate* is a green crystalline powder, insoluble in water, alcohol, and concentrated sulfuric acid, but soluble in dilute sulfuric acid and in hydrochloric acid, giving violet solutions.

Exchange of hydrogen in acid titanium(III) sulfate by metal ions gives rise to double or complex salts, with the composition $3Ti_2(SO_4)_3 \cdot M^I_2SO_4$, or $M^ITi_3(SO_4)_5$, (with variable water content)—e.g., an ammonium salt $NH_4Ti_3(SO_4)_5 \cdot 9H_2O$, and a rubidium salt $RbTi_3(SO_4)_5 \cdot 12H_2O$, both forming bright blue crystals, sparingly soluble in water. Another type of double sulfate of trivalent titanium corresponds, in composition and crystalline form, to the *alums*; the only known *titanium alums* are, however, the rubidium and cesium compounds—$RbTi(SO_4)_2 \cdot 12H_2O$ (red) and $CsTi(SO_4)_2 \cdot 12H_2O$ (bright red-violet). These salts can be recrystallized without decomposition from dilute sulfuric acid, but not from pure water. Yet another type is represented by the sodium double sulfate, a violet crystalline mass with the composition $NaTi(SO_4)_2 \cdot 2\frac{1}{2}H_2O$.

Double salts similar to the sulfates are formed by trivalent titanium also with oxalates: $M^ITi(C_2O_4)_2 \cdot 2H_2O$; $M^I = NH_4$, K, Rb; yellow-gold crystals.

---

* Such intensive colorations are characteristic of products in which one and the same element occurs in different states of oxidation. They often appear intermediarily in oxidations and reductions.

(*iii*) *Titanium(III) hydroxide* is formed by reaction of titanium(III) salt solutions with alkali hydroxides, as a deeply colored precipitate which acts as a most vigorous reducing agent, and is therefore difficult to obtain pure.

(*iv*) *Titanium(III)oxide*. Dititanium trioxide, Ti_2O_3 ($d = 4.49$, m.p. about $1900°$) may be obtained crystalline by heating titanium dioxide at $1000°$ in a stream of hydrogen and titanium tetrachloride. It has the same crystal structure as corundum; $a_0 = 5.42$ Å, $a = 56°32'$.

(*v*) *Titanium nitride*, TiN. If titanium compounds are reduced at high temperature in the presence of atmospheric nitrogen, *titanium nitride* is very readily formed, though generally in an impure state. Pure titanium nitride TiN is obtainable in the form of a bronze colored powder by strongly heating titanium tetrachloride or its ammoniate in a current of ammonia:

$$3TiCl_4 + 16NH_3 = 3TiN + \tfrac{1}{2}N_2 + 12NH_4Cl.$$

The nitride is obtained very pure, and in a compact state, by the filament growth process (as for zirconium nitride, cf. p. 75).

Titanium and zirconium nitrides can be prepared by the *filament* method because they possess a considerable electrical conductivity (cf. p. 50).

Although formally a compound of trivalent titanium, titanium nitride is so stable that it is not attacked by chlorine, even at $270°$. However, it dissolves in hydrofluoric acid in the presence of strong oxidizing agents, such as permanganate. It is also decomposed by hot caustic potash:

$$TiN + 2KOH + H_2O = K_2TiO_3 + NH_3 + \tfrac{1}{2}H_2.$$

A corresponding decomposition is brought about by superheated steam:

$$TiN + 2H_2O = TiO_2 + NH_3 + \tfrac{1}{2}H_2.$$

At one time an attempt was made to base a process for the synthesis of ammonia on this reaction.

Titanium nitride has a structure of rock salt type (cf. p. 46). Crystal lattices of the same type are formed also by ZrN, VN, NbN, and ScN.

(c) Titanium(IV) Compounds

(*i*) *Titanium tetrachloride*, $TiCl_4$, may be obtained by passing chlorine over heated titanium, titanium carbide, titanium-aluminum alloy (which is obtainable by the alumino-thermic method), or over a mixture of titanium dioxide and carbon. In the pure state, it forms a colorless liquid, boiling at $136.5°$, solidifying at $—23°$, and having a density of 1.76 at $0°$. It has a pungent smell, and fumes strongly in moist air. It is rapidly hydrolyzed by water:

$$TiCl_4 + 2H_2O = TiO_2 + 4HCl.$$

If the hydrolysis is repressed by addition of acid, or if only a little water enters into reaction, oxychlorides can appear as intermediate products.

Titanium tetrachloride readily forms addition compounds—e.g., with ammonia ($TiCl_4 \cdot 6NH_3$ and $TiCl_4 \cdot 8NH_3$) and also with pyridine; further, with phosphorus tri- and pentachloride, phosphorus oxychloride, nitrosyl chloride, sulfur tetrachloride, selenyl chloride, and many other compounds—especially those containing oxygen, sulfur or nitrogen.

If $TiCl_4$ is treated with alcohols and NH_3, titanium alkoxides $Ti(OR)_4$ are formed (R = alkyl or aryl group). As was shown by Meerwein (1927–29), these dissolve in alcohol to form *alkoxo-acids*, $H_2[Ti(OR)_6]$. It is possible to make use of the ability of titanium alkoxides to add on additional OR groups, as a means of preparing high-polymeric materials from polyhydroxy-organic compounds (Schmidt, 1952). The alcoholates of zirconium, tin, aluminum and trivalent iron behave similarly. Mixed titanium alkoxychlorides are also known—e.g., $TiCl_3 \cdot OC_2H_5$, which like titanium tetrachloride, forms a double compound with acetyl chloride; this is remarkable in that it can be distilled unchanged under reduced pressure (D. C. Bradley, 1952).

(*ii*) *Chlorotitanates.* Titanium tetrachloride adds on chloride ions, forming the complex chlorotitanate ion $[TiCl_6]^=$. Of the *chlorotitanates* (more precisely—*hexachlorotitanates(IV)*) the ammonium salt $(NH_4)_2[TiCl_6] \cdot 2H_2O$ (yellow crystals) and several salts of organic bases are known. The corresponding free acid can only exist in aqueous solution; its formation therein is demonstrated by the yellow coloration produced by the addition of concentrated hydrochloric acid to titanium tetrachloride. Much more intense colorations appear if hydrobromic acid or hydriodic acid are added.

(*iii*) *Titanium fluoride* and *Fluorotitanates.* Titanium fluoride, TiF_4, most conveniently prepared by reaction of $TiCl_4$ with HF (Ruff 1904), forms a white, loose powder of density 2.80. It displays a much stronger tendency than does the chloride to form acido salts. These correspond to the type $M^I_2[TiF_6]$ (hexafluorotitanates(IV), or more briefly—*fluorotitanates*). All the alkali- and alkaline earth salts of this type, and numerous salts of the heavy metals have been prepared. Potassium, rubidium, and cesium fluorotitanates are suitable for the detection of titanium, on account of their characteristic crystalline form.

(*iv*) *Titanium tetrabromide*, $TiBr_4$, obtainable in a similar manner to $TiCl_4$ (as amberyellow octahedral crystals, *d* 3.25, m.p. 40°, b.p. 230°), is very similar to the chloride in its chemical behavior. It is extremely hygroscopic, very soluble in alcohol (287 g in 100 ml)$_4$ and moderately soluble in ether. Like SiI_4 and GeI_4, $TiBr_4$ crystallizes with the SnI, structure (molecular lattice, see Vol. I, p. 531).

(*v*) *Titanium tetraiodide*, TiI_4, is most readily prepared by double decomposition of $TiCl^4$ with HI. It crystallizes in red-brown octahedra (crystal structure as for $TiBr_4$; m.p. 150°, b.p. 365°), but changes on standing, into a modification with a lower degree of symmetry.

(*iv*) *Titanium Sulfates and Sulfatotitanates.* When titanic acid or titanium dioxide are fumed down with concentrated sulfuric acid, *titanyl sulfate* $[TiO][SO_4]$ is formed, as a white powder, soluble in cold water. It is decomposed by hot water, with the deposition of gelatinous titanium dioxide:

$$[TiO][SO_4] + H_2O = TiO_2 + H_2SO_4.$$

In addition to titanyl sulfate, there are other titanium sulfates, both with greater and with smaller SO_3 contents. The neutral titanium sulfate, $Ti(SO_4)_2$, is not known with certainty in the free state. Double salts derived from it are, however, known; in particular the sulfatotitanates (trisulfatotitanate(IV) salts) of the type $M^I_2[Ti(SO_4)_3]$. Titanyl sulfate also forms double salts (disulfato-oxotitanates), e.g., $(NH_4)_2[TiO(SO_4)_2] \cdot H_2O$.

(*vii*) *Other Acidotitanates.* The double salts last mentioned belong to a type of acido-compounds of titanium which also includes complex salts with other acid radicals—for example the thiocyanato-oxotitanates $M^I_2[TiO(SCN)_4]$, and the dioxalato-oxotitanates $M^I_2[TiO(C_2O_4)_2]$. The latter are used as mordants in the dyeing of leather and in piece dyeing.

(*viii*) *Titanium dioxide*, TiO_2 occurs in Nature chiefly as *rutile*. This forms tetragonal, transparent to opaque, red, or frequently more yellowish, crystals, often of characteristic habit, ($d = 4.2$—4.3). More rarely, titanium dioxide occurs in the form of *anatase* ($d = 3.6$—3.95) which also crystallizes in the tetragonal system, but with a different axial ratio, and as the rhombic *brookite* ($d = 4.1$—4.2). *Edisonite*, which is found in the auriferous sands of North Carolina, is a variety of rutile.

The crystal structure of rutile is illustrated in Fig. 63, p. 265 of Vol. I ($a = 4.58$, $c = 2.95$, $d = 2.01$ Å). Each titanium atom in rutile is surrounded by two O-atoms at a distance of 2.01 Å, and by four at a distance of 1.92 Å; the six O-atoms form a somewhat distorted octahedron. Fig. 16 shows the crystal structure of *anatase*. In anatase, every titanium atom is surrounded, again in the form of a distorted octahedron, by two oxygen atoms at a distance of 1.95 Å (in the diagram these are joined by double lines to the corresponding Ti atom) and by four O-atoms at 1.91 Å. Whereas the rutile structure is frequently met with among the dioxides, and also the difluorides, the structure illustrated in Fig. 16 has hitherto been found only in anatase. In the somewhat more complicated rhombic structure of *brookite*, as in the two other structures, each Ti atom is surrounded, according to Pauling (1928), by six not exactly equidistant O-atoms making up a somewhat distorted octahedral arrangement. The mean Ti ↔ O distance in the brookite structure is practically the same as in the crystal lattices of rutile and anatase.

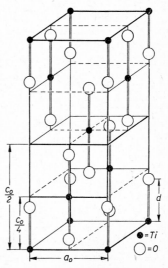

Fig. 16. Unit cell of anatase.
$a_0 = 3.73$ Å, $c_0 = 9.37$ Å,
$d = 1.95$ Å.

The different structures in which TiO_2 crystallizes occupy an interesting transitional position between pure coordination lattices and molecular lattices. One can regard them as coordination structures, in which Ti exerts the coordination number 6 towards O. In all structures, however, 2 of the 6 O-atoms surrounding the Ti differ from the remainder, namely those which are at a greater distance from the Ti atom than the rest. In the crystal lattice of anatase, the line joining these atoms lies in the direction of the *c* axis; in rutile, perpendicular to the *c* axis. In the brookite structure the two O-atoms furthest removed from the Ti atom (Ti ↔ O = 1.98 Å) are the only ones which are exactly equidistant from the Ti atom; the distances of the remaining four differ among themselves as well as from the first mentioned O-atoms. One can thus assign two of the O-atoms in each structure to a particular Ti atom, and can regard the structure as a molecular lattice instead of as a coordination lattice: the O-atoms belonging to the individual TiO_2 molecules are, indeed, repelled from the Ti atom by interaction with O-atoms of neighboring molecules, but they can nevertheless be recognized as belonging to it, (cf. in this connection the discussion of the Al_2O_3 structure in Vol. I, p. 351).

Anatase, according to Schröder (1928), undergoes an enantiotropic transformation at 642°. The two modifications are distinguished as α- and β-anatase. At about 915°, anatase changes monotropically into rutile. Brookite also changes monotropically into rutile. The velocity of transformation diminishes with decreasing temperature, and becomes immeasurably small below 600°.

TiO_2 melts at about 1800°. At higher temperatures it begins to lose oxygen with the formation of Ti_2O_3; the O_2-pressure reaches 1 atmosphere at 2230°. The heat of formation of TiO_2 from $\frac{1}{2}Ti_2O_3$ and $\frac{1}{4}O_2$ is calculated from the dissociation equilibrium to be about 99 kcal. (Junker 1936). Reduction to Ti_2O_3 by means of CO takes place at temperatures as low as 800°.

Titanium dioxide occurring naturally is seldom pure. As a rule it is more or less strongly ferruginous, and for that reason is almost invariably colored, often almost black (*nigrine*). Pure titanium dioxide is colorless when cold, yellowish when hot. It is insoluble in water and in dilute acids, even when prepared in amorphous form. It dissolves slowly in hot concentrated sulfuric acid, better in fused alkali hydrogen sulfates. The titanium sulfate or titanyl sulfate thereby formed is decomposed by boiling with water, even in dilute acid solution. The product of decomposition is called *b-titanic acid*, in contrast to ordinary titanic acid, *a-titanic acid*, which results from precipitation of a freshly prepared acid titanium sulfate solution with alkali- or ammonium hydroxide or carbonate, in the cold. The *b-titanic acid* differs from ordinary titanic acid just as b-stannic acid differs from ordinary stannic acid. It is much less reactive than ordinary titanic acid. Thus it is hardly dissolved by acids, except by hot concentrated sulfuric acid. As with the two sorts of stannic acid, there is a continuous transition between the two titanic acids.

Ordinary titanic acid behaves in a peculiar manner on heating. If it is not too slowly heated, it suddenly becomes incandescent at a certain temperature, without undergoing any change in weight. The b-titanic acid does not show this behavior. The phenomenon depends on the crystallization of the previously amorphous oxide, which takes place suddenly as the temperature is slowly raised, as was shown by Böhm from the X-ray diffraction patterns. The sudden evolution of heat, indicated by the glowing, is brought about by the considerable heat of crystallization which is liberated. A few other oxides display the same phenomenon—in particular ZrO_2, Nb_2O_5, Ta_2O_5, Sc_2O_3, Cr_2O_3 and Fe_2O_3. For most of these, the phenomenon was already noticed by Berzelius. The pre-requisites for its occurrence are that a substance must have a large heat of crystallization, a strong crystallizing tendency, and must be capable of preparation in the amorphous form; it is the superheating of the latter, which leads to the sudden onset of the crystalli-zation process. The 'glowing' is best observed in compact pressed masses, in which the temperature rise propagates itself more rapidly than through a loose powder.

Ordinary titanic acid is a gelatinous substance, of the nature of a hydro-gel. Older methods of investigation did not suffice to decide whether this was actually a gel of titanium dioxide, or of some stoichiometrically defined hydrate of titanium dioxide. Recently, however, by applying the acetone-drying method to a titanic acid gel, prepared by hydro-lysis of $TiCl_4$ at 0°, Schwarz has made it probable that titanium dioxide is present in such a gel in the form of a hydrate $Ti(OH)_4$, i.e.,—as orthotitanic acid.

It was found by Graham that *colloidal dispersions* (hydrosols) of titanium dioxide can be obtained by peptizing titanic acid, precipitated in the cold by ammonia, with hydrochloric acid. According to Wintgen (1936), considerably more concentrated hydrosols can be obtained by dialysis of 1–3% $TiCl_4$ solutions.

Like SiO_2, TiO_2 can also go into solution to a certain extent in *molecular disperse* form. The solutions are, however, much less stable than those of molecularly-disperse silicic acid (Brintzinger, 1931).

Uses of titanium dioxide: see p. 53.

(*ix*) *Titanates*. Neglecting their water of crystallization, the *alkali titanates* correspond mostly to the formulas $M^I_2TiO_3$ and $M^I_2Ti_2O_5$. They may be obtained in the wet way, by evaporating solutions of a-titanic acid in concentrated alkali hydroxide solutions (b-titanic acid is insoluble). They can be prepared in the anhydrous state by fusing titanium dioxide with alkali carbonates. The titanates of other metals can also be prepared by high temperature methods. Compounds of the type $M^I_4TiO_4$ (orthotitanates) are formed in this way by a few metals, as wel as those of the type $M^I_2TiO_3$ (metatitanates). Numerous *polytitanates* (compounds with more than 1 TiO_2 per 1 M^I_2O) have been obtained by fusion methods.

Alkaline earth titanates are known corresponding to the types $M^{II}TiO_3$, $M^{II}_2TiO_4$ and $M^{II}_3TiO_5$. Compounds of the type $M^{II}_3RO_5$ have relatively recently been recognized, through the work of Scholder (1953). He has shown that such compounds are formed by V^{IV}, Cr^{IV}, Fe^{IV} and Co^{IV}, as well as by Ti^{IV}.

Calcium titanate, $CaTiO_3$, occurs in Nature as *perowskite*. The still more commonly occurring mineral *ilmenite* (titaniferous iron ore) sometimes corresponds in composition to iron(II) titanate, $FeTiO_3$, which can also be prepared artificially. Usually, however, it contains more iron than is required by this formula. Its isomorphism with hematite, Fe_2O_3, is worthy of note. In consequence of this, red iron ore (hematite) often contains considerable amounts (up to 7%) of titanium dioxide.

The crystal structure of ilmenite is derived from that of corundum (Vol. I, Fig. 72), with which hematite is isomorphous, in such a way that the aluminum atoms are replaced alternately by titanium and iron atoms. The metatitanates of Co^{II}, Ni^{II}, Mn^{II}, Cd and Mg likewise crystallize with the ilmenite structure. $CaTiO_3$, $SrTiO_3$, and $BaTiO_3$, on the other hand, crystallize in the perowskite structure (see Vol. I, p. 398, Fig. 76). In the ilmenite structure, as also in the perowskite structure, Ti has the coordination number 6 towards oxygen. The same is true for the structure of the compound Fe_2TiO_5 (crystallizing rhombic), which occurs in Nature as *pseudobrookite*. It is built up of (somewhat distorted) TiO_6 octahedra, which are linked together by two opposite corners, so as to form chains running parallel to the c axis. The iron atoms are inserted into the spaces of the crystal lattice in such a way that every iron atom is surrounded, roughly tetrahedrally, by 4 O-atoms. Those titanates of the formula $M^{II}_2TiO_4$ for which the structures are known (M^{II} = Mg, Zn, Mn, Co), crystallize with the *spinel* structure (see Vol. I, p. 355). In these compounds, again, titanium is 6-coordinated; the 'tetrahedral' positions are filled by half the M^{II} atoms, so that these 'titanates' could be formulated as $M^{II}Ti[M^{II}O_4]$. They are, however, better thought of as double oxides, and not as salts.

X-ray structure analysis has shown that titanium is always 6-coordinated in oxide structures. 'Titanic acid' therefore does not replace silicic acid in minerals. *Titanite* or *sphene*, a monoclinic mineral of the composition $CaO \cdot TiO_2 \cdot SiO_2$, is therefore not to be regarded as a disilicate in which 1 Si is replaced by 1 Ti, as was formerly assumed. In view of its detailed structure, and the partial replaceability of Ti by Fe^{III}, it is best represented by the formula $Ca(Ti^{IV}, Fe^{III})(O,OH)[SiO_4]$.

X-ray structure analysis has shown that titanite is a silicate built up of isolated, somewhat distorted $[SiO_4]$ tetrahedra, in the structure of which the Ti atoms are arranged in such a way that each of them is surrounded by the vertices of 4 $[SiO_4]$ tetrahedra, and also by 2 O-atoms which link pairs of neighboring Ti atoms together. The Ti atoms can be partially replaced by Fe^{III}. A corresponding proportion of the O-atoms linking them together are then exchanged for OH groups.

The compound Li_2TiO_3 is not really a titanate, but is a double oxide with the rock salt structure, with a statistical distribution of the cations (see p. 275). The double oxide $2Li_2O \cdot 5TiO_2$ has the spinel structure, and can be formulated as $Li_{1/3}Ti_{5/3}O_4$—i.e., of the 3 cation sites in the spinel structure, $\frac{1}{3}$ are occupied by Li^+ ions, and $\frac{2}{3}$ by Ti^{4+} ions.

(*x*) *Peroxytitanic acid and Peroxytitanates.* Titanum salts, in neutral or acid solution, are colored an intense orange red by hydrogen peroxide. From sufficiently concentrated solutions, peroxytitanic acid, H_4TiO_5, can be thrown down by means of aqueous ammonia as a brownish-yellow precipitate.

Like titanic acid, peroxytitanic acid usually forms a gel with a variable water content. Schwarz was able to dehydrate this, however, by drying with acetone at $0°$. The residue had the composition $TiO_3 \cdot 2H_2O$, and contained 1 atom of 'active' oxygen for each atom of titanium, i.e., an oxygen atom which reduced 2 equivalents of permanganate in dilute sulfuric acid solution, like 1 molecule of hydrogen peroxide. The compound therefore contains 1 peroxygroup —O—O—. It was shown to be a true peroxyacid, and not an H_2O_2-addition product (peroxyhydrate) by Schwarz (1935) from the result of the Riesenfeld-Liebhafsky test (cf. Vol. I, p. 340 et seq.). On the other hand, the compound formerly referred to as potassium peroxytitanate is actually a peroxyhydrate of potassium orthotitanate $K_4TiO_4 \cdot 4H_2O_2 \cdot 2H_2O$.

True peroxytitanates exist, in which other radicals are bound to the titanium, as well as the peroxidic oxygen. To this type belong ammonium pentafluoroperoxotitanate,

$(NH_4)_3 \begin{bmatrix} O \\ | \quad TiF_5 \\ O \end{bmatrix}$ and potassium disulfatoperoxotitanate $K_2 \begin{bmatrix} O \\ | \quad Ti(SO_4)_2 \\ O \end{bmatrix} \cdot 3H_2O$.

The ion $[O_2=Ti(SO_4)_2]^=$ is only weakly complexed, and therefore decomposes in dilute solution to yield *peroxytitanyl ions*:

$$[O_2=Ti(SO_4)_2]^= \rightleftharpoons [O_2=Ti]^{++} + 2[SO_4]^=.$$

As has been shown by Jahr, peroxytitanyl ions are formed when any dilute, strongly acidified titanium or titanyl salt solutions are treated with hydrogen peroxide. This is made use of for the analytical detection of titanium (cf. p. 63). In aqueous solutions containing hydrogen peroxide, an equilibrium is established between the titanyl and the peroxytitanyl ions:

$$[TiO]^{++} + H_2O_2 \rightleftharpoons [Ti=O_2]^{++} + H_2O; \quad \frac{[TiO^{++}] \cdot [H_2O_2]}{[Ti\,O_2^{++}]} = 5.4 \cdot 10^{-5}.$$

Jahr was able to isolate the compound $[TiO][ClO_4]_2 \cdot H_2O_2$ (crystallizing in the hexagonal system) from concentrated solutions of peroxytitanyl perchlorate. This is very hygroscopic, but is colorless, unlike the intensely orange colored peroxytitanyl salt solutions.

(*xi*) *Titanium disulfide*, TiS_2, is formed by passing a mixture of titanium tetrachloride and hydrogen sulfide through a red-hot porcelain tube:

$$TiCl_4 + 2H_2S = TiS_2 + 4HCl.$$

A different reaction takes place between titanium tetrachloride and hydrogen sulfide in the cold, since the latter then acts as a reducing agent, according to the equation $TiCl_4 + H_2S = TiCl_2 + 2HCl + S$.

Titanium disulfide forms brassy-yellow scales, with a metallic luster (for crystal structure see p. 46). It is stable in air at ordinary temperature. When heated in air, however, it is converted into TiO_2, and on heating in a stream of hydrogen or of nitrogen it yields lower sulfides, Ti_2S_3 and TiS. It is not decomposed by water even on boiling and is likewise stable towards dilute sulfuric acid, hydrochloric acid, and ammonia. It is decomposed by nitric acid, however, as well as by hot concentrated sulfuric acid, with the deposition of sulfur. The disulfide is dissolved by boiling caustic potash, with the formation of potassium titanate and potassium sulfide. It reacts with dry carbon dioxide when heated, according to the equation $TiS_2 + 2CO_2 = TiO_2 + 2CO + 2S$.

Biltz (1937) has proved the existence of a higher sulfide of titanium, TiS_3, as well as at least one phase, structurally distinct from Ti and from TiS, with a lower sulfur content than the monosulfide.

(*xii*) *Titanium carbide*, TiC, is present in titaniferous cast iron. It was first prepared pure in an electric furnace by Moissan. Small amounts can most simply be obtained by the filament growth process (p. 65-6), based either upon the reaction $TiCl_4 + CO + 3H_2 = TiC + H_2O + 4HCl$, taking place on a highly heated wire, or on the heating of a carbon filament in $TiCl_4$ vapor (Burgers, 1934). For the preparation of larger quantities of the pure carbide in the compact form, the sintering process is best. In this, an intimate mixture of finely powdered Ti or TiO_2 and ignited carbon black is first heated in a graphite tube furnace, in an atmosphere of hydrogen, to about 2000°. The carbide so formed, after powdering, is compressed under high pressures (2000 kg/cm^2) into rods (pellets) and fired in a graphite tube furnace at 2500-3000°. The firing process is repeated, after fresh powdering and pressing, until the rod is sufficiently compacted and solidified by the 'presintering'. This is followed by a high temperature sintering process, in which a powerful electric current is passed through the rod—i.e., it is heated up almost to the melting point of the carbide, whereby the impurities still contained in the carbide evaporate. The procedure can also be applied to other high-melting carbides (e.f., ZrC, HfC, NbC, TaC) and to the preparation of nitrides, such as TiN, ZrN, TaN (Agte and Moers, 1931).

Titanium carbide is very similar to titanium metal in appearance and behavior, although less readily attacked by acids; melting point 3450°; crystal structure—see p. 46. According to Burgers, one g mol. of TiC is capable of taking up to about 0.1 g atom of Ti in solid solution. The possible existence of a second carbide, Ti_2C, is still uncertain.

On account of its great hardness, TiC finds applications for the preparation of sintered 'hard metals'.

(*xiii*) *Titanium Carbonitrides.* Two compounds of titanium with carbon and nitrogen are known: titanium dicyanide (titanium carbonitride, cochranite), $Ti(CN)_2$, blue crystals, harder than steel; and titanium cyanonitride, $Ti_{10}C_2N_8$ ($= Ti(CN)_2 \cdot 3Ti_3N_2$), copper red crystals. The latter is the compound occurring in the blast furnace slags from the smelting of titanium-rich iron ores, and was originally mistaken for elementary titanium.

(*xiv*) *Titanium Borides.* Titanium can take considerable amounts of boron into solid solution. It also forms two compounds with boron; TiB (cubic) crystallizes with the zinc blende structure ($a = 4.20$ Å), and TiB_2 (hexagonal) is isotypic with AlB_2 ($a = 3.03$, $c = 3.21$ Å). ZrB_2, VB_2, NbB_2, TaB_2, and CrB_2 crystallize with the same structure as the corresponding titanium compound; the structure of MoB_2 is related to the same type. TiB_2 is harder than any other known metallic boride. The borides of all the transition metals cited above are all notable for their great hardness and extremely high melting points; they are good electrical conductors and are very stable chemically. They can be alloyed with the metals of the iron group, and are used to some extent for high temperature materials and cutting tools. According to Kieffer, TiB_2 and ZrB_2 are conveniently prepared technically from boron carbide, B_4C, which is manufactured on a considerable scale as an abrasive. This is mixed with B_2O_3, and treated with metallic Ti:

$$7Ti + 3B_4C + B_2O_3 = 7TiB_2 + 3CO.$$

[*Z. anorg. Chem.*, 268 (1952), 191].

4. Analytical (Titanium)

The intense orange coloration produced in acidified solutions on addition of hydrogen peroxide is highly characteristic of titanium compounds. (see p. 61 *et seq.*)

The hydrogen peroxide reaction for titanium is very sensitive, but hydrofluoric acid, fluorosilicic acid, nitric acid, and large amounts of acetic acid interfere. Titanium also gives characteristically colored substances with many organic compounds, especially with alkaloids, phenols, and with many oxyacids. Thus chromotropic acid (1,8-dioxynaphthalene-3,6-disulfonic acid) is colored blood-red by even minimal amounts of titanium salts in aqueous solutions.

The microcosmic salt bead is not colored by titanium in the oxidizing flame, but is colored violet after prolonged heating in the reducing zone of the flame. The borax bead displays the same coloration, though only in the presence of larger amounts of titanium. The violet coloration characteristic of Ti(III) compounds also appears after the addition of zinc, tin, and other reducing agents to acid solutions of titanium salts. Similar color reactions are, however, given also by a few other substances (tungsten, vanadium, and molybdenum compounds).

Titanium salts are all readily hydrolyzed. However, titanium dioxide hydrate, precipitated in the cold and not aged, is readily dissolved and held in solution by acids, so that, in the course of wet-way analysis, titanium is found chiefly in the ammonium sulfide group. It is not difficult to separate it from other elements of this group, by utilizing the fact that titanium is precipitated as b-titanic acid from dilute acid solution on prolonged boiling.

In general, insoluble titanium compounds are best brought into solution by fusion with potassium hydrogen sulfate. However, the soda-potash melt is also to be recommended on occasion. In the former case, titanium goes into solution when the melt is dissolved in cold water; in the second case, after leaching the melt it is left as alkali titanate, sparingly soluble in cold water but readily soluble in acids.

Titanium can be determined by precipitation as b-titanic acid and weighing as TiO_2, or (more conveniently) titrimetrically, by oxidation of Ti^{3+} to Ti^{4+} by means of ferric alum solution, using KSCN as indicator. Small amounts of titanium are best determined colorimetrically, by means of the reaction with H_2O_2. For the determination of even smaller amounts, the color reaction of titanium(IV) sulfate with salicylic acid in concentrated sulfuric acid is suitable, according to Schenck [*Helv. Chim. Acta*, 19 (1936) 1127].

5. Zirconium (Zr) [*10, 11*]

(a) Occurrence

Compounds of zirconium, like those of titanium, are distributed in small amounts throughout the earth's surface. They seldom occur, however, in major deposits. The mineral which is most important technically (although not, as in the case of titanium, the most abundant) is the dioxide. This occurs as baddeleyite, ZrO_2, in considerable quantities (chiefly in South Brazil and in Ceylon). The silicate of zirconium, zircon, $ZrSiO_4$, is more frequently met with. This is at present less important then baddeleyite, since it is more difficult to work up.

Silicates of zirconium in which the silicic acid is partially replaced by titanic-, niobic, or tantalic acid are not infrequently found in Nature. Zirconium compounds are almost invariably found in the minerals which contain the rare earths. The mineral eudialyte, occurring in Greenland and Norway, contains minor amounts of cerium and manganese oxides, and also chlorine, as well as the oxides of zirconium, silicon, iron, calcium, and sodium.

According to Prandtl (1937), zirconium minerals almost invariably contain phosphoric acid, even though the phosphoric acid content is so small that it may be below the limit of detection, since the detection of PO_4^{3-} (like that of SO_4^{2-}) is hindered by a great excess of

[ZrO]$^{2+}$ ions. The presence of phosphate in zirconium dioxide is easily detected, however, since a dark coloration is produced on ignition in hydrogen, as a result of the formation of ZrP.

(b) History

Zirconium dioxide was first isolated by Klaproth in 1787, from zircon from the island of Ceylon. Berzelius, in 1824, succeeded in preparing the metal (in powder form) by reduction of potassium fluorozirconate with potassium.

All zirconium preparations made prior to 1924 were contaminated to a considerable degree with hafnium, the homologue of zirconium. Almost hafnium-free zirconium preparations were only recently obtained.

(c) Preparation

In a powdered and impure form, the metal is obtained by the method of Berzelius, by reduction of potassium fluorozirconate, $K_2[ZrF_6]$, with potassium or sodium:

$$K_2ZrF_6 + 4Na = 2KF + 4NaF + Zr.$$

Pure zirconium is now prepared industrially by a method similar to the Kroll process for titanium—i.e., by the reduction of zirconium tetrachloride by means of molten magnesium.

Molten zirconium can be obtained by the aluminothermic method, according to Weiss and Neumann; zirconium-aluminum is first produced by reduction of potassium fluorozirconate with aluminum. This is an alloy of zirconium with about 30% aluminum, which was formerly called 'crystallized zirconium'. If electrodes are fabricated from this, and an electric arc burned between them in an atmosphere of nitrogen or ammonia, under a pressure of less than 12 mm, the greater part of the aluminum evaporates while the zirconium melts. Fairly pure compact metallic zirconium can be obtained in this way, by repeated melting.

Analytically pure, but not compact, zirconium metal can be obtained by the method of Lely and Hamburger. In this process, a mixture of anhydrous zirconium tetrachloride with an excess of sodium is heated to 500°. Reaction ensues according to the equation $ZrCl_4 + 4Na = Zr + 4NaCl$. After washing out the product of reaction, first with water, then with hydrochloric acid, the zirconium remains in the form of a powder which can be pressed into rods. This cannot, however, be rolled. This is because each grain of the powder is covered with an extremely thin, analytically indetectable film of oxide or nitride.

Reduction of the dioxide with calcium, by Kroll's method (cf. p. 51) yields a compact metal, which can be worked when hot, but which is brittle when cold, on account of its ZrO content (cf. p. 67).

A process which furnishes quite pure, compact, and ductile zirconium metal has been described by Van Arkel and De Boer (1924) (filament growth process). It depends on the fact that a volatile zirconium compound (zirconium tetraiodide) is decomposed thermally in a vacuum, on an incandescent wire. The temperature of thermal decomposition of zirconium tetraiodide is far lower than the temperature of volatilization of metallic zirconium, so that the latter is deposited on the filament in crystalline form. Nearly pure zirconium in powder form, obtained by one of the processes given above, is first introduced together with some iodine through the opening A into a vessel, made of Pyrex glass, having the form show in Fig. 17. The opening A is then sealed up, and, after evacuating the vessel, B is sealed also. The tungsten filament F, stretched between the two tungsten rods S_1, S_2, is then heated to about 1800° by passing a current through it, while the whole vessel is heated to about 600° in an electric furnace. ZrI_4 first forms. This diffuses to the incandescent filament and is there decomposed, with deposition of the metal. The iodine set free can again transport zirconium, in the form of iodide, to the filament, and so on. As the filament grows in thickness through deposition of the metal, the heating current must be increased—for example, from $\frac{1}{4}$ ampere for a filament of 40 μ diameter to about 200 amperes when the zirconium rod has become 5 mm thick. The very thin tungsten filament

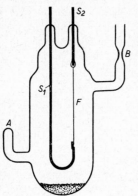

situated in the middle plays no significant role as impurity in such thick rods. In this way, it is possible to obtain zirconium which is as ductile as copper, so that it may be hammered and rolled cold without further ado. Other elements which are hard to prepare pure, such as boron, silicon, titanium, and hafnium, can also be prepared in the pure state and in compact form by the Van Arkel process. Use has also frequently been made recently of the filament growth process for the preparation of pure compounds of these elements (see, e.g., pp. 57 and 63).

Fig. 17. Apparatus for preparation of metallic zirconium by thermal decomposition of zirconium iodide, by the Van Arkel and De Boer method.

(d) Properties

Zirconium, in the pure compact state, is a lustrous metal with a steel-like appearance. In the finely divided state, it forms a black powder. The latter can readily be burned in air, whereas the melted, compact metal only tarnishes superficially on heating in air, and burns only at a very high temperature. In the powdered state, however, it burns in a current of oxygen at red heat. Thin ribbons or wires can be ignited in the flame of a match. The molecular heat of combustion exceeds even that of titanium (cf. Table 6, p. 45).

Metallic zirconium is not attacked by water or caustic alkalis, even when hot. Compact zirconium is also resistant to action by cold or hot nitric acid and hydrochloric acid at all concentrations. It is attacked, however, by hot 60% sulfuric acid. Aqua regia also reacts vigorously with the metal. Hydrofluoric acid is the best solvent both for zirconium metal, and for compounds of zirconium that are rather inert chemically, such as the nitride and carbide. Compact metallic zirconium is fairly resistant towards fused alkali hydroxides. It is attacked by gaseous hydrogen chloride and chlorine at a dark red heat. Zirconium, like titanium, combines vigorously with nitrogen at high temperatures (from about 1000° upwards), forming the refractory nitride ZrN. The carbide ZrC is also highly resistant, as also is the silicide $ZrSi_2$, obtainable by the aluminothermic method, or in the electric furnace, in the form of steel grey, brilliantly lustrous crystals. Hydrogen is also absorbed by powdered zirconium especially if the metal is first heated above its transition temperature in an atmosphere of hydrogen and then allowed to cool.

At room temperature, zirconium can incorporate more than 33 atom-% of hydrogen in its hexagonal crystal lattice. When heated however, a new phase begins to appear, even when the hydrogen content is small; this has the same structure as Mn_4N and Fe_4N, and accordingly is to be considered to be a compound Zr_4H. In the neighborhood of 50 atom-% hydrogen a further phase appears (the compound ZrH), and when the hydrogen content exceeds 65 atom-%, another new phase is formed—namely the compound ZrH_2 (Hägg, 1931). According to Sieverts, the heat of formation of this hydride is 20.2 kcal/g-atom of H.

Oxygen and nitrogen can also be dissolved homogeneously in zirconium in considerable amounts (oxygen up to 40 atom-%). The oxygen is remarkably freely mobile in the zirconium crystal lattice at high temperatures (e.g., 1000°). If an electric field is applied, the oxygen migrates to the anode (De Boer and Fast 1940). The transformation of the hexagonal modification (a-Zr) into cubic β-Zr no longer takes place isothermally with zirconium containing oxygen or nitrogen, but over a wide range of temperature, in which the oxygen-richer a-phase is in equilibrium with the oxygen-poorer β-phase (De Boer, 1936).

(e) Applications

Zirconium, like titanium, is employed as an additive in the casting of metals because of its power of combining with oxygen and nitrogen. It is commonly used in the form of the iron alloy, ferrozirconium. Added to aluminum, it is said to increase resistance towards corrosion by sea water. Filaments in discharge tubes, made from zirconium or zirconium alloys, have a good electron emissivity, and withstand overloading. Metallic zirconium is also used for the fabrication of crucibles, especially for alkaline fusions, in consequence of its excellent corrosion resistance. It is cheaper than tantalum, although the latter metal is in many respects superior in its resistance towards attack. A mixture of zirconium powder and zirconium nitrate forms a smokeless and odorless flashlight powder.

Zirconium dioxide finds many applications, and is increasingly of importance in the refractory ceramic industry, since it is extraordinarily resistant not only to heat and to fluctuations of temperature, but also to chemical effects at high temperatures. It is often mixed with other substances such as magnesium oxide (cf. p. 69). On account of its infusibility, it is suitable as a heat insulating material for furnaces in which the highest temperatures are to be attained. Mixed with graphite, it is used for the preparation of electric heating masses. Zirconium dioxide has also been recommended as a white pigment. It is used in the enamel industry, under the name of 'Terrar', as a turbidifier, in place of the more expensive stannic oxide, over which it also possesses a greater stability towards acids. It is also said to be a good grinding and polishing agent. Its peculiarity of emitting a brilliant incandescence when strongly heated ('thermoluminescence'), has been utilized in various ways (see below).

Zirconia has a high absorption coefficient for X-rays, and is absolutely non-poisonous; it is therefore suitable for use in radiological examinations, to render the digestive tract visible ('Kontrastine' preparation of the Polyphos-Gesell-schaft).

Zirconium dioxide hydrate has been suggested for clarifying sewage, as it has strong adsorptive properties. Various zirconium compounds (e.g., the sulfate) have been found suitable as weighting materials for silk. The extremely hard carbide, ZrC, finds application for cutting glass and as an abrasive.

6. Compounds of Zirconium

Zirconium almost invariably has the valence state +4 in its compounds. It was first proved by Ruff (1923) that, like titanium, it can exhibit other valence states. He reduced the colorless tetrachloride to the dark red-brown trichloride and the black dichloride, by heating it to $350°$ in a sealed, evacuated tube with aluminum powder containing aluminum chloride.

Zirconium dichloride is also obtainable by heating Zr in $ZrCl_4$ vapor, and the dibromide $ZrBr_2$ similarly (De Boer 1930). See also p. 72 for the lower halides of zirconium. ZrO also exists; it is soluble in metallic zirconium, and makes it cold-brittle.

Compounds derived from quadrivalent zirconium are usually colorless, as are the ions Zr^{4+}. These ions, however, hardly occur as such in appreciable amounts, since they have a strong tendency to add on negative ions. Thus, in the presence of F$^-$ ions, the complex ion $[ZrF_6]^=$ is formed; in the presence of sulfate ions, sulfato—or oxosulfato complexes—e.g., $[OZr(SO_4)_2]^=$—and so on. In other cases, the Zr^{4+} ions react with water

$$Zr^{4+} + H_2O \rightleftharpoons 2H^+ + [ZrO]^{++} \quad \text{(zirconyl radical or ion)}$$
$$Zr^{4+} + \tfrac{3}{2}H_2O \rightleftharpoons 3H^+ + \tfrac{1}{2}[Zr_2O_3]^{++} \quad \text{(dizirconyl radical or ion)}$$
$$Zr^{4+} + 2H_2O \rightleftharpoons 4H^+ + ZrO_2.$$

[See, however, Lister, *J. Chem. Soc.* (1952) 4315.]

In aqueous solutions, neutral zirconium salts accordingly undergo a more or less far-reaching hydrolysis, depending on the hydrogen ion concentration, with the formation of *zirconyl* salts $[ZrO]X_2$, or *dizirconyl* salts $[Zr_2O_3]X_2$, which are often obtainable in well crystallized form. On complete hydrolysis, zirconium dioxide, ZrO_2, or its hydrate, is thrown down as a white precipitate. In strongly alkaline solution, the latter, by adding on yet another oxygen ion, passes over into the zirconate ion $[ZrO_3]^=$:

$$ZrO_2 + 2OH^- \rightleftharpoons [ZrO_3]^= + H_2O.$$

Examples of well defined zirconyl and dizirconyl compounds are:

$ZrOCl_2$	zirconyl chloride	$Zr_2O_3Cl_2$	dizirconyl chloride
$ZrOBr_2$	zirconyl bromide	$Zr_2O_3Br_2$	dizirconyl bromide
$ZrO(NO_3)_2$	zirconyl nitrate	$Zr_2O_3(NO_3)_2$	dizirconyl nitrate
		$Zr_2O_3(SCN)_2 \cdot 5H_2O$	dizirconyl thiocyanate
$H_2[ZrO(SO_4)_2]$, zirconyl sul-		$K_2[Zr_2O_3(SO_4)_2]$	potassium dizirconyl sulfate
furic acid *or* disulfatooxozirconic		*or* potassium disulfatotrioxodizirconate	
acid			

The products of variable composition which are frequently deposited from solutions of zirconium salts,—for example, by interaction of zirconium salt solutions with alkali salts of weak acids—are generally only adsorption products of acid or salt on amorphous zirconium dioxide hydrates.

(a) Oxides and Acids

(*i*) *Zirconium dioxide* (*zirconia*), ZrO_2, is formed by the ignition of zirconium dioxide hydrate, or of salts of zirconium with volatile oxyacids, as a hard white powder, insoluble in water. Its melting point lies extremely high (2700°); nevertheless, by heating it with substances which lower the melting point or exert a solvent action (mineralizers) it can be converted into microscopic, tetragonal crystals. The naturally occurring zirconium dioxide, *baddeleyite* (brazilite), is monoclinic.

The monoclinic modification of zirconium dioxide is the stable one at ordinary temperature. According to Ruff and Ebert, a reversible transformation into the tetragonal modification takes place above about 1000°. The oxide prepared by igniting zirconium salts is the (metastable) tetragonal form if it is not heated above 600°. If certain oxides—for example magnesium oxide—are added when it is ignited, cubic mixed crystals (of the fluorspar type) may be obtained. The mixed crystal range ends, in the case of magnesium oxide, with a compound of the composition $2MgO \cdot 3ZrO_2$. The monoclinic modification of zirconium dioxide differs structurally from the cubic mixed crystals in that it has a somewhat distorted fluorspar structure.

Zirconium dioxide which has been recrystallized by fusion is not attacked by acids, except by hydrofluoric acid. Oxide that has been only weakly ignited, however, dissolves rather readily in mineral acid; after strong heating it is soluble only in concentrated sulfuric acid or hydrofluoric acid. Strongly ignited zirconium dioxide can be easily brought into solution by fusion with alkali hydroxide or carbonate; it thereby forms *zirconates*, which are soluble in acids. The density of crystallized, hafnium-free zirconium dioxide is 5.73.

When heated, zirconium dioxide emits an intense light. It therefore formerly found application for optical projection and similar purposes (zirconia light). It was also originally used as the incandescent material in the Auer gas mantle. It later attained importance again in the illuminating industry for the incandescent filament of the Nernst lamp. Nernst lamps contain rods (incandescent pencils) which consist of a mixture of zirconia with yttria. When hot—the heat of a match is sufficient—the mass conducts the electric current, and the heating effect of the current quickly heats it to a brilliant incandescence. The Nernst pencil was in turn displaced from the illuminating industry by the metal filament lamp, but was for a long time used as a conveniently operated intense light source for many laboratory purposes (e.g., for polarimetry). The conductivity of zirconia is electrolytic in nature. The conductivity of *absolutely pure* zirconium dioxide, however, even when hot, is very minute, probably even smaller than that of quartz.

As has been shown by Wagner (1943) and Hund (1952), the electrolytic conductivity of the Nernst filament arises from the fact that ZrO_2 forms 'anomalous mixed crystals' with Y_2O_3, in which there are vacant places distributed statistically at random throughout the anion lattice. These make it possible for the O^{2-} ions to jump from place to place, and therefore to migrate in an electric field. As has already been mentioned, ZrO_2 crystallizes with the fluorite structure if certain other metallic oxides are present. The cations thereby distribute themselves statistically over the appropriate lattice sites while vacancies are created in the anion lattice, in amount corresponding to the deficit of oxygen. Thus in a mixed crystal with the composition $1ZrO_2 + 1YO_{1.5}$, $\frac{1}{4}$ of the anion positions (in statistically random distribution) are unoccupied. The range of homogeneity of ZrO_2-Y_2O_3 mixed crystals extends from 8.5 to 63 mol-% $YO_{1.5}$. The cell dimension increases over this range from 5.12 to 5.23 Å. Mixed crystal formation between ZrO_2 and CaO or MgO also involves the formation of a fluorite-type structure with incompletely occupied anion lattice. *Thorium dioxide* is also capable of forming such anomalous mixed crystals. Pure thoria itself has the fluorite structure, and it is possible to incorporate up to 30 mol-% $YO_{1.5}$ or up to 52 mol-% $LaO_{1.5}$. As with the anomalous mixed crystals formed by ZrO_2, those derived from ThO_2 display an appreciable electrolytic conductivity at elevated temperatures. As is ususal, the temperature-variation of the electrical resistivity can be represented by an expression $\log \varrho_T = a + \dfrac{b}{T}$ or $\varkappa_T = \varkappa_0\, e^{-\dfrac{b}{T}}$ (ϱ_T, $\varkappa_T$ = specific resistance and specific conductivity, respectively, at T °K). Nernst heating elements made from ThO_2-Y_2O_3 mixed crystals have recently been employed for the construction of electric furnaces which can be used in the ordinary atmosphere up to 2000° (Geller, 1941). The interpretation of the connection between electrical conductivity and defects (vacancies) in the anion lattice is of some importance for the production of ceramics which retain their insulating properties at high temperatures. The only substances which can exhibit good insulating properties at high temperatures are those in which there are practically no vacant lattice positions.

The coefficient of expansion of zirconium dioxide is almost as small as that of silica glass. This, combined with the outstanding resistance towards acids and bases, and its high melting point, make zirconia a valuable material for high temperature refractory articles, such as crucibles.

Articles manufactured from pure zirconium dioxide have a tendency to crack. Ruff has shown that this is the result of the change of modification which occurs on strong heating and cooling down again. If magnesium oxide is added, however, the cubic mixed crystals mentioned above are formed, and their structure remains unaltered on cooling. Admixture with magnesium oxide, or with other suitable additions (e.g., CaO, Y_2O_3, CeO_2), therefore prevents the cracking of articles made from zirconium dioxide.

Naturally occurring zirconia is the usual starting point for the technical preparation of zirconium dioxide. It is extracted with boiling hydrochloric acid, in order to remove impurities; it is then converted into zirconium sulfate by fuming with concentrated sulfuric acid, or by fusion with potassium hydrogen sulfate. From the solution of zirconium sulfate, pure zirconium dioxide hydrate can be obtained by means of ammonia, and this is converted by ignition into zirconium dioxide.

Zirconium dioxide hydrate is a white, gelatinous precipitate, of variable water content. As precipitated in the cold, it is readily soluble in dilute acids. It readily forms colloidal dispersions. The gel is much inclined to adsorb acids, as well as alkali, and is peptized by both. It is also called 'zirconic acid', because of its ability to bind alkali. If precipitated hot, preparations are obtained with a smaller water content than is found, at the same temperature, in hydrates precipitated cold. Zirconium dioxide hydrate, when precipitated hot, is sparingly soluble in acids, and is known as b-zirconic acid.

Ordinary zirconic acid (a-zirconic acid), but not the b-zirconic acid displays the peculiar property of glowing vigorously when heated not too slowly to 300°, and after nearly all the water has been given off (cf. a-titanic acid). In this case also, the phenomenon arises from the sudden crystallization of the initally amorphous oxide.

(*ii*) *Zirconates*. Zirconium dioxide displays a weakly acidic, as well as a basic character. This shows itself chiefly in the formation of crystalline zirconates from melts. The marked adsorptive power of colloidal zirconium dioxide hydrate for alkalis is undoubtedly associated with the acidic character of the oxide.

The adsorptive power for alkalis goes so far that the formation of chemical compounds can be simulated. Such products, which in many ways resemble chemical compounds, except that their composition is continuously variable, can be obtained, for instance, by adding zirconium salt solutions to concentrated caustic alkalis. It may, however, be inferred that these are adsorption products of alkali hydroxide on the zirconium dioxide hydrate, formed by double decomposition of the salt with the alkali, and are not chemical compounds; this follows both from the variable (and usually not simple) composition of the precipitates, and because alkali is lost continuously when they are washed.

The zirconates obtainable from melts have been investigated chiefly by Ouvrard. Most correspond to the type $M^I_2ZrO_3$ (metazirconates); orthozirconates, $M^I_4ZrO_4$ also occur. Calcium zirconate $CaZrO_3$ (m.p. 2550°) is isomorphous with perowskite $CaTiO_3$ and with $CaSnO_3$.

The compound $CaO \cdot ZrO_2 \cdot SiO_2$ is isomorphous with titanite (cf. p. 61), and is accordingly to be considered as calcium zirconium oxysilicate $CaZrO[SiO_4]$, not as a mixed zirconate and silicate.

(*iii*) *Peroxyzirconic acid and Peroxyzirconates*. By the addition of hydrogen peroxide to a neutral solution of zirconium sulfate or acetate, a white gelatinous precipitate is obtained, which can be filtered and dehydrated only with difficulty. Schwarz was able to show, however, that after extraction with liquid ammonia it had a composition analogous to that of the corresponding titanium compound—$ZrO_3 \cdot 2H_2O$. It is thus a properly defined chemical compound, and since iodine alone is liberated in the Riesenfeld-Liebhafsky test, without evolution of oxygen, it is a true peroxy compound, *perzirconic acid*, $HOOZr(OH)_3$. The compound also results from the action of sodium hypochlorite on zirconium dioxide hydrate, or from the electrolysis of a sodium chloride solution in which zirconium dioxide hydrate is suspended. Hydrogen peroxide is liberated from it by the action of dilute sulfuric acid. It also loses oxygen, even on standing in warm air, but is more stable than peroxytitanic acid.

By the exchange of an O-atom for the peroxy-group —O—O—, or by combination with H_2O_2, the acidic nature of zirconium dioxide is strengthened to such an extent that true salts with alkalis can be obtained from aqueous solutions. The *alkali peroxyzirconates* and the *alkali zirconate peroxyhydrates* are soluble in water, but can be precipitated with alcohol. By working in the presence of an excess of hydrogen peroxide, and at low temperatures, Schwarz obtained the compound $K_4ZrO_4 \cdot 4H_2O_2 \cdot 2H_2O$, which is analogous to the peroxyhydrate of potassium titanate cited previously. Schwarz was also able to prepare potassium peroxyzirconyl sulfate (potassium disulfatoperoxozirconate)

$$K_2\left[\begin{array}{c} O \\ | \\ O \end{array} Zr(SO_4)_2 \right] \cdot 3H_2O,$$

the analogue of the corresponding titanium compound.

(b) Halogen Compounds

(*i*) *Zirconium tetrachloride and Zirconium oxychloride.* Zirconium tetrachloride can be obtained by passing chlorine over heated zirconium, zirconium carbide, or a mixture of zirconium dioxide (or silicate) and charcoal. It is a white crystalline powder ($d = 2.8$) which forms fumes of hydrochloric acid in moist air, and is vigorously decomposed by water, with the formation of the oxychloride:

$$ZrCl_4 + H_2O = ZrOCl_2 + 2HCl.$$

This hydrolysis is not markedly repressed by the addition of much acid.

Zirconium tetrachloride forms a molecular lattice of the SnI_4 type (see Vol. I, p. 531). Like titanium tetrachloride, it has a tendency to form addition compounds; thus it combines with up to 8 molecules of ammonia. Many organic compounds containing oxygen are also added on. Zirconium tetrachloride reacts with other organic compounds, liberating hydrogen chloride, as for example, with benzoic acid:

$$\begin{array}{c} Cl \\ Cl \end{array}\!\!\!\!Zr\!\!\!\!\begin{array}{c} Cl \\ Cl \end{array} + \begin{array}{c} H—O\cdot CO\cdot C_6H_5 \\ H—O\cdot CO\cdot C_6H_5 \end{array} = \begin{array}{c} Cl \\ Cl \end{array}\!\!\!\!Zr\!\!\!\!\begin{array}{c} O—CO\cdot C_6H_5 \\ O—CO\cdot C_6H_5 \end{array} + \quad 2HCl.$$

By addition of Cl^- ions to $ZrCl_4$, the weak complex ion $[ZrCl_6]^=$ is formed. The compounds of the type $M^I_2[ZrCl_6]$ (*hexachlorozirconates*) derived from this, are obtainable from alcoholic solution. They are immediately decomposed by water, like zirconium tetrachloride itself.

The oxychloride of zirconium formed by reaction of zirconium tetrachloride with water (*zirconyl chloride*) is readily soluble in water, but sparingly soluble in cold concentrated hydrochloric acid. It crystallizes in the form of characteristic tetragonal prisms or needles of the composition $ZrOCl_2 \cdot 8H_2O$; these are most perfectly formed by the slow cooling of a concentrated hydrochloric acid solution of the chloride. The formation of zirconyl chloride is used for the detection of zirconium, as well as for its purification.

Zirconyl chloride can be obtained with a lower water content or quite anhydrous under other conditions. Other zirconium oxychlorides are also known. Thus, by precipitating the alcoholic solution of zirconyl chloride, $ZrOCl_2$, with ether, *dizirconyl chloride*, $Zr_2O_3Cl_2$, is obtained. This compound is insoluble in water. It is also deposited slowly (on standing for months) from highly diluted aqueous zirconyl chloride solutions. These and other oxychlorides of well defined composition are completely different in nature from the products, formerly and erroneously called 'metazirconium chloride', which result from yet more extensive hydrolysis, or from the action of hydrochloric acid on zirconium dioxide hydrate. The latter are actually nothing but adsorption complexes of hydrogen chloride on zirconium dioxide, which readily pass into colloidal dispersion.

For the reduction of the tetrachloride to the trichloride—see p. 67. Zirconium tetra-bromide and zirconium tetraiodide, and the compounds derived from them, are almost exactly like the tetrachloride and its derivatives in behavior. They are also very similar in physical properties as the following table shows:

	$ZrCl_4$	$ZrBr_4$	ZrI_4
Melting point (under pressure)	437°	450°	499° C
Sublimation point	331°	357°	431° C
Heat of sublimation	26.0	26.5	29.5 kcal/mol.

By passing a mixture of $ZrBr_4$ and H_2 over an aluminum wire heated to 450°, Young (1931) obtained zirconium tribromide, $ZrBr_3$. $ZrBr_2$ is formed at the same time. The formation of ZrI_3 and ZrI_2 by heating ZrI_4 with an excess of Zr was observed by Fast (1938). ZrF_3 was isolated by Ehrlich (1952). Unlike TiF_3, ZrF_3 decomposes if the attempt is made to sublime it in a high vacuum.

(*ii*) *Zirconium fluoride and Fluorozirconates.* Zirconium fluoride forms white, highly refrigent monoclinic crystals ($d = 4.6$), hardly soluble in water.

From a solution of zirconium dioxide hydrate in hydrofluoric acid, a compound crystal-lizes which, from its composition, would appear to be simply a hydrate of zirconium fluoride, but which should probably be formulated as fluorozirconylic acid (tetrafluorooxozirconic acid), $H_2[ZrOF_4].2H_2O$. In the presence of alkali fluorides (or of the fluorides of various other elements), complex salts crystallize from the solution, most of them corresponding to the type $M^I_2[ZrF_6]$. The type $M^I_4[ZrF_8]$ is next in frequency of occurrence.

Potassium hexafluorozirconate, $K_2[ZrF_6]$, (colorless rhombic prisms) is par-ticularly stable. Since it is appreciably more soluble in hot water than in cold, it can easily be recrystallized. It can therefore serve for the purification of zirconium compounds. It is also a suitable starting material for the preparation of zirconium metal on the small scale.

Ammonium fluorozirconate, $(NH_4)_2[ZrF_6]$, decomposes into NH_3, HF and ZrF_4 on gentle heating, and is therefore convenient for the preparation of the last-mentioned compound. ZrF_4 begins to sublime above 600°. The sublimation of the fluoride offers a convenient way of separating zirconium from iron and other impurities.

(c) Salts of Oxy-Acids

(*i*) *Zirconium sulfate and Sulfatozirconic acids.* Anhydrous zirconium sulfate, $Zr(SO_4)_2$, may be obtained, as a white powder, by evaporating zirconium dioxide or zirconyl chloride with concentrated sulfuric acid*. It dissolves in water with considerable evolution of heat. Aqueous solutions containing a large excess of sulfuric acid yield a crystalline compound which, from its composition, $ZrO_2 \cdot 2SO_3 \cdot 4H_2O$, could be taken as a hydrate of the sulfate first mentioned. In reality, however, this is a complex acid, a disulfatooxozirconic acid, usually known as *zirconyl sulfuric acid*:

$$H_2\left[OZr\begin{matrix}SO_4\\SO_4\end{matrix}\right].3H_2O.$$

This inference follows from the observation that, in its solutions, the zirconium

* The sulfate so obtained always contains an excess of sulfuric acid, according to v. Hevesy (1931). This can be completely removed only at temperatures at which a per-ceptible decomposition of the sulfate occurs.

migrates to the anode, not to the cathode, and is thus contained in the anion. It accordingly does not exhibit the precipitation reactions characteristic of the Zr^{4+} or $[ZrO]^{2+}$ ions.

Zirconyl sulfuric acid is derived from *trisulfatozirconic acid* ,$H_2[Zr(SO_4)_3]$, by the exchange of one SO_4 group for oxygen. This complex separates in long colorless needles, with 3 molecules of water of crystallisation, from concentrated sulfuric acid solutions.

In neutral aqueous solution, zirconyl sulfuric acid undergoes hydrolytic decomposition, SO_4 groups being exchanged for O-atoms or OH-groups. The so-called basic zirconium sulfates resulting therefrom are mostly polynuclear complex compounds—oxo- and hydroxo-acids. An example of this is the tetrazirconyl sulfuric acid (octahydroxohexa-sulfatotetrazirconic acid), $H_4[Zr_4(OH)_8(SO_4)_6]\cdot 4H_2O$, isolated by Hauser (1907), of which the ammonium salt was subsequently prepared by Rosenheim (1919). These substances of high molecular weight may be obtained in well crystallized form, but their aqueous solutions display the reactions typical of colloids, (for example, they are unable to diffuse through a parchment paper membrane).

Sulfatozirconates. By evaporating mixed solutions of zirconium sulfate and alkali sulfates or hydrogen sulfates over sulfuric acid at ordinary temperature, sulfatozirconates are obtained: e.g., $K_4[Zr(SO_4)_4]\cdot 4H_2O^*$, potassium tetrasulfatozirconate (white silky needles); $(NH_4)_2[Zr(SO_4)_3]\cdot 3H_2O$, ammonium trisulfatozirconate. More frequently oxo- or hydroxo sulfatozirconates are obtained, especially from warm solutions. To this category belong the salts of the tetrazirconyl sulfuric acid already mentioned.

(*ii*) *Zirconyl oxalate and Oxalatozirconates.* Treatment of zirconium salt solutions with oxalic acid, yields a white granular precipitate, which consists of *zirconyl oxalate*, $ZrO(C_2O_4)$, containing some water. The compound is hydrolyzed by hot water.

Zirconium is able to form oxalato-complexes with excess of oxalic acid. For example, if zirconium dioxide hydrate is dissolved in a solution of sodium hydrogen oxalate, sodium tetraoxalatozirconate, $Na_4[Zr(C_2O_4)_4]\cdot 4H_2O$, crystallizes after some time. The acid corresponding to salts of this type is not known in the free state, although oxotrioxalato-zirconic acid (oxalatozirconylic acid), $H_4[ZrO(C_2O_4)_3]\cdot 7H_2O$, is known, and can be obtained crystalline by evaporating down an oxalic acid solution saturated with zirconium dioxide hydrate.

(*iii*) *Zirconium acetate.* If anhydrous zirconium tetrachloride is dissolved in hot glacial acetic acid, and allowed to cool, anhydrous neutral *zirconium acetate*, $Zr(C_2H_3O_2)_4$, crystal-lizes out. It is noteworthy that it should be just with the weak acetic acid that zirconium forms an undoubtedly neutral (normal) salt. With other salts of zirconium, which appear from their composition to be neutral but hydrated salts, it is always possible that they may in reality be oxo-acido compounds—as is demonstrably the case with the sulfate.

Zirconium tetracetate is very unstable. It loses acetic acid, even on standing over sulfuric acid, and is converted into zirconyl acetate, $ZrO(C_2H_3O_2)_2$. The latter is completely hydrolyzed by water, as would be expected for the salt of a very weak base with a weak acid. The zirconium dioxide hereby formed remains, however, in solution in colloidal form. The hydrosol, purified by dialysis, sets to a clear jelly on heating or on addition of salts.

(*iv*) *Nitrates of Zirconium.* By evaporating a solution of zirconium dioxide hydrate in nitric acid, colorless crystals are obtained, with the composition $ZrO(NO_3)_2\cdot 2H_2O$. It is customary to call the compound *zirconyl nitrate*, although the tight binding of the water points to the assumption that it is really an oxonitrato zirconic acid. If it is dissolved in absolute alcohol, and ether is added, *dizirconyl nitrate*, $Zr_2O_3(NO_3)_2\cdot 5H_2O$, is precipitated. A solution of freshly precipitated zirconium dioxide hydrate in nitric acid, prepared at ordinary tempera-ture, and evaporated *in vacuo* at about $15°$ over phosphorus pentoxide and sodium hy-droxide, furnishes large, water-clear prisms corresponding in composition to the *neutral nitrate*, $Zr(NO_3)_4\cdot 5H_2O$. This compound, which is extraordinarily hygroscopic and readily loses nitric acid, is probably to be regarded as a nitrato-acid, perhaps as oxotetranitrato-zirconic acid, $H_2[ZrO(NO_3)_4]\cdot 4H_2O$.

* Water content variable.

(v) *Zirconium phosphate and Phosphatozirconates.* Zirconium orthophosphate, $Zr_3(PO_4)_4$, should result from the precipitation of zirconium salt solutions with orthophosphoric acid or sodium phosphate. According to the experimental conditions, however, precipitates of variable composition are obtained. They generally contain more phosphoric acid than corresponds to the above formula. It has not yet been established whether these are acid phosphates, or whether the excess of phosphoric acid is retained by adsorption. In any case, zirconium dioxide hydrate in the freshly precipitated state has an extraordinarily great adsorptive power for phosphoric acid, as has stannic oxide hydrate. Compounds which are decomposed by water, depositing zirconium dioxide hydrate (e.g., zirconyl chloride or zirconyl nitrate) can accordingly be used in analytical chemistry, in place of stannic oxide hydrate or stannic chloride, for the removal of phosphoric acid.

A phosphate of the composition $2ZrO_2 \cdot P_2O_5$ has been obtained in the crystalline state from a solution of zirconium dioxide hydrate in hot syrupy phosphoric acid. The structure of this compound is not yet known. Crystal structure analysis has shown that the compound of composition $ZrO_2 \cdot P_2O_5$ obtainable in crystalline form by fusion of zirconium dioxide with glacial metaphosphoric acid is *zirconium pyrophosphate*, $Zr[P_2O_7]$, and not, as was formerly assumed, a zirconyl metaphosphate $ZrO(PO_3)_2$. The compound forms a cubic crystal lattice ($a = 8.20$ Å), which is derived from the rock salt structure (Vol. I, p. 209) in such a way that the Na^+ ions are replaced by Zr^{4+} ions, and the Cl^- ions by the central oxygen atoms of the $[O_3P.O.PO_3]^{4-}$ groups. The latter each consist of two tetrahedra of O-atoms, built around a P-atom, with one corner in common. The same crystal structure is possessed by $Hf[P_2O_7]$ ($a = 8.18$ Å), $Ti[P_2O_7]$ ($a = 7.80$ Å), $Si[P_2O_7]$ ($a = 7.46$ Å), and $Sn[P_2O_7]$ ($a = 7.89$ Å).

Some of the double salts of zirconium phosphate with other phosphates (phosphatozirconates) have the same composition as the double phosphates of thorium prepared in a similar manner, but are not isomorphous with the latter.

(vi) *Zirconium silicate.* $Zr[SiO_4]$ occurs in Nature as *zircon*; this constitutes the most widely distributed zirconium mineral. It is found, in the form of microscopically small crystals, in nearly all igneous rocks, but larger zircon crystals are

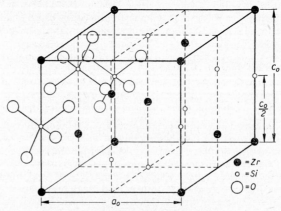

Fig. 18. Structure of zircon.

$a_0 = 6.58$ Å, $c_0 = 5.93$ Å, Si $\leftrightarrow$ O $= 1.62$ Å.

The only oxygen atoms shown are those surrounding the silicon atoms that lie in the left hand face of the unit parallelopiped. Certain of these O-atoms are situated outside the unit cell depicted in the figure.

by no means rare. They are almost invariably well shaped, in the forms of the
tetragonal system; they are generally colored brown by impurities and are more
or less opaque. Transparent colorless (Jargon) and colored zircons, expecially the
yellow-red hyacinth occurring in Ceylon, India, and New South Wales, are used
as gemstones. They have considerable hardness (7.5 on the Mohs scale), and possess
the highest specific gravity of all precious stones (up to 4.8; strangely enough,
zircons with a much lower specific gravity, such as 4.2, also occur).

Zircon is not isomorphous with rutile, as was formerly believed. It belongs, indeed, to the
same class (holohedral class) of the tetragonal system, has a similar axial ratio to that of
rutile, and generally occurs in quite similar forms. Nevertheless, it forms no mixed crystals
with rutile and has a different crystal structure.

The crystal structure of zircon is represented in Fig. 18, which shows the arrangement of
zirconium and silicon atoms. Each silicon atom must be considered as surrounded by 4
O-atoms in almost regular tetrahedral arrangement. The distance Si $\leftrightarrow$ O is 1.62 Å, the
shortest Zr $\leftrightarrow$ O distance, 2.05 Å. Y[PO$_4$], Y[AsO$_4$], and Y[VO$_4$] possess the same
crystal structure as zircon. At 862°, ZrSiO$_4$ changes reversibly into a cubic modification.

No compound other than zircon (m.p. 2430°) occurs in the SiO$_2$-ZrO$_2$ system. Up to
about 10 mol-% of SiO$_2$ is taken up by ZrO$_2$, with the formation of mixed crystals (Zirnowa,
1934).

Double silicates of zirconium, in addition to the compound CaZrO[SiO$_4$] mentioned on
p. 70, include the compounds Na$_2$ZrO[SiO$_4$] (m.p. 1477°, incongruent), Na$_4$Zr$_2$[SiO$_4$]$_3$
(m.p. 1540°), and Na$_2$Zr[Si$_2$O$_7$], which were found by D'Ans (1930) in the ternary system
Na$_2$O-ZrO$_2$-SiO$_2$.

(d) Other Compounds of Zirconium

(i) *Zirconium nitride.* Since zirconium, like titanium, combines, directly with
nitrogen at high temperatures, the metal always contains nitride if nitrogen has
access during the reduction process. If the Van Arkel and De Boer process for the
preparation of zirconium by thermal decomposition of the iodide is operated in the
presence of nitrogen, the nitride ZrN, analogous in composition to titanium nitride, is obtained in very hard, lustrous, metallic crystals (m.p. 2980°), resembling
silver though somewhat yellowish. This compound, which is extremely stable, has
a high metallic conductivity (cf. Table 10, p. 50), as does the boride ZrB
(m.p. 2990°).

A zirconium nitride of reputedly different composition is obtained by heating zirconium
tetrachloride, or its ammoniates, to a red heat in a stream of ammonia or hydrogen. Starting
from the higher ammoniates, ammonia is first split off on moderate heating, with the formation of the tetrammoniate, which is stable up to above 195°. On stronger heating, zirconium
amide is formed by elimination of hydrogen chloride, according to the equation
ZrCl$_4$·4NH$_3$ = Zr(NH$_2$)$_4$ + 4HCl. The amide is stable up to about 250°. If the temperature is finally raised to 350°, it is said that the nitride Zr$_3$N$_4$ is obtained according to the
equation 3Zr(NH$_2$)$_4$ = Zr$_3$N$_4$ + 8NH$_3$. The compound forms bronze-colored crystals,
insoluble in all mineral acids except hydrofluoric acid.

(ii) *Zirconium carbide.* ZrC is obtained in place of the pure metal by the reduction of zirconium compounds by means of carbon. It forms black, lustrous, very
hard crystals of metallic appearance, conducts the electric current very well, and
can be melted by this means (m.p. 3530°). In the fused state, it dissolves carbon,
which separates again in the form of graphite on cooling.

The carbide of zirconium is not nearly as stable as the nitride. It is, indeed, not
attacked by hydrochloric acid, but is attacked by nitric acid, sulfuric acid, and

aqua regia. It is also readily dissolved by molted caustic potash. It burns at a red
heat in a stream of oxygen, and under the same temperature condition is converted
into the nitride by nitrogen. It is much more readily attacked by chlorine, bromine, and iodine. It is thus very suitable as starting material for the preparation of
the halogen compounds of zirconium.

(*iii*) *Zirconium Borides*. In addition to the borides listed in Table 9, zirconium
combines with boron to form the compound ZrB_{12}. This was obtained by Post and
Glaser (1952), as a black powder. It has metallic properties, like ZrB, and is
exceedingly hard.

7. Analytical (Zirconium)

Zirconium has many reactions in common with the elements of the rare earths. It differs
from them in giving no precipitate with an *excess* of oxalic acid, because of the formation of
soluble oxalatosalts. It is distinguished from the elements of the cerium and yttrium earths,
and from thorium, in that it is not precipitated by an excess of hydrofluoric acid (formation
of soluble fluorozirconates). The formation of the oxychloride is also characteristic of
zirconium.

If turmeric paper is moistend with a zirconium salt solution, acidified with hydrochloric
or sulfuric acid, and dried on a water bath, a red-brown coloration is obtained. Solutions
of the rare earths or of thorium do not give this reaction. Titanium(IV) salts interfere,
however, since they similarly give a brown coloration; this interference can be avoided by
reduction of titanium to the +3 state. In contrast to titanium, zirconium gives a *white*
precipitate with hydrogen peroxide.

Zirconium is *determined quantitatively* by precipitation with ammonia as the dioxide
hydrate, or with hydrogen peroxide as peroxyzirconic acid, and weighing in the form of the
dioxide after ignition. *Microchemical detection* is possible by means of rubidium or caesium
fluorozirconates. The reaction with potassium hydrogen oxalate (formation of tetragonal
pyramids of potassium tetraoxalatozirconate) is still more sensitive. Zirconium can be
detected in great dilutions by means of drop reactions with organic reagents, which give
colored, very insoluble precipitates with zirconyl salts—e.g., with reagents such as
β-nitroso α-naphthol, alizarin, or *p*-dimethylamino-azo-phenyl arsenic acid.

8. Hafnium(Hf)

Hafnium was discovered in 1922 by von Hevesy and Coster [*12*] and named after
the city of Copenhagen (Latin, Hafniae). It constantly accompanies zirconium in its
minerals; most zircons contain over 1% of hafnium oxide. That hafnium remained long undetected in spite of this is attributable to the extraordinary similarity in chemical properties between these two elements.

Hafnium has the atomic number 72. It was first shown by X-ray spectroscopy,
on the basis of the Moseley law, that the element corresponding to this atomic
number was still unknown. The conclusion was reached that the element of atomic
number 72 did not, like the preceding elements, belong to the rare earth series, but
that it must be a homologue of zirconium. It was to be expected that such an element would also accompany zirconium in Nature, and indeed von Hevesy and
Coster were at once able to detect the new element by means of its characteristic
X-ray spectrum in every zirconium mineral that they investigated.

As long ago as 1895 Thomsen had argued from the Periodic System, that another element,
differing in character from the rare earths but a homologue of zirconium, must fill a place

between the elements of the rare earths and tantalum. This speculation was also expressed later by various others. When it became apparent from the Moseley law that the element of atomic number 72 was still undiscovered, most chemists nevertheless considered that this element would belong to the rare earth series. Urbain in 1911 had believed that he had discovered a new element of the rare earth series, which he called *Celtium*; Dauvilliers, in 1922, thought, on the grounds of some admittedly very weak X-ray lines, that the atomic number 72 could be assigned to Urbain's celtium. In contrast with this, von Hevesy and Coster arguing from the Bohr theory of atomic structure, expected to find that element 72 would be a homologue of zirconium in zirconium minerals, instead of a member of the rare earths. Bohr had not long previously succeeded in interpreting the Periodic System on the basis of the quantum theory of atomic structure. In so doing, and by considerations like those set out on p. IV, he came to the conclusion that the rare earth series must be concluded with element 71. Element 72 must accordingly have its outer shell constituted similarly to titanium, zirconium, and thorium, and in view of its atomic number, must be regarded as the next higher homologue of zirconium. It was readily identified by von Hevesy and Coster in the zirconium minerals that they examined, by means of a large number of lines of the L-series. The hafnium content could be inferred from the intensity of the lines and it was shown that an enrichment of hafnium could be achieved by fractional crystallization of the potassium and ammonium double fluorides of the hexafluoro-type, the hafnium compounds being more soluble. By sufficiently numerous repetitions of the separation, the pure hafnium salt can be obtained.

It has since been shown that hafnium invariably accompanies zirconium in Nature; the minerals cyrtolite, alvite, naegite, and malacone are particularly rich in hafnium. Cyrtolite contains 52.4% ZrO_2 and 5.5% HfO_2. The hafnium content is highest, as compared with the zirconium content, in thortveitite, a very rare mineral occurring in Norway and Madagascar, of which scandium oxide (combined with silicic acid, cf. p. 34) is the principal constituent. Thortveitite contains only 1 to 2% zirconium oxide, but almost the same amount of hafnium oxide with it. No mineral has been found in which hafnium is the predominant constituent. On the average, zirconium is accompanied in minerals by 2 to 2.5% hafnium. Hafnium is not found in association with any element other than zirconium—not even thorium.

The similarity of hafnium to zirconium, both in the metallic state and in its compounds, is greater than that between any other homologous elements. This is due partly to the fact that not only are the outer shells of the zirconium and hafnium atoms similarly constituted, but the atoms and also the ions are very closely similar in size (see Table 8, p. 47). This is connected with the filling of the intermediate $4f$ shell which leads to the occurrence of the lanthanide group.

The filling of the inner shell, which results in the appearance of the lanthanide group, has as its consequence not only a shrinkage in the atomic and ionic radii, but also a diminution in the polarizing action of the ions. Hafnium ions, in spite of their slightly smaller ionic radius are more weakly polarizing than zirconium ions, as follows from the lower stability of hafnium complex compounds, as compared with the corresponding compounds of zirconium. As a result of the weaker polarizing effect of hafnium ions, the *complex compounds* of this element are unusual, in that they mostly have a larger molecular volume than the corresponding zirconium complex compounds, whereas, according to more recent determinations (which disagree to some extent with older measurements) the molecular volumes of *simple compounds* of hafnium are uniformly *smaller* than those of zirconium compounds (see Table 11). Because of its stronger polarizing power, zirconium behaves in many respects as if it had a smaller radius than is deduced from its simple compounds. Thus the difference in radius between Zr and Hf is compensated (often, indeed a little overcompensated) by the difference in polarizing effects. The extraordinarily great similarity between the zirconium and hafnium compounds is thereby accounted for. It is much greater than that between niobium and tantalum compounds, even though the atomic and ionic radii of Nb and Ta differ from each other even less than do those of Hf and Zr (see Tables 8 and 12, pp. 47 and 88). Since the niobium ion polarizes more strongly than does the tantalum ion, it generally behaves as if it had an appreciably *smaller* radius than the tantalum ion. Proper allowance for differences in polarizing power does away with the discrepancies among the transition elements, which appear at first sight if the degree of similarity between the compounds of the different homologues is judged from atomic and ionic radii alone.

TABLE II

DENSITIES AND MOLAR VOLUMES OF SIMPLE ZIRCONIUM AND
HAFNIUM COMPOUNDS FROM X-RAY MEASUREMENTS

Compound	ZrO_2 cubic	HfO_2 cubic	ZrF_4	HfF_4	ZrC	HfC	ZrP_2O_7	HfP_2O_7
Density	6.27	10.43	4.66	7.13	6.613	14.19	3.195	4.278
Molar vol.	19.65	19.17	35.9	35.7	15.61	13.43	83.00	82.41

The compounds of zirconium and hafnium generally agree very closely in their solubilities, as well as in their melting points and volatilities. Even with the ammonium hexafluorosalts (the different solubilities of which have successfully been applied to the separation of Zr and Hf), the difference between the solubilities is only small. At 20°, the solubilities are, for $(NH_4)_2[ZrF_6]$, 1.050 mols, for $(NH_4)_2[HfF_6]$ 1.425 mols per liter. The melting point of hafnium oxide lies about 100° higher than that of zirconium oxide. The melting points of the carbides are: ZrC 3530°, HfC 3890°.

Larger differences in solubility often occur when the process of dissolution is accompanied by complex formation with some substance present in solution. In general, zirconium is more strongly inclined to form complexes than is hafnium. Thus, according to von Hevesy (1930) 0.017 g-mol. of ZrO_2 dissolve in 1 liter of 13-N HBr solution at 25°, but only 0.004 g-mol. of HfO_2. This relatively greatly different solubility is brought about by the difference in ability to form bromo complex ions, as is shown by the fact that it at once vanishes if the concentration of the hydrobromic acid is reduced. The solubility behavior of the oxides in hydrofluoric acid and hydrochloric acid parallels that in hydrobromic acid. The differences in solubility are nevertheless relatively smaller, but the solubilities considerably greater.

Use can advantageously be made of the different stability of zirconium and hafnium complexes for the separation of hafnium from zirconium. Thus, if a solution of zirconium and hafnium phosphates in concentrated sulfuric acid is diluted with water, the hafnium is preferentially precipitated as phosphate, since it is less firmly held in solution through the formation of sulfato complexes. In this instance, the separation is yet further favored in that hafnium phosphate is more sparingly soluble than zirconium phosphate; however, mere fractional precipitation of the phosphates from aqueous solution does not achieve nearly as rapid a separation as does the process previously described. It is also possible to employ a procedure in which zirconium and hafnium phosphates are dissolved in hydrofluoric acid, and the phosphatofluorozirconates and -hafnates are fractionated by the addition of boric acid (formation of fluoroborate ions) (De Boer, 1927).

Differences between the capacity of zirconium and hafnium for complex formation can be utilized to effect a separation of the elements by much more convenient methods than those used by the original investigators. One group of methods involves the preferential formation of complex compounds which are soluble in organic solvents—e.g., from thiocyanate solution (Fischer, 1939) or by means of complexing agents such as trifluoroacetylacetone (Schultz and Larsen, 1950) or thenoyltrifluoroacetone (TTA) (Huffmann and Beaufait, 1949). A second group of methods is based on selective adsorption from solutions in which the zirconium and hafnium are present as complex ions. Methods which are convenient on the laboratory scale and on a larger scale are adsorption on to silica gel from

concentrated hydrochloric acid solution in methyl alcohol, whereby hafnium is preferentially adsorbed (Beyer, Jacobs and Masteller, *J. Am. Chem. Soc.*, 74 (1952), 825), and the use of ion exchange resins to take up either the uncomplexed cations or the acido-complex anions present in solution (Huffmann and Lilley, 1949). The general procedure is to adsorb the mixed elements on to a column of the adsorbent or resin, and to pass a suitable complex-forming solution through the column to elute the adsorbed material. The effect of a deep bed of adsorbent is to multiply the separation which would be achieved in a single step of adsorption followed by elution, and thereby to achieve a great degree of fractionation in one operation.

Zirconium and hafnium compounds naturally often differ considerably in density. This difference can be used for the determination in a simple way of the hafnium content of zirconium preparations. The density of pure zirconium oxide, obtained by careful thermal decomposition of the sulfate, followed by ignition at 1000°, is 5.73 at 20°; that of hafnium oxide prepared in similar fashion is 9.68. The percentage hafnium content x of an oxide mixture with the density s at 20° may then be calculated by means of the formula

$$\frac{9.68x}{100} + \frac{5.73(100 - x)}{100} = s, \quad \text{or:} \quad x = \frac{100(s - 5.73)}{9.68}.$$

9. Thorium(Th)

(a) Occurrence

Thorium occurs principally in *monazite*, a mineral which is not rare, on the whole, although it is accumulated in only a few places. Monazite is an isomorphous mixture of the phosphates of the rare earth metals; besides these, it contains from 1–18% of thorium oxide, predominantly in the form of silicate, possibly in part as phosphate also.

Other rare earth minerals also almost invariably contain thorium. Pure thorium minerals, on the other hand, are quite rare. Thorium silicate, $ThSiO_4$, which is isomorphous with zircon, occurs in a more or less decomposed condition in Norway, as *orangeite* and *thorite*. The deposits there, however, are almost exhausted. *Thorianite*, occurring in Ceylon, consists chiefly of thorium oxide, ThO_2, but contains appreciable amounts of uranium oxide as well. Uranium is, in fact, almost always encountered in greater or less amounts in thorium minerals.

(b) History and Preparation

Thorium, in the form of its oxide, was discovered by Berzelius in 1828 in a Norwegian mineral. The element was named after the old Norse god of thunder, Thor, and the mineral was called thorite. Berzelius attempted to isolate the metal by reduction of the double fluoride or the double chloride with potassium or sodium, but because of the great affinity of thorium for oxygen, nitrogen, and other elements, and because of its high melting point, he was only very incompletely successful. The methods of preparation worked out later, which depend for the most part on the reduction of the chloride with sodium, have also almost always led to a more or less impure metal. Pure thorium has only been obtained quite recently, by the method of Van Arkel and De Boer, i.e., decomposition of the iodide on an incandescent filament, as described for zirconium.

(c) Properties

Thorium in powder form is grey; and in the compact state resembles platinum in appearance; the metal is soft and ductile, and its mechanical properties are not affected by a small oxide content. (For melting point and density, see p. 44). Except when extremely pure, the metal is hardly, or only slowly, attacked by dilute acids, including hydrofluoric acid. It dissolves readily, however, in fuming hydrochloric acid, and especially in aqua regia. It is stable towards caustic alkalis. At about 500°, thorium unites vigorously with the halogens, and at higher temperatures with nitrogen also. With the latter it forms a brown nitride Th_3N_4. Thorium burns in a stream of oxygen to the oxide ThO_2, with a very great evolution of heat (cf. p. 45). Thorium in powder form reacts with hydrogen at 300–400° with incandescence. The product, after cooling, is a black powder which contains 3 atoms of H to 1 of Th.

It has been shown that thorium forms two distinct compounds with hydrogen, ThH_2 and ThH_3. The former has been investigated by neutron diffraction, and shown to have the same structure as ZrH_2.

At elevated temperatures, thorium also unites directly with phosphorus and with sulfur. It forms the phosphides $\overline{ThP}$ and Th_3P_4 (Biltz and Meisel, 1938–9) and the sulfides $\overline{ThS}$, Th_2S_3, $\overline{ThS_2}$ and $\overline{Th_3S_7}$ (Biltz 1941).
Relatively little is yet known of the behavior of thorium towards other metals (cf. Table 9, p. 48–9). Only 0.014 atom-% of thorium dissolves in mercury at 25° (Parks 1936).

(d) Uses

Metallic thorium has, as yet, found but limited technical applications*. The oxide, on the other hand, is of great importance in the fabrication of incandescent gas mantles and as a high grade refractory. Thorium oxide is also used as an additive to tungsten for metal filament lamps to increase the life of the filament. (On the use of thorium oxide in the preparation of *single crystal filaments* from tungsten, see p. 174). In X-ray diagnostic work, thorium oxide is used, because of its high absorptive power for X-rays, as a contrast medium in the examination of the digestive organs; it also finds other uses in medicine. Recently, considerable quantities of thorium oxide have been consumed in preparing the mixed catalysts for the Fischer-Tropsch hydrocarbon synthesis. Thorium fluoride, mixed with other salts, is said to be a suitable filling material for cored carbons for arc lamps. Thorium nitrate, mixed with magnesium powder, is used for flash powders.

There has been an over-production of thorium because of the decline of the gas mantle industry, which accompanied the displacement of incandescent gas lighting by electric lighting. Therefore new possibilities of employing the metal or its compounds have been sought in recent years. Whereas the thorium compounds formerly constituted the main product in working up monazite sand, they now represent only a by-product because of the steeply rising application of compounds of cerium, lanthanum, neodymum, praseodymium, and other rare earth metals. As a consequence of the increasing technical use of the rare earths, the consumption of monazite has greatly increased in recent years, after a temporary decrease. In 1934, production in Brazil amounted to 580 tons, and in British India to 1025 tons (in 1935, to as much as 3880 tons). It is likely that the technology of nuclear energy will create a new demand for thorium as a starting material for the production of fissile material (^{233}U) in 'breeder' piles (see Chap. 13). ^{233}U is formed by the nuclear reactions

$$^{232}_{90}Th \xrightarrow{(n,\gamma)} {}^{233}_{90}Th \xrightarrow{\beta} {}^{233}_{91}Pa \xrightarrow{\beta} {}^{233}_{92}U.$$

---

* It is used chiefly as an additive to the alloys which find application for the heating elements of electric furnaces, to lower the rate of oxidation, and as a means for combining with residual gas (a 'getter') in high vacuum technology.

10. **Compounds of Thorium**

Thorium is electropositive and almost exclusively quadrivalent in its compounds. Thorium(IV) compounds are, in general, colorless; the simple compounds which are soluble in water dissociate to form colorless Th^{4+} ions. The hydroxide of thorium possesses purely basic character; the salts readily suffer hydrolysis in warm solutions. Most thorium salts are able, with the salts of more strongly basic metals, to enter into the formation of double (complex) salts (acido-thorates). The ability to form numerous double salts is particularly noteworthy in the case of thorium nitrate, since nitrates generally display little tendency for the formation of such compounds. The solubility of the double carbonates of thorium possesses significance for analytical and preparative purposes.

Compounds of trivalent and bivalent thorium include the iodides ThI_3, ThI_2 (D'Eye 1949, Hayek 1949) and the corresponding bromides (Hayek 1950). The monoxide ThO is known as a surface compound, and the sulfides ThS, Th_7S_3, and Th_7S_{12} are also known.

The starting material for the preparation of most thorium compounds is the nitrate, or the oxide obtained from it by ignition. The nitrate is prepared technically from monazite sand. This is a brightly colored mixture of gravity-concentrated heavy minerals of all sorts, the sandy weathering products of rocks containing monazite. It occurs in deposits of enormous size, in some places on the Brazilian sea coast, in Carolina in the United States, in New South Wales, in India, etc.

The crude monazite sand is first considerably enriched in monazite by mechanical ore dressing processes. Its commercial value is determined by the content of thorium oxide, which ordinarily amounts to 4–5% in the purified sand. The sand is opened up with concentrated sulfuric acid at a moderate temperature; after being cooled down, the solidified mass is dissolved in cold water, and the thorium is precipitated as phosphate, along with the rare earth metals, by neutralizing the solution. The phosphates are dissolved in concentrated hydrochloric acid and the oxalates precipitated with oxalic acid. When the well-washed precipitate is extracted with warm soda solution, the bulk of the rare earths remain behind, undissolved, while the thorium goes into solution as a complex carbonate. It is freed from the residue of rare earths still remaining by repeated crystallization as the sulfate, $Th(SO_4)_2 \cdot 8H_2O$, being at each stage precipitated as hydroxide by means of ammonia; this is then reconverted to the sulfate by dissolving it in sulfuric acid. The nitrate is obtained by finally dissolving the hydroxide in nitric acid. If the nitrate is to be obtained absolutely pure, special procedures must be employed.

(a) Oxide

(*i*) *Thorium oxide* (thoria), ThO_2, remains behind on the ignition of thorium hydroxide $Th(OH)_4$ (or thorium dioxide hydrate—see below), or of the thorium salts of volatile acids, as a white powder, which may be dense or voluminous according to the mode of preparation. Its density is 9.87. By melting it with a flux, such as borax, the oxide can be converted to a distinctly crystalline form; it crystallizes in the cubic system (fluorite structure, $a = 5.57$ Å). Crystallized or strongly ignited thorium oxide is practically insoluble in acids, but can be readily brought into solution by converting it to the sulfate by fusion with alkali hydrogen sulfates, or evaporation with concentrated sulfuric acid. It is not attacked by fused alkali carbonates, however, nor by molten caustic soda, since ThO_2, (contrast TiO_2, ZrO_2 and HfO_2) is not capable of salt formation with basic oxides.

If thorium oxide is prepared by cautious ignition of the oxalate (between 500° and 600°), it is obtained in the form of a very loose powder (so-called meta-thorium oxide) which, unlike ordinary thorium oxide, can apparently be brought into solution by treatment with acids—e.g., with dilute hydrochloric acid. This really involves a *peptization* process, however, similar to that for 'metastannic acid', except that the thorium oxide hydrosols are much more stable than the hydrosols of 'metastannic acid'. The residues remaining when they are evaporated down can be brought into colloidal dispersion once more by means of water. Thorium oxide is positively charged in the sols. The ease with which thorium oxide enters the colloidal state is important for its use as an additive to tungsten for electric light filaments.

When solutions of thorium salts are treated with ammonia or with alkali hydroxides, a white gelatinous precipitate is obtained, which is customarily called *thorium hydroxide*. Whether this is really such, or merely the hydrogel of a particularly reactive form of the oxide, is not yet known. In structure, the precipitate is amorphous. Only after heating for a long time under the solution from which it is precipitated does it give X-ray interference rings and these then correspond to those of crystalline thorium *oxide*. The fresh precipitate readily dissolves in acids, and avidly absorbs carbon dioxide from the air; it thus has fairly strong basic properties. It does not dissolve in caustic alkalis, but does, however, in alkali carbonate solutions, as a consequence of complex salt formation.

Pure thorium dioxide does not possess the property of emitting a brilliant light at the temperature of incandescence, as zirconium dioxide does. However, if it contains small amounts of colored rare earths—e.g., cerium oxide—it luminesces with extraordinary brilliance when brought into a flame. This is the basis of the use of thorium in the gas mantle industry, which was initiated by Auer von Welsbach, and which was at one time very important. In the year 1913, about 150,000 kg were utilized in the form of the nitrate for this purpose, and more than 3000 tons of monazite sand were used for the production of this quantity. It is true that incandescent gas lighting has been very largely displaced by the rivalry of the metal filament electric lamp; nevertheless, it is still used extensively in some parts of the world, especially for street lighting. Incandescent mantles are also widely used with vapor-type oil burners, where electricity supplies are not available.

The extraordinary luminosity of thorium oxide impregnated with a little cerium oxide is attributed to the fact that thorium oxide radiates but little heat, and so is heated to a high temperature in the Bunsen flame, whereas the admixed cerium oxide has the property of emitting a very intense light at high temperatures. Pure cerium oxide luminesces but little when introduced into the Bunsen flame, for the reason that is has the property of radiating not only much light, but also much heat simultaneously; as the emission of light increases strongly with temperature, and pure cerium oxide does not attain a very high temperature in the Bunsen flame because of its strong radiation of heat, its high emissivity for light can only come into play when the cerium oxide is embedded, in relatively small amount in thorium oxide, with its low emissivity for light and heat. Only in this way can it attain the high temperature necessary for strong luminosity in the Bunsen flame. In general, a mixture of 99% thorium oxide with 1% cerium oxide has shown itself to be the most satisfactory (cf. below). For a mantle prepared with this mixture, the maximum emission lies in the short-wave length part of the spectrum. For this reason, the incandescent gas light has a white color, tending towards greenish, as contrasted with the more reddish light of other incandescent bodies.

For the practicable utilization of incandescent bodies, two further properties are of importance. Thorium oxide possesses these in higher degree than all other oxides which might come into consideration from their thermal-optical properties; namely, extra-ordinary resistance to heat, and involatility. The last is also essential for the utilization of cerium oxide, since this must not show any tendency to sublime out of the mass in the course of time.

To produce the skeleton of the mantle, a 'stocking', woven from ramie fiber, is impregnated with a mixture of thorium and cerium nitrates in the right proportions. For ordinary gas mantles, solutions with 99.0–99.2% thorium nitrate and 1.0–0.8% cerium nitrate are used; for high pressure inverted gas mantles, up to 3% cerium nitrate is employed. After drying, the fabric is burned. The skeleton of ash remaining, consisting essentially of a mixture of the oxides of thorium and cerium, is stabilized for handling and dispatch by dipping it in a solution of collodion, which is easily removed by burning it off.

(ii) *Thorium oxide peroxyhydrate*. Hydrogen peroxide throws down a white, slimy precipitate from thorium salt solutions. Like the corresponding precipitates from titanium and zirconium salt solutions, this has the properties of a hydrogel and can be dehydrated only with difficulty. After extraction with liquid ammonia, it has the composition $Th_2O_7 \cdot 4H_2O$. It contains 3 g-atoms of active oxygen to 2 g-atoms of thorium, but on the basis of the Riesenfeld-Liebhafsky test it is to be considered not as a true peroxide, but as a peroxyhydrate of thorium oxide—i.e., as an addition compound of thoria with hydrogen peroxide $2ThO_2 \cdot 3H_2O_2 \cdot H_2O$ (Schwarz 1935). The compound is more stable than are the peroxyacids of titanium and zirconium. However, it cannot form peroxy salts with alkalis as they do. The inability of thorium to form a true peroxy acid is apparently conditioned by the more strongly basic nature of thorium. Corresponding to this and, unlike zirconium, thorium forms no peroxy-double sulfates.

(b) Halides

(i) *Thorium chloride*, $ThCl_4$, forms colorless needles of density 4.59, m.p. 765°, and b.p. 922°; the compound sublimes at about 750° in a vacuum. It may be obtained by passing a mixture of chlorine and carbon monoxide, or of carbon tetrachloride, over thorium oxide heated to a red heat:

$$ThO_2 + 2CO + 2Cl_2 = ThCl_4 + 2CO_2.$$

Thorium chloride dissolves in water with the evolution of much heat, undergoing extensive hydrolytic decomposition in the process. It is also readily soluble in alcohol. Hydrates of thorium chloride, and also under some conditions hydrated oxychlorides, crystallize when solutions of thorium hydroxide in dilute aqueous-alcoholic hydrochloric acid are evaporated.

Thorium chloride can add on other chlorides, with the formation of double salts of the type $2M^ICl \cdot ThCl_4$, or $M^I_2[ThCl_6]$. Like the chlorides of titanium and zirconium, it also forms addition compounds with ammonia ($ThCl_4 \cdot 6NH_3$ and $ThCl_4 \cdot 8NH_3$), with organic amines, and with numerous organic oxygen compounds. As with zirconium chloride, many of the compounds with the latter are not simply formed by addition, but involve simultaneous elimination of hydrogen chloride. Thus, with benzaldehyde:

$$\begin{array}{c} Cl \diagdown \diagup Cl \\ Th \\ Cl \diagup \diagdown Cl \end{array} + \begin{array}{c} H.CO.C_6H_5 \\ \\ H.CO.C_6H_5 \end{array} = \begin{array}{c} Cl \diagdown \diagup CO.C_6H_5 \\ Th \\ Cl \diagup \diagdown CO.C_6H_5 \end{array} + 2HCl.$$

thorium dibenzoyl
dichloride

(ii) *Thorium bromide* and *thorium iodide* closely resemble thorium chloride in their behavior. Oxyhalides are derived from these also. Thorium bromide is less inclined to form addition compounds than is thorium chloride, and no addition compounds of thorium iodide are known.

(iii) *Thorium fluoride and Fluorothorates*. Thorium fluoride, ThF_4, may be obtained in the anhydrous state as a white powder by passing hydrogen fluoride over anhydrous thorium chloride or bromide, heated to about 400°. Thorium fluoride containing water of crystallization, $ThF_4 \cdot 4H_2O$, also a white powder, results from the action of hydrofluoric acid on thorium hydroxide. It is insoluble in water and in excess hydrofluoric acid, and so is formed as a gelatinous precipitate, later becoming granular, when thorium salt solutions are treated with hydrogen fluoride or with ammonium fluoride.

The fact that only a few fluorosalts derived from thorium fluoride are known may well be connected with the insolubility of the parent compound; the concentration of the thorium fluoride present in aqueous solution does not suffice, in general, for the formation of the apparently weak fluorothorate complex. Only when the salts derived from the latter are likewise sparingly soluble can they be obtained from aqueous solution. The known compounds correspond in composition to the types $M^I[ThF_5]$ and $M^I_2[ThF_6]$.

(c) Salts of Oxy-Acids

(i) *Thorium nitrate and Nitratothorates.* The simple nitrate of thorium, $Th(NO_3)_4$, crystallizes from solutions of thorium hydroxide in nitric acid, and has a water content varying with the conditions of preparation. In the cold, it crystallizes from not too strongly acid solutions with 12 molecules of water. It is very readily soluble in water and in alcohol. The aqueous solution reacts acid, because of hydrolysis, and slowly deposits basic salts.

Thorium nitrate has been found particularly suitable for the preparation of the oxide skeleton of the gas mantle, since the oxide formed from it by ignition has a particularly fine state of subdivision, as is desirable for this purpose. The nitrate is, accordingly, prepared technically on a large scale. The commercial product is generally the 4-hydrate; it usually contains a little sulfate also.

Thorium nitrate combines very readily with the nitrates of univalent and bivalent metals, to form beautifully crystallized double (complex) salts. The majority of the double nitrates of thorium (nitratothorates) correspond to the type $M^I_2[Th(NO_3)_6]$. The alkali salts of this type are anhydrous, the alkaline earth salts all contain 8 molecules of water of crystallization. In addition, hydrated alkali nitratothorates of the type $M^I[Th(NO_3)_5]$ also exist.

(ii) *Thorium acetate*, $Th(C_2H_3O_2)_4$, separates in colorless needles when a solution of thorium hydroxide in acetic acid is evaporated. Precipitates of basic acetates are obtained when thorium salt solutions, treated with sodium acetate, are boiled.

(iii) *Thorium sulfate and Sulfatothorates.* Anhydrous thorium sulfate, $Th(SO_4)_2$, forms a white powder, readily soluble in water. It is obtained in a not quite homogeneous state by evaporating thorium dioxide with concentrated sulfuric acid. Hydrates crystallize from the aqueous solutions. According to Meyer, it may be obtained pure by the careful dehydration of the octahydrate. It is converted into the oxide by strong ignition in the blowpipe flame. The hydrates of thorium sulfate which crystallize from the solutions possess differing water contents, according to the conditions of preparations. The octahydrate $Th(SO_4)_2 \cdot 8H_2O$ is obtained by the evaporation, at 30–35°, of a thorium sulfate solution containing only a small excess of sulfuric acid. It is unstable, however; the forms stable, in contact with the solution, are the enneahydrate below about 45°, and the tetrahydrate at higher temperatures.

By the addition of alkali sulfates to thorium sulfate solutions, complex salts are formed, most being fairly sparingly soluble—the sulfatothorates correspond to the types

$$M^I_2[Th(SO_4)_3], \quad M^I_4[Th(SO_4)_4], \quad M^I_6[Th(SO_4)_5] \quad \text{and} \quad M^I_8[Th(SO_4)_6].$$

(iv) *Thorium oxalate and Oxalatothorates.* Thorium oxalate, $Th(C_2O_4)_2 \cdot 6H_2O$, results as a precipitate, insoluble in water and in dilute acids, when thorium salt solutions are treated with oxalic acid or alkali oxalates. Complex salts (oxalatothorates), principally of the type $M^I_4[Th(C_2O_4)_4]$, can be isolated from the solution of thorium salts in excess aqueous alkali oxalate.

(v) *Carbonatothorates.* Thorium hydroxide and many other sparingly soluble compounds of thorium can readily be brought into solution by means of water containing alkali carbonate or ammonium carbonate. This arises from the formation of complex salts—carbonatothorates, $M^I{}_6[Th(CO_3)_5]$—which are readily soluble in water. Their solubility permits of the separation of thorium from the rare earths accompanying it in monazite (cf. p. 81).

(vi) *Thorium phosphate and Thorium silicate.* Thorium metaphosphate, $Th(PO_3)_4$, can be isolated from the melt of the dioxide and phosphorus pentoxide; apparently no well defined compounds can be obtained by precipitating thorium salt solutions with alkali phosphates or phosphoric acid.

By melting thorium dioxide and quartz together, silicates are obtained—$ThSi_2O_6$ and $ThSiO_4$. The latter occurs in Nature as orangeite (greasy-lustrous, transparent or translucent orange colored crystals), or in a highly weathered state as black, opaque thorite.

(d) Other Compounds of Thorium

(i) *Thorium carbide and Silicide.* Thorium carbide, ThC_2, is formed by heating thorium oxide with carbon in an electric furnace. The pure compound forms yellowish crystals. It is decomposed not only by dilute acids, but even by water; various hydrocarbons, principally acetylene, together with hydrogen, are thereby evolved. *Thorium silicide,* $ThSi_2$, which is formed under similar circumstances when silicon is used for the reduction of the oxide, is less readily attacked.

According to Brauer, $ThSi_2$ has a tetragonal structure, with a highly symmetrical and regular arrangement of the atoms. Isotypic with it are $LaSi_2$, the silicides of the lanthanide elements Ce, Pr and Nd, and the silicides of the actinide elements, USi_2, $NpSi_2$ and $PuSi_2$.

Thorium nitride, Th_3N_4, is formed by heating metallic thorium, or a mixture of thorium oxide and Mg, in an atmosphere of nitrogen. It forms a yellow to yellow-brown powder, which is stable towards heat if air is excluded. If heated in air, it ignites and burns to ThO_2.

11. Analytical (Thorium)

In its analytical reactions, thorium resembles zirconium. Like the latter it is precipitated as the white oxide hydrate by ammonium sulfide, and by alkali- or ammonium hydroxide or by barium carbonate. It also resembles zirconium in that the precipitate thrown down by ammonium carbonate is soluble in excess of the precipitant. It may be differentiated from zirconium by its behavior towards oxalic acid and in dilute mineral acids. It dissolves, however, in ammonium oxalate, but is reprecipitated from the solution on the addition of hydrochloric acid. It is further distinguished from zirconium in that it is precipitated by hydrofluoric acid, even from dilute solutions.

Thorium is precipitated as the oxalate from acid solutions along with the rare earths, from which it can be separated by taking advantage of the solubility of its hydroxide in alkali- or ammonium-carbonate solutions, or by precipitation as the oxide peroxyhydrate. According to Meyer and Speter (1911), thorium, (even in very small amounts), can be detected in the presence of rare earths, by the precipitation of the iodate from strong nitric acid solution.

For quantitative determination, thorium is generally separated as the hydroxide or oxide peroxyhydrate, and converted by strong ignition to the oxide, in which form it is weighed. It can also be precipitated quantitatively as the iodate.

References

1 R. KIEFFER and P. SCHWARZKOPF, *Hartstoffe und Hartmetalle*, Vienna 1953, 717 pp.

2 W. M. THORNTON, *Titanium*, New York 1927, 262 pp.

3 M. DERIBÈRE, *Le Titane et ses Composés dans l'Industrie*, Paris 1936, 154 pp.

4 J. BARKSDALE, *Titanium, Its Occurrence, Chemistry, and Technology*, New York 1949, 591 pp.

5 G. F. COMSTOCK, S. F. URBAN and M. COHEN, *Titanium in Steel*, New York 1949, 320 pp.

6 J. L. EVERHART, *Titanium and Titanium Alloys*, New York 1954, 189 pp.

7 S. ABKOWITZ, J. J. BURKE and R. H. HILTZ, *Titanium in Industry*, New York 1955, 234 pp.

8 M. K. McQUILLAN and A. D. McQUILLAN, *Titanium* [H. M. FINNISTON (Editor), *Metallurgy of the Rarer Metals*, Vol. IV], New York 1955, 470 pp.

9 K. HEISE, *Titanweiss*, Dresden 1936, 96 pp.

10 F. P. VENABLE, *Zirconium and Its Compounds*, New York 1922, 173 pp.

11 G. L. MILLER, *Zirconium* [H. F. FINNISTON (Editor), *Metallurgy of the Rarer Metals*, Vol. II], New York 1954, 400 pp.

12 G. v. HEVESY, *Das Element Hafnium*, Berlin 1926, 49 pp.

CHAPTER 4

FIFTH SUB-GROUP OF THE PERIODIC SYSTEM: THE ACID EARTHS

Atomic numbers	Elements	Symbols	Atomic weights	Densi- ties	Melting points	Boiling points	Specific heats	Valence states
23	Vanadium	V	50.95	5.98	1715°	about 3500°	0.1203	II, III, IV, V
41	Niobium	Nb	92.91	8.58	1950°	about 5100°	0.0645	II, III, IV, V
73	Tantalum	Ta	180.95	16.69	3010°	about 6000°	0.0333	II, III, IV, V
91	Protactinium	Pa	231	15.37	—	about 4200°	—	IV, V

1. Introduction

(a) General

The fifth Sub-group of the Periodic System comprises the elements vanadium niobium, tantalum, and protactinium. In accordance with their group number, these elements are normally *pentapositive*. Although definitely metallic in character in the elementary state, they are decidedly acidic in their normal oxides, the pentoxides. This is true of vanadium, niobium, and tantalum at least—and for this reason their pentoxides are also known as the *acid earths* (i.e., acid-forming metal oxides), or as the earth acids. *Protactinium* is one of the natural radioelements, and is the parent element of actinium (cf. Chap. 11). It has been but little investigated chemically, as yet, because of its rarity and the difficulty of extracting it. In accordance with the general rule in the Periodic System, it appears that the basic character is more strongly developed in protactinium than in the first three members of the group, which are not capable of forming simple salts in aqueous solution even with the strongest acids.

Like the elements of the 5th Main Group, those of the 5th Sub-group can form compounds from lower valence states, as well as from the +5 state. However, although the tendency to assume a lower positive charge than 5 units *increases* with increasing atomic weight in the 5th Main Group, it *decreases* in that direction in the 5th Sub-group. Thus, treatment of solutions with zinc and acid reduces +5 vanadium to the bivalent state, and +5 niobium only to the trivalent state, whereas tantalum compounds are not reduced at all under these experimental conditions.

In contrast with the elements of the 5th Main Group, those of the 5th Sub-group—like the elements of the Sub-groups of the Periodic System generally—never function as electronegative constituents of salt-like compounds. This accords with their position, since they are not closely followed by the group of inert gases, which have electron configurations of special stability. Also connected with their electronic configuration is the fact that the elements of the 5th Sub-group do not form gaseous hydrides. They form *alloy-like* hydrides, however, like those found in the 4th Sub-group.

According to Sieverts, 1 g of vanadium at ordinary temperature can absorb 157 cc of hydrogen under atmospheric pressure, 1 g of niobium 104 cc, and 1 g of tantalum 55.6 cc. At 500°, the corresponding figures are 19.0, 47.4, and 14.8 cc; at 1000° 2.3, 2.8 and 1.4 cc. By variation of the hydrogen pressure, the amounts absorbed are approximately proportional to the square root of the pressure, at temperatures above 600°. At lower temperatures, because of the formation of compounds, the absorption of hydrogen increases much more strongly with increase of pressure. At room temperature and atmospheric pressure, 1 g-atom of vanadium takes up 0.71 g-atom of hydrogen, 1 g-atom of niobium takes up 0.87 g-atom, and 1 g-atom of tantalum, 0.90 g-atom (the corresponding quantities at 500° and 1000°, respectively, are 0.09, 0.39, 0.24, and 0.01, 0.02, 0.02 g-atom of hydrogen). Hägg (1930) has established by X-ray investigations that tantalum can take up to 12 atom-% of hydrogen into its crystal lattice; at higher hydrogen contents, Ta_2H is first formed, and at higher hydrogen contents still, TaH. The structure of the latter resembles that of zinc blende (distance $Ta \leftrightarrow H = 1.901$ Å).

Vanadium, niobium, and tantalum crystallize with the body-centered cubic structure (cf. Vol. I, p. 210, Fig 47) with $a = 3.034$, 3.299, and 3.298 Å, respectively. They thus resemble the alkali metals in crystal structure, but have considerably smaller atomic radii. The atomic radii of Nb and Ta are identical, within the error of measurement. The ionic radii also differ but little from one another (cf. Table 12). The compounds of niobium and tantalum are, accordingly, very similar in behavior; nevertheless, the similarity is not as great as that between the compounds of zirconium and hafnium (cf. p. 77).

TABLE 12
APPARENT ATOMIC AND IONIC RADII OF GROUP VB ELEMENTS

Element	Vanadium	Niobium	Tantalum
Atomic radius, Å	1.32	1.43	1.43
Radius of 4-valent ion, Å	0.61	0.69	—
Radius of 5-valent ion, Å	ca. 0.4	0.69	0.68

The elements of the 5th Sub-group are usually considered to be among the rarer elements. For vanadium this statement should be qualified with the proviso that minute amounts of this element are encountered in numerous minerals, although actual ores of vanadium are of rare occurrence. Protactinium, which always occurs in Nature along with radium, since both are formed by radioactive decay of the same element (i.e., from uranium or one of the isotopes of uranium; cf. pp. 536 *et seq.* for further details), is even rarer than radium.

The heats of formation of the pentoxides, in kcal/g equiv. of metal are: $[V_2O_5]$ 37.3, $[Nb_2O_5]$ 46.3, $[Ta_2O_5]$ 49.3. For the same quantity of metal they are thus considerably smaller than the heats of formation in the 4th Sub-group (see Table 6, p. 45). As with the latter, they increase with increasing atomic weight of the metal.

TABLE 13

MISCIBILITY AND COMPOUND FORMATION BETWEEN METALS OF SUB-GROUP V
AND OTHER METALS, INCLUDING CERTAIN NON-METALS

(Symbols have the same meaning as in Table 9, see p. 49)

Main Groups

	Be	Mg	B	Al	C	Si	N	P	As	S	Se
V	— Be_2V	—	— VB_2 2100°	$s > o$ $Al_3V?$ $AlV?$ $AlV_2?$	$s > o$ V_5C V_2C V_4C_3 $VC*$ 2750°	$s\ o$ V_2Si VSi_2 1654°	— $V_3N?$ V_2N VN $VN_2?$	— V_3P $V_2P?$ $V_3P_2?$ VP VP_2	— VAs VAs_2	$s\ o?$ VS V_2S_3 $\overline{VS_4?}$	— VSe V_2Se_3
Nb	—	—	— NbB_2 >3000°	— $Al_3Nb?$	— Nb_2C NbC 3500°	— $NbSi_2$	$s \sim o$ $\overline{Nb_2N}$ Nb_4N_3 NbN $Nb_3N_5?$	— NbP NbP_2	— $NbAs$	$s\ o$ $\overline{NbS}$ $\overline{Nb_2S_3}$	alloys
Ta	compound ?	alloys	— TaB_2 >3000°	— $Al_3Ta?$	— Ta_2C TaC 3900°	— $TaSi_2$	— TaN $Ta_3N_5?$ $TaN_2?$	$s\ o?$ TaP TaP_2	$s\ o$ —	$s\ o$ TaS $\overline{TaS_2}$ $\overline{TaS_3}$	—

Sub-groups

	Cr	Mo, W	Mn	Fe	Co	Ni	Cu	Ag
V	—	—	$s > O?$ $MnV?$	$s\ \infty$ $[FeV]$ 1234°	$s < \infty?$ o	$s\ \infty$ o	$liq \sim o$ $s\ o$ o	$liq \sim o$ $s\ o$ o
Nb	alloys	—	— Mn_2Nb	$s > o$ Fe_3Nb_2 1660° Fe_xNb_y	— Co_2Nb	$s > o$ $Ni_3Nb*?$ 1380° Ni_xNb_y	—	—
Ta	— Cr_2Ta	— alloys	— Mn_2Ta	$s > o$ Fe_2Ta	$s > o$ Co_2Ta	$s < \infty$ Ni_3Ta 1545° $NiTa*$ 1580°	—	$liq\ \infty$ $s\ o$

(b) Alloys

Our knowledge of the *alloys* of the 5th Sub-group metals is still fragmentary, as Table 13 shows. The reasons for this are the same as with the alloys of the metals of the 4th Sub-group. Precise knowledge is extant only for their behavior towards a few metals of the Sub-groups (Fe, Ni, Cu, Ag) which are of particular technical importance, as well as for a few systems, of alloy-like nature, in the formation of which non-metals are concerned.

Biltz (1938) has investigated the systems Nb-S and Ta-S, tensiometrically and with X-rays. It has been shown that neither Nb or Ta can incorporate appreciable amounts of sulfur in their crystal lattices, but that the sulfides of these elements may vary widely in composition. They thus have the character of non-Daltonide compounds. The majority of these sulfides can be obtained with the composition represented by the ideal formula, since the latter generally corresponds to one of the phase boundaries. The monosulfide of niobium, NbS, for example, can take up no excess of sulfur in its crystal structure, but can, however, incorporate up to about 1 atom of excess Nb. The sesquisulfide Nb_2S_3, on the other hand, can take up no excess of niobium, but up to as much as 5 atoms of excess sulfur ('NbS_4'). Tantalum forms a trisulfide TaS_3 which can take up excess sulfur but no excess of tantalum, and a disulfide TaS_2 which can likewise take up excess sulfur, and in such quantities, indeed, that its S-content can even exceed that of TaS_3. In addition, tantalum forms at least one lower sulfide (TaS?), likewise of variable composition.

It may be mentioned that V can apparently be alloyed with Pt. It has also been established that Ta and Nb can form mixed crystals with each other in all proportions. Nothing is known as yet of the systems V-Nb and V-Ta.

2. Vanadium (V)

(a) Occurrence

Vanadium is fairly widely distributed in the earth's crust, but nevertheless very rarely occurs in large quantities. Until a few years ago the most important ore for the technical extraction of vanadium was *patronite* (a vanadium sulfide), which occurs principally in a vast deposit in the Peruvian Andes, at an altitude of 5500 meters above sea level. At present, the South West African and Rhodesian vanadium ores (cf. p. 92) are far more important. The first vanadium mineral to be discovered was *vanadinite*, (vanadium lead ore), $3Pb_3(VO_4)_2 \cdot PbCl_2$; it occurs principally in Mexico and other South American countries, and also in Spain.

In Colorado and Utah, and also in South Australia, *carnotite* is found. This was formerly worked on account of its radium content, and is again being worked for uranium; it furnishes vanadium as a by-product. *Roscoelite* should also be mentioned; this is a vanadium-containing silicate of complex composition, which occurs in Colorado, California, New South Wales, and South Australia, chiefly in a sandstone of characteristic green color (vanadium sandstone). Before the discovery of patronite and carnotite, this was the most important vanadium ore.

Small amounts of vanadium are often found in lead ores, and occasionally in copper and iron ores also.

(b) History

In 1830 a new element was discovered by Sefström in an iron possessing notably good malleability and originating from the ores of Taberg in South Sweden. This he named vanadium after the Norse goddess Freyja, who also bore the name Vanadis (i.e., goddess of the Vanir*).

Twenty nine years before this, vanadium had already been discovered by the Mexican mineralogist del Rio, in a lead ore from Zimapan in Mexico, and had been first called

---

* The Vanir are a race of gods in Norse mythology, equivalent to the Aesir. The Saga recounts that Freyja, descended from the Vanir, came with her brother Freyr, the god of peace, to the Asen, by exchange of hostages.

Panchrom (from the Greek πᾶν and χρῶμα, 'the all colored') because of the varied color of its oxides, and then erythronium (ἐρυθρός, red) on account of the beautiful red color of its heated salts. Del Rio did not succeed, however, in definitely establishing the individuality of the substance he had discovered; in fact, he himself later concluded that there had been confusion with chromium. The lead ore of Zimapan was, nevertheless, in reality a vanadium ore, as was established by Wöhler as soon as he heard of Sefström's discovery. It received the name *vanadinite*. Vanadium was at first considered to be a homologue of chromium until the English chemist Roscoe, in 1867, recognized without doubt that it belonged to the same family as phosphorus—i.e., to the Vth group of the Periodic System.

(c) Preparation

In the technical extraction of vanadium from patronite, the rock containing this ore is first melted, with the addition of fluxes, in a reverberatory furnace. The specifically denser impurities are thereby deposited, while the vanadium passes into the slag. Since the preparation of pure vanadium metal is very difficult, alloys of vanadium are usually prepared for technical purposes, and especially the *iron alloy*, so-called *ferrovanadium*. Ferrovanadium is generally prepared by the alumino-thermic process, in which the vanadium-containing slag, mixed with powdered aluminum and iron, is heated to redness in crucibles, with the addition of fluxes (borax and fluorspar), and is then fired in the usual way. The process can also be carried out on a larger scale in *shaft furnaces*.

The preparation of almost pure (98–99%) vanadium metal was effected by Ruff and Martin, by aluminothermic reduction of the trioxide in a magnesia crucible (with the addition of carbon), the regulus so obtained being remelted in the vacuum arc furnace. Döring (1934) obtained pure vanadium in powder form by the prolonged heating of VCl_3 in a stream of hydrogen, a method already applied in principle by Roscoe, in 1870. Compact, completely pure and therefore ductile vanadium is obtainable by the filament growth process of De Boer and Van Arkel.

(d) Properties

When almost pure, vanadium is a steel grey metal, of relatively low density but very high melting point; it is extremely hard, but may be ground and polished.

According to Ruff, absolutely pure vanadium melts at 1715°, but even a small carbon content suffices to raise the melting point considerably. Thus Ruff found the melting point to be 2185° for vanadium with 2.7% carbon, and 2700° with 10% of carbon. Any oxygen or oxide content also leads to elevation of the melting point. Since the amount of the elevation is simply related to the content of carbon or oxygen, it is possible to extrapolate with certainty from the melting points of metal samples with varying amounts of impurity to the melting point of the pure metal.

In the compact state, vanadium is not affected at ordinary temperature by air, by water, or by caustic alkalis. It is also resistant to non-oxidizing acids, with the exception of hydrofluoric acid. It dissolves in strongly oxidizing acids such as aqua regia and nitric acid. When ignited, the finely divided metal burns in compressed oxygen, predominantly to the pentoxide V_2O_5. Lower oxides are also always present in the products of combustion.

The heats of formation (according to Siemonsen and Ulich, 1940) are, for V_2O_3 296; for VO_2 171; for V_2O_5 373 kcal/mol. Vanadium can incorporate up to 0.4 atom-% of oxygen in solid solution. Its crystal lattice is thereby deformed tetragonally (in the same way as is α-iron, through incorporation of carbon—cf. pp. 263–265).

When metallic vanadium is heated in a stream of nitrogen, vanadium nitride, VN, is obtained. Other nitrides are also said to exist (cf. Table 13). Vanadium combines when it is heated with arsenic, probably forming VAs and VAs$_2$. According to the conditions of experiment, the silicides V$_2$Si and VSi$_2$ are obtained by heating vanadium and silicon together in the electric furnace; these are characterized by hardness and resistance to attack by chemical reagents at low temperatures. In these properties they resemble the carbides V$_2$C and VC, which are formed by the union of vanadium and carbon at high temperatures (for other carbides see Table 13).

(e) Uses

The chief field of application of vanadium is in the iron and steel industry[1]. Vanadium has the property of making steel and cast iron stronger, tougher, more ductile, and more resistant to shock and impact. This is particularly important in the airplane-engine and automobile-engine industries. The addition of only 0.1 to 0.2% of vanadium serves to bring about the results indicated. They depend in part, like the effects of titanium, on the strong de-oxidizing action exerted by vanadium; in addition to this, there is a specific effect of vanadium on the texture of the alloy. Alloys of vanadium with other metals, as for example with copper, nickel, or aluminum, have proved useful for certain purposes.

Compounds of vanadium are suitable as oxygen carriers for many purposes. They can thus be used for the catalytic oxidation of organic compounds—e.g., for the preparation of aniline black by the oxidation of aniline by means of potassium chlorate or potassium dichromate. Their uses in the preparation of phthalic anhydride by the oxidation of naphthalene, and especially as catalysts for the manufacture of sulfuric acid by the contact process (cf. Vol. I, p. 707) are of very great industrial importance. It is also advantageous to use vanadium compounds as catalysts in the preparation of borate peroxyhydrates (so-called 'perborates') since they accelerate the addition of oxygen. Vanadic acid and sodium vanadate also find therapeutic use. Vanadium compounds are also used to some extent for photographic purposes; thus compounds of trivalent and quadrivalent vanadium may be employed as developers.

The world production of vanadium amounted in 1935 to about 630 tons. 58% of the ores consumed in this came from South West Africa, 27.5% from Northern Rhodesia, and only 10.6% from Peru, which in 1930 had been the principal source of vanadium ores.

3. Compounds of Vanadium

Vanadium forms four important oxides:

VO   Vanadium monoxide (vanadium(II) oxide), black, density 5.76;
V$_2$O$_3$ vanadium sesquioxide (vanadium(III) oxide), black, density 4.87;
VO$_2$ vanadium dioxide (vanadium(IV) oxide), dark blue, density 4.40;
V$_2$O$_5$ vanadium pentoxide (vanadium(V) oxide), yellow red, density 3.34.

In addition to these oxides which correspond to well defined valence states, Aebi has shown that there is an intermediate oxide V$_{12}$O$_{29}$. This oxide, like the complex oxides which are formed by molybdenum and tungsten also, is related in principle to the 'anion deficient' structures referred to (p. 69) in discussing the ZrO$_2$-Y$_2$O$_3$ mixed crystals. Whereas in the latter the vacant oxygen sites are distributed at random throughout the crystal lattice, in these complex intermediate oxides the 'defects' are ordered, to build up a regular repeating pattern in the crystal lattice.

Each of the four valence states of vanadium represented by the most important oxides gives rise to a series of vanadium compounds.

1. *Vanadium(II) Compounds.* Vanadium(II) hydroxide, $V(OH)_2$, functions as a base; its salts are compounds of the type VX_2. These are capable of forming double salts or weak complex salts [acidovanadates(II)]. Examples of simple vanadium(II) salts are: VCl_2 and $VSO_4 \cdot 7H_2O$ (isomorphous with $FeSO_4 \cdot 7H_2O$); complex salts: $K_4[V(CN)_6] \cdot 3H_2O$ (isomorphous with $K_4[Fe(CN)_6] \cdot 3H_2O$).

2. *Vanadium(III) Compounds.* Salts of the type VX_3 and the corresponding complex salts [acido vanadates(III)], are known. The tendency to form acido compounds is more strongly marked with trivalent vanadium than with bivalent, but it is weaker than, for example, with trivalent iron, chromium, and cobalt. The most stable of the vanadium(III) salts are the sulfate and the sulfatovanadic(III) acid, $HV(SO_4)_2 \cdot 6H_2O$ derived from it.

3. *Vanadium(IV) Compounds.* The oxide of quadrivalent vanadium is able to form salts with both acids and bases. Compounds of the latter type are called *hypo-vanadates*. Salts obtained from aqueous solution in which the vanadium dioxide functions as the basic constituent are all of the type $[VO]X_2$ (*vanadyl* salts), and not of the type VX_4 (where X is a univalent acid radical). In the compounds, the positive divalent radical $[VO]^{II}$ (vanadyl) plays a role corresponding to that of a bivalent heavy metal. It exists in aqueous solution as a bright blue ion. The acido-salts derived from the vanadyl salts are termed *oxovanadates(IV)*, in accordance with the general rule for naming coordination compounds (cf. Vol. I, p. 406 *et seq.*). Many vanadium(IV) compounds—e.g., the tetrafluoride VF_4, and also the alkali salts of the sulfatoxovanadic(IV) acid, $H_2[(VO)_2(SO_4)_3]$, undergo decomposition into vanadium(III) and vanadium(V) compounds when they are heated.

4. *Vanadium(V) Compounds.* Vanadium pentoxide, V_2O_5, is an acidic oxide. The salts derived from it are called *vanadates*; the simplest correspond to the type $M^I_3[VO_4]$ (orthovana\ates), and thus are analogous in composition to the salts of orthophosphoric acid.

Exchange of O^{2-} ions in orthovanadates by radicals of hydrogen peroxide, O_2^{2-}, produces *peroxyvanadates*, salts of the acids $H_3[VO_2(O_2)_2]$ and $H_3[V(O_2)_4]$. The oxygen atoms of the vanadates can also be exchanged by sulfur, thus forming the *thiovanadates*, e.g., $(NH_4)_3[VS_4]$, ammonium orthothiovanadate. Finally, halogeno salts and other acido salts [*acido-vanadates(V)*] are also derived from penta-valent vanadium: e.g., $K_2[VO_2F_3]$, potassium dioxotrifluorovanadate(V).

Vanadium(II) compounds were formerly designated as vanad*o* compounds, and vanadium(III) compounds as vanad*i* compounds. However, this terminology is inexpedient, for reasons given earlier (see Vol. I). According to the international rules for naming inorganic compounds (see supplement I) the valence should now quite uniformly be denoted by Roman numerals: e.g., vanadium(V) oxide, vanadium(III) chloride.

4. Vanadium(II), (III) and (IV) Compounds

(a) Vanadium(II) Compounds

(i) *Vanadium(II) oxide* (vanadium monoxide), $\overline{VO}$, (with 47–56 atom-% O) is obtained as a black powder (density 5.23), insoluble in water, and with a metallic luster in the compact state, when vanadium oxytrichloride, $VOCl_3$, or vanadium pentoxide is reduced by suitable means—e.g., by heating the latter compound to 1700° in hydrogen. It has a good electrical conductivity. It dissolves in acids with the formation of vanadium(II) salts.

$\overline{VO}$ forms a crystal lattice of rock salt type ($a = 4.081$ Å). It is stable only at high temperatures. When cooled slowly it decomposes into V-O mixed crystals and a higher oxide (Klemm, 1942).

(*ii*) *Vanadium(II) hydroxide*, $V(OH)_2$, is thrown down as a brown precipitate when alkal is added to vanadium(II) salt solutions. It has not yet been obtained pure, on account of its extraordinarily great tendency to be oxidized.

(*iii*) *Vanadium(II) chloride* (vanadium dichloride), VCl_2, is obtained in the form of bright green leaflets, with a micaceous luster, when a mixture of hydrogen and vanadium tetrachloride is passed through a red hot tube. It is very hygroscopic. The aqueous solution, fnitially violet, soon turns green because divalent vanadium is oxidized to the trivalent state, with the evolution of hydrogen. Vanadium(II) chloride dissolves in alcohol with a blue-, and in ether with a greenish yellow color. The different colors of the solutions indicate the formation of solvates.

(*iv*) *Vanadium(II) sulfate*, $VSO_4 \cdot 7H_2O$, results from the reduction of sulfuric acid solutions of vanadium(V) by means of zinc, sodium amalgam, and other metals. The preparation can best be done by electrolytic reduction at a mercury cathode. Although the tendency of vanadium(II) sulfate to be oxidized is also so great that it gradually decomposes water itself (i.e., the reaction $V^{++} + H_3O^+ \rightarrow V^{+++} + \frac{1}{2}H_2 + H_2O$ proceeds spontaneously), it can nevertheless be deposited in the crystalline state by suitable experimental technique. It then forms red-violet monoclinic crystals $VSO_4 \cdot 7H_2O$ which can form mixed crystals with other sulfates of the type $M^{II}SO_4 \cdot 7H_2O$ (the vitriols). It is strictly isomorphous with the vitriols, and depending on the other partner, it can crystallize not only in the monoclinic form, (as for example with $FeSO_4 \cdot 7H_2O$, iron vitriol), but also rhombic, (as for example in mixed crystals with magnesium sulfate, $MgSO_4 \cdot 7H_2O$).

(*v*) *Sulfatovanadates (II)*. Vanadium(II) sulfate readily forms double salts of the type $VSO_4 \cdot M^I_2SO_4 \cdot 6H_2O$ with alkali sulfates. These are sparingly soluble; although rather more stable than the simple vanadium(II) sulfate, they also are oxidized very readily.

(*vi*) *Potassium hexacyanovanadate(II)* crystallizes in the form of brownish yellow crystals, $K_4[V(CN)_6] \cdot 3H_2O$, when alcohol is added to a vanadium(II) salt solution treated with an excess of potassium cyanide. Like the analogously constituted potassium hexacyanoferrate(II), this compound gives precipitates with most bivalent metals, but it is much less stable than the iron compound.

(b) Vanadium(III) Compounds

(*i*) *Vanadium(III) oxide* (vanadium sesquioxide), $\overline{V_2O_3}$, may be obtained by the reduction of vanadium pentoxide at high temperatures by means of hydrogen, carbon, or potassium cyanide. It forms a black lustrous, crystalline, extremely infusible powder which has the corundum structure, $a = 5.43$ Å, $a = 53°53'$.

V_2O_3 forms a double compound with FeO, $FeO \cdot V_2O_3$, with spinel structure, which forms solid solutions with metallic iron (Mathewson, 1932). There is no double compound V_2O_3 with VO; however, the oxygen content of the compound can be lowered to 57 atom-% (corresponding to the composition $VO_{1.35}$) without any structural change ensuing.

V_2O_3 is said to be an excellent catalyst for hydrogenation reactions. (Komarewsky and Knaggs, 1951). It has the advantage of not being poisoned by organic sulfur compounds.

(*ii*) *Vanadium(III) hydroxide*, $V(OH)_3$, is obtained as a flocculent green precipitate by treating vanadium(III) salt solutions with alkali or ammonia. It absorbs oxygen with extreme avidity.

(*iii*) *Vanadium(III) chloride* (vanadium trichloride), VCl_3, can be obtained by heating V_2O_5 with S_2Cl_2 in a sealed tube at about 300° (Funk, 1940), or through thermal decomposition of VCl_4. It forms peach blossom colored, lustrous crystal leaflets, resembling anhydrous violet chromic chloride; it is not volatile, and is transformed into a mixture of oxychloride and pentoxide when it is heated in air.

In water, acidified to prevent hydrolytic decomposition, vanadium trichloride dissolves with a greenish color. Provided that oxidation by atmospheric oxygen is avoided, green hygroscopic crystals of the composition $VCl_3 \cdot 6H_2O$ are deposited from the solution on evaporation, a behavior corresponding to that of chromium and iron. Vanadium(III) chloride yields intense colorations with many organic hydroxy acids in aqueous solution, e.g., an amethyst violet with salicylic acid, and a deep red coloration with meconic acid. The latter can be used for the detection of opium (Woynoff, 1934).

The bromide, VBr_3, displays behavior similar to that of the chloride, except that it is appreciably less stable. The iodide, VI_3, may be obtained by direct union of the components above 150°, as a black-brown hygroscopic crystal powder (density 4.2), sparingly soluble in water and more soluble in absolute alcohol. It decomposes into I_2 and VI_2, when heated at 280°. The latter forms mica-like leaflets (density 5.0), insoluble in absolute alcohol. It may be sublimed at 800°, and begins to decompose into I_2 and V above 1000° (Morette 1938).

These vanadium(III) halides apparently have only a very small tendency to form double (complex) salts.

(iv) Vanadium(III) fluoride and Fluorovanadates(III). The *fluoride* of trivalent vanadium possesses a marked tendency to form double salts. Vanadium(III) fluoride itself is obtained, in the form of dark green rhombohedral crystals of the composition $VF_3\cdot3H_2O$, by the evaporation of a solution of vanadium(III) hydroxide in hydrofluoric acid.

The double fluorides of trivalent vanadium [fluorovanadates(III)] mostly correspond in composition to the type $M^I_2VF_5$; this formula takes no account of their water content. The double fluorides of this type formed with the fluorides of bivalent metals contain 7 molecules of water of crystallization and probably have the constitution $[M^{II}(H_2O)_6]$ $[VF_5(H_2O)]$.

(v) Vanadium(III) sulfate, Sulfatovanadic(III) acid and Sulfatovanadates(III). When a solution of vanadium pentoxide in sulfuric acid is reduced—e.g., electrolytically—a green solution is obtained from which green lustrous silky needles of the composition $V_2(SO_4)_3\cdot H_2SO_4\cdot12H_2O$ are deposited, if the concentrations of vanadium and of sulfuric acid are sufficient. By careful heating at 180°, this is changed into a substance having the composition of the neutral sulfate, $V_2(SO_4)_3$, a yellow powder insoluble in water, alcohol, and ether. The first-mentioned compound is probably to be regarded as a sulfatovanadic(III) acid $H[V(SO_4)_2\cdot6H_2O]$, from which are derived salts much as the green ammonium disulfatovanadate(III), $NH_4[V(SO_4)_2]\cdot6H_2O$. There are also numerous compounds of the type $M^IV(SO_4)_2\cdot12H_2O$—i.e., of the alum type—which can also be obtained from the solution of the compound mentioned, and which are distinguished from it and from the first-mentioned series of sulfatovanadates(III) only in that they contain an additional 6 molecules of water. The vanadium double salts of the alum type are mostly more or less distinctly violet in the crystalline state*. In (concentrated) aqueous solution, on the other hand, the salts of both series are *green*. (In more dilute solution a brown to yellow coloration appears, in consequence of hydrolysis.)

Trivalent vanadium is remarkably stable in the sulfato salts mentioned. Even in solution, these salts are oxidized only slowly by atmospheric oxygen.

(vi) Oxalatovanadates(III). Green monoclinic crystals of alkali trioxalatovanadates(III), $M^I_3[V(C_2O_4)_3\cdot3H_2O]$, separate out when a solution of vanadium pentoxide in oxalic acid, containing alkali oxalate, is evaporated in a vacuum after electrolytic reduction. These are only weak complex salts. Their solutions are therefore oxidized in air, whereas the solid salts are stable in air.

(vii) Potassium hexacyanovanadate(III), $K_3[V(CN)_6]$, is formed when vanadium(III) salt solutions are treated with potassium cyanide in excess. The salt is more weakly complexed than the corresponding compounds of trivalent chromium, iron, and cobalt. Nevertheless, the aqueous solutions behave substantially differently from solutions which contain free V^{3+} ions; they do not have the green color of the latter, but a wine-red color which is to be ascribed to the $[V(CN)_6]^{3-}$ ions.

(viii) Thiocyanatovanadates(III). These compounds of the type $M^I_3[V(SCN)_6]$ (with a varying content of water of crystallization) are only weakly complex, and are soon extensively decomposed on dissolution in water. The solid salts, such as $Na_3[V(SCN)_6]\cdot12H_2O$, $K_3[V(SCN)_6]\cdot4H_2O$, and $(NH_4)_3[V(SCN)_6]\cdot4H_2O$ are notable for lively and beautiful colors, and their dichroism or polychroism. The ammonium salt, for example, appears blackish-green in the compact state, but red in the powdered state.

---

* The rubidium and cesium salts each exist in a bluish-violet and a red modification.

(*ix*) *Vanadium(III) sulfide* (vanadium sesquisulfide), V_2S_3, may be obtained, in the form of a grey black powder or as graphite-like leaflets, by heating vanadium(III) oxide in a stream of hydrogen sulfide, or by heating vanadium pentoxide in carbon disulfide vapor at 700°. It is very resistant towards non-oxidizing acids, but dissolves in dilute nitric acid; with concentrated nitric acid it may even inflame. Ignition in a current of hydrogen converts V_2S_3 into the monosulfide VS. This crystallizes with the NiAs-type structure ($a = 3.34$, $c = 5.78$ Å), as also do the monoselenide VSe ($a = 3.58$, $c = 5.98$ Å) and monotelluride VTe ($a = 3.80$, $c = 6.12$ Å). The compounds VSe_2 and VTe_2 crystallize with the brucite structure.

The naturally occurring sulfide of vanadium, patronite, a lead-grey ore, with a metallic luster on a fresh cleavage surface, ($d = 2.46$, hardness 3.5) contains considerably more sulfur than corresponds to the composition of the sesquisulfide. According to Klemm (1936) and Biltz (1939), a higher sulfide of vanadium with a variable sulfur content, probably VS_4, corresponding to patronite, can be prepared in the laboratory.

(c) Vanadium(IV) Compounds

(*i*) *Vanadium(IV) oxide* (vanadium dioxide), VO_2, may be obtained by mild reduction of vanadium pentoxide V_2O_5—e.g., by fusing it with oxalic acid. It forms very deep blue, almost black lustrous crystals (density 4.654)—or, in some circumstances, a dark green powder. It readily dissolves in acids and alkalis, especially on warming, and thereby displays its character of an *amphoteric* oxide. It has no action on either red or blue litmus paper. It is oxidized to the pentoxide by concentrated nitric acid, or by heating it in air. VO_2 (also NbO_2) forms a rutile-type crystal lattice, $a = 4.54$, $c = 2.88$ Å (for $NbO_2 - a = 4.77$, $c = 2.96$ Å).

Scholder (1953) has obtained several types of vanadate(IV) compounds by heating VO_2 with the alkaline earth oxides in a high vacuum. These compounds correspond to the types $M^{II}VO_3$, $M^{II}{}_2VO_4$ and $M^{II}{}_3VO_5$.

(*ii*) *Vanadium Tetrachloride and Vanadyl Chloride.* Vanadium tetrachloride (vanadium(IV) chloride) forms a dark brownish-red, heavy oily liquid, which solidifies only when it is chilled with liquid air. The boiling point is 154°; its vapor density corresponds to the formula VCl_4. Above the boiling point, however, decomposition takes place fairly rapidly, into solid vanadium trichloride and chlorine; it already occurs slowly at ordinary temperature. Water immediately brings about partial hydrolysis according to the equation $VCl_4 + H_2O = VOCl_2 + 2HCl$. The preparation of pure vanadium tetrachloride is, therefore, not entirely simple. The compound can be prepared by the action of chlorine on metallic vanadium, vanadium nitride, or vanadium silicide, or by passing a mixture of vanadium oxytrichloride, $VOCl_3$, and excess of sulfur monochloride through a tube heated to dull redness. The tetrachloride is separated by fractional distillation from the by-products formed in this reaction. No double salts of vanadium tetrachloride are known.

The vanadyl dichloride (vanadium(IV) oxychloride), $VOCl_2$, resulting from the hydrolysis of the tetrachloride is also formed by the action of concentrated hydrochloric acid on vanadium pentoxide:

$$V_2O_5 + 6HCl = 2VOCl_2 + 3H_2O + Cl_2.$$

The elimination of chlorine is accelerated by the addition of mild reducing agents, such as alcohol or hydrogen sulfide. For the preparation of vanadyl dichloride in the pure state, it has been found most convenient to heat vanadyl trichloride with zinc in a sealed tube at 400°:

$$2VOCl_3 + Zn = 2VOCl_2 + ZnCl_2.$$

Vanadyl dichloride is hereby obtained in the form of lustrous, grass-green, hygroscopic crystal plates. The aqueous solutions are variously colored, mostly blue or brown.

With pyridinium chloride and quinolinium chloride, vanadyl dichloride forms double salts having the composition $VOCl_2.2RCl$ (blue) and $VOCl_2.4RCl$ (green)*.

* The salts crystallize in hydrated form.

Vanadyl dibromide, $VOBr_2$, resembles the dichloride in behavior. In the anhydrous state it is an ocher-brown powder; its aqueous solution is blue. Neither vanadium tetrabromide nor any tetraiodide or oxyiodide of vanadium is known.

(*iii*) *Vanadium tetrafluoride and Fluorovanadates(IV)*. Vanadium tetrafluoride (vanadium(IV) fluoride), VF_4, is obtained in the form of a very hygroscopic brown-yellow powder (Ruff) by the prolonged boiling of vanadium tetrachloride with anhydrous hydrofluoric acid. When heated above 300° in a current of nitrogen it decomposes (disproportionates) into vanadium trifluoride and pentafluoride. It is hydrolyzed by water. In the presence of other fluorides, double salts crystallize from the aqueous solution [*oxofluorovanadates(IV)*] which correspond chiefly to the type $M^I_2[VOF_4 \cdot H_2O]$—e.g.,

$$(NH_4)_2[VOF_4 \cdot H_2O] \text{ and } [Ni(H_2O)_6] [VOF_4 \cdot H_2O].$$

The potassium salt, however, crystallizes anhydrous, and the anhydrous ammonium salt has the composition corresponding to the formula $(NH_4)_3[VOF_5]$.

There is some evidence for the formation of the compound K_2VF_6 by dry methods, although the existence of this substance has not been established with certainty.

(*iv*) *Vanadyl Sulfate and Oxosulfatovanadates(IV)*. Anhydrous vanadyl sulfate, $VOSO_4$, exists in two modifications—one soluble and one insoluble in water. The latter is obtained by heating the acid vanadyl sulfate, $2VOSO_4 \cdot H_2SO_4$, to 260° with concentrated sulfuric acid. It forms a grey-green, finely crystalline powder, which changes into the blue soluble form when it is heated at 130° with a little water. Various *hydrates* of vanadyl sulfate have also been obtained, the colors of which vary from bright blue to dark blue. On prolonged standing, aqueous solutions of vanadyl sulfate deposit green vanadium dioxide, as a consequence of hydrolysis:

$$VOSO_4 + H_2O = VO_2 + H_2SO_4.$$

From solutions containing an excess of sulfuric acid, bright blue hydrates of the '*acid vanadyl sulfate*', $2VOSO_4 \cdot H_2SO_4$, crystallize on evaporation*. A few double salts (with alkali sulfates) are also derived from this compound—for example, $K_2SO_4 \cdot 2VOSO_4$, bright blue microscopically small quadratic plates. If these double salts are formulated as sulfato compounds: $M^I_2[V_2O_2(SO_4)_3]$, the '*acid vanadyl sulfate*' must correspondingly be regarded as dioxotrisulfato divanadic acid(IV), $H_2[V_2O_2(SO_4)_3]$. Another series of double salts of the vanadyl radical, (also known only in the form of hydrates of the alkali compounds) has the composition $M^I_2SO_4 \cdot VOSO_4$. These compounds, which separate in the form of fine dark crystals when concentrated solutions of salts of the first series are treated with alcohol, should be formulated as oxodisulfatovanadate(IV) salts, $M^I_2[VO(SO_4)_2]$.

If the double sulfates of the first type [dioxotrisulfato divanadates(IV)] are fumed down with concentrated sulfuric acid, disulfatovanadates(III), $M^I[V(SO_4)_2]$, are obtained; at the same time, an equivalent amount of V^{IV} is transformed into V^V (disproportionation). Oxidation to V^V does not take place if ammonium salts constitute the starting material, since the NH_4^+ ion reduces V^V to V^{IV}.

Boiling concentrated sulfuric acid partially transforms V^{IV} compounds into V^V compounds, with evolution of SO_2, so that equilibrium is established between V^{IV} and V^V. Conversely, a partial reduction to V^{IV} salt is brought about by heating V_2O_5 with boiling concentrated sulfuric acid, oxygen being evolved. (Auger 1921, Sieverts 1928).

Vanadyl sulfate can frequently be used with advantage for the potentiometric titration of weak oxidizing agents (del Fresno 1933).

(*v*) *Sulfitovanadates(IV)*. In addition to the oxodisulfitovanadates(IV), which correspond in type to the sulfato salts cited previously, there is also a second series of sulfitovanadates(IV):

$$M^I_2[V_3O_5(SO_3)_2].$$

* The range of existence of the 5-hydrate extends to about 100°. The 3-hydrate crystallizes at 120°, the 2-hydrate at 150°, the ½-hydrate at 175°. The compound crystallizes anhydrous at 190°.

The acid $H_2[V_3O_5(SO_3)_2]\cdot3\frac{1}{2}H_2O$, from which these are derived, is also known. It has been demonstrated, from the direction of migration of the vanadium during electrolysis of the solutions, that these compounds (which are blue in color, like most oxovanadium(IV) compounds) contain the vanadium in a complex anion (Koppel 1903).

(*vi*) *Oxalatovanadates(IV)* are readily obtained by treating vanadium pentoxide with oxalic acid. Two series of such complex oxalates of quadrivalent vanadium are known:

$$M^I{}_2[VO(C_2O_4)_2] \text{ and } M^I{}_2[V_2O_2(C_2O_4)_3].$$

They are blue or greenish blue. Oxalate ions cannot be detected in the solutions of these salts by precipitation reactions in the cold.

No compound of the first order between vanadium(IV) oxide and oxalic acid is known.

Other organic oxy compounds—e.g., tartaric acid, salicylic acid, catechol,—also form complex compounds with the vanadyl radical, of the same type as the oxalates of the first series mentioned above. In these compounds the H atoms both of the carbonyl groups, and of the alcoholic or phenolic hydroxyl groups are substituted by an equivalent amount of vanadium.

(*vii*) *Thiocyanatovanadates(IV)*. Vanadium(IV) ions similarly form no compounds of the first order with thiocyanate ions, but do form compounds of higher order—*oxotetrathio-cyanatovanadates(IV)*, $M^I{}_2[VO(SCN)_4]$. The tendency for their formation must be quite considerable, since the characteristic blue color of the thiocyanatovanadate(IV) ion appears at once on addition of SCN^- ions to a solution of vanadic acid. For the preparation of the salts, the reduction of V^V to V^{IV} can conveniently be promoted by adding sulfurous acid.

(*viii*) *Hypovanadates*. As has already been mentioned, vanadium dioxide dissolves in caustic alkalis. Salts are deposited from sufficiently concentrated, warm solutions; these compounds contain the vanadium in the anion and are called *hypovanadates*. The crystallized alkali hypovanadates correspond to the type $M^I{}_2[V_4O_9]\cdot7H_2O$.

(*in*) *Vanadium(IV) sulfide* and *Thiovanadates(IV)*. Soluble *thiohypovanadates* (thiovana-dates(IV)) of as yet unknown composition are formed by the action of hydrogen sulfide on solutions of alkali hypovanadates. If acid is added to their solutions a black sulfide, which later turns brown, is precipitated; the composition of this sulfide is not known with certainty.

5. Vanadium(V) Compounds

(a) Vanadium Pentoxide

Vanadium pentoxide (more strictly divanadium pentoxide), vanadium(V) oxide, V_2O_5, can be prepared by heating the so-called ammonium metavanadate, NH_4VO_3 (probably ammonium tetravanadate, $(NH_4)_4[V_4O_{12}]$, cf. p. 100 to a red) heat in platinum crucible:

$$2NH_4VO_3 = V_2O_5 + 2NH_3 + H_2O,$$

or by the action of water on vanadyl trichloride

$$2VOCl_3 + 3H_2O = V_2O_5 + 6HCl.$$

It forms an orange to cinnabar red, odorless, tasteless, poisonous powder, which melts at about 660°, and solidifies on cooling to yellow-red rhombic crystal needles (density 3.318). The heat of crystallization is so great that on rapid crystallization the compound, solidifying with considerable contraction, is again raised to in-candescence. Vanadium pentoxide is only slightly soluble in water (0.07 g in 100 g); its solution is yellow. It perceptibly reddens moist litmus paper. It dissolves easily in alkalis, forming *vanadates* (see below). It is also soluble, however, in strong acids, and thus displays a basic character, even though this is but weakly developed as compared with the acid character of the compound. Chlorine is evolved when vanadium pentoxide is dissolved in hydrochloric acid, since pentavalent vanadium is thereby partially reduced to the quadrivalent state. If the chlorine so formed is removed, the process goes to completion.

If a solution of V_2O_5 in dilute sulfuric acid (or an alkali vanadate solution acidified with sulfuric acid) is treated with arsenious acid and a few drops of a very dilute solution of OsO_4 (as catalyst), no reaction takes place at first. As soon, however, as potassium chlorate solution is added, the solution instantaneously turns blue, an indication that reduction of V^V to V^{IV} has now taken place. This experiment, described by Gleu in 1933, demonstrates in a most illuminating manner a phenomenon fairly often observed with induced reactions (called by Wilhelm Ostwald the 'topsy turvy world'), namely that a *reduction* can be initiated by an oxidizing agent (and conversely, an oxidation frequently by a reducing agent).

Vanadium pentoxide can easily be brought into solution in *colloidal form* (e.g., by peptization with nitric or hydrochloric acid). The dark blood red colloidal dispersion is remarkably stable, and can be evaporated; a velvety red mass remains behind, which re-forms the blood red colloid on addition of water. The particles in these dispersions are notable for their small size. They lie on the limit of ultramicroscopic visibility. During the aging of vanadium pentoxide hydrosols it is occasionally possible to observe the formation of minute rods, at first only detectable in the ultramicroscope, but gradually growing to microscopic dimensions. If the rodlets are orientated by the flowing motion of the liquid (e.g., even by stirring the sol with a glass rod) the sol becomes double-refracting. On X-ray investigation, the rodlets furnish the same diffraction pattern as does ordinary solid vanadium pentoxide, except that the lines are much broadened because of the small size of the crystallites.

The crystal structure of vanadium pentoxide provides an explanation for the tendency to form rod-shaped microcrystals. Each V atom in the crystal lattice is surrounded in a distorted tetrahedron by 4 O-atoms, of which 3 are also shared with 3 neighboring tetrahedra. In this manner, zigzag VO_4 chains of unlimited length are built up. These are, indeed, further linked together into sheets by bridges of O-atoms (cf. Fig. 19); however, only 1 O-atom of each VO_4 tetrahedron is utilized for this purpose, while 2 O-atoms are used in forming the chains. In Fig. 19, in which the linkage of the VO_4 tetrahedra is schematically illustrated, one

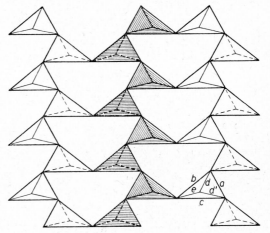

Fig. 19. Two-dimensional network of distorted VO_4 tetrahedra in vanadium pentoxide. Intersections of tetrahedron edges shown as thin lines lie above the plane of the drawing, those marked by dotted lines lie below that plane. The edges of the tetrahedra have the following lengths, in Ångström units: $a = 2.13$, $b = 3.06$, $c = 3.50$, $d = d' = 2.72$, $e = 2.36$ Å. The average V ↔ O distance is 1.72 Å.

of the chains is emphasized by shading. Those O-atoms which form the vertices of the tetrahedra lying above and below the plane of the drawing are not shared between two different tetrahedra; the cohesion between the individual lattice sheets is therefore not effected by principal valence forces. Vanadium pentoxide crystals accordingly exhibit an excellent cleavage parallel to the lattice sheets. There is also a distinct cleavage present in the direction perpendicular to this, but parallel to the chains. The preferential growth of the crystal in the direction of the chains can be understood in view of the higher density of atoms along the chain direction.

(b) Vanadic Acids and the Vanadates

Vanadic acids are formed by the dissolution of vanadium pentoxide in water, as is shown by the reaction of the solution towards litmus. *Metavanadic acid*, HVO_3, and *tetravanadic acid*, $H_2V_4O_{11}$, can be obtained in the free state; these compounds were recognized by Hüttig (1930) from thermal degradation and by X-ray studies. The salts of the vanadic acids, the *vanadates*, are of very manifold composition. According to the ratio of base anhydride: acid anhydride ($M^I_2O:V_2O_5$) contained in them, they are classified into the (4 : 1) vanadates, $4M^I_2O \cdot V_2O_5$; (3 : 1) vanadates, $3M^I_2O \cdot V_2O_5$; (2 : 1) vanadates $2M^I_2O \cdot V_2O_5$; (3 : 2) vanadates $3M^I_2O \cdot 2V_2O_5$; the (5 : 4) vanadates, $5M^I_2O \cdot 4V_2O_5$, etc. The (3 : 1) vanadates, $3M^I_2O \cdot V_2O_5$ or $M^I_3VO_4$ are also known as *orthovanadates*, the (2 : 1) vanadates, $2M^I_2O \cdot V_2O_5$ or $M^I_4V_2O_7$ as *pyrovanadates*, and the (1 : 1) vanadates, $M^I_2O \cdot V_2O_5$ or M^IVO_3 as *metavanadates*.

According to Jander and Jahr (1933 and later) the following equilibria are set up in aqueous solutions:

$$2[VO_4]^{3-} + 2H^+ \rightleftharpoons [V_2O_7]^{4-} + H_2O \quad (pH = 12\text{--}10.6) \tag{1}$$

$$2[V_2O_7]^{4-} + 4H^+ \rightleftharpoons H_2[V_4O_{13}]^{4-} + H_2O \quad (pH = 9.0\text{--}8.9) \tag{2}$$

$$5H_2[V_4O_{13}]^{4-} + 8H^+ \rightleftharpoons 4H_4[V_5O_{16}]^{3-} + H_2O \quad (pH = 7.0\text{--}6.8) \tag{3}$$

$$2H_4[V_5O_{16}]^{3-} + 6H^+ \rightleftharpoons 5[V_2O_5] + 7H_2O \quad (pH = 2.2) \tag{4}$$

$$[V_2O_5] + 2H^+ \rightleftharpoons 2[VO_2]^+ + H_2O \quad (pH < 1) \tag{5}$$

It may be seen from this that the more simple vanadate ions have a tendency to come together, with the elimination of water, to form higher-molecular ions. Acids of this kind, which may be derived from simple acids ('mono-acids') by *condensation*—that is by the linking up of two or more molecules and elimination of water—are termed, quite generally, *polyacids* (cf. p. 183 on this point). In the present instance, therefore, one speaks of *polyvanadic* acids, or of *polyvanadate* ions; the compounds are distinguished as *di*vanadates, *tetra*vanadates, *penta*vanadates, etc., according to the number of vanadium atoms in the ions.

Beside the equations given above are set down, on the right hand side, the pH values at which the individual changes take place in the solution. In the ranges between these, the substances written on the left hand side of the equations are present almost exclusively. Dissociation equilibria which do not involve condensation equilibria (for example $H[V_4O_{13}]^{5-} + H^+ \rightleftharpoons H_2[V_4O_{13}]^{4-}$) are also of importance in these intermediate ranges. For the sake of simplicity these have been left out of account here. The ions have also been formulated without taking account of the water molecules combined with them. The ionic weights (measured by Jander and Jahr, 1933, Brintzinger, 1935) have led to the conclusion that the VO_4^{3-} and $V_2O_7^{4-}$ ions are combined with $6H_2O$, whereas the $H_2V_4O_{13}^{4-}$ ions are not hydrated.

As the above equations show, simple vanadate ions (these being only VO_4^{3-} ions, not VO_3^- ions) exist only in *strongly alkaline solutions*. With increase of H^+ ion concentration, *divanadate* ions and *tetravanadate* ions are formed successively in the alkaline range and finally, on passing into the acid range, *pentavanadate* ions. The transformations (1) and (2) take place immeasurably fast; the change of tetravanadate into pentavanadate ions, however, takes place slowly, and probably proceeds by way of the formation of *octavanadic* acid $H_{10}[V_8O_{25}]$, which is not stable itself, but may be stabilized by added substances. The pentavanadate ions may be present until fairly far in the acid range. With increasing acid contents however, the solutions become unstable. and gradually begin to deposit hydrated vanadium pentoxide*. Simultaneously, the formation of positive $[VO_2]^+$ ions (or possibly $[VO]^{+++}$), according to equation (5), takes place to an increasing extent—i.e., vanadic acid begins to dissociate as a base, in accordance with its amphoteric nature.

It is a general rule that the condensation of acid ions is accompanied by a deepening of color (cf. Vol. I, p. 312). This also applies to the polyvanadate ions as shown by Jahr (1934). In the case of mono-, di-, and tetravanadate ions, the region of light absorption lies wholly in the ultraviolet. The mono-, di-, and tetravanadates therefore appear colorless to the eye (unless they contain colored cations). The pentavanadates are orange-yellow, the octa-vanadates are brown-red, and the simple-molecular oxyvanadium(V) cation has a pale yellow color.

On the basis of their color, mode of formation and other criteria, G. Jander has advanced an explanation of the constitution of the vanadates. This led to the view that the (1 : 3), (1 : 2) and (2 : 3) vanadates were not salts of a hexavanadic acid $H_4[V_6O_{17}]$ or $H_4[H_2V_6O_{18}]$, as had formerly been assumed, but were *oxyvanadium(V) pentavanadates*,

$$M^I_2[VO_2]H_4[V_5O_{16}], \quad M^I_3[VO_2]H_3[V_5O_{16}], \quad \text{and} \quad M^I_4[VO_2]H_2[V_5O_{16}]^{**}.$$

Under suitable experimental conditions, oxyvanadium(V) salts of octavanadic acid can be obtained—for example, $Na_2[VO_2]H_7[V_8O_{25}]$ (or $Na_2[VO]H_5[V_8O_{25}]$). The octavanadates are quite characteristically distinguished from the pentavanadates by their lower stability towards alkalis. The (4 : 5), (3 : 5), and (2 : 5) vanadates are to be considered as pentavanadates:

$$M^I_4H_3[V_5O_{16}], \quad M^I_3H_4[V_5O_{16}], \quad \text{and} \quad M^I_2H_5[V_5O_{16}].$$

The (3 : 2) and (5 : 4) vanadates are salts of tetravanadic acid, $M^I_6[V_4O_{13}]$, $M^I_5H[V_4O_{13}]$, as also are the (1 : 1) vanadates, $M^I_4H_2[V_4O_{13}]$, generally called 'meta-vanadates'. In the crystallization of the latter a molecule of water may be split off, whereby anhydrous metavanadates are obtained: these are also to be looked on as tetravanadates $M^I_4[V_4O_{12}]$. The anion of divanadic acid is present in the (2 : 1) vanadates (pyrovanadates), and that of monovanadic acid in the (3 : 1) (ortho-)vanadates. The (4 : 1) vanadates may perhaps represent basic salts, or oxy-orthovanadates $M^I_3O[VO_4]$.

The orthovanadates can be prepared by melting the component oxides together, as well as from aqueous solution. Most of the polyvanadates can also be obtained pure from aqueous solution without difficulty, as Jander has shown, if the range of existence of the individual polyvanadate ions is taken into account, and if time is given for the attainment of equilibrium. Of the so-called metavanadates (tetravanadates), the ammonium metavanadate NH_4VO_3 or $[NH_4]_4[V_4O_{12}]$, which is fairly sparingly soluble in the cold (unlike the alkali metavanadates), is readily prepared in the pure state. The vanadates of other metals can be prepared by double decomposition reactions between alkali vanadates and salts of the alkaline earths and heavy metals. The vanadates are peculiar in displaying a marked tendency to form alkali-alkaline earth or alkali-heavy metal mixed salts.

Vanadate ions not only have a marked tendency to undergo condensation processes, forming higher molecular ions, but they also tend to form complexes with the ions of other acids. Thus complex acids, known as *heteropolyacids*, are formed with the acids of silicon, tin, phosphorus, arsenic, molybdenum, and tungsten (see pp. 180 *et seq.*). The role of the central atom in these complex acids is played either by the elements named, or by vanadium.

* It has not yet been fully established whether V_2O_5 or HVO_3 is the stable solid phase in equilibrium with the solution. Some observations support the view that V_2O_5 is generally deposited first and that this subsequently changes into metavanadic acid, through chemical combination with water.

** If it is assumed that the compounds contain $[VO]^{3+}$ groups instead of $[VO_2]^+$ groups, the formulas become $M^I_2[VO]H_2[V_5O_{16}]$, $M^I_3[VO]H[V_5O_{16}]$, and $M^I_4[VO][V_5O_{16}]$. Since the oxyvanadium(V) cations are hydrated in aqueous solution, and as the compounds cited crystallize as hydrates, it cannot readily be determined which of the two modes of formulation is appropriate.

(c) Vanadinite Structure of Compounds of the Apatite Group

Another type of complex vanadates is represented by *vanadinite* (vanadium lead ore), $3Pb_3(VO_4)_2.PbCl_2$. This vanadium ore, occurring chiefly in Mexico (Zimapan), Argentina, and Arizona, and also in Scotland and Carinthia, crystallizes in hexagonal hemihedral columnar crystals, of yellow-brown to red-brown color, and is isomorphous with apatite, pyromorphite, mimetesite, and the other minerals of the *apatite* group.

The compounds of the apatite group have the general composition $3M^{II}_3(RO_4)_2.M^{II}X_2$, or $M^{II}_5(RO_4)_3X$: M^{II} can be Ca or Pb, R can be P, As, or V, and X can signify Cl or F. Elements denoted by the same symbol in this general formula can mutually replace each other continuously; Cl or F can also be partially replaced by the hydroxyl group OH, or by the equivalent amounts of O, CO_3 or SO_4; Ca is also replaceable by Sr. The compounds of the apatite group crystallize in the pyramidal-hemihedral class of the hexagonal system. They not only occur as minerals, but can also be obtained artificially in the crystalline state. That they are not simple double compounds of phosphates, arsenates, or vanadates with the chlorides, etc., is at once shown in that water leaches out no chloride from them. Werner accordingly regarded them as coordination compounds, and represented them as complex compounds of the type $[CaR_6]X_2$ which is typified, for example by calcium chloride hexahydrate $[Ca(H_2O)_6]Cl_2$. Werner derived apatite from this by exchange of each $2H_2O$ for 1 molecule of $Ca_3(PO_4)_2$, and so arrived at the structural formula $[Ca(Ca_3(PO_4)_2)_3]X_2$. X-ray structural analysis (Naray-Szabo, 1930, Schiebold 1931) has shown that the compounds of the apatite group are indeed coordination compounds. Werner's assumption must, however, be corrected in that in an apatite it is not a Ca-atom, but the univalent anion X which forms the central atom of the complex, or of the structural group corresponding to this in the crystal. $3(CaPO_4)^-$ groups are bound to this anion in the first sphere, and $12Ca^{2+}$ ions are arranged, in two superimposed regular hexagons, in the second sphere; each of the latter ions, however, is equidistant from $6[X(CaPO_4)_3]^{4-}$ groups, so that stoichiometrically $^{12}/_6 = 2Ca^{2+}$ ions belong to each $[X(CaPO_4)_3]^{4-}$ group. The structure of apatite can thus be roughly represented by the formula:

$$Ca \begin{bmatrix} CaPO_4 \\ F \quad CaPO_4 \\ CaPO_4 \end{bmatrix} Ca \quad \text{or} \quad Ca_2[F(CaPO_4)_3].$$

The general formula of the apatite group is thus $M^{II}_2[X(M^{II}RO_4)_3]$—e.g., vanadinite is $Pb_2[Cl(PbVO_4)_3]$. In some kinds of apatites the anion X^- is replaced by a group composed of several atoms—e.g., by the group $[CO_3]^{2-}$ in *carbonato-apatite*. Since this group is bivalent, half the coordination centers are simultaneously occupied by neutral molecules (H_2O). In *hydroxylapatite*, $Ca_5(OH)(PO_4)_3 (= Ca_2[(OH)(CaPO_4)_3])$ the anion X^- is replaced by the group $(OH)^-$. The formula of *oxyapatite* $Ca_{10}O(PO_4)_6$ is not yet explained; according to the investigation of Trömel (1932), its existence is questionable.

(d) Peroxy Compounds of Vanadium(V)

According to Jahr (1941) when hydrogen peroxide is added to alkali vanadate solutions, yellow *diperoxoorthovanadate* ions $[VO_2(O_2)_2]^{3-}$ are formed in alkaline or weakly acid solutions, whereas red-brown *peroxyvanadium(V) cations* $[V(O_2)]^{3+}$ are formed in strongly acid solutions. The diperoxoorthovanadate- and the peroxyvanadium(V) ions enter mutually into an equilibrium which depends on the hydrogen ion concentration and the hydrogen peroxide concentration of the solution:

$$[VO_2(O_2)_2]^{3-} + 6H^+ \rightleftharpoons [V(O_2)]^{3+} + H_2O_2 + 2H_2O.$$

Vanadium pentoxide hydrate dissolves in aqueous hydrogen peroxide, forming the free tribasic diperoxoorthovanadic acid $H_3[VO_2(O_2)_2]$; this readily decomposes, with evolution of oxygen, like all peroxy compounds of vanadium. Of the salts of this acid, only the primary potassium salt, $KH_2[VO_2(O_2)_2].H_2O$ (yellow rhombic leaflets), and the secondary ammonium salt, $(NH_4)_2H[VO_2(O_2)_2].aq$ (yellow needles), are as yet known; these were prepared by Jahr. The light-metal and heavy metal salts of a supposed peroxometavanadic acid $H[VO_2(O_2)]$, reported in the older literature, could not be obtained by Jahr. At temperatures around $0°$, in strongly alkaline solutions containing much hydrogen peroxide, the diperoxoorthovanadate ions are gradually transformed into blue-violet *tetraperoxoortho-vanadate* ions, $[V(O_2)_4]^{3-}$. The tetraperoxoorthovanadates $M^I{}_3[V(O_2)_4] \cdot aq.$ (M^I = Li, Na, K, NH_4) crystallize in blue-violet needles; their ease of decomposition runs parallel to increase in the radius of the cation.

(e) Vanadium Pentasulfide and Thiovanadates

Vanadium pentasulfide, V_2S_5, is said to be obtained by heating vanadium(III) oxide with sulfur at $400°$ in the absence of air. After extraction of the products of reaction with carbon disulfide, the reputed pentasulfide remains as a black powder. According to Klemm (1936), however, the existence of this compound is very doubtful. It cannot, in any case, be obtained by heating V_2S_3 with excess of sulfur; the polysulfide mentioned on p. 96 is thereby formed instead. The products described as V_2S_5 burn to vanadium pentoxide when they are heated in the air. Heated in an inert atmosphere, they lose sulfur and are changed into the sesqui-sulfide. They dissolve in alkali sulfide solutions, forming thio salts or oxythiosalts, of which two series are known. One series is derived from the orthovanadates, and other from the pyrovanadates, by exchange of oxygen atoms for sulfur. An example is *ammonium orthothio-vanadate*, $(NH_4)_3[VS_4]$, which is obtained, in the form of violet-black rhombic prisms, by evaporating vanadate solutions treated with ammonium sulfide. Like all alkali thio- and oxythiovanadates, it is easily soluble in water. Thiovanadates have also been obtained by fusion processes.

(f) Vanadium(V) Halides and Halogenovanadates(V)

The only oxygen-free halide as yet known is the pentafluoride VF_5 (solid white mass, readily soluble in water and alcohol, density 2.177, b.p. $111.2°$), which is obtained by thermal decomposition of the tetrafluoride. Oxygen-containing halides, such as VOF_3, $VOCl_3$ and $VOBr_3$, are more readily formed; the last mentioned, however, decomposes even on moderate heating, or slowly even at the ordinary temperature, according to the equation $VOBr_3 = VOBr_2 + \frac{1}{2}Br_2$. Iodine compounds of pentavalent vanadium cannot be prepared at all.

Vanadium oxytrichloride, $VOCl_3$, for example, may be obtained by passing dry hydrogen chloride over gently warmed vanadium pentoxide in the presence of phosphorus pentoxide (which serves to combine with the water set free in the reaction):

$$V_2O_5 + 6HCl = 2VOCl_3 + 3H_2O.$$

Vanadium oxytrichloride forms a yellow, mobile but fairly heavy liquid, of boiling point $127°$, with a vapor density corresponding to the simple molecular formula. It is hydrolyzed by water, so that the process represented by the above equation is reversible. With pyridi-nium chloride, vanadium oxytrichloride forms the double salt $[C_5H_5NH][VOCl_4]$. (pyridinium oxotetrachlorovanadate). This must be isolated from alcoholic solution, because of the hydrolysis occurring in water.

VF_5 combines with alkali fluorides to form *hexafluorovanadates*, $M^I[VF_6]$. These were first isolated by Eméleus (1949) by the action of BrF_3 on VCl_3 mixed with alkali halides.

A larger number of double salts of pentavalent vanadium are known, which are derived from the oxyfluorides. Several representatives of the following types are known:

$$3M^IF \cdot 2VOF_3; \quad 3M^IF \cdot 2VO_2F; \quad \text{and} \quad 2M^IF \cdot VO_2F.$$

Like the pyridinium double chloride cited above, these double salts can also be regarded as halogenovanadates—i.e., as vanadates in which the oxygen atoms are partially sub-stituted by halogen atoms. Radicals of oxyacids can also replace oxygen atoms in vanadates, in the same way as halogen atoms, e.g., the radicals of sulfuric acid, iodic acid, and oxalic acid. Thus *oxalatovanadates(V)* are known, and have been assigned the constitution $M^I{}_3[VO_2(C_2O_4)_2]$.

6. Analytical (Vanadium)

Vanadium is precipitated neither from acid solutions by hydrogen sulfide, nor from ammoniacal solutions by means of ammonium sulfide. Alkali vanadate solutions are colored cherry red by ammonium sulfide, however, by reason of the formation of thiovanadates. If they are subsequently acidified, most of the vanadium is then precipitated as the brown pentasulfide (or polysulfide? see p. 103). Acid solutions of vanadic acid are colored *blue* by hydrogen sulfide, through reduction to vanadyl salts. A blue coloration is also produced by numerous other reducing agents. In many cases (e.g., by the action of zinc and sulfuric acid), the blue color passes through green and finally to violet, as the reduction proceeds by way of vanadium(III) compounds to vanadium(II) compounds. The reaction with hydrogen peroxide is very useful. This colors acidified vanadic acid solutions an intense red-brown. The color is not extracted by ether, and this enables vanadium to be detected even in the presence of chromium (cf. p. 158). The test given by Ephraim [*Helv. chem. Acta*, 14 (1931) 1266] is considerably more sensitive, however; this depends on the reduction of Fe^{+++} ions by $[VO]^{++}$ ions in alkaline solution:

$$Fe^{+++} + [VO]^{++} + 6OH^- = Fe^{++} + [VO_4]^{3-} + 3H_2O.$$

Detection of the resulting Fe^{++} ions by means of dimethylglyoxime (intense cherry red color), enables as little as 2.5 γ of vanadium in 1 cc of solution to be recognized.

Gravimetrically, vanadium can be determined by precipitation as mercurous vanadate, which leaves a residue of pure vanadium pentoxide in ignition. It is more convenient to determine vanadium by a *volumetric* method—for example, by first reducing to a vanadyl salt by means of sulfurous acid, and then (after boiling off the excess of sulfur dioxide) titrating the hot solution, acidified with sulfuric acid, with permanganate:

$$5[VO]^{++} + MnO_4^- + 7H_2O = H_4[V_5O_{16}]^{3-} + Mn^{++} + 10H^+.$$

The reduction of V^V to V^{IV} by means of Fe^{++} ions, taking place in acid solution (the reverse of the process in alkaline solutions, see above) can also be used for the titrimetric determination of vanadium.

7. Niobium (Nb)

(a) Occurrence

Niobium minerals are found in many places, although generally in inconsiderable amounts. The most important niobium mineral is *niobite* or *columbite*, which is met with in various localities in the United States, and also in Greenland and at Tammela, Finland. In Germany, niobite is found at Bodenmais (Lower Bavaria). Its composition, is essentially that of a niobate of iron*, $Fe(NbO_3)_2$, which always contains iron tantalate, $Fe(TaO_3)_2$, in isomorphous admixture to a greater or less degree. If the tantalum predominates over the niobium, the mineral is called tantalite.

Niobium frequently occurs along with the rare earths; there is a whole series of rare earth minerals consisting of niobates, or isomorphous mixtures of niobates and tantalates. *Euxenite* is an isomorphous mixture of niobates and titanates of the rare earth metals; *polycrase* is of similar composition. A calcium niobate with a significant content of alkali fluoride is *pyrochlore*, $NaF \cdot CaO \cdot Nb_2O_5$, or $NaCaNb_2O_6F$.

(b) History

In 1801, a new metallic oxide was discovered by the English chemist Hatchett in a mineral from the British Museum, originally obtained from North America. He named the

* The iron is generally replaced partially by manganese.

element giving rise to this oxide *Columbium*, from the origin of the mineral: the latter he called *columbite*. It was later thought that this element was identical with the *tantalum* discovered by Ekeberg in 1802, especially on the basis of investigations by Wollaston and Berzelius. As we now know, neither of these chemists ever had pure tantalic acid, but only a mixture with niobic acid. In any case, it had already been noticed by Wollaston that there were significant variations in density not only between tantalites of different origin, but also between the oxides isolated. Rose, however, in 1844, was the first to conclude that these variations must be produced by another element admixed with the tantalum, and very similar to it, but different from it in weight. He called this element *niobium*, after Niobe, the daughter of Tantalus. Since the niobium oxide isolated by Hatchett from Columbite was certainly also contaminated with tantalum oxide, just as the tantalum oxide later isolated from tantalite was contaminated with niobium, Rose was the first to handle pure niobium oxide. The older name columbium, has been largely retained in the English (and especially in the American) literature.

(c) Preparation

Niobium minerals are best decomposed by fusion with potassium hydrogen sulfate. The melt is brought into solution with hydrofluoric acid, as is further described under tantalum; the two elements can then be separated by fractional crystallization of the potassium double fluorides. Separation from titanium, which may also be present, offers some difficulty, since it remains in the solution as fluorotitanate, together with the niobium. For this separation, Wernet (1952) has recommended precipitation of titanium as $(NH_4)_2[TiCl_6]$ from solutions saturated with HCl. If the titanium content is very high, however, niobium is also coprecipitated in the form of $(NH_4)_2[NbOCl_5]$, which forms mixed crystals with $[NH_4]_2[TiCl_6]$. In these circumstances, the method can be used as a means of separating titanium and niobium together from tantalum.

Moissan has described an elegant method of decomposition, which involves heating a mixture of the powdered mineral with sugar charcoal for a few minutes with a powerful current (1000 amps) in the short circuit furnace. Niobium and tantalum are thereby converted to their carbides, which are soluble in hydrofluoric acid, while the greater part of the other elements present is volatilized away.

For the preparation of the metal, according to von Bolton, the best starting material is the pure pentoxide. This can be reduced by the aluminothermic method, after which the resulting niobium-aluminum alloy is freed from aluminum by prolonged fusion in the electric vacuum furnace (whereby the aluminum evaporates). Alternatively, the pentoxide is plasticized by addition of paraffin wax, and moulded into filament form; this is converted to the dioxide NbO_2 by heating with carbon powder at a white heat. The dioxide, unlike the pentoxide, conducts the electric current (and is, indeed, an ionic conductor). Under the heating action of a strong alternating current it fairly rapidly gives off all its oxygen.

Attempts to deposit niobium electrolytically from solutions (Isgarischew, 1933 and later) have been unsuccessful.

(d) Properties

Niobium is a moderately hard, grey metal, with a white luster on polished surfaces; it is rather less ductile than tantalum, and its melting point is considerably lower. Its boiling point must also be much lower than that of tantalum, since, unlike the latter, it sputters strongly when heated in a vacuum.

Niobium is stable in air and the compact metal tarnishes only superficially when heated. It is only incompletely oxidized when heated to redness in oxygen, even when in the form of filings. In a stream of chlorine, however, it burns vigorously at a red heat, forming the chloride $NbCl_5$. It also unites directly when heated with sulfur or selenium. It alloys with difficulty with most metals, but nevertheless is said to form mixed crystals with iron in all proportions. With aluminum it forms the compound Al_3Nb.

Hydrofluoric acid slowly attacks niobium. All other acids, including aqua regia, have no action upon it. It is also insoluble in caustic alkalis. By contrast, it is attacked by fused alkali hydroxides and in the powdered state is even attacked with incandescence by fused potassium nitrate

The complete insolubility of niobium in oxidizing acids depends on the fact that it is extraordinarily strongly *passivated* by solutions of oxidizing agents. In contact with dilute sulfuric acid, it allows an electric current to flow only when the metal is the *cathode*. An alternating current rectifier constructed by Siemens and Halske depends upon this; it consists of two plates, one of platinum and one of niobium, dipping into a trough filled with 0.1-N sulfuric acid. This apparatus allows only that half-cycle of the alternating current to pass for which the platinum sheet is the anode, and the niobium sheet the cathode.

8. Compounds of Niobium

The well defined compounds of niobium are almost all derived from the element in its $+5$ state. Niobium can, indeed, function in lower valence states, but most of the compounds containing the element in these states are unstable.

(a) Niobium(IV) and (III) Compounds

Niobium is quadrivalent in the dioxide NbO_2 and in the tetrachloride $NbCl_4$. The dioxide may be obtained by heating niobium pentoxide to a white heat in hydrogen. It is formed as a bluish-black powder, insoluble in water and acids, which burns to the pentoxide again when heated to a dull red heat in air. By further abstraction of oxygen it can be transformed into the monoxide NbO, which has a crystal structure related to the rock salt type. The solubility of oxygen in niobium metal amounts to about 1.7 atom-%.

Niobium tetrachloride is obtained as brown needles by the reduction of $NbCl_5$ with Nb, Al, Fe, or H_2. It decomposes readily by the process

$$2NbCl_{4solid} \rightleftharpoons NbCl_{3solid} + NbCl_{5gas} - 28.3 \text{ kcal.}$$

For the vapor pressure of $NbCl_5$ over a mixture of the two solid compounds $NbCl_3$ and $NbCl_4$, Schäfer (1952) gives the values 1.2 mm. at $200°$, 230 mm. at $300°$, and 531 mm. at $320°$. The four phases $NbCl_{3solid}$, $NbCl_{4solid}$, $NbCl_{5liquid}$ and $NbCl_{5gas}$ can coexist in equilibrium at $420°$, under a pressure of 22 atm.

Niobium trichloride, $NbCl_3$, (black needles, with metallic luster), can be prepared by reducing niobium oxychloride, $NbOCl_3$, with heated magnesium, or from niobium penta-chloride $NbCl_5$, by thermal dissociation, or by reducing the vapor with hydrogen. Niobium(III) compounds are obtained in solution when niobium(V) compounds are treated with zinc in acid solution. When certain experimental conditions are adhered to, reduction proceeds quantitatively to the trivalent stage, as can be established by back-titration with permanganate, and re-oxidation to niobium(V). The color of the Nb^{+++} ions present in the reduced solution is blue; they have a very great tendency to be oxidized.

(b) Niobium(V) Compounds

(*i*) *Niobium pentoxide*, Nb_2O_5, is obtained as a white powder, (density 4.46, m.p. $1460°$), insoluble in water, by dehydration of its hydrate, *niobic acid*, or by ignition of niobium sulfide, nitride, or carbide in air. It can be brought into solution by fusing it either with alkali hydrogen sulfate or with alkali carbonate or hydroxide— an indication that it is an amphoteric oxide; its acidic character greatly pre-dominates, however.

Whereas Ta_2O_5 exists in only a single modification, Nb_2O_5 is trimorphous (Brauer, 1941). It is generally obtained in the low-temperature form, which is isotypic with Ta_2O_5, even after it has been ignited for a short time in the course of preparation. Nb_2O_5 and Ta_2O_5 form mixed crystals with each other. These can be obtained directly if the two oxides are precipitated together as gels from solution, and dehydrated by ignition. If the Nb_2O_5 content is high enough (more than 76 weight-per cent), the strongly ignited mixed crystals have the same crystal structure as the modification of Nb_2O_5 which is stable above 850°. With a lower Nb_2O_5 content, the Ta_2O_5 structure is obtained. When Nb_2O_5 is ignited in hydrogen, it can lose a certain amount of oxygen without undergoing any change in crystal structure. Only after the oxygen content has been reduced below that corresponding to the formula $NbO_{2.40}$ does a new phase appear—namely the dioxide, NbO_2.

(ii) *Niobates and Niobic Acid*. If niobium pentoxide is fused with sodium carbonate, it displaces carbon dioxide from the latter and enters as an acidic oxide in its place. If four parts of Na_2CO_3 are allowed to react with 1 part of Nb_2O_5 as much CO_2 is driven off as corresponds to the formation of an *orthoniobate*, according to the equation

$$Nb_2O_5 + 3Na_2CO_3 = 2Na_3NbO_3 + 3CO_2.$$

If the melt is leached with water, the residue remaining is not sodium *ortho*niobate, Na_3NbO_4, however, as would be expected, but sodium *meta*niobate, $NaNbO_3$. Sodium metaniobate forms a colorless, fine crystalline powder, sparingly soluble in cold water.

$NaNbO_3$ crystallizes with the *perowskite* type structure ($a = 3.89$ Å), as also does $KNbO_3$, ($a = 4.01$ Å), whereas $LiNbO_3$, crystallizes after the *ilmenite* type (see p. 61). In both structural types, Nb has the coordination number 6 with respect to oxygen. The same is true of the niobates of the *bivalent* metals, such as $Fe(NbO_3)_2$ and $Mn(NbO_3)_2$, which have the *columbite* type structure. The rhombic structure of $Fe(NbO_3)_2$ is built up from (distorted) FeO_6 and NbO_6 octahedra, which share corners in two spatial directions, and along the third direction share their edges in such a way that each oxygen atom belongs simultaneously to 3 octahedra. In this manner, each layer of FeO_6 octahedra, built up from

running in the direction of the crystallographic *a* axis, has a correspondingly constructed sheet of NbO_6 octahedra adjacent to it on each side. The heavy metal niobates as yet known have mostly been obtained from melts—as, for example, $Mn(NbO_3)_2$, which is isomorphous with $Fe(NbO_3)_2$, formed by heating a mixture of Nb_2O_5 and MnF_2 to a white heat with KCl (as flux).

The niobite or columbite occurring naturally (black orthorhombic crystals; structure, see above) is an isomorphous mixture of iron(II) and manganese niobates and the corresponding tantalates. The tetragonal *mossite*, of very rare occurrence, has the same composition.

Compounds of niobium pentoxide with basic oxides in other proportions, e.g., *pyroniobates*, $M^I_4Nb_2O_7$, may be mentioned. It has not yet been finally established whether the niobates, containing water of crystallization, that may be obtained from aqueous solutions, consist of *pentaniobates*, $M^I_7Nb_5O_{16}\cdot mH_2O$, or of *hexaniobates* $M^I_8Nb_6O_{19}\cdot nH_2O$, in which part of the water of crystallization can be replaced by M^IOH; the constitution of the hydrated tantalates is likewise uncertain. The nonexistence of ammonium niobates is noteworthy.

On treatment of niobate solutions with sulfuric acid, *niobium pentoxide hydrate* is thrown down as a white gelatinous precipitate, of continuously variable water content. Niobium pentoxide hydrate is also formed by hydrolysis of salt-like compounds of pentavalent niobium—e.g., of niobium pentachloride, or of the alkali hydrogen sulfate melts of niobium pentoxide. It is formed very slowly from the latter and is completely precipitated only on boiling the solutions. Niobium pentoxide hydrate, commonly known as 'niobic acid' is soluble both in caustic alkalis and in strong acids; in this it resembles stannic acid. It also shares with the latter the property of passing very easily into colloidal dispersion, and in this case also, (as with stannic acid), hydrochloric acid is an especially effective peptizing agent.

Whether or not niobium pentoxide forms stoichiometrically defined hydrates, so that it is possible to speak of a free niobic acid as a well defined chemical compound, it is difficult to decide; the hydrated preparations of Nb_2O_5 are, without exception, amorphous, and no clearly marked stages appear in the dehydration. The very firm binding of water in Nb_2O_5 gels is noteworthy however; some water is still firmly retained in the neighborhood of $500°$. Certain observations (Jander 1928, Hüttig 1930) support the assumption that amorphous pentaniobic acid, $H_7Nb_5O_{16}$, is obtained by the decomposition of niobate solutions. This is, however, apparently incapable of crystallizing and, as a typically colloidal material, it contains an excess of water which is mechanically so firmly bound that it cannot be fully removed without the loss, at the same time, of a portion of the chemically bound water.

(iii) *Acido-niobates.* Complex niobates, *acido-niobates*, are derived from niobium, by exchanging the oxygen atoms in the niobates with acid radicals—e.g., oxalatoniobates, formed by the entry of oxalate radicals, are colorless crystalline compounds; the majority have a composition corresponding to the formula $3M^I_2O \cdot Nb_2O_5 \cdot 6C_2O_3$, and must probably be formulated as *oxotrioxalatoniobates*, $M^I_3[ONb(C_2O_4)_3]$.

The radicals of other organic oxy-acids are also able to enter the niobate group by complex formation, as can the radicals of titanic, phosphoric, arsenic, chromic, and tungstic acids. Conversely, the niobic acid radical can enter into these. The formation of such complexes frequently shows itself in the mutual influence which these acids exert upon each others' reactions.

The *halogenoniobates*, discussed further below, also belong to the class of acidoniobates; the compounds are derived from niobates by exchange of oxygen atoms by halogen atoms.

Replacement of one or more oxygen atoms in the niobates by the equivalent amount of peroxy-radicals, —O——O—, leads to the formation of *peroxyniobates*.

(iv) *Peroxyniobates and Peroxyniobic Acid.* Hydrogen peroxide reacts with potassium niobate in the presence of excess alkali, to form a colorless or pale yellow salt. This salt, potassium peroxyniobate*

$$K_3NbO_8 = K_3\left[O_2^{O_2}Nb_{O_2}^{O_2}\right]$$

can be precipitated from aqueous solution by means of alcohol, and contains niobium and active oxygen in the ratio 1 : 4.

By adding dilute sulfuric acid to the concentrated solution of this peroxyniobate, a free peroxyniobic acid,

$$HNbO_4 = H\left[_O^O NbO_2\right]$$

(hydrated), containing niobium and active oxygen in the ratio 1 : 1, is obtained as a fine lemon-yellow powder. Peroxyniobic acid is remarkably stable. It is decomposed by dilute sulfuric acid (into hydrogen peroxide and niobic acid) only when heated.

(v) *Niobium pentachloride*, $NbCl_5$, can be obtained by heating metallic niobium in chlorine:

$$Nb + \tfrac{5}{2}Cl_2 = NbCl_5;$$

* The compound is obtained either anhydrous (Balke 1908) or containing $\tfrac{1}{2}H_2O$ (Sieverts 1928), according to the experimental conditions.

or by reaction of niobium pentoxide with certain chlorides—such as CCl_4:

$$Nb_2O_5 + 5CCl_4 = 2NbCl_5 + 5COCl_2.$$

Niobium pentoxide reacts with carbon tetrachloride appreciably more readily than does tantalum pentoxide. If a mixture of the two pentoxides is heated at 270° with excess carbon tetrachloride in a sealed tube from which air has been removed, the niobium compound reacts almost completely while tantalum pentoxide remains almost unaltered. This difference in reactivity can be utilized for the preparative separation of niobium and tantalum. When niobium pentachloride is prepared by the first-mentioned process it is possible to use niobium containing carbon, such as is obtained by melting the metal in the electric furnace, or the incompletely reduced product resulting from ignition of a mixture of niobium pentoxide and carbon. The niobium pentachloride can be separated by distillation from the oxychloride, $NbOCl_3$, which is admixed with the product of reaction in the latter case.

Niobium pentachloride forms yellow crystal needles, melting at 204.7° and boiling at 250°. The yellow vapor consists of $NbCl_5$ molecules, as shown by a vapor density determination (at 360°). It is soluble without decomposition in the usual organic solvents (alcohol, ether, chloroform, carbon tetrachloride), and also in sulfur monochloride, S_2Cl_2. It is decomposed hydrolytically by water, into niobic acid and hydrogen chloride. No niobic acid is precipitated, however, when the pentachloride reacts with concentrated strong acids (sulfuric acid or hydrochloric acid); the hydrolysis is probably suppressed by these strong acids. If the solution of niobium pentachloride in concentrated hydrochloric acid is boiled, precipitation no longer ensues even on subsequent dilution with water. The niobic acid resulting from hydrolysis on heating is peptized by the hydrochloric acid, and then remains in colloidal dispersion.

The heat of sublimation of $NbCl_5$ is 20 kcal per mol. (entropy of sublimation = 39.5 e.u.); heat of evaporation of the molten compound = 13.1 kcal per mol (entropy of evaporation 25.1 e.u.). $NbCl_5$ is completely miscible with $TaCl_5$ in both the liquid and the solid state.

Niobium pentachloride appears to have little tendency to form double salts. Addition products of niobium pentachloride with organic substances are well known, however,—e.g., a piperidine addition compound $NbCl_5 \cdot 6C_5H_{11}N$.

(vi) *Niobium oxychloride* (niobium oxytrichloride), $NbOCl_3$, forms white silky needles which volatilize at about 400°. The vapor density corresponds to the formula $NbOCl_3$. At very high temperatures, however, the oxychloride splits up into $NbCl_5$ and Nb_2O_5. It is also decomposed by water:

$$2NbOCl_3 + 3H_2O = Nb_2O_5 + 6HCl.$$

Hydrogen sulfide converts it into an oxysulfide on gentle heating. Niobium oxychloride forms double salts (*oxochloroniobates*) of the types $M^I[NbOCl_4]$ and $M^I_2[NbOCl_5]$. These are most readily obtained from a solution of niobic acid in concentrated hydrochloric acid, by addition of the corresponding chlorides (Weinland).

The purple-red pentabromide behaves similarly to the pentachloride. The yellow oxybromide of niobium is more easily decomposed than the oxychloride. It likewise forms double salts (*oxobromoniobates*). The pentaiodide, NbI_5, may be obtained by the action of I_2 vapor on an incandescent niobium wire (Körösy, 1939). TaI_5 is obtained in corresponding manner; it is more stable than NbI_5.

(vii) *Fluoroniobates and Niobium pentafluoride.* The tendency to from double salts is most marked with the fluorine compounds of niobium. If a solution of niobic acid in hydrofluoric acid is treated with metallic fluorides, either fluoroniobates or oxofluoroniobates are formed, according to the amount and concentration of the hydrofluoric acid.

The simplest examples of the fluoroniobates have compositions fitting the general formulas $NbF_5 \cdot M^IF$ or $M^I[NbF_6]$ and $NbF_5 \cdot 2M^IF$ or $M^I_2[NbF_7]$ (on this point see footnote on p. 115); most, however, have more complicated compositions. Among oxofluoroniobates, compounds of the compositions $NbOF_3 \cdot 2M^IF$ and $NbOF_3 \cdot 3M^IF$ are known, as well as those of more complex composition.

Niobium pentafluoride itself was prepared by Ruff in 1911, by boiling niobium pentachloride with anhydrous hydrogen fluoride, under a reflux condenser cooled with a freezing mixture. It forms colorless highly refringent crystals of the composition NbF_5, melts at 72–73°, and boils at 236°.

(viii) *Niobium Nitrides.* Niobium can take up only minimal amounts of nitrogen into its crystal lattice, if any at all. It forms several nitrides, however, (cf. Table 13, p. 89). According to Brauer and Jander (1952), NbN exists in three modifications which can be differentiated quite unambiguously by X-ray diffraction. In addition to the NaCl type (cubic) modification (with $a = 4.38$ Å), there are two different hexagonal modifications. One of these ($a = 2.95$, $c = 11.25$ Å) has a composition corresponding exactly to the formula NbN, whereas the composition of the cubic form may vary between NbN and $NbN_{0.89}$. The second hexagonal modification ($a = 2.93$, $c = 5.45$ Å) is only obtained under special conditions. The compound Nb_4N_3 has a structure derived from the NaCl type by a slight tetragonal distortion ($c/a = 0.987$), and covers a range of homogeneity between $NbN_{0.79}$ and $NbN_{0.75}$. The compound Nb_2N (with a range of homogeneity extending between $NbN_{0.50}$ and $NbN_{0.40}$) has a structure in which the Nb atoms are in hexagonal close packing (with $a = 3.05$, $c = 4.96$ Å), with the N atoms distributed statistically among the largest interstitial positions. All the niobium nitrides are very hard and brittle substances. The cubic NbN is yellow grey, but the others are pure grey in color.

9. Analytical (Niobium)

Niobium is readily precipitated as niobic acid both from acid and from alkaline solutions, and frequently in a form hard to filter. Ignited niobium pentoxide is insoluble in acids; it can, however, be brought into solution by fusion with potassium hydrogen sulfate or potassium pyrosulfate. A dirty blue coloration is produced in niobic acid solutions by zinc and dilute sulfuric acid, as a result of reduction of niobic acid. Tantalic acid does not give this reaction.

For the analytical separation of niobium from tantalum, see p. 116.

10. Tantalum (Ta)

(a) Occurrence

Tantalum is almost always found to accompany niobium in tantalites and niobites, which occur in almost all parts of the world, though mostly in only small amounts. The most important European sources of tantalite are in Finland and Scandinavia. Considerably larger deposits occur in North America and in Australia.

The *tantalites* are iron(II) metatantalates $Fe(TaO_3)_2$, in which manganese may also replace part of the iron. Part of the tantalum is almost invariably replaced by niobium. As has already been stated, niobium, predominates over tantalum in the niobites. Ordinary tantalite crystallizes in the rhombic system (structure as for niobite, see p. 107). The same compound can also occur tetragonal, as *tapiolite*, in forms which correspond to those of rutile. Tantalic acid also frequently occurs in combination with the rare earths. Thus *fergusonite* is substantially an yttrium orthotantalate (and niobate), $Y[TaO_4]$. *Yttro-tantalite* is an yttrium pyrotantalate, $Y_4[Ta_2O_7]_3$. Of similar composition, although with niobium predominating over tantalum, and notable for its uranium content, is *samarskite* (uranotantalite). *Microlite* is substantially a calcium pyrotantalate $Ca_2[Ta_2O_7]$.

More minerals are known with a predominant niobium content than with a preponderance of tantalum; the average ratio of abundance of the two elements is about 1.3 : 1. The fact that almost pure tantalum minerals occur, as well as pure niobium minerals, whereas hafnium never predominates over zirconium, can be explained from the relatively small difference in the abundance of niobium and tantalum, in conjunction with the fact that their chemical similarity is perceptibly less marked than that displayed by the pair zirconium-hafnium.

(b) History

Tantalum was discovered in 1802 by Ekeberg, in two then newly-found minerals (from Kimito in Finland, and from Ytterby in Sweden). Alluding to the inability of tantalum oxide, treated with an excess of acid, to utilize any of the acid for salt formation—a property remarkable in a metallic oxide—Ekeberg named the element after Tantalus, who thirsted in the underworld in the Greek legend. The two minerals accordingly received the names *tantalite* and *yttrotantalite*.

(c) Preparation

Tantalum ores are opened up technically by heating them with potassium hydrogen sulfate in iron vessels; the melt is leached with boiling water, and the tantalic acid remaining behind in powdered form (rendered impure by niobic acid) is dissolved in hydrofluoric acid. Separation from niobium is usually effected by fractional crystallization of potassium fluorotantalate, K_2TaF_7. Tantalum metal is obtained from this by reduction with sodium at high temperatures:

$$K_2TaF_7 + 5Na = Ta + 5NaF + 2KF.$$

The metal so obtained in the form of a black powder at first always contains some oxide. It can be freed from this, according to von Bolton, by heating it in an electric vacuum furnace, since tantalum pentoxide decomposes into the metal and oxygen at very high temperatures. Very pure tantalum can be obtained by the filament growth process, by thermal dissociation of $TaCl_5$ (Burgers 1934).

(d) Properties

Tantalum is a heavy, platinum-grey, lustrous metal. It is fairly hard, but at the same time is extremely ductile; its ductility increases with its degree of purity. It is distinguished by a very high melting point (3027 °C according to Worthing, 1926; 2996 °C according to Malter, 1939), and by extraordinarily great resistance towards attack by chemical agents at moderately high temperatures.

Apart from hydrofluoric acid, metallic tantalum is not attacked by any acids,—not even by aqua regia; aqueous alkali hydroxides are equally without action. It is attacked only with difficulty by molten caustic alkalis, but is eventually corroded. Compact tantalum is stable in air at ordinary temperature, and on gentle heating it merely assumes a blue-black tarnish. It is more extensively attacked, however, when strongly heated in the air, and burns with vigorous incandescence when heated in a finely divided state. Water is also vigorously decomposed by tantalum powder at a red heat. Chlorine attacks it strongly when hot, and fluorine does so even at ordinary temperature. Tantalum also unites with sulfur with inflammation. Tantalum absorbs hydrogen (cf. p. 88), and also nitrogen when heated. It thereby becomes hard and brittle.

(e) Uses

Since metallic tantalum combines excellent mechanical properties with outstanding chemical resistance, it has proved very suitable for the manufacture of surgical and dental instruments, such as forcep tips, canula, needles, etc. In many cases it can be used as a substitute for platinum. Pen nibs of tantalum are said to be practically as good as gold nibs tipped with iridium.

For some time, tantalum was used for the manufacture of incandescent lamp filaments, because it was possible to draw it into exceedingly fine wires. It was quickly displaced from this field, however, by the production of single-crystal wires of tungsten, which followed shortly afterwards.

11. Compounds of Tantalum

The compounds of tantalum are derived almost without exception from pentavalent tantalum. Well defined compounds of the lower valence states of tantalum are known, however. Thus the chlorides $TaCl_4$, $TaCl_3$, and $TaCl_2$ are obtained by heating tantalum pentachloride, $TaCl_5$, with aluminum (preferably in the presence of aluminum chloride) in a sealed tube in the absence of air. 'Nascent' hydrogen is not able to reduce tantalum compounds in aqueous solution, in contrast to the behavior of vanadium and niobium.

(a) Tantalum(V) Compounds

Tantalum pentoxide, Ta_2O_5, is best obtained in the pure state by strongly heating pure tantalum metal in a stream of oxygen. It is also obtained when compounds of tantalum with volatile or combustible elements are heated in air. It is generally prepared by dehydrating 'tantalic acid'. This tends, however, to be contaminated by substances adsorbed from the solution from which it was precipitated, and so does not furnish an absolutely pure oxide.

Tantalum pentoxide is a white powder, of density 8.02, insoluble in water and in acids other than hydrofluoric acid. It is unaltered by heating in air, or in an atmosphere of chlorine, hydrogen sulfide or sulfur vapor. In contrast with niobium pentoxide, it is not attacked when heated in gaseous hydrogen chloride or hydrogen bromide. On the other hand, it loses oxygen when heated to a white heat in a vacuum: $Ta_2O_5 = 2Ta + \frac{5}{2}O_2$. The elimination of oxygen proceeds at a lower temperature in the presence of carbon, since the latter reduces the oxygen pressure to a minimum by combining with it. In this case, however, a portion of the carbon combines with the tantalum, forming tantalum carbide (see p. 116).

Unlike niobium pentoxide, pure tantalum pentoxide is not reduced to the dioxide when it is heated in hydrogen, even when the powder is admixed with finely divided niobium pentoxide. The behavior is quite different, however, if the elements are present in the form of Nb_2O_5-Ta_2O_5 mixed crystals. If the Nb_2O_5 content is high enough, as has been shown by Schäfer, the whole of the Ta_2O_5 present in the mixed crystals can be reduced (i.e., all the Ta^{5+} converted to Ta^{4+}), giving homogeneous NbO_2-TaO_2 mixed crystals.

(*i*) *Tantalates and Tantalic acid.* *Tantalates* are obtained by melting tantalum pentoxide with alkali hydroxide or carbonate. These double compounds of basic oxides with tantalum pentoxide, which functions as an acid anhydride, generally have compositions analogous to those of the niobates. It does appear, however, that tantalum pentoxide tends even more strongly than niobium .pentoxide to form adsorption products of stoichiometrically ill defined composition, as well as the normal compounds with basic oxides.

The reaction of tantalum pentoxide with excess sodium carbonate, in the melt, proceeds much more sluggishly than with niobium pentoxide, but eventually leads likewise to the formation of the ortho-salt Na_3TaO_4, as may be deduced from the quantity of carbon dioxide eliminated from the melt. The properties of the metatantalates correspond fairly closely to those of the metaniobates; in particular, the two series of compounds are completely isomorphous. In aqueous solutions, even under strongly alkaline conditions, the ions present are exclusively those of *pentatantalic acid* $H_7[Ta_5O_{16}]$ according to the work of Jander (1925). Whether the hydrated crystalline tantalates are salts of this acid, and so are pentatantalates $M^I_7Ta_5O_{16}.mH_2O$, or are hexatantalates $M^I_8Ta_6O_{19}.nH_2O$, in which part of the water is replaced by M^IOH, is not yet settled. The potassium salt, crystallizing in six-sided prisms, and easily soluble in water (in contrast to the lithium and sodium salts), can be obtained according to Windmaisser (1941) with its composition, as expressed by the ratio $K_2O : Ta_2O_5$ corresponding either to 7 : 5 or to 8 : 6, without the variation in composition having any effect on the crystal structure. The alkali tantalates are strongly hydrolyzed in solution. When the hydrogen ion concentration is greater than 10^{-6} molar, complete decomposition occurs, with the deposition of gelatinous tantalum pentoxide, the so-called tantalic acid. The precipitation is hindered by some acids, which can unite with tantalum pentoxide, forming complex compounds—e.g., by arsenic acid, arsenious acid, and especially by organic oxyacids such as tartaric acid and citric acid. It has, however, not been possible to isolate any coordination compounds of tantalic acid (e.g., corresponding in type to apatite or vanadinite, p. 102).

It is very doubtful whether the so-called tantalic acid, precipitated from solution, represents a definite hydrate of tantalum pentoxide, such as the acid $H_7[Ta_5O_{16}]$ which Jander believes to be present, as its anion, in alkaline solutions. The problem is very similar to that of the Nb_2O_5 gels, except that the Ta_2O_5 gels give up water rather more readily than do the latter. If Ta_2O_5 gel is dehydrated by heating it not too slowly, a sudden incandescence occurs at the moment that practically all the water is given off. The phenomenon, which also occurs with niobium pentoxide, is due (as in the case of titanium dioxide), to the sudden crystallization of oxide, hitherto present in the quasi-amorphous state.

Freshly precipitated 'tantalic acid' is soluble in an excess of strong acid, such as hydrochloric, nitric, or sulfuric acid, probably with formation of acido-compounds, analogous to the behavior of niobic acid. These compounds are again decomposed, with the deposition of tantalic acid, on dilution with water.

Tantalic acid is prepared technically by sulfuric acid decomposition of potassium fluorotantalate $K_2[TaF_7]$, the tantalum compound most readily obtained in the pure state:

$$2K_2TaF_7 + 2H_2SO_4 + 5H_2O = Ta_2O_5 + 2K_2SO_4 + 14HF.$$

Tantalic acid prepared in this way is always contaminated, to a more or less considerable degree, with substances adsorbed from solution.

(*ii*) *Peroxytantalates and Peroxytantalic acid.* Pissarjewsky obtained potassium peroxytantalate, a compound of the composition $K_3TaO_8.\frac{1}{2}H_2O$* by the action of hydrogen peroxide on a melt of tantalum pentoxide and potassium hydroxide, dissolved in water. The salt was thrown down from the aqueous solution as a white, finely divided precipitate on addition of alcohol. Salts corresponding in composition to this (but with different water

* Under rather different experimental conditions, the anhydrous salt is obtained, as with the niobium compound (cf. p. 108).

content) could also be obtained with other cations. The compounds contain 4 active oxygen atoms. Their constitution is accordingly expressed by the formula

$$M^I_3 \begin{bmatrix} O_2 & & O_2 \\ & Ta & \\ O_2 & & O_2 \end{bmatrix}$$

—i.e., they are derived from the orthotantalates by exchanging the four O-atoms with four peroxy-radicals, just as the potassium peroxyniobate described on p. 108 is derived from the orthoniobates.

It was possible to prepare a free peroxytantalic acid by treating the potassium peroxytantalate with dilute sulfuric acid. It had the composition $HTaO_4$ (without inclusion of water content), and contained one active oxygen atom. It is thus analogous in its composition and constitution to the peroxyniobic acid described on p. 108.

(*iii*) *Tantalum pentachloride*, $TaCl_5$, results from the combustion of tantalum (or also of tantalum carbide, nitride, or sulfide) in a stream of chlorine. It can also be prepared by the reaction of tantalum pentoxide with certain chlorides (e.g., phosphorus pentachloride, carbon tetrachloride, carbonyl chloride, sulfur monochloride, or aluminum chloride). It is a yellow substance, usually vitreous, but obtainable in crystalline form by remelting or sublimation; it has a density of 3.68, melts at 216.5, and boils at 242°. The vapor density corresponds to the formula $TaCl_5$. If it is sublimed in an atmosphere containing oxygen, it decomposes with the formation of tantalum pentoxide. It is noteworthy that no oxychloride appears as an intermediate in this process, neither is such an intermediate product formed in the decomposition by water, which likewise leads immediately to the pentoxide (in this case in the gelatinous, hydrated form, the so-called tantalic acid), together with hydrogen chloride. At low temperatures, tantalum pentachloride adds on ammonia; Spacu (1937) has shown the existence of the 12-, 10- and 7-ammoniates. The last decomposes in the neighborhood of 0°, probably with the formation of $Ta(NH_2)_2Cl_3\cdot3NH_3$ (ammonolysis). It has not been possible to obtain double salts of tantalum pentachloride, although the pentafluoride has a great tendency to form such.

(*iv*) *Tantalum pentafluoride and Fluorotantalates.* According to Ruff, tantalum pentafluoride can be obtained like niobium pentafluoride, by halogen exchange between tantalum pentachloride and liquid hydrogen fluoride. It forms colorless prisms (m.p. 96.8°, b.p. 229°). According to Hahn it can also be prepared by heating barium tantalum fluoride, $3BaF_2\cdot2TaF_5$, in a platinum tube, while passing through a slow stream of air dried with phosphorus pentoxide.

Tantalum pentafluoride cannot be obtained from solutions of tantalum pentoxide in aqueous hydrofluoric acid, because of its great tendency to hydrolyze. When such solutions are evaporated, and also when the dried residues are ignited, there is invariably some loss, through volatilization of tantalum in the form of the fluoride, unless the tantalum pentoxide dissolved was *completely free from alkali*. If it does contain alkali, volatilization of tantalum fluoride can take place to a considerable extent, because the alkali fluorotantalate, formed under these circumstances, breaks up into its components when heated; one product is the relatively easily volatile tantalum pentafluoride. This volatilization can be prevented in practice by the addition of sulfuric acid which decomposes the tantalum fluoride.

In contrast to the free tantalum pentafluoride, its complex salts, the *fluoro-tantalates* can readily be obtained from the aqueous solution of tantalum or tantalic acid in hydrofluoric acid. The fluoro complexes separate from solution on the addition of the corresponding metallic fluorides. Most of the fluorotantalates correspond to the type $2M^IF\cdot TaF_5$. In addition to these compounds of the type $M^IF\cdot TaF_5$ and a sodium salt of the composition $3NaF\cdot TaF_5$ also exist. The potassium salt of the first mentioned type, $2KF \cdot TaF_5$*, which crystallizes anhy-drous, is used for the separation of tantalum from niobium and titanium, on ac-count of its sparing solubility. It forms at once if potassium ions are introduced into a solution of tantalum pentoxide in hydrofluoric acid. Since it is considerably more soluble in warm solutions than in cold, it can readily be obtained in a well-crystallized state. It forms fine rhombic needles, which can also serve for the microscopic detection of tantalum. Recrystallization from hot water must be carried out in the presence of excess hydrofluoric acid, since hydrolytic decompo-sition occurs otherwise.

It has not yet been fully established whether the hydrolysis of fluorotantalates leads immediately to tantalic acid, or whether oxofluorotantalates, in the form of definite com-pounds, appear as intermediate products. In any case, the tendency to form oxofluoro-compounds is much smaller with tantalum than with niobium. Hahn was able to isolate a fluorotantalic acid, with the composition $HF \cdot TaF_5 \cdot 6H_2O$, as feathery needles of m.p. $15°$, from a solution of pure tantalum pentoxide in hydrofluoric acid.

(b) Tantalum Compounds of Lower Valence States

(i) *Tantalum Chlorides*. Lower chlorides of tantalum can be obtained (Ruff, 1922 and 1925) by heating tantalum pentachloride at $300°$ in an air-free sealed tube with aluminum powder (in the presence of some aluminum chloride, to facilitate initiation of the reaction). Tantalum(III) chloride, $TaCl_3$, is the compound most readily obtained in this way. This reacts reversibly in the molten state with an excess of $TaCl_5$, forming $TaCl_4$:

$$TaCl_3 + TaCl_5 \rightleftharpoons 2TaCl_4.$$

If the tantalum pentachloride is distilled off at $350–400°$, pure tantalum trichloride remains behind. On raising the temperature further (to $500–600°$) the trichloride again splits off tantalum pentachloride:

$$3TaCl_3 = 2TaCl_2 + TaCl_5.$$

All these chlorides are green, and solid at ordinary temperature. They differ characteri-stically in their behavior towards water. The tetrachloride is decomposed according to the equation $2TaCl_4 + 5H_2O = TaCl_3 + Ta(OH)_5 + 5HCl$. The trichloride is soluble in cold water without change, forming a greenish solution. A green hydrogel is precipi-tated from the solution by OH^- ions, and is soluble not only in acids but also in excess of caustic alkali. It thus resembles aluminum hydroxide in its amphoteric behavior. It has a very energetic tendency to undergo oxidation, however, and decomposes water on boiling, forming tantalic acid:

$$Ta(OH)_3 + 2H_2O = Ta(OH)_5 + H_2.$$

A corresponding decomposition of water at $100°$ is also brought about by the trichloride. The dichloride of tantalum is practically insoluble in water, but is nevertheless oxidized to a tantalum(III) compound even by cold water, hydrogen being evolved:

$$TaCl_2 + H_2O = Ta^{+++} + 2Cl^- + OH^- + \tfrac{1}{2}H_2.$$

When a solution of tantalum(III) chloride, containing excess of hydrochloric acid, is evaporated in a vacuum, it deposits a particularly stable chlorine compound of trivalent tantalum, $Ta_3Cl_7O \cdot 3H_2O$. This compound was therefore actually obtained before the simple trichloride. From this, several derivatives may be prepared, which likewise contain 3 atoms of trivalent tantalum in the molecule.

* The compound is to be regarded structurally as a heptafluorotantalate, $K_2[TaF_7]$, with coordinatively 7-valent tantalum, as Hoard (1939) has shown by a complete X-ray structure determination, using Fourier analysis methods. The same applies to the niobium compound of analogous composition (cf. p. 110).

(*ii*) *Tantalum Carbides*. Tantalum forms two carbides. Like TiC and ZrC, the brassy-yellow TaC crystallizes with the rock salt structure ($a = 4.445$ Å). It resembles these other carbides in that it can take up a certain amount of metal in solid solution (up to 10 atom-%). (For conductivity, see p. 50.) Ta$_2$C is grey and forms a crystal lattice in which the tantalum atoms are arranged in hexagonal closest packing ($a = 3.091$ Å, $c = 4.93$ Å).

(*iii*) *Tantalum Nitride*, TaN, forms as a grey powder, when tantalum is heated in an atmosphere of nitrogen. The heat of formation is 58.1 kcal/mol. (Neumann, 1934).

12. Analytical (Tantalum)

Tantalum is extraordinarily similar to niobium from the analytical standpoint. However, potassium fluorotantalate, K$_2$TaF$_7$, which crystallizes from a concentrated hydrofluoric acid solution of tantalic acid on addition of potassium fluoride, is considerably more sparingly soluble than the analogous niobium compound. The latter, in particular, readily dissolves in warm water, being transformed into K$_2$[NbOF$_5$], whereas potassium fluoro-tantalate is changed into a very sparingly soluble basic salt on boiling with water. Contamination of niobium by tantalum can readily be detected by means of this reaction.

The method of Powell and Schoeller ('*Analysis of Minerals and Ores of the Rarer Elements*', Griffin, London, 1955) is generally employed for the analytical separation of tantalum from niobium. This depends on the fact that the oxalatotantalate complex is more readily decomposed than the oxalatoniobate complex. The tantalic acid liberated by hydrolysis is precipitated by the addition of tannin, with which it forms a pale yellow adsorption complex, whereas niobic acid, which is only precipitated at rather higher pH values, forms a bright red tannin adsorption compound.

13. Protactinium

Protactinium was discovered in 1918 by Hahn and Meitner and, independently, by Soddy and Cranston. The guiding idea in the search for the parent element of actinium was the supposition, based on the displacement law (p. 529) that the element sought must be a *homologue of tantalum*. When the trivalence of actinium had been established by von Hevesy's diffusion experiments and Fleck's investigation of its chemical behavior, so that actinium had been assigned to its place in the IIIrd Group of the Periodic System, it followed from the displacement law that actinium must necessarily be the daughter either of a β-emitting element of the IInd Group, or of an α-emitting element of the Vth Group. In the former case, the parent of actinium would have been an isotope of radium. However, since radium and actinium are both ultimately derived from uranium, the isotope in question must already be admixed with radium in the ores, and such a β-emitting isotope of radium could hardly have remained undiscovered. The second possibility was, accordingly, much more probable—that the element sought belonged to the Vth Group of the Periodic System, and would be an as yet unknown homologue of tantalum, disintegrating by emission of α-particles*. To isolate such an element, pitchblende residues were treated with tantalic acid, and the latter was separated again. Expectations that the homologue of tantalum sought for would be separated with the tantalum 'carrier' were fulfilled. The tantalum pentoxide recovered from the pitchblende residues proved to be radioactive, and was shown to contain the parent element of actinium by the fact that the short lived disinte-

* The only radioactive homologue of tantalum known up to that date, Uranium X$_2$, is a β-emitter.

gration products of actinium, which are easily identified by their characteristic decay curves, were formed from it. The name *protactinium* (Greek ὁ πρῶτος, first in the series), conferred on the element by Hahn and Meitner, expresses this relationship to actinium, as its immediate precursor.

Protactinium, or rather its pentoxide, is relatively easily obtained in the *radioactively pure state*—i.e., free from other radioactive substances. Its preparation in the *chemically pure* state, on the other hand, is extremely difficult, since it displays a very great similarity to tantalum pentoxide, from which it is distinguished only by its slightly stronger basic character. This difficulty is much increased by the minuteness of the quantities in which protactinium occurs in pitchblende. For 1 ton of uranium, in any uranium mineral, there is present in radioactive equilibrium only 314 mg of protactinium (cf. p. 537), as well as 332 mg of radium. However, in 1928, von Grosse succeeded in isolating 9 mg of the pure pentoxide Pa_2O_5* from about 500 kg of starting material. A few years later (1934) he prepared about 100 mg of pure Pa_2O_5, and at the same time Graue and Käding obtained 500 mg of Pa, in the form of K_2PaF_7, using $5\frac{1}{4}$ tons of starting material. The great similarity between protactinium and tantalum was confirmed by this preparative work. It was found, however, that Pa_2O_5 displayed perceptibly less acidic character than does Ta_2O_5; it approximates more closely to ZrO_2 in this respect. By applying zirconium and tantalum reactions alternately, such a far-reaching enrichment in protactinium was achieved that a solution of the potassium double fluorides ultimately furnished pure $PaF_5 \cdot 2KF$ by fractional crystallization.

Von Grosse (1934) also prepared minute amounts of protactinium in the metallic state, by two methods: (*i*) by decomposing the oxide by bombardement with cathode rays, and (*ii*) by thermal decomposition of $PaCl_5$ or PaI_5 on an electrically heated tungsten wire in a high vacuum. Metallic protactinium is grey-white and lustrous, and is not oxidized in air. The pentachloride $PaCl_5$ (pale yellow needles, easily sublimed, m.p. 301°) can be obtained from the oxide by heating it in a current of phosgene at 550°:

$$Pa_2O_5 + 5COCl_2 = 2PaCl_5 + 5CO_2.$$

The genetic relationship of protactinium to actinium, leaves no doubt that protactinium fills the place in the Periodic System indicated by the atomic number 91; its properties also show its general homology with tantalum. The value 231, found experimentally for its atomic weight by von Grosse (1935), agrees with that which may be deduced from the place of protactinium in the actinium disintegration series. The end product, actinium D, is a lead with the atomic weight 207 (cf. p. 537 *et seq.*). This is derived from protactinium, however, by a total of 6 α-ray transformations (cf. Table 60, p. 537). The atomic weight of protactinium must, accordingly, be about $6 \times 4 = 24$ units higher than that of actinium D (actinium lead)—i.e., about 231. For the radioactive properties of protactinium see p. 537.

Of the radioelements to which places must be assigned in the Periodic Table, protactinium ranks next to radium in the amounts in which it occurs in Nature. Although its half life (32,000 years) is much longer than that of radium (1590 years), it is less abundant than radium. This is because radium ^{226}Ra is a member of the decay series of the most abundant isotope of uranium, ^{238}U, hence 99.3% of the uranium follows the disintegration path through ionium and radium, and the natural abundance of radium is determined by radioactive equilibrium with the ^{238}U. Protactinium is formed by α-particle emission from actinouranium ^{235}U, which is present only to the extent of 0.7% in natural uranium.

---

* In this and the immediately following paragraph, the reactions are interpreted in terms of the chemical reactions postulated by von Grosse. It is not certain that he did, in fact, isolate Pa_2O_5 (cf. p. 118).

More recently it has been possible to extend the knowledge of protactinium chemistry by the use of the protactinium isotope ^{233}Pa, made from ^{232}Th by nuclear reactions in the atomic pile:

$$^{232}\text{Th} \xrightarrow{(n,\gamma)} {}^{233}\text{Th} \xrightarrow{\beta} {}^{233}\text{Pa} \xrightarrow{\beta} {}^{233}\text{U}.$$

^{233}Pa is a β-emitting radioelement (half life 27.4 days), and is convenient for the 'tracer' study of the chemistry of protactinium. Sellers, Fried, Elson and Zachariasen (1951) have identified a number of protactinium compounds on the microchemical scale. They state that the white oxide obtained by heating the oxalate or oxide hydrate is not Pa_2O_5, but is $PaO_{2.25}$ or Pa_4O_9 (cf. the uranium oxide system). This is reduced by hydrogen at 1550° to black PaO_2 (cubic, CaF_2 structure, with $a = 5.51$ Å), but can be oxidized to Pa_2O_5 (orthorhombic, isotypic with Ta_2O_5) by heating it in oxygen to 1100°. PaO was also observed (cubic, NaCl structure, $a = 4.97$ Å) on the surface of protactinium metal (cf. UO, p. 198).

Treatment of PaO_2 with a mixture of HF and hydrogen at 600° converted it to PaF_4 (red brown solid, isotypic with ZrF_4 and ThF_4). This tetrafluoride was reduced by barium vapor to metallic protactinium, which has a unique tetragonal structure ($a = 3.925$ Å, $c = 3.238$ Å) derived from the body-centered cubic structure. Metallic protactinium, like its neighbors in the Periodic System, thorium and uranium, reacted with hydrogen to give a cubic compound identified as protactinium hydride, PaH_3. A volatile chloride was formed by the action of CCl_4 on the oxide, and is presumed to be $PaCl_5$ although this has not been proved.

This more recent investigation of the chemistry of protactinium, in agreement with studies of the solution chemistry of protactinium (Bouissieres and Haissinsky, 1948–49) would seem to establish that the quadrivalent state is more important for the chemical behavior of protactinium than the pentavalent state, and that the element therefore has a closer horizontal relationship to the adjacent elements, thorium and uranium (in the uranium(IV) compounds) than it has to tantalum. The analogy with tantalum nevertheless proved a good guide in the discovery and first isolation of the element.

Reference

1 H. HOUGARDY, *Die Vanadinstähle; Aufbau, Eigenschaften und Verwendung von vanadinlegierten Stählen*, Berlin 1934, 224 pp.

CHAPTER 5

SIXTH SUB-GROUP OF THE PERIODIC SYSTEM: CHROMIUM, MOLYBDENUM, TUNGSTEN (WOLFRAM), AND URANIUM

Atomic numbers	Elements	Sym-bols	Atomic weights	Densi-ties	Melting points	Boiling points	Specific heats	Valence states
24	Chromium	Cr	52.01	7.2	1830°	ca. 2300°	0.1178	I, II, III IV, V, VI
42	Molybdenum	Mo	95.95	10.2	2600°	ca. 4800°	0.0589	II, III, IV, V, VI
74	Tungsten (Wolfram)	W	183.86	19.1	3400°	ca. 5700°	0.0324	II, III,IV, V, VI
92	Uranium	U	238.07	19.0	1130°	ca. 3500°	0.0274	II, III,IV, V, VI

1. Introduction

(a) General

The sixth Sub-group of the Periodic System contains the metals chromium, molybdenum, tungsten (or wolfram), and uranium. These are all found to function in several valence states; they have a maximum valence state of six in their normal compounds, as corresponds to the group number. In acid solutions, chromium is most stable in the $+3$ state, but is readily oxidized to the $+6$ state in the presence of alkalis. The homologues of chromium are most stable in the hexapositive state under ordinary conditions. For uranium alone, the tetrapositive state is comparable with the $+6$ state in importance, but the other valence states of this element are relatively unimportant. In general, these elements differ from those of the Main Group, in that odd valence states occur as well as even ones. The principal feature common to the elements of the VIth Sub-group and Main Group, apart from their maximum valence state of six, is the ability of their trioxides to form salts with basic oxides.

The simplest chromates, molybdates, tungstates, and uranates have compositions analogous to the sulfates $M^I_2SO_4$. The elements of the Sub-group differ from those of the Main Group in their strong tendency to form compounds which contain *more* than 1 equivalent of acid anhydride to 1 equivalent of basic anhydride. Thus, chromium, in addition to chromates $M^I_2CrO_4(= M^I_2O \cdot CrO_3)$ and dichromates $M^I_2Cr_2O_7(= M^I_2O \cdot 2CrO_3)$ (the latter corresponding in their composition to the pyrosulfates $M^I_2S_2O_7 = M^I_2O \cdot 2SO_3$), also forms trichromates

$M^I_2Cr_3O_{10}(= M^I_2O \cdot 3CrO_3)$ and tetrachromates $M^I_2Cr_4O_{13}(= M^I_2O \cdot 4CrO_3)$. In the molybdates and tungstates, the number of molecules of acid anhydride, MoO_3 or WO_3, per molecule of basic anhydrides M^I_2O, is frequently much greater. The acids underlying the formation of these types of salts are called *polyacids* (these are further discussed under *tungsten*). Uranium has a much smaller tendency to form such high molecular polyacids; nevertheless the ordinary uranates have formulas $M^I_2U_2O_7$, corresponding to the pyrosulfates, and not to the normal sulfates.

The strength of the acids derived from the trioxides decreases greatly from chromium to uranium. The compound of uranium, $UO_3 \cdot H_2O$ or $UO_2(OH)_2$, corresponding in composition to chromic acid, can be regarded as an amphoteric hydroxide, in which the acid character is quite subordinate, whilst the compound has a very marked capacity for salt formation with acids, forming the *uranyl* salts UO_2X_2 (X = univalent acid radical).

The hydroxide of quadrivalent uranium apparently has a purely basic character. Chromium(III) hydroxide is amphoteric, although its basic character greatly preponderates. Hexavalent chromium forms no salt-like compounds at all. Chromyl chloride, CrO_2Cl_2, is a typical acid chloride, whereas uranyl chloride UO_2Cl_2 possesses a very distinct salt-like character.

Molybdenum and tungsten do indeed form compounds with acid anions. Almost without exception, however, these are compounds in which the element in question combines with acid radicals, to form a more or less strongly bound complex anion. From the standpoint of the coordination theory, the constitution of the anions so formed—e.g., the complex ions $[Mo(CN)_8]^{4-}$ and $[W(CN)_8]^{4-}$ formed by molybdenum and tungsten—does not in any way differ in principle from that of the ions of the oxyacids, e.g., $[MoO_4]^{2-}$ and $[WO_4]^{2-}$.

As is the case almost throughout the Sub-groups of the Periodic System, the Sub-group VI elements display a very marked tendency to form acido-compounds (anionic complexes). With trivalent chromium, however, in addition to the capacity for binding *acid radicals* in the complex, there is also a strong tendency to combine firmly with *neutral molecules*, especially of ammonia and derivatives of ammonia. The complex compounds so formed are grouped together under the name of chromium ammines. A feature peculiar to the chemistry of molybdenum and tungsten is that the acid ions derived from them—e.g., $MoO_4^=$ or $WO_4^=$ and $Mo_3O_{10}^=$ or $W_3O_{10}^=$—readily attach themselves to certain elements as central atoms of complexes—e.g., to phosphorus in the form of phosphoric acid, to silicon in silicic acid, and to boron in boric acid. They thereby give rise to complex acids, often of quite complicated composition, though mostly based on relatively simple structural principles, the so-called *heteropolyacids*. The best known representative of these is the 12-molybdatophosphoric acid, which (neglecting its water content) has the analytical composition $H_3PO_4 \cdot 12MoO_3$*. The test for phosphoric acid by means of ammonium molybdate, already cited in Vol. I, Chap. 14, depends upon the formation of the sparingly soluble ammonium salt of this acid, when molybdate, phosphate, and ammonium ions are brought together in concentrated nitric acid solution.

The melting points and boiling points of the metals of the sixth Sub-group are for the most part very high. They rise extremely steeply from chromium, through

* For the constitution of the compound see p. 180 *et seq.*

molybdenum to tungsten, only to fall again strikingly with uranium. This is probably to be associated with the fact that uranium, at ordinary temperature, has a crystal structure differing from that of its homologues. Chromium, molybdenum, and tungsten crystallize with the body-centered cubic structure (see Vol. I, p. 210 Fig. 47), with the cell dimensions a = 2.878, 3.141, and 3.158 Å, whereas uranium is trimorphous, the form stable at ordinary temperature (a-uranium) being orthorhombic.

If tungsten is deposited electrolytically from a solution of WO_3 in fused alkali phosphates, it is obtained in another modification (β-W), which passes monotropically over into the ordinary modification (a-W) above 600°. In the cubic unit cell of β-W, the cube face centers, as well as the body center and cube corners are occupied by tungsten atoms (see Fig. 20). Chromium is also frequently deposited electrolytically in the form of unstable modifications, having either a structure of the Mg type (cf. Vol. I, p. 249) or of the a-Mn type (cf. p. 24). In this case also, a monotropic transformation into the ordinary form takes place on heating.

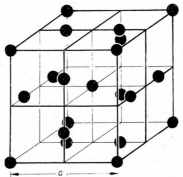

Uranium has a structure which is encountered with no other element. According to Warren (1937), it forms a rhombic crystal lattice which can be regarded as a deformed hexagonal packing. Whereas in the latter each atom is surrounded by 12 others at equal distances, in uranium 4 neighbors are drawn in closer (2 each at 2.76 and 2.85 Å, respectively) than the other 8 (4 each at distances of 3.27 and 3.36 Å).

This modification (a-uranium) changes above about 665° into β-uranium, and this in turn at 775° into γ-uranium. β-Uranium has a very complex, unique structure. Tucker (1951) and Thewlis (1951–52) have shown that it is tetragonal. It cannot be stabilized at

Fig. 20. β-Tungsten. $a = 5.038$ Å.

ordinary temperature by quenching, but Thewlis found the cell dimensions to be $a = 10.759$ Å, $c = 5.656$ Å at 720°. The uranium atoms fall into six crystallographically distinct sets, and the U $\leftrightarrow$ U distances range from 2.53 to 3.91 Å. γ-Uranium forms a body-centered cubic structure; at 805° the cell dimension is $a = 3.524$ Å (Thewlis, 1951). Although the γ-form of pure uranium cannot be obtained at ordinary temperature, it is stabilized by certain alloying additives (e.g., Cr and Mo). The hypothetical cell dimensions at room temperature have been deduced from those of the stabilized γ-phase in U-Mo alloys, by extrapolation to 0% Mo. A value of $a = 3.474$ Å was obtained in this way, corresponding to a U $\leftrightarrow$ U distance of 2.97 Å.

The *atomic radii* of the elements of the sixth Sub-group, based on the atomic distances in the crystal lattices of the metals, are collected in Table 14. The table also contains the radii of the M^{3+} and M^{4+} ions, based upon the lattice dimensions of the oxides. Adequate data are not yet available for calculating the radii of the ions in the 6-fold positive charged state.

TABLE 14

APPARENT ATOMIC AND IONIC RADII OF THE ELEMENTS OF SUB-GROUP VI

Element	Chromium	Molybdenum	Tungsten	Uranium
Atomic radius, Å	1.25	1.36	1.37	1.48
Radius of M^{3+} ion, Å	0.65	—	—	1.06
Radius of M^{4+} ion, Å	—	0.68	0.68	1.05

Uranium has the highest atomic number and the highest atomic weight of all the naturally occurring elements.

In Table 15 are collected the molar heats of formation of the principal oxides of the metals of the VIth Sub-group. Except for the case of tungsten, there is only a relatively small evolution of heat associated with transformation of the dioxides into the trioxides. The difference between the heats of formation of chromium sesquioxide and chromium trioxide is particularly small, however, when referred to the same quantity of metal. Whereas 89.6 kcal per g-atom of oxygen are liberated when chromium is oxidized from the metal to the sesquioxide, only 1.8 kcal per g-atom of oxygen is evolved in oxidizing the sesquioxide to the trioxide. Apart from this, the heat of formation even of chromium sesquioxide is relatively small, by comparison with the molecular heat of formation of aluminum oxide (393 kcal). This is of importance in connection with the *preparation* of chromium discussed below.

TABLE 15

HEATS OF FORMATION OF THE OXIDES OF THE ELEMENTS OF SUB-GROUP VI
(in kcal per mole of oxide)

Element	Chromium	Molybdenum	Tungsten	Uranium
Q_f for sesquioxide, M_2O_3	268.9	—	—	—
Q_f for dioxide, MO_2	—	142.8	131.4	269.7
Q_f for trioxide, MO_3	137.1	180.4	195.5	303.9

When heated, chromium, molybdenum, and tungsten readily unite not only with oxygen, halogens, phosphorus, sulfur and its homologues, but also with carbon, silicon, boron, and (less readily) with nitrogen.

These metals (except uranium) are notable for their ability to form compounds with carbon monoxide, namely carbonyls $M(CO)_6$ (colorless rhombic crystals). These may be obtained by the action of CO and Grignard's magnesium compounds on the halides of the metals in question (Job, cf. p. 354); the carbonyls of Mo and W can also be made directly, by the action of CO on the metals above 225° under high pressure (200 atm.). The metal carbonyls cited may be sublimed without decomposition under reduced pressure. They are insoluble in water, but soluble in indifferent organic solvents, such as benzene. They are remarkably stable towards chemical reagents, especially towards acids and alkalis. For further discussion of metal carbonyls see p. 351 *et seq.*

(b) Alloys

A fairly detailed knowledge is extant as to the alloys of the metals of the VIth Sub-group, other than uranium, as is shown by the data set out in Tables 16 and 17. It is evident that their behavior towards other metals depends on their place in the Periodic System. In general the tendency to form compounds with other metals is not very strong. Extensive mixed crystal formation is almost invariably the behavior displayed towards the transition elements.

Only in exceptional cases do the Sub-group VI metals form compounds with the *heavy metals of the Main Groups*, and mixed crystals are either not formed at all, or only within very narrow limits. As far as is known, there is also hardly any miscibility with the *light metals* in the solid state. The compounds with these latter mostly melt incongruently, indicating that the binding forces in the compounds are but weak. No compounds of the metals of the VIth Sub-group with alkali and alkaline earth metals are known. However, no exact metallographic investigations of these systems have been reported, and they could be carried

MISCIBILITY AND COMPOUND FORMATION BETWEEN ELEMENTS OF SUB-GROUP VI AND THE ELEMENTS OF THE MAIN GROUPS

(Symbols have the same meaning as in Table 9, see p. 49)

	II		III		IV				V					VI		
	Be	Mg	B	Al	C	Si	Sn	Pb	N	P	As	Sb	Bi	S	Se	Te
Cr (misc.)	—	—	$liq>0$; $s?$	$liq<\infty$; $s>0$	$s\sim0$	$s>0$	$liq<\infty$; $s>0$	$s\,0$	$s\sim0$	$s>0$	—	$s>0$	$liq\,0$; $s\,0$	$liq<\infty$; $s\,0$	—	—
Cr	$CrBe_2$	alloys	CrB 1550°; CrB_2 1850°	Al_7Cr^* 725°; $Al_{11}Cr_2^*$ 1000°; Al_3Cr^* 1160°; $Al_8Cr_5\square$ 890°; $AlCr_2\square$ 850°	Cr_4C^* 1530°; Cr_7C_3 1670°; Cr_3C_2 1820°; $CrC?$	Cr_3Si; Cr_2Si; $CrSi$; $CrSi_2$	o	o	Cr_2N; CrN	Cr_3P^* 1510°; Cr_2P; CrP; CrP_2	$CrAs$; $Cr_2As_3?$; $CrAs_2?$	$CrSb$ 1110°; $CrSb_2^*$ 675°	o	CrS 1565°; Cr_2S_3	$CrSe$; Cr_2Se_3	$CrTe$; Cr_2Te_3
Mo (misc.)	—	alloys	—	$s\,0$	$s>0$	$s\,0$	—	alloys	$s\sim0$	$s>0$		alloys	alloys	$s\,0$		
Mo	$MoBe_2$		Mo_2B 1850°; MoB_2 2250°	Al_4Mo^* 735°	Mo_2C 2407°; $MoC?$	$MoSi_2$; $Mo_2Si_3?$	—		Mo_2N; MoN	Mo_3P; MoP; MoP_2	$MoAs_2$			$MoS?$; $Mo_2S_3?$; MoS_2; MoS_3 MoS_4	$MoSe_3?$	Mo_2Te_3; $MoTe_2$
W (misc.)	—	—	—	—	$s>0$	$s\,0$	—	$liq<\infty$; $s\,0?$	$s\sim0$	$s\,0$						—
W	WBe_2		W_2B 2770°; WB 2860°	Al_4W; $Al_3W?$; $AlW_2?$	W_2C 2750°; WC^* 2600°	WSi_2 2170°; W_3Si_2 2360°		o	W_2N	$W_4P?$ unstable WP; WP_2	WAs_2	alloys	alloys	$WS?$; $W_2S_3?$; WS_2; WS_3	WSe_2; WSe_3	alloys

TABLE 17

MISCIBILITY AND COMPOUND FORMATION BETWEEN THE METALS OF SUB-GROUP VI AND THE ELEMENTS OF THE SUBSEQUENT SUB-GROUPS

(Symbols have the same meanings as in Table 9, see p. 49)

	Sub-group VI		Sub-group VII		Sub-group VIII					Sub-group I			Sub-group II		
	Mo	W	Mn	Re	Fe	Co	Ni	Pd	Pt	Cu	Ag	Au	Zn	Cd	Hg
Cr	$s\,\infty$	$s\,\infty$	$liq<\infty$ / s?	$s<\infty$	$s\,\infty$	$s\sim\infty$	$s\sim\infty$	$s<\infty$	$s<\infty$	$liq<\infty$ / $s\sim0$	$liq<\infty$ / $s\,0$	$liq<\infty$	$liq<\infty$	$liq<\infty$	—
	0	0	Mn₃Cr*	0	[FeCr] 930°	CoCr[] / Co₂Cr₃[] 1292°	0	Pd₂Cr₃ 1398°	0	0	0	0	0	0	CrHg? / CrHg₃?
Mo		$s\,\infty$	$liq<\infty$ / alloys	—	$s<\infty$	$s<\infty$	$s<\infty$	—	$liq<\infty$? / $s<\infty$	$liq\sim0$ / $s\,0$	$liq<\infty$ / $s\,0$	$liq<\infty$ / $s\,0$	—	—	$s\sim0$ / alloys
		0			Fe₃Mo₂* 1480° / Fe₇Mo₆* 1540°	CoMo* 1500° / Co₂Mo₃ 1550°	NiMo* 1345° / Ni₃Mo[] 925°			0	0	0			
W		W	—	$s\sim0$	$s>0$	$s<\infty$	$s<\infty$	—	$s<\infty$	$liq<\infty$	$liq<\infty$	$liq<\infty$ / $s\,0$	—	—	$s\sim0$ / alloys
				Re₃W₂ 3007°	Fe₇W₆* 1640° / Fe₂W[] 1040°	Co₇W₆* 1620° / Co₃W[] 1100°	Ni₆W 1525°		0	0	0	0			

out only with great difficulty, because of the wide disparity in melting points of the components, and the great chemical aggressiveness of the vapors of the alkali and alkaline earth
metals.

The compounds of the metals of the VIth Sub-group with the non-metals of the IIIrd
to VIth Main Groups closely resemble the intermetallic compounds in their structures and
in the compositions, and they are therefore also included in Table 16. In some cases,
non-Daltonide compounds are formed with the non-metals named. Apart from hydrogen, for
which chromium possesses a considerable solvent power (cf. p. 129), the metals of the VIth
Sub-group can not take up any considerable amounts of the non-metals without undergoing a change in crystal structure.

The behavior towards the *transition* metals is quite different from that towards metals of
the other Sub-groups. With the transition metals immediately following them*, i.e., with
the metals of the VIIth and VIIIth Sub-groups, the metals of the VIth Sub-group mostly
form mixed crystal systems of wide extent. In the liquid state, miscibility is complete, in all
cases, except perhaps in the system Mo-Pt. The excellent alloying properties of these
metals, especially with metals of the iron group, are of considerable technical importance
(see p. 268). In many cases, compounds between the metals of the VIth Sub-group and
those of the iron group are formed out of primarily deposited mixed crystals, as a result of
transformations taking place during cooling—frequently proceeding without change of
composition, i.e., simply characterized by a change of structure (compare pp. 15 and 339):
such compounds are denoted in Tables 16 and 17 by the sign []. Alternatively, superstructure phases appear on annealing, the formulas of these are enclosed in square brackets.
In systems where miscibility of the components in the solid state is limited, compounds of
incongruent melting point generally occur (denoted by asterisks).

The metals of the VIth Sub-group form no compounds among themselves. Chromium,
molybdenum, and tungsten form mixed crystals with each other in all proportions.

The metals of the VIth Sub-group are not completely miscible in the molten state with
those metals of the Sub-group which follow the transition elements (i.e., with the metals
of the Ist and IInd Sub-groups). The only exception known with certainty is formed by
the system Cr-Au; in any case this system displays a considerable degree of miscibility
even in the solid state: gold is able to incorporate up to 40 atom-% of chromium in its
crystal lattice. No compounds are formed between the metals of the VIth and those of
the Ist Sub-group, and the existence of compound with metals of the IInd Sub-group have
not been established with certainty.

2. Chromium (Cr)

(a) Occurrence

The most important chromium ore is *chromite*, a double compound of iron(II)
oxide and chromium(III) oxide, $FeCr_2O_4$. It is found in considerable quantities
in the Urals, in South Africa, New Caledonia, Greece, Asia Minor, in the
Scandinavian peninsula, in Hungary and in Bosnia, and in smaller amounts in
Upper Silesia, Moravia and Styria. In addition *crocoisite*, $PbCrO_4$, which occurs
especially in Brazil, as well as in the Urals, has some technical significance.

Chromium is frequently found to replace aluminum in aluminum minerals.
Thus chromium spinels, chromium tourmalines, chromium garnets, chromium
micas, and chromium chlorites are known**. The true emerald is a form of beryl,
colored dark green through incorporation of chromium in the place of aluminum.

---

* Very little is known of the behavior of the metals of the VIth Sub-group towards the
transition metals preceding them (cf. Tables 9 and 13).

** *Chlorites* in mineralogy comprise certain mica-like minerals (differing from the micas
in being alkali-free) which occur especially in many schists (chlorite schist).

(b) History

Chromium was discovered in 1797, by Vauquelin, in a Siberian mineral (crocoisite), and was named after the beautiful colors which characterized its compounds (Gr. χρῶμα = color). It was very soon identified in other minerals also, especially in chromite (by Tassaert in 1799). Vauquelin succeeded in reducing the oxide to the metal, by means of carbon. Starting from experiments on the production of metallic chromium, Goldschmidt, in 1894, devised the aluminothermic process, which he then immediately applied to the preparation of other metals of high melting point, as well as to the attainment of high temperatures within a small volume.

(c) Preparation

The starting point for the technical preparation of chromium and its compounds is almost always *chromite*. If pure chromium is not required, but only its (carbon-containing) iron alloy (*ferro-chromium*), the ore may be directly reduced with carbon (in the Siemens-Martin furnace, or in the electric furnace):

$$FeCr_2O_4 + 4C = 4CO + Fe + 2Cr.$$

For the preparation of pure chromium, on the other hand, the purest possible chromium oxide must first be obtained. From this, carbon-free chromium is obtained by aluminothermic reduction:

$$Cr_2O_3 + 2Al = Al_2O_3 + 2Cr + 130 \text{ kcal.}$$

The preparation of pure chromium oxide proceeds by way of the alkali chromates or dichromates. The chromite is first fused in the hearth of a reverberatory furnace, with the addition of alkali hydroxide or carbonate, and burned lime. By this means the chromium is transformed into chromate, through the oxidizing action of the furnace flame, in the presence of an excess of oxygen. The chromate leached out with water is converted to dichromate by the addition of acid. After the separation of more sparingly soluble compounds, the dichromate is deposited from the concentrated solution in a sufficiently pure state. Pure chromium(III) oxide is obtained from this by reduction with carbon; sawdust or starch can be used in place of wood-charcoal:

$$Na_2Cr_2O_7 + 2C = Cr_2O_3 + Na_2CO_3 + CO.$$

The aluminothermic preparation of metallic chromium from the oxide is carried out by igniting a mixture of the latter with aluminum powder. The addition of a little chromate is advantageous, so that the mixture burns uniformly, or the chromium oxide may be previously ignited with the addition of a little alkali (whereby chromate is formed). Rather less than the theoretical amount of aluminum is used, in order to obtain the chromium as free as possible from this agent. The economic superiority of the aluminothermic production of chromium over other processes is partly due to the fact that the fused aluminum oxide obtained as a by-product (artificial corundum) finds application as an abrasive and polishing agent. Chromium prepared by the aluminothermic method is, however, not absolutely pure. Absolutely pure chromium is malleable when hot, unlike ordinary aluminothermic chromium, although it is also brittle at ordinary temperature; it can be obtained by degassing electrolytic chromium by heating it in high vacuum, or by reducing $CrCl_3$ or Cr_2O_3 by means of Ca in a $CaCl_2$ or $BaCl_2$ melt (Kroll 1935).

The *electrolytic deposition* of chromium is employed for the production of *protective coatings* on other metals. The usual procedure is to deposit the chromium from warm chromic acid solutions, to which chromium(III) sulfate has been added. In accordance with the position of chromium in the electrochemical series, it is desposited as metal only at high current densities, when much hydrogen is evolved simultaneously. Since this alloys itself with the metal to some extent, and makes it porous and fissured, it must be subsequently removed by heating. In so doing the unstable modifications of chromium which are often formed first are transformed into the stable modification, and a considerable increase in hardness is brought about in consequence.

The technique of electrolytic chromium plating was originated in its fundamentals, by Foerster. The processes involved in it are not yet explained in all their details, although the essential features are clear, chiefly as a result of the investigations of Sargent (1920 and later), Liebreich (1921 and later), and Müller (1926 and later): [see Z. *Elektrochem.* 40 (1934) 326; 43 (1937), 361]. If a sufficiently strong current is passed through a pure chromic acid solution, between chemically inert electrodes, a vigorous evolution of hydrogen takes place at the cathode after reduction of $HCrO_4^-$ to Cr^{+++} ions has occurred initially to a slight extent. This occurs even though the potential at which H^+ ions are discharged lies about 1.3 volts lower than that required for the reduction of the $HCrO_4^-$ ion (cf. Vol. I, p. 765), and is a result of the formation on the cathode of a film of basic chromium(III) chromate, $Cr(OH)[CrO_4]$. This film is impermeable to $HCrO_4^-$ ions, and therefore keeps those ions from reaching the cathode. According to Müller, it must be supposed that the $[Cr(OH)]^{2+}$ groups are all directed towards the cathode and the $[CrO_4]^{2-}$ groups towards the anode; otherwise the chromium(III) chromate would be soluble in the chromic acid, as it is in all strong acids. The film is permeable to H^+ ions, as it is to certain other ions also—e.g., HSO_4^- and F^-. H^+ ions can accordingly pass through the film, and be discharged at the cathode. If foreign ions, such as HSO_4^- and F^-, are present simultaneously in the solution, a proportion of these are carried through the film by the H^+ ions. The acid which has reached the cathodic side attacks the membrane, making it porous. The pores are blocked again by the entry of $HCrO_4^-$ ions into them; these are at once reduced at the cathode to $[Cr(OH)]^{2+}$ ions, fresh ones are dissolved by the acid diffusing in and so on. In this way, Cr^{+++} ions are accumulated at the cathode surface, while H^+ ions are simultaneously used up, by reaction with $[Cr(OH)]^{2+}$ groups. The consequence is that at sufficiently high current densities, the discharge of H^+ ions is accompanied by the simultaneous discharge of Cr^{+++} ions and deposition of metallic chromium. A pre-requisite condition for this is, of course, that the acid content of the solution is not too high, otherwise only hydrogen evolution occurs. Accordingly, chromium sulfate is generally added to the chromic acid solution rather than sulfuric acid. It is considerably more difficult to obtain compact and well-adhering chromium platings by the electrolysis of chromium sulfate solutions, free from chromic acid, than by the electrolysis of solutions containing chromic acid. The very chromium chromate film that hinders the deposition of chromium when foreign ions are absent leads, in the presence of such ions, to the formation of an especially fine crystalline, and therefore dense, hard and lustrous chromium coating, since the rate of growth of the individual crystals is repressed, but the formation of crystallization nuclei is favored by the covering of the cathode surface with the film.

(d) Properties

Chromium is a white, brilliant, hard, brittle metal, of density 7.2. It has quite a high melting point (about 1800°) but its boiling point is not so high proportionately (about 2300° at atmospheric pressure).

Chromium is extraordinarily resistant chemically at ordinary temperature. It is not oxidized noticeably in air, even in the presence of moisture, and only tarnishes superficially even when it is warmed. It burns, however, with showers of sparks, in the oxy-blowpipe flame and also on heating with potassium chlorate or potassium nitrate. It also unites directly with the halogens, with sulfur, nitrogen, carbon, silicon, boron, and with some metals, although only when heated. Chromium containing carbon is appreciably more resistant still, and is also considerably harder than the pure metal. Chromium dissolves in aqueous hydrochloric acid or sulfuric acid,—quite vigorously in certain circumstances. Nitric acid, concentrated or dilute, on the other hand, and aqua regia in the cold, are absolutely without action. If the metal is boiled with these acids, a slight action occurs; the reaction at once comes to a standstill again when the heating is interrupted.

The peculiar stability of chromium towards the acids named arises from the fact that they put the metal into a particularly unreactive condition, which is termed a state of *passivity*.

The phenomenon of passivity has already been encountered with other metals; it has been thoroughly studied in the case of chromium. It is evident that chromium experiences an alteration of its chemical character by treatment with the oxidants mentioned, since after such treatment it is also no longer dissolved by dilute hydrochloric or sulfuric acid, even on boiling. If boiled for a long time, it can however be observed that the passive state eventually changes into the active state. This transition generally happens quite suddenly. At first there is hardly any observable evolution of hydrogen, but this sets suddenly, with such vigor that the whole liquid foams. If the hot acid is now replaced by cold acid, the evolution of hydrogen continues vigorously with this also. If, however, dilute or concentrated nitric acid is substituted for hydrochloric acid or sulfuric acid, the evolution of hydrogen ceases abruptly; the chromium has again become passive. The process may be repeated as often as desired.

In the *active* state, chromium has a great tendency to give up electrons, and go into solution as a positive ion. It stands between zinc and iron in the electrochemical series. (The standard potential of completely active chromium, in contact with a chromium(II) salt solution, referred to the normal hydrogen electrode, is + 0.56 volt). Chromium can therefore displace many other metals—e.g., copper, tin, and nickel—from solutions of their salts. In accordance with this, it can also discharge hydrogen ions and is accordingly soluble in acids in its normal state. By passivation, however, chromium is displaced completely to the noble end of the series, to so great an extent that strongly passive chromium actually has a lower solution tension than gold or platinum*. It behaves therefore, in the passive state, as a decidedly noble metal. The metal is generally neither completely active, nor completely passive. In practice the solution tension of chromium is therefore a very ill defined quantity.

The passivation of chromium is brought about not only by nitric acid and aqua regia, but by all other strong oxidants. Oxidizing agents of different strengths passivate chromium to various extents; a slight degree of passivation is even produced by the oxygen of the air. It is, accordingly, very difficult to obtain the metal in the pure active state. Acids which evolve hydrogen—e.g., hydrochloric acid—have an activating effect; activation is also caused to a lesser extent by many salt solutions, as also by some fused salts and also by many organic substances. The strongest activation is obtained when hydrogen is evolved electrolytically at a chromium surface, or when chromium is deposited electrolytically from its solutions.

A peculiar difference between active and passive chromium is observed during its anodic dissolution by means of the electric current. Whereas active chromium goes into solution as

---

* A normal potential of —1.2 volt, referred to the normal hydrogen electrode, has been measured for the most strongly passive chromium. In comparing this with the corresponding figures for gold and platinum it must be borne in mind that the solution tension of these metals is raised by their tendency to form complex ions. Gold and platinum, in contact with *analytically equimolecular* solutions show, accordingly, a greater tendency to go into solution than the most strongly passivated chromium, although if referred to the *same concentration of free elementary ions* their potentials would be rather lower than that of chromium.

the Cr^{++} ion—i.e., in the lowest common valence state that chromium displays*—, passive chromium dissolves with the formation of $CrO_4^=$ ions—i.e., with the highest valence state assumed by chromium. Except for their chemical or electrochemical behavior, active and passive chromium do not differ in any detectable way. There may perhaps be differences in their reflectivity for light, but no such differences have been established with certainty.

For theories to explain the phenomenon of passivity, see p. 750 *et seq.*

When chromium is deposited electrolytically from aqueous solutions of chromic salts, it can take up considerable quantities of hydrogen (up to 60 volumes per unit volume of chromium at ordinary temperature, and up to 300 volumes at —50°). The chromium probably thereby forms a sort of supersaturated solid solution with hydrogen. If it is heated in a vacuum, the greater part of the hydrogen is suddenly and irreversibly given off at about 60°; to achieve complete de-gassing, however, it is necessary to heat to nearly 600°. The crystal structure of chromium is not altered by the absorption of hydrogen, but the lattice dimensions are somewhat expanded (Hüttig 1925). According to Weichselfelder, a hydride of stoichiometric composition CrH_3 can be obtained by the action of hydrogen on an ether solution of chromium trichloride, $CrCl_3$, containing phenyl magnesium bromide. It has not yet been demonstrated, however, that this is indeed a chemical compound.

In the very finely divided state, as it is obtained, for example, from chromium amalgam by distilling off the mercury, chromium is pyrophoric in the air. *Colloidal chromium* can be obtained by electrical *dispersion* under isobutyl alcohol, or by treatment of the most finely powdered metal alternately with dilute acids and alkalis.

(e) Uses of Chromium and its Compounds

Metallic chromium finds its principal use in the *steel industry*. Chromium steels are characterized by hardness and strength, and are therefore employed for tools and for particularly heavily stressed portions of machines—e.g., the balls of ball bearings. Since too high a carbon content in these steels is deleterious, the ferro-chromium produced by reducing chromite with carbon, which is generally of high carbon content, cannot be used for the manufacture of the best products; chromium or ferrochromium produced by the aluminothermic process must be used. The chromium steels often contain a third metal—e.g., nickel or tungsten—to improve their quality further.

Other alloys, such as bronze and brass, are also frequently hardened by addition of chromium. A nickel-chromium-iron alloy in wire form ('Nichrome') is used for the winding of electric furnaces.

Since chromium is the hardest of the useful metals, it is very suitable for the *production of protective coatings*. Electrolytic chromium plating [1] has rapidly found wide applications in the automobile and cycle industries. For many other objects also, chromium plating is used increasingly, in place of the brass coating, formerly customary, or of nickel plating.

The compounds of chromium are even more widely used than metallic chromium. Chromium(III) salts and chromates are used in the dye and color industry as mordants. In the leather industry, they serve for the production of the durable chrome leather. Chromium(II) salts are used in vat dyeing because of their great reducing power. Chromates and dichromates are the most useful energetic oxidizing agents. They are used in large amounts as such, especially in the coal tar industry. They also possess the property of making gelatin and gum arabic insoluble in water after exposure to light, and find application in artistic photography (gum bichromate, bromoil, and similar processes). Numerous chromium

---

* Since passivation would set in as soon as even traces of oxygen were evolved at the anode, the current density in the process must be quite small.

compounds are employed as mineral pigments. Chrome yellow (lead chromate), chrome orange (basic lead chromate), chrome green (chromium(III) oxide), zinc yellow (a double salt of zinc chromate and potassium dichromate $3ZnCrO_4 \cdot K_2Cr_2O_7$), and zinc green (a mixture of zinc yellow and Paris blue) may be mentioned as such.

The chrome colors are distinguished by beauty, light-fastness, and stability towards air, and some of them possess particularly high covering power. They find application not only as building paints and artists' colors, but also in wall paper printing and color printing, in the manufacture of oil cloth and linoleum, and in colored glazes for porcelain. They are *poisonous*, and health precautions are necessary in the preparation and use of chrome colors (cf. p. 152 and 154).

In the production of chrome colors, and in chrome tanning, it is desirable to use chromium salts which are as free as possible from vanadium. This requirement is generally n ot fulfilled, since the chromite ore which constitutes the raw material for their preparation usually contains some vanadium. Vanadium can be removed from the chromate solutions obtained in opening up the chromite, by a method described by Perrin (1952). This involves stirring the solution with a suspension of lead sulfate, whereby the vanadium is precipitated in the form of lead vanadate (together with a certain amount of lead chromate). The precipitate can be worked up again for the recovery of vanadium.

The world production of chromium ores amounted in 1936 to 900 000 tons. About 40% of this came from the Union of South Africa and Rhodesia, 17% from Turkey, and about 15% from Russia.

3. Compounds of Chromium

(a) General

Whereas the compounds derived from *bivalent* chromium have a low stability, very stable compounds are formed from *trivalent* chromium. *Hexavalent* chromium also forms quite stable compounds. These are, however, reduced to chromium(III) compounds even by relatively weak reducing agents, such as hydrogen iodide, hydrogen sulfide, or alcohol, and generally decompose, with evolution of oxygen, at a red heat. Chromium is *quadrivalent* and *pentavalent* in but few compounds, and is still more rarely *univalent*.

Trivalent chromium has a *strong tendency to form complexes*. Atoms or radicals bound to it are, therefore, frequently not dissociated off in solution, so that solutions of chromium(III) compounds often fail to give the reactions characteristic of chromium(III) ions. The chromium(III) ions themselves are always present in aqueous solution in hydrated form; according to Brintzinger they are in the form of $[Cr(H_2O)_{15}]^{+++}$ in dilute solutions. Six of the molecules of water are especially tightly bound to the chromium, so that they pass over with it into the salts crystallizing from such solutions. The ordinary chromium(III) salts—i.e., salts of the general formula $[Cr(H_2O)_6]X_3$—have a violet color, as have their solutions. When the color of the salts of trivalent chromium, or of their solutions, differs appreciably from this violet hue, it invariably implies the presence, not of pure aquo-chromium(III) complexes, but of complexes involving the participation of other ligands—possibly with H_2O as well.

Except for their color, the ordinary hydrated chromium(III) salts closely resemble the salts of aluminum.

Thus chromium forms double sulfates with the character of alums. Like the aluminum salts, the chromium(III) salts are appreciably hydrolyzed in aqueous solution—to an even greater extent indeed, than the former. (On the course of the hydrolysis see p. 284). The analogy extends also to other chromium(III) and aluminum compounds. Thus chromium(III) oxide is isomorphous with corundum, and forms double compounds belonging to the spinel group with the oxides of bivalent elements. The similarity between the chromium(III) compounds and the aluminum compounds, which also shows itself in the associated occurrence of the two elements in Nature, has its origin in the close similarity between the aluminum ion and the trivalent chromium ion, both in valence, and in apparent ionic radius ($r_{Cr^{3+}}$ 0.65 Å; $r_{Al^{3+}}$ 0.57 Å).

The tendency of trivalent chromium to form *complex compounds* is very much greater than that of aluminum. The coordination number of chromium is 6 in these compounds, almost without exception. The complex compounds are variously, and often brightly, colored.

Those simple compounds of chromium which are insoluble in water are also for the most part deeply colored. Thus chromium(III) oxide, Cr_2O_3, is intense green; the sulfide Cr_2S_3 is black; the anhydrous trichloride, $CrCl_3$, which is likewise insoluble in water, is red violet, etc.

Chromium(III) oxide, Cr_2O_3, is amphoteric, and its salts with acids have a strong tendency to hydrolytic decomposition; the salts which it forms with strong alkalis, the *hydroxochromites*, are less stable than the corresponding salts of aluminum. Chromium(VI) oxide CrO_3 is a definite acid anhydride. The acid derived from it, *chromic acid*, H_2CrO_4, (known only in solution) tends to undergo condensation by elimination of water, forming dichromic acid, $H_2Cr_2O_7$, trichromic acid, $H_2Cr_3O_{10}$, and tetrachromic acid, $H_2Cr_4O_{13}$. The salts corresponding to these acids, and especially the (mono)*chromates* $M^I_2CrO_4$ and the *dichromates* $M^I_2[Cr_2O_7]$, are quite stable in the absence of reducing agents at ordinary temperature, and some are stable when not too strongly heated. Towards substances that can readily furnish electrons or accept oxygen, however, they act as energetic oxidizing agents, especially in acid solutions, being themselves converted into chromium(III) compounds (cf. the potential of the $HCrO_4^-$ ion shown in Vol. I, p. 765). Among the salts of the chromic acids, potassium dichromate is notable for its excellent power of crystallization.

'Intermediate oxides' of chromium are said to exist, e.g., an oxide with the composition Cr_5O_9. This is generally regarded as a double compound of chromium(III) and chromium(VI) oxide, or as chromium(III) chromate:

$$Cr_5O_9 = 2Cr_2O_3 \cdot CrO_3.$$

Its existence is, however, doubtful. It could not be detected by King (1939), at least, in the thermal degradation of CrO_3; he found, rather, that two non-Daltonide compounds appeared as intermediate steps, of which one had 2.6–2.2 and the other 1.9–1.7 atoms of O per atom of Cr.

Compounds of *pentavalent* chromium include the pentoxide Cr_2O_5 (cf. p. 153), the vivid red volatile pentafluoride, CrF_5 (von Wartenberg, 1941), the dioxyfluoride CrO_2F (Olsson, 1924), and the oxofluorochromate(V) salts, $KCrOF_4$, and $AgCrOF_4$, obtained by Sharpe and Woolf (1951). Weinland (1905) had earlier obtained oxochlorochromate(V) salts, $M^I_2[CrOCl_5]$, by the action of very concentrated hydrochloric acid on chromium trioxide at a low temperature, and addition of alkali chlorides. The cesium salt of this series forms mixed crystals with cesium oxochloroniobate, $Cs_2[NbOCl_5]$. Oxochlorochromate(V) salts of the type $M^I[CrOCl_4]$ (with M^I = pyridinium or quinolinium) had been prepared by Meyer (1899). The 'red peroxychromates' are also compounds of pentavalent chromium. According to Klemm (1936), the chromphenyl compounds, obtained by Hein (1919 and later) by means of Grignard compounds—e.g., $[Cr(C_6H_5)_3H]X$, (where X is a univalent acid radical), should also be regarded as chromium(V) compounds. Scholder (1953) has obtained chromate(V) salts by the thermal decomposition of chromate(VI) salts in the presence of basic oxides—e.g., Li_3CrO_4, $Sr_3(CrO_4)_2$ and $Ba_3(CrO_4)_2$. He also prepared chromate(V) salts of the same type as hydroxylapatite (e.g., $Sr_5(CrO_4)_3(OH)$). Fankuchen and Ward (1952) prepared compounds of the composition $M^ICr_3O_8$ by heating CrO_3 with alkali dichromates at 350° under a high pressure of oxygen, and there is some evidence that these should be regarded as alkali metal trichromate(V) salts.

From *quadrivalent* chromium, the fluoride CrF_4 and the chloride $CrCl_4$ are known, as well as the 'free radical' $Cr(C_6H_5)_4$ prepared by Hein. $CrCl_4$ is, however, capable of existence only in the gaseous state, and in presence of an excess of chlorine. Chromium is apparently also present in the +4 valence state in the oxide CrO_2, as is shown by the magnetic susceptibility of the compound (Bhatnagar 1939). Derivatives of this oxide include the alkaline earth chromate(IV) salts (e.g., Ba_2CrO_4), which have recently been prepared by Scholder (1953) by a variety of methods. Klemm isolated the compound $K_2[CrF_6]$ by the action of fluorine on a mixture of $CrCl_3$ and KCl. This also is undoubtedly derived from chromium(IV), since it has the same X-ray diffraction pattern as $K_2[MnF_6]$.

Univalent chromium is present in the compound $[Cr\ dipy_3]ClO_4$ (dipy = dipyridyl), prepared by Hein (1952). This was obtained by the reduction of $[Cr\ dipy_3](ClO_4)_2$ in aqueous solution, as a blue-black powder which is rapidly reoxidized in air. The compound was insoluble in water, but dissolved easily in a number of organic solvents—methanol, ethyl alcohol, acetone, dioxane and pyridine. The presence of chromium(I) in the compound was proved by its composition and its magnetic moment. The magnitude of the magnetic moment (2 Bohr magnetons) proves that the compound is a 'penetration complex', as also is $[Cr\ dipy_3](ClO_4)_2$ (observed moment 2.9, calculated 2.8 Bohr magnetons).

Organometallic compounds of chromium are also known. These are of the types R_3CrX, R_4CrX, and R_5CrX (R= phenyl, C_6H_5—, X = OH, halogen or 1 equivalent of acid radical). Hein found that these were quite readily prepared by the reaction between $CrCl_3$ and phenylmagnesium bromide. The tri-, tetra- and pentaphenylchromium hydroxides are strongly basic in character. The salts derived from them, and the tetrahydrate of pentaphenylchromium hydroxide itself—$(C_6H_5)_5CrOH \cdot 4H_2O$—can be obtained as well crystallized substances.

Chromium(III) salts can readily be oxidized to chromates in alkaline solution— for example by chlorine, bromine, hypochlorite, hydrogen peroxide, and also by lead dioxide or freshly precipitated manganese dioxide hydrate. The oxidation can also be effected in acid solution either electrolytically or by a mixture of nitric acid and potassium chlorate, or by potassium peroxysulfate in the presence of a silver salt as catalyst.

Chromates are reduced in acid solution to chromium(III) salts by reducing agents such as sulfurous acid, hydrogen sulfide, hydriodic acid, oxalic acid, or alcohol; also by hydrogen bromide or concentrated hydrochloric acid when warm.

Hydrogen peroxide also brings about reduction in acid solution, by way of chromium peroxide as intermediate product (cf. p. 158).

(b) Chromium(II) Compounds

When dissolved in acids, chromium passes at first predominantly into the *bivalent* state: $Cr + 2H^+ = Cr^{++} + H_2$. Since, however, chromium(II) ions are very easily transformed into the +3 state salts of *trivalent* chromium are obtained when chromium dissolves in acids with access of air.

Even in the absence of atmospheric oxygen, a portion of the chromium(II) ions takes up a further positive charge from the hydrogen ions, an equilibrium being thereby set up:

$$Cr^{++} + H^+ \leftrightarrows Cr^{+++} + \tfrac{1}{2}H_2 \qquad (1)$$

The position of the equilibrium depends upon the nature of the anions present in solution, as well as on the hydrogen ion concentration and the temperature. In general, it lies strongly over towards the right hand side. Pure chromium(II) compounds can therefore not be obtained directly by the dissolution of chromium in acids.

It is possible to get pure chromium(II) salts by the reduction of chromium(III) salts* with zinc in acid solution, provided that care is taken to secure a vigorous evolution of hydrogen. The interaction of chromium(II) ions with H^+ ions proceeds only slowly, especially in the absence of catalysts. Chromium(II) solutions can also be obtained by the electrolytic reduction of chromium(III) salts in aqueous solution at a lead cathode.

Anhydrous chromium(II) salts are most conveniently made by direct union of the components. The chloride and the bromide are readily prepared by the action of gaseous hydrogen chloride or hydrogen bromide on the heated metals.

Solutions of chromium(II) salts are sky-blue in color. The hydrated salts of strong acids are also mostly blue, whereas the anhydrous salts, and salts of the weak acids are variously colored.

In their reactions, chromium(II) salts show great similarity to the iron(II) salts, and yield precipitates with the same reagents. As with the iron(II) salts, their precipitation by ammonium hydroxide is prevented by ammonium salts.

Like iron(II) salt solutions, solutions of chromium(II) salts absorb nitric oxide; however, they do not simply form addition compounds with this oxide, but reduce it in acid solutions to hydroxylamine, and in alkaline solutions to ammonia. Chromium(II) solutions are very much stronger reducing agents, than iron(II) salt solutions, as is shown by the magnitude of the redox potentials.

The oxidation potential gives a measure of the tendency of chromium(II) ions to pass over into chromium(III) ions: for the couple Cr^{++}/Cr^{+++} this is about $+0.41$ volt. In contrast with this, a platinum electrode dipping in a Fe^{++}/Fe^{+++} solution has a potential in the opposite sense relative to the normal hydrogen electrode; for the couple Fe^{++}/Fe^{+++}, $E° = -0.77$ volts (cf. Vol. I, p. 765).

Bivalent chromium, in contrast with bivalent iron, is also able to form complex carbonato salts. One of the most stable chromium(II) salts is the oxalate, $CrC_2O_4 \cdot 2H_2O$, obtained as a yellow crystalline powder by boiling chromium(II) acetate with a solution of oxalic acid. A whole series of other chromium(II) salts of other organic acids is known. Among these, the red *chromium(II) acetate* is of importance for preparative purposes.

(i) *Chromium(II) oxide*, CrO, is formed from a solution of chromium in mercury by spontaneous oxidation in air. If it is heated with access of air, it is converted, with incandescence, into Cr_2O_3. Unlike chromium(III) oxide it can be reduced to the metal by heating it strongly in a current of hydrogen.

(ii) *Chromium(II) sulfide*, $\overline{CrS}$, obtainable directly from the components, is variable in composition within the limits 50 to 54 atom-% S. It has a structure like that of NiAs; nevertheless, in chromium(II) sulfide, at the composition corresponding to CrS, not all the lattice points are occupied by atoms. When all the positions in the sulfur lattice are occupied (at a content of 54 atom-% S) the compound becomes ferromagnetic (Haraldsen 1937). CrSe and CrTe also crystallize with the NiAs structure.

(iii) *Chromium(II) chloride*, $CrCl_2$ or Cr_2Cl_4, is best prepared by passing hydrogen chloride gas over chromium metal at a bright red heat (1), or by passing hydrogen over anhydrous chromium trichloride heated to red heat (2):

$$Cr + 2HCl = CrCl_2 + H_2 \qquad (1)$$
$$CrCl_3 + \tfrac{1}{2}H_2 = CrCl_2 + HCl \qquad (2)$$

* The violet chromium(III) salts are more easily reducible to chromium(II) salts than are the green.

It forms white lustrous silky needles, of density 2.88. It exists as double molecules, Cr_2Cl_4, to a considerable extent in the vapor state, even above $1500°$. It is fairly stable in dry air; in the presence of moisture, however, it absorbs oxygen avidly, forming Cr_2Cl_4O. Chromium(II) chloride is very hygroscopic. It dissolves in water with the evolution of considerable heat (18.6 kcal per g-mol.). From the cornflower blue solution, (which is most simply obtained by adding zinc to a strongly acid chromium(III) chloride solution) crystallized hydrates may be isolated; the most heavily hydrated of these has the composition $CrCl_2 \cdot 6H_2O$. It crystallizes at $0°$, and has a blue color. A blue tetrahydrate crystallizes at room temperature, and at about $50°$ a dark green hydrate isomeric with the latter. By dehydrating the dark blue tetrahydrate, a bright blue trihydrate and a bright green dihydrate can be obtained. The latter then passes over into the colorless anhydrous salt above $113°$.

According to Schlesinger (1933) and Ephraim (1934), chromium(II) chloride combines with ammonia to form the addition compounds $CrCl_2 \cdot 6NH_3$ (dark blue), $CrCl_2 \cdot 5NH_3$ (violet), $CrCl_2 \cdot 3NH_3$ (bright blue), $CrCl_2 \cdot 2NH_3$ (bright green), and $CrCl_2 \cdot NH_3$ (bright green). The removal of the last molecule of ammonia takes place at about $400°$. The hydrazine addition compound of chromium(II) chloride, $CrCl_2 \cdot 2N_2H_4$ which is precipitated from a chromium(II) chloride solution (or from a solution of chromium(II) acetate in dilute hydrochloric acid) on the addition of aqueous hydrazine hydrate, is noteworthy for its stability (Traube 1913).

The green *fluoride*, CrF_2, (m.p. $1100°$) and the white *bromide* $CrBr_2$ (density, 4.36) can be prepared by methods similar to those used for chromium(II) chloride. The red-brown iodide (density 5.02) can be obtained, according to Hein (1931/43) by direct union of Cr with I_2 at about $800°$. Except for the fluoride, these halides are also readily soluble in water.

(*iv*) *Chromium(II) sulfate and Double Sulfates*. The blue chromium(II) sulfate heptahydrate, $CrSO_4 \cdot 7H_2O$, which is isomorphous with iron vitriol, crystallizes on evaporation of solutions of chromium in dilute sulfuric acid, prepared with exclusion of air, or from a chromium(III) sulfate solution reduced with zinc. The colorless monohydrate, $CrSO_4 \cdot H_2O$, crystallizes at the cathode on the electrolytic reduction of concentrated sulfuric acid solutions of chromium(III).

Chromium(II) sulfate (like iron(II) sulfate) forms double salts of the type $M^I_2Cr(SO_4)_2 \cdot 6H_2O$ with the alkali sulfates.

A solution of chromium(II) sulfate in dilute sulfuric acid is an extremely efficient absorbent for oxygen, according to Stone (1936).

(*v*) *Chromium(II) acetate*, $Cr(C_2H_3O_2)_2$, is best prepared by allowing a solution of, chromium(II) chloride to run into a fairly concentrated solution of sodium acetate, saturated with carbon dioxide. It is thereby thrown down as a red precipitate, sparingly soluble in cold water, but readily soluble in dilute strong acids. Because of the ease with which it is prepared, it is a suitable starting material for the preparation of other chormium(II) salts.

(*vi*) *Chromium(II) Complex Compounds*. The chromium(II) ion may be stabilized by combination with neutral ligands. Thus the compounds $[Cr(N_2H_4)_n]I_2$ and $[Cr\ dipy_3](ClO_4)_2$, prepared by Hein (1943–52) are quite stable in air when they are dry, unlike the uncomplexed chromium(II) salts. The black-violet perchlorate, $[Cr\ dipy_3](ClO_4)_2$, is not hygroscopic. Its wine-red solution readily absorbs oxygen from the air, Cr^{II} being converted to Cr^{III}, as is general for chromium(II) salts in aqueous solution. Conversely, the perchlorate can also be quite easily reduced in solution, with the formation of the corresponding chromium(I) salt. The color thereby changes to a deep blue. $[Cr\ dipy_3]^+$ ions can also be formed from $[Cr\ dipy_3]^{++}$ ions by disproportionation:

$$2[Cr\ dipy_3]^{++} = [Cr\ dipy_3]^+ + [Cr\ dipy_3]^{+++}.$$

At high OH^- ion concentrations, this reaction goes to completion. The $[Cr\ dipy_3]^{+++}$ ions formed in the reaction are immediately converted into the red $[Cr(OH)(H_2O)\ dipy_2]^{++}$ ions (cf. p. 147).

(c) Chromium(III) Compounds

(*i*) *Chromium(III) oxide* (chromium sesquioxide), Cr_2O_3, in its usual form is a green powder, insoluble in water. In the crystallized state, such as is obtained by preparation at high temperatures, it is black and iridiscent, with a metallic luster; this oxide also forms a green powder when it is ground up, however. The crystallized chromium(III) oxide is isomorphous with corundum (see Vol. I, p. 351. Fig. 72). $a_0 = 5.33$ Å. $a = 55°17'$.

Chromium(III) oxide may be obtained in many different ways. The preparation generally starts from the chromates or dichromates, which can easily be obtained in the pure state. These are usually reduced by heating them with wood charcoal (see p. 126) or sulfur:

$$K_2Cr_2O_7 + S = Cr_2O_3 + K_2SO_4.$$

Reduction can also be effected by heating with ammonium chloride, since the ammonium dichromate which is first formed in this case decomposes, leaving a residue of chromium(III) oxide.

Chromium(III) oxide finds extensive uses as a pigment for oil and water paints, as well as for the preparation of metallic chromium, as has already been discussed. Since paints prepared with it show excellent resistance to heat and weathering, window frames, garden chairs, and machine parts, in particular, are frequently painted with chrome green. Chrome green is also much used in printing and lithography, in the glass* and clay industries, and also for porcelain glazing colors.

Ignited chromium(III) oxide can be dissolved by acids only with difficulty. It can readily be brought into solution, however, by warming it with an alkali bromate solution. It reacts therewith according to the equation:

$$5Cr_2O_3 + 6BrO_3^- + 2H_2O = 5Cr_2O_7^= + 4H^+ + 3Br_2.$$

The H^+ ions formed accelerate the reaction catalytically. Alkali *chlorates* dissolve the oxide much more slowly (Lydén, 1935).

If chromic oxide is prepared by dehydration of its hydrate, it may be observed that the expulsion of the last small portions of water is accompanied by a *glowing* (23 kcal per g-mol of Cr_2O_3 are liberated in the process). Böhm was able to show from the X-ray diffraction patterns, that this was due to a diminution in surface area, arising from the transformation of the amorphous oxide into the crystalline form, exactly as with titanium and zirconium dioxides.

The *hydrate of chromium(III) oxide*, $Cr_2O_3 \cdot xH_2O$, which although it has a variable water content is generally called chromium hydroxide, is thrown down by addition of OH^- ions to solutions of chromium(III) salts, as a precipitate which is light grey blue in the fresh state, and very insoluble in water. It has the typical properties of a *gel*, and can also be readily converted to a colloidal dispersion—e.g., by peptization with chromium(III) chloride.

Like other gels, chromium oxide hydrate is strongly surface-active. When precipitated from solutions containing alkalis, it is therefore apt to contain considerable quantities of adsorbed alkali. To obtain it in the pure state, it must be precipitated by substances which do not make the solution too strongly alkaline, but which repress the hydrogen ion concentration only far enough to disturb the hydrolysis equilibrium of the chromium salts:

$$2CrX_3 + 3HOH \rightleftharpoons Cr_2O_3 + 6H^+ + 6X^-,$$

* Glass acquires a beautiful emerald green color from chromium oxide. If the glass contains an excess of chromium oxide, this separates out on cooling in microscopically small green crystals.

e.g., by ammonium nitrite, zinc carbonate, zinc sulfide, or potassium iodide-iodate mixture (cf. Vol. I, p. 813). Precipitation with ammonia is less to be recommended, since an excess of this reagent can easily give rise to the formation of complex ammines, from which the hydroxide cannot be precipitated, especially when ammonium salts are also present in large amounts.

Like all gels, the chromium(III) oxide gels obtained by precipitation have the property of ageing—i.e., of changing into less reactive forms with lowers urface activity. Colloidal dispersions of chromium oxide age in the same way, especially in the presence of alkali hydroxide. Such dispersions therefore allow the greater part of the chromic oxide hydrate to precipitate out again when they are warmed. Aged chromic oxide hydrate is dissolved by acids, and particularly by alkalis, far less readily than is the fresh precipitate.

Chromic oxide hydrate dissolves in acids, forming the ordinary chromic salts, and in caustic alkali forming the *chromites*—i.e., salts containing the tripositive chromium in the anion. Chromic oxide hydrate is thus amphoteric, like aluminum hydroxide.

Hydroxochromate(III) ions, $[Cr(OH)_8]^{5-}$, $[Cr(OH)_7]^{4-}$, and $[Cr(OH)_6]^{3-}$ are present in concentrated alkali hydroxide solutions, as has been shown by the investigations of Muller (1922), Fricke (1924), and Scholder (1934). Salts of these can be obtained in crystalline form. Whereas the alkaline earth hydroxochromates(III) are stable towards water, the alkali metal compounds are readily hydrolyzed, with the deposition of chromic oxide hydrate. They therefore decompose when the solutions are diluted. As was shown by Herz and Fischer, the chromic oxide hydrate is present in dilute alkaline solutions in *colloidal* form; it does not diffuse through a parchment membrane, and the electrical conductivity of such a solution does not undergo any marked change if the chromium is precipitated from it by heating.

Barium hexahydroxochromate(III), when mixed with barium hydroxide and heated to about 800°C in a stream of water vapor, is oxidized to barium chromate(IV), Ba_2CrO_4 (Scholder, 1952). The presence of *tetra*-positive chromium in Ba_2CrO_4 was confirmed by magnetic measurements (Klemm). The formation of such an 'anomalous oxidation state' for chromium is undoubtedly made possible by the very great stability of the crystal lattice of the double oxide $2BaO \cdot CrO_2$. Scholder has shown that by suitable choice of reaction partners, other elements (e.g., Mn, Fe, Co) can also be obtained in double oxides in which these elements exhibit 'anomalous' oxidation states. The double oxides so obtained display considerable stability.

By fusing chromium(III) oxide with oxides of more strongly basic (mostly bivalent) metals, a number of well crystallized double compounds have been obtained, mostly of the composition $M^{II}O \cdot Cr_2O_3$. These were formerly regarded as chromites, and expressed by the general formula $M^{II}[CrO_2]_2$. Crystallographically, they belong to the class of *spinels*. Since they are identical with spinel, $MgO \cdot Al_2O_3$, in their detailed crystal structure, they should not be referred to as 'chromites'—i.e., salts of +3 chromium containing the chromium in a negative radical—any more than ordinary spinel should be called an aluminate (cf. Vol. I, p. 355). The most important of the spinel-type double compounds of chromium(III) oxide is chromite*, or chrome iron ore, $FeO \cdot Cr_2O_3$, which occurs native in large quantities.

(*ii*) *Guignet's Green.* When alkali dichromates are fused with boric acid, and the melt is leached with hot water, a chromic oxide hydrate of a particularly beautiful green color is obtained, which is named Guignet's green (also emerald green) after Guignet, who first described the preparation. It can also be obtained by heating the ordinary pale grey blue

---

* The mineralogical name *chromite* has nothing to do with the concept of a 'chromite' in the chemical sense.

chromic hydrate in an autoclave to 250°, in the presence of substances which promote agglomeration. It differs from the ordinary chromic oxide hydrate in that it is composed of coarser particles (Wöhler), but like the latter, it is amorphous. Guignet's green is used like chrome green, as an artist's color. To increase its otherwise rather small covering power, it is generally admixed with barium sulfate. A mixture of Guignet's green, barium sulfate, and zinc yellow (cf. p. 130) is used, as Victoria green in wall paper printing, in place of Schweinfurt green. Permanent green, Nurenberg green, and Mittler's green are similar mixtures.

(iii) *Chromium(III) Sulfides and Thiochromites.* Chromium(III) sulfide, Cr_2S_3, cannot be prepared from aqueous solution, but is formed by dry methods—e.g., by passing hydrogen sulfide over chromium trichloride at a red heat:

$$2CrCl_3 + 3H_2S = Cr_2S_3 + 6HCl,$$

or directly from its components. It forms black, hexagonal leaflets, which are very stable towards most acids but easily decomposed by agents such as nitric acid or fused potassium nitrate.

Direct union of the components gives rise to a phase of variable composition (55–60 atom-% of S) and monoclinic structure, which passes continuously into the hexagonal structure with increasing S-content (Haraldsen, 1937).

Well defined crystalline compounds of chromium(III) sulfide with the alkali sulfides are known—*alkali thiochromites.* For example, when sodium chromate is melted with sodium carbonate and sulfur, with the exclusion of air, sodium thiochromite $Na[CrS_2]$ remains as red hexagonal plates when the cooled melt is extracted with water. Potassium thiotetrachromite, $K_2[Cr_4S_7]$, can be prepared in an analogous way. The compounds have been chiefly studied by Schneider (1897). Schneider was able to prepare the thiochromites of the heavy metals also, by boiling the alkali thiochromites with heavy metal salt solutions. The heavy metal compounds form black or black-grey powdered substances, which are insoluble in water and in hydrochloric acid, but which are decomposed by nitric acid or aqua regia, and are pyrophoric in air.

(iv) *Chromium(III) chloride and Chlorochromium(III) Complexes. Chromic chloride* (chromium trichloride), $CrCl_3$, may be prepared anhydrous by passing dry chlorine over chromium metal, heated to redness (1), or over a mixture of chromium(III) oxide and carbon at a red heat (2). It can also be obtained by heating chromium(III) oxide with carbon tetrachloride (3) or with sulphur monochloride:

$$Cr + \tfrac{3}{2}Cl_2 = CrCl_3 + 133 \text{ kcal.} \tag{1}$$
$$Cr_2O_3 + 3C + 3Cl_2 = 2CrCl_3 + 3CO \tag{2}$$
$$Cr_2O_3 + 3CCl_4 = 2CrCl_3 + 3COCl_2 \tag{3}$$

Anhydrous chromium(III) chloride exists as red violet (peach blossom colored) lustrous leaflets. It can be sublimed in chlorine at red heat, but if heated in the absence of free chlorine, it undergoes partial decomposition into chromium(II) chloride and chlorine:

$$CrCl_3 \rightleftharpoons CrCl_2 + \tfrac{1}{2}Cl_2.$$

When ignited in air, chromium(III) chloride furnishes a fine green chromic oxide:

$$2CrCl_3 + \tfrac{3}{2}O_2 = Cr_2O_3 + 3Cl_2.$$

Chlorine can also be readily exchanged by sulfur, nitrogen, or phosphorus, by reaction with the corresponding hydrogen compounds.

$$2CrCl_3 + 3H_2S = Cr_2S_3 + 6HCl$$
$$CrCl_3 + NH_3 = CrN + 3HCl$$
$$CrCl_3 + PH_3 = CrP + 3HCl$$

Anhydrous chromium(III) chloride is insoluble in cold water and alcohol, and dissolves only extraordinarily slowly even when boiled. However, if a minute trace of chromium(II) chloride is added, rapid dissolution occurs, with considerable evolution of heat (21.3 kcal per mol of $CrCl_3$). The addition of substances which can bring about the transient reduction of a portion of the chromium(III) chloride to the chromium(II) compound has a similar affect. The *dark green* solution of chromium(III) chloride so obtained is identical in its reactions with the solution formed from a chromium(II) chloride solution, containing excess hydrochloric acid, by oxidation in air, and also with the solution obtained by dissolving chromium metal in acid, with access of air. It has a sweetish taste. When an aqueous solution is cooled, after evaporating it down, emerald green deliquescent crystals are deposited, which generally contain 6 molecules of water for each molecule of $CrCl_3$*. In a fresh solution of this salt, only *one third* of the chlorine present can be immediately precipitated as silver chloride: the remaining two thirds must be bound in a complex. In addition, 4 molecules of water must be bound in the complex, since the salt retains these when it is dried in a desiccator over sulfuric acid. The dark green chromium chloride hexahydrate is accordingly to be ascribed the formula $[CrCl_2(H_2O)_4]Cl \cdot 2H_2O$—*dichlorotetraquochromium(III) chloride dihydrate***. In this compound chromium has the *coordination number 6*, which is indeed the case in the very great majority of the complex compounds of $+3$ chromium.

Two other chromic chloride hexahydrates are also known. In the one, only one Cl is bound in the complex; both other chlorine atoms are dissociable. This is the *bright green* hexahydrate discovered by Bjerrum in 1906. This salt, which can be isolated from the solution only by adhering to certain specified conditions, is represented by the formula $[CrCl(H_2O)_5]Cl_2 \cdot H_2O$ (*chloropentaquochromium(III) chloride hydrate*). Two thirds of the chlorine are at once precipitated from its solution by silver nitrate.

There is, moreover, a *grey-blue* chromic chloride hexahydrate, which has long been known. *All* the chlorine can be immediately precipitated from the blue-violet solution of this salt, which is accordingly to be formulated as $[Cr(OH_2)_6]Cl_3$ (hexaquochromium(III) chloride). It can be prepared from the dark green compound. The dilute solution of the salt is boiled for a long time, and crystallization of the compound is brought about at temperatures below $0°$, by saturating the solution with hydrogen chloride. It also separates from solutions of other chromium(III) salts which contain violet chromium(III) ions—i.e., the ions $[Cr(H_2O)_6]^{+++}$, as for example, from a solution of chromic nitrate or chrome alum—if the solutions are saturated with hydrogen chloride at low temperature. Of the three chromic chloride hexahydrates, this is thus the one most readily obtained in crystalline form. The grey-blue hexahydrate contains all 6 water molecules bound in the complex: this salt loses no water on standing over sulfuric acid.

The three compounds just discussed, which have identical analytical compositions, but reveal different constitutions, provide an example of a type of isomerism which is fairly

* $CrCl_3 \cdot 10H_2O$ crystallizes if the solution is allowed to evaporate below 6° C.
** The groups bound within the complex in such compounds are always named in the following sequence: first come the names of the acid radicals; following these the names of the groups resembling ammonia in function (e.g., H_2O); and immediately before the name of the metal atom, the ammonia molecules.

often met with in the field of the complex compounds. When the differences between isomeric coordination compounds can be attributed to the different modes of binding of the water molecules, as in the foregoing example, they are said to display *hydrate isomerism*.

Equilibria are established in solution between the three isomeric chromic chloride hydrates, (or between their ions), the position of equilibrium being dependent upon the temperature and on the concentration and composition of the solutions. At the boiling point, attainment of equilibrium is reached in a few minutes, whereas at ordinary temperature months may be required under some circumstances. In the cold, and in dilute solutions, the grey-blue compound (colored blue-violet in solution) predominates; in highly concentrated solutions the green chlorides are predominant in the equilibrium state. The equilibrium also shifts in favor of the latter on warming. In dilute solutions, the blue-violet form is still predominant even in warm solutions.

The majority of the violet chromium(III) compounds can be prepared from the grey-blue (blue-violet in solution) chromic chloride $[Cr(H_2O)_6]Cl_3$, by exchanging the chlorine atoms with other acid radicals.

In Bjerrum's bright green chromic chloride $[CrCl(H_2O)_5]Cl_2 \cdot H_2O$ also, the ionically bound chlorine may be replaced by other acid radicals without significant change in the nature of the salts. Thus, the compound $[CrCl(H_2O)_5][PtCl_6] \cdot 5H_2O$ is obtained by exchanging the chlorine with the radical of chloroplatinic acid. Examples of the salts derived from the dark green chromic chloride, $[CrCl_2(H_2O)_4]Cl \cdot 2H_2O$, by exchange of the ionically bound chlorine by other acid groups are $[CrCl_2(H_2O)_4]Br$ (green crystals, very hygroscopic) and $[CrCl_2(H_2O)_4][SbCl_6] \cdot 6H_2O$ (green crystals).

The complex ions $[CrCl(H_2O)_5]^{2+}$ and $[CrCl_2(H_2O)_4]^+$, giving rise to the two green chromic chlorides and their derivatives, are themselves derived from the hexaquochromium(III) complex $[Cr(H_2O)_6]^{3+}$, in that 1 or 2 molecules of H_2O respectively are exchanged for Cl^- ions. It may be asked whether this exchange may not be extended in such a way that 3 or more Cl^- enter the complex in the place of H_2O. No *aquo* chromium(III) compound with 3 non-ionized chlorine atoms bound to chromium is known, but the corre-

sponding constitution $\begin{bmatrix} Cl & HOC_2H_5 \\ Cl\ Cr & HOC_2H_5 \\ Cl & HOC_2H_5 \end{bmatrix}$ can apparently be assigned to the re-

chromium(III) chloride alcoholate, which Koppel prepared by dissolving metallic chromium in alcoholic hydrochloric acid.

Numerous compounds are known, based upon a complex with *five* non-ionizable chlorine atoms bound to the chromium. Since, in this case, the sum of the negative charges on the chlorine (five) exceeds the positive charge on the chromium (three), the complex has the character of an anion. It is present, according to Werner, in the red double chlorides of chromium of the general composition $CrCl_3 \cdot 2M^ICl \cdot H_2O$, which crystallize from solutions of the constituents when they are evaporated, while passing in hydrogen chloride. The formulation of these compounds as pentachloroaquochromates(III), $[CrCl_5(H_2O)]M^I{}_2$, $(M^I = Li, Na, K, Rb, Cs, NH_4, Tl, \frac{1}{2}Ba, \frac{1}{2}Mg)$ is based on the observation that the very soluble lithium salt $Li_2[CrCl_5H_2O]$ does not give an immediate precipitate with silver nitrate. It must be concluded from this that *all the chlorine is bound in the complex*, and since chromium generally has the coordination number 6, it may be assumed that one molecule of water is also contained in the complex.

In aqueous solution, the red double chlorides rapidly change into *green* double chlorides, which can likewise be obtained crystalline. On addition of silver nitrate to nitric acid solutions of these green double chlorides, three fifths if the chlorine they contain is at once precipitated. It may be concluded from this that three chlorine atoms in these salts are combined in ionizable form. Werner gave them the following constitutional formula, in which, besides four molecules of water, two equivalents of alkali chloride are also coordinatively bound to the central chromium atom:

$$[Cr(H_2O)_4(ClM^I)_2]Cl_3 \quad (M^I = Li, Rb, Cs, NH_4).$$

Werner called such materials salt-substitution-compounds.

These green double chlorides would thus be derived from the type of the grey-blue chromic chloride, by exchanging two water molecules for 2 molecules of alkali chloride. This view can probably not be reconciled with our present knowledge, and the constitution of the compounds is not certain. They might be double compounds—e.g., $[Cr(H_2O)_4Cl_2]Cl \cdot 2M^ICl$. A double salt can be derived in a corresponding manner from Bjerrum's bright green chloride; it was prepared by Pfeiffer in the form of yellow brown needles by mixing chromic chloride and pyridinium chloride solutions:

$$[Cr(H_2O)_3Cl(ClPyH)_2]Cl_2 \quad (PyH = \text{pyridinium}, \ C_5H_5NH).$$

As the formula requires, only two fifths of the chlorine can be precipitated immediately from solutions of the salt.

The variously colored complex compounds derived from chromium(III) chloride exemplify very clearly the intimate relation between the color of complex compounds and their constitution.

(v) *Other Chromium(III) Halides.* Fluorides and bromides of positive chromium probably exist in the same, or similar, forms as the chloride, but have not been so thoroughly investigated. Chromium does not appear to form complex *iodides,* however. It has also not yet been possible to prepare pure anhydrous chromic iodide, though a hydrate $CrI_3 \cdot 9H_2O$ is known. This forms violet-black crystals, which furnish a green powder when ground up. It contains all the iodine in ionizable form, as has been concluded from measurements of the freezing point of its solutions.

(vi) *Chromium(III) cyanide and Cyanochromates(III).* If a chromium(III) salt solution is treated with potassium cyanide, a precipitate of chromium(III) cyanide, $Cr(CN)_3$, is obtained; this is soluble in an excess of the precipitant, and in acids. *Hexacyanochromates(III),* $M^I_3[Cr(CN)_6]$, may be obtained from solutions containing alkali cyanide in excess. The bright yellow potassium salt $K_3[Cr(CN)_6]$ is isomorphous with the red potassium ferricyanide [Potassium hexacyanoferrate(III)], $K_3[Fe(CN)_6]$.

(vii) *Chromium(III) thiocyanate and Thiocyanatochromates(III).* Chromic oxide hydrate dissolves in thiocyanic acid, forming chromium thiocyanate which can, however, be obtained crystalline only with great difficulty. The compound is a non-electrolyte. If the solution is treated with alkali thiocyanates, *hexathiocyanatochromates(III),* $M^I_3[Cr(CNS)_6]$, crystallize out on evaporation. These are strongly complexed, dark red salts; their fresh wine-red solutions give the characteristic reactions neither for the chromic ion nor for the thiocyanate ion; they do, however, display the conductivity, and the variation of conductivity upon dilution, which are characteristic of solutions of uni-trivalent salts. The solutions turn green on long standing, as a result of decomposition.

(viii) *Chromium(III) oxalate and Oxalatochromates(III).* Solutions of chromium(III) salts, treated with alkali oxalates yield the violet chromium(III) oxalate, $Cr_2(C_2O_4)_3$, insoluble in water and crystallizing in heavily hydrated form. The compound is unstable, however, and readily changes, with the expulsion of water, into a dark green, water-soluble amorphous product. *Alkali trioxalatochromates(III),* $M^I_3[Cr(C_2O_4)_3]$, separate in deep blue crystals* from solutions containing an excess of alkali oxalate. The corresponding alkaline earth and heavy metal oxalatochromates can be prepared from these by double decomposition.

It can be shown that solutions of the trioxalatochromates contain the strongly complexed ion $[Cr(C_2O_4)_3]^{3-}$. On the basis of the coordinative hexavalence of chromium, it may be suspected *a priori* that in these complex ions each of the three radicals occupies *two* coordination positions. If this is the case, the occurrence of *mirror image isomerism* is to be expected with the trioxalatochromates, since the oxalic acid radicals could be arranged about the chromium in either of the two ways represented schematically by Fig. 21. The two structural forms represented in Fig. 21 are related as object and mirror image. In fact, Werner in 1912 was able to bring proof that the trioxalatochromates could be resolved into optically active

* The alkali trioxalatochromates(III) are almost all pleochroic. According to the direction from which they are viewed, the crystals appear either deep cornflower blue, or blue-green or red.

components, thereby confirming the structure deduced from the principles of coordination chemistry.

By exchanging oxalic acid radicals for H_2O, for OH, or for other acid groups, several

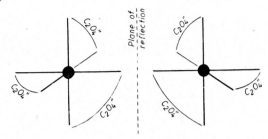

Fig. 21. Mirror image isomerism of trioxalatochromate(III) compounds.

other series of oxalatochromates(III) can be derived from the trioxalatochromates—e.g., the dioxalatodiaquochromates(III), $[Cr(C_2O_4)_2(H_2O)_2]M^I$; the dioxalatodihydroxochromates(III), $[Cr(C_2O_4)_2(OH)_2]M^I_3$; and the dioxalatohydroxoaquochromates(III) $[Cr(C_2O_4)_2(OH)(H_2O)]M^I_2$. The series of compounds named here all display the phenomenon of *cis-trans isomerism* discussed further below.

(*ix*) *Chromium(III) acetate and Acetatotrichromium(III) Salts.* Neutral chromium(III) acetate, $[Cr(H_2O)_6](C_2H_3O_2)_3$, can be obtained by dissolving freshly precipitated chromium oxide hydrate in glacial acetic acid. The salt crystallizes, with an additional 6 molecules of water of crystallization, in blue-violet needles. If chromium(III) salt solutions are treated with sodium acetate, complex compounds are formed which are based essentially upon acetatotrichromium(III) cations. Other methods are better suited for the preparation of the compounds in the pure state. Werner, and also Weinland, have prepared trichromi acetate salts of the following types in the pure state (Ac = acetic acid radical, $CH_3 \cdot CO_2^-$, X = any univalent acid anion):

$[Cr_3Ac_6(OH)_2]X$ (*hexacetatodihydroxotrichromium(III) salts*)
$[Cr_3Ac_6(OH)(H_2O)]X_2$ (*hexacetatohydroxoaquotrichromium(III) salts*)
$[Cr_3Ac_6(H_2O)_2]X_3$ (*hexacetatodiaquotrichromium(III) satls*.

Analogous salts can be obtained from other organic acids. It is noteworthy that 3 chromium atoms are always contained in these complexes; they are thus polynuclear complex compounds. The —OH groups are not the groups which function as bridges for the linkage of the chromium atoms, as is otherwise frequently the case, since the OH groups in these compounds can be transformed by addition of H^+ ions into H_2O molecules, and these latter never function as bridge groups since the maximum coordination number of oxygen is normally 3. It is therefore necessary to suppose that the acetato groups constitute the bridge groups.

From the diffusion coefficients, as determined by his dialysis method, Brintzinger (1936) concluded that the hexacetatodihydroxotrichromium(III) ions (and also the corresponding complex ions of Fe^{III} and Co^{III}) could add on certain anions in aqueous solution, forming 'double shelled' complex ions, of the type $[[Cr_3Ac_6(OH)_2]X_4]^{7-}$ (X = SO_4^{2-}, HPO_4^{2-}, $HAsO_4^{2-}$). Brintzinger considers that two-shelled complex ions of this type formed by adding acid radicals in a second coordination sphere around a coordinatively saturated complex cation, are of frequent occurrence in solution. Their existence is, however, open to dispute (cf. Spandau 1941).

Formation of readily soluble acetato-complexes, essentially of the type of those discussed above, inhibits the precipitation of chromic oxide hydrate when chromium(III) salt solutions are boiled with sodium acetate, although by analogy with aluminum it would be expected that precipitation should occur. After cooling, the solutions no longer give precipitates with other precipitants for chromium.

(*x*) *Chromium(III) nitrate*, $Cr(NO_3)_3$, is formed by dissolving chromic oxide hydrate in nitric acid. The solution is blue-violet by incident light, red by transmitted light. It turns green when heated, but the color rapidly reverts to violet on cooling. The salt crystallizing from solution has a water content that varies according to the conditions of preparation. It finds application as a mordant in piece dyeing.

(xi) *Chromium(III) sulfate and Sulfatochromium(III) Salts.* Anhydrous chromium(III) sulfate, $Cr_2(SO_4)_3$, forms a peach-blossom colored powder which, (like anhydrous chromium trichloride), will not dissolve in water and acids unless reducing agents are present. A hydrated *violet chromium sulfate*, $Cr_2(SO_4)_3 \cdot 18H_2O$, is also known, and several hydrated *green chromium sulfates*, formed from the violet salt on warming—e.g., by prolonged boiling.

The *hydrated violet chromium* sulfate—which may contain a few molecules of water less than is represented by the above formula, depending on the conditions of preparation—is readily soluble in water. Its solution shows the normal reactions for chromium(III) ions and sulfate ions, and the formula $[Cr(H_2O)_6]_2(SO_4)_3 \cdot nH_2O$ (n = 2 to 6) can be ascribed to it accordingly. The violet solution of this salt turns green on heating. The green chromium(III) sulfates are not only less highly hydrated and less soluble than the violet, but are also differently constituted. In them, sulfate groups are bound to the chromium in non-ionizable form, as follows from the reactions of their solutions. Hence the green salts are to be regarded as sulfatochromium(III) salts. Some of them represent also the products of hydrolytic decomposition, in which OH groups are non-ionizably bound, in addition to (H_2O) and (SO_4). The transformation of the violet into the green sulfates is therefore usually accompanied by a considerable increase in the conductivity of the solutions—arising from the liberation of H^+ ions. Erdmann (1952) has concluded from an analysis of the results of conductometric titrations that the following equilibria are set up in solutions of chromium(III) sulfate:

(1) $2[Cr(H_2O)_6]^{+++} + [SO_4]^= \rightleftharpoons$

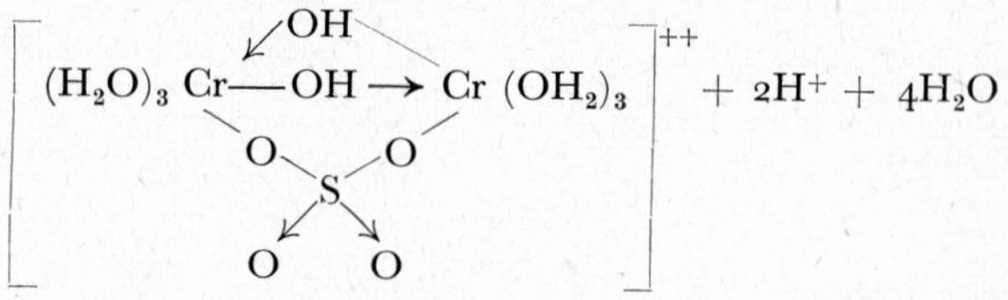

hexaquochromium(III) ion

(2) $2[Cr(H_2O)_6]^{+++} + 3[SO_4]^= \rightleftharpoons$

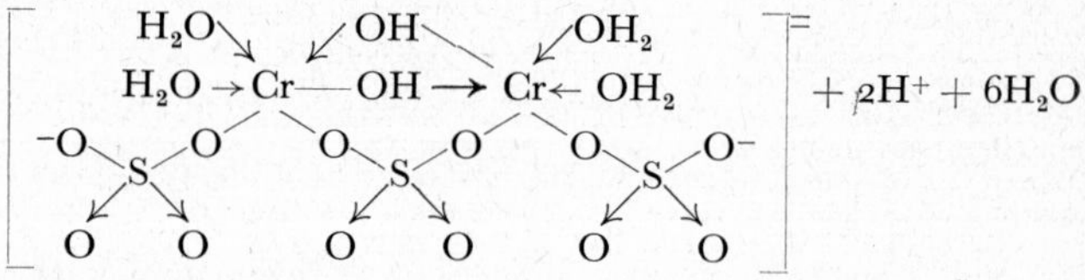

(3) $[(H_2O)_3Cr \genfrac{}{}{0pt}{}{(OH)_2}{(SO_4)} Cr(OH_2)_3]^{++} + 2[SO_4]^= \rightleftharpoons$

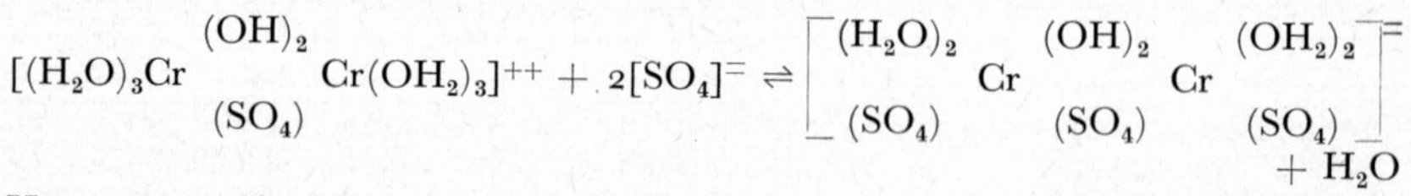

Hexaquo-μ-sulfato-diol chromium(III) ion

$+ H_2O$

Tetraquodisulfato-μ-sulfato-diol-dichromate(III) ion

Equilibrium (1) is of much importance only in highly dilute solutions of chromium sulfate. In moderately dilute solutions, equilibrium (2) predominates. Both equilibria are displaced from left to right with rise of temperature. Reactions 1 and 2 take place rapidly in a left-to-right direction, but only very slowly in the converse sense. This shows that the energy of activation involved in the degradation of the binuclear complex must be high. Equilibrium (3) depends on the concentration. It is established with similar speeds from both directions.

Recoura prepared a green sulfate which, in solution, gave the reactions neither for chromium(III) nor for sulfate, and which must accordingly have contained all the sulfate groups bound non-ionisably to chromium. By evaporation with sulfuric acid, he was able from it to obtain *sulfatochromium(III) acids*, namely disulfatochromium(III) acid, $H[Cr(SO_4)_2]$, and trisulfatochromium(III) acid, $H_3[Cr(SO_4)_3]$. They are easily soluble in water, giving solutions with a fine green color. The fresh solutions give precipitates neither with sodium phosphate (reagent for chromium(III) ions) nor with barium chloride (reagent for sulfate ions). The same applies to solutions of the salts of these acids, the sulfatochromates.

The hydrated violet chromic sulfates can also form addition compounds with sulfuric acid. The resulting acid chromic sulfate, $H[Cr(H_2O)_6](SO_4)_2 \cdot 2H_2O$* forms violet crystals, all the $SO_4^=$ is immediately precipitated from its solutions by Ba^{++} ions.

According to Krauss (1929) the violet 18-hydrate of Cr (III) sulfate first changes into the 9-hydrate when it is heated, without any color change, then into the 3-hydrate, becoming green; and finally into the anhydrous salt. The last three molecules of water are evolved continuously with rise of temperature (in the interval between 100° and 325°, under 10 mm pressure).

(*xii*) *Chrome Alum*. A mixed salt $KCr(SO_4)_2 \cdot 12H_2O$ crystallizes from mixed solutions of violet chromium(III) sulfate and potassium sulfate. It corresponds completely in composition and crystal form, to *alum*, $KAl(SO_4)_2 \cdot 12H_2O$, and its crystals continue to grow in a saturated solution of alum. The mixed salt is therefore called *chrome alum*.

Chrome alum crystallizes in large dark violet octahedra, of density 1.828. Its solubility in water at 25° is about 24.4 g in 100 g of water. The violet solution turns green when heated, as a result of the formation of the green sulfato-complex compounds discussed in the foregoing section; most of these, in contrast to the alums, have a very poor ability to crystallize. When the solution is cooled down chrome alum begins to crystallize out again only after standing for weeks.

Chrome alum is formed as an industrial by-product from many processes in which potassium dichromate is used as an oxidizing agent. It is used chiefly in tanning, and also in dyeing and calico printing.

In addition to ordinary chrome alum, (potassium chrome alum, $KCr(SO_4)_2 \cdot 12H_2O$) analogous mixed salts of violet chromium sulfate with other sulfates of univalent metals and radicals are known. These are chrome alums in a wider sense; they correspond exactly to ordinary chrome alum both in their general formula $M^ICr(SO_4)_2 \cdot 12H_2O$ (M^I = Na, Rb, Cs, Tl^I, NH_4, N_2H_5, NH_3OH) and in their crystal habit, and are known as sodium chrome alum, ammonium chrome alum, etc. according to the univalent metal or radical that they contain.

According to the results of X-ray structure analysis, both the *uni*valent atom (or radical) and the *tri*valent atom in the alums are surrounded by $6H_2O$. The chrome alums therefore contain exactly the same hexaquochromium(III) ion as is present in the other violet chromium(III) salts. The edge length a of the cubic unit cell of potassium chrome alum (12.17 Å), differs very little from that of potassium aluminum alum; the lattice points of the unit cell are occupied alternately by K and by Cr atoms, in the same arrangement as in the rock salt structure. Similar edge lengths have been found for the unit cells of other alums. The readiness with which the alums form mixed crystals with one another is to be explained by the close similarity of their cell dimensions.

According to Krauss (1929), potassium chrome alum behaves differently from potassium aluminum alum on heating, in that the latter forms only a 3-hydrate as an intermediate stage in the dehydration, whereas in the dehydration of potassium chrome alum a 6-hydrate and a 2-hydrate appear as intermediate stages.

* The compound may also crystallize with a higher water content, depending on the experimental conditions.

(d) Chromium Ammines

The marked ability of tripositive chromium to form complex compounds is particularly clearly shown in the numerous types of addition compounds formed with ammonia. These are called *chromium ammines*.

(*i*) *Mononuclear Chromium Ammines.* In accordance with its maximum coordination number 6, the chromium(III) ion can combine coordinatively with 6 ammonia molecules. The resulting complex ion $[Cr(NH_3)_6]^{+++}$ (*hexamminechromium(III) ion*) possesses the same charge as the chromium which functions as its central atom, since the ammonia molecules are uncharged. Although the ammonia molecules are only bound coordinatively to the chromium, they are nevertheless very strongly attached. Only a few other elements (in particular cobalt(III) and platinum(IV)) are able to bind six neutral molecules to their ions as strongly as ammonia is bound to the chromium(III) ion. The particularly strong binding in these complexes is due, as has been shown (cf. Vol. I, Chap. 11), to the fact that the neutral groups in them are not bound to the central atom by Van der Waals forces, as is ordinarily the case, but are united to it by covalent bonds—i.e., they share electrons with the central atom. Complexes of this kind have been called 'penetration complexes', to distinguish them from 'normal complexes'. Negative ions, which are generally bound by electrovalences, can also be linked to the central atom by covalences in exactly the same way as neutral groups.

The ammonia is so firmly bound that it undergoes no immediate dissociation when compounds containing the $[Cr(NH_3)_6]^{+++}$ complex are dissolved in water; even in the absence of ammonia, the complex goes into solution as a discrete ion*. Exchange of ammonia molecules for water molecules takes place only gradually. Such an exchange results in the formation of compounds in which 1, 2, 3, etc., and ultimately all the NH_3 molecules are replaced by H_2O molecules. An almost unbroken sequence of aquo-derivatives of the hexammine chromic ion exists—cf. Table 18. At the end of the series stands the hexaquochromic ion**, $[Cr(H_2O)_6]^{+++}$; mention has already been made of the exceptional stability of this, as compared with other hydrated ions. Substitution of H_2O for NH_3 involves no alteration in the charge of the complex. There also exists a series of complexes which can be regarded as derived from the hexammine chromic cation by successive replacement of NH_3 groups by *negative radicals* R (R = acid radical or hydroxyl). In this case, the (positive) charge on the complex at first decreases, according to the number of acid residues introduced, becomes zero, and then increases again, but with negative sign. If we suppose all six ammonia molecules to be replaced by univalent negative acid residues R, we have the negative trivalent *hexacidochromium(III)* complex anion $[CrR_6]^{3-}$.

---

* If the solutions contain anions which tend to form addition compounds, two-shelled complexes of the type $[[Cr(NH_3)_6]X_4]^{5-}$ (X = SO_4^{2-}, HPO_4^{2-}, $HAsO_4^{2-}$) may be formed, according to Brintzinger (1936). The anions constituting the second shell are, however, but loosely attached, and are probably bound by ordinary electrovalences. On this point see also p. 141.

** The hexaquochromic compounds cannot, of course, be included among the chromium ammines proper. They can, however, quite properly, be regarded as derivatives of the hexammines. The same is true of the hexacido- and the acido-aquo compounds.

TABLE 18

PRINCIPAL TYPES OF MONONUCLEAR AMMINO-, AQUO- AND ACIDO-CHROMIUM(III) COMPLEXES

	a	b	c	d	e	f	g
I	$[\mathrm{CrAm_6}]^{+++}$	$\left[\mathrm{Cr}\begin{smallmatrix}\mathrm{Am_5}\\(\mathrm{H_2O})\end{smallmatrix}\right]^{+++}$	$\left[\mathrm{Cr}\begin{smallmatrix}\mathrm{Am_4}\\(\mathrm{H_2O})_2\end{smallmatrix}\right]^{+++}$	$\left[\mathrm{Cr}\begin{smallmatrix}\mathrm{Am_3}\\(\mathrm{H_2O})_3\end{smallmatrix}\right]^{+++}$	$\left[\mathrm{Cr}\begin{smallmatrix}\mathrm{Am_2}\\(\mathrm{H_2O})_4\end{smallmatrix}\right]^{+++}$	—	$[\mathrm{Cr(H_2O)_6}]^{+++}$
II	$\left[\mathrm{Cr}\begin{smallmatrix}\mathrm{Am_5}\\\mathrm{R}\end{smallmatrix}\right]^{++}$	$\left[\mathrm{Cr}\begin{smallmatrix}\mathrm{Am_4}\\(\mathrm{H_2O})\\\mathrm{R}\end{smallmatrix}\right]^{++}$	$\left[\mathrm{Cr}\begin{smallmatrix}\mathrm{Am_3}\\(\mathrm{H_2O})_2\\\mathrm{R}\end{smallmatrix}\right]^{++}$	$\left[\mathrm{Cr}\begin{smallmatrix}\mathrm{Am_2}\\(\mathrm{H_2O})_3\\\mathrm{R}\end{smallmatrix}\right]^{++}$	—	$\left[\mathrm{Cr}\begin{smallmatrix}(\mathrm{H_2O})_5\\\mathrm{R}\end{smallmatrix}\right]^{++}$	
III	$\left[\mathrm{Cr}\begin{smallmatrix}\mathrm{Am_4}\\\mathrm{R_2}\end{smallmatrix}\right]^{+}$	$\left[\mathrm{Cr}\begin{smallmatrix}\mathrm{Am_3}\\(\mathrm{H_2O})\\\mathrm{R_2}\end{smallmatrix}\right]^{+}$	$\left[\mathrm{Cr}\begin{smallmatrix}\mathrm{Am_2}\\(\mathrm{H_2O})_2\\\mathrm{R_2}\end{smallmatrix}\right]^{+}$	—	$\left[\mathrm{Cr}\begin{smallmatrix}(\mathrm{H_2O})_4\\\mathrm{R_2}\end{smallmatrix}\right]^{+}$		
IV	$\left[\mathrm{Cr}\begin{smallmatrix}\mathrm{Am_3}\\\mathrm{R_3}\end{smallmatrix}\right]$	$\left[\mathrm{Cr}\begin{smallmatrix}\mathrm{Am_2}\\(\mathrm{H_2O})\\\mathrm{R_3}\end{smallmatrix}\right]$	—	$\left[\mathrm{Cr}\begin{smallmatrix}(\mathrm{H_2O})_3\\\mathrm{R_3}\end{smallmatrix}\right]$			
V	$\left[\mathrm{Cr}\begin{smallmatrix}\mathrm{Am_2}\\\mathrm{R_4}\end{smallmatrix}\right]^{-}$	—	$\left[\mathrm{Cr}\begin{smallmatrix}(\mathrm{H_2O})_2\\\mathrm{R_4}\end{smallmatrix}\right]^{-}$				
VI	—	$\left[\mathrm{Cr}\begin{smallmatrix}(\mathrm{H_2O})\\\mathrm{R_5}\end{smallmatrix}\right]^{=}$					
VII	$[\mathrm{CrR_6}]^{=}$						

Yet a further range of complexes can be derived by supposing a portion of the ammonia molecules in the hexammine chromic cation to be replaced by H_2O, and another portion by acid residues R. Table 18 provides a review of the multiplicity of ammino-, aquo- and acido-complexes obtained in this way. All the complexes indicated in this table are known in the form of compounds, and many of them are represented by very numerous derivatives. Not only may the negative radicals R in the formulas of Table 18 stand for various kinds of acid residues, or for the hydroxyl group OH^-, but in many instances other nitrogen compounds—e.g., hydroxylamine, organic amines, pyridine, etc.—can replace ammonia. Moreover, other oxygen compounds (alcohols, phenols, ethers, etc.) can in many cases be coordinatively bound in the place of water, so that a whole series of sub-types may be derived from most of the types represented in Table 18. Each of the complex cations and anions so formed gives rise to a series of individual compounds, originating from the union of the complex ion with various other ions of the opposite charge.

The principal types of compound derived from the complexes shown in Table 18 require a short discussion.

(Ia). *The hexamminechromic(III)* *compounds*: compounds of the general formula $[Cr(Am)_6]X_3$, in which Am = NH_3, $\frac{1}{2}$en, $\frac{1}{2}$pn*.
Ions bound outside the complex include Cl^-, Br^-, I^-, NO_3^-, SO_4^{2-}, PO_4^{3-}, etc. or the radicals of complex acids, such as $[PtCl_6]^{2-}$, $[Fe(CN)_6]^{3-}$, etc. They were discovered by Jörgensen about 1880. Because of their yellow, brown-yellow, or orange-yellow color, the compounds were formerly called *luteo-salts* (Latin *luteus*, orange yellow), as also were the analogously constituted ammonia complex salts of cobalt.

(Ib). *Aquopentamminechromium(III)* *compounds*, $[Cr(NH_3)_5(H_2O)]X_3$, (formerly called *roseo-salts*, because the analogous cobalt salts are rose-red in color) generally have a yellow to orange-yellow color, and are very similar to the hexammine compounds in mode of formation, crystal form, solubility, and chemical reactions. They differ from the hexammines in that they do not gradually undergo decomposition by dilute aqueous ammonia, with the deposition of a precipitate, but are transformed into the readily soluble hydroxopentammine salts, $[Cr(NH_3)_5OH]X_2$, (formerly called 'basic roseo-salts').

(Ic). *Diaquotetramminechromium(III)* *compounds*, $[Cr(Am)_4(H_2O)_2]X_3$ (Am = NH_3 or $\frac{1}{2}$en). These compounds resemble the preceding ones in their behavior. They are orange to vermilion in color. The ethylenediamine compounds of this type are known in both the isomeric forms expected according to theory ('cis-trans' isomerism, see below).

(Id). *Triaquotriamminechromium(III)* *compounds*, $[Cr(NH_3)_3(H_2O)_3]X_3$. Few of these are known; they closely resemble the compounds richer in ammonia, and are pale red in color.

(Ie). *Tetraquodiamminechromium(III)* *compounds*, $[Cr(NH_3)_2(H_2O)_4]X_3$ are red to violet-red in color, and approximate closely to the pure aquo-compounds in their properties. They have a strong tendency to undergo hydrolysis, and their hydroxides $[Cr(NH_3)_2(H_2O)_4](OH)_3$ are insoluble, in contrast to the readily water-soluble hexammine- and aquopentammine hydroxides.

(If). No pentaquomonammine compounds of chromium are known. The fact that only the monammine compound is missing from every series of Table 18 points to some underlying regularity, which shows up in the non-existence of the monammine compounds.

(Ig). *Hexaquochromium(III)* *compounds*, $[Cr(H_2O)_6]X_3$. The compounds of this type are violet in color, especially in solution. The ordinary hydrated salts of chromium, already described, belong in this class—e.g., the ordinary chromic chloride hexahydrate, the nitrate, acetate, sulfate, chrome alum, etc.

* en and pn are abbreviations for ethylenediamine, $NH_2 \cdot CH_2 \cdot CH_2 \cdot NH_2$, and (unsymmetrical) propylene diamine, $NH_2 \cdot CH_2 \cdot CH(CH_3) \cdot NH_2$. These compounds contain *two* groups which can replace NH_3. They can therefore occupy *two* coordination positions.

(IIa). *Acidopentamminechromium(III) compounds,* $\left[Cr\dfrac{Am_5}{R}\right]X_2,$ where

Am $=$ NH$_3$, CH$_3$NH$_2$, $\frac{1}{2}$en, etc.;   R $=$ Cl, Br, I, SCN, NO$_3$, NO$_2$, OH.

These compounds are closely related genetically to the roseo-(aquopentammine) compounds. The reaction $[Cr(Am)_5(H_2O)]^{3+} + R^- = [Cr(Am)_5R]^{++} + H_2O$ is often reversible. The chlorocompounds of this type are mostly dark red, and were formerly called *purpureo compounds.*

(IIb). *Acidoaquotetramminechromium(III) compounds,* $\left[CrH_2O\begin{smallmatrix}Am_4\\\\R\end{smallmatrix}\right]X_2.$

Am $=$ NH$_3$, $\frac{1}{2}$en, $\frac{1}{2}$ dipy (dipy $=$ α, α' dipyridyl);   R $=$ Cl, Br, I, OH.

(IIc). *Acidodiaquotriamminechromium(III) compounds,* $\left[Cr(H_2O)_2\begin{smallmatrix}Am_3\\\\R\end{smallmatrix}\right]X_2.$

Am $=$ NH$_3$;   R $=$ Cl, Br, OH, NO$_3$.

(IId). *Acidotriaquodiamminechromium(III) compounds,* $\left[Cr(H_2O)_3\begin{smallmatrix}Am_2\\\\R\end{smallmatrix}\right]X_2.$

Am $=$ NH$_3$, pyridine;   R $=$ OH.

(IIe). As for (If).

(IIf). *Acidopentaquochromium(III) compounds,* $\left[Cr\dfrac{(H_2O)_5}{R}\right]X_2.$

Bjerrum's bright green chromium(III) chloride falls into this type (R $=$ Cl).

(IIIa). *Diacidotetramminechromium(III) compounds,* $\left[Cr\dfrac{Am_4}{R_2}\right]X.$

The ethylenediamine compounds are the best known compounds in this series. They are interesting because of the phenomena of isomerism (*stereoisomerism*) that they display.

With an octahedral arragement of groups A and B, bound around the central atom Z,

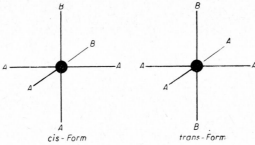

cis - Form   *trans - Form*

Fig. 22. *Cis-trans* isomerism in coordination compounds.

a complex of the general formula [ZA$_4$B$_2$] can occur in two forms, as shown by Fig. 22. These forms differ in that, in the one case, both of the groups B lie on the *same* side of the central atom, in adjacent positions, so that a line joining them forms one edge of the octahedron (e.g., a line joining two adjacent ends of the coordinate axes in Fig. 22). In the other case, the two groups are on *opposite* sides of the central atom, so that they are joined by a line passing through the central atom (a diagonal of the octahedron). The first form is said to be the *cis-* and the second the *trans-* isomer, this kind of isomerism is called *cis-trans isomerism.*

When a pair of compounds is found experimentally to stand in this sort of relationship it is possible to determine which should be regarded as the *cis-*, and which as the *trans-* compound, by testing which of them permits of replacement of the two groups B by a single 'bidentate' group, (that is, a group which can occupy two coordination positions). The isomer which does so must be the *cis* compound*. In the case of bis-ethylenediamine compounds there is the additional difference that the *cis* compound exists as a mixture of two optically active forms, whereas the trans isomer is not resolvable into optical isomers. As may be seen from Fig. 23, two *cis* forms occur, which are related to each other as an

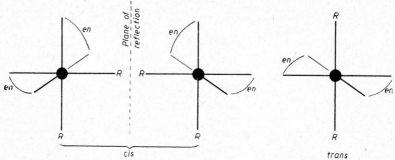

Fig. 23. *Cis-trans* and mirror image isomerism in bis-ethylenediamine compounds.

object and its mirror image. Pfeiffer, in 1904, succeeded in proving the occurrence of *cis-trans* isomerism in diacidobis(ethylenediamine)compounds. Werner, in 1911, succeeded in resolving the *cis*-dichlorobis(ethylenediamine)chromium(III) complex, [Cr en$_2$Cl$_2$]$^+$, into its two optically active components, by preparing from it the salts formed with optically active *a*-bromocamphorsulfonic acid. The salt of the *l*-dichlorobis(ethylenediamine) complex ion with the anion of the *d*-*a*-bromocamphorsulfonic acid (*l*-*d* salt) differs from the *d*-*d* salt, in being sparingly soluble. In the same way, the *d*-*l* salt is sparingly soluble, but the *l*-*l* salt is not. A separation of the *d* and *l*-cations can readily be brought about in this way. By replacing the bromocamphorsulfonic acid radical by other acid radicals X, the remaining salts of the *cis* series, of the type [Cr en$_2$Cl$_2$]X, may then be obtained in their optically active forms.

(IIIb). *Diacidoaquotriamminechromic(III) compounds,* $\left[\begin{array}{c} Am_3 \\ CrH_2O \\ R_2 \end{array} \right] X.$

Am = NH$_3$;   R = Cl and Br.

In this case, *three* isomers are predicted by theory. [cf. Fig. 24]. Riesenfeld did, indeed, obtain the chloride [Cr(NH$_3$)$_3$(H$_2$O)Cl$_2$]Cl in three different forms—red-violet, grey, and dark green.

(IIIc). *Diacidodiaquodiamminechromium(III) compounds,* $\left[\begin{array}{c} Am_2 \\ Cr(H_2O)_2 \\ R_2 \end{array} \right] X.$

Am = NH$_3$, pyr;   R = Cl, Br, OH, $\frac{1}{2}$C$_2$O$_4$.

Numerous stereoisomers can be foreseen from the theory, but they have not yet been prepared. The compounds of this type closely resemble the diacidotetraquochromium(III) compounds.

(IIId). cf. (If).

(IIIe). *Diacidotetraquochromium(III) compounds,* [Cr (H$_2$O)$_4$R$_2$]X. The best known compound of this type is the dark green chromium(III) chloride (cf. p. 138).

* As in all instances where *structure* is inferred from *reactivity*, it has to be assumed that if reaction takes place, it does so without any change of structure or configuration. During recent years a number of examples have been found which prove that this assumption is not always valid.

(IVa). *Triacidotriamminechromium(III) compounds,* $\left[\text{Cr}^{\text{Am}_3}_{\text{R}_3}\right]$.

Compounds of this type are non-electrolytes.

Am = NH_3, pyr;   R = F, Cl, SCN.

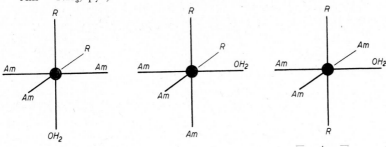

Fig. 24. Isomerism in compounds of the type $\left[\begin{array}{c}\text{Am}_3\\\text{CrOH}_2\\\text{R}_2\end{array}\right]$.

(IVb). *Triacidoaquodiamminechromium(III) compounds,* $\left[\begin{array}{c}\text{Am}_2\\\text{CrOH}_2\\\text{R}_3\end{array}\right]$.
Also non-electrolytes.

Am = NH_3, R = SCN, OH;   Am = pyr, R = F, Cl, OH.

(IVc). cf. type (If).

(IVd). *Triacidotriaquochromium(III) compounds.* Compounds of the general formula

$$\left[\text{Cr}^{(\text{H}_2\text{O})_3}_{\text{R}_3}\right]$$

are not known in the free state, but derivatives are known, in which H_2O is replaced by other oxygen- or sulfur-containing groups—e.g., the compound

$$\left[\text{Cr}^{(\text{C}_2\text{H}_5\text{OH})_3}_{\text{Cl}_3}\right],$$

already referred to.

(Va). *Tetracidodiamminechromate(III) salts,* $\left[\text{Cr}^{\text{Am}_2}_{\text{R}_4}\right]\text{M}^{\text{I}}$.

Am = NH_3, pyr, $\frac{1}{2}$en;   R = Cl, Br, SCN, $\frac{1}{2}C_2O_4$.

Since the complex contains 4 negative equivalents, it functions as an anion, and forms salts with metals.

To this class of compounds belongs the so-called Reinecke salt, $NH_4[Cr(SCN)_4(NH_3)_2]$ · H_2O; the anion of this complex has often proved useful in forming salts with organic cations, thereby enabling them to be isolated in a pure state. According to Mahr [*Z. anorg. Chem.* 225 (1935) 386], Reinecke's salt is a convenient precipitant for the quantitative determination of copper, which can be precipitated in the form of $Cu[Cr(SCN)_4(NH_3)_2]$ without separating other metals (except Ag, Hg, and Tl) from the solution. The reaction also provides a sensitive means of qualitatively detecting copper.

(Vb). See type (If).

(Vc). *Tetracidodiaquochromate(III) salts,* $\left[\text{Cr}^{(\text{H}_2\text{O})_2}_{\text{R}_4}\right]\text{M}^{\text{I}}$.

The dioxalatodiaquochromate(III) salts, already referred to, are of this type.

(VIa). See type (If).

(VIb). *Pentacidoaquochromate(III) salts,* $\left[Cr\genfrac{}{}{0pt}{}{H_2O}{R_5}\right]M^I{}_2.$

An example of this type is provided by the pentachloroaquochromates already discussed. The dioxalatohydroxoaquochromates, considered on p. 141, also belong here.

(VII). *Hexacidochromate(III) salts,* $[CrR_6]M^I{}_3.$

Examples of this class, already encountered, are the chloro-, cyano-, thiocyanato-, oxalato-, and sulfatochromate(III) salts.

(*ii*) *Polynuclear Chromium Ammines.* The polynuclear chromium ammines contain several central atoms of chromium. These are generally linked together by OH groups ('bridge' groups), which in this case are termed 'ol' groups (cf. Vol. I, p. 408).

The most important representatives of the *binuclear* chromium ammines are the so-called *rhodo-* and *erythro-chromium* salts and the *diol-dichromium(III) salts.* A *quadrinuclear* complex is probably present in the so-called *rhodoso-chromium salts* of Jörgensen and the ethylene-diamine compounds of analogous composition described by Pfeiffer.

The *rhodochromium salts* discovered by Jörgensen in 1882 are formed by the oxidation of chromium(II) salts by atmospheric oxygen in ammoniacal solution. They are readily converted into the *erythrochromium salts.* Thus the red neutral (normal) rhodochromium salts dissolve in aqueous ammonia with a blue color to give 'basic' rhodochromium salts, and the normal rhodo-salt is regenerated by the addition of acid. If the blue solution is allowed to stand, however, it turns carmine after some time, and then red 'basic' erythro-salts may be precipitated by adding alcohol, or 'normal' erythro-salts (also red) may be precipitated by acid. These are reconverted to rhodo-salts if they are kept for 24 hrs. at 100°.

Normal rhodo-salts   bases   'Basic' rhodo-salts
(red)   ⇌   (blue)
   acids

24 hrs heating ↑   │ standing in
at 100°   ↓ solution

Normal erythro-salts   acids   'Basic' erythro-salts
(red)   ←   (red)

These transformations, and the principal reactions of the salts, are explained by the following constitutional formulas (Jensen, 1937).

$[(NH_3)_5Cr—OH\cdots Cr(NH_3)_5]X_5$ NaOH ⟶ $[(NH_3)_5Cr—O—Cr(NH_3)_5]X_4$
                                            $+ NaX + H_2O$

Normal rhodosalt   'Basic' rhodosalt
decammine-ol-dichromium(III) salt   *decammine-μ-oxo-chromium(III) salt*

↑   ↓

$\left[(NH_3)_5Cr—NH_2\cdots Cr\genfrac{}{}{0pt}{}{(NH_3)_4}{H_2O}\right]X_5$ HX ← $\left[(NH_3)_5Cr—NH_2\cdots Cr\genfrac{}{}{0pt}{}{(NH_3)_4}{OH}\right]X_4$

Normal erythro-salt   'Basic' erythro-salt
aquo-enneammine-μ-amino   *hydroxo-enneammine-μ-amino*
dichromium(III) salt   *dichromium(III) salt*

These formulas accord not only with the chemical reactions of the salts but also with the observation that the rhodo- and erythro-salts give indistinguishable X-ray patterns. The complex cations of the normal erythro-salts are derived from those of the normal rhodo-salts by replacement of an OH-group (electron number = 9) by a NH_2-group (e.n. = 9), and an NH_3 molecule (e.n. = 10) by a H_2O molecule (e.n. = 10). Since the number of electrons at each point in the crystal lattice is not altered by the exchange, the scattering power for X-rays is unchanged. The 'basic' erythro-salts are derived from the 'basic' rhodo-salts by exchange of an O-atom (e.n. = 8) with an NH_2 group (e.n. = 9), and a NH_3 molecule (e.n. = 10) for a OH-group (e.n. = 9). In this case also, provided the structure of the crystal is the same, practically identical X-ray diagrams must be expected.

The *diol-dichromium(III)* *salts*, as their name implies, are complex salts in which two chromium atoms are joined through a pair of OH-groups. This class of compounds is represented chiefly by the tetraethylenediamine-diol-dichromium(III) compounds, $[Cr_2 en_4(OH)_2]X_4$, which were prepared by Pfeiffer by elimination of water (heating at 100–120°) from *cis*-hydroxo-aquo-diethylenediammine chromium salts*:

$$2\left[en_2Cr\begin{matrix}OH\\OH_2\end{matrix}\right]X_2 - 2H_2O = \left[en_2\ Cr\begin{matrix}OH\\OH\end{matrix}Cr\ en_2\right]X_4.$$

The salts of this series (of which the chloride, bromide, iodide, thiocyanate, nitrate, thio-sulfate, and chromate are known) are blue-violet in color.

The diol-chromium(III) complex

$$\left[\begin{matrix}en\\en\end{matrix}\ Cr\begin{matrix}OH\\OH\end{matrix}Cr\begin{matrix}en\\en\end{matrix}\right]^{4+}$$

can be formally derived from type Ic of Table 18,

$$\left[\begin{matrix}en\\en\end{matrix}\ Cr\begin{matrix}HOH\\HOH\end{matrix}\right]^{3+},$$

by replacement of the two HOH groups by the coordinatively bivalent group

$$\left[\begin{matrix}HO\\HO\end{matrix}\ Cr\begin{matrix}en\\en\end{matrix}\right]^{+}.$$

Three such groups could be introduced in a similar manner into the complex $[Cr (HOH)_6]^{3+}$, to form a *quadrinuclear* complex ion, with six-fold positive charge.

$$\left[Cr\left(\begin{matrix}HO\\HO\end{matrix}Cr\begin{matrix}en\\en\end{matrix}\right)_3\right]^{6+}.$$

The sulfate of this complex was obtained by Pfeiffer, by treating partially dehydrated chrome alum with ethylenediamine hydrochloride, and from it a series of other salts of the same type was prepared by double decomposition. These hexaethylenediamine-hexol-tetrachromium salts have a fine red color. The so-called 'rhodoso-chromium' salts of Jörgensen are probably of the same type, since they have the same composition except that ethylenediamine is replaced by a coordinatively equivalent amount of ammonia. They are thus probably dodecammine-hexol-tetrachromium salts,

$$\left[Cr\left(\begin{matrix}HO\\HO\end{matrix}Cr(NH_3)_4\right)_3\right]X_6.$$

(e) Chromium(VI) Compounds

(*i*) *Chromium trioxide* (chromium(VI) oxide, chromic anhydride), CrO_3, is precipitated from concentrated alkali chromate or dichromate solutions in the form of dark red crystal needles, on the addition of a large excess of concentrated sulfuric acid. These may be washed with pure very concentrated nitric acid (free from nitrogen oxides), and dried at about 70° on a porous plate. The oxide is

* The corresponding *trans* salts, unlike the *cis* salts, cannot change into diol salts without undergoing a change of configuration. They therefore remain unaltered under these experimental conditions.

odorless, has a sour taste, and is *very toxic* because of its destructive effect on organic substances. (It is particularly damaging to the kidneys: 0.6 g is fatal). It avidly absorbs moisture from the air, and is very soluble in water (166 g in 100 g of water at 15°), with a yellow color, to form chromic acid, H_2CrO_4, and dichromic acid, $H_2Cr_2O_7$, which are known only in solution.

Equilibrium is established in solution between chromic acid and dichromic acid:

$$2H_2CrO_4 \rightleftharpoons H_2Cr_2O_7 + H_2O.$$

This is shifted over, towards formation of chromic acid, by increasing dilution. The first stage dissociation of chromic and dichromic acids is almost complete; the second stage is only slight.

In the crystalline state, chromium trioxide has a space lattice built up from CrO_6 octahedra (see below). In solution, coordinative saturation of CrO_3 takes place through the addition of a water molecule. The binding of the hydrogen atoms in this H_2O molecule is so loosened by the repulsion of the 6-fold positively charged chromium atom, that they dissociate off as ions:

$$\begin{matrix} & O & \\ O & Cr + OH_2 \\ & O & \end{matrix} = \begin{bmatrix} & O & \\ O & Cr & O \\ & O & \end{bmatrix}^= + 2H^+.$$

Coordinative saturation of CrO_3 can also be achieved, however, by association with $[CrO_4]^=$ ions already existing in the solution, in which case the dichromate ion is formed:

$$\begin{matrix} & O & \\ O & Cr + \\ & O & \end{matrix} \begin{bmatrix} & O & \\ O & Cr & O \\ & O & \end{bmatrix}^= = \begin{bmatrix} & O & & O & \\ O & Cr & O & Cr & O \\ & O & & O & \end{bmatrix}^=.$$

Ions containing yet more CrO_3 units can be built up in the same manner—the *trichromate* ion, $[Cr_3O_{10}]^=$, and *tetrachromate* ion, $[Cr_4O_{13}]^=$. Formation of such polychromate ions takes place to an increasing extent with increase in the concentration of oxide and of H^+ ion.

Repulsion between the surrounding O^{2-} ions may be the reason that chromium has the coordination number 4 in the chromate and polychromate ions, whereas it has coordination number 6 in crystalline chromium trioxide. In the latter, the effect of the six Cr^{6+} ions which surround each Cr^{6+} ion in the second sphere may partly compensate the repulsions between O^{2-} ions.

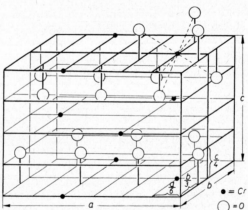

Fig. 25. Unit cell of chromium trioxide.
$a = 8.46.$ $b = 4.77,$ $c = 5.70$ Å.

Chromium trioxide crystallizes rhombic. Each Cr-atom in its crystal lattice (see Fig. 25) is surrounded by 6 O-atoms in a rather distorted octahedron, in such a way that every

O atom is common to two octahedra, and no two octahedra share more than one corner (Wyckoff, 1933, Megaw, 1935). Thus the CrO_6 octahedra make up a continuous network stretching throughout the entire crystal, in the same way as the SiO_4 tetrahedra in crystalline silicon dioxide.

Chromium trioxide has a density of 2.80, melts at about 190°, and begins to vaporize at a slightly higher temperature. At the same time it begins to lose oxygen. The thermal decomposition takes place in several stages: $CrO_3 \rightarrow Cr_3O_8 \rightarrow Cr_2O_5 \rightarrow CrO_2 \rightarrow Cr_2O_3$. It is possible to obtain the compounds Cr_3O_8 and Cr_2O_5 in a pure state if the degradation of CrO_3 is carried out in an atmosphere of oxygen at appropriate pressures. It has been proved by X-ray investigations (Ward, 1952) that the intermediate phases mentioned above are the only intermediate steps in the thermal decomposition. CrO_2 is ferromagnetic (Michel and Benard, 1935), and has the rutile structure.

Chromium trioxide is an extremely vigorous oxidizing agent, and reacts explosively with many oxidizable materials. Organic substances, in particular are generally attacked very energetically by chromium trioxide, and inflammation often takes place—e.g., if it comes into contact with alcohol. On the other hand, it can be boiled with acetic acid without attacking it.

Chromium trioxide is used in preparative chemistry as an oxidizing agent, and in medicine as a caustic. Dilute solutions are used for hardening microscopic preparations.

(ii) *Chromates*. The salts of the general formula $M^I_2[CrO_4]$ derived from chromium trioxide are known as *chromates* (strictly: monochromates(VI)). Unless their color is modified by the cationic component, they are all yellow. When the yellow chromate solutions are acidified, the color changes to the orange red of the *dichromates*, $M^I_2[Cr_2O_7]$.

The *trichromates*, $M^I_2[Cr_3O_{10}]$, and *tetrachromates*, $M^I_2[Cr_4O_{13}]$, obtained from very acid solutions containing an excess of chromium trioxide, are still deeper in color (deep red and brown red, respectively).

All the alkali chromates are soluble in water. The solubility of the alkaline earth chromates diminishes greatly from magnesium to barium (cf. Table 19). Of the heavy metal chromates, those of lead, bismuth, silver, and mercury (in both valence states) are practically insoluble. Those of the last three metals are dark red in color. Such dichromates as exist are mostly soluble, except for the dark red silver dichromate, $Ag_2Cr_2O_7$.

TABLE 19

SOLUBILITY OF SOME CHROMATES AND DICHROMATES IN WATER
(in g of anhydrous salt per 100 g of water)

	Potassium	Sodium	Magnesium	Calcium	Strontium	Barium
Chromate	62.9	76.6	72	2.3*	0.123	0.00035
Dichromate	12.7	180	—	very sol. deliquescent	very sol.	very sol.
Temperature	20°	20°	18°	19°	15°	18°

* This value is for the stable, anhydrous salt. The hydrated salt is metastable, and therefore more soluble. The solubility of calcium chromate decreases with rise of temperature, and is only 0.42 g in 100 g of water at 100°.

If a dichromate solution is treated with the salt of a metal that forms a soluble dichromate, but an insoluble chromate (e.g., with barium chloride), the chromate of the metal in question is precipitated (in this instance, barium chromate, $BaCrO_4$). This is because the equilibrium between chromate and dichromate ions may be temporarily disturbed:

$$2CrO_4^= + 2H^+ \rightleftharpoons Cr_2O_7^= + H_2O \tag{1}$$

Even though the dichromate ions predominate greatly in acid solutions, there are still sufficient chromate ions in equilibrium with them for the solubility product of a sparingly soluble chromate to be exceeded. The solubility product of barium chromate is, in fact, not attained in strongly acidic solutions, but is exceeded in weak acids (e.g., acetic acid) if a sufficient concentration of barium ions is introduced in the solution. If the acidity of the acetic acid is repressed yet further by adding sodium acetate, the chromium may be quantitatively precipitated as chromate, since the chromate ions present in equilibrium are continually removed by precipitation until there are practically no dichromate ions left to form them.

Whereas dichromates have an acid reaction, because of equilibrium (1), chromates have a basic reaction. The second stage dissociation of chromic acid takes place to such a small degree that the $CrO_4^=$ ions react with water to some extent, forming $HCrO_4^-$ ions:

$$CrO_4^= + H_2O \rightleftharpoons HCrO_4^- + OH^- \tag{2}$$

Salts of the formula M^IHCrO_4, i.e., acid chromates in the proper sense (the dichromates were formerly also often called 'acid chromates'), are not known. Under all experimental conditions when their formation might be expected, dichromates are obtained.

By heating barium chromate(VI) with barium carbonate in a nitrogen atmosphere to about 1200°, one obtains barium chromate(V) (Scholder, 1952):

$$2BaCrO_4 + BaCO_3 = Ba_3(CrO_4)_2 + CO_2 + \tfrac{1}{2}O_2.$$

By magnetic measurements, Klemm confirmed the presence of pentapositive chromium in this compound.

Chromates and dichromates are very important technically. They are used primarily as oxidizing agents—e.g., in the preparation of anthraquinone, benzoic acid, quinone, artifical camphor, etc. Their vigorous oxidizing properties also make them suitable as bleaching agents for oils and waxes, and for the purification of pyroligneous acid. They are also used in the production of explosives and igniter mixtures. Dichromate solutions mixed with concentrated sulfuric acid are also used for degreasing glassware ('cleaning solution'). Since chromates are converted into the chromium(III) state by reducing agents, they are frequently used as the starting materials for the preparation of chromium(III) compounds. They are also used in tanning and in the manufacture of inks. Their solutions are used to harden and preserve anatomical preparations, as a mordant in dye printing, and in photography; gelatin or gum arabic, when treated with chromate solutions, acquire the property of becoming insoluble in water when they are exposed to light (gum bichromate and bromoil printing processes, etc.).

The technical production of chromates starts from chromite. This is opened up by melting it with alkali, with free access of air ('oxidative fusion'). The process was formerly used chiefly to prepare potassium dichromate, which crystallizes extremely well, but it is now usual to manufacture sodium dichromate, which is made more cheaply and can also be readily purified by recrystallization.

The chromates are quite toxic, like chromic acid itself. They can give rise to deep sores if they come into contact with places where the skin is broken. Particular care has to be taken with their toxic properties in the manufactures and use of chrome colors.

(*ii*) *Sodium chromate and Sodium dichromate.* Sodium chromate, Na_2CrO_4, generally crystallizes from solution as a hydrate (with $10H_2O$ below $19.5°$, $6H_2O$ between $19.5°$ and $26.9°$, $4H_2O$ between $25.9°$ and $62.8°$, and anhydrous above $62.8°$). It forms yellow crystals, which are always contaminated with sodium sulfate if crystallization takes place from a solution containing this salt. Pure sodium chromate is therefore prepared from recrystallized sodium dichromate, by treatment with sodium carbonate:

$$Na_2Cr_2O_7 + Na_2CO_3 = 2Na_2CrO_4 + CO_2 \qquad (1)$$

Sodium dichromate generally crystallizes as the dihydrate, $Na_2Cr_2O_7 \cdot 2H_2O$, but is obtained anhydrous above $83°$. It is much more soluble than the monochromate. Nevertheless, it can readily be recrystallized since its solubility varies steeply with temperature (163 g dissolve in 100 g of water at $0°$, 433 g in 100 g at $98°$). The anhydrous salt is very deliquescent at ordinary temperature, and is also fairly soluble in alcohol. It melts at $320°$, and decomposes, with loss of oxygen, at about $400°$.

Sodium chromate is manufactured on a large scale by heating finely ground chrome iron ore (chromite), mixed with soda and lime, in air. The essential reaction follows the equation

$$2FeO \cdot Cr_2O_3 + 4Na_2CO_3 + \tfrac{7}{2}O_2 = 4Na_2CrO_4 + Fe_2O_3 + 4CO_2.$$

The chief purpose of adding lime is to keep the roasted mass porous in texture, so that the air has access to all parts. Roasting is generally carried out at $1000°$ to $1300°$. After the melt has cooled, the sodium chromate is leached out with water. It is generally converted into sodium dichromate, which is more valuable technically because of its higher chromium content, by addition of sulfuric acid to the solution, previously concentrated by evaporation. Anhydrous sodium sulfate first separates from the solution on addition of the acid. The solution decanted from this is either evaporated further, and allowed to crystallize at ordinary temperature, in which case the dihydrate is obtained, or is heated until all the water has evaporated. In the latter case, the mass solidifies on cooling as the anhydrous salt. If the process is worked so as to give the anhydrous salt, it is usual to add a further quantity of sodium hydroxide after acidification, so as to obtain a mixture of sodium chromate and sodium dichromate containing 35% Cr. This is done because potassium dichromate, which was formerly manufactured on a large scale, has this chromium content.

If carbon dioxide is passed under pressure into sodium chromate solution, dichromate is formed:

$$2Na_2CrO_4 + 2CO_2 + H_2O \rightleftharpoons Na_2Cr_2O_7 + 2NaHCO_3.$$

The lower the temperature, the higher the pressure, and the higher the concentration of chromate in the initial solution, the more this equilibrium is displaced towards the formation of dichromate (Agde, 1934). The formation of dichromate is considerably favored in that the sodium hydrogen carbonate formed separates out for the most part in the solid state. It has been suggested that this reaction should be used for the technical production of sodium dichromate, since it enables sulfuric acid to be replaced by the much cheaper carbon dioxide.

In addition to the mono- and dichromate, a tri- and tetrachromate of sodium are known, as for the other alkali and alkaline earth metals. A sodium chromate of the composition $Na_4CrO_5 \cdot 13H_2O$ is also known. This salt, which forms large sulfur-yellow rhombohedra can be crystallized unchanged from water. However, if sodium hydroxide (which would be expected to stabilize the compound, since the latter is more basic than the normal chromate) is added to the solution, the normal sodium chromate, Na_2CrO_4, separates.

(*iv*) *Potassium chromate and Potassium dichromate.* Potassium chromate always crystallizes anhydrous, in lemon yellow crystals which are isomorphous with those of potassium sulfate. It changes above about 670° into a red modification which melts at 970–980°. Potassium chromate is very soluble in water, the solubility being but little dependent upon the temperature.

Potassium dichromate, $K_2Cr_2O_7$, crystallizes from water in large orange-red triclinic plates, which are also anhydrous. The solubility in water is relatively small at room temperature, but increases greatly with rise in temperature (from 4.6 g in 100 g of H_2O at 0° to 94.1 g at 100°, and 263 g at 180°). The salt can therefore be very conveniently purified by recrystallization. Potassium dichromate is insoluble in alcohol. It melts at 396°, and crystallizes from the melt in a modification which is unstable at ordinary temperature, and which undergoes a transformation (with considerable change in volume) below 240°. Potassium dichromate loses oxygen only when it is strongly heated. It is completely stable in air, and is not hygroscopic. Since it can also readily be obtained in a high state of purity, it is used in volumetric analysis as a reference material for the standardization of thiosulfate solutions used in iodometry.

When potassium dichromate and dilute hydrochloric acid are added to a solution of potassium iodide, reaction takes place quantitatively:

$$Cr_2O_7^= + 6I^- + 14H^+ = 2Cr^{+++} + 3I_2 + 7H_2O.$$

Thus 6 atoms of iodine are liberated for every $Cr_2O_7^=$ ion.

Some potassium dichromate is manufactured by a method similar to that used for sodium dichromate, except that potash is used for the fusion of the chromite in place of soda. However, since considerable amounts of the relatively costly potassium oxide are lost by volatilization at the high temperature of the fusion, it is now generally preferred to prepare potassium dichromate from sodium dichromate, by double decomposition with sodium chloride.

(v) *Ammonium chromate and Ammonium dichromate*, $(NH_4)_2CrO_4$ and $(NH_4)_2Cr_2O_7$, crystallize anhydrous, like the potassium salts, but are much less stable than the latter. They deflagrate vigorously when they are heated to about 200°, leaving a residue of green chromic oxide in a very fine state of subdivision.

(vi) *Lead chromate*, $PbCrO_4$, is found native as *crocoite*. It is practically insoluble in water, and is thrown down as a yellow precipitate by mixing a solution of a chromate or dichromate with a solution of a lead salt. Lead dichromate is obtained only from solutions containing a very large excess of CrO_3.

Lead chromate generally crystallizes monoclinic, but a rhombic modification (stable between 707° and 783°) and a tetragonal form (stable above 783°) also exist. In accordance with Ostwald's law of stages, the rhombic modification is often formed by precipitation from solution, but on gentle warming, passes over into the monoclinic form, which is stable at ordinary temperature. Both the monoclinic and the rhombic lead chromate form mixed crystals with lead sulfate.

Lead chromate is distinctly soluble in caustic alkalis, since there is some formation of plumbites. However, if it is treated with a small amount of caustic alkali, a *basic lead chromate*, $PbCrO_4 \cdot Pb(OH)_2$, (which is also very insoluble) is formed in monoclinic plates. A ruby red lead chromate, still richer in lead oxide, with the composition $PbCrO_4 \cdot 2PbO$, is found in nature as *melanochroite* (phoenicite).

Large amounts of lead chromate are used as a pigment, under the name of *chrome yellow*. The basic chromate is also used as a pigment (*chrome red*).

(*vii*) *Halogen Substitution Products of the Chromates.* If one of the O^{2-} atoms in the chromate acid radical (I) is replaced by a chlorine atom, Cl^-, the univalent acid radical (II) is produced:

$$\begin{bmatrix} O & & O \\ & Cr & \\ O & & O \end{bmatrix}^{2-} \qquad \begin{bmatrix} O & & Cl \\ & Cr & \\ O & & O \end{bmatrix}^{1-} \qquad \begin{bmatrix} O & & Cl \\ & Cr & \\ O & & Cl \end{bmatrix}$$

$$\text{(I)} \qquad\qquad\qquad \text{(II)} \qquad\qquad\qquad \text{(III)}$$

The salts derived from (II), are known as *chlorochromates*, $M^I[CrO_3Cl]$. If two oxygen atoms are replaced by chlorine, the electroneutral molecule *chromyl chloride* (III) is obtained.

Compounds of the type $M^I[CrO_3X]$ are also known for the other halogens, but bromine and iodine form no compounds analogous to chromyl chloride.

(*viii*) *Chlorochromates.* Potassium chlorochromate, $K[CrO_3Cl]$, is obtained by adding hydrochloric acid to the solution of an equimolecular mixture of potassium chloride and chromium trioxide (1), or by boiling potassium dichromate solution with excess hydrochloric acid (2):

$$CrO_3 + KCl = K[CrO_3Cl] \tag{1}$$

$$K_2[Cr_2O_7] + 2HCl = 2K[CrO_3Cl] + H_2O \tag{2}$$

On cooling, potassium chlorochromate separates from the solution in orange, rectangular prisms of density 2.50. It breaks up into potassium dichromate and hydrochloric acid when it is dissolved in water, since the reaction represented by (2) is reversible. It can, however, be recrystallized from water acidified with hydrochloric acid or acetic acid.

Potassium fluorochromate, $K[CrO_3F]$, (ruby red octahedra), potassium *bromochromate*, $K[CrO_3Br]$, (dark brown crystals), and *potassium iodochromate*, $K[CrO_3I]$, (garnet red crystals) are obtained in a similar manner to potassium chlorochromate.

(*ix*) *Chromyl chloride*, CrO_2Cl_2, is a deep red liquid, boiling at $117°$ and forming a red brown vapor. It solidifies to form bright red needles (m.p. $-96.5°$), when it is strongly cooled. The density is 1.935 at $15°$. The compound is formed by the action of hydrogen chloride gas on dry chromium trioxide:

$$2HCl + CrO_3 = CrO_2Cl_2 + H_2O.$$

It can also be obtained by the action of phosphorus pentachloride, acetyl chloride, pyrosulfuryl chloride or chlorosulfonic acid on chromium trioxide or chromates. It is most simply prepared by warming a mixture of a chromate or dichromate with alkali chloride and concentrated sulfuric acid:

$$K_2Cr_2O_7 + 4KCl + 3H_2SO_4 = 2CrO_2Cl_2 + 3K_2SO_4 + 3H_2O.$$

If it is protected from light, chromyl chloride is fairly stable. In the undiluted state it reacts, usually very vigorously (often with inflammation) with oxidizable organic substances. It dissolves without decomposition in such substances as carbon tetrachloride, carbon disulfide, phosphorus oxychloride. In these solutions, and in the vapor state, its molecular weight corresponds to the formula CrO_2Cl_2. It is decomposed by water, with a considerable evolution of heat (about 17 kcal per mol):

$$CrO_2Cl_2 + 2H_2O = H_2CrO_4 + 2HCl.$$

It reacts in a corresponding manner with caustic alkalis, forming chromates.

For the detection of chlorides by means of the chromyl chloride reaction, the sample is mixed with finely powdered potassium dichromate and heated with concentrated sulfuric acid. The vapors evolved are passed into very dilute sodium hydroxide. If the solution thus obtained is found to contain chromate, the original sample must have contained *chloride*, since bromides and iodides form no volatile compounds with chromium, but are oxidized to free bromine and iodine under the given conditions. In carrying out the test, it has to be remembered that only those chlorides from which concentrated sulfuric acid will liberate HCl can give rise to chromyl chloride. Chlorides which are insoluble or very slightly dissociated—i.e., mercury(II) chloride, mercury(I) chloride, and silver chloride— give no chromyl chloride. Since fluorine gives a chromium compound very similar to chromyl chloride, fluorides must be removed before carrying out the chromyl chloride test.

(x) *Chromyl fluoride*, CrO_2F_2, can be obtained by the action of fluorine on CrO_2Cl_2, or by the reaction of CrO_3 with anhydrous HF, as a red-brown gas which can readily be condensed to deep violet-red crystalline needles (subl. temp., 30°, m.p. 31.6°, vap. press. at triple point 873 mm, heat of sublimation 14.6 kcal. per mol.). The compound is quite stable if kept in the dark. If exposed to light, it gradually changes into a dirty-white, non-volatile modification (von Wartenberg, 1941; Grosse, 1951).

(xi) *Peroxychromates and Chromium Peroxides.* The action of hydrogen peroxide on chromate solutions yields various *peroxychromates*, or a deep blue *chromium peroxide*, CrO_5, depending upon the conditions of experiment (H^+ ion concentration, amount of H_2O_2, temperature).

Two series of peroxychromates are known: blue peroxychromates, with the general formula $M^I_2[Cr_2O_{12}]$, and red peroxychromates, $M^I_3[CrO_8]$. The acids corresponding to these peroxychromates are not stable in the free state; a peroxide, CrO_5, not capable of forming salts, is formed in acid solution instead. This can be isolated in the form of crystalline addition compounds, as well as in solution, but does not exist in the free state.

The deep blue color which appears when acidified chromate solutions are treated with hydrogen peroxide is due to the formation of this peroxide. Since the latter decomposes, forming chromium (III) salts the blue color rapidly changes to green. However, if the solution is shaken with ether immediately after acidification, the chromium peroxide enters the ether layer, and persists longer. Since the blue color of chromium peroxide is extraordinarily intense, this reaction enables even quite small amounts of dissolved chromates to be detected.

By the action of aqueous ammonia on the ether solution of the blue peroxide, it is possible to obtain the ammoniate of a second chromium peroxide, CrO_4. This peroxide is also known only in the form of addition compounds. Its ammoniate, $CrO_4 \cdot 3NH_3$, is transformed into the compound $CrO_4 \cdot 3KCN$, of salt-like character, when it is gently warmed with potassium cyanide. Aqueous solutions of CrO_4 compounds are brown in color.

The addition compounds of CrO_4 were formulated by Werner as coordination compounds of electrochemically and coordinatively hexavalent chromium, containing one peroxy group,

$$\begin{bmatrix} O\diagdown\diagup NH_3 \\ O_2{=}Cr{\leftarrow}NH_3 \\ O\diagup\diagdown NH_3 \end{bmatrix} \quad \text{and} \quad \begin{bmatrix} O\diagdown\diagup CN \\ O_2{=}Cr{\leftarrow}CN \\ O\diagup\diagdown CN \end{bmatrix} K_3$$

It was first proved by Riesenfeld (1908) that CrO_5 and the blue and red perchromates are to be considered as true *peroxy*-compounds (i.e., hydrogen peroxide derivatives), since he showed that hydrogen peroxide was formed in the hydrolysis of these compounds. This can not be detected directly, since it is at once decomposed catalytically in acid solutions by chromium peroxide compounds. Its formation as an intermediate is shown, however, by its reducing action on permanganate. Riesenfeld considered that the peroxychromates were compounds of seven-valent chromium. Schwarz (1932) subsequently showed that the content of peroxidic oxygen is higher than had been found by Riesenfeld, in whose experiments some of the hydrogen peroxide liberated by hydrolysis had escaped reaction with the permanganate, by decomposing directly. Schwarz overcame this difficulty by catalytically accelerating the hydrogen peroxide-permanganate reaction, by the addition of small amounts of molybdate. He then found that the blue perchromates contain $2\frac{1}{2}$ peroxygroups per atom of chromium. Hence they are to be formulated as binuclear compounds, with 6-valent chromium. Boehm's observation (1926) that K_3CrO_8 forms mixed crystals with K_3NbO_8 and K_3TaO_8 (cf. pp. 108, 113) makes it reasonable to assume that chromium is pentavalent in the red peroxychromates, as are niobium and tantalum in their peroxysalts. This was proved by Tjabbes (1932) and Klemm (1933), who showed that the magnetic properties of the red peroxychromates corresponded to those expected for a compound of $+5$ chromium (cf. Vol. I, p. 305). The magnetic properties of the blue peroxychromates and of the pyridine addition compound of chromium pentoxide, on the other hand, correspond to those of the chromates and dichromates—i.e., to compounds of $+6$ chromium.

Chromium pentoxide contains 2 peroxygroups, as shown by its consumption of permanganate. Hence the following constitutional formulas may be written for these compounds.

$$\begin{array}{c} O\diagdown\quad\diagup O \\ |\ \ Cr\ \ | \\ O\diagup\ \|\ \diagdown O \\ O \end{array} \qquad \begin{bmatrix} O{-}O\quad\quad O{-}O \\ \diagdown\diagup\quad\quad\diagdown\diagup \\ O{-}Cr{-}O\quad O{-}Cr{-}O \\ \diagup\diagdown\quad\quad\diagup\diagdown \\ O{-}O\quad\quad O{-}O \end{bmatrix} M^I_2 \qquad \begin{bmatrix} O\diagdown\quad\diagup O{-}O \\ |\ \ Cr\ \ O{-}O \\ O\diagup\quad\diagdown O{-}O \end{bmatrix} M^I_3$$

Blue chromium   Blue peroxychromates   Red peroxychromates
peroxide, CrO_5

Now that the constitution of these compounds has been cleared up it removes the contradiction between the older assumption, that chromium is heptavalent in them, and our modern understanding of the relation between valence and atomic structure. The latter makes it excessively improbable that chromium should be 7-valent, since to do so it would have to be able to utilize an electron out of the argon shell. In view of the strong binding of electrons in the inert gas configurations, this would be hard to understand.

(*xii*) *Blue Chromium Peroxide*, CrO_5, is known only in solution and in the form of addition compounds. These correspond to the general formula $CrO_5 \cdot Am$ where Am represents one molecule of a neutral ligand (pyridine, aniline, quinoline, or ether). The instability of free chromium peroxide can be explained in that it is coordinatively unsaturated, and needs to add on a neutral molecule to achieve coordinative saturation. The pyridine addition compound of the blue chromium peroxide is the compound most readily obtained. It has the composition $CrO_5 \cdot C_5H_5N$, and crystallizes in deep blue leaflets insoluble in water. The methyl ether addition compound, $CrO_5 \cdot (CH_3)_2O$, which was formerly mistakenly thought to be an acid H_3CrO_8, can be made by the action of H_2O_2 on CrO_3 in methyl ether solution at low temperatures. It decomposes at about $-30°$ with explosive violence.

(*xiii*) *Blue potassium peroxychromate* can be obtained by cautious addition of 30% hydrogen peroxide to a solution of potassium dichromate at $0°$. It forms deep blue dichroic prisms, of composition $K_2Cr_2O_{12} \cdot 2H_2O$, which deflagrate on heating, through shock, or by contact with concentrated sulfuric acid. It also decomposes slowly at ordinary temperature, forming potassium dichromate.

The blue ammonium peroxychromate crystallizes with $2H_2O$, like the potassium salt, but the thallium salt, $Tl_2Cr_2O_{12}$, is anhydrous.

(xiv) *Red potassium peroxychromate*, K_3CrO_8, is formed by the action of 30% hydrogen peroxide on a strongly alkaline potassium chromate solution. It crystallizes in dark red-brown prisms. It is quite stable at ordinary temperature, but decomposes at 170°. Decomposition becomes explosive at higher temperatures.

The peroxychromates are not stable in solution at ordinary temperature. They decompose in alkaline solution, reforming chromates:

$$[Cr_2O_{12}]^= + 2OH^- = 2CrO_4^= + H_2O + \tfrac{5}{2}O_2$$

$$2[CrO_8]^\equiv + H_2O = 2CrO_4^= + 2OH^- + \tfrac{7}{2}O_2.$$

In acid solution they are transformed to compounds of chromium(III):

$$[Cr_2O_{12}]^= + 8H^+ = 2Cr^{+++} + 4H_2O + 4O_2$$

$$[CrO_8]^\equiv + 6H^+ = Cr^{+++} + 3H_2O + \tfrac{5}{2}O.$$

Hydrogen peroxide, if present, is decomposed catalytically at ordinary temperature. Preparation of the peroxychromates must therefore be carried out below 0°.

4. Analytical (Chromium)

Trivalent chromium behaves analytically like aluminum. Like the latter, it is precipitated from its solutions by ammonium sulfide, in the form of the *oxide hydrate*.

Chromates are gradually reduced by ammonium sulfide to chromium(III) salts. Hydrogen peroxide does so more rapidly. Conversely, chromium(III) compounds can readily be converted to chromates by fusion with soda and potassium nitrate (oxidative fusion). Chromates can then most simply be identified by the 'chromium peroxide' reaction.

The precipitation with silver nitrate from nitric acid solution as silver dichromate, $Ag_2Cr_2O_7$, or the precipitation of benzidinium chromate can be used as micro-reactions. The color reaction of the $Cr_2O_7^=$ ion with diphenylcarbazide (violet coloration), which can also be carried out as a drop reaction, is convenient for the detection of small amounts of chromium at high dilution. Chromium(III) salts are first oxidized by potassium peroxysulfate (in acid solution, with the addition of silver nitrate as catalyst) for the purpose of this test (Feigl, 1931).

The microcosmic bead is colored dark green by chromium, in both the oxidizing and reducing flame, since in the absence of alkalis the trivalent state (from which the green phosphate is derived) is the most stable state of chromium.

For quantitative determination, chromium is precipitated from chromium(III) salts as the oxide hydrate and is weighed as Cr_2O_3 after ignition. If chromium is present as chromate or dichromate it is most conveniently determined iodometrically. For gravimetric analysis, it can be precipitated as barium chromate, $BaCrO_4$, or as mercury(I) chromate, Hg_2CrO_4, which is converted to Cr_2O_3 on ignition:

$$2Hg_2CrO_4 = Cr_2O_3 + 4Hg + \tfrac{5}{2}O_2.$$

It is possible to determine chromium electrolytically, by deposition on a rotating mercury electrode, as described by Tutundžić [*Z. anorg. Chem.*, 202, (1931) 297; 215, (1933) 19].

5. Molybdenum (Mo)

(a) Occurrence

Molybdenum is found in Nature chiefly as *molybdenite*, MoS_2, and as *wulfenite* (yellow lead ore), $PbMoO_4$. Molybdenite is widely found, although seldom in large quantities. Wulfenite occurs as a secondary mineral in lead ore deposits, but is not very common.

A number of other molybdates are occasionally found is association with molybdenite— e.g., *powellite*, $CaMoO_4$, *pateraite*, $CoMoO_4$, *belonosite* $MgMoO_4$. More common, although invariably in only small quantities, is *molybdenum ocher*, MoO_3.

Molybdenum is widely distributed in Nature in minimal concentrations. According to Ter Meulen (1931) it is found in traces in all plants, and it has been stated (Bortels, 1930) to be essential for the life processes of azotobacter chroococcum, which is able to fix atmospheric nitrogen. Traces of molybdenum have been detected in the human and animal body, especially in the pancreas and liver. Cattle pastured in regions where the soil is lacking in molybdenum suffer from deficiency diseases which must be countered by the supply of small amounts in 'salt licks'.

(b) History

The old Greeks and Roman used the term *molybdaena* (μολυβδαινα) especially for galena, but for other lead ores also (Gk μόλυβδος = lead). The name was afterwards extended to other minerals which will make a dark mark, in the same way as galena,—especially to graphite and to the mineral now known as molybdenite. Molybdenite and graphite which are very similar in appearance, were for a long time thought to be the same substance. They were later named black lead or 'water lead' (latin, plumbago). Scheele in 1778, first recognized that they were different minerals. He decomposed molybdenite ('water lead') by means of nitric acid, and obtained from it a white oxide (molybdenum trioxide, MoO_3) which he named molybdic acid. The metal forming the oxide, *molybdenum*, was first isolated by Hjelm in 1782.

(c) Preparation

The most important starting material for the preparation of molybdenum is molybdenite. As found in Nature, this is generally very impure, so that the ores used for large scale production usually contain only a few tenths of a per cent of the pure sulfide, MoS_2. It is therefore necessary to subject it first to some concentration process. This is now generally carried out by *flotation*.

The flotation process (froth flotation) has attained great importance during recent years in ore dressing, especially for sulfide ores. [6–8]. In its original form, the flotation process depends on the fact that certain substances are not wetted by water, but are readily wetted by paraffin, petroleum, and other mineral oils. When such substances, which by themselves would sink in water, are agitated in a finely divided state with oil and water, those particles which are not wetted by water surround themselves with a layer of oil. This makes them float, since the bouyancy of the specifically light oil drop more than compensates for the weight of the small particle enclosed within it. If an ore contains large quantities of constituents which are wetted by water, together with small amounts of some material which is hydrophobic but wetted by oil—as is molybdenite, for example—then the latter stays in the froth, when subjected to flotation, while the impurities sink. Instead of using oil drops, gas bubbles may act as carriers for the particles to be floated, if suitable foam-forming substances are added. The flotation process is now used almost entirely in this form (foam flotation). The foam is produced by blowing air in, or by suction. Substances which would themselves be wetted by water can be made hydrophobic, or not readily wetted, by the addition of certain materials which are adsorbed at their surface. In froth flotation, these additives (known as 'collectors', since they bring about the collection in the foam of the mineral particles which are to be recovered) act by forming a coating on the surface of the particles of mineral, to which the air bubbles adhere by capillary forces. To understand the way in which the flotation process is now carried out, account must be taken of the strong capillary forces which act at points where three different phases meet (air, water, and the mineral particle or the adsorbed layer on its surface) (Ostwald, 1932). This view, that the mineral particles are not so much carried up to the surface by the buoyancy of the adherent air bubbles, as pulled into the foam boundaries by the three-phase boundary forces, accords with the observation that the foam is commonly stabilized by picking up the mineral particles. The action of the substances which produce the foam ('frothing agents') and the 'collectors' can be modified by so called 'modifying agents'. In this way it is possible to recover different minerals—e.g., lead sulfide and zinc sulfide—one after the other from the same ore by selective froth flotation.

By combining strongly acting 'collectors' with suitable 'regulators', it has even been found possible to separate *oxidic* ores (oxides, carbonates, etc.), which are themselves wetted very well by water, from impurities (gangue) which are themselves oxidic in nature, by direct flotation. It is generally preferable, however, to produce a sulfide layer on the surface of such mineral particles by suitable reactions, so they then behave in flotation like sulfide ores.

Molybdenum ore, concentrated to a content of 70% MoS_2 or over, is converted to molybdenum trioxide, either by roasting [eqn. (1)] or by fusion with soda in a reverberatory furnace, and decomposition of the resulting sodium molybdate [eqns (2) and (3)]. The metal is then obtained by reducing the oxide with hydrogen [eqn (4)] or with carbon or carbonaceous materials (e.g., colophonium).

$$MoS_2 + \tfrac{7}{2}O_2 = MoO_3 + 2SO_2 \tag{1}$$

$$MoS_2 + 3Na_2CO_3 + \tfrac{9}{2}O_2 = Na_2MoO_4 + 2Na_2SO_4 + 3CO_2 \tag{2}$$

$$Na_2MoO_4 + 2HCl = MoO_3 + H_2O + 2NaCl \tag{3}$$

$$MoO_3 + 3H_2 = Mo + 3H_2O \tag{4}$$

The molybdenum is first obtained in the form of powder, because of its infusibility. Compact, malleable molybdenum is obtained from this by pressing it into rods, and then heating it by low tension alternating current almost to its melting point, in an atmosphere of hydrogen.

Colloidal molybdenum can be made like colloidal chromium, by electrical dispersion under isobutyl alcohol, or by alternate treatment of the most finely divided molybdenum powder with dilute hydrochloric acid and dilute caustic soda. Molybdenum powder which peptizes readily by this treatment can be obtained by heating MoO_3 or MoO_2 with zinc dust.

(d) Properties

In the form of powder, molybdenum has a more or less matt dark grey color, but the compact metal is silver white and lustrous. Is is fairly hard, but can be polished and can be worked and welded at high temperatures. Its density is 10.2, and melting point around 2600°. The boiling point is very high indeed. Langmuir measured the rate of vaporization of molybdenum in a vacuum at temperatures up to the melting point, and calculated the corresponding sublimation pressures. From these data, Van Liempt estimated the boiling point at atmospheric pressure as around 3560°. According to later work this is rather higher—about 4800°. Molybdenum is quite a good electrical conductor; the specific conductivity at 0° is about 34% of that of silver.

Molybdenum is fairly stable in air at ordinary temperature. At a red heat, even the compact metal is oxidized fairly rapidly to the trioxide, MoO_3. It also reacts with chlorine and bromine at elevated temperatures, and with fluorine even in the cold. Iodine, however, is without action at red heat. Hydrogen is taken up only to a very small extent, even by finely divided molybdenum, and is given up again completely at $300°$. Molybdenum combines with carbon, when heated, forming a carbide. It can also combine directly with carbon monoxide, if the gas is treated under high pressure with finely divided molybdenum. The hexacarbonyl, $Mo(CO)_6$, is thereby formed (highly refractive, volatile crystals, density 1.96). Elementary nitrogen unites only with difficulty with molybdenum, but the nitrides Mo_2N and MoN are formed by heating molybdenum powder in ammonia gas.

The heat of formation of Mo_2N is 16.6 kcal per mol (Neumann, 1934). According to Hägg (1930), above $600°$, a third nitride is formed, with a nitrogen content of 28 atom-%. Molybdenum can hold small quantities of nitrogen in solid solution (Sieverts, 1936).

Direct combination of Mo with C generally produces the carbide Mo_2C, which is isotypic with Ta_2C and W_2C. In addition to this, there is another carbide, richer in carbon, and probably of the formula MoC. This appears to have a cubic structure, but gives rise very readily to mixed crystals with the hexagonal tungsten monocarbide, WC, (the structure of which is, admittedly, related to the NaCl type). It is much more readily obtained in the form of these mixed crystals than in the pure state (Weiss, 1946–48; Lander and Germer, 1947, Nowotny and Kieffer, 1952).

Molybdenum is hardly affected by dilute acids, nor by concentrated hydrochloric acid. It is attacked by concentrated nitric acid, but since this—like other oxidants—simultaneously *passivates* the molybdenum, the oxidation proceeds only slowly. Moderately concentrated nitric acid reacts more energetically with molybdenum than the very concentrated acid. Molybdenum is also oxidized by concentrated sulfuric acid when heated almost to boiling. Aqua regia, or mixtures of concentrated nitric acid with hydrofluoric acid or with sulfuric acid, react more vigorously. Molybdenum is practically insoluble in caustic alkalis, and is only slowly attacked by fused alkali hydroxides. Fusion with potassium nitrate, potassium chlorate, or sodium peroxide brings about rapid oxidation, however.

(e) Uses [3]

Molybdenum is used chiefly for the production of *special steels*. These are used, for example, for gun barrels, armor plate, rolling mills, etc. Even a small addition of molybdenum confers great strength and toughness on steel. With chromium, nickel, cobalt, and vanadium it is used in the manufacture of high speed steels. High alloy molybdenum steels retain their tensile strength up to very high temperatures, and molybdenum-nickel-chromium alloys have been important in the development of the gas turbine. Molybdenum is also used in making magnet steels and acid resistant alloys. For all these purposes it is usually used in the form of *ferromolybdenum*—i.e., an iron alloy of high molybdenum content, which is more fusible and less oxidizable than pure molybdenum.

In spite of its high affinity for oxygen, molybdenum cannot be used as a deoxidant, since the oxide formed is retained in solid solution. Conversely, therefore, particular care must be taken to ensure efficient deoxidation—preferably by adding some titanium or vanadium—when molybdenum is used as an alloy constituent.

Molybdenum is less suitable than tungsten for electric lamp filaments, since it sputters more quickly. However, molybdenum wires are used as filament supports in electric lamps, since the metal can readily be fused gas-tight through glass.

Of the compounds of molybdenum, ammonium molybdate is the most used, chiefly for the determination of phosphoric acid (e.g., in artificial fertilizers).

6. Compounds of Molybdenum [2]

Molybdenum forms compounds from a number of valence states. The maximum valence state, like that of chromium, is *six*. Apart from the mineral MoS_2, molybdenum is 6-valent in all the compounds which are themselves important—namely in molybdenum trioxide, MoO_3, and the molybdates derived from it.

SUMMARY

Chlorides	Fluorides	Sulfides	Oxides
	MoF_6	MoS_3	MoO_3
			Mo_9O_{26}
			Mo_8O_{23}
$MoCl_5$			Mo_2O_5?
$MoCl_4$	MoF_4?	MoS_2 and $Mo(S_2)_2$	MoO_2
$MoCl_3$	MoF_3?	Mo_2S_3?	Mo_2O_3?
$MoCl_2$ or Mo_6Cl_{12}			

a) Halides

(*i*) *Chlorides*. By heating molybdenum powder in chlorine, *molybdenum pentachloride* is obtained as a dark red vapor which condenses to a deep green almost black mass (m.p. 194°, b.p. 268°). According to Debray, the vapor density at 350° is 9.5 (relative to air), corresponding to the formula $MoCl_5$. Molybdenum pentachloride is a non-conductor of electricity, both in the solid and the fused states. It reacts with water, with evolution of much heat, forming the oxychloride:

$$MoCl_5 + H_2O = MoOCl_3 + 2HCl.$$

Molybdenum tetrachloride is most conveniently obtained by heating molybdenum dioxide with a solution of chlorine in carbon tetrachloride to 250°. It forms a brown powder which is readily volatilized to form an intensely yellow vapor. It partially decomposes into $MoCl_5$ and $MoCl_3$, when heated in a sealed tube.

Molybdenum trichloride, $MoCl_3$, can be obtained by passing $MoCl_5$ over heated molybdenum, or by heating $MoCl_5$ in hydrogen. It forms a dark red crystalline powder, insoluble in water and in hydrochloric acid. Purple-red aqueous solutions of molybdenum trichloride can be obtained, however, by electrolytic reduction at the mercury cathode of molybdenum trioxide, dissolved in concentrated hydrochloric acid. The molybdenum trichloride is present in these solutions in the form of chloromolybdate(III) ions. If alkali salts are added, either hexachloromolybdate(III) salts, $M^I_3[MoCl_6]$, or pentachloroaquomolybdate(III) salts, $M^I_2[MoCl_5(H_2O)]$, separate out, according to the concentration. Organic amines generally combine additively with $MoCl_3$, forming non-electrolytes, $[MoCl_3Am_3]$. However, it is possible for salts of complex molybdenum(III) cations to be formed by combination with neutral ligands—e.g., $[MoCl_2(NH_3)_4]Cl$ (Rosenheim, 1931). When anhydrous molybdenum trichloride is heated to a dull red heat in dry carbon dioxide, it decomposes into $MoCl_4$ and $MoCl_2$.

Molybdenum dichloride, $MoCl_2$ or Mo_6Cl_{12}, is a yellow powder, completely insoluble in water. It dissolves in alcohol and ether, however, and the molecular weight in solution was found to correspond with the formula Mo_3Cl_6. However, Brosset (1946) has determined the structure of molybdenum dichloride by means of X-rays, and has shown it to have the molecular formula Mo_6Cl_{12} in the solid state. When heated in hydrogen, it is reduced to metallic molybdenum.

(*ii*) *Fluorides*. Whereas molybdenum exhibits a maximum valence of *five* towards chlorine*, it can form a *hexafluoride*, MoF_6, with fluorine. Ruff obtained this compound by direct union of the elements, as a very hygroscopic, white, crystalline mass, which melted at $17°$ to a colorless liquid, and boiled at $35°$. It is very reactive, and is especially sensitive towards moisture. The *oxyfluorides* $MoOF_4$ and MoO_2F_2 (also colorless) are more stable. *Oxyfluoromolybdate(VI)* salts, (e.g., $K_2[MoO_2F_4] \cdot H_2O$, colorless glistening flakes) crystallize from their solutions when alkali fluorides are added. Oxofluoromolybdate(VI) salts of the type $M^I_3[MoO_3F_3]$ are formed by fusing MoO_3 with alkali fluorides. LiF differs from the other fluorides in that it reacts with MoO_3 to form MoO_2F_2 instead of a fluoromolybdate. The reason for this is that the stability of a complex salt depends not only on the energy of formation of the complex ion, but also on the energy liberated when the complex ion and the simple ions pack together to form a crystal lattice. It may be calculated that in the system $LiF-MoO_3$, although the free energy of formation of the complex ion $[MoO_3F_3]^{3-}$ is negative, the lattice energy is smaller than the sum of the lattice energies of the simple compounds. Similar calculations show that the trioxotrifluoromolybdates of the other alkali metals should be stable and (in agreement with experiment) should increase in stability from Na to Cs (Schmitz-Dumont, 1952). The alkali trioxotrifluoromolybdates are cubic in structure, and appear to be isotypic with the ammonium salt $(NH_4)_3[MoO_3F_3]$, which can be obtained from solution.

Oxyfluoro-salts derived from 5-valent molybdenum are also known (e.g., $K_2[MoOF_5]\cdot H_2O$, sky-blue glassy leaflets). The simple oxyfluoride from which the latter are derived has not yet been isolated, nor has molybdenum pentafluoride. The tetra-fluoride and trifluoride of molybdenum are also not known with certainty, but double salts of the latter exist (fluoromolybdate(III) salts)—e.g., $K[MoF_4] \cdot H_2O$.

(b) Sulfides

Two normal sulfides of molybdenum are known. The *trisulfide* is precipitated by hydrogen sulfide from acidified molybdate solutions; the disulfide occurs naturally as molybdenite.

According to older work, a sesquisulfide, Mo_2S_3, can be prepared by dry methods, but newer work makes the existence of this compound very dubious.

In addition to the normal sulfides, there is a higher sulfide of molybdenum, with the composition MoS_4. This is probably a polysulfide, $Mo(S_2)_2$, since salts derived from it are known and are regarded as *polythiomolybdates*—i.e., compounds derived from molybdates (but probably molybdate(IV) salts, not molybdate(VI) compounds) by exchange of polysulfide ions, S_2^{2-}, for O^{2-} ions.

(*i*) *Molybdenum trisulfide and Thiomolybdates*. Molybdenum trisulfide may be obtained by decomposing thiomolybdate solutions with acids, or by prolonged passage of hydrogen sulfide into warm molybdate solutions, acidified with hydrochloric acid.

Precipitation with hydrogen sulfide takes place only very incompletely in the cold, since partial reduction of molybdenum occurs, forming molybdenum blue (see below), which remains in colloidal dispersion. The trisulfide also readily forms colloidal dispersions. Only the first of the two methods cited is of preparative importance, since even the precipitate given by hydrogen sulfide from hot solutions is generally not pure molybdenum trisulfide.

Molybdenum trisulfide forms a deep brown precipitate, which dissolves readily in ammonium and alkali sulfides, and also in aqua regia. It loses sulfur, and is converted into molybdenum disulfide, when heated in the absence of air.

* *Oxychlorides* of 6-valent molybdenum are known, however.

The solubility of molybdenum trisulfide in ammonium sulfide is due to the formation of thiomolybdates. Most of these correspond to the general formula $M^I_2[MoS_4]$, but some thiomolybdates of more complex composition are also known. The normal thiomolybdates of the alkali and alkaline earth metals are soluble in water, and mostly crystallize well. They are intensely red in color. They are decomposed by acids, molybdenum trisulfide being precipitated:

$$(NH_4)_2[MoS_4] + 2HCl = MoS_3 + 2NH_4Cl + H_2S.$$

(*ii*) *Molybdenum disulfide*, MoS_2, occurs native as *molybdenite*. This usually exists as flat, thin, soft very flexible leaflets, which have a greasy feel, make a grey mark on paper, and are very similar to graphite. Density 4.7–4.8.

Molybdenite very rarely occurs in well-formed crystals. It belongs to the hexagonal system. Its crystal structure is shown in Fig. 26. The Mo atoms and the S atoms are all arranged in sheets, perpendicular to the *c*-axis, in such a manner that every sheet of Mo atoms has a sheet of S atoms (at a distance of $\frac{1}{8}c = 1.54$ Å) on each side of it. It thus forms a *layer lattice*, as in the case of graphite. This accounts for the excellent cleavage of molybdenite perpendicular to the *c*-axis, and its highly developed laminar habit—properties which it shares with graphite. The interatomic distances are as follows:

$$Mo \leftrightarrow S = 2.35 \text{ Å}, \qquad Mo \leftrightarrow Mo = a = 3.15 \text{ Å}, \qquad S \leftrightarrow S = \tfrac{1}{4}c = 3.08 \text{ Å}.$$

Molybdenite is found in many places in Europe, but usually in very small quantities. Norway has important deposits, but the principal sources of the ore are in Australia and North America.

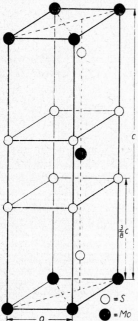

Fig. 26. Unit cell of molybdenite, MoS_2. $a = 3.15$, $c = 12.30$ Å.

Molybdenum disulfide can also be prepared artificially—e.g., it is formed by heating molybdenum dioxide, molybdenum trioxide, or ammonium molybdate in sulfur vapor. When it is heated to a very high temperature in the absence of air (in the Moissan furnace) it is said to lose part of its sulfur, and to form the sesquisulfide, Mo_2S_3. It readily burns in air, forming molybdenum trioxide, MoO_3.

(c) Oxides

The most stable oxide of molybdenum is the *trioxide*, MoO_3, which is therefore the end product of the oxidation of molybdenum by atmospheric oxygen. The dioxide, MoO_2, is also readily obtained. The oxides Mo_2O_5 and Mo_2O_3 are said to exist, but must be regarded as doubtful. The existence of the intermediate oxides Mo_9O_{26} and Mo_8O_{23} is noteworthy. In spite of their complex composition, these are well defined compounds with a characteristic structure (Magneli, 1949). Their structure is based on the existence of MoO_6 coordination octahedra as structural units in the crystal. Sharing of all corners between adjacent polyhedra would build up a crystal lattice of total composition MoO_3. If oxygen atoms are omitted from the structure, in such a way that a proportion of octahedra share an edge (i.e., have two O atoms in common), in an ordered arrangement, the structures of the intermediate oxides are obtained.

(*i*) *Molybdenum trioxide*, MoO_3, is obtained by igniting molybdenum, its sulfides, or other molybdenum compounds in air, or by treating them with nitric acid. It is usually prepared by prolonged heating of ammonium molybdate in air, or by repeatedly evaporating it down with nitric acid and extracting the ammonium nitrate so formed with water. Molybdenum trioxide forms a soft white powder of density 4.5. It turns yellow on heating, and melts without decomposition at 791°. It begins to sublime in the neighborhood of its melting point. Ignited molybdenum trioxide is only slightly soluble in water and in most of the common acids, although it dissolves in hydrofluoric acid and concentrated sulfuric acid. It dissolves readily in caustic alkalis, aqueous ammonia, and alkali carbonates, forming *molybdates*, salts with the general formula $M^I_2[MoO_4]$. Molybdenum trioxide can only be regarded as the anhydride of *molybdic acid*, H_2MoO_4, in a limited sense, since although it is readily obtained from the acid by heating, it cannot be reconverted to the acid by treatment with water.

If molybdenum trioxide is heated to 150–200° in dry hydrogen chloride, it forms a sublimate of pale yellow needles, of composition $MoO_3 \cdot 2HCl$. This is soluble in water, alcohol, ether, glacial acetic acid, and acetone. The aqueous solution leaves a residue of molybdenum trioxide when it is evaporated. This reaction enables molybdenum to be separated very simply from elements which do not form volatile oxides. MoO_3 unites with alkali fluorides (except LiF) to form colorless, cubic double compounds of the type $M^I_3[MoO_3F_3]$ (cf. p. 165).

(*ii*) *Molybdenum pentoxide* is said to be obtained as a violet-black powder when molybdenyl hydroxide, $MoO(OH)_3$, is heated in carbon dioxide. Molybdenyl hydroxide is thrown down as a rust-brown precipitate when ammonia is added to solutions of molybdenum(V) salts.

The intermediate oxides Mo_9O_{26} and Mo_8O_{23}, in which molybdenum has the average valence state of 5.78 and 5.75, respectively, are blue-black in color, and are interesting because of their relation to molybdenum blue. Their preparation and the elucidation of their structure is due to Hägg and Magneli (1944–48).

(*iii*) *Molybdenum dioxide*, MoO_2, is formed as an intermediate stage in the oxidation, when molybdenum is cautiously heated in air. It may also be obtained by passing steam over red hot molybdenum, and by treating molybdates with reducing agents (e.g., by fusion with zinc). Molybdenum dioxide is a brown-violet powder, with a coppery luster, and is a good electrical conductor. It is insoluble in caustic alkalis, unlike molybdenum trioxide, and is also insoluble in acids. It is oxidized to MoO_3 by nitric acid, and also by ammoniacal silver solutions, from which it precipitates silver. It combines with chlorine to form MoO_2Cl_2, but does not add on hydrogen chloride (cf. MoO_3). Molybdenum dioxide crystallizes with the rutile structure (Vol. I, Fig. 63, p. 265), $a = 4.86$, $c = 2.79$ Å.

Molybdenum dioxide was obtained by Paal in hydrated form, as a brown-black suspension, by reducing a solution of ammonium molybdate with hydrogen activated by colloidal palladium. After very careful drying, its water content corresponded roughly with the composition $MoO(OH)_2$. Drying in the warm gave a practically anhydrous oxide. Molybdenum dioxide was obtained in the form of a reversible colloid by use of sodium protalbinate as a protective colloid. The further action of activated hydrogen, with gentle warming and under excess pressure, led to a gel of the sesquioxide, Mo_2O_3 (black crusts, with a bluish luster).

Derived from molybdenum dioxide are the *molybdate(IV)* salts obtained by Scholder (1952) in the form of the alkaline earth compounds $M^{II}MoO_3$ and $M^{II}_2MoO_4$. These were prepared by reduction of the molybdate(VI) salts with hydrogen.

Molybdate(V) salts can also be prepared. Scholder found that they were formed by 'symproportionation' between molybdate(IV) and molybdate(VI) salts.

(d) Molybdic Acid and the Molybdates

As has already been mentioned, molybdenum trioxide dissolves easily in alkali hydroxides, forming *molybdates* (systematically: molybdate(VI) salts). The simplest molybdates have the general formula $M^I_2[MoO_4]$, but compounds of this type are obtained only from solutions containing a large excess of alkali hydroxide. Most molybdates—the so-called *polymolybdates*—contain more than one MoO_3 for each M^I_2O. The ordinary ammonium molybdate is such a polymolybdate.

(*i*) *Molybdic acid.* The sparingly soluble *molybdic acid*, monoclinic canary yellow crystals, with the composition $H_2MoO_4 \cdot H_2O$, frequently separates slowly from molybdate solutions containing nitric acid. It readily forms colloidal dispersions, as was observed by Graham. It changes into the unhydrated acid, H_2MoO_4, when gently heated, even in solution. This crystallizes in fine white needles, and exists in two modifications which differ but little in appearance, but are converted to the anhydride at quite different temperatures; one form also gives a milky suspension in water, which cannot be clarified by filtration, whereas the other does not.

The measurements of Jander and Jahr (1930 and later), Byé (1942), and Carpeni (1947) have shown that a variety of ions exist in molybdate solutions, depending upon the hydrogen ion concentration. In addition to $[MoO_4]^=$ ions in alkaline solutions, $[HMo_6O_{21}]^{5-}$ or $[Mo_7O_{24}]^{6-}$, $[Mo_8O_{26}]^{4-}$ and ions with 12 or 24 Mo atoms have been postulated in solution. Between them, equilibria are set up similar to those discussed on p. 100 for the polyvanadates, involving the H^+ ion concentration of the solution. Although there are only a few distinct types of polymolybdate ions—apart from the dimolybdates and trimolybdates, which are only rarely isolated from solution*—the salts which crystallize from the solutions may have wide variations in composition. This is in part because varying numbers of acidic hydrogen atoms may be replaced by metals in the polybasic acids from which they are derived. Although salts have been described with the ratio $M^I_2O : MoO_3 = 1 : 1$, $1 : 2$, $1 : 2.33$, $1 : 3$, $1 : 4$, $1 : 8$, $1 : 10$, it now appears that most of the polymolybdates are normal salts or acid salts of either the paramolybdate series, $M^I_6[Mo_7O_{24}]$, or the octamolybdate series, $M^I_4[Mo_8O_{26}]$.

(*ii*) *Ammonium molybdate.* Ordinary ammonium molybdate, a nitric acid solution of which is used as a reagent for phosphoric acid, is obtained in large colorless monoclinic crystals, by evaporating a solution of molybdenum trioxide in aqueous ammonia. The analytical composition of the crystals would correspond with either of the formulas $5(NH_4)_2O \cdot 12MoO_3 \cdot 7H_2O$ or $3(NH_4)_2O \cdot 7MoO_3 \cdot 4H_2O$. A decision between these alternatives was provided by Sturtevant (1937), who determined the weight of the unit cell of the monoclinic crystal lattice, from density and cell dimension measurements, and showed that it could be harmonized only with the presence of heptamolybdate anions, $[Mo_7O_{24}]^{6-}$, and has since been confirmed by the detailed structure determinations and precise analyses of Lindqvist (1948).

The normal salt $(NH_4)_2MoO_4$ can be obtained by dissolving molybdenum trioxide in warm ammonia. The ordinary ammonium molybdate of commerce is known as ammonium paramolybdate to distinguish it from the normal salt.

If a nitric acid solution of ammonium paramolybdate is added in large excess to a phosphate solution containing nitric acid, *ammonium molybdophosphate* is formed

---

* The anhydrous salt $Li_4Mo_3O_{11}$ is an example of a trimolybdate crystallizing anhydrous from solution. Jander and Jahr consider that the ion $[Mo_3O_{11}]^{4-}$ may be formed as an intermediate in the condensation of monomolybdates in solution, although it is not stable. Trimolybdates and dimolybdates (e.g., $K_2Mo_3O_{10}$, $Na_2Mo_2O_7$) can also be isolated from melts, but are of different types. Lindqvist (1950) has shown that the dimolybdates, at least, contain extended (infinite) anions (such as are found in the polysilicates), which can have no existence in solution.

as a fine yellow precipitate of the composition $(NH_4)_3PO_4 \cdot 12MoO_3 \cdot 6H_2O$*. The reaction is very convenient for the detection and determination of phosphoric acid, but it should be noted that arsenic acid, and in some circumstances silicic acid, react in similar fashion with ammonium molybdate.

(*iii*) *Molybdenum Blue*. If a solution of molybdic acid, or the acidified solution of a molybdate, is treated with a reducing agent (sulfur dioxide, hydrogen sulfide, hydriodic acid, hydrazine, glucose, zinc, or molybdenum metal), an intense deep blue coloration is produced, which is due to the formation of a (mostly colloidal) solution of *molybdenum blue*. Formation of this substance is utilized both as a qualitative analytical test for molybdenum. and for colorimetric determination.

The products comprised under the general name of molybdenum blue actually represent a variety of different substances. In addition to their deep blue color, it is common to all the substances that they contain molybdenum in an intermediate mean valence state equal to or higher than 5 but lower than 6. The products obtained by precipitation from aqueous solution are all amorphous to X-rays, have a variable water content, and pass more or less readily into colloidal dispersion. It is also possible to obtain molybdenum blue in the crystalline state—not only in the form of the intermediate oxides Mo_9O_{26} and Mo_8O_{23} referred to previously, but also in hydrated form. Glemser (1951) was able to prepare the blue compounds $Mo_8O_{23} \cdot 8H_2O$, $Mo_4O_{11} \cdot H_2O$ and $Mo_2O_5 \cdot H_2O$ in the crystalline state, and characterized them as definite compounds by analysis, isobaric degradation and X-ray analysis. The water in these compounds is constitutional, and Glemser therefore formulated them as hydroxides $(Mo_8O_{15}(OH)_{16}, Mo_4O_{10}(OH)_2$ and $Mo_2O_4(OH)_2)$. They are stable towards ammonia and caustic alkalis, and thereby differ from the amorphous blue preparations such as $Mo_8O_{23} \cdot$ xH_2O (Audrieth, 1942) and $Mo_4O_{11} \cdot xH_2O$ (Glemser, 1951). The compounds $Mo_4O_{10}(OH)_2$ and $Mo_2O_4(OH)_2$ can also be drepared by the action of atomic hydrogen on MoO_3. In this manner, or by reaction between $LiAlH_4$ and MoO_3, a Bordeaux red compound of the composition $Mo_5O_7(OH)_8$ is also obtained (Glemser, 1952). The substance $Mo_5O_7(OH)_8$ is not formed immediately when atomic hydrogen and MoO_3 react, but results from the elimination of hydrogen from an olive-green compound which is first produced. The latter, which has the empirical composition $MoO(OH)_2$, is, as Glemser concluded from its behavior, better formulated as $H_2Mo_5O_7(OH)_8$. Treadwell (1946) prepared a readily soluble molybdenum blue, Mo_3O_8OH, by the electrolytic reduction of ammonium paramolybdate in sulfuric acid solution. It is a monobasic acid, and has a molecular weight corresponding to the formula given. It is not yet known whether molybdenum is present in compounds of the type of molybdenum blue in the $+6$ and $+4$ valence states, or in the $+6$ and $+5$ states.

Colloidal molybdenum blue is readily adsorbed on surface-active materials, and especially on vegetable and animal fibers. It is therefore a typical dyestuff, and finds some use as such, especially for silk.

(*iv*) *Peroxymolybdic acid and Peroxymolybdates*. Hydrogen peroxide produces a yellow or orange coloration with acid solutions of molybdates. Compounds of similar color can be isolated from the solutions, which differ from the molybdates in that they contain an excess of oxygen, and give reactions typical of hydrogen peroxide (decolorization of permanganate, blue color with chromic acid, etc.). They are thus derived from the molybdates, by replacement of an oxygen atom by a peroxy group, and are therefore termed *peroxymolybdates*. The simplest peroxymolybdates have the formulas M^IHMoO_6 (pale yellow) and $M^I_2MoO_8$ (red, explosive), but like the molybdates they often have more complex compositions.

According to Jahr (1941), equilibria are set up in solution:

$$MoO_4^= + 2H_2O_2 \rightleftharpoons HMoO_6^- + OH^- + H_2O \qquad (1)$$

$$MoO_4^= + 4H_2O_2 \rightleftharpoons MoO_8^= + 4H_2O \qquad (2)$$

* For constitution, see p. 181–2.

If the OH^- ions formed according to (1) are removed by the addition of acid, the hydrogen diperoxymolybdate ion, $HMoO_6^-$, is formed quantitatively. Addition of much more acid brings about decomposition:

$$HMoO_6^- \rightarrow MoO_4^= + H^+ + O_2,$$

and subsequent condensation of the resulting $MoO_4^=$ ions, as in pure molybdate solutions. Formation of very stable condensed peroxymolybdate ions then takes place, by reaction with excess hydrogen peroxide, the existence of peroxypolymolybdate ions being shown by determinations of the rate of diffusion. The compounds derived from them are deep yellow or orange, and are usually of complex composition, but Hansson and Lindqvist (1949) have shown that it is probable that at least one series is related to the paramolybdate ion, $[Mo_7O_{24}]^{6-}$, by substitution of peroxy groups.

Free peroxymolybdic acid, $H_2MoO_5 \cdot \frac{3}{2}H_2O$, is an amorphous yellow powder. Unlike the blue chromium peroxide, CrO_5, it cannot be extracted by means of ether.

Allowance must be made for the formation of yellow peroxymolybdic acid if the molybdo-phosphate reaction is employed to test for the presence of phosphoric acid in hydrogen peroxide. The hydrogen peroxide must be removed by evaporation before carrying out the test.

(e) Compounds of Molybdenum with Acid Anions

Apart from its halides and sulfides, which in any case do not display well marked salt-like character, molybdenum forms practically no simple salts with acid anions, but forms numerous complex salts (acido-salts).

(i) Acido-salts of Hexavalent Molybdenum. Acido-salts of +6 molybdenum are usually obtained by dissolving molybdic acid in the corresponding acid, and adding the appropriate alkali salt, or by dissolving molybdates in the acids.

If hot concentrated sulfuric acid is saturated with molybdic acid, and allowed to cool, *molybdenum dioxysulfate* (molybdenyl sulfate), MoO_2SO_4, crystallizes out in brilliant, colorless, deliquescent, six-sided prisms. Weinland obtained oxosulfatomolybdate(VI) salts of the types $M^I_2[Mo_2O_6(SO_4)]$ and $M^I_2[Mo_2O_4(SO_4)_3]$ by dissolving this in alkali sulfate solutions, or by the addition of much sulfuric acid to alkali molybdates.

The oxofluoromolybdate(VI) salts are simpler in composition. They are formed by dissolving molybdates in hydrofluoric acid, or by adding fluorides to hydrofluoric acid solutions of molybdic acid. The principal types known are:

$$M^I_2[MoO_3F_2], \quad M^I[MoO_2F_3], \quad M^I_2[MoO_2F_4] \quad \text{and} \quad M^I[MoOF_5].$$

Oxochloromolybdate(VI) salts, mostly of the type $M^I_2[MoO_2Cl_4]$, were prepared by Weinland.

(ii) Acido-salts of Pentavalent Molybdenum. The oxohalogeno salts of molybdenum(V) are also the simplest acido-salts. The fluoro salts have already been mentioned (p. 165). The *chloro salts*, of general formula $M^I_2[MoOCl_5]$, can be obtained by adding alkali chlorides to solutions of molybdenum pentachloride, or to solutions of molybdenum trioxide or molybdates in hydrochloric acid, reduced by means of hydriodic acid. They are bright green in color, crystallize well, and are stable in dry air. The corresponding *bromo salts*, $M^I_2[MoOBr_5]$, are dark red, or in some cases olive green. Bromo salts of the type $M^I[MoOBr_4]$ are also known; they are similar in color. The thiocyanato salts of 5-valent molybdenum, which are deep purple-red in solution, are similar in composition, but the oxalato salts at present known are more complex.

(iii) Acido-salts of Quadrivalent Molybdenum. Of the acido salts of quadrivalent molybdenum, the *octacyanomolybdates*, $M^I_4[Mo(CN)_8]$, are notable for their stability. They are strongly complexed, and are obtained when compounds of either trivalent or 5-valent molybdenum are treated with a large excess of concentrated potassium cyanide. The potassium salt, $K_4[Mo(CN)_8] \cdot 2H_2O$, which is freely soluble in water, forms yellow rhombic plates. The free acid may be obtained from its concentrated solution, by adding fuming hydrochloric acid. It crystallizes in fine yellow needles, with the composition $H_4[MO(CN)_8] \cdot 6H_2O$. The octacyanomolybdate(IV) salts are of interest because, in them, molybdenum undoubtedly exerts the coordination number 8, which is otherwise very seldom encountered in complex compounds. If potassium cyanide is used in smaller excess for the preparation, the violet-red *dioxotetracyanomolybdate(IV)* salts are formed—$M^I_4[MoO_2(CN)_4]$.

(*iv*) *Acido-salts of Trivalent Molybdenum.* In addition to the fluoro- and chloro salts already mentioned, the best known acido-compounds of trivalent molybdenum are the *thiocyanato-molybdate(III)* salts, with the general formula $M^I_3[Mo(SCN)_6]$. They are usually yellow or red in color, and are in some cases isomorphous with the thiocyanatochromate(III) salts. This is true, e.g., of the rhombic ammonium salt $(NH_4)_3[Mo(SCN)_6] \cdot 4H_2O$, which is best obtained by electrolytic reduction of ammonium molybdate at a bright platinum cathode, in the presence of much ammonium thiocyanate. The sodium and potassium salts are similar. The cyanomolybdate(III), $K_4[Mo(CN)_7] \cdot 2H_2O$, prepared by Young (1932), may perhaps be an example of a compound in which molybdenum has the very rarely found coordination number 7.

(*v*) *Acido-salts of Bivalent Molybdenum.* The yellow addition compounds of alkali chlorides with molybdenum dichloride, first prepared by Blomstrand, are probably to be regarded as acido-salts of bivalent molybdenum. They correspond in composition to $Mo_3Cl_6 \cdot 2M^ICl \cdot 2H_2O$, and the work of Brosset (1946) makes it probable that they have the true complexity $M^I_4[Mo_6Cl_{16}(H_2O)_4]$. The corresponding bromine and iodine compounds are also known. Blomstrand made the interesting observation that when Mo_6Cl_{12} or Mo_6Br_{12}, or their yellow solutions in caustic alkalis, are treated with acids, four of the twelve halogen atoms can be replaced by other acid radicals.

7. Analytical (Molybdenum)

The presence of molybdenum can generally be inferred during the course of the systematic separation of the cations, from the fact that the filtrate from the hydrogen sulfide precipitation is deep blue in color. Molybdenum is partially precipitated as its sulfide by the passage of hydrogen sulfide; it goes into solution again on treatment with ammonium sulfide.

When a molybdenum compound is heated in a tube open at both ends, a sublimate of molybdenum trioxide is obtained. This dissolves readily in ammonia, forming ammonium molybdate, and if the solution is evaporated to dryness and dissolved in hydrochloric or sulfuric acid, the molybdenum blue reaction is obtained on adding zinc or tin(II) chloride. A blue coloration is also obtained if a molybdenum compound is fumed down not quite to dryness with concentrated sulfuric acid, and allowed to stand some time in the air. An intensely deep blue incrustation is obtained when molybdenum compounds are reduced before the blowpipe. This substance is similar in nature to the molybdenum blues, and is considered by Glemser (1951) to be probably a definite crystalline hydroxide.

Molybdenum can be detected microanalytically by the intense reddish blue color given by the addition of potassium ethyl xanthate, $SC(OC_2H_5)SK$, to a molybdate solution. This reaction permits the detection of 0.04 γ of molybdenum in a dilution of 1 : 1,000,000. The reaction for molybdates with potassium thiocyanate and tin(II) chloride (red coloration, due to formation of $K_3[Mo(SCN)_6]$) is also very sensitive, and is conveniently carried out as a drop reaction.

Molybdenum can be determined quantitatively in molybdate solutions by precipitation as mercury(I) molybdate with mercury(I) nitrate, or by saturating with hydrogen sulfide and acidifying to precipitate the sulfide. In either case, the precipitate is ignited in air to form molybdenum trioxide, which is weighed. Precipitation of the compound $MoO_2(C_9H_6ON)_2$ with 8-hydroxyquinoline from molybdate solutions containing acetic acid is more convenient (Balanescu, 1930). Molybdenum is also conveniently precipitated with benzoin oxime.

8. Tungsten (W)

(a) Occurrence

Tungsten occurs in Nature chiefly in the form of tungstates—most usually as *wolframite* (an isomorphous mixture of $FeWO_4$, and $MnWO_4$), *scheelite*, $CaWO_4$, and *stolzite*, $PbWO_4$. It is also occasionally found as the free trioxide, *tungstite* (wolfram ocher), WO_3.

Pure ferrous tungstate, $FeWO_4$, occurs tetragonal as *reinite*, and monoclinic as *ferberite*; pure manganous tungstate always monoclinic as *hübnerite*. The isomorphous mixture of $FeWO_4$ and $MnWO_4$ is invariably monoclinic. Lead tungstate is usually found as the tetragonal stolzite (isomorphous with scheelite); a monoclinic modification is known as *raspite*.

The most important deposits of tungsten ores are in China, Korea, Indo-China, Burma and Siam, Malacca, Bolivia, Argentine, Australia and the United States. The principal European occurrences are in Portugal. World production of tungsten (as ore) in the years 1930, 1935 and 1936 amounted to 7950, 10800 and 11900 tons, the contributions of the producing zones in these years being: China 56.6%, 35.6% and 30.9%; Burma 16.2%, 20.2% and 21.4%, Malaya 7.4%, 9.1%, 8.2%, the United States 3.8%, 9.7%, 9.6%, and Europe (chiefly from Portugal and Spain) 5.9%, 6.2%, 6.7%.

(b) History

Tungsten, or its oxide WO_3, was discovered in 1781 by Scheele, in the mineral then known as 'tungstein', and now called scheelite. Soon afterwards, two Spanish chemists, the brothers d'Elhuyar, found that the same oxide was present in wolframite, but combined with iron and manganese oxides instead of with lime, as in 'tungstein'. They also succeeded in reducing the oxide to the metal. Because of its occurrence in 'tungstein' and in 'wolfram' (the term then used for wolframite*), the element was given both the names tungsten and wolfram. In German nomenclature, the element has retained the name of wolfram. English and French usage has adopted the name tungsten.

(c) Preparation

Tungsten ores, after a preliminary concentration by mechanical and electro-magnetic ore dressing processes, are attacked by fusion with soda in a reverberatory furnace [Eqn (1)]. The sodium tungstate so formed is leached out with water, and then decomposed by hot concentrated hydrochloric acid [Eqn (2)]. Tungstic acid thus precipitated is heated to convert it to tungsten trioxide (3), and this is reduced with carbon or (especially when tungsten of high purity is required) with hydrogen to the metal [Eqn (4)].

$$2FeWO_4 + 2Na_2CO_3 + \tfrac{1}{2}O_2 = 2Na_2WO_4 + Fe_2O_3 + 2CO_2 \tag{1}$$

$$Na_2WO_4 + 2HCl = H_2WO_4 + 2NaCl \tag{2}$$

$$H_2WO_4 = WO_3 + H_2O \tag{3}$$

$$WO_3 + 3H_2 = W + 3H_2O \tag{4}$$

The metal is obtained in the form of powder, because of its high melting point, and can be converted into the compact form by sintering and swaging (cf. p. 174). Fused tungsten can be made by reducing the oxide with carbon, or by reaction between the sulfide and

---

* The mineral was already known by this name to Agricola: lupi spuma, Wolfrahm, = wolf's foam. It was so called because the mineral, which often accompanies tin ores, caused difficulty in tin smelting by slagging or 'eating up' the tin.

calcium oxide in the electric arc furnace. Colloidal tungsten can be obtained by methods similar to those used for molybdenum (cf. p. 162).

(d) Properties

Tungsten powder is dull grey, the fused metal white and lustrous. Its density is 19.1 and hardness 7. The melting point of tungsten is extremely high—about 3400°. Its boiling point has been estimated as about 5700°.

In accordance with its high atomic weight, and the Dulong and Petit law, the specific heat (heat capacity) of tungsten is very small (0.0324 cal per degree at room temperature). The electrical conductivity at 0° is about 28.3% of that of silver at the same temperature. The resistance increases about 14 fold when a tungsten wire is heated to 2000°.

Tungsten is stable in air at ordinary temperature. It is oxidized to the trioxide, WO_3, on heating. Red hot tungsten is oxidized by steam to the dioxide. Nitrogen does not react perceptibly with tungsten, even at 1500°. Hydrogen is absorbed only to a very small degree. Of the halogens, fluorine reacts vigorously with tungsten powder even at ordinary temperature, whereas chlorine reacts only at a red heat.

Tungsten combines with nitrogen to form a 'non-Daltonide' compound, which is formed by heating tungsten powder in NH_3 gas. It corresponds in structure to the compound Mo_2N formed by molybdenum, but can be obtained with only up to 18.2 atom-% N, instead of 33.3 atom-%, as would be the case if all the positions in the nitrogen lattice were occupied.

Tungsten is very resistant towards acids; this is in part due to the ease with which the metal is passivated. Compact tungsten is only superficially attacked even by concentrated nitric acid and aqua regia. It dissolves slowly in a mixture of nitric acid and hydrofluoric acid. Soda-potassium nitrate melts and other oxidizing alkaline melts attack tungsten energetically, however.

(e) Uses [4, 5]

Tungsten is used extensively in the production of special steels, which require particular hardness, elasticity and tensile strength. Tungsten and chromium together alloyed with iron produce the so-called 'high speed steels', which retain their hardness and sharpness even when red hot. Tungsten is generally used for these purposes in the form ferrotungsten, a tungsten-iron alloy usually containing 81–83% tungsten, which is made from wolframite by reducing it with carbon in the electric furnace.

High speed steel usually contains about 19% W, 4% Cr, 0.5% V, and often some Co also. A double carbide Fe_3W_3C is present, with a crystal lattice built up from a network of Fe and W octahedra. Up to one third of the tungsten atoms may be replaced by Fe atoms, so that the carbide saturated with iron has the composition Fe_4W_2C. Cobalt and nickel form similar double carbides but their stability diminishes in the direction Fe → Ni (Westgren, 1933).

Tungsten has become of great importance in the electric lamp and electronic industries. Since it has been possible to produce durable filaments it has displaced all other metals as the material for electric lamp filaments, because of its high melting point and low volatility, combined with its small heat capacity.

Stringent requirements are laid upon the purity of tungsten to be used for filaments. It is therefore usually prepared by reducing the trioxide in hydrogen. It is thereby obtained as a powder, which is very brittle and not readily worked. The original method of making lamp filaments from it was by 'squirting', the finely divided metal being mixed into a plastic mass in a suitable binding medium, and extruded through a diamond die with a very fine aperture. The threads produced in this way had such poor mechanical properties, however, that they were of but little use. A method was subsequently devised whereby the powder tungsten could be converted into a ductile metal, from which wires could be made by drawing. This is achieved by first making bars, by pressing and sintering; these are initially very brittle. By prolonged alternate annealing and intensive hammering ('swaging'), these are converted into rods, which are already relatively ductile. If these are carefully drawn into wires, their ductility increases yet further, and by repeated drawing it is ultimately possible to get very flexible, strong filaments, with diameters down to 0.01 mm.

It is also possible to obtain flexible filaments—which are single crystal wires—from the 'squirted' wires. If a little thorium oxide* is mixed in with the mass from which the filament is to be extruded, and the initially brittle wire is passed through a short high temperature zone at about 2400–2600° (produced by means of a tungsten wire spiral in a hydrogen atmosphere), it becomes converted into single crystals which are often several meters long. To achieve this, the wire must clearly not be passed through the 'forming' zone faster than the rate of crystal growth. Single crystal wires have the same flexibility and bending strength as drawn tungsten wires, but they do not lose these properties when they are heated to high temperatures, whereas the drawn wires become rather brittle again as a result of re-crystallization.

Tungsten compounds find use to some extent as pigments—e.g., lead tungstate which possesses brilliant whiteness and excellent covering power as also does zinc tungstate. Tungstic acid is employed in ceramics to produce fiery yellow under-glaze colors. Sodium tungstate is used to impregnate textiles and combustible materials, to make them non-inflammable. The so-called 'tungsten bronzes' (see p. 178) are now but little used. *Tungsten carbide*, W_2C, is the principal constituent of the hard metal 'Widia alloy', which is used for the fabrication of metal cutting tools and drills, and in place of diamond for wire drawing dies for making electric lamp filaments.

9. Compounds of Tungsten

Like molybdenum, tungsten forms compounds from all valence states from 2 to 6. It has, however, a stronger tendency than molybdenum to function as 6-valent. This shows itself, among other evidence, in that it forms a hexachloride, as well as the hexafluoride, and—although it is not very stable—a hexabromide. The most important compounds of tungsten are the *tungstates*, the acid, *tungstic acid*, from which they are derived, and its anhydride, *tungsten trioxide*.

SUMMARY OF SOME BINARY COMPOUNDS

Chlorides	Sulfides	Oxides
WCl_6	WS_3	WO_3
WCl_5		$W_{10}O_{29}$
WCl_4	WS_2	W_4O_{11}
WCl_3 (known only in		WO_2
the form of double chlorides)		
WCl_2		

* It has not yet been fully explained why thorium oxide should promote the formation of single crystals of tungsten. It has been shown, however, that a part of the thorium oxide is reduced to metallic thorium on heating.

(a) Halides

(*i*) *Hexahalides*. Tungsten freshly reduced in hydrogen combines with chlorine at a red heat forming *tungsten hexachloride*, WCl_6. This exists as dark violet to blue black crystals (density 3.52, m.p. 275°, b.p. 347°). Tungsten hexachloride is practically insoluble in cold water, and is decomposed by warm water. It is soluble in alcohol, ether, and other organic solvents. It has a strong tendency to form oxychlorides—e.g., $WOCl_4$ (red, m.p. 211°, b.p. 228°) and WO_2Cl_2 (yellow, m.p. 265°, volatile when heated). Tungsten hexachloride which contains such oxychlorides is at once completely hydrolyzed by water to tungstic acid and hydrogen chloride.

The *hexabromide*, WBr_6, resembles the hexachloride, but is more readily decomposed. The gaseous *hexafluoride* (m.p. 2.5°, b.p. 19.5°) is also very reactive.

(*ii*) *Pentahalides*. *Tungsten pentachloride*, WCl_5, is formed in the thermal decomposition of the hexachloride. As vapor density measurements have shown, decomposition begins immediately above the boiling point (the bromide decomposes still more readily). The pentachloride is prepared by distilling the hexachloride repeatedly in hydrogen. The pentachloride forms brilliant black-green needles (m.p. 248°, b.p. 276°). The greenish yellow vapor has a density corresponding to the formula WCl_5. WCl_5 burns in oxygen, forming $WOCl_4$. It dissolves in water, with partial decomposition. The pentabromide (m.p. 276°, b.p. 333° with partial decomposition) behaves similarly.

(*iii*) *Tetrahalides*. When tungsten hexachloride is distilled in hydrogen, the tetrachloride and dichloride are formed as well as the pentachloride, especially if the temperature is high. Unlike the higher chlorides, these are not volatile. *Tungsten tetrachloride*, WCl_4, is a loose grey-brown hygroscopic crystalline mass, which is readily hydrolyzed by water. The black *tetraiodide*, WI_4, made by heating tungsten hexachloride with liquid hydrogen iodide in a sealed tube, is similar in properties.

(*iv*) *Dihalides*. *Tungsten dichloride*, WCl_2, is formed by reducing tungsten hexachloride with hydrogen (at not too high a temperature, as reduction otherwise proceeds to the metal), and by the thermal dissociation of the tetrachloride. It is a grey material which is not stable in air. It reacts with water, with the liberation of hydrogen, and has a strongly reducing action on other materials. The *dibromide*, WBr_2, is similar. The brown *diiodide*, WI_2, formed by passing iodine vapor over freshly reduced metallic tungsten, is insoluble in cold water. It is decomposed by hot water, undergoing hydrolysis and oxidation simultaneously.

(b) Sulfides

If hydrogen sulfide is passed into a tungstate solution, the oxygen of the tungstic acid undergoes first partial, and ultimately complete replacement by sulfur, and first *oxothiotungstates* and finally *thiotungstates* are formed:

$$M^I_2[WO_4] \qquad M^I_2[WO_3S] \qquad M^I_2[WO_2S_2] \qquad M^I_2[WOS_3] \qquad M^I_2[WS_4].$$

According to Brintzinger (1934), the solution of the latter contains the ions $[WS_4(H_2O)_2]^=$. The addition of acids to the solution precipitates chocolate brown *tungsten trisulfide*, WS_3.

Tungsten trisulfide has an extraordinary tendency to go into colloidal dispersion—e.g., when it is merely washed with water. It is converted to tungsten trioxide when it is heated in air. Heated in the absence of air, it loses sulfur and forms the *disulfide*, WS_2. The latter, which is also formed by melting tungsten trioxide with potassium carbonate and sulfur, is a soft, insoluble powder (density 7.5). Further loss of sulfur (leading to formation of metal) occurs in the absence of air only at very high temperatures. When it is heated in air, tungsten disulfide readily burns to the trioxide. It also reacts with fluorine, chlorine, and bromine, forming the corresponding hexahalides. Its crystal structure corresponds exactly to that of molybdenum disulfide (cf. Fig. 26, p. 166); $a = 3.18$, $c = 12.5$ Å, $W \leftrightarrow S = 2.48$ Å.

(c) Oxides

As with the case of molybdenum, the *trioxide* is the end-product of oxidation of tungsten or its compounds when heated in air. The *dioxide* is formed from this by treatment with reducing agents.

It has been shown by Glemser (1943), by X-ray diffraction studies, that there are two other tungsten oxides, in addition to WO_3 and WO_2. These, like WO_2, were obtained by reducing WO_3, with hydrogen, and also by the reaction of WO_3 with metallic tungsten. One of these oxides, the blue-violet $\overline{W_{10}O_{29}}$, has a homogeneity-range from $WO_{2.92}$ to $WO_{2.88}$ [for the structure see Magnéli, *Arkiv Kemi*, 1 (1950) 513]. The other, $\overline{W_4O_{11}}$ (extending over the homogeneity range $WO_{2.76}$ to $WO_{2.65}$) is red-violet. The dioxide $\overline{WO_2}$ is stated to have the composition range $WO_{2.03}$ to $WO_{2.00}$.

(*i*) *Tungsten trioxide*, WO_3, is a soft lemon yellow powder, which turns orange when heated. It becomes distinctly crystalline when it is strongly ignited, or if it is fused with borax. The brilliantly lustrous, wine-yellow crystals have a density 7.23, and belong to the rhombic (or possibly triclinic, pseudo-rhombic) system. Tungsten trioxide melts at $1473°$ and volatilizes above $1750°$. It is completely volatilized at about $500°$, however, in a current of hydrogen chloride gas, probably as an oxychloride. Tungsten trioxide is completely insoluble in water. It is dissolved by caustic alkalis, however, forming *tungstates*, the salts of *tungstic acid*, of which tungstic oxide is only in a restricted sense the anhydride. It is formed from the acid by ignition, but cannot be reconverted to the acid directly (e.g., by addition of water).

The crystal structure of WO_3 is very similar to that of ReO_3 (cf. p. 237). The O^{2-} ions are not exactly mid-way between the W^{6+} ions, however. This lowers the symmetry, so that the cubic lattice of ReO_3 is replaced by a rhombic (or possibly triclinic) lattice.

WO_3 can lose a certain fraction of its oxygen (down to the limit $WO_{2.95}$) without any significant change in crystal structure. However, the smallest diminution in oxygen content brings about a change in the color, and the oxide is distinctly blue at the composition $WO_{2.98}$.

(*ii*) *Tungsten dioxide*, WO_2, is formed by reduction of tungsten trioxide—e.g., by mild ingition in hydrogen. Tungsten dioxide is a brown powder, (density 12.1, m.p. about $1300°$). Prolonged heating in hydrogen at not too low a temperature reduces it to metallic tungsten. The crystal structure of WO_2 resembles that of MoO_2 (rutile type, $a = 4.86$, $c = 2.77$ Å).

(*iii*) *Intermediate Oxides*. The oxides already referred to above ($\overline{W_{10}O_{29}}$ and $\overline{W_4O_{11}}$) are formed as intermediates in the reduction of tungsten trioxide to metal.

The intermediate oxide $\overline{W_4O_{11}}$ is also formed in the reduction of WO_3 by other methods—e.g., electrolytic reduction in aqueous solution. Older statements, that the blue color observed in these processes was due to the formation of an oxide W_2O_5, which could also be obtained in the reduction of WO_3 with H_2, can now be regarded as superseded through the investigations of Van Liempt (1931), Ebert (1935), and Glemser (1943).

(d) Tungstic Acid and the Tungstates

Like molybdenum trioxide, tungstic oxide can unite in very varied proportions with basic oxides. The compounds are termed *tungstates*. The simplest (so-called *normal*) tungstates have the general composition $M^I_2O \cdot WO_3$, or $M^I_2[WO_4]$. The so-called *metatungstates* contain $4WO_3$ per M^I_2O; they also contain water which, from its behavior during dehydration, must be regarded as constitutionally bound. The '*paratungstates*', which generally contain $12WO_3$ for $5M^I_2O$, have more complex compositions, and also invariably contain water.

Confirmation that the paratungstates have the ratio $WO_3 : M^I_2O = 12 : 5$, and do not belong to the same structural type as the paramolybdates (q.v.), has been afforded by precise analytical work and X-ray methods (Saddington, 1949, Lindqvist, 1950). G. Jander (1927 and later) had concluded from measurements of diffusion coefficients in tungstate solutions that the paratungstates are acid hexatungstates, $M^I_5[HW_6O_{21}]$. Lindqvist has found, however, from X-ray structure determinations that in the crystalline paratungstates the anion contains 12 tungsten atoms, i.e., that the formula is to be regarded as $M_{10}[H_2W_{12}O_{42}]$. The (1 : 4) tungstates, which are grouped together without distinction as 'metatungstates', on the basis of their analytical composition, are probably in some cases derived from a 'hexatungstic' acid. Free, crystalline *metatungstic acid*, however, is not a hexatungstic acid, but a *dodecatungstic* acid quite distinct from that postulated by Lindqvist as the parent substance of the paratungstates. On the basis of its structure, it is not to be regarded as an *isopoly-acid*, but as a *heteropoly-acid* (cf. p. 182). The salts of this acid have the formula $M^I_6[H_2W_{12}O_{40}]$. Since all the (1 : 4) tungstates are hydrated, it is not possible to distinguish by any analytical data whether they are dodecatungstates of this type, or are of the type $M^I_3H_2[HW_6O_{21}]$ or $M^I_3[H_3W_6O_{21}]$, derived from paratungstic acid (whether from the hexatungstate form that appears to exist in solution, or the dodecatungstate form represented by crystalline sodium paratungstate). However, in the case of a series of crystalline (1 : 4) tungstates it has been proved that they are isomorphous with the salts of certain heteropoly-acids (p. 182). Those (1 : 4) tungstates of which this is true are undoubtedly salts of metatungstic acid, $M^I_6[H_2W_{12}O_{40}]$, but it is not impossible that there are also (1 : 4) tungstates which are salts of hexa-tungstic acid, e.g., $M^I_3H_3[W_6O_{21}]$ or $M^I_6H_6[W_{12}O_{42}]$.

In alkaline solutions, down to a pH value 8, *monotungstate* ions, $[WO_4]^=$ are present. Formation of hexa-tungstate ions takes place between pH 8 and pH 6:

$$6[WO_4]^= + 7H^+ \rightleftharpoons [HW_6O_{21}]^{5-} + 3H_2O.$$

Between pH 6 and pH 4, the solutions contain practically exclusively the hydrogen hexatungstate ion, $[HW_6O_{21}]^{5-}$. With further increase in hydrogen ion concentration, Jander considers that this may be transformed either into more acid ions—$H_3[W_6O_{21}]^{3-}$—by combination with hydrogen ion, a process favored in more dilute solutions, or by further condensation, into dodecatungstate ions:

$$2[HW_6O_{21}]^{5-} + 4H^+ \rightleftharpoons [H_2W_{12}O_{40}]^{6-} + 2H_2O.$$

The latter process is favored in concentrated solutions (Jander, 1925, 1940).*

The complex ion $[HW_6O_{21}]^{5-}$ can apparently exist in two forms, with differing water contents and different behavior. Only one of these is converted, with increase in hydrogen ion concentration, into the 'pseudo-metatungstate' ion which is formulated as $[H_3W_6O_{21}]^{3-}$ (which also contains water of constitution). This transformation takes place by way of a third hexatungstate ion, formally a dibasic acid:

$$[HW_6O_{21}]^{5-} \rightleftharpoons H_2[HW_6O_{21}]^{3-} \rightleftharpoons [H_3W_6O_{21}]^{3-},$$

as has been shown by conductometric titrations and absorption photometry (Souchay, 1943–44, G. Jander, 1951).

There is a marked contrast between the condensation processes of the tungstates, which lead directly to hexatungstate ions without any stable intermediate stages, and the successive formation of di-, tri- and tetrachromate ions in the condensation of the chromates. In principle, although not in detail, the condensation reactions in molybdate solutions are related to those of the tungstates.

* Our understanding of these condensation processes is by no means complete. Other workers have not found evidence for the dodecatungstate ions postulated in solution by Jander. See, for example Souchay, *Ann. Chim.*, 18 (1943) 61, 169.

(i) *Tungstic Acid.* The addition of strong acids to hot solutions of most tungstates (but not metatungstates) precipitates *yellow tungstic acid*, which is practically insoluble in water and in most acids.

If tungstate solutions are treated in the cold with a sufficient excess of acids (see below), a white voluminous precipitate is formed, the so-called *white tungstic acid*. Its composition and properties indicate that this is a gel of tungsten trioxide, with a continuously variable water content. Treated with acids (and especially if warmed), it can be converted into the ordinary yellow form. White tungstic acid is perceptibly soluble in water, with an acid reaction. It has a strong tendency to remain in collodial dispersion, especially if no excess of acid, or only a small excess, was employed in its precipitation. Its hydrosols are less stable, however, than those of molybdic acid. High concentrations of hydrogen ions bring about immediate flocculation.

If solutions of the so-called metatungstates are treated with acid, no precipitation takes place at all, since metatungstic acid is very soluble. It can be isolated by double decomposition of (e.g.,) barium metatungstate with sulfuric acid, or lead metatungstate with hydrogen sulfide, and is obtained on evaporation in the form of large rhombohedra. Its composition corresponds to the formula $4WO_3 \cdot 9H_2O$. Its constitution will be discussed in conjunction with that of tungstosilicic acid.

(ii) *Sodium tungstate,* Na_2WO_4, the technical preparation of which has already been discussed (p. 172), is the principal starting material for the preparation of other tungsten compounds and of pure tungsten metal. It is very soluble in water (73.2 g in 100 g of water at 21°), and crystallizes in thin, colorless, glistening leaflets of the rhombic system, with the composition $Na_2WO_4 \cdot 2H_2O^*$. The salt loses water when it is heated in air, or on standing over sulfuric acid. It is decomposed by boiling with strong acid, and yellow tungstic acid is precipitated.

In addition to the normal sodium tungstate, various sodium polytungstates exist. Of these, *sodium metatungstate*, with the analytical composition $Na_2O \cdot 4WO_3 \cdot 10H_2O$, may be noted. Its solution gives no precipitate with acid, since metatungstic acid is soluble.

(iii) *Calcium tungstate,* $CaWO_4$, occurs native as scheelite, often in association with cassiterite. It crystallizes tetragonal (yellow pyramids, density 6.0). Its crystal structure is represented in Fig. 27. Stolzite, $PbWO_4$, and wulfenite, $PbMoO_4$, have the same structure.

Ferberite, $FeWO_4$, hübnerite, $MnWO_4$, wolframite, $(Fe, Mn)WO_4$, and a series of other tungstates of bivalent metals that crystallize monoclinic ($CoWO_4$, $NiWO_4$, $ZnWO_4$, $MgWO_4$) have a crystal structure of characteristic type (wolframite structure) hitherto found only among the tungstates.

(iv) *Tungsten Bronzes.* The name 'tungsten bronze' is applied to a group of variously colored (blue, purple red, golden yellow), excellently crystallized, chemically extremely resistant substances, which have a bright metallic luster and also metallic conductivity. They are formed by treating alkali or alkaline earth tungstates with reducing agents at high temperatures. Analytically, the tungsten bronzes contain equivalent amounts of alkali or alkaline earth oxide, and W_2O_5, with a variable, more or less great excess of WO_3. Analogous substances are formed by molybdenum (molybdenum bronzes). The nature of the molybdenum and tungsten bronzes has been interpreted as a result of X-ray investigations (de Jong, 1932, Hägg, 1935). Over the range in which they are cubic, the sodium tungsten bronzes, for example, have crystal lattices of perovskite type in which not all the places available as sites for Na^+ ions need be occupied. The composition of these bronzes varies, according to the number of vacant sites, between the limits $Na_{0.3}WO_3$ and $NaWO_3$. At the composition $NaWO_3$, all the lattice points are occupied; the tungsten is then present exclusively in the +5 state. $NaWO_3$ is golden yellow in color; with a decrease in the sodium content, the color changes through red and violet to deep blue violet. This deepening of color is associated with the fact that as the number of Na^+ ions decreases, a corresponding

* The decahydrate crystallizes below 6°.

number of W^{5+} ions are oxidized to W^{6+} ions. There are also sodium tungsten bronzes with a sodium content even smaller than that of $Na_{0.3}WO_3$. These are blue, and crystallize tetragonal, since the perovskite structure is no longer stable with further increase in the number of vacant sites. The cubic sodium tungsten bronzes are electronic semi-conductors, with high conductivities ($\varkappa = 1$ to 10 ohm^{-1} at ordinary temperature).

(v) *Tungsten Blue.* Intense blue products, which are usually not uniform in composition, are obtained by the mild reduction of tungsten trioxide or tungstic acid in solution, or in contact with water. Depending upon the method of preparation, several different compounds may be present in these substances, which are referred to by the general name of 'tungsten blues'. Properties common to all the substances are their blue color, and the fact that they have compositions which correspond to the presence of tungsten in a mean oxidation state lower than 6, but higher than 5. Thus tungsten blue resembles molybdenum blue in many respects. Tungsten trioxide takes on a bluish tint merely by exposure to sunlight, under water. The formation of tungsten blue, by the addition of zinc or tin(II) chloride to tungstate solutions in the presence of hydrochloric acid, is a very sensitive reaction for tungstates.

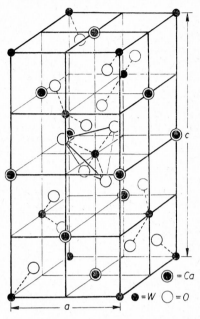

Fig. 27. Scheelite, $CaWO_4$ ($a = 5.24$, $c = 11.38$ Å). $BaWO_4$, $PbWO_4$, $CaMoO_4$, $PbMoO_4$ are isostructural with scheelite, as also are $NaIO_4$, KIO_4, NH_4IO_4, $AgIO_4$, $KReO_4$ and $KOsO_3N$.

The blue precipitate thrown down from tungstate solutions by the action of 'nascent hydrogen' (Zn + HCl) is stable in air. The precipitate formed by means of $SnCl_2$ and HCl is gradually decolorized on standing in air, being transformed into yellow tungstic acid, H_2WO_4. This different behavior in itself indicates that there may be different kinds of tungsten blue. According to Glemser (1943–51), the products in the first case are hydrogen analogues of the tungsten bronzes—i.e., compounds H_xWO_3 (with x = 0.1 to 0.5), whereas in the other case the compound which separates out is closely related in structure to tungstic acid, H_2WO_4; it differs from the latter in having a rather lower oxygen content and a variable water content. The crystalline compound $H_{0.1}WO_3$ can also be obtained by the action of atomic hydrogen or of $LiAlH_4$ on WO_3 (Glemser, 1952). As in the case of molybdenum blue, tungsten blue is frequently obtained in the X-ray amorphous state, rather than crystalline. It also has a strong tendency to go into colloidal dispersion.

(vi) *Peroxytungstic acid and Peroxytungstates.* Tungstic acid and tungsten trioxide both dissolve in hydrogen peroxide solutions at 100°, forming a compound which is regarded as *peroxytungstic acid*, H_2WO_5*. *Peroxytungstates* are obtained by treating tungstates with hydrogen peroxide.

Yellow *tetraperoxomonotungstates*, $M^I_2[W(O_2)_4]$**, crystallise from weakly alkaline solutions of alkali tungstates, containing much hydrogen peroxide. From weakly acid solutions rich

* It has not been proved that this is really a peroxy-acid, and not merely a tungsten oxide-peroxyhydrate, $WO_3 \cdot H_2O_2$.
** (O_2) represents the peroxy group. The compounds crystallize as hydrates.

in hydrogen peroxide, or weakly alkaline solutions with little hydrogen peroxide, colorless *alkali tetraperoxoditungstates*, $M^I_2[W_2O_3(O_2)_4]$, crystallize. The formation of these two types of compounds, depending on the concentrations of H^+ ions and H_2O_2, can be explained in terms of the following equilibria in solution, as shown by Jahr (1938, 1952):

$$2[WO_4]^= + 4H_2O_2 \rightleftharpoons [W_2O_3(O_2)_4]^= + 2OH^- + 3H_2O$$

$$[W_2O_3(O_2)_4]^= + 2OH^- + 4H_2O_2 = 2[W(O_2)_4]^= + 5H_2O$$

$$[HW_6O_{21}]^{5-} + 12H_2O_2 \rightleftharpoons 3[W_2O_3(O_2)_4]^= + H^+ + 12H_2O.$$

Whereas a sufficient excess of H_2O_2 converts the hexatungstate ions $[HW_6O_{21}]^{5-}$, present in acid solution, to tetraperoxoditungstate ions, $[W_2O_3(O_2)_4]^=$, yellow *peroxopolytungstates* are formed in acid solutions if smaller amounts of H_2O_2 are present—i.e., peroxy-compounds derived from polytungstic acids. The composition and constitution of these compounds are as yet not fully explained.

(e) Tungstosilicic acid. Heteropolyacids

As was first observed by Marignac (1862), silicic acid is soluble in hot solutions of acid alkali tungstates. Very well crystallized salts are so formed, which display the peculiar feature that they contain a large number (up to 12) WO_3 for each SiO_2, in addition to alkali oxide and water. These salts have hitherto generally been called *silicotungstates*; the name *polytungstosilicates* is more strictly correct, since they are derived from silicic acid or its salts by the addition of tungstic acid radicals (tungstatogroups). The potassium salt of this group of compounds with the highest tungstic acid content has the analytical composition $SiO_2 \cdot 12WO_3 \cdot 8KOH \cdot 10H_2O$, or $H_4SiO_4 \cdot 12WO_3 \cdot 8KOH \cdot 8H_2O$. If the tungstosilicates are treated with acids, the alkali alone is lost, and the ratio $SiO_2 : WO_3$ remains unaltered. The tungstosilicates are immediately decomposed by alkalis.

It has been found that tungstic acid forms compounds exactly analogous to the silicotungstates with other acids than silicic acid—especially with boric acid (*tungstoborates*), phosphoric acid (*tungstophosphates*) and arsenic acid (*tungstoarsenates*).

Examples:

$H_3BO_3 \cdot 12WO_3 \cdot 5KOH \cdot 14H_2O$ potassium (dodeca)tungstoborate

$H_3PO_4 \cdot 12WO_3 \cdot 3KOH \cdot 2\frac{1}{2}H_2O$ potassium (dodeca)tungstophosphate

$H_3AsO_4 \cdot 9WO_3 \cdot 3KOH \cdot 4H_2O$ potassium (ennea)tungstoarsenate.

As shown by the last example, the number of WO_3 radicals associated with each molecule of the other acid may be smaller than 12. The ratio 12 : 1 is an upper limit, which is never exceeded.

The complex acids from which such salts as those shown above are derived are grouped together under the name of *heteropolyacids*. The characteristic property of these is that they contain at least two different acid radicals, and as a rule contain one of the radicals in large numbers.

It was formerly assumed that the numerous acid radicals in the heteropolyacids were linked together in some form of chains. The great stability, of the high-molecular species in particular, was not easily reconciled with such chain formulas, which furthermore offered no explanation for the existence of an upper limit to the molecular ratio of the two sorts of acid radical. A more satisfactory theory of the heteropolyacids was developed by Miolati (1908), on the basis of Werner's coordination theory. Miolati's ideas were extended and substantiated experimentally by Rosenheim. The application of X-ray structure determinations to this class of compounds (Keggin, 1933; Signer, 1934; Kraus, 1936) has finally clarified the matter. It has confirmed the general principles underlying the theory of Miolati and Rosenheim, even although it has been necessary to make considerable modifications in the ideas as originally developed.

Miolati regarded the tungstosilicates of the limiting series—i.e., that in which the ratio $WO_3 : SiO_2 = 12 : 1$—as the derivatives of a particular form of silicic acid in which the silicon atom was surrounded by six oxygen atoms (in conformity with the importance assigned to the coordination number 6 in Werner's theory). If it may be supposed that each of the six oxygen atoms is replaced by a W_2O_7 residue, then the hypothetical silicic acid $H_8[SiO_6]$ (hexaoxosilicic acid) leads at once to the formulation of the tungstosilicic acid of the limiting series as $H_8[Si(W_2O_7)_6]$.

Thus tungstosilicic acid would be regarded as a hexaoxosilicic acid in which all the oxygen atoms were replaced by W_2O_7 groups. The tungstoboric acid can similarly be derived from a boric acid $H_9[BO_6]$, and tungstophosphoric acid from the acid $H_7[PO_6]$, so that they would be formulated as $H_9[B(W_2O_7)_6]$ and $H_7[P(W_2O_7)_6]$. Analogous formulas follow for the heteropolyacids containing molybdic acid radicals. The Rosenheim-Miolati theory naturally allows also for the existence of acids which can be regarded as arising from the replacement of only a part of the oxygen atoms by W_2O_7 or other radicals. Such acids would be termed unsaturated acids of the limiting series.

The large number of ionizable (replaceable) hydrogen atoms which must be assigned to these acids in terms of the Miolati theory is striking. The discovery of salts with compositions corresponding to the number of replaceable hydrogen atoms was later regarded as confirmation of the theory. Indeed, the potassium tungstosilicate referred to at the outset contains 8 potassium atoms per atom of Si, as required by the theory. Rosenheim found that tungstophosphoric acid (and molybdophosphoric acid), which should be 7-basic according to the theory, did in fact form a compound with 7 molecules of guanidine, CN_3H_5,— $[CN_3H_5 \cdot H]_7[P(W_2O_7)_6]$. The mercury(I) salt of tungstoboric acid, corresponding to the theoretical 'basicity', could also be prepared, and had the composition $Hg_9[B(W_2O_7)_6] \cdot 12H_2O$ (Copaux, 1909).

The evidence of X-ray structure analysis leads to the following model for the structure of heteropolyacids of the type of dodecatungstosilicic acid. 4 oxygen atoms are arranged tetrahedrally around the Si atom, which is the central atom of the complex, and each of these oxygens forms the common vertex of three WO_6 octahedra (consisting of 6 O atoms arranged in an octahedron around a W atom) (see Fig. 28). Each of these three octahedra also shares another O atom with each of the two neighboring octahedra of the same group, in such a way that each pair of octahedra shares an edge with each of its neighbors. In addition, each octahedron shares yet another oxygen atom with one octahedron of an adjacent group attached to the same Si atom, while the sixth O atom of each octahedron is unshared. Thus, $3 \times (\frac{1}{3} + \frac{2}{2} + \frac{2}{2} + 1) = 10$ oxygen atoms in all can be assigned to each group of three WO_6 octahedra, so that each of

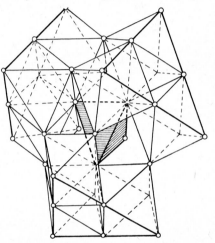

Fig. 28. Arrangement of WO_6 octahedra around the central atom in heteropolyacids of the type of 12-tungstophosphoric acid. The fourth W_3O_{10} group, which would lie behind the groups shown, has been omitted for the sake of clarity.

the four groups surrounding the central Si atom has the composition W_3O_{10}. The structure of the complex ion of dodecatungstosilicic acid can thus be represented by the formula $[Si(W_3O_{10})_4]^{4-}$. The acid can be regarded structurally as a tetra(tritungstato)-silicic acid, $H_4[Si(W_3O_{10})_4]$.

Analogous in structure are

dodecatungstoboric acid	$H_5[B(W_3O_{10})_4]$
dodecatungstogermanic acid	$H_4[Ge(W_3O_{10})_4]$
dodecatungstophosphoric acid	$H_3[P(W_3O_{10})_4]$
dodecatungstoarsenic acid	$H_3[As(W_3O_{10})_4]$

and the corresponding molybdo-acids. The same complex structural groups have also been found in salts of these acids, where their structures have been determined. Thus ammonium molybdophosphate, which is important for the determination of phosphoric acid, is the normal ammonium salt of dodecamolybdophosphoric acid,

$$(NH_4)_3[P(Mo_3O_{10})_4] \cdot 6H_2O.$$

The heteropolyacids (and, as a rule, their salts also) always crystallize as hydrates, with a variable content of water, according to the conditions of preparation. The H_2O molecules are arranged in a different way in the crystal lattices of the various hydrates; thus the 5-hydrates (for example) crystallize cubic, the 14-hydrates triclinic, the 24-hydrates trigonal. The complex anionic groups persist throughout with unaltered structure. The differing crystal structures of the 29-hydrates (cubic) and the 30-hydrates (tetragonal) is noteworthy. Such a relatively small difference in water content is thus by no means unimportant for the crystal structure*, even though *exchange of the central atom* in the anionic complex has *no effect* on the crystal structure, as long as the water content is unaltered. Hence all the corresponding hydrates of these polyacids are isomorphous with one another.

The essential difference between the structural formulas based on the X-ray crystallographic measurements and those proposed by Miolati lies in the recognition that the central atom of the anionic complex has a different coordination number from that assumed in the older theory. The polyacids under consideration are derived from *tetraoxo-acids* (coordination number = 4), and not from *hexaoxo-acids* (coordination number = 6). They therefore contain, not six ditungstato groups, but *four* tritungstato-groups (or trimolybdato-groups)**. Since these heteropolyacids are derived from tetraoxo-acids—i.e., from acids of lower 'basicity' than formerly assumed, and should themselves also display this smaller 'basicity', a certain difficulty still arises, in accounting for compounds, such as those cited on p. 181, which contain more equivalents of basic metal than there are H-atoms in the free acid. X-ray investigations (Kraus, 1939) seem to indicate that these compounds are a sort of basic salt, which can be considered to arise from the hydrates of the normal salts by exchange of HOH molecules for metal hydroxide. Which hydroxides can be introduced in this manner, and to what extent, would depend, on this hypothesis, essentially on spatial considerations and on the particular properties of the ions to be incorporated. This probably explains why such compounds, which simulate in composition a higher 'basicity' of the acids, can only be obtained in exceptional cases, and are invariably poorly crystallized.

The so-called *metatungstic acid* must also be included, on structural grounds, in the class of heteropolyacids. The isomorphism which exists between metatungstic acid and its salts, and the corresponding compounds of typical heteropolyacids, had already led Copaux to the conclusion that they must be structurally similar. Thus the following structural formulas are obtained for the metatungstates and metatungstic acid:

$K_6[H_2(W_3O_{10})_4] \cdot 18H_2O$ potassium metatungstate	isomorphous with	$K_4[Si(W_3O_{10})_4] \cdot 18H_2O$ potassium tungstosilicate
	(hexagonal)	
$Ba_3[H_2(W_3O_{10})_4] \cdot 27H_2O$ barium metatungstate	isomorphous with	$Ba_{2.5}[B(W_3O_{10})_4] \cdot 27H_2O$ barium tungstoborate
	(tetragonal)	
$H_6[H_2(W_3O_{10})_4] \cdot 24H_2O$ metatungstic acid	isomorphous with	$H_3[P(W_3O_{10})_4] \cdot 24H_2O$ tungstophosphoric acid
	(trigonal)	

X-ray investigations have confirmed that metatungstic acid exactly corresponds in structure to the tetra(tritungsto) silicic, boric and phosphoric acids. The two hydrogen atoms written

* The contradictory data often found in the literature for the crystallographic properties of the hydrates of the heteropoly acids and their salts are to be explained by insufficient attention to this effect.

** Even before X-ray structure determinations had been carried out, G. Jander rejected the hypothesis that dimolybdato- or ditungstato-groups were involved in the structure of heteropolyacid complexes, on the grounds that it has never been possible to detect them in solution.

within the bracket thus play, in metatungstic acid, the same role as the silicon atom in tungstosilicic acid, or the boron atom in tungstoboric acid, etc. Hence these two hydrogen atoms can be formally regarded as the central atoms of the complex in metatungstic acid, which can be considered as a *tetra(tritungsto)dihydrogen acid*. In the same way, metatungstates are to be regarded as heteropoly salts, in so far as they are isomorphous with tungstosilicates or molybdosilicates. Thus the sodium metatungstate referred to on p. 178 is

$$Na_6[H_2(W_3O_{10})_4] \cdot 29H_2O.$$

Those heteropolyacids which contain less than $12WO_3$ (or MoO_3)* per molecule of the non-metal acid (so called 'unsaturated' heteropolyacids) are probably substantially similar in structure to the 'saturated' heteropolyacids already considered, but more precise knowledge of their structure is lacking. In the case of the heteropolysalts formed by combination of WO_3 with tellurates, $M^I_6[TeO_6]$, and periodates, $M^I_5[IO_6]$, the generally assumed formulation as hexa(monotungsto)tellurates or periodates, $M^I_6[Te(WO_4)_6]$ or $M^I_5[I(WO_4)_6]$, is instructive: it derives the heteropolysalts from the simple salts by replacement of O-atoms by WO_4-groups. However, Jander has shown that salts of this type cannot be obtained from solutions which contain only $WO_4^=$ ions. They are obtained, rather, from acid solutions, in which tungstate is present almost entirely as $[HW_6O_{21}]^{5-}$ ions. This observation would accord better with a formulation as di(tritungsto) salts, $M^I_6[TeO_4(W_3O_{10})_2]$, etc. More recent considerations, and X-ray study of the corresponding molybdotellurate, have shown that these complex anions are cage structures, based on the sharing of corners between WO_6 (and MoO_6) octahedra, in such a way that the central $[TeO_6]$- or $[IO_6]$-group shares each of its oxygen atoms with two $[WO_6]$ octahedra, which also share edges with their neighbors so as to form a flat ring of $[WO_6]$ octahedra (Anderson, 1937, Evans, 1948). Salts of the types $M^I_2TeO_4 \cdot M^I_2WO_4$, or $M^IIO_4 \cdot M^I_2WO_4$, crystallize from alkaline solutions. These can probably be regarded as ordinary double salts. It is by no means impossible that salts of higher-molecular condensed acids (isopolyacids) may also be able to combine with salts of simple acids, forming compounds with the structure of double salts.

(f) Isopolyacids

If, in any oxyacid, the O-atoms are replaced by radicals of the same acid (e.g., introduction of one or more CrO_4 radicals in H_2CrO_4), the resulting compounds are termed *isopolyacids*, as distinct from the heteropolyacids formed by the introduction of radicals of a different acid. The isopolyacids are generally referred to merely as 'polyacids', except when it is wished to emphasise the distinction from the heteropolyacids. They can also be defined as compounds derived from a simple acid by the elimination of water between two or more molecules of the acid

E.g., $\qquad 2H_2CrO_4 - H_2O = H_2Cr_2O_7, \qquad 3H_2CrO_4 - 2H_2O = H_2Cr_3O_{10}$, etc.

The ions of the isopolyacids are similarly derived from those of the simple acids by elimination of O^{2-} ions—e.g.,

$$2[CrO_4]^{2-} - O^{2-} = [Cr_2O_7]^{2-}, \text{ etc.}$$

The O^{2-} ion eliminated from the one acid ion is replaced by the other acid radical (or rather by one of its O^{2-} ions).

$$\begin{bmatrix} & O & \\ O & Cr & O \\ & O & \end{bmatrix}^{2-} + \begin{bmatrix} & O & \\ O & Cr & O \\ & O & \end{bmatrix}^{2-} = \begin{bmatrix} & O & & O & \\ O & Cr & O & Cr & O \\ & O & & O & \end{bmatrix}^{2-} + O^{2-}.$$

Reactions of this kind, which lead to the linking of radicals, with the elimination of H_2O or of O^{2-} ions (also of NH_3, H_2S, etc.) are generally referred to as *condensations*.

The ions of the oxyacids can be divided into five groups, on the basis of the ease with which they can undergo condensation:

* V_2O_5 can also combine with non-metal acids, to form heteropoly acids.

[i] Ions which cannot condense with one another at all.
Examples — $[CO_3]^{2-}$, $[NO_3]^-$, $[ClO_4]^-$, $[MnO_4]^-$.

[ii] Ions which can undergo condensation only to a limited extent, but in all degrees up to that limit.
Examples — $[CrO_4]^{2-}$ and $[SO_4]^{2-}$. These form only anionic radicals of finite size by their condensation — $[Cr_2O_7]^{2-}$, $[Cr_3O_{10}]^{2-}$, $[Cr_4O_{13}]^{2-}$, $[S_2O_7]^{2-}$, $[S_3O_{10}]^{2-}$. The stability of the radicals derived from any one acid decreases with increase in the degree of condensation. In most of the ions of this type, the capacity for condensation does not extend beyond the first stage (formation of pyro-acid anions).

[iii] Ions which can condense together in both finite and infinite numbers.
Example — $[SiO_4]^{4-}$. The condensation of a finite number of $[SiO_4]^{4-}$ ions gives rise to anionic radicals such as $[Si_2O_7]^{6-}$, $[Si_3O_9]^{6-}$, $[Si_4O_{13}]^{10-}$ (?), $[Si_6O_{18}]^{12-}$*. The condensation of an indefinite number of ions produces anionic chains or bands, sheets or networks, with unlimited extension in one, two, or three dimensions.

[iv] Ions which can condense in indefinite (essentially infinite) numbers.
Example — $[SbO_8]^{11-}$. The anionic network present in the anhydrous antimonates can be regarded structurally as the condensation product of an indefinite ('infinitely great') number of $[SbO_8]^{11-}$ ions (cf. Vol. I, p. 668). This anion does not exist either in the uncondensed form, or in the form of anionic radicals fromed by the condensation of a finite number (2, 3, 4, etc.) of $[SbO_8]^{11-}$ groups.

[v] Ions which can condense in finite numbers, but only in certain definite proportions below the maximum degree of condensation.
Examples. $[VO_4]^{3-}$, $[MoO_4]^{2-}$, $[WO_4]^{2-}$ (cf. pp. 100, 168, 177) and $[GeO_3]^{2-}$ (cf. Vol. I, p. 521).

The difference in capacity for condensation between the ions of group [i] and those of group [ii] is only gradual. With ions of group [ii], this capacity generally falls off with increasing charge and decreasing radius of the central atom**. The mutual effect of both influences can reduce the tendency for condensation to zero (group [i]). The relation of group [iii] to group [ii] is the converse of that between group [i] and group [ii], the mutual effect of charge and radius leading to an optimum tendency for condensation. Moreover, it may be assumed that stereochemical factors also enter into the particularly strongly developed tendency of the SiO_4^{4-} ion to condense. A further result of steric influences may also be that an ion such as the Sb^{5+} ion seeks to surround itself by more oxygen ions than can be bound by its positive charge. The groups which would be so built up—in the present instance the $[SbO_8]$ group—can only be formed if they can share as many O^{2-} ions as possible with neighboring groups—i.e., if condensation is carried to the highest possible degree***. This is achieved if an indefinite number of $[SbO_8]^{11-}$ groups are linked together to form a single anionic network, such as has been found in the crystal structures of the antimonates. The ions of group [iv] exist *only* in the form of networks, in crystal lattices, derived in this manner. They stand in direct formal contrast with the ions of group [i], which exist only in the uncondensed form. In fact, however, the forces which inhibit condensation in the one case, and which produce it to its fullest extent in the other, are the same.

The ions of group [v] occupy a special place, as compared with those of groups [i]–[iv]. It is peculiar to these ions that the only detectable condensation products involve discontinuous jumps in the number of acid radicals. Thus, in the case of the tungstates, condensation

* It is not known whether these radicals, which exist as structural groups in crystals, can exist also in solution.
** Where the polarizing power of the central atom is strong (elements of the Sub-groups), the effect of its charge upon the capacity for condensation may be compensated or overcompensated.
*** The group $[SbO_6]$, which is also possible stereochemically, becomes stable only if the mutual repulsion of the O^{2-} ions is reduced—e.g., conversion of O^{2-} ions to OH^- ions in aqueous solution, formation of $[Sb(OH)_6]^-$.

of the monotungstate leads directly* to a hexatungstate, and from this directly to a dodeca-tungstate anion. None of the structures found for the condensation products of groups [i] to [iv] can satisfactorily explain this peculiarity. The fact that just those acids belong to group [v] which are capable of forming heteropolyacid complexes suggests that the isopolyacid complexes of this group must be structurally related to the heteropoly-acid complexes. Indeed, the dodecatungstate complex has already been considered as a heteropolyacid complex, on structural grounds (cf. p. 182). Classification of dode-catungstic acid among the heteropolyacids, instead of among the isopolyacids would no longer seem arbitrary and puzzling if it proved that the condensed acid ions of group [v] were based, quite generally, upon structures related to those of the heteropolyacids. Until recently, no X-ray evidence was available to bear upon this question, but the recent work of Lindqvist (1950) has shown that this is the case. Paratungstic acid involves a second dodecatungstate complex, formed by condensation of two hexatungstate ions, whereas the paramolybdate and octamolybdate ions are both cage structures, formed by condensation of $[MoO_6]$ octahedra, although they do not involve any 'hetero'-atoms similar to the central hydrogen atoms in the metatungstates.

The ions $[NbO_6]^{7-}$ and $[TaO_6]^{7-}$ may also be assigned to group [v]. These stand in the same relation to the group [v] ions already mentioned as do the ions of group [iv] to those of the preceding groups. They do not exist in the uncondensed state, but are found either condensed in infinite numbers, in anions stretching throughout the crystal, such as are present in the anhydrous niobates mentioned on p. 107, or else condensed in finite, but not small numbers, giving the $[Nb_5O_{16}]^{7-}$ and $[Ta_5O_{16}]^{7-}$ or $[Nb_6O_{19}]^{8-}$ and $[Ta_6O_{19}]^{8-}$ ions (cf. p. 113), although lower condensation steps do not exist.

Anionic networks formed by condensation of an indefinite number of ions, extending throughout a crystal, are not usually included among the polyacid anions. However, oxyacid salts containing such anionic networks are often thought of as salts of polyacids, on the grounds of their analytical composition**. By way of differentiation from these, the acids or salts with *finite* anions, derived from oxyacids by condensation, can be designated isopolyacids or isopolysalts in the stricter sense. The class of isopolyacids and their salts may then be divided into two groups, typified by the polychromates, on the one hand, and the polyvanadates, polymolybdates, and polytungstates on the other.

(g) Compounds of Tungsten with Acid Anions

Apart from the simple halides (including the oxyhalides) and sulfides, compounds of tungsten with acid anions are known only in the form of complex salts (acido-salts). The number of such compounds, other than halogeno- and oxohalogeno-compounds, is only small.

(i) *Acido-salts of Hexavalent Tungsten.* A number of *oxohalogeno-salts* of +6 tungsten are known. Oxofluorotungstate(VI) salts—usually colorless and well crystallized and mostly of the type $M^I_2[WO_2F_4]$—can be obtained by dissolving tungstates in aqueous hydrofluoric acid. The rhombic copper salt, $Cu[WO_2F_4] \cdot 4H_2O$, is isomorphous with $Cu[TiF_6] \cdot 4H_2O$. The salt $(NH_4)_3[WO_3F_3]$ is isomorphous with $(NH_4)_3ZrF_7$ and with $(NH_4)_3NbOF_6$. Double salts of the oxychloride WO_2Cl_2 are also known.

(ii) *Acido-salts of Pentavalent Tungsten.* By dissolving alkali tungstates in a concentrated, acid alkali oxalate solution, and reducing with tin foil, Olsson obtained fine red oxalato salts of +5 tungsten, with the general formula $M^I_3[WO_2(C_2O_4)_2]$. These dissolve with a deep blue color in concentrated hydrochloric acid, and by saturating the solution with hydrogen chloride deep-colored oxochlorotungstate(V) salts can be isolated—$M^I_2[WOCl_5]$ (green), $M^I[WOCl_4(H_2O)]$ (bright blue) and $M^I[WOCl_4]$ (brown-yellow). The corresponding bromo salts have also been prepared. Complex thiocyanato salts of pentapositive tungsten are also known.

(iii) *Acido-salts of Quadrivalent and Trivalent Tungsten.* The most important acido-compounds of +4 tungsten are the *octacyanotungstate(IV)* salts, $M^I_4[W(CN)_8]$, the analogues of

---

* It is quite possible, of course, that lower unstable intermediate stages take part in the condensation process.

** E.g., compounds containing several equivalents of SiO_2 per 1 M_2O are generally called 'polysilicates', irrespective of whether their structures involve finite anionic radicals or chains, networks, etc., of infinite extent.

the octacyanomolybdenum(IV) compounds discussed earlier. They can be prepared by various methods, and form yellow crystals which are mostly quite stable in air, and their aqueous solutions are neutral towards litmus. The free acid is also known, and exists as yellow deliquescent needle crystals with the composition $H_4[W(CN)_8] \cdot 6H_2O$. Potassium permanganate in sulfuric acid solution oxidizes these compounds to the *octacyanotungstate(V)* salts, $M^I_3[W(CN)_8]$ which are also yellow, whereas in other compounds this oxidant in variably converts the tungsten to the $+6$ stage*. The free acid $H_3[W(CN)_8] \cdot 6H_2O$ forms orange yellow needles, which are not stable in air.

A solution of potassium tungstate in concentrated hydrochloric acid, obtained by a particular method given by Ollson or by Rosenheim, yields potassium hydroxopenta-chlorotungstate(IV), $K_2[W(OH)Cl_5]$, when it is reduced with tin; this compound is a dark green powder which is red by transmitted light or in solution. If the reduction is carried further, it eventually produces chloro salts of tripositive tungsten, of the type $M_3[W_2Cl_9]$— e.g., $K_3[W_2Cl_9]$, yellow green hexagonal platelets.

10. Analytical (Tungsten)

Yellow tungstic acid, H_2WO_4, is precipitated by boiling tungstate solutions with strong acids. This precipitation does not occur from metatungstate solutions, or only after prolonged boiling (whereby the metatungstic acid is ultimately decomposed). Hydrogen sulfide also gives no precipitate. However, if alkali tungstate solutions are saturated with hydrogen sulfide, and then acidified, light brown tungsten trisulfide, WS_3, is precipitated; it is insoluble in acids and soluble in ammonium sulfide.

If zinc is added to a tungstate solution containing hydrochloric acid, the slimy white precipitate of tungstic acid first formed turns a fine blue. A blue coloration or blue precipitate, depending upon the concentration, is obtained by heating with tin(II) chloride and hydrochloric acid (tungsten blue, see p. 179). Tungsten blue is also formed when a tungsten compound or metallic tungsten is fumed down with concentrated sulfuric acid. It should be noted, however, that molybdenum gives a similar reaction. Nb_2O_5 also turns blue if it is subjected to the action of 'nascent' hydrogen (cf. p. 110).

The microcosmic salt bead, which is colorless in the oxidizing flame, turns blue in the reducing flame, and blood red on addition of a little iron sulfate.

Tungsten is usually determined gravimetrically as tungsten trioxide, after precipitation as yellow tungstic acid or as mercury(I) tungstate, Hg_2WO_4. Separation from molybdenum can be effected by volatilizing the latter as $MoO_3 \cdot 2HCl$, or by utilizing the insolubility of tungstic acid, but not of molybdic acid in 50% sulfuric acid. Belcher and Nutten (1951) recommend the use of 1-amino 4 (*p*-aminophenyl) naphthalene for the separation of tungsten from molybdenum. If the HCl concentration is sufficiently high, this reagent precipitates only tungsten, without carrying down any molybdenum.

Quite small quantities of tungsten (down to 10γ and less) can be accurately determined by means of the intense yellow coloration produced when tin(II) chloride is added to a weakly alkaline tungstate solution containing potassium thiocyanate (Feigl, 1932).

11. Uranium (U) [9]

(a) Occurrence

Uranium occurs in Nature chiefly in the form of *pitchblende*. This mineral is a compound of either uranium(IV) or uranium(V) oxide and uranium(VI) oxide (cf. p. 200), approximating U_3O_8 in formula, but is invariably contaminated with varying amounts of other metal oxides (iron-, lead-, calcium oxides, thoria and the

* The same applies to the corresponding molybdenum compounds.

rare earths). The first deposit of pitchblende to be exploited was the deposit of St. Joachimsthal, in Bohemia, but the richest and most productive deposits are those of the Belgian Congo (Katanga), after which may be ranked those of the far North West of Canada (Great Bear Lake). Pitchblende has also been discovered in Portugal, Australia (Northern Territory), and in the United States (Colorado); and in smaller quantities in England (Cornwall), Norway, and Sweden.

A uranium mineral which is met with in certain primeval rocks (the so-called pegmatites), and not enriched in secondary ore deposits as is pitchblende, is *uraninite*, UO_2. This invariably contains thorium oxide, since UO_2 and ThO_2 are isomorphous with each other. Certain varieties of uraninite contain considerable quantities of rare earths, as well as thoria—e.g., the minerals *cleveite*, *bröggerite*, and *nivenite*, which are found in Norway. *Thorianite* is a mixture of thorium and uranium oxides, with ThO_2 predominating. *Thorite*, $ThSiO_4$, also invariably contains some uranium oxide, but usually only in small amount. An important ore of uranium, occurring in the United States (Colorado and Utah), is *carnotite*, $K_2(UO_2)_2[VO_4]_2 \cdot 3H_2O$. Other minerals which may be mentioned are *autunite* (uranite, uranium mica), $Ca(UO_2)_2[PO_4]_2 \cdot 8H_2O$, *uranocirite* (barium uranite), $Ba(UO_2)_2[PO_4]_2 \cdot 8H_2O$, *torbernite* (chalcolite, copper uranite), $Cu(UO_2)_2[PO_4]_2 \cdot 8H_2O$ and *zeunerite*, $Cu(UO_2)_2[AsO_4]_2$. Euxenite and polycrase also always contain some uranium.

(b) History

Uranium was discovered by Klaproth in 1789 in pitchblende, and received its name from the planet uranus, discovered a few years previously by Herschel (1781).

Klaproth obtained uranium dioxide by reduction of uranium(VI) oxide hydrate. The dioxide was for long regarded as the metal, until Peligot in 1841 discovered the means of preparing the true metal. This he obtained by the reduction of UCl_4 with metallic potassium—a method which, in principle, still remains one of the most suitable ways of preparing the metal in a pure state.

Until the inception of projects for the utilization of nuclear energy (see p. 603), during the World War of 1939–45, the principal objective in working up uranium ores was the extraction of radium. Uranium oxide was obtained only as a by-product; the metal was of no technical interest, and had, indeed, never been obtained in a high state of purity. With the construction of nuclear reactors, and the development of atomic energy projects in the United States, Canada, Britain, France and other countries the production of uranium metal has become of greater importance than the extraction of radium, and is now carried out on a considerable scale.

Mining for pichblende was begun in the middle of the 19th century, at the mines of St. Joachimsthal in the Bohemian Erzgebirge. From the 16th to the beginning of the 19th centuries, these mines were at the height of their prosperity for the production of silver*, but the silver workings were later abandoned because the ore was exhausted, and finally the production of lead, which was continued for some time, declined also. The uranium compounds extracted from the pitchblende found only limited uses for a long while—chiefly for the manufacture of glasses with a fine fluorescence, and for ceramic colors. Only after the discovery of radium (1898, cf. Chap. 11) did the uranium minerals become important. The extraction of radium was started on a large scale in almost every industrialized country, and especially in the United States. Even so, the uses of uranium compounds remained very restricted, and the metal itself was of no technical interest at all. With the discovery of nuclear fission by Hahn (1939), metallic uranium has become a material of the utmost technical importance for the production of plutonium and for nuclear power reactors.

(c) Preparation

The uranium ore containing pitchblende is first leached with sulfuric, nitric, or hydrochloric acid. The solution is treated with $Ca(OH)_2$ and an excess of Na_2CO_3,

* The silver coins from Joachimsthal were briefly known as 'Joachimsthaler', and later simply as 'thaler'. This in turn was corrupted to the modern 'dollar'.

in order to precipitate aluminum, iron, cobalt, and manganese. The soluble carbonatouranate which is thereby formed is decomposed by the addition of HCl, and uranium is precipitated from the resulting uranyl salt solution in the form of 'ammonium uranate' (nominally $(NH_4)_2U_2O_7$, but often approximating hydrated UO_3 in composition), by passing in ammonia. The ammonium uranate is converted to the oxide U_3O_8 by igniting it in air.

If the ores contain copper and arsenic, uranium is generally precipitated as $Na_2U_2O_7$ by means of NaOH, from the carbonatouranate solution, after acidification with HCl. The $Na_2U_2O_7$ is dissolved once more in hydrochloric acid, and H_2S is passed through the solution, precipitating CuS and As_2S_3. The filtrate is freed from H_2S by boiling, and the uranium is finally precipitated as $(NH_4)_2U_2O_7$ by the addition of ammonia.

In working up carnotite, it is necessary to use methods which enable the vanadium and the phosphoric acid to be separated from the uranium. Special processes must also be used for low-grade uranium ores—e.g., heating in a current of chlorine or of other chlorinating agents (SCl_2, $SOCl_2$, $COCl_2$, CCl_4), whereby the uranium is sublimed out as UCl_4.

Reduction of the oxide U_3O_8 to the metal can be effected by heating it with carbon in the electric arc furnace, as was shown by Moissan. Metal prepared in this way, however, contains the carbide.

The preparation of the pure metal is rendered very difficult both by the strong tendency of uranium to form the carbide, and by its high affinity for oxygen and nitrogen.

On a laboratory scale, it is possible to prepare the very pure metal by the filament growth process of Van Arkel and De Boer, adapted in a way described by Foote (1944–45, cf. also Prescott and Holmes, 1944). The conditions for deposition require much more exacting control than is the case when the method is used for the preparation of titanium, zirconium, etc., both because of the low melting point of uranium (1130°), and because equilibrium is established with the non-volatile UI_3, according to the equation $2UI_4 \rightleftharpoons UI_3 + I_2$. A feasible technical method for the preparation of metallic uranium is the *Westinghouse process*, which makes use of the electrolytic decomposition of UF_4 or KUF_5, dissolved in a $CaCl_2$-KCl melt. Pure metal can also be produced on a manufacturing scale by the reduction of UCl_4 with metallic calcium (James, 1926) or with metallic potassium (which can be produced by the reaction of CaC_2 with KCl) (Lautié, 1947):

$$UCl_4 + 2Ca = U + 2CaCl_2 + 131 \text{ kcal.}$$

$$UCl_4 + 4K = U + 4KCl + 167 \text{ kcal.}$$

The process in general use for the manufacture of pure metallic uranium for use in nuclear reactors consists of the reduction of uranium tetrafluoride with metallic calcium:

$$UF_4 + 2Ca = U + 2CaF_2 + 134 \text{ kcal.}$$

The starting point for this process is the pure U_3O_8 obtained by heating ammonium diuranate, $(NH_4)_2U_2O_7$. This can be converted, into uranium dioxide, UO_2, by heating it in hydrogen, and the dioxide in turn to the tetrafluoride, either by treatment with 40 per cent hydrofluoric acid, and dehydrating the product, by heating it under reduced pressure, or directly, by the action of gaseous hydrogen fluoride at elevated temperatures. The pure calcium metal needed for the process can be made by the reaction of calcium oxide on aluminum, followed by sublimation in a vacuum. The reaction between uranium tetrafluoride and calcium yields the metal in the molten state. It is allowed to solidify in moulds of sintered fluorite, is re-melted in a vacuum in crucibles of the same material, and is finally obtained 99.98% pure.

(d) Properties

Pure uranium is a silver-white, lustrous metal, which gradually tarnishes in air. It is not very hard, and can be deformed at ordinary temperature by hammering and rolling. When it is heated, it first becomes brittle, and at still higher temperatures becomes distinctly plastic. The reason for these changes in properties lies in the allotropic transformations a-U $\xrightarrow{660°}$ β-U $\xrightarrow{770°}$ γ-U (cf. p. 121). The density of pure uranium is 19.05, and its specific heat 0.0274 (at room temperature). It has a relatively low melting point, (1130°) as compared with its homologues. The latent heat of fusion is about 2.8 kcal per g-atom.

From data given by Moore and Kelley (1947), the temperature-variation of the atomic heat (C_p) of a-uranium above 0° C can be represented by $C_p = 3.15 + 8.44 \cdot 10^{-3} T + 0.80 \cdot 10^5 T^{-2}$ (T = temp., in °K). The atomic heats of β- and γ-uranium are independent of the temperature (C_p = 10.38 and 9.10, respectively). Uranium is perceptibly more volatile than molybdenum and tungsten at high temperatures. The vapor pressure is 0.026 mm at 1900° and 2.4 mm at 2300° C, and the boiling point may be calculated from the temperature-variation of the vapor pressure as about 3500°. The specific electrical conductivity of the pure metal is about $4 \cdot 10^4$ ohm^{-1}cm^{-1}, and the thermal conductivity about $65 \cdot 10^{-3}$ cal $\cdot$ cm^{-1}sec^{-1}degree^{-1} (at ordinary temperature). Uranium is weakly paramagnetic. Its specific susceptibility was found by Owen (1912) to be $\chi = 2.6 \cdot 10^{-6}$ at 18° C.

Uranium is a highly reactive metal, especially at elevated temperatures. Hydrogen at 250–300° converts massive uranium to the finely divided *uranium hydride*, UH_3. The metal burns, with showers of sparks, when gently heated in air, forming uranium(IV,VI) oxide, U_3O_8. It also combines directly with the halogens and at once inflames in fluorine. Dry hydrogen chloride attacks it at a dull red heat. Sulfur combines slowly with uranium above 250°, and vigorous combustion occurs in sulfur vapor above 500°. Uranium combines with nitrogen at 450° and above. Rapid reaction with phosphorus, arsenic, and carbon takes place only at about 1000°. Uranium combines with boron only at the temperatures attained in the electric arc. It unites directly with many metals, as well as with the non-metals.

Uranium dissolves readily in dilute acids, evolving hydrogen and forming uranium(IV) salts. It reacts with water even at ordinary temperature if it is finely divided.

Uranium is *radioactive*, and is the element for which the phenomena of radioactivity were first observed (cf. Chap. 11). The radioactive disintegration of uranium leads to the formation of radium, by way of the intermediate steps set out in Table 59, p. 533, and this in turn disintegrates further in the manner shown in Table 59. The isotope ^{235}U, *actinouranium*, which is present to the extent of about 0.7% in natural uranium, is the parent element of the actinium disintegration series (cf. Table 60, p. 537).

(e) Uses

The principal use of uranium is as the fissionable material in nuclear reactors ('atomic piles'; cf. Chap. 13), as a direct source of nuclear energy, and for the production of plutonium.

Many of the *compounds* of uranium play a part in the preparation of the pure metal or in the separation of the uranium isotopes. Apart from this, uranium compounds find only limited applications. Uranyl salts are used in photography for intensifying negatives, and occasionally in toning baths. Sodium uranate is used under the name of *uranium yellow* as a yellow pigment for glass and ceramic glazes. Uranyl acetate is a valuable analytical reagent, for the determination of sodium by means of zinc uranyl acetate. Uranium, containing some uranium carbide, is a possible catalyst for the synthesis of ammonia by the Haber process.

12. Compounds of Uranium

(a) Survey

Like molybdenum and tungsten, uranium appears to be capable of forming compounds from every valence state from (II) to (VI). It is *bivalent* in the monoxide, UO, and *trivalent* in a number of compounds including the very stable anhydrous trihalides, UX_3. *Quadrivalent* uranium is present in a large number of compounds; these include the tetrahalides, UX_4 and the double salts derived from them, the dioxide UO_2, and the salts with oxyacids derived from the dioxide. Compounds derived with certainty from the +5 state are the two pentahalides, UF_5 and UCl_5. There are numerous derivatives of hexapositive uranium, including the two hexahalides UF_6 and UCl_6, the trioxide UO_3, and the two principal types of derivatives of the latter—the *uranyl* compounds, of the type UO_2X_2, and the *uranates*, compounds with metallic oxides formally derived from UO_3 in its function as an acid anhydride.

With a few exceptions, only uranium(IV) and uranium(VI) compounds are capable of existing in aqueous solution. In general, the uranium(VI) compounds are by far the most stable in aqueous solution. Unless they are stabilized by complex formation, uranium(IV) ions have a strong tendency to undergo oxidation, with the formation of uranyl cations, UO_2^{++}, or uranate ions, $UO_4^{=}$ or $U_2O_7^{=}$.

The tendency for the ions U^{++++} to change into uranyl ions, UO_2^{++}, can be expressed quantitatively by the oxidation potential. For the conditions under which the values of Table 103, p. 765 of Vol. I, are valid this has the value —0.41 volt for the reaction $U^{++++} + 2H_2O \rightarrow UO_2^{++} + 4H^+ + 2e$. Comparison with other values in the Table shows that, in a solution 1-normal in hydrogen ions, the U^{++++} ion has a considerably stronger tendency to be oxidized than the ion Fe^{++}, but not so strong as the ion Sn^{++}. If the hydrogen ion concentration is lowered, the tendency of the U^{++++} ion to be oxidized is considerably increased. The oxidation potential of the U^{++++} ion is also dependent upon *exposure to light*; the value quoted is valid in the dark. Irradiation raises the oxidation potential—i.e., the tendency of U^{++++} ions to undergo oxidation increases upon irradiation. At the same time, it may frequently be observed that the reduction of uranyl salts is promoted by sunlight or ultraviolet light. This is due to a different effect—a catalytic acceleration of the corresponding reactions by light.

Many compounds of uranium(IV) are isomorphous with the corresponding compounds of thorium—e.g., UO_2 with ThO_2, U_3P_4 with Th_3P_4, UP with ThP, and also the hydrates of $U(SO_4)_2$ with the corresponding hydrates of $Th(SO_4)_2$.

Uranium combines with silicon to form the compounds U_3Si, U_3Si_2, (m.p. 1665°), USi, USi_2 (*a*- and *β*-forms, m.p. 1700°) and USi_3 (incongruent m.p. at 1510°). The tetragonal *a*-USi_2, is isotypic with $ThSi_2$. *β*-USi_2 crystallizes in the hexagonal system.

Uranium compounds generally confer to glass melts a yellow color with a fine yellow green fluorescence, characteristic of the uranyl radical. Soluble uranium compounds are *very poisonous*.

(b) Classification of Uranium Compounds

The compounds of uranium can be subdivided into

[i] the compounds which are *not salt-like* in character,
[ii] *salts* derived from uranium.

The salts formed by uranium fall into two groups—

(a) *uranates* and *peroxyuranates*
(b) compounds of uranium with acid anions.

    1. uranium(VI) salts. These, without exception, are *uranyl* salts, i.e., salts of the radical $[UO_2]^{II}$

    2. uranium(IV) salts

    3. uranium(III) salts.

It is not possible to draw an absolutely sharp distinction between those binary compounds of uranium which are salt-like in character and those which are not. The former will therefore be treated in the following sections along with the other binary compounds of uranium. Compounds of uranium containing more than two elements will not be considered here, except for those which are definitely salt-like in character.

The *uranates* are derived from *uranium trioxide*, UO_3, by salt formation with other *basic* oxides. The simplest uranates correspond to the general formula $M^I_2[UO_4]$, but most of the compounds of uranium trioxide with basic oxides are *diuranates*, of the type $M^I_2[U_2O_7]$.

The *peroxyuranates* are derived from the uranates, by replacement of O^{2-} by peroxy groups, O_2^{2-}. Peroxyuranates are represented e.g., by the compounds $M^I_4[UO_6]$ (or $M^I_4[O_2=UO_4]$).

The *uranyl salts* are also derived from *uranium trioxide*. They result from salt formation between uranium trioxides and *acids*. Only one O^{2-} ion of the UO_3 is thereby replaceable by acid radicals, giving rise to salts containing the uranyl radical, $[UO_2]^{II}$:

$$UO_3 + 2HX = UO_2X_2 + H_2O.$$

The *uranium(IV) salts* are derived from uranium dioxide, UO_2, which acts as a base anhydride. The salts correspond to the general formula $U^{IV}X_4$ and, as already stated, they are less stable in aqueous solution than the uranyl salts.

The *uranium(III) salts*—i.e., salts of the type UX_3—are in general quite exceptionally unstable in aqueous solution. One relatively stable simple uranium(III) salt is the iodide, UI_3.

(c) Binary Compounds of Uranium

Table 20 gives a summary of the binary compounds of uranium with metals and with non-metals.

The data concerning solubility in the solid state, given in the table, all relate to α-uranium. The β-modification has a rather better solvent power for certain metals, and γ-uranium tends towards still greater miscibility. Thus the latter is able to incorporate in its crystal lattice up to 4–5 atom-% of aluminum, 3–4 atom-% of chromium or manganese, about 2 atom-% of iron or nickel, and as much as 36 atom-% of molybdenum. Niobium, which can be taken up only to the extent of 0.25 atom-% in α-uranium, will dissolve to the extent of 85 atom-% in γ-uranium at 1350° (although the solubility is only 3.6 atom-% at 660°). As far as is known, uranium can be incorporated only to a small extent in the crystal lattice of other metals. Thus bismuth, chromium, and copper can take up no uranium at all to form mixed crystals. The solvent power of molybdenum, tungsten, and thorium for uranium is also very small. Mercury dissolves less than 0.01 atom-% of uranium at 18°, and 1.25 atom-% at 350°.

Among the intermetallic compounds of uranium, UBi is isotypic with sodium chloride. UCo has a more complicated, body-centered cubic structure. UAl_2, UMn_2, UFe_2, and UCo_2 form Laves-phases of the $MgCu_2$ type, whereas UNi_2 forms one of the $MgZn_2$ type (cf. p. 24). UHg_2 crystallizes hexagonal, like UNi_2, but has a different structure. UAl_3 and USn_3 crystallize with the cubic $CaSn_3$ structure (cf. Vol. I). UHg_3 forms a hexagonal structure, UHg_4 a rather complicated crystal lattice, related to the body-centered cubic type. UNi_5 and UCu_5 (isotypic with $PdBe_5$ and $AuBe_5$) have a face-centered cubic structure, related to that of the $MgCu_2$ type. The compounds U_6Mn, U_6Fe, U_6Co and U_6Ni form a series of isotypic body-centered tetragonal structures.

Table 21 provides a conspectus of the fine structure of the binary compounds of uranium with non-metals. In addition to the modification shown in Table 21, there are several other forms of UO_3. The structure of these forms has not been worked out, however.

Table 22 records the *heats of formation* and the *free energies* of a series of uranium compounds. These values relate to the formation of the compounds from the elements in their standard states at 25°. In the case of uranium trioxide hydrate alone, the heat of reaction is for the formation from UO_3 and H_2O_{gas}.

(*i*) *Uranium hydride*, UH_3. Massive uranium metal reacts directly with hydrogen at temperatures above about 230° (Driggs, 1929, Spedding 1943 and later), being changed into a black, very finely divided powder of uranium hydride, UH_3, which has an appreciable dissociation pressure of hydrogen at higher temperatures. Equilibrium pressures of hydrogen over a mixture of metallic uranium and uranium hydride are at 200°, 0.6 mm; 300°, 24.8 mm; 400°, 345 mm; and 436°, 760 mm. Uranium hydride has semi-metallic properties and is an electrical conductor. Its crystal structure is cubic ($a = 6.632$ Å), and quite distinct from that of uranium itself. It is thus a definite compound, and not a hydride of the interstitial type (cf. p. 45). Its density is 11.4, and heat of formation 30.4 kcal per mol. The finely divided hydride is highly reactive, and combines not only with oxygen (to form uranium oxides), but also with the halogens, and with nitrogen at slightly elevated temperatures. It also reacts at 200–400° with water vapor, H_2S, PH_3, the halogen hydrides etc., to give the oxides, sulfides, phosphides, lower halides, etc. of uranium. The formation and decomposition of uranium hydride provides a convenient means of preparing pyrophoric and highly reactive uranium metal.

Above the decomposition temperature of UH_3, uranium can take up a certain amount of hydrogen in solid solution. α-uranium can dissolve only minimal amounts—less than iron, for example. The solubility of hydrogen in β- and γ-uranium is rather greater, and it is quite considerable in the molten metal.

(*ii*) *Fluorides*. Uranium combines with fluorine to form fluorides from all the valence states between (VI) and (III).

(*a*) *Uranium hexafluoride*, UF_6, is formed by the direct action of excess fluorine upon metallic uranium, or upon uranium tetrafluoride, UF_4. It can also be prepared by the action of fluorine on heated uranium carbide, UC_2, or on UCl_5. It is a colorless volatile compound, forming orthorhombic crystals which sublime at 56.5° without melting (melting point 64.05° under 1134 mm pressure); the density of the solid at the melting point is about 4.8, and that of the liquid 3.62, the increase of volume upon liquefaction being unusually great. The crystals of the compound are made up of UF_6 molecules, in which the 6 F atoms are arranged about the U atom in the form of a somewhat distorted octahedron (Hoard, 1944). In the gaseous state, the UF_6 molecules are quite undistorted regular octahedra (Bauer, 1943, Smyth 1944). The dipole moment of UF_6 in the gaseous state is therefore practically zero. Uranium hexafluoride is very hygroscopic and highly reactive. It is hydrolyzed by water to *uranyl fluoride*, UO_2F_2, and is readily reduced to the tetrafluoride, UF_4.

TABLE 20

MISCIBILITY OF URANIUM WITH OTHER ELEMENTS AND COMPOUNDS FORMED WITH ELEMENTS OF THE MAIN GROUPS AND SUB-GROUPS

(See Table 9 for meaning of symbols)

ELEMENTS OF THE MAIN-GROUPS

III		IV			V					VI				VII			
B	Al	C	Si	Sn	N	P	As	Sb	Bi	O	S	Se	Te	F	Cl	Br	I
—	$s\ o$	$s > 0$	$s \sim 0$	o	$s \sim 0$	—	—	—	o	$s > 0$	—	—	—	—	—	—	—
UB_2	UAl_2 1590°	UC 2250°	$U_3Si[]$	U_5Sn_4 1500°	UN	UP	U_2As	U_3Sb_8?	UBi	UO	US	U_2Se_3	UTe	UF_3	UCl_3	UBr_3	UI_3
UB_4	UAl_3* 1350°	U_2C_3 only stable above 2000°	U_3Si_2 1665°	USn_3* 1350°	U_2N_3	U_3P_4	UAs			UO_2	U_2S_3	USe_2	U_2Te_3	UF_4	UCl_4	UBr_4	UI_4
UB_{12}	UAl_5* 730°		USi* 1580°		UN_2		U_3As_4?			U_4O_9	US_2		UTe_2?	U_4F_{17}			
		UC_2 2400°	USi_2 1700°							U_2O_5?				U_2F_9			
			USi_3 1510°							U_3O_8	US_3			UF_5	UCl_5		
										UO_3				UF_6	UCl_6		

ELEMENTS OF THE SUB-GROUPS

IV			V			VI			VII	VIII			I			II	
Ti	Zr	Th	V	Nb	Ta	Cr	Mo	W	Mn	Fe	Co	Ni	Cu	Ag	Au	Zn	Hg
—	—	$s\ o$	$s\ o$	$s > 0$	$s\ o$	$s\ o$	$s\ o$	$s > 0$		$s\ o$				$s\ o$	$s\ o$		$s\ o$
—	—	—	—	—	—	—	—	—	U_6Mn* 726°	U_6Fe* 815°	U_6Co	U_6Ni* 754°	UCu_5* 1052°	—	—	—	UHg_2* 450°
									UMn_2 1120°	UFe_2 1235°	UCo	UNi_2* 810°					UHg_3* 390°
											UCo_2	UNi_5 1295°					UHg_4 360°

TABLE 21

CRYSTAL STRUCTURES OF BINARY COMPOUNDS OF URANIUM WITH NON-METALS

Compounds	Structure	Edge lengths of the unit cell, Å			Formula weights per unit cell	Density, X-ray
		a	b	c		
Borides						
UB_4	tetragonal	7.07		3.97	4	9.425
UB_{12}	cubic	7.47	—	—	4	5.854
Carbides						
UC	cubic, NaCl type	4.95	—	—	4	13.65
UC_2	tetragonal face-centered, related to CaC_2 type	3.517	—	5.987	2	11.68
Silicides						
U_3Si	tetragonal face-centered	6.02	—	8.68	4	15.58
U_3Si_2	tetragonal	7.315	—	3.982	2	12.20
USi	orthorhombic	5.65	7.65	3.90	4	10.40
a-USi_2	tetragonal face-centered	3.97	—	13.71	4	8.98
β-USi_2	hexagonal	3.85	—	4.06	1	9.25
USi_3	cubic	4.03	—	—	—	—
Nitrides						
UN	cubic, NaCl type	4.880	—	—	4	14.31
U_2N_3	cubic, Sc_2O_3 type	10.67	—	—	16	11.24
UN_2	cubic, CaF_2 type	5.31	—	—	4	11.73
Phosphide arsenides						
UP	cubic, NaCl type	5.59	—	—	4	10.23
U_2As	different from UAs and UO_2					
UAs	cubic, NaCl type	5.77	—	—	4	10.77
Oxides						
UO	cubic, NaCl type	4.92	—	—	4	14.1
UO_2	cubic, CaF_2 type	5.458	—	—	4	10.97
U_4O_9	cubic, based on CaF_2 type	5.430	—	—	6	11.2
U_2O_5	orthorhombic	8.27	31.65	6.72	16	8.35
U_3O_8	orthorhombic	6.703	11.94	4.140	2	8.39
a-UO_3	hexagonal	3.963	—	4.16	1	8.34

(Continued)

TABLE 21   (*Continued*)

CRYSTAL STRUCTURES OF BINARY COMPOUNDS OF URANIUM WITH NON-METALS

Compounds	Structure	Edge lengths of the unit cell, Å			Formula weights per unit cell	Density, X-ray
		a	*b*	*c*		
Sulphides						
US	cubic, NaCl type	5.473	—	—	4	10.87
U_2S_3	orthorhombic, related to Sb_2S_3 type	10.63	10.39	3.88	8	
a-US_2	tetragonal	10.25	—	6.30	10	7.54
β-US_2	orthorhombic, isotypic with ThS_2	4.22	7.08	8.45	4	7.90
Hydride						
UH_3	cubic, U atoms as in β-W	6.631	—	—	8	10.92
Fluorides						
UF_3	hexagonal, LaF_3 type	4.138	—	7.333	6	8.95
UF_4	triclinic (pseudo-monoclinic), ThF_4 type	12.79	10.72 $a = 126°$	8.39	12	6.70
a-UF_5	tetragonal chain structure	6.512	—	4.463	2	5.81
β-UF_5	tetragonal network structure	11.45	—	5.198	8	6.45
UF_6	orthorhombic molecular lattice	9.900	8.962	5.270	4	5.06
Chlorides						
UCl_3	hexagonal, $La(OH)_3$ type	7.428	—	4.312	2	5.51
UCl_4	tetragonal	8.30	—	7.49	4	4.87
UCl_6	hexagonal, molecular lattice	10.90	—	6.03	3	3.59
Bromides and Iodides						
UBr_3	hexagonal, $La(OH)_3$ type	7.926	—	4.432	2	6.53
UI_3	orthorhombic, LaI_3 type	13.98	4.33	9.99	4	6.76

TABLE 22

HEATS OF FORMATION, $-\Delta H_{298}$, AND FREE ENERGIES OF FORMATION, $-\Delta F_{298}$, OF URANIUM COMPOUNDS AT $25°$, IN KCAL PER G ATOM OF URANIUM

	Carbides		Nitrides			Oxides				Hydride
	UC	UC$_2$	UN	U$_2$N$_3$	UN$_2$	UO	UO$_2$	U$_3$O$_8$	UO$_3$	UH$_3$
$-\Delta H_{298}$	—	3.92	106	128	—	135	259.8	284.6	287.7	30.4
$-\Delta F_{298}$	—	9.83	—	—	—	—	247.3	266.6	269.1	17.7

	Fluorides						Chlorides			
	UF$_4$	U$_4$F$_{17}$	U$_2$F$_9$	α-UF$_5$	β-UF$_5$	UF$_6$	UCl$_3$	UCl$_4$	UCl$_5$	UCl$_6$
$-\Delta H_{298}$	444	455	467	484	485	505	212.0	251.0	262.1	272.3
$-\Delta F_{298}$	422	432	442	458	459	485	196.7	229.6	235.7	241.4

	Bromides		Iodides		Oxyhalides		Uranium trioxide hydrates			
	UBr$_3$	UBr$_4$	UI$_3$	UI$_4$	UOCl$_2$	UOBr$_2$	UO$_3 \cdot \frac{1}{2}$H$_2$O	UO$_3 \cdot$H$_2$O	UO$_3 \cdot 1\frac{1}{2}$H$_2$O	UO$_3 \cdot 2$H$_2$O
$-\Delta H_{298}$	181.6	211.3	129.4	149.2	261.7	246.9	13.3	23.4	31.7	39.2
$-\Delta F_{298}$	167	190	121	140	246.3	231.3	—	—	—	—

(β) *Uranium pentafluoride*, UF_5, and intermediate fluorides. Ruff and Heinzelmann (1911) found that the reaction of uranium pentachloride with anhydrous hydrogen fluoride yielded *uranium pentafluoride*:

$$UCl_5 + 5HF = UF_5 + 5HCl.$$

The same product can be obtained by reaction between UF_4 and UF_6 at 125° (Grosse, 1941), but in this reaction at least two other intermediate fluorides may also be formed, with the compositions U_2F_9 (= $UF_5 \cdot UF_4$) and U_4F_{17} (= $UF_5 \cdot 3UF_4$), respectively. UF_5 is a colorless compound, existing in two modifications, both tetragonal, with a transition point at 125°. U_2F_9 is cubic, and U_4F_{17} is a black compound of unknown structure. At 300°, the dissociation pressures of UF_6 are approximately 200 mm over UF_5, 8 mm over U_2F_9, and 1.5 mm over U_4F_{17}.

(γ) *Uranium tetrafluoride* is formed in the reaction of metallic uranium with fluorine (Ruff), but is readily fluorinated further. It can be obtained as a hydrate (with $2\frac{1}{2}H_2O$ or $1\ H_2O$) by precipitation from uranium(IV) solutions with fluoride ions, or by the reaction of uranium dioxide with hydrogen fluoride gas at elevated temperatures. The hydrated tetrafluoride can be dehydrated to anhydrous UF_4, but readily undergoes partial hydrolysis to UO_2, and oxidation, if heated in the presence of moisture. Uranium tetrafluoride forms green triclinic (pseudo-monoclinic) crystals, which are insoluble in water (m.p. 960°, density 6.70). The heat of formation is 446 kcal per mol. It is not very reactive, but can be reduced by dry hydrogen at 1000° to UF_3. Double fluorides—e.g., $NaUF_5$, Na_2UF_6, Na_3UF_7, KU_2F_9—are formed from melts of UF_4 with the fluorides of the alkali and alkaline earth metals (Zachariasen, 1948).

(δ) *Uranium trifluoride*, UF_3, is formed by reducing UF_4 with hydrogen, but can be prepared in the pure state by this means only if the uranium tetrafluoride employed is completely free from oxygen. (Skinner, 1944). It is also formed by the action of metallic uranium on UF_4, at high temperatures. The reaction $3UF_4 + U \rightleftharpoons 4UF_3$ is reversible, and proceeds from right to left if the temperature is raised much above 1000°, since UF_4 sublimes out of the system. UF_3 forms deep violet-red crystals, which are insoluble in water and in dilute non-oxidizing acids. It has the same hexagonal crystal structure as is found in the fluorides of the rare earth metals.

(*iii*) *Chlorides*. When uranium burns in chlorine, a mixture of *uranium tetrachloride*, UCl_4, and *uranium pentachloride*, UCl_5, is obtained. The chloride of +6 uranium has only recently been isolated.

(a) *Uranium hexachloride*, UCl_6, can be prepared by the thermal decomposition of the pentachloride:

$$2UCl_5 = UCl_4 + UCl_6$$

or by volatilization of lower chlorides in chlorine. It is a volatile compound, with a vapor pressure of about 2 mm at 140°, and can be purified by sublimation in chlorine. It then forms dark green-black needles, which decompose and melt at about 177°. The heat of formation is 272 kcal per mol.

(β) *Uranium pentachloride*, UCl_5, was obtained by Roscoe (1874) as a by-product in the preparation of UCl_4, but was first prepared in the pure state by Ruff (1911). It is formed by the action of chlorine on UCl_4 at 520°, or by the reaction of UO_3 with liquid CCl_4, under pressure, at 100–250°. It exists as red-black crystals, with a green reflex (density 3.81, heat of formation 262 kcal per mol.). The molecular weight in CCl_4 solution agrees with the dimeric formula U_2Cl_{10}. Uranium pentachloride is a reactive compound, and readily loses chlorine even at ordinary temperature.

(γ) *Uranium tetrachloride*, UCl_4, can be prepared by a variety of methods—e.g., by the action of chlorine on a mixture of uranium dioxide and carbon at a dull red heat (Peligot, 1842), or by the liquid-phase chlorination of UO_3 with CCl_4 at 115–180°. UCl_5 is formed initially in the latter reaction, and is decomposed by further heating. Uranium tetrachloride forms dark green tetragonal crystals of octahedral habit ($a = 8.296$, $c = 7.487$ Å). The heat of formation is 251 kcal per mol, m.p. 590°, b.p. 792°. It dissolves in water with the evolution of much heat, giving a green solution in which it is extensively hydrolyzed.

At low temperatures, UCl_4 adds on up to 12 molecules of NH_3 per molecule UCl_4; this is given up in stages when the compounds are degraded isobarically. The 5-ammoniate is stable at ordinary temperature.

(δ) *Uranium trichloride* is formed by reducing the tetrachloride with hydrogen at 575° (Peligot, 1842), or by the action of hydrogen chloride gas on uranium hydride. It forms dark red, hygroscopic needles, which dissolve in water to give a purple solution; this solution evolves hydrogen, and rapidly reverts to the green color characteristic of uranium(IV).

(*iv*) *Bromides.* The highest bromide of uranium is UBr_4, which closely resembles UCl_4 in its properties and in the reactions whereby it may be prepared. Uranium tribromide, UBr_3, corresponds closely to UCl_3.

(*v*) *Uranium tetraiodide* is much less stable than the chloride or bromide, and the dissociation pressure of iodine in the equilibrium $UI_4 \rightleftharpoons UI_3 + \frac{1}{2}I_2$ becomes appreciable at high temperatures. Mixed halides of uranium(IV) have been prepared—e.g., $UClF_3$, UBr_2I_2.

(*vi*) *Uranium Sulfides. Uranium disulfide*, US_2, can be prepared by the direct combination of uranium with sulfur, by reaction between UCl_4 and hydrogen sulfide at 500°, or by the action of hydrogen sulfide on a mixture of uranium dioxide and carbon at 1200°. It forms grey-black lustrous tetragonal crystals of density 7.96. It is slowly attacked by water, more readily by hydrochloric acid, and very rapidly by nitric acid. According to Biltz (1940), it is converted to *uranium trisulfide*, US_3, by heating it under pressure with sulfur. *Diuranium trisulfide*, U_2S_3, is formed by the reduction of US_2 with hydrogen, by the thermal decomposition of US_2 in a vacuum at 1600°, or by the reaction of uranium metal with the calculated amount of sulfur or hydrogen sulfide. U_2S_3 is orthorhombic, and its structure is related to that of Sb_2S_3 (Brewer, 1948). A lower sulfide US (or possibly U_4S_3,—cf. Biltz, 1940) can be obtained by the direct union of the calculated quantities of uranium and sulfur. This phase, which may exist over a range of composition, is quasi-metallic and cubic, having the NaCl structure. It is isomorphous with ThS and CeS.

(*vii*) *Uranium Nitrides.* Uranium combines with nitrogen to form the compounds UN, U_2N_3, and UN_2. The mononitride (light grey powder) is the most stable; it can be heated to over 1900° in a vacuum without decomposition. U_2N_3 begins to lose nitrogen in a vacuum at 750°, whereas UN_2 can only be obtained under pressure of nitrogen. In UN the uranium atoms lie on a face-centered cubic lattice; the same is probably true of the nitrogen atoms, so that a crystal lattice of NaCl type is formed. U_2N_3 crystallizes with the Sc_2O_3 structure ('rare earth type C' structure), and UN_2 with the fluorite structure. The sesquinitride can be converted continuously into the dinitride by uptake of nitrogen, as is understandable from the relation between the Sc_2O_3 and CaF_2 type structures, explained on p. 37–8. For cell dimensions, see Table 21.

(*viii*) *Uranium Oxides.* Uranium functions principally as a 4- or 6-valent element in its compounds with oxygen, forming a brown-black *dioxide*, UO_2 and an orange *trioxide*, UO_3. In addition, it forms at least two intermediate oxides—*triuranooctoxide*, U_3O_8 (the most stable of the uranium oxides), and U_4O_9 (obtained by the oxidation of UO_2). U_3O_8 is to be regarded as a double oxide of UO_3 with either UO_2 ($U_3O_8 = U^{IV}O_2, 2U^{VI}O_3$) or U_2O_5 ($U_3O_8 = U^V_2O_5, U^{VI}O_3$) (cf. p. 200). Uranium is found in Nature in secondary ore deposits largely in the form of U_3O_8.

(a) The *monoxide*, UO, which has the sodium chloride structure, has also been unambiguously identified by X-ray methods. This compound is exceedingly difficult to obtain pure, and has indeed probably never been obtained as a bulk specimen. It is formed as a surface layer on metallic uranium by carefully controlled oxidation—e.g., by superficially oxidizing the metal to UO_2 and annealing the sample in a vacuum, but is only observed under specified conditions. A similar oxide (ThO) is formed on the surface of thorium; this has also not been isolated in weighable amounts.

(β) The *dioxide*, UO_2, can take up an excess of oxygen in its crystal lattice, without any change of crystal structure. It is oxidized to a limiting composition of about $UO_{2.38}$ at temperaures below 200°. The cell dimensions contract with increasing oxygen content, and it has been shown from the accompanying increase in density that the excess oxygen must be present in interstitial positions in the crystal lattice. The cubic phase of variable composition, $\overline{UO_2}$, is not stable at high temperaures, but breaks up into a well defined intermediate oxide U_4O_9 + either $UO_{2.0}$ or U_3O_8, according to the total composition. U_4O_9 has a structure derived from the CaF_2-type structure of UO_2, with one additional oxygen atom per unit cell in an interstitial position. Two other intermediate oxides have also been observed in the oxidation of UO_2. These have tetragonal structures, derived from that of UO_2 by slight distortion, and their compositions approximate $U_{16}O_{37}$ and $U_{16}O_{38}$, respectively (compare intermediate oxides in the molybdenum and tungsten oxide systems) (Perio 1952–53, Anderson 1953–54). It is not entirely certain whether the homogeneity range of UO_2 extends to lower oxygen contents also; Zachariasen (1945) observed a cubic phase believed to have the composition $UO_{1.75}$ ($= U_4O_7$).

According to work of Rundle (1944), the phase $UO_{2.5}$ is probably to be considered as a definite compound, U_2O_5, although its crystal structure differs only very little from that of U_3O_8; it is stated to undergo a continuous transformation to U_3O_8 by upstake of oxygen.

(γ) *Uranium trioxide*, UO_3, is obtained as an orange or brick red powder by the cautious heating of uranyl nitrate, ammonium uranyl carbonate, or ammonium uranate. The purest product is prepared by heating uranyl peroxide dihydrate to 400° in a stream of oxygen. When more strongly heated, uranium trioxide loses oxygen and is converted into U_3O_8. Uranium trioxide has the formal semblance of an amphoteric oxide, since it reacts with acids to form *uranyl salts*, and with basic oxides to form *uranates*; according to the most recent work (Zachariasen 1949), however, the uranates are not salts based upon a 'uranate' radical, but are of the nature of double oxides.

Uranium trioxide is converted to the hydrates $UO_3 \cdot H_2O$ (uranyl hydroxide, $UO_2(OH)_2$ or 'uranic acid') and $UO_3 \cdot 2H_2O$ by boiling it with water. Both hydrates are yellow in color, and exist in several crystalline modifications, but are generally amorphous in appearance.

Hüttig (1922) found that two further hydrates appear as definite stages in the isobaric degradation of these hydrates—namely $UO_3 \cdot \frac{3}{2}H_2O$ and $UO_3 \cdot \frac{1}{2}H_2O$. $UO_3 \cdot 2H_2O$ initially loses water continuously until it reaches the composition $UO_3 \cdot \frac{3}{2}H_2O$; a further $\frac{1}{2}H_2O$ is then lost at constant temperature. With further increase of temperature, the hydrate $UO_3 \cdot H_2O$ loses another $\frac{1}{2}H_2O$ continuously. The hemihydrate is finally converted at constant temperature into the anhydrous oxide.

(δ) *Uranium dioxide* is obtained in the form of a brown to black powder (density 10.95, m.p. well above 2500°) when uranium trioxide or U_3O_8 is reduced in hydrogen. It is practically insoluble in water and alkalis. Except in nitric acid, which oxidizes it to form uranyl nitrate, it dissolves only with difficulty in acids, producing uranium(IV) salts. It is converted to U_3O_8 when it is heated in air.

Uranium dioxide crystallizes cubic, with the CaF_2 structure, and forms mixed crystals with ThO_2 in all proportions. Selwood (1951) and Dawson (1951) have shown that the magnetic properties of dilute solid solutions of UO_2 in ThO_2 accord with the 'spin only' susceptibility for two unpaired valence electrons in the U^{4+} ion. In UO_2 itself (as in the oxides of other paramagnetic metallic ions) the paramagnetic ions are not magnetically 'dilute', and interaction effects between ions interfere with the unambiguous measurement of magnetic moments (cf. Haraldsen, 1940).

(ε) *Uranium(IV, VI) oxide*, U_3O_8, is formed when either the dioxide or the trioxide is ignited in air, or by the ignition of ammonium uranate or the uranyl salts of volatile acids (e.g., uranyl nitrate). It begins to lose oxygen when it is heated to high temperatures (above 900° in air, perceptibly at 600° in a vacuum), forming U_4O_9 rather than pure UO_2. It is completely reduced to UO_2 by hydrogen or carbon monoxide.

U_3O_8 is dark green to black in color and has a density of 8.30. It is orthorhombic in crystal structure, $a = 6.70$, $b = 11.94$, $c = 4.14$ Å In its naturally occurring form, as pitchblende, the density of U_3O_8 varies considerably, depending upon the purity. The mineral is amorphous.

Uranium trioxide is also obtained generally in amorphous form, but can exist in the crystalline state. At least three crystalline modifications have been described, differing in their X-ray diffraction patterns and in their color (orange, brick red, and yellow, respectively). The orange modification crystallizes with hexagonal symmetry (cf. Table 21) and according to Zachariasen has a structure in which there are –O–U–O–U–O–U–O–, etc. chains running parallel to the c-axis (distance U—O $= 2.08$ Å). In addition to these two O atoms which are closely associated with the uranium, each U atom is surrounded by six other O-atoms (U—O distance $= 2.39$ Å) which lie in sheets perpendicular to the 'uranyl chains'.

UO_2 has the properties of a base anhydride. UO_3 is amphoteric. It dissolves in acids to form *uranyl salts*, and in caustic alkalis to form *uranates* (see p. 199). When U_3O_8 is dissolved in acids, it forms a mixture of uranium(IV) salts and uranyl salts. This is commonly taken as evidence that it should be regarded as uranium(IV,VI) oxide, $U^{IV}U^{VI}_2O_8$. However, Haraldsen (1940, see also Dawson and Lister, 1950) found that the magnetic properties of U_3O_8 did not accord well with this formulation. The magnetic moment was found to be 1.39 Bohr magnetons, which is considerably smaller than the moment calculated for $U^{IV}U^{VI}_2O_8$ ($= 1.63$ Bohr magnetons), but close to that expected for $U^{V}_2U^{VI}O_8$ ($= 1.42$). On this basis, U_3O_8 should be considered to be uranium(V,VI) oxide. According to Zachariasen (1944) and Rundle (1944–45), all the U atoms in U_3O_8 occupy equivalent positions in the crystal lattice, whereas one quarter of the oxygen atoms are arranged differently from the remainder.

Uranium dioxide forms anomalous mixed crystals with a number of other metallic oxides in which the cations are of appropriate radius—e.g., with calcium oxide, and with rare earth oxides such as erbium oxide and yttrium oxide. The cubic forms of the rare earth oxides are related in type to the CaF_2-type structure of uranium dioxide. Thus UO_2 will incorporate up to about 40 mol-per cent $YO_{1.5}$, or 45 mol-per cent CaO. The replacement of U^{4+} ions by ions of lower valence necessitates a corresponding omission of O^{2-} ions, i.e., the creation of vacancies in the anion lattice. The anomalous mixed crystals react with oxygen far more readily than does UO_2 itself; U^{4+} ions are converted to ions of higher valence (U^{5+} or U^{6+}) while the oxygen vacancies are progressively filled up and possibly interstitial oxygen atoms introduced (Johnson, 1953, Anderson, 1953, Hund, 1952). Such anomalous mixed crystals have been described as mixed crystals between U_3O_8 and rare earth oxides, etc. However, they may be obtained with compositions corresponding to continuously variable average uranium valence numbers from 4 upwards, and no unique significance attaches to the average valence number exhibited by uranium in U_3O_8 ($= 5.33$).

(d) Uranates

The compounds of uranium trioxide with basic oxides are known as *uranates*. The simplest uranates, (monouranates) correspond to the formula $M^I_2[UO_4]$, and are generally prepared by high temperature processes—e.g., by heating U_3O_8 in fused metallic chlorides in the presence of air. The *diuranates*, $M^I_2[U_2O_7]$, which correspond formally in composition to the dichromates are more commonly formed, and are precipitated by adding the appropriate base to uranyl salt solutions. All uranates, including those of the alkali metals, are insoluble in water. They are very readily hydrolyzed, however, the free base being leached out and uranyl hydroxide (i.e., UO_3 hydrate) remaining. For this reason, much of the reported work on metal uranates is unreliable.

(*i*) *Sodium uranate* or *sodium diuranate*, $Na_2U_2O_7$, is obtained as a yellow precipitate of the hexahydrate when uranyl salts are treated with sodium hydroxide. It may be dehydrated by heating, and is used as a coloring matter (*uranium yellow*) for glass and procelain glazes.

(*ii*) *Ammonium uranate*, $(NH_4)_2U_2O_7$, is precipitated by ammonia from uranyl salt solutions as a yellow flocculent precipitate, practically insoluble in water, but readily soluble in ammonium carbonate solution. The compound provides a very convenient form for the precipitation of uranium in gravimetric analysis, but tends to become colloidal as the precipitate is washed.

(*iii*) *Calcium uranate*, $CaUO_4$, is obtained in the form of yellow rhombohedral leaflets when U_3O_8 is heated in fused calcium chloride. Zachariasen (1951) has shown from X-ray structure determinations that there is no anionic $[UO_4]^{2-}$ radical present in this compound, or in $BaUO_4$ and K_2UO_4, but rather cationic $[UO_2]^{2+}$ groups. It is therefore likely that all the uranates should be formulated as double oxides, $M^I_2(UO_2)O_2$, rather than as true salts, $M^I_2[UO_4]$.

(*iv*) *Uranium peroxide dihydrate* and *Peroxyuranates*. If hydrogen peroxide is added to a concentrated solution of uranyl nitrate or uranyl acetate, a light yellow precipitate is obtained, which has the composition $UO_4 \cdot 2H_2O$ after drying at 100°. The compound furnishes 1 molecule of H_2O_2 per atom of uranium when it is treated with dilute sulfuric acid, and is therefore generally assumed to be the dihydrate of a hypothetical uranium peroxide, UO_4, which is derived from the trioxide, UO_3, by replacement of an oxygen atom by a peroxy group. Since uranium trioxide has been shown to contain UO_2^{2+} radicals as structural units, the peroxide would thus be a uranyl peroxide, $[UO_2](O_2)$. This constitution has not been conclusively proved, however, and the hydrated compound could alternatively be a peroxyhydrate-monohydrate of UO_3, i.e., $UO_3 \cdot H_2O_2 \cdot H_2O$. It displays no acidic character, and the old name of 'peruranic acid' is incorrect. Peruranates cannot be obtained by the action of alkali on the peroxide dihydrate, but the joint action of alkalis and hydrogen peroxide does lead to the formation of *peroxyuranates*. These may also be obtained directly by the action of hydrogen peroxide and alkali hydroxide on uranyl compounds—e.g., the potassium salt $K_4[UO_6] \cdot 6H_2O_2$ (crystallizing with variable water content also), which was shown by Schwarz (1938) to be a monoperoxyuranate 6-peroxyhydrate. The compound $UO_4 \cdot 2H_2O$ loses water and oxygen when it is heated, and is thereby converted into UO_3. This reaction is employed for the preparation of pure UO_3. If the heating is carried out carefully, it is possible for only one half of the peroxidic oxygen to be lost, with the formation of anhydrous peroxide, $U_2O_7 (= O_2U_2O_5)$ (Hüttig, 1922, Kraus, 1942). This same peroxide can also be obtained by heating ammonium uranate in a stream of oxygen (Kraus, 1944).

(e) Compounds of Uranium with Acid Radicals

Compounds of +6 uranium with acid radicals are known in large numbers, and are for the most part definitely salt-like in nature. Almost all of them are derived from the positive, bivalent uranyl radical, $[UO_2]^{2+}$. In many ways this behaves like the elementary ion of a bivalent metal. Most uranyl salts have a distinct tendency to form double or complex salts (acido-salts).

In the case of $+4$ uranium, only the compounds formed with acids are known in well defined form, and not those with alkalis. Uranium(IV) hydroxide is thus predominantly, if not wholly, basic in character. The uranium(IV) salts are much less stable than the uranium(VI) (uranyl) salts.

Trivalent uranium is present in the readily prepared trihalides, UX_3. *Disulfatouranium(III) acid*, $H[U(SO_4)_2]$, is an example of a compound of trivalent uranium with oxyacids. It was prepared by Rosenheim, in the form of deep brown crystal leaflets, by adding sulfuric acid to an ice-cold, electrolytically reduced solution of uranyl chloride.

(f) Uranyl Salts

The uranyl salts correspond to the general formula UO_2X_2 (X = univalent acid radical). They generally crystallize well, and most of them are soluble in water. As a rule they are yellow in color, with a yellow-green fluorescence, the absorption spectrum and fluorescence spectrum being characteristic of the $[UO_2]^{2+}$ group. The commonest uranium compounds of commerce are *uranyl nitrate*, $UO_2(NO_3)_2 \cdot 6H_2O$, and *uranyl acetate*, $UO_2(C_2H_3O_2)_2 \cdot 2H_2O$.

(*i*) *Uranyl chloride*, UO_2Cl_2, (m.p. 578°) can be obtained anhydrous by the action of chlorine on uranium dioxide at 500°, or by the reaction of oxygen with uranium tetrachloride at 300°:

$$UO_2 + Cl_2 = UO_2Cl_2; \quad UCl_4 + O_2 = UO_2Cl_2 + Cl_2.$$

A trihydrate, $UO_2Cl_2 \cdot 3H_2O$, crystallizes in yellow-green birefringent needles from a solution of uranium trioxide in hydrochloric acid. The monohydrate, $UO_2Cl_2 \cdot H_2O$, is obtained by drying the trihydrate in a desiccator over phosphorus pentoxide. Uranyl chloride is very soluble in water, in which it is perceptibly hydrolyzed. It combines with alkali chlorides, and with the chlorides of organic bases, to form double salts of the type $M^I_2[UO_2Cl_4]$ (dioxotetrachlorouranate(VI) salts). *Uranyl fluoride*, UO_2F_2, also tends to form double salts with alkali fluorides, those of the type $M^I_3[UO_2F_5]$ probably being the most stable.

At low temperatures, UO_2Cl_2 forms addition compounds with up to $10NH_3$ (Spacu, 1936). Ammoniates with 5, 4, 3, 2 and 1 NH_3 are obtained during the progressive degradation of this compound.

(*ii*) *Uranyl nitrate* is obtained when the oxides of uranium are dissolved in nitric acid. It crystallizes from concentrated solutions as the hexahydrate, $UO_2(NO_3)_2 \cdot 6H_2O$, a deliquescent lemon yellow material with a yellow green fluorescence (m.p. 59.5°). Uranyl nitrate hexahydrate is freely soluble in water, alcohol, and ether, and can be recrystallized well. 100 g of water dissolve 127 g of the anhydrous salt at 21°. A dihydrate, which melts at 179°, crystallizes from solutions of uranyl nitrate in concentrated nitric acid. The anhydrous nitrate can also be prepared.

Uranyl nitrate combines with alkali nitrates, forming complex salts of the type $M^I[UO_2(NO_3)_3]$, which crystallize anhydrous, but are very hygroscopic.

(*iii*) *Uranyl acetate*, $UO_2(C_2H_3O_2)_2 \cdot 2H_2O$, can be prepared by dissolving uranyl hydroxide or uranium trioxide in acetic acid. Since the uranium trioxide used for this purpose is prepared by heating uranyl nitrate, the commercial salt is often contaminated with nitrate. Uranyl acetate crystallizes from solution as the dihydrate, in rhombic prisms with a strong fluorescence. It can be dehydrated by heating to $110°$, and decomposes at $275°$, leaving a residue of uranium trioxide. It is relatively sparingly soluble in water (7.69 g of the dihydrate in 100 g of water at $15°$). A basic salt separates from the solution on prolonged standing, as a result of hydrolysis.

Uranyl acetate has a tendency to add on an additional acetate group, to form the (weakly complexed) anion $[UO_2(C_2H_3O_2)_3]^-$. It therefore readily forms complex salts with other acetates. Of these, the salts of the types $M^I[UO_2(C_2H_3O_2)_3]$ and $M^IM^{II}[UO_2(C_2H_3O_2)_3]_3$ are noteworthy for their excellent crystallizing power and strong greenish fluorescence. Formation of the sodium salt $Na[UO_2(C_2H_3O_2)_3]$, which crystallizes in tetrahedra, is widely used as a test for sodium. The rather insoluble double salts $NaZn[UO_2(C_2H_3O_2)_3]_3 \cdot 6H_2O$, and $NaMg[UO_2(C_2H_3O_2)_3]_3 \cdot 9H_2O$, are also important in analysis. The former is widely used for the gravimetric determination of sodium.

(*iv*) *Uranyl carbonate* and *Carbonatouranates*. Pure *uranyl carbonate*, UO_2CO_3, occurs native (as *rutherfordine*) in East Africa, as an alteration product of pitchblende. The precipitates obtained by adding alkali carbonates to uranyl salt solutions are, however, generally inhomogeneous. The double- or complex salts derived from uranyl carbonate are, however, readily obtained pure.

If a uranyl salt solution is treated with ammonia and ammonium carbonate in excess, or if freshly precipitated ammonium uranate is treated with ammonium carbonate, a clear yellow solution is obtained. The complex salt $(NH_4)_4[UO_2(CO_3)_3] \cdot 2H_2O$ (ammonium dioxotricarbonatouranate) separates from this solution in yellow monoclinic crystals upon evaporation (solubility about 5 g in 100 g of water at $5°$). The double salt is decomposed by prolonged boiling with water, ammonium uranate being precipitated. Analogous soluble carbonato salts are formed with the other alkali metals. The calcium salt, $Ca_2[UO_2(CO_3)_3] \cdot 10H_2O$, is found native as *uranothallite*.

The ability of uranium to form soluble carbonato salts provides a convenient method for separating uranium from iron, aluminum, etc., and the rare earth metals which generally accompany it in its ores.

(*v*) *Uranyl sulfate*, $UO_2SO_4 \cdot 3H_2O$, separates in yellowish green crystals from the solution which is obtained when uranyl nitrate is fumed down with sulfuric acid, and taken up again with water. The crystals slowly effloresce in air, with partial loss of water. Uranyl sulfate is soluble in water (17.4 g of the anhydrous salt in 100 g of water). The hydrate can be completely dehydrated by heating to $175°$. The anhydrous salt is amber colored, and (unlike the hydrate) not fluorescent. Conductivity measurements have shown that complex ions, $[UO_2(SO_4)_2]^=$, are present to some extent in uranyl sulfate solutions. Double or complex salts, chiefly of the types $M^I_2[UO_2(SO_4)_2(H_2O)_2]$ and $M^I_4[UO_2(SO_4)_3]$, can be isolated from solutions of uranyl sulfate to which alkali sulfates have been added.

(*vi*) *Uranyl sulfide*, UO_2S, is deposited as a brown precipitate, soluble in dilute acids (including acetic acid), when ammonium sulfide is added to a uranyl salt solution. When moist, it readily decomposes in air, forming uranyl hydroxide, $UO_2(OH)_2$. Uranyl sulfide (uranium oxysulfide) prepared by dry methods forms black glistening needles, which are much more resistant to acids than the sulfide precipitated from solution.

(g) Uranium(IV) Salts

The composition of the simple uranium(IV) salts conforms to the general formula UX_4. They are almost all green in color, and readily soluble in water. The oxalate, however, is practically insoluble in water and in acids; it is also quite stable, whereas the soluble uranium(IV) salts, which are hydrolyzed to a considerable extent in solution, have a strong tendency to be oxidized. Alkali hydroxides or ammonia give a voluminous red-brown precipitate of uranium(IV) hydroxide with uranium(IV) salt solutions; this is also very readily oxidized.

Uranium(IV) salts can be prepared, in general, by the reduction of the corresponding uranium(VI) salts (uranyl salts)—e.g., electrolytically, or with 'nascent' hydrogen.

Only in exceptional cases is it possible to prepare uranium(IV) salts by treating UO_2 with the corresponding acids. Uranium(IV) fluoride can be prepared in this way, but not the other uranium(IV) halides (see pp. 197-8).

In general, it seems that the uranium(IV) salts have a smaller tendency than the uranyl salts to form clearly defined complexes in solution. Uranium(IV) oxalate, however, has a very strong tendency for complex formation.

(i) *Uranium(IV) Halides.* UF_4 has also a strong tendency to form complex salts. Complex salts of the type $M^{II}UF_6$—e.g., $BaUF_6$, $PbUF_6$—can be obtained by precipitation from aqueous solution; they have the LaF_3 structure. In addition to these, UF_4 forms numerous complexes with the alkali fluorides. Thus by melting UF_4 with KF, the following compounds may be formed; their existence has been proved by X-ray structural studies (Zachariasen, 1948): K_3UF_7 (two forms), K_2UF_6 (trimorphous), KUF_5, KU_2F_9, KU_3F_{13}, KU_6F_{25}. K_2UF_6 can crystallize with the CaF_2 structure (i.e., as $K_{2/3}U_{1/3}F_2$, with statistical distribution of the cations) and with the AlB_2 structure, as well as in a third type of lower symmetry. Na_2UF_6 is similar.

UCl_4 gives complex salts which are mostly of the types $M^I_2\,UCl_6$ and $M^{II}UCl_6$; however, it forms a salt of the composition Ba_2UCl_8 with barium chloride. UBr_4 (m.p. 518°) and UI_4 (m.p. 506°) which resemble the tetrachloride in their chemical properties, but are less stable, have little tendency to form complex halides.

(ii) *Uranium(IV) sulfate.* If a solution of UO_3 or U_3O_8 in concentrated sulfuric acid is mixed with alcohol, and exposed to sunlight or ultraviolet light, *uranium(IV) sulfate*, $U(SO_4)_2 \cdot 4H_2O$, separates slowly in dark green rhombic plates. The salt dissolves in water to give a solution which is clear at first, but soon becomes turbid, through the deposition of a basic sulfate. The octahydrate frequently crystallizes from solution in place of the tetrahydrate, but is stable only below 19.5°. The solubility of the tetrahydrate is: at 24°, 10.9 g; at 63°, 6.7 g of anhydrous salt in 100 g of water. Corresponding figures for the octahydrate are 11.3 g at 18°, and 58.2 g at 62° (at which temperature it is metastable). The enneahydrate, $U(SO_4)_2 \cdot 9H_2O$, can also be obtained. All the hydrates are isomorphous with the corresponding hydrates of thorium sulfate.

(iii) *Uranium(IV) oxalate* may be prepared by reducing uranyl salts with oxalic acid in sunlight, or by precipitating uranium(IV) salt solutions with oxalic acid. It crystallizes in dark green cubes or prisms, is insoluble in water and dilute acids, but dissolves in alkali oxalate solutions. It thereby forms complex salts (oxalatouranate(IV) salts) which differ markedly in color from other uranium(IV) salts. Potassium tetraoxalatouranate, $K_4[U(C_2O_4)_4] \cdot 5H_2O$, which is very soluble in water, crystallizes in grey hexagonal plates. The barium salt, $Ba_2[U(C_2O_4)_4] \cdot 6H_2O$, forms red violet crystals.

13. Analytical (Uranium)

Uranium follows the course taken by iron in the usual systematic separation of the cations, and is precipitated by ammonium sulfide as the oxysulfide, UO_2S, which is soluble in acids. Addition of ammonia or alkali hydroxide to uranyl salt solutions precipitates uranium as ammonium uranate or alkali uranate, which can readily be separated from the oxide hydrates precipitated by the same reagents, since the uranate is soluble in ammonium carbonate.

With potassium hexacyanoferrate(II), uranyl salts give a brown precipitate of uranyl hexacyanoferrate(II), $(UO_2)_2[Fe(CN)_6]$. This reaction is very sensitive. It is also used, as a drop reaction, for the detection of free hexacyanoferrate(II) ions, in 'ferrocyanide' titrations.

Uranium gives a yellow microcosmic salt bead in the oxidizing flame (pale greenish yellow after cooling), and a green bead in the reducing flame.

The *microanalytical* detection of uranium can be effected by means of the double sodium uranyl acetate, $Na[UO_2(C_2H_3O_2)_3]$.

Uranium is generally determined *gravimetrically* by precipitation as ammonium uranate by means of ammonia. This is subsequently ignited in oxygen, or in air at 800°, and weighed as U_3O_8.

References

1 W. Birett, *Die Praxis der Verchromung*, Berlin 1935, 76 pp.

2 D. M. Killeffer and A. Linz, *Molybdenum Compounds*, New York 1952, 408 pp.

3 J. L. Gregg, *The Alloys of Iron and Molybdenum*, New York 1932, 507 pp.

4 C. J. Smithells, *Tungsten, Its Metallurgy, Properties and Applications*, 3rd. Ed., London 1952, 326 pp.

5 J. L. Gregg, *The Alloys of Iron and Tungsten*, New York 1934, 511 pp.

6 W. Petersen, *Schwimmaufbereitung* (Wissenschaftliche Fortschrittsberichte, Band 36), Dresden 1936, 337 pp.

7 A. M. Gaudin, *Flotation*, London 1932, 562 pp.

8 H. Havre, *Concentration des Minerais par Flotation*, Paris 1938, 464 pp.

9 J. J. Katz and E. Rabinowitch, *The Chemistry of Uranium*, New York 1951, Part I, 609 pp.

CHAPTER 6

SEVENTH SUB-GROUP OF THE PERIODIC SYSTEM: MANGANESE GROUP

Atomic numbers	Elements	Symbols	Atomic weights	Densi- ties	Melting points	Boiling points	Specific heats	Valence states
25	Manganese	Mn	54.94	7.21	1247°	2030°	0.1214	I, II, III, IV, V, VI, VII
43	Technetium	Tc	98.91*	11.50				VII
75	Rhenium	Re	186.22	20.9	3150°	—	0.0327	I, II, III, IV, V, VI, VII

* Atomic weight of the isotope of longest known half-life.

1. Introduction

(a) General

The seventh Sub-group of the Periodic System contains the elements *manganese*, *technetium*, and *rhenium*. No element of atomic number higher than that of uranium (92) occurs in Nature in substantial amounts, and the artifically prepared *trans*-*uranic* elements belong to the so-called *actinide* (or $5f$) transition series, so that neptunium (atomic number 93) bears a horizontal relationship to uranium, but is not a homologue of rhenium.

The principal member of group VIIB is manganese. Of its two homologues, the element of atomic number 43 is *unstable*, and does not exist in Nature. It can be obtained, however, as a product of nuclear transmutation processes, and has been given the name of *technetium* (from τέχνη, art), as being the first new element to be prepared artificially (Segré, 1937). *Rhenium*, the homologue of manganese with atomic number 75, which was discovered by Noddack and Tacke in 1925, is a stable element, but belongs to the very rarest of the chemical elements. In the range of valence states that they can assume, technetium and rhenium resemble manganese, but are more closely related in other respects to their neighbors in the same periods—i.e., to molybdenum and ruthenium or tungsten and osmium, respectively.

In accordance with its group number, manganese has a maximum oxidation state of seven, but also exists in numerous lower valence states. Valence states of two, four, and seven are the most common, but compounds of $+3$ and $+6$ manganese can also readily be obtained. As a rule, manganese forms compounds of normal composition only with elements of strongly electronegative character, and it may be inferred that in its normal compounds it generally bears a positive charge. The basic character of the hydroxides of manganese diminishes, and their acidic character increases, in a very marked manner as the positive charge on the atom increases (cf. p. 212 *et seq.*). Almost all manganese compounds are colored— those of bivalent manganese weakly so (pale pink), but the rest mostly strongly, and in some cases (e.g., permanganates) the color is extraordinarily intense.

The relationship between manganese and the elements of the seventh Main Group (the halogens) is very loose, and is most marked in the highest oxidation state. Thus manganese heptoxide, Mn_2O_7, may be compared with chlorine heptoxide, Cl_2O_7, and permanganic acid, $HMnO_4$, with perchloric acid, $HClO_4$. The relations are much closer between manganese and its horizontal neighbors in the same Period, chromium and iron. It shares with these elements the ability to form salts of the type $M^I_2R^{VI}O_4$—i.e., the chromates, $M^I_2CrO_4$, manganates, $M^I_2MnO_4$, and ferrates, $M^I_2FeO_4$—and likewise forms insoluble oxides from its lower valence states. The closest resemblance is found between manganese and its right hand neighbor, iron. This shows itself not only in their analytical reactions, but also in the almost invariable association of the two elements in their natural occurrence.

From the standpoint of the Kossel theory, manganese has a maximum valence of seven, because it follows seven places after an inert gas (argon). It always functions as electropositive because it is not closely followed by an inert gas, so that a stable configuration cannot be built up by the acquisition of a small number of electrons. From the more modern standpoint, the electron configuration of the manganese atom is $KL3s^23p^63d^54s^2$, so that the 'outer' electron levels are completely different from those of the halogen atoms (ns^2np^5), but the Mn^{7+} and Cl^{7+} ions would both have the inert gas configuration. Quantum mechanical considerations indicate that manganese can form a set of tetrahedral bonds, and that the $[MnO_4]^-$ ion is strictly comparable with the $[ClO_4]^-$ ion if, as is very probable, the binding forces are quantum mechanical exchange forces. The binary compounds of the highest valence states of manganese are likely to be covalent, rather than ionic, in character.

Even although rhenium displays, in general, the same valence states as manganese, it differs in a very characteristic manner in the marked predominance of the $+7$ state. This accords with the general rule that in the Sub-groups formed by the transition elements, the tendency to exercise the highest valences is much stronger for the heaviest elements than for the lighter elements. The tendency of rhenium to function in the heptapositive state is so strong that it dominates the whole chemistry of rhenium.

From the position of technetium in the Periodic System, it may be assumed that its chemical properties must resemble those of manganese and rhenium, but will be more closely allied to the latter. As yet, however, but little is known experimentally of its properties.

The heats of formation of some simple compounds of manganese and rhenium are collected in Table 23.

TABLE 23

HEATS OF FORMATION OF MANGANESE AND RHENIUM COMPOUNDS

MnO	Mn_3O_4	Mn_2O_3	MnO_2	Mn_2O_7	ReO_3	Re_2O_7	
93.1	336.5	232.7	125.4	ca. 165	82.5	297.5	kcal per mol of compound
93.1	112.2	116.4	125.4	82	82.5	148.7	kcal per g atom of metal

$MnCl_2$	MnI_2	MnS	MnSe	Mn_5N_2	ReS_2	
112.7	49.8	44.6	26.3	57.8	70.5	kcal per mol of compound
112.7	49.8	44.6	26.3	11.6	70.5	kcal per g atom of metal

(b) Alloys

A brief survey of the behavior of manganese towards other metals, and towards non-metals which form alloy-like systems, is given by Table 24 (p. 209). Rhenium is not included in the table since, except for those quoted in Table 17 (p. 124), only a few rhenium alloys have as yet been studied.

A comparison of the data of Table 24 with those of Tables 16 and 17 will show that the behavior of manganese towards other metals in its broad features is substantially similar to that of chromium. This similarity shows itself not only with respect to the miscibility with other metals and capacity for forming compounds, but also frequently in the existence of compounds of analogous composition in the corresponding systems. This is even true to some extent for the compounds with non-metals (C and Si). In other respects, however, a comparison of these tables shows some pronounced differences in behavior between manganese and chromium. Thus, unlike chromium, manganese forms compounds with most of the heavy metals of the Main Groups of the Periodic Table. Further points of difference emerge when the individual chromium and manganese alloy systems are compared with due regard to the temperature-dependence of the relations between the components as summarized by the equilibrium diagrams. The technically important systems Cr-Fe and Mn-Fe provide illustrations of this (cf. Figs. 40 and 42, on p. 266). Tables 17 and 24 record complete or almost complete miscibility between the components in the solid state for both systems. This statement is valid at temperatures which lie not very much below the melting point. In the system Cr-Fe, this miscibility persists upon cooling down to ordinary temperature, but in the system Mn-Fe, two fairly broad miscibility gaps appear on cooling. At 500°, one of these extends from about 2 to 31 atom-% Mn, and the other from about 48 to 63 atom-% Mn. The reason for this unmixing is that only the two γ-modifications of manganese and iron are (almost) completely miscible. α-iron, on the other hand, can form mixed crystals only with very small amounts of Mn, and α-Mn can incorporate Fe only to a limited extent. Hence the transformation of the γ-modifications, which are stable at high temperatures, into the α-modifications stable at lower temperatures (cf. pp. 211, 262), results in the creation of a wide miscibility gap at lower temperatures. The transition temperature of γ-iron is lowered more and more as manganese is progressively added, so that γ-Fe is stable even at 500° if the manganese content exceeds 31 atom-%. Hence a third region of homogeneity, consisting of solid solutions of Mn in γ-iron, interposes itself in the miscibility gap between α-Fe crystals saturated with Mn (with about 2 atom-% Mn), and α-Mn crystals saturated with iron (with 37 atom-% Fe, or 63 atom-% Mn). (See Fig. 42, p. 266). At 500°, this third homogeneous range extends from 31 atom-% Mn to about 48 atom-% Mn, the limit of solubility of γ-Fe for Mn at that temperature. Since the ability to form mixed crystals is dependent upon the crystal structure, it is not surprising that the existence of *three modifications* of manganese, stable at different temperatures, as compared with the *one* stable modification of chromium, should occasion substantial differences in the way the capacity to form mixed crystals varies with temperature.

2. Manganese (Mn)

(a) Occurrence

Except for iron, manganese is the most abundant of the heavy metals. It is found in small amounts almost everywhere, but actual ores of manganese are widely distributed. The most important of these is *pyrolusite*, MnO_2. It is found in vast quantities in Russia (Caucasus), India, West Africa (the Gold Coast), South Africa, and South America (Brazil and Chile). Other oxides of manganese which are important as ores are *braunite*, Mn_2O_3, *manganite*, $MnO(OH)$, and *hausmannite*, Mn_3O_4. The carbonate, $MnCO_3$, *manganese spar*, is also of some importance as an ore. The ores of iron, the neighbor of manganese in the Periodic System, almost

TABLE 24

MISCIBILITY AND COMPOUND FORMATION BETWEEN MANGANESE AND THE ELEMENTS OF THE MAIN GROUPS AND SUB-GROUPS

(Symbols have the same meanings as in Table 9)

Main Group II

Be	Mg	Ca
—	$s<\infty$	—
	Leg.	
$MnBe_2$	—	

Main Group III

B	Al	Ga	In	Tl
	$s>0$			$liq<\infty$ $s>0$
MnB	Al_6Mn* 710°	—	$InMn_3$	0
MnB_2	Al_4Mn* 820°			
	Al_3Mn* 990°			
	$AlMn*$ 1160°			

Main Group IV

C	Si	Ge	Sn	Pb
$s<\infty$	$s>0$	—	$s>0$	$liq<\infty$ $s o$
Mn_4C	Mn_3Si* 1084°	$Mn_{13}Ge_4$ 900°	Mn_4Sn 990°	0
Mn_3C 1245°	Mn_5Si_3 1300°	Mn_5Ge_2 920°	Mn_2Sn* 897°	
Mn_7C_3	$MnSi$ 1270°	Mn_5Ge_3 932°	$MnSn*?$ 548°	
	$MnSi_2*$ 1144°	Mn_3Ge_2 745°		

Main Group V

N	P	As	Sb	Bi
$s<\infty$	$s o$	$s>0$	$s>0$	$liq<\infty$
Mn_4N	Mn_4P	Mn_2As 1029°	Mn_2Sb 948°	$MnBi?$
Mn_5N_2	Mn_2P 1390°	$MnAs$	Mn_3Sb_2* 872°	
Mn_3N_2	MnP		$MnSb$ 809°	
	MnP_3			

Main Group VI

S	Se	Te
$liq<\infty$ $s o$	—	—
MnS 1580°	$MnSe$	$MnTe$
MnS_2	$MnSe_2$	$MnTe_2$

Sub-group VIII

Fe	Co	Ni	Ru	Rh	Pd	Os	Ir	Pt
$s\sim\infty$	$s\infty$	$s\infty$	—	—	$s\sim\infty$	—	—	—
0	0	0	—	—	$Mn_2Pd_3\square$ 1175°	—	—	Leg.
					$MnPd$ 1515°			

Sub-group I

Cu	Ag	Au
$s\infty$	$liq<\infty$ $s<\infty$	$liq>0$ $s>0$
0	0	$Mn_3Au\square$ 1237°
		$MnAu$
		$MnAu_2\square$
		$MnAu_3\square$

Sub-group II

Zn	Cd	Hg
$s<0$	$s>0$	$liq<\infty$
Zn_7Mn*	—	—
Zn_3Mn*		

always have a more or less considerable manganese content. Siderite (spathic iron ore) and brown hematite, in particular, are often relatively rich in manganese. In certain countries (e.g., Germany), the manganiferous iron ores represent the principal source of manganese.

Manganese is also present in minimal amounts in plants and in the animal body. It is an indispensable constituent, because of its catalytic role in the chemical processes which occur within the cell of the living organism.

(b) History

Pyrolusite, the dioxide of manganese, has been known since ancient times, and its use as 'glass maker's soap' was then already known. It was considered to be a variety of magnetic oxide of iron (*magnes*). Pyrolusite, which is black and shows no magnetic forces, was termed 'the female lodestone', by Pliny, to distinguish it from the brown magnetite. In medieval times a distinction was drawn between *magnes* or *magnesia lapis* (= magnetic iron ore) and *magnesia* or *pseudomagnes*, the false lodestone, (= pyrolusite). The German name 'Braunstein' dates back to the alchemist Basil Valentine, who gave this name to the (generally grey-black) mineral because of its property of giving brown glazes on earthenware. The glassmakers, on the other hand, called it 'glass-maker's soap', because of its property of decolorizing iron-containing glass, and altered its older name of *magnes* into *manganes* or *lapis manganensis*, probably to accord with the Greek μανγανίζειν = to purify.

Until about the middle of the 18th century it was held that pyrolusite was an iron ore, but it was eventually concluded that it must contain some other, as yet unknown metal. Clear proof that this was so was furnished by Scheele, in a treatise on pyrolusite laid before the Academy of Sciences in Stockholm, in 1774. In the same year Gahn succeeded in isolating the 'metal of pyrolusite' as a regulus, by strongly heating a mixture of pyrolusite and charcoal. It then received the name *manganesium*. To avoid confusion with *magnesium*, which had been discovered in the meanwhile, this name was later altered to the German Mangan, or British (and French) manganese (latin manganium).

(c) Preparation

Metallic manganese can most simply be obtained from pyrolusite by the aluminothermic process. Since aluminum reacts too violently with pyrolusite itself, the latter is first converted to the red oxide of manganese, Mn_3O_4, by strong heating:

$$3MnO_2 = Mn_3O_4 + O_2.$$

Determination of the loss in weight indicates when the expulsion of oxygen has proceeded far enough. The red manganese oxide is then mixed with aluminum and ignited:

$$3Mn_3O_4 + 8Al = 9Mn + 4Al_2O_3 + 602 \text{ kcal.}$$

It is advantageous to mix in slightly less than the theoretical quantity of aluminum. Slagging of the excess manganese oxide is facilitated by adding some fluorspar.

The older process of preparation, i.e., the reduction of the oxide with carbon, is no longer of any importance for the preparation of manganese itself, since it furnishes a manganese which is more or less rich in carbon, which renders it useless for many purposes. However, the process is often used for *iron-manganese alloys* (ferromanganese and spiegeleisen), in which the carbon content is not harmful. In such cases it is usual simply to add the manganese ore directly in appropriate amounts to the iron ore being smelted.

(d) Properties

Manganese resembles metallic iron in appearance, but differs from iron in being hard and very brittle. It is silvery white, like iron, when it is pure, but is grey, like cast iron, if it contains carbon. Its density (7.2) is also close to that of iron but its melting point (1247°) is much lower than that of pure iron. As prepared in a regulus, by the aluminothermic method, the metal takes a rather characteristic tarnish color in the air, but is not then oxidized any further even when it is heated. In a fine state of subdivision, however, manganese is readily oxidized, and may even be pyrophoric.

Finely divided manganese can decompose water, especially if ammonium chloride is added (to hinder the formation of insoluble $Mn(OH)_2$). The normal potential of manganese in contact with manganese(II) salt solutions is $+1.1$ volt. Manganese is thus the most strongly electropositive metal after the alkalis, alkaline earth metals, rare earth metals and aluminum (cf. Vol. I, Table 4, p. 30).

As prepared by the aluminothermic process, manganese is usually a mixture of two modifications, α- and β-Mn, of which the former (density 7.21), is stable at ordinary temperature and the latter (density 7.29) at higher temperatures (between 742° and 1070°). Both are hard and brittle, and have complicated cubic crystal structures. Electrolytically deposited manganese, however, which is fairly soft and ductile, forms a face-centered tetragonal lattice $a = 3.77$ Å, $c = 3.53$ Å, Mn $\leftrightarrow$ Mn $= 2.67$ Å (γ-Mn, density 7.21, stable between 1070° and 1160°). It gradually changes into brittle α-Mn. Above 1160°, a fourth modification, δ-Mn, appears, as was shown by Grube (1938).

As corresponds with its position far above hydrogen in the electrochemical series, manganese dissolves readily in dilute acids, forming the bivalent ion Mn^{++} and liberating hydrogen:

$$Mn + 2H^+ = Mn^{++} + H_2.$$

It dissolves in concentrated sulfuric acid, with the evolution of sulfur dioxide*. and in nitric acid with the evolution of nitric oxide.

Manganese burns in chlorine to the dichloride, $MnCl_2$. It reacts very vigorously with fluorine, forming MnF_2 and MnF_3. It catches fire in nitrogen above 1200°, and then burns with a very smoky flame to the nitride Mn_3N_2. (In addition to this compound it forms two other nitrides—cf. Table 24, p. 209). It also inflames with phosphorus. It unites directly with sulfur, carbon, silicon, and boron also, but not with hydrogen.

Heusler's alloys. It was discovered by Heusler in 1898 that manganese formed alloys with many metals—e.g., with aluminum, tin, or antimony—which had *ferromagnetic* properties, although they contained no ferromagnetic metal. The ferromagnetism of these alloys can be considerably augmented by the addition of copper. The ferromagnetism appears to be associated with the presence of certain intermetallic compounds in these alloys, but often attains its maximum value when these compounds are not present in the pure form, but in the form of mixed crystals. Alloys of chromium are also known which display ferromagnetism.

(e) Uses of Manganese and its Compounds

The principal use of manganese is as a deoxidant for iron and steel. For this purpose it is used chiefly in the form of iron-manganese alloys (*ferromanganese* and *spiegeleisen*). It is also used as a deoxidizing additive for other alloys also, and especially for bronzes (manganese bronzes). True alloys of manganese and copper also find applications. *Manganin,* an alloy of 84 parts copper, 12 parts manganese

---

* Concentrated sulfuric acid attacks compact manganese only very slowly in the cold, but rapidly when it is heated.

and 4 parts nickel, is used for precision resistances, because of the very low temperature coefficient of its electrical conductivity.

The consumption of manganese ores (and highly manganiferous iron ores) in the steel
industry is enormous. The use of manganese ores in other branches of industry is insignificant in comparison, although quite considerable in actual amount.

The world production of manganese ores (not including manganiferous iron ores, etc.)
in 1936 amounted to 5.1 million tons, with a manganese content of 2.2 million tons. Of this
production, 54.9% (based on the manganese content) came from Russia, 18.9% from
British India, 9.9% from the Gold Coast, 5.5% from the Union of South Africa, 3.2% from
Brazil and 0.64% from the United States.

Pyrolusite is used in glass making as a decolorizing agent for glass which
contains iron (glass maker's soap). It is also used to give a violet color to glasses,
and for brown glazes on earthenware and tiles, etc. Its use as an oxidizing agent
in Leclanché cells, and especially in dry cells, is important. It is also used in the
production of varnishes as a 'drying agent', because it has the property of catalytically accelerating the oxidation ('drying') of linseed oil by atmospheric oxygen.
It now occupies only an unimportant role in the chlorine industry (see the Weldon
process, Vol. I, p. 779 et seq.).

The technically most important compounds of manganese are *potassium permanganate*, $KMnO_4$ (generally referred to simply as 'permanganate'), which finds a
multiplicity of uses as an oxidant, bleaching agent, and disinfectant, and also
manganese(II) chloride, $MnCl_2$ and *manganese(II) sulfate*, $MnSO_4$. These and other
manganese salts are used in dyeing and printing, and also as a growth stimulant
for seed. Manganese compounds are said to have the property of stimulating
certain plant ferments, which are of importance for the growth of the seed, into
enhanced activity. Various manganese compounds, especially manganese(II)
chloride, are also used like manganese dioxide, for the preparation of siccatives
(drying agents) for varnishes and paints. Many manganese compounds, especially
the naturally occurring *umbers*, are used as artists' pigments; the most important
artificial pigments are manganese brown (mineral bister), and occasionally
manganese white, $MnCO_3$, manganese green (Kassel green) and manganese
violet (permanent violet).

3. Compounds of Manganese

Manganese exhibits oxidation states of $+1$, $+2$, $+3$, $+4$, $+5$, $+6$ and $+7$ in
its compounds. Oxides corresponding to most of these valence states exist.

The lowest oxide, MnO, has a well defined basic character. The next higher
oxide, Mn_2O_3, is also basic, but the dioxide, MnO_2, is amphoteric. The existence
of manganese trioxide, MnO_3, in the free state is doubtful. Compounds derived
from this oxide, the manganates, $M^I_2[MnO_4]$, are salts, in which manganese is
present within the acid radical. From the highest oxide of manganese, Mn_2O_7, is
derived permanganic acid, $HMnO_4$, which is one of the strongest acids known.
The rule that the basic character of any element diminishes, and the acidic
character increases, with increase in oxidation state, is thus particularly well
exemplified by the oxides of manganese.

The various *oxides* and *hydroxides* of manganese will first be considered together.

The individual oxides, and the oxidation states of manganese which they define, will then be discussed individually.

As far as is known, the compounds of manganese with S, Se, and Te correspond in composition to the lower oxides of manganese. The compounds with the non-metals of the IIIrd, IVth and Vth Main Groups are quite unrelated in composition to the oxides of manganese or the manganese compounds of salt-like character. They are therefore listed in Table 24, p. 209, together with the intermetallic compounds of manganese, with which they may be compared in composition.

The most important compounds of manganese are derived from Mn^{II}, Mn^{IV} and Mn^{VII}. Mn^{III} and Mn^{VI} can also be prepared readily, and it is now known that even manganese(V) compounds can be obtained quite easily in a pure state. Manganese(I) compounds are known in the form of the cyano salts, $M^I{}_5[Mn(CN)_6]$.

(a) Oxides and Hydroxides

The following simple oxides of manganese are known:

MnO   Manganese monoxide, manganese(II) oxide, green manganese oxide.
Mn_2O_3 Manganese sequioxide, manganese(III) oxide, black manganese oxide.
MnO_2 Manganese dioxide, manganese(IV) oxide, pyrolusite.
[MnO_3 Manganese trioxide, manganese(VI) oxide. The existence of this compound in the free state is doubtful.]
Mn_2O_7 Manganese heptoxide, manganese(VII) oxide.

In addition to these compounds, an intermediate oxide Mn_3O_4 is known, which is called red manganese oxide, after its color. It crystallizes according to the spinel type (cf. Vol. I, p. 355), but the crystal lattice is distorted so as to be tetragonal ($a = 5.75$ Å, $c = 9.42$ Å). It can be formally regarded as a manganese manganite, $Mn^{II}{}_2[Mn^{IV}O_4]$. Mn_3O_4 occurs native as *hausmannite* (brilliant pyrolusite), generally in association with braunite but in larger quantities than the latter, from which it is not readily distinguished by its external properties.

Of the two possible formulations $Mn^{III}{}_2[Mn^{II}O_4]$ and $Mn^{II}{}_2[Mn^{IV}O_4]$, permitted by the structure determination, the latter is the correct one, as Verwey and De Boer (1936) showed, from measurements of the X-ray scattering power of the manganese ions belonging to and not belonging to the MnO_4 radical, respectively (cf. Vol. I, p. 213). From the observation that Mn_3O_4, like Mn_2O_3, dissolves in many concentrated acids (e.g., phosphoric acid) with a violet color (i.e., forming Mn^{+++} ions), whereas MnO_2 is insoluble, it was formerly inferred that Mn_3O_4 should be regarded as a compound of 1 MnO with 1 Mn_2O_3, rather than of 2 MnO with 1 MnO_2. However, the formation of Mn^{+++} ions can be explained as arising from the reduction of Mn^{++++} ion by Mn^{++} ions during dissolution, and it is not permissible to use the insolubility of MnO_2 as a basis for any conclusions about the solubility of compounds derived from it, but having different structures.

MnO crystallizes cubic, with the rock salt structure (a = 4.43 Å), and MnO_2 tetragonal, with the rutile structure (a = 4.40 Å, c = 2.87 Å). Mn_2O_3 usually crystallizes cubic, with the scandium oxide structure (cf. p. 37–8). In addition to the stable cubic modification, (a-Mn_2O_3), an unstable tetragonal modification is also known (cf. p. 214) In the structures of MnO, a-Mn_2O_3 and MnO_2, manganese has the same coordination number, *six*, with respect to oxygen. The distance Mn ↔ O is 2.21 Å in the MnO structure, 2.00 to 2.03 Å in Mn_2O_3, and about 1.89 Å in MnO_2. It thus decreases as the charge on the manganese ion increases.

According to Le Blanc (1934), the oxides MnO, Mn_3O_4, and Mn_2O_3 can take up an excess of oxygen in solid solution, up to the compositions $MnO_{1.13}$ in the case of MnO,

$Mn_3O_{4\cdot26}$ for Mn_3O_4, and up to $Mn_2O_{3\cdot16}$ for Mn_2O_3. Unlike PbO_2, MnO_2 cannot give up any oxygen without the destruction of its crystal lattice.

(*i*) *Manganese(II) oxide*, MnO, is obtained in the form of a greenish grey to dark green powder by the reduction of the higher oxides of manganese with hydrogen or carbon monoxide, or by igniting manganese carbonate in hydrogen or nitrogen. In the finely divided state it is oxidized very readily. It can be reduced only with difficulty, however,—e.g., only at very high temperatures by hydrogen. Manganese(II) oxide is practically insoluble in water. The oxide is rarely found native, as *manganosite*.

(*ii*) *Manganese(II) hydroxide*, $Mn(OH)_2$, is thrown down as a white precipitate which rapidly turns brown in air, when manganese(II) salt solutions are treated with alkali hydroxides. It is only incompletely precipitated by ammonia, and may not be precipitated at all in the presence of ammonium salts. The reason for this is firstly that, as in the case of magnesium hydroxide, the concentration of hydroxyl ions under these conditions is only small in relation to the solubility product ($K_{sp} = [Mn^{++}] \cdot [OH^-]^2 =$ about 10^{-12}). In addition, the concentration of manganese ions is significantly reduced through complex formation with the ammonia (Weitz, 1925).

Manganese(II) hydroxide is occasionally found in Nature, in the form of white transparent leaflets of *pyrochroite*. These also gradually turn brown upon exposure to air. The brown coloration is due to oxidation, which can lead, through the intermediate oxides or their hydrates, to the ultimate formation of manganese dioxide. A variety of oxidation products of $Mn(OH)_2$ has been obtained, with different structures and compositions; they are mostly non-Daltonide compounds. For example, a proportion of the OH^- ions of $Mn(OH)_2$ (which has a layer lattice structure) may be replaced by O^{2-} ions. The formation of such oxidation products involves typical *topochemical processes* (Feitknecht, 1944).

The crystal structure of manganese(II) hydroxide resembles that of magnesium hydroxide (brucite), $a = 3.34$ Å, $c = 4.68$ Å, $Mn \leftrightarrow O = 2.30$ Å. The heat of formation of $Mn(OH)_2$ from MnO and liquid H_2O is about 4.2 kcal per mol.

(*iii*) *Manganese(III) oxide* (manganese sesquioxide), Mn_2O_3, occurs in Nature as *braunite*, in brownish black masses, generally accompanying other manganese ores. It is obtained artificially as a black amorphous powder by heating manganese dioxide in air at 530–940°, or by igniting manganese(II) salts in air or oxygen.

Mn_2O_3 loses oxygen when it is heated above 940° in air, or above 1090° in oxygen, and is converted into the oxygen-poorer manganese(II,IV) oxide, Mn_3O_4. Provided that the latter has been heated sufficiently long, it does not take up oxygen again on cooling. This fact is of importance in the gravimetric determination of manganese.

If Mn_2O_3 is prepared by cautious heating of MnO (or Mn_3O_4 which has not been deactivated by strong ignition) in oxygen (Le Blanc, 1934), or by dehydrating MnO(OH) in a vacuum at 250° (Dubois, 1934), an unstable tetragonal modification is first obtained. This is converted monotropically into the stable cubic modification by prolonged heating. The cubic modification is obtained directly by the thermal decomposition of MnO_2. The cubic α-Mn_2O_3, as already stated, is isotypic with Sc_2O_3. Its cubic unit cell, containing $16Mn_2O_3$ (which thus corresponds to a cell built up from 8 unit cubes of fluorite), has an edge of 9.41 Å. According to Verwey and De Boer (1936), the structure of the tetragonal β-Mn_2O_3 can be described as a Mn_3O_4 lattice, but containing a number of vacant positions to correspond with the smaller Mn content. Unlike Mn_3O_4, however, the Mn ions in this structure are either all trivalent or else, if Mn^{2+} and Mn^{4+} ions are present, they must be distributed at random over all the positions.

When manganese(III) oxide is heated in hydrogen, it is first reduced to manganese(II,IV) oxide, Mn_3O_4, at about 230°, and this in turn to the green monoxide above 300°.

Dissolution of manganese(III) oxide in acids either produces manganese(III) salts (cf. manganese(III) sulfate and manganese(III) chloride), or leads to formation of manganese(II) salts and manganese dioxide, depending upon the nature of the acid and the temperature.

The hydrate of manganese(III) oxide, $Mn_2O_3 \cdot H_2O$ (or manganese(III) metahydroxide, $MnO(OH)$), occurs native as *manganite* (brown manganese ore). It is black-brown, and occasionally is found in well formed crystals which are isomorphous with goethite, $FeO(OH)$, and diaspore, $AlO(OH)$. The hydrates obtained by precipitation from aqueous solution also correspond in composition to manganite after drying at 100°.

(*iv*) *Manganese Brown.* Artificially prepared manganese(III) oxide hydrate is used as a brown-black pigment (manganese brown, mineral bister). It is prepared by treating manganese(II) chloride solution with bleaching powder and lime water. The pigment is used in cloth dyeing and printing, and for that purpose formed directly on the fiber by the oxidation of a fabric soaked in manganese(II) chloride. A chestnut brown pigment, obtained by grinding and firing a naturally occurring mixture of manganese oxide hydrate with the hydrated oxides, of iron and aluminum, is known as *umber*.

(*v*) *Manganese dioxide*, MnO_2, in the form of pyrolusite, constitutes the most important ore of manganese. For crystal structure see p. 213.

Several varieties of native manganese dioxide are recognized, according to their crystallographic properties,—*polianite*, (= recognizably crystallized manganese dioxide), *pyrolusite* or soft manganese ore (manganese dioxide derived from other manganese minerals, and retaining their morphological structure), grey pyrolusite (fine grained or quite dense), *psilomelane* (manganese dioxide in the form of dense nodules, and generally very impure), *wad* or *bog manganese* (also nodular, but loose), and *manganese black* (very finely divided and of sooty appearance). Recent investigations of artificial manganese dioxide have shown that at least three different structures may be formed, which are differentiated as α-, β- and γ-MnO_2 (Dubois 1936, Glemser, 1939, Walkley and Wadsley 1949). β-MnO_2 has the structure of natural manganese dioxide (polianite). In γ-MnO_2, a proportion of the O^{2-} ions is replaced by OH^- ions (with the simultaneous substitution of Mn^{3+} ion for Mn^{4+} ions). Preparations of γ-MnO_2 also frequently contain foreign metals, replacing part of the Mn^{4+} ions (Feitknecht, 1944).

Manganese dioxide is grey to grey black in color, both in its naturally occurring forms (except wad, which also occurs in brownish tones) and when prepared artifically in the anhydrous state (e.g., by the mild ignition of manganese(II) nitrate). Manganese dioxide begins to lose oxygen when heated above 530° in air, but loss of oxygen occurs at a considerably lower temperature in a vacuum or in the presence of reducing substances.

Manganese dioxide is readily attacked by strong reducing agents in the presence of dilute acids, in spite of its extremely low solubility. With sulfurous acid it forms manganese(II) dithionate (1), and with hydrogen peroxide and sulfuric acid it gives manganese(II) sulfate and oxygen (2):

$$MnO_2 + 2H_2SO_3 = MnS_2O_6 + 2H_2O \tag{1}$$

$$MnO_2 + H_2O_2 + H_2SO_4 = MnSO_4 + 2H_2O + O_2 \tag{2}$$

Hydrogen peroxide is also decomposed in the absence of acids, but in this case only one half as much oxygen is evolved as in acid solution, since manganese dioxide is not reduced but merely acts catalytically:

$$H_2O_2 = H_2O + \tfrac{1}{2}O_2.$$

The evolution of oxygen from potassium chlorate is also catalytically accelerated by manganese dioxide (cf. Vol. I, p. 806).

The use of manganese dioxide as an oxidizing agent in Leclanché cells and dry batteries is based upon the ease with which it gives up its oxygen to other substances. Manganese dioxide mixed with copper oxide is used for the catalytic combustion of carbon monoxide—e.g., in breathing and rescue apparatus (Vol. I, p. 443).

If concentrated hydrochloric acid is added to manganese dioxide, chlorine is evolved, especially on warming. This reaction is widely employed for the generation of chlorine on a small scale.

The mechanism of the reaction involves the initial formation of manganese tetrachloride, which readily decomposes into manganese trichloride and free chlorine. The trichloride also decomposes on warming:

$$MnO_2 + 4HCl = MnCl_4 + 2H_2O$$

$$MnCl_4 = MnCl_3 + \tfrac{1}{2}Cl_2; \quad MnCl_3 = MnCl_2 + \tfrac{1}{2}Cl_2.$$

Manganese dioxide is not attacked by cold concentrated sulfuric acid, but when it is warmed to $110°$, oxygen is evolved and manganese(III) sulfate is formed. Further evolution of oxygen takes place on stronger heating, and manganese(III) sulfate is converted to the manganese(II) compound.

(vi) *Manganese dioxide hydrate* is formed as a brown, or sometimes blackish precipitate by the oxidation of manganese salts, or by the reduction of manganates or permanganates in alkaline solution. The oxidation of manganese(II) salts to manganese dioxide hydrate can be carried out in alkaline solution—e.g., by bleaching powder or by atmospheric oxygen; see p. 218 for the oxidation in acid solution. Manganese can be deposited electrolytically upon the anode, as manganese dioxide hydrate, from solutions rich in nitric acid.

The products obtained from alkaline solution invariably contain alkali (or alkaline earth). The preparation of pure manganese dioxide hydrate must therefore always be carried out from acid or neutral solutions. In this case, however, the oxygen content of the compound does not as a rule correspond exactly to that required for a hydrate of the dioxide. It is doubtful whether manganese dioxide hydrate has a well defined water content—e.g., corresponding to the formula $Mn(OH)_4$.

Manganese dioxide hydrate is important because it is much more reactive than the anhydrous, grey manganese dioxide. It is therefore more suitable for use as an oxidant than natural pyrolusite—e.g., in the dyestuffs industry and in varnish manufacture.

Alkali hydroxides and other bases are not merely adsorbed by manganese dioxide hydrate, but are bound chemically. Compounds of manganese dioxide with basic oxides have been obtained in a well crystallized state, and with stoichiometrically well defined composition. Thus manganese dioxide hydrate displays in this respect the properties of an acid, even if a weak one. It salts are known as *manganites* [cf. Feitknecht, *Helv. Chim. Acta,*

28 (1945), 129, 149]. On the other hand, manganese dioxide hydrate can also form salts with strong acids, although these are rather unstable because of the strong tendency of manganese to pass over to the bivalent state in acid solutions. Thus manganese dioxide is amphoteric in character, although it is so extremely insoluble that neither its acidic nor its basic character appears very strongly pronounced.

(*vii*) *Manganese heptoxide and Permanganic acid. Manganese heptoxide*, Mn_2O_7, may be obtained by the action of concentrated sulfuric acid on potassium permanganate:

$$2KMnO_4 + H_2SO_4 = K_2SO_4 + Mn_2O_7 + H_2O.$$

It forms a dark, heavy oil, with a greenish brown luster (density 2.4). It is quite stable in dry air at ordinary temperature. It decomposes explosively into manganese dioxide and oxygen when it is warmed.

Manganese heptoxide also decomposes when it is added to a small amount of water, because of the rise of temperature brought about by its considerable heat of solution (about 12 kcal per mol. of Mn_2O_7). It dissolves without decomposition, however, in a large excess of cold water. The solution is violet, and has the same absorption spectrum as a permanganate solution (cf. p. 229). From this it may be concluded that it contains permanganate ions, MnO_4^-, formed by the process

$$Mn_2O_7 + H_2O = 2MnO_4^- + 2H^+.$$

The *permanganic acid* which is thus present in solution in dissociated form cannot be isolated in the anhydrous state. The solution can be concentrated only up to a content of about 20% $HMnO_4$ without decomposition occurring. Conductivity measurements have shown that permanganic acid is a very strong acid. It is almost completely dissociated in solution, the apparent degree of dissociation determined from conductivity coefficients being 93% in 0.1 molar solution at 25°.

Permanganic acid is also formed by the reaction of lead dioxide on manganese salts, in the presence of concentrated sulfuric or nitric acid:

$$2Mn^{++} + 5PbO_2 + 4H^+ = 2MnO_4^- + 5Pb^{++} + 2H_2O.$$

This reaction is of importance in analysis, since even traces of manganese may be detected from the intensive violet color of the MnO_4^- ions. The test fails if Cl^- ions are present; this is because hydrochloric acid reduces permanganic acid, being oxidized to chlorine.

Mn^{++} ions are very readily oxidized to MnO_4^- ions by peroxymonophosphoric acid, H_3PO_5 (Schmidlin, 1910). If a solution of peroxymonophosphoric acid (free from hydrogen peroxide) is available, the detection of manganese by means of this reaction is more convenient than with PbO_2 and HNO_3 or H_2SO_4.

The oxidizing properties of permanganic acid are very powerful and free manganese heptoxide reacts even more vigorously; combustible substances are attacked by the latter, and inflame. However, the heptoxide dissolves without decomposition in acetic anhydride, forming a violet solution.

(b) Manganese(II) Salts

The manganese(II) salts are derived from MnO, the lowest oxide of manganese, by replacing the oxygen by acid radicals. They are mostly pale pink in color, as also are their concentrated aqueous solutions which contain the Mn^{++} ion.

Manganese(II) salts are quite stable in the dry state, and also in solutions containing excess acid. In alkaline solution, however, manganese(II) hydroxide is formed, and this has a strong tendency to become oxidized up to brown manganese dioxide hydrate.

Neutral manganese(II) salt solutions also gradually assume a brownish color, and ultimately a flocculent brown deposit forms. This is because manganese(II) hydroxide is formed in traces in the solution, and is then oxidized by atmospheric oxygen to the extremely insoluble manganese dioxide hydrate. It is not certain whether formation of manganese(II) hydroxide is due to a slight hydrolysis, or to the action of alkali dissolved out of the glass.

Powerful oxidants, such as ozone and peroxysulphates, can oxidize manganese(II) salts even in solutions which contain a moderate amount of free acid. The ability of peroxysulphates to precipitate manganese quantitatively as the hydrated dioxide, from solutions containing a little free sulfuric acid, is utilized in quantitative analysis.

Manganese(II) salts have a catalytic action on the progress of many oxidations, especially those involving atmospheric oxygen. This underlies the use of manganese compounds as *siccatives*. These are preparations which, when dissolved or suspended in linseed oil, make the oil dry out quickly, by accelerating its oxidation by atmospheric oxygen. Linseed oil (or other drying oils, such as tung oil) containing such a siccative is known as a *varnish*. Manganese salts, in particular, are used as siccatives, especially the manganese salts of resinic and linoleic acids (manganese resinate and manganese linoleate), since these dissolve more readily in linseed oil than the manganese salts of inorganic acids. Lead compounds may also be employed as siccatives, but are inferior in efficiency to manganese driers. Varnishes containing both manganese and lead have the best drying properties.

Most of the ordinary manganese(II) salts are soluble in water, especially the chloride, nitrate, sulfate, acetate, and thiocyanate. The sulfide, phosphate, and carbonate are sparingly soluble.

Double salts of manganese(II) with alkali salts also exist. In solution these are almost completely dissociated into their components. Thus the tendency to form complex anions is much less strongly developed for bivalent manganese than for trivalent.

Mn^{2+} occupies a special position in the series of ions from Ca^{2+} to Zn^{2+}, in the same way as the Gd^{3+} ion in the series of lanthanide ions (cf. p. 480). This is because the Mn^{2+} ion with the electron configuration $KL3s^23p^63d^5$ contains a half-completed 3d level. This special position shows itself in various properties of the manganese(II) compounds (e.g., molecular volumes), when they are compared with the compounds of the neighboring elements (Klemm, 1942).

(*i*) *Manganese(II) chloride* (manganous chloride), $MnCl_2$, can be obtained in the anhydrous state by the action of dry hydrogen chloride on manganous oxide, manganous carbonate or metallic manganese. It may also be prepared by burning manganese in chlorine (heat of formation—p. 207). The anhydrous chloride forms pink crystalline leaflets, isomorphous with calcium chloride and cadmium chloride. It melts at 650°, and begins to volatilize at a red heat (b.p. 1190°) in a current of hydrogen chloride. It is not reduced by hydrogen, but is converted into Mn_2O_3 by heating in oxygen or water vapor. Manganous chloride is very soluble in water (72.3 g in 100 g of water at 25°).

The tetrahydrate, $MnCl_2 \cdot 4H_2O$, generally crystallizes from aqueous solutions. This exists in two modifications, the unstable form being isomorphous with $FeCl_2 \cdot 4H_2O$.

Manganese(II) chloride also forms a 6-hydrate (isomorphous with $CoCl_2 \cdot 6H_2O$) and a dihydrate. All the water may be driven off by heating the compound in a stream of hydrogen chloride. The anhydrous chloride is very hygroscopic, and evolves 16 kcal per mol, when it is dissolved in much water. The anhydrous chloride is also soluble in absolute ethanol, and crystallizes from the solutions with 3 molecules of alcohol of crystallization.

Anhydrous manganese(II) chloride can add on 6 molecules of ammonia, and also many other nitrogen compounds. Thus with hydroxylamine it forms a quite stable addition compound, $MnCl_2 \cdot 2NH_2OH$.

Manganese(II) chloride can also form double salts with other chlorides, especially those of the alkali metals. These double salts are mostly of the types $M^ICl \cdot MnCl_2$ and $2M^ICl \cdot MnCl_2$. They are decomposed by water.

Hydrated manganese chloride, $MnCl_2 \cdot 4H_2O$, is most conveniently prepared by dissolving manganese carbonate in hydrochloric acid and evaporating the solution (below 58°).

In so far as chlorine is still prepared by the Weldon process, manganese chloride can be prepared technically from the manganese chloride liquors. The excess chlorine and hydrogen chloride are first driven off, and the chief impurities (iron and aluminum) are precipitated by adding manganese carbonate. The manganese(II) chloride which crystallizes out when the resulting solution is concentrated is best purified by way of the carbonate.

Manganese(II) chloride gives rise to three basic salts, with different structures, according to Feitknecht, (1951): $Mn(OH)Cl$, $Mn_2(OH)_3Cl$, and $Mn_3(OH)_5Cl$, respectively. The latter is not very stable however. It is variable in composition, and has the brucite type structure, like $Mn(OH)_2$ itself. The OH- and Cl-ions are distributed statistically, and the lattice is markedly expanded in the direction of the c-axis.

(ii) *Manganese(II) bromide* and *iodide* are very similar to the chloride in their appearance and properties, and form almost the same hydrates, ammoniates, etc. The capacity for forming double salts is, however, much weaker in the case of the bromide, and appears to be lacking in the iodide.

(ii) *Manganese(II) fluoride*, MnF_2, pink quadratic prisms, differs from the other manganese(II) halides in that it is much less soluble (1.06 g in 100 g of water). It forms a rather unstable tetrahydrate and a very unstable ammoniate, $3MnF_2 \cdot 2NH_3$, as well as double salts of the type $M^I[MnF_3]$.

(iv) *Manganese(II) sulfate*, $MnSO_4$, is one of the most stable compounds of manganese. It is formed when practically all manganese compounds, except those containing involatile acids, are evaporated down with sulfuric acid. The anhydrous salt is almost pure white. It forms rose-pink crystals from aqueous solution, with a water content depending on the conditions (hepta-, penta-, tetra- and monohydrates exist).

The heptahydrate, $MnSO_4 \cdot 7H_2O$, which is stable below 9°, is occasionally found as a mineral (mallardite). It generally crystallizes monoclinic, but can also form rhombic mixed crystals with zinc vitriol.

The hydrate of manganous sulfate stable at ordinary temperature (in contact with the solution between 9° and 26°) is the pentahydrate, $MnSO_4 \cdot 5H_2O$, known as *manganese vitriol*, which is isomorphous with copper vitriol. Ordinary commercial manganese(II) sulfate is the 4-hydrate, however. The rombic tetrahydrate is, indeed, stable only over an exceedingly narrow range of temperature*. However, a monoclinic form of the tetrahydrate exists, which is actually less stable, but which nevertheless crystallizes from solution over a wider range of temperature. Thus it is obtained when a manganese(II) sulfate solution is evaporated between 35° and 40°. This is the manganese sulfate of commerce.

* Between about 26° and 27°. Above 27°, only the monohydrate is stable in contact with the solution.

Manganese(II) sulfate is manufactured from pyrolusite, which is either dissolved in hot concentrated sulfuric acid (1), or heated strongly with dehydrated iron vitriol (2):

$$MnO_2 + H_2SO_4 = MnSO_4 + H_2O + \tfrac{1}{2}O_2 \tag{1}$$

$$2MnO_2 + 2FeSO_4 = 2MnSO_4 + Fe_2O_3 + \tfrac{1}{2}O_2 \tag{2}$$

Like the chloride, manganese(II) sulfate serves as the starting material for the preparation of many other manganese compounds. It is also used in dyeing; material used for this purpose must be largely freed from iron.

With the exception of the monohydrate, which is also occasionally found as a mineral (*szmikite*, monoclinic), the hydrates of manganese sulfate gradually effloresce when they are left exposed to air. All the water may be driven off by heating, especially in the presence of concentrated sulfuric acid. The anhydrous sulfate can add on 6 molecules of ammonia, and the monohydrate 5 molecules NH_3.

Double salts, of the general formula $M^I_2SO_4 \cdot MnSO_4$, crystallize from mixed solutions of manganese(II) sulfate with alkali sulfates (anhydrous salts, or hydrates with 2, 4 or 6 molecules of H_2O exist).

(*v*) *Manganese(II) carbonate*, $MnCO_3$, is found native as *manganese spar*, and is important as a manganese ore for the production of spiegeleisen and ferromanganese. It crystallizes in the hexagonal-rhombohedral system, and is isomorphous with calcite, magnesite, and siderite. The naturally occurring manganese spar is often strongly contaminated with these substances, and is therefore generally grey, greenish, or brown in color, whereas it is raspberry red or pink in the pure state. The mineral is generally found in fine-grained or dense masses, and its crystalline texture is recognizable only under the microscope.

Precipitation from aqueous solution of soluble manganese(II) salts with soluble carbonates usually yields *basic manganese carbonate*, but if sodium hydrogen carbonate is employed as precipitant, in a solution saturated with carbon dioxide, the monohydrate of the neutral carbonate, $MnCO_3 \cdot H_2O$, is obtained as a white precipitate. This can be converted into the anhydrous carbonate by heating it with the solution (containing carbon dioxide) under pressure and in the absence of atmospheric oxygen.

The solubility of manganese carbonate in water is very low. It is therefore not decomposed by water at ordinary temperature, although it is hydrolyzed on boiling. The dry carbonate also decomposes fairly readily when it is heated, dissociation being perceptible even below 100°, according to:

$$MnCO_3 + 23.2 \text{ kcal} \rightleftharpoons MnO + CO_2.$$

At higher temperatures, the carbon dioxide oxidizes the manganese(II) oxide to some extent, and therefore becomes admixed with carbon monoxide above 330°. If manganese carbonate is heated in air, it is transformed into Mn_3O_4, and in oxygen forms Mn_2O_3.

Artificially prepared manganous carbonate is occasionally used as an artists' pigment (manganese white).

$MnCO_3$ has the calcite crystal structure, with $a = 5.84$ Å, $a = 47°2'$.

(*vi*) *Manganese(II) nitrate* crystallizes from a solution of manganese carbonate in dilute nitric acid. Evaporation at ordinary temperature yields the hexahydrate,

$Mn(NO_3)_2 \cdot 6H_2O$, whereas above 25°, or from concentrated nitric acid solutions, the trihydrate, $Mn(NO_3)_2 \cdot 3H_2O$, is obtained in colorless monoclinic prisms. The monohydrate also exists. It can be completely dehydrated by warming it gently with nitric acid anhydride. The anhydrous salt adds on up to 9 molecules of ammonia. It dissolves in water with considerable evolution of heat (12.9 kcal per mol). The solubility is 134 g of $Mn(NO_3)_2$ in 100 g of water at 18°.

Manganese(II) nitrate readily forms double nitrates with the rare earth metals. These salts have proved very useful for the separation of the rare earths by fractional crystallization.

(vii) *Manganese(II) acetate*, $Mn(C_2H_3O_2)_2 \cdot 4H_2O$, crystallizes from a solution of manganese carbonate in acetic acid, in pale red needles or plates, which are stable in air (1 part is soluble in 3 parts of cold water). It is used as a stimulating fertilizer, a siccative, and in other ways as an oxygen carrier. Anhydrous manganese(II) acetate can be prepared by the action of acetic anhydride upon the nitrate.

(viii) *Manganese(II) Phosphates* and *Arsenates*. Excess of disodium phosphate throws down hydrated manganese orthophosphate, $Mn_3(PO_4)_2 \cdot 7H_2O$, from neutral manganese(II) salt solutions, as a flocculent white precipitate. Other phosphates, pyrophosphates, and metaphosphates, as well as acid phosphates, can be obtained under different conditions. The beautifully crystalline double salt, *ammonium manganese phosphate*, $Mn(NH_4)PO_4 \cdot H_2O$ forming fine silky crystals, can be precipitated from manganese(II) salt solutions by the addition of ammonium chloride, ammonium phosphate (or disodium phosphate), and a little ammonia. This reaction is utilized in gravimetric analysis. Like the analogous magnesium salt, this compound is transformed into the pyrophosphate when it is ignited:

$$2Mn(NH_4)PO_4 \cdot H_2O = Mn_2P_2O_7 + 2NH_3 + 3H_2O.$$

The (weakly) complex acid $H_4[Mn(PO_4)_2] \cdot 3H_2O$ (diphosphatomanganese(II) acid) crystallizes from manganese(II) phosphate solutions containing a large excess of phosphoric acid. It has been shown by transport measurements that the manganese is present in the anion of this acid. The same holds for the *diarsenatomanganese(II) acid*, $H_4[Mn(AsO_4)_2] \cdot 3H_2O$ (rose pink hexagonal needles). This reacts with alkali carbonates to form the salts $Na_4[Mn(AsO_4)_2]$ and $K_4[Mn(AsO_4)_2]$, which crystallize anhydrous, or with ammonium carbonate, to form the acid salt $(NH_4)_2H_2[Mn(AsO_4)_2]$ (Grube, 1936).

Of the simple arsenates of manganese, the salts $Mn_3(AsO_4)_2 \cdot H_2O$, $MnHAsO_4 \cdot H_2O$, and $Mn(H_2AsO_4)_2$ may be mentioned; also the double salt $Mn(NH_4)AsO_4 \cdot 6H_2O$. Basic manganese arsenates, or double compounds of $Mn_3(AsO_4)_2$ with $Mn(OH)_2$, occur in Nature (sarkinite, hemafibrite, allactite, arsenoclasite).

(ix) *Manganese sulfide*, MnS. When a manganese(II) salt solution is treated with ammonium or alkali sulfide solution, a flesh pink precipitate is usually obtained, although under some conditions it may be more yellowish or reddish in tint. It consists of manganese(II) sulfide, generally in a hydrated form. On prolonged standing in the solution from which it was precipitated, and especially if it is boiled, it is converted into the more stable green form. The transformation is not conditional upon the presence of ammonium salts, but takes place particularly rapidly if precipitation is carried out with a large excess of ammonium sulfide. The green modification of manganese sulfide occurs native as *manganese blende* (*alabandite*), in the form of sparkling black regular crystals, which grind up to a green powder. Distinct crystallites of the green manganese sulfide can also be fairly readily obtained by artificial means. Manganese sulfide is not very stable in air.

The more reactive, flesh colored form, in particular, turns brown fairly rapidly on exposure to air, as a result of decomposition (hydrolysis and simultaneous oxidation to the dioxide hydrate).

The crystal structure of manganese blende and the green form of the artificial sulfide (a-MnS), like that of the alkaline earth sulfides, is of the rock salt type, with $a = 5.21$ Å, Mn $\leftrightarrow$ S $= 2.61$ Å. The unstable red form exists both in a cubic modification, with the zinc blende structure, ($a = 5.60$ Å), and a hexagonal modification ($a = 3.98$ Å, $c = 6.43$ Å). Both of the unstable modifications are generally present together in the flesh colored sulfide precipitated from solution, and are transformed monotropically into the green modification on warming.

Like a-MnS, the selenide MnSe crystallizes with a structure of the NaCl type. This is noteworthy in that the monosulfides and monoselenides of the adjacent elements in the Periodic System (Cr and V; also Fe, Co, and Ni) all crystallize with the NiAs structure. MnTe crystallizes with the NiAs structure, as do all the tellurides from TiTe to NiTe.

(x) *Manganese disulfide*. In addition to manganese blende, a *disulfide* of manganese is also found native (*hauerite*, MnS$_2$). This has the same crystal structure as pyrite (cf. p. 277), $a = 6.097$ Å, S $\leftrightarrow$ S $= 2.09$ Å, Mn $\leftrightarrow$ S $= 2.58$ Å. Manganese disulfide can also be obtained from aqueous solution, though generally only at higher temperatures (about 160°). See p. 277 for a discussion of the valence of manganese in this compound. MnS$_2$ readily loses sulfur when it is heated, and at 304° has the same sulfur vapor pressure (55 mm) as elementary sulfur (Biltz, 1936). MnSe$_2$ and MnTe$_2$ have the pyrite structure, like MnS$_2$ (whereas FeSe$_2$ and FeTe$_2$ have the marcasite structure).

(xi) *Manganese(II) borate* is used as a siccative. It is obtained by double decomposition between manganese(II) salt solutions and borax. The composition of the preparations obtained in this way corresponds roughly to that of a hydrated tetraborate. It is difficult to obtain material of uniform composition by precipitation from solution, but such compounds can be prepared by fusing manganese(II) oxide or manganese carbonate with boric acid or borax.

(xii) *Manganese(II) oxalate*. If hot oxalic acid solution is added to a hot dilute solution of a manganese(II) salt, or to a suspension of manganese carbonate in hot water, a white crystalline powder with the composition MnC$_2$O$_4 \cdot$ 2H$_2$O separates on slow cooling. If the reaction is carried out in the cold, the trihydrate MnC$_2$O$_4 \cdot$ 3H$_2$O is first obtained, crystallizing in fine pink triclinic needles. Formation of this compound can be used as a microchemical reaction for the identification of manganese. The trihydrate is not very stable in air, and changes to the dihydrate on standing. The latter does not lose water at ordinary

MnC$_2$O$_4$, above 100°.

Manganese(II) oxalate is sparingly soluble in water. Its solubility is considerably enhanced by strong acids and also by oxalic acid. Hydrated double salts, of the general formula M$^I_2$C$_2$O$_4 \cdot$ MnC$_2$O$_4$, can be isolated from solutions containing alkali oxalates.

($xiii$) *Manganese(II) sulfite*. Manganese carbonate reacts with an aqueous solution of sulfur dioxide, to form *manganese(II) sulfite*, MnSO$_3$. This salt is rather sparingly soluble, and crystallizes from the solution—with 3H$_2$O below 70°, and with 1H$_2$O at higher temperatures—in pink needles, which are oxidized easily in the air. With alkali sulfites it forms double salts, mostly of the type M$^I_2$SO$_3 \cdot$ MnSO$_3$.

(xiv) *Manganese(II) cyanide and Cyanomanganate(II) Salts*. A precipitate is formed when cyanide ions are added to manganese(II) salt solutions. This apparently consists of manganese(II) cyanide, Mn(CN)$_2$, but has never been obtained pure, since it rapidly decomposes in air. The complex salts formed between manganese(II) cyanide, and other cyanides are more stable (*cyanomanganate(II) salts*). These conform to the type M$^I_4$[Mn(CN)$_6$]—e.g., potassium hexacyanomanganate(II), K$_4$[Mn(CN)$_6] \cdot$ 3H$_2$O, dark blue to steel blue plates; sodium hexacyanomanganate(II), Na$_4$[Mn(CN)$_6] \cdot$ 8H$_2$O, amethyst octahedra; calcium hexacyanomanganate(II), Ca$_2$[Mn(CN)$_6$], dark blue crystals. The aqueous solutions of cyanomanganate(II) salts are almost colorless. These compounds cannot be kept in the air for long without undergoing oxidation, and also decompose fairly quickly in solution.

Double decomposition reactions with heavy metal salts can be carried out with freshly prepared solutions, however, and the radical $[Mn(CN)_6]^{4-}$ thereby passes unchanged from one salt to another. The free acid, $H_4[Mn(CN)_6]$ can be obtained by the action of hydrogen sulfide on the lead salt. It crystallizes in colorless scales, is very soluble in water, and decomposes in solution.

By treating solutions of cyanomanganate(II) salts with aluminum turnings, or Devarda's alloy, in the absence of air, Manchot (1926) succeeded in preparing *cyanomanganate(I)* *salts* $M^I_5[Mn(CN)_6]$. These cyano salts of *uni*positive manganese are intensely yellow in solution, and colorless in the solid state; they are extraordinarily easily oxidized. They can also be prepared by electrolytic reduction (Grube, 1927). It is curious that all these complex salts are paramagnetic (Bhatnagar, 1939; Goldenberg, 1940), whereas it would be expected that they should display diamagnetism, like $K_4[Fe(CN)_6]$.

(*xv*) *Manganese(II)* *thiocyanate* and *Thiocyanatomanganate(II)* *Salts*. Manganese(II) thiocyanate, $Mn(SCN)_2$, crystallizes when a solution of manganese carbonate in thiocyanic acid is evaporated. The degree of hydration of the salt depends upon the conditions of preparation, the tetrahydrate being most stable at ordinary temperature. It exists as large, bright green crystals, which are exceedingly soluble in water. The concentrated solution is green, and the dilute solution pink. The anhydrous salt, obtained from the hydrates at 100° (a trihydrate and dihydrate exist, in addition to the tetrahydrate) is yellow. With alkali thiocyanates, it forms complex salts of the type $M^I_4[Mn(SCN)_6]$ (hexathiocyanatomanganate(II) salts).

(*xvi*) *Manganese silicates*. Manganese metasilicate, $MnSiO_3$, occurs as the mineral *rhodonite*. When pure, this forms rose-red triclinic needles (density 3.5). It is generally found in dull masses, however, with a dirty brown or grey color due to impurities present.

Manganese orthosilicate, *tephroite*, Mn_2SiO_4, is much rarer than rhodonite. Both of these minerals are often found in beautifully crystalline form, in blast furnace slags, etc.

(c) Manganese(III) Salts

The manganese(III) salts are derived from the oxide Mn_2O_3, oxygen being replaced by acid radicals. Most of the salts are dark in color, have a strong tendency to form complex salts (acido-compounds), and in many cases, indeed, can only be obtained pure in such a form. All the manganese(III) salts have a low stability. In acid solution, they are readily reduced to manganese(II) compounds. In neutral solution, the simple salts in particular are hydrolyzed, initially to manganese(III) hydroxide; on exposure to air, however, the latter is rapidly converted to manganese dioxide hydrate.

Proof that the manganese(III) salts represent a definite oxidation state of manganese, and not perhaps a mixture of manganese(II) and manganese(IV), is afforded, among other arguments, by determinations of molecular weight. It has been possible to make such measurements on manganese(III) acetylacetonate, $Mn[CH(COCH_3)_2]_3$ (black, lustrous crystals, melting at 172°), which is soluble in benzene.

(*i*) *Manganese(III)* *chloride* *and* *Chloromanganate(III)* *Salts*. The dark brown, solution obtained by the action of concentrated hydrochloric acid on manganese dioxide probably contains manganese(III) chloride, $MnCl_3$. It is not possible to obtain the simple chloride from this solution, but it is possible to isolate double (or complex) salts derived from it, the *chloromanganate(III)* salts. They crystallize out when the solution is treated with alkali chlorides, and saturated with hydrogen chloride at 0°. The formulas of the chloromanganate(III) salts correspond to the general type $M^I_2[MnCl_5]$. They are dark red in color, and all crystallize anhydrous, except the ammonium salt, which contains 1 molecule of water of crystallization.

(*ii*) No bromide or iodide of trivalent manganese is known, nor have the corresponding complex salts been prepared. *Manganese(III) fluoride*, MnF_3, however, can be obtained by the action of free fluorine on manganese(II) iodide:

$$2MnI_2 + 3F_2 = 2MnF_3 + 2I_2.$$

It decomposes into fluorine and manganese(II) fluoride at elevated temperatures, and is converted into Mn_2O_3 when heated in oxygen. It dissolves in a small amount of water, to form a red brown solution, which decomposes rapidly when it is diluted. The same red-brown solution is obtained by dissolving Mn_2O_3 in hydrofluoric acid (1), or by the reaction between permanganate and a manganese(II) salt in the presence of hydrofluoric acid (2):

$$Mn_2O_3 + 6HF = 2MnF_3 + 3H_2O \tag{1}$$

$$MnO_4^- + 4Mn^{++} + 15HF = 5MnF_3 + 4H_2O + 7H^+ \tag{2}$$

Hydrated manganese(III) fluoride, $MnF_3 \cdot 2H_2O$, crystallizes out, in ruby red prisms, when the solution containing excess hydrofluoric acid is evaporated. Dark red complex salts, of the types $M^I[MnF_4]$ and $M^I_2[MnF_5]$ crystallize out in the presence of alkali fluorides.

(*iii*) *Manganese(III) sulfate, Sulfatomanganate(III) Salts and Manganese Alums.* Manganese(III) sulfate is formed by dissolving manganese(III) oxide or hydroxide in cold, moderately concentrated sulfuric acid*. The compound $Mn_2(SO_4)_3 \cdot H_2SO_4$ $\cdot 4H_2O$ crystallizes from the red solution on cooling; the same compound can also be prepared by the action of concentrated sulfuric acid on potassium permanganate (the manganese heptoxide which is first formed in this reaction dissolves in the sulfuric acid when it is gently warmed, giving off oxygen). On stronger heating, the red manganese(III) sulfate is converted to green manganese(III) sulfate. This is also formed, according to Carius, by treating manganese dioxide hydrate with hot concentrated sulfuric acid.

Manganese(III) sulfate is also obtained by the electrolytic oxidation of warm manganese(II) sulfate solutions containing much sulfuric acid. The solutions turn red, and the compound $Mn_2(SO_4)_3 \cdot H_2SO_4 \cdot 4H_2O$ can be isolated from them.

Manganese(III) sulfate forms two series of double salts with the alkali sulfates. Those of the first series are either anhydrous, or contain only a few molecules of water of crystallization. They correspond in composition to the formula $M^I_2SO_4 \cdot Mn_2(SO_4)_3$, and can be formulated as *disulfatomanganates(III)*, $M^I[Mn(SO_4)_2]$. The 'red manganese(III) sulfate' is probably the parent acid of this series, $H[Mn(SO_4)_2] \cdot 2H_2O$, disulfatomanganese(III) acid.

The second series of alkali double sulfates of trivalent manganese belongs, in composition and in crystal form, to the class of the *alums*. The most stable is cesium manganese alum, $CsMn(SO_4)_2 \cdot 12H_2O$, garnet red crystals. The analogous rubidium compound loses some water even at ordinary temperature, and the potassium and ammonium compounds are even more readily decomposed.

Double salts of manganese(III) sulfate with iron(III)-, chromium(III)- and aluminum sulfate also exist.

* The oxide will not dissolve unless a little manganese(II) oxide is also present.

(*iv*) *Manganese(III) phosphate.* Several phosphates of trivalent manganese are known, of which the orthophosphate $MnPO_4 \cdot H_2O$ is green, and the meta-phosphate $Mn(PO_3)_3$ is red. The others are violet. All are insoluble in water. The formation of manganese(III) phosphate is responsible for the violet micro-cosmic salt bead given by manganese in the oxidizing flame.

A violet manganese(III) phosphate, prepared in a particular way, is used as an artist's pigment (manganese violet, permanent violet), which is notable for its pure tone and stability in air.

A complex acid, $H_3[Mn(PO_4)_2]$, is derived from manganese(III) phosphate. Its salts, of the type $M^I H_2[Mn(PO_4)_2]$ (diphosphatomanganate(III) salts, mostly $+3H_2O$), have been investigated by Meyer.

(*v*) The *triarsenatomanganese(III) acid*, $H_6[Mn(AsO_4)_3] \cdot 3H_2O$, obtained by Deiss (1925) is derived from manganese(III) arsenate.

(*vi*) *Cyanomanganate(III) Salts.* Manganese(III) cyanide is not known, but the complex salts derived from it, the dark red *hexacyanomanganate(III) salts*, $M^I_3[Mn(CN)_6]$, which are isomorphous with the hexacyanoferrate(III) salts (ferricyanides), can readily be obtained crystalline. They are quite stable in dry air. They may be prepared either by the oxidation of hexacyanomanganate(II) salts, or by the reaction between manganese(III) acetate and cyanides.

The existence of the hexacyano complex, $[Mn(CN)_6]^{3-}$, is shown by double decom-position of the soluble hexacyanomanganate(III) salts with heavy metal salts. Precipitates are thereby formed, in which the hexacyano complex persists unchanged. Only in the double decomposition with iron(II) salts does tripositive manganese become exchanged for bivalent iron, since the hexacyanoferrate(II) complex is more stable than the hexacyanomanga-nate(III) complex.

(*vii*) *Manganese(III) acetate* may be obtained by oxidizing manganese(II) acetate, dissolved in hot glacial acetic acid, by means of permanganate or chlorine. The dihydrate, $Mn(C_2H_3O_2)_3 \cdot 2H_2O$, is soon deposited, in the form of cinnamon brown, silky crystals, when a little water is added to the solution thus obtained. It is stable in dry air, but is decomposed by water, although it may be recrystallized from glacial acetic acid. Manga-nese(III) acetate has often been found to be a convenient starting material for the prepa-ration of other manganese(III) compounds.

(*viii*) *Oxalatomanganate(III) Salts.* Manganese(III) oxalate is known only in the form of the oxalatomanganate(III) complex. *Potassium trioxalatomanganate(III)*, $K_3[Mn(C_2O_4)_3] \cdot 3H_2O$, forms deep red, almost black, monoclinic prisms, which are isomorphous with potassium trioxalatoferrate(III). It is fairly stable at ordinary temperature, when kept in the dark, but is decomposed in the light, or on heating; Mn^{III} is thereby reduced to Mn^{II}, and the oxalate radical is oxidized to CO_2:

$$2K_3[Mn(C_2O_4)_3] = 2MnC_2O_4 + 3K_2C_2O_4 + 2CO_2.$$

(d) Manganese(IV) Compounds

Simple salts of quadrivalent manganese are very unstable, and in general only the complex salts derived from them (*acidomanganate(IV) salts*) are known. Even these are all more or less readily hydrolyzed by water, and the exchange of acido-groups with oxygen takes place yet more readily in the presence of alkali. Oxo-manganate(IV) salts (usually known as *manganites*) are thus formed from acido-manganate(IV) salts. However, the acid from which these are derived (H_2MnO_3, manganous acid) is an extremely weak acid, and its anhydride, MnO_2, is charac-terized by its great insolubility. This oxide, or its hydrate, is therefore generally precipitated. Even when manganites are formed, in solutions of sufficiently high hydroxyl ion concentration, they are almost invariably contaminated to a variable extent by admixed manganese dioxide hydrate.

The great stability of manganese dioxide, which is shown in that it occurs naturally in such large amounts as pyrolusite, is due in part to its extremely low solubility. Manganese(IV) compounds are not stable in solution. In acid solutions they have a strong tendency to undergo reduction, although they can also be oxidized to permanganates by powerful oxidants. In alkaline solution they are oxidized to manganates even by atmospheric oxygen.

Of the *acidomanganate(IV)* salts (general formula $M^I_2[MnX_6]$), the chloro and fluoro-compounds are among the most important. Double iodates of this same type are also known —$M^I_2[Mn(IO_3)_6]$—and the so-called glyceryl manganites (diglycerylomanganate(IV) salts) belong to the same class. The latter are obtained by heating freshly precipitated manganese dioxide hydrate with glycerol, $C_3H_5(OH)_3$, and strong bases—e.g., sodium diglycerylomanganate(IV), $Na_2[Mn(C_3H_5O_3)_2]$, a yellow-red powder which dissolves with blood red color in a mixture of alcohol and glycerol.

(*i*) *Manganese(IV) chloride and Chloromanganate(IV) Salts.* Manganese(IV) chloride is probably the primary product formed when manganese dioxide is dissolved in concentrated hydrochloric acid. It at once decomposes, however, losing chlorine, and cannot be isolated. The chloromangananate(IV) salts, which it forms by combining with alkali chlorides, are more stable.

Potassium hexachloromanganate(IV), $K_2[MnCl_6]$, very deep red crystals, is most readily obtained, according to Weinland, by running calcium permanganate and potassium chloride solutions into well cooled 40% hydrochloric acid. The ammonium and rubidium salts can be obtained in a similar manner. The fluoromanganate(IV) salts (e.g., $K_2[MnF_6]$, golden yellow hexagonal plates and $K[MnF_5]$, pale red powder) can be obtained similarly.

(*ii*) *Manganese(IV) sulfate.* If manganese(II) sulfate, in sulfuric acid solution, is oxidized at 50–60° with permanganate, black crystals of manganese(IV) sulfate, $Mn(SO_4)_2$, are obtained on cooling if the solution is sufficiently concentrated. These dissolve with a deep brown color in 50–80% sulfuric acid, but are hydro-lyzed by more dilute acid, and especially by water, with the deposition of hydrated manganese dioxide. It has been found that manganese(IV) sulfate is a particularly convenient oxidant for certain purposes, and finds some technical applications

Manganese(IV) sulfate is capable of forming double salts. These do not conform in composition to simple types, however.

(*iii*) *Manganites* (manganate(IV) salts). Manganese dioxide unites in very varied proportions with other metallic oxides. The compounds are known as *manganites*, or manganate(IV) salts.

Thus with calcium oxide the following compounds are formed:

$$2CaO \cdot MnO_2, \quad CaO \cdot MnO_2, \quad CaO \cdot 2MnO_2, \quad CaO \cdot 3MnO_2, \quad CaO \cdot 5MnO_2.$$

These are obtained from melts. Under similar conditions, analogous compounds are obtained with the other alkaline earth metals. Compounds considerably richer in manganese are usually obtained from melts with the alkali oxides. The products obtained from aqueous solutions are for the most part very ill defined stoichiometrically, and consist to some extent merely of adsorption products of alkali or alkaline earth oxides on hydrated manganese dioxide. Many of the naturally occurring varieties of manganese dioxide are also highly contaminated with adsorbed metallic oxides (especially psilomelane and wad). The (usually hydrated) manganese oxides, with a content of between 3 and 4 equivalents of oxygen per atom of manganese, are also to be regarded as consisting in part of manganese(II) manganite and in part of adsorption products.

(e) Manganate(V) salts

It is often found that a blue coloration in place of the usual green is obtained when MnO_2 is fused with soda and potassium nitrate. It was shown by Lux (1946) that this was due to the formation of alkali manganate(V) salts. He was able to isolate the compound $Na_3[MnO_4] \cdot 10H_2O$ in a pure state, and Klemm showed that its magnetic susceptibility accorded with that expected for a manganese(V) compound. Auger and Billy had earlier (1904) described a compound with the composition of a manganate(V) salt. Schmal (1950) observed that any manganese oxide was converted into a manganese(V) compound when it was heated to about 900° in air with BaO, although $Ba_3[MnO_4]_2$ was first isolated by Scholder, who subsequently prepared a series of other manganate(V) salts. A pigment, manganese blue, which is now manufactured industrially has been shown by Klemm, by magnetic susceptibility methods, to contain pentapositive manganese.

(f) Manganate(VI) salts

When manganese dioxide is heated with potassium nitrate, a green melt is obtained. This contains manganate(VI) salts—i.e., salts of the general formula $M^I_2[MnO_4]$. This reaction was already known to Scheele. Other oxidants, in conjunction with alkali, can be used in place of niter. Solutions of the alkali manganate(VI) salts (usually known as *manganates*) contain the ions $[MnO_4]^{2-}$ and are dark green in color.

Industrially, an intimate mixture of pyrolusite and potassium hydroxide is heated with free access of air, which acts as oxidant. The process must be carried out in such a way that the water formed in the reaction:

$$MnO_2 + 2KOH + \tfrac{1}{2}O_2 = K_2MnO_4 + H_2O$$

brings about as little decomposition of the resulting manganate(VI) as possible. Yields up to 60–70% can be obtained under suitable conditions; the formation of manganate(IV) or manganate(V), apparently in a definite relation to the amount of manganate(VI) formed, prevents the achievement of higher yields.

Only the *alkali* manganates(VI) have as yet been obtained in a state of purity. They form deep green, almost black, crystals. Potassium manganate(VI), $K_2[MnO_4]$, crystallizes anhydrous. A tetra-, hexa- and a decahydrate of sodium manganate(VI) are known.

The alkali manganate(VI) salts are very soluble in dilute caustic alkalis, giving solution with a green color. They are at once hydrolyzed by pure water or dilute acids, however, forming manganese dioxide and permanganate:

$$3MnO_4^= + 2H_2O = MnO_2 + 2MnO_4^- + 4OH^-.$$

This reaction is apparently the consequence of the instability of free manganic acid, H_2MnO_4, and is favored by the extreme insolubility of the manganese dioxide which is formed as a product of decomposition. This decomposition occurs even in alkaline solutions if they are heated.

When heated above 500°, potassium manganate(VI) decomposes into potassium manganate(IV) (so-called potassium manganite) and oxygen, although the latter is not readily evolved quantitatively, owing to the formation of solid solutions:

$$K_2MnO_4 = K_2MnO_3 + \tfrac{1}{2}O_2.$$

The pigment manganese green (Kassel green), sometimes recommended for fresco painting, as being non-toxic, is an impure barium manganate(VI).

(g) Permanganates (Manganate(VII) salts)

The permanganates are derived from manganese heptoxide, Mn_2O_7, which functions as an acid anhydride. They have the general formula $M^I[MnO_4]$, and the violet permanganate ion, $[MnO_4]^-$, is present in their solutions.

By far the most important of these salts is *potassium permanganate*, $KMnO_4$. Sodium permanganate (which crystallizes as the hydrate $NaMnO_4 \cdot 3H_2O$) is far less important industrially, because of its excessive solubility and corresponding difficulty of crystallization. The other alkali permanganates are only of academic interest. The solubilities of the alkali permanganates are shown in Table 25.

TABLE 25

SOLUBILITY OF THE ALKALI PERMANGANATES IN WATER
(g of anhydrous salt in 100 g of water)

Formula	$LiMnO_4 \cdot 3H_2O$	$NaMnO_4 \cdot 3H_2O$	$KMnO_4$	$RbMnO_4$	$CsMnO_4$
Solubility	about 71	V. soluble deliquescent	6.34	1.1	0.23
Temperature	16°	20°	20°	19°	19°

Of the alkaline earth permanganates, the *calcium* salt, $Ca(MnO_4)_2 \cdot 5H_2O$, may be mentioned. It is very soluble in water, and finds some application for the sterilization of drinking water. Permanganates of the heavy metals are also known.

Potassium permanganate, $KMnO_4$, the most important salt of permanganic oxid, is manufactured on a large scale, usually by the electrolytic oxidation of potassium manganate(VI):

$$MnO_4^= - e = MnO_4^-.$$

Older processes either employed chlorine as oxidant:

$$MnO_4^= + \tfrac{1}{2}Cl_2 = MnO_4^- + Cl^-, \quad \text{or} \quad K_2MnO_4 + \tfrac{1}{2}Cl_2 = KMnO_4 + KCl,$$

or utilized the spontaneous decomposition of potassium manganate(VI) in solution, and neutralized the liberated alkali by means of CO_2:

$$3K_2MnO_4 + 2CO_2 = 2KMnO_4 + MnO_2 + 2K_2CO_3.$$

In the latter case, one third of the original manganate becomes reconverted to manganese dioxide. However, this process has the advantage over that involving oxidation with chlorine, in that the potassium carbonate formed as a by product can be returned to the operations, and used for the oxidative fusion of the pyrolusite. In the chlorine process, the by-product potassium chloride is lost.

Potassium permanganate exists as deep purple, almost black, prisms, with a brownish reflex. It is isomorphous with potassium perchlorate, with which it forms a complete range of mixed crystals*. Potassium permanganate is moderately

* $NH_4[MnO_4]$ likewise forms a complete range of mixed crystals with $NH_4[ClO_4]$. $KMnO_4$ and $RbMnO_4$, however, can only form mixed crystals with one another over quite a restricted range.

soluble in water (just over $\frac{1}{3}$ mol per liter at room temperature; cf. Table 25). Its intensely red violet solution has a characteristic absorption spectrum. It is widely used as an oxidant both in analytical and preparative chemistry.

In acid solutions, permanganate reacts with sufficiently strong reducing agents according to the equation:

$$MnO_4^- + 8H^+ + 5e = Mn^{++} + 4H_2O.$$

This reaction takes place quantitatively with iron(II) salts, hydrogen peroxide, and oxalic, formic and nitrous acids and it thus enables these substances to be determined by a very simple titrimetric procedure. Since a permanganate solution is perceptibly colored even in very great dilution, and its color disappears when it is converted to a manganese(II) salt, the end point of the titration when permanganate is used as an oxidant is indicated when the reagent is no longer decolorized. It is usual to employ for volumetric analysis a permanganate solution which is $\frac{1}{10}$ normal with respect to the oxygen furnished to the reducing agent. Such a solution contains $\frac{1}{50}$ mol (or 3.1605 g) of potassium permanganate per liter. Each milliliter of such a solution corresponds to $\frac{1}{10}$ milliequivalent of the reducing agent in question. Oxidizing agents can be determined indirectly by means of permanganate. They are first treated with an excess of a suitable reducing agent (e.g., oxalic acid or iron(II) sulfate), the total amount of which is known. The unconsumed portion of the reducing agent is then titrated back with permanganate. *Manganometric* methods (as titrimetric determinations based on the use of permanganate are termed) can occasionally be employed even for the determination of substances which have neither oxidizing nor reducing properties. Thus calcium may be determined by precipitating it with oxalic acid. The precipitate is dissolved in dilute sulfuric acid, and the oxalic acid in this solution, which is equivalent in amount to the calcium, is titrated with permanganate.

In alkaline, neutral, or weakly acid solutions, permanganate is reduced to manganese dioxide:

$$MnO_4^- + 4H^+ + 3e = MnO_2 + 2H_2O.$$

This reaction can be employed for the determination of manganese in manganese(II) salts (Volhard's method):

$$2MnO_4^- + 3Mn^{++} + 2H_2O = 5MnO_2 + 4H^+.$$

A permanganate solution which is $\frac{1}{10}$ normal with respect to active oxygen for oxidations in strongly acid solution (i.e., a $\frac{1}{50}$ molar solution) is only $\frac{3}{50}$ normal for the determination of manganese by Volhard's method. In carrying out titrations of manganese by Volhard's method, steps must be taken to prevent manganese(II) ions from escaping oxidation through the formation of insoluble manganese(II) manganites. For this purpose, zinc ions are added in high concentration to the solution, since the precipitated manganese dioxide then binds these preferentially, rather than manganese(II) ions.

If concentrated potassium hydroxide solution is added to potassium permanganate, manganate(VI) is formed and oxygen is evolved:

$$4MnO_4^- + 4OH^- = 4MnO_4^= + O_2 + 2H_2O.$$

Decomposition proceeds further on prolonged standing, MnO_2 also being formed. Oxygen is also evolved when dry permanganate is heated to dull red heat. In this case manganite (i.e., manganate(IV)) is formed as well as manganate(VI):

$$10KMnO_4 = 3K_2MnO_4 + 2K_2O \cdot 7MnO_2 + 6O_2.$$

Since MnO_4^- ions are far less stable in alkaline solution than in acid solution, it is possible to titrate many substances with permanganate in alkaline solution, even though they are only very slowly attacked in acid solution—e.g., hypophosphites, methanol, erythritol, and numerous other organic substances. The spontaneous decomposition of the MnO_4^- ions which takes place in alkaline solution (although only very slowly at low concentrations of alkali) can easily introduce errors, unless the reactions take place very rapidly. It is therefore preferable, in determining such substances titrimetrically, to utilize only the *first* stage of the reduction of the MnO_4^- ions, which leads to the formation of $MnO_4^=$ ions; this probably takes place according to the equation:

$$2MnO_4^- + 2OH^- = 2MnO_4^= + H_2O + O,$$

giving rise to the formation of atomic oxygen as primary product, which then reacts with the reducing agent. This reaction is much faster than the second stage of the reduction ($MnO_4^= \rightarrow MnO_2$). Stamm has shown that if a soluble barium salt is present, the reaction is checked at the first stage, since the $MnO_4^=$ ions thus formed are at once removed, through formation of the insoluble barium manganate, $BaMnO_4$. It is desirable to add some heavy metal salts also (e.g., a few drops of $Co(NO_3)_2$ solution), which acts as a catalyst to accelerate the transfer of the oxygen. In this form, titration with alkaline permanganate solution is suitable not only for the determination of many organic substances (e.g., in testing drinking or process water), but also for the titrimetric estimation of iodates ($[IO_3]^- \rightarrow [IO_4]^-$), iodides ($I^- \rightarrow [IO_4]^-$), cyanides ($CN^- \rightarrow CNO^-$), thiocyanates ($SCN^- \rightarrow CNO^- + SO_4^=$), phosphites ($[HPO_3]^= \rightarrow [PO_4]^{3-}$), and hypophosphites ($[H_2PO_2]^- \rightarrow [PO_4]^{3-}$).

Potassium permanganate is a favorite oxidizing agent in preparative organic chemistry. Its uses for the bleaching of wool, cotton, silk, and other textile fibers, and for the decolorization of oils, are also dependent on its strong oxidizing properties. It is also employed to free materials of organic impurities—e.g., for the removal of empyreumatic constituents from crude pyroligneous acid. Potassium permanganate is also widely used as a disinfectant. Various varieties of wood are stained with permanganate to give them a nut brown color. Permanganate also finds use in dyeing. It is frequently used in the laboratory and industrially for washing gases—e.g., for purifying carbon dioxide in the manufacture of mineral waters.

Potassium permanganate does not react at a perceptible rate with hydrogen under ordinary conditions. However, if silver nitrate is added to a permanganate solution, it is able to absorb hydrogen at room temperature and ordinary pressure. This observation can be utilized for the gas-analytical determination of hydrogen in the presence of CH_4, C_2H_6, and N_2, since these gases are not attacked (Hein, 1931).

4. Analytical (Manganese)

Manganese is not precipitated by hydrogen sulfide, but is precipitated by ammonium sulfide as the flesh colored (hydrated) sulfide MnS. If manganese is present in a higher valence state, it is reduced to a manganese(II) salt by boiling it with ammonium sulfide or much more quickly, with hydrogen sulfide.

Manganese is readily separated from the other elements of the ammonium sulfide group. Unlike cobalt and nickel sulfides, its sulfide is readily soluble in dilute hydrochloric acid. Manganese differs from zinc and aluminum, in that its oxides MnO, Mn_2O_3, and MnO_2 are insoluble in sodium hydroxide. Iron and chromium can be separated from bivalent manganese by precipitation with ammonia in the presence of ammonium chloride*.

* For this purpose, the iron must be converted to the trivalent state by the addition of a little concentrated nitric acid, as it is otherwise incompletely precipitated. The manganese is not oxidized by the nitric acid.

When manganese compounds are fused with potassium nitrate (oxidizing fusion), a green manganate(VI) is formed, and its solution turns violet when it is acidified, manganese dioxide being silmultaneously precipitated. Oxidation to permanganic acid by means of lead dioxide in the presence of concentrated nitric or sulfuric acid constitutes a very sensitive identification reaction (cf. p. 217).

Manganese may be identified microanalytically as the oxalate (cf. p. 222), or by means of a drop reaction based on the blue coloration of benzidine produced by auto-oxidizing $Mn(OH)_2$. The drop reaction with potassium periodate and Arnold's reagent is considerably more sensitive still (limit of detection 0.001γ manganese). Mn^{++} ions can be oxidized by periodates in acid solution to MnO_4^- ions, which give an intense blue color with Arnold's reagent (cf. Vol. I, p. 557).

In the oxidizing flame, manganese confers a violet color on the microcosmic or borax beads, due to the formation of manganese(III) phosphate or borate. The coloration disappears in the reducing flame, since the corresponding manganese(II) compounds are formed.

Suitable methods for the gravimetric determination of manganese are precipitation and weighing as sulfide, or precipitation as the hydrated dioxide, which is converted by ignition to manganese(II, III) oxide, Mn_3O_4. An excellent method for determination of manganese in manganese(II) salts is the precipitation as manganese ammonium phosphate, $MnNH_4PO_4 \cdot H_2O$, and weighing as pyrophosphate, $Mn_2P_2O_7$, after ignition. Electrolytic decomposition at the anode, as the hydrated dioxide from concentrated nitric acid solutions, often provides a good method for the separation of manganese from other metals.

For the titrimetric determination of manganese, and the use of potassium permanganate in titrimetric analysis, see p. 229 *et seq.*

5. Technetium (Tc)

(a) Occurrence and History

Technetium is an unstable element, which can be obtained only as the product of artificial nuclear transmutations (cf. p. 232). As mentioned earlier (p. 206), the name of *technetium* was given to element 43. This follows the proposal of Segré, who first obtained it in 1937 as the product of a *nuclear reaction* (bombardment of molybdenum with deuterons, cf. p. 563), and investigated its most characteristic chemical properties. Several isotopes of technetium have since been obtained by nuclear reactions. All are unstable, in accordance with the rules concerning nuclear stability (see p. 592). The longest lived isotope of technetium (^{99}Tc), obtained by Segré in 1947 (by neutron bombardment of molybdenum and also as a product of uranium fission) has a half-life of about $2 \cdot 10^5$ years. The age of the earth is more than a thousand times as great as this, so that even if technetium were originally present in the earth's crust, it must have completely disappeared in the intervening time (cf. p. 514).

(b) Preparation

Although no stable isotope of technetium exists, considerable quantities of the element should become available in course of time, since it is formed to the extent of 6.2% of the total fission products from ^{235}U in nuclear reactors. Owing to the enormous β-ray and γ-ray activity of the fission products, it is not readily extracted. However, ^{99}Tc is the longest-lived of the fission products (half-life $2.12 \cdot 10^5$ years, according to Fried, 1951). After a sufficiently long period of storage, the other fission product activities have largely died away, and it becomes practicable to recover technetium from the residues. Small quantities (of a few milligrams) can conveniently be prepared by irradiation of molybdenum oxide with neutrons, in a nuclear reactor:

$$^{98}Mo \xrightarrow{\ ^{1}_{0}n\ } {}^{99}Mo \longrightarrow {}^{99}Tc + \beta^-.$$

The technetium can be sublimed off quantitatively as the volatile oxide, Tc_2O_7. It can be isolated from solutions (e.g., of old fission products) by precipitation as sulfide (see below).

(c) Chemical Properties

Technetium is very similar to rhenium in its chemical behavior. The resemblance to rhenium is much closer than to manganese, just as molybdenum resembles tungsten rather than chromium, or ruthenium resembles osmium rather than iron. Thus technetium heptasulfide is insoluble in dilute hydrochloric acid, as is the corresponding rhenium sulfide, whereas manganese sulfide is soluble. Metallic technetium also volatilizes as the heptoxide when it is heated in oxygen, as rhenium does, whereas manganese heptoxide cannot be formed by the direct combination of the elements. Technetium can be separated from rhenium by passing a stream of moist hydrogen chloride through a solution of the oxide in 80% sulfuric acid at 200°. Rhenium distils over as a volatile chloride, whereas technetium remains behind.

In many reactions, technetium also resembles molybdenum, its neighbor in the Periodic System. Advantage can be taken of this resemblance, by using molybdenum compounds as 'carriers' for the corresponding technetium compounds when working on a 'tracer'

The chemical properties of technetium could at first be studied only with unweighably small quantities, using the 'tracer' methods of radiochemistry (cf. Chap. 11). Weighable amounts of the weakly radioactive isotope ^{99}Tc are now available, however (see above). Thus Boyd and his coworkers (1952) have isolated about 0.6 g of spectroscopically pure metallic technetium from fission products. The sulfide Tc_2S_7 was precipitated by means of H_2S from a solution acidified with HCl. The black precipitate was dissolved in ammoniacal hydrogen peroxide. *Ammonium pertechnetate*, NH_4TcO_4, was thereby obtained, and was reduced to the metal by heating to about 600° in hydrogen. Very small quantities of technetium can be precipitated as the sulfide by use of PtS_2 as a carrier. The sulfide in this case is dissolved in ammoniacal H_2O_2 and distilled from concentrated sulfuric acid. Under these conditions, Tc_2O_7 distils over, and is absorbed in ammonia.

The atomic weight of technetium, determined by chemical methods, agrees closely with the atomic weight as determined by means of the mass spectrograph, by Inghram ($= 98.91$). Technetium metal crystallizes with the hexagonal close packed structure (isomorphous with Re, Ru and Os), $a = 2.735$ Å, $c = 4.388$ Å; atomic radius $= 1.36$ Å, density (by X-ray method) $= 11.50$. The metal does not dissolve either in hydrochloric acid or in alkaline hydrogen peroxide, but is dissolved by nitric acid and aqua regia. It burns when it is heated in oxygen, forming the bright yellow volatile heptoxide, Tc_2O_7. Technetium heptoxide is hygroscopic, and dissolves in water to form *pertechnetic acid*, $HTcO_4$, which can be obtained in the form of deep red needles when the solution is evaporated. $HTcO_4$ is a strong monobasic acid. The deep red color of its concentrated solutions rapidly pales when the solution is diluted, and a 1-molar solution is quite colorless. Ammonium pertechnetate, NH_4TcO_4, is colorless and is not hygroscopic when pure. Its solutions show an intense absorption in the ultraviolet, at 2470 Å and 2890 Å [see Boyd, *J. Am. Chem. Soc.* 74 (1952), 556; Rulfs and Meinke, *ibid.*, 74 (1952), 235].

Metallic technetium can be deposited electrolytically from acid solutions. The standard potential for the formation of TcO_4^- ions from the metal, according to the equation

$$Tc + 4H_2O = TcO_4^- + 8H^+ + 7e,$$

was found by Flagg and Bleidner (1945) to be $E° = -0.41$ volts. This value lies between the standard potentials of the MnO_4^- and ReO_4^- ions. From the standard potential, the free energy of formation of the TcO_4^- ion can be calculated as $\Delta F° = -160$ kcal per g-ion TcO_4^-. Technetium is displaced, in the metallic form, from solutions of the TcO_4^- ion by means of Mg, Zn, Fe, Ni, Sn, Pb, and Cu. Tin(II) chloride and hydrochloric acid reduce the TcO_4^- ion to a cation derived from some lower valence state of technetium. It has been shown by Rogers (1949) that, in principle, it should be possible to separate technetium from both rhenium and molybdenum by electrolytic deposition from alkaline solution. If technetium is heated in dry hydrogen chloride, it volatilizes as a chloride (cf. rhenium).

6. Rhenium (Re)

(a) Occurrence

Rhenium is an extremely rare element indeed, and is exceeded in rarity by none but the inert gases krypton and xenon, and the strongly radioactive elements.

The average content of rhenium in the portion of the earth's crust which is accessible to us has been estimated as about 10^{-7} per cent. Rhenium is not uniformly dispersed throughout the crust, however, but is relatively strongly concentrated in certain minerals—usually to the greatest extent in molybdenite, and often in platinum ores also. Depending upon its source, molybdenite may contain $6 \cdot 10^{-5}$ to $2 \cdot 10^{-3}\%$ rhenium. A content of $10^{-4}\%$ rhenium was detected in a Russian platinum ore—i.e., 1 mg of rhenium in 1 kg of platinum ore. Oxidic ores rarely attain such a 'high' rhenium content; in columbites, for example, it lies between $0.5 \cdot 10^{-5}$ and $2 \cdot 10^{-5}\%$. Even among the sulfide ores, however, molybdenite seems to be the only one in which the rhenium content may rise above $10^{-4}\%$. Rhenium becomes relatively highly enriched in the smelter residues from the Mansfeld copper shales (cf. p. 235).

(b) History

Rhenium was discovered in 1925 by Noddack and Tacke, as the result of deliberately planned experiments. It had been inferred long before this, from the Periodic system, that two unknown homologues of manganese ('ekamanganese' and 'dvimanganese') ought to exist, and this inference was strengthened when it became possible to state with certainty, on the basis of Moseley's law, that the atomic numbers 43 and 75 could not be assigned to any of the known elements. Unsuccessful attempts to discover these elements had already often been made. From the failure of such attempts, and from a comparison of the abundances of the various elements in the earth's crust, Noddack and Tacke drew the conclusion that 'eka-' and 'dvi-manganese' must be exceedingly rare elements. If this were so, there could be no possibility of detecting them directly in minerals, by means of their characteristic X-ray spectra. A preliminary enrichment to a content of at least 0.1% would be essential, since this was at that time the limit of sensitivity of X-ray spectroscopy. In order to carry out such an enrichment, Noddack and Tacke tried to predict the chemical properties of the elements from the Periodic System. To do so, they assumed that in the VIIth Sub-group the properties would change with atomic number in the same way as in the two neighboring groups— i.e., in the VIth Sub-group and the first column of the VIIIth Sub-group. They reached the conclusion that, like manganese, elements 43 and 75 would be capable of existing in a large number of valence states, but that the higher valence states would be much more stable than with manganese; this statement corresponds to the fact that in the neighboring columns the stability of the highest valence state rises in a very marked degree from chromium to tungsten and from iron to osmium. They deduced the most important properties of elements 43 and 75 and their compounds by a comparison with neighboring elements and, with these as a basis, applied chemical separation processes to various minerals, in order to bring the concentrations of the hypothetical elements within the range of detection by X-ray spectroscopy. For this purpose it was again necessary to select those minerals in which, according to the laws governing the distribution of the elements in the earth's crust, (Chap. 15), it might be expected that elements 43 and 75 should be present, if they existed. For this purpose, the first choice fell upon platinum ores. Since, however, only small quantities of such ores were available to the investigators, the search was extended to oxidic ores, and especially to those which contained both molybdenum and ruthenium, or tungsten and osmium together, in appreciable amounts. This is the case, for example, with columbite and tantalite. The estimates of abundance suggested that element 75 should be about 500,000 times as rare as niobium in the earth's crust*. If it were assumed that the same ration between niobium and element 75 obtained in columbite, then it would be necessary to enrich element 75 by a factor of at least 500 in order to make spectroscopic detection possible. Since, however, it might be that in columbite the actual amount of the sought for element would be even smaller than this, the enrichment process was carried yet further, by a factor of ten—i.e., 0.2 g of concentrate was prepared from 1 kg of mineral, by a process which was so chosen that the total quantity of the elements adjacent to nos. 43 and 75 (i.e., Mo, W, Ru and Os), present in traces in the mineral, remained together. By this means a quantity of element 75 which was detectable by X-ray spectroscopy was, in fact, obtained in the enriched product. The element which was thus detected by its X-ray spectrum (cf. Table 26) was given the name *rhenium* (after the river Rhine) by its discoverers. By a similar process, Noddack and Tacke believed that they had detected the element of atomic number 43, which they called *masurium* (after Masuria, in East Prussia). However, it has later been shown that this element does not exist in Nature as already mentioned (see pp. 206, 231).

TABLE 26

PREDICTED AND MEASURED LINES IN THE CHARACTERISTIC L-SERIES
X-RAY SPECTRUM OF RHENIUM

Symbol of line	$L\alpha_1$	$L\alpha_2$	$L\beta$	$L\gamma$
Wave length, calculated from Moseley's law	1.431	1.441	1.235	1.204 Å
Wave length, measured	1.430	1.441	1.235	1.205 Å

A determination of the rhenium content of numerous minerals (after an initial enrichment process, in each case) revealed the fact that molybdenite, especially that from Norway, had a relatively high rhenium content. By starting with 660 kg of almost pure molybdenite Noddack and Tacke were able, in 1928, to isolate for the first time 1 g of almost pure rhenium. The industrial preparation of larger quantities soon followed, and the metal is now commercially available. It now finds technical applications on an appreciable scale (see below). Hönigschmid, in 1930, determined the atomic weight of rhenium as 186.31. Noddack and Tacke, in 1925, had predicted, from the regularities of the Periodic System, that rhenium would have an atomic weight of 187 to 188. The chemical properties have also been found to agree, in their essentials, with those predicted.

* Noddack and Tacke adopted the working hypothesis that the average abundance of elements 43 and 75 would bear to that of ruthenium and osmium about the same ratio as that of manganese to the abundance of iron—i.e., about 1 : 50. Noddack and Tacke derived their estimates from considerably lower values for the abundance of the platinum metals than those shown in Table 84 (p. 644), which are based on more recent investigations. Cf. the discussion of this matter on p. 647.

(c) Preparation

At the present time, the raw materials used for the technical extraction of rhenium are certain residues obtained in smelting the Mansfeld copper shales, in which rhenium is present in a relatively highly enriched state. These residues, which contain on the average about 0.005% rhenium, are subjected to an oxidation process and are then leached. The rhenium then passes into solution in the form of the ReO_4^- ion, together with large quantities of heavy metal sulfates. The greater part of the latter is first deposited by evaporation, and then the rhenium is precipitated as potassium perrhenate, $KReO_4$, by the addition of potassium chloride. This is redissolved, freed from heavy metals which may have been carried down with it, by the addition of potassium hydroxide, and is finally purified by repeated recrystallization. By this process, an annual production of at least 250 kg of potassium perrhenate, or 160 kg of rhenium, can be obtained from the smelter residues.

Metallic rhenium is obtained, as a grey powder, by heating potassium perrhenate in hydrogen. It is freed from the accompanying potassium hydroxide by washing it in hot water, and then heating again in hydrogen. If ammonium perrhenate is heated in hydrogen, pure rhenium is obtained directly, in the form of a brilliant metallic mirror. Single crystal wires of rhenium can be obtained by depositing them according to the Van Arkel and De Boer process.

(d) Properties [1]

Metallic rhenium resembles platinum in appearance. It is fairly soft and quite malleable in the compact state when it is pure. Lumps obtained by sintering or melting the grey powdered metal which is produced by reduction in hydrogen are only moderately ductile, and are very hard (hardness 8 on Mohs' scale). Rhenium melts at 3150°. It crystallizes with the close packed hexagonal structure (Vol. I, p. 249, Fig. 57), $a = 2.76$ Å, $c = 4.47$ Å, atomic radius 1.38 Å. Its density in the compact state is 20.9, and its specific electrical resistance $0.20 \cdot 10^{-4}$ ohm cm^{-1} at 20°. Rhenium is fairly resistant to air oxidation at ordinary temperature. The compact metal is perceptibly attacked by oxygen only above 1000°. Powdered rhenium, however, slowly volatilizes when it is heated gently in oxygen, giving the oxide Re_2O_7. It gradually is oxidized in moist air, even at ordinary temperature, forming perrhenic acid $HReO_4$.

Rhenium does not unite with hydrogen to form hydrides, and nitrides are also at present unknown. With phosphorus it forms the compounds Re_2P, ReP, ReP_2 and ReP_3. It also combines with arsenic forming a compound corresponding approximately with the formula Re_3As_7, and with silicon to form $ReSi_2$. Rhenium combines with tungsten, yielding the compound Re_3W_2 (cf. Table 17, p. 124). It forms a wide range of mixed crystals, but no compound, with chromium. The behavior of rhenium towards the other metals is but little known.

Rhenium is practically unattacked by hydrochloric and hydrofluoric acids. It dissolves rapidly in nitric acid, however, and more slowly in sulfuric acid, giving perrhenic acid. The salts of perrhenic acid are also obtained as the final products when rhenium is fused with alkali hydroxides, with adequate access of air (cf. p. 238).

Rhenium is a relatively poor hydrogenation catalyst, but is better for dehydrogenation. It is less suitable for catalytic oxidations, because of the ease with which the volatile heptoxide is formed.

(e) Uses

Rhenium has been found very useful for the manufacture of tips for fountain pen nibs, since when alloyed with small amounts of other metals it combines great hardness with resistance towards corrosion by dried ink. It is also used for making platinum alloys with excellent resistance towards mechanical wear—e.g., for electrodes used in electro-analysis. The hardness and tensile strength of Pt-Re alloys with 5% Re excels not only that of pure platinum, but also that of the Pt-Ir alloys (with 10% Ir) formerly used for making electrodes. The volatility of rhenium when it is heated in the air is an objection to its use for crucibles. The volatility of rhenium when heated in an oxidizing atmosphere has to be borne in mind when platinum-platinum rhenium thermoelements are employed, since this loss leads to a gradual decrease in the thermo-e.m.f. These thermocouples have advantages over platinum-platinum rhodium thermocouples in that the thermo-e.m.f. is 3 to 4 times as large.

7. Compounds of Rhenium

The $+4$ and $+7$ valence states are distinctly favored by rhenium in its compounds.

Rhenium can only function as heptapositive towards the halogens when it is combined with oxygen as well as with the halogens. Its maximum valence is otherwise *six* towards fluorine and *five* towards chlorine. Among the halogeno-salts, those of the type $M^I{}_2[ReX_6]$, with *tetra*positive rhenium, are particularly readily formed. Compounds of *tri*positive rhenium are also known. There are some indications that rhenium can also exist in the *bi*valent and *uni*valent state, but pure compounds corresponding to these valence states have not yet been isolated. Rhenium can also exist in the *uninegative* state which is produced when potassium perrhenate is passed through a Jones reductor (Lundell and Knowles, 1937). The compound $KRe \cdot 4H_2O$ has recently been isolated (Bravo, Greswold and Kleinberg, 1954).

A survey of the principal classes of rhenium compounds is given in Table 27.

(a) Rhenium Oxides

Besides the oxides given in Table 27, rhenium probably forms an oxide of character similar to molybdenum blue and tungsten blue, since a blue deposit is frequently obtained when rhenium is heated in a current of oxygen, in addition to the yellow oxide Re_2O_7 and

TABLE 27

MOST IMPORTANT TYPES OF COMPOUNDS FORMED BY RHENIUM

Valence state	Oxides	Oxy salts	Sulfides	Fluorides	Chlorides	Chloro salts
III	$Re_2O_3 \cdot xH_2O$	—	—	—	Re_2Cl_6	$M^I[ReCl_4]$
IV	ReO_2	$M^I{}_2[ReO_3]$	ReS_2	ReF_4	—	$M^I{}_2[ReCl_6]$
V	—	$M^I[ReO_3]$	—	—	$ReCl_5$	$M^I{}_2[ReOCl_5]$
VI	ReO_3	$M^I{}_2[ReO_4]$	—	ReF_6	$ReOCl_4$	$M^I{}_2[ReOCl_6]$
VII	Re_2O_7	$M^I[ReO_4]$	Re_2S_7	—	ReO_3Cl	—

the white oxide (see below). In the preparation of ReO_2 by the reduction of Re_2O_7, or of compounds derived from this oxide, violet or blue substances are again often formed as intermediate products. These have compositions intermediate between ReO_3 and ReO_2, and may perhaps be identical in nature with the blue sublimate. It has not yet been possible to determine the composition of the latter.

(i) *Rhenium heptoxide*, Re_2O_7, is produced when rhenium powder is heated in a current of oxygen. It forms yellow crystals which melt at 220°, sublime even below their melting point, and dissolve very readily in water to form *perrhenic acid*. Rhenium heptoxide is also very soluble in alcohol, but not very soluble in ether. For heat of formation, see p. 207.

When the yellow rhenium heptoxide is gently heated in oxygen, and in the course of preparing the heptoxide by the combustion of rhenium, a *white* oxide of rhenium is frequently obtained as a by-product, in the form of a mist which is not readily deposited. In the condensed state, this forms a snow white, highly refractive mass, which is not perceptibly crystalline. It is uncertain whether this is a second modification of rhenium heptoxide, Re_2O_7, or whether it may be a peroxide, Re_2O_8.

(*ii*) *Rhenium trioxide*, ReO_3, was prepared by Biltz (1931), by heating Re_2O_7 with Re in a sealed capillary tube at 300°. It is red, and has a density of 7. The X-ray investigation of its structure (Meisel, 1932) has shown that the Re^{6+} ions form a simple cubic lattice, with the O^{2-} ions occupying the mid-points of the 12 edges of the cubic unit cell (Vol. I, p. 209, Fig. 45).

(*iii*) *Rhenium dioxide*, ReO_2, can be obtained by heating rhenium heptoxide in hydrogen at 300°, or by dehydrating rhenium dioxide hydrate, $ReO_2 \cdot xH_2O$, in a vacuum at 300°. It is a brown-black powder, which burns to rhenium heptoxide (often inflaming) if it is heated in oxygen. Rhenium dioxide is reduced to the metal when it is heated to 800° in hydrogen. If strongly heated in a vacuum it decomposes, according to the equation $7ReO_2 = 3Re + 2Re_2O_7$ (Biltz, 1933). ReO_2 dissolves with a green color in hydrochloric acid, forming $H_2[ReCl_6]$. The salts of this acid (see below) have a tendency to hydrolyze, with the deposition of the brown-black rhenium dioxide hydrate.

(*iv*) *Dirhenium trioxide* (rhenium sesquioxide), Re_2O_3, has as yet been obtained only in the hydrated form, $Re_2O_3 \cdot xH_2O$, by the addition of alkali hydroxide to a solution of Re_2Cl_6. It is unstable, and gradually decomposes water with the evolution of hydrogen (Geilmann, 1933).

(b) Oxy-Salts of Rhenium

Rhenium forms the following oxy salts.

Re(VII)	$\begin{cases} M^I[ReO_4] \\ M^I_3[ReO_5] \end{cases}$	(1 : 1) perrhenates, ordinary perrhenates, colorless. (3 : 1) perrhenates, yellow to red	
Re(VI)	$M^I_2[ReO_4]$	rhenates, green	
Re(V)	$\begin{cases} M^I[ReO_3] \\ M^I_4[Re_2O_7] \\ M^I_3[ReO_4] \end{cases}$	(1 : 1) hyporhenates, metahyporhenates, yellow (2 : 1) hyporhenates, pyrohyporhenates, yellow (3 : 1) hyporhenates, orthohyporhenates, yellow	
Re(IV)	$M^I_2[ReO_3]$	rhenites, brown.	

The *perrhenates* are by far the most stable of these salts. The *rhenites* may also be obtained relatively easily. The preparation of pure *rhenates* and *hyporhenates*, however, presents great difficulties, since they are present in melts, in temperature-dependent equilibrium with the rhenites and the perrhenates (see below).

(*i*) *Perrhenic acid* and the *Perrhenates*. Rhenium heptoxide is the anhydride of *perrhenic acid*, $HReO_4$.

The reaction $Re_2O_7 + H_2O = 2H^+ + 2ReO_4^-$ takes place at great dilution with considerable evolution of heat (6.3 kcal per mol of $HReO_4$). Nevertheless, when the solution is evaporated it is not the free acid $HReO_4$, which separates out, but its anhydride, Re_2O_7. Perrhenic acid is a strong monobasic acid. Its aqueous solutions are colorless. It is most readily obtained by dissolving metallic rhenium in 30 per cent nitric acid.

When perrhenic acid is neutralized with bases, its salts, the *perrhenates*, $M^I[ReO_4]$, are obtained. Unless they contain colored cations, these are colorless, and generally crystallize anhydrous.

Most perrhenates are soluble in water, and many of them can be melted without decomposition. In so far as their structures have been determined, the perrhenates of univalent metals crystallize in the scheelite structure (Fig. 27, p. 179).

(*ii*) *Potassium perrhenate*, $KReO_4$, is the starting material for the preparation of most other rhenium compounds. It crystallizes anhydrous, in colorless tetragonal bipyramids (scheelite type, $a = 5.615$ Å, $c = 12.50$ Å). Its solubility is:

at	0°	10°	20°	25°	50°	100°
	0.474	0.573	0.990	1.235	3.18	10.44 g of $KReO_4$ in 100 g of H_2O.

Unlike potassium permanganate, potassium perrhenate is quite stable, even in strongly alkaline solutions.

Perrhenates may be obtained directly, by fusing rhenium powder with alkali hydroxides in the presence of oxygen. When an amount of oxygen equivalent to the oxidation of the metal to perrhenate has been absorbed, the melt assumes a fiery red color. This is because the great excess of alkali hydroxide in such a melt leads to the formation of the red (3 : 1) perrhenates, $M^I_3ReO_5$, and not the colorless ordinary perrhenates, M^IReO_4.

Before the formation of perrhenate is completed, the alkali melt turns first dark brown, then yellow and finally green, as the oxygen content increases. As was shown by I. and W. Noddack (1933), this is due to the occurrence of the oxidation in stages, by way of the brown rhenites, yellow hyporhenates, and green rhenates. The oxy salts corresponding to the intermediate stages of oxidation are difficult to isolate from the melt in a pure state, however, since equilibria are set up between the various oxidation states, depending in part upon reversible disproportionation processes (e.g., $2Re(V) \rightleftharpoons Re(IV) + Re(VI)$, $2Re(VI) \rightleftharpoons Re(V) + Re(VII)$), which are shifted by changes in temperature and oxygen content. Pure *rhenites*, $M^I_2[ReO_3]$, are obtained most simply by fusing ReO_2 with alkali hydroxide. Thus sodium rhenite is formed by the reaction

$$ReO_2 + 2NaOH = Na_2[ReO_3] + H_2O,$$

as a brown powder, insoluble in water and caustic alkalis.

(c) Sulfur Compounds of Rhenium

Unlike manganese, rhenium forms no monosulfide, but a *disulfide* ReS_2 (with tetrapositive rhenium), and a *heptasulfide*, Re_2S_7. Whereas the disulfide is a very stable compound (heat of formation 40 kcal), the latter is an endothermic compound. Hence only the disulfide can be obtained by the direct union of its constituents, whereas the heptasulfide can be prepared only by precipitation from solution.

(*i*) *Rhenium heptasulfide* is obtained as a black precipitate by the action of H_2S on strongly acidified solutions of perrhenates. It tends to stay in colloidal dispersion if the concentration of acid is too high or too low, the optimum conditions for precipitation being 12 g of HCl per 100 ml of solution. Rhenium heptasulfide is practically insoluble in hydrochloric acid and in alkali sulfide solutions. It dissolves in nitric acid, forming perrhenic acid. It loses sulfur irreversibly when heated, giving the disulfide, (Biltz, 1931).

(*ii*) *Thioperrhenates.* If hydrogen sulfide is passed into a neutral perrhenate solution, or into dilute perrhenic acid, the solution turns yellow as a result of the formation of monothioperrhenate ions, $[ReO_3S]^-$. The alkali monothioperrhenates are very soluble in water. For example, the potassium salt $K[ReO_3S]$, which crystallizes in yellowish green needles, dissolves to the extent of 1 part in 1.5 parts of water at 20°. The yellow thallium(I) thioperrhenate, $Tl[ReO_3S]$, is sparingly soluble. The monothioperrhenates gradually decompose in solution, apparently to form compounds with higher sulfur content—e.g., $Tl[ReS_4]$ thallium(I) tetrathioperrhenate, a dark brown precipitate.

(*iii*) *Rhenium disulfide* can be prepared both by the direct combination of the components and by heating rhenium heptasulfide (e.g., in nitrogen). It forms black, trigonal leaflets (density 7.51), insoluble in hydrochloric acid and in alkali sulfide solutions. It decomposes into its elements when sufficiently strongly heated ($ReS_2 = Re + S_2$, dissociation pressure

13 mm of sulfur vapor at 1110°, 99 mm at 1225°), without the intervention of any lower sulfides as intermediate stages. Whereas Re forms mixed crystals with ReS_2 only to a very small extent, if at all, ReS_2 can take up an excess of sulfur in solid solution. It is therefore not easy to obtain the sulfide with a composition corresponding exactly to the formula ReS_2.

(d) Halogen Compounds of Rhenium

Table 27, p. 236, gives a summary of the *fluorides* and *chlorides* of rhenium. Most of the halogen compounds of rhenium have a tendency to form acido-complexes, and some are stable only in such a form. This is true of the *bromides* and *iodides* in particular, which are themselves but little known in the free state, although their acido salts correspond in composition to the chloro salts.

(*i*) *Rhenium Fluorides*. Rhenium powder reacts with fluorine at about 125°, yielding *rhenium hexafluoride*, ReF_6. This is formed as a colorless gas which condenses on cooling to a yellow crystalline mass (m.p. 18.8°, b.p. 47.6°). The gas density corresponds to the formula ReF_6. Oxyfluorides, $ReOF_4$ and ReO_2F_2, have also been prepared, but it has not been possible to isolate a fluoride containing heptapositive rhenium (Ruff, 1932 and onwards). If ReF_6 mixed with hydrogen is passed through a tube heated to 200°, *rhenium tetrafluoride*, ReF_4, is deposited as a greenish black mass (density 5.38, m.p. 124.5°). ReF_4 is soluble in water, with decomposition, but it dissolves in 40% hydrofluoric acid to give a green solution, from which *potassium hexafluororhenate(IV)*, $K_2[ReF_6]$, crystallizes when KF is added. Potassium hexafluororhenate is more simply prepared by reducing $KReO_4$ by means of KI, in hydrofluoric acid solution. It forms small green octahedra, isotypic with $K_2[PtCl_6]$.

(*ii*) *Rhenium Chlorides and Chlororhenates*. Rhenium combines with chlorine when heated, giving *rhenium pentachloride*, $ReCl_5$. $ReCl_3$ (or Re_2Cl_6) is formed at the same time, but the pentachloride can readily be separated by sublimation in a high vacuum. Rhenium pentachloride is deep brown-black in color. Although it can be sublimed in high vacuum, it loses chlorine when it is heated in an atmosphere of nitrogen, and forms Re_2Cl_6. Heated in oxygen, $ReCl_5$ loses chlorine and forms the oxychlorides $ReOCl_4$ and ReO_3Cl. When $ReCl_5$ is treated with dilute hydrochloric acid it dissolves, in part with the production of the acid $H_2[ReOCl_5]$, but also to some extent with the formation of $HReO_4 + H_2[ReCl_6]$. It reacts with concentrated hydrochloric acid according to the equation:

$$ReCl_5 + 2HCl = H_2[ReCl_6] + \tfrac{1}{2}Cl_2$$

(Geilmann and Biltz, 1933).

The acid $H_2[ReOCl_5]$ is not very stable, and the *oxopentachlororhenate(V) salts*, $M^I_2[ReOCl_5]$, derived from it also readily decompose in solution, undergoing hydrolysis and disproportionation. These salts resemble the hexachlororhenate(IV) salts in appearance, but differ in that they are birefringent and give different X-ray diffraction patterns (Jakob, 1933, Hölemann, 1937).

Chlororhenic acid (more correctly hexachlororhenium(IV) acid), $H_2[ReCl_6]$, derived from the chloride $ReCl_4$, which is itself not known in the free state, is very similar to chloroplatinic acid, $H_2[PtCl_6]$, in its reactions, although less strongly complexed than the latter. In the form of its aqueous solution or of its salts, it is the most readily accessible chlorine compound of rhenium, and can be prepared in various ways,—e.g., by dissolving ReO_2 in hydrochloric acid or, starting with perrhenates, by reducing them with HI in concentrated hydrochloric acid solution. *Potassium chlororhenate* (potassium hexachlororhenate(IV)), $K_2[ReCl_6]$, is most readily obtained pure by warming an intimate mixture of potassium perrhenate and potassium iodide with concentrated hydrochloric acid (Enk, 1931). It is isotypic with $K_2[PtCl_6]$, and is rather sparingly soluble in hydrochloric acid; water alone, without addition of acid, brings about hydrolytic decomposition. *Cesium chlororhenate*, $Cs_2[ReCl_6]$, is far less soluble still.

Rhenium trichloride, $ReCl_3$ or Re_2Cl_6, obtained by the thermal decomposition of $ReCl_5$, is red. It is soluble in water, but is rapidly hydrolyzed in pure aqueous solution, with deposition of the hydrated sesquioxide, $Re_2O_3 \cdot xH_2O$. Rhenium trichloride is notably stable in hydrochloric acid solution, however, and under these conditions is not attacked at ordinary temperature by strong oxidizing agents such as permanganate or chlorine water. Rhenium trichloride is present in hydrochloric acid solution in the form of the complex acid $H[ReCl_4]$; *chlororhenate(III)* salts, $M^I[ReCl_4]$, crystallize out when alkali chlorides are added to the hydrochloric acid solution. These salts undergo decomposition when they are heated:

$$6M^I[ReCl_4] = 3M^I_2[ReCl_6] + Re + Re_2Cl_6$$

(Geilmann and Biltz, 1932, I. and W. Noddack, 1933).

A freshly prepared solution of rhenium trichloride does not react immediately with silver nitrate. Hence the trichloride is not present in the solution in a dissociated form. In accordance with this, it can be extracted from solution by means of ether. Molecular weight determinations in glacial acetic acid point to the formula Re_2Cl_6. According to Biltz, the similarity in absorption spectrum between the free chloride and $H[ReCl_4]$ suggests that the former is a complex compound:

$$\begin{bmatrix} Cl & & Cl & & Cl \\ & Re & & Re & \\ Cl & & Cl & & Cl \end{bmatrix}$$

It differs from other complex non electrolytes in being readily soluble. This may be due to the fact that rhenium in this compound is not coordinatively saturated, and is able to add on H_2O molecules.

(*iii*) *Rhenium oxychlorides* are obtained by the action of a mixture of chlorine and oxygen upon metallic rhenium. The compounds are more readily obtained pure, however, by other methods—e.g., by heating $ReCl_5$ or Re_2Cl_6 in oxygen, or by the action of $ReCl_5$ on Re_2O_7. *Rhenium trioxychloride*, ReO_3Cl, was first prepared in this way by Brukl (1932), as a colorless liquid, m.p. 4.5°, b.p. 131°. If an excess of $ReCl_5$ is employed, *rhenium oxytetrachloride*, $ReOCl_4$ (m.p. 29.3°, b.p. 223°) is obtained as brown-red radiating needles. Whereas ReO_3Cl is smoothly hydrolyzed by water, to give $HReO_4$ and HCl, the hydrolysis of $ReOCl_4$ is accompanied by a self-oxidation-reduction, so that $HReO_4$ and HCl are formed and black ReO_2 is deposited simultaneously. $ReOCl_4$ dissolves in concentrated hydrochloric acid, forming the complex acid $H_2[ReOCl_6]$, the salts of which are very unstable. The potassium salt, $K_2[ReOCl_6]$, obtained by I. and W. Noddack, decomposes as soon as it is separated from its hydrochloric acid solution, and undergoes decomposition according to the equation

$$3K_2[ReOCl_6] + 5H_2O = 2K[ReO_4] + K_2[ReCl_6] + 2KCl + 10HCl.$$

$ReOCl_4$ reacts with ammonia, forming $ReO(NH_2)_2Cl_2$, which decomposes above 400°, leaving a residue of Re and ReO_2. If the oxydiamidodichloride is brought into contact with ice water, it exchanges Cl for OH, and forms *diamidorhenic acid*, $H_2[ReO_3(NH_2)_2]$. This loses water in a vacuum at 100°, and is transformed into *diamidorhenium dioxide*, $Re(NH_2)_2O_2$ (Brukl, 1933).

(e) Carbon Monoxide Compounds of Rhenium

Rhenium shares with the elements of the VIth, VIIIth and Ist Sub-groups the capacity to form compounds with carbon monoxide (*carbonyl* compounds); it forms both a compound with true metal carbonyl character, namely the *pentacarbonyl*, $[Re(CO)_5]_2$, and also *carbonyl halides*, $Re(CO)_5X$. These compounds were discovered by Hieber, 1939.

(*i*) *Rhenium pentacarbonyl*, $[Re(CO)_5]_2$, may be obtained by the action of carbon monoxide under high pressure (300–400 atm.) upon rhenium heptoxide at 250°:

$$Re_2O_7 + 17CO = [Re(CO)_5]_2 + 7CO_2.$$

It exists as colorless pseudo-cubic crystals, which are but slightly volatile, and not very soluble in organic solvents. It is quite stable at ordinary temperature, but undergoes decomposition above 250°.

(*ii*) *Rhenium Carbonyl Halides. Rhenium chloropentacarbonyl*, $Re(CO)_5Cl$, can be obtained, e.g., by heating $ReCl_5$ with CO and Cu, or $KReO_4$ with CO and CCl_4:

$$ReCl_5 + 9CO + 4Cu = Re(CO)_5Cl + 4Cu(CO)Cl$$

$$KReO_4 + 8CO + CCl_4 = Re(CO)_5Cl + COCl_2 + 3CO_2 + KCl.$$

If $K_2[ReBr_6]$ or $K_2[ReI_6]$ are used, in place of $ReCl_5$, the corresponding bromo- or iodopentacarbonyl is obtained. The rhenium pentacarbonyl halides form colorless crystals, which are not very soluble in organic solvents. They are quite stable at ordinary temperatures.

Both in rhenium pentacarbonyl and in the pentacarbonyl halides, a portion of the carbon monoxide can be replaced by certain organic nitrogen compounds (e.g., pyridine (pyr) or orthophenanthroline (o-phen). In this way such compounds as $Re(CO)_3(pyr)_2$, $Re(CO)_3(pyr)_2X$, (pale yellow-green), $Re(CO)_3(o\text{-phen})$, $Re(CO)_3(o\text{-phen})Cl$ (yellow) are formed.

8. Analytical (Rhenium)

$+7$ Rhenium is precipitated by hydrogen sulfide from strongly acid solutions as the heptasulfide, Re_2S_7, which is insoluble in ammonium sulfide. The formation of the volatile oxide Re_2O_7 when materials containing rhenium are heated in oxygen, is a very characteristic reaction. Basic oxides, which would interfere with the volatilization of rhenium heptoxide (by forming perrhenates) can be rendered ineffective by adding H_3PO_4 or SiO_2. Perrhenic acid, formed by dissolving Re_2O_7, may be identified by converting it into salts of characteristic crystal habit.

In addition to the perrhenates, the hexachlororhenates(IV), $M^I_2[ReCl_6]$, are suitable salts for the *microanalytical* detection of rhenium.

Rhenium is detected with the greatest certainty by means of its arc spectrum (characteristic lines at 3452, 3461 and 3465 Å), or by its X-ray spectrum (cf. p. 234). Volatile rhenium compounds impart a dull green color to the oxidizing flame of the Bunsen burner.

Rhenium is most conveniently determined gravimetrically by precipitating and weighing *nitron perrhenate*, $C_{20}H_{16}N_4 \cdot HReO_4$. If necessary, the rhenium may first be separated by precipitating it as sulfide. This is then oxidized by means of hydrogen peroxide [Geilmann, *Z, anorg. Chem.*, 195 (1931)].

Rhenium can be separated quantitatively from molybdenum by precipitating the latter element by means of 8-hydroxyquinoline. The rhenium can then be determined in the filtrate as nitron perrhenate (Geilmann, 1931).

Reference

1 I. and W. NODDACK, *Das Rhenium*, Leipzig 1933, 86 pp.

CHAPTER 7

EIGHTH SUB-GROUP OF THE PERIODIC SYSTEM: METALS OF THE IRON GROUP AND PLATINUM METALS

	Atomic numbers	Elements	Symbols	Atomic weights	Densities	Melting points	Boiling points	Specific heats	Valence states
Metals of the iron group	26	Iron	Fe	55.85	7.86	1528°	2735°	0.1077	I, II, III, IV, V
	27	Cobalt	Co	58.94	8.83	1490°	3100°	0.0928	I, II, III, IV
	28	Nickel	Ni	58.71	8.90	1452°	2840°	0.107	I, II, III, IV
Light platinum metals	44	Ruthenium	Ru	101.1	12.30	ca. 2400°	ca. 4200°	0.0553	II, III, IV, V, VI, VII, VIII
	45	Rhodium	Rh	102.91	12.42	1966°	ca. 3900°	0.0591	I, II, III, IV, V
	46	Palladium	Pd	106.4	12.03	1555°	3170°	0.0543	II, III, IV
Heavy platinum metals	76	Osmium	Os	190.2	22.7	ca. 2700°	ca. 4600°	0.0311	II, III, IV, VI, VIII
	77	Iridium	Ir	192.2	22.65	2454°	ca. 4500°	0.0309	I, II, III, IV, V
	78	Platinum	Pt	195.09	21.45	1774°	ca. 3800°	0.0318	I, II, III, IV, V

1. Introduction

(a) General

Sub-group VIII of the Periodic System comprises three series, each made up of three consecutive elements with the atomic numbers 26 to 28 (iron, cobalt, and nickel), 22 to 46 (ruthenium, rhodium, and palladium) and 76 to 78 (osmium, iridium, and platinum). The first three resemble each other especially closely, and together make up the metals of the *iron group*. The other six are known collectively as the *platinum metals*, after the most abundant and most important member, which the rest closely resemble in properties, and with which they are usually associated in Nature. The platinum metals can be subdivided into the *light* and *heavy* platinum metals, according to their density (cf. Table, above).

The nine elements of Group VIIIB are arranged in the Periodic Table as follows (the names are prefixed by the atomic numbers):

	a	b	c
1.	$_{26}$iron	$_{27}$cobalt	$_{28}$nickel
2.	$_{44}$ruthenium	$_{45}$rhodium	$_{46}$palladium
3.	$_{76}$osmium	$_{77}$iridium	$_{78}$platinum

Similarity between *adjacent* elements is hardly shown to the same degree anywhere else in the Periodic Table as in these three series. In addition, there is also, a considerable similarity between the elements standing one above another in the three columns *a, b, c.*

A comparison of the most important physical and chemical properties shows that the metals of the *iron series* display closer resemblances to one another than any of them does to the elements standing under them, whereas in the platinum metals the closest resemblances are between elements in the same column. In discussing the chemical properties it is therefore convenient to subdivide Group VIIIB, with iron, cobalt and nickel constituting one subdivision (the *iron series*), and the platinum metals being divided into three *dyads*, each consisting of the pair of vertically related elements. This subdivision is set out in the scheme of Table 28.

TABLE 28

EIGHTH SUB-GROUP OF THE PERIODIC SYSTEM

A. IRON SERIES		
Iron	Cobalt	Nickel

B. PLATINUM METALS		
1st dyad	2nd dyad	3rd dyad
Ruthenium Osmium	Rhodium Iridium	Palladium Platinum

Of the three elements standing on the extreme left of Group VIIIB—iron, ruthenium and osmium—, the first two occupy positions 8 places after the preceding inert gas (argon and krypton, respectively). According to the Bohr theory, the outer electron shell of osmium should have essentially the same configuration as that of iron and ruthenium (see below). It might therefore be expected that these three elements would display a maximum valence state of *eight*. This is, in fact, true of ruthenium and osmium, whereas iron exhibits a maximum valence state of six. In general, the tendency to exercise high valence states increases from top to bottom within each of the three columns of vertically related elements (cf. the data in the last column of the summary Table). Within each horizontal series, however, the tendency to exert high valence states diminishes from left to right.

A similar tendency emerges when instead of the *maximum*, the *preferred* valence states are compared within Sub-group VIII. Iron is usually bi- or trivalent, ruthenium is preferentially quadrivalent, whereas osmium has a marked tendency to function in the $+6$ or $+8$ valence state. Cobalt is predominantly bivalent in its simple compounds, and trivalent in complex compounds; rhodium is predominantly trivalent in its simple compounds also, whereas in the case of iridium some of the most important compounds are derived from the quadrivalent state. Nickel, as a rule, is definitely bivalent; although the bivalent state is the most important for palladium also, the distinction is not so marked, and platinum exhibits a $+4$ valence state in some of its most important compounds—hexachloroplatinic acid and its salts.

	Be	Mg	Ca	B	Al	Ga	In	Tl	C	Si
Fe	$s < \infty$ Be_xFe $x > 5$ Be_5Fe Be_2Fe	$s\ o?$ —	$liq < \infty$ $s\ o?$ o	$s > o$ Fe_2B^* 1389° FeB 1540°	$s < \infty$ Al_3Fe^* Al_5Fe_2 Al_2Fe^* $AlFe[]$ $[AlFe_2]$	$s > o$ o	—	$liq\ o?$ $s\ o$ o	$s < \infty$ Fe_3C $Fe_2C?$ meta- stable	$s <$ Fe_3Si 103c FeS 141c FeS 122c
Co	$s > o$ Be_xCo $BeCo$	$s\ o?$ —	$Alloys?$ —	— Co_2B CoB CoB_2	$s > o$ Al_3Co^* 943° $Al_5Co_2^*$ 1170° $AlCo$ 1628°	—	— $InCo_2$	$liq < \infty$ $s \sim o$ o	$s > o$ Co_3C $Co_2C[]?$ 225°	$s <$ Co_3S 121c Co_2 1327 Co_3Si 1214 CoS 144c CoS 1277 CoS 130c
Ni	$s > o$ $Be_{21}Ni_5$ 1262° $BeNi$ 1492°	$s\ o$ Mg_2Ni^* 760° $MgNi_2$ 1145°	—	$s\ o?$ Ni_2B 1230° Ni_3B_2 1160° NiB^* 1020° Ni_2B_3 1280°	$s > o$ Al_3Ni^* 842° $Al_3Ni_2^*$ 1132° $AlNi$ 1640° $AlNi_3^*$ 1396°	$s < \infty$ Ga_4Ni Ga_3Ni_2 $GaNi$ $GaNi_2$ 1220°	— In_2Ni In_3Ni_2 $InNi_2$ $InNi_3$	$liq < \infty$ $s > o$ o	$s \sim o$ Ni_3C meta- stable	$s >$ Ni_3S 1163 Ni_5S 1255 Ni_2S 1290 Ni_3Si 845 NiS 995° $NiSi_$ 1000°
Pd	$s \sim o$ $BePd$ 1465° $Be_4Pd_3^*$ 1200° $BePd_2^*$ 1090° $BePd_3^*$ 960°	—	—	$s\ o$ —	— Al_3Pd $AlPd$ $AlPd_2$	—	— In_3Pd $InPd$ $InPd_2$	—	$s\ o$ o	$s\ o$ Pd_2S $\sim$ 14c PdS 900°
Pt	$s > o$ compd.	— compd.	— compd.	— compd.	$s\ o$ Al_3Pt^* 780° Al_2Pt $AlPt$	— Ga_2Pt Ga_3Pt_2 $GaPt$	— In_3Pt In_2Pt In_3Pt_2	$s\ o$ $TlPt$ 685°	$s\ o$ —	$alloy$ Pt_2Si $PtSi_2$

F SUB-GROUP VIII AND THE ELEMENTS OF THE MAIN GROUPS

eaning as in Table 9)

Ge	Sn	Pb	N	P	As	Sb	Bi	S	Se	Te
—	*s > 0* Fe$_2$Sn* 900° FeSn[] 800° FeSn$_2$* 496°	*liq o* *s o* o	*s > 0* Fe$_4$N[] $\overline{Fe_2N}$	*s > 0* Fe$_3$P* 1166° Fe$_2$P 1365° FeP FeP$_2$	*s > 0* Fe$_2$As 919° Fe$_3$As$_2$ FeAs 1030° FeAs$_2$	*s > 0* Fe$_3$Sb$_2$ 1010° $\overline{FeSb}$ FeSb$_2$* 728°	*liq o* *s o* o	*s ~ 0* $\overline{FeS}$ 1208° FeS$_2$ 689° (decomp. to FeS + S vap.)	*s < ∞* $\overline{FeSe}$ FeSe$_2$	— FeTe FeTe$_2$
Co$_2$Ge 1200° CoGe[] 982° o$_2$Ge$_3$[] 750° oGe$_2$[] 842°	*s > 0* $\overline{Co_3Sn_2}$ 1151° CoSn* 943° CoSn$_2$* 525°	*liq < ∞* *s o* o	*s ~ 0* Co$_3$N Co$_2$N? Co$_3$N$_2$	*liq < ∞* *s o* Co$_2$P 1386° CoP CoP$_3$	*s ~ 0* Co$_5$As$_2$* Co$_2$As* Co$_3$As$_2$* CoAs 1180°	*s > 0* CoSb 1191° CoSb$_2$* 897°	*liq < ∞* *s o* o	*s ~ 0* $\overline{Co_4S_3}$* 930° Co$_9$S$_8$* 832° CoS Co$_3$S$_4$ CoS$_2$	— $\overline{CoSe}$ CoSe$_2$	CoTe CoTe$_2$
—	*s > 0* Ni$_3$Sn* 1174° Ni$_3$Sn$_2$ 1264° Ni$_3$Sn$_4$ 794°	*liq < ∞* *s > 0* o	*s > 0* Ni$_3$N Ni$_3$N$_2$?	*s o* Ni$_3$P* 970° Ni$_5$P$_2$ 1175° Ni$_2$P 1110° Ni$_6$P$_5$? NiP$_2$ NiP$_3$	*s > 0* Ni$_5$As$_2$ 998° Ni$_3$As$_2$*? NiAs 970° NiAs$_2$	*s > 0* Ni$_3$Sb[] 698° Ni$_5$Sb$_2$ 1162° Ni$_7$Sb$_3$[] 588° NiSb 1153° Ni$_2$Sb$_3$[] Ni$_3$Sb$_5$* 626°	*s ~ 0* NiBi$_3$* 469° NiBi* 655°	*s > 0* Ni$_3$S$_2$* Ni$_6$S$_5$* NiS Ni$_3$S$_4$ NiS$_2$	— $\overline{NiSe}$ NiSe$_2$	— NiTe NiTe$_2$
—	— Probably several compds.	*s < ∞* Pd$_3$Pb 1220° Pd$_3$Pb$_2$* 830° PdPb* 596° PdPb$_2$ 454°	*s o?* o	*s o* $\overline{Pd_5P}$* 807° Pd$_3$P 1047° Pd$_5$P$_2$* 860° PdP$_2$ ~1150°	Pd$_3$As$_2$? PdAs$_2$	*s > 0* Pd$_3$Sb 1182° $\overline{Pd_2Sb}$[] 850° PdSb 800° PdSb$_2$ 680°	compd.	*s o* Pd$_4$S[] 761° $\overline{Pd_5S_2}$? PdS 970° PdS$_2$	PdSe? PdSe$_2$?	PdTe PdTe$_2$
—	*s > 0* Pt$_3$Sn* 1365° PtSn 1300° PtSn$_2$* 848° PtSn$_4$* 745°	*s o* Pt$_3$Pb* 915° PtPb* 795° PtPb$_2$* 360°	— o?	*liq < ∞* *s o* Pt$_{20}$P$_7$[] 590° PtP$_2$ >1500°	*s o?* Pt$_2$As$_3$? PtAs$_2$	*s > 0* Pt$_4$Sb*? 752° Pt$_5$Sb$_2$*? ~637° PtSb* 1050° PtSb$_2$ 1226°	*s ~ 0* PtBi PtBi$_2$?	PtS Pt$_2$S$_3$? PtS$_2$	PtSe? PtSe$_2$	PtTe PtTe$_2$ 1250°

An increase in the tendency to function in the higher valence states, with increase in atomic number, has been noticed among the vertically related elements of each of the Sub-groups of the Periodic System to be discussed so far. It shows itself clearly in Sub-groups IV, V, VI, VII and VIII, but *only in these* Sub-groups. It cannot be detected in groups IB, and IIB, and is, in fact, reversed in group IIB, the tendency being for *lower* valence states to be preferred as the atomic number increases.

Osmium is closely related to ruthenium in its chemical properties, and especially in its ability to exist in a valence state of $+8$, even although osmium does not follow eight places after an inert gas. This is because the series of lanthanides has been interposed between osmium and the inert gas preceding it (xenon). According to the Bohr theory, the occurrence of the lanthanide series is due to the entry of electrons into the $4f$ shell, which lies 'inside the xenon configuration' (cf. the Table in Appendix II). Osmium possesses 22 ($= 14 + 8$) electrons more than xenon, but of this number only 8 are bound 'outside' the xenon configuration—namely in $5d$ and $6s$ orbits. Thus the neutral osmium atom, like ruthenium and iron, has an electronic configuration with 8 electrons bound 'outside' the preceding inert gas configuration.

Properties common to all the elements of group VIIIB are their metallic character, grey to grey-white color, high melting points, extremely high boiling points, and very small atomic volume (only about $\frac{1}{4}$ to $\frac{1}{5}$ of that of potassium). They all display to some extent the property of occluding and activating hydrogen. Even apart from this, they all show well marked catalytic properties. The ability to form compounds from several valence states, and to pass readily from one state to another is common to all the Group VIIIB metals, as also is their distinct tendency to form complex compounds. In this respect it is characteristic that they can combine not only with NH_3, but also with CO and often with NO, and have a particular affinity for CN radicals. They also all have the property of forming colored compounds. Even in the form of so-called 'free electrolytic ions' (that is to say, only sheathed with water molecules), those cations that exist are invariably colored. The hydroxides of the sub-group VIII elements are in some cases weakly basic, in some cases weakly acidic, and some are amphoteric. In each of the three series, the affinity for oxygen diminishes from left to right. The elements of the last column (nickel, palladium and platinum) constitute a transition to the elements of Group IB, which are all noble metals. The elements of sub-group VIII all have a high affinity for sulfur, which shows itself not only in their simple compounds, but also in compounds of higher order (complex compounds)—e.g., in the numerous addition compounds formed by certain platinum salts with organic sulfides. In each of the three series, this affinity for sulfur increases from left to right. In this respect also, the elements of group VIII form a transition to those of Group IB for which—especially in the case of copper and silver—the ease with which they combine with sulfur, as compared with their low affinity for oxygen, is very characteristic.

In each of the three series of Sub-group VIII, the melting points decrease from left to right, and they rise on passing down each column:

$$
\begin{array}{ccc}
\text{Fe} & \text{Co} & \text{Ni} \\
1528° > & 1490° > & 1452° \\
\wedge & \wedge & \wedge \\
\text{Ru} & \text{Rh} & \text{Pd} \\
\text{ca. } 2400° > & 1966° > & 1555° \\
\wedge & \wedge & \wedge \\
\text{Os} & \text{Ir} & \text{Pt} \\
\text{ca. } 2700° > & 2454° > & 1774°
\end{array}
$$

The values of the 'atomic volumes', and the atomic radii derived from interatomic distances in the crystals of the elements follow a similar regular trend:

Atomic volumes in cc.				Atomic radii in Å.		
Fe	Co	Ni		Fe	Co	Ni
7.10 >	6.67 >	6.59		1.24 <	1.26 >	1.24
∧	∧	∧		∧	∧	∧
Ru	Rh	Pd		Ru	Rh	Pd
8.27 <	8.29 <	8.87		1.32 <	1.34 <	1.37
∧	∧	∧		∧	∧	∧
Os	Ir	Pt		Os	Ir	Pt
8.38 <	8.53 <	9.10		1.33 <	1.35 <	1.38

The Sub-group VIII metals have the following hardness values, on Mohs' scale.

Fe	Co	Ni	Ru	Rh	Pd	Os	Ir	Pt
4.5	5.5	3.8	6.5	—	4.8	7.0	6.5	4.3

CRYSTAL STRUCTURES OF GROUP VIIIB METALS

Fe*	Co*	Ni*
Body-centered cubic	Hexagonal close packed	Face-centered cubic
$a = 2.86$ Å	$a = 2.51 \quad c = 4.10$ Å	$a = 3.52$ Å
Ru	Rh	Pd
Hexagonal close packed	Face-centered cubic	Face-centered cubic
$a = 2.69 \quad c = 4.27$ Å	$a = 3.80$ Å	$a = 3.88$ Å
Os	Ir	Pt
Hexagonal close packed	Face-centered cubic	Face-centered cubic
$a = 2.71 \quad c = 4.31$ Å	$a = 3.82$ Å	$a = 3.91$ Å

(b) Alloys

As may be seen from the data listed in Tables 29 and 30 (pp. 244–5 and 248), the metals of Sub-group VIII have a notable capacity for forming alloys. So far as is known, the metals of the alkalis and alkaline earths are the only ones with which they form mixed crystals nor compounds. With the other metals of the Main Groups they form numerous compounds, which almost invariably are well defined in composition. This is associated with the fact that there is no miscibility (or at least very little) in the solid state between the Sub-group VIII metals and those of the Main Groups (cf. Table 29). As shown in Table 29, similar compounds are formed with the non-metallic elements of the Main Groups, especially with B, C, Si, N, P, As, and to some extent with S.

The elements Ru, Os, Rh, and Ir are not included in Table 29, since very little is known of their behavior towards the metals of the Main Groups. According to Rode (1929), Rh forms with Bi the compounds Bi_4Rh, Bi_2Rh and $BiRh$, which melt incongruently, but no mixed crystals. Rh and Ir are stated to be completely miscible with Pb in the liquid state, but quite immiscible in the solid state.

The behavior of the Sub-group VIII metals towards the metals of the other Sub-groups differs according to whether the other component is a metal of the transition series or of the later Sub-groups. They form extensive mixed crystals both among themselves, and with the other metals of the transition series (cf. Tables 13, 17, 21). It is only with the metals standing furthest from them in the transition series (the transition metals of Groups III and IV, also Nb and Ta) that they display just restricted miscibility (cf. Tables 9 and 13). On the other hand, they usually show very little or no ability to form mixed crystals with the elements of

* These data are for the modification stable at ordinary temperature. Crystal structures of the other modifications are given on p. 262 (see also Fig. 36, p. 263), pp. 291 and 309.

TABLE 30

MISCIBILITY AND COMPOUND FORMATION BETWEEN METALS OF SUB-GROUP VIII AND OTHER METALS OF THE VIIITH, IST, AND IIND SUB-GROUPS

(Symbols have the same meaning as in Table 9)

	Co	Ni	Ir	Pd	Pt	Cu	Ag	Au	Zn	Cd	Hg
Fe	$s \sim \infty$ / 0	$s \sim \infty$ $[FeNi_3]$ 600° / ∞ 0	alloys / —	$s \sim \infty$ $[FePd][FePd_3]$ / ∞ 0?	$s\,\infty$ $[FePt]$ / ∞ 0	$s > 0$ / 0	liq 0 / s 0 , 0	$s < \infty$ $Fe_3Au[]$ ~850°	$s > 0$ / Zn_7Fe^* $Zn_{21}Fe_5^*$	liq 0? / s 0 , 0?	$s \sim 0$ / 0
Co		$s\,\infty$ / 0	— / —	$s\,\infty$ / 0	$s\,\infty$ $Co_3Pt[]$ $[CoPt]$ 825°	$s\,\infty$ / 0	liq 0 / s 0 , 0	$s > 0$ / 0	$s < \infty$ $Zn_{21}Co_5^*$ $ZnCo^*$	s 0? / —	alloys Co_3Hg_{10}?
Ni			— / —	$s\,\infty$ / 0	$s\,\infty$ $[Ni_3Pt]$	$s\,\infty$ / 0	liq $< \infty$ / $s > 0$, 0	$s\,\infty$ / 0	$s < \infty$ $Zn_{15}Ni_2$? $Zn_{21}Ni_5$ 882° $ZnNi^*$ 1043°	s 0? $Cd_{21}Ni_5^*$ 502°	— / —
Ru			alloys / —	— / —	— / —	$s \sim$ / —	— / —	$s \sim$ / —	alloys / —	— / —	— / —
Os			alloys $[OsIr]$	— / —	— / —	$s \sim$ / —	— / —	$s \sim$ / —	alloys o?	— / —	— / —
Rh				$s\,\infty$ / 0	$s\,\infty$ / 0	$s\,\infty$ $[Cu_3Rh][CuRh][CuRh_3]$	$s > 0$ / 0	$s \wedge$ / —	$Zn_{21}Rh_5$	$Cd_{21}Rh_5$	alloys / —
Ir				— / —	$s\,\infty$ / 0	alloys / —	liq 0? / alloys?	$s > 0$ / —	alloys / —	— / —	alloys / —
Pd					$s\,\infty$ / 0	$s\,\infty$ $[Cu_3Pd][CuPd]$	$s\,\infty$ / 0	$s\,\infty$ / 0	$Zn_{21}Pd_5$	alloys $Cd_{21}Pd_5$	alloys / —
Pt						$s\,\infty$ $[Cu_3Pt]$ 500° $[CuPt]$ 800°	$s < \infty$ $Ag_3Pt[]$ 960° $AgPt[]$ $AgPt_2[]$	$s\,\infty$ / 0	$Zn_{21}Pt_5$	s 0 Cd_xPt^* $(x > 4)$ 615° Cd_2Pt	Hg_2Pt?

Groups IB and IIB (cf. Table 30). An exception to this generalization is provided by the behavior of the IB metals, and probably Zn, towards Ni, Pd, and Pt.

As far as the ability to form *compounds* is concerned, the metals of Sub-group VIII behave towards those of Sub-groups III, IV, V and VI (except vanadium and chromium) in the same way as they do to the elements of Main Groups III to VI. Almost all the Sub-group VIII metals combine with the metals of Group IIB, and especially with zinc, to form Hume-Rothery phases (cf. p. 22). In general, no compounds are formed with the metals of the other Sub-groups, or with vanadium or chromium. However, *super-structure phases* occur in a few cases in alloys with these metals.

On pages 264, 265 and 266 are shown a few typical phase diagrams for binary alloys of *iron*, as being technically the most important of the metals of Group VIIIB.

A. THE METALS OF THE IRON SERIES

2. General

Iron, cobalt, and nickel display a considerable similarity to each other, both in the metallic state and in their compounds. Iron is also related in its behavior to manganese, which precedes it in the Periodic Table, and which generally occurs with it in Nature. The properties of nickel lead naturally to those of copper, the following element in the Periodic Table. The metals of the iron series share the property of being usually bivalent in their simple salts. Iron can also be tripositive, and this valence state is preferred in some compounds. Cobalt can also be converted to the tripositive state, and is in fact especially stable in this state in its complex compounds. In *simple* salts, however, unlike iron, it very rarely functions as trivalent, and it appears that nickel cannot be tripositive in salts at all, except in certain complex salts. Conversely, the capacity for acting as *unipositive* increases from iron to nickel, corresponding to increasing proximity to copper, which readily exhibits a $+1$ state.

Whereas iron, like manganese, can relatively easily be converted into compounds in which it is present in the $+6$ state [*ferrates(VI)*], cobalt and nickel can, at most, be *tetrapositive*. They form dioxides which have a strong tendency to decompose with loss of oxygen, and which combine with strongly basic oxides to form the rather unstable *cobaltites* [cobaltates(IV)] and *nickelites* [nickelates(IV)].

The *oxides* FeO, CoO and NiO, like MnO, crystallize with the rock salt structure (Vol. I, p. 209, Fig. 44). The hydroxides $Fe(OH)_2$, $Co(OH)_2$, and $Ni(OH)_2$, like $Mn(OH)_2$, have layer lattice structures of the brucite teyp (Vol. I, p. 261, Fig. 61).

	FeO	CoO	NiO		Fe(OH)$_2$	Co(OH)$_2$	Ni(OH)$_2$
$a =$	4.29	4.24	4.17	$a =$	3.24	3.19	3.07 Å
				$c =$	4.47	4.66	4.60 Å

The *dichlorides* of the iron group metals form layer lattices very similar to those of the hydroxides. They differ from the brucite structure only in that the negative ions are not arranged in hexagonal closest packing, but are arranged approximately in cubic closest packing. This structure ('cadmium chloride type') has been found not only for $FeCl_2$, $CoCl_2$, and $NiCl_2$, but also for $MnCl_2$, $ZnCl_2$, $CdCl_2$, and $MgCl_2$, as well as for $NiBr_2$ and NiI_2.

The *monosulfides* of Fe, Co, and Ni (as also the mono-selenides, tellurides and antimonides) crystallize with the NiAs type structure (cf. p. 313). The cell dimensions of these compounds have the values given in Table 31 (p. 250). Some of the compounds listed in Table 31 also exist in other modifications (cf. pp. 295, 313).

Of the *disulfides*, FeS_2 is dimorphous (pyrite and marcasite structures). CoS_2 and NiS_2, and the diselenides $CoSe_2$ and $NiSe_2$, all crystallize with the pyrite structure, whereas $FeSe_2$ and $FeTe_2$ have the marcasite structure. $NiTe_2$ has a crystal structure of the brucite type, and $CoTe_2$ can exist with both the marcasite and the brucite structures. There is a continuous transition from the monotellurides of Co and Ni to the ditellurides, just as there is between TiTe and $TiTe_2$.

A property common to the metals iron, cobalt, and nickel, when present in the finely divided state, is their considerable capacity for absorbing hydrogen. Schlenk and Weichselfelder have claimed that hydrides with stoichiometric compositions—FeH_2, CoH_2, NiH_2—

could be obtained by the action of hydrogen on ethereal solutions of $FeCl_2$, $CoCl_2$, and $NiCl_2$, in the presence of phenyl magnesium bromide. It was stated that a compound still richer in hydrogen, namely FeH_6, was formed with iron trichloride. However, the existence of these hydrides has not been substantiated.

TABLE 31

CELL DIMENSIONS OF SULFIDES, ANTIMONIDES, SELENIDES, ETC. OF THE
IRON SERIES METALS

(s = shortest interatomic distance M $\leftrightarrow$ S)

	FeS	β-CoS	β-NiS	FeSe	CoSe	β-NiSe	FeSb	CoSb	NiSb	NiAs
$a =$	3.43	3.37	3.42	3.61	3.59	3.66	4.06	3.87	3.92	3.61 Å
$c =$	5.86	5.14	5.30	5.87	5.27	5.33	5.13	5.19	5.11	5.03 Å
$s =$	2.45	2.33	2.38	2.55	2.46	2.50	2.67	2.58	2.60	2.43 Å

The relationships between the stability of the strong complex compounds of the iron series metals and the oxidation states of the central atoms are noteworthy. Whereas with *iron*, for example, the complex $[Fe(CN)_6]^{4-}$, with a dipositive central atom, is more stable than the $[Fe(CN)_6]^{3-}$ complex which contains tripositive iron, with *cobalt*, on the other hand, the complexes that are especially stable are those containing the element in the *tri*positive state. In the presence of substances with which it can form strong complexes cobalt is extremely readily oxidized to the $+3$ state. Dipositive *nickel*, when it is present in the form of certain strong complex compounds, is converted immediately to the *tetra*positive state even by the action of oxidizing agents such as O_3 and H_2O_2, which are not extremely powerful.

These relationships between the oxidation states of the central atoms and the stability of the complexes can be understood in terms of the theory developed by Pauling for complex compounds of the transition elements. The complexes here considered are of the *penetration* type. In the penetration complex formed by the union of six CN^- ions with one Fe^{2+} ion,

$$\begin{bmatrix} NC & & CN \\ NC & \!\!\!-Fe-\!\!\! & CN \\ NC & & CN \end{bmatrix}^{4-}$$

the central atom possesses 18 electrons ($3d^{10}\ 4s^2\ 4p^6$) beyond its argon shell. The central atom in the complex thus possesses a krypton-like electronic configuration. The difference between the actual krypton configuration and that of the complex is that in the latter the three $4p$-levels and the $4s$-level are hybridized with each other and with two of the $3d$-levels (cf. Vol. I, Chapter 9). The hybridization is accompanied by the pairing of d-electrons which were originally unpaired in the free Fe^{2+} ion, the energy required for this pairing being more than compensated by the extra stabilization energy resulting from the hybridization process. A situation like that encountered in the $[Fe(CN)_6]^{4-}$ complex exists for the penetration complexes of Co^{III} and Ni^{IV} in which there are present six ligands each of which shares two electrons with the central atom (see Table 32). Complex compounds of tetrapositive nickel are obtained, as Hieber [*Z. anorg. Chem.* 269 (1952), 12] has shown, only with ligands which have an especially great affinity for nickel. The energy required for the oxidation of Ni^{II} to Ni^{IV} is considerably greater than that necessary for the conversion of Fe^{II} to Fe^{III} and of Co^{II} to Co^{III}. This is brought out by a comparison of the oxidation potentials for the elementary ions in aqueous solution:

TABLE 32

ELECTRONIC CONFIGURATIONS IN ELEMENTARY AND COMPLEX IONS OF THE METALS OF THE IRON SERIES

The electronic spins are shown by means of arrows, and the hybrid levels by boxes.

Ion	Electronic configuration
Fe^{2+}, Co^{3+}, Ni^{4+}	Argon configuration + $3d^6$: ↑↓ ↑ ↑ ↑ ↑
$[Fe^{II}(CN)_6]^{4-}$	Argon configuration + $3d^{10}$ $4s^2$ $4p^6$: ↑↓ ↑↓ ↑↓ | ↑↓ ↑↓ ↑↓ ↑↓ ↑↓ ↑↓ |
$[Co^{III}(CN)_6]^{4-}$	Argon configuration + $3d^{10}$ $4s^2$ $4p^6$: ↑↓ ↑↓ ↑↓ | ↑↓ ↑↓ ↑↓ ↑↓ ↑↓ ↑↓ |
$\left[\begin{smallmatrix} & S & \\ S & \vert & S \\ & Ni^{IV} & \\ R & \vert & R \\ & S & \end{smallmatrix}\right]^{2-}$	Argon configuration + $3d^{10}$ $4s^2$ $4p^6$: ↑↓ ↑↓ ↑↓ | ↑↓ ↑↓ ↑↓ ↑↓ ↑↓ ↑↓ |
Fe^{3+}	Argon configuration + $3d^5$: ↑ ↑ ↑ ↑ ↑
$[Fe^{III}(CN)_6]^{3-}$	Argon configuration + $3d^9$ $4s^2$ $4p^6$: ↑↓ ↑↓ ↑ | ↑↓ ↑↓ ↑↓ ↑↓ ↑↓ ↑↓ |
Co^{2+}	Argon configuration + $3d^7$: ↑↓ ↑↓ ↑ ↑ ↑
$[Co^{II}(CN)_6]^{4-}$	Argon configuration + $3d^{10}$ $4s^2$ $4p^6$ $4d$: ↑↓ ↑↓ ↑↓ | ↑↓ ↑↓ ↑↓ ↑↓ ↑↓ ↑↓ | ↑
Ni^{2+}	Argon configuration + $3d^8$: ↑↓ ↑↓ ↑↓ ↑ ↑
$\left[\begin{smallmatrix} & S & \\ S & \downarrow & S \\ & Ni^{II} & \\ R & \uparrow & R \\ & S & \end{smallmatrix}\right]^{4-}$	Argon configuration + $3d^{10}$ $4s^2$ $4p^6$ $4d^2$: ↑↓ ↑↓ ↑↓ | ↑↓ ↑↓ ↑↓ ↑↓ ↑↓ ↑↓ | ↑ ↑
Kr	Argon configuration + $3d^{10}$ $4s^2$ $4p^6$: ↑↓ ↑↓ ↑↓ ↑↓ ↑↓ ↑↓ ↑↓ ↑↓ ↑↓

$$Fe^{++} = Fe^{+++} + e \qquad Co^{++} = Co^{+++} + e \qquad Ni^{++} = Ni^{++++} + 2e*$$
$$\text{—0.77 volt} \qquad\qquad \text{—1.84 volts} \qquad\qquad \text{—2.1 volts}$$

The greater stability of the $[Fe(CN)_6]^{4-}$ ion as compared with the $[Fe(CN)_6]^{3-}$ ion can be understood directly from the oxidation potential for $Fe^{++} \rightarrow Fe^{+++}$, in as much as the energy which must be expended to free two d-levels of the central atom for entry by the binding electrons supplied by the ligands is practically the same for both complex ions. Two d-electrons which are unpaired in the elementary ions must be paired in both Fe^{II} and Fe^{III} in order to form the penetration complexes (see Table 32). Since the $[Fe(CN)_6]^{4-}$

* The oxidation potential for the reaction $Ni^{++} = Ni^{++++} + 2e$ is estimated from the measured potential of —1.75 volts for the reaction $Ni^{++} + 2H_2O = NiO_2 + 4H^+ + 2e$, the assumption being made that the difference between this measured potential and that for the elementary ions is equal to that between the corresponding potentials for lead (cf. Table 103, Vol. I).

ion contains only paired electrons, it is diamagnetic. The $[Fe(CN)_6]^{3-}$ ion, however, contains one unpaired electron and is therefore paramagnetic.

Whereas energy is liberated in reaction $Fe^{+++} \rightarrow Fe^{++}$, energy is required to convert Co^{++} to Co^{+++}. The amount of energy required for the latter conversion is considerable, as the value for the oxidation potential shows. That the $[Co^{III}(CN)_6]^{3-}$ ion is, however, considerably more stable than the $[Co^{II}(CN)_6]^{4-}$ ion is attributed by Pauling to the fact that in order to free two d-levels of Co^{II} for entry by electron pairs of the ligands, one electron from the $3d$-levels must be promoted to a $4d$-level. For Ni^{II} to form a $[Ni^{II}(CN)_6]^{4-}$ complex, two electrons from $3d$-levels must be promoted to $4d$-levels. However, Ni^{II} can form a $[Ni(CN)_4]^{2-}$ penetration complex directly. Through the pairing of both unpaired electrons in the Ni^{2+} ion there is made available a $3d$-level, which admits the two binding electrons of a CN^- group, the other three cyanide groups being disposed in $4s$- and $4p$-levels. Therefore, in the $[Ni(CN)_4]^{2-}$ complex the eight binding electrons have the configuration $3d^2\,4s^2\,4p^4$, and in this case, as was stated in Chap. 9, Vol. I, the ligands have a *planar* arrangement, whereas the complexes described in Table 32 have an octahedral configuration. The complex $[Ni(CN)_4]^{2-}$, like the $[Fe(CN)_6]^{4-}$ and $[Co(CN)_6]^{3-}$ ions, is diamagnetic, the electrons in the three complexes being all paired.

What has been said regarding the cyano complexes applies also to other penetration complexes of the metals of the iron series which possess six ligands, each of which has a pair of electrons available for binding. The stability of the various complexes is naturally influenced by the nature of the ligands; for example, the reduction potential of the $[Co(CN)_6]^{3-}$ ion is —0.83 volt in comparison with +0.1 volt for the $[Co(NH_3)_6]^{3+}$ ion. The tendency for the formation of the $[Co(CN)_6]^{3-}$ complex is so great, that the cobalt(II) cyanide complex decomposes water with the evolution of hydrogen. On the other hand, the complex $[Co(NH_3)_6]^{2+}$ is perfectly stable in aqueous solution in the absence of atmospheric oxygen. However, in contact with oxygen, in concordance with the low value of its oxidation potential (—0.1 volt), it reverts directly to the $[Co(NH_3)_6]^{3+}$ ion.

3. Iron (Ferrum, Fe)

(a) Occurrence

Iron is the most widely distributed of the heavy metals in the earth's substance. In the crust of the earth it occurs almost exclusively in the form of its compounds. It is very rarely found in the native state in the rocks which form the surface of the earth, although it is occasionally found in small amounts, as inclusions in basalt. On the other hand, metallic iron forms the chief constituent of many *meteorites* (*siderites*), and it has been assumed that the core of the earth consists essentially of metallic iron. In the meteorites iron is alloyed with nickel, which usually makes up 5.5 to 20 per cent by weight of the alloy. Meteoric iron is not infrequently accompanied by iron(II) sulfide (troilite), iron phosphide, and iron oxides.

The most important ores for the extraction of iron are the oxides—*hematite*, Fe_2O_3, *magnetite*, Fe_3O_4, *brown iron stone*, *limonite* $Fe_2O_3 \cdot H_2O$ or $FeO(OH)$, and the carbonate, $FeCO_3$, *spathic iron ore* or *siderite*.

These ores are found in almost every country in the world, although not always in a form which can be smelted.

The British iron industry was originally based largely on the ironstone and limonites of the Wealden beds in Southern England, the forsets of the Weald providing the charcoal used for smelting. At present, the principal British ores are probably the black-band ironstones (siderites) of the Cleveland District of Yorkshire, and the ironstones of the North Midlands. The steel industry in Britain, as in Germany and other Western European countries, draws much of its ore from Sweden and other external sources. The Swedish bog iron ore, which contains a certain amount of phosphoric acid, has an iron content of about

46% Fe after drying at 100°. Similar ores found along the Dutch and North German coastal plains generally have too high a content of sand and of organic matter to be smelted conveniently, but they possess excellent absorptive properties for hydrogen sulfide and find extensive uses as purifying materials for town's gas. In North America, about 85% of all the ore smelted comes from the Lake Superior District (which includes Minnesota, Michigan, Wisconsin, and Ontario)—largely soft hematites and limonites. The annual production of ore from the Mesabi range, shipped across the Great Lakes from Duluth, is more than 100 million tons. There are also great reserves of ore in the form of *taconites*—siliceous ironstones, containing about 27% Fe_2O_3 and 50% SiO_2, but these cannot be employed without beneficiation.

The iron compound most widely distributed in nature is *pyrite*, FeS_2. The ore is not directly suitable for smelting, however, both because of its high sulfur content and because it contains the sulfides of copper and other metals which are undesirable as impurities in metallic iron. However, the roasted pyrite ('burned pyrite') which is a residue from the manufacture of sulfuric acid is smelted for iron, after the impurities which are undesirable for this purpose, although often valuable in themselves, have been removed in special refineries: silver and gold are present as well as copper and zinc.

Many silicates also contain considerable amounts of iron. The products of their weathering are carried away in suspension by the rivers, and deposited in the form of *marl* (i.e., clay contaminated with iron oxide and sand).

(b) History

The use of iron has been known since the earliest historical times, and seems to be as ancient as the use of bronze. In early times, and at the present time also among peoples in a low stage of cultural development, it was prepared by the so-called bloomery hearth, or Catalan forge. Iron ores were heated in a shallow trench with a large excess of wood charcoal, fanned by a bellows. By this means, more or less coherent lumps (blooms) of wrought iron were obtained, and were welded together by strong hammering. As technology advanced during the Middle Ages, the trench or flat hearth was replaced by a small shaft furnace, and from this the present day *blast furnace* has developed. The use of water power to operate the blast was introduced during the 14th century. The consequent very considerable increase in furnace temperature resulted in the production of iron with a much higher carbon content than formerly, namely *cast iron*. At first, indeed, this was not malleable but it was soon discovered how this might be converted into malleable iron by a second heating in an ample supply of air (refining). The iron industry received a great impetus at the end of the 18th century, when the demand for iron began to increase tremendously, as a result of the invention of the steam engine and the railway. The shortage of wood charcoal led to the introduction of coal and coke, both as fuel and as reducing agent. Coke was first used in the blast furnace by Abraham Darby, in 1732. The refining process underwent fundamental improvements during the 19th century, through the introduction of the blast refining method (Bessemer process 1855, Thomas-Gilchrist process 1878) and of regenerative heating (Siemens-Martin process 1865). In more recent times, smelting in the electric furnace has been introduced for the production of certain high grade steels.

(c) Production of Iron

Chemically pure iron can be prepared either by the reduction of pure iron oxide (best obtained for this purpose by heating the oxalate in air) with hydrogen, or by the electrolysis of aqueous solutions of iron(II) salts—e.g., of iron(II) ammonium oxalate. On the technical scale, pure iron is prepared chiefly by the thermal decomposition of iron pentacarbonyl.

The so-called 'carbonyl iron' prepared in this way initially contains some carbon and oxygen in solid solution. These impurities can be removed by suitable after treatment. It has not been possible to detect the presence of S, P, Cu, Mn, Ni, Co, Cr, Mo, Zn, or Si in carbonyl iron, even by the most rigorous analytical procedure. According to the conditions under which the decomposition of the iron carbonyl is carried out, the iron may be obtained in the form of a uniform mirror or layer, in a porous form (similar to that which is deposited electrolytically from aqueous solutions), or as a dense fine-grained powder (from which extraordinarily homogeneous molded forms may be produced, by pressing and heating—e.g., to 1000°: 'powder metallurgical' or 'sintering' process), or finally as extremely voluminous cotton-wool like flocks, of which a liter weighs only about 10 g. In this last form the iron is extremely reactive, so that it can be ignited by a spark.

Although chemically pure iron [8] is growing in technical importance (cf. p. 262), incomparably greater importance attaches to the production of ordinary technical iron, which is not the pure metal, but is essentially an alloy of iron with carbon. In addition to carbon, however, technical iron [9] contains a whole range of other elements, which are either alloying components added intentionally, or impurities which are more or less deleterious. The properties of the iron are influenced in a high degree by the carbon and other constituents, even if they are present in quite small amount. The properties also depend to a considerable extent on the previous treatment of the alloys concerned.

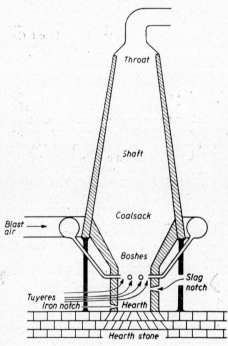

The preparation of technical varieties of iron [1–7] can be divided into two sections, the first involving the reduction of the ore to an impure metal (crude iron, pig iron), whereas the second covers the production of sorts of iron with specified properties, such as are required technically, by removing the impurities and adding valuable alloying components if necessary.— e.g., wrought iron and steels. The first process, *iron smelting*, is carried out in the *blast furnace*. The second, refining i.e., conversion to wrought iron, steel, etc., was formerly effected by *puddling* or hearth refining, but is now carried out mostly by the blast-refining (Bessemer) and Siemens-Martin processes, whereas crucible melting and electric furnace methods are used for special steels.

Fig. 29. Schematic diagram of blast furnace.

(i) *The Blast Furnace Process.* The blast furnaces ordinarily used for the production of pig iron or cast iron* are shaft furnaces, usually 60 to 100 feet high, built from refractory bricks or Dinas bricks in the shape sketched in Fig. 29. Oxidic iron ore, broken up into lumps the size of a fist, is fed in from above in

---

* Pig iron can be produced in the electric furnace instead of in the blast furnace. In this case, carbon is used only as the reducing agent, and not as fuel, since heating is effected by means of the electric current. Electric furnaces for this purposes are best constructed as short shaft furnaces, and not as blast furnaces [cf. *Z. Elektrochem.*, 42 (1936) 337]. In general, the use of electric furnaces comes into question only in countries which are poor in coal but rich in water power (e.g., Sweden). The world production of iron by the electric furnace is less than 0.1% of the total production.

layers, alternating with layers of coke, while pre-heated compressed air is blown in at the base through so-called *tuyeres*. The air burns the coke, and the carbon monoxide thus formed liberates the iron from its oxides:

$$3CO + Fe_2O_3 = 2Fe + 3CO_2 + 5.7 \text{ kcal*} \tag{1}$$

Reaction (1) is reversible. In accordance with Le Chatelier's principle, the equilibrium is displaced from right to left with rise of temperature. Hence reaction (1) takes place predominantly in the upper, cooler portion of the blast furnace. Reduction to iron(II) oxide proceeds to some extent in this zone of the furnace:

$$Fe_2O_3 + CO = 2FeO + CO_2$$

(see below). In the lower, very hot zone, reduction takes place directly by carbon:

$$FeO + C + 34.5 \text{ kcal} = Fe + CO \tag{2}$$

In the lowest zone of the furnace the heat is so intense that the iron melts and runs down. The space which is emptied through this, and through the combustion of the coke, is continually filled up again as the burden sinks down.

If iron is present in the ore as carbonate, this is converted to the oxide by roasting, before introducing it into the blast furnace:

$$2FeCO_3 + \tfrac{1}{2}O_2 = Fe_2O_3 + 2CO_2.$$

Powdered ores are briquetted into lumps of a suitable size, by the addition of some binding agent. In order to remove the gangue, with which the iron ores are almost invariably contaminated, basic materials (usually limestone) are added if the gangue is acidic (silica and alumina), and acidic substances (slate or granite) are added if the gangue is basic (limestone). These so-called fluxes combine with the gangue to form fusible slags, and thereby effect a separation from the iron, which sinks through. At the same time, the slag covers the iron and protects it from oxidation by the air blast. The molten iron is tapped every 4–6 hours from an opening—the 'iron notch'—close to the base of the furnace. The slag is also run off from time to time through an opening ('slag notch') at a suitable level. Above the level where the tuyeres are introduced (the 'tuyere line'), the blast furnace first widens out—this part is known as the boshes—and then gradually narrows again in a long shaft. The direct reduction of iron oxide by carbon in the lower part of the furnace involves the absorption of heat (cf. eqn. 2), and the temperature in the blast furnace therefore decreases rapidly in a vertical direction in the neighborhood of the boshes. At the tuyere level it is over 1600°. At the junction of boshes and shaft, the 'coalsack', it has dropped to 800°. The fall of temperature along the shaft then becomes relatively gradual. In the lower two thirds of the shaft it drops from 800° to about 600°. In the upper part of the shaft, it falls rather more steeply, and the temperature of the gases issuing from the upper opening, the *throat* is about 200°. The peculiar shape of the blast furnace is intended to allow both for the rapid expansion of the blast gases, through heating and the formation of carbon monoxide, in the region of the boshes, and for the unhindered spreading of the burden as it is introduced from above, and slips down into the increasingly hotter zones of the furnace. In the portion of the furnace below the tuyeres, the 'hearth', the molten iron collects. The sole of the furnace which is continually covered by the molten iron, the 'hearthstone', is shaped so as to key it in position, so that it cannot be torn from its fastenings by the buoyant action of the specifically heavier iron.

The ore, introduced into the blast furnace through the 'cup and cone' at the throat, is pre-heated and simultaneously dehydrated in the upper part of the shaft, by the ascending hot gases. Reduction of iron(III) oxide by carbon monoxide begins below 400°. It leads,

---

* The heats of reaction shown in this and subsequent equations are the values holding at ordinary temperature.

in the first place, to iron(II,III) oxide, Fe_3O_4, and then, in the lower part of the shaft, above 700°, to metallic iron, which is at first still solid and porous:

$$3Fe_2O_3 + CO = 2Fe_3O_4 + CO_2 + 8.4 \text{ kcal*} \tag{3}$$

$$Fe_3O_4 + 4CO = 3Fe + 4CO_2 + 4.3 \text{ kcal} \tag{4}$$

In order that reduction should proceed practically to completion in the time available, a fairly considerable excess of carbon monoxide must be present. The gases issuing from the throat are therefore always rich in carbon monoxide. A portion of the carbon monoxide decomposes, in contact with the porous iron, according to

$$2CO = C + CO_2 + 38.6 \text{ kcal} \tag{5}$$

since carbon monoxide is not stable by itself below 1000° (Vol. I, p. 441). The solid carbon formed by this decomposition is deposited upon the iron, and as the iron sinks down, and its temperature rises, it takes up more and more carbon. The melting point of the iron is thereby lowered, and it liquefies. In the course of its passage through the lowest layer of coke, through which it trickles, the molten iron has the opportunity to pick up still more carbon.

Reaction of the fluxes with the gangue takes place concurrently with the reduction of the iron oxide. Some of the iron(II) oxide, formed by partial reduction, at first enters the resulting slag, in the form of a silicate, and is thereby removed from further reaction with the carbon monoxide. However, while the molten slag runs through the layer of coke below it, the iron is reduced from the silicate (cf. eqn. (2), p. 255), together with some of the manganese present, phosphorus and, to a lesser extent, silicon.

Once a blast furnace has been 'blown', it must be kept in uninterrupted operation. When external circumstances make a break in its working unavoidable, all the openings in the furnace are luted over, to protect the contents as far as possible from the entry of air and the loss of heat. Under favorable conditions, the furnace can be started up again even after more than a month's interruption.

The stack gases, which contain about two thirds of the carbon burned in the form of carbon monoxide, are freed from dust and burned in so-called Cowper's stoves, to pre-heat the blast air (which enters the furnace through the tuyeres at about 800°), and to supply power to drive the blowers. The *slag* is widely used as road metal, molded with clay to make building stones ('breeze' blocks), and also for the manufacture of cement ('iron portland cement'). Some of the *pig iron* obtained from the blast furnace is used directly for *cast iron*, but by far the greater part is converted into *steel* or *wrought iron*.

(*ii*) *Refining* In addition to other impurities, such as silicon, manganese, phosphorus (and often sulfur also), the pig iron produced from the blast furnace contains considerable amounts of carbon. This makes it brittle, and not malleable. It melts sharply when it is heated, without softening previously, as wrought iron and steel do. The conversion of pig iron to malleable iron is known as refining.

(*iii*) *Charcoal Hearth Refining.* The oldest method of producing malleable iron from pig iron is the charcoal hearth refining process. In this, the pig iron is melted with wood charcoal in a container (the hearth) made of cast iron plates, and blown with compressed air (Fig. 30). To convert grey cast iron into wrought iron, three fusions are necessary. The first time (called 'fining'), only the most easily oxidized impurities in the iron (silicon and manganese) are burned. Oxidation of the silicon, which is present in the iron as silicide, takes place essentially according to the equation:

$$FeSi + \tfrac{3}{2}O_2 = FeSiO_3 \text{ (iron metasilicate or 'bisilicate')} \tag{6}$$

Iron orthosilicate, Fe_2SiO_4 ('singulosilicate') is formed at the same time, and manganese

* See footnote on preceding page.

silicate also if manganese is present. In the second fusion ('first refining'), oxidation of the phosphorus contained in the iron takes place, roughly according to the equation:

$$2Fe_3P + 6O_2 = Fe_3(PO_4)_2 + Fe_3O_4 \qquad (7)$$

At the same time, some of the iron is oxidized to iron(II,III) oxide, and this brings about

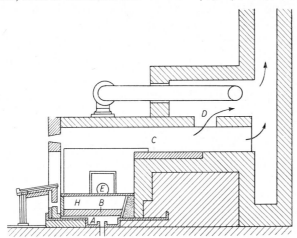

Fig. 30. Hearth refining furnace.

The hearth H is encased in cast iron plates. The base plate B may be cooled by means of water in the container A. The hot flame gases pass over the pre-heating hearth C, on which is stacked the crude iron for the next charge. Before entering the stack, a portion of the flue gases is passed through the chamber D, where it pre-heats the blast air which is injected into the furnace at E.

the oxidation of a portion of the carbon which is dissolved in the iron chiefly in the form of iron carbide, Fe_3C:

$$3Fe + 2O_2 = Fe_3O_4 \qquad (8)$$

$$Fe_3O_4 + 4Fe_3C = 15Fe + 4CO \qquad (9)$$

The iron thus obtained still contains 1 to 2% of carbon—i.e., is a *steel*. To convert this into true wrought iron, a further refining is necessary. In this last stage, the 'thorough refining', in which the iron is no longer completely molten, because of its greatly reduced carbon content, iron(II,III) oxide is again produced (refining slag). This then reduces the carbon content to less than 0.5% according to eqn. (9). It is usual to expedite the decarbonization by adding Fe_3O_4 from a previous operation.

In the hearth refining process, the sulfur contained in the iron is volatilized as sulfur dioxide, and the more prolonged the refining, the more complete is the removal of sulfur. The charcoal hearth refining process yields a very pure iron, with a relatively low slag content. It has therefore been able to persist in a few regions which are very rich in wood. (Sweden, Russia).

(*iv*) *Reverberatory Furnace Refining* (*Puddling*). The reverberatory furnace or puddling process differs from the foregoing in that the iron is not melted in a hearth which also contains the fuel, but in a reverberatory furnace—i.e., a furnace such as that shown in Fig. 31, in which the flame from the fuel (usually coal), burning on the grate of the combustion chamber R, passes over the iron, which is contained in the flat trough T (the hearth, lined with hematite, Fe_2O_3). The essential feature of the puddling process is that the molten iron is stirred, and thereby brought continually into fresh contact with the air and with the iron oxide of the hearth lining. This enables the removal of carbon to proceed so far that wrought iron is produced in a single operation. The stirring is spoken of as 'puddling'. As the carbon is lost, the iron becomes increasingly pasty, until ultimately the furnace temperature is no longer high enough to keep the iron completely melted, and forms large balls or 'blooms'. These still contain considerable inclusions of slag, which is removed by hammering and rolling.

It is also possible to produce steel by a puddling process. In this case, as soon as the 'balling' stage is reached, the melt is heated with a reducing flame. In order to ensure that the impurities (silicon and phosphorus) are nevertheless completely oxidized, it is necessary to heat to a higher temperature and for a longer period than when wrought iron is produced.

(*v*) *Mild Steel and Ingot Steel. The Crucible Steel Process.* Iron obtained by melting the blooms produced by puddling or by the charcoal hearth refining process is known as wrought iron. It is always more or less inhomogeneous, by reason of the particles of slag which cannot be completely removed, and is apt to split in consequence. In the middle of the 18th

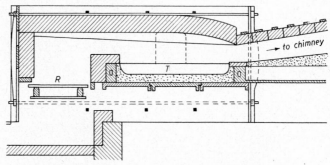

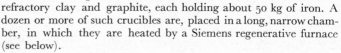

Fig. 31. Reverberatory puddling furnace.

century, Benjamin Huntsman, a Sheffield watchmaker, had the idea of making the steel used for producing clock springs more uniform, by remelting it in a crucible. The process was subsequently developed by Krupp, in Essen, for large scale operation, and is still used for the production of high grade steels, under the name of the *crucible steel process*. The crucible process can be used not only to homogenize already refined iron, but can be made to effect refining simultaneously. In this case, the pig iron, which is to be melted with the addition of iron oxide, must be free from phosphorus and sulfur, since these substances are not removed by the crucible fusion. Melting is carried out in crucibles made from a mixture of

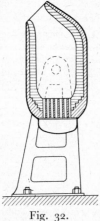

refractory clay and graphite, each holding about 50 kg of iron. A dozen or more of such crucibles are, placed in a long, narrow chamber, in which they are heated by a Siemens regenerative furnace (see below).

(*vi*) *Air Blast Decarburization Processes. Bessemer and Gilchrist-Thomas Processes.* The air blast processes depend upon blowing compressed air (blast) through the molten iron, and this burns the carbon and other impurities out of the iron. The oldest process, still used for decarburizing pig iron of very low phosphorus content, is the Bessemer process, which was invented in 1855 by the English steelmaker Henry Bessemer. The process is carried out in a pear-shaped vessel, the *Bessemer converter*, Fig 32, which is furnished with a refractory lining, and can be tilted roughly about its centre line. It has an opening at the top. Pig iron is run into the converter—most conveniently, straight from the blast furnace—while it is in the

Fig. 32.
Bessemer converter.

horizontal position. It is then raised to the upright position, and a blast of air is blown through the molten iron from the perforated base of the converter. This oxidizes the impurities present in the iron, and the heat of oxidation is so great that the iron remains molten in spite of the considerable rise in melting point which accompanies decarburization.

As in charcoal hearth refining, it is possible in principle to distinguish three stages in the Bessemer process, corresponding to the fining, first refining, and thorough refining, although these stages are not nearly so clearly differentiated. This is due in part to the high temperature, which results in increasing the oxidizability of the carbon, and in part to the much greater rapidity of the whole process of blast refining, which is usually complete within 10 to 20 minutes. For this reason it is possible for almost completely decarburized Bessemer iron still to have a relatively high silicon content. For many purposes this is advantageous. It is possible to control the course of blast refining by observation of the flame at the mouth of the converter.

The original Bessemer process could produce satisfactory wrought iron or steel only from pig irons with a *phosphorus content of less than* 0.1%. This is because the continuous, intimate intermixing of the slag with the carbon-containing iron perpetually regenerates iron phosphide from the primarily oxidized phosphorus:

$$Fe_3(PO_4)_2 + 2Fe_3C + 3Fe \rightleftharpoons 2Fe_3P + 6FeO + 2CO \tag{10}$$

The silica present in the refractory lining of the converter favors reaction to the right, since it reacts with iron(II) oxide to form silicate. Hence iron cannot be freed from phosphorus by the Bessemer process in its original form—the so called 'acid Bessemer process'. Phosphorus in amounts exceeding 0.1% is a most undesirable impurity in iron, however, since it renders the metal brittle at ordinary temperature ('cold short').

It first became possible to decarburize pig iron with an appreciable phosphorus content in the Bessemer converter when Thomas and Gilchrist in England, in 1878, provided the converter with a *basic lining* (of calcined dolomite). This combines with phosphoric acid, and protects it from reduction. The action of the basic lining is promoted by adding lumps of quick lime. The so-called '*basic Bessemer process*' of Thomas and Gilchrist produces slags with a high phosphorus content ('basic slag'), which constitute a valuable by-product. Finely ground basic slag is used as a fertilizer ('Thomas meal').

The greater part of the transfer of phosphorus to the slag takes place, when the carbon has been practically completely burned. Figs. 33 and 34 give examples illustrating the course of the acid and basic Bessemer processes. The ordinates represent the percentages of carbon, silicon, manganese, etc. in the iron, as they vary with the duration of blowing. It may be seen that, in the example represented by Fig. 33 (acid Bessemer process), silicon mostly undergoes oxidation in the earliest stages, and then the carbon is oxidized, while the phosphorus and sulfur contents remain practically unchanged. Fig. 34 shows the course of the oxidation of phosphorus in the basic Bessemer process.

Removal of sulfur is not effected by the acid Bessemer process, and must be carried out by adding manganese before decarburization (cf. p. 276). The basic Bessemer process is also unsuitable for pig irons which contain more than about 0.15% of sulfur. In general, the Bessemer process can be applied only to pig irons of certain specified compositions.

Decarburizing with the air blast reduces the carbon content of iron to a lower figure than is desired for technical purposes, especially when the basic process is employed. Hence before the converter is emptied, its contents are brought to some higher content by the addition of high-carbon ferromanganese ('spiegeleisen'). The manganese has the property of de-oxidizing such iron oxide as is dissolved in the iron; manganese oxide which is thus formed is readily rejected as slag. Numerous other substances can be used as de-oxidants in place of manganese—especially silicon, and also aluminum, vanadium, titanium, and other metals with a high affinity for oxygen. Beryllium has also been used more recently. In the basic Bessemer process it is necessary to remove the slag, which contains phosphoric acid, by cautiously tilting the converter, before adding the de-oxidant.

(*vii*) *The Siemens-Martin Process.* In principle, the Siemens-Martin process is only a reverberatory furnace decarburizing process which, through the use of higher temperatures, yields low-carbon steel (mild steel) in place of the wrought iron

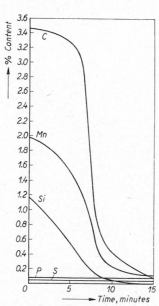

Fig. 33. Course of decarburization, etc., during 'acid' blast refining.

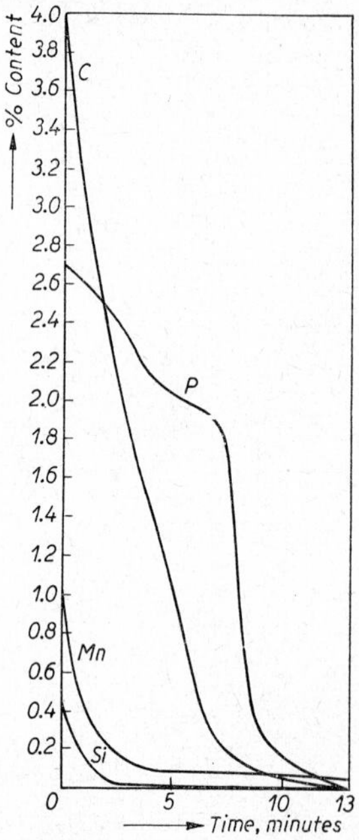

Fig. 34. Course of decarburization, etc., during 'basic' blast refining.

obtained from the puddling process. The possibility of attaining the requisite temperatures was provided by the principle of *regenerative firing*, introduced by the brothers Siemens in 1860. This is a gas firing process, in which the heat of the flue gases is employed to pre-heat the fuel gas and the air. For this purpose they are passed through a pair of chambers packed with refractory bricks ('regenerators') in which they give up their waste heat. After some time, the flow of flue gases is switched so as to heat up another pair of chambers, while the producer gas and the air for its combustion, respectively, are passed through the now incandescent regenerators. The flow of gases is reversed at regular intervals. The high temperature which can be attained in such a furnace was first used by the brothers Martin, in France, in 1865, for the production of ingot steel.

Fig. 35 represents a Siemens-Martin furnace, such as is employed for the production of low carbon steel and cast steel. Mild steel was originally made in the Siemens-Martin furnace by melting pig iron with wrought iron—i.e., with iron which had already been extensively decarburized. It was soon found, however, that it was, possible to obtain steel and even wrought iron directly in the open hearth furnace. This can be achieved, in the first place, by so controlling the air supply that there is an excess of oxygen in the flames passing over the hearth which contains the iron, so that they exert an oxidizing action, and further by adding oxidic iron ores (e.g., hematite), hammer scale, or rusty scrap to the pig iron melt. The hearth is provided with an 'acid' or a 'basic' lining, according as the pig iron is free from phosphorus or not. In the latter case, burned lime is also added, as in the basic Bessemer process. In this case, the extensively decarburized iron which is first obtained must be carburized again, if it is desired to make steel, and must be simultaneously de-oxidized, as in the Bessemer process.

Unlike the Bessemer process, the Siemens-Martin process is not restricted to pig irons of any particular composition, since in it the iron is heated from without, and not only by the heat of combustion of the impurities contained in the iron itself. It is also slower in operation

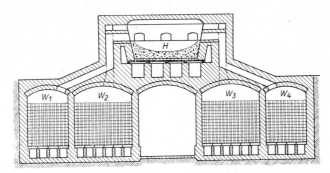

Fig. 35. Siemens-Martin open hearth furnace.

The iron is situated in the hearth H. The smaller of each of the pairs of regenerative heat exchangers W_1, W_2, W_3, W_4, are used to pre-heat the producer gas, and the larger pair to pre-heat the combustion air. During one phase of the process, producer gas and air pass separately through the previously heated chambers W_1 and W_2, while the products of combustion flow through the checker work in W_3 and W_4. The flow is then changed over, so that the flue gases are passed through W_1 and W_2, while air and producer gas are being pre-heated in W_3 and W_4, respectively.

than the Bessemer process, and this is beneficial to the quality of the iron since it permits of a more complete separation of the slag from the iron. A disadvantage attending the open hearth process is that it is difficult to avoid oxidation of the iron by the flame gases during the carburization process that follows refining.

(viii) Electro-steel Making. Furnaces with electric heating are now used to an increasing degree for the production of high-grade steels. They are employed, in particular, for the improvement of steels made by the Bessemer process, by remelting them, and also for carburizing and de-oxidizing iron after refining by the Siemens-Martin open hearth method. Electricity is costly as a means of heating, but it has the advantage that there can be no action of flame gases on the molten iron, and it is possible to obtain iron alloys of exactly the compositions desired.

(ix) Malleablizing. Small objects, such as bolts, nuts, etc., are commonly cast in white cast iron, and are subsequently converted into low-carbon iron by embedding them in iron oxide and heating them for a long time (about a week) to 850–1000°—i.e., below the melting point. The iron carbide, Fe_3C, contained in white cast iron thereby undergoes decomposition, with the separation of finely divided carbon. This diffuses outwards and is burned by the iron oxide. The process is known as malleablizing or *tempering*, and the articles produced as malleable cast iron.

(x) Cementation. In the same way as cast iron can be decarburized by malleablizing, it is possible to raise the carbon content of wrought iron by prolonged heating with powdered wood charcoal in the absence of air. The process is known as *cementation*, and the steel so made is called *blister steel*. The famous Damascene steel was produced from direct charcoal iron (i.e., from wrought iron made directly from the ore in the bloomery furnace) by cementation. In the 18th and 19th centuries, high quality tool steels were often made by cementation from thin wrought iron bars, which were subsequently welded together to a 'faggot' ('faggot steel' or 'shear steel').

At the present time, it is usual to employ cementation only to convert a layer near the surface of shaped mild steel objects into high-carbon steel. This process, used to produce a surface hardening, is known as 'case hardening'.

(d) Properties of Pure and Technical Iron

(*i*) *Pure Iron. Pure iron* [*8*] is a white, lustrous, rather soft metal (hardness 4. 5). Its specific gravity is 7.86, and melting point 1528°. It exists in two modifications in the solid state, depending upon the temperature. At ordinary temperature, and again above 1401°, it forms a *body-centered cubic* lattice ($a = 2.8605$ Å at 25°), but between 906° and 1401° the structure is *face-centered cubic* ($a = 3.63$ Å at 1100°). At temperatures below 768°, iron is *ferromagnetic* (i.e., becomes strongly magnetized when it is placed in a magnetic field). The temperature (768°) at which it loses this property—that is, the *Curie point* of iron—also shows up as a halt in the heating and cooling curves, in the same way as do the points (906° and 1401°) at which structural rearrangements take place. It is therefore usual to distinguish iron in its magnetizable state as α-iron, from the structurally identical, but not magnetizable form, which is known as δ-iron.* The face-centered modification, which is also not magnetizable, is known as γ-iron.

In the temperature interval from 768° to 906°, iron was formerly known as β-iron. However, no distinction can be drawn between the β- and δ-forms, and in phase diagrams of alloys where the γ-range in mixed crystals is restricted, it is quite impossible draw any boundary between the β- and δ-areas. (Cf. Figs. 39, 40, pp. 265, 266).

Unlike iron which contains carbon, pure iron has a very low remanence, i.e., it instantly loses its magnetization when the applied electric field is removed. For this reason it finds certain applications in electrotechnology—e.g., for electric motors and transformers, in which rapid fluctuations must occur in the magnetism of an iron core.

The coefficient of thermal expansion (linear) of pure iron between 0° and 100° is $1.1 \cdot 10^{-5}$, the thermal conductivity at 18° is 0.17 cal.cm^{-1}.sec^{-1}.degrees^{-1}. The specific heat is 0.1055 at 0°, 0.1467 at 725°, 0.1571 between 785° and 919°, and 0.1637 at the melting point. It changes but little further above the melting point.

Iron has a great affinity for oxygen. It *rusts* in moist air—i.e., the surface gradually becomes converted into iron oxide hydrate. Compact iron reacts with dry air only above 150°. When heated in air it forms the intermediate oxide, Fe_3O_4, which is also formed during the forging of red hot wrought iron ('hammer scale'). In a very finely divided state—such as is obtained, for example, when iron oxalate is gently heated in hydrogen—iron is pyrophoric.

Iron objects are protected from rusting by covering them with coatings of other metals (e.g., zinc, tin, chromium, nickel) or with paint (red lead). A particularly effective protection from rust can be achieved by converting the iron superficially to iron(II) phosphate ('phosphatizing'). [*22*] This is done by treatment with a solution of acid manganese or zinc phosphate, $Mn(H_2PO_4)_2$ or $Zn(H_2PO_4)_2$ ('Parkerizing process'). The phosphatizing process can be accelerated by suitable additions ('Bonderizing').

In accordance with its place in the electrochemical series, iron dissolves in dilute acids with the evolution of hydrogen and the formation of iron(II) salts. The normal potential of iron in contact with iron(II) salt solutions is $+0.440$ volt at 25°, relative to the normal hydrogen electrode (Randall, 1932). If iron is dipped into a copper sulfate solution, it becomes covered with metallic copper:

$$Fe + Cu^{++} = Fe^{++} + Cu.$$

* With other substances it is not customary to characterize the states above and below the Curie temperature by different letters. The designation 'α- or δ-iron' is therefore often used without distinction for the body-centered cubic form, both above and below the Curie point.

Iron loses this property of discharging hydrogen ions and copper ions if it is immersed in fuming nitric acid. It does not dissolve in the acid, but is converted to a state of *passivity*, in the same way as chromium (cf. pp. 128, 750 *et seq.*).

At ordinary temperature, iron is hardly attacked by air-free water, since a covering film of iron(II) hydroxide is formed on it, and this has a protective action even when it is very thin. If air has access, however, the porous iron(III) oxide hydrate is formed, and corrosion goes on continuously as a result. In the absence of air, iron is attacked by dilute sodium hydroxide still less than by water, chiefly because the OH^- ions repress the solubility of $Fe(OH)_2$ (furthermore, the deposition potential of the H^+ ions is raised, as follows from the increase in pH). *Concentrated* sodium hydroxide, however, attacks iron, fairly strongly, even in the absence of air, especially at high temperatures, since the $Fe(OH)_2$ goes into solution through the formation of hydroxo-salts when the OH^- concentration is high (cf. p. 272). Since the formation of hydroxo-ions at the same time depresses the concentration of Fe^{++} ions in the solution to an infinitesimal value, iron has quite a strongly positive potential $(+0.86$ volts$)$ in concentrated (e.g., 40%) sodium hydroxide solution. Under some conditions, however, this can also be changed by passivation even in concentrated caustic soda, to a strongly negative value $(-0.65$ volt$)$. As soon as this occurs, the iron no longer dissolves spontaneously, with evolution of hydrogen and formation of hydroxo-ferrate(II) ions, but dissolves only under the influence of an applied potential, and then forms ferrate(VI) ions.

Iron combines very energetically with chlorine when heated, and also with sulfur and phosphorus, but not directly with nitrogen. It has a strong tendency, however, to unite or alloy with carbon and silicon. These alloys are most important for the properties of *technical iron*.

● = Fe   ○ = Available for C atoms

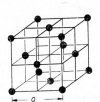

Fig. 36. Detailed structure of the textural constituents of iron-carbon alloys.

γ-iron	austenite	martensite	$\alpha(\delta)$-iron
$a = 3.59$ Å (extrapolated to room temperature)	$a = 3.63$ Å (for 8 atom-per cent C)	$a_0 = 2.84, c_0 = 3.00$ Å (for 6 atom-per cent C)	$a = 2.86$ Å

(ii) Iron-Carbon Alloys [9–14]. A variety of textural constituents can be distinguished in solidified iron-carbon alloys. The most important are the following:

1. *Ferrite*:   pure iron—more strictly, α-iron.
2. *Graphite*:   hexagonal crystalline carbon.
3. *Cementite*:   a compound of iron and carbon, Fe_3C (contains 6.68% C).
4. *Austenite*:   a solid solution of carbon in iron (more precisely, in γ-iron).
5. *Martensite*:   a metastable conversion product of austenite, formed by rapid cooling.
6. *Ledeburite*:   a eutectic mixture of cementite with austenite (saturated with carbon).
7. *Pearlite*:   a eutectoid* mixture of ferrite and cementite.

The formation of *austenite* depends on the fact that carbon atoms can take up positions at the center and in the middle of the cell edges of the unit cube of γ-iron (see Fig. 36).

* The term *eutectoid* is applied to a mixture which is formed by an unmixing process *in the solid state*, in the same way as a eutectic is formed by the cooling of a melt.

If these sites were occupied in every unit cell, a compound FeC, with 50 atom-% C, would be produced. However, the crystal lattice of austenite is stable only when not more than a small proportion of the available positions is occupied by carbon atoms. This proportion may vary between 0 and 8 atom-% C. The C atoms thus built into the crystal lattice distribute themselves statistically between the cell centers and cell edges, and austenite has the characteristics of a 'solid solution'. The crystal lattice of γ-iron is expanded uniformly in all directions by the incorporation of the C atoms, but the expansion is small.

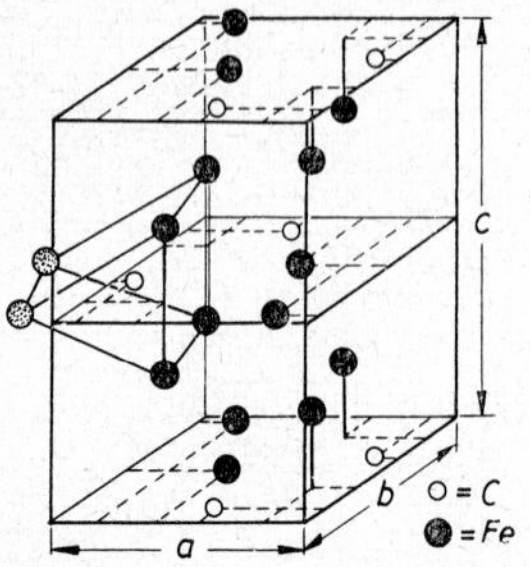

Martensite can be similarly regarded as a (supersaturated) solid solution of carbon in α-iron. Like all supersaturated solutions, it is unstable (or metastable). As shown in Fig. 36, only a fraction of the sites available for C atoms is occupied in the martensite structure also. If all lattice positions were occupied, a compound FeC would result in this case also. As compared with the α-iron structure, the crystal lattice of martensite has undergone tetragonal distortion; it is stretched in one direction, and contracted a little in the two others*.

Cementite, Fe₃C, as is shown in Fig. 37, has a considerably more complex structure (Shimura, 1931, Westgren, 1932). Another iron carbide, probably with the formula Fe₂C, exists in addition to cementite. It is formed when iron is heated with CO at 225°, but decomposes at higher temperatures. It is apparently not obtainable by the direct union of iron and carbon.

Fig. 37. Structure of cementite, Fe₃C.

$a = 4.52$, $b = 5.08$, $c = 6.73$ Å. Fe ↔ C = 2.01 Å. The Fe atoms form prisms, with the C atoms located at their mid-points. One such prism is shown in the diagram; two of its Fe atoms lie outside the unit cell represented in the figure.

Fig. 38 represents the phase diagram of the *iron-carbon alloys*, up to a content of 5% by weight of C. The melting point of iron is first lowered by the addition of carbon, from *A* to *E*, and is then raised

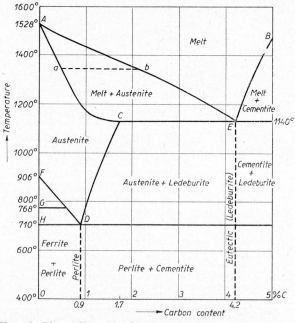

Fig. 38. Phase diagram of iron-carbon alloys (simplified).

* In addition to the ordinary tetragonal martensite, there is also a cubic modification. This is produced from the tetragonal form by a slight shift in the position of the C atoms.

again with further increase of carbon content. The eutectic point E corresponds to 4.2 weight-% C, and a temperature of 1140°. A melt of iron containing 4.2 weight-% of carbon solidifies in the form of *ledeburite*. From melts with a lower carbon content, mixed crystals of γ-iron with carbon (*austenite*) first separate. The mixed crystals have a lower carbon content than the melt, so that a melt having the composition corresponding to the point b on the liquidus curve AE is in equilibrium with mixed crystals having the composition given by the point a on the *solidus* curve AC. The melt is thus enriched in carbon through the deposition of austenite. At the same time, the concentration of carbon

in the austenite deposited rises continuously* until, at the point C, with a carbon content of 1.7 weight-% C, this phase is saturated with carbon. The melt has then reached the point E, and the remaining melt solidifies as ledeburite. Melts containing up to 1.7% of carbon therefore yield only austenite, and those with a higher content (up to 4.2%) give ledeburite as well. Austenite, however, is only stable at high temperatures. When the alloy is slowly cooled, it undergoes transformation into a mixture of *ferrite* and *pearlite*, at temperatures between F (the transformation temperature of γ-iron into δ-iron) and H (or D), depending on its carbon content (G is the transformation temperature of δ-iron into α-iron). *Martensite*, which is characterized by its extreme hardness, is formed as an intermediate product in this transformation, and persists at ordinary temperature if cooling is effected rapidly ('quenching'). Under microscopic examination on a polished section, martensite stands out clearly in the form of dark needles from the light, not yet transformed austenite.

On passing to a melt with more than 4.2 weight-% of carbon, *cementite*, Fe_3C, first separates on rapid cooling. When the carbon content of the residual melt has thereby been reduced to 4.2%, the rest solidifies as ledeburite. If cooling is allowed to take place slowly, however, *graphite* crystallizes out, for the most part, in place of cementite. This is because, at temperatures below the melting point, a mixture of iron and graphite is more stable than a mixture of iron and cementite. Hence the latter is slowly converted into the former at about 1000°. As a result of this transformation within the solid alloy, the graphite is deposited in an extremely finely divided state ('temper carbon'). This fact, as already mentioned, is taken advantage of in malleablizing.

(*iii*) *Other Iron Alloys*. Examples of other iron alloys are given in Figs. 39–42, which represent the phase diagrams of the binary alloys of iron with silicon, chromium, nickel, and manganese.

The phase diagram of the *iron-silicon alloys* (Fig. 39), resembles that of the iron-carbon system, in that compounds as well as mixed crystals, appear in both systems. There is, however a fundamental difference between silicon and carbon, in their effect on the range of existence of the γ- and δ-phases of iron. In the iron-carbon diagram, the γ-iron mixed crystals (austenite) occupy a fairly extensive field, whereas δ-iron mixed crystals are not

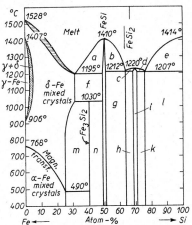

Fig. 39. Phase diagram of the iron-silicon system.

a and b = melt + FeSi
c and d = melt + $\overline{FeSi_2}$
e = melt + Si
f = FeSi + eutectic I
g = FeSi + eutectic II
h = $\overline{FeSi_2}$ + eutectic II
i = $\overline{FeSi_2}$
k = $\overline{FeSi_2}$ + eutectic III
l = Si + eutectic III
m = Fe_3Si_2 + eutectoid I
n = Fe_3Si_2 + eutectoid II

Eutectic I = δ-Fe mixed crystals + FeSi

Eutectic II = FeSi + $\overline{FeSi_2}$

Eutectic III = $\overline{FeSi_2}$ + Si

Eutectoid I = δ-Fe mixed crystals (α-Fe mixed crystals below 490°) + Fe_3Si_2

Eutectoid II = Fe_3Si_2 + FeSi

---

* The austenite already deposited also takes up more carbon.

stable except for very low carbon concentrations*. Conversely in the iron-silicon system, the γ-iron mixed crystals are restricted to a quite narrow range, whereas the δ-iron mixed crystal field is of considerable extent.

The varying influence of the other alloy component upon the existence range of the γ- and

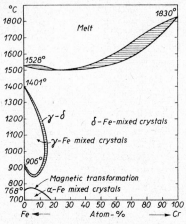

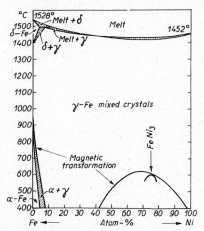

Fig. 40. Phase diagram of the iron-chromium system.

Fig. 41. Phase diagram of the iron-nickel system.

δ-phases is shown even more clearly in the phase diagrams of the *iron-chromium* and *iron-nickel alloys* (Figs. 40 and 41). In both systems, except for a narrow region of inhomogeneity at the junction between the γ-and the δ- or α -phase, there is complete miscibility in the solid state. In the iron-chromium system, the range of stability of the γ-iron mixed crystals is rather restricted, and that of the δ-iron mixed crystals is of very wide extent. By contrast, in the iron-nickel system, δ-iron mixed crystals exist only in a very small field, whereas the range of existence of the γ-iron mixed crystals extends over practically the entire diagram.

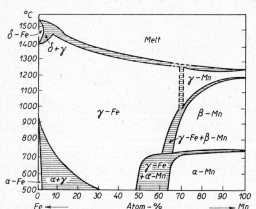

The γ-iron mixed crystal range is also by far the most important in the phase diagram of the iron-manganese alloys (Fig. 42), which was discussed on p. 208. A noteworthy feature in this system is the strong decrease in the capacity of the components to form mixed crystals, as the temperature is lowered. (Regions of inhomogeneity are indicated by shading in Figs. 40–42.)

Fig 42. Phase diagram of the iron-manganese system (according to Westgren and Ohmann).

Whether the stability range of γ-iron is broadened or contracted by any element which forms mixed crystals with iron is systematically related to the position of the element in the Periodic System. The atomic radius appears to be the prime determining factor [cf. Wever, *Naturwissenschaften*, 17 (1929) 304].

* In the field *FHD* of Fig. 38, the phases present are austenite and δ- or α-iron of very low carbon content. Only a very narrow strip at the edge of this field (not shown in the diagram) is occupied by homogeneous δ- or α-iron mixed crystals. In the same way, immediately beneath the point *A* in Fig. 38 there is a very restricted range (also not shown) in which δ-iron mixed crystals, containing carbon, are present, similar to the nickel-containing δ-iron mixed crystals of Fig. 41.

Except for carbon and silicon, iron can take *non-metals* into solid solution only in very minute amounts (cf. Table 29, p. 244–5). *Oxygen* is dissolved by molten iron, as FeO; some of this is retained in solid solution on cooling (δ-iron can dissolve up to 0.12 weight-%, α-iron only up to 0.04 weight-% of O in solid solution). The oxygen content of iron reduces its workability when hot—produces 'red shortness'. *Nitrogen* is absorbed from the air by molten iron only in minimal amounts. However, if iron is heated in ammonia gas, an iron-nitrogen compound Fe_2N is formed, which displays a considerable solubility in solid iron, and confers great hardness. This property is utilized in the surface-hardening of iron articles, by heating them in ammonia (*nitriding* process)[17]. In addition to the foregoing, there is a second compound Fe_4N with a rather narrow range of homogeneity. Above 660° this is transformed without change of composition into crystals of the compound $\overline{Fe_2N}$, in which only one half of the available lattice positions are occupied by nitrogen atoms.

Hydrogen is also absorbed by iron at a red heat. The amount absorbed is small, and is proportional to the square root of the pressure. Electrolytic iron, however, may contain larger amounts of hydrogen, which make it hard and brittle. The hydrogen is driven off on heating, and the iron then becomes ductile.

According to Frankenburger (1931), very considerable quantities of hydrogen and nitrogen (up to 6 H_2 or N_2 per atom of iron) are taken up by iron in a state of *atomic* dispersion, such as is obtained when iron atoms (from vaporized iron) are condensed with a large excess of some inhibitor (e.g., sodium chloride) so as to prevent the union of the condensing iron atoms into larger aggregates. It is probable that the attachment of the gas molecules to the Fe atoms can be attributed to the action of Van der Waals forces, in the same way as, e.g., in the formation of inert gas hydrates. (Vol. I, p. 102). Atomic-dispersed nickel behaves similarly. Neither nickel nor iron, in atomic dispersion, is able to take up the inert gases.

(*iv*) *Technical Varieties of Iron.* Ordinary technical iron contains silicon, manganese, and other impurities, as well as carbon, and the properties of the iron-carbon alloys may be modified to a considerable extent by these other constituents. Thus silicon represses the formation of cementite and favors the deposition of graphite, whereas manganese acts in the opposite manner.

The influence of silicon, manganese, etc. on the range of existence of the various modifications of iron, as discussed on p. 265–6, becomes practically important only when these elements are present in considerable concentration, as in the special steels (see p. 268).

The following varieties of iron are recognized, according to their carbon content (and to some extent, their content of other elements also).

Pig iron or *cast iron* is iron which contains more than 2.3%, and usually 5–10% of foreign constituents, with a carbon content of 2–5%. It melts without any previous softening, and therefore cannot be forged, but it casts well, since it fills the molds sharply. The melting point of pig iron lies between 1100° and 1200°. Pig iron is brittle at ordinary temperature. Ordinary *grey cast iron* (cast iron in the narrower sense) contains the carbon chiefly in the form of graphite (e.g., typically 0.90% 'combined carbon', 2.8% graphite). *White cast iron*, on the other hand, contains its carbon essentially as cementite (e.g. 3.0% 'combined carbon', 0.10% graphite). Since it is harder and more brittle than grey cast iron, it is less suitable for castings (except for malleable cast iron), and is used almost exclusively as raw material for the production of malleable iron.

Malleable iron has a lower carbon content than pig iron, and contains fewer impurities altogether. The carbon content is usually between 0.04 and 1.5%. Malleable iron melts at a higher temperature than pig iron. It softens gradually at high temperatures, and can therefore be forged and welded, as well as rolled or drawn. A distinction is drawn between *wrought iron* and *mild steel*, according to the method of production (cf p. 257–8). Depending on whether the iron can be hardened or not, a further distinction is also made between *steel* and *soft iron*.

Soft iron contains 0.5% of carbon at the most. It is tough and relatively soft, and can therefore be worked particularly well. It approaches pure iron in its properties, but differs from the latter in that, when magnetized, it loses its magnetism with a greater or less delay (hysteresis).

Steel has a higher carbon content than soft iron—usually between 0.5 and 1%. It can be forged and welded less readily than soft iron. It is also harder than the latter, and not tough but elastic at ordinary temperature. The most important characteristic of steel is that it may be *hardened* [*15, 16*]. If it is heated to bright redness, and suddenly cooled (by plunging it into water or oil), it becomes extraordinarily hard and brittle. The brittleness can be removed without any reduction in hardness, by 'tempering' the steel—that is, by heating it carefully for a short time to a moderately high temperature (250—300°).

The possibility of hardening steel is based on the fact that iron-carbon alloys with a carbon content below 1.7% can be converted into austenite by heating them. When this is suddenly cooled ('quenched') it passes over, partially or completely, into the very hard martensite. As follows from the phase diagram (Fig. 38), the temperature necessary for hardening varies, according to the carbon content of the steel. It is usually about 900°. The effect of tempering is to release or diminish the internal strains that result from quenching.

If, in course of tempering, the temperature is raised higher, so that pearlite begins to separate out in a state of very fine subdivision, the hardness decreases to some extent but the tensile strength is increased. It is essential, however, that the heat treatment should not be continued so long that the martensite decomposes completely into pearlite, as the hardness and strength would then be lost once more.

As follows from Fig. 38, there is no intrinsic distinction between soft iron and the iron-carbon alloys which are known as steel. For this reason, soft iron also displays the property of hardening, although usually only to a negligibly small degree. It is not possible, however, to draw any sharp limit between soft irons and steels, and it is now a common technical practice to include under the heading of 'steel' all iron alloys which can be forged, including soft iron.

(*v*) *Special Steels* [*18–21*]. As already mentioned, steel generally contains silicon and manganese, as well as carbon, although manganese is usually present only in small amounts (a few tenths of a per cent). If the silicon content is raised above 1% (e.g., to about 2.5%), particularly hard and elastic steels are obtained, such as are used, e.g., for springs (*silicon steels*). If the manganese content is raised, *manganese steels* are produced, which are also highly elastic, and have particularly high tensile strength. They are employed for axles and other highly stressed machine components. Varieties of steel of particularly high quality, suited for specific purposes, can be obtained by alloying with nickel, chromium, vanadium, molybdenum, tungsten, etc. Thus steel with a high nickel content (36%) has a very low coefficient of expansion (*invar steel*), and is especially suitable for clock pendulums, etc. Steel with 46% nickel has the same coefficient of expansion as glass and platinum, so that wires made from this alloy can be sealed through glass, in the same way as platinum wire. By the addition of a high proportion of chromium, alloys are obtained which resist corrosion by furnace gases, water vapor, etc. at high temperatures (*heat resistant steels*). Steels containing tungsten and vanadium, as well as (usually 4%) chromium are notable for their property of retaining their hardness at a red heat. Tools fashioned from these steels therefore permit operation at high cutting speeds (*high speed steels*). *Rustless* or *stainless steels* are produced by alloying with 12–15% chromium (VM steel) or with 18% chromium and 8% nickel (V2A steel, 'nirosta'); such steels are frequently stabilized by the addition of other alloying components, such as niobium. These steels are now used in increasing quantities, not only for the production of industrial chemical equipment, but also for the fabrication of ordinary objects—especially cutlery. A similar alloy (*wipla* metal) has been used for making artificial dentures.

The properties of special steels depend to a considerable degree upon the influence exerted by the various alloying components upon the range of existence of the structurally different phases of iron. If Ni, Mn, Cr, or W are added in sufficient quantity to the Fe-C alloys, the product of slow cooling is not pearlite, but martensite or, with still larger additions, austenite. Thus iron with 0.4% C and less than 8 weight-% Ni solidifies with pearlitic structure, whereas with 8–22 weight-% Ni (and the same C content) it is martensitic, and with higher Ni content, austenitic. Steels which soldify with the martensitic structure even when they cool slowly are known as *self-hardening* steels, since it is not necessary to subject them to a special hardening process. Steels which solidify with an austenitic structure are called *naturally hard*, since in them the austenite structure is stable at ordinary temperature.

Certain iron-nickel alloys are characterized by a very high magnetic permeability, and are therefore used in electromagnetic devices (Permalloys). [*23*] The material recently developed by Boothby and Bozorth [*J. Applied Phys.*, 18 (1947), 173] has an especially high permeability and very small hysteresis (i.e., very small remanent magnetism). It contains 79% nickel, 15% iron, 5% molybdenum, and 0.5% manganese. It must be subjected to a special heat treatment after it has been rolled into sheet.

(*vi*) *Silicon Cast Iron.* Iron-silicon alloys with a relatively high silicon content (15–18 weight-%, or 26–30 atom-%) are characterized by a high resistance towards attack by acids. They are therefore used for acid pumps, pipe lines, etc., and since these alloys cannot be rolled or forged, such objects are made by casting. The usual procedure is to heat a mixture of (low-carbon) iron and silicon to about 1200°. The reaction which then sets in raises the mixture to its melting point. Silicon cast iron (tantiron) is harder than cast iron, but has a lower bending strength. Its thermal conductivity is about 50%, and electrical conductivity about 65% of that of cast iron. As follows from the phase diagram (Fig. 39, p. 265), silicon cast iron consists essentially of iron-silicon mixed crystals. Between these there is enclosed, at the most, a very small amount of iron silicide, together with carbon, precipitated from the iron on cooling, in the form of graphite. In order to exert its protective action, it is therefore sufficient that the silicon should be incorporated in the crystal lattice of the iron, without the occurrence of any chemical combination in the narrower sense. There are numerous other examples, also, in which mixed crystals formed between elements of very different electroaffinity display properties (brittleness, enhanced resistance towards chemical attack, lower electrical conductivity) which are otherwise regarded as typical of intermetallic compounds (cf. Vol. I, p. 571 ff). In the case of the silicon-iron alloys, resistance towards acids is probably enhanced by the formation of a protective layer, consisting of SiO_2, on the surface. This view is supported by the fact that these alloys are generally particularly resistant towards strongly oxidizing acids.

4. Compounds of Iron

Iron is predominantly *bivalent* and *trivalent* in its compounds. The oxide FeO and a hydroxide $Fe(OH)_2$ are derived from bivalent iron, as also are many salts. The most important of these is the sulfate, crystallizing with 7 molecules of water, $FeSO_4 \cdot 7H_2O$ (*iron vitriol*). From trivalent iron are derived the oxide Fe_2O_3 and a hydroxide FeO(OH). This also forms salts with many acids—e.g., iron(III) chloride, $FeCl_3 \cdot 6H_2O$ and iron alum, $NH_4Fe(SO_4)_2 \cdot 12H_2O$. Iron is present in the +6 state in the rather unstable *ferrates(VI)*, $M_2[FeO_4]$.

In addition to the oxy-acid salts of +6 iron (ferrate(VI) salts), similar derivatives of tetrapositive iron (ferrate(IV) salts) and tripositive iron (ferrate(III) salts) are known. To distinguish them from the ordinary ferrates (ferrate(VI) salts), ferrate(IV) and ferrate(III) salts are commonly known as perferrites and ferrites, respectively. Both bi- and tri-positive iron can also form *hydroxo-salts* (hydroxoferrate(II) salts and hydroxoferrate(III) salts), although these are not very stable. For compounds of *uni*positive iron see pp. 270, 289, 356.

Iron(II) compounds are commonly known as ferrous compounds, and iron(III) compounds as ferric compounds; but it is advisable to discontinue this terminology in favor of the more systematic nomenclature.

Salts of both bivalent and trivalent iron with the oxyacid anions of chlorine, sulfur and nitrogen are soluble in water, as also are the halides (except for iron(II) fluoride). The phosphates, carbonates, and sulfides are insoluble, as are the oxides and hydroxides.

In both its common valence states, the salts of iron readily form complex salts (acido-salts) with salts of more electropositive elements. This tendency is particularly strong in the salts of iron with weak acids. In many instances, such as the cyanide compounds, the complex salts can readily be obtained pure, whereas it is impossible to isolate in the pure state the simple salts from which they are derived.

Many iron compounds are also capable of adding on ammonia. Thus iron(II) and iron(III) chlorides, in the dry state, both combine with 6 molecules of NH_3. The ammoniates undergo decomposition when they are dissolved in water. However, it is possible to crystallize iron(II) salt ammoniates from solutions containing a high concentration of ammonium salts, by saturation with ammonia (Weitz, 1925).

The complexes formed by iron with certain hydroxylated organic compounds are much more stable in solution than the ammoniates. The precipitation of iron from its solutions (as hydroxide) is therefore inhibited when such substances are present in high concentration.

Many compounds of iron combine additively with carbon monoxide, and numerous others with nitric oxide. The 'brown ring' test for nitric acid, discovered by Des-bassins de Richmond in 1835, depends on the fact that iron(II) sulfate absorbs nitric oxide, producing a deep brown coloration. Solutions of iron(II) chloride and other simple iron(II) salts also take up nitric oxide.

The amount of nitric oxide absorbed by iron salt solutions depends strongly upon the temperature and pressure. The absorption does not obey Henry's law, however, and under varied experimental conditions always tends towards the limiting value 1 NO for each Fe^{++}. Manchot was able to isolate NO-compounds of this kind, in the crystalline state, in spite of their instability,—e.g., the compound $[Fe^{II}(NO)]SO_4$ (nitrosoiron(II) sulfate). Transport experiments prove that the nitric oxide in compounds of this type is bound to the iron. Manchot was able to show from other investigations that the nitric oxide was combined as a neutral group, without changing the electrochemical valence of the iron.

Nitric oxide is more firmly bound in the nitroprusside compounds (compounds of the general formula $M^I_2[Fe(CN)_5(NO)]$) than in compounds such as nitrosoiron(II) sulfate. Nitroprussides can readily be isolated, and may be recrystallized from hot solutions without decomposition. Other nitroso compounds, which also contain firmly bound nitric oxide, are the Roussin's salts, of which there are two principal types, with the general formulas $M^I[SFe(NO)_2]$ ('red salts of Roussin') and $M^I[S_3Fe_4(NO)_7]$ ('black salts of Roussin'), respectively. According to Manchot, the NO-group is probably a neutral ligand in these salts also, and he therefore regards the iron in Roussin's salts (or 3 of the 4 Fe atoms in the black salt) as unipositive.

Among the prusside compounds*, complexes containing carbon monoxide of the type $M^I_2[Fe(CN)_5(CO)]$ (carbonyl prusside salts), are known. Carbon monoxide can also combine directly with (finely divided) iron to form iron pentacarbonyl, $Fe(CO)_5$. According to Manchot, an iron tetranitrosyl, $Fe(NO)_4$, is formed by the reaction of iron pentacarbonyl with nitric oxide (see further, p. 289 et seq.).

(a) Oxides and Hydroxides

Iron forms the oxides FeO, Fe_2O_3, and Fe_3O_4. The hydroxides $Fe(OH)_2$ and FeO(OH) are known in the crystalline state.

Heats of formation are, for FeO 64.0 kcal, for Fe_2O_3 197.6 kcal, and for Fe_3O_4 266.7 kcal per mol.

* For definition of the prusside compounds, see p. 287–8.

It is probable that an iron(III) peroxide or peroxyhydrate exists, which can, perhaps be formulated as:

$$\begin{array}{ccc} HO.O{\diagdown} & & {\diagup}O.OH \\ & Fe{\rule{1em}{0.4pt}}O{\rule{1em}{0.4pt}}O{\rule{1em}{0.4pt}}Fe & \\ HO{\diagup} & & {\diagdown}OH \end{array}$$

It can be obtained as a red, very unstable powder, according to Pellini (1909) and Wieland (1938), by adding H_2O_2 and KOH at a low temperature ($-79°$) to $FeCl_2$ or $FeCl_3$ dissolved in absolute alcohol.

(*i*) *Iron(II) oxide (ferrous oxide)*, FeO, is obtained as a black, pyrophoric powder when iron(II) oxalate is heated in the absence of air. It decomposes water, especially when warmed. It loses its great reactivity when it is heated.

Crystalline FeO (wüstite) has the density 5.9; for crystal structure, see p. 249. The compound invariably contains rather less iron than corresponds to the ideal formula FeO, since its crystal lattice is stable even when the lattice positions available for Fe atoms are not all fully occupied. However, FeO is only stable at high temperatures, and it decomposes into metallic iron and Fe_3O_4 when it is cooled slowly. The researches of Schenck and Dingmann (1929), and of Darken and Gurry [*J. Am. Chem. Soc.*, 67 (1945), 1398] have shown that the wüstite phase exists over a certain composition range. Although the oxygen in iron(II) oxide is very firmly bound, it is possible by indirect means to measure the dissociation pressure of oxygen exerted at high temperatures. This may be determined from the equilibrium pressure ratio p_{H_2}/p_{H_2O} which is established when hydrogen or steam are passed over a mixture of FeO and Fe, or of FeO and Fe_3O_4, at high temperatures (cf. Vol. I, p. 37). At $1500°$ K, according to Nernst and von Wartenberg, the equilibrium constant for the thermal dissociation of water, $K = p_{H_2}^2 \cdot p_{O_2}/p_{H_2O}^2$ has the value $8.15 \cdot 10^{-14}$. This may be combined with equilibrium ratios of H_2 and H_2O over a metallic oxide (e.g., FeO). Such measurements were made by Schenck and Dingmann, and by Kapustinsky, but the best data are those of Darken and Gurry (1945). They found

	$\dfrac{p_{CO_2}}{p_{CO}}$	$\dfrac{p_{H_2O}}{p_{H_2}}$	$\dfrac{p_{CO_2}}{p_{CO}}$	$\dfrac{p_{H_2O}}{p_{H_2}}$	p_{O_2}, atm.
1100° C	0.355	0.738	6.12	12.7	1.49×10^{-11}
1200°	0.322	0.802	8.02	20.0	7.21×10^{-10}
1300°	0.297	0.861	10.79	31.3	2.40×10^{-8}
1400°	0.285	0.888			
	over $\overline{FeO} + Fe$		over $\overline{FeO} + Fe_3O_4$		

The same authors determined the composition of the wüstite phase at the upper and lower ends of the composition range:

	Lower limit		Upper limit	
1100°	$FeO_{1.048}$	$(Fe_{0.95}O)$	$FeO_{1.153}$	$(Fe_{0.87}O)$
1200°	$FeO_{1.049}$	$(Fe_{0.95}O)$	$FeO_{1.167}$	$(Fe_{0.86}O)$
1300°	$FeO_{1.050}$	$(Fe_{0.95}O)$	$FeO_{1.180}$	$(Fe_{0.85}O)$
1350°	$FeO_{1.052}$	$(Fe_{0.95}O)$		
1400°			$FeO_{1.196}$	$(Fe_{0.84}O)$

The ideal compound, $FeO_{1.000}$ appears from their experiments not to exist, although Benard has given evidence that it is formed as the equilibrium phase at much lower temperatures (lower stable limit of existence about 700–800° C).

Jette and Foote [*J. Chem. Phys.*, 1 (1932) 29] showed that the departure of the wüstite phase from the ideal composition was the result of the replacement of a proportion of Fe^{2+} ions by Fe^{3+} ions in the structure, and a simultaneous creation of vacant sites in the iron-atom lattice of the crystal. The change of composition is accompanied by a change in cell dimensions. If the Fe : O ratio changes through incorporation of extra (interstitial) oxygen atoms, the mean weight per unit cell increases. If iron atoms are omitted from the structure, the weight per unit cell diminishes. Comparison of the pycnometrically determined density with that calculated on both models shows quite clearly that the iron-vacancy hypothesis is correct. The non-Daltonian iron(II) oxide is thus more correctly formulated as $FeO_{1-x}O$ than as FeO_{1+y} (see table above).

(*ii*) *Iron(II) hydroxide*, $Fe(OH)_2$, is formed by treating iron(II) salts in absolutely air-free solution with alkali; it precipitates as a white, flocculent material which absorbs oxygen with the greatest avidity. It thereby turns a dirty green, which gets progressively darker and eventually passes into the red brown color of iron(III) oxide hydrate. The very deep colored intermediate product formed in the reaction contains both bivalent and trivalent iron. Its intense color is typical of the fact that compounds which contain the same element in two different valence states are apt to be deeply colored.

Iron(II) hydroxide is only incompletely precipitated by ammonia. If the solution also contains a high concentration of ammonium salts, there is no precipitation at all, since the OH^- ion concentration cannot, in these circumstances, rise above a very low value. Since the concentration of Fe^{++} ions is also repressed, through the formation of the iron-ammonia complexes identified by Weitz, it is not possible for the solubility product of iron(II) hydroxide, $K_{SP} = [Fe^{++}] \cdot [OH^-]^2$, to be exceeded under these conditions.

The density of crystalline $Fe(OH)_2$ is 3.40 (for crystal structure—cf. p. 249). Its heat of formation from FeO and H_2O is 4.0 kcal.

$Fe(OH)_2$ is soluble in concentrated sodium hydroxide, forming sodium hydroxoferrate(II), $Na_2[Fe(OH)_4]$. According to Scholder (1936), this may best be prepared by dissolving finely divided iron in boiling 50% sodium hydroxide. The compound separates out on cooling, in the form of fine blue-green crystals. If $Sr(OH)_2$ or $Ba(OH)_2$ are added, the alkaline earth salts $Sr_2[Fe(OH)_6]$ and $Ba_2[Fe(OH)_6]$ may also be obtained (as greenish white microcrystalline precipitates).

(*iii*) *Iron(III) oxide (ferric oxide)*, Fe_2O_3, is obtained by igniting the iron(III) oxide hydrates as a red-brown powder or, after stronger ignition, as a dark grey, lustrous crystalline mass. When it has been strongly ignited, it is insoluble in acids, as are the corresponding oxides of chromium and aluminum, with which it is isomorphous. It has the corundum structure (Vol. I, Fig. 72, p. 351), $a_0 = 5.414$ Å, $a = 55°17'$. The density of ignited iron(III) oxide is 5.20. Iron(III) oxide is found native as *hematite* (red ore), usually in dense, granular, earthy or flaky masses with a steel-grey, black, or red-brown color. The various forms in which it occurs are distinguished by the names of iron mica (flaky), porous hematite, ocher (earthy), bloodstone (dense), and specular iron ore. The last is perceptibly crystalline, and steel-grey to black, but like the other varieties it gives a red streak on an unglazed porcelain plate. Hematite, which occurs in enormous deposits in some places (e.g., the Mesabi Range of Minnesota) is an important ore for the smelting of iron. A few varieties of natural iron oxide, especially ocher (red bolus, terra sigillata), and some artificially prepared varieties (venetian red, English red, colcothar, 'caput mortuum') are used as red pigments and as polishing agents (rouge), and also as catalysts for various reactions.

Like aluminum oxide, iron oxide can occur in a second modification, This does not apparently exist native, but can be obtained by dehydrating *lepidocrocite* (ruby mica). It was detected by X-ray methods by Haber and Böhm, who called it γ-iron(III) oxide. γ-Iron oxide crystallizes cubic, and its structure can be derived from that of magnetite (cf. p. 275) by taking out $\frac{1}{9}$ of the iron atoms. Like magnetite, γ-iron oxide is ferromagnetic, in contrast to α-iron oxide which is paramagnetic. The γ-oxide is converted monotropically into the α-oxide when it is heated above 400°.

A third modification (δ-Fe_2O_3) can be obtained by the simultaneous hydrolysis and oxidation of iron(II) salts (Chevallier, 1927, Glemser, 1939). This is hexagonal, like α-Fe_2O_3, but is ferromagnetic. It is transformed into γ-Fe_2O_3 by prolonged heating above 110°.

(*iv*) *Iron(III) oxide hydrate* occurs in nature as *limonite* or brown iron ore. The varieties of brown iron ore which are most important for smelting are reniform hematite (xanthosiderite), pitticite or triplite, pisolite ironstone, iron oolite, the Scandinavian lake ore, and the bog iron ores. Goethite (needle iron ore), which forms needle crystals, and lepidocrocite (ruby mica), which generally forms thin plates, are well crystallized varieties of brown iron ore. Limonite is amorphous. Very impure bog iron ores, such as are found in various localities, are used as gas purification masses and as catalysts (e.g., for the recovery of sulfur by the Chance-Claus process) (Vol. I, p. 700). They are also used for the preparation of red pigments (red ocher).

Only the oxide hydrate $Fe_2O_3 \cdot H_2O$, or $FeO(OH)$, occurs in Nature as a compound of well defined water content. It is found in two crystalline modifications—as *goethite** and as *lepidocrocite*, as well as in the amorphous form, as limonite**. The water content of the brown iron ores found in Nature is frequently rather higher than corresponds with the formula $FeO(OH)$, since the compound can take up some water in solid solution. Iron(III) oxide (hematite) can also take up water in solid solution (up to 8%), forming *hydrohematite****. This gives the red streak characteristic of hematite, whereas the other iron oxide hydrates give a more brownish streak (Posnjak, 1919, Haber and Böhm, 1925, Kurnakow, 1926).

Both goethite (α-$FeO(OH)$) and lepidocrocite (γ-$FeO(OH)$) have orthorhombic crystal structures. The former is isomorphous with diaspore, the latter forms a well marked layer lattice. According to Weiser (1935) and Kratky (1938) there is also a third, metastable modification [β-$FeO(OH)$], which is formed by the slow hydrolysis of $FeCl_3$. Other iron(III) salts yield α-$FeO(OH)$ on slow hydrolysis, and α-Fe_2O_3 on rapid hydrolysis.

The mutual stability relations between the various modifications of the oxide and hydroxide may be summarized by the following scheme:

$$\gamma\text{-FeO(OH)} \xrightarrow{\quad 6 \text{ kcal} \quad} \alpha\text{-FeO(OH)}$$
$$\uparrow 4 \text{ kcal} \qquad\qquad\qquad \uparrow 2 \text{ kcal}$$
$$\gamma\text{-Fe}_2\text{O}_3 \xrightarrow{\quad 8 \text{ kcal} \quad} \alpha\text{-Fe}_2\text{O}_3$$

The energy quantities indicated apply to the compounds with undistorted crystal structure, and sufficiently great particle size. However, since their power of crystallization is very poor, they are often obtained in such a finely divided state that their energy content is markedly affected by their great specific surface (cf. p. 677.) The lattice distortions which are often present in high concentration have an even greater effect (cf. p. 18). Thus Fricke was able to prepare γ-$FeO(OH)$ specimens which differed by as much as 6 kcal in their energy content for these reasons. Preparations of α-Fe_2O_3, made by dehydrating slimy amorphous iron(III) oxide hydrates at different temperatures, may exhibit energy differences of up to 13 kcal per mol. of Fe_2O_3 (Fricke, 1935). The energy rich states of the iron oxides are characterized by a very high catalytic efficiency; they are therefore said to be 'active' states****. Other properties may also be modified to a marked extent by lattice imper-

---

* The name goethite was originally proposed for lepidocrocite (ruby mica), but later came into general use for needle iron ore.

** The amorphous character of limonite is only apparent. It gives an X-ray diffraction pattern of weak and broadened interference rings, which agree in position with the (sharp) diffraction rings of goethite.

*** Hydrohematite and hematite give the same X-ray diffraction pattern.

**** Similar active states have been observed with other compounds that have a tendency for lattice imperfections.

fections and by the surface development—e.g., the thermal dissociation pressure and electrolytic dissociation. As is illustrated by the figures given above, it is possible for the energy differences exhibited by one and the same modification in states of different activity to be greater than the energy differences between the normal states of the several modifications.

If iron(III) oxide hydrate is precipitated by adding ammonia to iron(III) salt solutions, it is obtained as a red-brown, slimy amorphous precipitate, which yields a gel of variable water content when it is dried.*

The precipitate has a strong tendency to carry down with it other substances from the solution, since it has a high surface activity. Freshly precipitated iron oxide hydrate is a specific adsorbent for arsenic trioxide, and can therefore be used as an antidote for acute arsenic poisoning. Iron oxide hydrate can also be obtained in colloidal dispersion, and is used in medicine in this form.

Freshly precipitated iron(III) oxide hydrate dissolves readily in dilute strong acids. It is also perceptibly soluble in hot concentrated caustic soda or potash, forming sodium or potassium ferrite (ferrate(III)), $NaFeO_2$ or $KFeO_2$. Iron(III) oxide hydrate is thus amphoteric in behavior, although its ability to function as an acid is only very weakly developed.

Ferrites (ferrate(III) salts) are more readily obtained by melting iron(III) oxide with alkali oxides, hydroxides, or carbonates. In addition to the 1 : 1 compounds, it is also possible to prepare 1 : 2 ferrites in this way—e.g., $K_2Fe_4O_7$. *Hydroxoferrites* (hydroxoferrate(III) salts) are also known. The sodium salt, $Na_5[Fe(OH)_8] \cdot 5-6H_2O$, was obtained by Scholder (1936) by the oxidation of a sodium hydroxoferrate(II) solution with atmospheric oxygen at ordinary temperature. It forms colorless crystals, which are rather sparingly soluble in concentrated caustic soda. It is immediately decomposed by water or by dilute sodium hydroxide, with the deposition of iron(III) oxide hydrate. If oxidation of the hydroxoferrate(II) salt is carried out at 40–60°, the salt which separates is a heptahydroxo-compound, $Na_4[Fe(OH)_7(H_2O)] \cdot 1-2H_2O$, which is also colorless. At still higher temperatures (100–130°), hydroxo-salts are not obtained, but the anhydrous oxo-salt $NaFeO_2$ is formed, in olive green crystals, whereas the same salt is obtained in red-brown crystals from 60 per cent sodium hydroxide above 130°. The latter salt is also obtained when sodium ferrite is prepared by the method first cited. The nature of the difference between the green and red forms of sodium ferrite is not yet known.

Red sodium ferrite is isotypic with $Na[HF_2]$, although the anionic radicals $[OFeO]^-$ do not stand out as structural groups quite so clearly as the $[FHF]^-$ groups in the structure of $Na[HF_2]$. $Cu[FeO_2]$ has the same structure. In $KFeO_2$, however, there is an infinite continuous network of FeO_4 tetrahedra (which corresponds structurally in every way with the network structure of β-cristobalite), in the meshes of which are located the K ions (Barth, 1935). $RbFeO_2$ and $Pb[FeO_2]_2$ have the same structure (Hilpert, 1933), as also have the ferrites of a few other uni- and bivalent metals. Oxides of many bivalent metals combine with Fe_2O_3, to form not ferrites, but compounds with the *spinel structure*—i.e., double oxides. Such double oxides without salt-like character are formed, for example, by the oxides of Mg, Zn, Cd, Cu^{II}, and Mn^{II}. Iron(II, III) oxide, Fe_3O_4, has the same structure. The ferromagnetism which is displayed by many of these oxides is not unambiguously associated with their spinel structure, since cadmium iron spinel, for example, is only

---

* Iron(III) oxide hydrate, freshly precipitated from iron(III) salt solutions, is amorphous to X-rays, and gives no interference pattern. As aging takes place, the diffraction pattern characteristic of hematite (or hydrohematite) develops. However, if precipitates obtained by the action of OH^- ions on iron(II) salts are allowed subsequently to be oxidized in air, they furnish preparations which display the diffraction pattern of goethite or limonite. Hüttig states that the stability of the compounds commonly found in nature increases in the order: hydrogel of amorphous iron(III) oxide hydrate → hematite → hydrohematite → limonite → goethite.

paramagnetic, whereas another modification of the compound $CdO \cdot Fe_2O_3$, which does not have the spinel structure, is ferromagnetic.

The compound $LiFeO_2$ is to be regarded structurally as a double oxide, and not as a ferrite. It has the rock salt structure ($a = 4.14$ Å), in which the ions Li^+ and Fe^{3+} are distributed statistically (i.e., quite at random) over the positions of the cation lattice (Barth, 1931). The same structure has been found for some other double oxides—namely Li_2TiO_3 ($a = 4.10$ Å), Na_2CeO_3 ($a = 4.82$ Å) and Na_2PrO_3 ($a = 4.84$ Å).

There is a noteworthy difference in structure between the iron spinels and ordinary spinel. In ordinary spinel, the Mg^{2+} ion is at the center of a *tetrahedron* of oxygen atoms, while every Al^{3+} ion is at the mid-point of an oxygen octahedron. In magnesium iron spinel, $MgO \cdot Fe_2O_3$, however, it is not the Mg^{2+} ions which are located at the centers of the oxygen tetrahedra, but Fe^{3+} ions. The Mg^{2+} ions and the remaining half of the Fe^{3+} ions occupy the positions of octahedral coordination. The difference between the structures can be expressed by writing the formula of ordinary spinel as $Al_2[MgO_4]$, and that of magnesium iron spinel as $MgFe[FeO_4]$. According to Verwey (1953), this is not strictly correct; most, but not all of the Mg^{2+} ions in magnesium iron spinel are in the positions of octahedral coordination, and the formula schould be written as $Mg_{1-x}Fe_{1+x}[Mg_xFe_{1-x}O_4]$.

(*v*) *Iron(II, III) oxide*, Fe_3O_4 (density 5.1, m.p. 1538°), is a double compound of FeO and Fe_2O_3, belonging to the spinel class. It is obtained (in an impure form) by burning iron filings in air. The 'hammer scale' referred to on p. 262 also consists essentially of iron(II, III) oxide. It forms a black powder, insoluble in acids. It is found in very large quantities in Nature as magnetite (lodestone, magnetic iron ore), and is an important ore of iron. Magnetite is strongly ferromagnetic, and has a fairly high electrical conductivity. It is therefore used as a material for electrodes, especially for the electrolysis of alkali chlorides, and also to some extent for arc light electrodes. In the Siemens-Martin process, magnetite and hammer scale are used for decarburizing iron. Hammer scale is also used for the preparation of thermite.

It has been proved by X-ray diffraction that magnetite has the spinel structure. However, the iron(II) ions do not occupy the mid-points of the oxygen tetrahedra, these positions being taken by one half of the iron(III) ions present. The other iron(III) ions and the iron(II) ions occupy (probably in random distribution) the positions which, in the lattice of ordinary spinel, Al_2MgO_4, are taken by the aluminum ions. The constitution of magnetite can therefore be represented by the formula $Fe^{II}Fe^{III}[Fe^{III}O_4]$. The high electrical conductivity of magnetite can be explained by a ready exchange of charge between the Fe^{2+} and Fe^{3+} ions which occupy crystallographically identical positions outside the $[FeO_4]$ groups (Verwey and De Boer, 1936). The unit cell of magnetite has the edge length $a = 8.41$ Å. When moderately heated in air, powdered magnetite is oxidized to Fe_2O_3. This property was utilized at a very early period for preparing red iron oxide from magnetite. Conversely, Fe_3O_4 is formed from Fe_2O_3 by very strong ignition (above 1400°). At still higher temperatures it passes into FeO by the further loss of oxygen.

(b) Sulfur Compounds

Direct combination of iron with sulfur can give rise to the two compounds: FeS, *iron sulfide*, iron monosulfide, iron(II) sulfide, and FeS_2, *iron disulfide*, iron(II) disulfide.

These are the only compounds which appear in the phase diagram of the iron-sulfur system; i.e., these are the only sulfides obtainable by direct combination of the elements. However, it is not to be completely excluded that another sulfide—namely iron(III) sulfide, Fe_2S_3—might be obtained by other methods, such as by precipitation from solution. No definite evidence for the existence of such a compound has as yet been obtained (cf. p. 276).

(*i*) *Iron monosulfide* (*ferrous sulfide*), FeS, is formed by melting iron and sulfur together, and is prepared technically by this method since it is used for the generation of hydrogen sulfide. The technical product is usually contaminated with a considerable excess of uncombined iron. It is sold in grey-black lumps, but the pure crystalline iron(II) sulfide is a light brassy brown.

Iron(II) sulfide is obtained as a black precipitate when $S^=$ ions are added to solutions of iron(II) salts. The precipitate is practically insoluble in water, but soluble in dilute acids. When moist it undergoes partial oxidation by the air, forming the sulfate.

Iron sulfide, crystallized in the hexagonal system, is found in Nature as *pyrrhotite* (magnetic pyrites). The iron sulfide occurring in meteorites, with the same crystal structure, is called *troilite*. Pyrrhotite almost invariably contains nickel, and is therefore of importance as a nickel ore. The sulfur content of magnetic pyrites is invariably 1 to 2% higher than corresponds with the formula FeS, the excess sulfur being built into the crystal lattice.

FeS is isotypic with NiAs (cf. p. 249). Its ability to take up a certain excess of sulfur (up to 3.3 atom per cent) arises from the fact that a proportion of the positions which should be occupied by Fe atoms may remain vacant. The density of FeS varies between 4.5 and 5.0. Its heat of formation is 22.8 kcal per mol.

FeS is completely miscible with iron in the molten state, but quite immiscible in the solid state. Hence FeS separates out during the solidification of iron with even a very low sulfur content, and (like FeO) gives rise to 'red shortness'—i.e., cracks develop in the iron when it is forged at a red heat. If a sufficient quantity of manganese is added to the iron, this removes the sulfur quantitatively, forming MnS, which is very sparingly soluble in the iron-manganese melt. This makes it feasible to desulfurize pig iron almost completely by the addition of manganese. The other impurities present in pig iron—C, P, Si—promote the desulfurization, since they still further reduce the solubility of MnS in the melt. Iron-manganese alloys are now produced, to which sulfur is actually added—the so-called *sulfur steels*. When these solidify, the plastic MnS separates out, in extremely fine subdivision, instead of the brittle and deleterious FeS. The sulfur steels possess excellent tensile strength, for this reason.

(*ii*) *Iron sesquisulfide.* A black precipitate is obtained by the action of $S^=$ ions on iron(III) salt solutions. This is practically insoluble in water, but is soluble in dilute acids, and has a composition corresponding to the formula Fe_2S_3. It decomposes readily in air when it is moist, with the formation of iron oxide hydrate and the deposition of sulfur. Sulfur is also deposited when the precipitate is treated with hydrochloric acid:

$$Fe_2S_3 + 4HCl = 2FeCl_2 + 2H_2S + S.$$

It has not yet been established whether, as is commonly assumed, this is a definite compound (iron(III) sulfide), or merely a mixture of iron(II) sulfide and sulfur formed by the reaction:

$$2Fe^{+++} + 3S^= = 2FeS + S.$$

Iron(III) sulfide occurs in Nature in the form of double sulfides, especially with copper(I) sulfide—e.g., *chalcopyrite*, $CuFeS_2$ or $Cu_2S \cdot Fe_2S_3$, and *bornite*, Cu_3FeS_3 or $3Cu_2S \cdot Fe_2S_3$.

(*iii*) *Iron disulfide*, FeS_2, is very widely distributed in the form of *pyrite*. The compound also occurs as *marcasite*. It readily loses sulfur when it is heated, and burns to Fe_2O_3 and SO_2 when it is heated in air. The ease with which sulfur dioxide is produced by roasting pyrite makes pyrite the most important raw material for the manufacture of sulfuric acid. When once initiated, the combustion continues spontaneously in the pyrite roasting oven, in consequence of the high heat of combustion and ease with which sulfur is lost. The use of the burned pyrites, remaining as a residue after roasting has been discussed in Vol. I.

Pyrite and marcasite have a brassy yellow color and metallic luster. They differ in their crystal structures. Pyrite, which is not uncommonly found in well formed large crystals (usually cubes or pentagonal dodecahedra, or combinations of these forms), belongs to the pentagonal hemihedral class of the cubic system. Marcasite is orthorhombic.

The iron atoms in the crystal lattice of pyrite are arranged in the same way as the sodium atoms in the rock salt structure. In the pyrite structure, however, the position corresponding to each Cl atom is occupied by the center of gravity of a pair of S atoms, 2.14 Å apart, which are inclined at an angle to the axes joining the Fe atoms. The unit cell of pyrite has an edge of length $a = 5.40$ Å. The structure of marcasite is related to that of rutile but is of lower symmetry. The S $\leftrightarrow$ S distance in marcasite is 2.21 Å.

Magnetic measurements (Klemm, 1935) have shown that the metal is not quadrivalent in pyrite and other disulfides of the same type (MnS_2, CoS_2, NiS_2). MnS_2 must be regarded, from its magnetic behavior, as a manganese(II) sulfide, $Mn^{2+}(S_2)^{2-}$, but of only weakly ionic character. The other disulfides listed have even weaker ionic character, FeS_2 being the least ionic. The binding forces involved in them are probably closely related in nature to the metallic bond. (Cf. Chap. 9, Vol. I).

The sulfides of iron form *double sulfides* with some other heavy metal sulfides. They are not capable of forming true *thio-salts*, however, and are therefore insoluble in alkali sulfides. However, sulfur can enter the hydroxoferrate(II) and-(III) complexes as a substituent, replacing a proportion of the hydroxyl groups, and forming *thiohydroxo*-salts (Scholder, 1936)— e.g., $Na_3[SFe^{II}(OH)_3] \cdot 2H_2O$ (black-green) and $Na_8[(H_2O)(OH)_6Fe^{III} \cdot S \cdot Fe^{III}(OH)_6(H_2O)] \cdot 2H_2O$ (dark brown). The green or deep red colorations, formed when Na_2S is added to strongly alkaline solutions containing iron(II) or iron(III) salts, are due to the formation of such compounds.

(c) Iron(II) Salts

Soluble iron(II) salts, such the sulfate and chloride, are most conveniently obtained by dissolving the metal in the corresponding acids. In solution, and in the hydrated crystalline state, the compounds are pale green in color. This is thus the color of the hydrated iron(II) ion. The solutions have an acid reaction, because of hydrolysis. Insoluble iron(II) salts may be obtained by metathesis. Most iron(II) salts can form double or complex salts with the salts of the strongly electropositive elements, especially with the alkali and ammonium salts. Most iron(II) salts are not completely stable in air; the double salts are usually more stable than the simple salts. Strong oxidants convert iron(II) quantitatively into iron(III).

It is noteworthy that when iron is dissolved in acids of definitely oxidizing properties— e.g., in nitric, perchloric, chloric, bromic, and iodic acids—the products are salts of bivalent iron, provided that the acids are dilute, and used at ordinary temperature. The Fe^{++} ion in itself has but a small tendency to become oxidized to the $+3$ state. A platinum foil dipping into a solution which containes the ions Fe^{++} and Fe^{+++} in equal concentrations assumes a potential 0.77 volts lower than the normal hydrogen electrode. Thus the spontaneous conversion of iron(III) ions into iron(II) ions would take place in the closed circuit, hydrogen simultaneously going into solution as hydrogen ions. The fact that iron(II) compounds usually have a rather strong tendency to undergo oxidation is really to be attributed to the extreme insolubility of the oxide, which is very readily formed by hydrolysis in aqueous solution or in the presence of moisture. The stability of iron(II) salts in solution can therefore be raised to a considerable extent by adding an excess of acid to repress the hydrolysis.

If metallic iron is treated with aqueous sulfur dioxide solution, *iron(II) thiosulfate* is formed as well as the *sulfite*:

$$2Fe + 3SO_2 = FeSO_3 + FeS_2O_3.$$

The formation of *iron(II) dithionate*, $FeS_2O_6 \cdot 5H_2O$, by the action of sulfurous acid on a dilute iron(III) sulfate solution at 0°, is also of some interest. Oxidation of iron(II) thiosulfate by means of iron(III) chloride yields *iron(II) tetrathionate*:

$$2FeS_2O_3 + 2FeCl_3 = FeS_4O_6 + 3FeCl_2.$$

(d) Iron(II) Halides

(*i*) *Iron(II) fluoride and Fluoroferrate(II) Salts.* The colorless relatively insoluble hydrated *iron(II) fluoride*, $FeF_2 \cdot 8H_2O$, separates out when a solution of iron in hydrofluoric acid is evaporated. The anhydrous compound, which is also colorless, is obtained by heating the hydrate in a stream of dry hydrogen fluoride, or by the direct-action of hydrogen fluoride on iron at high temperatures. Complex salts (*fluoroferrate(II) salts*), of the types $M^I[FeF_3]$ and $M^I_2[FeF_4]$, crystallize from mixed solutions of iron(II) fluoride and alkali fluorides.

(*ii*) *Iron(II) chloride and Chloroferrate(II) Salts.* Iron(II) chloride, $FeCl_2$, can be obtained in the anydrous state by passing dry hydrogen chloride over iron filings at a red heat, by heating iron(III) chloride in hydrogen, or by heating the hydrated iron(II) chloride (see below) in the absence of air. It is volatile at a yellow heat, and its vapor density at 1000° corresponds to the formula $FeCl_2$. It is deliquescent in air, and very soluble in water and in alcohol. The tetrahydrate, $FeCl_2 \cdot 4H_2O$, separates from the aqueous solution, which is most conveniently obtained by dissolving iron in dilute hydrochloric acid in the absence of air (in a flask provided with a Bunsen valve). It forms blue-green, monoclinic deliquescent needles. 68.5 g of $FeCl_2$ dissolve in 100 g of water at 20°. With the chlorides of strongly electropositive metals, iron(II) chloride forms complex salts, mostly of the type $M^I_2[FeCl_4]$ (tetrachloroferrate(II) salts). Ammonia also combines avidly with iron(II) chloride.

(*iii*) *Iron(II) bromide*, $FeBr_2$, is the immediate product of burning iron in bromine vapor, as long as iron is in excess (contrast iron(II) chloride). $FeBr_2 \cdot 6H_2O$ separates as a pale green crystalline powder when a solution of iron in hydrobromic acid is strongly cooled. Above 45°, the hexahydrate is converted to the tetrahydrate.

(*iv*) *Iron(II) iodide*, FeI_2, can be prepared by direct union of the components. Iron dissolves gradually at ordinary temperature in water containing iodine, and $FeI_2 \cdot 4H_2O$ is deposited in green crystals when the solution is evaporated in a desiccator. Higher hydrates are formed at low temperatures. The hydrated salts and their solutions turn black when they are heated to 50°, but recover their original color again on cooling.

See pp. 353 and 356 for CO addition compounds of iron(II) halides.

(e) Other Iron(II) Salts

(*i*) *Iron(II) thiocyanate* crystallizes in green prisms of the composition $Fe(SCN)_2 \cdot 3H_2O$ when solutions of iron in aqueous thiocyanic acid are evaporated in the absence of air.

(*ii*) *Iron(II) nitrate* is formed when iron is dissolved in cold dilute nitric acid:

$$8Fe + 20HNO_3 = 8Fe(NO_3)_2 + 2NH_4NO_3 + 6H_2O.$$

It is better obtained in the pure state by double decomposition between iron(II) sulfate and lead nitrate. It usually crystallizes as the hexahydrate, $Fe(NO_3)_2 \cdot 6H_2O$, in light green rhombic plates. The 9-hydrate is stable below —10°. The aqueous solution decomposes when it is heated, and basic iron(III) nitrate is formed.

(*iii*) *Iron(II) perchlorate*. Iron dissolves in dilute perchloric acid at low temperatures (0°), yielding iron(II) perchlorate. According to Lindstrand (1936), the compound is better prepared by dissolving FeS in perchloric acid, since this enables a higher concentration of acid to be used at ordinary temperature, without oxidation of the iron(II) occurring. Iron(II) perchlorate crystallizes from solution as the hexahydrate, $Fe(ClO_4)_2 \cdot 6H_2O$, in light green hexagonal prisms, which are unchanged in air dried by $CaCl_2$, but which deliquesce and simultaneously undergo oxidation in moist air. The salt is extremely soluble in water (80.12 g of $Fe(ClO_4)_2$ in 100 g of water at 0°, 106.6 g in 100 g of water at 60°), and also in alcohol. The solubility is greatly repressed by the addition of $HClO_4$.

(*iv*) *Iron(II) sulfate* usually crystallizes from solution as the heptahydrate 'ferrous sulfate', $FeSO_4 \cdot 7H_2O$, which is technically the most important salt of iron. It was already known as 'green vitriol' to Albertus Magnus in the 13th century. It forms light green monoclinic crystals, of density 1.88. These slowly effloresce in air, and undergo superficial oxidation to yellow-brown basic iron(III) sulfate. Ferrous sulfate readily loses $6H_2O$ when it is heated, but the last molecule of water of crystallization is lost with much more difficulty. The anhydrous salt is a white powder of density 3.0. When strongly heated, it decomposes with the loss of sulfur dioxide:

$$2Fe^{II}SO_4 = (Fe^{III}O)_2SO_4 + SO_2.$$

Ferrous sulfate is prepared technically by dissolving iron in dilute sulfuric acid, or by allowing pyrrhotite to weather in air, moistening it frequently. It is obtained as a by-product in the manufacture of chrome alum. It finds uses in the manufacture of ink and of prussian blue, in dyeing (for black dyeing of wool), for the preservation of wood, as a fungicide, and occasionally in medicine. It may be purified by precipitation from aqueous solution by means of alcohol.

100 g of water dissolve at 20° 26.6; at 56° 54.4; at 64° 54.9, and at 90° 37.3 g of $FeSO_4$. Above 56°, $FeSO_4$ crystallizes from solution with $4H_2O$, and above 64° with $1H_2O$. Iron vitriol may be formed as rhombic or triclinic crystals (as well as the monoclinic modification), namely when crystallization is initiated from supersaturated solutions by seeding with rhombic zinc sulfate or triclinic copper sulfate. In the latter case it crystallizes as the pentahydrate.

Iron(II) sulfate is found native in small amounts, as an efflorescence on pyrrhotite.

Iron(II) sulfate forms double salts with alkali sulfates, principally of the composition $M^I_2SO_4 \cdot FeSO_4 \cdot 6H_2O$. The best known of these is the ammonium double salt, $(NH_4)_2SO_4 \cdot FeSO_4 \cdot 6H_2O$, which is stable in air, and which was introduced into volumetric analysis by Mohr.

It is not impossible that the double salts should be regarded as disulfatoferrates, $M^I_2[Fe(SO_4)_2] \cdot 6H_2O$. The corresponding parent acid, $H_2[Fe(SO_4)_2]$ (disulfatoiron(II) acid), is known, both in the anhydrous state and with 6, 5, and $3H_2O$. Double sulfates with the sulfates of bivalent metals are also known (especially with Be, Mg, Zn, and Cd). With chromium(II) and manganese(II) sulfates, iron(II) sulfate forms mixed crystals, but not double salts.

(*v*) *Iron(II) carbonate*, $FeCO_3$, is found in Nature as *siderite* or *spathic iron ore*. This, like calcite, crystallizes in rhombohedra, is insoluble in water, but dissolves in water containing carbonic acid, with the formation of iron(II) hydrogen carbonate, $Fe(HCO_3)_2$. Such waters rapidly deposit iron(III) oxide hydrate when exposed to air, since the excess carbon dioxide escapes, the carbonate deposited is hydrolyzed, and is oxidized by atmospheric oxygen. The precipitate of iron(II) carbonate, obtained by adding soluble carbonates to iron(II) salt solutions, which is initially white, rapidly darkens and ultimately goes brown through the same decomposition.

When heated, $FeCO_3$ is transformed (via several intermediate stages) into FeO and CO_2. The dissociation pressure reaches 1 atm. at 282°. The heat of dissociation is 20.9 kcal per mol.

(vi) *Iron(II) oxalate and Oxalatoferrate(II) Salts.* Iron dissolves in aqueous oxalic acid, with the evolution of hydrogen, and the oxalate $2FeC_2O_4 \cdot 3H_2O$ can be isolated in lemon yellow crystals from the resulting solution. From solutions which also contain alkali oxalates, complex salts (oxalatoferrate(II) salts) can be precipitated by the addition of alcohol. These have the general composition $M^I_2[Fe(C_2O_4)_2]$ with an amount of hydration water depending upon the nature of the cation.

A mixture of iron(II) sulfate and potassium oxalate can be used as a photographic developer.

(vii) *Iron(II) silicates* may be obtained by fusing iron(II) oxide with silica sand. Iron(II) silicates are also very widely distributed in Nature, although almost invariably in the form of isomorphous mixtures with the silicates of other metals (especially magnesium). *Olivine* $(Mg,Fe)_2[SiO_4]$, *hypersthene* $(Fe,Mg)_2[Si_2O_6]$ and *bronzite*, $(Mg,Fe)_2[Si_2O_6]$ are of this type. An example of a double silicate of bivalent iron, of stoichiometric composition, is *hedenbergite*, $CaFe[Si_2O_6]$.

(viii) *Iron(II) phosphate.* Hydrated iron(II) orthophosphate, $Fe_3(PO_4)_2 \cdot 8H_2O$ occurs in Nature as *vivianite*, a mineral which is white on a freshly broken surface, but which at once turns blue on exposure to air.

When iron(II) solutions are treated with sodium phosphate, white insoluble precipitates of iron(II) phosphates are obtained, the composition depending upon the experimental conditions.

(ix) *Cyanoferrate(II) Salts (Ferrocyanides).* Pure iron(II) cyanide is unknown. The interaction of Fe^{++} and CN^- ions invariably gives rise to complex ions, and in presence of a sufficiency of CN^- ions the product is always the *hexacyanoferrate(II)* complex, $[Fe(CN)_6]^{4-}$, which is characterized by particular stability. Many examples are known of the salts derived from this complex, the hexacyanoferrate (II) salts or *ferrocyanides*, $M^I_4[Fe(CN)_6]$ (usually hydrated). The most important of these is potassium hexacyanoferrate(II) ('yellow prussate of potash', see below). Except for barium hexacyanoferrate(II), all the alkali and alkaline earth salts are soluble in water. The $[Fe(CN)_6]^{4-}$ ions present in the solution are strongly complexed, and give reactions neither for Fe^{++} ions nor for CN^- ions. With the ions of bivalent heavy metals, $[Fe(CN)_6]^{4-}$ ions usually give insoluble compounds, many of which provide an analytical detection for the corresponding heavy metal, by reason of their characteristic color or crystal form.

According to Brintzinger (1934), the hexacyanoferrate(II) ion is highly hydrated in solution. It has been deduced from its rate of diffusion that it is associated with $12H_2O$, unlike the hexacyanoferrate(III) ion, which has a rate of diffusion corresponding to the ionic weight of the unhydrated ion $[Fe(CN)_6]^{3-}$.

(x) *Hydroferrocyanic acid* (hexacyanoiron(II) acid), $H_4[Fe(CN)_6]$, can be precipitated in the form of an ether addition compound, by adding ether to a solution of a hexacyano-iron(II) salt, treated with concentrated sulfuric acid. The pure acid can be obtained from this by evaporating off the ether. It is a white powder, which is stable when dry, but gradually turns blue in moist air; it is soluble in water and still more soluble in alcohol.

Hexacyanoiron(II) acid is a fairly strong tetrabasic acid. It has the property of adding on organic oxygen compounds (ethers, alcohols, aldehydes, ketones, esters, etc.), forming oxonium salts, and with some restrictions can be considered as a reagent for 'coordinatively unsaturated' oxygen.

Hexacyanoiron(II) acid forms peculiar addition compounds with sulfuric acid—e.g., $H_4[Fe(CN)_6] \cdot 7H_2SO_4$, rhombic plates; $H_4[Fe(CN)_6] \cdot 5H_2SO_4$, needles. Double salts of peculiar composition, derived from the cyanoferrate(II) salts, are also known—e.g., $(NH_4)_4[Fe(CN)_6] \cdot 2NH_4Cl \cdot 3H_2O$, $K(NH_4)_3[Fe(CN)_6] \cdot 2NH_4Cl$, $K_2Na_2[Fe(CN)_6] \cdot 4KNO_3$. Esters of hexacyanoiron(II) acid are also known.

(xi) *Potassium hexacyanoferrate(II)* has been known since the middle of the 18th century. It was formerly prepared by adding nitrogenous organic substances (blood, leather, horn, etc), mixed with iron scrap, to molten potassium carbonate, and leaching the melt with water. The popular German name 'gelbes Blutlaugensalz' (yellow prussate of potash) is derived from this method of manufacture.

The compound is now generally prepared from the 'spent oxide' from the purifiers in gas works, which has absorbed the hydrogen cyanide formed in the destructive distillation of coal, and contains it chiefly in the form of prussian blue (see below). Most of this is first converted into the very soluble calcium cyanoferrate(II), by heating it with slaked lime. Leaching with water and treatment with potassium chloride furnishes a precipitate of the mixed salt $K_2Ca[Fe(CN)_6]$, which can be converted into pure potassium cyanoferrate(II) by metathesis with the calculated quantity of potassium carbonate:

$$K_2Ca[Fe(CN)_6] + K_2CO_3 = CaCO_3 + K_4[Fe(CN)_6].$$

Potassium cyanoferrate(II) crystallizes from solution as the hydrate $K_4[Fe(CN)_6] \cdot 3H_2O$, in rather soft monoclinic crystals of density 1.85. The water is lost at 100°, and there is formed a white hygroscopic powder, which decomposes with the evolution of nitrogen when it is more strongly heated.

At 0° 14.5 g, and at 98° 74 g of anhydrous potassium cyanoferrate(II) dissolve in 100 g of water. The salt is *not poisonous*, because the hexacyanoferrate(II) radical is so strongly complexed. It is a suitable antidote against poisoning by corrosive salts of copper and iron. It is used in analysis as a reagent for iron, copper, and other heavy metals.

(f) Iron(III) Salts

Iron(III) salts are obtainable by the oxidation of the corresponding iron(II) compounds (e.g., with nitric acid or hydrogen peroxide), or by the dissolution of freshly precipitated iron(III) oxide hydrate in acids. They are similar in solubilities to the iron(II) salts. Their solutions are yellow-brown to dark brown in the absence of any excess acid. This coloration is not that characteristic of the hydrated iron(III) ion, however, but is due to basic salts or iron oxide hydrate present in the solution in the colloidal state, as a result of hydrolysis. If the hydrolysis is repressed by the addition of acid, formation of acido-complexes generally occurs at the same time. In this case, the resultant color of solutions containing an excess of acid depends upon the nature of the acid added; e.g., iron(III) fluoride solutions become pink on treatment with hydrochloric acid, whereas iron(III) chloride solutions become yellow. In general, however, solutions of iron(III) salts with an excess of acid are colorless. Iron(III) may be quantitatively converted into iron(II) by treatment with reducing agents (e.g., tin(II) chloride, or zinc and hydrochloric acid) in acid solution. Quantitative reaction also takes place with potassium iodide in acid solution, according to:

$$Fe^{+++} + I^- = Fe^{++} + \tfrac{1}{2}I_2. \tag{1}$$

These reactions make it possible to determine iron volumetrically, when it is present in the trivalent state. In the first case, the iron(II) salts obtained by reduction are titrated with potassium permanganate, and in the latter case, iodine liberated by reaction (1) is titrated with thiosulfate.

(g) Iron(III) Halides

(i) *Iron(III) fluoride and Fluoroferrate(III) Salts.* Iron(III) fluoride separates in pale pink crystals from solutions of iron oxide hydrate in hydrofluoric acid—as the $4\frac{1}{2}$ hydrate at ordinary temperature (evaporation over calcium chloride), and as the 3-hydrate at rather higher temperatures (evaporation on a water bath). According to Nielsen (1940), the trihydrate exists in two modifications. The anhydrous compound, which is greenish, forms hexagonal crystals of density 3.52. Complex salts (fluoroferrate(III) salts) crystallize from solutions containing alkali or ammonium fluorides. These compounds are mostly of the type $M^I_2[FeF_5]$ (commonly crystallizing with H_2O or HF), but salts of the types $M^I_3[FeF_6]$ and $M^I[FeF_4]$ also are known (Remy 1933).

(ii) *Iron(III) chloride and Chloroferrate(III) Salts.* Anhydrous iron(III) chloride (ferric chloride) is most conveniently obtained by heating metallic iron in a current of dry chlorine. The chloride formed sublimes away, and condenses in brown black flakes (density 2.90, m.p. about 300°, b.p. 317°). It is converted into iron(III) oxide when heated in air, or on treatment of the heated compound with steam:

$$2FeCl_3 + \tfrac{3}{2}O_2 \;\; = Fe_2O_3 + 3Cl_2$$

$$2FeCl_3 + 3H_2O = Fe_2O_3 + 6HCl.$$

Iron(III) chloride crystallizes in hexagonal plates, which have a greenish metallic luster by reflected light, but are garnet red by transmitted light. When it is heated in a vacuum above 500°, iron(III) chloride undergoes partial decomposition, according to:

$$2FeCl_3 \;\rightleftharpoons\; 2FeCl_2 + Cl_2 \qquad\qquad (1)$$

As is shown by vapor density determinations (in which the dissociation reaction (1) is repressed by the addition of chlorine), iron(III) chloride is present almost entirely in the form of dimeric molecules, Fe_2Cl_6, in the vapor state at relatively low temperatures (around 400°). Above 750°, however, it is split up almost completely into single molecules $FeCl_3$. It is also monomolecular in alcoholic and ethereal solutions, as shown by ebullioscopic measurements.

Iron(III) chloride absorbs water avidly in moist air, and deliquesces to a dark brown liquid (*oleum martis*). It is extremely soluble in water and in alcohol, with an appreciable evolution of heat. It forms a number of hydrates from aqueous solution. The ordinary yellow ferric chloride of commerce is the hexahydrate, $FeCl_3 \cdot 6H_2O$. This hydrate often separates out in masses of characteristic shape (hemispherical lumps). It is usually prepared technically by dissolving iron in hydrochloric acid, passing in chlorine to oxidize the iron(II) chloride, and evaporating the solution on a steam bath in earthenware pans.

Solutions of iron(III) chloride have a strongly acid reaction, as a result of hydrolysis. They rapidly coagulate albumen, and this reaction is the basis for the use of such solutions, or of cotton wool impregnated with iron(III) chloride, as styptics. Iron(III) chloride finds its chief technical uses in the organic dyestuffs industry as an oxidant, and occasionally as a condensing agent and chlorine carrier. It is also used as a mordant in piece dyeing.

Complex salts (*chloroferrate(III) salts*) crystallize from solutions containing the chlorides of the alkalis or of other strongly electropositive metals. The most common of these are of the types $M^I[FeCl_4]$ and $M^I_2[FeCl_5]$ (Remy, 1925).

X-ray examination has shown that α-FeOOH is always obtained at ordinary temperature as the stable end product, when the hydrolysis of iron(III) chloride solutions is carried so far that a solid product is deposited. The substances at first precipitated are basic chlorides, which are converted by way of γ-FeOOH into α-FeOOH. This transformation usually takes place very slowly, so that precipitates formed by slow hydrolysis in the cold often still contain basic iron(III) chloride, even after standing for months in contact with the solution. Even when the basic chlorides remain in colloidal dispersion, their complete conversion into FeOOH may take months. Starting from solutions containing an excess of hydrochloric acid, or of soluble chlorides, in which chloroferrate(III) ions (e.g., $[FeCl_4]^-$) are therefore present, the gradual reduction of the hydrogen ion concentration leads at first to the partial replacement of Cl by OH, according to the equations:

$$[FeCl_4]^- + OH^- \rightleftharpoons [FeCl_3(OH)]^- + Cl^-$$
$$[FeCl_3OH]^- + OH^- \rightleftharpoons [FeCl_2(OH)_2]^- + Cl^-.$$

The occurrence of these replacement reactions was demonstrated, by G. Jander and Jahr, by measurements of diffusion velocities and absorption spectra. At the same time, deposition of basic chlorides or hydroxide takes place. The iron remaining in solution is, however, present for the most part in monomolecular form, as chlorohydroxo ions, which are directly in equilibrium with the precipitate. The precipitation does not take place, therefore, through the formation of higher molecular hydrolysis products in solution, as it does with iron(III) perchlorate or nitrate solutions (cf. below and p. 284).

If hydrolysis is carried out in warm solutions, α-Fe$_2$O$_3$ is formed. If the hydrogen ion concentration is suddenly and drastically reduced, as by the addition of caustic alkali or ammonia in excess, amorphous iron(III) oxide hydrate is thrown out of solution.

(*iii*) *Iron oxychloride* (basic iron chloride), FeOCl, is best obtained in the pure state by heating a mixture of Fe$_2$O$_3$ and FeCl$_3$ in a sealed tube to 350°. It forms rhombic crystal leaflets (density 3.55) with a good cleavage, which have a metallic luster and are red by transmitted light. Iron oxychloride has a well defined layer lattice, similar to that of γ-FeOOH.

(*iv*) *Iron(III) bromide* (iron tribromide, ferric bromide), FeBr$_3$, brown-red plates, is very similar to the chloride.

(*v*) *Iron(III) iodide*, however, exists only in equilibrium with a large excess of iron(II) iodide:

$$FeI_3 \rightleftharpoons FeI_2 + \tfrac{1}{2}I_2.$$

(h) Other Iron(III) Salts

(*i*) *Iron(III) perchlorate*, Fe(ClO$_4$)$_3$, can be obtained by fuming FeCl$_3$ down with perchloric acid. It crystallizes from solution at ordinary temperature as the decahydrate, Fe(ClO$_4$)$_3 \cdot$ 10H$_2$O, in the form of pale pink doubly refracting crystals, which do not lose water when dried over calcium chloride and are extremely hygroscopic. 4H$_2$O are lost by drying over concentrated sulfuric acid or P$_2$O$_5$ (Lindstrand, 1936). The enneahydrate crystallizes from solution above 42°. The solubility of iron(III) perchlorate (96.99 g in 100 g of water at 0°, 142.0 g in 100 g at 60°) is even greater than that of iron(II) perchlorate. The salt is also very soluble in alcohol.

As has been shown by G. Jander (1930 and later), the hydrolysis of iron(III) perchlorate follows a different course from that of iron(III) chloride, in that the exchange of acid radicals for OH groups is associated with a simultaneous condensation process, which leads to products of continuously increasing molecular complexity—e.g., as in the scheme:

$$2Fe(ClO_4)_3 + H_2O \rightleftharpoons (ClO_4)_2Fe-O-Fe(ClO_4)_2 + 2HClO_4$$

$$2(ClO_4)_2Fe-O-Fe(ClO_4)_2 + H_2O \rightleftharpoons (ClO_4)_2Fe-O-Fe-O-Fe-O-Fe(ClO_4)_2 + 2HClO_4, \text{ etc.}$$
$$\qquad\qquad\qquad\qquad\qquad\qquad\qquad\qquad \underset{ClO_4}{|} \quad \underset{ClO_4}{|}$$

or, since ferric perchlorate may be assumed to be ionized,

$$2Fe^{3+} + H_2O \ \rightleftharpoons \ Fe^{2+}\!\!-\!\!O\!\!-\!\!Fe^{2+} + 2H^+$$

$$2Fe^{2+}\!\!-\!\!O\!\!-\!\!Fe^{2+} + H_2O \ \rightleftharpoons \ Fe^{2+}\!\!-\!\!O\!\!-\!\!Fe^+\!\!-\!\!O\!\!-\!\!Fe^+\!\!-\!\!Fe^{2+} + 2H^+, \qquad etc.$$

Hydrolysis products of this kind exist in equilibrium with the normal salt (or its ions) even in relatively strongly acidic solutions. With decreasing H^+ ion concentration (e.g., on the gradual addition of sodium hydroxide), the equilibria undergo a *continuous* displacement, in the direction of a further increase in the particle size (as contrasted with the *sudden* condensation which is usual for the acid anions). It appears, indeed, that the particles at first grow preferentially in one direction, and only when they have built up long linear poly-cations do these aggregate into rods or platelets. Even when the particles have in this way attained a complexity corresponding to molecular weights of 40,000 or more, they usually remain in colloidal dispersion, unless they undergo conversion into pure iron hydroxide, FeOOH, by elimination of the last acidic groups. Ultimately, however, unless the solutions are strongly acid, iron hydroxide is deposited as the stable end-product of hydrolysis, in the form of α-FeO(OH), goethite. An equilibrium state is often reached, in which fairly high-molecular basic salts are present in solution, in contact with α-FeO(OH) as solid phase. As befits the complicated nature of the reactions, the hydrolysis equilibria are usually attained in solution only very slowly, often in the course of weeks or months.

Jander and Jahr have found that the hydrolysis of iron(III) nitrate follows a similar course (except that, in this case, nitrato ions are also present in equilibrium with the hydrolysis products in solutions containing excess of nitric acid). The hydrolysis of iron(III) perchlorate appears to be typical of the course taken by the hydrolysis of the iron(III) salts of strong acids which do not form complexes with Fe^{+++} ions, or have only a weak tendency to do so. The hydrolysis of chromium and aluminum salts in aqueous solutions takes place in an essentially similar manner. Here also, the average particle size of the hydrolysis products does not increase suddenly, but increases gradually as the H^+ ion concentration is lowered. The solutions never contain hydrolysis products of a uniform molecular complexity, but always of a range of ionic weights, which shifts continuously towards greater particle sizes as hydrolysis proceeds.

(*ii*) *Iron(III) nitrate* is obtained by dissolving iron in 20–30 per cent nitric acid. Depending upon the acid content and concentration of the solution, it crystallizes at ordinary temperature either as almost colorless cubes of $Fe(NO_3)_3 \cdot 6H_2O$, or in monoclinic crystals with the composition $Fe(NO_3)_3 \cdot 9H_2O$*. It dissolves in water with a brown color, which is attributable to hydrolysis (see above), and which disappears upon the addition of nitric acid. Iron(III) nitrate is used as a mordant in dyeing, and is also employed in medicine.

(*iii*) *Iron(III) sulfate* is used as a mordant, and for the manufacture of iron alum and of prussian blue. It also finds medicinal applications. As a rule, it is prepared only in aqueous solution, either by the oxidation of iron(II) sulfate with nitric acid, or by dissolving iron oxide in concentrated sulfuric acid. In the latter case, the anhydrous salt which is first formed is brought into solution by boiling with water.

Iron(III) sulfate forms several hydrates, which are not easily isolated in the pure crystalline state. *Coquimbite*, $Fe_2(SO_4)_3 \cdot 9H_2O$ (hexagonal) and *quenstedtite* $Fe_2(SO_4)_3 \cdot 10H_2O$ (monoclinic) are found native. If the hydrates are cautiously heated anhydrous iron(III) sulfate remains as a white powder, which dissolves only slowly in water. Dissolution takes place rapidly, however, if a little iron(II) sulfate is present. If iron(III) sulfate is more strongly heated, it breaks up into iron(III) oxide and sulfur trioxide:

$$Fe_2(SO_4)_3 = Fe_2O_3 + 3SO_3.$$

Hydrolysis takes place to a considerable degree in solution, and a whole series of *basic*

* Other hydrates may exist as well as these.

sulfates can be isolated. Various basic sulfates have also been found in Nature—e.g., amaranthite (hohmannite) $Fe_2O_3 \cdot 2SO_3 \cdot 7H_2O$, glockerite (vitriol ocher), $2Fe_2O_3 \cdot SO_3 \cdot 6H_2O$, raimondite, $2Fe_2O_3 \cdot 3SO_3 \cdot 7H_2O$, and others.

(*iv*) *Double Salts of Iron(III) Sulfate. Iron Alums.* The most important of the double salts of iron(III) sulfate are those belonging to the class of alums, with the general formula $M^IFe(SO_4)_2 \cdot 12H_2O$. *Ammonium iron alum*, $NH_4Fe(SO_4)_2 \cdot 12H_2O$, and *potash iron alum*, $KFe(SO_4)_2 \cdot 12H_2O$, are of technical importance, chiefly as mordants in dyeing. They are prepared by oxidizing iron vitriol in solution by means of nitric acid, and crystallizing after the addition of ammonium or potassium sulfate. When quite pure, the iron alums are colorless, but they often have a pale violet color, due to traces of manganese(III) sulfate.

The mineral *voltaite*, which is occasionally found native, is a double sulfate of apparently complex constitution. Its composition approximately fits the formula

$$K_2Fe^{II}_5(Fe^{III}, Al)_4(SO_4)_{12} \cdot 32\text{-}36H_2O.$$

A number of double sulfates of analogous composition, but containing NH_4, Rb, or Tl^I in place of K, and Mn^{II}, Co^{II}, Mg, Zn, or Cd in place of Fe^{II}, have been prepared by Gossner (1930).

(*v*) *Iron(III) oxalate and Oxalatoferrate(III) Salts* ('Ferrioxalates'). Addition of ammonium oxalate to iron(III) salt solutions yields brown precipitates of variable composition. The complex salt $(NH_4)_3[Fe(C_2O_4)_3] \cdot 3H_2O$ (ammonium trioxalatoferrate(III)) can be isolated as green crystals which are stable in air, from solutions which contain an excess of ammonium oxalate. Other salts of the same type are also known.

When a solution of potassium trioxalatoferrate(III) is exposed to light, the trivalent iron oxidizes the oxalic acid to carbon dioxide (being itself reduced to the bivalent state), to an extent which is proportional to the amount of light absorbed. The reaction can be utilized for the measurement of quantities of light (actinometry).

(*vi*) *Iron(III) silicate* generally occurs in Nature in the form of double silicates—e.g., acmite (aegirine), $NaFe^{III}[Si_2O_6]$, andradite (lime iron garnet), $Ca_3Fe^{III}_2[SiO_4]_3$,—and very commonly in isomorphous admixture with double sulfates of aluminum, with corresponding compositions; aluminum usually preponderates in these.

(*vii*) *Iron(III) phosphate*, $FePO_4$, is precipitated as a yellowish white substance when disodium phosphate is added to a solution of iron(III) chloride. If a yellow solution of iron(III) chloride is treated with phosphoric acid, it is practically decolorized, as a consequence of the formation of colorless phosphatoferrate(III) complex ions—e.g., $[Fe(PO_4)_3]^{6-}$. It was proved by Jensen that pure phosphato complexes are formed, and not chlorophosphato complexes, from the fact that the solubility of $FePO_4$ in acid solutions containing chloride and phosphate increases with the concentration of phosphate ion, but is independent of the chloride ion concentration.

(*viii*) *Acetatotriiron(III) Salts.* The action of concentrated acetic acid upon freshly precipitated iron(III) oxide hydrate yields a dark red solution. On evaporation, this leaves a glassy red mass, which was formerly regarded as iron(III) acetate, $Fe(CH_3CO_2)_3$. Actually, as was shown by Weinland, products obtained by this and similar methods (although usually inhomogeneous) consist, like the chromium acetates, of compounds containing an electropositive *hexaacetatotriiron(III)* complex. This complex usually exists in solution as a univalent cation, $[Fe_3(CH_3CO_2)_6(OH)_2]^{+*}$. Weinland was able to isolate, in the crystalline state, a number of salts derived from this cation, of the type $[Fe_3(CH_3CO_2)_6(OH)_2]X$. Among the salts isolated was the acetate, $[Fe_3(CH_3CO_2)_6(OH)_2]CH_3CO_2 \cdot H_2O$ (hexaacetatodihydroxotriiron(III) acetate). According to Weinland, the blood red color produced by adding sodium acetate to solutions of iron(III) salts is also due to the formation

---

* The formation of this cation has **also** be confirmed by electrometric titration (Treadwell, 1930).

of compounds containing an acetatotriiron(III) complex. When solutions containing sodium acetate are boiled, the whole of the iron is thrown out of solution as a red brown precipitate, which contains iron and acetate ion in the approximate ratio 3 : 1, and which probably consists essentially of the insoluble compound

$$\left[Fe_3 \begin{array}{l} CH_3CO_2 \\ O_3 \\ (OH)_2 \end{array} \right]$$

which is derived from the hexaacetatodihydroxoiron(III) complex by exchange of $5CH_3CO_2^-$ with $3O^=$ (or for $6OH^-$), whereby the complex loses its cationic charge. Since chromium forms similar compounds, which do not, however, undergo decomposition and precipitation when they are boiled, and since iron and chromium can mutually replace one another to some extent in these substances*, some iron is left in solution on boiling with sodium acetate if much chromium is present. Conversely, if much iron is present it may carry down a portion of the chromium with it.

(ix) *Iron(III) thiocyanate and Thiocyanatoferrates.* When thiocyanate ions are added to an iron(III) salt solution, a blood red coloration is produced. The colored compound can be extracted with ether. The principal constituent of the solutions is iron(III) thiocyanate, $Fe(SCN)_3$. If an excess of SCN^- ions is present, hexathiocyanatoferrate(III) ions, $[Fe(SCN)_6]^{3-}$, are formed as well**. If the concentration of thiocyanate ions is high enough, the red coloration appears even at extremely low Fe^{+++} ion concentrations. The reaction therefore lends itself to the detection and determination of iron(III) salts.

The *hexathiocyanatoferrates(III)* (e.g., potassium hexathiocyanatoferrate(III), $K_3[Fe(SCN)_6] \cdot 2H_2O$, deliquescent, dark red hexagonal prisms) are only weakly complexed. They undergo extensive dissociation when they are dissolved in water, unless an excess of SCN^- ions is present. The existence of $[Fe(SCN)_6]^{3-}$ ions in the solution has been proved, however, by transport experiments in alcohol solution. Iron(III) thiocyanate, like other heavy metal thiocyanates, is decomposed by alkali hydroxides.

(x) *Cyanoferrate(III) Salts (Ferricyanides).* With an excess of CN^- ions the Fe^{+++} ion forms the very strongly complexed ion $[Fe(CN)_6]^{3-}$. Salts derived from this, $M^I{}_3[Fe(CN)_6]$, are the hexacyanoferrate(III) salts or ferricyanides, the most important being the potassium salt, $K_3[Fe(CN)_6]$, (German 'rot Blutlaugensalz'— see above), which was discovered by Gmelin. It is prepared by oxidation of potassium ferrocyanide (hexacyanoferrate(II)) in hydrochloric acid solution, by means of chlorine, permanganate, or other strong oxidizing agents:

$$[Fe(CN)_6]^{4-} + \tfrac{1}{2}Cl_2 = Cl^- + [Fe(CN)_6]^{3-}.$$

It exists as dark red monoclinic crystals, which form a yellow powder when they are crushed, and dissolve in water to give a yellow solution. It is less stable than potassium ferrocyanide, and is therefore *poisonous*, unlike the latter. It has strong oxidizing properties, especially in alkaline solution. Thus it reacts with hydrogen peroxide, liberating oxygen and re-forming potassium ferrocyanide, which is stable in alkaline solution. The redox potential of the couple $[Fe(CN)_6]^{4-}$ / $[Fe(CN)_6]^{3-}$, relative to the normal hydrogen electrode, is —0.44 volts.

* Weinland was able to prepare crystalline compounds of the general formulas $[Fe_2CrAc_6(OH)_2]X$ and $[FeCr_2Ac_6(OH)_2]X$ (Ac = CH_3CO_2—).
** The $[FeSCN]^{++}$ ion has also been identified in solutions containing Fe^{+++} and SCN^- ions.

The free acid $H_3[Fe(CN)_6]$ can be prepared by the action of fuming hydrochloric acid on a pure concentrated solution of potassium ferricyanide, but is much less readily obtained pure than is the hexacyanoiron(II) acid (hydroferrocyanic acid), $H_4[Fe(CN)_6]$.

(*xi*) *Prussian Blue and Turnbull's Blue*. If the solution of an iron(III) salt is treated with a solution of potassium ferrocyanide, a dark blue precipitate of *prussian blue* or *berlin blue* is obtained. A dark blue precipitate—*Turnbull's blue*—is also obtained if the solution of an iron(II) salt is treated with potassium ferricyanide. Both reactions are of analytical importance, for the detection of iron and for identifying the state of oxidation in which it is present. Prussian blue is also used as a pigment, and is manufactured for this purpose.

The formation of prussian blue and of Turnbull's blue can be represented by the following equations:

Prussian blue reaction, $4Fe^{+++} + 3[Fe^{II}(CN)_6]^{4-} = Fe^{III}_4[Fe^{II}(CN)_6]_3$

Turnbull's blue reaction, $Fe^{++} + [Fe^{III}(CN)_6]^{3-} = Fe^{+++} + [Fe^{II}(CN)_6]^{4-}$
$4Fe^{+++} + 3[Fe^{II}(CN)_6]^{4-} = Fe^{III}_4[Fe^{II}(CN)_6]_3$

Proof that a transfer of charge takes place between the Fe(II) present in the form of free ions and the Fe(III) bound in the complex, is afforded by the red coloration which the Fe^{+++} ions give with SCN^- ions in the instant of their formation (Simon, 1936). The precipitates known as 'prussian blue' and 'Turnbull's blue' are thus substantially identical. They nevertheless vary more or less considerably from the composition required by the formula $Fe_4[Fe(CN)_6]_3$. In particular, the precipitates always contain potassium. If an iron(III) salt solution is added cautiously, drop by drop, to excess of potassium cyanoferrate(II), it is possible to obtain a blue salt, of the composition $KFe[Fe(CN)_6]$ ('*soluble prussian blue*'), which forms colloidal dispersions in pure water, but is coagulated by salts. Ordinary prussian blue is insoluble in water and dilute acids. It dissolves only in oxalic acid, to give a deep blue color. This is utilized in the manufacture of inks. Although prussian blue is very stable towards dilute acids, it is very sensitive towards even very dilute alkalis. It is decomposed, with the deposition of iron(III) oxide hydrate, and the formation of alkali hexacyanoferrate(II).

(*xii*) *Prusside Compounds*. A group of complex iron cyanides containing only *five* cyanide groups is known as the *prusside* compounds. The best known of these is *sodium nitroprusside*, which can be obtained by the action of nitric acid on potassium ferrocyanide. Potassium nitrate first crystallizes from solution, and then, after neutralization of the excess nitric acid with sodium carbonate, sodium nitroprusside, $Na_2[Fe(CN)_5(NO)] \cdot 2H_2O$, is deposited in ruby red rhombic crystals, which are stable in air. It is easily soluble in water and alcohol, but the solutions are not very stable. A few other nitroprussides, of the general formula $M^I_2[Fe(CN)_5(NO)]$, are also known.

Sodium nitroprusside is used as a reagent for $S^=$ or SH^- ions, with which it gives an intense violet coloration. Free hydrogen sulfide does not give the reaction. It also reacts with $SO_3^=$ ions to give a red coloration, especially in the presence of zinc sulfate or zinc nitrate. Sulfites can thereby be distinguished from thiosulfates, which do not give this reaction.

The iron in the nitroprusside complex is probably to be regarded as tripositive, the NO group acting as an electroneutral group. As long as there is any uncertainty about the valence state of the iron it is best not to designate the complex as a pentacyanoferrate(III), but to state explicitly the number of atoms bound outside the complex—e.g., the sodium salt is *di*sodium pentacyanonitrosoferrate.

If sodium nitroprusside is treated with concentrated ammonia solution, the NO group is replaced by ammonia. The iron at the same time undergoes reduction to the dipositive state, while the nitric oxide is converted to nitrous acid:

$$[(CN)_5Fe(NO)]^{2-} + NH_3 + OH^- = [(CN)_5Fe(NH_3)]^{3-} + O{=}N{-}OH.$$

The sodium 'amminoferrocyanide' (sodium pentacyanoammineferrate(II)) $Na_3[Fe^{II}(CN)_5$ $(NH_3)]$, thus obtained can be oxidized by nitrous acid to sodium amminoferricyanide, (sodium pentacyanoammineferrate(III)), $Na_2[Fe^{III}(CN)_5(NH_3)] \cdot H_2O$ (dark yellow powder). Other examples of prusside compounds are:

$$M^I{}_2[Fe^{III}(CN)_5(H_2O)] \qquad\qquad M^I{}_3[Fe^{II}(CN)_5(H_2O)]$$
$$M^I{}_2[Fe^{III}(CN)_5(NO_2)] \qquad\qquad M^I{}_4[Fe^{II}(CN)_5(NO_2)]$$
$$M^I{}_4[Fe^{II}(CN)_5(AsO_2)]$$
$$M^I{}_5[Fe^{II}(CN)_5(SO_3)]$$
$$M^I{}_3[Fe^{II}(CN)_5(CO)]$$

Most of these compounds were prepared by Hofmann. The potassium pentacyanocarbonyl-ferrate(II), $K_3[Fe(CN)_5(CO)] \cdot 3\frac{1}{2}H_2O$, which can be obtained by the action of carbon monoxide on a warm solution of potassium ferrocyanide, is generally present as an impurity in crude potassium ferrocyanide. The corresponding free acid is also known.

The compound $Na_4[Fe(CN)_5(NO)]$, prepared in 1925 by Ungarelli, may also be mentioned in this connection. It was obtained by the action of sodium hyponitrite on a solution of sodium pentacyanoaquoferrate(II), $Na_3[Fe(CN)_5(H_2O)]$. Cryoscopic measurements show that it possesses the monomeric formula given, so that it must contain the radical NO, and not such a radical as N_2O_2. The question whether this is a neutral ligand or a negative group remains open, and the (electrochemical) valence of the iron in this compound cannot be specified.

As with the cyanoferrates, Brintzinger found that complexes containing Fe^{III} had diffusion velocities corresponding to the unhydrated ions, whereas those with Fe^{II} as the central atom appeared, from their diffusion rates to be heavily hydrated.

(i) Ferrates(VI)

If iron filings are heated with potassium nitrate, they undergo oxidation with a lively incandescence. The cooled melt gives a bright violet solution in water, and on addition of barium chloride, a carmine red precipitate is produced which has the composition $BaFeO_4 \cdot H_2O$ after drying at 100°. This belongs to a class of iron compounds which correspond in composition to the chromates and sulfates, and contain iron in the hexapositive state. They are known as *ferrates* (strictly, ferrate(VI) compounds). The deep red potassium ferrate, K_2FeO_4, which is very soluble in water, and is therefore less readily isolated than the barium salt, is isomorphous with potassium sulfate. Ferrates(VI) can be prepared not only by the above method, but also by oxidation of a suspension of freshly precipitated iron oxide hydrate, by means of chlorine or bromine, or by anodic dissolution of iron (preferably cast iron) in warm caustic potash or (better) soda:

$$Fe + 8OH^- - 6e = [FeO_4]^= + 4H_2O.$$

Ferrates are even stronger oxidizing agents than the permanganates. Thus they oxidize ammonia to nitrogen in the cold. Neither the acid corresponding to the ferrates, nor the parent oxide has yet been isolated. When ferrate solutions are treated with dilute acids, oxygen is evolved, and iron is transferred from the $+6$ to the $+3$ state:

$$2FeO_4^= + 10H^+ = 2Fe^{+++} + \tfrac{3}{2}O_2 + 5H_2O.$$

On treatment with concentrated potassium hydroxide solution, many ferrates(VI), e.g., $BaFeO_4$, gradually change to *ferrates(IV)* with evolution of oxygen. Scholder (1952) obtained pure strontium ferrate(IV), Sr_2FeO_4, by heating a mixture of $Sr(OH)_2$ and $Sr_3[Fe(OH)_6]_2$ in an oxygen stream to 500–800°. After removal of excess SrO, the compound Sr_2FeO_4 is left as a deep-black crystalline powder. According to X-ray evidence this compound is isotypic with Sr_2MnO_4 and Ba_2MnO_4, whereas pure Ba_2FeO_4, also first prepared by Scholder, is isotypic with Ba_2TiO_4, Ba_2CrO_4 and Ba_2CoO_4.

(j) Iron Carbonyls and Iron Nitrosyl Compounds

(i) *Iron Carbonyls.* Finely divided iron combines directly with carbon monoxide—slowly at ordinary temperature and pressure, more rapidly at high pressures and elevated temperatures—forming a highly refractive, pale yellow liquid, of the composition $Fe(CO)_5$, *iron pentacarbonyl*. The compound has a density of 1.494 (at 0°), freezes at —20° and boils at 102.7°. It dissolves readily in benzene, paraffin hydrocarbons, and ether, but is insoluble in water. It is not attacked by gaseous hydrogen chloride, hydrogen bromide, or hydrogen sulfide, but is attacked by hydrogen iodide. The first mentioned acids attack iron carbonyl dissolved in ether or alcohol, but not in aqueous solution. In ethereal solution, nitric and sulfuric acids also decompose iron carbonyl, the latter reacting smoothly according to the equation:

$$H_2SO_4 + Fe(CO)_5 = FeSO_4 + H_2 + 5CO.$$

Iron carbonyl is prepared on a technical scale by heating finely divided iron with carbon monoxide at 180–220° under pressure (150–250 atm.). The procedure is to condense liquid iron pentacarbonyl out of the carbon monoxide issuing from the autoclave, by cooling, and to return the gas to the reaction vessel. For some time, iron pentacarbonyl was added to gasoline as an 'antiknock', and this requirement originally led to its large scale production. It is now used principally for the production of very pure iron (cf. p. 253), but also for the preparation of iron(III) oxide suitable for use as a pigment and polishing agent, which can be obtained by burning iron pentacarbonyl. The presence of iron carbonyl in illuminating gas is undesirable, since it leads to the formation of brown oxide deposits on incandescent mantles, which reduces their luminosity. It may be removed from coal gas by passage over porous charcoal impregnated with chromic acid or chromate solution.

Iron pentacarbonyl gradually undergoes decomposition in sunlight*, with the deposition of a solid with the composition $Fe_2(CO)_9$, *iron enneacarbonyl* (hexagonal yellow leaflets, structure cf. p. 355). This compound is very sensitive to air. When heated in the absence of air it decomposes at about 100° into Fe and CO, with some re-formation of $Fe(CO)_5$. More stable than iron enneacarbonyl is the so-called iron *tetracarbonyl* (more correctly *trisiron dodecacarbonyl*), $Fe_3(CO)_{12}$, which may be obtained by treating iron pentacarbonyl with sodium alcoholate and mild oxidizing agents (see also pp. 355, 358 for the preparation). This forms deep green monoclinic prismatic crystals (density 2.0), which are very sparingly soluble in the usual solvents. It can be dissolved in iron pentacarbonyl, however, and in this solvent has a molecular weight corresponding to the formula $Fe_3(CO)_{12}$. The crystal structure of the solid compound is also built up from $Fe_3(CO)_{12}$ molecules (cf. p. 355).

(ii) *Iron tetranitrosyl*, $Fe(NO)_4$, was prepared in 1929 by Manchot, by warming iron pentacarbonyl with nitric oxide at 44–45° in an autoclave. It forms black crystal needles, and is very reactive. It decomposes, giving nitrosoiron(II) sulfate, $[Fe(NO)]SO_4$, when it is added to dilute sulfuric acid. With potassium hydrogen sulfide it yields Roussin's black salt, (cf. p. 270), $K[Fe_4S_3(NO)_7]$. It forms $(NO)_2Fe$—S—SO_3K with potassium thiosulfate, $(NO)_2Fe$—SC_2H_5 with ethyl mercaptan, C_2H_5SH, and $(NO)_2Fe$—$C\overset{S}{\underset{OC_2H_5}{}}$ with potassium xanthate, $K\left[S—C\overset{S}{\underset{OC_2H_5}{}}\right]$. These compounds, some of which had previously been isolated by Hofmann and some, in other ways, by Manchot, can be regarded as derivatives of Roussin's red salt (p. 270), and like the latter are considered by Manchot to contain *uni*positive iron.

---

* Photo-decomposition of the CO compounds of *other metals* has not been observed. It is quite characterisitic of the CO compounds of iron, and has been shown to occur, for example, with carbon monoxide-hemoglobin, as was first demonstrated by J. Haldane, 1897.

For further discussion of metal carbonyls and nitrosyls see p. 351 *et seq.*

Hieber found that NO reacted with iron(II) iodide, one halogen atom being replaced by two NO groups. *Iron dinitrosyl iodide*, $Fe(NO)_2I$, is thus obtained, as a deep brown-black compound, which may be sublimed unchanged but is extremely sensitive to air and moisture. The bromide, $Fe(NO)_2Br$, is less readily obtained than the iodide. A chloride, with the composition $Fe(NO)_3Cl$, is still more difficult to prepare; it is more volatile than the bromide or iodide. A corresponding fluoride does not seem to be formed at all.

5. Analytical (Iron)

In the course of analysis, iron comes into the *ammonium sulfide group*, being precipitated as a black precipitate of FeS or $2FeS + S$, according to the valence state present in the solution. If hydrogen sulfide is previously passed through the solution, the iron is invariably present in the *bivalent* state. It can be separated from chromium, aluminum, and zinc by precipitation as the oxide hydrate by means of a mixture of sodium hydroxide and hydrogen peroxide; from manganese, by precipitation (also as oxide hydrate) with ammonia, in the presence of ammonium chloride and hydroxylammonium chloride (to hinder precipitation of manganese). For the latter purpose, iron must be present in the trivalent state, being previously oxidized with nitric acid if necessary.

Iron is usually identified by the red coloration produced by adding thiocyanate to the solution of an iron(III) salt; this can be extracted with ether.

The state of oxidation of iron in a substance being analyzed can be distinguished by the reaction with potassium hexacyanoferrate(II) and hexacyanoferrate(III), respectively. Iron(II) salts give with hexacyanoferrate(II) a precipitate which is white at the moment of its formation, and with hexacyanoferrate(III) a deep blue precipitate (Turnbull's blue). Iron(III) salts give a deep blue precipitate (prussian blue) with hexacyanoferrate(II), but no precipitate with hexacyanoferrate(III). If a deep blue precipitate is obtained with both reagents, iron is present in both oxidation states.

For *gravimetric determination*, iron is generally precipitated with ammonia, as the oxide hydrate, and weighed as the oxide after ignition. Iron is frequently determined *volumetrically*, by titration with permanganate or dichromate:

$$5Fe^{++} + MnO_4^- + 8H^+ = 5Fe^{+++} + Mn^{++} + 4H_2O$$

$$6Fe^{++} + Cr_2O_7^= + 14H^+ = 6Fe^{+++} + 2Cr^{+++} + 7H_2O$$

or iodometrically:

$$Fe^{+++} + I^- = Fe^{++} + \tfrac{1}{2}I_2.$$

The solubility of iron(III) chloride in organic solvents makes it possible to separate iron quantitatively from all other elements (except, under certain conditions, Ga) by extraction with isopropyl ether from a 6N-hydrochloric acid solution containing an iron(III) salt.

6. Cobalt (Co)

(a) Occurrence

Cobalt always occurs in Nature in association with nickel, and usually in combination with arsenic. The most important cobalt minerals are *smaltite*, $CoAs_2$, and *cobalt glance* (cobaltite), CoAsS.

Large quantities of a manganese black containing cobalt (*black cobalt earth, asbolane*) are found in New Caledonia and Canada. Cobalt is found in the native state (in quantities between 0.5 and 2.5%) in meteoric iron.

(b) History

In former times, minerals which could not be smelted for metal, in spite of their metallic appearance, were commonly called 'cobalts' in miners' slang, since they deceived the miners, like the impudent mountain spirits ('Kobolds', also formerly called 'Cobalts'). The term was subsequently restricted to those ores which were hard to smelt, and which gave a blue coloration to glass. The metal from which they were derived, the present day cobalt, was first isolated and recognized as an new element by the Swedish chemist Brandt in 1735.

(c) Preparation

The raw material for the technical extraction of cobalt is furnished chiefly by the 'speisses' obtained in smelting arsenical nickel, copper, and lead ores, in which nickel and cobalt are present as arsenides. Cobalt oxide, which is used for the preparation of cobalt pigments, and which does not have to be especially pure for this purpose, is obtained chiefly from these. Metallic cobalt is also produced technically in increasing quantities.

The preparation of pure metallic cobalt is somewhat troublesome, the complete separation from nickel causing especial difficulties. The usual procedure is to convert the speisses or the arsenical cobalt ore, by roasting, into a mixture of oxides and arsenates, called 'zaffre, after its reddish color. This is then dissolved in hydrochloric acid. Copper, lead, bismuth' etc., are first precipitated with hydrogen sulfide; arsenic and iron (after oxidation with chlorine) are then precipitated with calcium carbonate, as calcium arsenate and iron oxide hydrate, respectively; finally, bleaching powder is added in amount exactly sufficient to throw down the cobalt. Provided that not too large an excess of bleaching powder is used, the nickel remains substantially in solution. When the oxide has been sufficiently purified by repeated precipitation, it may be reduced to the metal by heating with suitable reducing agents.

(d) Properties

Cobalt is a lustrous metal, resembling iron, of density 8.8. Its melting point is a little lower than that of iron. Cobalt is very tough, and excels steel in hardness and tensile strength. Like iron, it is ferromagnetic, and undergoes transformation into a non-magnetic modification above 1000°.

Cobalt ordinarily crystallizes with the magnesium-type structure (hexagonal close packed), $a = 2.510$, $c = 4.06$ Å. It has been found, however, that the finely divided metal, obtained by gently heating the oxide in hydrogen, has the face-centered cubic structure ($a = 3.554$ Å) with which nickel usually crystallizes. This modification is stable above about 480°, and is apparently stabilized at lower temperatures by the absorption of hydrogen. The shortest interatomic distance is the same in both modifications (Co $\leftrightarrow$ Co = 2.514 Å).

Cobalt has a rather smaller capacity for absorbing hydrogen than has iron. Sieverts (1934) found that 100 g of cobalt absorbed 0.08 mg of H_2 at 600°, and 0.49 mg of H_2 at 1200°, under 1 atm. pressure. The quantity absorbed is proportional to the square root of the hydrogen pressure. Nitrogen is practically insoluble in cobalt up to 1200°.

Compact cobalt is not attacked by air or water at ordinary temperature, but is pyrophoric, like iron, when finely divided. It dissolved much less readily than iron in dilute acids (e.g., sulfuric and hydrochloric acids), as accords with its position below iron in the electrochemical series. (Standard potential = +0.28 volt, relative to the normal hydrogen electrode). Cobalt is readily dissolved by dilute nitric acid, but is passivated, like iron, by concentrated nitric acid.

Cobalt is oxidized when it is heated in air, and burns at a white heat to Co_3O_4. It also combines with many other elements when it is heated, often with incandescence—e.g., with S, P, As, Sb, Sn, and Zn. It forms several compounds when melted with silicon (cf. p. 244–5, Table 29). I talso combines directly with boron at high temperatures, but not with nitrogen. It combines readily with the halogens. Cobalt forms mixed crystals in all proportions with iron and nickel, and also with manganese and chromium. In its behavior towards carbon, cobalt resembles iron; however, the carbide Co_3C never separates out during cooling (although, according to Ruff, the existence of the compound is probable in the melt), but graphite is always formed when the carbon content exceeds the saturation limit of the mixed crystals. The action of CH_4 or CO on metallic cobalt at slightly raised temperatures (below 225°) gives the compound Co_2C, which decomposes at higher temperatures (Bahr, 1930). Cobalt can bring about the catalytic decomposition of CO and CH_4 at temperatures at which the carbide is unstable (cf. p. 709).

(e) Uses

The chief use of cobalt was formerly in the form of the potassium double silicate (cobalt glass) as a blue pigment (*smalt*).

Smalt, which was already known to the ancient Egyptians and Romans, is obtained by fusing cobalt oxide (or roasted cobalt ores) with silica sand and potash. The fine dark blue melt is ground up, after it has cooled, and the powder is used for coloring glass melts in ceramic work and in the glass and enamel industry.

It is only during recent years that metallic cobalt has found extensive uses, but considerable quantities are now employed for the production of alloys for certain high speed steels and for cutting alloys.

High speed steels permit the use of high working speeds when they are used for tools in metal turning, since they retain their hardness at a red heat. Still higher working speeds can be achieved by the use of the so-called 'hard alloys', which posses very great hardness—e.g., widia metal and stellite. These alloys can be used in place of diamonds in rock drills, etc., as well as for cutting tools. *Widia metal* (German 'wie Diamant') consists of tungsten carbide with about 10% cobalt. *Stellite* contains about 50% cobalt, 27% chromium, 12% tungsten, 2.5% carbon, and up to 5% iron, with some manganese and silicon.

7. Cobalt Compounds

The majority of simple cobalt compounds are derived from *dipositive* cobalt, and most complex compounds from the *tripositive* state. Oxides derived from +2, +3 and +4 cobalt are known. The oxide of tetrapositive cobalt, CoO_2, is not known in the pure state, however; it can combine with strongly basic oxides to form salts (*cobaltites* or *cobaltates(IV)*) which, like the parent oxide, are not very stable. Cobalt(III) oxide, Co_2O_3, and especially cobalt(II) oxide, CoO, act as basic anhydrides, and form salts of the general formula CoX_3 (cobalt(III) salts) and CoX_2 (cobalt(II) salts), respectively. Simple cobalt(III) salts are known in only a few cases; nearly all simple cobalt salts are derived from cobalt(II) oxide. The most important of them are the chloride, $CoCl_2 \cdot 6H_2O$; nitrate, $Co(NO_3)_2 \cdot 6H_2O$; and sulfate $CoSO_4 \cdot 7H_2O$. These are soluble in water, as is the acetate, $Co(CH_3CO_2)_2 \cdot 4H_2O$. Most simple cobalt salts of weak acids are sparingly soluble. Hydrated cobalt(II) salts are pink to red at ordinary temperature, as are their solutions. They turn deep blue when they are warmed.

Strictly considered, the hydrated compounds mentioned are not really *simple* salts, but are *aquo salts*—i.e., are coordination compounds. The anhydrous salts can readily be obtained from them by heating, however (except for the nitrate, which decomposes). Ammonia can generally be bound in place of water, and is also but loosely bound in the cobalt(II) salts. Nearly all cobalt(II) salts are also capable of forming acido-salts, most of which are only weakly complexed, however—i.e., have definite double-salt character.

The acido-compounds of tripositive cobalt are much more strongly complexed. Complex compounds formed by combination of *neutral ligands* with tripositive cobalt are also characterized by relatively great stability, as in the case of chromium. Tripositive cobalt can, in fact, form salts with many acid radicals only through the formation of such complex ions—e.g., $[Co(NH_3)_6]^{+++}$, $[Co(NH_3)_5(OH)]^{++}$, $[Co(NH_3)_4(NO_2)_2]^+$. Numerous salts are known, in which cobalt is present in such complexes—e.g., nitrates, nitrites, chlorides, bromide, iodides, cyanides, thiocyanates, carbonates, oxalates, acetates, whereas the corresponding simple salts (including, in the cases cited, the aquo compounds $[Co(H_2O)_6]X_3$) are all unobtainable.

Tripositive cobalt almost invariably has the coordination number six in its complex compounds, irrespective of the nature of the combined atoms or groups. The coordination number of bivalent cobalt varies with the nature of the ligands and the experimental conditions, and is not infrequently lower than six.

Electrochemically *quadrivalent* cobalt is present in the oxide CoO_2 (although this has not been prepared in a state of purity), and in the double oxides derived from CoO_2—e.g.. Sr_2CoO_4 and Ba_2CoO_4, which were obtained by Schmahl (1950) and Scholder (1952). The double oxide Co_3O_4 (or $2CoO \cdot CoO_2$) contains cobalt in the $+4$ and $+2$ valence states simultaneously, whereas certain complex compounds (e.g., those mentioned on p. 306) contain cobalt in both the $+4$ and $+3$ valence states.

Electrochemically univalent cobalt may be present in a thiosulfate complex containing nitric oxide, related in type to the Roussin's salts of iron; this was prepared by Manchot (1926). The compound forms brassy crystal leaflets, dissolves in water with a brown-black color, and can be precipitated with alcohol. It has the constitution

$$\left[\begin{array}{c} NO \\ \\ NO \end{array} \; Co \; \begin{array}{c} S-SO_3 \\ \\ S-SO_3 \end{array} \right] K_3.$$

The compounds $Co(NO)_2Cl$, $Co(NO)_2Br$, and $Co(NO)_2I$, obtained by Hieber (1939) may also be regarded as compounds of unipositive cobalt if it is permissible to regard the NO groups in these molecules as neutral ligands. Grube (1926) obtained cyanocobaltate(I) complex ions, in aqueous solution, by the electrolytic reduction of hexacyanocobaltate(II) ions.

(a) Oxides and Hydroxides

Cobalt(II) oxide is prepared as an olive-green powder by heating cobalt(II) hydroxide or carbonate in the absence of air. Cobalt(II) hydroxide, $Co(OH)_2$, is obtained as a precipitate which is blue at first, but turns pale pink on standing, when the solution of a cobalt(II) salt is treated with potassium hydroxide. In air, the precipitate is slowly oxidized to brown cobalt(III) oxide hydrate, as is iron(II) hydroxide. This oxidation occurs more rapidly in the presence of strong oxidants such as NaOCl, Cl_2, Br_2, H_2O_2, and then proceeds as far as the partial formation of black cobalt dioxide hydrate, $CoO_2 \cdot xH_2O$. The cobalt(III) oxide hydrate*

---

* Preparations made by precipitation from aqueous solution have a composition corresponding fairly exactly to the formula $Co_2O_3 \cdot 3H_2O$ after drying. They are generally in a high state of dispersion, but in so far as they yield X-ray diffraction patterns, they show the structure of anhydrous Co_2O_3. It is curious that the monohydrate $Co_2O_3 \cdot H_2O$, which was recognized as a definite compound by Hüttig (1929), on the basis of its dehydration curve, and which occurs native (as *stainierite*), does not differ in structure (according to Natta, 1928) from anhydrous Co_2O_3.

can only be dehydrated to anhydrous brown *cobalt(III) oxide* under certain experimental conditions. As a rule, loss of oxygen occurs before all the water has been given up, with the formation of the black cobalt(II, IV) oxide, Co_3O_4.

Co_3O_4 is isotypic with the spinels and with magnetite ($a = 8.07$ Å). Unlike Fe_3O_4, however, the metal ions in Co_3O_4 are not dipositive and tripositive, but di- and tetra-positive, as in Mn_3O_4. The Co^{4+} ions occupy those lattice positions which, in the ordinary spinel structure, are taken up by Mg^{2+} ions. (Verwey and De Boer, 1936).

All the cobalt oxides pass into the monoxide, CoO, when they are very strongly ignited. They are all reduced to metallic cobalt when they are heated in hydrogen.

CoO crystallizes with the rock salt structure ($a = 4.24$ Å). The pink $Co(OH)_2$ has the brucite structure ($a = 3.17$, $c = 4.64$ Å), and forms mixed crystals with other hydroxides of the same structure—e.g., $Mg(OH)_2$, $Zn(OH)_2$, $Ni(OH)_2$. The blue cobalt(II) hydroxide differs from the pink form by its higher degree of dispersity, and by the disordered arrangement of its atoms. It is open to question, whether it is a special modification, with its own crystal structure*. Feitknecht (1936) inferred from its X-ray diffraction pattern that the blue hydroxide, like the green basic chloride of cobalt, had a 'double layer structure', in which the 'principal layers' had the same structure as in the pink hydroxide, but were pushed apart, to a distance of about 8 Å, and were mutually displaced, by intercalated 'intermediate layers', of a disordered hydroxide structure. This constitution can be schematically represented as $[Co(OH)_2]_4 \cdot Co(OH)_2$, which implies that 80% of the cobalt atoms are in the 'principal layers', and 20% in the 'intermediate' layers. The action of atmospheric oxygen on the freshly precipitated compound leads first to the oxidation of the cobalt of the intermediate layers, giving $[Co(OH)_2]_4 \cdot CoO(OH)$, which is analogous in structure to the green basic chloride. Oxidation of the cobalt of the principal layers takes place much more slowly, and furnishes CoO(OH) or $Co_2O_3 \cdot H_2O$ as the end product.

The color of anhydrous CoO also depends upon the degree of dispersion and the perfection of ordering of the atoms. It may be yellow, grey, brown, reddish, bluish or black, as well as olive green, but it has not yet been proved that any other modification than the cubic form exists.

The heat of formation of crystallized CoO is 57.2 kcal per mol (Roth, 1931). The heat of formation of $Co(OH)_2$ from CoO_{cryst} and H_2O_{liq} is about 6.2 kcal per mol. The heat liberated by the combination of very finely disperse ('amorphous') CoO with H_2O may be twice as great as this, since, according to Mixter (1909), the energy content of 'amorphous' cobalt(II) oxide is about 7 kcal per mol higher than that of distinctly crystalline oxide.

Cobalt(II) hydroxide dissolves in boiling concentrated alkali hydroxide, with a blue-violet color. Scholder (1933) showed that *hydroxocobaltate(II) ions* are present in the solution. He isolated the compounds $Na_2[Co(OH)_4]$, $Sr_2[Co(OH)_6]$ and $Ba_2[Co(OH)_6]$ in crystalline form. These are red-violet, but rapidly blacken in air as a result of oxidation. They are immediately decomposed by water. Solutions of $Co(OH)_2$ in hot caustic alkali of lower concentrations therefore yield red crystalline $Co(OH)_2$, and not hydroxosalt.

The true nature of the anhydrous compounds obtained by fusion of cobalt oxides with basic oxides, in the presence of air, and commonly known as 'cobaltites', has not yet been cleared up**. Some of them are of rather complex constitution, and contain the cobalt in more than one valence state. Thus, according to Belucci, steel grey needles of the composition $K_2O \cdot CoO \cdot 3CoO_2$ are obtained when any oxide of cobalt is fused in air with caustic potash. They are insoluble in water and in dilute hydrochloric acid, but are slowly hydrolyzed by water. They liberate chlorine from concentrated hydrochloric acid, the +4 cobalt being reduced to the dipositive state. 'Cobaltites' of simpler composition, $BaO \cdot CoO_2$ and $BaO \cdot 2CoO_2$, and also $MgO \cdot CoO_2$, are obtainable according to Rousseau and Dufau, by heating cobalt(III) oxide in air with barium oxide and barium chloride, or with magnesium oxide. The existence of these compounds is open to some doubt. However, Scholder has recently isolated the compound $2BaO \cdot CoO_2$ ($= Ba_2CoO_4$) (dibarium cobaltate(IV)) in a state of purity.

* The 'green cobalt hydroxide' which has been described by several workers is actually a basic cobalt chloride, according to Feitknecht (1935) (cf. p. 298).

** The name 'cobaltite' is used both for compounds regarded as oxy-salts of tripositive cobalt (cobaltate(III) salts) and for those of +4 cobalt (cobaltate(IV) salts).

By gently igniting cobalt nitrate with the nitrates of dipositive metals, e.g., zinc, magnesium, Holgersson and Karlsson (1929) were able to prepare double oxides of the general composition $M^{II}O \cdot Co_2O_3$, which were shown by their X-ray diffraction patterns to be spinels.

Blue, vitreous potassium cobalt(II) orthosilicate is obtained by heating cobalt(II) oxide with silica and potassium carbonate. If cobalt(II) oxide is ignited with aluminum oxide, a blue double oxide of spinel type is also formed. The Thenard's blue reaction for aluminum (Vol. I, p. 366) is based on the formation of this double compound. Ignition of cobalt(II) oxide with zinc oxide, or of zinc oxide moistened with cobalt nitrate solution, (Rinmann's green reaction) can lead to the formation either of spinel-type double oxides, or simply of mixed crystals (cf. p. 444), according to the temperature of ignition. If magnesium oxide is ignited after moistening it with cobalt nitrate solution, it turns pale pink. This reaction is given by many minerals which contain magnesium oxide.

(b) Sulfides

(i) *Cobalt(II) sulfide.* Cobalt forms several sulfides (cf. Table 29, p. 244–5), of which the most important is *cobalt(II) sulfide*, CoS. It is formed as a black amorphous precipitate when ammonium sulfide is added to cobalt(II) salt solutions. The precipitate is practically insoluble in water, but dissolves in dilute acids, including acetic acid, when it is freshly precipitated (α-CoS). Crystalline cobalt sulfide is precipitated by hydrogen sulfide from acetic acid solutions (β-CoS) and is practically insoluble in dilute hydrochloric acid. If the cobalt sulfide precipitated by ammonium sulfide from ammoniacal solution (i.e., α-CoS) is filtered with access of air, it becomes practically insoluble in dilute hydrochloric acid, although it does not become crystalline, but remains amorphous to X-rays. According to Fricke and Dönges, the precipitate becomes insoluble because, in presence of air, sulfur from adhering ammonium sulfide solution is taken up to form a cobalt sulfide richer in sulfur. Only when precipitation is carried out with colorless ammonium sulfide solution (i.e., free from sulfur) is the precipitate completely soluble in dilute hydrochloric acid, as long as it has not oxidized. Precipitation with polysulfide solution yields a sulfide with higher sulfur content from the outset, which is almost insoluble in dilute hydrochloric acid even although it is X-ray amorphous. If atmospheric oxygen is allowed to react with sulfide precipitated by colorless ammonium sulfide, and subsequently washed in the absence of air, a basic sulfide of tripositive cobalt is formed. About half of this dissolves in cold dilute hydrochloric acid, with the deposition of sulfur. The insolubility of the residue, which is not the hexagonal crystalline sulfide, is due to its enhanced sulfur content. The behavior of nickel sulfide is very similar. Cobalt sulfide (like the nickel compound) has a strong tendency to go into colloidal dispersion. The sols are brown in color, and may be flocculated by boiling with acetic acid.

Whereas α-CoS is X-ray amorphous, β-CoS has a hexagonal structure. β-CoS is most readily obtained by dry methods—by heating cobalt powder to 700° in a current of H_2S, and subsequent heating to 750–800° in high vacuum. The compound Co_9S_8, which differs but little in composition from CoS, crystallizes cubic. It may be obtained by thermal degradation of CoS in hydrogen sulfide, or from melts. Its structure involves a face-centered cubic lattice of S atoms, into which the Co atoms are so inserted that eight-ninths of them are tetrahedrally surrounded by 4 S atoms, and one ninth octahedrally surrounded by 6 S atoms.

According to Weibke (1936), cobalt sulfide, CoS, is only homogeneous when it contains a certain stoichiometric excess of sulfur (S : Co = at least 1.05 : 1). Its crystal lattice is only stable when a certain proportion of the lattice sites belonging to Co atoms remains un-occupied. This is shown by the fact that the cell dimensions found by X-rays would lead to a higher density than is observed pycnometrically. If excess S atoms were inserted in the crystal structure, the density found pycnometrically would be greater than that calculated from the cell dimensions.

Cobalt(II) sulfide occurs only extremely rarely as a mineral, but has occasionally been found in India (syepoorite).

(*ii*) *Tricobalt tetrasulfide*, Co_3S_4, is found native as *linnaeite*, but can be prepared artificially also—best by heating cobalt powder to 400° in hydrogen sulfide (Schenck, 1942). The naturally occurring sulfide commonly has a part of the cobalt replaced by nickel (nickeli-ferous linnaeite). Linnaeite is often found in well formed crystals (regular octahedra), reddish-grey to copper red in color, with a metallic luster. Its crystal structure, like that of Co_3O_4, is of the spinel type ($a = 9.41$ Å).

(*iii*) *Cobalt disulfide*, CoS_2, can be obtained as a black powder, by the prolonged action of molten sulfur on cobalt monoxide. It loses sulfur when strongly heated. Its structure resembles that of pyrite ($a = 5.64$ Å).

(c) Arsenides

Cobalt also forms several compounds with arsenic, and from the phase diagram of the arsenic-cobalt alloy system (investigated up to 53.5 weight-per cent cobalt), the existence of Co_5As_2, Co_2As, Co_3As_2, and CoAs may be inferred. The arsenides $CoAs_2$ and $CoAs_3$ are found native, the former (smaltite) being the most abundant cobalt ore. It is isomorphous with pyrite, but is rarely found in well formed crystals. It usually occurs in dull grey lumps, and always in association with nickel arsenides. $CoAs_2$ also occurs occasionally in a rhombic form, as *safflorite*. Cobaltite (cobalt glance), CoAsS, which is also important as a cobalt ore, is also isomorphous with pyrite, and is derived from smaltite by the exchange of one arsenic atom with sulfur. The arsenide $CoAs_3$ is also occasionally found in Nature, as *skutterudite*.

(d) Carbonyl and Nitrosyl Compounds

When finely divided cobalt is heated in carbon monoxide to 150–200° under a high pressure (100 atm.), cobalt tetracarbonyl, $[Co(CO)_4]_2$, is obtained in the form of orange crystals which begin to decompose a little above 50°. The compound is insoluble in water, and fairly resistant towards non-oxidizing acids. It is soluble in organic solvents (alcohol, ether, benzene, carbon disulfide), and molecular weight determinations have shown it to be present in solution as dimeric molecules $Co_2(CO)_8$. Loss of carbon monoxide yields black cobalt tricarbonyl, $[Co(CO)_3]_4$ or $Co_4(CO)_{12}$, which may be recrystallized from benzene.

For cobalt carbonyl hydride and cobalt nitrosyl carbonyl, see pp. 356, 358.

As with iron halides, the action of nitric oxide on cobalt(II) halides produces *cobalt nitrosyl halides*, $Co(NO)_2X$ (Hieber). Their stability is greater than that of the corresponding iron compounds, and falls from the iodide to the chloride: the fluoride cannot be prepared. The iodide melts at 131°, the bromide at 116°, the chloride at 101°. The compounds can be sublimed without decomposition in air, but are hydrolyzed by water. Hieber (1941) assumes that in this type of compound, and in the metal-nitric oxide, complexes generally, the NO group is present in the form of the ion $(NO)^+$, or $(:N :: O:)^+$, which is bound to the central atom through the electron pair on the N atom. The binding is thus similar to that of the CO group present in carbonyl compounds, with which $(NO)^+$ is isoelectronic; it is covalent in character, as is typical of the binding of ligands in strong complexes ('pene-tration complexes', cf. Chap. 9, Vol. I).

The number of nitrosyl groups which can be bound in place of a chlorine atom decreases regularly along the series $_{26}Fe$—$_{27}Co$—$_{28}Ni$—$_{29}Cu$, the compounds obtained being $Fe(NO)_3Cl$; $Co(NO)_2Cl$; $Ni(NO)Cl$; CuCl, respectively. The iron compound is fairly volatile, whereas the volatility decreases steadily along the series of compounds.

In aqueous solution, Co^{++} ions, unlike Fe^{++} ions, cannot combine with NO, even in the presence of metals which can combine with halogens (e.g., silver powder). Such conditions favor the reaction of the anhydrous cobalt(II) halides with NO, and the chloride reacts only in presence of some metal which acts as a halogen acceptor.

(e) Cobalt(II) Halides

(i) *Cobalt(II) chloride* (cobalt chloride), $CoCl_2$, is obtained in the anhydrous state by burning cobalt in chlorine, or by dehydrating the hydrated chloride. It is a pale blue powder, which sublimes without melting. It dissolves very readily in water (46.3 g of $CoCl_2$ in 100 g of water at 0°, according to Foote); the aqueous solution is pink at ordinary temperature. The anhydrous chloride is very hygroscopic, and soon turns pink in air.

It is also soluble in various organic solvents—e.g., in alcohol, acetone, quinoline, and benzonitrile. These solutions have a fine blue color, and the chloride crystallizes out from them in blue needles when they are evaporated or cooled. 56.2 g of $CoCl_2$ dissolve in 100 g of absolute ethanol. The solubility in ether, however, is very slight (0.02 g in 100 g).

With water, cobalt chloride forms hydrates of various colors

$CoCl_2 \cdot H_2O$	blue violet	$CoCl_2 \cdot 4H_2O$	peach blossom red
$CoCl_2 \cdot 1\frac{1}{2}H_2O$	dark blue violet	$CoCl_2 \cdot 6H_2O$	pink
$CoCl_2 \cdot 2H_2O$	violet-pink		

The hexahydrate turns blue when it is gently warmed. The aqueous solution also turns violet to blue, depending on the concentration, when it is warmed. A blue coloration is also produced by the addition of concentrated hydrochloric acid or sulfuric acid, or on the addition of chlorides of strongly electropositive metals (especially calcium and magnesium chlorides).

The color changes of the solutions are due, essentially, to the dehydration brought about on warming. Addition compounds of cobalt(II) chloride (and of the other cobalt(II) halides) with the smallest number of ligands are violet to blue, whereas those with the largest number of ligands—addition compounds containing 6 molecules of the attached neutral substance—are pink. This rule is followed not only by the hydrates, but by addition compounds of other oxygen-containing solvents (e.g., with methyl alcohol), where such are known in the solid state. Cobalt(II) chloride can also add on many nitrogen-containing solvents, with the formation of crystalline compounds, and in these cases also the 'saturated' compounds are pink, and the 'unsaturated' are violet or blue (Hantzsch, 1927).

The *double chlorides* formed with the chlorides of the alkalis and other strongly electropositive metals (*chlorocobaltate(II) salts*) are also mostly deep blue. The composition of these salts may be exemplified by the double chlorides with cesium chloride—$CsCl \cdot CoCl_2 \cdot 2H_2O$, $2CsCl \cdot CoCl_2$ (or $Cs_2[CoCl_4]$), and $3CsCl \cdot CoCl_2$. The apparent formation of hexahalogeno anions is rarely observed—e.g., in $4LiCl \cdot CoCl_2 \cdot 10H_2O$—and it is not certain that this is correctly represented as $Li_4[CoCl_6] \cdot 10H_2O$. X-ray structure determination has shown that the crystal lattice of $3CsCl \cdot CoCl_2$ is actually built up from Cs^+ cations, and $[CoCl_4]^{2-}$ and Cl^- anions—i.e., the compound is $Cs_2[CoCl_4] \cdot CsCl$.

Paper impregnated with cobalt chloride turns blue in dry air, but the pink color is restored in moist air. Such paper is therefore used as a 'weather indicator', or hygrometer. Cobalt chloride finds applications chiefly as a laboratory reagent and as starting material for the preparation of other cobalt compounds.

(ii) *Cobalt(II) bromide*, $CoBr_2$, forms lustrous green leaflets when it is anhydrous. These are very soluble in water, and deliquesce to give a red liquid in air. The hexahydrate, $CoBr_2 \cdot 6H_2O$ (red crystals) is also deliquescent, but loses water on standing over concentrated sulfuric acid. It melts at 100°, simultaneously losing water and forming the purple dihydrate, $CoBr_2 \cdot 2H_2O$. Transition to the anhydrous salt occurs at 130°. The principal type of double bromide appears to be $M^I_2[CoBr_4(H_2O)_2]$; an apparent hexahalogeno salt (see above), $Li_4[CoBr_6] \cdot 12H_2O$ (dark blue deliquescent crystals) was prepared by Meyer (1935).

(iii) *Cobalt(II) iodide*, CoI_2, is obtained anhydrous by the direct union of the components, or by dehydrating the hydrated salt at 130°. It forms a grey-green mass, which is soluble in alcohol and acetone, as well as in water. The hexahydrate, $CoI_2 \cdot 6H_2O$, is dark red. Its aqueous solution is olive green at 20°, and a pure chrome-green at 35°. At very low temperatures an enneahydrate (light red) is also stable. The known double iodides are mostly of the type $M^I_2[CoI_4]$—e.g., $Cs_2[CoI_4]$. Hexaiodo salts, $M^I_4[CoI_6]$, are known only in isolated cases.

(iv) *Cobalt(II) fluoride* separates in rose red crystals of $CoF_2 \cdot 2H_2O$, when solutions of cobalt carbonate in aqueous hydrofluoric acid are evaporated. The anhydrous salt CoF_2 is also reddish in color, as are the double fluorides—e.g., $K_2[CoF_4]$ and $(NH_4)_2[CoF_4]$.

The color change from pink through blue to green in the series of anhydrous cobalt(II) halides, CoF_2, $CoCl_2$, $CoBr_2$, and CoI_2, arises from a shift in the light absorption from blue green towards the red. It is assumed that this displacement is associated in some manner with the increase in covalent character of the metal-halogen bond, in going from CoF_2 to CoI_2. The color changes which occur as dehydration takes place step by step can be explained on the same basis. The greater the number of H_2O molecules (or other neutral ligands) bound to the cobalt(II) ion, the more the halogen atom is displaced from the cation, and the smaller is the polarizing effect of the cobalt cation upon the halogen anion (cf. Vol. I, p. 312). It is found with other cobalt(II) salts also that the stronger the polarizing action of the cobalt ion on the attached anion—i.e., the more the bond character tends towards the covalent type—the more the color is shifted from red to green.

(v) *Basic Cobalt Halides*. The basic cobalt halides are of interest, as representatives of two groups of basic salts, of which the composition and properties were first explained as a result of the determination of their crystal structures (Feitknecht, 1933 and later).

Basic salts of the first group, exemplified by the 'green basic cobalt chloride', $CoCl(OH) \cdot 4Co(OH)_2 \cdot 4H_2O$, have so-called 'double layer lattices'. There may be considerable variations in the composition of these salts, although in the structural sense they are well defined compounds. The basic salts of the other group, of which 'pink basic cobalt chloride', $Co_2Cl(OH)_3$, is a member, possess 'simple layer lattices'. Salts having this structure are invariably much simpler in composition than the double layer lattice salts, and their composition is either invariable, or varies only within narrow limits.

(vi) *Green Basic Cobalt Chloride. Double Layer Lattices*. If a solution of cobalt(II) chloride is treated with a quantity of alkali insufficient to precipitate the hydroxide completely, the blue $Co(OH)_2$, which is at first incompletely precipitated, is subsequently transformed, by reaction with the $CoCl_2$ remaining in solution, into *green basic cobalt chloride**. In the ideal case, this corresponds in composition to $4Co(OH)_2 \cdot CoCl(OH) \cdot 4H_2O$, although the composition is variable. The structure of this compound is built up from layers of $Co(OH)_2$, exactly as occur in crystalline $Co(OH)_2$. Between these $Co(OH)_2$ layers ('principal layers') are others ('intermediate layers'), built up from $CoCl(OH)$ and H_2O. Principal layers and intermediate layers are stacked alternately. The intercalated $CoCl(OH)$ layers push the principal layers apart to a distance of 8.2 Å, as compared with 4.65 Å in $Co(OH)_2$ itself. At the same time, a displacement in the plane brings the OH groups of one principal layer vertically above an OH group of the next principal layer. This stacking is not found in the hydroxide. The constituents of the intermediate layers differ from those in the principal layers, in that they are distributed quite at random over the lattice sites. Feitknecht concluded from the color (or absorption spectrum) of green cobalt chloride that the cobalt was predominantly covalently bound in the intermediate layers, and ionic in the principal

---

* This compound was formerly, but erroneously, spoken of as a second modification of cobalt(II) hydroxide.

layers. The difference in binding shows itself in the chemical behavior, in that the cobalt of the intermediate layers is more readily oxidized than that of the principal layers.

Feitknecht, who first described the essentials of this type of structure, called them *double layer lattices*. They are found in the basic salts of other bivalent metals, as well as of cobalt—e.g., with nickel, iron(II), manganese(II), cadmium, and magnesium. Salts which have this structure form foliated crystals, and can lose water continuously, since it is not bound at definite lattice positions. The same may be true of other constituents of the intermediate layers (as in the case of green cobalt basic chloride), and such constituents may be present in variable amount. However, X-ray diffraction studies have shown that in many of these basic salts the cations and anions are regularly arranged in the intermediate layers (as they invariably are in the principal layers). Even so, the over-all ratio of hydroxide to salt can vary between certain limits, in that hydroxyl ions and acid anions can mutually replace one another in the intermediate layers. Alternatively, there may be zones in the crystal in which several hydroxide layers follow one another consecutively, the intermediate layers being omitted. Substances having double layer lattice structures have a particular tendency to undergo topochemical reactions (cf. Chap. 19). If they are brought into reaction with dissolved substances, the attack takes place only along the layer planes, and never at right angles to the layers.

(*vii*) *Pink Basic Cobalt Chloride. Simple Layer Lattices.* On prolonged standing, in contact with a not too dilute solution of cobalt chloride, the green basic chloride gradually changes into the pink basic cobalt chloride, $Co_2Cl(OH)_3$. This has a so-called 'simple layer lattice structure', i.e., a structure built up from sheets all of the same kind. This kind of structure was first found for a basic salt by Hoard (1934), in the case of basic cadmium chloride, $Cd(OH)Cl$ (cf. Fig. 50, p. 449). Pink basic cobalt chloride consists, according to Feitknecht, of single sheets of cobalt ions, on either side of which the hydroxyl and chloride ions are distributed in an ordered arrangement. This arrangement is probably very similar to that of the individual ions in the crystal lattice of cobalt(II) hydroxide (i.e., the brucite structure, Fig. 61, Vol. I, p. 261), except in so far as the partial replacement of hydroxyl ions by chloride ions leads to an expansion of the distance between the sheets. From the color of the compound Feitknecht inferred that the binding between cobalt and chlorine is essentially ionic.

It has been established that simple layer lattices are present in basic salts of nickel and zinc, and in the basic cadmium chlorides, as well as in some of the basic cobalt salts.

(*viii*) *Basic Cobalt Bromides.* Four basic bromides of cobalt are known: a *green* salt isomorphous with the green basic chloride; a *violet-red* salt, $Co_2Br(OH)_3$, which forms a simple layer lattice, like the chloride of analogous composition, although with a different arrangement of ions within the individual layers; a *blue-violet* salt with double layer lattice; and a *pink* basic chloride of still unknown structure. Whereas up to one third of the chloride ions in the pink basic chloride may be exchanged for bromide ions, none of the bromine in the violet-red basic bromide can be exchanged for chlorine.

(f) Other Cobalt(II) Salts

(*i*) *Cobalt(II) cyanide and Cyanocobaltate(II) Salts.* With CN^- ions, cobalt(II) salts give a precipitate which is insoluble in water and dilute acids, but soluble with a yellow color in ammonia and ammonium carbonate solutions; after drying, it has the composition $Co(CN)_2 \cdot 2H_2O$. Treatment of the fresh precipitate with potassium cyanide solution yields a solution which is at first green, and from which a green precipitate, $K_2Co[Co(CN)_6]$, is often deposited. It is essential that no rise of temperature should occur, as the cobalt is otherwise converted to the trivalent state, with evolution of hydrogen. On standing, the solution gradually turns red, and if it is sufficiently concentrated, violet crystal leaflets of potassium *pentacyanocobaltate(II)* (Adamson, 1951) separate out. These may be washed with alcohol and ether (in which they are insoluble), and may be kept without changing, in an atmosphere of hydrogen. The compound is very soluble in water, with a deep red color, and is converted into the hexacyanocobaltate(III) on boiling; in the absence of oxygen the reaction proceeds by the discharge of hydrogen ions and evolution of hydrogen gas; in the presence of excess CN^- ions the reaction occurs even on gentle warming. Other cyanocobaltate(II) salts have been prepared, in addition to the potassium salt.

(ii) *Cobalt(II) thiocyanate and Thiocyanatocobaltate(II) Salts.* Cobalt thiocyanate, $Co(SCN)_2 \cdot 4H_2O$, exists as deep red-violet, rhombic, very deliquescent crystals, which dissolve in a little water to give a dark blue solution. This turns red when it is diluted, but the deep blue color is restored on the addition of acetone. Cobalt thiocyanate also dissolves with deep blue color in alcohol. The pure salt is best obtained by neutralizing cobalt carbonate with a solution of thiocyanic acid, or by metathesis between cobalt sulfate and barium thiocyanate; the solution may be evaporated over sulfuric acid.

Most of the *thiocyanatocobaltate(II)* salts conform to the type $M^I_2[Co(SCN)_4]$. Thus the ammonium salt is obtained by treating a cobalt(II) salt solution with ammonium thiocyanate, extracting with amyl alcohol, (or with a mixture of equal volumes of amyl alcohol and ether), and evaporating the solution. It forms blue, highly refractive needles, which may be recrystallized from acetone. It crystallizes from water as a hydrate, $(NH_4)_2[Co(SCN)_4] \cdot 4H_2O$. The potassium salt is obtained by a similar method. Other tetrathiocyanatocobaltates, and a few penta- and hexathiocyanato salts, are also known.

The reaction of cobalt salts with ammonium or potassium thiocyanate is often used for the analytical detection of cobalt. It is usual to shake up the red aqueous solution with a mixture of amyl alcohol and ether. The test can be carried out with greater sensitivity by producing the blue color in the solution itself, by the addition of acetone.

(iii) *Cobalt(II) nitrate,* $Co(NO_3)_2 \cdot 6H_2O$, forms deliquescent, carmine red, monoclinic prisms or plates, which are very soluble in water (to give a carmine solution). It is used in analytical chemistry (especially in blowpipe analysis) as a reagent, and finds applications in the production of cobalt pigments and in ceramics. It is prepared by dissolving cobalt, cobalt oxide, or cobalt carbonate in dilute nitric acid. The hexahydrate first loses $3H_2O$ when it is heated, and then decomposes on stronger heating, with the evolution of nitrous fumes and the formation of the black-brown oxide.

When a little sodium hydroxide is added to a cobalt(II) nitrate solution, *green basic cobalt nitrate* is obtained. This has a composition which is usually very close to $Co(NO_3)_2 \cdot 6Co(OH)_2$, and possesses a double layer lattice structure; the distribution of the components between the principal and intermediate sheets can be represented by the formulation $4Co(OH)_2 \cdot Co_{1.25}(NO_3)_{1.5}(OH)_{1.0}$, analogous to that of the green basic chloride. In contact with concentrated cobalt nitrate solution, the green salt is converted to the pink basic nitrate, $Co_2(NO_3)(OH)_3$, having a simple layer lattice (Feitknecht, 1934).

(iv) *Cobalt(II) Nitrite and Double Nitrites.* Cobalt(II) nitrite has never been isolated, although it may exist in solution. Double nitrites have been obtained by mixing hot concentrated solutions of cobalt(II) chloride and alkali nitrites—e.g., $2KNO_2 \cdot Co(NO_2)_2 \cdot H_2O$, yellow powder.

(v) *Cobalt(II) sulfate* crystallizes at room temperature from solutions of cobalt, cobalt oxide, or cobalt carbonate in dilute sulfuric acid, as 'cobalt vitriol', $CoSO_4 \cdot 7H_2O$—red monoclinic crystals, isomorphous with iron vitriol, stable in air. The salt is readily soluble in water (36 g of $CoSO_4$ in 100 g of H_2O at 20°), but insoluble in alcohol.

At 40–50°, the hexahydrate $CoSO_4 \cdot 6H_2O$ crystallizes from the solutions, and the action of dehydrating agents leads to hydrates with a smaller water content. Anhydrous cobalt sulfate is obtained as a red powder which dissolves readily in water, by heating the hydrated materials with concentrated sulfuric acid. Double sulfates crystallize from solutions containing alkali sulfates—e.g., $(NH_4)_2SO_4 \cdot CoSO_4 \cdot 6H_2O$, red crystals isomorphous with ammonium magnesium sulfate and with Mohr's salt. On the basis of diffusion measurements, Brintzinger considers that solutions with a high sulfate ion concentration contain the binuclear complex ions $[Co_2(SO_4)_4]^{4-}$.

There are two basic cobalt sulfates—a blue salt with the composition $CoSO_4 \cdot 3Co(OH)_2$ (with variable water content), and a violet salt, $2CoSO_4 \cdot 3Co(OH)_2 \cdot 5H_2O$. Both have double layer lattice structures. Both salts can lose water, and reabsorb it, continuously; however, the violet salt turns blue in the course of dehydration, and the color change is not reversed when it is hydrated again. X-ray studies show that the initially ordered structure of the intermediate layers is destroyed during dehydration, and the distance between the principal layers ($Co(OH)_2$ layers) is diminished. During rehydration this dimensional change is reversed, but the disordered structure of the intermediate layers remains (Feitknecht, 1934).

Cobalt vitriol is found in Nature in small amounts, as a pink bloom on cobalt ores (*bieberite*).

(*vi*) *Cobalt(II) sulfite and Sulfitocobaltate(II) Salts*. When sulfur dioxide is passed through an aqueous suspension of cobalt hydroxide, *cobalt(II) sulfite*, $CoSO_3 \cdot 5H_2O$, is obtained in reddish granular crystals, sparingly soluble in water. Treatment of a solution of a hot solution of cobalt(II) sulfite or chloride with potassium sulfite produces a pale red crystalline precipitate of potassium disulfitocobaltate(II), which is not very stable in air. The same compound is obtained by boiling cobalt hydroxide with potassium hydrogen sulfite. Other sulfitocobaltates (e.g., the sodium and ammonium salts) have been obtained similarly.

(*vii*) *Cobalt(II) carbonate and Carbonatocobaltate(II) Salts*. Blue precipitates of *basic carbonates* are generally obtained by the addition of alkali carbonates to cobalt(II) salt solutions. The hexahydrate of the neutral carbonate, $CoCO_3 \cdot 6H_2O$, violet red microscopic crystals, can be prepared by carrying out the reaction in a solution saturated with carbon dioxide. The anhydrous salt is formed at 140° in a sealed tube, as a light red powder (consisting of microscopic rhombohedra).

The precipitate obtained by treating a cobalt(II) salt with ammonium carbonate solution turns crystalline when left standing for some time under excess of the precipitant, the double salt $(NH_4)_2CO_3 \cdot CoCO_3 \cdot 4H_2O$ being formed in bright red prisms. The sodium and potassium double carbonates (carbonatocobaltates) are similar in composition.

(*viii*) *Cobalt(II) oxalate and Oxalatocobaltate(II) Salts*. Cobalt oxalate is formed as a pink precipitate, when oxalate ions are added to a cobalt(II) salt solution. It is best obtained pure by dissolving cobalt(II) carbonate in aqueous oxalic acid, and then separates as a pink powder of the composition $CoC_2O_4 \cdot 2H_2O$. It is almost insoluble in water and aqueous oxalic acid, but dissolves in warm concentrated ammonia and especially in ammonium carbonate. Double oxalates (oxalatocobaltate(II) salts) can be obtained from solutions containing excess alkali oxalate—e.g., $K_2C_2O_4 \cdot CoC_2O_4 \cdot 6H_2O$—red rhombic prisms, soluble in water.

(*ix*) *Cobalt(II) acetate*, $Co(CH_3CO_2)_2 \cdot 4H_2O$, may be prepared by dissolving cobalt carbonate in acetic acid. It forms very soluble red crystals, isomorphous with the acetates of magnesium, zinc, manganese, and nickel. It finds application as a bleaching and drying agent for varnishes and lacquers.

(*x*) *Ammoniates of Cobalt(II) Salts*. Anhydrous cobalt(II) salts of strong acids— e.g., the halides and sulfate, and also the thiocyanate—can usually combine with up to 6 molecules of ammonia. Thus *hexamminecobalt(II) sulfate*, $[Co(NH_3)_6]SO_4$, is formed as a reddish-white powder when ammonia gas is passed over the anhydrous sulfate. The compound can also be formed by passing ammonia into a solution of cobalt(II) sulfate. It is very soluble in aqueous ammonia, but may be precipitated by addition of alcohol. It is decomposed by pure water, and is thus to be regarded as a *very weakly complexed compound*.

Other cobalt(II) ammoniates are similar in behavior. Thus the pale pink *hexammine-cobalt(II) chloride* is also obtainable by passing a large excess of ammonia into concentrated cobalt chloride solution. It is very soluble in dilute ammonia, much less so in concentrated ammonia, and is decomposed by water. $[Co(NH_3)_6]Br_2$ and $[Co(NH_3)_6]I_2$ also are red powders. Thus the hexammine compounds of bivalent cobalt resemble the hexaquo compounds in their color. In many cases, hydroxylamine, hydrazine, and organic nitrogen anhydrobases can be combined in place of ammonia. The hydrazine addition compound of cobalt oxalate, $CoC_2O_4 \cdot 2N_2H_4$, (raspberry red) may be specially mentioned since, unlike the hydrazine compounds of the chloride, bromide, and sulfate, it is not markedly decomposed by water.

Cobalt(II) halides, and cobalt(II) salts with strongly polarizable acid groups, can usually combine with 4 neutral ligands without displacement of the anion from coordination with the central atom. Compounds of the type $CoX_2 \cdot 4Am$ (X = Cl, Br, I, CN, SCN, etc., Am = organic amine) are thus usually non-electrolytes, $[CoX_2Am_4]$. It is the exception for the anion to be displaced from combination with the central atom when 4 neutral ligands are bound, so that the resultant compound can dissociate electrolytically. Thus *o*-phenylene-diamine forms $[Co(C_6H_4(NH_2)_2)_2]Cl_2$, probably because the large size of the neutral ligand sterically hinders the simultaneous attachment of the halogen atoms to the central atom.

The compounds $[CoX_2Am_4]$ are not very stable, and this is generally true of substances of the type $[M^{II}X_2Am_4]$. This is probably why the existence of *cis-trans* isomerism has never been established among such compounds [Hieber, *Z. Elektrochem.*, 39 (1933) 24]. In the few cases where the stability is great enough for a compound of this type to exist, only the energetically favored configuration of the ligands can be obtained, since the ligands, which are weakly bound, can readily interchange their places.

(g) Cobalt(III) Salts

(*i*) *Cobalt(III) fluoride* was first obtained by Barbieri, as the hydrated salt $CoF_3 \cdot 3\frac{1}{2}H_2O$, by the electrolysis of a saturated solution of cobalt(II) fluoride in 40% hydrofluoric acid in a cooled platinum dish which served as anode. It is a chrome green powder, which is decomposed by water, with the deposition of cobalt(III) oxide hydrate. The anhydrous fluoride, CoF_3, obtained in 1929 by Ruff by the action of fluorine on cobalt(II) chloride, is a light brown powder (density 3.88), crystallizing in the hexagonal system. It decomposes into cobalt(II) fluoride and fluorine when it is heated to about 300° in CO_2, and is a valuable fluorinating agent.

Potassium fluorocobaltate(III), $K_3[CoF_6]$, formed by the action of fluorine on a mixture of $CoCl_2$ and KCl, is the only known cobalt(III) compound with the magnetic susceptibility of the Co^{3+} ion (corresponding to 3 unpaired electrons, and indicative of a 'normal complex', with ionic bonds between central atom and ligands). All other cobalt(III) complex compounds are so-called 'penetration complexes'; they are diamagnetic, indicating the rearrangement of electron levels involved in the formation of covalent bonds between central atom and ligands.

Cobalt(III) chloride has not yet been obtained. Solutions of cobalt(III) oxide hydrate in hydrochloric acid evolve chlorine. $CoCl_3$ may well be stable in the gaseous state at higher temperatures, however. According to Schäfer (1951), $CoCl_2$ is considerably more volatile at 800° in a stream of chlorine than in a stream of hydrogen chloride. It was inferred from this that a volatile higher chloride is formed when the cobalt chloride is heated in chlorine.

(*ii*) *Cobalt(III) sulfate*, $Co_2(SO_4)_3 \cdot 18H_2O$, can be prepared by the anodic oxidation of a cold, concentrated cobalt(II) sulfate solution containing sulfuric acid in a diaphragm cell. It exists as blue needles, and is at once decomposed by water, but dissolves without appreciable decomposition in dilute sulfuric acid.

Cobalt(III) sulfate forms alums with the alkali sulfates—e.g., potassium cobalt alum, $KAl(SO_4)_2 \cdot 12H_2O$, made by mixing equivalent amounts of cooled solutions of cobalt(III) sulfate and potassium sulfate. The alum forms dark blue octahedral crystals which are decomposed by water, with the evolution of oxygen, but can be washed with glacial acetic acid and acetone. The deep blue cesium and rubidium alums are much less soluble than the potassium salt.

The hydrate and alums of cobalt(III) sulfate undoubtedly contain the hexaquocobalt(III) ion, $[Co(H_2O)_6]^{+++}$, and are thus to be included among the complex salts of tripositive cobalt (see below, cobaltammines).

(iii) Oxalatocobalt(III) Salts. A green solution is obtained by dissolving cobalt(III) oxide hydrate in concentrated oxalic acid solution. With calcium carbonate this gives a precipitate having the composition $Ca_3[Co(C_2O_4)_3]_2$, and thus contains the complex acid $H_3[Co(C_2O_4)_3]$. Neither *trioxalatocobalt(III) acid*, nor the cobalt(III) oxalate from which it is derived, is known in the free state, however, and the acid is not very stable in solution. It decomposes, with deposition of cobalt(II) oxalate. The salts $M^I_3[Co(C_2O_4)_3]$ (trioxalato-cobaltate(III) salts) are more stable than the free acid—e.g., the deep green monoclinic ammonium salt, $(NH_4)_3[Co(C_2O_4)_3] \cdot 3H_2O$, which can be prepared either by shaking ammonium oxalatocobaltate(II) solution with lead dioxide and acetic acid, or by electro-lytic oxidation of mixed ammonium oxalate and cobalt(II) salt solutions. Only a few complex cobalt(III) oxalates, other than the trioxalato salts, are known, in contrast to chromium. 'Durrant's salt', according to Werner (1914) and Spacu (1934), is potassium dioxalatohydroxoaquocobaltate(III), $K_2[Co(C_2O_4)_2(OH)(H_2O)]$.

(iv) Hexacyanocobaltate(III) Salts and Hexacyanocobalt(III) acid. The action of excess of potassium cyanide on cobalt(II) salts in solution at first produces potassium pentacyano-cobaltate(II), $K_2[Co(CN)_5]$, and this is converted, by access of oxygen or by discharge of hydrogen ions, into the *hexacyanocobaltate(III)* salt. *Potassium hexacyanocobaltate(III)*, $K_3[Co(CN)_6]$, forms transparent pale yellow crystals, isomorphous with those of potassium hexacyanoferrate(III). Its solution yields precipitates of the hexacyanocobaltate(III) salts of the heavy metals, e.g., by double decomposition with lead or copper(II) salts. The free hexacyanocobalt(III) acid is obtained by decomposing these heavy metal salts with hydrogen sulfide. From it, other soluble hexacyanocobaltate(III) salts may be prepared—e.g., the sodium, ammonium, barium, and strontium salts. Hexacyanocobalt(III) acid can also be obtained directly from the potassium salt, by evaporation with nitric or sulfuric acid, and extraction of the residue with water. It forms colorless feathery crystals, $H_3[Co(CN)_6] \cdot H_2O$, which are soluble in water and alcohol, but not in ether.

$K_3[Co(CN)_6]$ reacts with metallic potassium, dissolved in liquid ammonia, to form a complex of electrochemically zerovalent cobalt:

$$K_3[Co(CN)_6] + 3K = 2KCN + K_4[Co(CN)_4]$$

(Hieber, 1952). $K_4[Co(CN)_4]$ is obtained as a brown-violet, finely crystalline, pyrophoric substance. It dissolves in water with the vigorous evolution of hydrogen, and is attacked even by dry CO_2.

(v) Hexanitrocobaltate(III) Salts. Although cobalt(III) nitrite is not known, its complex salts, the hexanitrocobaltate(III) salts ('cobaltinitrites') $M^I_3[Co(NO_2)_6]$, are readily prepared. Sodium hexanitrocobaltate, $Na_3[Co(NO_2)_6]$, which is used as a reagent for potassium ions, is prepared by mixing cobalt nitrate with sodium nitrite solution, adding acetic acid, and blowing air through. After the solution has been allowed to stand for some time, to permit potassium or ammonium hexanitrocobaltate arising from impurities to settle out, the sodium salt can be precipitated by the addition of alcohol. It may be recrystallized by redissolving and adding alcohol again, and forms a yellow powder which readily gives a yellow brown solution in water.

Potassium hexanitrocobaltate, $K_3[Co(NO_2)_6]$, (Fischer's salt), is thrown down as a sparkling deep yellow powder by the addition of sodium hexanitrocobaltate to solutions of potassium salts. Under microscopic examination it exhibits well formed prisms and double pyramids. It is very sparingly soluble in cold water (1 part in 1120 at 17°), and is decomposed by hot water, with the evolution of NO.

It is almost insoluble in alcohol and ether. Under the name of cobalt yellow or aureolin it is used in oil painting and water color painting, as a substitute for the true Indian yellow. It is also employed in glass and porcelain painting, since unlike the cobalt oxide ordinarily used, it is free from nickel and iron, and therefore confers a purer blue tone to silicate melts.

Ammonium hexanitrocobaltate resembles the potassium salt in solubility. The rubidium and cesium salts are particularly insoluble, their solubilities being 1 part in 19,800 and 1 in 20,100, respectively at 17°. Both crystallize with $1H_2O$.

(h) Cobalt Ammines

Ammonia is much more firmly bound by tripositive cobalt than by the element in its +2 state. Ammoniacal solutions of cobalt(II) salts also have a great tendency to be oxidized to the +3 state, and this oxidation can be effected even by atmospheric oxygen.

If air is passed for some time through a solution of a cobalt(II) salt containing ammonium chloride and ammonia, it is possible to isolate the compound $[Co(NH_3)_6]Cl_3$ (*hexamminecobalt(III) chloride*) from the solution. It forms orange yellow crystals, and was formerly known as *luteocobaltic* chloride. The constitution of this compound follows from its analytical reactions (all the chlorine is immediately precipitated by silver nitrate), and from the conductivity of its solutions, which has a magnitude typical of the values found for salts which dissociate into 4 ions.

In addition to the yellow salt, it is possible to isolate a red salt from the solution. This was formerly called *roseocobaltic* chloride, and has the composition $CoCl_3 \cdot 5NH_3 \cdot H_2O$. All the chlorine is ionized in solution, and if the chlorine is replaced by other acid radicals, the resulting salts invariably contain at least $1H_2O$ in addition to $5NH_3$. It is therefore assumed that the H_2O molecule is bound to the cobalt inside the complex, together with the $5NH_3$ molecules, and the salt is accordingly formulated as $[Co(H_2O)(NH_3)_5]Cl_3$ (*aquopentamminecobalt(III) chloride*).

Yet another purple red anhydrous compound can be obtained from the solution; this has the composition $CoCl_3 \cdot 5NH_3$, and was formerly called *purpureocobaltic* chloride. Its reactions indicate that 1 Cl atom is complex-bound, in addition to $5NH_3$; as was first established by Krock in 1870, only 2 Cl atoms can be immediately precipitated by silver nitrate from a freshly prepared solution. It is accordingly to be formulated as $[CoCl(NH_3)_5]Cl_2$, *chloropentamminecobalt(III) chloride*.

Another compound exists, with the composition $CoCl_3 \cdot 4NH_3$, which contains 2Cl bound within the complex, and ionizes only 1 Cl atom in solution, except in so far as it undergoes decomposition. This compound, formulated as *dichlorotetramminecobalt(III) chloride*, $[CoCl_2(NH_3)_4]Cl$, like the chromium compounds of the same type, exists in two isomeric forms (p. 147)—*cis* and *trans* isomers. The *cis* compounds are blue-violet, and the trans compounds *green*. They were originally known as *violeo-* and *praseo-* cobalt chlorides, respectively.

The compounds mentioned are examples of the first members of the most important series of cobalt(III) salt-ammonia compounds, known briefly as *cobaltammines*. The known types of mononuclear cobaltammines are summarized in Table 33. They exist in as great a multiplicity as do the chromium ammines*.

As with chromium ammines, ammonia may often be replaced by the molecules of other nitrogen anhydrobases. Negative groups R within the complex may be either halogen atoms

---

* It is noteworthy that complexes containing a single NH_3 molecule are unknown, among both cobalt(III) and chromium(III) complexes. Such compounds exist, however, in the case of +2 and +3 iron, and also for tripositive iridium.

or the radicals of weak acids, whereas the groups X outside the complex may frequently be the radicals of strong acids.

The aquoamminocobalt(III) ions are acidic in nature—i.e., they can split off protons, and pass over into the corresponding hydroxo ions—e.g.,

$$[Co(H_2O)(NH_3)_5]^{+++} + H_2O \rightleftharpoons [Co(OH)(NH_3)_5]^{++} + H_3O^+.$$

The position of equilibrium between the complementary aquo- and hydroxo ion pairs

$$[Co(H_2O)(NH_3)_5]^{+++} \rightleftharpoons [Co(OH)(NH_3)_5]^{++},$$

$$[Co(H_2O)_2(NH_3)_4]^{+++} \rightleftharpoons [Co(OH)(H_2O)(NH_3)_4]^{++},$$

etc., depends in a systematic fashion on the number of aquo molecules in the complex. The greater the number of aquo molecules in the complex, the more the equilibrium tends to favor formation of the hydroxo ion, at any given hydrogen ion concentration (Brönsted, 1928). Parallel with the tendency to lose a proton runs the tendency to acquire an electron—i.e., to pass into a complex ion of *di*positive cobalt.

TABLE 33

PRINCIPAL TYPES OF MONONUCLEAR AMMINES, AQUO- AND ACIDO-COMPLEXES
FORMED BY TRIPOSITIVE COBALT

	a	b	c	d	e
I	$[CoAm_6]^{+++}$	$\left[Co{Am_5 \atop (H_2O)}\right]^{+++}$	$\left[Co{Am_4 \atop (H_2O)_2}\right]^{+++}$	$\left[Co{Am_3 \atop (H_2O)_3}\right]^{+++}$	$\left[Co{Am_2 \atop (H_2O)_4}\right]^{+++}$
II	$\left[Co{Am_5 \atop R}\right]^{++}$	$\left[Co{Am_4 \atop (H_2O) \atop R}\right]^{++}$	$\left[Co{Am_3 \atop (H_2O)_2 \atop R}\right]^{++}$	$\left[Co{Am_2 \atop (H_2O)_3 \atop R}\right]^{++}$	—
III	$\left[Co{Am_4 \atop R_2}\right]^{+}$	$\left[Co{Am_3 \atop (H_2O) \atop R_2}\right]^{+}$	$\left[Co{Am_2 \atop (H_2O)_2 \atop R_2}\right]^{+}$	—	
IV	$\left[Co{Am_3 \atop R_3}\right]$	?	—	?	
V	$\left[Co{Am_2 \atop R_4}\right]^{-}$	—	?		
VI	—	$\left[Co{(H_2O) \atop R_5}\right]^{=}$			
VII	$[CoR_6]^{3-}$				

Am = neutral ligand, such as NH_3, pyridine, etc.

R = complex-bound acidic group

R represents one univalent acidic group, or the equivalent amount of multivalent groups.

The radical of nitrous acid can exist in *two forms* inside certain cobalt complexes. Thus two series of the compounds of the type $[Co(NO_2)(NH_3)_5]X_2$ are known. The compounds of one series are all yellow, and are more stable than the salts of nitrous acid—e.g., towards acetic acid. The other series is red, and much less stable, so that they can only be isolated in relatively few cases. It is assumed that, in the yellow compounds, as in organic nitro compounds which are also yellow (e.g., nitroethane, C_2H_5—NO_2), the NO_2 radical is bound to the metal (cobalt) through the nitrogen atom, whereas in the red compounds, as in the alkali nitrites $M^+[O$—$N{=}O]^-$, it is linked through the oxygen. Compounds of the yellow series are called *nitro* compounds, and those of the red series *nitrito* compounds

$$\text{e.g.,}\quad \begin{bmatrix} NH_3 & NH_3 \\ NH_3 \ Co \ NH_3 \\ NH_3 & NO_2 \end{bmatrix} Cl_2 \qquad \begin{bmatrix} NH_3 & NH_3 \\ NH_3 \ Co \ NH_3 \\ NH_3 & O-N=O \end{bmatrix} Cl_2$$

Nitropentamminecobalt(III)   Nitritopentamminecobalt(III)
chloride   chloride.

Werner gave the name *salt isomerism* to this type of isomerism. In addition to this, the cobalt-ammines display the same phenomena of isomerism as have been described for the chrom-ammines.

Among the dinitrotetrammine cobalt(III) salts, $[Co(NO_2)_2(NH_3)_4]X$ (and also the dinitrodiethylenediamine cobalt(III) salts, $[Co(NO_2)_2 \ en_2]X$), the *cis* compounds are yellow brown, and the *trans* compounds bright yellow or reddish yellow. The former are accordingly known as *flavo*-salts, and the latter as *croceo*-salts.

The cobaltammines were the compounds for which the existence of mirror image isomerism (optical isomerism) was first detected among inorganic compounds by Werner in 1911.

As examples of *polynuclear complexes*, there may be cited two compounds which (with other similar substances) appear as intermediates in the autoxidation of cobalt(II) salt solutions. The first product obtained by the oxidation of ammoniacal cobalt(II) salt solutions by atmospheric oxygen is the compound $[(NH_3)_5Co^{III}-O-O-Co^{III}(NH_3)_5](NO_3)_4 \cdot 2H_2O$ (*decammine-μ-peroxodicobalt(III) nitrate*), crystallizing in brown-black prisms. Oxidation of ammoniacal cobalt(II) chloride solutions by oxygen gives, as first product, a very deep colored mixture of various polynuclear ammino and peroxo compounds (the so-called *melano-chloride*, μέλας = black, dark colored). One characteristic constituent of these mixtures is the so-called 'pure melanochloride', a sparingly soluble violet-black salt of the formula

$$\begin{bmatrix} (NH_3)_3 & & (NH_3)_3 \\ & Co^{III}\cdots NH_2-Co^{III} & (H_2O) \\ Cl_2 & & Cl \end{bmatrix} Cl_2$$

(*trichloroaquohexammine-μ-amidodicobalt(III) chloride*).

The brown-black peroxo salt first mentioned is converted, by the action of nitric acid or by decomposition in the dry state, into an intensely green salt, considered to be *decammine-μ-peroxocobalt(III, IV) nitrate* $[(NH_3)_5Co^{III}-O-O-Co^{IV}(NH_3)_5](NO_3)_5$. This compound, and other salts derived from the same cation, are interesting in that they contain cobalt in both the +3 and +4 states, as was first suspected by Werner, and more recently proved experimentally by Gleu (1938). Magnetic measurements (Malatesta, 1942) have also confirmed this conclusion.

$$\text{The compounds}\quad \begin{bmatrix} Co\Big(\overset{HO}{\underset{HO}{}} Co(NH_3)_4\Big)_3 \end{bmatrix} X_6 \qquad (\textit{dodecamminehexoltetracobalt(III) salts})$$

are the analogues of the *tetranuclear* type of complex already described for chromium. The salts of this type are brown black in color. It may be mentioned that there is also a series of red-brown cobalt salts, with the same empirical composition as the above, but of only half the molecular weight. These salts have the constitution

$$\begin{bmatrix} NH_3 & OH & NH_3 \\ NH_3 {\rightarrow} Co-OH{\rightarrow}Co{\leftarrow}NH_3 \\ NH_3 & OH & NH_3 \end{bmatrix} X_3$$

(hexamminetrioldicobalt(III) salts). This sort of polymerism is termed *nuclear polymerism*.

8. Analytical (Cobalt)

Cobalt falls within the ammonium sulfide group in the course of analysis. The sulfide precipitated by ammonium sulfide is no longer soluble in dilute hydrochloric acid after filtration in the presence of air, for the reasons discussed on p. 295.

It can therefore readily be separated (along with nickel sulfide which behaves similarly) from the other elements of the ammonium sulfide group, by digestion with very dilute hydrochloric acid. Cobalt can be separated from nickel by conversion to the sparingly soluble potassium hexanitrocobaltate(III), $K_3[Co(NO_2)_6]$, or by treating the solution with alkali cyanide and hypobromite (caustic alkali and bromine). Whereas nickel is precipitated as the black oxide hydrate under these conditions, cobalt remains in solution as the strongly complexed cyanide, $M^I_3[Co(CN)_6]$. A quantitative separation is possible by both methods. The *gravimetric* determination is best effected by electrolytic deposition of the metals from a sulfate solution, containing ammonium sulfate and an excess of ammonia. Cobalt can also be precipitated as the oxide, and converted to the metal by ignition in hydrogen.

Cobalt compounds give a blue coloration to the borax and metaphosphate beads, both in the oxidizing and reducing flames. This reaction provides a very convenient recognition of the presence of cobalt. Cobalt may be recognized microchemically, either as potassium hexanitrocobaltate, $K_3[Co(NO_2)_6]$, or in the form of the double salt with mercury thiocyanate, $Co(SCN)_2 \cdot Hg(SCN)_2$, which forms fine blue aggregates of needle-like crystals.

The reaction with thiocyanate and acetone, referred to on p. 300, is also very sensitive. The presence of Fe^{+++} ions interferes however, since the resulting red color masks the blue coloration. Kolthoff recommends adding ammonium fluoride, to bind the Fe^{+++} ions by complex formation and so avoid the interference. According to Ditz (1935), however, the ammonium fluoride reduces the sensitivity of the reaction for cobalt, so that it is better to remove the iron by precipitation with calcium carbonate.

9. Nickel (Ni)

(a) Occurrence

Nickel is found in Nature chiefly in combination with sulfur, arsenic, and antimony—as yellow nickel ore, *millerite*, NiS, red nickel ore ('Kupfernickel'), NiAs, *breithauptite*, NiSb, and also as *chloanthite*, white nickel ore or *niccolite*, NiAs₂, which is isomorphous with smaltite. Other minerals are *gersdorffite*, NiAsS, and *ullmannite*, NiSbS.

More important for the extraction of nickel than the foregoing minerals are *garnierite*, a magnesium-nickel silicate of variable composition formed by weathering, and certain varieties of pyrrhotite—especially *pentlandite*, which occurs in large amounts at Sudbury, Ontario—which can contain up to 3% nickel in isomorphous admixture.

Nickel is also found native, alloyed with iron, in many meteorites.

(b) History

Nickel was recognized as a new metal by Cronstedt, in 1751, and named by him after its occurrence in red nickel ore ('Kupfernickel'). The word 'Nickel' was at that time a term of abuse in miners' slang, and the miners applied the name 'kupfernickel' to any ore which, from its appearance, they judged to contain copper, although they were not able to extract any copper from it in spite of their efforts. Even after the discovery of nickel, the view that 'Kupfernickel' was a copper ore was maintained by many chemists for some time, until Bergmann (1775) described the properties of nickel more accurately, and showed how it could be prepared pure. Bergmann also recognized the close similarity between nickel and iron.

(c) Preparation

The most important nickel ores are the garnierite of New Caledonia and the Canadian pentlandite. The latter contains considerable amounts of copper, as well as nickel, and important quantities of the platinum metals also. The speisses which are by-products of copper and lead smelting are also often used as raw materials for the extraction of nickel.

In extracting nickel from garnierite, use is made of the high affinity of nickel for sulfur. The ore is melted with substances which can furnish sulfur, whereby the nickel is converted into Ni_3S_2, while most of the impurities are slagged, by formation of silicates. The product is partially roasted, and melted again with the addition of quartz sand, and by blowing in a converter; the 'crude metal', which at first usually has a very high iron content, is converted into 'nickel matte', consisting essentially of nickel sulfide. This is roasted to nickel(II) oxide(1), which is briquetted into cubes with wood charcoal, water, and a little flour (as binder). These yield cubes of crude, porous nickel when they are heated (2).

$$Ni_3S_2 + \tfrac{7}{2}O_2 = 3NiO + 2SO_2 \tag{1}$$

$$NiO + C = Ni + CO \tag{2}$$

Nickeliferous pyrrhotite is worked up in a similar manner, except that no sulfur is added in this case, but the sulfur content of the ore is reduced to some extent by roasting. Further, the high copper content of the ore leads to the production of a copper-nickel alloy, in place of crude nickel. This can be resolved into its constituents by electrolytic methods, but it is generally preferred to effect as complete a separation as possible between copper and nickel, before carrying out the reduction to metal. In Canada it is usual to employ the Orford or 'tops and bottoms' process (of Thompson). The copper-nickel matte is melted with sodium hydrogen sulfate and coke in a shaft furnace. Sodium sulfide, formed by reduction of the 'salt cake', forms double compounds with the copper(I) sulfide present in the matte, and these pass into the melt which collects in the upper part of the furnace, whereas the nickel sinks to the bottom. After cooling, the 'tops', enriched in copper, can be separated from the 'bottoms' which are enriched in nickel. In this way, by repeating the process if necessary, it is possible to obtain a product containing 70% nickel with only 2–3% copper. The crude nickel is refined either by electrolysis or by the nickel carbonyl process.

The *carbonyl process* is based on the preparation of volatile nickel carbonyl, $Ni(CO)_4$, by the action of carbon monoxide on nickel, and the subsequent thermal decomposition of the carbonyl. Spongy crude nickel, obtained by reducing copper-poor nickel oxide from the Orford process by means of water gas, can be treated with carbon monoxide at 50° and at ordinary pressure (Mond-Langer process). Nickel carbonyl can, however, also be prepared directly from the nickel-copper matte by the action of carbon monoxide at 200–250° and 200 atm. (I.G. Farbenindustrie process). Decomposition of the nickel carbonyl is carried out at about 200° under ordinary pressure, the carbon monoxide thereby liberated being returned to the cycle of operations. Nickel of very high purity is obtained by the carbonyl process (99.9–99.99%). A considerable proportion of the world's pure nickel is prepared by this method.

Electrolytic refining is used chiefly for platinum-bearing crude nickels, since the platinum metals can readily be recovered from the anode slimes obtained in the process.

(d) Properties

Nickel is a silver white metal of high luster, density 8.85–8.90, hardness 3.8 on Mohs' scale, m.p. 1452°. It takes an excellent polish, is very ductile, can be forged, welded, rolled into sheet, and drawn into wire. It is ferromagnetic, although to a smaller extent than iron. The electrical conductivity at 18° is 13.8% of that of silver (14.9% of that of copper). Certain nickel alloys have a considerably lower conductivity—e.g., constantan (p. 310). The thermal conductivity of nickel is about 15% of that of silver.

Compact nickel is very resistant towards water and air at ordinary temperature, but the finely divided metal may be pyrophoric under certain conditions. A heated nickel wire will burn in oxygen, with showers of sparks. Nickel sheet tarnishes like steel when it is heated in air. The attack of dilute acids on nickel is perceptibly slower than on iron; it dissolves readily in dilute nitric acid but, like iron, it is passivated by concentrated nitric acid.

Nickel (like palladium and platinum) ordinarily has a face-centered cubic lattice, $a = 3.517$ Å. Bredig (1927–31) found, however, that nickel cathodically sputtered in hydrogen had hexagonal closest packing ($a = 2.65$, $c = 4.32$ Å)—i.e., the structure usual for cobalt. According to Leclerc and Michel, hexagonal nickel is also formed when ordinary, finely divided nickel is kept for a long time in an atmosphere of CO. It would accordingly appear to be the stable modification at ordinary temperature. It changes at $250°$ into the cubic form. Hexagonal nickel is not ferromagnetic.

Nickel stands next to cobalt in the electrochemical potential series. Its normal potential is $+0.25$ volts, relative to the normal hydrogen electrode.

Heated nickel catches fire in chlorine and bromine. It also unites with phosphorus, arsenic, and antimony (cf. Table 29, p. 244–5). Nickel is brittle if it contains much phosphorus, but a small phosphorus content (about 0.3%) improves the casting and working qualities of the metal, because of its deoxidizing action. In the molten state, nickel also takes up carbon very readily (up to about 6.25%.) On solidification, most of the carbon separates out as graphite (nickel can hold 0.5 weight % C in solid solution at $1315°$, and only 0.15 wt.% at room temperature). A compound between nickel and carbon is not stable in the solid state. However, the metastable compound Ni_3C—the existence of which has been established by X-ray methods—can be obtained by the thermal decomposition of CO on finely divided nickel (Bahr, 1928). Nickel combines extremely vigorously with aluminum. An equi-atomic mixture of nickel and aluminum combines with explosive violence at $1300°$, forming the compound NiAl (Al_2Ni and Al_3Ni also exist, but decompose when they are melted). Nickel is completely miscible with cobalt, both in the solid and liquid states, but no compounds are formed. Its behavior with manganese, and also with chromium, is similar, at least at high temperatures. It is also practically completely miscible with iron in the solid state, but forms with iron a 'super-structure compound' $FeNi_3$ (Fig. 41, p. 266). See Tables 29 (p. 244–5) and 30 (p. 248) for its behavior with other metals.

Metallic nickel decomposes gaseous ammonia into hydrogen and nitrogen at moderate temperatures. It does not combine directly with nitrogen, and does not dissolve it to any appreciable extent. Hydrogen, however, can be absorbed in fairly large quantities by finely divided nickel, especially at high temperatures. Even at ordinary temperature, a considerable occlusion of hydrogen may be observed when hydrogen ions are discharged at a porous nickel surface. It is still open to question whether definite *hydrides* of nickel are formed (in weighable amounts). The ability of nickel to take up hydrogen, and to activate it by converting it to the atomic state, is the basis of its action as a hydrogen carrier for unsaturated compounds, and of its use as a catalyst in hydrogenations.

(e) Uses

Nickel was used as a coinage metal in very old times. Bactrian coins of the pre-Christian period are known, which consist of 78% copper and 21% nickel. Typical modern coinage alloys may contain 75% copper, 25% nickel, and pure nickel has also been used. The use of nickel for alloys such as 'nickel silver', widely used for utensils, is also old. Articles made of such alloys, then called *packfong*, came to Europe from China in the 18th century. New silver, or nickel-silver, is the name applied to alloys containing about 10–20% nickel, 40–70% copper, 5–40% zinc, which are silver-white, fairly resistant chemically, and capable of taking a good polish. Pure nickel is now also widely used for domestic appliances and for laboratory equipment (spatulas, tongs, crucibles, etc.). Iron objects were formerly commonly electrolytically plated with nickel, as a protection against corrosion and to achieve a pleasing appearance: the use of nickel for this purpose, has largely been replaced by chromium plating, especially, e.g., in the automobile industry. Large quantities of nickel are used in the steel industry, since the addition of nickel makes it possible to produce steels of great tensile strength and toughness (cf. p. 268). Certain nickel alloys are used as electrical resistance wires, because of their low conductivity—e.g., *constantan*, an alloy of 40% Ni, 60% Cu, which also has a very low temperature coefficient of conductivity; *nickeline*, with about 31% Ni, 56% Cu, and 13% Zn, which also has a high electrical resistivity and low temperature coefficient; *manganin* (4% Ni, 12% Mn, 84% Cu), which also has a small thermoelectric power towards copper, and is therefore chiefly employed for making precision resistances. *Nichrome* wire (chromium-nickel, 60% Ni, 40% Cr) is widely used for winding electric furnaces and heaters. Finely divided nickel or nickel oxide is used extensively as a catalyst in the hardening of fats.

The salts of nickel are used in the preparation of nickel plating baths, and to some extent in ceramics, to produce certain colors.

10. Nickel Compounds

Nickel is predominantly electropositively bivalent in its compounds, and *all the simple nickel salts* are derived from that valence state. Hydrated simple nickel salts and their solutions are light green in color. This is therefore the color of the hydrated nickel ion Ni^{++}. The anhydrous salts are generally different in color.

Almost all nickel salts can form addition compounds or coordination compounds—both of the type of the *ammines* (to which, in principle, the hydrated simple salts belong) and of the type of acido compounds. The coordination compounds of nickel are less strongly complexed than those of cobalt—especially those of $+3$ cobalt.

The oxides of nickel correspond in composition and in general properties with those of cobalt. The monoxide, NiO, is basic in character. The higher oxides Ni_2O_3 and NiO_2 (which are only known in the hydrated state) have definite, although weak, acidic properties. This is shown in their ability to form double oxides with the more strongly basic oxides. In so doing, the higher oxidation state of the nickel is simultaneously stabilized.

In addition to the higher oxides of nickel, several complex compounds are known in which nickel is present in the tripositive and tetrapositive states. Tripositive nickel is present in the complexes $[Ni(P(C_2H_5)_3)_2Br_3]$ and $[Ni((As(CH_3)_2)_2C_6H_4)_2Cl_2]^+$, as has been proved by magnetic measurements (Jensen, 1949; Nyholm, 1950). Both these complexes contain neutral ligands with highly polarizable atoms (P and As, respectively); the elevation of the nickel to the higher oxidation state is thereby favored. Simpler complex compounds of $+4$ nickel are known—namely $K_2[NiF_6]$ (Klemm, 1949), and $Na[NiIO_6] \cdot xH_2O$ and $K[NiIO_6] \cdot xH_2O$ (Rây, 1946). The molybdatocomplexes of tetrapositive nickel (Rây, 1948) are of somewhat complicated composition. Feigl (1924) and Hieber (1949) have obtained nickel(IV) complexes of organic thio-compounds.

See also p. 314 for the fluoro complexes of $+3$ and $+4$ nickel.

In a few exceptional compounds, nickel can also function in the *unipositive state*. Thus the complex cyanide of $+1$ nickel, $K_2Ni(CN)_3$, is definitely characterized. This compound is diamagnetic, and is therefore probably to be regarded as a binuclear compound, $K_4[Ni_2(CN)_6]$. It can be converted into $K_2[Ni(CN)_3CO]$ by additive combination with CO (Nast and Goehring, 1946). If it is permissible to regard NO as a neutral ligand in the compounds $Ni(NO)Cl$, $Ni(NO)Br$ and $Ni(NO)I$, obtained by Hieber and Nast (1940), these can also be included among the compounds of unipositive nickel. The substances which were formerly considered to be nickel(I) oxide and sulfide are not homogeneous chemical compounds, according to the work of Levi and Bornemann.

By treating $K_2[Ni(CN)_4]$ with metallic potassium, dissolved in liquid ammonia, Eastes and Burgess (1942) prepared the compound $K_4[Ni(CN)_4]$; this contains nickel in the zero oxidation state. The complex present in this compound is apparently related to $Ni(CO)_4$ (involving covalent bonds formed from the $4s4p^3$ hybrid orbitals). Chemical evidence supporting this is provided by the compounds containing both CO and CN^- in the complex simultaneously, such as $K[Ni(CO)_3(CN)]$, prepared by Burg and Dayton (1949).

(a) Oxides and Hydroxides

(*i*) *Nickel(II) oxide*, NiO, remains as a residue when nickel(II) hydroxide, carbonate, or nitrate is strongly ignited. It is a green powder, practically insoluble in water but readily soluble in acids. It is readily reduced to the metal by heating it in hydrogen. When very strongly ignited, it is transformed into grey-black regular octahedra with a metallic luster, and thereby loses its solubility in acids. Nickel(II) oxide occurs native, as *bunsenite*, in pistachio-green crystals, of density 6.4. It finds applications in ceramics, for colors and enamels, and is used to give a grey color to glass. It is also used as a catalyst. The heat of formation of nickel(II) oxide is 58.6 kcal per mol.

(*ii*) *Nickel(II) hydroxide* is precipitated when alkali hydroxides are added to nickel salt solutions; it is formed as a voluminous apple green precipitate, which changes to a green crystalline powder on prolonged standing under the solution. Nickel(II) hydroxide is readily soluble in acids, and also in aqueous ammonia and ammonium salt solutions. It is converted to the oxide by ignition.

The decomposition of $Ni(OH)_2$ to form the oxide takes place at a temperature as low as $230°$ (at 10 mm water vapor pressure), as was found by Hüttig (1930). However, a small proportion of the water formed in the decomposition of the hydroxide is firmly retained by the oxide, and is fully given up only at a red heat.

NiO crystallizes with the NaCl structure ($a = 4.17$ Å); $Ni(OH)_2$ forms a layer lattice of brucite type ($a = 3.08$ Å, $c = 4.61$ Å).

(*iii*) *Higher Nickel Oxides*. When nickel carbonate or nickel nitrate is cautiously heated to about $300°$ in air, grey to black powders are obtained, which dissolve in hydrochloric acid with the evolution of chlorine, and in oxyacids with the evolution of oxygen. It was formerly assumed that the product was nickel(III) oxide, Ni_2O_3, although the composition of the material is subject to considerable variations, and the active oxygen is generally far below that required by the formula Ni_2O_3 (Le Blanc, 1926). This was originally overlooked, because the products assumed to be Ni_2O_3 always contained water. 'Nickel(III) oxide' can be obtained in still more highly hydrated form, but also with a higher content of active oxygen, by treating an aqueous suspension of nickel(II) hydroxide with chlorine, or by the electrolytic oxidation of nickel(II) hydroxide in potassium hydroxide.

Foerster states that an electrode of Ni_2O_3, dipping into 2.8-N potassium hydroxide, has a potential $+0.48$ volts, relative to the normal hydrogen electrode. However, the potential displayed by $Ni(OH)_2$ immediately after it has been electrolytically oxidized is at first considerably higher, +0.6 volt. Foerster attributes this to the initial formation of some nickel(IV) oxide, NiO_2, which gradually decomposes, over the course of a few days, to give nickel(III) oxide, Ni_2O_3. Since the two compounds form solid solutions with one another, the fall in potential takes place gradually, and not discontinuously as in other cases. The formation of nickel dioxide hydrate is also the primary reaction in the addition of sodium hydroxide and chlorine to nickel salt solutions.

Hofmann obtained the compound $Ni_2O_3 \cdot 2H_2O$ in the crystalline state by burning potassium metal on nickel sheet. The same compound can also be obtained by precipitation from solution (Ott, 1933). Ott found that the mechanically held water (i.e., that which is not chemically combined) could be removed from the precipitate by drying over calcium chloride, whereas more complete dehydration was accompanied by decomposition and loss of oxygen. According to Glemser, there are several distinct phases present in the hydrated higher oxides—$NiO_{1.07-1.22} \cdot xH_2O$, $Ni_3O_4 \cdot 2H_2O$, $Ni_2O_3 \cdot xH_2O$ and $NiO_2 \cdot xH_2O$. These are not peroxides, but are compounds containing nickel in the higher oxidation states, in amounts corresponding to the analytically found composition.

The formation of nickel(III) oxide hydrate by the anodic oxidation of nickel(II) hydroxide plays a part in the working of the Edison accumulator. The electrodes of this accumulator consist of highly active iron powder and of 'nickel(III) oxide', respectively, the latter being mixed with finely flaked nickel to improve its conductivity. The electrolyte is potassium hydroxide. During discharge, the nickel(III) oxide hydrate is reduced to nickel(II) hydroxide, and the iron is oxidized, substantially to iron(II) hydroxide. The converse processes take place during charging:

$$Fe + Ni_2O_3 + 3H_2O \xrightleftharpoons[\text{Charge}]{\text{Discharge}} Fe(OH)_2 + 2Ni(OH)_2.$$

The Edison or nickel-iron ('NIFE') accumulator withstands heavy overloading, or prolonged standing in the discharged state, without suffering damage, unlike the lead accumulator. The terminal voltage is about 1.3 volts during discharge, and 1.7 volts or more during charging. The considerable difference between on-charge and discharge potentials involves a low over-all efficiency of energy conversion. Nevertheless, the Edison accumulator finds important applications—e.g., in electric traction.

If freshly prepared nickel(III) oxide hydrate is treated in the cold with concentrated acetic acid, a black solution is obtained, from which black nickel(III) oxide hydrate can once more be precipitated by caustic soda. The solution at once turns green on the addition of reducing agents, and decomposes on standing or when it is warmed. It is not certain whether nickel(III) acetate is present in the solution, or whether it is merely a colloidal dispersion of nickel(III) oxide hydrate.

Double oxides containing the nickel in higher oxidation states can be obtained by heating NiO with BaO or SrO in an atmosphere of oxygen. Schmahl obtained the compounds $Ba_2Ni^{III}_2O_5$ and $Sr_2Ni^{III}_2O_5$ in this way; $BaNi^{IV}O_3$ was prepared in a similar manner by Lander and Wooten. It is a black powder (X-ray density 6.22), which is insoluble in caustic alkalis, but soluble in dilute hydrochloric acid with the evolution of chlorine. When heated to 730° in oxygen (at 730 mm press.), it loses oxygen to form a phase which is structurally distinct from $BaNiO_3$. Oxygen is lost continuously on further heating, until the composition is $Ba_2Ni_2O_5$. $Ba_2Ni_2O_5$ can be melted in an atmosphere of oxygen at 1200°, without decomposition.

(b) Sulfides

Nickel(II) sulfide is precipitated from nickel salt solutions by ammonium sulfide—initially in an acid-soluble form (α-NiS), which like cobalt(II) sulfide rapidly changes when exposed to air in contact with the solution, into a sulfur-richer compound which is no longer soluble in very dilute hydrochloric acid. Nickel sulfide remains to some extent in colloidal dispersion (as may be perceived from the brown color of the medium) when it is precipitated with ammonium sulfide. It can be flocculated by adding acetic acid and boiling. Hydrogen sulfide precipitates a crystalline nickel sulfide from acetic acid solution (β-NiS); this is sparingly soluble in cold dilute hydrochloric acid.

a-NiS is amorphous. β-NiS crystallizes with the NiAs structure (Fig. 43, cf. also p. 249). A third (rhombohedral) modification γ-NiS is found native as *millerite*, but can also be obtained from solution under certain conditions. Artificially prepared γ-NiS gradually changes into β-NiS in contact with the solution.

Millerite forms lance-like radiating or sometimes hair-like crystals, of bright brassy yellow to brown black color, density 5.3–5.9. It has a rhombohedral structure (Ni $\leftrightarrow$ S = 2.18 Å). Considerable quantities of nickel sulfide are often present in pyrrhotite. *Pentlandite*, (Ni, Fe)$_9$S$_8$, is isotypic with Co$_9$S$_8$. It also occurs as a mineral and is an important constituent of the nickeliferous ores of Ontario.

Nickel disulfide, NiS$_2$, can be obtained as a dark iron-grey product by strongly heating nickel carbonate with potassium carbonate and sulfur. It has the same crystal structure as CoS$_2$, FeS$_2$, and MnS$_2$ (a = 5.74 Å). A sulfide of the composition Ni$_3$S$_4$ occurs native as *polydymite*, with the same crystal structure as linnaeite (Co$_3$S$_4$) and the spinels (a = 9.5 Å). In addition to these, the incongruently melting compounds Ni$_3$S$_2$, Ni$_6$S$_5$, and Ni$_7$S$_6$ can be obtained from melts, and from the thermal degradation of NiS.

Double sulfides of nickel are also known. Thus when nickel(II) oxide is fused with sulfur and an excess of potassium carbonate at 600–1100°, yellow leaflets with a metallic luster are obtained, with the composition K$_2$S · 3NiS. Belucci obtained the compound BaS · 4NiS (dark red crystals) in a similar manner from nickel chloride, sulfur and barium oxide.

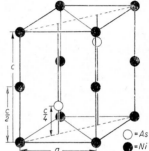

Fig. 43. Unit cell of nickel arsenide, NiAs.
a = 3.61 Å, c = 5.03 Å.

(c) Arsenides and Antimonides

Nickel combines very readily with arsenic and antimony, as it does with sulfur. It has already been mentioned that the arsenides NiAs and NiAs$_2$ and the antimonide NiSb$_2$ occur as minerals. Mixed crystals of NiAs and NiSb (*arite*) are also found native. The compounds formed from melts are listed in Table 29, p. 244–5.

The crystal structure of ordinary nickel arsenide (red nickel ore) is represented in Fig. 43. The same structure is found for NiSb and for β-NiS, β-NiSe and NiTe (cf. Table 31, p. 250). The diarsenide, NiAs$_2$, exists in two modifications—commonly as the cubic *niccolite* (chloanthite), and more rarely rhombic, as *rammelsbergite*. The latter has the same crystal structure as löllingite, FeAs$_2$, and safflorite, CoAs$_2$. NiAsS (gersdorffite) and NiSbS (ullmannite) are isotypic with cobalt glance (cf. p. 296).

(d) Nickel Tetracarbonyl

Nickel tetracarbonyl, Ni(CO)$_4$, discovered in 1888 by Mond and Langer, is formed by passing carbon monoxide at 50–100° over finely divided nickel (obtained by reduction of NiO with hydrogen below 400°).

In the laboratory, nickel carbonyl is most conveniently prepared, according to Hieber [*Z. anorg. Chem.* 269 (1952), 28], by the action of carbon monoxide on the readily obtainable nickel(II) phenyldithiocarbamate, Ni(S · SC · NHC$_8$H$_5$)$_2$. As Hieber has shown, this reaction depends upon the fact that nickel compounds of certain organic thio acids on treatment with carbon monoxide undergo disproportionation in the following manner:

$$2\text{Ni}^{\text{II}} + 4\text{CO} \longrightarrow \text{Ni}^{\text{IV}} \text{ (complex)} + \text{Ni}°(\text{CO})_4.$$

If the conditions for the formation of a nickel(IV) compound are not provided, carbonyl formation does not occur. (Cf. p. 250, 318.)

The carbonyl is a colorless liquid which boils at 43°, solidifies at —25°, and has a density of 1.310 at 20°. Its critical temperature is about 200° and critical pressure about 30 atm. The vapor density corresponds to the molecular weight 173 (calculated 170.7). The vapor is *poisonous*.

For the industrial application of nickel carbonyl, for the preparation of pure nickel, see p. 308.

Unlike most other nickel compounds, which are paramagnetic, nickel carbonyl is *diamagnetic*. It has a high dispersion. The compound is perfectly stable at ordinary temperature when kept in a sealed tube, but is oxidized gradually in air. It burns with a bright luminous flame when it is ignited: mixtures of nickel carbonyl vapor and air are explosive. It is insoluble in water, and is therefore not attacked by dilute acids and alkalis. It dissolves in ether, benzene, and chloroform. Nickel carbonyl vapor dissociates into its components when it is passed through a tube heated to 180–200°, depositing a brilliant mirror of metallic nickel. Strong oxidizing agents (e.g., concentrated nitric acid or chlorine) decompose nickel carbonyl, forming nickel(II) salts and carbon dioxide or carbonyl chloride (phosgene).

In the presence of moisture, nickel carbonyl reacts with NO, forming deep blue nickel nitrosyl hydroxide, $Ni(NO)OH$.

(e) Nickel Halides

(*i*) *Nickel chloride*, $NiCl_2$, is obtained in the anhydrous state by burning nickel in chlorine, or by heating the hydrate or ammoniate. It forms sparkling golden yellow crystal scales, which are soft, resembling talc in feel, and are readily sublimed (density 2.56, crystal structure—cf. p. 249). It is converted into nickel(II) oxide when it is gently heated in air. It is readily soluble in water and alcohol (100 g of water dissolve 64.0 g of $NiCl_2$ at 20°, and 81.2 g at 60°). From the light green aqueous solution, it crystallizes as the *hexahydrate*, $NiCl_2 \cdot 6H_2O$, in grass-green granular prismatic monoclinic crystals, isomorphous with those of cobalt chloride hexahydrate. It is gradually converted to the dihydrate on standing over concentrated sulfuric acid. A monohydrate is also known.

Nickel chloride forms double salts with the alkali chlorides—e.g., $NH_4Cl \cdot NiCl_2 \cdot 6H_2O$, green rhombic prisms; $CsCl \cdot NiCl_2$, yellow crystals.

(*ii*) *Nickel bromide*, $NiBr_2$, and *nickel iodide*, NiI_2, are very similar to the chloride. They generally crystallize from solutions as the hexahydrates. The trihydrate and enneahydrate of the bromide are known, as also are double bromides.

(*iii*) *Nickel fluoride*, NiF_2, is best obtained in the anhydrous state by heating the ammonium fluoride double salt. It exists as long prisms, which are light brown to green when dry, very sparingly soluble in water, and insoluble in alcohol and ether. The trihydrate, $NiF_2 \cdot 3H_2O$, is obtained by evaporating a solution of nickel hydroxide in hydrofluoric acid, in the form of pale green granular crystals. Double salts have been obtained by melting the appropriate salts together, or by mixing their solutions—e.g., $KF \cdot NiF_2 \cdot H_2O$, a sulfur-yellow precipitate.

The compounds $K_3[Ni^{III}F_6]$ and $K_2[Ni^{IV}F_6]$ are formed by the action of F_2 on a mixture of $NiCl_2$ and KCl. $K_2[NiF_6]$ is the only known fluoro complex which must be classed as a 'penetration' complex; it is diamagnetic, and therefore involves d^2sp^3 hybrid covalent bonds. All other fluoro complexes have the magnetic susceptibility appropriate to the central ion of the complex; they thus apparently involve ionic binding forces, and are classified as 'normal' complexes.

(*iv*) *Basic Nickel Halides*. When a warm, moderately concentrated nickel chloride solution is incompletely precipitated with sodium hydroxide, a *basic nickel chloride* is obtained, of the average composition $NiCl_2 \cdot 3Ni(OH)_2$, but with a rather wide range of existence (Cl : OH varying between 1 : 1.9 and 1 : 4.3). This forms a 'simple layer lattice', derived from that of nickel hydroxide by the partial, random substitution of Cl^- ions for OH^- ions; a certain expansion of the lattice takes place simultaneously ($a = 3.18$ Å, $c = 5.48$ Å). The considerable dimensional change accounts for the fact that the hydroxide lattice does not pass by a continuous transition into that of the basic salt. The *basic bromide* has the same structure and average composition as the basic chloride, but the range of homogeneity is much narrower (Br : OH varying from 1 : 2.8 to 1 : 3.3). There is also a second basic nickel chloride, Ni(OH)Cl, of constant, or nearly constant, composition. This is formed when a warm *concentrated* solution of nickel chloride is incompletely precipitated with sodium hydroxide. It also possesses a 'simple layer lattice', derived from that of $NiCl_2$ by the substitution of OH^- ions for one half of the Cl^- ions. This probably occurs, as in Cd(OH)Cl, with all the chloride ions being on one side of the nickel ion sheet, and all the hydroxyl ions on the other (cf. Fig. 50, p. 449) (Feitknecht, 1936).

(f) Other Nickel Salts

(*i*) *Nickel cyanide and Complex Cyanides.* Nickel cyanide is precipitated in a green hydrated form when CN^- ions are added to solutions of nickel salts. The yellow-brown anhydrous salt, $Ni(CN)_2$, is obtained by heating the precipitate to 180–200°. Freshly precipitated nickel cyanide dissolves with a golden yellow color in an excess of alkali cyanide solution, and complex salts (*cyanonickelates*) can be isolated by evaporating the resulting solutions— e.g., $Na_2[Ni(CN)_4] \cdot 3H_2O$ (long yellow six sided prisms which lose their water when gently heated). The cyanonickelates are decomposed by strong acids, nickel cyanide being precipitated.

The action of 'nascent' hydrogen (e.g., from sodium amalgam and water) upon an alkali cyanonickelate(II) turns the solution blood red in color. Belucci (1914) showed that this reaction was due to the reduction of the nickel to the +1 state. He was able to isolate the red potassium cyanonickelate(I), $K_2[Ni(CN)_3]$, and to establish its composition by analysis. This compound is diamagnetic, both in solution and in the crystalline state. It may be inferred from this that it must be a *binuclear* complex. This inference has been confirmed by Nast (1952), by an X-ray structure determination, which showed the compound to be

potassium tetracyano-μ-dicyanonickelate(I), $K_4[(CN)_2Ni\underset{CN}{\overset{CN}{\diagup\diagdown}}Ni(CN)_2]$.

(*ii*) *Nickel thiocyanate and Complex Thiocyanates.* Nickel thiocyanate, $Ni(SCN)_2$, is obtained as a (generally somewhat hydrated) yellow brown precipitate, by dissolving nickel carbonate in aqueous thiocyanic acid, or by double decomposition of nickel sulfate with barium thiocyanate. It dissolves fairly easily in water to give a green solution. Well crystallized complex salts (thiocyanatonickelates) can be obtained by evaporation of solutions containing alkali thiocyanates. These can be recrystallized from alcohol, and in some cases from water, without decomposition—e.g., $Na_2[Ni(SCN)_4] \cdot 8H_2O$, green crystals, and $K_4[Ni(SCN)_6] \cdot 4H_2O$, blue crystals.

(*iii*) *Nickel nitrate* usually crystallizes from aqueous solutions as the hexahydrate, $Ni(NO_3)_2 \cdot 6H_2O$, in emerald green monoclinic crystals, isomorphous with those of the corresponding cobalt salt. The tetrahydrate crystallizes above 54.0°, and the dihydrate above 85.4°. An enneahydrate is stable below —3° (Sieverts, 1934). Nickel nitrate is used in ceramics to produce brown colors.

Of the *basic nitrates* of nickel, that with the composition $Ni(NO_3)_2 \cdot Ni(OH)_2 \cdot 6H_2O$ is noteworthy because of its structure. According to Feitknecht it probably has a complex ionic lattice, $[Ni(H_2O)_6][Ni(OH)_2(NO_3)_2]$, in contrast with the majority of basic nickel salts, which have layer lattice structures.

(*iv*) *Nickel nitrite and Complex Nitrites.* Double decomposition between barium nitrite and nickel sulfate yields a solution from which, on evaporation, *nickel nitrite*, $Ni(NO_2)_2$, is obtained in orange crystals, which are stable in air and soluble in water, giving green solutions. Complex nitrites crystallize from solutions containing the nitrites of the alkalis and alkaline earths; these are of the type $M^I_4[Ni(NO_2)_6]$ (hexanitronickelates)—e.g., $K_4[Ni(NO_2)_6]$, brownish red octahedra which are insoluble in alcohol, but can be recrystallized from water.

(v) *Nickel sulfate* can be obtained by dissolving the metal, its oxide, or carbonate, in dilute sulfuric acid. In the technical preparation, nickel is generally dissolved in nitric acid or aqua regia, and evaporated down with sulfuric acid. The diluted solution is treated with barium carbonate to precipitate impurities, and is evaporated. The *hexahydrate*, $NiSO_4 \cdot 6H_2O$, crystallizes out on cooling, in two modifications—bluish green tetragonal crystals formed between 31.5° and 53.3°, and green monoclinic crystals above 53.3°. The latter are isomorphous with magnesium sulfate hexahydrate. The emerald green *heptahydrate*, $NiSO_4 \cdot 7H_2O$ ('nickel vitriol') generally crystallizes at ordinary temperature from pure aqueous solutions.

Nickel vitriol is occasionally found native, as *morenosite*. Unlike iron- and cobalt vitriols, nickel vitriol is rhombic, and has the same crystal form as epsom salt, with which it forms a complete range of mixed crystals. Iron can replace nickel in rhombic nickel vitriol only to the extent of 21%, and nickel can replace iron in monoclinic iron vitriol up to about 50%.

Double salts crystallize from solutions of nickel sulfate containing also the alkali sulfates or ammonium sulfates. The most important of these are the salts of the general composition $M^I_2SO_4 \cdot NiSO_4 \cdot 6H_2O$, which are isomorphous with the corresponding salts of iron, cobalt, magnesium, and other bivalent metals ('Tutton's double salts'). The blue green ammonium nickel sulfate, $(NH_4)_2SO_4 \cdot NiSO_4 \cdot 6H_2O$, is used in the preparation of electroplating baths.

(vi) *Nickel carbonate and Double Carbonates.* Hydrated nickel carbonate, $NiCO_3 \cdot 6H_2O$, is obtained as a pale green, finely crystalline precipitate on the addition of alkali hydrogen carbonates to nickel salt solutions. Precipitation with neutral alkali carbonates yields basic nickel carbonate.

With ammonium hydrogen carbonate as precipitant, the double salt $(NH_4)HCO_3 \cdot NiCO_3 \cdot 4H_2O$ is obtained in apple green crystals, either admixed with $NiCO_3 \cdot 6H_2O$ or alone. It is possible to obtain anhydrous nickel carbonate also, by precipitation from hot solution under certain conditions—e.g., by heating a solution of nickel chloride with calcium carbonate in a sealed tube at 150°. In addition to the ammonium double salt mentioned, other double carbonates of nickel are known—e.g., $K_2CO_3 \cdot NiCO_3 \cdot 4H_2O$, apple green needles, and $Na_2CO_3 \cdot NiCO_3 \cdot 10H_2O$, grass green microscopic rhombohedra.

Nickel carbonate is used in the preparation of ceramic pigments, and as the starting material for the preparation of other nickel salts.

(vii) *Nickel oxalate* may be obtained by dissolving nickel hydroxide or carbonate in aqueous oxalic acid solution (metallic nickel does not dissolve in oxalic acid). The compound is also formed as a flocculent green precipitate by adding oxalic acid to nickel salt solutions. It has the composition $NiC_2O_4 \cdot 2H_2O$ after drying at 100°. It is practically insoluble in water, but dissolves readily in strong acids and in ammonia solution (forming complex salts in the latter case). Nickel oxalate is also somewhat soluble in boiling potassium oxalate solution; the light green sparingly soluble double salt $K_2C_2O_4 \cdot NiC_2O_4 \cdot 6H_2O$ crystallizes from the solution on cooling.

(viii) *Nickel acetate*, $Ni(CH_3 \cdot CO_2)_2 \cdot 4H_2O$, separates out in apple green crystals when a solution of nickel hydroxide in acetic acid is evaporated in the cold. It is readily soluble in water (1 part in 6) but insoluble in alcohol. The aqueous solution has a sweet taste. Hydrolysis, with deposition of nickel hydroxide, takes place rapidly in hot solutions.

(ix) *Nickel phosphates. Nickel orthophosphate* may be taken as an example of the rather large number of compounds formed by nickel with the phosphoric acids. It is obtained as a pale apple green flocculent precipitate, insoluble in water but soluble in acids, when secondary sodium phosphate (Na_2HPO_4) is added to a nickel salt solution. The air dried precipitate corresponds roughly in composition to $Ni_3(PO_4)_2 \cdot 7H_2O$. If ammonium phosphate is used as precipitant, the compound $Ni(NH_4)PO_4$ is formed as a precipitate which is at first flocculent, but which turns crystalline on standing in contact with the mother liquor.

(x) *Nickel silicate*. *Nickel orthosilicate*, $Ni_2[SiO_4]$, isotypic with olivine (Vol. I, Fig. 90, p. 499), is obtained by heating NiO with SiO_2. A hydrated nickel silicate (reputedly of the composition $Ni_2Si_3O_8 \cdot 2H_2O$ but probably closely related in structure to meerschaum, q.v.) is found in Nature, as *konarite*, in the form of small green crystal grains or leaflets. An earthy variety is called *röttisite* (after its occurrence at Röttis in Vogtland). More abundant than this pure silicate of nickel, which occurs only in small amounts, is *garnierite*, an apple green to emerald green hydrated magnesium-nickel silicate, of very variable nickel content, formed by the weathering of nickeliferous olivine rocks. It is found in very large quantities at Noumea, in New Caledonia, and is worked as a nickel ore both there and in Oregon.

(g) Nickel Ammoniates

Most nickel salts can combine with ammonia, either when the gas is passed over the anhydrous salt or when ammonia is added to their solutions. Thus hexamminenickel(II) chloride, $NiCl_2 \cdot 6NH_3$, is formed from nickel chloride and ammonia, as a faintly violet, almost white powder, which dissolves without decomposition in water containing ammonia. When ammonia is passed over anhydrous nickel sulfate, $NiSO_4 \cdot 6NH_3$ is formed as a pale violet powder, and a solution of nickel sulfate in concentrated aqueous ammonia yields $NiSO_4 \cdot 4NH_3 \cdot 2H_2O$ in dark blue rectangular prisms. $Ni(NO_3)_2 \cdot 4NH_3 \cdot 2H_2O$ can be similarly obtained in large, sapphire blue octahedral crystals.

It is to be assumed that ammonia is structurally combined in these compounds, in the same manner as with cobalt—i.e., the ammoniates cited contain the complex ions $[Ni(NH_3)_6]^{2+}$ and $[Ni(H_2O)_2(NH_3)_4]^{2+}$. The compounds derived from these and from other nickel ammoniate ions are readily soluble in water, and formation of such ions accounts for the solubility in ammonia solution of many nickel compounds which are insoluble in water—e.g., the hydroxide and the phosphate. Ammoniacal solutions of nickel salts, like corresponding solutions of copper(II) are usually blue in color. This has at times led not only miners but chemists to the mistaken view that certain nickel ores (such as nickel arsenide) contained copper.

In addition to the ammoniates of the types mentioned above, in which nickel has the coordination number 6, there are also ammoniates in which this maximum coordination number is not attained. Numerous compounds of the same type as the ammoniates are also known, but containing other nitrogen bases aniline, pyridine, quinoline, ethylenediamine, etc.

As with cobalt(II) salts, many nickel salts can combine with up to 4 such neutral ligands (or with 2 'bidentate' ligands) without displacement of the acid radical from coordination with the central atom. Compounds of the type $[NiX_2Am_4]$ are generally not very stable, and some show a tendency to undergo decomposition. Thus Hieber found that the compounds $[NiBr_2\ en_2]$ and $[Ni(SCN)_2\ en_2]$ react in aqueous solution according to $3[NiX_2\ en_2] + 6H_2O = 2[Ni\ en_3]X_2 + [Ni(H_2O)_6]X_2$. However, there are, many compounds of the type $NiX_2 \cdot 4Am$ in which the nickel only exercises the coordination number 4, so that the acid radicals are ionically bound—$[Ni\ Am_4]X_2$. This is true especially of those compounds in which the acid ions are only weakly polarizable.

Nickel also has a strong tendency to form *inner complex salts*. These are salts in which the metal that replaces the hydrogen ion is, at the same time, coordinatively bound at some other point to the group which functions as the acidic radical. Inner complex salts are often characterized by extreme insolubility in water, although many of them are soluble in non-ionizing organic solvents. They have, accordingly, become very important in modern analytical chemistry. One of the best known compounds of this type is *nickel dimethylglyoxime*, widely used in the determination of nickel.

The reaction of nickel with dimethylglyoxime (diacetyl dioxime) to form the scarlet inner complex salt follows the equation:

$$2 \begin{array}{c} CH_3-C=NOH \\ | \\ CH_3-C=NOH \end{array} + Ni^{++} = \begin{array}{c} CH_3-C=N-O \diagdown \quad \diagup O-N=C-CH_3 \\ | \qquad\qquad Ni \qquad\qquad | \\ CH_3-C=\!=\!=N \diagup \quad \diagdown N=\!=\!=C-CH_3 \\ \quad | \qquad | \\ \quad HO \qquad OH \end{array} + 2H^+.$$

Nickel has the coordination number four in the resultant inner complex salt. The compound separates only from ammoniacal, neutral, or weakly acidic (acetic acid) solutions; it has the properties of an anhydrobase and forms salts with strong or fairly strong acids. Thus, representing nickel dimethylglyoxime schematically by the formula [DHNiDH],

$$[DHNiDH] + 2HCl = [DH_2NiDH_2]Cl_2 \text{ (dark blue crystals)}.$$

Nickel can also form other types of dimethylglyoxime compounds under other conditions— e.g., $[DH_2NiCl_2]$ (yellow green), $[DNi(NH_3)_2]$ (carmine) and $[DNi(H_2O)_2]$ (carmine). Nickel has the coordination number 4 in all of these compounds, whereas cobalt has the coordination number 6 in all its stable compounds with dimethylglyoxime, even when it is dipositive. This is why cobalt forms no compound corresponding to [DH Ni DH], and therefore why cobalt is not precipitated by the reagent.

It is likely that the yellow, insoluble, amorphous compound $Ni(S \cdot C_6H_4 \cdot NH_2)_2$, prepared by Hieber (1952) by reaction of o-aminothiophenol with nickel salts in ammoniacal solution, is an *inner complex salt*. This compound is probably built up of a large number of $Ni(S \cdot C_6H_4 \cdot NH_2)_2$ groups which are joined to each other in the manner shown in constitutional formula I. This material is readily oxidized to a deep-blue, crystalline compound containing *tetra*positive nickel, which on the basis of composition and behavior probably has the constitution given in formula II. By reaction of an alcoholic solution of nickel chloride with an ethereal solution of dithiobenzoic acid, $C_6H_5CS(SH)$, Hieber obtained the blue nickel(II) dithiobenzoate, $Ni(C_6H_5CS_2)_2$, which also is easily oxidized to a nickel(IV) complex compound. The latter possesses a deep-violet color and has the constitution shown in formula III. The dimeric nature of the compound was demonstrated by Hieber by means of the freezing point lowering of its benzene solution.

Nickel(II)-o-aminothiophenolate   Tetra-o-aminothiophenolo-µ-dioxo-dinickel(IV)

Tetradithiobenzoato-µ-dithio-dinickel(IV)

11. Analytical (Nickel)

Nickel accompanies cobalt in the systematic analysis of cations. It is precipitated by ammonium sulfide as a black sulfide, which is no longer soluble in dilute, hydrochloric acid after it has been filtered and exposed to air. Nickel can be separated from cobalt by precipitating it from the solution of the complex cyanides, by means of alkali hydroxide and bromine or chlorine. It is also quantitatively separated from cobalt by precipitation with dimethylglyoxime, which was introduced as a reagent for nickel by Tschugaeff, and which is particularly suitable for the microanalytical detection and the gravimetric determination of nickel.

Although cobalt forms no precipitate with dimethylglyoxime, it reduces the sensitivity of the reaction if it is present in great excess, since the reagent is consumed in the formation of soluble complex compounds of cobalt. The solubility of the nickel dimethylglyoxime is also enhanced in the presence of these complexes. The interfering effect of $+3$ cobalt is appreciably less than that of the $+2$ state, and it is therefore advisable to oxidize the cobalt in ammoniacal solution by means of hydrogen peroxide before adding dimethylglyoxime. The reagent permits the detection of nickel in the presence of a 200 fold excess of cobalt. Dipositive iron also forms a red, but soluble, complex with dimethylglyoxime. Palladium and platinum form insoluble yellow complexes, corresponding exactly to the nickel salt in constitution.

Nickel may be determined quantitatively by means of dimethylglyoxime, and also by means of Grossmann's reagent (dicyandiamide sulfate) $(C_2H_6N_4O)_2 \cdot H_2SO_4$. This also involves the formation of an inner complex salt. A yellow precipitate is formed in ammoniacal solution on the addition of potassium hydroxide; this has the constitution

$$
\begin{array}{ccc}
H_2N-C-O & & O-C-NH_2 \\
\| & & \| \\
N & Ni & N \\
| & \nearrow \nwarrow & \| \\
H_2N-C=NH & HN=C-NH_2 \, .
\end{array}
$$

Nickel can also be very conveniently and quantitatively deposited electrolytically from ammoniacal solution.

It may also be precipitated as 'nickel(III) oxide hydrate' by means of potassium hydroxide and bromine water, and weighed as the monoxide after ignition.

B. THE PLATINUM METALS

12. General

The platinum metals are characterized by their inertness towards acids, their relative infusibility, and in some cases by their great hardness. They also display excellent *catalytic* properties.

In their *compounds*, *ruthenium* and *osmium* show a number of resemblances to iron, which stands vertically above them in the Periodic Table, and also to iron's neighbour manganese. *Rhodium* and *iridium* have a clear relationship to cobalt, and in particular they very readily form complex compounds of the type $[M^{III}Am_6]X_3$. *Palladium* forms a transition to silver in its properties as does *platinum* to gold, in the same way as nickel links iron and cobalt with copper. The tendency to exercise the higher valence states decreases from left to right among both the light and the heavy platinum metals, as it does in the iron group.

Preferred valence states are four for ruthenium, three with rhodium, and two for palladium. Osmium has a strong tendency to form the tetroxide, in which it is octapositive, and is also much more stable in the $+6$ state than is ruthenium. Iridium is chiefly tri- and tetrapositive, and platinum exhibits valence states of $+2$ and $+4$.

(a) Occurrence

In Nature, the platinum metals almost invariably occur in association with one another. They are usually found as the native metals, often in the form of alloys with one another (e.g., *osmiridium*), in sands and gravels ('placer' deposits), associated with the heavier minerals, such as magnetite, ilmenite, and chromite, and with gold. A few compounds also occur as minerals—e.g., *sperrylite*, $PtAs_2$, in the nickel-iferous pyrrhotite of Ontario, and *braggite*, PdS. Platinum is usually the principal constituent of those platinum metal ores which consist of the native metals, but the Sudbury nickel ores, and some related platinum ores of South Africa, are notable in that platinum and palladium are present in approximately equal amounts. The platinum usually contains copper and gold, as well as considerable amounts of iron, in addition to the minor platinum metals.

The following are typical compositions of platinum ores (after Péchard):

	Pt	Ir	Rh	Pd	Osmiridium	Cu	Au	Fe	Sand
Russian platinum ore	77.5	1.45	2.8	0.85	2.35	2.15	—	9.6	1.0 %
Colombian platinum ore	80.0	1.55	2.50	1.0	1.40	0.65	1.5	7.2	4.35%

(b) History

The first reliable account of platinum is due to Antonio de Ulloa (1748), in the course of a report on the French expedition to the west coast of South America in 1735. He mentioned that in Colombia gold occurs mixed with other metals, in rocks which were not exploited because they were difficult to work; these rocks are platinum-bearing, and Colombia is still one of the principal producers of platinum. At the time of Ulloa's visit the name 'platina del Pinto' was already applied to a white metal, resembling silver, discovered in Colombia. It was brought to Europe, and its properties were described as those of a peculiar metal by Watson in 1750, shortly after the publication of Ulloa's report. Doubt was subsequently cast upon the elementary nature of platinum until this was established by Blondeau in 1774. Extensive deposits of the platinum metals were found in the Urals in the opening years of the 19th century, and the finding soon led to the discovery of the minor platinum metals; palladium, rhodium, iridium, and osmium were all discovered in 1803–1804, leaving only ruthenium to be discovered in 1845.

(c) Preparation

Various methods are in use for the extraction of platinum and the allied elements from their ores, and some of the processes are kept secret. The essential steps involve first a separation by gravity concentration and elutriation of the platinum ore from the sand in which it is contained. After grinding as finely as possible, the ore is boiled with aqua regia, whereby *osmiridium*, which is essentially an alloy of osmium and iridium, remains as an insoluble residue. Iron, copper, iridium, rhodium, and a part of the palladium are precipitated from the solution by the addition of milk of lime, and the solution, which now contains principally platinum with some palladium, is evaporated to dryness and ignited. The platinum is thereby converted to the spongy metal, which is washed first with water, and then with hydrochloric acid, and is finally hot-pressed. It is not yet pure, and must be refined by a further series of operations. There are also small amounts of platinum in the precipitate thrown down by the milk of lime. These are recovered by treating the precipitate with dilute sulfuric acid, and ultimately precipitating the platinum as ammonium chloroplatinate, $(NH_4)_2[PtCl_6]$.

A considerable proportion of the world's output of platinum metals is now derived from the Sudbury nickel ore, in the form of residues either from the nickel carbonyl refining of nickel, or from the anode slimes of the electrolytic refining process. A measure of separation from baser metals can be effected by smelting with litharge and charcoal. The resulting lead alloy is cupelled, and treated with aqua regia, after 'parting' with sulfuric acid to remove the greater part of the silver. Platinum metals from the Sudbury ore contain about equal proportions of platinum and palladium as the predominant constituents. These go into solution in aqua regia, and they are separated and purified by taking advantage of the difference in stability of the $[M^{IV}Cl_6]^=$ complexes: platinum is precipitated as $(NH_4)_2[PtCl_6]$, and palladium is recovered as the insoluble diamminepalladium(II) chloride. [See 'Platinum and Allied Metals'. Imperial Institute, London, 1936; also Atkinson and Raper, *J. Inst. Metals*, 3 (1936) 207].

(d) Uses

Platinum [24] finds widespread technical applications, because of its excellent properties—its extreme resistance to chemical attack, durability, infusibility, and excellent working properties, as well as its outstanding properties as catalyst for a wide range of reactions. It is used for the manufacture of crucibles, dishes, boats, spatulas, etc., for chemical work, on both the laboratory and the technical scale, and in making electrodes and resistance wires for electric furnaces. It is also suitable for surgical instruments. Metals are often plated with platinum to protect them from corrosion. It is also favored for jewellery, especially for the mounting of precious stones. Large quantities of platinum were formerly used (often as platinized asbestos) as a contact catalyst in the sulfuric acid industry. The use of platinum as catalyst, in the manufacture of sulfuric acid at least, has been largely replaced by the use of promoted catalysts (cf. p. 715 *et seq.*), which are equally efficient and far less sensitive to catalytic poisons. Considerable quantities are still used in other branches of chemical industry, however, and it is widely employed as a laboratory catalyst. Platinum compounds are used in photography, in radiology (for barium cyanoplatinate screens), and in glass and porcelain painting.

Iridium is often alloyed with platinum to increase its hardness, especially for physical and surgical instruments. Platinum-iridium alloys find a particularly wide use for electrical contacts. The standard meter bar is made from 90% platinum-10% iridium alloy. Iridium is also used to give a black color to porcelain.

Palladium is used for jewellery, since its white luster resembles that of silver*. It has the advantage over silver that it is not blackened by hydrogen sulfide. Palladi-

---

* Silver-palladium alloys are more commonly used than pure palladium. Quite small additions of palladium to silver (17%) suffice to give protection against the black discoloration produced by H_2S, and the protective action is complete if the palladium content is 30%.

um is also used in place of gold for dental fillings. It is employed in the chemical laboratory in the form of palladium black—i.e., very finely divided palladium—or, still more efficiently, in the form of colloidal palladium as a catalyst for organic hydrogenations (cf. p. 339).

Rhodium is used chiefly in the form of a 90% Pt–10% Rh alloy for thermocouples [25]. The platinum-platinum rhodium thermoelement has the advantage over the platinum-platinum iridium couple formerly used of being more permanent, since rhodium is less volatile than iridium*. For extremely high temperatures (1600–2000°), thermoelements are used having one limb of iridium and the other of an alloy of 40% iridium with 60% rhodium. Platinum-rhodium alloys are also employed on a fairly large scale as catalysts for the oxidation of ammonia. Rhodium alloys are also used for the divided circles of astronomical instruments. Rhodium crucibles are suitable for use at very high temperatures (above 1500°), at which platinum becomes quite soft and commences to volatilize too rapidly. However, rhodium crucibles cannot be used for analytical work, since they do not remain constant in weight, owing to superficial oxidation of the metal. Electrolytically deposited rhodium platings have been increasingly used in the jewellery industry, in place of silver plating, since they have excellent protective properties, differ very little from silver in their color, and do not tarnish in air containing hydrogen sulfide. Rhodium is especially suitable for surface-coated mirrors. For the use of colloidal rhodium as a hydrogenation catalyst see p. 331.

Osmium was for some time of importance in the incandescent lamp industry ('Osram' lamps). One of its most important fields of application was for a long time for the tips of fountain pen nibs, but rhenium is now frequently used in its place (cf. p. 235). Osmium is occasionally used as a catalyst—e.g., in Hofmann's chlorate pipette (cf. p. 328). For the use of osmium tetroxide as a microscopic stain see p. 328.

Ruthenium is used to some extent as an alloy constituent for jewellery and is also suitable as a catalyst for many purposes. '*Ruthenium red*' obtained by treating ruthenium trichloride with aqueous ammonia, can be used in histology as a stain for tendons, and may also be employed to distinguish between natural and artificial fibers.

13-16. Dyad I: Ruthenium and Osmium

General

Ruthenium and osmium are hard, brittle, extremely infusible metals. They are very resistant towards acids, but combine rather easily with oxygen. A characteristic property of ruthenium and osmium is their ability to function as octa-positive, forming the volatile *tetroxides*, RuO_4 and OsO_4. They are also noteworthy for the large number of valence states from which they form compounds (cf. the summary table at the beginning of the chapter), and in many of their compounds they readily undergo changes of valence. This may be associated with their exceptional catalytic properties for a variety of reactions.

The oxides derived from the lowest valence states of ruthenium and osmium are basic in nature. With increase in the charge borne by the metal, there is an increase in the ability of the oxides to form acids, and the tendency for acid formation is completely lacking only in the highest oxides, the *tetroxides*, MO_4. The ability of many of the oxides to combine with water, to form oxyacids (and, correspondingly, with basic oxides to form oxyacid salts), can be explained in terms of the coordination theory, on the assumption that in these oxides the atoms are not exerting their maximum covalence towards oxygen. For most elements, the maximum coordination number towards oxygen, in radicals which exist as free ions or self-contained structural groups in crystals, is 4. For a few elements of the later series of the Periodic Table in their highest oxidation states—in particular for antimony, tellurium, and iodine—, the maximum coordination number towards oxygen is such radicals is 6. It may

* See p. 334 regarding the volatility of iridium.

be concluded that osmium, in the tetroxide, is not coordinatively saturated, from the fact that the compound can form addition compounds with basic hydroxides (p. 329). However, the capacity to form coordination compounds is only feebly developed in osmium tetroxide, and the compounds concerned, such as $K_2[OsO_4(OH)_2]$, have the properties of loose addition compounds. It is still uncertain whether the coordination number of ruthenium towards oxygen can be higher than 4. However, both ruthenium and osmium display a very strong tendency to form coordination compounds—stronger than that of iron.

13. Ruthenium (Ru)

As a constituent of platinum ores, ruthenium is the rarest of the platinum metals.* For this reason, it was discovered considerably later than the others, by the Russian chemist Claus in 1845. It was named in honor of Russia (Ruthenia is the old name for Little Russia)

In native platinum, ruthenium is contained chiefly in the osmiridium which remains as a residue when the platinum ore is dissolved in aqua regia. It is found very rarely indeed as a mineral of its own, e.g., *laurite*, a ruthenium disulfide RuS_2 (isotypic with pyrite), containing osmium.

(a) Physical Properties

Depending on the mode of preparation, ruthenium is a dull grey or silver white lustrous metal, of very great hardness, but so brittle that it can easily be pulverized. It is much less fusible than platinum. It can be melted in the electric arc, and vaporizes simultaneously. It also volatilizes slowly when it is strongly heated in air, but in this case it does not vaporize as the metal, but as the tetroxide, which is stable at high temperatures.

Ruthenium solidified from the melt has a density of 12.30. The density of metal obtained by reducing the oxide or chloride in hydrogen may be considerably lower. Its specific heat is 0.0553. Molten ruthenium dissolves 4–5% of carbon, according to Moissan. The metal can also absorb considerable amounts of hydrogen or oxygen, especially when charged with these gases electrolytically. In the finely divided state it has very highly developed ability to act as a carrier of oxygen or hydrogen to other substances—e.g., to catalyze the union of hydrogen and nitrogen to form ammonia, or the oxidation of alcohol to aldehyde and acetic acid by atmospheric oxygen.

Colloidal ruthenium can be prepared by reducing it from its salts by means of hydrazine in the presence of gum arabic (Gutbier), or by means of acrolein (Castoro).

(b) Chemical Behavior

In the absence of oxygen, ruthenium is not attacked by any acids—even by aqua regia. Hydrochloric acid containing dissolved air or oxygen, however, slowly attacks ruthenium, and does so rapidly at 125° (in a sealed tube). Ruthenium turns black when it is heated in air, as a result of surface oxidation.

Ruthenium burns in the oxidizing blowpipe flame, forming a black smoke of ruthenium dioxide. The characteristic smell of ruthenium tetroxide is evident at the same time. Powdered ruthenium is completely converted to the blue black dioxide, RuO_2, when it is ignited in oxygen. Powdered ruthenium is attacked by fluorine below a red heat, and by chlorine at a red heat or above. It combines with sulfur directly only under suitable conditions (see below). The compounds RuP_2, RuP and Ru_2P are formed with phosphorus (Biltz, 1939), and with arsenic a diarsenide, $RuAs_2$, corresponding to that of platinum (Wöhler, 1931).

* As a minor element in sulfide and oxidic ores, in very minute traces, ruthenium is about as abundant as platinum.

Ruthenium is vigorously attacked when it is heated with alkalis in the presence of oxygen or of oxidizing agents,—e.g., with a mixture of KOH and KNO_3, or of K_2CO_3 and $KClO_3$—or with peroxides such as Na_2O_2 and BaO_2. *Ruthenates*, $M^I_2RuO_4$, are thereby formed. Powdered ruthenium is dissolved by alkali hypochlorite solutions, forming the tetroxide, the reaction proceeding slowly even at ordinary temperature.

14. Ruthenium Compounds

(a) Oxides

The usual product of the direct union of oxygen and ruthenium is the blue black *dioxide*, RuO_2. At high temperatures (above 600°) *ruthenium tetroxide* is formed in traces, but decomposes in course of cooling. The tetroxide is metastable at ordinary temperature.

Lower oxides of ruthenium are known only in the form of oxide hydrates or hydroxides. *Ruthenium(III) hydroxide* is obtained as a black precipitate by the addition of alkali hydroxide to ruthenium(III) salt solutions, but is difficult to prepare in the pure state. A brown precipitate of *ruthenium(II) hydroxide* is formed on the addition of alkali hydroxide to ruthenium(II) salt solutions, but cannot be isolated, since it is oxidized with excessive ease. It gradually undergoes oxidation even under air-free water, from which it liberates hydrogen.

(*i*) *Ruthenium dioxide*, RuO_2, is obtained as a blue black powder by heating the powdered metal, or its chloride or sulfide, in oxygen. It has the rutile structure ($a = 4.51$ Å, $c = 3.11$ Å). It is not attacked by acids. It is readily reduced by hydrogen at slightly elevated temperatures. In general, it is unchanged when heated in air, but begins to decompose into ruthenium and oxygen when it is very strongly heated. The oxygen pressure is about 36 mm at 930°. No lower oxides are formed in the course of the thermal decomposition.

(*ii*) *Ruthenium tetroxide*, RuO_4, is most readily obtained by oxidizing an alkali ruthenate solution with warm periodic acid, or by passing chlorine through a ruthenate solution or a solution of a ruthenium salt in excess alkali hydroxide. It crystallizes in yellow needles, which melt at 25° to an orange liquid, and is fairly soluble in water (20.3 g in 1 liter at 0°). It is much more soluble in inert organic solvents such as carbon tetrachloride. Ruthenium tetroxide is quite volatile even at room temperature. Its boiling point cannot be measured at atmospheric pressure, since it decomposes explosively at about 108° into RuO_2 and O_2. Debray and Joly found that at 100° and 106 mm pressure the vapor density corresponded to that calculated from the molecular weight ($RuO_4 = 165.1$). Ruthenium tetroxide is stable at very high temperatures, and metastable at room temperature, although it undergoes no change if it is kept in the dark and complete absence of moisture.

Ruthenium tetroxide reacts very vigorously with organic substances, giving a black coloration as a result of the formation of RuO_2. It attacks rubber vigorously, and reacts explosively with alcohol. It has a characteristic smell, irritates the mucous membranes, and blackens the skin. Ruthenium tetroxide reacts with concentrated hydrochloric acid, evolving chlorine and yielding an oxychloride of hexapositive ruthenium, RuO_2Cl_2, as the initial product. This is rapidly reduced further, to $RuCl_4$, by excess of hydrochloric acid. The tetrachloride is gradually hydrolyzed to $Ru(OH)Cl_3$, even in concentrated hydrochloric acid, but under certain conditions may alternatively break up into $RuCl_3$ and chlorine, if hydrolysis has not occurred.

(b) Sulfides

Ruthenium disulfide, RuS_2, is obtained as a black precipitate (density 6.14) by the action of H_2S on aqueous solutions, or as a light grey crystalline mass by heating $RuCl_3$ in a current of H_2S. It is possible to bring the elements into direct combination with one another by means of a device dating back to Faraday's researches: powdered ruthenium and sulfur are heated in an evacuated quartz tube, in such a manner that one part of the tube, containing the ruthenium, is heated to 1200°, while the end of the tube projecting out of the furnace is maintained at 450°. Ruthenium disulfide prepared by dry methods is not attacked by acids—even by aqua regia. Thermal decomposition sets in above 1000°, yielding metal and sulfur. The heat of formation of the compound is 77 kcal per mol. (Juza 1933, Wöhler, 1933). Other sulfides of ruthenium are said to be formed by precipitation from solution, but these are unstable if, indeed, well defined compounds are formed at all.

The compounds $RuSe_2$ and $RuTe_2$ are very similar to ruthenium disulfide, but are less stable (Wöhler, 1933).

The sulfides, selenides, and tellurides of ruthenium and osmium crystallize with the pyrite structure. Cell dimensions are:

	RuS_2	$RuSe_2$	$RuTe_2$	OsS_2	$OsSe_2$	$OsTe_2$
$a =$	5.57	5.92	6.36	5.61	5.93	6.37 Å.

(c) Halides

(i) *Fluorides*. Ruff (1924) obtained *ruthenium pentafluoride*, RuF_5, by heating finely powdered ruthenium to about 300° in fluorine, in a platinum tube; small amounts of lower ruthenium fluorides (and considerable amounts of platinum fluoride) were formed at the same time. The *trifluoride*, RuF_3, also exists (Robinson, 1952). The pentafluoride forms a dark green transparent mass, which melts at 101° and boils above 250°. The fluoro salt $K[RuF_6]$, derived from RuF_5, has also been prepared. The existence of ruthenium pentafluoride is of some interest, as being the only compound known with certainty in which ruthenium is present in the pentapositive state.

(ii) *Chlorides. Ruthenium trichloride* (brown-black crystal leaflets, insoluble in water and acids) is formed by direct combination of ruthenium and chlorine at a red heat: reaction is facilitated by the presence of carbon monoxide. The compound can be prepared in a water-soluble state by treating ruthenium tetroxide with hydrochloric acid and evaporating the resulting solution to dryness in a current of hydrogen chloride gas (Remy); it is thereby obtained as the monohydrate, $RuCl_3 \cdot H_2O$ (Grube and Fromm, 1939). Products obtained from solution which, in the older literature, were considered to be ruthenium(III) chloride, were however for the most part more or less pure preparations of *ruthenium hydroxytrichloride*, $Ru(OH)Cl_3$. *Ruthenium tetrachloride*, $RuCl_4$, is formed as an intermediate product of low stability in the reduction of RuO_4 with HCl. It decomposes readily, with loss of chlorine. Ruthenium trichloride also decomposes into ruthenium and chlorine when it is heated, the decomposition pressure attaining 1 atm. at 850° (Remy). Lower chlorides do not appear as stages in the thermal decomposition. On the other hand, aqueous solutions of ruthenium chlorides turn deep blue when they are treated with strong reducing agents (sodium amalgam, titanium(III) chloride, cathodic reduction), and the ruthenium is reduced to the +2 state. Ruthenium tetrachloride readily undergoes reduction to the metal when it is heated in hydrogen, and is converted to the dioxide when it is ignited in oxygen. The chlorides of ruthenium have a strong tendency to form acido-compounds. The most important types of these are $M^I_2[Ru^{IV}Cl_6]$ (hexachlororuthenate(IV) salts), $M^I_4[Ru^{IV}_2Cl_{10}O]$ (decachloro-oxodiruthenate(IV) salts) and $M^I_2[Ru^{III}Cl_5(H_2O)]$ pentachloroaquoruthenate(III) salts).

(d) Coordination Compounds of Ruthenium

Ruthenium forms a wide variety of coordination compounds. Among derivatives of *dipositive* ruthenium, the *hexacyanoruthenate(II)* salts, $M^I_4[Ru(CN)_6]$, are noteworthy for their stability. The colorless potassium salt $K_4[Ru(CN)_6] \cdot 3H_2O$, which is not very soluble in cold water, is isomorphous with the analogous compound of iron, potassium ferrocyanide. Compounds of *tripositive* ruthenium include not only the pure acido-compounds of the types $M^I_2[RuX_5]$ and $M^I_3[RuX_6]$, but also compounds containing neutral ligands bound to the central atom. The *pentachloroaquoruthenate(III)* salts, $M^I_2[RuCl_5(H_2O)]$ and the *pentachloronitrosoruthenate(III)* salts, $M^I_2[RuCl_5(NO)]$, are of this type, as are the *ammines* such as $[Ru(OH)(NO)(NH_3)_4]Cl_2$ (hydroxonitrosotetrammineruthenium(III) chloride), etc. Examples have already been given of coordination compounds of +4 ruthenium. Coordination compounds of +6 ruthenium are represented by the dioxotetrachloro-ruthenate(VI) salts, $M^I_2[RuO_2Cl_4]$, prepared by Howe, and especially by the *ruthenates*,

$M^I{}_2[RuO_4]$, discussed below; the *perruthenates*, $M^I[RuO_4]$, are derivatives of heptapositive ruthenium. It is not finally established whether $+8$ ruthenium forms coordination compounds—i.e., whether ruthenium tetroxide forms addition compounds with other substances.

(e) Ruthenium Ammines

As was shown by Gleu, ruthenium is capable of forming strongly complexed ammoniates, in the same way as cobalt and chromium. Whereas the latter elements form their strongly complexed ammines exclusively from the $+3$ state, ruthenium ammines are derived both from the $+2$ and the $+3$ state.

Gleu has prepared ammines of the following series (where R stands for a complex-bound acid group, and X for an ionized acid radical).

Ruthenium(III) ammines

$[Ru(NH_3)_6]X_3$	hexammineruthenium(III) salts ('luteo salts')
$[Ru(OH)(NH_3)_5]X_2$	hydroxopentammineruthenium(III) salts ('roseo')
$[RuR(NH_3)_5]X_2$	acidopentammineruthenium(III) salts ('purpureo')
$[RuR_2(NH_3)_4]X$	diacidotetrammineruthenium(III) salts, of *cis* and *trans* configurations
$[RuCl_3(NH_3)_3]$	trichlorotriammineruthenium.

The hydroxonitrosotetrammineruthenium(III) chloride, which has long been known (see above) belongs to this class.

Ruthenium(II) ammines

$[Ru(SO_2)(NH_3)_5]X_2$	thioxopentammineruthenium(II) salts
$[Ru(SO_2)(H_2O)(NH_3)_4]X_2$	thioxoaquotetrammineruthenium(II) salts
$[RuR(SO_2)(NH_3)_4]X$	acidothioxotetrammineruthenium(II) salts
$[Ru(SO_3)(NH_3)_5] \cdot 2H_2O$	sulfitopentammineruthenium(II) (dihydrate)
$[Ru(SO_3H)_2(NH_3)_4]$	di(hydrogensulfito)-tetrammine ruthenium
$(NH_4)_2[Ru(SO_3)_2(NH_3)_4] \cdot 4H_2O$	ammonium disulfitotetrammineruthenate(II) (tetrahydrate)
$Na_4[Ru(SO_3)_2(SO_3H)_2(NH_3)_2] \cdot 6H_2O$	sodium disulfito-di(hydrogen sulfito)-diammine ruthenate(II).

Most of the ruthenium(III) ammines are but weakly colored, those of the so-called 'luteo', 'roseo', and 'purpureo' series being usually colorless or light yellow. The intensity of color deepens as the number of acido groups is increased: dichlorosalts are orange and trichlorotriammine ruthenium red.

The ruthenium(II) ammines listed are all colorless, except for the thioxo compounds, containing the neutral SO_2 group within the complex*. These are mostly yellow, brown, or red. These ruthenium complexes containing sulfur dioxide are noteworthy compounds, since although many complex compounds are known in which the ions of sulfurous acid, $SO_3{}^{2-}$ and $HSO_3{}^-$, are present, (sulfito- and hydrogen sulfito-complexes), there are no other known compounds in which the neutral SO_2 molecule is present within a coordination complex.

The '*ruthenium red*' discovered by Joly in 1892 seems to be related to the ammines of tripositive ruthenium. It has the composition $2RuCl_2(OH) \cdot 7NH_3 \cdot 3H_2O$. Its constitution is still unexplained, but Gleu found that the red coloration characteristic of $+3$ ruthenium was obtained when hexammine- or pentammineruthenium(III) salts were warmed in weakly alkaline solution.

---

* The suggested designation 'thioxo' group, for the complex-bound SO_2 molecule (as abbreviation for 'thiodioxo' group) is based on the use of 'thionyl' for the radical SO, derived from SO_2.

(f) Ruthenates and Perruthenates

When powdered ruthenium, or ruthenium dioxide, is heated with potassium hydroxide and potassium chlorate (or nitrate), a deep green melt is obtained, which is very hygroscopic and dissolves in water with a deep orange red color. On evaporation of the solution, which is rapidly blackened in contact with organic matter, RuO_2 being deposited, *potassium ruthenate(VI)*, $K_2RuO_4 \cdot H_2O$, is obtained in lustrous green prisms, which appear red in thin layers by transmitted light. If chlorine is passed into the red solution it initially turns green, due to formation of perruthenate, and on continued passage of chlorine, ruthenium tetroxide is evolved. If the passage of chlorine is stopped before this stage, *potassium perruthenate* (potassium ruthenate(VII)), $KRuO_4$, crystallizes in black tetragonal crystals when the solution cools down again. Unlike the ruthenates, the perruthenates are not stable above 200°. They are also decomposed by alkalis with the formation of ruthenates. Conversely, the ruthenates are converted to perruthenates on treatment with dilute acids (ruthenium dioxide hydrate being deposited simultaneously). The ruthenates and perruthenates are thus very similar in behavior to the manganates and permanganates.

(g) Carbonyls and Nitrosyls

Like iron and nickel, ruthenium can combine directly with carbon monoxide, but does so only under high pressures. *Ruthenium pentacarbonyl* can be obtained by heating ruthenium powder with carbon monoxide at 180° under 200 atm. pressure. It is more readily prepared by heating ruthenium triiodide, RuI_3, mixed with silver powder, in carbon monoxide, or by the action of carbon monoxide under high pressure upon ruthenium disulfide (Hieber). It is a water-clear, volatile liquid, which solidifies at —22° to give colorless crystals. It is soluble in benzene, alcohol, and similar solvents, but gradually decomposes in solution, or more rapidly in the liquid state, to give the orange *enneacarbonyl* $Ru_2(CO)_9$ (hexagonal leaflets). A third solid green ruthenium carbonyl is thereby formed as a decomposition product—probably with the composition $Ru(CO)_4$ (or $Ru_3(CO)_{12}$).

In the reaction of CO on RuI_3, the compound $RuI_2(CO)$ is formed as the initial product of reaction. The analogous bromine and chlorine compounds, and the compound $RuBr(CO)$, are also known (Manchot 1924–1930).

According to Manchot, the action of NO on $Ru_2(CO)_9$ at moderately elevated temperatures (best under pressure) yields a red, well crystallized *ruthenium nitrosyl*, the composition of which has not been finally established. It may be $Ru(NO)_4$ or $Ru(NO)_5$ (but see p. 358 for comparison with the corresponding reaction of iron carbonyl).

15. Osmium (Os)

Osmium is present in platinum ores in the form of an alloy with iridium, known as *osmiridium*, which is insoluble in aqua regia. Its osmium content varies from 17–80%. Russian platinum ore usually contains between 0.5 and 2.5% of osmiridium, which is usually present to the extent of a few per cent in Colombian ore also. Platinum ores have been found in Oregon and Australia (Tasmania), however, containing up to 37% osmiridium, and almost pure osmiridium has been found in California.

Osmium was discovered in 1804 by Tennant, whose work followed the earlier observations of two French scientists, Fourcroy and Vauquelin. In the course of the analysis of platinum minerals, they had discovered a residue insoluble in aqua regia, and this, when attacked by fusion with potash and treatment with nitric acid, evolved vapors which attacked the eyes and mucous membranes, and blackened organic matter. These were obviously osmium tetroxide. Tennant isolated the metal from the volatile tetroxide, and gave it the name of osmium from its characteristic smell (Greek ὀσμή = smell).

(a) Physical Properties

Osmium is a blue grey metal, or blue-black in the powdered state. It is very hard (7.5 on Mohs' scale), brittle, and easily powdered. It has the highest melting point of all the platinum metals (2700°). Metal solidified from the melt has the density 22.5, as compared, with the value 22.7 calculated from X-ray data. No other metal excels osmium in catalytic activity, but it is readily poisoned. Colloidal osmium can be prepared in the same way as colloidal ruthenium, although less readily because of the ease with which osmium is oxidized.

(b) Chemical Behavior

A characteristic property is the ease with which osmium forms the *tetroxide*, OsO_4. The powdered metal always possesses the characteristic smell of the tetroxide, which is formed in traces even at ordinary temperature on exposure to air. Since the tetroxide is reduced by traces of grease, dust, etc., to the black, non-volatile dioxide, the walls, and especially the necks, of bottles in which osmium is stored are always blackened with this compound.

Osmium is oxidized very vigorously when it is heated. The compact metal can be stored without change indefinitely, however, and there is no detectable oxidation. Because of its strong tendency to combine with oxygen, powdered osmium is also oxidized by nitric acid (especially the concentrated acid), by hot concentrated sulfuric acid, by molten alkali hydrogen sulfates, and by sodium hypochlorite solution. Osmium also combines vigorously with sulfur when it is heated. It is not attacked, however, by oxygen-free acids. It combines with fluorine above 100°, and also with chlorine when heated, but not with bromine or iodine. It forms neither mixed crystals nor compounds with carbon, and combines much less readily with phosphorus than do ruthenium and iron. It forms only the one compound OsP_2, which was obtained by Biltz, by heating powdered osmium with phosphorus in the 'Faraday tube' arrangement.

On fusion with potash under oxidizing conditions (access of air or addition of potassium chlorate or nitrate), osmium is converted to potassium osmate(VI), K_2OsO_4. If the solution of this salt is acidified, disproportionation occurs, as with a manganate melt, in this case with the formation of OsO_4 and OsO_2.

With carbon monoxide, osmium forms carbonyls which correspond closely in composition, properties and mode of formation to those of ruthenium. See Table 34, p. 352 for osmium carbonyl halides.

16. Osmium Compounds

(a) Oxides, Sulfides etc.

Osmium forms the pale yellow volatile tetroxide OsO_4 and a brown dioxide OsO_2 (black in powdered form). It is questionable whether any lower oxides exist.

(i) *Osmium tetroxide*, OsO_4, is always formed when osmium or one of its compounds is oxidized by air, or treated with strong oxidants. It is most simply prepared by heating finely divided osmium to redness in a current of oxygen. The tetroxide is thereby condensed in an ice-cooled trap in the form of transparent, almost colorless, lustrous monoclinic crystals. It melts at 40° to an oily liquid, and boils without decomposition at about 100°. The vapor density (8.88 at 0° and 760 mm) corresponds to the molecular weight required by the formula OsO_4. The tetroxide is appreciably volatile at ordinary temperature, and has a penetrating, characteristic smell. The vapors attack the mucous membranes, especially of the eyes, and are very dangerous for the respiratory organs. Osmium tetroxide, although more stable than ruthenium tetroxide, is nevertheless a strong oxidant, since it is readily reduced to the dioxide, and also to the metal. It is used in microscopy as a stain, under the name of 'osmic acid'. Its aqueous solution is not acid in reaction towards litmus. However, as was shown by Tschugaieff (1918), osmium tetroxide can form loose addition compounds with strong alkalis, e.g., $OsO_4 \cdot 2KOH$ or $K_2[OsO_4(OH)_2]$, potassium tetroxo-dihydroxo osmate(VIII). Kraus (1925) prepared the ammonium, cesium, and barium salts of the same series, and also the rubidium and cesium salts of the type $M^I_2[OsO_4F_2]$ (tetroxo-difluoro-osmate(VIII) salts), in which two F atoms replace the OH groups. Osmium tetroxide is soluble in alcohol and ether, as well as in water, but the solutions soon undergo decomposition. It is also decomposed by concentrated hydrochloric acid ($d = 1.16$), forming $OsCl_4$, but (unlike ruthenium tetroxide) is not attacked by more dilute hydrochloric acid.

(*ii*) *Osmium dioxide*, OsO_2, is obtained by the reduction of osmium tetroxide under conditions where conversion to metal does not occur. It is a black powder, insoluble in water and acids. It can also be prepared by other methods—e.g., by fusing potassium hexachlorosmate, $K_2[OsCl_6]$, with soda. In the crystalline state, it is brown with a yellow luster. It has the same structure as RuO_2 ($a = 4.51$ Å, $c = 3.19$ Å). Osmium dioxide is converted to the tetroxide when it is heated in air, and is also very readily reduced to the metal by means of hydrogen. Osmium dioxide is used in gas analysis, in Hofmann's chlorate pipette, as a catalyst for the oxidation of elementary hydrogen.

(*iii*) *Osmium disulfide*, OsS_2 (heat of formation 62 kcal per mol) is obtained by methods similar to those for RuS_2 which it closely resembles. It is slowly attacked by aqua regia and fuming nitric acid.

The stability of the compounds increases in the series OsS_2, $OsSe_2$, $OsTe_2$, with increase in atomic number of the non-metal, as also with the ruthenium chalcogenides. (Structure—see p. 325.)

(b) Halides

Our present knowledge of the osmium halides is largely due to Ruff.

(*i*) *Fluorides*. Ruff prepared *osmium octafluoride*, OsF_8, by direct combination of the elements above 250°. The octafluoride can be condensed, in a trap cooled with solid carbon dioxide, to a lemon-yellow sublimate, m.p. 34.4°, b.p. 47.5°. *Osmium hexafluoride*, OsF_6, is formed at the same time, in bright green crystals. Whereas the octafluoride dissolves in water giving a clear solution, the hexafluoride is decomposed forming osmium tetroxide, osmium dioxide, and hydrofluoric acid. Incomplete fluorination yields the *tetrafluoride* OsF_4 also.

(*ii*) *Chlorides*. The treatment of osmium with chlorine at elevated temperatures usually gives a mixture of chlorides. Ruff obtained pure *osmium tetrachloride*, $OsCl_4$, by heating osmium in chlorine at 650–700°, and slowly cooling the resulting yellow brown vapors. The tetrachloride forms black crusts, with a metallic luster, which evaporate without melting when heated in a vacuum or in chlorine. It is only slowly dissolved by water, forming a yellow solution which rapidly deposits osmium dioxide as a result of hydrolysis. By chlorination at higher temperatures, and rapid cooling of the products, Ruff obtained *osmium trichloride*, $OsCl_3$, as a brown-black hygroscopic powder, soluble in water and alcohol. Its deep brown aqueous solution has a weakly acid reaction, and is stable both in the cold and on boiling; it is not attacked by mild reducing agents (e.g., $FeSO_4$, sulfurous acid, formaldehyde) but forms osmium tetroxide on prolonged heating with concentrated nitric acid. If the trichloride is heated in a vacuum to 500°, it is converted to the dark brown, insoluble, hydrophobic *dichloride*, $OsCl_2$. Osmium trichloride reacts with carbon monoxide above 220°, forming carbonyl chlorides such as $OsCl_2(CO)_3$ (colorless prisms) (Manchot, 1925).

(c) Coordination Compounds of Osmium

Like ruthenium, osmium forms a variety of coordination compounds. *Hexacyanoosmium(II) acid*, $H_4[Os(CN)_6]$, derived from dispositive osmium, was already known to Claus. Its potassium salt, $K_4[Os(CN)_6] \cdot 3H_2O$ (light yellow, tabular monoclinic crystals) is obtained, for example, by adding potassium cyanide to a solution of osmium tetroxide. Its reactions with heavy metal salts resemble those of potassium ferrocyanide, and it was shown by Dufet (1895) to be isomorphous with the latter and with the corresponding ruthenium compound. As examples of coordination compounds of tripositive osmium, mention may be made of potassium pentanitroosmate(III), $K_2[Os(NO_2)_5]$ (amber yellow triclinic crystals) and *potassium hexachloroosmate(III)*, $K_3[OsCl_6]$. Osmium can probably also be regarded as tripositive in the salts of the type $M^I_2[OsX_5(NO)]$. When osmium is heated to dull redness with potassium chloride in a current of chlorine, or when potassium chloride and alcohol are added to a solution of osmium tetroxide, *potassium hexachloroosmate(IV)*, $K_2[OsCl_6]$, is obtained in dark red octahedra, isomorphous with the analogous platinum compound. A number of *sulfito* compounds of +4 osmium are also known, of the same type as the hexachloro-compound—e.g., $Na_2[Os(SO_3Na)_6]$, $Na_2[Os(SO_3Na)_4(OH)_2]$. Derivatives of +6 osmium include the ordinary *osmates*, $M^I_2[OsO_4]$ (*tetroxate(VI)* salts), and the so-called *osmyl* compounds; in Werner's nomenclature the latter would be termed dioxoosmate(VI) compounds. They have the general formula $M^I_2[OsO_2X_4]$, and can be obtained by the following reactions:

$$OsO_4 + 4HCl + 2KCl = K_2[OsO_2Cl_4] + Cl_2 + 2H_2O$$

$$OsO_4 + 3H_2C_2O_4 + 2KOH = K_2[OsO_2(C_2O_4)_2] + 2CO_2 + 4H_2O$$

$$OsO_4 + 2NO + 2KNO_2 = K_2[OsO_2(NO_2)_4]$$

$$K_2OsO_4 + 4HX \rightleftharpoons K_2[OsO_2X_4] + 2H_2O.$$

The last reaction is reversible, and the dioxoosmates are decomposed by pure water to give ordinary osmates. A cationic complex of +6 osmium is present in the dioxotetrammine-osmium(VI) compounds $[OsO_2(NH_3)_4]X_2$. The action of nitrites on OsO_4 at ordinary temperature yields *trioxoosmate(VI)* salts, $M^I_2[OsO_3X_2]$, formerly called 'osmyl oxysalts':

$$OsO_4 + 3KNO_2 = K_2[OsO_3(NO_2)_2] + KNO_3.$$

The coordination compounds of hexapositive osmium also include the peculiar salts of the type $M^I_2[OsNX_5]$, containing nitrogen. However, in *osmiamic acid*, OsO_3NH, and its salts, osmium is in the +8 state.

(d) Osmates

The osmates, $M^I_2[OsO_4]$, correspond in composition to the ruthenates, but are less stable. They readily pick up oxygen, and undergo conversion to the tetroxide, which is formed in perceptible amounts, for example, when they are exposed to moist air. *Potassium osmate*, $K_2[OsO_4] \cdot 2H_2O$ may be obtained by reducing osmium tetroxide, dissolved in caustic potash, by means of alcohol or with potassium nitrite. It crystallizes in pale red-violet regular octahedra. It loses its crystal water when heated to 200° in an inert atmosphere, but forms the tetroxide when it is heated in air.

17-20. Dyad II: Rhodium and Iridium

General

Rhodium and iridium are hard metals, more infusible than platinum; they are remarkably resistant towards acids, but combine with oxygen when heated. Whereas rhodium has a very marked preference for a single valence state (the tripositive state), iridium, like ruthenium and osmium, readily changes its valence. Neither element, however, ever exhibits a higher valence state than six, and in general the +4 state is not exceeded. Rhodium and iridium are very active catalytically, especially when in the finely divided state. In addition to the simple compounds, a relatively large number of well defined coordination compounds is known. Many of these involve complex cations, such as are rarely encountered among the compounds of ruthenium and osmium.

17. Rhodium (Rh)

Rhodium is found in platinum ores, and also in certain auriferous sands from South America. As much as 43% rhodium has been found in Mexican gold. Apart from this, however, the rhodium content of even relatively rich ores is generally small—usually between 0.5 and 4.5%.

Rhodium was discovered by Wollaston in 1803, at the same time as palladium. It received its name from the rose pink color of many of its compounds (ῥόδεος = rose red).

(a) Physical Properties

Rhodium is a white ductile metal (density 12.42, m.p. 1966°). It is oxidized superficially when it is melted in air. It can be distilled in the electric arc.

Molten rhodium dissolves up to 7% of carbon, which separates out again as graphite on cooling. Carbon dioxide, nitrogen, and hydrogen are not absorbed to a perceptible extent by rhodium between 400° and 1000°. Rhodium can be obtained in colloidal dispersion by the reduction of rhodium trichloride with hydrazine, or by sputtering the metal. According to Zenghelis (1938), colloidal rhodium is an extremely efficient catalyst for hydrogenations, making it possible to carry out at ordinary pressure and in neutral solution hydrogenations which, with colloidal palladium, can be carried out only under high pressures and in acid solution. Colloidal rhodium also possesses bactericidal properties, while being relatively free from danger to the human organism, so that it may have some therapeutic value in certain circumstances. Rhodium can also be obtained in a very active form when it is reduced from solutions of its salts by means of ammonium formate in the presence of ammonium acetate and ammonia, or by alcohol or formaldehyde in the presence of potassium carbonate. The very finely divided metal obtained in this way (known as *rhodium black*) is able to decompose formic acid into carbon dioxide and hydrogen at ordinary temperature. It is remarkable that this activity is not at once destroyed by sulfur compounds (as is usually the case with other catalysts), but is conserved or even increased (Bredig). In alkaline solution, alcohol is converted to acetate by rhodium black, with the evolution of hydrogen; oxygen is evolved from chlorine water or hypochlorite solutions; and O_3 is converted to O_2.

(b) Chemical Behavior

At a bright red heat, rhodium is gradually oxidized in air to the sequioxide, Rh_2O_3. This in turn undergoes decomposition at much higher temperatures. Rhodium can take up to 2.3 weight % (= 13 atom %) of oxygen in solid solution. The last traces of this are given up again only with great difficulty in the absence of reducing gases. Rhodium is converted to the trichloride $RhCl_3$ by chlorine at a red heat, but is extraordinarily resistant towards fluorine, according to Ruff. Pure, compact rhodium is *completely insoluble in all acids*, including aqua regia. Rhodium black, however, is soluble in aqua regia and in hot concentrated sulfuric acid, and also in hydrochloric acid in the presence of air. Massive rhodium is attacked by a mixture of oxygen and hydrochloric acid at 150°, and it also dissolves to some extent in fused metaphosphoric acid. Rhodium is completely dissolved at a red heat by potassium pyrosulfate, and is thereby converted to the water-soluble rhodium sulfate, $Rh_2(SO_4)_3$. If, however, it is fused with soda and potassium nitrate, or with barium peroxide, insoluble oxides are obtained. Rhodium is readily obtained in the metallic state by heating its compounds in hydrogen; it is necessary to cool the metal in an atmosphere of carbon dioxide, as the 'activation' of the hydrogen by the rhodium may otherwise lead to inflammation and reoxidation.

Rhodium and arsenic unite directly only with some difficulty. $RhAs_2$ was prepared by Wöhler in the same way as $IrAs_2$ (cf. p. 335). With P, rhodium forms the compounds Rh_2P, Rh_5P_4, RhP_2, and RhP_3, which were obtained by Biltz (1940) by heating the components in sealed quartz tubes. Rh_2P is anti-isotypic with fluorspar ($a = 5.50$ Å) and isotypic with Ir_2P ($a = 5.53$ Å).

Among the intermetallic compounds of rhodium (cf. p. 248), Bi_4Rh has some technical importance, since the ready solubility of this compound in nitric acid enables rhodium to be separated simply from other platinum metals.

18. Rhodium Compounds

Most of the known compounds of rhodium are derived from the $+3$ oxidation state, although the present knowledge of the chemistry of rhodium is rather incomplete. It has been established that an oxide hydrate of $+4$ rhodium exists, and rhodium is unquestionably tetrapositive in the complex compound $K_2[RhF_6]$.

The valence of rhodium towards fluorine may exceed three in simple compounds also, but it is not certain whether the highest fluoride has the formula RhF_4 or RhF_5. The principal product of the reaction of rhodium with fluorine is the trifluoride, RhF_3, a red finely crystalline powder of density 5.38.

It has been assumed, but not definitely established, that compounds of $+6$ rhodium exist. Grube (1937) detected the existence of rhodium(VI) ions in solution. Solutions containing rhodium(VI) ions yield deep blue precipitates, soluble in excess caustic alkali to give blue solutions, whereas solutions containing rhodium(IV) ions yield dark green precipitates, soluble in caustic alkali to give green solutions. Dipositive rhodium exists in the monoxide RhO, detected tensimetrically, and in the complex sulfito salt $Na_2[Rh(SO_3)_2] \cdot 2H_2O$ prepared by Reihlen (1933).

The unstable oxide Rh_2O is the only known compound of unipositive rhodium.

(a) Oxides, Sulfides etc.

(i) *Rhodium dioxide* (rhodium(IV) oxide), RhO_2, cannot be obtained completely anhydrous, according to Wöhler, It can be obtained in the hydrated state by the action of powerful oxidants on Rh(III). In the electrolysis of sodium hexachlorohodate(III), $Na_3[RhCl_6]$, the oxide is obtained at the anode as a black green precipitate of variable water content, usually contaminated with admixed Rh_2O_3. Dehydration of the hydrated dioxide is accompanied by simultaneous loss of oxygen, and formation of the anhydrous sesquioxide Rh_2O_3. Rhodium dioxide hydrate has a strong tendency to adsorb alkali- or alkaline earth hydroxides. The adsorption products were formerly considered to be salts of the types, e.g., $K_2O \cdot 6RhO_2$, $Na_2O \cdot 8RhO_2$, $BaO \cdot 12RhO_2$, but is very doubtful whether they are definite compounds at all.

(ii) *Rhodium sesquioxide (rhodium(III) oxide)*, Rh_2O_3, can be obtained as a black-grey powder both by heating the metal in air and by igniting the nitrate. It has the corundum structure (cf. Vol. I, Fig. 72, p. 351) ($a = 5.47$ Å, $a = 55°40'$). Alkali hydroxides precipitate the pentahydrate, $Rh_2O_3 \cdot 5H_2O$, as a lemon yellow precipitate, insoluble in water but soluble in acids. This furnishes the anhydrous sesquioxide, insoluble in acids, when it is heated.

Rh_2O_3 can be degraded to lower oxides, RhO and Rh_2O, when it is heated under reduced pressure. The lower oxides appear to be metastable (Wöhler, 1925, Schenck 1937–38).

(iii) *Sulfides*. Wöhler obtained *rhodium pentasulfide*, Rh_2S_5, by heating $RhCl_3$ with sulfur in a sealed tube; it is a dark grey crystalline mass, insoluble in acids and even in aqua regia. The compounds Rh_2Se_5 and Rh_2Te_5 have been obtained in the same way (Wöhler 1933, Biltz 1937). It is curious that these three pentachalcogenides have a structure similar to that of pyrite ('pseudopyrite' type). According to Biltz, the sulfides Rh_2S_3, Rh_3S_4, and Rh_9S_8 also exist.

(b) Simple Salts of Rhodium

The anhydrous chloride $RhCl_3$ is obtained by heating the metal to dull red heat in a stream of chlorine. It is a red powder, insoluble in water and acids. It decomposes into the metal and chlorine above 440°. The hydrated chloride $RhCl_3 \cdot 4H_2O$, which is very soluble in water (and also in alcohol), is obtained as a red deliquescent mass, by dissolving rhodium(III) oxide hydrate in hydrochloric acid. It loses its water of hydration, without becoming insoluble, when it is heated to 180° in hydrogen chloride, but is converted to the insoluble form of the trichloride when heated at higher temperatures.

Hydrated rhodium chloride (but not the anhydrous compound) reacts at moderate temperatures with carbon monoxide, forming $Rh_2OCl_2(CO)_3$ (ruby red needles, m.p. 125.5°—Manchot, 1925).

The bromide and iodide of rhodium resemble the chloride. Besides the halides, rhodium can form simple salts with nitric acid, sulfuric acid, sulfurous acid, hydrogen sulfide, hydrocyanic acid, etc.*

Rhodium also forms alums $M^IRh(SO_4)_2 \cdot 12H_2O$**. These lead to a consideration of the true coordination compounds of rhodium.

(c) Coordination Compounds of Rhodium

Rhodium forms rather strongly complexed coordination compounds. These can be divided into two classes, involving complex cations and complex anions, respectively. Rhodium invariably has the coordination number 6 in the former, and is usually 6-coordinated in the latter.

Rhodium salts with complex cations ('rhodammines') have been studied principally by Jörgensen. They correspond closely in properties to the cobalt and chromium compounds of analogous composition, and are typically strongly complexed. Representatives of the following types are known:

$[Rh(NH_3)_6]X_3$, hexamminerhodium(III) salts
$[Rh en_3]X_3$, tris-ethylenediaminerhodium(III) salts
$[Rh(H_2O)(NH_3)_5]X_3$, aquopentamminerhodium(III) salts
$[RhR(NH_3)_5]X_2$, acidopentamminerhodium(III) salts (R = univalent
                           acido group or hydroxyl group)
$[RhCl_2 pyr_4]X$, dichlorotetrapyridinorhodium(III) salts.

The dimethylglyoxime compounds of rhodium, prepared by Tshugaieff, are to be included in the last mentioned type. They have the constitution

$$\left[\left(\begin{array}{c} CH_3 \cdot C = N - O \\ | \\ CH_3 \cdot C = N \\ HO \end{array} \right)_2 Rh \ (NH_3)_2 \right] X.$$

* As in the case of chromium, some of the rhodium compounds, which appear from the analytical data to be simple salts, really possess a more complicated structure. They accordingly often occur in different forms, which are apparently isomeric except for their water content. Thus, according to Krauss, rhodium(III) sulfate, $Rh_2(SO_4)_3$, exists in a yellow and a red form. The former, in freshly prepared aqueous solutions, gives the reactions for Rh^{+++} and $SO_4^=$, whereas these reactions are not observed in fresh solutions of the red form. Similar observations have been made for rhodium selenate (Meyer, 1936). Rhodium chloride also exists in a brown red form and a yellow form—$[RhCl_3(H_2O)_3]$, soluble without any ionization of Cl^- ions, and $[Rh(H_2O)_3]Cl_3$, apparently existing only in solution.

** The alums are derived from the *yellow* rhodium sulfate, and cannot be obtained from the red form.

The tris-ethylenediamine compounds belong to the type of complex which, according to Werner's theory, should be capable of existing in two enantiomorphous forms, and they have in fact been resolved into their two optically active components. Their optical rotation is found to be opposite in sense to that of the corresponding cobalt and chromium compounds, thereby demonstrating the decisive importance of the central atom for the direction of optical rotation.

Rhodium salts with complex anions are represented chiefly by acido salts of the type $M^I_3[Rh\,X_6]$, where $X = Cl, NO_2, CN, \frac{1}{2}SO_4, \frac{1}{2}SO_3, \frac{1}{2}C_2O_4$, etc. A few compounds of the general formula $M^I_2[Rh\,X_5]$ are also known, however, $(X = Cl, Br)$, and also dimethylglyoxime compounds of the formula $M^I[RhCl_2(C_4H_6N_2O_2H)_2]$. This complex anion is derived from the cation depicted above, by exchange of the two neutral NH_3 groups for Cl^- ions. Iridium compounds of the same type are also known.

Reference has been made to the hexafluororhodium(IV) complex, $K_2[RhF_6]$ (p. 332). For rhodium carbonyl compounds (Hieber and Lagally, 1943) see Table 34, p. 352.

19. Iridium (Ir)

Iridium is invariably found as the native metal, alloyed sometimes with platinum and sometimes with osmium. Iridium-osmium alloys ('osmiridium') are found in smaller or larger amounts in all platinum ores—usually in very fine subdivision, but occasionally in larger grains. Traces of iridium are also found in a few occurrences of gold. It is often distinctly to be detected in meteoric iron, and the presence of iridium in the sun's photosphere has been established.

Iridium was discovered by Tennant in 1804, at the same time as osmium. Its name (ἰρίδιος = rainbow colored) symbolizes the range of colors found among its compounds.

(a) Physical Properties

Iridium is a silver white, very hard and rather brittle metal, with distinctly crystalline texture. Its ductility is low even at red heat, but it can be filed and polished. The density of iridium is 22.65, and it has the highest melting point (2454°) of any platinum metal but osmium.

Iridium is, after osmium, the least volatile of the platinum metals in a vacuum. However, volatilization becomes perceptible even at a bright red heat when the metal is heated in air. This volatility, which is often an inconvenience in using apparatus made of platinum-iridium alloys, arises from the formation of the dioxide.

Molten iridium dissolves a certain amount of carbon, without combining with it, and the carbon separates out as graphite on cooling. Fine iridium powder can absorb a considerable amount of hydrogen, especially when charged with the gas electrolytically. Colloidal iridium was obtained by Paal, by mixing iridium chloride, lysalbumin, and strong sodium carbonate solution, and by Gutbier, from the reduction of the chloride with hydrazine.

Iridium is inferior to platinum and palladium in its catalytic properties. Iridium finds technical applications not only because of its chemical inertness, but especially because it possesses great hardness.

(b) Chemical Behavior

Iridium is not attacked by any of the usual acids, even including aqua regia. It is, however, attacked by hydrochloric acid in the presence of air, when heated under pressure at 125°. If powdered iridium is heated to a dull red heat in air or oxygen, it is converted to the dioxide, IrO_2. This decomposes again at higher temperatures, so that no oxidation of iridium occurs above 1140°. Iridium is relatively easily attacked by chlorine at red heat—especially if mixed with sodium chloride, with which iridium chloride forms a double salt.

Iridium mixed with sodium chloride also reacts at high temperatures with bromine and iodine, which otherwise scarcely attack it. It also combines with sulfur when heated, and forms with phosphorus a fusible compound which decomposes again at higher temperatures. A method for the technical preparation of fused iridium is based on this fact. Iridium combines directly with arsenic only with difficulty, although the compound $IrAs_2$ can be prepared by heating $IrCl_3$ with arsenic in hydrogen (Wöhler, 1931). With phosphorus, iridium forms the compounds Ir_2P and IrP_2 (cf. p. 332 for structure). Molten alkali hydroxides have no action on iridium, although Claus obtained a mixture of soluble and insoluble 'iridates' (see below) by adding powdered iridium to a soda-potassium nitrate melt. Fused potassium pyrosulfate forms only insoluble Ir_2O_3.

The redox potential $[IrCl_6]^{3-}/[IrCl_6]^{2-}$ in a solution 1-molar with respect to HCl is —1.02 volts, relative to the normal hydrogen electrode (Woo, 1931). Thus the tendency of iridium(IV) ions to revert to a lower valence state is greater than that of iron(III) ions, for example (cf. Vol. I, p. 765, Table 103). It should be noted, however, that the redox potential may be shifted considerably through complex formation.

20. Iridium Compounds

(a) Oxides, Sulfides, etc.

(*i*) *Iridium dioxide*, IrO_2, is obtained as a black powder by heating finely powdered iridium in air or oxygen. According to Wöhler, the optimum temperature is about 1070°, since dissociation occurs at higher temperatures, and leads directly to metallic iridium. Since the metal can dissolve oxygen, it is not possible to specify a definite dissociation pressure. Iridium dioxide, like the dioxides of ruthenium and osmium, has the rutile structure, with $a = 4.49$ Å, $c = 3.14$ A. *Iridium dioxide hydrate*, $IrO_2 \cdot 2H_2O$ or $Ir(OH)_4$, may be prepared by reaction between a solution of sodium chloroiridate(III), $Na_3[IrCl_6]$,* and potassium hydroxide, accompanied by passage of oxygen through the solution. Violet and blue colloidal dispersions are first obtained thereby, and the dioxide hydrate separates out from these after some time. According to the conditions of precipitation, the product may be indigo blue, violet, dark blue, or black in color. The water content corresponds to that of the formula given, after drying over concentrated sulfuric acid. The dioxide hydrate is completely insoluble in caustic alkalis, even when freshly precipitated. It is fairly readily soluble in acids, however. The anhydrous dioxide is insoluble in nitric acid and sulfuric acid, but soluble in hydrochloric acid with which it forms the complex acid $H_2[IrCl_6]$ (Wöhler).

It was formerly assumed that iridium dioxide could form well defined compounds with strongly basic oxides. These were termed 'iridates', but these substances are apparently merely adsorption products. According to Claus, +6 iridium can also exercise acid-forming functions, and forms, e.g., a potassium iridate $K_2[Ir_2O_7]$ which is insoluble in water. However, this statement also needs confirmation.

(*ii*) *Iridium sesquioxide* (*iridium(III) oxide*), Ir_2O_3, is obtained in hydrated form by adding potassium hydroxide to a solution of sodium chloroiridate(III); to avoid oxidation, precipitation should be carried out in an atmosphere of carbon

* The salt crystallizes with $12H_2O$.

dioxide. The iridium sesquioxide is green if precipitation is effected with dilute potassium hydroxide, but black if concentrated potash is used. The green hydrated oxide is converted to the black form by heating it with concentrated caustic potash. It invariably adsorbs some alkali, which cannot be removed completely by washing (Wöhler).

(*iii*) *Sulfides*. Hydrogen sulfide precipitates *iridium disulfide*, IrS_2, from iridium(IV) solutions. This can also be obtained in the dry way, by heating $IrCl_3$ in a stream of H_2S to 630° (Wöhler, 1933). Thermal degradation of this compound yields Ir_2S_3 (Biltz, 1937), which degrades further directly into the metal and free sulfur. There is also a sulfide richer in sulfur. This has a definite dissociation pressure at a composition corresponding to the formula Ir_3S_8, but can take up a considerable amount of sulfur in solid solution. Heats of formation of the sulfides, calculated from the dissociation pressures are: Ir_2S_3 25.5; IrS_2 30; Ir_3S_8 about 35 kcal per g-atom of Ir. The sulfide Ir_3S_8 has the same structure as the triselenide $IrSe_3$ and the tritelluride $IrTe_3$ (pseudo-pyrite type, cf. p. 332).

$IrSe_3$ and $IrTe_3$ are converted to $IrSe_2$ and $IrTe_2$, respectively, on heating; further dissociation of the latter leads directly to the metal.

(b) Halides

(*i*) *Fluorides*. Direct combination of iridium with fluorine leads to *iridium hexafluoride*, IrF_6. This was obtained by Ruff (1927) by gently heating powdered iridium in a current of fluorine, in a fluorspar tube. At ordinary temperature, iridium hexafluoride forms a very volatile, yellow, vitreous mass, which liquefies at 44°. Its boiling point, from the vapor pressure curve, is 53°. It is a very reactive compound, and not only combines with water according to the equation:

$$IrF_6 + 5H_2O = Ir(OH)_4 + 6HF + \tfrac{1}{2}O_2,*$$

but is reduced even by chlorine, to form IrF_4 and ClF.

Iridium tetrafluoride, which is most simply obtained by heating iridium hexafluoride with iridium powder, is a viscous, yellow-brown, non-volatile oil, which is immediately hydrolyzed by water to iridium(IV) hydroxide and hydrofluoric acid.

(*ii*) *Chlorides*. The chlorides of iridium were the object of an exhaustive investigation by Wöhler (1913). The most readily obtained is the *trichloride*, $IrCl_3$, which is best made by heating powdered iridium in chlorine to 600–620°. It is an olive green powder, which becomes yellow and crystalline, without change in composition, when it is sintered at 650°. It is stable under 1 atmosphere of chlorine pressure up to 763°. By loss of chlorine, it passes into *iridium dichloride*, which is stable between 763° and 773°. Between 773° and 798°, under 1 atm. of chlorine pressure, the *monochloride*, $IrCl$, is stable (copper red crystals, density 10.18). All these chlorides are insoluble in water, and also in acids and in caustic alkalis. When iridium trichloride is heated with chlorine under high pressure, *iridium tetrachloride*, $IrCl_4$, is slowly formed. This can be obtained in hydrated form by treatment of ammonium hexachloroiridate(IV), $(NH_4)_2[IrCl_6]$, with aqua regia. The hydrated tetrachloride forms a brown-black deliquescent mass. It is not very stable, and evolves chlorine on heating, even under a chlorine pressure of 1 atm. Hydrated iridium trichloride forms dark green, water-soluble crystals. These water-soluble chlorides are, however, probably not simple salts, but complex chloro-hydroxo compounds.

(c) Iridium Salts

The chlorides of iridium do not have the typical properties of salts, and very little is known at all of *simple* salts of iridium. Both the composition and the constitution of some of the compounds mentioned in the literature must be regarded as questionable. On the other hand, iridium forms numerous well defined *complex*

---

* Some O_3 is formed, as well as O_2.

salts. The well characterized *alums,* $M^IIr(SO_4)_2 \cdot 12H_2O$ (M^I = NH_4, K, Rb, Cs, Tl^I) represent a transition between the simple salt and complex salt types.

Whereas tetrapositive iridium appears to form only salts with complex anions (acido-salts), both salts with complex anions and with complex cations are known as derivatives of tripositive iridium.

Iridium(III) salts with complex cations ('iridammines'), like the corresponding cobalt and chromium compounds, are typical strongly complexed compounds. They conform essentially to the types known also for rhodium:

$[Ir(NH_3)_6]X_3$, hexammineiridium(III) salts
$[Ir(H_2O)(NH_3)_5]X_3$, aquopentammineiridium(III) salts
$[IrR(NH_3)_5]X_2$, acidopentammineiridium(III) salts
$[IrR_2(NH_3)_4]X$, diacidotetrammineiridium(III) salts.

The known salts of tripositive iridium with complex anions are almost without exception *hexacidoiridates*—i.e., of the type $M^I_3[IrX_6]$. Numerous representatives of this type are known (X = Cl, Br, I, NO_2, CN, $\frac{1}{2}SO_4$, $\frac{1}{2}SO_3$, SO_3K, etc., M^I = Na, K, NH_4, Ag, Hg^I, and often H).

Sodium hexachloroiridate(III), $Na_3[IrCl_6] \cdot 12H_2O$, often used as the starting material for the preparation of other iridium(III) compounds, is prepared from the corresponding chloroiridate(IV) salt. The latter is obtained by heating an intimate mixture of powdered iridium and sodium chloride in chlorine, and may be reduced by gentle means—e.g., by heating in hydrogen chloride at 400–500°, or by treatment in solution at room temperature with hydrogen sulfide. Pentachloroammineiridate(III) salts $M^I_2[IrCl_5(NH_3)]$ were prepared by Lebedinski (1938). For dimethyl glyoxime compounds of iridium see p. 334.

Complex compounds of +4 iridium conform to the type $M^I_2[IrX_6]$ (hexacidoiridate(IV) salts), and comprise chiefly the halogen compounds. *Ammonium hexachloroiridate(IV),* $(NH_4)_2[IrCl_6]$, isomorphous with ammonium hexachloroplatinate, is precipitated by adding ammonium chloride to a solution of sodium hexachloroiridate(IV) (see above), as deep red, octahedral crystals. It is very sparingly soluble in cold water, but considerably more soluble in hot water, and can therefore conveniently be recrystallized. Hydrated iridium tetrachloride can be prepared from it by treatment with aqua regia, and it is converted to the trichloride by heating it in chlorine. The pure metal is readily obtained by heating ammonium chloroiridate in hydrogen.

(d) Carbonyl Compounds

The *iridium carbonyls,* $[Ir(CO)_4]_2$ and $[Ir(CO)_3]x$ (cf. Table 34, p. 352) correspond in composition to the cobalt carbonyls. They may be readily prepared by the action of CO on the iridium halides or halogeno-complex salts under pressure. *Iridium carbonyl halides* are formed as intermediates in this process. These have the compositions $Ir(CO)_2X_2$ and $Ir(CO)_3X$, and do not correspond to the cobalt carbonyl halides; their properties—e.g., color and volatility—show a regular gradation with the halogen and CO content. These compounds appear to be monomeric, covalent compounds, with the coordination number 4. (Hieber, 1940 and later).

Hieber has also shown the existence of an *iridium carbonyl hydride,* $H[Ir(CO)_4]$.

21-24. Dyad III: Palladium and Platinum

General

Palladium and platinum are ductile, not very hard metals. Their melting points are higher than that of iron, but lower than the melting points of the other platinum metals. Palladium in particular is attacked by acids, but platinum is also more readily attacked than other platinum metals; palladium and platinum (especially the latter) have a smaller affinity for oxygen than do their congeners. Catalytic properties are very highly developed in both palladium and platinum.

Both elements are predominantly *di-* and *tetra*-positive in their compounds. Very few simple compounds are formed, but both elements have a very strong tendency to form complex compounds.

21. Palladium (Pd)

Palladium is found native in the form of individual grains in platinum ores, and in certain auriferous sands of Brazil, Colombia and the Caucasus. It frequently occurs alloyed with gold or silver (or both). In these ores, the palladium is usually about 1 to 2 per cent of the platinum content. The proportion of palladium to platinum is very much higher in the primary deposits of platinum, such as are found in ultrabasic rocks in the Transvaal, and in the nickeliferous pyrrhotite of Ontario. In the Transvaal, palladium is present as *braggite*, (Pd,Pt,Ni)S; as *stibiopalladinite*, Pd_3Sb, and as a mercuride PdHg. In the Sudbury deposits, palladium is present in amount comparable with the platinum, probably largely as the selenide. Lockyer showed that the lines of palladium were observable in the spectrum of the sun.

Palladium was discovered by Wollaston in 1803. He named the element in honor of the minor planet Pallas, which had newly been discovered.

(a) Physical Properties

In appearance, palladium stands between silver and platinum. Its density is 12.03 and it is the most fusible of the platinum metals (m.p. 1555°). It softens before melting, and can therefore be worked and welded. It is a little harder and tougher than pure platinum, but its elasticity is lower.

Palladium has a considerable absorptive capacity for many gases, and especially for hydrogen. Palladium consequently manifests a highly specific *permeability* for hydrogen. This is distinctly measurable at 240°, and increases further with rise of temperature.

According to Graham, a palladium sheet 1 mm thick allows the passage of 42.3 cu. mm of hydrogen per min. per cm^2 at 240°, or 400 cu. mm per min. per cm^2 at 1060°. At ordinary temperature, palladium can take up 350–850 times its own volume of hydrogen. It swells markedly in so doing, and becomes brittle and cracked. The hydrogen is given off again on gentle warming in a vacuum. It is dissolved in the metal in the *atomic* state. Hydrogen is strongly activated by palladium, accordingly.

The absorption of hydrogen is accompanied by a decrease in the electrical conductivity of the palladium. The passage of current through a palladium wire charged with hydrogen is accompanied by some transport of hydrogen, and it follows that the H atoms in palladium are dissociated (at least to some degree) into protons and electrons. Hydrogen is thus present in palladium in the 'metallic state'.

Depending upon the pretreatment of the metal, the velocity of absorption of hydrogen by palladium may be proportional to the gas pressure, or to the square root of the gas pressure. In the former case, the rate determining step is the dissociation of H_2 molecules, adsorbed at the metallic surface, and in the latter case it is the rate of penetration of H atoms, formed by dissociation, into the interior (Wagner, 1932).

X-ray investigations have shown that the absorption of hydrogen is not attended with any change in the crystal structure of palladium. However, if the hydrogen content exceeds a certain amount (which depends on the temperature), the lattice undergoes a discontinuous expansion (from $a = 3.884$ Å to 4.020 Å at room temperature). Further increase in hydrogen content brings about a further continuous expansion (to 4.07 Å at a hydrogen content of 0.80 atom-%). Other properties also undergo a discontinuous change with the expansion of the lattice. The palladium-hydrogen system with an expanded lattice is therefore regarded as distinct from the solid solution phase, which shows no lattice expansion— i.e., it is considered to be a *compound*, which has a rather broad range of homogeneity. Its ideal formula may be Pd_2H, but there are some reasons, based on the electronic theory of metals, for considering that it may approximate $PdH_{0.6}$.

In contact with palladium, hydrogen can react with chlorine, bromine, iodine, and oxygen in the dark and at ordinary temperature.

Palladium sponge charged with hydrogen ignites in air. The occluded gas also reduces $HgCl_2$ to Hg_2Cl_2, $FeCl_3$ to $FeCl_2$, $[Fe(CN)_6]^{3-}$ to $[Fe(CN)_6]^{4-}$, $[AsO_4]^{3-}$ to $[AsO_3]^{3-}$, $[ClO_3]^-$ to Cl^-; it converts SO_2 to H_2S, $[NO_3]^-$ to $[NO_2]^-$ and NH_3, CH_3NO_2 to CH_3NH_2, $C_6H_5NO_2$ to $C_6H_5NH_2$, etc. Palladium charged with hydrogen, in contact with oxygen and water, can also bring about interesting *oxidation* reactions—e.g., it converts CO (which is not oxidized by ozone at its decomposition temperature) into CO_2 at ordinary temperature, N_2 into NH_4NO_2, benzene into phenol, and toluene to benzoic acid.

Palladium is especially active catalytically in the colloidal state.

Palladium can be obtained as a reversible colloid (according to Paal) by the reduction of aqueous palladium(II) chloride, to which sodium protalbinate (as protective colloid) and caustic soda (rather more than necessary to precipitate the hydroxide) are added, by means of hydrazine hydrate. Electrolytes are removed by dialysis, and the solution is evaporated at 60–70°; after drying over sulfuric acid, black lamellae are obtained, from which the sol may be reconstituted simply by the addition of water.

Colloidal palladium prepared by Paal's method, which can absorb 3–8 times as much hydrogen as ordinary palladium, has proved particularly suitable for the hydrogenation of unsaturated organic compounds. Hydrogenation can be carried out at room temperature, with the minimum risk of interference with the molecule, as is necessary, for example, in determinations of constitution.

(b) Chemical Behavior

Palladium is oxidized to palladium monoxide, PdO, by oxygen at a dull red heat. The oxide decomposes at higher temperatures. Palladium which has been melted in an atmosphere of oxygen spurts on solidification, like silver, since in the molten state it dissolves more oxygen than in the solid state.

At a red heat, fluorine converts palladium to the difluoride PdF_2, and chlorine forms the dichloride $PdCl_2$. Sulfur and selenium attack the metal at a rather higher temperature, with vigorous evolution of heat. Phosphorus and arsenic react rather less energetically, and silicon reacts only at a white heat. Carbon is dissolved by molten palladium, but is redeposited, as graphite, upon solidification. Palladium forms alloys with most metals; it forms unbroken series of mixed crystals with cobalt, nickel, copper, silver, and gold. For further data on the behavior of palladium towards the metals and non-metals see Table 29, p. 244–5.

The system Mn-Pd, referred to in Table 24 (p. 209) (Grube, 1936), is noteworthy in that a compound Mn_2Pd_3 appears, which is formed from the mixed crystals primarily deposited by a transformation occurring in the solid state *without any change of composition*. In the formation of this compound the primary mixed crystals, which have the same crystal structure as Pd, undergo a structural transformation below 1175°, to form a crystal structure peculiar to the compound Mn_2Pd_3. In this structure, the Mn and Pd atoms are initially distributed completely at random, but below 530° an ordering process sets in (formation of a 'superstructure' phase), without change of crystal structure. The second compound which appears in this system, MnPd, corresponds to a maximum (at $1515°$) in the melting point curve. It does not differ in any respect in structure, above $608°$, from Pd and its face-centered cubic mixed crystals. Below $608°$ it undergoes a structural transformation, forming a tetragonal face-centered structure with an ordered distribution of Pd and Mn atoms.

Palladium is slowly attacked by dilute nitric acid. It dissolves rapidly in concentrated nitric acid, especially in the presence of oxides of nitrogen. The best solvent for palladium is aqua regia. Hydrochloric acid is without action on compact palladium, even when the acid is concentrated, as long as it is free from dissolved oxygen or free chlorine.

Boiling concentrated sulfuric acid dissolves palladium, forming $PdSO_4$ and SO_2. The same reaction occurs on fusion with potassium pyrosulfate. Palladium is not oxidized by fusion with soda and potassium nitrate, but is converted to the monoxide, PdO, if it is heated with sodium peroxide.

The standard potential of palladium is about -0.82 volt—i.e., palladium is rather more noble than silver.

22. Palladium Compounds

Palladium is usually dipositive and less often tetrapositive in its compounds. Known compounds of $+2$ palladium include the monoxide, PdO, the monosulfide, PdS, and a few simple salts, among which are the halides PdX_2, as well as a very large number of complex salts (coordination compounds). The dioxide, PdO_2 (which is known only in the hydrated state), the disulfide, PdS_2, and coordination compounds (especially the *hexahalogenopalladates* $M^I_2[PdX_6]$) are known as derivatives of $+4$ palladium.

Palladium is predominantly *trivalent* towards fluorine. Ruff (1929) obtained the *trifluoride*, PdF_3, with relative ease, by treatment of $PdCl_2$ with fluorine at $200–250°$. The trifluoride is a black, finely crystalline, hygroscopic powder (density 5.06). It was not possible to obtain the difluoride, PdF_2, in a state of purity by dry methods.

(a) Oxides, Sulfides, etc.

(*i*) *Palladium monoxide*, PdO, may be obtained as a black powder, insoluble in all acids (including aqua regia), by heating palladium powder in oxygen. It decomposes into palladium and oxygen at high temperatures, the decomposition pressure reaching 1 atm. at $875°$. A hydrated, acid-soluble form of the oxide is obtained by precipitation from solution—e.g., by hydrolysis of palladium nitrate. The hydrated monoxide is brown when air dried, and black when it is dried on a water bath. Its water content is variable, and water is not lost completely when it is heated to $500–600°$, although loss of oxygen then commences. Its solubility diminishes as its water content is lowered. It has the properties of a weak oxidant. Hydrogen reacts even at ordinary temperature, both with the hydrated and the anhydrous oxide, reduction to the metal occurring with incandescence.

(*ii*) *Palladium dioxide*. It has been possible to prepare *palladium dioxide*, PdO_2, only by wet methods, and in a hydrous condition. Thus it is obtained by the addition of alkali hydroxide to the solution of a chloropalladate, $M^I_2[PdCl_6]$, as a dark red precipitate which turns black when any attempt is made to dry it. It is completely converted to palladium monoxide at 200°, and decomposes slowly even at ordinary temperature. It has strongly oxidizing properties. When freshly prepared, it is soluble in cold dilute acids and in concentrated sodium hydroxide solution. The solubility diminishes as its water content is reduced.

(*iii*) *Sulfides*. The sulfides of palladium have been studied chiefly by Wöhler (1933), Biltz (1936) and Weibke (1935)—cf. Table 29, p. 244–5. *Palladium monosulfide*, PdS, is precipitated by hydrogen sulfide from solutions of palladium(II) compounds, as a brown-black precipitate, insoluble in dilute hydrochloric acid and in ammonium sulfide. Prepared in the dry way (e.g., by heating ammonium chloropalladate with sulfur, or by thermal decomposition of the disulfide) it forms hard bluish or silvery lustrous crystals, insoluble in nitric acid or aqua regia. *Palladium disulfide*, PdS_2, can be obtained by various methods—e.g., by heating $PdCl_2$ with sulfur to 450–500°. In the finely disperse state it is brown black; in the crystalline state grey black. It is insoluble in strong acids, but is readily dissolved by aqua regia. Above 600° it loses sulfur, and is converted to the monosulfide.

The crystal structure of PdS_2 is not known. $PdSe_2$ and $PdTe_2$ form layer lattices of brucite (or CdI_2) type. The naturally occurring (Pd,Pt,Ni)S, braggite, has a quite different structure from NiS. The palladium atoms are located at the center of a square planar arrangement of sulfur atoms; thus the configuration is the same as in the coordination compounds of +2 palladium. It has not been established whether the synthetic pure PdS has the same structure.

Double compounds of the palladium sulfides with alkali sulfides have been prepared by dry methods.

(b) Simple Palladium Salts

These, e.g., the dark red-brown, very hygroscopic dichloride, $PdCl_2 \cdot 2H_2O$, the red brown, deliquescent sulfate $PdSO_4 \cdot 2H_2O$ and the nitrate $Pd(NO_3)_2$, which crystallizes in yellow brown deliquescent prisms—may obtained by dissolving palladium(II) oxide hydrate in the appropriate acid. The anhydrous dichloride, $PdCl_2$, which may be prepared by heating palladium sponge to a red heat in chlorine, is also freely soluble and deliquescent. The aqueous solution is decolorized by carbon monoxide, ethylene, methane, and other reducing gases, with deposition of metallic palladium. Traces of carbon monoxide or coal gas can thus readily be detected by filter paper impregnated with very dilute palladium chloride solution.

X-ray studies have shown that in anhydrous palladium(II) chloride each palladium atom is at the center of a square of co-planar chlorine atoms, so that the crystal might be formulated as a 'linear polymer' of $PdCl_2$:

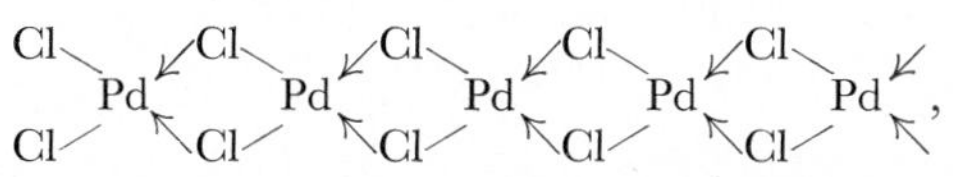

The square planar, 4-coordinated arrangement is characteristic of many compounds of dipositive palladium (e.g., the chloropalladites or tetrachloropalladate(II) salts, see below), both in crystal structures, and in complex ions in solution. It is not improbable that the so-called 'simple' salts mentioned above are all coordination compounds—e.g., $[PdX_2(H_2O)_2]$ in the case of the solid state, whereas the ion $[Pd(H_2O)_4]^{++}$ may possibly be present in their solutions.

When anhydrous palladium dichloride is heated in a stream of dry carbon monoxide that has been charged with methyl alcohol vapor, it combines to form the compound $PdCl_2(CO)$. This is instantly decomposed by water, with the formation of metallic palladium. (Manchot, 1926).

In hydrochloric acid solution, palladium(II) chloride forms complex ions $[PdCl_4]^=$. On evaporation solutions containing palladium(II) chloride and alkali chlorides furnish crystals of *tetrachloropalladate*(II) salts (chloropalladites), of the formula $M^I{}_2[PdCl_4]$—e.g., potassium tetrachloropalladate(II), $K_2[PdCl_4]$, red brown dichroic prisms, isomorphous with the corresponding platinum salt (p. 347). The $[PdCl_4]^=$ ion has been shown by X-ray studies to have a coplanar square configuration of the chlorine atoms. (Dickinson).

(c) Complex Salts of Palladium

A wide variety of palladium compounds of the type $M^I{}_2[Pd\ X_4]$ is known, in addition to the tetrachloropalladates, and including cases where the corresponding 'simple' palladium salt cannot be isolated. There are also numerous *ammines* of $+2$ palladium. Nearly all of these are of the types $[Pd(NH_3)_4]X_2$ (*tetramminepalladium(II) salts*) and $[Pd(NH_3)_2X_2]$ (*diamminepalladium(II) compounds*, of non-electrolyte type). Tetrammine compounds are in general obtained by passing ammonia gas over anhydrous simple palladium salts; they may also be obtained from solution. If a deficiency of ammonia is used, or if tetrammine salts are treated with acid, the diammine compounds are obtained.

For the configuration of diammine and tetrammine palladium(II) salts, see p. 349.

In the tetrapositive state, palladium forms no complex cations (unlike platinum), but is able to form complex anions. Thus the action of chlorine or aqua regia on a solution of potassium tetrachloropalladate(II) produces *potassium hexachloropalladate(IV)*, which is sparingly soluble in cold water and crystallizes in vermilion regular octahedra. It is also obtained by dissolving palladium in aqua regia (or hydrochloric acid-chlorine mixture) and adding potassium chloride. It readily loses chlorine, and undergoes conversion to the tetrachloropalladate(II)—e.g., merely on boiling the aqueous solution. When it is treated with aqueous ammonia, it forms diamminepalladium(II) chloride, $[Pd(NH_3)_2Cl_2]$, with evolution of nitrogen.

According to Wellmann, the equilibrium constant $\dfrac{[PdCl_6^=]}{[PdCl_4^=][Cl_2]}$ has a value 4160, in a solution 1-normal with respect to Cl^- ions. In such a solution, the oxidation potential for the half-reaction $2Cl^- + PdCl_4^= = PdCl_6^= + 2e$ is -1.29 volts at 25°, relative to the normal hydrogen electrode.

23. Platinum (Pt)

Platinum is most commonly found in alluvial deposits as the native metal. It is generally associated with the other metals of the platinum group, and also with iron, lead, copper, gold, and silver. A few compounds occur as minerals—e.g., *sperrylite*, $PtAs_2$ and *cooperite*, PtS.

Unlike gold, which is found in *acid* (i.e., quartzitic) rocks, platinum ores are found almost exclusively in ultrabasic rocks, and in particular, in secondary alluvial deposits derived from them. Primary deposits of platinum (which, except for those of the Transvaal, are very little worked), contain the native metal or platinum minerals very finely dispersed in silicates of the olivine and pyroxene types. Detritus from the primary deposits, and water borne gravels, have carried the platinum with them in the course of geological time, and have laid down deposits, enriched in the heavier minerals, in the places where the secondary deposits of ore are now found.

The most productive source of platinum was formerly the Urals. Outside Russia, platinum is found in Europe only in very small amounts in France and Spain. Outside Europe, platinum is found in Canada, in the nickeliferous pyrrhotite of Sudbury, where it occurs chiefly as sperrylite, dispersed in pyrrhotite, chalcopyrite, and pentlandite. These deposits now represent the greatest single source of platinum and (as already mentioned) of palladium. Platinum also occurs as sperrylite in the ultrabasic rocks of the Transvaal, and as native metal in Colombia, Brazil, Haiti, Borneo, California, New Caledonia, and Lapland.

In the Sudbury ore, the platinum content is about $\frac{1}{2}$ g per ton—i.e., 1 part in 2,000,000. The platinum metals become concentrated in the copper-nickel matte of the Orford process, and can be recovered from residues of the nickel carbonyl process of nickel refining, and from anode slimes in electrolytic nickel refining. Such residues may contain about 2% platinum and as much palladium, with 0.2% rhodium, 0.2% ruthenium, 0.04% iridium, and traces of osmium, together with silver and gold.

The world production of platinum, in the form of ores, amounted in 1936 to about 10 tons, of which 41% came from Canada, 31% from Russia, 12% from Colombia and 10% from the Union of South Africa.

(a) Physical Properties

Platinum is a grey-white, lustrous, metal, which is not very hard, is fairly ductile, and can readily be fashioned and welded when hot. Its density is 21.45, m.p. 1774°; it can be vaporized in the electric arc.

From measurements at low pressures, Van Liempt extrapolated the boiling point to be 3800°. The specific heat of platinum is 0.318, electrical conductivity $\varkappa = 9.1 \cdot 10^4$ at 0°, and thermal conductivity $\lambda = 0.165$ (at 17°). Its coefficient of thermal expansion $(9.1 \cdot 10^{-6})$ is close to that of soda glass, so that platinum can be sealed into soda glass.*

Platinum is markedly permeable to hydrogen at red heat, but not to other gases such as O_2, N_2, Cl_2, HCl, CO, CH_4, H_2O, H_2S, or NH_3. The two latter are decomposed in contact with hot platinum, and the hydrogen so formed diffuses through the metal.

Platinum can absorb considerable quantities of hydrogen. In the finely divided state (as platinum sponge, or platinum black), it can take up more than 100 times its own volume of hydrogen. It can most readily be saturated with hydrogen by generating the gas at its surface electrolytically. Platinum gives the occluded gas up again less readily than does palladium, but the gas can be driven out completely by heating the metal in a vacuum.

Very small amounts of helium are also absorbed by platinum, and also oxygen in rather large amounts. Platinum black can absorb up to 100 times its own volume of oxygen. Hydrogen and oxygen, dissolved in platinum, are highly activated, This is the basis of the efficacy of platinum as a hydrogen- or oxygen-carrier. *Colloidal platinum*, prepared by reduction of hexachloroplatinic acid by means of hydrazine in weakly alkaline solution, and in presence of a protective colloid, is particularly effective for the hydrogenation of unsaturated organic compounds. Paal, who introduced the use of protected colloidal platinum prepared in this

---

* But not, however, into such glasses as Pyrex. Vacuum-tight glass-to-metal seals in borosilicate glass apparatus can be made by using tungsten and certain special alloys, such as Kovar or Fernico, which have low coefficients of expansion.

way, used sodium lysalbuminate as protective agent; Skita found that gum arabic or the purest gelatin could often be used with advantage. Bredig had already previously obtained irreversible (because unprotected) colloidal dispersions of platinum, by vaporization of the metal in an electric arc, struck under water. Platinum sols prepared by Bredig were noted for their catalytic acceleration of the decomposition of hydrogen peroxide, which was still quite perceptible even at the highest dilutions (0.03 mg per liter). This hydrogen peroxide catalysis is inhibited by protective colloids, such as gelatin.

(b) Chemical Behavior

Platinum has the properties of a decidedly noble metal. It is practically without reaction with oxygen, and is not attacked by strong acids, except by aqua regia, in which it dissolves readily to form hexachloroplatinic acid, $H_2[PtCl_6]$.

According to Wöhler, platinum not only dissolves oxygen when it is moderately heated in an oxygen atmosphere, but can also combine to form the monoxide PtO. Platinum sponge can undergo partial oxidation even at ordinary temperature. The oxide is no longer stable above 560°. Platinum is also not quite completely insoluble in concentrated sulfuric acid, especially at high temperatures.

Platinum is rapidly attacked by hot alkali peroxides. Unless air is rigorously excluded, traces of peroxide are also formed when alkali hydroxides are melted; hence these melts also attack platinum. Carbon monoxide does not directly attack platinum. The risk of heating platinum crucibles in flame gases containing carbon monoxide or hydrogen (see p. 345) is probably chiefly due to the reducing action on the contents of the crucible of the gases which diffuse through the platinum.

Above 250°, platinum unites with dry chlorine to form the dichloride, $PtCl_2$. It is slowly attacked by chlorine water even at ordinary temperature. It combines with fluorine at a dull red heat, forming chiefly platinum tetrafluoride, PtF_4, with small amounts of the difluoride, PtF_2. Platinum is also attacked by sulfur under certain conditions, and still more easily by selenium, tellurium, and especially by phosphorus, with which it forms very fusible alloys. Platinum also alloys readily with arsenic, antimony, bismuth, tin, lead, and silver. It forms a complete range of mixed crystals with iron, cobalt, nickel, copper, and gold. Platinum (unlike palladium) has only a limited miscibility with silver in the solid state. For further details of alloys see Table 29 and 30 (pp. 244-5, 248) and Table 17 (p. 124).

Platinum-chromium alloys with up to 5 weight-% chromium have higher melting points and higher electrical resistivity than pure platinum. They are therefore very suitable for the windings of electric furnaces.

Platinum stands at the bottom of the electrochemical series. However, the tendency of platinum to form complex ions is so great that it is not possible to specify its real normal potential more closely.

(c) Treatment of Platinum Apparatus

In using platinum apparatus in the laboratory, it is necessary to bear in mind the ease with which it may be corroded by certain substances. Particular heed must be given to the possible formation of alloys at high temperatures.

Platinum crucibles should be heated only on triangles of silica, platinum, or fireclay.

Heated platinum can decompose hydrocarbons, with the deposition of soot; this carbon may diffuse into the metal which, when the carbon is burned off again, is left in a roughened, brittle state. Platinum crucibles should therefore never be heated in a luminous (sooty) flame. When a non-luminous Bunsen flame is used for heating, care should be taken that the crucible is high enough up in the flame to ensure that the platinum does not come in contact with the blue cone in the interior of the flame. A blow pipe or Meker burner is more suitable for the purpose than a Bunsen burner, but the use of an electric furnace is to be preferred. Neither the compounds of easily reducible metals (such as silver, copper, lead, tin, and bismuth) not the metals themselves, may be heated in a platinum crucible. Ignition of boron, borides, silicon, silicides, phosphorus, phosphides, arsenic, arsenides, antimony, antimonides, sulfides, sulfites, or dithionites in platinum must also be avoided.

Phosphates, arsenates, and antimonates must also not be ignited in platinum, except in an electric furnace, since they are so readily reduced by flame gases. It is possible for attack on platinum apparatus to take place during the ashing of organic substances containing phosphorus, even if carried out in an oxidizing atmosphere.*

Fusions of alkali hydroxide or peroxide must never be carried out in platinum.

Platinum crucibles are best cleaned by careful scouring with the finest silver sand (not with ordinary coarse-grained sand). The crucible may also be boiled with concentrated hydrochloric acid or concentrated nitric acid, but *not* with a mixture of the two, or may be cleaned by fusing potassium hydrogen sulfate or pyrosulfate in it. Platinum crucibles are not to be recommended, however, for carrying out pyrosulfate fusions in analysis, as the attack is perceptible; fused quartz crucibles are superior for this purpose.

The same rules apply to the use of other platinum apparatus as for crucibles.

24. Platinum Compounds

Platinum is predominantly di- and tetrapositive in its compounds. It has a strong tendency to form complex compounds, and exists exclusively in that form in solutions. Platinum forms *acido*-compounds in both the +2 and +4 states, those derived from the latter state being much more stable than those of +4 palladium. Platinum differs from palladium in that in the tetrapositive state it can also form *complex cations*.

A few types of complex compounds derived from tripositive platinum have been described. In the monochloride, PtCl, which is stable only within a narrow range of temperatures, it is unipositive, and in the very unstable trioxide PtO_3 it is hexapositive.

Platinum is the only member of the VIIIth Sub-group to form alkyl compounds. The *trialkyl platinum halides*—e.g., $(CH_3)_3PtI$—were first prepared by Pope and Peachey (1909) by the action of the Grignard reagent on platinum(IV) chloride. $(CH_3)_3PtI$ forms orange crystals (decomp. about 220°), which are soluble in organic solvents but insoluble in water. The molecular weight in benzene solution shows that the compound is highly associated. When treated with moist silver oxide, the iodide yields the base $(CH_3)_3PtOH$. This is soluble in organic solvents, but insoluble in water, and does not react with cold mineral acids. *Tetramethyl platinum*, $(CH_3)_4Pt$, was obtained by Gilman (1938) by the action of sodium methyl on trimethylplatinum iodide. It is a solid, fairly stable in air and soluble in organic solvents. Both $(CH_3)_3PtCl$ and $(CH_3)_4Pt$ have been shown by X-ray analysis to be associated in the solid state as well as in solution. They form tetrameric molecules $[(CH_3)_3PtX]_4$ (where $X = CH_3$ or Cl), of essentially similar structure. The Pt atoms occupy four alternate corners of a cubic framework, with Cl (or CH_3) at the other four corners. Each Pt atom is thus octahedrally coordinated with 3 unshared CH_3 groups and 3 shared Cl atoms or CH_3 groups. $Pt(CH_3)_4$ is remarkable in that the 'bridge' methyl groups form bonds with 3 Pt atoms. The present theoretical understanding of such structures is very imperfect.

---

* Apparatus made from relatively phosphorus-resistant platinum alloys (Pt-Ru-Nb, or Pt-Ta) has recently been introduced for such purposes.

(a) Oxides and Hydroxides

Although, as shown by Wöhler, platinum is able to combine directly with oxygen at moderate temperatures to form an oxide, the pure oxides and their hydrates can only be obtained by indirect means.

(i) The lowest oxide of platinum is *platinum monoxide*, PtO; this is not itself known, however, except in solid solutions with platinum or with platinum dioxide. However, the corresponding *hydroxide* Pt(OH)$_2$ can be obtained as a black precipitate by the action of alkali hydroxide on a solution of potassium tetrachloroplatinate(II), K$_2$PtCl$_4$ (potassium chloroplatinite). Air must be excluded, as the moist hydroxide is oxidized easily. After drying in carbon dioxide at 120–150° its composition corresponds to the formula Pt(OH)$_2$. It decomposes partially, to give platinum dioxide and the metal, if more strongly heated.

(ii) *Platinum sesquioxide*, Pt$_2$O$_3$, is also known only in the hydrous state. It can be obtained by the action of sodium hydroxide on *disulfatoplatinum(III) acid*, H[Pt(SO$_4$)$_2$]*, first isolated by Blondel. The latter compound is obtained by reducing a solution of platinum dioxide hydrate in concentrated sulfuric acid, by means of oxalic acid; it crystallizes in orange plates. The sesquioxide is readily oxidized when warmed in oxygen, and simultaneously gives up water.

(iii) *Platinum dioxide* (platinum(IV) oxide), PtO$_2$, is relatively the most stable oxide of platinum. Is is obtained in the hydrated state, as a red brown precipitate, by boiling a solution of platinum tetrachloride with soda. The dioxide hydrate is soluble both in acids and in strong alkalis. It becomes insoluble in acids on prolonged heating to about 200°, and begins to lose oxygen at rather higher temperatures, before it has lost all its water. It thereby decomposes directly into platinum and oxygen, the lower oxides of platinum not being formed as intermediates in the dissociation. It has not been possible to measure definite dissociation pressures over the platinum oxides, owing to the formation of solid solutions, but it can be stated that no oxide of platinum is stable in oxygen under 1 atmosphere pressure above 500°.

(iv) The very unstable *platinum(VI) oxide*, PtO$_3$, was obtained by Wöhler by the anodic oxidation of a solution of the dioxide in caustic potash at temperatures below 0°. The anode became coated with a lustrous golden deposit, having the composition 3PtO$_3$ · K$_2$O, which could readily be freed from alkali by treatment with dilute acetic acid. Platinum trioxide slowly decomposes at ordinary temperature. Oxygen is rapidly and completely lost, down to the composition PtO$_2$, when PtO$_3$ is gently heated. Chlorine is at once evolved if the trioxide is treated with dilute hydrochloric acid, but there is no evolution of oxygen if hydrogen peroxide is added.

(b) Sulfides

(i) *Platinum monosulfide*, PtS, can be obtained either directly from the elements (by heating an intimate mixture of very finely divided platinum and sulfur powder), or by the action of hydrogen sulfide on a solution of a chloroplatinate(II) salt or tetrachloroplatinum(II) acid: H$_2$[PtCl$_4$] + H$_2$S = 4HCl + PtS. When crystalline, it forms grey needles, but is precipitated from solution as a black precipitate. The crystalline sulfide, in particular, is very resistant towards both acids and alkalis, and is insoluble even in aqua regia.

(ii) *Platinum disulfide*, PtS$_2$, is formed as a black precipitate when hydrogen sulfide is passed into an almost boiling solution of a chloroplatinate(IV) salt or of hexachloroplatinic acid:**

$$H_2[PtCl_6] + 2H_2S = 6HCl + PtS_2.$$

* This compound crystallizes with water of hydration.
** Reduction of a part of the platinum to the metallic state often occurs simultaneously.

Platinum disulfide is also very resistant towards acids, but dissolves slowly in hot concentrated nitric acid and in aqua regia. It is almost insoluble in alkali solutions, unless it is mixed with other sulfides which are themselves soluble. It dissolves in polysulfide solutions, however, and does so with especial ease if a sulfide-soluble sulfide such as arsenic, antimony, tin, or gold sulfide is present. The platinum disulfide is precipitated again, with the other sulfide, when the solutions are acidified.

(*iii*) By heating finely divided platinum with alkali carbonate and sulfur, Schneider obtained double compounds, of the nature of thiosalts, having the general composition $M^I_4Pt_3S_6$ and $M^I_2Pt_4S_6$, and he stated that the free acid corresponding to the second type of thiosalt was oxidized on exposure to air, forming *diplatinum trisulfide*, Pt_2S_3. This was a steel grey powder (density 5.52), which smoldered when it was heated in air, leaving a residue of metallic platinum.

With selenium and tellurium, platinum forms the compounds $PtSe_2$ and $PtTe_2$, which are very resistant towards acids. Like PtS_2, they have the brucite structure.

PtS has a tetragonal structure in which the Pt atoms are arranged like the Zn atoms in zinc blende. Each Pt atom is surrounded by four S atoms, each of which is equidistant from 4 Pt atoms ($Pt \leftrightarrow S = 2.32$ Å). The S atoms do not form tetrahedra about the metallic atoms, as in the ZnS structure, but are in *square planar* arrangement. This type of coordination recurs frequently in the stereochemistry of platinum.

(c) Chlorides

Platinum forms four chlorides; their ranges of existence were determined by Wöhler in 1913. *Platinum monochloride* is only stable over a very narrow temperature range (from $581-583°$, under 1 atm. pressure of chlorine). The stability range of platinum dichloride extends from $435-581°$, that of *platinum trichloride* from $370-435°$, and *platinum tetrachloride* is only stable below $370°$.

(*i*) *Platinum dichloride*, $PtCl_2$, may be obtained by heating spongy platinum to about 500° in chlorine, or (better) by the thermal decomposition of platinum tetrachloride or chloroplatinic acid. It is a greenish grey or brownish powder, (density 5.87), which is insoluble in water. It dissolves in hydrochloric acid, to form the complex acid $H_2[PtCl_4]$, but the dissolution is accompanied by a certain decomposition into platinum and hexachloroplatinic acid.

The most familiar salt of tetrachloroplatinum(II) acid is the potassium salt, $K_2[PtCl_4]$ (potassium chloroplatinite), which is best obtained by boiling potassium hexachloroplatinate with the theoretically necessary quantity of potassium oxalate. Potassium tetrachloroplatinate(II) forms dark red, freely soluble tetragonal prisms. As was first proved by Dickinson (1922) from X ray studies, it has the structure represented in Fig 44; the $[PtCl_4]^{2-}$ ions constitute square planar structural groups. $K_2[PdCl_4]$ and $(NH_4)_2[PdCl_4]$ have similar

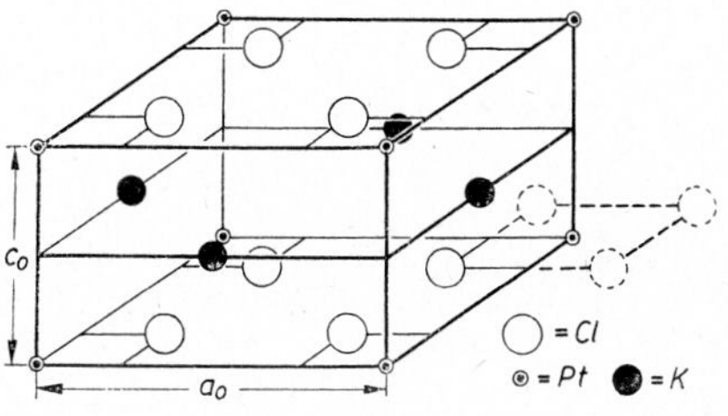

Fig. 44. Crystal structure of $K_2[PtCl_4]$. The diagram shows one unit cell. $a_0 = 6.99$ Å, $c_0 = 4.13$ Å, $Pt \leftrightarrow Cl = 2.32$ Å. The Cl atoms belonging to one of the $PtCl_4$ groups, but located outside the unit cell shown, are drawn with dotted lines.

Compounds with the same structure are $K_2[PdCl_4]$ ($a_0 = 7.04$ Å, $c_0 = 4.10$ Å, $Pd \leftrightarrow Cl = 2.29$ Å) and $(NH_4)_2[PdCl_4]$ ($a_0 = 7.21$ Å, $c_0 = 4.26$ Å, $Pd \leftrightarrow Cl = 2.35$ Å).

structures. The *planar* coordination of the four Cl atoms about the central Pt atom is noteworthy (cf. p 349).

As was discovered by Schützenberger (1870), $PtCl_2$ adds on CO at moderate temperatures, forming the compounds $[PtCl_2(CO)]_2$ (m.p. 194°), $[PtCl_2(CO)_2]$ (m.p. 142°), and $[(PtCl_2)_2(CO)_3]$ (m.p. 130°). It also forms very stable addition compounds with phosphorus trichloride, ammonia, and other anhydrobases (see below).

(*ii*) *Platinum tetrachloride* (platinum(IV) chloride), $PtCl_4$, can most conveniently be obtained by heating hexachloroplatinic acid to about 300° in chlorine. It forms a red-brown, slightly hygroscopic mass, which is readily soluble in water (and in acetone), but not very soluble in alcohol. It crystallizes from aqueous solutions with variable amounts of water of crystallization, depending upon the temperature. It is present in solution, not as a salt, but as the complex acid $H_2[PtCl_4(OH)_2]$ (dihydroxotetrachloroplatinum(IV) acid), which can be isolated by heating the more heavily hydrated compounds to 100°. When platinum tetrachloride is dissolved in hydrochloric acid, it forms hexachloroplatinic acid (hexachloroplatinum(IV) acid), $H_2[PtCl_6]$, with considerable evolution of heat.

The other halides of platinum are very similar to the chlorides. The bromides decompose at lower temperatures than the chlorides, and the iodides are still less stable (Wöhler, 1925). Both the free acid $H_2[PtI_6]$ and the tetraiodide PtI_4 are known, whereas the corresponding palladium compounds have not been obtained. The greater stability of platinum in the tetrapositive state, as compared with palladium, is thereby exemplified.

(d) Complex Compounds of Platinum

A very large number of *coordination* compounds of platinum is known. Most of these can be assigned to one or other of a few simple structural types.

SURVEY OF THE MOST IMPORTANT TYPES OF PLATINUM COMPLEXES

Valence state of platinum = II		III	IV
Acido-compounds	$M^I_2[PtX_4]$	$M^I[PtX_4]$	$M^I_2[PtX_6]$
	$M^I[PtX_3 Am]$	$M^I_2[PtOX_3]$	$M^I_2[PtO_2X_2]$
Platinum ammines	$[Pt Am_4]X_2$		$[Pt Am_6]X_4$
	$[PtR Am_3]X$		$[PtR_2 Am_4]X_2$
	$[PtR_2 Am_2]$		

As shown in the summary table, dipositive platinum usually exercises the coordination number 4, and tetrapositive platinum has coordination number 6.

The best known of the acido-compounds of platinum are the hexachloroplatinates(IV), $M^I_2[PtCl_6]$. Treatment of these salts with caustic alkalis yields the hexahydroxoplatinates(IV), $M^I_2[Pt(OH)_6]$ (golden yellow crystals) and hexahydroxoplatinic acid, $H_2[Pt(OH)_6]$, is thrown down as a white precipitate when acids are added to solutions of these salts.

Among acido-compounds of +2 platinum, tetrachloroplatinates(II) have already been mentioned. The corresponding acid, $H_2[PtCl_4]$ is conveniently obtained by reduction of hexachloroplatinic acid solutions by means of sulfur dioxide. It cannot be isolated, and is known only in its dark red solution and as its salts. The cyanoplatinates described below, and other salts, are derived from complex acids of the same structural type.

Complex compounds of ammonia and other nitrogen bases with platinum are known as *platinum ammines*, and exist in very large number. Among those which have been longest known are the two Reiset's chlorides, $[Pt(NH_3)_4]Cl_2$ and $[PtCl_2(NH_3)_2]$. The former is obtained by boiling a solution of tetrachloroplatinum(II) acid with a large excess of ammonia. It crystallizes from solution in long colorless needles (with 1 molecule of water, which is expelled at 110°). When it is heated to 250°, $2NH_3$ are expelled, and it is converted into 'Reiset's second chloride', a sulfur-yellow powder which is not very soluble in cold water, but soluble in hot, and which crystallizes in rhombohedra. A fresh solution gives no immediate precipitate with silver nitrate, indicating that the chlorine is complex-bound.

Isomeric with Reiset's second chloride is 'Peyrone's chloride', which is obtained as a greenish yellow precipitate when ammonia is added to a cold solution of tetrachloroplatinum(II) acid. This compound is more soluble in cold water than in hot. It also contains the chlorine bound in non-ionized form. The isomerism between this compound and Reiset's second chloride can be explained only on the hypothesis that the four ligands in compounds of this type are arranged about the central atom in a *planar configuration*. This makes it possible for *cis-trans* isomerism to occur.

$$NH_3\diagdown \quad \diagup Cl \qquad\qquad NH_3\diagdown \quad \diagup Cl$$
$$Pt \qquad\qquad\qquad\qquad Pt$$
$$NH_3\diagup \quad \diagdown Cl \qquad\qquad Cl\diagup \quad \diagdown NH_3$$

$$\text{\textit{cis} form} \qquad\qquad\qquad \text{\textit{trans} form}$$

Exhaustive investigations have proved the validity of this hypothesis, and both chemical reasoning and the evidence of dipole moment determinations (Jensen, 1936) have shown that Peyrone's chloride is the *cis*-compound.

The occurrence of *cis-trans* isomerism has also been proved for other platinum(II) complexes of the type $[PtX_2 Am_2]$ (e.g., the sulfine derivatives, with $Am = (C_2H_5)_2S$, etc.), and also for complex cations involving two different ammines as ligands—e.g., $[Pt\,pyr_2(NH_3)_2]Cl_2$, (dipyridinediammineplatinum(II) chloride). Hantzsch showed cryoscopically, for the last mentioned compound, that both forms were monomeric in phenol solution. Hence the isomerism cannot be attributed to *polymerism*.

The *planar configuration* of platinum(II) and palladium(II) compounds with coordination number 4 is very important, since with most elements of the Periodic System it is usual to find that four ligands are bound in a *tetrahedral* arrangement about the central atom.* Doubts cast upon the correctness of the hypothesis have been settled by very thorough studies carried out between 1930 and 1940 by a number of workers [see especially a review by Mellor, *Chemical Reviews*, 33 (1943) 137], using both chemical evidence, arguments based upon optical enantiomorphism (Mills and Quibell, 1935; Drew, 1934), and physical measurements (dipole moments, Jensen). For the acido-complexes $[PtCl_4]^{2-}$ and $[PdCl_4]^{2-}$,

$$\left[M\left(\begin{matrix} \diagup SC.CO \\ \diagdown SC.CO \end{matrix} \right)_2 \right]^{2-}$$, the planar structure has been proved by complete X-ray investiga-

tions (Cox, Dickinson). X-ray evidence also supports the hypothesis of planar arrangement in the ammines, but complete structure analyses have not been carried out.

Many compounds are known which arise from the combination of platinum ammine cations with complex anions of platinum. An example is "Magnus' green salt", which has the same empirical formula as Reiset's second chloride and Peyrone's chloride, but is shown by its mode of formation to be tetrammineplatinum(II) tetrachloro-platinate(II), $[Pt(NH_3)_4][PtCl_4]$. Yet another salt with the same empirical formula is $[Pt(NH_3)_4]$ $[PtCl_3(NH_3)]_2$. Magnus' salt was described by Magnus in 1828, and was the first known ammonia complex.

In addition to the ammines proper, there are many complex platinum compounds containing organic sulfides R_2S, tertiary phosphines R_3P, tertiary arsines R_3As, etc.; these correspond closely to the ammines.

* The planar configuration is not restricted to Pd^II and Pt^II alone. It has been proved to occur in complex compounds of Ni^II, Cu^II, Ag^II, and Au^III, and possibly Ru. The operation of purely electrostatic forces in complex formation (cf. Vol. I, p. 401 *et seq.*) necessarily results in a tetrahedral arrangement, with the highest possible spatial symmetry. Where *covalent* binding is responsible for complex formation, four coplanar bonds can be formed only when the appropriate electronic orbits (dsp^2 hybrid orbits) are available for bond formation. Complexes with this steric arrangement are therefore formed in one section of the Periodic System only—namely in the transition series, by metals with incompletely filled d levels. See Vol. I, p. 406.

(e) Chloroplatinic Acid and the Chloroplatinates

Chloroplatinic acid (hexachloroplatinum(IV) acid), $H_2[PtCl_6]$, is one of the most important platinum compounds. It is generally prepared by dissolving platinum in aqua regia, made up for this purpose with a high content of hydrochloric acid, in order to minimize the formation of compounds of the nitrogen oxides (e.g., $PtCl_4 \cdot 2NOCl$). The solution is evaporated to dryness with hydrochloric acid several times, to destroy such compounds; it is desirable to pass chlorine in during evaporation, as partial reduction to tetrachloroplatinum(II) acid often occurs. Pure hexachloroplatinic acid can be obtained directly by treating platinum sponge with hydrochloric acid containing chlorine. The acid crystallizes with $6H_2O$ from its fine yellow solution, in brownish red deliquescent prisms of density 2.43. It is freely soluble in water, alcohol, and ether. Of its salts, $M^I{}_2[PtCl_6]$, the hexachloroplatinates, the ammonium salt, $(NH_4)_2[PtCl_6]$, is the best known. It forms as a lemon yellow precipitate, when chloroplatinic acid is added to the solution of an ammonium salt, and consists of minute regular octahedra. It is sparingly soluble in water (0.67 g in 100 g at $15°$); the solubility in water containing ammonium chloride is considerably smaller, and can be reduced practically to zero by the addition of alcohol. It is therefore a very suitable form in which to precipitate platinum, both in analysis and for preparative purposes. It decomposes completely when it is heated, leaving a residue of spongy platinum. The potassium salt, $K_2[PtCl_6]$, is isomorphous with ammonium chloroplatinate (for crystal structure, see Vol. I, p. 398), which it resembles in solubility. The sodium salt, however, crystallizes with $6H_2O$, like the free acid, and is very soluble in both water and in alcohol. When silver nitrate is added to a solution of a chloroplatinate, the practically insoluble yellow silver salt, $Ag_2[PtCl_6]$, is precipitated.

(f) Cyanoplatinum(II) Acid and the Cyanoplatinates

The potassium salt of *tetracyanoplatinum(II) acid*, $H_2[Pt(CN)_4]$, is formed when platinum sponge is heated with potassium ferrocyanide, and crystallizes as the hydrate, $K_2[Pt(CN)_4] \cdot 3H_2O$, from the solution obtained by leaching the cooled melt. The salt was first prepared in this manner by Gmelin. Potassium tetracyanoplatinate(II) forms yellow, pleochroic rhombic needles (appearing blue along the axial direction), of density 2.45. It is very soluble in hot water, but far less soluble in cold, and can therefore be recrystallized well. *Barium tetracyanoplatinate(II)* ('barium platinocyanide'), $Ba[Pt(CN)_4] \cdot 4H_2O$, (which is also pleochroic) has the property of fluorescing brightly when it is exposed to X-rays or cathode rays. Fluorescent screens, consisting of a thin layer of barium tetracyanoplatinate coated on cardboard, are therefore often used to make visible the track of X-rays or cathode rays, and to show a visible image of objects undergoing radiographic examination.

Tetracyanoplatinum(II) acid is also known in the free state. It is most simply obtained by the action of hydrogen sulfide on its silver or copper salts. It is a fairly strong dibasic acid. Its alkali and alkaline earth salts are soluble in water, and crystallize well. They do not evolve any hydrogen cyanide when they are treated with strong acids—an indication of the strongly complexed nature of the tetracyanoplatinate(II) anion.

25. Analytical (Platinum Metals)

Platinum compounds are readily reduced to the metal, although less readily than those of gold. The metal can be converted to chloroplatinic acid by treatment with aqua regia and evaporation with hydrochloric acid, and from this the sparingly soluble and characteristically crystalline ammonium chloroplatinate is precipitated in yellow octahedra, on the addition of ammonium chloride.

Hydrogen sulfide precipitates platinum from solutions of hexachloroplatinic acid, as the disulfide. This is soluble in aqua regia, and also in ammonium polysulfide (especially if mixed with other sulfides which are soluble in that reagent), forming thio-salts which are decomposed on the addition of acid.

Hexachloroiridium(IV) acid, $H_2[IrCl_6]$, gives with ammonium chloride a precipitate exactly analogous to the foregoing, except that it is deep red in color. Any substantial amount of iridium, present in platinum as an impurity is therefore indicated by the darker color of the precipitate obtained with ammonium chloride. Palladium, if present in the form of hexachloropalladate(IV) ion, is also precipitated by ammonium chloride as red ammonium hexachloropalladate(IV). Hexachloropalladic acid is not stable in hot solution, however, and cannot be present after evaporation with hydrochloric acid. Ammonium tetrachloropalladate(II), $(NH_4)_2[PdCl_4]$, is soluble. Rhodium is also not precipitated by ammonium chloride. If a not too dilute rhodium solution, freed from excess hydrochloric acid by evaporating it to dryness, is treated with ammonia, yellow chloropentamminerhodium(III) chloride, $[RhCl(NH_3)_5]Cl_2$, is precipitated after a little time. Solutions containing osmium evolve osmium tetroxide (recognizable by its smell) when they are heated with dilute nitric acid. Ruthenium can also be distilled off as the tetroxide, by oxidation of an alkaline (ruthenate) solution with chlorine or periodic acid. The ruthenium tetroxide can be collected in concentrated hydrochloric acid, giving a deep red-brown solution which at once turns dark blue when zinc is added.

Very minute quantities of rhodium, iridium, palladium, and platinum can be detected by means of their great catalytic activity for the combination of hydrogen and oxygen. If a small drop of a solution containing any of these metals is placed on asbestos paper and ignited, a distinct incandescence of the resulting fleck can be observed when a jet of hydrogen is directed on to the paper (Hahn, 1930). A specific, and very sensitive, reaction for palladium is the catalytic acceleration of the reduction of molybdophosphoric acid (to molybdenum blue) by means of carbon monoxide (Zenghelis, 1910; Feigl, 1930).

The detection and determination of any one of the platinum metals, when present alone, is not difficult. However, the platinum metals mutually modify each other's reactions to an extent which depends upon the proportions present, so that when several, or the whole group, are present together an exact analysis may present severe difficulties*.

C. METAL CARBONYLS

(a) General

The compounds of carbon monoxide with the heavy metals (iron, cobalt, nickel, chromium, tungsten, rhenium, and some of the platinum metals), known generally as *metal carbonyls*, are quite different in structure and constitution from those formed by carbon monoxide with the light metals such as potassium. Whereas 'potassium carbonyl' (cf. Vol. I, p. 444) is really a benzene derivative, the metal carbonyls are substances in which carbon monoxide is bound as such to the metals in question. They are therefore a group of compounds of particular interest in relation to valence theory. Their technical importance has already been mentioned in connection with the most important of the group (pp. 289, 308).

The metal carbonyls at present known are summarized in a systematic manner in

* For a fuller discussion see Schoeller and Powell, *Analysis of Minerals and Ores of the Rarer Elements*, Griffin, London 1955.

TABLE 34

SURVEY OF THE KNOWN METAL CARBONYLS, CARBONYL HYDRIDES, AND
METAL CARBONYL HALIDES

METAL CARBONYLS

		$_{24}Cr$	$_{25}Mn$	$_{26}Fe$	$_{27}Co$	$_{28}Ni$
First 18 Period	I	$Cr(CO)_6$ $d = 1.77.$ Colorless rhombic cryst. Sublimes readily		$Fe(CO)_5$ $d = 1.49.$ Yellow liq. m.p. —20° b.p. 103°		$Ni(CO)_4$ $d = 1.31.$ Liq., Colorless, m.p. —25° b.p. 43°
	II		$[Mn_2(CO)_{10}]$ Golden monocl. cryst. m.p. 155°	$[Fe_2(CO)_9]$ $d = 2.09.$ Golden pseudo- hexag. cryst. Decomp. 100°	$[Co_2(CO)_8]$ $d = 1.73.$ Orange crystals. M.p. 51°	
				$[Fe_3(CO)_{12}]$ $d = 2.00.$ Green monoclinic. Decomp. 140°	$[Co_4(CO)_{12}]$ Black crystals. Decomp. 60°	
		$_{42}Mo$	$_{43}Tc$	$_{44}Ru$	$_{45}Rh$	$_{46}Pd$
Second 18 Period	I	$Mo(CO)_6$ $d = 1.96.$ Colorless rhombic cryst. Sublimes readily		$Ru(CO)_5$ Colorless liq. m.p. —22°		
	II			$[Ru_2(CO)_9]$ Orange monoclin. Sublimes	$[Rh_2(CO)_8]$ Orange, m.p. 76° (decomp.)	
				$[Ru_3(CO)_{12}]$ Green needles	$[Rh(CO)_3]x$ $[Rh_4(CO)_{11}]$	
		$_{74}W$	$_{75}Re$	$_{76}Os$	$_{77}Ir$	$_{78}Pt$
32 Period	I	$W(CO)_6$ $d = 2.65.$ Colorless rhombic cryst. Sublimes readily		$Os(CO)_5$		
	II		$[Re_2(CO)_{10}]$ Colorless	$[Os_2(CO)_9]$ Yellow cryst. Sublimes	$[Ir_2(CO)_8]$ Green yellow cryst. Sublimes	
				$[Os_3(CO)_{12}]$	$[Ir(CO)_3]x$ Canary yellow rhombohed. cryst. Decomp. 210°	

I = Mononuclear, volatile compounds, readily soluble in organic solvents. The central atom
in these compounds probably has an inert gas electronic configuration.

II = Polynuclear compounds, either non-volatile or only slightly volatile, and generally
insoluble or sparingly soluble in organic solvents.

	$_{24}$Cr	$_{25}$Mn	$_{26}$Fe	$_{27}$Co	Second 18 Period
First 18 Period	?	HMn(CO)$_5$?	H$_2$Fe(CO)$_4$ m.p. —70° Colorless liq.	HCo(CO)$_4$ m.p. —26.2° Yellow liq.	HRh(CO)$_4$ Yellow liq. m.p. —10°
	$_{74}$W	$_{75}$Re	$_{76}$Os	$_{77}$Ir	Only stable in solid state
32 Period		HRe(CO)$_5$?	H$_2$Os(CO)$_4$?	HIr(CO)$_4$	

METAL CARBONYL HALOGEN AND CYANO-COMPOUNDS

Mols. CO per atom metal	5CO	4CO	3CO	2CO	1CO
$_{25}$Mn	Mn(CO)$_5$X				
$_{26}$Fe	Fe(CO)$_5$X$_2$	Fe(CO)$_4$X$_2$ Fe(CO)$_4$en$_2$I$_2$		Fe(CO)$_2$I$_2$ Fe(CO)$_2$I Fe(CO)$_2$A$_2$I$_2$ Fe(CO)$_2$(CN)$_2$pyr	Fe(CO)pyr$_2$I$_2$ M$^{I}_3$[Fe(CN)$_5$(CO)]
$_{27}$Co					Co(CO)I$_2$ K$_3$[Co(CN)$_5$(CO)]
$_{28}$Ni					K$_2$[Ni(CN)$_3$(CO)]
$_{29}$Cu					Cu(CO)Cl
$_{43}$Tc					
$_{44}$Ru				Ru(CO)$_2$X$_2$	Ru(CO)Br
$_{45}$Rh				Rh$_2$(CO)$_4$X$_2$	
$_{46}$Pd					[Pd(CO)Cl$_3$] M^{I}[Pd(CO)Cl$_3$] [Pd(CO)Cl$_2$]x [PdCl(CO)en]
$_{47}$Ag				½CO: Ag$_2$(CO)SO$_4$	
$_{75}$Re	Re(CO)$_5$X				
$_{76}$Os		Os(CO)$_4$X$_2$ Os$_2$(CO)$_8$X$_2$	Os(CO)$_3$X$_2$	Os(CO)$_2$X$_2$	
$_{77}$Ir			Ir(CO)$_3$X	Ir(CO)$_2$X$_2$	
$_{78}$Pt				Pt(CO)$_2$X$_2$ [Pt(CO)$_2$(NH$_3$)$_2$]Cl$_2$	Pt$_2$(CO)$_2$X$_4$ M^{I}[PtX$_3$(CO)] [PtCl(CO)(NH$_3$)$_2$]Cl
				Pt$_2$(CO)$_3$Cl$_4$	
$_{79}$Au					Au(CO)Cl

X = Cl, Br, I.
A = o-phenylenediamine, pyridine or ½o-phenanthroline.
en = ethylenediamine.   pyr = pyridine.

Table 34. The *pure metal carbonyls*—i.e., compounds containing only one metal and carbon monoxide—are all derived from the metals of Sub-groups VI to VIII of the Periodic System. The elements of the succeeding Sub-group (copper group) can combine with carbon monoxide only when present in the form of their compounds. All the metals of Groups VIII and IB can form carbonyl halides or related compounds. So far as is now known, the metals of Group VI can form only pure metal carbonyls, and no other types of carbonyl compounds. In Sub-group VII, manganese, like rhenium, forms both the pure carbonyl and carbonyl halides.

(b) Historical

In 1888, Langer, in Mond's laboratory, attempted to free technically produced hydrogen from admixed carbon monoxide, by passing it over heated nickel in order to catalyze the reaction $2CO \rightleftharpoons C + CO_2$. The resulting gas burned with a brightly luminous flame, and gave a deposit of nickel, when it was heated. By cooling it he was able to condense out a liquid which, from its analysis, appeared to be a compound of carbon monoxide with nickel, $Ni(CO)_4$. The reaction was at once turned to practical account by Mond, as a means of preparing very pure nickel (p. 308). The search for similar compounds of related metals led to the discovery of iron pentacarbonyl in 1891, simultaneously by Berthelot in Paris and by Mond and Quincke in London. Attempts to prepare other metal carbonyls were not successful until the reaction of carbon monoxide with finely divided metals was investigated at high pressures. By this means Mond, in 1908, obtained the carbonyls of cobalt and molybdenum, and showed the existence of a ruthenium carbonyl, although the latter compound was first prepared in a pure state by Manchot in 1936. In 1926, Job obtained chromium carbonyl, $Cr(CO)_6$, by the action of carbon monoxide on a mixture of anhydrous chromium chloride and phenyl magnesium bromide. He also prepared tungsten carbonyl, $W(CO)_6$, by the same method in 1928, although it was subsequently found (I.G. Farbenindustrie, 1931) that tungsten and molybdenum carbonyls could both be conveniently prepared directly from the metals by the high pressure synthesis. Iron carbonyl has been manufactured on a large scale since 1924.

A considerable number of other metal carbonyls, and of compounds derived from them, has been discovered by Hieber (since 1928), who has devised new methods of preparation for the compounds and has made extensive investigations of their reactions.

(c) Constitution and Properties

According to Hieber, the metal carbonyls can be divided into two groups, as shown in Table 34. Those of the first group can be vaporized without decomposition, and are readily soluble in inert organic solvents such as benzene, ether, chloroform, etc. They have molecular weights corresponding to their empirical formulas. The carbonyls of the second group are sparingly soluble or practically insoluble in indifferent organic solvents, and in most cases cannot be melted without decomposition. They are all polynuclear compounds, as has been proved by determinations of the molecular weight of $Fe_3(CO)_{12}$, $Co_2(CO)_8$, and $Co_4(CO)_{12}$.

The CO content of the compounds of the first group—i.e., the mononuclear carbonyls— is related in a systematic way to the position of the metal in the Periodic System. As was first pointed out by Sidgwick, the CO content could be explained if the pair of electrons on the carbon atom which are not involved in the C—O bond (the 'lone pair') are shared with the outer electrons of the metal atom to form a common electron system which, in every case, contains 18 electrons. The 'effective atomic number' of the metal atom is thereby made up to the electron number of the succeeding inert gas. The physical properties of the compounds clearly indicate that they are covalent compounds, in harmony with this idea. In order to bring their outer levels up to a total of 18 electrons, Cr, Mo, and W need 12 electrons, and therefore add on 6 molecules of CO. Fe and Ru, needing 10 electrons, combine with 5 CO, and Ni, needing 8 electrons, combines with 4CO. Co cannot form a mononuclear compound with CO, since it needs an odd number (9) electrons to bring the oute levelr up to 18 electrons; it therefore forms only polynuclear compounds. This view of the constitution of the carbonyls, according to which CO groups differing little from CO molecules are covalently bound to the metal atoms, is concordant not only with the composition of the compounds, but also with their physical properties and chemical behavior*.

In the hexacarbonyls of Cr, Mo, and W, the crystal lattice in each case is built up from $M(CO)_6$ molecules, according to Hofmann (1935). In these, the metallic atom M is surrounded octahedrally by CO groups, which are probably so oriented that the C atoms are adjacent to the metal. The distances $M \leftrightarrow C$ accord with those expected for a single covalent bond. A regular arrangement of CO groups around the metal has been found in iron pentacarbonyl also, the configuration probably being that of a trigonal bipyramid. Iron enneacarbonyl, from the structure investigations of Brill (1927) and Powell and Ewens (1939), is built up from molecules of the following configuration:

$$
\begin{array}{ccccc}
\text{CO} & & \text{CO} & & \text{CO} \\
\text{CO} & \text{Fe} & \text{CO} & \text{Fe} & \text{CO} \\
\text{CO} & & \text{CO} & & \text{CO}
\end{array}
$$

The so-called iron tetracarbonyl is trimeric, both in solution and in the crystalline state, but the structure of the $Fe_3(CO)_{12}$ molecule has not been fully worked out. The CO groups in nickel carbonyl are arranged tetrahedrally around the nickel atom, as follows from the electron diffraction investigations of Brockway (1935) and from the analysis of the Raman and infrared spectra of the molecule. The $Ni \leftrightarrow C$ bond length (1.82 Å) is rather shorter than the sum of the covalent radii of the atoms (2.01 Å), and it is probable that the bond is intermediate between a single covalent bond of the type discussed above, and a double bond, $Ni=C$, involving the electrons of the uppermost level (d level) of the nickel atom. The diamagnetism of the metal carbonyls (Klemm, 1931) indicates that the electron levels in the central atoms form completed groups of high symmetry.

(d) Chemical Behavior

The *chemical behavior* of the metal carbonyls provides further evidence for the constitution considered. Their *substitution reactions* are of particular interest. It has been shown that the CO groups in the carbonyls may be partially replaced both by neutral molecules (amines or alcohols) and by electronegative atoms or radicals (Hieber).

If neutral molecules are introduced in place of CO groups, such compounds are obtained as $Mo(CO)_3Am_3$; $Fe_2(CO)_4Am_3$; $Co_2(CO)_5Am_4$; $Ni_2(CO)_3Am_2$ (Am = NH_3 or organic amine, especially pyridine and o-phenanthroline). These are notable in that they contain *nothing but neutral ligands* bound to the metallic atoms, and have therefore been called by Hieber 'pure coordination compounds'. Examples of alcohol compounds of this kind are $Fe(CO)_3(CH_3OH)$, $Fe_2(CO)_6(C_2H_5OH)$, and $Co_2(CO)_5(CH_3OH)$. Provided that oxygen and moisture are rigorously excluded, these may be relatively easily obtained, by warming $Fe_3(CO)_{12}$ or $Co_2(CO)_8$ with methanol or ethanol. The compound $Fe(CO)_3(CH_3OH)$ is monomeric in solution, both in water and in benzene, whereas when $Fe_2(CO)_6(C_2H_5OH)$ is dissolved in water it decomposes into $Fe(CO)_3$ (which is known only in solution) and $Fe(CO)_3(C_2H_5OH)$. Both the amine-substituted and the alcohol-substituted carbonyls are decomposed by acids, in such a way that a portion of the electroneutral iron is oxidized by H^+ ions to Fe^{++} ions. Thus:

$$6Fe(CO)_3pyr + 12H^+ = Fe_3(CO)_{12} + 3Fe^{++} + 3H_2 + 6CO + 6[pyrH]^+ \quad (1)$$

$$4Fe(CO)_3(CH_3OH) + 2H^+ = Fe_3(CO)_{12} + Fe^{++} + H_2 + 4CH_3OH. \quad (2)$$

Oxidation by the atmosphere follows a similar course:

$$4Fe(CO)_3(CH_3OH) + \tfrac{1}{2}O_2 = Fe_3(CO)_{12} + FeO + 4CH_3OH. \quad (3)$$

---

* Evidence based on bond lengths and on the infrared and Raman vibrational spectra of the carbonyls indicates that this hypothesis is only approximately correct. In addition to the purely coordinate covalent bond $M \leftarrow C{\equiv}O$, there is a certain measure of double bond character, which may be represented as $M=C{\doteq}O$. This becomes of increasing importance in the sequence $Cr \to Fe \to Ni$.

Substitution of *halogens* for CO groups leads to compounds which are substantially of the same type as those produced by introduction of neutral ligands. In the metal carbonyl halides, the halogen atoms are bound directly to the metal atoms, and are strongly polarized, whereas in most complex metallic halides—e.g., $[Fe(NH_3)_6]X_2$—the metal-halogen binding is almost purely ionic, with very little polarization. Hieber (1934) concluded from measurements of the lattice energies that in the metal carbonyl halides the metal-halogen binding is similar to that in the anhydrous halides of the same metals—i.e., strongly polarized, and approximating a covalent bond. Klemm (1931) found that the metal carbonyl halides, like the metal carbonyls, were diamagnetic, indicating completed, symmetrical electron configurations. There is a difference between the binding of the halogen atoms and the CO groups, in that the free halogen atom contains an electron with unpaired spin, which can form an electron pair bond with one electron from the levels of the metallic atom. Even although the bond approximates a pure covalence, it is permissible to consider that there is a mutual polarity between metal atom and halogen atom in the carbonyl halides in the same way as there is in the simple halides from which they are derived—e.g., FeI_2. Examples of metal carbonyl halides are

$$Fe(CO)_4I_2 \qquad\qquad Fe(CO)_2(pyr)_2I_2$$
$$Fe(CO)_4Br_2 \qquad\qquad Fe(CO)(pyr)_2I_2.$$

These may be regarded equally well as substitution products of iron pentacarbonyl or as addition compounds of iron(II) halides, and their mode of formation accords with either viewpoint. They can be made either by the action of halogens on the metal carbonyls, or by the action of CO (under pressure) on metallic halides.

Hieber (1940) found that the thermal decomposition of $Fe(CO)_4I_2$, in which the iron is dipositive, yielded the compounds $Fe^{II}(CO)_2I_2$, $Fe^{I}(CO)_2I$ and $Fe^{I}I$. The existence of the two latter compounds is of particular interest in that they contain *unipositive* iron. This oxidation state has not been observed in any other compounds except the so-called Roussin salts.

(e) Metal Carbonyl Hydrides

Carbon monoxide in metallic carbonyls is much more readily oxidized than it is in the free state. If it is oxidized (in the absence of air) by *hydroxyl ions*, the CO which is split off by oxidation to $CO_3^=$ is replaced by hydrogen, and metal *carbonyl hydrides* are obtained— e.g., *iron carbonyl hydride*, by the reaction

$$Fe(CO)_5 + 2OH^- = H_2Fe(CO)_4 + CO_3^= \tag{4}$$

Cobalt carbonyl hydride, $HCo(CO)_4$, has been obtained in an analogous manner by the action of OH^- ions on $Co_2(CO)_8$. It is also possible for the oxidation of the central metallic atom by H^+ ions to take place in such a way that metal carbonyl hydride is formed*. Thus (A = alcohol or amine):

$$2Fe(CO)_3A + 2H^+ = H_2Fe(CO)_4 + Fe^{++} + 2CO + 2A \tag{5}$$

$$Co_2(CO)_5A + 2H^+ = HCo(CO)_4 + Co^{++} + \tfrac{1}{2}H_2 + CO + A \tag{6}$$

The metal carbonyl hydrides are very unstable compounds, and undergo decomposition, with loss of hydrogen, at temperatures above —20°:

$$2H_2Fe(CO)_4 = 2H_2 + [Fe(CO)_3]x + Fe(CO)_5 \tag{7}$$

$$2HCo(CO)_4 = H_2 + Co_2(CO)_8 \tag{8}$$

In their physical properties, they are intermediate between iron pentacarbonyl and nickel carbonyl. Thus, in the series $Fe(CO)_5 — H_2Fe(CO)_4 — HCo(CO)_4 — Ni(CO)_4$ there is a continuous decrease in the dipolar character of the molecules. It must be concluded that hydrogen in these compounds is also covalently bound. It has also been shown (Ewens and

* It is not uncommon for reactions of type (1) and type (5) to occur concurrently.

Lister, 1939) that iron and cobalt carbonyl hydrides have a tetrahedral molecular configuration, and they can be regarded as 'pseudo-nickel carbonyls', although their complete
molecular structure (and especially the precise mode of binding of the hydrogen) has not
been worked out.

In other respects, there are relations between the metal carbonyl hydrides and the
volatile hydrides of the Main Groups of the Periodic System, which also have completed
electron shells of high symmetry, in which the hydrogen nuclei are 'embedded' (cf. Vol. I,
p. 788). This relationship is shown, e.g., by the ability to form salts. Thus:

$$2H_2Fe(CO)_4 + [Ni(NH_3)_6]^{++} + 2NH_3 = [Ni(NH_3)_6][HFe(CO)_4]_2 + 2NH_4^+.$$

True salts of this kind are usually formed only with complex cations. However, by reaction,
of metal carbonyls or metal carbonyl halides with alkali metals dissolved in liquid ammonia
one can obtain alkali metal salts of the metal carbonyl hydrides, as Behrens (1952) has
shown. The metal carbonyl hydrides can be liberated from their salts by treatment with
acids.

A carbonyl hydride, the hydrogen of which *cannot* be replaced by metals or by positive
complex groups, is *nickel carbonyl hydride*, $[Ni(CO)_3H]_2$, which was obtained by Behrens by
the reaction:

$$Ni(CO)_4 + Na + NH_3 = NaNH_2 + CO + \tfrac{1}{2}[Ni(CO)_3H]_2.$$

This substance dissolves in liquid ammonia to give a deep-red solution and forms a vermilion ammoniate, $[Ni(CO)_3H]_2 \cdot 4NH_3$, from which the ammonia is easily split off.
Nickel carbonyl hydride is decomposed by acids with the liberation of hydrogen:

$$2[Ni(CO)_3H]_2 + 2H^+ = 3Ni(CO)_4 + Ni^{++} + 3H_2.$$

That the compound is a dimer is shown from its molecular weight, which was determined
from the vapor pressure lowering caused in liquid ammonia. In the $Ni(CO)_3H$ radical the
nickel is surrounded by the same number of electrons as cobalt in the $Co(CO)_4$ radical.
Therefore, the dimerization of the former is to be expected on the same grounds as that of
the latter.

The metal carbonyl hydrides more usually give rise to metallic derivatives which have
the character of non-electrolytes. These are formed, for example, with the ammoniates of
metals such as Zn, Cd, Cu, and Ag, which have an especially great tendency to form
covalent bonds. Examples are—$Fe(CO)_4Zn(NH_3)_3$ (colorless prisms), $Fe(CO)_4Cd(NH_3)_2$
(colorless prisms), $Fe(CO)_4Cu_2(NH_3)_2$ (yellow needles). It can probably be assumed that
in these compounds, the metal atoms which replace the hydrogen of the carbonyl hydride
are covalently bound to CO groups (Hieber). In accordance with this idea, they contain
less ammonia* than could be bound by the same metallic atoms in ionic form.

The so-called mixed metallic carbonyls can also be regarded as derivatives of the metal
carbonyl hydrides. E.g.,

$$2HCo(CO)_4 + HgCl_2 = Hg[Co(CO)_4]_2 + 2HCl$$

$$Fe(CO)_4Cd(NH_3)_2 \overset{100°}{=} Fe(CO)_4Cd + 2NH_3$$

$$Fe(CO)_5 + HgSO_4 + H_2O = Fe(CO)_4Hg + H_2SO_4 + CO_2$$

(Hock and Stuhlmann). The cobalt compounds can also be prepared directly from their
components by high pressure synthesis. Mixed carbonyl compounds of iron containing the
metals of Groups IIIA and IVA can also be prepared, as well as those of the Group IIB
metals (Hieber). The mixed carbonyls of iron are insoluble and cannot be sublimed,
whereas those of cobalt can be sublimed in a current of CO, and are soluble in inert organic
solvents. It may be assumed that the bonds between the metal atoms and CO groups in
these compounds are covalences.

* Organic amines, such as pyridine, or *o*-phenanthroline, may be present in place of
ammonia.

(f) Metal Nitrosyl Carbonyls and Metal Nitrosyls

CO groups in the metal carbonyls may also be replaced by NO groups. The resulting compounds are also of interest in relation to valence theory. It is probable that nitric oxide is present in these substances in a form which is isosteric with CO, i.e., as an electropositive NO$^+$ group, bound by a coordinate covalent bond (cf. p. 296). Since each NO group thereby gives up one electron to the metal atom, fewer NO groups than CO groups are needed in order that the central metal atom shall acquire the stable electronic configuration found in this class of compound. Thus the nitrosocarbonyl $Fe(NO)_2(CO)_2$ (Anderson, 1932) corresponds to the carbonyl $Fe(CO)_5$, and for the same reason the $Co(CO)_4$ radical can be stabilized either by the addition of 1 H or by the exchange of one CO group for an NO group, giving $Co(NO)(CO)_3$. The compounds $Co(NO)(CO)_3$ and $Fe(NO)_2(CO)_2$ are red, volatile liquids. It has been shown that their molecules have a tetrahedral configuration, with CO and NO groups of similar dimensions (Brockway and Anderson, 1938), and like the metal carbonyl hydrides, they may be considered as 'pseudo-nickel carbonyls'. In the metal nitrosyl carbonyls, only the CO groups (and not the NO groups) can be replaced by amines or other ligands. Hieber has pointed out that this is to be expected from the assumptions made above as to the metal-NO binding, which should be stronger than the coordinate covalence between metal and CO groups.

The constitution of the *pure metal nitrosyls*, which exist only in the solid state, is as yet unknown. Examples of these substances are iron tetranitrosyl, $Fe(NO)_4$, and ruthenium nitrosyl (p. 327).

(g) Formation and Preparation of Metal Carbonyls

In addition to the direct combination of metals with carbon monoxide, several methods are now known whereby metal carbonyls may be formed, and some of these are important as preparative methods. In some cases, Job's method, involving the use of Grignard reagents, may be used. It is considered by Hieber (1935) that unstable organo-metallic carbonyls are thereby formed as intermediate products, and that these decompose in acid solution to give the pure carbonyls

$$3Cr(CO)_2R_4 + 6H^+ = Cr(CO)_6 + 2Cr^{+++} + 12R + 3H_2$$

(R = organic radical). It has not hitherto been possible to prepare chromium hexacarbonyl by any other method.

The metal carbonyl halides may often more readily be prepared from their components (metal halide + carbon monoxide) than the metal carbonyls themselves. By the action of suitable metals (copper or silver) it is possible to abstract the halogen from these substances, and thus obtain the pure metal carbonyls. Manchot (1936) obtained ruthenium carbonyl in this way:

$$RuI_3 \underset{\xrightarrow{\hspace{1.2cm}}}{\overset{CO\ under\ press.}{\rightleftharpoons}} RuI_2(CO)_2 \overset{Ag}{\rightleftharpoons} Ru(CO)_5.$$

The action of carbon monoxide under pressure on the halides of carbonyl-forming metals is often a convenient method of obtaining metal carbonyls in other cases also, as Hieber has shown. Thus the iridium carbonyls mentioned on p. 337, and the rhodium carbonyls and rhodium carbonyl hydride listed in Table 34, were obtained by Hieber in this manner.

Other methods involve the reactions of sulfur compounds. Thus Manchot first observed that $Ni(CO)_4$ can readily be obtained by the action of CO on a suspension of NiS in alkali hydroxide solution. It has, in fact, been observed quite generally that the reaction of carbon monoxide with the carbonyl-forming metals is promoted by the presence of sulfides.

Finally, it has been shown that in salts of the type of $[Ni(NH_3)_6][HFe(CO)_4]_2$, the NH_3 can be smoothly replaced by CO. In the instance cited, $Ni(CO)_4$ is obtained, while the $[HFe(CO)_4]$ radical undergoes disproportionation into $H_2Fe(CO)_4$ and $[Fe(CO)_4]_3$. If the same reaction is applied to the compound $[Co(NH_3)_6][HFe(CO)_4]_2$, cobalt carbonyl hydride is obtained, one half of the hydrogen migrating to the cobalt atom:

$$[Co(NH_3)_6][HFe(CO)_4]_2 + 4CO = [Co(CO)_4][HFe(CO)_4]_2 + 6NH_3$$

$$2[Co(CO)_4][HFe(CO)_4]_2 = 2HCo(CO)_4 + H_2Fe(CO)_4 + [Fe(CO)_4]_3.$$

References

1 R. Durrer, *Erzeugung von Eisen und Stahl*, Dresden 1936, 159 pp.

2 B. Osann, *Kurzgefasste Eisenhüttenkunde*, 2nd Ed., Leipzig 1939, 188 pp.

3 Ver. deutscher Eisenhüttenleute, *Gemeinfassliche Darstellung des Eisenhüttenwesens*, 14th Ed., Düsseldorf 1937, 591 pp.

4 H. M. Boylston, *Introduction to the Metallurgy of Iron and Steel*, 2nd Ed., New York 1936, 563 pp.

5 B. Stoughton, *The Metallurgy of Iron and Steel*, 4th Ed., London 1934, 559 pp.

6 R. Durrer, *Die Metallurgie des Eisens*, 3rd Ed., Berlin 1943, 997 pp.

7 H. Schenck, *Einführung in die physikalische Chemie der Eisenhüttenprozesse*, 2 vols., Berlin 1932 and 1934, 306 and 274 pp.

8 H. E. Cleaves and J. G. Thompson, *The Metal Iron*, New York 1935, 574 pp.

9 P. Oberhoffer, *Das technische Eisen; Konstitution und Eigenschaften*, 3rd Ed., by W. Eilender and H. Esser, Berlin 1936, 642 pp.

10 E. Heyn, *Theorie der Eisen-Kohlenstoff-Legierungen*, Berlin 1924, 185 pp.

11 M. C. Neuburger, *Röntgenographie des Eisens und seiner Legierungen*, Stuttgart 1928, 124 pp.

12 S. Epstein and F. T. Sisco, *The Alloys of Iron and Carbon*, 2 vols., New York 1936 and 1937, 476 and 777 pp. [Monographs on other iron alloy systems—*e.g.*, with silicon, chromium, manganese, nickel etc.—are also to be found in the same series, edited by F. T. Sisco.]

13 C. H. Plant, *The Metallography of Iron and Steel*, New York 1933, 220 pp.

14 A. Sauveur, *Metallography and Heat Treatment of Iron and Steel*, 4th Ed., London 1935, 550 pp.

15 F. Reiser, *Das Härten des Stahls*, 8th Ed., Leipzig 1932, 200 pp.

16 H. Brearley, *Die Einsatzhärtung von Eisen und Stahl*, Berlin 1926, 249 pp.

17 F. Giolitti, *La Nitrurazione dell'Acciaio*, Milan 1933, 453 pp.

18 E. Houdremont, *Einführung in die Sonderstahlkunde*, Berlin 1935, 566 pp.

19 F. Rapatz, *Die Edelstähle*, 3rd Ed., Berlin 1942, 482 pp.

20 O. Pattermann, *Werkzeugstähle; Eigenschaften und Verfahren zu ihrer Wärmebehandlung*, Kladno 1937, 496 pp.

21 J. H. G. Monypenny, *Stainless Iron and Steel*, 2 vols., 3rd Edition, London 1951 and 1954, 524 and 330 pp.

22 O. Macchia, *Der Phosphatrostschutz*, Berlin 1940, 250 pp.; Supplement Berlin 1942, 304 pp.

23 W. S. Messkin and A. Kussmann, *Die ferromagnetischen Legierungen und ihre gewerbliche Verwendung*, Berlin 1932, 418 pp.

24 G. Münzer, *Das Platin*, Leipzig 1929, 136 pp.

25 A. Schulze, *Metallische Werkstoffe für Thermoelemente*, Berlin 1940, 100 pp.

CHAPTER 8

FIRST SUB-GROUP OF THE PERIODIC SYSTEM: COPPER, SILVER, AND GOLD

Atomic numbers	Elements	Symbols	Atomic weights	Densities	Melting points	Boiling points	Specific heats	Valence States
29	Copper	Cu	63.54	8.92	1083°	2350°	0.0916	I, II, III
47	Silver	Ag	107.880	10.50	960.5°	1980°	0.0556	I, II, III
79	Gold	Au	197.0	19.3	1063°	2700°	0.0313	I, III, IV

1. Introduction

(a) General

The elements of Sub-group I, *copper, silver* and *gold* follow immediately after the right hand column of Sub-group VIII (nickel, palladium, and platinum). In many ways they rather resemble these preceding elements. However, they have many properties which also reveal an inner relation to the elements of the first Main Group, the alkali metals. Like the latter, they can function as unipositive in their compounds. However, whereas the alkali metals are exclusively unipositive, the elements of the first Sub-group, (the copper group), and especially copper and gold*, can also form compounds in higher valence states—copper chiefly in the dipositive, and gold chiefly in the tripositive state. In fact, in general the +2 state of copper and the +3 state of gold are the preferred states in salt-like compounds.

The compounds in which copper, silver, and gold are unipositive are not strictly analogous in properties to the alkali compounds, but mostly differ strongly from these. The compounds of +1 copper and gold are all either very insoluble, as for example the chlorides CuCl and AuCl, or are strongly complexed. Most compounds of the alkali metals are readily soluble, are stable, and highly dissociated in aqueous solution, and these elements are, therefore, almost always present as free elementary ions. By contrast, copper and gold in the unipositive state are only present in vanishingly small amounts as elementary ions in solutions. Unipositive silver resembles +1 copper and gold in its insoluble and complex compounds. However, it can also form some soluble salts of a non-complex nature. Many of these, e.g., the nitrate and the sulfate, are isomorphous with the corresponding sodium salts.

The elements of Main Group I and Sub-group I stand in marked contrast with each other in their electrochemical character. Whereas the free alkali metals are

---

* For compounds of dipositive and tripositive silver see p. 407–8.

among the most active of the elements, the metals of the first Sub-group are noble metals. *The alkali metals stand at the top, copper, silver and gold at the bottom, of the electrochemical potential series.*

The alkali ions have the smallest tendency for complex formation of all the elementary ions. Copper, silver, and gold, on the other hand are marked complex formers, in all the valence states in which they occur.

The oxides of the alkali metals combine avidly with water, forming strongly basic hydroxides, which are very soluble, with great evolution of heat. The oxides of copper, silver, and gold have a very low solubility in water. The hydroxides of the copper group, present in extremely low concentration in the solutions, display a more or less marked amphoteric character.

The alkali metals are typical light metals, whereas copper, silver, and gold are typical heavy metals.

The alkali metals have a great affinity for oxygen. Copper, silver, and gold, like the metals of the VIIIth Sub-group, have a strong tendency to combine with sulfur.

The alkali salts are all colorless (unless they contain colored acid radicals). Some of the salts of unipositive copper, silver, and gold are colorless; many of them are weakly, but characteristically, colored. In the salts with colored acid radicals, a distinct deepening of color is often to be observed. Nearly all the salts of $+2$ copper display definite colors, mostly blue or greenish. Salts derived from $+3$ gold are also colored, although more weakly so.

The analogy between the metals of the Main Group and Sub-group of the first family is shown much more strongly in the optical spectrum than in chemical behavior.

In the arc spectra of the metals of the copper group, as in those of the alkali metals, the p and d terms occur doubled, i.e., lines formed by means of these terms, when examined with sufficiently strong dispersion, are found to be doublets, like, e.g., the yellow sodium-line (cf. Vol. I, p. 171). As in the spectra of the alkali metals, the ground term in the spectra of copper, silver, and gold, that is the numerically largest term, is an s term. According to what was said on page 117, Vol. I, this means that the outermost electron is bound in an orbit with the subsidiary quantum number $l = 0$. Like the atoms of the alkali metals, those of the metals of the first Sub-group contain only one electron bound in this way. The univalence of the elements can be explained in this manner, although, for reasons discussed below, they also have the ability to exercise higher valences. The ionization potentials, calculated from the absorption series limits and in some cases measured directly (cf. Vol. I, p. 115, Table 22) show that the electrons are considerably more strongly bound than in the alkali metal atoms. This explains to some extent why the copper group metals are considerably more noble than the alkali metals. In addition, the very high heats of evaporation of the metals of the copper group, as compared with those of the alkali metals, (calculated from the high boiling points by means of Trouton's rule) are important in determining their noble character. (cf. the summary table at the beginning of this chapter, and Vol. I, p. 149).

The electronic configurations in the free atoms, as deduced from the Bohr theory, are shown in Table 35.

It can be seen from this that the metals of the copper group differ from the alkali metals in that the outermost shell, with its single s electron, does not immediately follow a core having the inert gas configuration (i.e., with two s and six p electrons). Instead there is a shell containing ten d electrons (subsidiary quantum number $l = 2$). The electrons in this shell are considerably more loosely bound than in the inert gas shell. This is shown by the fact that atoms of lower atomic number cannot build up this shell by accepting electrons, thereby converting the atom into a free negative ion. More direct evidence is provided by

TABLE 35

ELECTRONIC CONFIGURATION IN THE ATOMS OF COPPER, SILVER, AND GOLD

(The numbers indicate the number of electrons in the various energy states, in the ground state of the atom)

Energy states	1s	2s	2p	3s	3p	3d	4s	4p	4d	4f	5s	5p	5d	6s
Copper	2	2	6	2	6	10	1							
Silver	2	2	6	2	6	10	2	6	10		1			
Gold	2	2	6	2	6	10	2	6	10	14	2	6	10	1

the spectra of copper, silver, and gold, which involve other terms, in addition to the alkali-like terms. These arise from the fact that when the atoms are excited to emit light, not only the most loosely bound electron, but also one or more of the ten d electrons can be in an excited state. These term systems are much more prominent for copper and for gold than for silver, with which the corresponding lines are less numerous and of lower intensity. The comparatively loose binding of the d electrons, which manifests itself in the occurrence of these term systems, makes possible the loss of more than one electron from the atomic structure of the elements of the copper group. The considerably more marked tendency of copper and gold to be multivalent as compared with silver, is thus in complete accord with the spectral behavior. The atom models deduced from the spectra very satisfactorily express not only the similarity between the elements of the Main Group and the Sub-group of the first family, but also, on closer examination, the contrast in behavior which is so apparent in their chemical properties.

Some physical constants of the elements of the copper group are set out for comparison in Table 36.

TABLE 36

MOST IMPORTANT PHYSICAL CONSTANTS OF ELEMENTS OF THE COPPER GROUP

	Specific electrical conductivity	Thermal conductivity at 18° cal/cm sec deg.	Atomic heat, cal	Heat of fusion (at the m.p.)		Heat of sublimation at 0°K (kcal per g-atom)
Copper	$57.2 \cdot 10^4$	0.989	5.82	42 cal/g	2.65 kcal/g-atom	81.2
Silver	$61.4 \cdot 10^4$	1.006	6.01	25	2.66	68.0
Gold	$41.3 \cdot 10^4$	0.700	6.17	16	3.16	about 92

Copper, silver and gold are diamagnetic. Their atomic magnetic susceptibilities at room temperature are $-5.4 \cdot 10^{-6}$, $-22 \cdot 10^{-6}$, and $-27.3 \cdot 10^{-6}$, respectively.

The metals of the copper group crystallize with the face-centered cubic structure (Vol. I., p. 210, Fig. 46) and not, like the alkali metals, with the body-centered cubic structure. The edge lengths of the unit cells at 20° are $a = 3.6075$ Å for Cu, 4.0774 Å for Ag, 4.0700 Å for Au. The apparent atomic and ionic radii of copper, silver and gold are collected in Table 37. Whereas silver and gold form mixed crystals with one another in all proportions, as also do copper and gold, copper and silver are capable of mixed crystal formation only in very restricted proportions (see pp. 11 and 16).

The occurrence of the superstructure phases discussed on page 13 et seq. was first observed in the copper-gold mixed crystal series. According to the results of X-ray structure determinations, an ordered arrangement of the atoms in the crystal lattice normally occurs at ordinary temperature in the neighborhood of the compositions CuAu and Cu₃Au; at temperatures above about 385° the arrangement is disordered. If alloys of the composition CuAu or Cu₃Au are allowed to cool slowly from temperatures above 385°, or if the rapidly

TABLE 37

APPARENT ATOMIC AND IONIC RADII OF ELEMENTS OF SUB-GROUP I IN Å

(Ionic radii in this table are those of the univalent ions)

	Copper	Silver	Gold
Atomic radius	1.27	1.442	1.441
Ionic radius	about 1.0	1.13	—

cooled alloys are annealed for a long time (e.g., at 370°), so that they can attain the arrangement stable at low temperatures, these alloys furnish diffraction patterns characteristic of an ordered atomic arrangement. These show that with alloys of the composition Cu_3Au, the corners of the unit cube (cf. Vol. I, p. 210, Fig. 46) are all occupied by gold atoms and all the face centers by copper atoms. For alloys of the composition CuAu, X-ray analysis shows that the corners of the unit cube and also the centers of two opposite faces are occupied by gold atoms, the remaining face centers being occupied by copper atoms (Borelius, 1925–28). Copper-gold alloys cooled slowly, or annealed below 385°, give sharp peaks at the same composition in the curves representing the dependence of electrical conductivity upon the proportions of the alloy components (Kurnakow, 1916). Alloys cooled rapidly, or heated above 385° and then quenched, do not show these phenomena. They give the diffraction pattern characteristic of the ordinary face-centered cubic lattice, and of mixed crystals with completely random distribution of the atoms. According to Kurnakow the transformation at 385° is also shown by a halt in the cooling curve in the case of the alloy corresponding in composition to CuAu.

TABLE 38

SOLUBILITIES OF SOME COMPOUNDS OF COPPER, SILVER, AND GOLD

(in moles per liter at room temperature)

	Copper(I)	Silver(I)	Gold(I)	Copper(II)	Gold(III)
Chloride	$1.1 \cdot 10^{-3}$	$1.31 \cdot 10^{-5}$	decomp.	very sol.	very sol.
Oxide	almost insol.	$1.08 \cdot 10^{-4}$	ca. $2 \cdot 10^{-11}$	$6.8 \cdot 10^{-5}$	almost insol.
Sulfide	$3.1 \cdot 10^{-7}$	$1.8 \cdot 10^{-17}$	$< 10^{-7}$	$3.5 \cdot 10^{-7}$	decomp.

In consequence of their smaller atomic radii, the ions of the metals of Sub-group I exert a considerably stronger polarizing effect than do the alkali ions. Their compounds with negative ions, in the anhydrous state, therefore have a much less marked salt-like character than the corresponding alkali compounds. Many of them approximate covalent compounds in properties and behavior, and should indeed probably be regarded as purely covalent (cf. p. 381). On the basis of the Heitler-London theory (cf. Vol. I, p. 131 et seq.) it would be expected that these metals should always be univalent in purely covalent compounds. This conclusion needs to be modified, however, since the presence of electrons in available d levels presents the further possibility of forming covalent bonds from hybridized d-s-p wave functions (see Vol. I, Chap. 5 and 11). This is in fact, found to take place with gold especially. The organometallic compounds of gold are mostly derived from trivalent gold (e.g., $[(C_2H_5)_2AuBr]_2$). They have the properties of typical covalent compounds, and it has been shown by X-ray structure analysis that in all such compounds the gold lies at the center of a square planar arrangement of attached groups. It may be inferred that the wave functions used in forming the covalences are not only the $6s$ wave function, as might be inferred from the simple Heitler-London theory, but a set of d^2sp or dsp^2 hybrid wave functions.

The volatile hydrides CuH, AgH, and AuH, which are formed in minute amounts at high temperatures as has been detected spectrographically, are probably to be regarded as pure covalent compounds. The formation of gold hydride is also shown by the observation

TA|

MISCIBILITY AND COMPOUND FORMATION BETWEEN THE ME|

(Meaning of sym|

	Li	Na	Be	Mg	Ca	Sr	Ba	B	Al	Ga	In
Cu	s o o	alloys o	s < ∞ Be₂Cu 1206° BeCu	s > 0 Mg₂Cu 568° MgCu₂ 820°	s < ∞ Cu₄Ca 935° CuCa₄[]? 480°	—	alloys —	liq o s o o	s < ∞ Al₂Cu* 590° AlCu* 626° Al₄Cu₉* 1016° AlCu₃ 1047°	s < ∞ GaCu₀.₈* 249° GaCu₁.₃* 467° Ga₄Cu₉ 837° GaCu₃* 909°	s < InCu 310 InCu 671 InCu 683 In₄C 685 InC 715
Ag	s o Li₃Ag 450° LiAg 955°	s o o	s < ∞ Be₂Ag* 960°	s < ∞ Mg₃Ag* 492° MgAg 820°	s o Ca₂Ag? 555° CaAg 665° CaAg₂* 595° CaAg₃ 725° CaAg₄* 683°	s o Sr₃Ag₂ 666° SrAg 680° Sr₃Ag₅ 757° SrAg₄ 781°	liq < ∞ s o? Ba₃Ag? 584° Ba₄Ag₃? 585° Ba₂Ag₃ 846° Ba₃Ag₅* 797° BaAg₄ 729°	liq ∼ o s o o	s < ∞ Al₃Ag₅* 698° AlAg₂* 728° AlAg₃* 772°	s < ∞ Ga₃Ag₂?* 326° Ga₂Ag₅* 619° GaAg₃	s < 3 int meta phas
Au	alloys com- pound	s o Na₂Au NaAu₂ 990° K s o KAu₂ KAu₄	s ∼ o Be₅Au Be₂Au BeAu 730° BeAu₂* 645°	s > 0 Mg₃Au 830° Mg₅Au₂*? 796° Mg₂Au 796° MgAu 1150°	s > 0 Ca₂Au* 798° Ca₄Au₃* 849° CaAu 1014° CaAu₂ 864° CaAu₃* 853° CaAu₄ 880°	—	— Ba₂Au₃ BaAu₂	—	s > 0 Al₂Au 1060° AlAu* 625° AlAu₂ 624° Al₃Au₅?* AlAu₄*	s > 0 Ga₂Au 492° GaAu 468° GaAu₂.₃[] 286° GaAu₂.₆* 352°	s > Au₅I 647 Au₅I 488 Au₇I 484 AuI 506 AuI 544

of Farkas (1929), that gold volatilizes substantially more rapidly when heated to 1400° in a stream of hydrogen than in N_2 or He. Silver behaves similarly, but the effect is less marked.

The solid hydrides (cf. pp. 383, 395, and 411.), obtained by the action of atomic hydrogen on Cu, Ag, or Au, are probably salt-like in nature. In contrast to the gaseous hydrides, which can exist at low concentration in equilibrium with the metal and H_2 at high temperatures, they are unstable (metastable at ordinary temperature). Of the solid hydrides, the silver compound seems to be the most stable, whereas gold hydride is the most stable of the volatile hydrides.

Table 38 summarizes the solubilities of some compounds of copper, silver and gold.

SUB-GROUP I AND THE ELEMENTS OF THE MAIN GROUPS

in Table 9, p. 49)

Tl	Si	Ge	Sn	Pb	P	As	Sb	Bi	S	Se	Te
$< \infty$ $s\ 0$ 0	$s < \infty$ $CuSi_{0.17}$* 852° $Cu_5Si[]$ 726° $CuSi_{0.22}$ 824° $Cu_{15}Si_4[]$ 610° Cu_3Si 859°	$s < \infty$ Cu_4Ge? 828° Cu_3Ge* 700°	$s < \infty$ Cu_5Sn* $Cu_{31}Sn_8$ Cu_3Sn Cu_6Sn_5* 410°	$liq < \infty$ $s > 0$ 0	$liq < \infty$ $s > 0$ Cu_3P 1018° CuP_2	$s > 0$ Cu_3As 835° Cu_5As_2?* 710°	$s > 0$ $Cu_9Sb_2[]$ 469° Cu_3Sb 681° Cu_2Sb* 585°	$s \sim 0$ 0	$liq < \infty$ $s\ 0$ Cu_2S 1130° CuS	$liq < \infty$ $s\ 0$ Cu_2Se 1113° $CuSe$ unstable Cu_2Se_3	$liq < \infty$ $s\ 0$ Cu_3Te* 623° Cu_2Te 855°
> 0 0	$s > 0$ 0	$s > 0$ 0	$s < \infty$ Ag_6Sn* 724° Ag_3Sn* 480°	$s > 0$ 0	$liq < \infty$ $s \sim 0$ Ag_3P unstab. AgP_2 Ag_2P_5 unstab. AgP_3	$s > 0$ Ag_9As?* 595°	$s < \infty$ $AgSb_{0.16}$* 562° Ag_3Sb* 705°	$s > 0$ 0	$liq < \infty$ $s\ 0$? Ag_2S 841°	$liq < \infty$ $s\ 0$? Ag_2Se	$liq < \infty$ $s\ 0$ Ag_4Te unstable Ag_2Te 958° Ag_3Te_2* 465° $AgTe$ unstable
$s\ 0$ 0	$s > 0$ 0	$s\ 0$ 0	$s > 0$ $Au_x Sn$* $x > 5$ $AuSn$ 418° $AuSn_2$* 309° $AuSn_4$* 252°	$s\ 0$ Au_2Pb* 418° $AuPb_2$* 254°	$s\ 0$ Au_2P_3	*alloys* Au_3As? Au_4As_3? $AuAs$?	$s \sim 0$ $AuSb_2$* 460°	$s > 0$ Au_2Bi 373°	— Au_2S AuS Au_2S_3	— Au_2Se_3	— $AuTe_2$ 464°

(b) Alloys

The metals of the first Sub-group are notable for their specially marked ability for forming alloys, as is evident from the data of Tables 39 and 40. A large number of compounds appear among these alloys. For the most part these have the character of non-Daltonide compounds, in some cases with very wide ranges of homogeneity. Many of the compounds are of the Hume-Rothery type (cf. p. 21. *et seq.*), The metals of the first Sub-group invariably have the valence electron number 1 in these compounds. There are also numerous other compounds however, and the regularities underlying their formation are still unexplained.

There is no clear difference in the behavior of Sub-group I elements towards the metals

TABLE 40

MISCIBILITY AND COMPOUND FORMATION BETWEEN THE ELEMENTS OF
SUB-GROUP I AND THE ELEMENTS OF SUB-GROUPS I TO III

(Meaning of symbols as in Table 9, p. 49)

	Ag	Au	Zn	Cd	Hg	La
Cu	$s > 0$ 0	$s \infty$ [Cu$_3$Au] [CuAu]	$s < \infty$ Zn$_3$Cu* 594° ZnCu$_{0.39}$* 700° Zn$_8$Cu$_5$ 833° ZnCu* 905°	$s > 0$ Cd$_3$Cu* 397° Cd$_8$Cu$_5$ 563° Cd$_3$Cu$_4$ CdCu$_2$* 549°	$s < \infty$ HgCu 96.2° and two other compounds	$s\ 0$ LaCu* 551° LaCu$_2$ 834° LaCu$_3$* 793° LaCu$_4$ 902°
Ag		$s \infty$ 0	$s < \infty$ Zn$_3$Ag 635° Zn$_8$Ag$_5$* 665° ZnAg 710°	$s < \infty$ Cd$_3$Ag* 592° Cd$_8$Ag$_5$* 640° CdAg* 722°	$s < \infty$ Hg$_8$Ag$_5$* 127° Hg$_3$Ag$_4$?* 276°	$s\ 0$ LaAg 885° LaAg$_2$* 864° LaAg$_3$ 955°
Au			$s < \infty$ Zn$_3$Au* 475° Zn$_8$Au$_5$ 644° ZnAu 725° [ZnAu$_3$] 425°	$s < \infty$ Cd$_3$Au Cd$_8$Au$_5$ CdAu 627° [CdAu$_3$]	$s < \infty$ Hg$_2$Au* 124° Hg$_3$Au$_2$* 310° HgAu$_3$* 421°	$s\ 0$ La$_2$Au* 665° LaAu 1360° LaAu$_2$ 1214° LaAu$_3$ 1204°

of the Main Groups, on the one hand, and towards those of the second and third Sub-groups (Tables 39 and 40). With the metals of Sub-groups IV–VIII their behavior is different, however. As may be seen from Tables 9 (p. 48–9), 13 (p. 89), 17 (p. 124), 24 (p. 209), and 30 (p. 248), compounds are formed with these elements in only a few cases. The tables contain only four certain examples of compounds, apart from superstructure phases. Among the transition elements, only those of Sub-groups VII and VIII appear to be capable of extensive miscibility in the solid state with the elements of the first Sub-group. Miscibility is frequently unrestricted in these cases.

Towards the metals of Sub-group VIII, the metals of the first Sub-group display the same behavior as they do among themselves (p. 362). In both cases, unlimited miscibility can occur even with a fairly large difference of atomic radii. There are several examples of superstructure phases, as in the Cu-Au alloys. The relationship between the metals of the eighth and the first Sub-groups is also shown in their joint ability to function as metals of type I in Hume-Rothery phases.

For bronzes and related alloys see p. 372–3 *et seq.*; cf. also p. 16.

(c) Historical

Copper, silver and gold are among the metals which have been longest known. Copper, in the form of bronze, served in prehistoric times (the Bronze Age) for the manufacture of weapons, articles, and jewellery. Silver and gold were already generally known and treasured in the times from which our oldest records date. From the books of the law of the Egyptian king Menes, about 3600 B.C., we gather that at that time the relative value of silver and gold, with the ratio 1 : 2½, lay considerably more in favor of silver than it does today. We find the three metals mentioned in the oldest parts of the Bible and also in Homer. The Phoenicians possessed gold washings on the Isle of Thasos, silver mines in Spain, and copper quarries in Cyprus. For this reason the Romans called copper *aes cyprium*, from which *cuprum* was later derived.

In Germany, after the Dark Ages, during which the small amount of mining activity previously carried on was completely destroyed, the smelting of silver and copper ores on a large scale began again in the 10th Century, in Saxony and in the Harz. The Mansfeld copper mine which is still the leader in German copper production, dates from the beginning of the 13th century. In the production of silver, the mines of St. Andreasberg, in the Harz, and at Freiberg in Saxony played for a time the principal role. Gold was formerly found in some German rivers, especially in the Rhine, the ancient golden riches of which are described in mythical symbolism in the Nibelung Saga.

2. Copper (Cuprum, Cu)

(a) Occurrence

Copper is found in many places, in great quantities, in the form of its compounds, but large deposits of *native copper* are more rare (e.g., by Lake Superior in North America). It is most commonly found as *sulfide*. Cupric sulfide, CuS (*covellite*), is of but small importance. Cuprous sulfide, Cu_2S (*copper glance* or *chalcocite*), is an important copper mineral, however. By far the most abundant and important copper ore is *chalcopyrite*, $CuFeS_2$, which very frequently occurs together with *bornite* Cu_3FeS_3 (e.g., in America). Double sulfides of copper with other heavy metals are, indeed, very widely distributed in Nature. Thus, *tetrahedrite* (and other double compounds of Cu_2S, chiefly with Sb_2S_3 and As_2S_3) and *bournonite* $(Cu_2, Pb)_3[SbS_3]_2$ are quite important for the American production of copper.

Other copper ores, which are also important occasionally are the basic carbonates, *malachite*, $CuCO_3 \cdot Cu(OH)_2$, and *azurite*, $2CuCO_3 \cdot Cu(OH)_2$, and also the oxide Cu_2O, *cuprite*.

The silicates of copper are less important; of these *chrysocolla*, $CuSiO_3 \cdot 2H_2O$, and *dioptase*, (copper-emerald), $CuSiO_3 \cdot H_2O$, which is occasionally used as a gem stone, may be mentioned.

Enargite, Cu_3AsS_4, *famatinite* Cu_3SbS_4, *tenorite* (black copper ore) CuO, and *atacamite*, $CuCl_2 \cdot 3Cu(OH)_2$, have only mineralogical interest.

Copper minerals occur in many cases in very fine particles, interspersed in other rocks— as for example in the very thinly foliated marls, colored black by a high bitumen content, on the south east edge of the Harz (the Mansfeld copper shales).

Copper also occurs in traces in the human and animal body, especially in bones and teeth (Tiede, 1934). The copper content of the tooth substance amounts to about 0.001%. The red to pale violet coloring which bone ashes display after very strong ignition, as was first observed by Gabriel (1894), depends on their copper content. The coloration originates from Cu_2O formed above 800° by thermal dissociation of CuO, and deposited in colloidal dispersion. The question whether copper has biological importance as a catalyst in the animal body is not yet settled. On the poisonous action of copper compounds see p. 375.

(b) Preparation

The sulfide ores are particularly important for the extraction of copper. [1] They are worked up by converting the copper sulfide first into the oxide, which is then reduced to the metal by means of carbon. However, since copper ores generally contain a considerable proportion of foreign impurities, these operations must generally be preceded by others which serve essentially for the enrichment of the copper.

Oxide ores, which generally occur only as minor associates of the sulfide ores (from which they have been formed by chemical change), are usually not worked up by themselves, but are fed into the smelting process of the other ores after their conversion into copper oxide.

The smelting of sulfide ores can be divided into the following stages:

(i) *The roasting process.* The purpose of this is to remove part of the sulfur. The longer the duration of the roasting, the greater is the extent to which the sulfides present are converted into the oxides. The oxides of arsenic and antimony volatilize during roasting.

When the sulfur content has been so far decreased through roasting that the ore as a whole contains only about one atom of sulfur for each atom of copper, stage two follows.

(ii) *Smelting for matte (smelting process).* In this, the roasted product, which generally contains much ferrous sulfide among other materials, is melted in a shaft furnace or reverberatory furnace. Coke is added, along with slag-forming materials when necessary. Copper, converted to the oxide during the roasting process, is thereby changed to cuprous sulfide, according to the equation:

$$2CuO + FeS + C + SiO_2 = Cu_2S + FeSiO_3 + CO*.$$

Iron oxide (ferrous oxide), which is formed simultaneously, goes as silicate into the slag. The oxides of other baser metals present in the reactants behave like the iron oxide, and are also slagged through formation of silicates. The resulting cuprous sulfide combines with the remaining ferrous sulfide forming $Cu_2S \cdot FeS$. This double compound separates out under the slag, as *coarse metal* or *matte*. By a repetition of the process, nearly pure cuprous sulfide, known as *white metal*, or *fine metal* can be obtained.

If the furnace charge still contains arsenic or antimony, these elements separate in the form of *speisses*, compounds analogous to the matte. If nickel and cobalt are present, these also may pass into any *speisses* formed; otherwise they enter the matte.

With copper ores rich in pyrite, the roasting and smelting processes are frequently combined in one (pyrite smelting). In this case, the heat necessary for the smelting is furnished by the combustion of the iron disulfide present in the ore.

(iii) *Conversion of copper matte to black copper.* This was formerly carried out in the German process by the *roast reduction* method, and later by the *roast reaction* process

---

* This is only one of numerous concurrent reactions.

(the English process). Both these processes have now practically dropped out of use. Extraction of metallic copper from the copper matte is now effected almost universally by air-blowing, in a converter. This process is known as *air blast roasting*, or the *copper Bessemer process*.

(1) *The roast reduction process*. The matte is roasted until the sulfides have been converted almost completely into the oxides (CuO and Fe$_2$O$_3$). Coke and quartz are added and smelting is performed in a shaft furnace, whereby the copper is reduced to the metal and iron mostly to the monoxide FeO, which passes as silicate into the slag.

(2) *Roast reaction process*. In the original form of this process, the matte is first slowly melted in a reverberatory furnace with free access of air, whereby, as in ordinary roasting, the sulfides are converted into the oxides. However, the air is excluded before this process has reached completion, so that the oxide of the most easily reducible metal (copper) enters into reaction with the remaining sulfides. By this means, metallic copper is formed, with elimination of SO$_2$, e.g., according to:

$$2Cu_2S + 3O_2 \quad = 2Cu_2O + 2SO_2 \qquad \text{(first phase—\textit{roasting})};\qquad\qquad (1)$$

$$2Cu_2O + Cu_2S = SO_2 + 6Cu \qquad \text{(second phase—\textit{reaction stage})}.\qquad (2)$$

The iron monoxide, being more difficult to reduce to the metal, combines with the added quartz to form the silicate:

$$FeO + SiO_2 = FeSiO_3.$$

Repetition of the process yields 'black copper', with about 98% Cu.

(3) *Converter roasting process (copper Bessemer process)*.

This process, which is now in general use, has developed from the two earlier processes just considered. In converter roasting, the molten matte or white metal, like the pig iron in the Bessemer process, is introduced into a *converter*. This is a vessel furnished with an acid or basic lining (the latter is used if acidic additions are to be made to the melt, and is more resistant), through which compressed air is blown. The FeS then first undergoes oxidation, and the resulting iron oxide is slagged. Oxidation of a portion of the copper sulfide then follows. The resulting oxide at once reacts vigorously with the still unchanged sulfide, until all the copper is transformed into the metal in accordance with equations (1) and (2).

An essential step in extending the Bessemer process to the production of copper was introduced by the French metallurgist Manhés (1880) who had the idea of introducing the air from the side, at some distance above the bottom, instead of at the bottom of the converter. This ensures that the air at once enters the reactant mixture, which is also maintained in a white hot molten state by the exothermic oxidation processes occurring within it, without having to pass through the copper, which collects at the bottom, and which readily cools and sets solid. Drum converters are widely used in the copper converter process, in place of the pear-shaped converter usual in steel smelting.

(iv) *Extraction of copper by autogenous smelting*. It has been seen that the preparation of copper can be subdivided into three stages (this being true of the copper Bessemer process also). Two of these stages—the roasting process and the converter stage—take place exothermically whereas heat is absorbed in the smelting stage. It is an obvious idea that the necessary heat should be provided by the two exothermic stages. To a certain extent this aim had already been achieved in the so-called *pyrite smelting* process, which came into use in a few places at the end of the last century. It was applicable, however, only to certain copper ores. A process which can be applied to almost any sulfidic copper concentrate, and which provides an excellent material and energy balance sheet, is the so-called *autogenous smelting process* worked out by the Outokumpu Oy (Finland).

The finely pulverized copper ore, concentrated by flotation, is mixed with slag-forming substances, and fed from above into a shaft furnace. This has a 'burner', from which a blast of heated air emerges. This maintains the finely pulverized charge in a state of suspension for some time, and simultaneously brings about some elimination of sulfur, which burns to SO_2. In practice, copper ores always contain iron sulfide. This is oxidized to FeO at the same time, whereas the Cu_2S remains unattacked if the air supply is correctly regulated. The heat of combustion raises the dust particles to the temperature of fusion, so that the iron oxide is slagged by formation of the silicate. The molten droplets settle out, and collect at the base of the furnace, with a lower layer of molten copper matte and an upper layer of fused iron silicate. The matte and slag are tapped off from time to time. The former is run directly into a converter, in which it is 'blown' to crude copper, while the slag is run into an electric furnace. The melt is covered with a mixture of quick lime and coke, which is converted by an electric arc into calcium carbide. This reacts with the silicate slag, forming a low-carbon iron (steel), and a basic calcium silicate which can be used as a fertilizer. CO is formed as a by-product, and this is burned to provide the heat required for lime burning. The hot roaster gases emerging from the shaft furnace in which the copper matte is smelted provide energy for electric power generation (for the electric furnace), and are used also to pre-heat the air blast of the shaft furnace. The gases are finally freed from dust and used for the manufacture of sulfuric acid.

(*v*) *Extraction of copper by wet processes*. Copper is obtained by *leaching* from ores poor in copper and from copper-containing residues (e.g., burned pyrites). This is generally done by treatment with dilute sulfuric acid, if necessary after a previous roasting to convert copper sulfide into the oxide and sulfate. The copper can be allowed to crystallize from the leaching solutions as copper vitriol, or it can be precipitated by metallic iron (occasionally by other precipitants also, such as H_2S). As the present time it is generally deposited from solution electrolytically.

In the year 1938 the total production of copper, as ore, was 1.992 million tons. Of this production the United States contributed 25.4%, Chile 17.6%, Canada 13.3%, Rhodesia 10.9%, and Belgian-Congo 6.2%. Only 12% of the total copper ore production came from all European countries (1.5% from Germany). The smelter production of copper amounted to 1.979 million tons, and the production of refinery copper to 2.246 million tons.* Of this 13.6% and 32.8%, respectively, were in Europe. Europe accounted for 63.3% of the world consumption of refined copper, which amounted in the same year to 2.144 million tons.

(c) Purification (Refining) of Copper [2]

As a rule, the copper obtained directly by the various methods is very impure. It can be refined by dry or wet methods, in the latter case electrolytically. Although in itself it is more costly, this method of operation is especially profitable when the copper contains noble metals as impurities. Moreover, it has the advantage of furnishing copper of a very high degree of purity, and at the present time it is therefore used almost exclusively for the copper required for electrical machinery or conductors.

(*i*) *Refining by the dry method* is carried out by melting the copper in a reverberatory furnace. A portion of it is thereby oxidized to Cu_2O. This dissolves in the molten copper, and can pass its oxygen on to the baser elements contained therein. When this has happened, the Cu_2O which still remains in excess is deoxidized, by the introduction of wood or wood charcoal into the melt ('poling' of copper). The process furnishes refined copper of about 99.5% purity.

---

* The production capacity of the copper refineries is higher than that of the smelters, because in addition to crude copper the former also work up scrap and residues for fine copper.

(*ii*) *Refining by the wet method* is carried out electrolytically. The blister (unrefined) is suspended as anode in a copper sulfate solution acidified with sulfuric acid. On electrolysis, copper goes into solution at the anode, and pure copper is deposited at the cathode. The voltage required for the electrolysis is very small, since no decomposition potential has to be overcome, and the potential is determined almost entirely by the ohmic resistance of the solution. With cells of the usual dimensions and a current strength of 1000 amps. a few tenths of a volt suffice for this. A few hundred cells are therefore usually connected in series.

The energy consumption per kg of pure electrolytic copper produced can be calculated from the fact that a current of one amp. deposits 19.76 mg of copper per minute, so that 843.4 amp. hours are needed to deposit one kg. Assuming a terminal voltage of 0.3 volt per cell, this represents an energy consumption of $843.4 \cdot 0.3 : 1000 = 0.253$ kilowatt hours. The voltage losses in the current carrying conductors are not thereby taken into account.

The metals standing below copper in the electrochemical potential series, (i.e., silver, gold, and platinum in particular) do not go into solution with the copper when the anode dissolves, but settle on the bottom as an anode slime. The possibility of recovering the precious metals from the anode slimes is very important for the economics of the electrolytic process.

The electrolysis refining of copper was the first electrochemical process to find application for industrial purposes. The first installation was erected by Elkington in England, only two years after the invention of the dynamo.

The electrolytic process is also of considerable importance for the recovery of copper from residues and from scrap. The usual procedure resembles that used for the refining of crude copper. For the extraction of copper from copper-plated iron scrap, a process was worked out by Schaarwächter (1939), and depends on anodic dissolution of the copper in a solution containing sodium cyanide, and its redeposition cathodically (cf. p. 382). The iron remains practically unattacked in this process. Brass can be recovered directly from brass-plated scrap by the same process (cf. page 374).

(d) Properties

Copper is a metal of characteristic red color; very thin layers appear greenish blue by transmitted light. It crystallizes in the cubic system (cf. also page 362) and frequently occurs native in well formed octahedra and cubes. Pure copper is rather soft (hardness 3 on Mohs' scale). It is very tough and ductile. It is notable for its very high electrical and thermal conductivity (cf. Table 36 p. 362.). The conductivity is considerably diminished by traces of most impurities (which also increase the hardness), and especially by the brittle metals arsenic and antimony. It is a general rule in metallurgy that the further apart any alloying element and the main constituent stand in the Periodic System, the greater is the effect of the alloying element on all properties.

For further physical properties of copper compare the summary table at the beginning of this chapter.

Compact copper is not markedly attacked by dry atmospheric oxygen at ordinary temperature. It tarnishes on heating, with the superficial formation of oxide. When strongly heated, it is ultimately completely converted into cuprous oxide or, at higher oxygen pressures, into cupric oxide.

Moist chlorine rapidly attacks copper, even at ordinary temperatures. The metal also combines readily with the other halogens, and has a very marked affinity for sulfur and also for selenium.*

Gaseous nitrogen has no perceptible action even at high temperatures. However, if ammonia gas is passed over copper at a red heat, nitrogen is absorbed.

---

* Of the three selenides cited in Table 39, only one can be obtained by direct combination of the constituents. All three are found in Nature however; Cu_2Se as *berzelianite*, $CuSe$ as *klockmannite*, and Cu_2Se_3 as *umangite*.

Copper is attacked by the oxides of nitrogen even below a red heat; by N_2O and NO with the formation of Cu_2O, and by NO_2 with the formation of CuO. Finely divided, freshly reduced copper combines directly with NO_2 at about 25–30°, with the formation of *nitrocopper*, Cu_2NO_2, a brown mass stable in dry air, which is decomposed vigorously by water with the evolution of NO. Copper does not combine directly with carbon. When present in the form of its compounds however, it can replace the hydrogen atoms of acetylene C_2H_2. Either Cu_2C_2 or CuC_2 can be obtained, depending on whether compounds of +1 or +2 copper are employed.

Copper does not combine directly with molecular hydrogen, apart from the formation of a volatile hydride in minimal amounts at very high temperatures; this has only been detected spectroscopically (cf. p. 363). It does however combine with atomic hydrogen (Pietsch, 1931), with the formation of a very unstable solid hydride*. The solvent power of copper for hydrogen is only small. The quantity dissolved is proportional to the square root of the hydrogen pressure, and increases with rise of temperature. According to the measurements of Sieverts, at 409° 100 g of copper absorb only 0.006 mg of hydrogen at atmospheric pressure. At the melting point (1083°), the same amount of copper in the solid state can take up 0.19 mg hydrogen, and in the liquid state 0.54 mg at 1083° and 1.24 mg at 1550°.

Copper is able to alloy with many metals. With a number of them, for example aluminum, combination occurs with the evolution of considerable heat. As has already been discussed, there are many compounds among the alloys of copper.

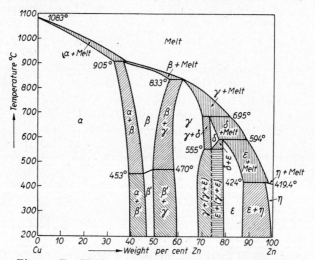

Fig. 45. Equilibrium diagram of the copper-zinc alloys.

a = Cu-Zn mixed crystals (f.c. cubic)
β = $\overline{ZnCu}$ (b.c. cubic); β' = ZnCu phase with ordered arrangement of atoms
γ = $\overline{Zn_8Cu_5}$ (cub., very large cell)
δ = high temperature phase (hexag.), structure not yet known
ε = $\overline{Zn_3Cu}$ (hexag. close packed)
η = Zn-Cu mixed crystals (hexag. close packed, but with axial ratio different from ε-phase. Ratio c/a rises with increasing Cu content of ε-phase, but falls with increase in Cu content of η-phase).
$(\gamma+\varepsilon)$ = eutectoid mixture of γ and ε formed by decomposition of δ-phase.

* A hydrated solid hydride, differing from this, is obtained by the reaction of copper sulfate with hypophosphorous acid (cf. p. 383.).

Some of the phase diagrams of the copper alloys are rather complex. For example, not less than eight different intermetallic phases appear in the system copper-aluminum. Of these, four are stable at ordinary temperature (cf. Fig. 6 p. 17.). Furthermore, copper can take up to 10.5 weight-% of aluminum without change in its crystal structure, and aluminum up to 0.5 weight-% of copper*. Four intermetallic phases (or five, including the super-structure phase of β-brass) are involved in the system copper-zinc; three of these are stable at ordinary temperature. They all have fairly broad ranges of homogeneity. Copper can dissolve up to 39.5 weight-% of zinc without change of crystal structure; zinc can dissolve only 0.5 weight-% of copper without change of structure. The areas of the different phases in the copper-zinc alloys are represented in Fig. 45. The ranges in which the different phases coexist (regions of inhomogeneity) are indicated by shading. Six or seven intermetallic phases occur in the copper-tin alloys. Three of these are stable at ordinary temperature, namely the compounds $Cu_{31}Sn_8$, Cu_3Sn, and $\overline{Cu_6Sn_5}$. All three have quite narrow ranges of homogeneity. The last mentioned, however changes markedly in composition with rise of temperature; it melts incongruently. The two others are likewise unstable at high temperatures, and undergo transformations in the solid state. Copper can dissolve up to 14.2 weight-% of tin without change of crystal structure. The (white) tin however does not incorporate copper into its crystal lattice in appreciable amounts. An alloy of 94.5 weight-% tin and 5.5 weight-% copper solidifies from the melt as a eutectic mixture of Cu_6Sn_5 and tin crystallites. As has already been discussed in Chap. 1, (p. 16 *et seq.*), a close relationship exist between the alloys of copper with tin, zinc, and aluminum (on this point see also p. 375.) The phase diagram of the copper-nickel alloys is very simple. It corresponds exactly to the type represented by Fig. 103, Vol. I, p. 569; that is to say copper and nickel are completely miscible with one another both in the solid and in the liquid state, and form no chemical compounds with each other.

Copper dissolves in dilute nitric acid, with the evolution of nitric oxide, forming copper(II) nitrate; in hot sulfuric acid yielding copper(II) sulfate**. On the other hand, as follows from its position in the electrochemical potential series, it is in general unable to discharge hydrogen ions. In accordance with this, it is not attacked by dilute sulfuric acid, dilute hydrochloric acid, acetic acid, etc., in the absence of air.

However, gaseous hydrogen chloride reacts with heated copper, forming copper(I) chloride:

$$Cu + HCl \rightleftharpoons CuCl + \tfrac{1}{2}H_2.$$

Metallic copper also dissolves in concentrated alkali cyanide solutions, with the evolution of hydrogen:

$$Cu + 2CN^- + H_2O = [Cu(CN)_2]^- + OH^- + \tfrac{1}{2}H_2.$$

This reaction depends on the fact that the oxidation of the copper is facilitated by complex formation.

---

* The saturation limits given here and in the following apply to ordinary temperature.
** The reaction proceeds substantially according to:

$$Cu + H_2SO_4 \quad = CuO + SO_2 + H_2O$$
$$CuO + H_2SO_4 = CuSO_4 + H_2O.$$

This main reaction is however accompanied by side reactions, which are formulated by Fowles (1928) as follows:

$$4Cu + SO_2 = Cu_2S + 2CuO$$
$$Cu_2S + H_2SO_4 = Cu_2SO_4 + H_2S$$
$$H_2S + H_2SO_4 = SO_2 + S + 2H_2O$$
$$S + 2H_2SO_4 = 2H_2O + 3SO_2$$
$$Cu_2SO_4 + 2H_2SO_4 = 2CuSO_4 + 2H_2O + SO_2.$$

It must be assumed that an equilibrium is set up in the vicinity of the metal:

$$Cu + H^+ \rightleftharpoons Cu^+ + \tfrac{1}{2}H_2. \tag{3}$$

This generally lies so far over to the left, however, that there is never any perceptible dissolution of copper and evolution of hydrogen. It is possible to displace the equilibrium from
left to right to a considerable extent by creating conditions such that either the concentration of Cu^+ ions—which are ordinarily present in solution only in minimal amounts—is still
further substantially lowered, or that the dissolved hydrogen (also present only in minimal
amounts) is removed from the solution. The first can happen if the Cu^+ ions are captured by
strongly complex-forming reagents, such as the CN^- ions in the example cited. If halide
ions (which have a much weaker tendency to form complexes with copper than the cyanide
ions) are used to take up the Cu^+ ions, it is not possible to displace the equilibrium (3)
towards the right, unless the H^+ ion concentration is simultaneously raised to a very high
value. Copper does, in fact, dissolve in *highly concentrated* halogen acids (with the exception
of hydrofluoric acid) with the evolution of hydrogen, although it stands below hydrogen in the electrochemical potential series. The other possibility of shifting equilibrium
(3) towards the right, the removal of the elementary hydrogen, can also be achieved by
oxidants. However, these may also react directly with the copper, by accepting electrons
from it so that they transform it directly into the ionic state.

The initial rate of dissolution of copper in dilute nitric acid is very slow. However, when a certain amount of nitrite ion has been formed in the solution,
according to the equation

$$Cu + 2H^+ + NO_3^- = Cu^{++} + H_2O + NO_2^-,$$

the reaction becomes very vigorous. If some nitrite is previously added to the
dilute nitric acid used for dissolving the copper, reaction takes place from the
beginning at its full speed. Such a case as this, when one of the reaction products
itself accelerates the reaction, is spoken of as *autocatalysis* (cf. Chap. 17).

Copper is precipitated from its solutions by less noble metals such as iron. Use
is frequently made of this in the technical preparation of copper (cf. p. 370.).
Precipitation in this way is called cementation, and the powdered (and generally
very impure) copper so obtained is called cementation copper.

The normal potential of copper has already been cited as an electrochemical measure
of the noble character of the metal (cf. Vol. I, p. 30, Table 4). The potential of a copper
foil dipping into a solution which contains one mole of cupric ions per liter, is 0.34 volts
below that of the normal hydrogen electrode. A copper foil dipping into a solution which
contained the same concentration of cuprous ions would exhibit a still lower potential, namely 0.52 volts below the normal hydrogen electrode; at the same ionic concentration,
therefore, cuprous ions would have a greater tendency than cupric ions to pass over into the
metallic state, giving up their charge—in other words stronger oxidizing properties. In
reality, however, cuprous ions are always present only in quite minimal concentrations,
and since the electrode potential depends upon the ionic concentration, the situation is
actually reversed. In practice, the cupric compounds are the ones with the greater oxidizing
power.

It is possible to effect the simultaneous electrodeposition of copper with much more
strongly electropositive metals, such as zinc, if the copper ion concentration is strongly
repressed through complex formation. Its solution potential is thereby displaced (according
to eq. 1b, Vol. I, p. 32) towards the side of the baser metal. Both metals are deposited
simultaneously, in the form of brass, by electrolysis of a solution of copper and zinc salts
containing much sodium cyanide, since the deposition potential of copper in such a solution is raised in magnitude by formation of the strongly complexed ion $[Cu(CN)_2]^-$ to a
far greater extent than is that of zinc; the cyano ions $[Zn(CN)_3]^-$ and $[Zn(CN)_4]^=$ are only
weakly complexed. This effect is utilized for the production of brass coatings on iron objects, as protection against corrosion.

Copper is strongly toxic towards lower organisms. Small amounts of copper (e.g., average daily doses of less than 0.1 g) are not harmful to man, even with continued ingestion. Larger amounts of soluble copper salts can bring about serious disorders; 2–3 g in a single dose can be fatal. Copper poisoning from food-stuffs is rare, since the presence of copper in foodstuffs confers a distinct and un-pleasant taste which makes them unpalatable.

(e) Uses of Copper and its Compounds

Except for iron and aluminum, copper is probably the most important tech-nical metal, even though within recent years other cheaper metals have replaced copper to a considerable extent in many of its applications. Its alloys, however, still find the most manifold applications for machine parts, tools, instruments and equipment of all sorts, and for ornamental objects. Copper sheet is used for roof covering, and for coating ships' bottoms. In the form of wire, it is used in very great quantities for electrical conductors.

Among the alloys of copper, [3–7] the bronzes call for particular mention. Bronzes are alloys of copper and tin, to which zinc and lead are frequently added to lower the cost. The bron-zes are characterized by their hardness and strength. The best are gunmetal and bell-metal, which contain only copper and tin. A general property of the bronzes is their ability to make sound castings. Moreover, bronzes containing up to about 10% tin are ductile, malleable, and can be die-stamped. In accordance with what was stated on p. 17, such bronzes con-sist only of Cu-Sn mixed crystals*. To avoid oxidation of copper and tin during the melting and casting of bronzes, or to reduce any oxides formed again, phosphorus, in the form of phosphor-tin or phosphor-copper, is frequently added to the bronze as a de-oxidant, before casting. The *phosphor-bronzes* so obtained possess especially great strength and durability, because of their freedom from oxide**. Silicon is frequently added as de-oxidant in casting pure copper. The oxide-free copper so produced, containing traces of excess silicon, is generally known as *silicon-bronze*, although it is not a bronze in the proper sense, i.e., it is not a copper-tin alloy (cf. also Vol. I, p. 254 et seq.).

Aluminum-bronzes are coming increasingly into use in place of the tin-bronzes. As a rule they contain 55–75% Cu, 22–42% Zn and 2–20% Al. When correctly worked, many of them display properties superior to the ordinary bronzes. *Tombak* and *brass* are other tech-nical metals in wide use. They are alloys containing copper and zinc as principal consti-tuents, often with the addition of tin, lead, and iron. They are cheaper, softer, and more easily worked than the bronzes. With a zinc content of less than 18%, the color is still reddish (tombak or red brass); with a higher zinc content they are yellow (up to 50% zinc) and are called brass. Brasses with a zinc content up to about 45% can not only be cast but can also be worked by cold-rolling, pressing, and drawing. Copper-zinc alloys with a zinc content between 38 and 45 % (*Muntz metal*) may be readily hot-rolled and forged. Accord-ing to the equilibrium diagram of the zinc-copper alloys, the intermetallic phase known as β-brass occurs at red-heat in this region of concentrations***. This changes below 470° into the so-called β'-brass which differs from β-brass in the ordered arrangement of the atoms on

---

* In practice the limits for cold working lie rather lower than the miscibility limits at ordinary temperature (14.2%). This may be due to the fact that at higher temperatures the limit is displaced towards the region of alloys poorer in tin—at 792° this is at 7.5%. The equilibrium corresponding to that obtaining at ordinary temperature is not attained in alloys which have solidified relatively quickly from the melt.

** The finished phosphor-bronzes usually contain only traces of phosphorus, since no more is usually added than is required for the reduction of the oxides.

*** At 800°, the β-brass region extends from about 38–53% Zn. At 700°, it extends from 42–49% Zn. The boundaries of the region approach increasingly closely with decrease of temperature.

the lattice points of its structure*. It appears that the hot-working properties and malleability observed for Muntz-metal are closely associated with the occurrence of β-brass. An alloy with about 30% copper and 65–70% nickel (*Monel metal*, m.p. 1410°), which is obtained directly by the reduction of nickel-copper matte (cf. p. 308. *et seq.*), is distinguished by its resistance to acids, great strength, and good working qualities. *German silver*, much used for the manufacture of cutlery and small articles of jewellery, is an alloy with 50–70% copper, 13–25% nickel, 13–25% zinc, and frequently small amounts of lead, tin, or iron. [7]

The coinage alloys must also be mentioned. Most copper coins contain 95% copper, 4% tin, and 1% zinc. Nickel coins consist of copper and nickel (cupronickel) or pure nickel (cf. p. 309).

Devarda's alloy, consisting of 50% copper, 45% aluminum, and 5% zinc, which is so brittle that it can easily be powdered, is often used in the laboratory as a reducing agent.

Copper finds many applications as a catalyst in preparative inorganic and organic chemistry and in technology.

Both metallic copper and the compounds of copper possess catalytic properties, accelerating oxidations in particular. The best known example of this is the oxidation of hydrogen chloride by atmospheric oxygen, in the presence of copper compounds, in the Deacon process. Copper compounds are also frequently used as oxygen carriers in other ways, as for example in the oxidation of organic dye-stuffs. If methyl alcohol mixed with air is passed over a heated copper spiral, the latter continues to glow, since formaldehyde is formed with considerable evolution of heat, according to the equation:

$$CH_3OH + \tfrac{1}{2}O_2 = CH_2O + H_2O.$$

Copper can also transfer atmospheric oxygen to numerous other substances. Oxidation with combined oxygen is also frequently accelerated by copper salts.

Use is made of this in the laboratory—for example in Kjeldahl nitrogen determinations in which a little copper sulfate is generally added, as a catalyst, to the sulfuric acid used for the destructive oxidation of the organic substances. Reduction processes can also be accelerated by copper. The widespread use of cuprous halides and cyano compounds in preparative organic chemistry for the Sandmeyer reaction (conversion of diazo compounds into halogen compounds, etc.) may also be mentioned in this connection.

Many compounds of copper are used as pigments. Ground up malachite (cf. p. 389), or artificial preparations of similar composition are used under various names in distemper and water paints. "Mountain-blue" is ground up azurite (p. 389), a beautiful blue pigment, but unstable. More frequently used is Bremen blue, a voluminous copper(II) hydroxide prepared by a particular process. Schweinfurt green (cf. p. 390) is a color highly valued in many places. It is very poisonous, however, because of its arsenic content, and is therefore prohibited in certain countries, as also is Scheele's green, a copper(II) arsenite of variable composition.

Other copper compounds find application in medicine or to combat plant pests. Thus Bordeaux mixture is used to combat Peronospora and potato disease (cf. p. 387). Copper borate or copper phosphate sprays, prepared by dissolving copper vitriol and borax or sodium phosphate in water, are used against diseases of wheat, such as rust.

Copper(I) chloride, dissolved in concentrated hydrochloric acid or in aqueous ammonia, is used in gas analysis for the absorption of carbon monoxide. It is also used for the purification of acetylene. Copper(II) chloride, copper(I) oxide, and copper(II) oxide are used for coloring glasses and enamels (Cu_2O gives a red

* The transformation of the one type into the other shows up in the cooling curve, and manifests itself further in anomalous changes in electrical conductivity, specific volume, and other properties during cooling or heating.

color, CuO a green or bluish tint). Copper oxide-ammonia solutions are used in the manufacture of artificial silk. One of the most widely used copper salts is copper vitriol, $CuSO_4 \cdot 5H_2O$.

3. Compounds of Copper

The familiar compounds of copper are derived both from the $+1$ and $+2$ oxidation states of the metal. The copper(II) compounds are generally the more stable.

In addition to a black oxide CuO, and a blue hydroxide $Cu(OH)_2$, dipositive copper forms numerous salts, CuX_2, of which those formed with strong acids are easily soluble in water, and are almost completely dissociated in dilute aqueous solution, being also hydrolyzed to a slight extent. The (hydrated) Cu^{++} ions have a sky-blue color. They have a fairly strong tendency to form complex ions, this being generally manifested by a change of color. Such complexes are formed both with negative ions (acido-complexes), e.g., $Cu^{++} + 4Cl^- = [CuCl_4]^=$, and with neutral molecules, especially ammonia—$Cu^{++} + 4NH_3 = [Cu(NH_3)_4]^{++}$.

The inhibition of many precipitation reactions of copper (e.g., precipitation by alkali hydroxides) by certain organic substances, such as glycerol, sugar, or tartaric acid, is due to the formation of complexes. A factor common to these substances is the presence of several hydroxyl groups. Copper can replace the hydrogen atoms of such hydroxyl groups; thus in alkaline solution it can react with the ion of tartaric acid $[C_4H_4O_6]^=$ according to:

$$
\begin{array}{ccc}
\overset{\text{O}}{\overset{\|}{-\text{O}-\text{C}}}-\text{CH}-\text{O}-\text{H} & & \overset{\text{O}}{\overset{\|}{-\text{O}-\text{C}}}-\text{CH}-\text{O} \\
\phantom{-\text{O}-\text{C}-}| & \text{HO} & \phantom{-\text{O}-\text{C}-}| \text{Cu} + 2\text{H}_2\text{O.}\\
-\text{O}-\text{C}-\text{CH}-\text{O}-\text{H} & + \text{Cu} = & -\text{O}-\text{C}-\text{CH}-\text{O} \\
\overset{\|}{\text{O}} & \text{HO} & \overset{\|}{\text{O}}
\end{array}
$$

$$
\begin{array}{c}
\overset{\text{O}}{\overset{\|}{-\text{O}-\text{C}}}-\text{CH}-\text{O}-\text{H} + \text{HO}-\text{Cu}-\text{OH} + \text{H}-\text{O}-\text{CH}-\overset{\text{O}}{\overset{\|}{\text{C}}}-\text{O}^- \\
-\text{O}-\text{C}-\text{CH}-\text{O}-\text{H} + \text{HONa} \quad \text{NaOH} + \text{H}-\text{O}-\text{CH}-\text{C}-\text{O}^- \\
\overset{\|}{\text{O}} \overset{\|}{\text{O}}
\end{array} \quad =
$$

$$
\begin{array}{c}
\overset{\text{O}}{\overset{\|}{-\text{O}-\text{C}}}-\text{CH}-\text{O}-\text{Cu}-\text{O}-\text{CH}-\overset{\text{O}}{\overset{\|}{\text{C}}}-\text{O}^- \\
-\text{O}-\text{C}-\text{CH}-\text{ONa} \quad \text{NaO}-\text{CH}-\text{C}-\text{O}^- + 4\text{H}_2\text{O.} \\
\overset{\|}{\text{O}} \overset{\|}{\text{O}}
\end{array}
$$

In these and similar complex ions the copper is so firmly bound that free Cu^{++} ions are not present in such solutions in sufficient concentration for the solubility product, e.g., of $Cu(OH)_2$ to be attained. Precipitation of this compound is therefore prevented.

The sodium salts, $Na_2[C_4H_2O_6Cu] \cdot 2H_2O$ and $Na_4[C_8H_4O_{12}CuNa_2] \cdot 13H_2O$ have been isolated in the crystalline state from tartrate solutions; their constitution can however not yet be considered to be fully elucidated. *Fehling's solution*, prepared by mixing copper sulfate solution with Seignette salt ($KNaC_4H_4O_6 \cdot 4H_2O$) and potassium hydroxide, which is used as an analytical reagent for many organic substances (e.g., glucose) contains such copper-tartaric acid complexes. The copper in them is reduced to the $+1$ state by substances such as glucose, and is thereby precipitated as cuprous oxide.

The compounds of unipositive copper, e.g., the red cuprous oxide Cu_2O, gray cuprous sulfide Cu_2S, and the simple salts CuX are all practically insoluble in water, but in many cases can be brought into solution through complex formation. Most of the complex compounds and also the simple salts of $+1$ copper are colorless, or nearly so. The ability of copper(I) salt solutions to absorb carbon monoxide with the formation of a complex is noteworthy.

The conditions governing the stability of copper(I) salts are defined by the alternative electrode processes (1) and (2):

$$Cu \rightleftharpoons Cu^+ + e \tag{1}$$

$$Cu \rightleftharpoons Cu^{++} + 2e \tag{2}$$

At equilibrium, in the presence of metallic copper, $\dfrac{[Cu^{++}]}{[Cu^+]^2} = K$. At 20°, K has the magnitude $1 \cdot 10^6$. Hence copper will dissolve in a solution of a copper(II) salt until enough copper(I) ion has been formed to establish equilibria in the system; conversely, Cu^+ ions would undergo disproportionation, $2Cu^+ \rightleftharpoons Cu^{++} + Cu$, until equilibrium was reached. The numerical magnitude of K implies that the concentrations of copper(I) ion coexisting with copper(II) ion would be

$$[Cu^{++}] = \quad 1 \quad\quad 10^{-2} \quad 10^{-4} \quad 10^{-6} \quad \text{molar}$$

$$[Cu^+] \;\; = \quad 10^{-3} \quad 10^{-4} \quad 10^{-5} \quad 10^{-6} \quad \text{molar}$$

It follows that appreciable concentrations of Cu^+ ions cannot exist in solution at all, and the known copper(I) compounds are limited to those which are extremely insoluble (the oxide, sulfide, and halides), so that the Cu^+ ion concentration in equilibrium with the solid cannot rise to a value involving appreciable disproportionation, and those which are complexed, so that the Cu^+ ion concentration is held at a low figure. Other copper(I) salts are unstable. Thus Cu_2SO_4 is formed by the action of dimethyl sulfate on Cu_2O, but is decomposed by water, but complex sulfates (e.g., $[Cu(NH_3)_2]_2SO_4$) and nitrates (e.g., $[Cu(CH_3CN)_4]NO_3$) exist.

The corresponding hypothetical equilibria for the Ag^+ and Ag^{++} ions lie far over to the left, the Ag(II) compounds being powerful oxidants unless the free Ag^{++} ions concentration is held at a very low figure by complex formation. The corresponding reactions with gold involve the gold(I) and gold(III) valence states — $3Au(I) \rightleftharpoons Au(III) + 2Au(metal)$ — and greatly favor the existence of gold(III) compounds. The equilibrium is still further displaced in this direction, however, by the strong tendency for complex formation by gold(III) compounds.

With many non-metals, e.g., with Si, P, Te, copper forms compounds which differ completely in composition from the salt-like compounds, and appear to be related to the intermetallic compounds (cf. Table 39, p. 364–5.).

It is also possible to obtain compounds in which copper is *tripositive*. Scagliarini and Torelli (1921) obtained a garnet red powder, said to contain the oxide Cu_2O_3, by treating freshly precipitated copper(II) hydroxide with potassium peroxysulfate. This evolves chlorine from hydrochloric acid, but gives no hydrogen peroxide with dilute acids. It is thus not a derivative of the latter, i.e., is not a peroxide such as the brown-black copper peroxide

$$CuO_2 \cdot H_2O = Cu\underset{\diagdown O}{\overset{\diagup O}{\big|}} \cdot H_2O \text{ or } Cu\underset{\diagdown OH}{\overset{\diagup OOH}{}}$$

which Moser has prepared by the action of H_2O_2 on finely divided $Cu(OH)_2$. (A peroxide or peroxyhydrate of copper with a still higher oxygen content, $HOOCuOOCuOOH$, has been prepared by Wieland, 1938.) Scagliarini and Torelli therefore formulate the oxide Cu_2O_3 with $+3$ copper. The oxide Cu_2O_3 is acidic in character, and forms red very unstable salts with alkalis of the type $M^I[Cu(OH)_4]$, the tetrahydroxocuprate(III) salts.

Oxocuprate(III) salts, $KCuO_2$ and $Ba(CuO_2)_2 \cdot H_2O$ have recently been isolated by Klemm and by R. Scholder (1951). Klemm (1949) also succeeded in preparing a fluoro complex of $+3$ copper, $K_3[CuF_6]$. Other examples of copper(III) compounds are $K_7[Cu(IO_6)_2] \cdot 7H_2O$ and $Na_5H_4[Cu(TeO_6)_2] \cdot 18H_2O$ (Malatesta, 1941). Except for the fluoro complex referred to above, all these Cu(III) compounds are diamagnetic. They are thus compounds involving covalent bonds within the complex (probably dsp^2 hybrid, square planar bonds).

(a) Copper(I) Compounds

(*i*) *Cuprous oxide*, Cu_2O, occurs in Nature as red copper ore (*cuprite*), in the form of dense red to brown-black masses, occasionally also as well formed regular crystals (usually octahedra) of density 5.75–6.09. It is obtained artificially by the addition of caustic soda and a not too strong reducing agent, such as glucose, hydrazine, or hydroxylamine, to a cupric sulfate solution or to Fehling's solution. On gentle heating a yellow precipitate is first formed. From its X-ray diffraction pattern this is to be regarded as a gel of finely dispersed cuprous oxide. The precipitate slowly turns red, (faster on stronger heating) as a consequence of growth of particle size.

Cuprous oxide is practically insoluble in water. It readily dissolves in aqueous ammonia however, and also in concentrated solutions of halogen acids, forming colorless complex compounds:

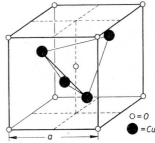

$$[Cu(NH_3)_2]OH \text{ and } [CuX_2]H \quad (X = \text{halogen}).$$

O = O ● = Cu

Fig. 46. Unit cell of cuprite, Cu_2O. $a = 4.26$ Å

Cuprous oxide is also perceptibly soluble in caustic alkalis. It is transformed by dilute halogen acids into cuprous halides, which are likewise insoluble in water. It is dissolved by dilute oxy-acids, e.g. sulfuric acid, but thereby undergoes simultaneous decomposition into a copper(II) salt and free metal:

$$Cu_2O + H_2SO_4 = CuSO_4 + H_2O + Cu.$$

The crystal structure of cuprite is illustrated in Fig. 46. The oxygen atoms form a body-centered cube of side $a = 4.257$ Å, the copper atoms a tetrahedron with the edge length $\dfrac{a}{\sqrt{2}}$. The shortest distance Cu$\leftrightarrow$O is $\dfrac{\sqrt{3}}{4}a = 1.85$ Å. The heat of formation of cuprous oxide, according to Wöhler, is 43.0 kcal per mole of Cu_2O.

Cuprous oxide finds technical application as a red pigment for glass and enamel and as a paint for ships' bottoms. It is also used as a pesticide.

Cuprous oxide, containing cupric oxide as an impurity, is formed in the rolling and forging of copper, as 'copper hammer scale'. At the present time the technical preparation of pure cuprous oxide is generally carried out electrolytically. If an electric current is passed through an alkali chloride solution in a cell with a copper sheet as anode, the copper passes into solution in the form of chlorocuprate(I) ions, $[CuCl_2]^-$, which react with the hydroxyl ions formed at the cathode to give cuprous oxide:

$$2[CuCl_2]^- + 2OH^- = Cu_2O + H_2O + 4Cl^-.$$

(*ii*) *Copper(I) Halides. Cuprous chloride* (copper(I) chloride) can be prepared by warming copper turnings with concentrated hydrochloric acid, to which a little potassium chlorate has been added. The solution is at first black, but ultimately becomes completely colorless. On dilution, the salt, dissolved in complex form (as the addition compound with HCl), is precipitated as a powder which is snow-white in the pure state, but rapidly becomes greenish-colored in moist air. The preparation can also be carried out by the reduction of copper(II) chloride solution, or of mixed solutions of copper(II) sulfate and sodium chloride, with sulfur dioxide. Cuprous chloride is very sparingly soluble in water (cf. Table 38, p. 363), but is readily soluble in hot concentrated sulfuric acid. It also dissolves in concentrated alkali chloride solutions, as well as in concentrates aqueous ammonia.

The solutions of cuprous chloride in concentrated hydrochloric acid absorb carbon monoxide, but give it up again on heating. It is possible to isolate the addition compound $CuCl \cdot CO$ in the crystalline form. The compound $H[CuCl_2]$ can be thrown down in pearl grey needles by cooling a solution of cuprous chloride in hydrochloric acid (saturated when hot), and passing in HCl gas. Alkali double chlorides are also known, e.g., $KCl \cdot CuCl = K[CuCl_2]$. Complex ammoniates are present in the ammoniacal solutions of cuprous chloride, e.g., $[Cu(NH_3)_3]Cl$, but they can be isolated only with difficulty. Ammoniates of the composition $2CuCl \cdot 3NH_3$ and $2CuCl \cdot NH_3$, as well as the ammoniate mentioned, have been isolated by dry methods.

Cuprous chloride can also form addition compounds with acetylene. The compound $C_2H_2 \cdot 6CuCl$ is precipitated by acetylene from a solution of copper(II) chloride in anhydrous alcohol. The compound $C_2H_2 \cdot 2CuCl$ crystallizes from a solution of CuCl in concentrated hydrochloric acid after the prolonged passage of acetylene.

Cuprous chloride forms double molecules $(CuCl)_2$ in the vapor state. As against this however, molecular weight determinations in many solvents, e.g., pyridine, have given values lying closer to the unimolecular formula CuCl. Cuprous chloride crystallizes in regular tetrahedra, as can at once be perceived by microscopic examination of the crystals. X-ray measurements have shown that its structure is of the lattice type known as the zinc blende structure, since it was first found for zinc blende. The unit cell of this structure consists of a face-centered cube formed by the atoms of one kind (e.g., the Cu atoms), into which a regular tetrahedron of atoms of the other kind (e.g., the Cl atoms) is inserted (see Fig. 47). If all the lattice points in a crystal lattice of the zinc blende type were occupied by atoms of the same kind, the resulting crystal lattice would be of the diamond type (Vol. I, Fig. 79, p. 421).

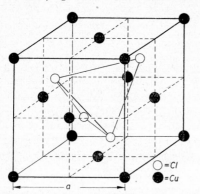

$\bigcirc = Cl$
$\bullet = Cu$

Fig. 47. Unit cubic cell of copper(I) chloride (zinc blende type).
$a = 5.41$ Å

Copper(I) bromide and copper(I) iodide have the same structure as copper(I) chloride. The edge length a of the unit cube is 5.41Å for CuCl, 5.68 Å for CuBr, and 6.05Å for CuI. The shortest interatomic distances $(\frac{\sqrt{3}}{4}a)$ are Cu↔Cl 2.34, Cu↔Br 2.46, and Cu↔I 2.62Å. These distances unlike those in typical ionic structures, e.g., in the crystal lattice of the alkali halides, are not equal to the sums of the apparent ionic radii (2.81, 2.96, and 3.20Å, respectively). This is true not only of the copper(I) halides, but of all compounds crystallizing with the zinc blende structure. It is also true of substances crystallizing with the Wurtzite structure (an example of which is beryllium oxide, Vol. I, p. 258), and also with the cuprite structure (see above p. 379). The crystal lattices of the zinc blende, wurtzite, and cuprite types are, 'incommensurable' with the typical ionic lattices (Goldschmidt). They are, however, commensurable among themselves and with the crystal lattices of most free elements. Grimm and Sommerfeld (1926) have advanced the hypothesis that the structural units in the zinc blende and wurtzite types are un-ionized atoms. If un-ionized atoms are involved in the structures of copper(I) chloride, bromide, and iodide, we might expect to obtain the apparent atomic (covalent) radii of Cl, Br, and I by deducting the apparent (covalent) radius of the copper atom, obtained from the crystal lattice of the metal, from the shortest interatomic distance in the halide structures. We then obtain

$$r_{Cl} = (2.34—1.27) = 1.07\text{Å}, \quad r_{Br} = (2.46—1.27) = 1.19\text{Å}, \quad r_{I} = (2.62—1.27) = 1.35\text{Å}.$$

Since silver iodide can also crystallize with the zinc blende structure, the atomic radius of iodine can also be calculated from its crystal lattice. The value so obtained, $r_I = (2.81—1.44) = 1.37\text{Å}$ is in satisfactory agreement with that deduced from the cuprous iodide structure. It must be mentioned however that there are many arguments against the assumption that zinc sulfide, cuprous oxide, cuprous chloride, etc., have purely covalent structures. It is therefore still open to dispute whether incommensurability of the zinc blende and other structures with the typical ionic lattices should be explained on this basis.

The ability of the cuprous halides to form addition compounds with PH_3 can also be correlated with their assumption of the zinc blende structure. PH_3 has a much smaller dipole moment than has NH_3, and in consequence can only add on to substances with particularly high polarizing power. In particular, it combines with compounds forming molecular lattices, such as the halides of Ti(IV), Sn(IV), and Al (Höltje, 1930), and also with halides crystallizing with the zinc blende structure, such as CuCl, CuBr, and CuI and AgI precipitated from aqueous solution. The compounds $CuCl \cdot PH_3$, $CuBr \cdot PH_3$, $CuI \cdot 2PH_3$ and $2AgI \cdot PH_3$ were obtained by Scholder (1934) by the action of PH_3 at low temperatures on solutions of the metal halides in concentrated alcoholic solutions of the halogen hydrides. The compounds readily decompose forming the corresponding phosphides. The silver halides AgCl and AgBr crystallizing with the rock salt structure, are unable to add on PH_3.

Copper(I) iodide (as also silver iodide) combines with tertiary alkyl phosphines to give compounds $R_3P \rightarrow CuI$ of a different type. These are reasonably stable, do not have the properties of salts, but are soluble in inert organic solvents. Exactly analogous compounds, $R_3As \rightarrow CuI$ are formed by trialkyl arsines. These are tetrameric in organic solvents (molecular weights determined ebullioscopically in benzene, cryoscopically in benzene and ethylene dibromide). The structure of the tetrameric molecular complex has been determined by X-ray analysis. The molecule is essentially tetrahedral, with a copper atom at each vertex and one R_3As (or R_3P) group linked to each Cu, and lying on the lines joining the center of the tetrahedron to each vertex. The iodine atoms occupy positions above the center of each face of the tetrahedron of Cu atoms. Every copper atom therefore forms four tetrahedrally disposed covalent bonds (by the use of the sp^3 hybrid orbital set which is available for the Cu^+ ion)—one to the R_3As group, and three to I atoms.

Whereas the silver compounds $R_3P \rightarrow AgI$ and $R_3As \rightarrow AgI$ have the same structure as the copper compounds, the gold(I) compounds $(CH_3O)_3P \rightarrow AuCl$ and $(C_6H_5)_3P \rightarrow AuCl$ are known to be monomeric in benzene solution [See Mann, Wells and Purdie, *J. Chem. Soc.*, (1936) 1503, (1937) 1828].

Copper(I) bromide and iodide correspond completely to the chloride, both in chemical behavior and in crystal structure. Copper(I) bromide forms pale greenish-yellow crystals, melting at about 500°. Copper(I) iodide may be prepared by heating copper with iodine and concentrated hydriodic acid, or by the reaction of copper(II) sulfate solution with alkali iodides, preferably with the addition of a reducing agent such as SO_2 or $Na_2S_2O_3$. It is a white powder, which generally has a somewhat brownish color, due to admixed iodine. Like CuCl, CuBr and CuI readily form complexes both with ammonia and with halogen acids and alkali halides. They are therefore soluble in the concentrated solutions of these substances. Their solubility in water is even smaller than that of CuCl; it amounts at room temperature to $2.0 \cdot 10^{-4}$ mole of CuBr or $2.2 \cdot 10^{-6}$ mole of CuI per liter. Copper(I) fluoride, CuF, is formed by the action of HF on CuCl at temperatures above 1000°; it is a ruby red, transparent, crystalline mass, which is insoluble in water. It was found by Ebert (1933) to crystallize also with the zinc blende structure, $a = 4.255\text{Å}$, $Cu \leftrightarrow F = 1.84\text{Å}$.

The cyanide and the thiocyanate of $+1$ copper are closely related in properties to the halides.

(*iii*) *Copper*(I) *cyanide*, CuCN, forms a white powder (consisting of monoclinic prisms) insoluble in water and dilute acids. It is soluble in concentrated acids, and also in aqueous ammonia and in ammonium salt solutions, forming complex ammoniates, and dissolves especially easily in alkali cyanide solutions, with the formation of complex cyanides. Numerous copper(I) complex cyanides are known. Solutions of alkali copper(I) cyanides are used in electrodepositing, as baths for copper-plating.

(*iv*) *Copper*(I) *thiocyanate*, CuSCN, is most readily formed by addition of SCN^- ions to a solution of a copper(II) salt containing a reducing agent. It is a white powder, practically insoluble in water and in cold non-oxidizing acids. It dissolves in concentrated thiocyanate solutions, yielding soluble complex salts. Dry copper(I) thiocyanate absorbs ammonia in the cold, forming a black powder from which ammonia is readily evolved again; this compound has the composition $2CuSCN \cdot 5NH_3$. $CuSCN \cdot NH_3$ has been obtained as a white crystalline powder from aqueous solution. This substance also readily loses ammonia.

(*v*) *Copper*(I) *acetate*. If hydroxylamine hydrogen sulfate is added dropwise to a hot mixture of ammoniacal copper(II) acetate solution, containing large excess of ammonium acetate, the subsequent addition of an excess of acetic acid precipitates copper(I) acetate, $CuC_2H_3O_2$, in the form of fine white needles. The compound is fairly stable in air in the dry state, but it is decomposed by water.

(*vi*) *Copper*(I) *oxalate*. Oxalic acid or potassium oxalate throws down a white precipitate from a solution of cuprous chloride in concentrated hydrochloric acid. It has been assumed that this is the oxalate of unipositive copper.

(*vii*) *Copper*(I) *sulfite* is precipitated by passing sulfur dioxide into a concentrated acetic acid solution of copper(II) acetate. It exists as a white crystalline powder, which has the composition $Cu_2SO_3 \cdot H_2O$, and is soluble in an aqueous solution of sulfurous acid. With alkali sulfites, it forms rather unstable double compounds, some of which are readily soluble in water.

(*viii*) *Copper*(I) *sulfide*, Cu_2S, is formed in almost quantitative yield, as a blackish lead-grey crystalline powder, by heating copper(II) sulfide, CuS, together with some sulfur, in a stream of hydrogen*. It occurs in Nature in rhombic crystals as *copper glance* or *chalcocite*; density 5.785, m.p. 1130°. It solidifies from the melt in cubic crystals. It is also found in this form in smelter products. Copper(I) sulfide conducts the electric current fairly well, though not so well as the copper(II) compound. The cubic copper(I) sulfide obtained by fusion is a much poorer conductor than the rhombic form.

Cubic copper(I) sulfide, like the alkali sulfides Li_2S and Na_2S, is anti-isotypic with fluorspar (cf. Fig. 62, Vol. I, p. 265.); $a = 5.56$Å at 170°. However, according to Rahlfs (1936), only a fraction of the lattice points belonging to Cu atoms are occupied, so that the cubic sulfide is deficient in copper, and has the composition $Cu_{1.8}S$. The rhombic sulfide has the normal composition Cu_2S. Its crystal structure is more complicated, and is retained at higher temperatures.

Copper(I) sulfide is practically insoluble both in water and in ammonium sulfide solution. Complex double sulfides can be obtained by other methods, however, and may be regarded as addition products of copper(I) sulfide and alkali polysulfides. Thus by dropping an ammoniacal copper sulfate solution into am-

---

* This reaction was formerly frequently used for the gravimetric determination of copper. However, copper(I) sulfide begins to lose sulfur as soon as the added sulfur is consumed, so that it cannot be obtained in the perfectly pure state in this way. It is obtained pure by heating copper, either in a mixture of hydrogen and hydrogen sulfide, or in a stream of carbon dioxide which is first passed through a wash bottle filled with methyl alcohol, and then charged with sulfur vapor by a simple procedure described by Hahn [*Ber.* 63 (1930), 1616].

monium polysulfide solution, the compound $NH_4[CuS_4]$ can be obtained as garnet red needles.

(ix) *Copper nitride*, Cu_3N, can be prepared by passing dry ammonia gas at 250° over precipitated copper(I) oxide or copper(II) oxide, or over copper(II) fluoride. It is a dark olive-green powder, with density 5.84, and is anti-isotypic with ReO_3. The heat of formation is —17.8 kcal per mol of Cu_3N (Juza 1938). Above 300°, copper nitride decomposes into copper and nitrogen. It is decomposed by chlorine, with the formation of $CuCl_2$, and by hydrogen chloride to give CuCl.

(x) *Ammoniates of Copper(I) Salts.* Copper(I) salts generally have a strong tendency to form addition compounds (ammoniates) with ammonia, as has already been shown by various examples. The ammoniates obtained from aqueous solutions generally correspond to the type $[Cu(NH_3)_2]X$. These complex compounds are often more stable than the simple salts from which they are derived. Thus the ammoniate of copper(I) sulfate $[Cu(NH_3)_2]_2SO_4$ is obtained well crystallized, in white hexagonal lamellae, by dissolving copper(I) oxide and ammonium sulfate in aqueous ammonia in the absence of air, or by reducing a solution of copper(II) sulfate, supersaturated with ammonia, by means of hydrazine sulfate; by contrast, it has not been possible to prepare the simple copper(I) sulfate in the uncomplexed state.

If metallic copper (or some other metal such as magnesium or zinc) is allowed to react with an aqueous solution of copper(II) sulfate, a considerable proportion of the copper present in solution may be converted to the unipositive state. Nevertheless, the solution still contains no appreciable amount of Cu^+ ions; instead, the +1 copper is present almost exclusively in the form of an anionic complex.

Solutions of copper(I) ammoniates are completely colorless if no dipositive copper is present.

As Bjerrum (1934) has shown, $[Cu(NH_3)]^+$ and $[Cu(NH_3)_2]^+$ ions are present in the solutions in equilibrium with one another. Both are considerably more strongly complexed than the copper(II) ammine ions (cf. p. 390). The solutions do not contain copper(I) ammine ions with more than $2NH_3$ per Cu atom.

Numerous complex salts of the above mentioned type are also known containing other nitrogen compounds in place of ammonia—e.g., aniline, pyridine, quinoline, phenylhydrazine, thiourea. As an example, the thiourea addition compound $[Cu\{CS(NH_2)_2\}_2]NO_3 \cdot H_2O$ may be mentioned; this compound is noteworthy in that the simple copper(I) nitrate is unknown.

(xi) *Copper hydride.* When an aqueous solution containing two molecules of hypophosphorous acid for each molecule of copper sulfate is warmed to 40 or 50°, it slowly deposits a red-brown precipitate, consisting of a very unstable compound of copper and hydrogen. If the compound is filtered off and dried, it evolves hydrogen when heated. If heated in air, it catches fire and the hydrogen burns. The precipitate is also extremely easily oxidized in the moist state, even in the cold. The compound of which it consists has therefore not been obtained in the pure state, but is generally given the formula CuH. This Cu : H ratio is in agreement with the experiments of Hüttig. Hüttig, however, thought that the water (which cannot be removed without some decomposition of the compound) might be in some way essential to its structure. According to Hüttig the compound has a face-centered cubic structure with the cube edge length $a = 4.33$Å, i.e., an expanded copper lattice. According to Bradley (1926), however, the copper atoms in it form a lattice of the magnesium type (hexagonal closest packing) with $a = 2.89$Å, $c = 4.61$Å. Hägg (1931) considered that in this structure the H atoms should be arranged like the S atoms in the wurtzite lattice; the radius so obtained for the H atom is the same as it possesses in the compounds TiH, ZrH, and $\overline{Ta}$H. The solid copper hydride is thermodynamically unstable. Its heat of formation (Sieverts, 1928) is —5.1 kcal per mole of CuH.

Another (likewise unstable) anhydrous solid copper hydride, probably of saltlike character, was obtained by Pietsch (1931) by the action of atomic hydrogen on superficially roughened copper.

(b) Copper(II) Compounds

(i) *Copper(II) oxide (cupric oxide)*, CuO, occurs native in the form of *melaconite*, a black earthy weathering product of copper ores. The oxide also has been found crystallized, in black triclinic tablets (*tenorite*), in the lava of Vesuvius. It is obtained artificially by heating copper, in the form of turnings or wire, to a red heat in air, or by ignition of the nitrate or carbonate. Copper oxide obtained in this way is amorphous, and possesses a marked adsorptive power for gases. The density of copper oxide is 6.3 to 6.4. Copper oxide exerts an appreciable dissociation pressure of oxygen even below 1000°. This dissociation pressure is considerably diminished through the dissolution of the resulting copper(I) oxide in the copper(II) oxide. CuO can be reduced by hydrogen to copper even below 250°, or more readily by carbon monoxide.

Cupric oxide is used in the glass and enamel industry for green and blue tints, and also for the production of copper ruby glass. For the manufacture of the latter, small amounts of copper oxide, together with a reducing agent (generally tin(II) oxide) are dissolved in the molten glass. A colorless glass is first obtained, but this takes on a fine red color after the subsequent re-heating. This 'temper color' is due to metallic copper, distributed through the glass in colloidal form (cf. also ruby-gold glass).

Copper(II) oxide acts as an oxidizing agent in the Cupron cells. It is also used as an oxidizing agent in organic analysis. It finds application in medicine, principally in the form of salves.

The crystal lattice of copper(II) oxide (tenorite) can be called a triclinically deformed rock salt structure. It would be obtained from the rock salt lattice (Fig. 44, Vol. I, p. 209.) if the edges of the cell were set obliquely to each other, ($a = 85°21'$, $\beta = 86°25'$, $\gamma = 93°35'$), and their lengths made unequal ($a = 3.74$Å, $b = c = 4.67$Å). The tenorite lattice is commensurable with the typical ionic structures, whereas the cuprite lattice is not. The heat of formation of CuO is 33.0 kcal per mol (Wöhler, 1933).

(ii) *Copper(II) hydroxide*, $Cu(OH)_2$, is deposited as a voluminous blue precipitate, when copper(II) salt solutions are treated with alkali. It is generally obtained in the form of a gel, which dries out to a powder of variable water content, and changes into black cupric oxide on warming, even in contact with the aqueous solution. It can also be obtained in crystallized form, and can then be dried at 100° without undergoing any decomposition; it has a composition corresponding to the formula $Cu(OH)_2$.

Freshly precipitated copper hydroxide is markedly soluble in caustic alkalis. Violet solutions are obtained at moderately high alkali concentrations; the copper hydroxide is present in these predominantly in the colloidal state. The solutions are deep blue at very high alkali concentrations, and contain the copper hydroxide in true solution, in the form of hydroxo-anions. Copper hydroxide is thus amphoteric; nevertheless its character as an anhydro-acid is only very weakly developed.

The salts derived from copper hydroxide, functioning as an anhydro-acid, are called *hydroxocuprates(II)* or *cuprites*. The sodium salt $Na_2[Cu(OH)_4]$, was first obtained in the crystalline state by Müller (1923). Scholder (1933) showed that it loses one molecule of H_2O above 180°, becoming black at the same time through decomposition into CuO and 2NaOH. The second H_2O molecule is not given off even at 500°, but only on melting with $K_2Cr_2O_7$, according to:

$$K_2Cr_2O_7 + 2NaOH = K_2CrO_4 + Na_2CrO_4 + H_2O.$$

It follows from this that the compound is not the dihydrate of an oxo-salt $Na_2[CuO_2]\cdot 2H_2O$, but is a hydroxo-salt. Alkaline earth hydroxocuprates have also been prepared by Scholder: $Sr[Cu(OH)_4]\cdot H_2O$ (blue violet), $Sr_2[Cu(OH)_6]$ (bright blue) and $Ba_2[Cu(OH)_6]$ (sky blue), as well as halogeno-hydroxocuprates, e.g., $Na_5[CuCl(OH)_6(H_2O)]\cdot 6H_2O$. A pyrocatechato-hydroxocuprate, $Na_3[Cu(C_6H_4O_2)_2(OH)(H_2O)]\cdot 6H_2O$ was earlier described by Weinland (1923).

Copper hydroxide dissolves in aqueous ammonia with a deep blue color: $Cu(OH)_2 + 4NH_3 = [Cu(NH_3)_4](OH)_2$. The solution (Schweizer's reagent) has the property of dissolving cellulose. The cellulose is precipitated again from the solution by addition of much water or by acids. Use is made of this in the manufacture of artificial silk.

Schweizer's reagent is most simply obtained by treating copper, in free access of air, with aqueous ammonia solution, preferably containing some ammonium chloride. The compound $[Cu(NH_3)_4](OH)_2\cdot 3H_2O$ can be isolated in long azure blue deliquescent needles by evaporating the solution in a stream of dry ammonia.

(iii) Copper(II) fluoride. Copper(II) fluoride separates from solutions of the hydroxide or carbonate in an excess of aqueous hydrofluoric acid, as light blue small crystals with the composition $CuF_2\cdot 2H_2O$. The water can be eliminated by heating in a current of HF; anhydrous CuF_2 remains behind as a white crystalline powder (m.p. 950°), which can also be obtained directly by dry methods. It crystallizes with the fluorspar structure $a = 5.41$Å. Its heat of formation from the elements is 129 kcal. [von Wartenberg (1939)]. Copper(II) fluoride is sparingly soluble in cold water (4.7 g in 100 g of H_2O). It is decomposed by hot water, with the deposition of a basic fluoride $Cu(OH)F$.

(iv) Copper(II) chloride, $CuCl_2$, forms a dark brown deliquescent mass in the anhydrous state (m.p. 498°). It crystallizes from water between 26° and 42° C as the dihydrate in sky blue rhombic needles (density 2.47). These turn dark green on exposure to traces of water. Copper(II) chloride not only has an extraordinarily high solubility in water, but is also easily soluble in many organic solvents such as alcohol, acetone, and pyridine.

Above 42°, the monohydrate is stable in contact with the solution, between 25.7° and 15.0° the trihydrate, and below 15° the tetrahydrate (Boye 1933). Solubility, in g of $CuCl_2$ per 100 g of solution: at 0°, 40.7; at 20°, 42.2; at 25°, 43.3; at 50°, 45.0; and at 100°, 52.6.

X-ray analysis of the crystal structure of $CuCl_2$ and $CuBr_2$ shows that the crystals contain extended chains,

in which (as in most copper(II) compounds of known crystal structure) the Cu^{2+} ion is surrounded by a square of coplanar linked atoms or groups. The distance $Cu\leftrightarrow Br =$

2.40 Å; each Cu^{2+} also has two more distant Br neighbors, at 3.18 Å. The dihydrated chloride also has a structure based on square complex groups,

$$\begin{array}{ccc} Cl\diagdown & & \diagup OH_2 \\ & Cu & \\ H_2O\diagup & & \diagdown Cl \end{array}$$

The same complex groups are present in the salt $2KCl \cdot CuCl_2 \cdot 2H_2O$, which appears in the solid state to consist of $[(H_2O)_2CuCl_2]$ neutral complexes, K^+ ions, and Cl^- ions, even though complex anions exist in solution. The complex salts $M^I[CuCl_3]$ have been shown to consist of infinite chains of $[CuCl_4]$ groups, linked by the sharing of Cl atoms [Wells, *J. Chem. Soc.*, (1947) 1662].

The solution of copper(II) chloride in a very small amount of water is dark brown; on dilution it first turns green, and with further dilution pale blue. On addition of concentrated hydrochloric acid, even the highly dilute solutions become greenish again. A fairly concentrated copper(II) chloride solution in concentrated hydrochloric acid is yellowish green at ordinary temperature and brown when hot. The very dilute solutions show the color of copper(II) ions (sky blue). Auto-complexes are present in the concentrated solutions, and the solutions treated with hydrochloric acid contain complex compounds with hydrochloric acid, which can be isolated in the crystalline state from the solution. Thus Engel obtained the compound $CuCl_2 \cdot HCl \cdot 3H_2O$ in dark garnet red needles; Sabatier obtained hyacinth red needles of the composition $CuCl_2 \cdot 2HCl \cdot 5H_2O$. Numerous complex chlorides are also formed by copper(II) chloride, particularly those of the types $M^I[CuCl_3]$ and $M^I_2[CuCl_4]$, whereas CuCl predominantly forms such compounds of the types $M^I[CuCl_2]$ and $M^I_2[CuCl_3]$ (Remy, 1933). The chlorocuprates(I) are colorless; the anhydrous chlorocuprates(II) are yellow to brown as a rule, and those containing water of crystallization are mostly pale blue to blue-green*.

A concentrated aqueous solution of copper(II) chloride can absorb considerable quantities of nitric oxide, at the same time turning a black-brown color. When the solution is diluted the nitric oxide is evolved again with effervescence.

Ammonia is added on still more readily. An aqueous cupric chloride solution, saturated with ammonia, yields a pentammine containing water of crystallization when it is evaporated at $0°$. The hexammine, $CuCl_2 \cdot 6NH_3$, can be prepared by passing well dried ammonia gas into a solution of copper(II) chloride in ethyl acetate.

Copper(II) chloride was formerly used as an oxygen carrier in the Deacon process. It is at present chiefly used as an oxygen carrier in the preparation of organic dyestuffs. It is employed in pyrotechnics to produce green flames. It also finds medical application. It is prepared technically by dissolving cupric oxide or copper carbonate in hydrochloric acid, or often also by the reaction of copper vitriol with barium chloride.

(v) *Copper(II) bromide*, $CuBr_2$, forms brilliant, almost black, crystals when anhydrous. It evolves half its bromine when it is heated to a red heat. It separates from solution with 2 or $4H_2O$, depending on the temperature, in brownish green crystals which deliquesce in air. The anhydrous salt is converted into the hexammine, $CuBr_2 \cdot 6NH_3$, by a current of dry ammonia below $20°$.

When a solution of hydrobromic acid or an alkali bromide is treated with copper(II) bromide it turns a purple red color. The color change indicates that complex formation takes place in this case also. Weinland was able to isolate a complex acid of the composition $CuBr_2 \cdot HBr \cdot 10H_2O$, in the form of brilliant black needles. Complex bromides of the types $M^I CuBr_3$ and $M^I_2 CuBr_4$ have also been obtained crystalline.

* The lithium salt $LiCuCl_3 \cdot 2H_2O$ is garnet red, like the HCl addition compound.

(*vi*) *Copper(II) iodide.* When a copper sulfate solution is treated with potassium iodide, copper(II) iodide is formed initially. This readily decomposes, however, into copper(I) iodide and iodine:

$$Cu^{++} + 2I^- = CuI_2; \qquad CuI_2 = CuI + \tfrac{1}{2} I_2 \qquad (1)$$

Double iodides exist, e.g., $CuI_2 \cdot 2NH_4I \cdot 2NH_3 \cdot 4H_2O$, deep green needles, insoluble in water.

(*vii*) *Copper(II) sulfate*, $CuSO_4$, is a white powder when anhydrous (density 3.64); it is turned blue by the absorption of water, and is therefore used for the detection of water in organic liquids, and also as a drying agent. When heated it is decomposed into $CuO + SO_2 + \tfrac{1}{2}O_2$.

Copper sulfate crystallizes from aqueous solutions as copper (blue) vitriol, $CuSO_4 \cdot 5H_2O$, in azure blue transparent triclinic crystals, which effloresce superficially in the air. The density is 2.286. Blue vitriol is the most important copper salt technically. It finds application for killing algal growth in water, in the preparation of mineral pigments, for the impregnation of wood against dry rot, for the preservation of bird skins, and for the treatment of cereal seeds. It is also employed to make Bordeaux mixture (a mixture of blue vitriol solution and milk of lime) which is used as a fungicide for fruit trees and grape vines. Copper vitriol is also used for making up copper plating baths, and in Daniell and Meidinger cells. It also finds application in the organic dyestuff industry and in leather dyeing. It is used in medicine as an astringent, a caustic, and emetic.

Blue vitriol is manufactured by dissolving scrap copper in hot concentrated sulfuric acid (eqn 2) or better and more economically, by treating copper with warm dilute sulfuric acid in an ample supply of air (eqn 3). It is also frequently prepared by dissolving copper oxide, which may in some cases be obtained by roasting sulfide ores, in sulfuric acid (eqn 4).

$$Cu + 2H_2SO_4 = CuSO_4 + 2H_2O + SO_2 \qquad (2)$$

$$Cu + H_2SO_4 + \tfrac{1}{2}O_2 = CuSO_4 + H_2O \qquad (3)$$

$$CuO + H_2SO_4 = CuSO_4 + H_2O \qquad (4)$$

Copper vitriol also occurs native in many places, especially in Chile. The native copper vitriol is also called *chalcanthite*. In addition, a basic copper sulfate, $CuSO_4 \cdot 3Cu(OH)_2$ occurs as *brochantite*. Copper vitriol is also prepared by dissolving these natural sulfate ores in dilute sulfuric acid.

When copper vitriol is heated, the water is eliminated in successive steps. The trihydrate is first formed, then the monohydrate, and complete dehydration takes place only above 200°.

The crystal lattice of the pentahydrate is that of a distorted octahedron, each Cu^{++} ion being surrounded by four H_2O molecules and by two O atoms belonging to different $SO_4^=$ groups. The fifth H_2O molecule is not bound coordinatively to the copper ion, but is surrounded by two H_2O molecules and by two O atoms of the $SO_4^=$ groups (Lipson, 1934).

Copper sulfate forms well crystallized double salts with the sulfates of strongly electropositive metals and radicals (sulfato salts). Most of them correspond to the type $M^I_2SO_4 \cdot CuSO_4 \cdot 6H_2O$, and are pale greenish blue in color. Rieger showed by means of transference experiments that complex sulfato anions of copper are present in the solution of the monoclinic potassium double salt $K_2SO_4 \cdot CuSO_4 \cdot 6H_2O$ (which is isomorphous with $(NH_4)_2SO_4 \cdot MgSO_4 \cdot 6H_2O$).

The solubility of copper sulfate in water is: at 0° 14.8, at 15° 19.3, at 25° 23.05, at 50° 33.5, and at 100° 73.6 g of $CuSO_4$ in 100 g of water. The saturated solution boils at 104.0°. Above about 105°, the trihydrate is stable in contact with the solution (under pressure).

The action of dry ammonia gas on anhydrous copper sulfate yields the pentammine $CuSO_4 \cdot 5NH_3$, as a blue-violet powder. It is interesting that this contains as many NH_3 molecules as copper vitriol does H_2O molecules. In moist air, it exchanges the ammonia step by step for water molecules, forming first $CuSO_4 \cdot 4NH_3 \cdot H_2O$ and then $CuSO_4 \cdot 2NH_3 \cdot 3H_2O$. The pentammine dissociates into $CuSO_4 \cdot 4NH_3$ and NH_3 on gentle warming; the dissociation pressure is one atm. at 90°.

If an aqueous solution of copper sulfate is saturated with ammonia, the compound $CuSO \cdot 4NH_3 \cdot H_2O$ crystallizes out on evaporation. The same compound may be obtained by the action of ammonia gas on solid blue vitriol. It forms dark azure blue, transparent, long rhombic prisms, and is very soluble in water (1 : 1.5). The aqueous solution is not stable when diluted, but gradually deposits basic copper sulfate.

(*viii*) *Copper nitrate*, $Cu(NO_3)_2$, is very soluble in water, and crystallizes from the solution with a water content which depends on the temperature. It can also separate out anhydrous as a white or somewhat greenish crystal meal, from solutions rich in nitric acid.

The trihydrate forms deep blue prisms, stable in air (m.p. 114.5°). The hexahydrate has the form of blue tablets. When heated, it begins to liquefy above 26°, with the elimination of water. The enneahydrate is stable below —20°.

A so-called basic nitrate crystallizes from a copper nitrate solution to which copper hydroxide is added, and has the composition of a double compound of copper nitrate and copper hydroxide, $Cu(NO_3)_2 \cdot 3Cu(OH)_2$. It occurs native as *gerhardtite*. Werner regarded it as a hex-ol salt (cf. p. 390). Its structure is probably similar, however, to those of the basic salts of cobalt and nickel discussed on pp. 298 and 314 *et seq.*

(*ix*) *Copper(II) nitrite*. A blue-green solution is first formed when copper dissolves in nitric acid. The color is considered to be that of copper(II) nitrite, although the nitrite cannot be obtained in the free state. The nitrate is formed with evolution of nitric oxide when a solution which contains only Cu^{++} and NO_2^- ions (obtained for example by double decomposition of $CuSO_4$ with $Pb(NO_2)_2$) is evaporated. The so-called basic copper nitrite $Cu(NO_2)_2 \cdot 3Cu(OH)_2$, which was regarded by Werner as a hex-ol salt, can readily be isolated. It crystallizes in fine needles, sparingly soluble in water and in alcohol. The nitrites derived from the ammines of the copper(II) ion are likewise stable, as also are various triple nitrites corresponding in composition to the general formula

$$2M^INO_2 \cdot M^{II}(NO)_2)_2 \cdot Cu(NO_2)_2.$$

(*x*) *Copper(II) acetate*. The monohydrate of cupric acetate generally crystallizes from solutions of copper oxide in acetic acid in dark blue green monoclinic prisms. It can also be obtained by recrystallization of verdigris from acetic acid, and when prepared in this manner it is also known as purified or crystallized verdigris. The rhombic pentahydrate crystallizes at lower temperatures.

Ordinary verdigris is a mixture of basic cupric acetates. It is prepared artificially by repeatedly moistening copper plates with vinegar and exposing them to the action of the air. According to the conditions green or blue verdigris is so obtained. It finds application as an oil and water pigment and for the preparation of other copper pigments; also in pharmacy, and in combating mildew. It serves for the preparation of the "gilder's wax" for fire gilding.

Cupric acetate can form readily soluble double salts with other acetates, e.g., with ammonium acetate.

(*xi*) *copper carbonate*. A simple neutral carbonate of copper has never yet been obtained. Well crystallized basic carbonates are well known, however, and are frequently found native. *Malachite*, a basic carbonate of the composition $CuCO_3 \cdot Cu(OH)_2$, is widely distributed, though seldom found in large amounts. It often occurs in emerald green monoclinic needle-shaped crystals, mostly growing in bundles, but more often in dense or fibrous green masses. The likewise monoclinic blue *azurite* has the composition $2CuCO_3 \cdot Cu(OH)_2$. It has been found in particularly beautiful crystals at Chessy, near Lyons, and is therefore also called *chessylite*. It is used as a painters' pigment under the name of azurite, and also for blue flare mixtures in pyrotechnics. In moist air it gradually changes into green malachite. This adversely affects its use as a pigment.

A basic carbonate of the composition $CuCO_3 \cdot Cu(OH)_2 \cdot 0.5H_2O$ is obtained by the action of Na_2CO_3 on $CuSO_4$ in aqueous solution. It changes into malachite when heated to 200°. According to Binder (1937) it gives the same X-ray interference pattern as malachite, and is to be regarded structurally as a malachite into the crystal lattice of which H_2O molecules are inserted.

Of the double carbonates of copper, the potassium salt $K_2CO_3 \cdot CuCO_3$ may be mentioned; according to the experimental conditions it crystallizes either anhydrous as a dark blue powder, with $1H_2O$ in bright blue silky needles, or with $4H_2O$ in large greenish-blue plates.

(*xii*) *Copper(II) oxalate* is precipitated by oxalic acid, or better by alkali oxalates, from cupric salt solutions as a bright blue precipitate of the composition $CuC_2O_4 \cdot H_2O$. It is insoluble in water, soluble in strong acids, but insoluble in an excess of oxalic acid.

Copper(II) oxalate dissolves in a sufficient excess of alkali oxalates forming soluble double salts which may be also obtained in the solid state. Most of the double salts correspond in composition to the formula $M^I_2C_2O_4 \cdot CuC_2O_4 \cdot 2H_2O$. Cupric oxalate also dissolves in aqueous ammonia. The pentammoniate, $CuC_2O_4 \cdot 5NH_3$, crystallizes when a hot saturated solution of CuC_2O_4 in ammonia is allowed to cool.

(*xiii*) *Copper(II) cyanide*, $Cu(CN)_2$, is obtained by reaction of cupric salt solutions with small amounts of CN^- ions, as a brownish yellow precipitate which rapidly changes at ordinary temperature into cuprous-cupric cyanide $2CuCN \cdot Cu(CN)_2$, and completely into cuprous cyanide on warming: $Cu(CN)_2 = CuCN + \frac{1}{2}(CN)_2$. Complex cyanides may be obtained from solutions containing excess alkali cyanide, e.g., $K_2[Cu(CN)_4]$, white hexagonal crystals, very soluble in water.

(*xiv*) *Copper(II) thiocyanate*, $Cu(SCN)_2$, can be prepared by the dissolution of copper(II) hydroxide or carbonate in concentrated aqueous thiocyanic acid, or by precipitation from a concentrated cupric salt solution by potassium thiocyanate. It is a velvety black powder, which is decomposed by water, or more quickly by solutions of alkali thiocyanates, with the formation of copper(I) thiocyanate. In the presence of water it acts as a strong oxidizing agent towards certain organic substances such as indigo. Both the solubility in water and the stability of cupric thiocyanate are enhanced in the presence of ammonia, through the formation of complex ammoniates. Depending upon the temperature and the ammonia concentration, the products may be either $[Cu(NH_3)_2][SCN]_2$, fairly sparingly soluble light blue needles, or $[Cu(NH_3)_4][SCN]_2$, readily soluble dark blue plates. A monohydrate of the latter in the form of violet needles has also been obtained. A pentammoniate is obtained from liquid ammonia.

(*xv*) *Ammoniates of Copper(II) Salts*. Dipositive copper forms a still greater range of ammoniates or ammines* than unipositive copper. The two types, $[Cu(NH_3)_2]X_2$ and $[Cu(NH_3)_4]X_2$, are most frequently encountered.

* The distinction between ammoniates and ammines corresponds to that between hydrates and aquo compounds. Ammoniates are addition compounds with ammonia, quite generally. Compounds specially designated ammines are those in which the NH_3 groups are attached to particular atoms or ions, forming strong or weak complexes, i.e., not merely included in lattice cavities.

Equilibria are established in solutions containing ammonia and ammonium chloride, between the ions $[CuNH_3]^{++}$, $[Cu(NH_3)_2]^{++}$, $[Cu(NH_3)_3]^{++}$, $[Cu(NH_3)_4]^{++}$ and $[Cu(NH_3)_5]^{++}$. As Bjerrum (1931) has shown, from determinations of equilibrium constants, the tetrammine ions always strongly predominate, except at quite small ammonia concentrations. At high ammonia concentrations, pentammine ions are also formed, but hexammine ions do not exist in the solutions. In agreement with this, Brintzinger (1935) found by the dialysis method (see Vol. I, p. 299.) that the cations in ammoniacal solutions of most copper salts had an ionic weight corresponding to the formula $[Cu(NH_3)_4]^{++}$. It is noteworthy, however, that in ammoniacal copper sulfate and selenate solutions he found ionic weights corresponding to the formulas $[Cu_2(SO_4)(NH_3)_4]^{++}$ and $[Cu_2(SeO_4)(NH_3)_4]^{++}$.

Organic nitrogen compounds such as aniline, pyridine, and quinoline, may be combined in place of ammonia. Examples of such ammines have already been mentioned, among them compounds derived from simple salts which are not themselves stable.

(xvi) *So-called Hex-ol Salts of Copper.* Dipositive copper forms a whole series of salts with the general composition $CuX_2 \cdot 3Cu(OH)_2$. Werner regarded these as hex-ol compounds, i.e., as compounds in which six hydroxyl groups, attached to the metal, were bound by coordinate covalences to the central atom as represented by the general formula (I).

$$\left[Cu\left(\begin{smallmatrix} H \\ O \\ \diagup \diagdown \\ Cu \\ \diagdown \diagup \\ O \\ H \end{smallmatrix} \right)_3 \right] X_2 \quad (I) \qquad \left[Cu\left(\begin{smallmatrix} AsO \\ O \diagup \\ \diagup \diagdown \\ Cu \\ \diagdown \diagup \\ O \diagdown \\ AsO \end{smallmatrix} \right)_3 \right] X_2 \quad (II)$$

The structure of these compounds has not yet been worked out by X-rays; nevertheless, on the basis of X-ray structure determinations of other basic salts, which Werner also regarded as hex-ol salts—e.g., the basic cobalt and nickel halides of analogous composition (cf. p. 298 and 314)—it appears that Werner's idea must be considerably modified. Examples of basic copper salts of this type are provided by the minerals atacamite, (p. 367), brochantite (p. 387), gerhardtite (p. 388), and langite ($CuSO_4 \cdot 3Cu(OH)_2 \cdot H_2O$). Further, basic salts of similar composition have been prepared artificially, in which $X = Br$, (NO_2), (ClO_3), (IO_4), or $\frac{1}{2}(S_2O_6)$. As far as their crystal form has been investigated they all display the rhombic-bipyramidal habit.

Werner regarded the pigment Schweinfurt green as related to the type of 'hexol' salts. This compound, formed by adding Scheele's green, $CuHAsO_3$, gradually to boiling acetic acid, contains 3 molecules of copper arsenite to each molecule of copper acetate. Werner formulated it as (II), analogous to the 'hexol' salts, but as for the case of apatite, which Werner also assigned to this group, this formula must be modified (cf. p. 102).

Other compounds analogous to Schweinfurt green are known, in which the ions of trichloracetic acid, propionic acid or formic acid replace the acetate ion.

(xvii) *Copper(II) sulfide*, CuS, is obtained as a black precipitate when hydrogen sulfide is passed into solutions of copper(II) salts. It is insoluble in water and in dilute strong acids. In the absence of acid it forms colloidal dispersions rather readily. When it is moist, it also undergoes partial oxidation to copper sulfate in air. When it is intended to separate the precipitate by filtration from the solution in which it was precipitated, it must therefore be washed with water containing hydrogen sulfide and a little acetic acid.

Copper(II) sulfide occurs in nature as *covellite* (copper indigo) in dull, rather soft, blue-black lumps, which furnish an indigo-blue powder (density 4.68) when they are ground up. Large, well formed crystals are but rarely found. They belong to the hexagonal system. The crystal structure of copper(II) sulfide is rather complicated.

Copper(II) sulfide has a fairly good electrical conductivity. It is converted into copper(I) sulfide* by strong heating in the absence of air, or when it is gently heated (below 600°) in hydrogen. It dissolves readily in hot, moderately dilute nitric acid, but only with great difficulty in boiling dilute sulfuric acid. Copper sulfide is practically insoluble in sodium and potassium sulfide solutions. It is somewhat soluble, however, in ammonium sulfide, and still more soluble in alkali polysulphide solutions. Polysulfides of copper—e.g., CuS_3 and Cu_2S_5—have been prepared by mixing copper(II) salt solutions with alkali polysulfide solutions, and by other methods. They are difficult to obtain pure.

The solubility of CuS in alkali polysulfide solutions increases considerably with the content of polysulfide sulfur, and also with increase in concentration and in temperature. Potassium polysulfide dissolves more CuS than the corresponding sodium salt (Höltje, 1935).

4. Analytical (Copper)

Copper compounds yield red metallic flakes, or a ductile red bead, soluble in nitric acid, when they are heated with soda on the charcoal block before the blowpipe. The solution turns deep blue on addition of ammonia. Hydrogen sulfide produces a black precipitate in the acid solution, and potassium hexacyanoferrate(II) (ferrocyanide) a red-brown precipitate of copper(II) hexacyanoferrate(II), $Cu_2[Fe(CN)_6]$. The latter reaction is very sensitive.

In the course of systematic analysis, copper comes into that part of the hydrogen sulfide precipitate which is insoluble in ammonium polysulfide, although copper(II) sulfide is slightly soluble in ammonium sulfide, and more so in the polysulfide.

In the oxidizing flame, copper colors microcosmic salt yellow when hot and greenish blue when cold. The bead becomes opaque and red brown in the reducing flame, through formation of copper(I) oxide or metallic copper.

Formation of $K_2PbCu(NO_2)_6$, which exists as brown-black tetragonal prisms, is a very convenient microanalytical test (Limit of detection 0.03 γ Cu). The drop reaction with rubeanic acid $[(S=C-NH_2)_2]$, with which copper forms an inner complex salt, is still more sensitive (limit 0.006 γ Cu). Suitably applied (Feigl, 1930), this reaction enables a minimal concentration of copper to be directly detected in alloys.

Copper may be *determined gravimetrically* by precipitating copper(II) sulfide, which is weighed as copper(I) sulfide after ignition in a current of hydrogen and hydrogen sulfide or a current of carbon dioxide charged with methyl alcohol and sulfur vapor. It may also be weighed as copper(II) oxide after ignition in air. Alternatively, copper is electrodeposited as metal and weighed as such. The latter method is the most accurate. Determination as copper(I) thiocyanate, by precipitation with ammonium thiocyanate in the presence of sulfurous acid, is also very reliable (cf. p. 382).

Copper may also be determined volumetrically—e.g., iodometrically, by treating copper sulfate solution with potassium iodide and titrating the iodine liberated (cf. p. 387).

* See footnote to p. 382.

Ephraim (1930) recommends salicylaldoxime, $C_6H_4(OH)CH=NOH$, as a reagent for the detection, and especially for the determination of copper, since it enables copper to be precipitated quantitatively from acid solutions in the presence of other metals. It furnishes with Cu^{++} ions in acetic acid solution a light greenish yellow precipitate of $Cu(C_7H_6O_2N)_2$. Very small amounts of copper can conveniently be determined colorimetrically with sodium diethyl dithiocarbamate (Ensslin, 1941). Copper can also be accurately determined polarographically (Remy and Michaelsen, 1941). For the detection and determination of copper with Reinecke's salt see p. 149.

5. Silver (Argentum, Ag)

(a) Occurrence

Silver occurs in Nature chiefly as the *sulfide*, very often in association with other sulfides—e.g., those of lead, copper, antimony, and arsenic. With the two latter, double sulfides are formed, which may be regarded as thio salts—e.g., the 'dark red silver ore' (antimonial silver blende, *pyrargyrite*) is a silver thioantimonite, Ag_3SbS_3. With this, in lesser amounts, occurs 'light red silver ore' (arsenical silver blende, *proustite*), a silver thioarsenite, Ag_3AsS_3. Both crystallize hexagonal, but the same compounds are also very rarely found as monoclinic crystals of *fire blende* and *xanthocone*. Another monoclinic mineral, *silver antimony glance* (miargyrite, hypargyrite, also occasionally known as dull silver ore) can be considered to be silver metathioantimonite, $AgSbS_2$; it is rather rare. Silver glance (*argentite*), Ag_2S, is an important silver mineral, occurring particularly in Mexico and South America. Silver sulfide is almost invariably present in small amounts (up to 1%) as an isomorphous admixture in galena. It also forms mixed crystals with copper glance, Cu_2S; varieties poor in silver are known as *stromeyerite* or silver copper glance, those richer in silver as *jalpaite*.

Discrasite (antimonial silver), Ag_2Sb, and *polybasite*, a mixed sulfide of silver, copper, and antimony, are also occasionally found with silver ores. Considerable quantities (up to 32%) of silver may also be present in argentiferous tetrahedrite. The cubic *cerargyrite* (horn silver), AgCl, is frequently found in Central and South America, but is otherwise rare, as also is *bromargyrite*, AgBr (also cubic) and the hexagonal *iodargyrite*, AgI. Silver also often occurs native, sometimes in well formed cubes and octahedra.

The most important silver producing countries are at present Mexico, the United States, and Canada, each of which produces more silver than the entire output of Europe. In 1937, out of a world production of 8734 tons of silver, Mexico contributed 30.2%, the United States 25.4%, and Canada 8.2%, while Europe produced 7.7%.

(b) Preparation

In addition to the silver ores proper, the argentiferous ores of lead, zinc, and copper provide the raw materials for the extraction of silver to a large extent, as also frequently do the residues from the roasting of pyrites.

Large quantities of silver are obtained as a by-product of the smelting of argentiferous galena. Two processes are in use for the recovery of the—generally very small—amount of silver in the lead so obtained: the *Pattinson process* and the *Parkes process* (zinc desilverization).

In the Pattinson process (invented 1833), the molten silver-lead is allowed to cool slowly, and the pure lead which then crystallizes out is ladled out of the melt. The silver-lead system has a eutectic at 2.6% silver, melting at 303°, whereas pure lead melts at 326°. Hence, when an alloy poor in silver is cooled, lead must first separate—in theory, until the silver content of the melt has risen to 2.6%. In practice, it is not difficult to obtain a lead with more than 2% of silver. This is then melted in a cupelling furnace. The lead is thereby oxidized to litharge, which runs off the surface, and the silver remains behind.

In the process devised by Karsten and Parkes, the lead of low silver content is heated to the melting point of zinc, and zinc is stirred in. The affinity of zinc for silver is greater than that of lead, so that the zinc extracts the silver (with some lead also) and separates as an emulsion on top of the melt. If the temperature is slightly lowered, this solidifies, and can readily be removed from the melt. The zinc is first distilled from the zinc froth, and the silver is then separated from the remaining lead by cupellation.

If the lead is initially relatively rich in silver (e.g., lead with more than 0.1% silver), it can be cupelled directly.

Cupellation is often effected in two stages, the first being carried only to the stage of a silver-lead alloy with 50–80% silver (so called 'Bleileder'). Cupellation is then completed in a smaller furnace. Often, however, cupellation is carried out so as to give 'Blicksilber' with 90% silver, or even further, in a single process.

The argentiferous copper ores are also of great importance for the extraction of silver. After these have been smelted for matte or black copper, the copper can be leached out by treatment with moderately concentrated hot sulfuric acid, leaving the silver as a residue. However, as has already been mentioned, argentiferous black copper is usually refined electrolytically, whereby the silver may be recovered from the anode slimes.

Since 1841, argentiferous copper ores have been treated at the Mansfeld smelters by Ziervogel's method. The silver sulfide present in them is converted to silver sulfate by careful roasting at 700–800°, and this is then leached out with water.

The processes described are often used also for smelting silver ores which are free from lead or copper. In such cases, galena or copper ore is added before smelting. This is done especially with ores rich in silver. Ores poor in silver are treated by processes which enable the silver to be extracted at ordinary temperature, if for any reason it is not possible to smelt them with copper or lead ores. The *amalgamation* process, and various processes based on the dissolution of silver chloride by thiosulfate solutions* were formerly used. The *Augustin process*, which attempted to utilize the solubility of silver chloride in concentrated sodium chloride solution, may also be mentioned. All these methods have now been superseded, however, by cyanide leaching.

Prior to 1914, the amalgamation process was in use especially in Mexico and South America. It was based on the ease with which silver amalgamates with mercury. Even when present as the chloride, silver is taken up by the mercury, the chloride being decomposed. Other silver ores must first be converted to silver chloride by a chloridizing roast.

At the present time, ores with a low silver content, which are not suitable for smelting for copper or lead, are almost invariably treated by *cyanide leaching*. This process can be applied directly to silver sulfide, as well as to silver chloride. It is based on the fact that silver compounds are dissolved by alkali cyanides, with the formation of complex cyanides.

Where silver is present as the sulfide, the following reaction occurs:

$$Ag_2S + 4CN^- \rightleftharpoons 2[Ag(CN)_2]^- + S^=.$$

Since an equilibrium is set up, the $S^=$ ions must be removed in order that the reaction shall proceed as far as possible to the right. This is effected by blowing air in, whereby the $S^=$ ions are converted into $S_2O_3^=$, $SO_4^=$ and SCN^- ions. The silver can be thrown out of the cyanide solution as metal by the addition of zinc dust.

* Silver sulfide can readily be converted to the chloride by a chloridizing roast.

(c) Purification of Silver

As extracted by the methods described, silver almost always contains a little gold, and generally copper also. It is purified either by affination or by electrolytic refining.

The term affination is applied to the purification of silver by treatment with hot concentrated sulfuric acid. Silver thereby goes into solution as silver sulfate, and gold remains as a metallic residue. Copper is also converted to its sulfate, which is, however, only very slightly soluble in concentrated sulfuric acid. The silver may be precipitated from the solution as the metal by adding sheets of copper or iron.

Electrolytic refining is more widely employed. [10] In this process, plates of the silver to be refined, which should have only a low content of copper and lead, are suspended as anodes in very dilute nitric acid or silver nitrate solution. Lead and copper also go into solution during the electrolysis, but as long as the concentration of the less noble metals is not very high compared with that of silver, only silver is deposited at the cathode. Gold falls to the bottom at the anode, undissolved. The 'electrolytic silver' obtained in this way has a purity of 99.95 or better.

For the electrodeposition of silver from argentiferous gold, see p. 410.

(d) Properties

Silver is a noble metal, which exhibits a beautiful white luster, capable of taking a high polish. In hardness it is intermediate between copper and gold. Next to gold, it is the most ductile of all metals, and it has the highest electrical and thermal conductivities of all the metals (Specific electrical conductivity of silver, in reciprocal ohms: $\varkappa = 6.73 \cdot 10^5$ at $0°$, $6.14 \cdot 10^5$ at $18°$, $6.07 \cdot 10^5$ at $20°$. Thermal conductivity: $\lambda = 1.006$ cal. cm^{-1} sec^{-1} degrees^{-1} at $18°$). Silver crystallizes in the cubic system (for structure, see p. 362). The crystal structure is retained even in the small particles of colloidal silver, as Scherrer showed by X-ray diffraction method.

Silver forms an unbroken series of mixed crystals with gold and also with palladium, which immediately precedes it in the Periodic Table. Its miscibility with copper in the solid state is only very limited, however (cf. p. 11). This is connected with the fact that gold and palladium not only have the same lattice type (face-centered cubic) as silver, but also differ very little in cell dimensions ($a =$ for Ag 4.077, Au 4.070, Pd 3.88 Å), whereas although copper has the same crystal structure, its cell constant is considerably smaller ($a = 3.61$ Å).

In the liquid state, silver is completely miscible with copper and with many other metals. Its miscibility with chromium, manganese and nickel is very limited, in both the liquid and the solid states*, and it is immiscible with iron and cobalt. It also forms no compounds with these elements, and so far as is known, combines with none of the metals of Sub-groups IV to VIII, except platinum. It also forms no compounds with its homologues copper and gold, but forms several compounds, all of variable composition, with the metals of the IInd and IIIrd Sub-groups. It forms many compounds with the metals of the Main Groups, but none with Na, Tl, Ge, Pb, and Bi, or with the non-metals B and Si (cf. Tables 39, 40, pp. 364–5 and 366).

Silver forms amalgams with mercury. These are most readily obtained by shaking metallic mercury with silver nitrate solution. Mercury thereby passes into solution—$Hg + Ag^+ = Ag + \frac{1}{2}Hg_2^{++}$—and the displaced silver forms crystalline compounds with the remaining metallic mercury. These form long lustrous needles ('arbor Dianae'), and are stated to have the composition Ag_3Hg_4, Ag_3Hg_2, and Ag_3Hg. Compounds of different composition have been obtained by crystallization from molten alloys (cf. Table 40, p. 366).

* Silver is completely immiscible with chromium in the solid state.

As accords with its character as a noble metal, compact silver does not unite directly with oxygen*; it can, however, in the molten state *dissolve* considerable quantities of oxygen. The greater part of this is given off again upon solidification. The outer surface of the solidifying melt is thereby torn apart by the gases evolved from within, often with such violence that the silver spurts. This phenomenon has long been known as the 'spitting' of silver. Silver oxide, Ag_2O, which can be obtained indirectly, is decomposed by even gentle heating. Ozone, however, reacts directly with silver, especially on gentle warming (240°). It blackens the silver, forming silver(II) oxide or silver peroxide. The reaction can be used for the detection of ozone, and if a freshly polished silver sheet is used (traces of iron oxide remaining on the surface have a catalytic action), the test is very sensitive.

White phosphorus also dissolves freely in molten silver, and a portion of the phosphorus is given off again as vapor, and spurting occurs, when the melt solidifies. However, solid silver can also contain quite appreciable amounts of phosphorus. Of the compounds listed in Table 39, p. 364–5 only AgP_2 and AgP_3 are stable. The compounds Ag_3P and Ag_2P_5 can be obtained from solution, by double decomposition reactions, but not by the direct union of their componenets.

Silver has a great affinity for sulfur. Moist hydrogen sulfide blackens silver, forming Ag_2S. The tarnishing of silver objects when they are exposed to air is due to the superficial formation of this compound. The free halogens also slowly combine with silver to some extent, even at ordinary temperature.

Pietsch (1931) found that silver reacted with atomic hydrogen at ordinary temperature, forming a solid, colorless, salt-like hydride, AgH. It can also react with ordinary hydrogen at high temperatures, forming traces of a gaseous hydride (cf. p. 363).

Silver is not attacked by aqueous solutions of non-oxidizing acids in the absence of air, as accords with its position in the electrochemical series. When it is heated to red heat in hydrogen chloride gas, however, it reacts, with the liberation of hydrogen:

$$2Ag + 2HCl \rightleftharpoons 2AgCl + H_2 + 17.2 \text{ kcal.} \tag{1}$$

As indicated by this equation, the reaction leads to an equilibrium. According to Jouniaux (1898), if one starts with one atmosphere pressure of either hydrogen chloride or hydrogen at ordinary temperature, and heats in a sealed tube, 92.8 volume–% of hydrogen chloride gas and 7.2 volume–% of hydrogen are in equilibrium with Ag + AgCl as solid phase at 600°**. At 700°, the corresponding figures are 95% and 5.0% by volume, respectively. Thus the equilibrium (1) is displaced to the left with rising temperature, as required by the Le Chatelier principle. It follows further from the Le Chatelier principle that the equilibrium will also be displaced to the left by a reduction of the pressure. This was confirmed by Jouniaux.

In contact with water containing dissolved air, silver goes into solution according to (2), even though to only a minute extent:

$$2Ag + H_2O + O_2 = 2Ag^+ + 2OH^-. \tag{2}$$

The increase in pH, of the solution, due to reaction (2), has been detected.

* Presumably a very thin oxide film is formed on silver at ordinary temperature, and protects the metal from further attack. The oxide film cannot grow to any appreciable thickness on compact silver, since at ordinary or slightly elevated temperature the rate of diffusion is too small (cf. p. 754). At higher temperatures, however, silver oxide is no longer stable. Finely divided silver can absorb appreciable quantities of oxygen at 170–180°, with the formation of oxide.

** Equilibrium is established only very slowly below 600°.

Oxidizing acids (e.g., nitric acid, concentrated sulfuric acid) readily dissolve silver. This is in contrast with the behavior of gold, which can therefore be separated from silver by treatment with these acids (cf. pp. 394, 410). If air has access, or if hydrogen peroxide is present, silver also readily dissolves in alkali cyanide solutions; unlike copper, however, it does not dissolve with the liberation of hydrogen. Silver is only slightly attacked by fused caustic alkalis; the attack is greater if potassium nitrate is added to the melt.

(e) Uses

Metallic silver is used especially for jewellery and utensils, [*8,9*] but for other purposes also—e.g., for chemical apparatus (crucibles for alkali fusions), for medical instruments, and in the electrical industry (fuses). [*11*]

The unpleasant smell and taste often acquired by silver utensils after long use are due, according to Raub (1934) to the superficial formation of compounds of silver with organic sulfur compounds. Copper and copper alloys behave like silver in this respect, but other metals do not. The smell and taste can readily be removed by boiling with very dilute hydrochloric acid.

The value of currency was formerly based upon the price of silver. Even today, very large quantities of silver—possibly as much as half the annual production—are used in minting coins. For coinage, and for most other purposes, silver is alloyed with other metals (generally copper), since pure silver ('fine silver') is too soft. English Sterling Silver is 92.5% silver. Silver jewellery and utensils usually contain 80% silver. The eutectic in the silver-copper alloys contains 28% copper, and melts at 778°; it thus melts almost 200° lower than pure silver. The admixture of copper with silver first becomes clearly evident from the color when the copper content exceeds 50%. Silver solder usually contains 10–15% of zinc in addition to 20–35% of copper. Objects made of base metals may be silvered either by applying a coating (rolled silver), or in a furnace (Sheffield plate), or galvanically.

Fine silver was formerly generally used for chemical equipment (e.g., crucibles for alkali fusions). Silver alloyed with a small quantity (0.1–0.2%) of nickel is more suited for this purpose, however (Fröhlich, 1939). The nickel hinders the recrystallization of the silver, which leads to brittleness and cracking of the metal after long heating, and does not significantly increase its susceptibility to chemical attack.

Silver salts and other silver preparations have wide application. The most important salt is silver nitrate, which is the usual starting material for the preparation of other silver compounds. Very large quantities are consumed by the photographic industry. Colloidal silver is frequently used therapeutically, since it has excellent bactericidal properties while being relatively non-toxic towards the bodily organs. One of the oldest preparations of this kind is *Collargol* (argentum colloidale), green or blue-black metallic lustrous leaflets, which disperse in a little water to a turbid, but in much water to a clear, colloidal dispersion. *Protargol*, (argentum proteinicum), a yellow or brown powder, is also widely used. These colloidal silver preparations contain varying amounts of organic substances—mostly albumen or its degradation products—as protective colloids.

The bactericidal properties of silver ions (which do not, however, extend to spores also) may be observed at extraordinarily great dilutions—down to about $2 \cdot 10^{-11}$ g-ions per liter (the 'oligodynamic action of silver', discovered by von Naegeli, 1893). Metallic silver displays this action only when it is contaminated at some spot, so that local corrosion currents are formed whereby traces of silver pass into solution. In preparations of colloidal silver, this condition is invariably fulfilled, because of the impurities present.

6. Compounds of Silver

The compounds of silver are almost all derived from the *unipositive state*, so that the familiar compounds correspond to the general formula AgX. In a few compounds, silver functions as *dipositive* (cf. p. 407 *et seq.*), and even *tripositive* silver is known in the form of some derivatives.

(a) Silver(I) Compounds

Many compounds of silver are practically insoluble in water and, in many cases, in acids also (cf. pp. 363, 399). The salts of silver which are *readily soluble* in water include the fluoride, nitrate, chlorate, and perchlorate.*

The acetate and sulfate also dissolve fairly well, especially when heated, as also does the nitrite.

The Ag^+ ions are colorless, as also are the salts mentioned above, with the exception of the pale yellowish nitrite. The silver ion not infrequently forms colored salts, however, with colorless anions; these are usually yellow—e.g., Ag_3PO_4, Ag_3AsO_3, Ag_2CO_3, AgI. $AgBr$ is pale greenish yellow, Ag_3AsO_4 is chocolate brown, Ag_2S black. Combination of the silver ion with an anion which is already colored often leads to an intensification of the color— e.g., Ag_2CrO_4 is red brown (K_2CrO_4 yellow), $Ag_2Cr_2O_7$ deep red ($K_2Cr_2O_7$ orange). The color of the silver periodates varies from yellow ($Ag_2H_3IO_6$) to black (Ag_5IO_6).

The brown black oxide of silver, Ag_2O, is extremely sparingly soluble in water. It imparts a perceptibly alkaline reaction to water, however, since the dissolved portion is hydrated to $AgOH$, which is almost completely ionized to Ag^+ and OH^- ions.

Many silver salts—e.g., the nitrate and sulfate—can form mixed crystals with the corresponding sodium salts.

Most of the silver salts which are insoluble in water dissolve immediately on the addition of certain substances, such as ammonia, sodium thiosulfate, potassium cyanide, etc. This is because silver has a strong tendency to form *complex ions*.

The following are the most important silver complex ions detected in solution by equilibrium measurements:

$$[Ag(NH_3)_2]^+ \quad [Ag(CN)_2]^{1-} \quad [Ag(SCN)_2]^{1-} \quad [Ag(S_2O_3)_2]^{3-} \quad [Ag_2I_4]^{2-}$$
$$[Ag(CN)_3]^{2-} \quad [Ag(SCN)_4]^{3-} \quad [Ag(S_2O_3)_3]^{5-} \quad [AgI_4]^{3-**}$$
$$[Ag(NO_2)_2]^{1-} \quad \text{(less strongly complexed)}$$

The species of complex ion which predominate in a solution cannot be directly inferred from the composition of the salts which crystallize from it; the composition of the solid phase is determined also by the other ion which builds up the crystal lattice, jointly with the complex ion, during crystallization.

Hein (1935) obtained two optically active, isomeric forms of a complex compound of silver with hydroxyquinoline, in which silver had the coordination number 4. It may be inferred from this that the ligands are arranged in tetrahedral configuration in the com-

---

* Silver perchlorate is not only readily soluble, but is actually hygroscopic. On the other hand, $[Ag(pyr)_2]ClO_4$ (obtained by treating a silver nitrate solution containing pyridine with $NaClO_4$) is sparingly soluble.

** Various transition forms, such as $[Ag_2I_5]^{3-}$, $[Ag_2I_6]^{4-}$, probably exist between $[Ag_2I_4]^{2-}$ and $[AgI_4]^{3-}$.

plexes of 4-coordinate univalent silver. The silver(I) compounds thus differ from those of platinum, palladium, and other metals of Sub-group VIII (see p. 349), but resemble the complexes of copper(I).

(b) Silver(I) Oxide

Silver oxide, Ag_2O, is formed as a brown-black precipitate when OH^- ions are added to solutions of silver salts. It may be assumed that silver hydroxide, AgOH, is first formed, and that this at once passes into silver oxide by loss of water:

$$Ag^+ + OH^- = AgOH; \quad 2AgOH = Ag_2O + H_2O.$$

The precipitate obstinately occludes some of the precipitant (alkali or alkaline earth hydroxide). If the precipitant is not completely halogen-free, silver halide is precipitated with the hydroxide.

A suspension of silver oxide in water has a distinctly alkaline reaction, indicating that the reactions shown by the equations above can be reversed, even if only to a slight extent. If the silver ions present in the solution of silver oxide are removed by reaction with halide ions, more silver oxide must pass into solution in order to maintain equilibrium. The solution may become strongly enriched in hydroxyl ions by this means. This property of exchanging halide ions (or other ions forming insoluble silver compounds) for hydroxyl ions is often utilized in preparative chemistry. Silver oxide prepared for this purpose is customarily stored under water, since the preparation is not readily wetted again properly by water if it is once dried. Silver oxide also finds medicinal applications; it is frequently used for this purpose in colloidal form, as prepared by precipitation in the presence of protective colloids (albumen, etc.) ('Syrgol').

In water with a CO_2 concentration in equilibrium with the air, the solubility of silver oxide is $1.1 \cdot 10^{-4}$ mol per liter at $18°$ (Remy). In absolutely CO_2-free water, it is less ($0.57 \cdot 10^{-4}$ mol. per liter at $18°$, Laue, 1927). The solubility is diminished by addition of small amounts of OH^- ions, but increased by higher concentrations. According to Laue, the increase in solubility is due to the formation of $[AgO]^-$ ions in the strongly alkaline solutions. This implies that silver hydroxide has not only basic properties but also, in very slight degree, acidic character. On the assumption that the dissolved silver oxide is present almost exclusively as dissociated silver hydroxide, AgOH, it follows from Laue's measurements that the proportion undergoing acidic dissociation (i.e., forming $[AgO]^-$ ions) is about 10^{-4} of that dissociating as a base. In a solution containing 0.01-M alkali hydroxide, the extents of acidic and basic dissociation are about equal.

The heat of formation of silver oxide from its elements is 7.02 kcal, the free energy of formation $\Delta F°$ is -2.45 kcal at $25°$ (Benton, 1932). In accordance with its small free energy of formation, silver oxide begins to lose oxygen when only gently heated (just detectably at $160°$; the oxygen pressure reaches 1 atmosphere at 185 to $190°$). Decomposition takes place even at ordinary temperature on exposure to light. Silver oxide is reduced by hydrogen at a temperature as low as $100°$. It reacts vigorously with hydrogen peroxide at ordinary temperature:

$$Ag_2O + H_2O_2 \; = \; 2Ag + H_2O + O_2.$$

Silver oxide is readily soluble in ammonia solution, forming complex compounds. As was first noticed by Berthollet, such solutions on long standing deposit dark precipitates, which are *extraordinarily explosive*, even in the moist state—e.g., 'fulminating silver', Ag_3N (not to be confused with silver fulminate, $AgONC$, and silver azide, AgN_3, which are also explosive).

If a compact crust of fulminating silver has formed upon the surface of an ammoniacal silver oxide solution, it is impossible to pour the liquid out without an explosion occurring.

The crystal structure of Ag_2O resembles that of Cu_2O (body-centered cubic): $a = 4.74$Å, $Ag \leftrightarrow O = 2.05$ Å.

No oxide of silver lower than Ag_2O exists. The products formerly believed to be such lower oxides are mixtures of Ag_2O and Ag, as has been proved by X-ray diffraction (Levi, 1924).

(b) Silver(I) Halides

Of the halides of silver, the fluoride is very soluble, and the others sparingly soluble. The iodide is the least soluble. The most important properties of the silver halides are summarized in Table 41.

TABLE 41

HALIDES OF SILVER

	AgF	AgCl	AgBr	AgI
Formula weight	126.88	143.34	187.80	234.81
Density	5.85	5.56	6.47	5.67
Color	white	white	pale yellow	yellow
Melting point	435°	450°	419°	552°
Solubility, mols per liter at 25°	very sol.	$1.31 \cdot 10^{-5}$	$0.725 \cdot 10^{-6}$	$1.00 \cdot 10^{-8}$
Crystal system and structure type	cubic NaCl $a = 4.92$ Å	cubic NaCl $a = 5.54$ Å	cubic NaCl $a = 5.76$ Å	α-AgI cubic zinc blende $a = 6.48$ Å β-AgI hexag. wurtzite, $a = 4.59$, $c = 7.52$ Å γ-AgI cubic, random struct. see p. 402
Heat of formation, kcal per mol	46.7	30.13	23.70	15.34

(i) *Silver fluoride*, AgF, obtained by dissolving silver oxide in hydrofluoric acid, is very soluble (1 part in 0.55 parts of water at 15°), and crystallizes as the hydrate (with 1 or $2H_2O$, according to conditions) from solution. The anhydrous salt forms a flaky crystalline mass. One volume of the anhydrous compound can absorb 844 volumes of ammonia. Double compounds $AgF \cdot HF$ and $AgF \cdot 3HF$ have been obtained from solutions containing excess hydrofluoric acid. It is not known whether these are fluoro acids, or addition compounds analogous to the hydrates ('fluorohydrates').

(ii) *Disilver fluoride*. A concentrated solution of AgF is able to dissolve silver, as was first observed by Guntz (1890). Disilver fluoride, Ag_2F, is thereby formed, and separates from the solution in bronze colored hexagonal crystals, with a greenish luster (density 8.6). It is of interest because it deviates in composition from the type of the normal valence compounds. Its structure is anti-isotypic with CdI_2 (brucite structure, Fig. 61, p. 261, Vol. I), and it thus has a layer lattice structure, unlike AgF, but its constitution is otherwise as yet unknown.

(*iii*) *Silver chloride*, AgCl, is thrown out of solution as a white, curdy precipitate when Ag^+ and Cl^- ions are mixed. It occurs native as *horn silver* (cerargyrite). This silver ore, which is found in quantity only in America, forms soft transparent masses of various colors (as the result of impurities), but is ordinarily pearl grey and is made up of usually very minute cubic crystals, with density 5.5–5.6. Silver chloride is practically insoluble in water. It dissolves in ammonia, forming complexes, and also in concentrated hydrochloric acid. On evaporation of these solutions, it is left as a residue of very small cubes, easily recognizable as such under the microscope. Precipitated silver chloride melts at 450° to an orange yellow liquid, which solidifies on cooling to a horny mass, with considerable increase in volume. Silver chloride begins to vaporize appreciably above 1000°, without any decomposition. It is polymerized to some extent in the vapor state. The boiling point is 1554° (von Wartenberg).

Dry silver chloride absorbs ammonia to give the compounds $AgCl \cdot NH_3$, $2AgCl \cdot 3NH_3$, and $AgCl \cdot 3NH_3$. These can also be obtained from aqueous solutions (the last mentioned, however, only when crystallization is carried out under high pressure). Equilibrium measurements have shown that aqueous ammoniacal solutions of silver chloride contain the complex ion $[Ag(NH_3)_2]^+$. Silver chloride also forms very soluble complex salts with sodium thiosulfate and potassium cyanide. It is also appreciably soluble in concentrated solutions of other chlorides, also as the result of complex formation. Its solubility in concentrated hydrochloric acid arises from the formation of complex acids, such as $H[AgCl_2]$.

Silver chloride is also quite soluble in hot concentrated nitric acid, as also are silver bromide, silver iodide, silver cyanide, and silver thiocyanate. Double salts crystallize from the solutions—$AgNO_3 \cdot AgCl$, colorless prisms, m.p. 160°; $AgNO_3 \cdot AgBr$, m.p. 182°; $AgNO_3 \cdot AgI$, m.p. 94°; $2AgNO_3 \cdot AgI$, m.p. 105°; $2AgNO_3 \cdot AgCN$; $2AgNO_3 \cdot AgSCN$.

The *sensitivity of silver chloride towards light* is of very great practical importance. The compound turns violet and ultimately blue-green on exposure to light, chlorine being lost and finely divided metallic silver being deposited. The use of silver chloride in photography—chiefly in photographic papers and for lantern slides—is based on this photosensitivity.

If silver chloride is covered with water or very dilute sulfuric acid, and zinc is added, reduction takes place according to:

$$2AgCl + Zn = 2Ag + Zn^{++} + 2Cl^-.$$

Wislicenus observed that this decomposition enables silver to be obtained in a state of extremely fine subdivision, so called 'molecular silver'. This is conveniently prepared by a method described by Gomberg. Well washed silver chloride is covered with water, in a glass battery jar, and a clay pot of fine porosity, containing zinc rod, is dipped in it. A platinum foil is immersed in the silver chloride, and placed in metallic contact with the zinc by means of a platinum wire welded to it. To accelerate the beginning of the reaction, a few drops of hydrochloric acid are added to the water in the clay pot, and then the level of the liquid in the clay pot is maintained always at a rather lower level than that in the outer jar, by means of a siphon, so that diffusion of zinc chloride out of the clay pot is minimized. The grey, powdery silver thus obtained is washed, first with water and then (to remove residual AgCl) with ammonia, again with water, and finally with alcohol and ether. After drying over sulfuric acid in a vacuum desiccator it is heated to 150°, and finally sieved to give a uniform product. 'Molecular silver' is used, for example, in organic combustion analysis, by Dennstedt's method, to combine with halogens.

(*iv*) *Silver bromide*, AgBr, is very similar to silver chloride, but is still less soluble, is more easily reduced, and (of great technical importance) is more photosensitive. It is obtained from solutions as a white to pale greenish yellow precipitate, according to conditions of preparation. The white form has the smallest particle size, and is therefore relatively the most soluble. It soon passes into the greenish yellow, less soluble form on standing in contact with potassium bromide solution. Silver bromide also occurs native, especially in Mexico, in the form of soft aggregates of regular crystals, of olive green to yellow color (*bromargyrite* or *bromite*), but almost always admixed with silver chloride. Silver bromide is used in large quantities for the production of photographic plates, films, and papers. It is prepared technically by the reaction of potassium bromide with silver nitrate in aqueous solution.

In the photographic industry, silver bromide-gelatin emulsions are prepared by mixing gelatin solutions, containing potassium bromide, with silver nitrate. Silver bromide is initially formed in a state of extremely fine subdivision. It is then made more photosensitive by a process of 'ripening', i.e., by prolonged standing in a warm place, so that grain growth occurs.

(*v*) *Silver iodide*, AgI, is even more insoluble than silver bromide. Its solubility product is so small that it is not dissolved to an appreciable extent even by concentrated ammonia.

The diamminesilver cations are dissociated to a slight extent, even in concentrated ammonia solution, according to

$$[Ag(NH_3)_2]^+ \rightleftharpoons Ag^+ + 2NH_3 \qquad (1)$$

In a concentrated ammonia solution, the silver ion concentration due to this equilibrium is considerably less than the silver ion concentration in solutions of silver chloride or silver bromide. Hence, in presence of these salts, reaction (1) proceeds from right to left. In an aqueous solution of silver iodide, however, so few Ag^+ ions are present that silver ammine complexes can be formed only to a negligible extent. However, if silver iodide subjected to reagents which complex the silver ion more strongly than ammonia does—e.g., the CN^- ion—the solution is so impoverished in silver ions that the ionic product falls below even the small solubility product of silver iodide ($[Ag^+][I^-] = 10^{-16}$), and even this particularly insoluble salt passes into solution.

Silver iodide is appreciably soluble in concentrated hydriodic acid and in concentrated alkali iodide solution, especially if warm. Complex salts, such as $K[AgI_2]$ and $K_2[AgI_3]$, have been isolated from such solutions, in the form of colorless crystalline compounds.

Silver iodide precipitated from solution is yellow. It gradually turns greenish-grey-black on exposure to light. On treatment with concentrated ammonia, silver iodide turns pure white, as a result of the formation of the compound $AgI \cdot \frac{1}{2}NH_3$.

Dry silver iodide absorbs 0.5, 1, 1.5 2, or 3 mole of NH_3 per formula weight, depending on the pressure, whereas silver bromide forms the same ammoniates as silver chloride. The stability of the ammoniates decreases from the chloride to the iodide.

Silver iodide exists in three modifications (cf. Table 41, p. 399). Precipitation of silver iodide from aqueous solution, in presence of excess Ag^+ ions, yields the α-modification, crystallizing with the zinc blende structure and stable at ordinary temperature.* This changes at 137° into the β-modification, crystallizing in the wurtzite type structure. This

* The modification stable above 146° is often referred to as α-AgI, and that stable at room temperature as γ-AgI.

modification is obtained in the metastable state at ordinary temperature when precipitation is effected in the presence of excess of I^- ions; conversion to the α-modification
takes place when the precipitate is rubbed. The γ-modification is stable above 146°. It
forms a cubic crystal lattice, in which the I atoms occupy the corners and mid-point of the
unit cell (cube of edge 5.03 Å), while the Ag atoms are distributed at random between 42
lattice 'holes'. The distance Ag$\leftrightarrow$I is 2.80 Å in α-AgI and 2.78 Å in β-AgI (at room temperature). In γ-AgI the distances vary between 2.52 and 2.86 Å (at 146°).

The modification stable above 146° is particularly interesting in that, as first shown by
Bruni (1913), its conductivity is purely electrolytic, and quite large. Tubandt (1920)
proved that the current is transported exclusively by the Ag^+ ions, the I^- ions playing no
part. The specific conductivity of γ-AgI at the transition point is 1.31 ohm^{-1}, whereas
that of β-AgI at the same temperature is only 0.00034. For comparison, it may be noted that
the specific conductivity of sulfuric acid of maximum conductivity is 0.739 ohm^{-1} at 18°.
At the melting point (552°), the specific conductivity of γ-AgI is 2.64 ohm^{-1}, and is thus
actually greater than that of the melt at the same temperature, which is only 2.36 ohm^{-1}.
On the other hand, with silver chloride and silver bromide, which are also pure electrolytic
conductors*, the specific conductivity increases on melting from 0.12 to 3.76, and from 0.53
to 2.76, respectively. The particularly high electrolytic conductivity of γ-AgI can be
understood in terms of the crystal structure found by Strock. The fact that the Ag^+ ions
do not occupy definite places in the crystal lattice must considerably facilitate their migration. However, the occurrence of electrolytic conduction does not necessarily presuppose
that one species of ion should be distributed at random in the crystal lattice. The same property is also displayed, although to a lesser degree, by compounds, e.g., AgBr and AgCl, in
which all the ions are bound in definite lattice positions. The possibility of ionic migration
is then bound up with the lattice imperfections and disorder present in the crystal lattice
(cf. p. 18 et seq.).

Silver iodide is used in photography for the preparation of collodion emulsion plates.
These are exposed while still moist, immediately after preparation, are developed with
iron(II) sulfate in the presence of silver nitrate, and are fixed with potassium cyanide
solution. Silver iodide-collodion plates furnish particularly sharp images, but are little
used because of the inconvenient technique involved. The oldest form of photography employing development, the *Daguerreotype*, depends on the photosensitivity of silver iodide.
Daguerre's procedure was first to subject a silver plate to the action of iodine vapor (thereby
producing a thin film of silver iodide), and then, after illumination, to expose it to mercury
vapor, which condensed preferentially on the exposed portions of such a plate.

(vi) Use of Silver Halides in Photography. [15,16]

The photographic plates and films in customary use today consist of glass plates, or thin
strips of celluloid or cellite,** which are coated with a layer of colloidal silver halide. This
is usually a colloidal dispersion (emulsion) of silver bromide in gelatin, or less often in
collodion. For special purposes (reproduction of drawings, and for transparencies and slides), silver chloride-gelatin plates, or plates with silver chloride and silver bromide, are
used. These are less sensitive to light than the bromide plates, but are more 'contrasty'
('harder'). Under the action of light, decomposition of the silver halide takes place in
traces, where it is illuminated, forming extremely minute particles or nuclei of silver, which
can be detected only in the ultramicroscope. For weaker or shorter exposures, the number
of such nuclei is roughly proportional to the quantity of light falling on the emulsion. The
emulsion bearing such a 'latent image' has undergone no change detectable to the eye,
but if treated with suitable reducing agents ('developers'),*** more of the silver bromide is

---

* Ag^+ ions are the sole current carriers in these salts also. An example of a salt in which
only the anions transport current in the solid state is provided by lead chloride.

** Cellite, (secondary) acetylcellulose, has the advantage of being far less inflammable
than celluloid, which is prepared from nitrocellulose and camphor. As against this, it has
the disadvantage of undergoing dimensional changes on wetting and drying.

*** Various organic substances, such as hydroquinone HO—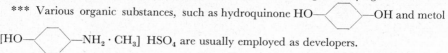—OH and metol

[HO—⟨ ⟩—$NH_2 \cdot CH_3$] HSO_4 are usually employed as developers.

reduced to metallic silver. In this process, the very fine particles of silver already produced by the light act as crystallization centers ('nuclei') for the metallic silver, and facilitate deposition to such a great extent that reduction occurs almost exclusively in the immediate neighborhood of these particles—i.e., at the exposed places. The emulsion is therefore blackened at these points. When the blackening has reached a sufficient depth, the silver halide still remaining unaltered is dissolved out by means of a 'fixing bath', since it would also ultimately be blackened if not removed. The usual fixing salt is sodium thiosulfate, $Na_2S_2O_3$, but other compounds which will convert silver halides to soluble complexes (e.g., potassium cyanide) may also be used. The picture thus first obtained is a *negative*. In it, the areas with the deepest blackening are those which have received the most light, i.e., which correspond to the brightest parts of the original. To obtain a '*positive*' with the proper distribution of light and dark, the process must be repeated in principle. This is effected by *printing*. The negative is laid on a paper coated with a light-sensitive emulsion (or, to produce a 'diapositive', on a plate or film). On exposure to light, the blackened areas on the negative protect the underlying emulsion from illumination, whereas the parts that are not blackened allow the light through. The distribution of light and dark is thus once more reversed, to correspond with that of the original object. In making printing papers, silver chloride, or silver chloride-silver bromide emulsions, are usually employed in place of silver bromide, since the sensitivity to light need not be so high. Use is often made, also, of emulsions in which blackening takes place during the exposure, so that development is not necessary, and the paper requires only to be fixed ('printing out paper'). When such papers are used, the color of the image is unsatisfactory, and the paper must be 'toned' during or before fixing. Toning depends upon the replacement of silver by the more noble gold, by immersion in a bath of a gold salt solution: $3Ag + Au^{+++} = 3Ag^+ + Au$. Selenium compounds may also be used for toning, in place of gold. Especially fine pictures may be obtained by toning with platinum salts. The following papers are the most widely used for the production of prints.

Developing papers	*Bromide paper*	Coating—silver bromide-gelatin emulsion
	Gaslight paper	Coating—silver chloride-silver bromide-gelatin emulsion
Printing out papers	*Celloidin paper*	Coating—silver chloride-collodion emulsion
	Aristo paper	Coating—silver chloride-gelatin emulsion
	Albumen paper	Coating—silver chloride-egg albumen emulsion.

(d) Other Silver(I) Salts

(*i*) *Silver nitrate*, $AgNO_3$, which is technically the most important compound of silver, is usually prepared by dissolving metallic silver in nitric acid. If the metal contains copper, it is purified by precipitating the dissolved silver as chloride, by the addition of common salt; the chloride is reduced to the metal by means of zinc and dilute sulfuric acid, or by boiling with grape sugar or formaldehyde and caustic soda, and is redissolved in nitric acid.

Silver nitrate forms colorless rhombic crystals (density 4.35) and is not hygroscopic. Above $159.6°$ a hexagonal-rhombohedral form (density 4.19) is stable. It melts at $208.5°$. Silver nitrate is very soluble in water. The solution is neutral in reaction and has a bitter metallic taste. The solubility increases considerably with rise of temperature (at $20°$, 215 g of $AgNO_3$, and at $100°$, 910 g of $AgNO_3$ dissolve in 100 g of water).

Silver nitrate is partly precipitated from its saturated aqueous solution by the addition of concentrated nitric acid. It is only moderately soluble in dilute ethanol (3.8 g in 100 g of 92.5% alcohol at $15°$) and in methanol. It is still less soluble in pure acetone (0.35 g in 100 g at $15°$) and in benzene.

Silver nitrate is readily reduced to the metal by many organic compounds—e.g., glucose, tartaric acid, paper. Tartaric acid deposits the metal upon glass in the form of a brilliant mirror from ammoniacal solution. This property is used both for the detection of tartaric acid and especially for the manufacture of mirrors. Albumen is coagulated by silver nitrate, which accordingly has a caustic and destructive effect on organic tissues. This property, and its disinfectant action, are the basis for its medicinal uses (lunar caustic). Great quantities of silver nitrate are consumed by the photographic industry for the production of light-sensitive plates, films, and papers.

Silver nitrate, together with barium nitrate (or chloride), is extensively used in analytical chemistry as a group reagent for acid anions.

If any one of the ions Cl^-, Br^-, I^-, CN^-, SCN^-, ClO^-, IO_3^-, $[Fe(CN)_6]^{3-}$, $[Fe(CN)_6]^{4-}$, or $S^=$ is present in solution, a precipitate is obtained *in the presence of nitric acid* when Ag^+ ions are added, usually in the form of silver nitrate. Anions precipitated by Ag^+ ions from *acetic acid solutions*, but not from nitric acid, are: $C_2O_4^=$, $CrO_4^=$, $Cr_2O_7^=$, $SO_3^=$, $S_2O_3^=$*, PO_3^-, $P_2O_7^{4-}$, $HPO_3^=$ and BrO_3^-. Precipitated by Ag^+ ions from *neutral solution only* are: PO_4^{3-}, AsO_4^{3-}, AsO_3^{3-} (or AsO_2^-), BO_3^{3-} (or BO_2^-), $SiO_3^=$ (or $Si_2O_5^=$), and many organic anions, such as $C_4H_4O_6^=$. $CO_3^=$ also gives a precipitate with Ag^+ in neutral solution. If tests for anions are to be carried out on a soda extract (as is usually the case), it is therefore necessary to remove carbonic acid by acidifiying and boiling, before cautiously neutralizing with ammonia and testing for the last named group of anions.

Silver nitrate is also used as a reagent for arsine in the Gutzeit test (see Vol. I. p. 661).

Silver nitrate forms double salts with potassium nitrate and thallium nitrate, and forms mixed crystals with sodium nitrate. Whereas there is a miscibility gap in the system sodium nitrate-potassium nitrate, extending from 15–76 mol-% $NaNO_3$ at 218°, the miscibility gap between sodium nitrate and silver nitrate in the solid state at the same temperature extends only between 26 and 38 mol-% $NaNO_3$. In both cases the miscibility gap becomes broader as the temperature is reduced.

(*ii*) *Silver nitrite*, $AgNO_2$, is obtained as a pale yellow precipitate when NO_2^- ions are added to silver salt solutions. The solubility is 0.332 g in 100 g of water at 18°, and increases considerably with rise of temperature. The salt crystallizes from hot solutions in very fine, almost colorless, rhombic needles. It combines with excess of nitrites to form complex salts of the type $M^I[Ag(NO_2)_2]$. Abegg proved the presence of the ion $[Ag(NO_2)_2]^-$ in solutions of these complex salts by means of electrometric measurements.

Silver nitrite blackens in light. It also decomposes slowly in solution, especially on boiling, according to the equation: $2AgNO_2 = AgNO_3 + NO + Ag$. The dry salt decomposes into silver and nitrogen dioxide when it is heated; $AgNO_3$ and NO are formed as the first intermediate products in this case also.

Silver nitrite is used in preparative organic chemistry for the preparation of aliphatic nitro compounds and nitrous acid esters.

(*iii*) *Silver sulfate*, Ag_2SO_4, forms colorless rhombic crystals (density 5.45, m.p. 660°). It is rather sparingly soluble in water (0.80 g in 100 g of water at 25°), and more soluble in dilute sulfuric acid. *Acid sulfates*—e.g., $AgHSO_4$—crystallize from the solution in sulfuric acid, and are reconverted by water to the normal salts and free sulfuric acid. The sulfate is completely converted into silver chloride by fusion in dry hydrogen chloride: $Ag_2SO_4 + 2HCl = 2AgCl + H_2SO_4$. The same double decomposition takes place in solution, in consequence of the insolubility of silver chloride.

* $S_2O_3^=$ ions are not precipitated if they are present in excess as compared with the Ag^+ ions.

Silver sulfate is prepared either by dissolving silver powder or filings in concentrated sulfuric acid—$2Ag + 2H_2SO_4 = Ag_2SO_4 + SO_2 + 2H_2O$—,or by adding alkali sulfate or dilute sulfuric acid to a concentrated solution of silver nitrate—$2AgNO_3 + Na_2SO_4 = Ag_2SO_4 + 2NaNO_3$.

Manchot (1924) found that a solution of silver nitrate in concentrated sulfuric acid would absorb considerable quantities of CO, especially at low temperatures. No absorption of CO takes place in dilute sulfuric acid solutions.

(iv) *Silver sulfite*, Ag_2SO_3, is precipitated by adding sodium sulfite or sulfurous acid to silver nitrate solution. It is a white substance, sparingly soluble in water, which turns purple and finally black when exposed to light. It is decomposed on boiling with water: $2Ag_2SO_3 = Ag_2SO_4 + SO_2 + 2Ag$. Silver sulfite dissolves in dilute strong acids, with decomposition. It also dissolves in solutions containing excess alkali sulfite, forming complex salts, chiefly of the composition $M^I[Ag(SO_3)]$.

(v) *Silver thiosulfate*, $Ag_2S_2O_3$, is precipitated when sodium thiosulfate is added to a silver nitrate solution. It is a white substance, insoluble in water but very soluble in excess precipitant. It is decomposed by dilute strong acids, and slowly by water; in the latter case silver sulfide is deposited and sulfuric acid is formed.

Complex salts (thiosulfatoargentates) of varying composition have been isolated from solutions of silver thiosulfate in excess alkali thiosulfates—e.g., $Na_2S_2O_3 \cdot 3Ag_2S_2O_3 \cdot 2H_2O$; $Na_2S_2O_3 \cdot Ag_2S_2O_3 \cdot 2H_2O$; $5Na_2S_2O_3 \cdot 3Ag_2S_2O_3 \cdot 4H_2O$; $3Na_2S_2O_3 \cdot Ag_2S_2O_3 \cdot 4H_2O$; $3K_2S_2O_3 \cdot Ag_2S_2O_3 \cdot 2H_2O$; $5K_2S_2O_3 \cdot 3Ag_2S_2O_2$. Except for the first mentioned sodium salt, they are extremely soluble in water. The formation of such compounds, which also takes place through the action of sodium thiosulfate solution on such sparingly soluble silver compounds as the chloride and bromide, is of fundamental importance in the photographic 'fixing' process. In order to avoid formation of the less soluble sodium salt mentioned it is best to avoid the use of too highly dilute thiosulfate solutions. None of the complex salts cited dissolves unchanged. According to Schmitz-Dumont (1941), the trithiosulfato anion, $[Ag(S_2O_3)_3]^{5-}$, is the principal species in the solutions.

(vi) *Silver carbonate*, Ag_2CO_3, is deposited as a light yellow powder when silver nitrate solutions are treated with alkali carbonate or hydrogen carbonate. If carbonate is employed, an excess of the precipitant is to be avoided, as the precipitate is otherwise readily contaminated with oxide. The solubility of silver carbonate in water is extremely small, but is greater in concentrated alkali carbonate solutions, since soluble complex salts such as $K[Ag(CO_3)]$ are formed, which can also be isolated in the crystalline state.

Silver carbonate loses carbon dioxide on very gentle heating. The CO_2 pressure attains 6 mm at $132°$, and 752 mm at $218°$.

(vii) *Silver acetate*, $AgC_2H_3O_2$, crystallizes in colorless, lustrous, flat, flexible needles, from solutions of silver carbonate in hot acetic acid. It is rather more soluble than the sulfate (solubility at $20°$, 1.04 g; at $80°$, 2.52 g in 100 g of water).

(viii) *Silver oxalate*, $Ag_2C_2O_4$, is obtained as a white precipitate, insoluble in water and in dilute acetic acid, when soluble oxalates are added to silver nitrate solution. The precipitate is also insoluble in excess of the precipitant, since complex formation with excess $C_2O_4^=$ ions does not occur. Silver oxalate decomposes when heated to $110°$, and can detonate when heated rapidly. Its solubility is 0.004 g of $Ag_2C_2O_4$ in 100 g of water at $18°$.

(ix) *Silver cyanide*, AgCN, is precipitated by the addition of CN^- ions to solutions of silver salts. It is practically insoluble in water and dilute acids. (1 liter of water dissolves about $2 \cdot 10^{-6}$ mol. of AgCN). It dissolves very readily in alkali cyanide solutions, however, forming complex ions.

Silver cyanide is also present in the form of complex ions in its aqueous solution:

$$2AgCN = Ag^+ + [Ag(CN)_2]^-.$$

Polynuclear complex ions, such as $[Ag_2(CN)_3]^-$ and $[Ag_3(CN)_4]^-$, are present to some extent in other dissociating solvents (such as pyridine, in which silver cyanide is much more soluble than in water).

Silver cyanide is also considerably more soluble in hot concentrated potassium carbonate solution than in water, and separates out on cooling in the form of fine needles of density 3.943. It is only sparingly soluble in aqueous ammonia (0.52 g of AgCN in 100 cc of 10% ammonia at 18°). When it is heated, silver cyanide loses half its cyanogen, and 'silver paracyanide' remains.

The solubility of silver cyanide in alkali cyanide solutions depends upon the formation of soluble complex salts (cyanoargentates), of which potassium argen-tocyanide (potassium dicyanoargentate), $K[Ag(CN)_2]$, is typical. This crystal-lizes in colorless six-sided plates; 1 part dissolves in 4 parts of water at 20°. This salt is also obtained when potassium cyanide reacts with metallic silver in the presence of air. It is even more strongly complexed than the thiosulfate complex salts, and the concentration of silver ions in its solution is therefore infinitesimally small. This manifests itself, for example, in the very considerably increased deposi-tion potential of silver from such solutions, which is so far raised that metals such as zinc, which stand far above silver in the electrochemical series, are not capable of displacing silver from cyanide solutions. The use of alkali silver cyanide baths in electroplating is based upon this fact. If baser metals were suspended, for silvering, in silver nitrate solution silver would be deposited as a result of chemical displacement, and this would lead, in general, to powdery deposits, with poor adhesion. Deposit of silver from the complex cyanide solutions takes place, how-ever, only when a sufficiently high potential is applied. The metal is then obtained in a compact, adherent form.

Silver cyanide crystallizes hexagonal-rhombohedral. Contrary to older assumptions, there is not a second modification, but according to Natta it can form a considerable range of cubic mixed crystals with silver bromide.

(x) *Silver thiocyanate*, AgSCN, is formed as a white curdy precipitate by the action of SCN^- ions on silver salt solutions; it is practically insoluble in water $(0.8 \cdot 10^{-6}$ mol per liter at room temperature) and in dilute strong acids, but dissolves readily in excess of the precipitant. Soluble complex salts (thiocyanato-argentates) are formed in this case also. The potassium salts $K[Ag(SCN)_2]$, $K_2[Ag(SCN)_3]$, and $K_3[Ag(SCN)_4]$ have been obtained crystalline.

Silver thiocyanate forms a very restricted range of mixed crystals with silver bromide, the miscibility gap extending from 3% to 90% by weight of AgBr.

(xi) *Silver phosphates. Silver orthophosphate*, Ag_3PO_4, is formed as a yellow precipi-tate, insoluble in water but soluble in dilute acids, when silver nitrate is added to solutions of orthophosphates. It may be obtained in the form of cubic crystals (density 7.32) from acetic acid or dilute phosphoric acid. It gradually blackens in light. It finds some applications in medicine.

Silver nitrate gives white precipitates from solutions of pyro- and metaphosphates— $Ag_4P_2O_7$ and $(AgPO_3)_x$. The latter exists in three forms, which differ in their solubility and their behavior on fusion.

(*xii*) *Silver sulfide*, Ag_2S, is deposited as a black precipitate when hydrogen sulfide is passed into silver salt solutions. It is also formed by the action of hydrogen sulfide or metal sulfides on metallic silver. It is the *most insoluble* salt of silver. From potentiometric measurements of the equilibrium, Jellinek deduced a value of $5.7 \cdot 10^{-51}$ for the solubility product, $K_{sp} = [Ag^+]^2 \cdot [S^=]$. The theoretical solubility of silver sulfide—i.e., its solubility in pure water, if the salt were almost completely dissociated and not hydrolyzed at all—therefore would be $1.8 \cdot 10^{-17}$ mol per liter at $10°$.

Silver sulfide is obtained in crystalline form by passing sulfur vapor over red hot silver. It occurs native as silver glance (argentite), in the form of dark leaden grey crystals, usually with a cubic habit (density 7.2–7.4, hardness 2–2½). An acicular form of silver glance, known as *acanthite*, that is occasionally found, is identical in crystal structure with argentite.

Ag_2S crystallizes rhombic, but appears to be related in crystal structure to the cubic Cu_2O. It undergoes a transition at $180°$, forming a cubic structure.

The heat of formation of silver sulfide is 6.6 kcal per mol at $20°$ (Roth, 1935).

When silver sulfide is treated with concentrated alkali sulfide solutions, it is converted into red crystalline double sulfides—e.g., $Na_2S \cdot 3Ag_2S \cdot 2H_2O$ and $K_2S \cdot 4Ag_2S \cdot 2H_2O$. Double sulfides of silver sulfide with sulfides of acidic character (especially the sulfides of arsenic and antimony) are widely distributed in Nature. Of this type are *proustite*, $3Ag_2S \cdot As_2S_3$ (or $Ag_3[AsS_3]$); *xanthoconite*, $9Ag_2S \cdot 2As_2S_3 \cdot As_2S_5$, which is rare; *polyargyrite*, $12Ag_2S \cdot Sb_2S_3$; *stephanite*, $5Ag_2S \cdot Sb_2S_3$ or $Ag_5[SbS_4]$; *pyrargyrite*, $3Ag_2S \cdot Sb_2S_3$ or $Ag_3[SbS_3]$; *miargyrite*, $Ag_2S \cdot Sb_2S_3$ or $Ag[SbS_2]$; and *silver bismuth glance*, $Ag_2S \cdot Bi_2S_3$ or $Ag[BiS_2]$.

(e) Silver(II) Compounds

Compounds containing dipositive silver are known in the form of the difluoride, AgF_2, and probably also the oxide AgO, obtained by the anodic oxidation of silver. There is also a series of complex compounds of the type $[Ag^{II} Am_4]X_2$, where Am stands for a neutral ligand, such as an organic nitrogen anhydrobase.

Barbieri (1912) obtained the compound $[Ag^{II}(C_5H_5N)_4]S_2O_8$ (tetrapyridinesilver(II) peroxysulfate) by adding a cold saturated solution of potassium peroxysulfate to a solution of silver nitrate and pyridine. It forms beautiful orange prisms, which give mixed crystals with the corresponding compound of $+2$ copper. Barbieri later (1927) obtained the corresponding nitrate $[Ag^{II}(C_5H_5N)_4](NO_3)_2$, by electrolytic oxidation of a solution containing silver nitrate and pyridine; this also exists as orange crystals. Hieber shortly afterwards (1928) prepared a series of silver(II) salts containing the nitrogen anhydrobase *o*-phenanthroline bound to silver in the complex:

X_2; $X = ClO_4, ClO_3, NO_3, SO_4H, \frac{1}{2}S_2O_8.$

He first prepared the chocolate-brown peroxysulfate, by adding ammonium peroxysulfate to a solution containing silver nitrate and *o*-phenananthroline in the molecular proportion 1 : 2. The other salts (also brown) were obtained from this by double decomposition. They form mixed crystals in all proportions with the analogous salts of dipositive copper and cadmium, and must therefore be similar in structure. Hence the silver must be present in the $+2$ state. The dipositive character of silver in these compounds was confirmed by

Klemm (1931), who found them to display the same paramagnetism as the corresponding copper(II) salts.

A long known compound which, according to Barbieri, is also derived from +2 silver, is the substance AgO, formed by anodic oxidation of silver salts or silver in aqueous solution. The formation of this was first observed by Ritter in 1799, and it was formerly regarded as a peroxide (i.e., as a derivative of hydrogen peroxide). However, the constitution of this compound has not yet been established with certainty.*. An unambiguous example of simple compound of +2 silver is the silver difluoride prepared by Ruff.

Silver difluoride, AgF_2, is formed directly by the action of fluorine on metallic silver at not too high a temperature. (Ruff, 1934). Fluorine reacts only superficially with *compact* silver below a red heat; above 450°, the monofluoride AgF is formed. With 'molecular silver', however, reaction takes place even at ordinary temperature, and the *difluoride* is formed. According to Rochow [*J. Am. Chem. Soc.*, 74 (1952) 1615], AgF_2 is more conveniently obtained by passing ClF_3 over AgCl. Silver difluoride (density 4.7, m.p. 690°) is deep brown in color, and strongly paramagnetic (actually ferromagnetic below —110°). It is at once hydrolyzed by water; Ag(OH)F is apparently first formed as an intermediate product, and this rapidly decomposes, yielding AgF and ozonized oxygen. Silver difluoride readily gives up its fluorine to other substances, and is therefore an excellent fluorinating agent. Its heat of formation from its elements is 84.5 kcal per mol. (von Wartenberg, 1939).

(f) Silver(III) Compounds

A compound containing silver in the +3 valence state is the fluorosalt $K[AgF_4]$ prepared by Klemm (1953). In addition to this, there are several silver(III)-anionic complexes, of more complicated composition—e.g., some tellurato and periodato compounds, such as $Na_6H_3[Ag(TeO_6)_2]$ and $K_6H[Ag(IO_6)_2]$ prepared by Malatesta (1941), and some organic complex silver(III) compounds described by Rây (1944). The latter contain complex cations. Compounds of both types are diamagnetic, and involve covalent (probably dsp^2 hybrid, square planar) bonds.

7. Analytical (Silver) [12]

Two reactions characteristic of silver are its precipitation by means of hydrochloric acid from solutions of its simple salts, in the form of curdy white silver chloride, soluble in aqueous ammonia, and the ease with which reduction to the metal takes place—e.g., by heating the salts on charcoal with fusion mixture before the blowpipe. The smooth, white, ductile metallic bead is readily dissolved in dilute nitric acid. The resulting solution gives a white precipitate with hydrochloric acid, and a black precipitate (silver sulfide) with hydrogen sulfide; after neutralization, it gives a red-brown precipitate with potassium chromate (silver chromate, Ag_2CrO_4), and a yellow precipitate with secondary sodium phosphate (silver phosphate, Ag_3PO_4).

For microanalytical detection, suitable forms are silver chromate and, especially, the crystallization of silver chloride in characteristic form when its ammoniacal solution is evaporated (limit of detection, 0.1 γ Ag). Rather more sensitive still is the reaction with dithizone, $C_6H_5—N=N—CS—NH—NH—C_6H_5$, given by Fischer [*Z. angew. Chem.*, 42, (1929) 1025]. This cannot be applied, however, if salts of mercury and the platinum metals are also present.

In *gravimetric analysis*, silver is precipitated as chloride or deposited electrolytically as metal, and weighed as such. From solutions in which it is present in

---

* The nature of the compound formed by the action of ozone on silver (cf. p. 395), which is generally regarded as a peroxide of silver, is also not fully explained.

complex form, it is sometimes advantageous to reduce it to the metal by means of sodium dithionite (Steigmann, 1923). Silver is often determined *titrimetrically*.

In titrating silver by Gay Lussac's method, a sodium chloride solution of known concentration is added to a nitric acid solution containing silver, as long as any precipitate is formed. In Mohr's method, a neutral silver nitrate solution is run into a neutral sodium chloride solution (of known concentration, if the silver is to be determined), and potassium chromate is used as the indicator for the end-point of the reaction. The red silver chromate, being more soluble than silver chloride, persists only when practically all the chloride ions have been removed from the solution. Volhard's method of determining silver consists in titration with ammonium or potassium thiocyanate, and utilizes an iron(III) solution as indicator. As long as Ag+ ions are present in the solution being titrated, white silver thiocyanate is precipitated. When essentially all the Ag$^+$ ions have disappeared, a red coloration indicates the formation of iron(III) thiocyanate.

Small amounts of silver can be titrated conductometrically with sodium chloride, even in the presence of a large excess of lead salts (Jander, 1937).

8. Gold (Aurum, Au)

(a) Occurrence

In accordance with its noble character, gold is generally found free in Nature, in two forms—as reef gold and alluvial gold. [*13*]

Gold which is found still in its primary ore deposit is known as reef gold. As such, it occurs partly in the form of very fine metallic particles dispersed through quartz, to some extent associated with sulfides such as pyrite, chalcopyrite, arsenopyrite, and stibnite, which are also found either finely dispersed or as thicker veins and reefs, predominantly in quarzitic rock. Gold often occurs, mixed with these sulfides, in the form of chemical compounds, and especially as tellurides—e.g., *calaverite*, $AuTe_2$, *sylvanite*, $AgAuTe_4$, and *nagyagite* (foliated tellurium), a double telluride and sulfide of lead, gold, and antimony. Compounds of gold with selenium are also occasionally found. Native gold is always found along with its compounds, as might be expected from the ease with which reduction to metal takes place, but the gold is finely dispersed.

In the course of the history of the earth's crust, some of the gold has been removed from its original ore deposits, and laid down again in secondary deposits. The *alluvial* or *placer* gold which is now found in this form is invariably native, usually in the form of fine or very fine grains, but occasionally as large aggregates or nuggets.

In former times, gold was extracted only from the placer deposits. With the perfection of methods of extraction, however, reef gold has been increasingly important in world production, and by 1912 upwards of 90% of the world's production came from reef gold.

(b) Extraction [*14*]

The oldest method of extracting gold is by 'washing'. In this process, auriferous sand is subjected to a sedimentation, whereby the specifically heavier grains of gold settle out faster and are enriched.

This process, much improved technically, is still in use in the form of 'hydraulic mining', in which auriferous loose sedimentary rocks (sands and gravels) are broken up and elutriated by powerful jets of water. To make the process economic on a large scale, it must be combined with the amalgamation process.

The amalgamation process depends upon the capacity of mercury to dissolve gold, forming an amalgam, from which the gold can be recovered by distilling off the mercury. A requisite condition for the method to be applicable is that the particles of gold should not be too small, as they are otherwise not sufficiently wetted by the mercury.

The process which is now most important for the extraction of gold is the *cyanide process* introduced in 1886. This is used to some extent in conjunction with the amalgamation process, for the recovery of the last portion of the gold, ('tailings'). Cyaniding enables even the finest particles, which cannot be amalgamated, to be leached out. It is therefore suitable not only for working up the still auriferous residues of the amalgamation process, but especially for those ores which contain gold in the finest state of subdivision—i.e., for reef gold.

Cyanide leaching, which was first tried by MacArthur and Forrest in 1886, has practically completely displaced the chlorination process (invented by Plattner in 1850) which was formerly much used, especially in America. In the latter process, gold was converted to the soluble chloride, by treating the moist ore with chlorine; the chloride was leached out, and reduced by ferrous sulfate, hydrogen sulfide, or by adsorption on wood charcoal. Attempts have also been made to utilize the volatility of gold chloride for the extraction of the metal from certain ores, by heating them with common salt, whereupon gold chloride sublimes.

In cyanide leaching, the finely ground gold ore is treated with a dilute (0.1– 0.2%) solution of potassium or sodium cyanide. With the aid of air (dissolved in the solution), the gold is converted to a soluble complex cyanide, and is subsequently deposited from solution by reduction with zinc shavings* or (very rarely) by electrolysis.

The precipitate is more or less strongly contaminated with zinc, which is dissolved out by treatment with dilute sulfuric acid. The dried residue is then fused under borax. The melted down gold (bullion) still contains larger or smaller amounts of silver, which is separated by the process of *refining*. On a large scale, this is now generally carried out electrolytically. In smaller refineries, it is frequently still effected by treatment with concentrated sulfuric acid, or often also by means of nitric acid.

In *electrolytic refining* of gold, [10] which is used especially when the gold contains platinum, the gold plate to be refined is suspended as an anode in a solution of chloroauric acid (tetrachlorogold(III) acid, $HAuCl_4$). When the current is passed, some of the metals alloyed with the gold do, indeed, enter the solution; others (Ir, Rh, Ru) fall to the bottom unattacked; Ag is precipitated as AgCl; and Pb is deposited as $PbSO_4$ by the addition of sulfuric acid. Only gold is deposited on the cathode, since it has by far the lowest tendency to remain in solution. Even platinum remains dissolved, provided that care is taken that it does not build up to too high a concentration.

Sulfuric acid refining (*parting*) depends upon the solubility of silver, and the insolubility of gold, in boiling concentrated sulfuric acid. It is applicable to all gold-silver alloys, irrespective of their gold content, but it is not suitable for alloys containing platinum, since platinum may make a portion of the silver insoluble.

The parting of gold and silver can also be effected by means of moderately concentrated nitric acid (preferably about 60%). For this process, it is necessary that the alloy should consist of about $\frac{1}{4}$ gold and $\frac{3}{4}$ silver; if the gold content is higher, an alloy of appropriate composition is made up by melting the mixture with added silver. The production of an alloy with the suitable composition is known as '*quartation*', and this term has been extended to apply to the whole process of refining. Quartation was the usual process of gold refining from the 15th to the 19th century, but with the growth of the sulfuric acid industry, it has been largely displaced by sulfuric acid parting.

---

* The zinc goes partly into solution as a complex alkali zinc cyanide, and is partly precipitated as $Zn(CN)_2$.

(c) **Properties**

From ancient times, gold has been the 'king of metals', excelling all others in its beautiful yellow color, bright luster, stability in air, and ductility. Its hardness is only $2\frac{1}{2}$–3 on Mohs' scale, and it is among the heaviest of the metals (density 19.3).

The melting point of gold (1063°) is only about 100° higher than that of silver, and below that of copper. Its boiling point (about 2700°) is higher than that of copper, however. Nevertheless, the volatility of gold becomes quite appreciable at much lower temperatures —from about 1000°. Platinum containing gold becomes quite gold-free after prolonged fusion. Gold expands considerably when it melts.

The electrical conductivity and thermal conductivity of gold are very high. Its specific electrical conductivity $\varkappa$ (at 18°) is 67% of that of silver, its thermal conductivity λ is 70% of that of silver. ($\varkappa_{Au} = 4.13 \cdot 10^5$ ohm^{-1}; $\lambda_{Au} = 0.705$ cal. cm^{-1}sec^{-1}degree^{-1} at 18°).

Gold crystallizes in the cubic system (for crystal structure see p. 362). Gold crystals have been found native, with cube, octahedron, and rhombic dodecahedron faces developed. It forms mixed crystals in all proportions with platinum, palladium, silver, and also with copper (in spite of the considerable difference in cell dimensions – cf. p. 22).

Gold unites with atomic hydrogen, forming a colorless solid hydride, which is probably salt-like in character (Pietsch, 1931). This compound is very unstable. A gold hydride is stable in the gaseous state, however, at high temperatures (cf. p. 363).

As a decidedly noble metal, gold remains practically unchanged in air or in contact with water*. It is also very resistant to other forms of chemical attack. Even the halogens, in the form of the dry gases, are without action, or react only superficially, with gold at ordinary temperature. Thus gold combines with fluorine only between 300° and 400° (the fluoride decomposes again at higher temperatures). It is rapidly dissolved, however, by an aqueous solution of chlorine at ordinary temperature. The tendency of gold to combine with chlorine:

$$Au + \tfrac{3}{2}Cl_2 = AuCl_3,$$

is considerably augmented by the fact that complex ions are formed—tetrachloroaurate ions:

$$AuCl_3 + Cl^- = [AuCl_4]^-$$

or, if no excess of chloride ions is present, trichlorooxoaurate ions:

$$Au + \tfrac{3}{2}Cl_2 + H_2O = [AuCl_3O]^= + 2H^+.$$

* According to Müller (1935), gold becomes covered in air with a very thin oxide film. If this assumption is correct, the complete stability of gold in air arises not from its inability to combine directly with atmospheric oxygen, but because a protective coating is formed upon it. In that case, the only difference in behavior towards oxygen between gold and other metals such as copper and nickel, which also form protective films, is that at temperatures at which the protective film would be capable of growing (cf. pp. 739, 754), the oxide forming the protective film on gold undergoes decomposition. The heats of formation of the gold oxides are not very reliably known. According to Thomsen, gold(III) hydroxide has a fairly large exothermic heat of formation (about 100 kcal per mol).

Gold dissolves particularly readily in aqua regia which produces 'nascent' chlorine*.

The dissolution of gold is yet more strongly facilitated by CN^- ions than by the presence of Cl^- ions. Its tendency to dissolve is so far augmented that atmospheric oxygen can oxidize gold in the presence of CN^- ions**:

$$2Au + \tfrac{1}{2}O_2 + H_2O + 4CN^- = 2[Au(CN)_2]^- + 2OH^-.$$

This reaction is the basis of the cyanide process for extracting gold.

Gold is not attacked by molten caustic alkalis. It is also unattacked by acids such as sulfuric, hydrochloric, nitric, phosphoric or arsenic acid, whether dilute or concentrated, unless powerful oxidizing agents are also present.

Gold dissolves in concentrated sulfuric acid in the presence of iodic acid, nitric acid, manganese dioxide, etc. Yellow solutions are thereby obtained, from which gold(III) hydroxide is precipitated by the addition of water. Selenic acid, which is itself a very strong oxidant, can dissolve gold directly.

The insolubility of gold in the less strongly oxidizing acids is the consequence of its position in the electrochemical series, in which it stands far below hydrogen. Not only, therefore, is it unable to go into solution by displacement of hydrogen, but it may be deduced from the numerical magnitude of its standard potential (cf. Table 4 on p. 30 of Vol. I) that in general it cannot even enter solution as an ion under the oxidizing action of atmospheric oxygen. Unlike copper, for example, it is therefore insoluble in acids in the presence of air, as well as in air-free acids.

A gold electrode in contact with a 1-molar gold(III) chloride solution has a potential 1.08 volts lower than that of the normal hydrogen electrode. However, the Au^{+++} ion concentration in such a solution is far less than 1 mol per liter, as a result of complex formation with the Cl^- ions. The standard *potential* of gold therefore must be considerably lower than this value. It was estimated as -1.37 to -1.39 volts by Jirsa and Jellinek, from measurements of gold(III) sulfate and nitrate solutions. This low value evidences the difficulty with which the gold atom assumes an ionic charge, and the ease with which the gold ions are discharged.

The standard potential of the gold(I) ion is even lower than that of the gold(III) ion. However, free gold(I) ions are never present except in infinitesimal concentrations, so that the relations are reversed in practice, just as with copper.

The discharge potential of gold is lower than for any other metal, so that provided the current density is not too high, gold alone is deposited cathodically from gold chloride solutions of sufficient strength, even when the solutions contain appreciable amounts of other noble metals.

The situation is different if the gold is present in the form of its extraordinarily strongly complexed cyanide. Its ionic concentration is then reduced to so low a value that its solution potential becomes greater than that of silver or mercury. Not only are mercury and silver unable to displace gold from gold(I) cyanide-alkali cyanide solutions, but they are themselves displaced by gold, which dissolves, from their alkali cyanide complex salt solutions.

It follows from the position of gold in the electrochemical series, and accords with experiment, that gold can be displaced from its solutions by almost all other

* Nitrosyl chloride, NOCl, which is evolved simultaneously from aqua regia, is quite without action on gold even at 100°.
** Cf. also p. 417].

metals, provided it is not present in the form of compounds which are exceedingly strongly complexed.

It should be noted that in the displacement of noble metals from solution by means of base metals, mixed crystals or compounds of the noble metal with the displacing metal are often obtained, instead of the pure noble metal. Thus cadmium precipitates the compound Cd_3Au from gold salt solutions.

Gold is also precipitated from solutions of its chloride by ions with reducing properties, such as Fe^{++}, Sn^{++}, $SO_3^=$, NO_2^-. Mercury(I) ions reduce it chiefly to the singly charged state:

$$Au^{+++} + Hg_2^{++} = Au^+ + 2Hg^{++}.$$

The resulting gold(I) ions are at once precipitated as the hydroxide by reaction with water:

$$Au^+ + HOH = AuOH + H^+$$

before they can be formed in appreciable amount. At the same time they react mutually:

$$3Au^+ = Au^{+++} + 2Au,$$

so that metallic gold separates out in addition to gold(I) hydroxide.

Reduction with $SO_3^=$ ions also frequently proceeds only as far as unipositive gold, particularly when the conditions of concentration favor the formation of sulfito complexes, which protect the gold from further reduction.

Many organic substances can also precipitate gold as the metal—e.g., oxalic acid, formaldehyde, sugar, tartaric, citric, and acetic acids, and their salts. Carbon monoxide also precipitates the metal from gold(III) chloride solutions, as does acetylene. The latter is oxidized to glyoxal, CHO—CHO, according to Kindler. Gold is often obtained in colloidal form when precipitated by these means. As such, it forms hydrosols which are usually intensely colored (e.g., purple red, blue, violet, brown or black), and notably stable even in the absence of protective colloids.

It is an essential condition for stability that the sols should be quite free from substances causing coagulation—e.g., electrolytes. The flocculating action of electrolytes is largely determined by the nature of the species going into solution as cation, since the particles of colloidal gold are negatively charged. In some circumstances, the charge on the particles may be reversed by larger concentrations of electrolyte. It is therefore also possible for large amounts of electrolyte to have a stabilizing effect.

Pure gold hydrosols are best prepared by two processes worked out by Zsigmondy*. The first depends on the reduction of chloroauric acid with formaldehyde, in very dilute solution made just alkaline by addition of potassium carbonate. In the second, reduction is effected by adding a few drops of an ether solution of phosphorus. The two methods may be combined. Such sols, with a well defined particle size—in the red gold sols prepared by reduction with phosphorus, the particles are usually between 1 and 6 · 10^{-7} cm in diameter —have found applications in medicine for diagnostic purposes (e.g., in investigations of the cerebrospinal fluid) and in biology (for the characterization of albumenoids and their decomposition products). These applications depend upon the fact that different proteins give rise to agglomeration processes, which show up in different ways through color changes of the sol.

Hydrazine, carbon monoxide, and hydrogen peroxide have also been proposed as reducing agents for the preparation of colloidal gold sols. Donau, in 1913, described an

* See Zsigmondy's *Kolloidchemie* for further details.

interesting process, in which a burning jet of hydrogen is directed against the surface of a very dilute gold chloride solution (1 in 50,000 or 1 in 100,000). At the point touched by the flame, red wisps of colloidal gold appear. As was shown by Halle and Pribram, the reducing action of the hydrogen flame itself is only a partial cause of the separation of gold; nitrogen oxides produced in the flame also play an essential part.

(d) Uses

Gold is highly valued as a material for the manufacture both of jewellery and of useful objects. Since pure gold is too soft for this purpose, it is alloyed with other metals, generally copper or silver. [11]

The gold content ('fineness') of these alloys is generally expressed either in parts per thousand or in carats. 24 carat represents pure gold. An 18 carat alloy contains 18 parts by weight of gold in 24 parts, i.e., 750 in 1000 parts. 14 carat gold contains 14 parts of gold in 24 parts, or roughly 585 parts per thousand.

Considerable quantities of gold are used for gilding objects made of baser metals—principally copper and silver objects. In dentistry, gold is used for fillings, bridge work, etc. Gold is also occasionally used for chemical apparatus (fusion crucibles, retorts, condensers, etc.).

Of the compounds of gold, chloroauric acid, $HAuCl_4 \cdot 4H_2O$, the 'gold chloride' of commerce, is the most widely employed. It is used for making up baths (cyanide solutions) for electroplating with gold, in glass and porcelain painting, in photography (for toning baths) and occasionally in medicine. Its sodium salt, $Na[AuCl_4] \cdot 2H_2O$, known in commerce as 'gold salt', is often used instead of chloroauric acid.

Gold ruby glass, which is valued on account of its deep red color, contains metallic gold in the form of colloidal particles which can be rendered visible under the ultramicroscope. It is prepared by adding a little gold, in the form of any of its compounds, to the glass melt. A pale yellow, greenish or even quite colorless glass is first obtained, but when this is reheated it suddenly turns blood red when the gold particles attain such a size as to impart this color to the sol. If the glass is heated too long, continued particle growth leads to a poor and turbid color.

Almost half of the annual production of gold is stored away, in the form of gold bar in the central reserve banks of the various countries, as cover for the currency notes which are circulated in place of gold. Through the accumulation of gold in the reserve banks of a few countries (and especially in the United States), the significance of gold as a cover for bank notes, and as a monetary standard, has rather diminished.

In those few countries where gold coins are still used as currency, they generally consist of 900 parts gold and 100 parts silver.

Around 1800, the average annual production of gold in the whole world was only 17,784 kg. By 1901 this had risen to 392,705 kg, and in 1912 it reached a temporary peak production of 701,370 kg. The average annual gold production rose to 652,290 kg during 1906–1910, 691,409 kg during 1911–1915, and 541,778 kg in 1921–1925. Since 1929 it has followed a continuous upward trend (1929 597,420 kg; 1932 749,550 kg; 1935 933,080 kg; 1937 1,128,000 kg).

Of the total world production of gold in 1912, about 44.7% came from Africa (39.8% from the Transvaal alone), 27.8% from North America, 2.6% from South America, 11.9% from Australia, 5.9% from Asia other than Siberia, and 7% from Europe (chiefly from

Russia, including Siberia). The proportions produced in the various countries has shifted from time to time in favor of Africa, (57.7% in 1929) at the expense of American and Australian production. The share due to Europe has tended to increase recently as a result of the rapid rise in Russian output. Of the gold production in 1937, 38.8% was attributable to Africa (32.5% from the Union of South Africa), 26.9% to North America, 4.4% to South and Central America, 5.0% to Australia, 7.6% to Asia without Russia, and 17.3% to Europe (15.7% to Russia with its Asiatic territories).

9. Compounds of Gold

Gold ordinarily exhibits valence states of $+3$ and $+1$ in its compounds. In general, it is most stable in the tripositive state. All gold compounds are easily decomposed, however. They break up, with the separation of free gold, under the action of reducing agents, and often when only gently heated.

The most important compounds of $+1$ gold are gold(I) sulfide, Au_2S, the halides AuX, and various complex anions. Simple gold(I) compounds are not capable of existing in aqueous solution, except at infinitesimal concentrations. Hence there are no soluble simple gold(I) salts. Even the very insoluble gold(I) chloride, $AuCl$, is decomposed by water forming gold(III) chloride and free gold. Free gold(I) ions can only exist in solution in minimal concentrations, which are not directly detectable. On the other hand, strongly complexed gold(I) compounds, such as the cyanoaurate(I) salts, $M^I[Au(CN)_2]$, are remarkably stable in solution.

A very strong tendency to form complex salts is also characteristic of the salts derived from $+3$ gold. Strong complexes are formed especially with the halogen, cyano, and thiocyano groups. The salts of those acids which are not capable of combining with gold(III) ions to form acido-complexes are stable only in concentrated solutions of the corresponding acids. They are at once hydrolyzed on the addition of water, with the deposition of gold(III) hydroxide. Thus Au^{+++} ions are also incapable of existing in directly measurable concentration in pure aqueous solution.

Gold has a much weaker capacity for forming complex cations than complex anions. However, Weitz (1915) was able to prepare a number of compounds of the type $[Au(NH_3)_4]X_3$ (tetramminegold(III) salts). The nitrate, $[Au(NH_3)_4](NO_3)_3$, may be obtained by the action of ammonia on a solution of chloroauric acid saturated with ammonium nitrate. Other salts, such as the oxalate-nitrate, $[Au(NH_3)_4](C_2O_4)(NO_3)$, and the sparingly soluble chromate, $[Au(NH_3)_4]_2(CrO_4)_3$, can be obtained by metathesis. The ammonia is very firmly bound in the complex from which these salts are derived, and is not eliminated by the action of concentrated acids. Alkalis bring about decomposition, whereby explosive compounds containing nitrogen are formed, similar in nature to those obtained when chloroauric acid solution is treated with excess ammonia, or by the prolonged action of ammonia or ammonium salts on gold oxide ('fulminating gold'). According to Weitz, these consist chiefly of the compounds $Au_2O_3 \cdot 2NH_3$ and $Au_2O_3 \cdot 3NH_3$, mixed with diamido-imido-gold(III) compounds, $\begin{matrix} X \diagdown & & \diagup X \\ & Au-NH-Au \\ NH_2 \diagup & & \diagdown NH_2 \end{matrix}$, where X is a univalent acid radical (Cl, NO_3). The action of ammonia on chloroauric acid, in solutions saturated with ammonium chloride, leads to diamido-gold(III) chloride, $Au(NH_2)_2Cl$, which is not explosive, and deflagrates only feebly when heated.

Gold(III) hydroxide dissolves only in quite concentrated acids, except when complex formation can occur. It also dissolves in strong caustic alkalis. It thus has the character of an amphoteric substance, and is commonly known as *auric acid*.

A number of compounds are known which, on the basis of their formulas, could be regarded as gold(II) compounds—e.g., $AuCl_2$, $AuBr_2$, AuO, AuS, $AuSO_4$. It appears, however, that all these represent double compounds of unipositive with tripositive gold—e.g., $AuCl \cdot AuCl_3$ or $Au^I[Au^{III}Cl_4]$.

Hoppe (1950) has obtained some evidence that fluoro-salts of *tetrapositive* gold may be formed. The existence of such compounds is by no means established, however.

(a) Gold(I) Compounds

(*i*) *Gold(I) oxide*, Au_2O, is said to be formed by the action of hydroxyl ions on gold(I) compounds, or on gold(III) compounds in the presence of a reducing agent. Its existence is questionable, however.

Krüss (1887) claimed to have obtained gold(I) oxide in the pure state by decomposing potassium dibromoaurate(I), $K[AuBr_2]$, by means of caustic potash. (The potassium dibromoaurate(I) was prepared by the cautious reduction of potassium tetrabromoaurate(III), $K[AuBr_4]$, with sulfurous acid in ice-cold solution.) A dark violet precipitate, thought to be gold(I) hydroxide or gold(I) oxide gel, was first formed; this changed to a light grey-violet powder when dried over phosphorus pentoxide. More recent investigations make it probable that the substance considered to be gold(I) oxide is merely a mixture of Au_2O_3 with finely divided gold. It is possible that gold(I) oxide or hydroxide is initially formed in the precipitation, but at once decomposes. In any case, gold(I) oxide is extremely unstable if it exists at all.

(*ii*) *Gold(I) Halides. Gold(I) chloride*, $AuCl$, is formed as a pale yellow powder when anhydrous gold(III) chloride is moderately heated (to 185°):

$$AuCl_3 = AuCl + Cl_2.$$

Gold(I) chloride is difficult to obtain pure. When more strongly heated, it decomposes into gold and free chlorine. It is rapidly decomposed by water at ordinary temperature:

$$3AuCl = AuCl_3 + 2Au.$$

It dissolves in alkali chloride solutions, forming complex ions, $[AuCl_2]^-$ (chloroaurate(I) ions). These also soon undergo decomposition in solution, however, depositing metallic gold and forming complex ions of +3 gold.

The chloroaurate(I) salts are stable in the solid state, however. Thus the yellow potassium chloroaurate(I) can be obtained by melting potassium chloroaurate(III):

$$K[AuCl_4] = K[AuCl_2] + Cl_2.$$

$AuCl$ also forms complex compounds with other substances—e.g., with PCl_3 and especially with ammonia. It can combine with up to 12 molecules of NH_3 (on treatment with liquid ammonia).

$AuCl$ combines with CO on gentle warming, forming *gold carbonyl chloride*, $Au(CO)Cl$—colorless, highly refractive tabular crystals. The compound is more readily prepared by warming $AuCl_3$ in a current of CO. It is at once decomposed by traces of water, with deposition of gold and the formation of CO_2 and HCl (Manchot, 1925).

Gold(I) bromide, $AuBr$, obtained by cautiously heating gold(III) bromide, is still more easily decomposed than the monochloride. It dissolves in alkali bromide solutions, with the formation of complexes.

Gold(I) iodide, AuI, is obtained as the product of reactions which would be expected to yield gold(III) iodide—e.g., from the dissolution of gold(III) oxide in hydriodic acid, or by adding iodide ions to gold(III) salt solutions. It can also be obtained, however, by the direct union of gold with iodine at moderately elevated temperatures. Gold(I) iodide is a lemon yellow powder. It is even more readily decomposed by heat than are AuCl and AuBr, but is much less rapidly decomposed by water. This is apparently due to its much smaller solubility. It dissolves in potassium iodide solution (forming a complex), but does so without decomposition only if the solution contains free iodine (or potassium triiodide) in amount corresponding to the iodine pressure of the compound. Solvents such as chloroform or carbon disulfide, which readily take up iodine, decompose it.

Heats of formation, according to Biltz, are: for AuCl $+8.4$, for AuBr $+3.4$ and for AuI -0.2 kcal per mol.

(iii) Gold(I) cyanide and Cyanoaurate(I) Salts. Gold(I) cyanide, AuCN, (yellow microscopic six-sided plates) is often formed, like the iodide, in processes which would be expected to yield gold(III) cyanide. It is also obtained by warming complex gold(I) cyanides with hydrochloric acid at 50°:

$$Na[Au(CN)_2] + HCl = AuCN + NaCl + HCN.$$

Gold(I) cyanide is not decomposed by water, apparently because it is too insoluble, nor by dilute acids or even by hydrogen sulfide. It readily dissolves in alkali cyanide solutions, yielding strongly complexed salts of the type $M^I[Au(CN)_2]$, (*cyanoaurate(I)* salts). It is also soluble in sodium thiosulfate, ammonium sulfide, caustic potash, and ammonia.

Potassium dicyanoaurate(I), $K[Au(CN)_2]$, is obtained when gold dissolves in potassium cyanide solution in the presence of air:

$$2Au + 4KCN + H_2O + \tfrac{1}{2}O_2 = 2K[Au(CN)_2] + 2KOH.$$

As was shown by Bodländer (1896), the process takes place through the intermediary formation of hydrogen peroxide*:

$$2Au + 4KCN + 2H_2O + O_2 = 2K[Au(CN)_2] + 2KOH + H_2O_2$$

$$2Au + 4KCN + H_2O_2 \qquad = 2K[Au(CN)_2] + 2KOH.$$

Potassium dicyanoaurate(I) is very soluble in water (about $\tfrac{1}{2}$ mol per liter in the cold, 7 mols per liter at the boiling point). It is sparingly soluble in alcohol and quite insoluble in ether. The sodium salt has similar properties. The importance of these compounds in the extraction of gold has already been mentioned.

(iv) Thiosulfatoaurate(I) Salts. Sodium dithiosulfatoaurate(I), $Na_3[Au(S_2O_3)_2] \cdot \tfrac{1}{2}H_2O$, is precipitated by adding alcohol to a solution of gold chloride treated with sodium thiosulfate. It forms colorless, needle-like crystals, freely soluble in water and having a sweetish taste. Reducing agents such as iron(II) chloride, tin(II) chloride, and oxalic acid do not deposit gold from its solution, and no sulfur is precipitated by the addition of acid. However, sodium tetrathionate and gold(I) iodide are formed if iodine is added. The free acid (dithiosulfatogold(I) acid), $H_3[Au(S_2O_3)_2] \cdot \tfrac{1}{2}H_2O$, can be obtained by the action of dilute sulfuric acid on the barium salt, $Ba_3[Au(S_2O_3)_2]_2$.

(v) Sulfitoaurate(I) Salts. If gold chloride solution is added gradually to an alkaline solution of potassium sulfite, potassium disulfitoaurate(I), $K_3[Au(SO_3)_2] \cdot H_2O$, is obtained in the form of white needles. It is soluble in water and strongly complexed: no gold is precipitated from the solution by hydrogen sulfide. Purification can be effected by means of the sparingly soluble barium salt. The corresponding sodium salt may also be prepared.

* The dissolution of metallic silver in alkali cyanide solutions takes place similarly (Simon, 1935).

(*vi*) *Gold(I) sulfide*, Au_2S. If hydrogen sulfide is passed into a hot solution of gold(III) chloride, metallic gold is deposited. In the cold however, a precipitate with the composition AuS separates, which is probably a double compound of Au_2S and Au_2S_3. Au_2S, gold(I) sulfide, is best obtained by saturating a $K[Au(CN)_2]$ solution with hydrogen sulfide, and adding hydrochloric acid. It is steel grey when moist, brown black when dry; is practically insoluble in water and dilute acids, but readily forms colloidal dispersions, especially in the presence of hydrogen sulfide. It is decomposed and dissolved by strong oxidants (chlorine, aqua regia). It dissolves in alkali cyanide solutions, forming cyanoaurate(I) salts, and also is soluble in excess alkali sulfide solutions, with which it forms complex salts—the thio-aurate(I) salts: monothioaurates, $M^I[AuS]$, and dithioaurates, $M^I_3[AuS_2]$. Such thio-aurates are also obtained when the gold sulfides richer in sulfur, AuS and Au_2S_3, are dis-solved in colorless alkali sulfide solutions; the gold sulfides give up sulfur to the excess alkali sulfide, forming polysulfide. The thioaurates are decomposed by acids, gold(I) sulfide being thereby re-formed:

$$2[AuS]^- + 2H^+ = Au_2S + H_2S.$$

(b) Gold(III) Compounds

(*i*) *Gold(III) oxide and Gold(III) hydroxide (Auric acid)*. When a solution of gold(III) chloride is treated with alkali or alkaline earth hydroxide, or is boiled after addition of alkali carbonate, a yellow brown precipitate is obtained, consist-ing essentially of gold(III) hydroxide, but usually strongly contaminated with the precipitant. It is possible to remove the impurities by extraction with acids. Drying over phosphorus pentoxide yields an orange or ocherous powder, with the compo-sition AuO(OH). It is soluble in hydrochloric acid and other acids, if these are sufficiently concentrated, and also in hot caustic potash, and is thus amphoteric. Since its acidic character is the more strongly developed, gold(III) hydroxide is usually known as auric acid, and the salts derived from it as *aurates* (more strictly, as aurate(III) salts)—e.g., potassium aurate(III), $K[AuO_2] \cdot 3H_2O$.

From the values found by Jirsa and Jellinek (1924) for the solubility of gold hydroxide in sulfuric and nitric acids, the solubility product $K_{sp} = [Au^{+++}] \cdot [OH^-]^3$ is about $9 \cdot 10^{-46}$.

Gold(III) hydroxide may be dehydrated by cautious heating (to 140–150°). The oxide Au_2O_3 begins to lose oxygen to a marked extent at only slightly higher temperatures, however (from about 160°).

(*ii*) *Gold(III) chloride and Tetrachlorogold(III) acid (Chloroauric acid)*. *Gold(III) chloride* (auric chloride), $AuCl_3$ (or probably Au_2Cl_6), is best obtained anhydrous by the action of chlorine on gold leaf at a little above 200°, or on gold powder reduced by ferrous sulfate and well dried. It sublimes in chlorine at 200°, and condenses in the form of red needles (density 3.9). The decomposition pressure of gold(III) chloride reaches 1 atmosphere at 251°. The heat of formation is 28.3 kcal per mol (Biltz). It melts at 287–288° in chlorine at elevated pressures. It dissolves readily in water, with a brown red color, forming the complex trichloro-oxogold(III) acid, $H_2[AuCl_3O]$:

$$AuCl_3 + H_2O = H_2[AuCl_3O].$$

This acid forms an insoluble yellow silver salt, $Ag_2[AuCl_3O]$.

If hydrochloric acid is added to the brown-red solution of gold(III) chloride, it turns lemon yellow through the formation of *chloroauric acid* (tetrachlorogold(III) acid), $H[AuCl_4]$:

$$H_2[AuCl_3O] + HCl = H[AuCl_4] + H_2O.$$

The 'gold chloride' of commerce is usually chloroauric acid, and this is also the compound usually signified by the name of auric chloride in the older scientific literature.

Chloroauric acid is prepared by dissolving gold in aqua regia, and evaporating down with hydrochloric acid. Chloroauric acid crystallizes in long, bright yellow needles with the composition $H[AuCl_4] \cdot 4H_2O$, which effloresce in moist air. One molecule of water is lost in dry air. The compound is also soluble in alcohol and ether; it crystallizes anhydrous from alcohol.

Numerous salts, $M^I[AuCl_4]$, are derived from chloroauric acid, the most important being sodium chloroaurate, $Na[AuCl_4] \cdot 2H_2O$, briefly known as 'gold salt'. This crystallizes in large rhombic prisms or plates. It is soluble in ether, as well as in water, and also crystallizes from ether with $2H_2O$. When it is heated, it loses its water only at temperatures at which chlorine is evolved also. In the case of potassium chloroaurate, which is insoluble in ether, a dihydrate, $K[AuCl_4] \cdot 2H_2O$, is known, crystallizing in bright yellow rhombic plates from dilute hydrochloric acid, as well as a hemihydrate, $K[AuCl_4] \cdot \frac{1}{2}H_2O$, which separates in small yellow monoclinic needles from concentrated hydrochloric acid solutions. Both hydrates lose their water completely at 100°.

(*iii*) *Gold(III) bromide*, Au_2Br_6, and the *tetrabromoaurates(III)*, $M^I[AuBr_4]$ resemble the corresponding chloro compounds in their properties. Burawoy has shown by ebullioscopic measurements in benzene that gold(III) bromide has the dimeric formula Au_2Br_6. Potassium tetrabromoaurate(III), $K[AuBr_4] \cdot 2H_2O$, (purple red monoclinic crystals) may be prepared by treating gold with bromine water and potassium bromide. An X-ray structure determination (Cox, 1936) has shown that the Br atoms are arranged in square planar arrangement about the Au atom as a center; the K atom and H_2O molecules are built into interstices in the resulting structure. The H_2O cannot be regarded as coordinatively bound to the Au atom, since the distance $Au \leftrightarrow H_2O$ is 3.57 Å, whereas the $Au \leftrightarrow Br$ distances are only 2.50 and 2.57 Å. Cox finds that the Cl atoms are also probably disposed in planar arrangement in $KAuCl_4$.

(*iv*) *Gold(III) iodide*, AuI_3, cannot be obtained by the direct union of its components, nor by addition of soluble iodide to $HAuCl_4$ solution, since the monoiodide is formed in each case. However, if a solution of $HAuCl_4$, neutralized with potassium carbonate, is slowly added to potassium iodide solution, complex $[AuI_4]^-$ ions are first formed, and when all free iodide ions have been removed by this means, the further addition of $[AuCl_4]^-$ results in the formation of gold(III) iodide as a dark green precipitate:

$$[AuCl_4]^- + 4I^- = [AuI_4]^- + 4Cl^-$$

$$[AuCl_4]^- + 3[AuI_4]^- = 4AuI_3 + 4Cl^-.$$

When this is dried it loses iodine and is converted to gold(I) iodide. The complex acid $H[AuI_4]$ (tetraiodogold(III) acid), which is formed by dissolving gold(III) iodide in aqueous hydriodic acid, and by the action of hydriodic acid, containing iodine, on finely divided gold, is more stable than the simple triiodide, as also are its salts, the *tetraiodoaurates(III)*, $M^I[AuI_4]$ (e.g., $K[AuI_4]$, thin black prisms). These complex salts also readily lose iodine, however.

(*v*) *Gold(III) cyanide and Cyanoaurate(III) Salts*. No precipitate is obtained when potassium cyanide is added to a solution of gold(III) chloride. If the solution is evaporated, the very soluble *potassium tetracyanoaurate(III)* $K[Au(CN)_4] \cdot \frac{3}{2}H_2O$, crystallizes out in colorless plates. It may be dehydrated by heating to 200°, but loses cyanogen when more strongly heated, yielding potassium dicyanoaurate(I), $K[Au(CN)_2]$.

Other cyanoaurate(III) salts are also known—e.g., the ammonium salt $[NH_4][Au(CN)_4]$·H_2O, which is soluble in alcohol. The corresponding acid cannot be obtained, however. When strong acids are added, hydrocyanic acid is evolved, and, by evaporation over sulfuric acid, the trihydrate of gold(III) cyanide, $Au(CN)_3$·$3H_2O$, may be crystallized from the resulting solution in large colorless plates.

The reaction of the free halogens with the dicyanoaurate(I) salts yields dicyanodihalogenoaurates(III), $M^I[Au(CN)_2X_2]$, where X = Cl, Br, or I. Many salts of this type are known.

(vi) *Thiocyanatoaurate(III) Salts.* Potassium tetrathiocyanatoaurate(III) is obtained as a bulky orange precipitate, when gold chloride (neutralized with potassium carbonate) is added to an excess of cold potassium thiocyanate solution. Other thiocyanatoaurate(III) salts are also known, but the free gold(III) thiocyanate is not stable.

(vii) *Gold(III) salts of oxyacids*—e.g., gold(III) sulfate or nitrate—are only stable in the concentrated solutions of the corresponding acids. If the solutions are diluted with water, hydrolysis occurs immediately, with the deposition of gold(III) hydroxide (auric acid).

It is probable that the compounds referred to exist in solution as complex compounds, and not as simple salts. By evaporating a solution of gold(III) hydroxide in very concentrated nitric acid, over soda lime, it has been found possible to isolate the complex acid $H[Au(NO_3)_4]$·$3H_2O$ (*tetranitratogold(III) acid*) in octahedral crystals (density 2.84, m.p. 72–73°). Various salts of this acid are also known—e.g., $K[Au(NO_3)_4]$, golden yellow, lustrous rhombohedral crystals. If crystallization is carried out in the presence of more alkali nitrate, salts of hexanitratogold(III) acid are formed—e.g., dipotassium hexanitratoaurate(III), $K_2H[Au(NO_3)_6]$, tabular crystals.

The *acetatoaurates*, $M^I[Au(C_2H_3O_2)_4]$, which have been obtained from solutions of the corresponding aurates, $M^I[AuO_2]$, in glacial acetic acid, are analogous to the tetranitratoaurates.

The *sulfatoaurate(III) salts* also belong to the same type—e.g., potassium disulfatoaurate(III), $K[Au(SO_4)_2]$, a bright yellow crystalline powder obtained by dissolving gold(III) hydroxide in concentrated sulfuric acid, adding potassium hydrogen sulfate, and evaporating at 200°.

Evaporation of a solution of gold in selenic acid yields *gold(III) selenate*, $Au_2(SeO_4)_3$, in small yellow crystals.

(viii) *Gold(III) sulfide*, Au_2S_3, cannot be obtained from aqueous solution, since it is decomposed, but is formed, according to Antony and Lucchesi, by the action of hydrogen sulfide on dry lithium tetrachloroaurate(III) $Li[AuCl_4]$·$2H_2O$, at —10°. LiCl is extracted by means of alcohol from the resulting mixture of LiCl and Au_2S_3, and gold(III) sulfide remains as a black amorphous powder. It is decomposed into gold and sulfur when it is heated to 200°.

If gold(III) sulfide is treated with colorless sodium sulfide solution at 3–4°, it rapidly dissolves with a red-brown color, probably as a result of the initial formation of $Na_3[AuS_3]$. This thioaurate(III) at once decomposes, however, to give the thioaurate(I):

$$Na_3[Au^{III}S_3] + Na_2S = Na_3[Au^IS_2] + Na_2S_2.$$

(c) Organometallic Compounds of Gold

All three metals of Sub-group I can form unstable alkyl or aryl compounds in which the elements are univalent. *Copper(I) ethyl*, Cu · C_2H_5, is believed to be formed by the action of ethyl magnesium bromide on CuI, but decomposes above —50°. *Copper(I) phenyl*, Cu · C_6H_5, was obtained by Gilman (1936) by the action of phenyl magnesium bromide on CuI. It is an unstable solid (decomp. 86°), insoluble in most organic solvents, and slowly hydrolyzed by water to C_6H_6 and Cu_2O. *Silver phenyl*, Ag · C_6H_5, was first isolated in the form of the addition compound 2Ag · C_6H_5 · $AgNO_3$, by the action of $AgNO_3$ on $Bi(C_6H_5)_3$ or C_2H_5 · $Pb(C_6H_5)_3$. Ag · C_6H_5 is formed by the reaction of AgCl with the Grignard reagent. When dry, it decomposes vigorously at ordinary temperature, forming metallic silver and diphenyl, $(C_6H_5)_2$. Gold(I) alkyls may be formed by the action of the Grignard reagent on $AuCl \cdot CO$, but are very unstable.

Organometallic compounds of trivalent gold are far more stable than the foregoing, and several series of organogold(III) derivatives are known. They were first prepared by Pope and Gibson (1907). *Trialkyls*, R_3Au, have been prepared at low temperatures, in the form of their ether addition compounds (e.g., $(CH_3)_3Au \cdot O(CH_3)_2$, decomp. —35°), by the reaction of lithium alkyls with gold(III) bromide. They react with hydrogen chloride, or with gold(III) bromide, to form the *dialkyl gold(III) halides*, R_2AuX, which are also obtained directly by the action of Grignard reagents on gold(III) halides. The diakyl gold(III) halides and the *monoalkyl gold(III) dihalides* (obtained, e.g., by the reaction of bromine with the dialkyl gold(III) compounds) are covalent compounds, soluble in organic solvents. They react with Ag_2SO_4 to give the sulfates $[R_2Au]_2SO_4$, and with AgCN to give cyanides R_2AuCN.

All these compounds are polymerized, and have structures in which the gold attains its stable coordination number (4) and stable stereochemical configuration (square coplanar arrangement). The dialkyl gold(III) halides are dimeric in solution, forming molecules (I) which are also present in the crystalline compound, as is proved by X-ray analysis [Burawoy, Gibson, Hampson, and Powell, *J. Chem. Soc.*, (1937) 1690]. The dialkyl gold(III) cyanides are

$$
\begin{array}{ccc}
R & Br & R \\
\diagdown & \diagdown & \diagup \\
 & Au \quad\quad Au & \\
\diagup & \diagup\diagdown & \diagdown \\
R & Br & R
\end{array} \qquad (I)
$$

tetrameric in solution, $[R_2AuCN]_4$. The only structure compatible with the coordinating properties of gold, and the collinearity of groups bound to the —CN-radical is the 12-membered cyclic structure (II). This unusual molecular structure has been confirmed by the full X-ray analysis of the propyl compound $[(C_3H_7)_2AuCN]_4$ [Phillips and Powell, *Proc. Roy. Soc.*, 173A, (1939) 147]. The dialkylgold cyanide readily decomposes, losing the hydrocarbon R_2 and forming first a more highly polymerized monoalkyl gold cyanide, $[RAuCN]_x$, and ultimately gold(I) cyanide, which has the structure of a linear polymer, $[Au—C{\equiv}N{\rightarrow}Au—C{\equiv}N{\rightarrow}....]_x$.

$[(C_2H_5)_2Au]_2SO_4$ is soluble in benzene, cyclohexane, acetone, and water. It is dimeric in inert organic solvents, but is electrolytically dissociated in aqueous solution, and

$$
\begin{array}{ccc}
R & & R \\
| & & | \\
R—Au—C{\equiv}N{\rightarrow}Au—R \\
\uparrow & & | \\
N & & C \\
{\parallel}{\parallel}{\parallel} & & {\parallel}{\parallel}{\parallel} \\
C & & N \\
| & & | \\
R—Au{\leftarrow}N{\equiv}C—Au—R \\
| & & | \\
R & & R
\end{array} \qquad (II)
$$

gives the reactions of $SO_4^=$ ions. These are probably formed by the reaction:

$$[(C_2H_5)_2Au]_2SO_4 + 4H_2O = 2[(C_2H_5)_2Au(OH_2)_2]^+ + SO_4^=.$$

Anhydrous gold(III) chloride reacts directly with aromatic hydrocarbons. If reaction is allowed to go to completion, the products are AuCl, HCl, and chlorinated hydrocarbons, but the addition of ether checks the reaction at an intermediate stage. Monoaryl gold(III) dihalides—e.g., $(C_6H_5)AuCl_2$—may then be isolated. These dissolve unchanged in solutions of potassium chloride, probably forming complex salts (Kharasch, and Isbell 1931). The reaction is known as *auration* of the hydrocarbons, and takes place readily with a variety of aromatic compounds.

10. Analytical (Gold) [12]

Gold is generally detected by means of reducing agents, which deposit *metallic gold*. In the compact state, gold is immediately recognizable by its fine yellow color. If reduction is carried out, in very dilute solution, the gold remains in colloidal dispersion, and, provided that suitable reducing agents have been used (e.g., iron(II) sulfate), is then detectable even in minimal amounts by the color of the colloidal sol. If tin(II) chloride is used for the reduction, an intensely colored and particularly stable colloidal dispersion is obtained ('purple of Cassius'). In this reaction the tin dioxide hydrate, formed in the reaction, serves as a protective colloid (cf. p. 677).

The color reaction with dithizone (cf. p. 408)is also a very sensitive test for gold, as also is the benzidine blue reaction of Tananaev (1930), which is best carried out as a drop reaction on filter paper. The latter reaction depends on the oxidation of benzidine, NH_2—C_6H_4—C_6H_4—NH_2, by gold(III) ions to a quinonoid dyestuff. It must be noted, however, that other oxidants (e.g., manganese dioxide hydrate) can also bring about formation of benzidine blue.

For quantitative determination, gold is also usually separated as the metal and weighed in that form. Even very small quantities of gold can be determined in this manner.

Haber worked out a method for the determination of quite minimal quantities of gold (down to 10^{-10} g). It is based on the fact that the gold can first be precipitated together with lead, which may, if necessary, be used in a very large excess*. The lead is then cupelled off, the remaining grain of gold fused into a sphere within a borax bead, and its diameter measured under a microscope. The greatest difficulty in determining minimal quantities of gold consists in excluding the accidental introduction of gold into the analysis, from traces of gold present in the reagents.

* Even if the gold is present in extreme dilution in solution, or in suspension in water— e.g., for such purposes as the determination of the gold content of sea or river water—it is possible to carry it down quantitatively by producing a precipitate of lead sulfide in the sample.

References

1 N. E. CRUMP, *Copper; A Survey of Sources of the Metal, Methods of Manufacture, Uses and Conditions of Trade*, New York 1926, 253 pp.
2 M. WAEHLERT, *Die Kupferraffination*, Halle 1927, 142 pp.
3 A. SCHIMMEL, *Metallographie der technischen Kupferlegierungen*, Berlin 1930, 134 pp.
4 O. BAUER and M. HANSEN, *Der Aufbau der Kupfer-Zink-Legierungen*, Berlin 1927, 150 pp.
5 H. C. DEWS, *The Metallurgy of Bronze*, London 1930, 147 pp.
6 R. HINZMANN, *Nichteisenmetalle*, Vol. I, *Kupfer, Bronze, Messing, Rotguss*, 2nd Ed., Berlin 1941, 62 pp.
7 R. KRULLA, *Neusilber; Eigenschaften, Verwendung, etc.*, Munich 1935, 63 pp.
8 W. LAATSCH, *Die Edelmetalle – eine Übersicht*, Berlin 1925, 91 pp.
9 E. RAUB, *Die Edelmetalle und ihre Legierungen*, Berlin 1940, 323 pp.
10 G. EGER, *Das Scheiden der Edelmetalle durch Elektrolyse*, Halle 1929, 120 pp.
11 E. A. SMITH, *Working in Precious Metals*, London 1935, 399 pp.
12 A. WOGRING, *Analytische Chemie der Edelmetalle* [Vol. 36 of *Die chemische Analyse*, edited by W. BÖTTGER], Stuttgart 1936, 141 pp.
13 G. BERG and F. FRIEDENSBURG, *Das Gold* [*Die metallischen Rohstoffe*, Vol. 3], Stuttgart 1940, 248 pp.
14 T. K. ROSE and W. A. C. NEWMAN, *The Metallurgy of Gold*, 7th Ed., London 1937, 561 pp.
15 M. ANDRESEN, *Photo-Handbuch*, Berlin 1930, 376 pp.
16 W. MEIDINGER, *Die theoretischen Grundlagen der photographischen Prozesse* [Vol. 5 of *Handbuch der wissenschaftlichen und angewandten Photographie*, edited by A. HAY and M. VON ROHR], Vienna 1932, 513 pp.

CHAPTER 9

SECOND SUB-GROUP OF THE PERIODIC SYSTEM: ZINC, CADMIUM, AND MERCURY

Atomic numbers	Elements	Symbols	Atomic weights	Densities	Melting points	Boiling points	Specific heats	Valence states
30	Zinc	Zn	65.38	7.13	419.4°	906°	0.0924	II
48	Cadmium	Cd	112.41	8.64	320.9°	767°	0.0553	II
80	Mercury	Hg	200.61	13.595	—38.84°	356.95°	0.0334	(I) II

1. Introduction

(a) General

Sub-group II of the Periodic System comprises the elements zinc, cadmium and mercury. The two former are almost invariably *dipositive* in their compounds as is typical of elements of Group II. Mercury can not only have a valence state of +2, but can also be unipositive in many of its compounds. However, the mercury(I) compounds are invariably polymerized, in that they always contain a pair of linked mercury atoms—as, for example, in mercury(I) chloride, Cl—Hg—Hg—Cl—so that mercury is only univalent in the electrochemical sense, but is formally bivalent.

The resemblances between the Main Group and Sub-group elements are greater in Group II than in Group I. In particular, zinc and cadmium are closely related to magnesium. Thus the sulfates of all these elements are isomorphous (when similarly hydrated), and their double salts and complex salts often have analogous compositions. The double sulfates of mercury, of the type $M^I_2SO_4 \cdot HgSO_4 \cdot 6H_2O$, correspond to schönite, $K_2SO_4 \cdot MgSO_4 \cdot 6H_2O$. The elements of Sub-group II also share with magnesium the ability to form alkyl compounds readily.

Zinc and cadmium have the same crystal structure as beryllium and magnesium (cf. Fig. 57, p. 249 of Vol. I). For Zn, $a = 2.6595$ Å, $c = 4.9368$ Å; for Cd, $a = 2.9731$ Å, $c = 5.6069$ Å. The length of the cell edge in the basal plane, a, is here equal to the shortest interatomic distance. The crystal structures of zinc and cadmium deviate from perfect hexagonal close packing, in the sense that the c axis is somewhat stretched. Axial ratios are, for hexagonal close packing, $c/a = 1.633$, for Zn $c/a = 1.86$, for Cd $c/a = 1.89$. For the crystal structure of Hg, see p. 457.

Table 42 shows the atomic and ionic radii deduced from the cell dimensions of the metals and their crystalline compounds. Ionic radii are those for the bivalent ions.

TABLE 42

APPARENT ATOMIC AND IONIC RADII OF ELEMENTS OF THE ZINC GROUP

Element	Zinc	Cadmium	Mercury
Atomic radius, Å	1.33	1.49	1.50
Ionic radius, Å	0.83	1.03	1.12

The metals of Groups IIA and IIB all have relatively low melting points and relatively high volatility. These properties are more marked in the Sub-group than in the Main Group, and mercury has the lowest melting point of all the metals.

As follows from Trouton's Rule (Vol. I, p. 322), and their low boiling points, the metals of the zinc group have low latent heats of vaporization (cf. Table 43)—considerably smaller than those of the metals of Group IIA (cf. Vol. I, Table 46, p. 235), and very much smaller than those of the metals of the copper group (cf. Table 36, p. 362).

Like the metals of the copper group, the Group IIB metals are diamagnetic. Their magnetic susceptibilities are given in Table 43.

TABLE 43

MOST IMPORTANT PHYSICAL CONSTANTS OF THE ELEMENTS OF THE ZINC GROUP

	Heat of fusion		Heat of vaporization (at the b.p.)		Heat of sublimation (at 0° K) in	Spec. electrical conductivity at 18°	Spec. magnetic susceptibility at 18°
	cal/g	kcal per g-atom	cal/g	kcal per g-atom	kcal per g-atom		
Zinc	26.4	1.73	419	27.4	31.4	$16.5 \cdot 10^4$	$-0.15 \cdot 10^{-6}$
Cadmium	12.9	1.46	212	23.9	27.0	$13.2 \cdot 10^4$	$-0.17 \cdot 10^{-6}$
Mercury	2.74	0.55	69.5	14.0	15.45	$1.044 \cdot 10^4$	$-0.168 \cdot 10^{-6}$

The metals of Group IIB, like those of Group IB, are *heavy metals*; in this they differ from the elements of Group IIA. The electrochemical character also alters in the sequence zinc—cadmium—mercury in the same way as it does in the sequence copper—silver—gold, but in the opposite sense to the gradation in the Main Group: the affinity for the positively charged state diminishes markedly from zinc to mercury, so that zinc and cadmium stand above hydrogen in the electrochemical series, but mercury below. Mercury, indeed approaches the noble metals in properties. It does form an oxide by direct combination with oxygen, but this breaks up again into the free metal and oxygen if it is rather more strongly heated.

As in Sub-group I, the oxides are colored, although zinc oxide is colored only when heated. The hydroxides are very sparingly soluble, and only weakly basic in character. Zinc hydroxide is definitely amphoteric*. As with silver, the oxide of mercury is precipitated instead of the expected hydroxide (cf. p. 464), when hydroxyl ions are added to mercury salt solutions.

* $Cd(OH)_2$ and HgO also possess a certain measure of amphoteric character, although far less strongly developed than for $Zn(OH)_2$. It shows itself in that $Cd(OH)_2$ and HgO are rather more soluble in concentrated caustic alkali than in dilute.

The sulfides of the Sub-group II elements are insoluble in water, like the sulfides of the copper group, whereas the sulfides of the alkaline earths are soluble. Mercury(II) sulfide is black, like the Group IB sulfides; cadmium sulfide is also colored (yellow), but the zinc compound is white.

The spectra of the elements of Sub-group II are closely related in structure to those of the Main Group (especially magnesium)—i.e., in the arrangement of energy levels from which the spectra arise. As with the alkaline earth metals, the arc spectra of zinc, cadmium, and mercury involve *two* systems of principal, subsidiary, and fundamental series—namely a singlet system and a triplet system. The spark spectra, like those of the alkaline earth

TABLE 44

ELECTRONIC CONFIGURATION IN ATOMS OF THE ALKALINE EARTH METALS AND ATOMS OF THE ZINC GROUP

Element	Atomic number	1s	2s	2p	3s	3p	3d	4s	4p	4d	4f	5s	5p	5d	6s
Calcium	20	2	2	6	2	6		2							
Zinc	30	2	2	6	2	6	10	2							
Strontium	38	2	2	6	2	6	10	2	6			2			
Cadmium	48	2	2	6	2	6	10	2	6	10		2			
Barium	56	2	2	6	2	6	10	2	6	10		2	6		2
Mercury	80	2	2	6	2	6	10	2	6	10	14	2	6	10	2

metals, are similar in structure to the arc spectra of the alkali metals. The ground term, both in the arc spectra and in the spark spectra, is an *s* term in each case—i.e., both the most loosely bound and the second most loosely bound electron occupy orbits with the subsidiary quantum number $l = 0$ in the ground state of the atom. The configuration of the most loosely bound electrons—the 'valence electrons'—is thus the same in the elements of Sub-group II and in the elements of the Main Group. Table 44 sets out the electronic configuration of the elements of Sub-group II, as deduced from the principles of the Bohr theory, with the configurations of the alkaline earth elements for comparison. It may be seen that the essential difference in electronic structure between the zinc atom and the calcium atom lies in the fact that, in zinc, 10 additional electrons in the 3d level are interposed between the 'shell' containing the valence electrons, characterized by the quantum symbol 4s, and the 3s and 3p levels of the 'argon shell'. In zinc (as contrasted with copper), these are so firmly bound that they have no influence on the valence *number* of zinc. However, the effect of this intermediate shell of 10 electrons is not confined simply to screening a corresponding amount of the nuclear charge: if this were so, the resemblance between the Zn^{++} and Ca^{++} ions would be much closer than it is. The difference between the atoms of cadmium and strontium is similar to that between zinc and calcium, as also is the difference between the mercury atom and the barium atom, except that in the latter case there is interposed not merely the 10 electrons of the 5d shell, but also another intermediate shell of 14 electrons, in the 4f level. The latter levels lie so deep, however, that their existence is practically without effect upon the valence properties of the atom.

It is probable that the *polarizing effect* of the ions of the elements of Sub-group II, which is distinctly greater than would be expected from their ionic radii, can be related to the presence of the intermediate levels. This property shows itself in a marked tendency to form layer lattices and structures which are 'incommensurable' with the ordinary ionic lattice types (cf. p. 380). In those compounds which do have structures commensurable with ordinary ionic lattices, all the elements of Sub-group II have larger ionic radii than magnesium (cf. Fig. 3, p. 16 of Vol. I). In compounds which are incommensurable with ordinary ionic structures, however, the interatomic distances are considerably smaller than would be predicted, using the ionic radii deduced from structures of the first type. This contraction of radii can be ascribed to the effect of polarization. It is often assumed (as in the case of

TAI

MISCIBILITY AND COMPOUND FORMATION BETWEEN ELEME

Symbols have the same mea

	Li	Na	K	Rb	Cs	Mg	Ca	Sr	Ba	B
Zn	$s > 0$ $[\overline{LiZn}]$? $\overline{Li_2Zn_3}$ 520° $\overline{LiZn[]}$ 93° $\overline{LiZn_{2.5}}$* 502° $\overline{LiZn_4}$* 481°	$liq < \infty$ $s\ 0$ $NaZn_4$(a) unstab. $NaZn_{13}$ 557°	$liq < \infty$ $s\ 0$ KZn_{13} 585°	0	0	$s > 0$ $\overline{MgZn}$* 354° $\overline{MgZn_2}$ 590° $\overline{MgZn_5}$* 381°	$s\ 0$ Ca_4Zn* 450° $\overline{CaZn}$* 431° $\overline{Ca_2Zn_3}$ 688° $CaZn_4$ 680° $CaZn_{10}$ 717°	alloys	alloys	liq s 0
Cd	$s > 0$ Li_3Cd 272° $LiCd$ 549° $LiCd_3[]$ 370° also 'β-phase'	$s\ 0$ $NaCd_2$ 385° $NaCd_5$ 360°	$liq < \infty$ $s\ 0$ KCd_7? KCd_{13} 487°	$liq < \infty$ $RbCd_{13}$	$liq < \infty$ $CsCd_{13}$	$s < \infty$ $[Mg_3Cd]$ 150° $[MgCd]$ 251° $[MgCd_3]$ 89°	$liq < \infty$? $s < \infty$ Ca_3Cd_2* 510° $\overline{CaCd}$ 685° $CaCd_3$ 612°	$s\ 0$? $SrCd$ $SrCd_{12}$	alloys	—
Hg	$s \sim 0$ Li_6Hg* 164° Li_3Hg 375° Li_2Hg* 375° $LiHg$ 590° $LiHg_2$* 340° $LiHg_3$* 235°	$s\ 0$ Na_3Hg* 35° Na_5Hg_2* 66° Na_3Hg_2* 119° $NaHg$ 212° Na_7Hg_8* 222° $NaHg_2$ 354° $NaHg_4$* 156°	$s\ 0$ KHg* 178° KHg_2 279° KHg_3* 204° K_2Hg_9* 173° KHg_9* 70°	$s\ 0$ Rb_7Hg_8* 157° Rb_3Hg_4* 170° $RbHg_2$ 256° Rb_2Hg_7 197° Rb_5Hg_{18}* 194° Rb_2Hg_9* 162° $RbHg_6$* 132° $RbHg_9$* 67°	$s\ 0$ Cs_2Hg_3?* 171° $CsHg_2$ 208° $CsHg_4$ 164° $CsHg_6$ 158° $CsHg_{10}$* 73°	$s\ 0$ Mg_3Hg* 509° Mg_5Hg_2? Mg_2Hg Mg_5Hg_3 $MgHg$ $MgHg_2$ 170°	$s\ 0$? $CaHg_3$ $CaHg_5$* 265° $CaHg_{10}$* 84°	— Sr_2Hg_5? $SrHg_8$? $SrHg_{12}$?	— $BaHg_{10}$* 165°	—

(a)   This compound is obtained from liquid ammonia solutions and not from melts.

many compounds of Sub-group I) that this mutual distortion of the components of such compounds goes so far that the atoms quite lose their ionic character, and the binding forces become *covalent*.

The *alkyl compounds*, which are formed by all the elements of Sub-group II, are undoubtedly covalent compounds. The most readily obtained of these are the zinc alkyls

SUB-GROUP II AND THE ELEMENTS OF THE MAIN GROUPS

Table 9, p. 49.

Al	Ga	In	Tl	Sn	Pb	As	Sb	Bi	S	Se	Te
< ∞ 0	s 0 0	s 0 0	liq < ∞ s 0 0	s 0 0	liq < ∞ s 0 0	s 0 Zn_3As_2 1015° $ZnAs_2$ 771°	s 0 Zn_3Sb_2 566° Zn_4Sb_3* 563° ZnSb 546°	liq < ∞ s > 0 0	— ZnS	liq < ∞ ZnSe	s 0 ZnTe 1239°
< ∞ s 0 0	liq < ∞ s 0 0	s > 0 0	s ~ 0 0	s > 0 0?	s > 0 0	s 0? Cd_3As_2 721° $CdAs_2$ 621°	s 0 Cd_3Sb_2 unstab. 423° CdSb 456°	s 0 0	— CdS	liq < ∞ CdSe	s 0 CdTe 1045°
> 0 0	liq < ∞ s 0 0	— —	s < ∞ Tl_2Hg_5 14.5°	s ~ 0 Hg_3Sn? $HgSn_{12}$?	s < ∞ 0	— Hg_3As_2	— Hg_3Sb_2	s 0 0	— HgS	— HgSe	s 0? HgTe

(p. 444). Their formation can be interpreted in the same way as for the elements of Main Group II (cf. Vol. I, p. 239).

(b) Alloys

The metals of the zinc group, like those of Sub-group I, have great alloying power. Their behavior towards the metals of the Main Groups is summarized in Table 45. Tables given in earlier chapters have included the alloys formed with the metals of the Sub-groups already discussed. The Sub-group II metals also form numerous intermetallic compounds with other metals, but do not combine either among themselves or (with very few exceptions) with the metals of Groups IIIA and IVA (cf. Table 45).

Compounds of the Sub-group II elements with other metals can be divided into four groups. The first of these comprises the compounds formed with strongly electropositive metals (alkali and alkaline earth metals); as is shown in Table 45, these are extraordinarily numerous,—especially the compounds of the alkali metals with mercury. These conform to a rule stated by Biltz and Weibke:—as judged by the mercury content and thermal stability of the various compounds, the capacity to combine with mercury increases amongst the alkalis from lithium to cesium. The existence and stability of lower mercurides is typical of the lighter alkali metals, whereas with the heavier alkalis the higher mercurides are more stable. The second group of compounds includes the compounds of the Sub-group II metals with elements of only weakly metallic character, such as arsenic and antimony. Compounds of this group constitute a transition towards the normal valence type, formed with non-metals, and they generally have compositions corresponding to normal valence compounds. It is noteworthy that between these two classes—compounds of Sub-group II metals (a) with the strongly electropositive metals and (b) with the very weakly electropositive elements—there is interposed a section of the Periodic Table (Main Groups III and IV) in which practically no compounds are formed with the elements of Sub-group II. The third group of compounds are the 'electron compounds', of the Hume-Rothery type. It may be seen from Table 1 (p. 21) that zinc and cadmium are particularly prone to form compounds of this type, and invariably enter into them as 'metals of the IInd type'. It follows that the formation of this group of compounds is also restricted to a definite part of the Periodic System, namely that in which the 'metals of the first type' are found. The metals of Sub-group II appear to have a very small capacity for forming compounds with the metals of Sub-groups IV to VI and none have as yet been definitely characterized. If any exist, they can be assigned to a fourth class of compounds, which includes the compounds between the Sub-group II elements and the metals of Sub-groups VII, VIII and I, in so far as they are not of Hume-Rothery type. Only a few compounds of this class are known, and the factors which determine their formation and composition have not been worked out.

Compounds of the elements of Sub-group II with *non-metals* are almost invariably of the normal valence type. This is true even for those non-metals which, with elements of the Sub-groups previously considered, tend to form compounds which approach the intermetallic type. This corresponds with the fact that, in atomic structures, the metals of Sub-group II are rather similar to those of Main Group II; as is usual with metals in the Main Groups, these form normal valence compounds with the non-metals.

2. Zinc (Zn)

(a) Occurrence [1]

Zinc is widely distributed in the earth, in the form of its compounds. The most common ores are *calamine* or zinc spar, $ZnCO_3$, and *zinc blende* (sphalerite, marmatite, or 'black jack'), ZnS. A few silicates are also of importance as zinc ores, namely *siliceous zinc ore*, $Zn_2SiO_4 \cdot H_2O$, *willemite*, Zn_2SiO_4, and *troostite*, $(Zn, Mn)_2SiO_4$.

Ores which are almost peculiar to America include *zincite* or red zinc ore, ZnO, and *franklinite*, $(Zn, Mn)O \cdot Fe_2O_3$. Related to the latter is *zinc spinel* or gahnite, $ZnAl_2O_4$, which occurs in well formed crystals, but usually only in very small amounts. It therefore has no significance as an ore mineral.

Iron ores often contain small amounts of zinc. When they are smelted in the iron blast furnace, the zinc is enriched in the dust which is deposited from the stack gases (especially in the finest fractions). These can contain as much as 30 % of zinc (as oxide), and are therefore valuable as a raw material for the extraction of the metal, to supplement the zinc ores proper.

(b) History

As a pure metal, zinc became known only in relatively recent times, at least in Europe, since the technique of extraction from its ores is not a simple one. Its alloy with copper, *brass*, has been known since Homeric times, however, and was obtained by melting copper with an ore which was known to the Greeks as καδμεία (referred to as cadmia by Pliny). It is possible that the name calamine for the earliest recognized zinc ore was derived from this name.

Pure zinc seems to have become known in Europe towards the end of the Middle Ages, and was first clearly recognized as a metal by Paracelsus. It was obtained from China and India, where it had long been known, until in 1721 Henkel found that zinc could be obtained from calamine. A works for the production of zinc was erected about 1740 in Bristol (England), by Champion. The smelting of zinc in Silesia began in 1798. This region became the principal producer of zinc until 1907, when it was exceeded by the United States.

(c) Preparation [*1, 3*]

Until recently, zinc was extracted almost exclusively by purely chemical methods (by the so-called distillation process). In recent years, the electrolytic process has come into use on a rapidly increasing scale, and the electrolytic extraction of zinc now ranks after the production of copper as the most important electrometallurgical process carried out in solution. The distillation process is now chiefly used for working up those zinc ores which present special difficulty for the application of electrolytic methods—e.g., in smelting zinc blendes which contain certain silicates as gangue minerals; if these are decomposed by acids, the resulting leaching solutions are excessively difficult to filter.

The purely chemical extraction by the *distillation process* [*2*] is based on the ease with which zinc oxide is reduced by carbon (coke):

$$ZnO + C = Zn + CO. \tag{1}$$

Zincite, franklinite, and the anhydrous silicate can be subjected directly to reduction with coke. Calamine is first converted to zinc oxide, by heating it in a shaft furnace (2), and blende is roasted in a reverberatory furnace or muffle (3):

$$ZnCO_3 = ZnO + CO_2 \tag{2}$$

$$ZnS + \tfrac{3}{2}O_2 = ZnO + SO_2 \ (+114.5 \text{ kcal}). \tag{3}$$

The SO_2 formed by reaction (3) is ordinarily converted to sulfuric acid.

The old 'muffle process' is still used to some extent for the reduction. In this, the zinc oxide, mixed with coke, is heated in relatively small, inclined, fireclay retorts or muffles, a considerable number of which are usually combined to make a single furnace. The zinc distils out of the muffles, and is carried over by the escaping carbon monoxide into small receivers in which it condenses as a liquid. Any zinc which does not condense there is caught as 'zinc dust' (a mixture of powdered metal and zinc oxide) in sheet iron cans (known as adapters or assay crucibles) attached to the receivers. The crude zinc or *spelter* which is produced in this way usually contains a considerable amount of lead as impurity, and some iron as well. It is purified by remelting. The *refined zinc* usually still contains about 1% of lead. *Fine zinc*, with a content of at least 99.8% Zn, can be obtained from this by redistillation*.

* Spring and Romanoff give the solubility of lead in zinc, at the melting point of the latter, as 1.5% by weight. However, the vapor pressures of lead and zinc at 1000° are roughly in the ratio 1 : 2000. It follows that a far more complete separation between zinc and lead can be acheived by distillation than by remelting. By close control of the process, it is indeed possible to prepare zinc of 99.99% purity on the technical scale.

The *zinc dust* obtained as a by-product in this process cannot be melted down directly, since the individual particles are covered by an oxide coating. Much of it comes directly into commerce (for uses, see p. 433); otherwise it is mixed in with the zinc ores to be smelted, and re-cycled.

The *New Jersey process* is considerably more economic in operation than the old muffle process; it permits continuous operation, and yields directly a crude zinc of relatively high purity (99.5–99.8%). It utilizes vertical retorts built of silicon carbide bricks, which are externally heated to 1200–1300° by means of producer gas. The charge is introduced into these retorts in the form of briquettes, which are made from a mixture of finely ground zinc oxide (or roasted blende) and coke, with a suitable binder, by pressing at a moderate temperature. The zinc, which collects in a condenser, can be refined by a further process of the New Jersey Zinc Company, which involves heating it to 1100°; the vapors are fractionally condensed in a rectification column (similar in principal to that used in the rectification of alcohol). Zinc rich in cadmium passes into the upper end of the column, fine zinc with a purity of 99.995% collects in the middle, while the so-called 'wash zinc', which contains all the lead, copper, and iron present in the original crude zinc, is obtained from the bottom of the column.

Zinc is now (since 1951) extracted from ores with a high iron content by the *Sterling process*, which was also developed by the New Jersey Zinc Company. This employs an electric arc for heating, and the radiant heat brings about a reduction of the ZnO, by the carbon in the furnace charge, before any molten slag has been formed. Reduction of the Fe_2O_3 to Fe takes place only after this stage is reached, and the iron collects in the molten state below the slag. The slag and the iron are tapped off at intervals, while the zinc vapors leaving the upper part of the furnace are condensed to liquid zinc in a condenser.

The *electrolytic preparation of zinc* [3] is generally carried out by leaching zinc from its ores as the sulfate, by means of sulfuric acid, and depositing the metal electrolytically from the acid sulfate solution, at a high current density.

The fact that zinc can be electrodeposited from *acid* solution, although it stands well above hydrogen in the electrochemical series, depends upon the high over-voltage of hydrogen when liberated cathodically at a zinc surface. This factor is operative only as long as the zinc is not contaminated by substances which have a smaller over-voltage for hydrogen. The hydrogen will otherwise be discharged and liberated at these impurities; this results not only in a reduction of the current efficiency, through liberation of hydrogen, but may lead to the dissolution of zinc which has already been deposited, as a consequence of the establishment of local couples (p. 432). This effect can be brought about by even quite small traces of impurities with low over-voltage, which may be present in the electrolyte, if they are more noble than zinc, and are therefore deposited at the same time. For example, germanium, which is present in traces in many zinc ores, has occasionally caused great difficulty in the electrolytic separation of zinc. The more noble impurities are now generally removed by adding zinc dust to the electrolyte solution. The manufacture of anodes which were sufficiently resistant towards electrochemical attack also presented great difficulties at first. For these reasons, the electrolytic extraction of zinc only developed into a large scale industry during 1920–30, although it had been introduced in 1894 (by Höpfner at Fürfurth-on-Lahn), and is in general preferable to the distillation process. When the technical difficulties had once been overcome, the electrolytic process expanded rapidly, and by 1934 a third of the world's zinc production came from this process. It not only furnishes zinc of a very high degree of purity, but the recovery of zinc content from the ores used is much more efficient than is usual with the distillation process, and especially as compared with the old muffle process (Belgian or Silesian process). It can also be applied to low-grade ores, which are not suitable for smelting with coke, and it permits the recovery of other valuable metals which are frequently present in the zinc ores. However, it requires the expenditure of a relatively high consumption of electrical energy. Cell voltages of 3.3 to 3.7 volts are customary, which—with a current efficiency of 90%—corresponds to 3000–4000 kWh of direct current energy per ton of zinc, without making allowance for current losses in conductors and converters.

The electrolytic preparation from zinc sulfate also has the advantage, as compared with distillation processes, of being more widely applicable, and imposing less limitation on the type of ore used. Thus it makes it possible to work up highly siliceous ores, or ores of relatively poor zinc content, while recovery of the other metals present in the ores can readily be effected. The economics of the electrolytic process, however, depend greatly on the cost of generation of electric current. About 40% of the present world zinc production comes from the electrolysis of zinc sulfate.

Fused salt electrolysis (from $ZnCl_2$-KCl mixtures) has occasionally been used for the preparation of zinc. It has some advantages for ores from which $ZnCl_2$ can readily be obtained by chlorination and distillation.

Zinc is also obtained in the form of $ZnCl_2$ from pyritic ores which contain zinc, by leaching the products of a chloridizing roasting process. $ZnCl_2$ solutions with a high NaCl content are formed, and zinc may be deposited electrolytically from these by the *amalgam process*, developed (1937) by the Duisburg copper smelters. This is based on the electrolysis of alkali chlorides. Graphite is used for the anode, and the cathode is formed by mercury flowing in a thin layer over the bottom of the cell. The resulting zinc amalgam runs through a second cell in which it forms the anode. If is to be converted into zinc white, a solution of $NaHCO_3$ is used as electrolyte. As the zinc goes into solution, it is precipitated as carbonate, which is ignited to the oxide. If the zinc is to be recovered as the metal, a $ZnSO_4$ solution is employed as electrolyte in the second cell, and the zinc is deposited on sheet aluminum cathodes.

(d) **Properties**

Zinc is a bluish-white metal with a strong luster, which gradually disappears in moist air as a result of superficial oxidation. It is fairly brittle at ordinary temperature, but is sufficiently ductile between 100° and 150° to be rolled into sheet or drawn into wire. Above 200°, however, it becomes so brittle that it can be pulverized by grinding. Zinc is very fusible (m.p. 419.4°), and is one of the most volatile metals (b.p. 905.7°). The specific gravity of the purest cast zinc is 7.13 (at 18°), that of rolled zinc being a little higher. The density of liquid zinc is 6.92 at the melting point. The hardness of zinc is 2.5 on Mohs' scale. The mechanical properties of zinc are considerably modified by impurities. Thus the presence of 0.12% of iron markedly impairs the rolling properties.

The thermal conductivity of zinc is 61—64% and the electrical conductivity 27% of that of silver (cf. Table 43 p. 424).

The *standard potential* of zinc is +0.762 volts, relative to the normal hydrogen electrode (La Mer, 1934). Zinc accordingly stands well above hydrogen in the electrochemical series.

In accordance with the fairly strong tendency of zinc to become positively charged, which is evident from the position of zinc in the electrochemical series, zinc dissolves vigorously in dilute acids, with the evolution of hydrogen:

$$Zn + 2H^+ = Zn^{++} + H_2.$$

It is not perceptibly attacked by pure water, but it dissolves in strong caustic alkalis, and also in aqueous ammonia and in ammonium chloride solution, especially when these are warm. If the zinc is extremely pure, however, the dissolution is inhibited, or takes place only very slowly. The same applies to dissolution in acids. When very pure zinc is to be dissolved in acids, it is therefore usual to add a few drops of a very dilute solution of copper sulfate, to accelerate the reaction. Copper, which stands far below zinc in the electrochemical series, is at once precipitated by displacement, contaminates the surface of the zinc, and accelerates the dissolution.

The action of impurities in the zinc or on its surface, in speeding up the dissolution, arises from the formation of galvanic couples between the zinc and the impurities. Those portions of the zinc which are not covered by the impurities then go into solution, while the hydrogen is liberated on the impurities, which generally consist of more noble metals. By a process of charge transfer within the metal, *local currents* come into operation, such as are shown (by the arrows) in Fig. 48.* In the absence of impurities, the hydrogen can be discharged only at the surface of the zinc itself. It is generally assumed that it remains on the surface, to form an exceedingly thin film, which hinders the entry of further portions of zinc into the solution; the further liberation of hydrogen is thereby inhibited, so that bubbles cannot grow and break away. In view of the high overvoltage which must be established before hydrogen can be evolved from a zinc surface, it is likely that the hydrogen ions cannot be discharged, or at least that the combination of the resulting H atoms to form H_2 molecules cannot take place on an absolutely pure zinc surface. (Vol. I, p. 31 et seq). The effect is so marked that really pure zinc is practically insoluble in dilute acids.

By amalgamating the zinc surface, the effect of the impurities is diminished, since local currents can no longer be formed, as indicated in Fig. 48, on the uniformly amalgamated surface. Dissolution of the zinc then takes place only when some other metal, such as copper,

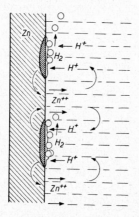

Fig. 48. Effect of impurities in the surface of zinc, in accelerating dissolution through formation of local couples (schematic).

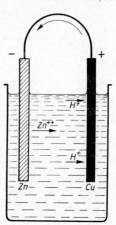

Fig. 49. Galvanic cell.

dips into the solution and is joined to the zinc by a conducting path as in the galvanic cell shown in Fig 49 [6]; the zinc dissolved is then equivalent to the hydrogen liberated on the copper. Zinc plates used in galvanic cells are customarily amalgamated before making them up into cells, in order to give them a long life.

According to the Nernst formula (eqn. (1) on p. 29 of Vol. I), the deposition potential of hydrogen from pure water is only 0.41 volt greater in magnitude than from a solution 1-normal with respect to H^+ ions. A zinc electrode therefore assumes a negative potential, with respect to the hydrogen electrode, if it dips into pure water or a neutral salt solution. In spite of this, detectable amounts of zinc do not dissolve in water, even when the zinc is impure so that local currents could be set up. This is because zinc hydroxide, formed by reaction with water, is practically insoluble, and as soon as it is formed in traces at the surface, the zinc is protected from further reaction with the water. If the protective film is removed by the addition of substances in which zinc hydroxide is soluble (e.g., caustic alkali or ammonia), the zinc dissolves with evolution of hydrogen. Ammonium salts can also facilitate the dissolution of zinc, since some ammonia is always present in their solutions through traces of hydrolysis. The use of ammonium chloride as electrolyte in dry cells and Leclanché cells depends on this fact.

* The formation of *local currents* in this way plays an important role quite generally in the dissolution of metals in acids.

In a solution 1-normal with respect to hydroxyl ions, the solution potential of hydrogen ($+0.81$ volts relative to the normal hydrogen electrode) lies above the normal potential of zinc. However, the solution potential of zinc is also displaced considerably towards more positive values in strongly alkaline solutions, since the concentration of the free zinc ion, Zn^{++}, becomes very small through the action of the equilibrium

$$Zn^{++} + 3OH^- \rightleftharpoons [Zn(OH)_3]^-.$$

In consequence of this, the solution potential of zinc lies above that of hydrogen in strongly alkaline solutions also. The action of ammonia is similar, since it dissolves zinc hydroxide to form tetramminezinc hydroxide, $[Zn(NH_3)_4](OH)_2$, and accordingly holds the zinc ion concentration at a low level through this complex formation. Neutral ammonium salts have a much weaker complex-forming action, but the hydrogen potential is also considerably lower in their neutral solutions.

The heat of solution of zinc in dilute hydrochloric acid is 18.2 kcal, in dilute sulfuric acid 18.9 kcal per g equivalent (at constant pressure).

Zinc does not dissolve with evolution of hydrogen in dilute nitric acid, but the hydrogen is consumed *in statu nascendi* in the reduction of the acid. Very concentrated nitric acid is thereby reduced chiefly to nitrogen oxides. At lower concentration, ammonia is formed as well. Sulfur dioxide is liberated, in place of hydrogen, from concentrated sulfuric acid. The hydrogen which is liberated from dilute sulfuric acid is also often contaminated with sulfur dioxide and hydrogen sulfide.

Zinc burns with a brightly luminous, blue-green flame when it is heated in air, forming the oxide ZnO. It is also oxidized at a red heat by steam and by carbon dioxide, the latter being thereby reduced to carbon monoxide.

Zinc reacts with the halogens rather sluggishly at ordinary temperature, and only in presence of moisture. Hydrogen sulfide reacts with zinc even at ordinary temperature, but the sulfide formed superficially protects the metal from further attack. Powdered zinc combines directly with sulfur. Zinc does not react to a detectable extent with nitrogen, even in the vapor state, but the metal reacts with ammonia at red heat, forming the nitride Zn_3N_2. The reaction is generally incomplete, however, even with zinc powder.

In the molten state, zinc is completely miscible with many metals—e.g., with Cu, Ag, Au, Cd, Hg, Mg, Ca, Mn, Fe, Co, Ni, Al, Sn, Sb, As. In the solid state, however, it forms homogeneous mixtures only over a very restricted range. It can enter to the extent of about 1% by weight into the crystal lattice of cadmium, and also forms mixed crystals over a very narrow range with magnesium. Zinc forms compounds with many metals (p. 426–7)—e.g., with Cu, Ag, Au, Mn, Fe, Co, Ni, Rh, Pd, Pt, As, Sb, Mg, Ca, Li, Na, K, but not with Cd, Ga, Tl, Cr, Bi, Sn, and Pb. Zn and Pb are imcompletely miscible the molten state.

(e) Uses

The metal finds extensive uses, especially in the form of sheet zinc—e.g., for roofing and wall coverings, rain water gutters, the lining of troughs, etc. Cast zinc is also a suitable material for many objects. Large amounts are also used in galvanic cells (notably in dry batteries). It is used in the laboratory, generally in the form of *zinc rod*, for the generation of hydrogen. *Zinc dust* (usually containing much oxide) has extensive uses as a reducing agent in the laboratory and in industry. Metallurgical uses of zinc are for the desilverization of lead (Parkes process, p. 393) and for the precipitation of gold after cyanide leaching.

Among alloys if zinc [4], those with copper (the *brasses*) are the most important. During recent years these have been replaced for many purposes by alloys of low copper content or copper-free alloys (especially aluminum alloys), of which zinc is often the principal constituent. Examples of these are white brass, (85% Zn, 5% Al, 10% Cu), lumen bronze (88% Zn, 6% Al 6% Cu) and Zelco (83% Zn, 15% Al, 2% Cu).

Large quantities of zinc are used for protective coatings on other metals, and especially on iron. These may be applied by dipping in molten zinc (hot galvanizing, pot galvanizing) [5], by spraying (metal spraying process), by the action of vapor (vapor galvanizing), or by galvanizing in the proper sense. Zinc-coated iron sheet is commonly known as 'galvanized iron', irrespective of the method used for coating.

Many of the compounds of zinc are used as pigments—in particular *lithopone* (a mixture of ZnS with BaSO$_4$, produced by double decomposition of BaS with ZnSO$_4$) and *zinc white*, ZnO. Very fine zinc dust ('zinc grey') or finely powdered zinc blende ('zinc sulfide grey') are much used for rust-preventive paints on iron— e.g., on bridges, and for machine parts. A number of zinc compounds, such as the chloride and sulfate, find medicinal uses as antiseptics. Basic zinc borate is used as a dusting powder.

Zinc compounds are relatively low in toxicity, but large quantities may have injurious effects. In view of the ease with which zinc is dissolved, zinc vessels are therefore unsuitable for the storage of food stuffs, and their use for this purpose is prohibited by law.

3. Compounds of Zinc

Zinc is always positively bivalent in its compounds. The colorless ion Zn^{++} is present in solutions. The compounds are invariably colorless unless the other component imparts some selective absorption of light. Even the sulfide is colorless, as also is the oxide at ordinary temperature; the latter turns yellow when it is heated.

In general, the tendency of zinc to form complexes with most anions is smaller than that exhibited by copper. However, the tendency to combine with additional hydroxide ions—i.e., the tendency of the hydroxide to form salts with strong alkalis —is much greater for zinc than for copper. Zinc hydroxide is defintely amphoteric. Like aluminum hydroxide, it acts in alkaline solution as an anhydro-acid; i.e., it goes into solution not by splitting off H$^+$ ions, but by adding on OH$^-$ ions to form hydroxozincate ions (commonly called *zincate* ions)—e.g.,

$$Zn(OH)_2 + OH^- = [Zn(OH)_3]^-; \quad Zn(OH)_2 + 2OH^- = [Zn(OH_4)]^=, \text{ etc.}$$

The *zincates* derived from them are of the types

$$M^I[Zn(OH)_3], \; M^I_2[Zn(OH)_4] \text{ or } M^{II}[Zn(OH)_4], \text{ and } M^{II}_2[Zn(OH)_6].$$

The view that the zincates obtained from solution were actually hydroxosalts was first suggested by Pfeiffer in 1908, on the basis of Werner's theory of anhydracids. This conception found support in the formation of hydroxosalts by other metallic hydroxides (hydroxostannates and hydroxoplumbates, Vol I, pp. 532, 547), but was first proved by Scholder (1933), when he showed that the water present in the compounds is not water of crystallization, but is constitutively bound. Thus the sodium salt of composition Na$_2$O · ZnO · 2H$_2$O loses only one molecule of water between 190° and 200°. The second molecule is retained even after long heating to 465° and can be driven off completely only by fusion with K$_2$Cr$_2$O$_7$. If this compound were really a dihydrate, Na$_2$ZnO$_2$ · 2H$_2$O, it would be expected that it would lose all its water on moderate heating. If, however, it is a hydroxo salt, $Na_2[Zn(OH)_4]$, the observations can be interpreted on the hypothesis that this decomposes a little below 200°, with the loss of $1H_2O$ since $Zn(OH)_2$ is not stable above 100°. $Na_2[Zn(OH)_4] = ZnO + 2NaOH + H_2O$. Fusion with potassium dichromate brings about expulsion of water from the NaOH:

$$2NaOH + K_2Cr_2O_7 = K_2CrO_4 + Na_2CrO_4 + H_2O.$$

NaOH does not react with ZnO, to form Na_2ZnO_2 with liberation of water, even upon prolonged heating.

The following compounds, isolated by Scholder, are examples of hydroxozincates:

$$Na[Zn(OH)_3], \ Na[Zn(OH)_3] \cdot 3H_2O, \ Na_2[Zn(OH)_4], \ Na_2[Zn(OH)_4] \cdot 2H_2O,$$

$$Sr[Zn(OH)_4] \cdot H_2O, \ Ba[Zn(OH)_4] \cdot H_2O, \ Sr_2[Zn(OH)_6], \ Ba_2[Zn(OH)_6].$$

Some of these contain water of crystallization as well as constitutional water. The former is usually lost easily—e.g., from $Ba[Zn(OH)_4] \cdot H_2O$ at 87°, and from $Na[Zn(OH)_3] \cdot 3H_2O$ at room temperature over sulfuric acid.

Zinc also forms complex cations, by association with neutral ligands, and especially with ammonia. Compounds (such as the hydroxide) which are insoluble in water may dissolve in aqueous ammonia through the formation of such complex ions.

Although they are very soluble, it has been possible to isolate a series of ammoniates in the crystalline state. Most of them conform to the type $[Zn(NH_3)_2]X_2$ (diamminezinc salts), but some crystallize with $4NH_3$ (tetramminezinc salts). It has not been possible to prepare ammoniates with $6NH_3$ from solution, but they have been obtained by dry methods. The series $[Zn(NH_3)]Cl_2$, $[Zn(NH_3)_2]Cl_2$, $[Zn(NH_3)_4]Cl_2$, $[Zn(NH_3)_5]Cl_2$, $[Zn(NH_3)_6]Cl_2$ is known for the chlorides, and the series $[Zn(NH_3)_2]SO_4 \cdot H_2O$, $[Zn(NH_3)_4]SO_4$, and $[Zn(NH_3)_5]SO_4$ for the sulfates.

Zinc salts can combine with many organic nitrogen compounds in the same way as with ammonia—e.g., with aniline, phenylhydrazine, pyridine, quinoline, strychnine, etc. Most of these are also of the type $[Zn\ Am_2]X_2$, with two molecules of the organic anhydrobase as ligands, although compounds of the type $[Zn\ Am_6]X_2$ are also known. Among these are the tris-ethylene-diamine zinc bromide described by Werner, $[Zn\ en_3]Br_2$.

Table 46 summarizes some properties of the most important simple compounds of zinc.

TABLE 46

HEATS OF FORMATION, DENSITIES, MELTING POINTS, AND BOILING POINTS
OF SOME SIMPLE ZINC COMPOUNDS

Compound	Heat of formation kcal/g-equivalent	Density at 25°	Melting point	Boiling point
ZnO	41.6	5.78	ca. 2000° (under press.)	ca. 1725° (sublimation temp.)
ZnS	20.7	4.06	1800–1900° (under press.)	1180° (sublimation temp.)
ZnF_2	96.4	4.84	872°	—
$ZnCl_2$	49.4	2.904	318°	730°
$ZnBr_2$	—	4.20	394°	650°
ZnI_2	24.9	4.74	446°	—

Among the simple salts, the chloride (and the other halides, except the fluoride), the nitrate, the sulfate and the acetate are very soluble in water. Sparingly soluble compounds are the oxide, hydroxide, and sulfide, and also the carbonate, fluoride, phosphate, silicate, cyanide, and cyanoferrate(II). Potassium zinc cyano-ferrate(II), $K_2Zn_3[Fe(CN)_6]_2$, which is also sparingly soluble, is of importance in analysis.

Many *basic salts* of zinc are known, most of them, according to Feitknecht (1933) being of the 'double layer lattice' type (cf. p. 298). The principal layers in these are built up from $Zn(OH)_2$; in the intermediate layers the ions of the salt are usually arranged in a regular manner on definite lattice positions, although in some of them the ions in the intermediate layers are not bound in fixed positions. As in the case of the green basic cobalt chloride (p. 298), this results in a variable composition for the compound, although it may be well crystallized. '*Simple* layer lattice' structures are very rarely found in basic zinc salts (e.g., in $Zn(OH)Cl$, whereas this structural type is common among basic cadmium salts.

On the basis of Werner's ideas, the basic zinc salts were formerly regarded as 'ol-compounds' (cf. p. 390). Although X-ray studies have shown that this idea is not valid for the solid state, it appears from the work of Hayek (1934) that it is probably true of the dissolved state. It is assumed that, in solution, hydroxide molecules (of zinc or other metal), by virtue of their dipolar character, may add on to a zinc ion, to form a complex, in the same way as water molecules add on to form aquocomplexes:

$$Zn^{++} + 6HOH \quad = [Zn\,(HOH)_6]^{++}$$

$$Zn^{++} + 3Zn(OH)_2 = \left[Zn\left(\begin{matrix}HO\\HO\end{matrix}{>}Zn\right)_3\right]^{++}.$$

The appreciable solubility of heavy metal oxides and hydroxides in solutions of zinc salts is attributed to the formation of complexes of this kind. This property is utilized in the action of zinc chloride solutions as soldering fluxes.

(a) Oxygen Compounds

(*i*) *Zinc oxide*, ZnO, occurs native in large amounts only in one place, in New Jersey, as *zincite* or red zinc ore. The red color of the mineral is due to its manganese content, which may amount to 9%. Pure zinc oxide is colorless at ordinary temperature, but is yellow when hot. Its structure is hexagonal (wurtzite structure, Fig. 60, p. 258 of Vol. I); $a = 3.243$ Å, $c = 5.195$ Å, $Zn{\leftrightarrow}O = 1.94$ Å). The density of the pure oxide is 5.78, and the hardness 4–5 on Mohs' scale. For technical purposes the oxide is usually prepared by burning metallic zinc. It is obtained by this means in the form of a pure white, loose powder, the 'philosophers' wool' of the alchemists. It is used chiefly as an artist's pigment (zinc white, chinese white). It is also used as a filler for soft rubber, for the preparation of cements and hard resins, for cleaning powders and for the preparation of other zinc compounds. Very pure zinc oxide, obtained by ignition of the basic carbonate, is used for cosmetic powders and creams, and in medicine (as a dusting powder and for the preparation of ointments; it is also a constituent of plasters). Zinc oxide is also employed as a catalyst for the synthesis of methanol, either by itself or admixed with other oxides.

The catalytic activity of zinc oxide depends to a considerable degree on the mode of preparation. Thus a catalytically active oxide is obtained by the careful thermal decomposition of zinc hydroxide or carbonate, but the oxide prepared by decomposition of the nitrate has only a low catalytic activity. The lower the temperature at which the oxide is prepared from the carbonate or hydroxide, the greater is its catalytic efficiency. The various

preparations have the same crystal structure, but they differ in the extent of their lattice imperfections (p. 18), and these are of importance in determining the catalytic properties.
The most active preparations are those in the so-called 'intermediate states', characterized
in that the crystal lattice of the original material (e.g., $ZnCO_3$) is already destroyed, but
that of the reaction product (ZnO) is not formed, or only very imperfectly formed. Such
active intermediate states are found to occur in many other reactions which take place in
the solid state (cf. p. 746).

The chemical reactivity, like the catalytic efficiency, is also dependent on the degree of
lattice imperfection. Thus Hüttig (1937) showed that ZnO made by thermal decomposition of basic zinc carbonate would react with CoO to form Rinmann's green with increasing ease as the decomposition was effected at lower and lower temperatures. The heat of
solution of zinc oxide preparations also depends to a marked extent on the mode of preparation and the starting material. Fricke (1932) showed that active preparations might
have a heat of solution up to 1.2 kcal per mol higher than that of 'normal' ZnO (i.e., ZnO
with no significant amount of lattice imperfections). He showed by X-ray investigations
that the differences in energy content and activity were not due to the presence of different
modifications, nor to considerable differences in particle size, but could be explained only
in terms of incomplete ordering of the crystal lattice.

Zinc oxide reacts with BaO at high temperatures (1100°), to form a double oxide or oxo
salt, $BaZnO_2$. A corresponding compound is formed between BaO and CdO (Scholder,
1953).

Zinc oxide is practically insoluble in water (in water in equilibrium with the
CO_2 of the atmosphere, the solubility is 3.0 mg ZnO per liter).

(*ii*) *Zinc hydroxide*, $Zn(OH)_2$, is thrown down as a white voluminous precipitate,
which is generally contaminated with adsorbed ions, when zinc salt solutions are
treated with alkali hydroxide. It is soluble in excess of the precipitant, as well as
in acids, and also in aqueous ammonia.

It is soluble in acids because neutralization occurs, with the formation of soluble salts (1),
in alkalis because of the formation of zincates (2), and in ammonia by reason of complex
formation (3):*

$$Zn(OH)_2 + H_2SO_4 = ZnSO_4 + 2H_2O; \ Zn(OH)_2 + 2H^+, = Zn^{++} + 2H_2O \qquad (1)$$

$$Zn(OH)_2 + NaOH = Na[Zn(OH)_3]; \ Zn(OH)_2 + OH^- = [Zn(OH)_3]^- \qquad (2)$$

$$Zn(OH)_2 + 4NH_3 = [Zn(NH_3)_4](OH)_2; \ Zn(OH)_2 + 4NH_3 = [Zn(NH_3)_4]^{++} + 2OH^- \ (3)$$

Zinc hydroxide is also soluble to a slight extent in solutions of ammonium salts. In such
solutions, there is the equilibrium:

$$[NH_4]^+ \rightleftharpoons NH_3 + H^+.$$

The dissociation of the ammonium ion in neutral solutions is very slight indeed, but the
equilibrium is displaced considerably towards the right in the presence of zinc hydroxide,
which can react both with the H^+ ion and with the NH_3:

$$2NH_3 + Zn(OH)_2 + 2H^+ = [Zn(NH_3)_2]^{++} + 2H_2O \quad or$$

$$Zn(OH)_2 + 2NH_4^+ \ = [Zn(NH_3)_2]^{++} + 2H_2O.$$

If zinc salt solutions are treated in the cold with ammonia, a gelatinous amorphous precipitate is first obtained ('zinc oxide hydrate'). This becomes flocculent
after a time, and then gives an X-ray diffraction pattern which reveals the inci-
pient ordering of the structure—probably to form zinc oxide in the first instance.
The diffraction pattern characteristic of zinc hydroxide appears only after further

* This is demonstrated, e.g., by transport measurements. Zinc migrates towards the
cathode in solutions containing an excess of ammonia.

aging, and it may take several months of standing in contact with water before complete transformation into the stable form of the crystalline hydroxide has taken place (Fricke, 1924). Formation of the crystalline hydroxide takes place more rapidly when alkali hydroxide is used as precipitant. The rate of crystallization also depends upon the salt from which the zinc is precipitated—thus crystalline zinc hydroxide is obtained considerably more rapidly by precipitation from zinc chloride solution than from zinc nitrate solution.

$Zn(OH)_2$ can exist in five distinct crystalline modifications. Their properties, and the conditions under which they are formed, have been investigated chiefly by Fricke (1924) and Feitknecht (1930). The forms are designated a-, β-, γ-, δ-, and ε-$Zn(OH)_2$, in order of decreasing energy content. Only the last modification, ε-$Zn(OH)_2$, which is rhombic, is stable. The other forms, obtained under certain specified conditions, are metastable. Some are produced as intermediates in the transformation of the amorphous zinc oxide hydrate into the rhombic zinc hydroxide; all of them are eventually transformed, in contact with water, into ε-$Zn(OH)_2$. The latter is stable in contact with water below 39° (Hüttig, 1933). At higher temperatures, ZnO is the *stable phase* in contact with water. However, equilibria are often established so sluggishly in this system that it is possible to heat ε-$Zn(OH)_2$ in vacuum to 60° for a long time without any detectable loss of water. The least stable modification of zinc hydroxide, a-$Zn(OH)_2$, can undergo conversion into ZnO, even at ordinary temperature and under water, the latter is subsequently hydrated once more to form $Zn(OH)_2$ in the stable ε-modification.

According to Feitknecht (1938), the very unstable a-$Zn(OH)_2$ forms a hexagonal 'double layer lattice', of which the principal layers have the same structure as in $Mg(OH)_2$ and $Cd(OH)_2$.* The stable ε-hydroxide, however, has an orthorhombic structure not found for any other hydroxide, built up from a network of $Zn(OH)_4$ tetrahedra similar to the SiO_4 tetrahedra in the various modifications of SiO_2. The $Zn(OH)_4$ tetrahedra are slightly distorted, so that the $Zn \leftrightarrow OH$ distances vary between 1.94 and 1.96 Å. The shortest distance is the same as the $Zn \leftrightarrow O$ distance in the crystal lattice of zinc oxide (Wyckoff, 1933; Megaw, 1935).

The heat of formation of crystalline ε-$Zn(OH)_2$ from normal crystalline ZnO and H_2O is 2.28 kcal. per mol (Fricke, 1932).

(b) Nitrogen Compounds

Zinc amide and Zinc nitride. Zinc amide, $Zn(NH_2)_2$, formed by the reaction of zinc ethyl with ammonia in ether solution—$Zn(C_2H_5)_2 + 2NH_3 = Zn(NH_2)_2 + 2C_2H_6$—is a white powder, which reacts vigorously with water to form $Zn(OH)_2$ and NH_3. Its density is 2.13, and heat of formation from the elements 35 kcal per mol (Juza, 1937). It decomposes at red heat to form *zinc nitride*:—$3Zn(NH_2)_2 = Zn_3N_2 + 4NH_3$. The nitride, a grey powder ($d = 6.2$), is thermally stable in the absence of air, but is decomposed by water:—$Zn_3N_2 + 6H_2O = 3Zn(OH)_2 + 2NH_3$. It can also be formed by the action of ammonia gas on powdered zinc. It is diamagnetic. The heat of formation from the elements is 5 kcal per mol (Juza, 1938).

(c) Sulfur Compounds

(*i*) *Zinc sulfide*, ZnS, is found native in many places in the cubic form, as *zinc blende* or *sphalerite*. The hexagonal form, *wurtzite*, is rare. The naturally occurring material is often more or less dark in color, as a result of the presence of impurities (especially FeS). Zinc sulfide is precipitated as an amorphous white material from zinc salt solutions on the addition of ammonium sulfide, and also by the action of hydrogen sulfide in the presence of a sufficient concentration of acetate ions.

* The crystal lattice of a-$Zn(OH)_2$ is only stable when the intermediate layers contain a certain proportion of a salt, such as zinc chloride or carbonate, in addition to (randomly arranged) hydroxide.

Freshly precipitated zinc sulfide is readily soluble in dilute acids, but gradually changes into a less soluble modification. It also forms colloidal dispersions easily— e.g., on prolonged treament with hydrogen sulfide solution. It is also formed in the colloidal state by passing hydrogen sulfide into an ammoniacal zinc solution.

Precipitated, amorphous zinc sulfide becomes crystalline when it is heated with hydrogen sulfide solution under pressure. In these circumstances, zinc blende is formed, but if the dry sulfide is heated in hydrogen or hydrogen sulfide gas, the wurtzite modification is obtained. The latter is the modification stable at high temperatures; the transition temperature is 1020°, according to Allen. If wurtzite is ground up in a mortar, it is converted into blende, as is proved by X-ray examination.

Zinc blende possesses a cubic structure, with a unit edge 5.42 Å (cf. Fig. 47, p. 380), and a shortest Zn↔S distance of 2.35 Å. Wurtzite forms the hexagonal structure represented in Fig. 60, Vol. I, p. 258, with $a = 3.84$ Å, $c = 6.28$ Å. The shortest Zn↔S distance in wurtzite, 2.36 Å, is practically the same as in blende, showing that the two structures are 'commensurable' (cf. p. 380).

ZnSe and ZnTe are isotypic with zinc blende ($a = 5.66$ Å and 6.09 Å, respectively). HgSe, HgTe, and CdTe have the same structure. Like ZnS, CdS and CdSe can crystallize both with the zinc blende structure and the wurtzite structure.

The density of crystalline zinc sulfide is 4.09, and hardness 3.5 to 4. Blende and wurtzite have the same density within the limits of experimental error. The solubility of the crystalline sulfide is about $7 \cdot 10^{-6}$ mol per liter, that of the freshly precipitated sulfide being about ten times as great. Biltz found that zinc sulfide sublimed at about 1180°, and Tiede and Schlede were able to melt it at about 1800–1900°, under a pressure of 100–150 atm.

Artificially prepared zinc sulfide is used as a white paint—rarely by itself (as metallic white), but generally as *lithopone* (p. 434). Lithopone is much cheaper than lead white, and is, moreover, not poisonous nor sensitive to hydrogen sulfide. However, it is considerably inferior to white lead and to zinc white (ZnO) in covering power and durability. The principal disadvantage is the instability of many varieties of lithopone towards light.

Zinc sulfide turns grey when it is exposed to ultraviolet light, as a result of the deposition of elementary zinc. This takes place most rapidly with wurtzite, and more slowly with blende. Added substances may either increase or decrease the photosensitivity. It has been found possible to obtain lithopones of adequate stability in light by adding a very small amount of cobalt salt to the zinc sulfate solution before reaction with barium sulfide. The reason for the action of the cobalt salt is not yet fully understood.

(*ii*) *Sidot's Blende*. When it is rendered crystalline by fusion, by sublimation, or by heating in the presence of a mineralizing agent, zinc sulfide, like the alkaline earth phosphors, has the property of phosphorescing brightly after exposure to light. For this purpose it must be freed as completely as possible from impurities, but must contain traces of a heavy metal such as manganese or copper. Zinc sulfide prepared in this manner is known as Sidot's blende, after the chemist who first observed the phenomenon. It differs from the alkaline earth phosphors in that it is strongly excited to luminescence by the radiations from radioactive substances and by X-rays. It is therefore used for the preparation of X-ray viewing screens. The 'radium paint' used for luminous clock and instrument dials, etc., contains about 1 mg of radium (or the equivalent amount of mesothorium) to 10 g of zinc sulfide.

(d) Zinc Phosphides

When a current of hydrogen or nitrogen, charged with phosphorus vapor, is passed over molten zinc, *trizinc diphosphide*, Zn_3P_2, is formed in lustrous, silvery, opaque, tetragonal needles and leaflets ($d = 4.54$), insoluble in water. Orange transparent tetragonal crystals of *zinc diphosphide*, ZnP_2, are obtained if phosphorus is present in excess.

The structure of Zn_3P_2, according to von Stackelberg, is similar to that of Sc_2O_3 (cf. p. 37 *et seq.*), with the P atoms occupying the positions of cubic close packing in the same way as do the O atoms in Sc_2O_3. The Zn atoms occupy the interstices of the structure, in such a way that each Zn atom is surrounded by 4 P atoms in the form of a distorted tetra-hedron, and each P atom by 6 Zn atoms located at six of the eight corners of a distorted cube. Zn_3As_2, Cd_3P_2, and Cd_3As_2 have the same structure.

(e) Halides

(*i*) *Zinc chloride*, $ZnCl_2$, is prepared by dissolving zinc blende, zinc oxide, or scrap zinc in hydrochloric acid. If metallic zinc is used, the solution will contain no heavy metals except iron, since the electrolytic solution pressure of all the heavy metals concerned is much smaller than that of zinc. The iron is precipi-tated as iron(III) oxide hydrate by the addition of zinc oxide, after oxidation with chlorine. When the solution is evaporated down with an excess of hydro-chloric acid, the zinc chloride is obtained as a white granular powder, consisting of small hexagonal (rhombohedral) crystals. It is very fusible, and solidifies to a translucent mass, like porcelain. It is sold commercially both as powder and as fused lumps or sticks.

Molten zinc chloride is a fairly good electrical conductor. Schultze found the specific conductivity to be $\varkappa = 0.03$ ohm^{-1} at 400°, and $\varkappa = 0.46$ ohm^{-1} at 700°. Zinc chloride vaporizes at red heat (b.p. 730°), and solidifies in white needles. It is extremely hygroscopic, and deliquesces in moist air. Nevertheless, above 28° it can crystallize from water in the anhydrous state; various hydrates are formed below this temperature, although the anhy-drous salt can be obtained from very concentrated solutions at temperatures as low as 10°.

Zinc chloride dissolves in water with the evolution of considerable heat (15.6 kcal per mol). The solution is acid in reaction, as a result of hydrolysis (to the extent of about 0.1% in $^1/_{200}$-normal solution). Zinc chloride is often used to split off water or bring about condensations in organic compounds, because of its great affinity for water. Apart from its uses in preparative organic chemistry, zinc chlo-ride is used in large quantities in dye printing, in the manufacture of organic dyestuffs (which often are sold as their zinc chloride double salts), and for the impregnation of wood. It is also used as an etchant for metals—e.g., in soldering—and has medicinal applications. Zinc chloride has an unpleasant metallic taste, and a strongly corrosive action.

Zinc chloride is highly dissociated in aqueous solution. In concentrated solutions, how-ever, as was shown by Hittorf by transport measurements, it is to some extent associated in the form of autocomplexes, although to a much smaller extent than cadmium chloride.
Zinc chloride is very soluble in absolute methyl and ethyl alcohols, as well as in water, and also in ether, acetone, glycerol, ethyl acetate, and other organic compounds containing oxygen; also in nitrogen compounds, such as pyridine and aniline.

Zinc chloride tends to form double complex salts, especially with the chlorides of the alkalis and alkaline earths. Most of these complexes are of the type $M^I_2[ZnCl_4]$ (tetrachlorozincates), and some crystallize as hydrates. The compound $2ZnCl_2 \cdot HCl \cdot 2H_2O$ crystallizes at 0° from a zinc chloride solution saturated with hydrogen chloride gas. The compound $ZnCl_2 \cdot HCl \cdot 2H_2O$ (long needles) can also be prepared.

The (rhombic) ammonium chloride double salt (ammonium tetrachlorozincate), $(NH_4)_2[ZnCl_4]$ ($d = 1.88$, m.p. 150°) is used as 'soldering salt'. The flux prepared by tin-smiths, by dissolving scrap zinc in crude hydrochloric acid and adding ammonium chloride, contains this compound although all the zinc chloride double salts are almost completely dissociated into their components in solution. The action of the solution as a flux depends on the ability of zinc chloride to combine with metallic oxides to form complex compounds (ol-compounds, cf. p. 436), which exist only in solution.

(ii) *Basic Zinc Chlorides*. Of the known, crystalline basic zinc chlorides, the simplest in composition is *zinc hydroxy-chloride*, $Zn(OH)Cl$. This crystallizes in long hexagonal prisms ($d = 4.57$) and forms a simple layer lattice like that of cadmium hydroxychloride (Fig. 50, p. 449).

Another basic salt which can be obtained as a compound of well defined composition, $ZnCl_2 \cdot 4Zn(OH)_2$, crystallizes in microscopic hexagonal leaflets ($d = 3.29$), and forms a double layer lattice, based on hydroxide layers similar in structure to those of α-$Zn(OH)_2$, with regularly ordered $ZnCl_2$ layers between them (Feitknecht, 1930).

Partial exchange of zinc ions for other metallic ions leads to *mixed basic chlorides*. Partial replacement of chloride ions by hydroxyl ions in the salt layers can also take place. Substances with a variable, and not stoichiometrically simple, ratio of hydroxide to salt are thereby formed. However, only the stoichiometrically simple salts are stable in equilibrium with the solution, although equilibrium is often extremely slow of attainment.

Feitknecht finds that $ZnBr_2 \cdot 4Zn(OH)_2$, $Zn(NO_3)_2 \cdot 4Zn(OH)_2 \cdot 2H_2O$, $ZnSO_4 \cdot 3Zn(OH)_2 \cdot 4H_2O$, and some basic salts of cadmium, cobalt, and nickel have structures like that of $ZnCl_2 \cdot 4Zn(OH)_2$. However, most of the basic salts of cobalt and nickel either form single layer lattices, or form double layer lattices in which the atoms of the intermediate layers are randomly disposed. This type (discussed on p. 299) is represented by a third zinc chloride which differs from the two already mentioned in that it has a higher hydroxide content and generally does not have the stoichiometrically simple composition. Its composition and structure may be ideally represented by the formula $Zn(OH,Cl)_2 \cdot 4Zn(OH)_2$. The principal layers consist of $Zn(OH)_2$ layers, with the $CdCl_2$ structure (p. 451). In contact with a very dilute zinc chloride solution, it is transformed into yet another basic chloride. The composition and structure of this last salt have not yet been interpreted.

(iii) *Other Halides*. *Zinc bromide* and *iodide* are very similar to the chloride, although not isomorphous with it. The anhydrous bromide is orthorhombic, and the iodide (which is yellow) is cubic. Fromherz (1934) showed from absorption spectrum measurements that halogeno complexes are present in zinc iodide solutions of high concentration, or containing alkali iodide, to a far greater degree than with zinc chloride. The bromide is intermediate in tendency for complex formation. *Zinc fluoride*, which is said to be a very good wood preservative, differs from the other halides not only in its crystalline form (monoclinic), but in being sparingly soluble (about $5 \cdot 10^{-5}$ mol per liter). All the zinc halides form double salts with alkali halides, with the general formulas M^IZnX_3, $M^I_2ZnX_4$ and $M^I_3ZnX_5$.

(f) Other Zinc Salts

(i) *Zinc nitrate*, $Zn(NO_3)_2$, exists as the anhydrous salt and as hydrates with 2, 4, 6 and $9H_2O$ (cf. nickel and magnesium nitrates). The hexahydrate ($d = 2.07$) is the stable form at ordinary temperature. It crystallizes from solutions above $-17.6°$ and melts (with decomposition into tetrahydrate and water) at $36.5°$. 115 g of $Zn(NO_3)_2$ dissolve in 100 g of water at $18°$.

In addition to the basic nitrate $Zn(NO_3)_2(H_2O)_2 \cdot 4Zn(OH)_2$, with double layer lattice and ordered atomic arrangement, there is a more highly basic salt with double layer lattice and random atomic arrangement of the intermediate layers. This has hitherto been obtained only in highly disperse form and (unlike the former compound, which is usually constant in composition) its composition is variable. It usually contains about $9Zn(OH)_2$ for each $Zn(NO_3)_2$ (Feitknecht 1933–38). There is also another basic nitrate which has not been fully investigated.

(ii) *Zinc nitrite and Double Nitrites*. Zinc nitrite, $Zn(NO_2)_2$ is precipitated when sodium nitrite is added to zinc salt solutions. It is readily hydrolyzed, and forms double nitrites with the alkali nitrites—e.g., $K_2Zn(NO_2)_4 \cdot H_2O$, yellow hygroscopic prisms, readily decomposed.

(*iii*) *Zinc sulfate*, $ZnSO_4$, is manufactured in large quantities by dissolving scrap zinc in dilute sulfuric acid (1) or by the cautious roasting of zinc blende (2).

$$Zn + H_2SO_4 = ZnSO_4 + H_2 \text{ (1)}; \quad ZnS + 2O_2 = ZnSO_4 \text{ (2)*.}$$

Zinc blende containing iron lends itself particularly to the sulfatizing roast (2). The ions of nobler metals present as impurities in the solution obtained by leaching the roasted ore are precipitated as the metals by suspending strips of zinc in the solution. Iron and manganese are precipitated as hydroxides by means of bleaching powder.

Zinc vitriol, $ZnSO_4 \cdot 7H_2O$, crystallizes below 39° from the evaporated solution, in colorless, glassy, columnar, rhombic crystals, isomorphous with epsom salt. The solubility is: at 0°, 41.7; 18°, 52.7, 100°, 78.6 g of $ZnSO_4$ in 100 g of water. Zinc vitriol melts in its own water of crystallization when it is heated rapidly. It loses water and effloresces on exposure to air.

Zinc vitriol forms mixed crystals with the other vitriols $MSO_4 \cdot 7H_2O$ (M = Mg, Fe, Mn, Co, and Ni), but except with epsom salt the range of mixed crystal formation is restricted. With copper sulfate, the following mixed crystals are formed: 0 to 3 mol-% Cu— colorless rhombic crystals; 16 to 33 mol-% Cu—pale blue monoclinic crystals; 84 to 100 mol-% Cu—blue triclinic crystals.

Zinc sulfate crystallizes from aqueous solutions above 39° with $6H_2O$, the hydrate again being isomorphous with the corresponding hydrate of magnesium sulfate. Crystallization above 70° yields the monohydrate. The heptahydrate can exist in a monoclinic form as well as in the rhombic form. It is metastable in the former modification, which is appreciably more soluble than the rhombic heptahydrate (e.g., solubility 58.7 g in 100 g of water at 18°).

Zinc sulfate first loses 6 molecules of water of crystallization when it is cautiously heated, and the last molecule is not lost below 240°. On strong ignition, it is decomposed into zinc oxide, sulfur trioxide, sulfur dioxide and oxygen. Decomposition is more readily effected in the presence of reducing agents such as carbon.

Zinc vitriol dissolves in water with the absorption of heat (—4.3 kcal per mol at 18°). The monohydrate dissolves with evolution of heat (+5.6 kcal per mol).

The addition compound $ZnSO_4 \cdot H_2SO_4$ crystallizes in silky needles from a solution in hot concentrated sulfuric acid. Double salts $M^I_2SO_4 \cdot ZnSO_4 \cdot 6H_2O$ are obtained from solutions containing the alkali sulfates.

Zinc sulfate is occasionally found native—anhydrous as *zincosite* (isomorphous with barytes), and in the hydrated form as *native zinc vitriol* (white vitriol or goslarite).

Zinc sulfate is technically the most important zinc salt. It is used in dyeing and dye printing, in the preparation of lithopone, for the impregnation of wood, and in galvanostegy, for the preparation of zinc plating baths. Very dilute solutions are used in medicine as an astringent disinfectant, which reduces the secretions, in washings and dressings, and occasionally as an emetic. Zinc sulfate is also the starting material for the preparation of other zinc compounds and for the extraction of absolutely pure zinc by electrolytic refining.

The only *basic sulfate* of zinc hitherto investigated is that mentioned on p. 441. There is also another more highly basic salt, with double layer lattice structure.

(*iv*) *Zinc sulfite*, $ZnSO_3$, is formed as a hydrated crystalline precipitate, by treatment of zinc sulfate solution with sodium sulfite. It can also be obtained by the action of SO_2 on an aqueous suspension of ZnO or $ZnCO_3$. Berglund has described double sulfides, some of which have complex compositions.

---

* Reaction (2) does not take place quantitatively.

(v) *Zinc thiosulfate*, ZnS_2O_3, which is very soluble in water, is not very stable. It decomposes when its solution is evaporated. Well defined double salts of zinc thiosulfate have been isolated from solution by Rosenheim—e.g.,

$$K_2Zn(S_2O_3)_2 \cdot H_2O \text{ and } (NH_4)_2Zn(S_2O_3)_2 \cdot H_2O$$

(long white prisms).

(vi) *Zinc cyanide*, $Zn(CN)_2$, is obtained as a white precipitate by the addition of alkali cyanides to zinc salt solutions. It is insoluble in water and alcohol, but soluble in excess of precipitant. It can be obtained crystalline, in the form of rhombic prisms. It has a tendency to form colloidal dispersions. Zinc cyanide is tasteless, but poisonous. It is occasionally used in medicine.

Zinc cyanide dissolves in excess alkali cyanide solution with the formation of soluble complex salts (*cyanozincates*). Most of these are of the type $M^I_2[Zn(CN)_4]$, and some of the composition $M^I[Zn(CN)_3]$. Mixed solutions of zinc and copper(I) cyanides in excess alkali cyanide are used as plating baths for the electrodeposition of brass.

(vii) *Zinc thiocyanate*, $Zn(SCN)_2$, can be obtained by the action of thiocyanic acid on zinc carbonate. It is very soluble in water and alcohol.

(viii) *Zinc acetate*, $Zn(C_2H_3O_2)_2$, prepared by dissolving zinc oxide in acetic acid, crystallizes from solution as the dihydrate in brilliant monoclinic scales ($d = 1.73$). It can be recrystallized from glacial acetic acid as the anhydrous salt (octahedra, $d = 1.84$, m.p. $242°$). It can be sublimed without decomposition at low pressures. It is very soluble in water (about 1 in 3 parts), and is partially hydrolyzed in solution—to the extent of about 1% in 1-normal solution (Löfmann). Zinc acetate is used for fireproofing. It is also used in medicine as a gargle, as a wash for infections of the skin, and also internally.

(ix) *Zinc carbonate*, $ZnCO_3$, is found in Nature as *calamine* or *zinc spar*, crystallized in the rhombohedral system. It is rarely found in large crystals, but commonly forms dull masses which are colored yellowish, brown, grey, or greenish by the impurities present. Pure zinc carbonate, obtained by precipitation of a zinc salt with a solution of an alkali hydrogen carbonate, saturated with carbon dioxide, or by the action of carbon dioxide on the aquous suspension of freshly precipitated zinc hydroxide, is pure white. Its density is 4.44. $ZnCO_3$ has the calcite structure, with $a = 5.62$ Å, $a = 48°2'$. It is practically insoluble in water, but is slowly decomposed with the formation of *basic zinc carbonate*. Even in the dry state it begins to lose CO_2 at $150°$.

Basic zinc carbonate is obtained as a white amorphous precipitate when zinc salt solutions are treated with ordinary alkali carbonates. The precipitate is variable in composition, depending on the conditions of precipitation, but generally corresponds roughly to the formula $2ZnCO_3 \cdot 3Zn(OH)_2$. Basic zinc carbonate of this composition also occurs native, as a weathering product of calamine (*hydrozincite*, flowers of zinc). Its structure is similar to that of other basic zinc salts.

Feitknecht (1952) has recognized three distinct zinc hydroxycarbonates: $ZnCO_3 \cdot Zn(OH)_2 \cdot H_2O$ (unstable); $2ZnCO_3 \cdot 3Zn(OH)_2 \cdot H_2O$ (which forms extremely minute crystalline plates, recognizable only by means of the electron microscope); and $ZnCO_3 \cdot 3Zn(OH)_2 \cdot nH_2O$ (very fine needles or fibers). The two last-mentioned compounds are variable in composition. They can be derived structurally from the brucite type. If some of the OH^- ions in the hydroxide sheets of this structure are replaced by CO_3^{2-} ions, zinc hydroxycarbonate sheets are formed, which bear an excess negative charge. This is compensated by intercalation of a corresponding number of Zn^{2+} ions (together with H_2O molecules) between the sheets. If every fourth OH^- ion is replaced by a CO_3^{2-} ion, the resulting structure can be represented by the formula $[Zn_4(OH)_6(CO_3)_2]^{2-}-[Zn(H_2O)_n]^{2+}$, corresponding to the analytical composition $2ZnCO_3 \cdot 3Zn(OH)_2 \cdot nH_2O$.

(x) *Zinc oxalate* is precipitated as the dihydrate, $ZnC_2O_4 \cdot 2H_2O$, when oxalate ions are added to zinc salt solutions. It is a white, sparingly soluble salt (6.4 mg of ZnC_2O_4 dissolve in 1 liter of H_2O at 18°), but dissolves in excess of alkali oxalate, with the formation of complex ions, $[Zn(C_2O_4)_2]^{2-}$ and $[Zn(C_2O_4)_3]^{4-}$.

(xi) *Zinc Silicates*. The orthosilicate Zn_2SiO_4 occurs native as *willemite* and, in isomorphous mixture with Mn_2SiO_4 as *troostite* (both rhombohedral). It has been prepared artificially. A hydrated zinc orthosilicate, $Zn_2SiO_4 \cdot H_2O$, occurs in large amounts as siliceous zinc ore, usually in fibrous or nodular aggregates of rhombic crystals ($d = 3.3$ to 3.5), which may be colorless or colored by impurities. The water is given up only at a red heat.

Zinc metasilicate, $ZnSiO_3$, has been prepared artifically by fusion methods. If zinc salt solutions are treated with alkali silicate, slimy precipitates of variable composition are formed.

(g) Zinc Alkyls

Zinc reacts with methyl iodide at ordinary temperature to form methyl zinc iodide, CH_3ZnI. When this is heated it disproportionates to form *zinc methyl*, $Zn(CH_3)_2$. This compound was discovered by Frankland in 1849. It forms a colorless, evil smelling liquid ($d = 1.38$, b.p. 46°), which fumes in air, and inflames very readily. *Zinc ethyl*, $Zn(C_2H_5)_2$ ($d = 1.18$, b.p. 118°) and other zinc alkyls have been obtained similarly. The zinc alkyls are extremely reactive, and therefore find uses in organic synthetic work. They are decomposed by water, to give zinc hydroxide and hydrocarbons—$Zn(CH_3)_2 + 2H_2O = Zn(OH)_2 + 2CH_4$.

The *cadmium alkyls* (e.g., cadmium methyl, $Cd(CH_3)_2$, b.p. 104°) are very similar to the zinc alkyls in properties. For mercury alkyls, see p. 476.

4. Analytical (Zinc)

Zinc comes into the ammonium sulfide group in the systematic separation of cations, since it is precipitated as sulfide by ammonium sulfide from ammoniacal solution, but not by hydrogen sulfide from acid solution. It can be distinguished from cobalt and nickel by the solubility of its sulfide in dilute acids, from iron and manganese in the solubility of its hydroxide in sodium hydroxide, and from aluminum and chromium in the ready solubility of its hydroxide in aqueous ammonia. The white color of the precipitated sulfide is also characteristic.

In blowpipe reactions on the charcoal block, zinc gives an incrustation (zinc oxide) which is yellow when hot, but white when cold. If this is moistened with cobalt nitrate solution and re-heated, a characteristic green product is obtained (Rinman's green).

As was shown by Hedvall (1914, 1932) and Natta (1929), the composition and color of Rinman's green depend upon the conditions of preparation. If it is very strongly ignited (above 1000°), mixed crystals of zinc oxide and cobalt(II) oxide are obtained. These are pink, not green, if they contain more than 70% of CoO. Ignition at lower temperatures (800–900°), in the presence of sufficient oxygen, produces a double oxide, of spinel type, between zinc oxide and cobalt(III) oxide, $ZnCo_2O_4$. This double oxide is formed by a reaction proceeding in the solid phase (cf. Chap. 19). Co_3O_4, the primary product of reaction, exchanges cobalt(IV) ions for zinc ions by a process of diffusion and simultaneous redistribution of charge:

$$Co^{II}_2[Co^{IV}O_4] + ZnO = Co^{III}_2[ZnO_4] + Co^{II}O.$$

Formation of the zinc mercury(II) thiocyanate, $ZnHg(SCN)_4$, provides a suitable microchemical reaction for zinc, as also does the sodium zinc carbonate, $3Na_2CO_3 \cdot 8ZnCO_3 \cdot 8H_2O$, which crystallizes in tetrahedra (limit of detection 0.01 γ Zn). The color reaction with dithizone (cf. p. 408) is a sensitive drop reaction, and the intense red color of the resulting inner complex salt enables 0.05 γ of zinc to be detected in the presence of 2000 times as much aluminum.

Zinc is determined gravimetrically by precipitation and weighing as the sulfide ZnS, or by precipitation as carbonate and conversion to the oxide by ignition.

Zinc may also be precipitated as ammonium zinc phosphate $(NH_4)ZnPO_4$, and converted to the pyrophosphate, $Zn_2P_2O_7$, for weighing. It may also be deposited electrolytically as the metal preferably from the alkali double cyanide solution. A volumetric method involves precipitation from dilute hydrochloric acid solution as $K_2Zn_3[Fe(CN)_6]_2$, by means of a potassium cyanoferrate(II) solution of known concentration, the end point being detected by 'spotting' with uranyl acetate.

5. Cadmium (Cd)

(a) Occurrence

Cadmium is commonly found to accompany zinc in its ores, especially in calamine and in zinc blende. These minerals almost invariably contain more or less cadmium.

Pure cadmium compounds are very rare in Nature. The sulfide, CdS, occurs very occasionally as *greenockite*, and the oxide, CdO is still rarer. A basic carbonate (*otavite*) has been reported from Otavi (South West Africa).

(b) History

Cadmium was discovered in 1817 by Stromeyer, in Göttingen, in an investigation of a zinc carbonate bought from a Hildesheim apothecary. This unexpectedly yielded a brownish oxide when it was ignited, although it contained no iron. Almost simultaneously Hermann found cadmium in a zinc oxide which gave a yellow precipitate, in place of white, when its solution was treated with hydrogen sulfide; it had accordingly been confiscated in Magdeburg as suspected of containing arsenic. The new element was given the name of cadmium, because of its frequent occurrence in zinc oxide (which was known to the ancient Greeks under the name of καδμεία).

(c) Preparation

Cadmium was formerly prepared almost invariably from the zinc dust obtained in the extraction of zinc by the distillation process. With the change over to electrolytic extraction of zinc, cadmium is now obtained in increasing measure by electrolytic methods. Extraction of cadmium from zinc dust, and from various flue-dusts, etc., containing cadmium, is still important, however.

(*i*) *Preparation from Zinc Dust.* Cadmium is more volatile than zinc, and is therefore considerably enriched in the zinc dust obtained in the Belgian or Silesian distillation process. The zinc dust is mixed with coke, to effect reduction of the oxide, and subjected to repeated fractional distillation, whereby the cadmium is progressively enriched in the first fractions to pass over each time.

The losses are often considerable in the distillations, because of the difficulty of condensing the cadmium vapor. These losses can be avoided by wet extraction methods. The zinc dust containing cadmium is dissolved in hydrochloric or sulfuric acid, and the cadmium is deposited from solution by means of zinc ($Cd^{++} + Zn = Cd + Zn^{++}$), or electrolytically. The dry and wet methods are occasionally combined, the cadmium deposited from solution (having a small zinc content) being redistilled to free it almost completely from zinc (down to about 0.1% Zn).

(*ii*) *Electrolytic Extraction.* There is no enrichment of cadmium in the zinc sulfate solutions which are prepared for the electrolytic extraction of zinc. Since the deposition potential of cadmium is considerably lower in magnitude than that of zinc (cf. Table 4, p. 30 of Vol. I), the cadmium can easily be deposited selectively. The tendency of cadmium to form spongy and dendritic deposits from acid solutions originally caused considerable difficulty, but this can be overcome by suitable procedures (e.g., by the addition of colloids such as gelatin or glue to the baths). The process is essentially similar to the electrolysis of zinc.

Extremely pure cadmium, such as is required for standard cells, can be produced by repeated electrolytic refining, and by sublimation in hydrogen or in a vacuum.

(d) Properties

Cadmium is a white, lustrous metal, but the surface rapidly dulls on exposure to air, through the formation of a thin oxide layer. Cadmium is rather soft, and it is possible to cut off shavings with a knife. It is very ductile, and can be rolled into sheet or drawn into wire. The strength of pure cadmium is very low, but may be enhanced by alloying with zinc. The metal can be welded very readily. It melts at 320.9°, and boils at 767°. It can be sublimed in a vacuum at a temperature as low as 164°.

Deville found the vapor density of cadmium at 1040° to be 3.94 (relative to air = 1), from which the molecular weight can be calculated as 114.1*. The vapor of cadmium, like that of most metals, is therefore almost completely monatomic.

Cohen (1914 and later) reported that cadmium exists in three distinct modifications, which differ in their coefficients of expansion and in their electrolytic solution potentials (though only by a few millivolts). The interconversion of these modifications is said to be extremely sluggish, so that they are ordinarily all present together. However, X-ray investigation shows that there are no structurally distinct modifications.

The electrical conductivity of cadmium at 18° is 21.5%, and the thermal conductivity 22.0% of that of silver. Cadmium exists as hexagonal crystals, with the same structure as those of zinc and magnesium. It forms mixed crystals in all proportions with magnesium, but not with zinc.

Cadmium can be obtained in colloidal dispersion in water by means of the under-water electric arc (Bredig). In the absence of air, the deep brown hydrosol is quite stable.

Cadmium burns with a red flame when it is strongly heated in air, forming a brown smoke of cadmium oxide, CdO**. The heat of combustion is 65.2 kcal per

---

* The molecular weight of any gas is equal to the vapor density relative to oxygen = 32.000. The liter weights of air and of oxygen are 1 293 g and 1.4289 g, respectively, and the 'mean molecular weight' of air is 28.957. Hence molecular weight = 28.957 × vapor density relative to air = 1.

** A trace of the peroxide, CdO_2, is formed at the same time.

g-atom of Cd. Cadmium also combines readily with the halogens when heated, but not with nitrogen and hydrogen.

Cadmium is above hydrogen in the electrochemical series, but is nobler than zinc, and is therefore displaced from it solutions by the latter.

The standard potential of cadmium is $+0.402$ volt at $25°$, relative to the normal hydrogen electrode (La Mer, 1934 and Harned, 1936). Cadmium therefore dissolves in dilute, non-oxidizing acids, with the evolution of hydrogen. Pure cadmium, like absolutely pure zinc, does not dissolve in non-oxidizing acids.

(e) Uses

Cadmium is chiefly used for metallic coatings, as a protection against corrosion. The growth in this field of application of cadmium during 1925–1935, together with the improvements in extraction of the metal resulting from the introduction of the electrolytic process, led to a great increase in world production and a considerable fall in the price of cadmium. Protective cadmium plating is used on automobile parts, aircraft, machine components, electrical and photographic apparatus, piano wires, bolts, and precision instruments. It must not be employed, however, for any objects which may come into contact with foodstuffs, since it is readily corroded by acids, and soluble cadmium compounds are *highly toxic*.

Protective cadmium coatings are usually applied electrolytically, and ordinarily by deposition from a cyanide bath. Spray coating with the molten metal is sometimes employed: heed must then be paid to the toxicity of cadmium vapor. Very thin coatings suffice to exert a protective action. Thus cadmium plating on iron achieves, with a film thickness of 5μ, the same rust protection as is given by zinc 12μ or nickel plating 25μ thick. The coatings are not very resistant towards mechanical abrasion, because of the softness of cadmium, but have the advantage that they do not flake off when plated objects are subjected to bending, tension, or compression strains. Cadmium is therefore particularly suited for plating metal parts which must withstand strongly corrosive conditions without being subjected to mechanical wear.

In assessing the value of rust-preventive coatings, it is important to consider whether the cadmium plating on a iron surface retains its protective action when it has been damaged. It is well known that a coating of *tin* not merely fails to inhibit the corrosion of iron at points where the coating is broken, but actually promotes it, since the iron then acts as the anode in the local couples that are thereby formed. A *zinc* coating, on the other hand, still acts protectively even when it has been damaged, since in this case the iron becomes cathodic. It might be expected from the relative positions of iron and cadmium in the electrochemical series that cadmium would resemble tin in its behavior towards iron. However, it has been shown by various workers that in a galvanic cell made up from an iron sheet dipping in iron sulfate solution, and cadmium in cadmium sulfate solution, the iron becomes the cathode, just as it does in combination with zinc. In agreement with this, cadmium plated on to iron still exercises a protective action when it has been allowed to diffuse completely into the iron surface, by heating it, so that the original deposit has completely disappeared. It is, however, not clear how this should be interpreted in terms of the normal potentials. The discrepancy may be connected with the ease with which the iron becomes passivated. Since iron is only a little baser than cadmium even in its active state, a very small degree of passivation would suffice to shift the potential of the iron so that it behaved as if it were nobler than cadmium.

Apart from its use in protective coatings, metallic cadmium is used chiefly form the production of fusible alloys such as Wood's metal and Lipowitz's metal (cf. Vol. I, bismuth), rapid solder (50% Sn, 25% Pb, and 25% Cd melting at

149°), amalgams for dental fillings, and bearing metals. An iron-platinum alloy, containing cadmium and having an extremely small coefficient of expansion, is used in watchmaking. Cadmium is also employed in the manufacture of Weston standard cells and of Jungner accumulators.

The Jungner accumulator (alkaline nickel-cadmium accumulator) is a modified version of the Edison accumulator (p. 312). In place of an iron electrode, it has an electrode of metallic cadmium mixed with iron(II, III) oxide. The Cd is oxidized to CdO during discharge, and reduced again during charging. Whereas in the Edison cell the evolution of hydrogen takes place from the outset during charging, this occurs only towards the end of the charging process at the cadmium electrode. The onset of hydrogen evolution results in a considerable increase in the terminal voltage, so that avoidance of the effect markedly improves the current efficiency.

Of the compounds of cadmium, the sulfide is used as an artists' pigment, the sulfate in Weston cells, and the bromide and iodide in photography.

6. Compounds of Cadmium

Cadmium compounds are similar to zinc compounds in all respects. They have the general formula CdX_2, and cadmium is invariably bivalent.* Unlike the colorless oxide and sulfide of zinc, cadmium oxide is brown and the sulfide is yellow. Cadmium sulfide differs from the zinc compound in being insoluble in dilute strong acids.

The salts of cadmium (except those derived from colored anions) are colorless. Cadmium salts of strong acids are soluble in water, and the solutions contain colorless Cd^{++} ions.

Cadmium salts have a repulsive metallic taste and are very poisonous. They have an astringent action in dilute solution, and are caustic when more concentrated They are more efficient emetics than zinc salts.

Cadmium salts have a fairly marked tendency to form *complexes*, and the binding between cadmium and halogen, in particular, is strikingly strong. Unlike other metallic salts, the chloride, bromide, and especially the iodide of cadmium are only incompletely dissociated in solution. In addition, *autocomplex* formation takes place in such solutions—e.g.,

$$3CdI_2 = Cd^{++} + 2[CdI_3]^-.$$

If excess iodide ions (e.g., as KI) are added to such solutions, complex anions, containing up to 6 halogen atoms, are formed. Evaporation of such solutions furnishes crystalline complex salts of the general compositions $M^I[CdX_3]$, $M^I_2[CdX_4]$, or $M^I_4[CdX_6]$, or occasionally $M^I_3[CdX_5]$. The other complex or double salts conform to the same general formulas. The products depend upon the proportions of the components in the solution and on the other experimental conditions.

Cadmium ions can also combine with neutral ligands such as ammonia and its organic derivatives, and the formation of such complexes is responsible for the fact that most cadmium compounds which are sparingly soluble in water are

* The existence of cadmium(I) compounds has often been assumed, but has never been conclusively proved.

soluble in aqueous ammonia. However, the sulfide, which is extremely insoluble, does not dissolve in ammonia. The ions $[Cd(NH_3)_4]^{++}$ appear to predominate in ammoniacal cadmium solutions. When ammonia gas is passed over dry salts, the maximum uptake is generally much more than $4NH_3$ per formula of compound— e.g., $6\ NH_3$ by the sulfate, up to $10\ NH_3$ by the chloride, and up to $12NH_3$ by the bromide.

Cadmium forms a considerable number of basic salts, most of which have been shown by X-ray investigations to possess simple layer lattices. Unlike zinc, very few basic cadmium salts are known with double layer lattices. Cadmium resembles zinc, however, in forming well crystallized *mixed basic salts*, such as $Cd(NO_3)_2 \cdot Cu(OH)_2 \cdot 4H_2O$ (green crystals) or $CdBr_2 \cdot 3Cu(OH)_2$ (green hexagonal platelets). Werner regarded these as 'ol' compounds, but there is little doubt that—like the solid simple basic salts of cadmium—they are really layer lattice compounds.

The *basic chlorides* may be taken as examples of the basic salts of cadmium. The simplest of these has the composition $Cd(OH)Cl$ which (according to Hoard, 1934) has the hexagonal layer lattice structure shown in Fig. 50, with fully ordered atomic arrangement. Another

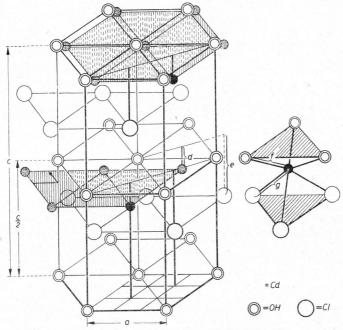

= Cd
◎ = OH   ◯ = Cl

Fig. 50. Structure of cadmium hydroxychloride, Cd(OH)Cl. Hexagonal layer lattice, $a = 3.66$, $c = 10.27$ Å. Spacing between sheets, $d = 1.03$, $e = 2.70$ Å. Distances between centers of ions, Cd $\leftrightarrow$ OH $= f = 2.34$ Å, Cd $\leftrightarrow$ Cl $= g = 2.69$ Å. The unit cell is marked out by rather heavier lines.

basic chloride, with composition corresponding to the formula $CdCl_2 \cdot 4Cd(OH)_2$ also has a fully ordered simple layer lattice structure, according to Feitknecht (1937). Two other basic chlorides, intermediate in composition between the foregoing, have structures with disordered atomic arrangement. One of them is structurally derived from anhydrous cadmium chloride (p. 451), by replacement of five-eighths of the chlorine ions in the crystal lattice by hydroxide ions. The other is essentially similar in structure to cadmium hydroxide. It may contain 2 to 2.6 $Cd(OH)_2$ for each $CdCl_2$, and is thus variable in composition, but there is no continuous transition into cadmium hydroxide. The latter can, indeed, take up some $CdCl_2$, forming mixed crystals, although only to the extent of 0.13 mol %. The particular basic salt formed depends upon the concentration and pH of the solutions from which crystallization occurs, and also on the age of the precipitate; the immediate products of precipitation are often not really stable under the conditions of formation, but subsequently undergo transformation into the stable substances.

(a) Oxygen Compounds

(i) Cadmium oxide, CdO, formed by burning cadmium, is a brown powder ($d =$ 8.15) which may be transformed into deep red cubic crystals by ignition in oxygen. It has also been found native, in the form of a lustrous black deposit of octahedral crystals. When it is heated, it begins to sublime at about 700° without melting, and loses oxygen at higher temperatures. It is very readily reduced, accordingly— by hydrogen at temperatures as low as 270–300°, and by carbon or carbon monoxide at rather higher temperatures.

Cadmium oxide crystallizes with the rock salt structure, $a =$ 4.70 Å, Cd $\leftrightarrow$ O = 2.35 Å. It thus has the same structure as the alkaline earth oxides, but not as zinc oxide.

The color of cadmium oxide depends very markedly on the temperature to which it is heated. Thus by heating $Cd(OH)_2$ at 350°, greenish-yellow CdO is obtained, whereas heating a 800° gives a deep blue-black oxide. The blue black oxide was also obtained by Scholder (1941) by prolonged boiling of $Cd(OH)_2$ with very concentrated caustic potash.

It was formerly assumed that when CdO was strongly heated or reduced at a high temperature, a 'suboxide', Cd_2O, was formed. This is contrary to the evidence of X-ray investigations, and there is, in fact, no change of structure. According to Fricke (1940), the greenish-yellow and the blue-black oxides differ only in particle size. Others consider that there is, indeed, some expulsion of oxygen on strong ignition, but without change in crystal structure except for the vacant sites distributed at random throughout the oxygen lattice.

Cadmium oxide is converted to cadmium chloride by heating it in a current of chlorine.

(ii) Cadmium hydroxide is obtained as a white precipitate on addition of alkali hydroxides to cadmium salt solutions. It is practically insoluble in excess of precipitant, but soluble in aqueous ammonia.

Cadmium hydroxide has the same crystal structure as magnesium hydroxide (brucite structure, Fig. 61, p. 261 of Vol. I), with $a =$ 3.47, $c =$ 4.64 Å. Cadmium iodide has the same structure, which is often referred to as the CdI_2 type.

The heat of formation of cadmium hydroxide from CdO and H_2O is about 5 kcal per mol (Fricke). Its solubility in water amounts to $1.15 \cdot 10^{-5}$ mol per liter at 25°. This is at first diminished by the addition of NaOH, but on further increase in the caustic soda concentration, it rises again (to $9.0 \cdot 10^{-5}$ mol per liter in 5-normal NaOH). Piater (1928) inferred from this evidence that cadmium hydroxide is amphoteric (though only weakly so). More recently, Scholder (1941) has succeeded in isolating *hydroxocadmates*, i.e. salts derived from $Cd(OH)_2$ acting as an anhydro-acid—e.g., $Na_2[Cd(OH)_4]$, $Sr_2[Cd(OH)_6]$, $Ba_2[Cd(OH)_6]$. He obtained the sodium salt by heating $Cd(OH)_2$ or CdO with very concentrated sodium hydroxide solution. $Cd(OH)_2$, instead of a hydroxo-salt, crystallizes from very concentrated potassium hydroxide solution.

(b) Amide

Cadmium amide, $Cd(NH_2)_2$, is formed by double decomposition between $Cd(SCN)_2$ and KNH_2 in liquid ammonia (Bohart, 1915). It is a yellowish-white powder ($d =$ 3.05) which turns brown in air. The heat of formation (from the elements) is 13 kcal per mol (Juza, 1937).

(c) Sulfide

Cadmium sulfide, CdS, is formed as a yellow precipitate by passing hydrogen sulfide into cadmium salt solutions. It is very occasionally found native (as *greenockite*), as an earthy incrustation on zinc ores. A highly crystalline form is obtained by heating cadmium oxide with sulfur. The density of the crystalline form, and of the native sulfide, is 4.8; the precipitated material is rather less dense.

Ignited cadmium sulfide (and greenockite) has the wurtzite structure, $a = 4.14$ Å, $c = 6.72$ Å, Cd $\leftrightarrow$ S $= 2.52$ Å. Cadmium sulfide precipitated from solution has the zinc blende structure, $a = 5.82$ Å, Cd $\leftrightarrow$ S $= 2.52$ Å.

Cadmium sulfide is insoluble in dilute hydrochloric acid, but soluble in warm dilute nitric acid, in boiling dilute (1:5) sulfuric acid, and in concentrated acids. The color of cadmium sulfide is dependent on the conditions of precipitation, and may vary from lemon-yellow to orange. It is a valuable artists pigment (cadmium yellow), being both brilliant and permanent. It is manufactured either by precipitation from sulfuric acid solution (usually with sodium sulfide), or by heating a mixture of cadmium carbonate and flowers of sulfur.

Freshly precipitated cadmium sulfide is very slightly soluble in ammonium sulfide, but not in alkali sulfide solutions. It also forms colloidal dispersions very readily; these have a fine yellow color. Thus hydrogen sulfide acts as a peptizing agent, in the absence of strong acids.

(d) Halides

(*i*) *Cadmium chloride*, $CdCl_2$, may be prepared anhydrous by heating cadmium or cadmium oxide in chlorine, or by dehydrating the hydrated salt. It forms colorless transparent masses, which may be converted into rhombohedral leaflets by sublimation ($d = 4.05$, m.p. about 565°, b.p. 964°).

The structure of cadmium chloride is similar to that of brucite and of cadmium iodide— i.e., is of layer lattice type. The difference lies in that, in the CdI_2-type, the sheets of negative ions are stacked one on another in the manner of hexagonal closest packing, whereas in cadmium chloride they are arranged in cubic closest packing (cf. Fig. 10, p. 25).

Cadmium chloride is very soluble in water (110.6 g in 100 g of water at 18°); the heat of solution is 3.2 kcal per mol of $CdCl_2$. The solution—which is usually prepared by dissolving metallic cadmium in hydrochloric acid—yields various hydrates (with 1 to 4 H_2O), according to the conditions of crystallization. The hydrates may be dehydrated completely at 120°.

Cadmium chloride is also somewhat soluble in many organic solvents—e.g., in ethyl alcohol (1.5 g in 100 g), in methyl alcohol (1.7 g in 100 g), in ethyl acetate and acetone. It is appreciably dissociated in the two solvents first mentioned, but has the normal molecular weight in urethane.

See p. 449 for basic cadmium chlorides.

(*ii*) *Cadmium bromide*, $CdBr_2$, obtained by heating cadmium in bromine vapor or by treating the metal with boiling bromine water, is similar to the chloride in every respect. The anhydrous salt forms white, nacreous leaflets ($d = 5.2$, m.p. about 570°, b.p. 863°). It is rather more strongly complexed in solution than the chloride.

(*iii*) *Cadmium iodide*, CdI_2, unlike the chloride and bromide, is only known in the anhydrous state. It may be obtained by the action of iodine on cadmium, in the presence of water, or by dissolving cadmium oxide or carbonate in dilute hydriodic acid, and forms lustrous, colorless trigonal plates ($d = 5.7$, m.p. about 385°, b.p. 708–719°).

Cadmium iodide has a brucite-type layer lattice, $a = 4.24$ Å, $c = 6.84$ Å, Cd $\leftrightarrow$ I $=$ 2.99 Å. Its heat of solution is negative (-0.96 kcal per mol), in accordance with the very small tendency for hydration of the compound. The solubility (85.2 g in 100 g of water at 18°) increases but little with rise of temperature.

Cadmium iodide crystallizes not only with the brucite structure (C6 type) but also with another structure (C27 type). This is intermediate in type between the brucite and the nickel chloride structures (C19 type). The metal ions occupy corresponding positions in all three structures, but in the C27 type CdI_2 structure the halogen ions in the individual anionic sheets occupy the positions of the brucite structure and the nickel chloride structure alternately. Starting from the structure of Cd(OH)Cl (Fig. 50), the C27 structure is obtained by replacing both OH⁻ ions and Cl⁻ ions by I⁻ ions.

(*iv*) *Cadmium fluoride*, CdF_2, ($d = 6.6$, m.p. 1110°) is only slightly soluble in water (0.29 mol per liter at 25°), and, unlike the other cadmium halides, is not volatile.

(*v*) *Autocomplex Formation by the Cadmium Halides*. It has been concluded, from measurements of conductivity, osmotic pressure, electrode potentials, and transport numbers, that cadmium iodide exists in solution very largely in the form of *autocomplexes*, such as $Cd[CdI_3]_2$. For example, in a solution which is 0.1 molar, or stronger, with respect to CdI_2, the transport number of the anion is apparently greater than 1. This implies that more cadmium migrates to the anode than to the cathode, and to do so it must be present as *complex anions*, such as $[CdI_3]^-$. The formation of such anions—e.g., according to $Cd^{++} + 3I^- = [CdI_3]^-$—brings about a reduction in the total number of ions in solution, and therefore lowers the conductivity considerably. It is, in fact, found that moderately dilute solutions of cadmium iodide have conductivities appreciably lower than are ordinarily observed for bi-univalent electrolytes. This peculiarity is exhibited in lesser degree by solutions of cadmium bromide and chloride. Table 47 affords a comparison between the equivalent conductivities Λ and the conductivity quotients Λ_c/Λ_0 for various cadmium salts with those found for a typical bi-univalent electrolyte, $MgCl_2$. The anomalous dissociation of the cadmium halides is particularly apparent in the values for the conductivity quotient, whereas the conductivity quotients of cadmium nitrate, and also of zinc chloride, hardly differ from that of magnesium chloride.

TABLE 47

EQUIVALENT CONDUCTIVITIES, Λ_c, AND CONDUCTIVITY QUOTIENTS Λ_c/Λ_0 OF CADMIUM HALIDES, COMPARED WITH THOSE FOR NORMAL ELECTROLYTES

Λ_c = equivalent conductivity at the concentration c.

		$CdCl_2$	$CdBr_2$	CdI_2	$MgCl_2$	$Cd(NO_3)_2$	$ZnCl_2$
Λ_1	=	22.4	18.3	15.4	61.5	54.3	55
$\Lambda_{0.1}$	=	50.0	44.6	31.0	83.4	80.8	82
$\Lambda_{0.01}$	=	83	76.3	65.6	98.1	96	98
Λ_0	=	115	116	120	110.5	112	112
Λ_1/Λ_0	=	0.19	0.16	0.13	0.55	0.48	0.49
$\Lambda_{0.1}/\Lambda_0$	=	0.43	0.38	0.26	0.75	0.72	0.73
$\Lambda_{0.01}/\Lambda_0$	=	0.72	0.66	0.55	0.89	0.86	0.87

Cryoscopic and ebullioscopic measurements on solutions of cadmium halides have confirmed the relatively small concentration of ions revealed by the conductivity quotients. According to McBain, it is necessary to assume—especially for cadmium iodide solutions—that *undissociated molecules* are present in considerable concentration, in addition to complex ions:

$$Cd^{++} + 3I^- \rightleftharpoons CdI_2 + I^- \rightleftharpoons [CdI_3]^-.$$

It is interesting in this connection that Bruns (1925) found the electrical conductivity of cadmium iodide solution to be *increased* by the addition of elementary iodine, whereas that of other iodide solutions (e.g., potassium iodide) was diminished. I_2 combines with I⁻ to form the less mobile I_3^- ions. With cadmium iodide, however, the reaction with iodine leads to an appreciable increase in the *number* of ions—$[CdI_3]^- + 3I_2 = Cd^{++} + 3I_3^-$—and thereby to an increase in the conductivity.

(*vi*) *Cadmium Double Halides.* The formation of complex anions, which occurs even in ordinary cadmium halide solutions, is considerably augmented by the addition of other halides. Double or complex salts therefore crystallize from mixed solutions of cadmium halides and other halides. These compounds are mostly of the type $M^I[CdX_3]$, although salts of the types $M^I_2[CdX_4]$ and $M^I_4[CdX_6]$ are also known.

Clear evidence for the formation of complex ions, when other halides are added to solutions of cadmium halides, is afforded by the fact that the conductivities of the mixed solutions are not additively related to the conductivities of the constituents. Thus the equivalent conductivity of 1-normal KI at 18° is 103.6, that of CdI_2 in 1-normal solution is 15.4, whereas a solution 1-normal with respect to both KI and CdI_2 has an equivalent conductivity of only 82, instead of $103.6 + 15.4 = 119.0$.

(e) Cyanides and Thiocyanates

(*i*) *Cadmium cyanide and Complex Cyanides.* Cadmium cyanide, $Cd(CN)_2$, is formed as a white precipitate, soluble in strong acids, by the addition of alkali cyanide to cadmium salt solutions. It dissolves in excess of precipitant, forming complex ions:

$$Cd^{++} + 2CN^- = Cd(CN)_2; \quad Cd(CN)_2 + 2CN^- = [Cd(CN)_4]^=.$$

Crystalline complex cyanides of the types $M^I_2[Cd(CN)_4]$ and $M^I_4[Cd(CN)_6]$ can be isolated from the solutions.

Transport experiments and other measurements have shown that the complex cyanides of cadmium are much more strongly complexed than the corresponding iodides, but hydrogen sulfide nevertheless effects quantitative precipitation of cadmium, as sulfide, from such solutions.

Solutions of the complex cyanides are useful for the electrolytic deposition of cadmium.

(*ii*) *Cadmium thiocyanate and Complex Thiocyanates.* Cadmium thiocyanate, $Cd(SCN)_2$, can be obtained by dissolving cadmium carbonate in thiocyanic acid, or by metathesis between cadmium sulfate and barium thiocyanate, and separates in colorless crusts when its solution is evaporated. Complex thiocyanates, mostly of the type $M^I_2[Cd(SCN)_4(H_2O)_2]$, are formed in the presence of other thiocyanates; they crystallize well—e.g., $K_2[Cd(SCN)_4 (H_2O)_2]$, regular octahedra. Crystalline thiocyanates of cadmium ammines can also be obtained from solution—e.g., $[Cd(NH_3)_2](SCN)_2$. The compound $[Cd\ pyr_2](SCN)_2$ is precipitated when alkali thiocyanates are added to cadmium salt solutions in the presence of pyridine (Spacu).

(f) Salts of Oxyacids

(*i*) *Cadmium nitrate*, most readily obtained pure by dissolving cadmium carbonate in dilute nitric acid, crystallizes at ordinary temperature as the tetrahydrate (hydrates with $2H_2O$ and $9H_2O$ are also known). The tetrahydrate forms radiating clusters of deliquescent needles ($d = 2.45$), which melt at 59.3° in their water of crystallization. (Solubility—127 g of $Cd(NO_3)_2$ in 100 g of water at 18°.).

Solutions of cadmium nitrate display a fairly normal conductivity (cf. Table 47), and thus do not contain complex ions to any considerable extent. Double compounds of cadmium nitrate with the alkali or alkaline earth nitrates are not known in the crystalline state.

Several basic cadmium nitrates are known. Of these, the salt represented by the formula $Cd(NO_3)_2 \cdot 4Cd(OH)_2$ forms a double layer lattice with ordered atomic arrangement, like the corresponding basic zinc nitrate.

(*ii*) *Cadmium nitrite and Complex Nitrites.* Unlike the nitrate, cadmium nitrite $Cd(NO_2)_2$ is very unstable as a simple salt (it is readily hydrolyzed), whereas it forms relatively stable and well crystallized complex salts—e.g., $K[Cd(NO_2)_3]$, pale yellow or colorless cubes; $K_2[Cd(NO_2)_4]$, well formed pale yellow crystals.

(*iii*) *Cadmium sulfate*, $CdSO_4$, is formed by dissolving cadmium or cadmium carbonate in dilute sulfuric acid. It is the starting material for the preparation of other cadmium compounds (e.g., cadmium yellow), and is used in Weston cells.

Cadmium sulfate ordinarily crystallizes from solution as the hydrate $CdSO_4 \cdot \frac{8}{3}H_2O$, in colorless monoclinic prisms ($d = 3.09$). Hydrates with $7H_2O$ or 1 H_2O can also be obtained; the latter exists in two forms with a transition point at $74.5°$. On addition of much sulfuric acid, anhydrous cadmium sulfate is precipitated in the form of rhombic prisms ($d = 4.7$). The solubility (76.2 g of $CdSO_4$ in 100 g of water at $18°$) at first rises slightly with increase in temperature, but suddenly drops off above $41.5°$ (the stability range of the monohydrate); solubility at $100°$, 60.8 g of $CdSO_4$ in 100 g of water. Heats of solution are— for $CdSO_4$ 10.69 kcal, for $CdSO_4 \cdot H_2O$ 6.05 kcal, and for $CdSO_4 \cdot \frac{8}{3}H_2O$ 2.54 kcal per mol.

Cadmium sulfate forms an isodimorphous series of mixed crystals with iron(II) sulfate and with copper sulfate. When the cadmium content is high, the crystal form and water content of ordinary cadmium sulfate, $CdSO_4 \cdot \frac{8}{3}H_2O$, determine the properties; with a large excess of iron or copper, the mixed crystals are of the types $MSO_4 \cdot 7H_2O$ or $MSO_4 \cdot 5H_2O$, respectively. There are broad miscibility gaps in both series—between 0.26 and 51.08 mol % iron, and between 0.55 and 98.29 mol % copper. With cobalt(II) sulfate, there are mixed crystal series with 7, $\frac{8}{3}$, 1, and also with $4H_2O$ (Bassett, 1934), and mixed crystals are also formed with $MnSO_4 \cdot 4H_2O$. A double salt is formed with magnesium sulfate; $[Mg(H_2O)_6]SO_4 \cdot [Cd(H_2O)_6]SO_4$.

Cadmium sulfate also forms double sulfates with the alkali sulfates—mostly of the type $M^I_2Cd(SO_4)_2 \cdot 6H_2O$, but there are also salts of much more complex composition. Thus Benrath found that the following double salts were formed with K_2SO_4: $K_2Cd(SO_4)_2$ (with $1\frac{1}{2}$ and $4H_2O$), $K_2Cd_3(SO_4)_4$ ($+2$ and $5H_2O$), and $K_4Cd_3(SO_4)_5 \cdot H_2O$. In solution, these salts are largely, but not quite completely, split up into their components; Rouyer has shown that the boiling point elevation of mixed solutions of ammonium sulfate and cadmium sulfate is not given additively by the boiling point elevations of the constituent solutions, and the deviation is greatest when the two components are present in a 1 : 1 ratio. The existence of sulfato ions $[M^{II}(SO_4)_2]^=$ in mixed solutions of ammonium sulfate and the sulfates of Fe^{II}, Co^{II}, Zn, Mg, and Cu^{II}, has been inferred from similar evidence. Their stability decreases along the series from cadmium to copper.

It appears that sulfato ions are also present, to a very limited extent, in solutions of cadmium sulfate itself. This is indicated by the values for the conductivity quotients of cadmium sulfate solutions, which (as also the values for zinc and copper sulfate solutions) are significantly smaller than the values found for magnesium sulfate solutions (cf. Table 48).

Cadmium sulfate crystallizes from ammoniacal solutions as the ammoniate, $[Cd(NH_3)_4(H_2O)_2]SO_4$. The hexammine, $[Cd(NH_3)_6]SO_4$, is obtained by passing ammonia gas over the anhydrous salt.

Cadmium sulfate begins to volatilize when heated above $700°$, and undergoes partial decomposition at the same time. The total pressure is 330 mm at $1000°$, the SO_3 partial pressure being 14 mm.

Little is known of the basic cadmium sulfates.

TABLE 48

EQUIVALENT CONDUCTIVITIES Λ_c AND CONDUCTIVITY QUOTIENTS Λ_c/Λ_0 OF
SOLUTIONS OF BI-BIVALENT ELECTROLYTES

Λ_c = equivalent conductivity at the concentration c.

		$CdSO_4$	$ZnSO_4$	$CuSO_4$	$MgSO_4$
Λ_1	=	23.58	26.6	25.8	28.9
$\Lambda_{0\cdot1}$	=	42.21	46.2	45.0	50.1
$\Lambda_{0\cdot01}$	=	70.34	73.4	72.2	76.7
Λ_0	=	115	114	114	113
Λ_1/Λ_0	=	0.20	0.23	0.23	0.26
$\Lambda_{0\cdot1}/\Lambda_0$	=	0.37	0.41	0.39	0.44
$\Lambda_{0\cdot01}/\Lambda_0$	=	0.61	0.64	0.63	0.68

(iv) *Cadmium carbonate*, $CdCO_3$, is thrown down as a white precipitate, on the addition of $CO_3^=$ ions to cadmium salt solutions; it is usually admixed with some hydroxide. It can be obtained in crystalline form by heating the precipitate with ammonium carbonate solution to about 170° in a sealed tube, and cooling slowly. It decomposes into CdO and CO_2 when heated; the CO_2 pressure attains 1 atm. at 327°.

(v) *Cadmium oxalate*, CdC_2O_4, is formed as a white precipitate when oxalic acid or ammonium oxalate is added to cadmium salt solutions. Leaflets of the trihydrate crystallize in the cold, and prisms of the anhydrous salt from hot solutions. It is sparingly soluble in water (1 part in 13,000), but is considerably more soluble in concentrated alkali oxalate solutions. Complex oxalates—e.g., $K_2Cd(C_2O_4)_2 \cdot 2H_2O$—crystallize from the latter solutions. Cadmium also dissolves in concentrated alkali chloride solutions, forming complex salts— e.g. $K_4Cd_2(C_2O_4)_3Cl_2 \cdot 6H_2O$.

7. Analytical (Cadmium)

Cadmium follows copper in the systematic analysis for cations. It is precipitated by hydrogen sulfide from acid solutions. The sulfide is not soluble in ammonium sulfide solution, but is dissolved by warm, fairly dilute nitric acid. Cadmium forms complexes with ammonia, as does copper, and is not precipitated by ammonia. It differs from copper in that the cyanide complex is less strongly complexed. Hence yellow cadmium sulfide alone is precipitated by the action of hydrogen sulfide on a solution containing copper and cadmium as complex cyanides.

Metallic cadmium alone is deposited when an electric current is passed (at room temperature) through a solution of the complex cyanides, provided that the concentration of alkali cyanide is high enough and the voltage not too high, copper remaining in solution. If the current is passed through an acidified solution of the simple salts, copper alone is deposited, whereas cadmium, standing above hydrogen in the electrochemical series, remains quantitatively in solution.

Cadmium compounds yield a brown oxide incrustation on the charcoal block, before the blowpipe. The yellow color of the sulfide, together with its insolubility in ammonium polysulfide solution, are characteristic.

For microanalytical detection, cadmium is conveniently converted into the well crystallized cadmium mercury thiocyanate, $CdHg(SCN)_4$. The rubidium double chloride, $Rb_4[CdCl_6]$, affords a still more sensitive means of detection (limit of detection, 0.01 γ Cd).

Cadmium is usually determined by precipitation as sulfide and weighing as sulfate. It can also be electrodeposited as metal and weighed as such. Deposition is best effected from cyanide solution.

8. Mercury (Hydrargyrum, Hg)

(a) Occurrence

Mercury occurs in Nature chiefly in the form of its sulfide, HgS, as *cinnabar*. Associated with cinnabar are smaller quantities of native mercury, distributed in droplets through the rock. The principal sources of mercury are at Almaden (Spain), Idria (Italy), Monte Amiata (Tuscany), and various places in California, Mexico, Oregon, Nevada, and in Russia.

Mercury minerals of no technical importance include *tiemannite*, HgSe, *coloradoite*, HgTe, native calomel, Hg_2Cl_2, and *coccinite*, HgI_2.

Many tetrahedrites contain mercury, as also do some zinc blendes. When these are roasted, the resulting flue dusts contain metallic mercury.

(b) History

Mercury, and its extraction from cinnabar, were already known to the ancient Greeks and Romans. The first reliable reports date from Theophrastus (about 300 B.C.). The mines of Almaden, which are still productive, were worked by the Romans. The second oldest mercury mines in Europe are those of Idria, which date from the end of the 15th century.

Use was made of mercury for the extraction of gold from its ores as early as the end of the 6th century B.C. Bartholomew of Medina introduced the use of the amalgamation process for silver ores, in Mexico, just after the middle of the 16th century. For some time this consumed a considerable proportion of the mercury production of Almaden.

Mercury was a particularly favored object of study by the alchemists, who considered it to be the common principle of all the metals. They maintained that it was possible to bring about the conversion of one metal into another (*transmutation*) by changing the mercury content. Mercury preparations were introduced into medicine chiefly by the Iatrochemists. Hitherto the poisonous nature of mercury had made physicians hesitant to use it in therapy, although the curative virtues of mercury compounds—e.g., of cinnabar—had been known in ancient times.

(c) Preparation

Mercury is prepared by heating its sulfides, either in a current of air or with the addition of iron or quick lime:

$$HgS + O_2 = Hg + SO_2$$
$$HgS + Fe = Hg + FeS$$
$$4HgS + 4CaO = 4Hg + 3CaS + CaSO_4.$$

The mercury vapors which distil off were formerly condensed in so-called *aludels*—small clay flasks, with necks top and bottom, which were joined together in long strings. Today it is more usual to use water-cooled, flattened earthenware tubes as condensers.

From mercury-containing flue dusts, the bulk of the mercury is pressed out, and the remainder is recovered by distillation.

Mechanical impurities, such as small amounts of dust, grease, etc., are separated from mercury by running it through a filter paper perforated with small holes. The presence of dissolved heavy metals may be detected by the grey 'tail' left by otherwise quite pure mercury when it runs over the surface of smooth paper. Such impurities are removed by running the mercury several times in a fine stream down a tall column of dilute nitric acid or mercury(I) nitrate solution. Final purification may be effected by vacuum distillation.

(d) Properties

At ordinary temperature, mercury is a silver-white lustrous liquid metal. It boils at 356.95° under 1 atm. pressure, and solidifies at —38.84°.

Solid mercury forms octahedra, which appear to be regular in crystal structure. However, mercury actually belongs to the hexagonal system and crystallizes in a unique structural type. Each Hg atom is surrounded by six equidistant atoms, at 3.01 Å distance.
Solid mercury is soft and ductile. The density of mercury is at

0°	15°	18°	20°	100°
13.59546	13.55846	13.55108	13.54616	13.35166 g/cc.

According to Scheel and Blankenstein, the density d_t^4 at temperatures between 0° and 100° can be represented by the formula

$$d_t^4 = \frac{13.59546}{1 + \left[18.182\left(\dfrac{t}{100}\right) + 0.078\left(\dfrac{t}{100}\right)^2\right] \cdot 10^{-3}}.$$

The thermal expansion of mercury is quite considerable, and between 0° and 100° is almost exactly proportional to the thermal expansion of gases. The thermal conductivity of mercury at 0° is only 2.2% of that of silver, and the specific electrical conductivity at 0° is 1.58% of that of silver.

The electrical conductivity of mercury is used in the definition of the statutory unit of electrical resistance, the *international ohm*, which is defined as the resistance at 0° of a column of mercury 1 sq. mm in cross section, and 106.3 cm long.

Mercury is perceptibly volatile at ordinary temperature. Its vapor pressure is:

at	0°	20°	100°
	0.00021	0.0013	0.279 mm of Hg.

1 cubic meter of air, saturated with mercury vapor, contains

at	0°	20°	100°
	0.002	0.014	2.42 g of Hg.

The vapor density of mercury has been found to be close to 200 over a wide range of temperatures (relative to $H_2 = 2$). The vapor is therefore monatomic.

Mercury vapor is *highly toxic*[7]. Even very small traces can produce serious lesions if exposure is very prolonged.

Symptoms characteristic of mercury poisoning are excessive salivation, a peculiar reddening of the gums, and loosening of the teeth. There are serious nervous disturbances, such as giddiness, digestive disorders, and trembling of the hands and head. However, it is possible for general nervous disorders—such as depression, dullness, deafness, sleeplessness, or loss of memory—to occur before there are any of the typical symptoms of chronic mercury poisoning. Different persons differ widely in their susceptibility to the toxic action of mercury. It is possible for sensitive individuals to sustain serious harm from dental fillings which give off traces of mercury over long periods, as is often the case with copper amalgam fillings. The risk is far greater from mercury scattered round the typical laboratory such as collects in cracks in bench tops, joins in linoleum, etc. This evaporates slowly, and continuously poisons the air. Evaporation is particularly favored when the mercury is finely subdivided through wax polishing the floor. Attention has been drawn by Stock in particular (1926 and later) to the dangers to which workers in such rooms are exposed, but the health hazard is still far too often under-rated.

Mercury can readily be emulsified by rubbing it up with fat (grey ointment). It can be dispersed to a black powder by vigorous agitation. It is difficult, however, to obtain pure aqueous colloidal dispersions, although mercury sols are stable in the presence of a protective colloid. Colloidal mercury is used in medicine for injections, since it has a strong bactericidal action even at very great dilutions, and is far less toxic than soluble mercury salts.

Pure mercury is not oxidized by dry air at ordinary temperature. When heated almost to the boiling point in the presence of air, it is gradually converted to oxide, but this decomposes again if heated to a higher temperature. In moist air, or if impure, mercury soon becomes covered at ordinary temperature with a skin of oxide. Mercury is also very vigorously attacked by chlorine at room temperature, and also combines with sulfur when rubbed up with it. It does not combine directly with phosphorus, however, even when heated, but merely dissolves in molten phophorus. It is dissolved by warm concentrated sulfuric acid, or by concentrated (or even dilute) nitric acid. Salts of either unipositive or dipositive mercury are thereby formed, depending upon whether the mercury or the acid is present in excess. Mercury dissolves very readily in aqua regia, forming the chloride. It does not dissolve, however, in air-free hydrochloric acid or dilute sulfuric acid, because mercury stands well below hydrogen in the electrochemical series, and cannot discharge hydrogen ions.

The standard potential of mercury in contact with a mercury(I) salt solution is —0.795 volt relative to the normal hydrogen electrode; for mercury in contact with a mercury(II) salt solution the potential is —0.86 volts. The oxidation potential $Hg_2^{++} \rightarrow 2Hg^{++} + 2e$ is —0.911 volt.

The contact potential between mercury and a 0.1-normal solution of potassium chloride saturated with mercury(I) chloride (calomel) is 0.613 volt at 18°. Such a 'calomel electrode', which is very simply constructed, (cf. Fig. 51), is often used, instead of a hydrogen electrode,

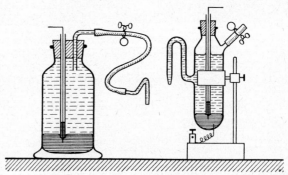

Fig. 51. Forms of calomel electrode.

as the reference electrode for potential measurements (cf. Vol. I, p. 30). The potential of an electrode of the type mentioned is 0.338 volt lower than that of the normal hydrogen electrode at 18°. If the electrode is constructed with 1-normal potassium chloride (instead of

0.1 normal KCl)—a 'normal calomel electrode'—its potential is 0.286 volts lower than that of the normal hydrogen electrode, and if saturated (about 3.5-normal)KCl is used—the 'saturated calomel electrode'—the potential is 0.254 volts lower.

(e) Amalgams

Mercury forms alloys, known as *amalgams*, with many metals. It amalgamates especially easily with sodium, potassium, silver, and gold; with zinc, cadmium, tin, and lead also, and with copper when the latter is finely divided. It does not alloy at all with manganese, iron, cobalt, or nickel.

For the preparation of amalgans, it often suffices merely to introduce the metal in lumps into mercury, which is warmed if necessary. Metals which are not dissolved to any considerable extent when they are in compact form, such as copper, are rubbed up with mercury in the finely divided state. Amalgams can often best be prepared by electrolysis. The metal in question is then deposited by the current on a mercury cathode; if desired, the mercury can be run in a fine stream through the solution.

The amalgams occupy a special place among metallic alloys, in that many of them are liquid or pasty at ordinary temperature. This is often of some importance for their practical applications. Thus the use of amalgams for dental fillings depends on the fact that they are soft and plastic at temperatures close to that of boiling water, but are quite hard at the body temperature.

In their *chemical nature* the amalgams differ in no respect from other metallic alloys. Some of them are simple solutions of the metals concerned, in mercury, and may be solid or liquid, depending on the temperature. Chemical compounds are often formed, however. Mercury forms numerous compounds with the elements of Main Group I, in particular (cf. Table 45, p. 426–7). Many important amalgams, such as those of zinc and cadmium, contain no compounds but are merely mixtures.

(f) Uses

Metallic mercury finds many applications in technical appliances and in scientific instruments. Thus it is used in mercury vapor lamps and mercury vapor rectifiers, automatic electric switches, pressure regulators, and non-return valves, and in the construction of thermometers and barometers for scientific purposes; also for the working liquid of mercury vapor pumps and many laboratory purposes.

Considerable amounts of mercury are employed in the preparation of artificial cinnabar (the pigment vermilion). Mercury fulminate is used as a detonator in the munitions and explosives industries. Mercury is used as an ointment (grey ointment) for skin infections, and many other mercury compounds have pharmaceutical applications. The nitrate and chloride are the most important salts technically, and are used, among other purposes for carrotting the hair in the manufacture of felt hats. The use of cerium compounds (cf. p. 496) for this purpose is to be preferred, however, on account of the great toxicity of mercury.

The catalytic properties of mercury compounds, are important especially for the reactions of carbon compounds. Thus mercury(I) sulfate, like copper sulfate, accelerates the destructive oxidation of organic compounds by concentrated sulfuric acid in the Kjeldahl estimation of nitrogen. Mercury(II) sulfate is used industrially as catalyst for the hydration of acetylene to acetaldehyde (Vol. I, p. 437).

Amalgams are now less important than formerly for many purposes—e.g., for the extraction of gold and silver, for fire gilding, and for coating mirrors. Considerable quantities of mercury are still employed for dental fillings*. The amalgamation of zinc plates is important in the manufacture of primary batteries. Sodium amalgam (which is readily prepared by immersing lumps of sodium in warm mercury) is often used in the laboratory as a reducing agent.

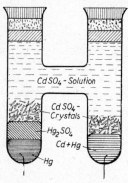

Fig. 52.
Weston standard cell.

(g) Standard Cells

Since chemically pure mercury is easily obtained, and its surface remains completely pure in contact with aqueous solutions, it is particularly well suited for the production of cells with an accurately defined potential (*standard cells*). The most important of these is the *Weston cell*, which consists of an H-shaped vessel (Fig. 52), one limb of which contains mercury, covered with a layer of Hg_2SO_4. This in turn is covered with a saturated solution of $CdSO_4$, which fills the other limb also. The bottom of this limb contains some 12.5 % cadmium amalgam, which is run while molten and allowed to solidify. In order to ensure that the cadmium sulfate solution remains saturated at all temperatures, crystals of $CdSO_4 \cdot \frac{8}{3}H_2O$ are added as solid phase. Platinum wires sealed through each limb serve as electrical connections. The electromotive force of such a cell varies very little with the temperature, being

at   0°   10°   15°   16°   17°   18°   19°   20°   25°
    1.0187   1.0186   1.0185   1.0184   1.0184   1.0184   1.0183   1.0183   1.0181 volts.

The electromotive force is still less dependent upon the temperature if a cadmium sulfate solution saturated at 4° is used, instead of a solution which is kept saturated at all temperatures. The temperature variation of the older *Clark cell* was much more serious. This contained zinc and zinc sulfate solution, in the presence of an excess of crystalline zinc sulfate, in place of cadmium with cadmium sulfate.

9. Compounds of Mercury

Mercury forms two series of normal simple compounds. These correspond in empirical composition to the general formulas HgR and HgR_2, in which R is a univalent radical. So far as it is permissible to regarded mercury as ionically charged in its compounds, it is electro-positively univalent in the first series, and bivalent in the second. The two series are known as mercury(I) (or mercurous) compounds and mercury(II) (or mercuric) compounds accordingly.

(a) Mercury(I) Compounds

(*i*) *Constitution of Mercury(I) Compounds.* In every mercury(I) compound of which the constitution has been determined, there are two mercury atoms linked together. The formulas must be represented therefore by R—Hg—Hg—R (or Hg_2R_2), instead of by the empirical formula HgR. The mercury is thus formally

---

* Because of the hazards to health which can arise from the use of mercurial dental fillings in certain circumstances, attempts have been made to replace them by other alloys with similar properties—e.g., by an alloy of Sn, Bi, and Ga. These alternatives have not yet proved completely satisfactory.

bivalent, although electrochemically unipositive. The solutions of the salts contain the ions Hg_2^{++}.

The presence of Hg_2^{++} ions in solutions of electrochemically unipositive mercury has been proved in the following manner. If mercury is shaken with a solution of silver nitrate, an equilibrium is established and, depending upon whether this is represented by process (1) or process (2), one or other of the mass action expressions (1a) or (2a) must be constant:

$$Hg + Ag^+ \rightleftharpoons Ag + Hg^+ \tag{1}$$

$$2Hg + 2Ag^+ \rightleftharpoons 2Ag + Hg_2^{++} \tag{2}$$

$$\frac{[Ag^+]}{[Ag] \cdot [Hg^+]} = K \;(1a) \qquad \text{or} \qquad \frac{[Ag^+]}{[Ag] \cdot [Hg_2^{++}]^{1/2}} = K \;(2a)^*.$$

Experiments by Ogg (1898) showed that only (2a) was constant. Ogg obtained a similar result for the equilibrium between mercury(I) nitrate, mercury(II) nitrate, and metallic mercury. If mercury(II) nitrate is shaken with mercury, the metal is dissolved and the mercury(II) nitrate reduced to mercury(I) nitrate. This reaction does not proceed to completion, but an equilibrium is reached, represented either by (3) or by (4):

$$Hg^{++} + Hg \rightleftharpoons 2Hg^+ \tag{3}$$

$$Hg^{++} + Hg \rightleftharpoons Hg_2^{++} \tag{4}$$

The experiments showed that in presence of an excess of metallic mercury, the ratio (4a) was constant, but not (3a):

$$\frac{[Hg^{++}]}{[Hg^+]^2} \;(3a); \qquad \frac{[Hg^{++}]}{[Hg_2^{++}]} \;(4a).$$

(4a) has the value $\frac{1}{120}$ at $25°$. In each case, therefore, the mass action law is obeyed on the assumption that electropositively univalent mercury is present in solution as Hg_2^{++} ions.

Among other observations which point to the formulation of mercury(I) ions as Hg_2^{++}, it may be mentioned that, when the influence of hydrolysis is eliminated, the electrical conductivity of solutions of soluble mercury(I) salts corresponds both in its magnitude and in its dependence on concentration to that of typical bi-univalent electrolytes (e.g., barium nitrate), and not to uni-univalent salts (such as potassium nitrate).

It was until recently open to doubt whether electrochemically unipositive mercury was already dimeric in its crystalline compounds, or whether the formation of the Hg_2^{++} ion was limited to aqueous solutions. The results of vapor density determinations on mercury(I) chloride and bromide could be interpreted as supporting the monomeric formulation, since they indicated the molecular weights required by the monomolecular formula. However, they could also be explained in terms of a dissociation of Hg_2Cl_2 into $HgCl_2$ and Hg, in the course of evaporation, since this would also lead to an average molecular weight corresponding to the formula $HgCl$. A gold leaf exposed to the vapor of mercury(I) chloride is, in fact, immediately amalgamated, but this does not justify the deduction that the vapor is completely dissociated into $HgCl_2$ and Hg.

The crystal structure of the mercury(I) halides has, however, been determined by X-ray methods, and has shown that the mercury atoms in them are always associated in pairs, in the arrangement X—Hg—Hg—X. The existence of the Hg—Hg bond is thereby proved for the solid state as well as for solutions, and there can be no doubt that it is a peculiar and characteristic property of electrochemically unipositive mercury to exist always in the dimeric form, Hg_2^{++}.

---

* Symbols in square brackets represent the molar concentrations (more precisely, activities) of the ions in the solution, or of the silver in the resulting amalgam. The concentration of metallic mercury can be considered constant.

(ii) *Preparation of Mercury(I) Compounds.* Soluble mercury(I) salts can generally be prepared by treating the corresponding mercury(II) salts with mercury, or by dissolving mercury(I) carbonate in acids. The sparingly soluble salts and other sparingly soluble compounds can be prepared from the soluble salts by precipitation. However, this method not infrequently leads to a mixture of the corresponding mercury(II) compound with metallic mercury, instead of the desired mercury(I) compound.

This invariably happens when the mercury(II) compound concerned is so strongly complexed, or is so little dissociated, that the concentration of Hg^{++} ions in the solution becomes vanishingly small. In that event it follows from eqn. (4) that the concentration of Hg_2^{++} ions must also become vanishingly small, so that it is no longer possible to exceed the solubility product of a normally dissociated mercury(I) compound, however insoluble it may be. Thus the non-existence of mercury(I) cyanide is occasioned by the extraordinarily small degree of dissociation of $Hg(CN)_2$, which results in the equilibrium

$$Hg_2^{++} + 2CN^- \rightleftharpoons Hg + Hg(CN)_2$$

being displaced completely to the right.

(iii) *Properties of Mercury(I) Compounds.* Most mercury(I) compounds are sparingly soluble, but a few salts (in particular the nitrate, chlorate, and perchlorate) are quite soluble. These salts are normally dissociated in solution, in the same way as salts of the barium nitrate type. They are hydrolyzed to a slight extent, and their solutions have an acidic reaction. Mercury(I) compounds are generally fairly strong reducing agents.

A few mercury(I) compounds are characteristically colored. The carbonate is yellow, as is the (very unstable) iodide. The arsenite is also yellowish and the arsenate red-orange.

b) Mercury(II) Compounds

(i) *Preparation of Mercury(II) Compounds.* Mercury(II) nitrate is obtained by dissolving mercury in excess of strong nitric acid, and mercury(II) oxide is formed by heating the nitrate. Most mercury(II) salts can be prepared by dissolving the oxide in the corresponding acid.

As has already been mentioned, mercury(II) compounds are often obtained, mixed with metallic mercury, by precipitation reactions, when it would be expected that mercury(I) compounds should be formed. This is the case with the sulfide and cyanide, for example. These reactions may be contrasted with those of copper, where the extreme insolubility of some copper(I) compounds (e.g., CuI) leads to their formation in place of the expected copper(II) compounds.

(ii) *Properties of Mercury(II) Compounds.* Mercury(II) salts of strong, colorless acids are colorless, as are the mercury(I) compounds. Electrochemically dipositive mercury tends to form insoluble basic salts, however, and these are mostly yellow to orange.

Mercury(II) salts of weak acids are also often colored. The nitrite is bright yellow, the orthoarsenate lemon yellow, and the sulfide exists in a red and a black modification.

Many mercury(II) salts are freely soluble. The nitrate, perchlorate, and similar salts are quite normal in their dissociation. They are much more extensively hydrolyzed than the mercury(I) salts, and their solutions are strongly acidic in reaction. Precipitation of basic salts generally takes place when they are strongly diluted, and only the salts of very strong acids, such as mercury(II) perchlorate, are exempt from basic salt formation.

In addition to the normally dissociated salts, there are also freely soluble mercury(II) salts which are only slightly dissociated. This is the case with the chloride, bromide, cyanide, and thiocyanate (the iodide is also very slightly dissociated, but is very sparingly soluble). The salts which dissociate only to a slight extent are hydrolyzed to a correspondingly small degree; it is impossible to detect any hydrolysis of the cyanide, which is ionized to an especially small extent.

The mercury(II) salts of moderately strong organic acids, such as the acetate, have a degree of dissociation similar to that of the acid itself, rather than resembling the typical salts.

Mercury(II) ions differ strongly from mercury(I) ions in that they have a *very strong tendency to form complex ions.*

Salts of low solubility include the iodide, oxalate, and phosphate. The sulfide is extremely insoluble, and the thiocyanate and sulfate are fairly insoluble.

Many mercury(II) salts have an unusually high solubility in organic solvents such as alcohol, ether, and benzene.

It is rare that ammonia merely forms addition compounds with mercury compounds. More often, mercury *replaces hydrogen* in ammonia (or its derivatives), forming amido compounds or even ammonium compounds containing substituted mercury. The neighboring element, gold, shows the same peculiarity, although to a much slighter extent (cf. p. 415).

As an example of the ease with which hydrogen bound to nitrogen undergoes substitution, by reaction with mercury compounds, it may be mentioned that mercury(II) oxide dissolves smoothly in aqueous solutions of acid amides, which are in other respects quite indifferent towards bases. Thus

$$HgO + 2H_2N \cdot CO(CH_3) = Hg\begin{matrix} NH \cdot CO(CH_3) \\ NH \cdot CO(CH_3) \end{matrix} + H_2O.$$

acetamide   mercury acetamide

The reaction corresponds formally to the formation of a salt. The resulting compound— like many other mercury compounds which would be formulated as salts—is only very slightly dissociated in solution.

Mercury can also be combined very readily with carbon, or with organic radicals. The univalent radical —HgX, in particular, can be very readily substituted for hydrogen in organic compounds. Thus the reaction of mercury(II) oxide on ethanol, in the presence of caustic soda leads to the formation of the compound:

$$O\begin{matrix} Hg \\ Hg \end{matrix}C - C\begin{matrix} Hg \\ Hg \end{matrix}O \qquad \text{(Hofmann).}$$
$$HO-Hg \quad HgOH$$

Dry mercury(II) acetate, heated with benzene to 110° for a few hours, forms phenyl mercury acetate, C_6H_5—Hg—O $\cdot$ CO(CH$_3$). It reacts in analogous manner with toluene and phenol, and other mercury(II) compounds (e.g., the sulfate and nitrate) behave similarly (Dimroth). See p. 476 for organo-mercury compounds of the type HgR$_2$ (mercury alkyls).

Except for the fluoride, the halides of mercury(II) are only slightly ionized in solution. The binding of Cl, Br, or I with mercury is thus not truly ionic, and radicals linked to mercury through C, N and S are essentially covalently bound.

The close relationship between mercury compounds derived from the two valence states makes it desirable to consider both series of compounds together.

(c) Oxides

Mercury(II) oxide, (mercuric oxide), HgO, is thrown down as a yellow precipitate, on the addition of excess alkali hydroxide to a solution of mercury(II) chloride or nitrate, and is obtained as a red crystalline powder by gently heating mercury(II) nitrate or mercury(I) nitrate:

$$Hg(NO_3)_2 = HgO + 2NO_2 + \tfrac{1}{2}O_2; \quad Hg_2(NO_3)_2 = 2HgO + 2NO_2.$$

These methods of preparation were known to the alchemists, who called the product 'red precipitate'. Mercury(II) oxide is decomposed into mercury and oxygen if it is heated to a higher temperature. The red and yellow forms differ only in their particle size, as was long supposed and ultimately proved by determination of the size of the crystallites from the width of their X-ray diffraction lines. The more finely divided yellow oxide is slightly less insoluble in water than the red form (solubility 1 part in 19,500 parts of water at 25°). The solution is weakly basic in reaction:

$$HgO + H_2O \rightleftharpoons Hg^{++} + 2OH^-.$$

Mercury(II) oxide crystallizes in the orthorhombic system. It has been found native as the mineral *montroydite*.

Mercury(II) oxide is used in preparative chemistry (e.g., for the preparation of hypochlorous acid, Vol. I, p. 803) and in medicine. It also serves as starting material for the preparation of other mercury compounds.

Colloidal dispersions of mercury(II) oxide can be prepared in the presence of protective colloids, such as protalbinic acid or lysalbinic acid.

Addition of alkali hydroxide to mercury(I) salt solutions gives a black precipitate, which was formerly taken to be 'mercurous oxide', Hg_2O. It has been shown by X-ray examination, however, to be a mixture of finely divided mercury and mercury(II) oxide (Fricke and Ackermann, 1933):

$$Hg_2^{++} + 2OH^- = Hg + HgO + H_2O.$$

Mercury(II) oxide is formed directly from its elements at temperatures just below the boiling point of mercury. The heat of formation from the elements is 21.6 kcal per mol and in accordance with the low affinity of mercury for oxygen, dissociation sets in at temperatures just above the boiling point of mercury. The dissociation pressure is a few mm at 440°, 1250 mm at 610°.

(d) Mercury Sulfide and Thiosalts

(i) *Mercury(II) sulfide*, HgS, exists in two modifications, black and red respectively. Both occur native.

Red mercury sulfide occurs as *cinnabar*, the principal ore of mercury. It rarely occurs in well formed crystals, but usually exists as dense aggregates of brick red to dark brown crystals, which may be colored blue black by impurities. Artificially prepared cinnabar, however, is a vivid scarlet, and is used as an artists' pigment (vermilion). It is usually prepared for this purpose by heating mercury with a solution of potassium pentasulfide (Döbereiner's method):

$$Hg + K_2S_5 = HgS + K_2S_4.$$

Vermilion gradually blackens on exposure to light. It is not certain whether this is due to decomposition, or to conversion into the black modification.

Black mercury(II) sulfide is found native in small amounts, with cinnabar, as *metacinnabarite*, a black powder ($d = 7.7-7.8$), apparently amorphous, but actually composed of minute regular tetrahedral crystals.

Cinnabar can be described as having a deformed sodium chloride structure, of hexagona symmetry. The structure is derived from that of sodium chloride by displacement of the non-metals atoms (in this case sulfur atoms) somewhat from their ideal positions, thereby changing the symmetry. The shortest distance Hg↔S is 2.52 Å, and the binding between mercury and sulfur is probably ionic in character. Black mercury sulfide has the zinc blende structure (Fig 47, p. 380), with $a = 5.84$ Å. The shortest Hg↔S distance (2.53 Å) is the same as in cinnabar.

Black mercury(II) sulfide is always formed by precipitation from solution with hydrogen sulfide, and under other conditions. Thus it is prepared technically by treating mercury with molten sulfur (Dutch process) or powdered sulfur (Irish process). It is converted into the red form by sublimation at ordinary pressure, or by digestion with alkali polysulfide solution.

The solubility product of the sulfide is so extraordinarily small ($3 \cdot 10^{-54}$ at 26°) that all mercury compounds, including the most stable complexes, are decomposed by hydrogen sulfide, with the deposition of HgS. For this reason hydrogen sulfide precipitates a mixture of mercury(II) sulfide and mercury from mercury(I) salt solutions:

$$Hg_2^{++} + H_2S = Hg + HgS + 2H^+.$$

The sulfide is attacked very slowly, if at all, even by concentrated acids, but is easily dissolved by aqua regia.

(*ii*) *Thiosalts.* Mercury(II) sulfide is insoluble in ammonium sulfide solution and in caustic alkalis, but it dissolves in concentrated solutions of alkali or alkaline earth sulfides. *Thiosalts* are thereby formed, and may also be obtained by fusing mercury sulfide with sulfur and alkali hydroxides—e.g., $K_2[HgS_2] \cdot 5H_2O$, bright, very deliquescent needles.

The thiosalts of mercury are stable in solution only in the presence of an excess of alkali hydroxide. If hydrogen sulfide is passed into a solution of mercury sulfide in alkali sulfide solution, all the mercury is reprecipitated since the H_2S converts the $S^=$ ions of the solution into HS^- ions:

$$S^= + H_2S \rightleftharpoons 2HS^-. \qquad (1)$$

The reaction

$$HgS + S^= \rightleftharpoons [HgS_2]^= \qquad (2)$$

must therefore proceed from right to left. Conversely, addition of alkali hydroxide augments the solvent properties of alkali sulfides for mercury sulfide:

$$SH^- + OH^- \rightleftharpoons S^= + H_2O. \qquad (3)$$

Mercury sulfide is insoluble in solutions of ammonium sulfide, even though it is capable of forming thiosalts; in a solution of ammonium sulfide the sulfur is present almost exclusively in the form of HS⁻ ions (Vol. I, p. 736). Hence reaction (2) cannot proceed to any perceptible extent from left to right in such a solution.

If hydrogen sulfide is passed slowly into a mercury(II) salt solution, white, yellow, or brownish precipitates may be transiently formed. These consist of mixed salts, in which the ion originally combined with the mercury is only partially replaced by sulfur. Such mixed salts are obtained especially readily from solutions of weakly dissociated mercury salts—e.g.,

$$3HgCl_2 + 2H_2S = Hg_3S_2Cl_2 + 4HCl.$$

(e) Mercury(I) Halides

(*i*) *Mercury(I) chloride* (calomel), Hg_2Cl_2, is occasionally found native in small tetragonal crystals. It is made technically by both dry and wet methods. The dry process involves heating an intimate mixture of mercury(II) chloride (corrosive sublimate) and mercury (1a), or a mixture of mercury(II) sulfate, mercury, and sodium chloride (1b). In the wet process, mercury(II) hloride is reduced with sulfur dioxide (2), or mercury(I) nitrate solution is treated with dilute hydrochloric acid (3):

$$HgCl_2 + Hg = Hg_2Cl_2 \tag{1a}$$

$$HgSO_4 + Hg + 2NaCl = Hg_2Cl_2 + Na_2SO_4 \tag{1b}$$

$$2HgCl_2 + SO_2 + 2H_2O = Hg_2Cl_2 + H_2SO_4 + 2HCl \tag{2}$$

$$Hg_2^{++} + 2Cl^- = Hg_2Cl_2. \tag{3}$$

The calomel which sublimes off in the dry process forms a lustrous white crystalline mass. The precipitated material is a powder. It received its name 'calomel' from the deep black coloration obtained when it is treated with ammonia (greek καλόν μέλας, beautiful black).

It is generally assumed that this black color is due to the formation of a mixture of (colorless) mercury(II) amidochloride, $ClHgNH_2$ (cf. p. 474 *et seq.*) and finely divided metallic mercury:

$$Hg_2Cl_2 + 2NH_3 = ClHgNH_2 + Hg + NH_4Cl.$$

There is some evidence, however, which seems to point to the existence of a black mercury(I) amidochloride, $ClHg_2NH_2$, a compound which readily decomposes into metallic mercury and mercury(II) amidochloride, especially in the presence of excess precipitant.

$$Hg_2Cl_2 + 2NH_3 = ClHg_2NH_2 + NH_4Cl; ClHg_2NH_2 = ClHgNH_2 + Hg.$$

Mercury(I) chloride is very sparingly soluble in water (2.1 mg per liter at 18°), but is much more soluble in chloride solutions. It is practically insoluble in ethanol, acetone, and ether. The density of mercury(I) chloride is 7.16. It turns yellowish when it is gently heated (or rubbed), and darkens on exposure to light—probably as a result of partial decomposition into mercury(II) chloride and mercury. Mercury(I) chloride sublimes at 383° without melting, but Blitz and Klemm found that it could be melted to a red-brown liquid at 525° in a sealed tube (with some decomposition into $Hg + HgCl_2$).

The principal use of calomel is as a purgative in medicine.

(ii) Mercury(I) bromide, Hg_2Br_2, is obtained as a white precipitate by the addition of potassium bromide to a solution of mercury(I) nitrate. It can be recrystallized from hot mercury(I) nitrate solution, and forms pearly tetragonal leaflets.

(iii) Mercury(I) iodide, Hg_2I_2, is formed as a green precipitate by the addition of a little alkali iodide to mercury(I) nitrate solution, or as a greenish yellow powder by rubbing up mercury with iodine in the presence of a little ethanol. It is practically insoluble in water ($0.37 \cdot 10^{-9}$ mol per liter at $25°$) and ethanol. It has some therapeutic uses. The pure compound is bright yellow, but readily suffers partial decomposition into $HgI_2 + Hg$. If mercury(I) iodide is treated with a solution of alkali iodide, half the mercury goes into solution as an iodo complex of dipositive mercury, and half the mercury is reduced to the metal:

$$Hg_2I_2 + 2I^- = [HgI_4]^= + Hg.$$

(iv) Mercury(I) fluoride, Hg_2F_2, obtained by dissolving freshly precipitated mercury(I) carbonate in hydrofluoric acid, is a yellowish material ($d = 8.7$) which blackens on exposure to light. It is more soluble than mercury(I) chloride, and is hydrolyzed by water.

(v) Crystal Structure of the Mercury(I) Halides. X-ray investigations have shown that Hg_2Cl_2, Hg_2Br_2, and Hg_2I_2 have structures of the type represented diagrammatically in Fig. 53. It may be seen that the structure is that of a tetragonal layer lattice, built up from Hg_2X_2 molecules.

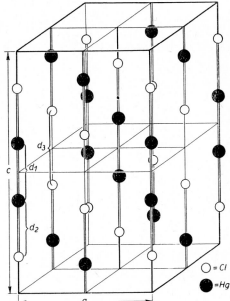

Fig. 53. Crystal structure of mercury(I) chloride, Hg_2Cl_2.

$a = 6.30, c = 10.88$ Å. $Hg\leftrightarrow Hg = d_1 = 2.54$ Å, $Hg\leftrightarrow Cl = d_2 = 2.51$ Å, $Cl\leftrightarrow Cl = d_3 = 3.32$ Å. Hg_2Br_2 (with $a = 6.61, c = 11.16$ Å) and Hg_2I_2 (with $a = 6.95$, $c = 11.57$ Å) have similar structure. If the $Hg\leftrightarrow Hg$ distance in these is taken as 2.54 Å, other distances are $Hg\leftrightarrow Br = 2.58$, $Hg\leftrightarrow I = 2.68$, $Br\leftrightarrow Br = 3.46$, and $I\leftrightarrow I = 3.67$ Å.

(f) Mercury(II) Halides

(i) Mercury(II) chloride, $HgCl_2$, (corrosive sublimate) is usually made either by heating a mixture of mercury(II) sulfate and sodium chloride:

$$HgSO_4 + 2NaCl = HgCl_2 + Na_2SO_4,$$

or by dissolving mercuric oxide in hydrochloric acid.

As prepared in the dry process, by sublimation (hence its trivial name), it forms a white transparent crystalline mass ($d = 5.44$). It crystallizes from aqueous solution in colorless rhombic bipyramids with m.p. $277–280°$ and b.p. $302°$. Corrosive sublimate is thus more volatile than calomel. The solid substance is stable towards light.

Mercury(II) chloride has a definite molecular lattice according to Braekken (1934)—cf. Fig. 54. The four mercury(II) halides differ quite characteristically in this respect, in that mercury(II) fluoride possesses a true ionic (coordination) lattice of fluorite type ($a =$ 5.54 Å), whereas HgBr$_2$ and HgI$_2$ have layer lattices, that of HgBr$_2$ being a transition between the molecular type and the true layer lattice type (see below).

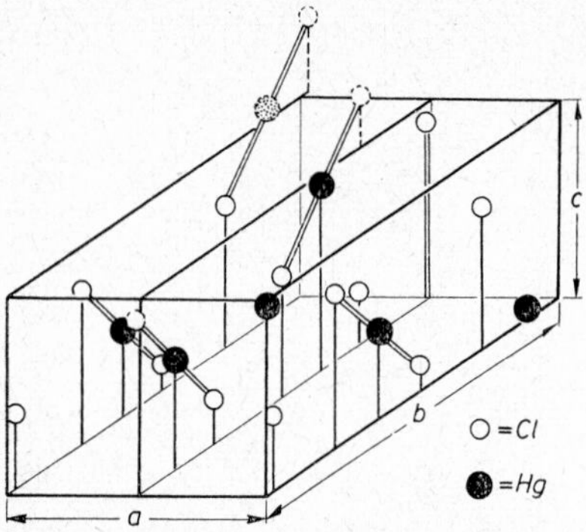

Fig. 54. Crystal structure of mercury(II) chloride. Orthorhombic unit cell, $a = 5.963$, $b = 12.753$, $c = 4.325$Å. Distance Hg$\leftrightarrow$Cl $= 2.25$Å. Atoms belonging to the same molecule are joined by double lines. (Atoms shown dotted are located in adjacent unit cells.)

Mercury(II) chloride is fairly soluble in water (100 g of water dissolve 4.3 g at 0°, 7.4 g at 20° and 55 g at 100°). The solubility is raised by the presence of HCl or NH$_4$Cl. It is also freely soluble in ethanol (50.5 g in 100 g of absolute ethanol at 25°), and quite soluble in ether (6.45 g in 100 g of absolute ether at 18°). It is also distinctly soluble in many other organic solvents, such as benzene (0.5 g in 100 g at 25°). Dilute aqueous solutions gradually decompose, with deposition of calomel. Aqueous solutions are distinctly acid in reaction.

Corrosive sublimate is widely used in medicine, especially as a very efficient antiseptic and disinfectant, and also for the treatment of syphilis and other infectious diseases. It is a powerful internal poison—lethal dose 0.2 to 0.4 g. Tablets of sublimate, for making up disinfectant solutions, are often colored red with eosin, to avoid risk of confusion. Wood (e.g., telegraph poles) may be protected from decay by prolonged soaking in a solution of mercury(II) chloride, a process introduced by Kyan in 1823. The chloride is also used as an intensifier for photographic negatives. Its property of forming well crystallized double chlorides is often utilized in preparative chemistry for the identification or purification of organic bases.

The molar conductivity of aqueous solutions of HgCl$_2$ is very small, as compared with that of other salts, even though the solutions give reactions indicative of an appreciable H$^+$ ion concentration, and therefore contain some free hydrochloric acid. The small conductivity arises from the weak dissociation of the mercury chloride. Various methods—e.g. potential measurements—indicate that the concentration of Hg^{++} ions in a saturated solution of HgCl$_2$ (i.e., about 0.25-molar) is of the order 10^{-8} g ions per liter—i.e., is actually only about one tenth of the hydrogen ion concentration in pure water. The chloride ion concentration, although very small, is much greater than the Hg^{++} ion concentration, because of the hydrolysis, which probably takes place according to the equation:*

$$2HgCl_2 + H_2O \rightleftharpoons (HgCl)_2O + 2H^+ + 2Cl^- \qquad (1)$$

Thus by far the greater part of the mercuric chloride is present in solution in the undissociated state. From the conductivity measurements of Ley, the Cl$^-$ ion concentration in a $1/_{32}$ molar solution of HgCl$_2$ at 25° is $1.6 \cdot 10^{-4}$ g ions per liter, on the assumption that it is present entirely as hydrochloric acid formed by hydrolysis. If it were regarded as due exclusively to the dissociation

$$HgCl_2 \rightleftharpoons [HgCl]^+ + Cl^- \qquad (2)$$

it would still amount to only $6 \cdot 10^{-4}$ g ions per liter at the most. It must, in fact, lie between these limits, since processes (1) and (2) occur concurrently. In any case, it may be

* Account must be also be taken of the process HgCl$_2$ + 2Cl$^- \rightleftharpoons$ [HgCl$_4$]$^=$. However, at the very low Cl$^-$ ion concentration involved, this equilibrium is displaced far to the left.

stated that in a $^1/_{32}$ molar mercuric chloride solution, not more than 0.5% is hydrolyzed according to eqn. (1), less than 2% undergoes primary dissociation, and that the extent of the secondary dissociation $[HgCl]^+ \rightleftharpoons Hg^{++} + Cl^-$ is vanishingly small. Luther (1904) found 0.47% hydrolyzed and 0.14% primarily dissociated, at the temperature and concentration cited, with 2.7% hydrolysis and 0.5% dissociation in $^1/_{512}$ molar solution.

Mercury(II) chloride provides a good example of a peculiarity of substances which are only slightly dissociated both electrolytically and hydrolytically. If a nearly saturated solution of the salt, which has an acid reaction towards methyl orange, is treated with a neutral solution of NaCl or KCl, a color change takes place and the solution shows a neutral reaction. The addition of Cl^- ions lowers the hydrogen ion concentration, in accordance with equilibrium (1).* The behavior of the solution is thus analogous to that of a weak acid on the addition of one of its neutral salts, and is in fact that which would be shown by hydrochloric acid solutions if HCl were a weak acid. It may also be perceived from eqn. (1) that increase of the H^+ ion concentration must yet further diminish the Cl^- ion concentration. One consequence of this is that concentrated sulfuric acid is unable to liberate hydrogen chloride from mercury(II) chloride, even on boiling.

A further consequence of the very low chloride ion concentration in mercuric chloride solutions is that sparingly soluble, but normally dissociated chlorides, such as silver chloride, are far more soluble in solutions of mercuric salts (e.g., mercury(II) nitrate) than in water, provided that the mercury ion concentration is high. Since the Hg^{++} ions bind the Cl^- ions to form undissociated $HgCl_2$, the solubility product of silver chloride cannot be exceeded until a considerable amount of Ag^+ ions has entered the solution. Silver is thus only incompletely precipitated by Cl^- ions from nitrate solutions in the presence of mercury(II) salts, and chloride is only incompletely precipitated by silver nitrate from solutions of mercury(II) salts. These phenomena are of importance in analytical chemistry.

(*ii*) *Complex Salts of Mercury(II) chloride.* Mercury(II) chloride has a strong tendency to add on to other chlorides, forming well crystallized complex salts, mostly of the types $M^I[HgCl_3]$ and $M^I_2[HgCl_4]$. There are also compounds of more complex formulas. Addition compounds of HCl with $HgCl_2$ have also been isolated—e.g., $HgCl_2 \cdot 2HCl \cdot 7H_2O$ or $H_2[HgCl_4] \cdot 7H_2O$. Mixed double salts are also frequently formed, resulting from the addition of mercury(II) chloride to the salt of another acid—e.g., $HgCl_2 \cdot K_2Cr_2O_7$ or $K_2\left[Hg\genfrac{}{}{0pt}{}{Cl_2,}{Cr_2O_7}\right]$, orange rhombic crystals. $HgCl_2$ also forms a number of crystalline double compounds with HgO.

The considerable increase in the solubility of mercury(II) chloride, brought about by hydrogen chloride, ammonium chloride, or other chlorides, affords evidence that complex ions, such as $[HgCl_3]^-$ or $[HgCl_4]^=$, are formed in considerable amount in solutions of mercuric chloride, containing excess Cl^- ions. The acidic reaction of mercury(II) chloride solutions disappears on addition of alkali chloride, as already stated, Furthermore, no precipitate of basic salt is obtained when solutions of $HgCl_2$, containing alkali chlorides, are mixed with solutions of low OH^- ion concentration (e.g., $NaHCO_3$ solutions), as is the case with pure mercury(II) salt solutions. This property is utilized in manufacturing sublimate pastilles, which are made from a mixture of equal weights of $HgCl_2$ and NaCl, so that their solutions will be neutral in reaction.

(*iii*) *Mercury(II) bromide*, $HgBr_2$, crystallizes from aqueous solution in silvery leaflets, and from ethanolic solution in colorless rhombic crystals. It is rather less soluble in water than the chloride (0.62 g of $HgBr_2$ in 100 g of water at 25°), and is even more weakly dissociated

---

* Combining eqn. (1) with the equation in the footnote, p. 468, the resulting Mass Action Law expression is $\dfrac{[(HgCl_4)^=]^2 \cdot [H_2O]}{[(HgCl)_2O] \cdot [H^+]^2 \cdot [Cl^-]^6} =$ constant. This expression brings out the way in which the H^+ ion concentration is strongly influenced by the Cl^- ions concentration.

in solution. In other respects it is very similar to mercury(II) chloride, and forms addition compounds with HgO and complex salts with other halides, in the same way.

The crystal structure of mercury(II) bromide is intermediate in character between the molecular type and the true layer lattice type. As in the layer lattices of brucite, cadmium chloride, and the mercury(II) iodide type, the metal atoms in the $HgBr_2$ structure are arranged in sheets, with non-metal atoms in layers on either side of them. However, in mercury(II) bromide, one Br atom in each sheet can always be assigned to a particular mercury atom, so that each Hg atom, with its two nearest-neighbor Br atoms (one from each sheet) can be recognized as a discrete molecule of $HgBr_2$ (distance $Hg \leftrightarrow Br = 2.48$ Å, as compared with a distance of 3.23 Å from each of the next nearest Br atoms). No delimitation of a molecule is possible in the true layer lattice structures. Thus, in the CdI_2 or $CdCl_2$ structures, each Cd atom is equidistant from 3 I or Cl atoms in each sheet, and in HgI_2 each Hg atom is equidistant from two I atoms in each sheet (i.e., is coordinated tetrahedrally with 4 I atoms in all).

(iv) *Mercury(II) iodide*, HgI_2, is formed as a red powder by rubbing mercury with iodine, moistened with a little alcohol, or by precipitation from a mercury(II) nitrate solution by addition of a small amount of alkali iodide. It forms a tetragonal layer lattice, with mercury atoms 4-coordinated (distance $Hg \leftrightarrow I = 2.78$ Å from each of four neighbors). It is sparingly soluble in water (4.4 mg in 100 g of water at 25°), but more soluble in warm ethanol. It also dissolves in many other organic solvents, and is more soluble in benzene than in water. It is transformed reversibly into another modification, the rhombic yellow form, when it is heated above 126°. This is also formed as an unstable intermediate product when mercury(II) iodide is precipitated from mercuric chloride solution by adding alkali iodide, or when its alcoholic solution is suddenly diluted with water. In the latter case it may persist for some time before conversion to the red form. Indeed, close investigation has shown that the yellow form is always produced in the first moment when mercury(II) iodide is forced to crystallize from solution. This exemplifies Ostwald's law of successive transformations (Vol. I, p. 489), whereby a substance which exists in several modifications often crystallizes initially in the less stable form.

Tammann, in about 1910, described a third, more unstable, colorless modification of HgI_2, which is obtained by the rapid condensation of the vapor. It reverts within a few minutes to the red form.

Mercuric iodide is used, in the form of ointment, for the treatment of skin infections.

Mercury(II) iodide is dissociated to a still smaller extent than the other mercury halides. Its solutions give no precipitate of silver iodide, on addition of silver nitrate, neither do they form mercury oxide or basic salts on the addition of alkali. It readily forms complex salts, like the chloride and bromide. An addition compound with mercuric oxide, $HgI_2 \cdot 3HgO$, can be obtained as a yellow brown powder, by heating mercury(II) iodide with dilute caustic potash.

Potassium tetraiodomercurate(II), $K_2[HgI_4] \cdot 2H_2O$, is typical of the complex iodides (light yellow crystals, freely soluble in water and ethanol). A solution of this compound, to which alkali hydroxide is added, is used as a sensitive reagent, for the detection and determination of very small traces of ammonia (Nessler's reagent, cf. p. 476 and Vol. I, p. 624).

The red copper(I) complex iodide, Cu_2HgI_4, and the yellow silver complex iodide, Ag_2HgI_4, undergo color change when they are warmed to a moderate temperature—e.g., the copper salt turns chocolate brown at 70°. They can be used as 'thermoscopes', when painted on machine parts, to reveal excessive rise of temperature.

(*v*) *Mercury(II) fluoride*, HgF_2, can be made by heating mercury(I) fluoride to 275° in chlorine (Ruff, 1918). It forms colorless octahedra (m.p. 645°, fluorite structure). A hydrate, $HgF_2 \cdot 2H_2O$ crystallizes from solutions of mercury oxide in hydrofluoric acid. The ability of the fluoride to form a hydrate contrasts strongly with the properties of the other mercury(II) halides. It also differs in being extensively hydrolyzed in aqueous solution, an indication that it is normally ionized. No complex salts derived from mercury(II) fluoride have been obtained.

(g) Cyanides and Thiocyanates

(*i*) *Mercury(II) cyanide*, $Hg(CN)_2$, is obtained by the action of aqueous hydrocyanic acid on mercuric oxide, or by reaction between alkali cyanides and mercury(II) salts in solution. If mercury(I) compounds are taken as starting material, mercury(II) cyanide is also formed, together with metallic mercury.

Mercury cyanide is ionized to such an extraordinarily small extent that even minimal concentrations of CN^- ions in any solution suffice to bring about the general reaction

$$Hg^{++} + 2CN^- = Hg(CN)_2.$$

Thus mercury cyanide can be formed by heating mercury(II) oxide with an aqueous suspension of Prussian blue.

The cyanide forms colorless, columnar crystals. It is fairly soluble in water (8.0 g in 100 g at 0°), but not in ethanol. Its aqueous solutions give no precipitates with potassium hydroxide or potassium iodide, since the compound is practically undissociated, but hydrogen sulfide precipitates mercury(II) sulfide.

A solution saturated with $Hg(CN)_2$ can dissolve a considerable amount of HgO, to form the less soluble mercury oxycyanide, $Hg(CN)_2 \cdot HgO$ (possibly $(CN \cdot Hg)_2O$). Solutions of this substance are basic in reaction.

If mercury cyanide is heated to 320°, it decomposes into mercury and cyanogen, which is partially polymerized to paracyanogen.

Mercury cyanide has a very strong tendency to form complex cyanides. Most of these are of the types $M^I[Hg(CN)_3]$ and $M^I_2[Hg(CN)_4]$; there are also more complicated compounds, and numerous examples of mixed complexes such as $M^I[Hg(CN)_2X]$, where X may be Cl, Br, I, or some radical such as the NO_3^- ion, with a weaker tendency to form complexes. In the latter case, the complex salts are largely decomposed by dissolution in water.

Mercury cyanide is completely odorless, but has a repulsive taste. It is *extremely poisonous*, but has therapeutic uses and is also used as a disinfectant. For the latter purpose, the oxycyanide mentioned above is still more widely used. Mixtures of zinc and mercury cyanides, of varying composition, are used for the impregnation of surgical dressings.

(*ii*) *Mercury(I) thiocyanate*, $Hg_2(SCN)_2$, is formed as an insoluble, white crystalline precipitate, when potassium thiocyanate is added to a solution of mercury(I) nitrate.

(*iii*) *Mercury(II) thiocyanate*, $Hg(SCN)_2$, is similarly precipitated from mercury(II) salt solutions, as a sparingly soluble (1 part in 1440 parts of water at 25°) white precipitate. It is rather more soluble in hot water and also in ethanol, from which it crystallizes in fine needles. The electrical conductivity of its aqueous solution is vanishingly small, showing that the salt is practically un-dissociated.

Mercury(II) thiocyanate swells up to an extraordinary bulk when it is heated ('Pharaoh's serpents').

In the presence of an excess of SCN^- ions, complex salts of the types $M^I[Hg(SCN)_3]$ and $M^I_2[Hg(SCN)_4]$ crystallize from mercury(II) thiocyanate solutions. Salts of the first type are usually sparingly soluble, and the latter are mostly very soluble. The free acid $H_2[Hg(SCN)_4]$ has also been isolated in yellow crystals.

Ions of heavy metals, such as Zn^{++}, Ni^{++}, Co^{++}, Fe^{++}, give insoluble precipitates with the $[Hg(SCN)_4]^=$ ion. The very soluble potassium salt $K_2[Hg(SCN)_4]$ shows, in solution, the molar conductivity and change of conductivity on dilution ($\mu_{1024} - \mu_{32} = 20.7$) typical of uni-bivalent electrolytes.

(h) Nitrates and Nitrites

(i) *Mercury(I) nitrate*, $Hg_2(NO_3)_2$, can most conveniently be prepared by treating an excess of metallic mercury with moderately dilute nitric acid. It is also formed by the action of mercury on mercuric nitrate solution. It crystallizes from solution as the hydrate $Hg_2(NO_3)_2 \cdot 2H_2O$ in short, colorless, monoclinic prisms, which effloresce in air, with loss of water, and melt at 70° with partial decomposition. Mercury(I) nitrate is very soluble in water (30 g in 100 g at 25°). The solution is acid in reaction, because of hydrolysis, and basic salts such as $Hg_2(NO_3)(OH)$ (lemon yellow) are precipitated on dilution—slowly in the cold, and more rapidly on warming. If sufficient nitric acid is present in the solution, no decomposition occurs on heating.

Mercury(I) nitrate solutions have strong reducing properties. To prevent partial oxidation by atmospheric oxygen, they are best stored over metallic mercury.

The nitrate is the only compound of electrochemically unipositive mercury to form double salts—mostly of the type $2Hg_2(NO_3)_2 \cdot M^{II}(NO_3)_2$, involving the nitrates of bivalent metals.

(ii) *Mercury(II) nitrate*, $Hg(NO_3)_2$, is usually prepared by dissolving mercury in excess nitric acid. When the solution is evaporated it crystallizes in large colorless deliquescent needles of the composition $Hg(NO_3)_2 \cdot H_2O$. It is the starting material for the preparation of most other mercury(II) compounds.

Mercury(II) nitrate is stable in solution only if sufficient nitric acid is present. It is strongly hydrolyzed in dilute solution, and can be completely decomposed into mercury oxide and nitric acid by a large excess of water.

(iii) *Mercury(I) nitrite*, $Hg_2(NO_2)_2$, which can be prepared by the action of dilute nitric acid ($d = 1.04$) on excess of mercury, reacts with organic compounds in the same way as silver nitrite.

(iv) *Mercury(II) nitrite*, $Hg(NO_2)_2$, was obtained by Rây, by the double decomposition of mercuric chloride solution with the calculated amount of silver nitrite. Very deliquescent, unstable needle-like crystals were obtained on evaporation in a desiccator. The complex salts formed with alkali nitrites are much more stable than the simple salt. They are conveniently prepared by the action of mercury(II) nitrate on alkali nitrite solutions. Examples—$K[Hg(NO_2)_3]$, bright yellow crystals, $Na_2[Hg(NO_2)_4]$, yellow hygroscopic prisms, $K_3[Hg(NO_2)_5(H_2O)]$, yellow rhombic crystals.

(i) Other Mercury Compounds

(i) *Mercury(I) sulfate*, Hg_2SO_4, forms colorless monoclinic prisms ($d = 7.12$), sparingly soluble in water and dilute sulfuric acid. It is formed by precipitation from mercury(I) nitrate solution with dilute sulfuric acid, by anodic dissolution of mercury in dilute sulfuric acid, or by heating an excess of mercury with concentrated sulfuric acid. The sulfate is gradually decomposed by water, to give still more sparingly soluble greenish yellow basic salts. It darkens on exposure to light. It catalyses the oxidation of organic substances by fuming sulfuric acid, and is employed industrially in the oxidation of napthalene to phthalic acid. For its use in Kjeldahl nitrogen determinations, see p. 459.

(ii) *Mercury(II) sulfate*, $HgSO_4$, may be obtained by repeated evaporation of mercury with an excess of concentrated sulfuric acid (with the addition of nitric acid initially, if required), or by dissolving mercuric oxide in sulfuric acid. The anhydrous salt forms clusters of white leaflets. With a small amount of water, the hydrate $HgSO_4 \cdot H_2O$ is formed, crystallizing in colorless, rhombic prisms. Large amounts of water bring about hydrolysis and the formation of basic mercuric sulfate. The latter is also formed by treating mercury(II) nitrate solution with alkali sulfate. This basic salt was known to Basil Valentine, and was called 'Turpeth mineral' by Paracelsus. In the pure state it has the composition $HgSO_4 \cdot 2HgO$, and is a brilliant lemon yellow powder ($d = 6.44$), which reversibly turns red when it is heated. It is very sparingly soluble in water. The neutral sulfate also becomes colored—first yellow and then red-brown—when heated. It decomposes into Hg, SO_2, and O_2 when it is heated to a red heat.

Mercury(II) sulfate is used industrially as a catalyst for the conversion of acetylene to acetaldehyde (cf. Vol I, p. 437), and also as a starting material for the preparation of other mercury compounds.

Mercury(II) sulfate can form double compounds with the alkali sulfates—e.g., $K_2SO_4 \cdot 3HgSO_4 \cdot 2H_2O$. It can also add on 1 or 2 molecules of HCl or HBr.

(iii) *Mercury(II) sulfite and Sulfitomercurates*. Mercury(II) sulfite, $HgSO_3$ is formed as a white, unstable precipitate when mercury(II) nitrate solution is treated with alkali sulfite. The basic sulfite, $HgSO_3 \cdot HgO$, is more readily obtained pure than is the simple compound.

Mercury(II) oxide dissolves easily in neutral alkali sulfite solutions, by reaction with the hydrogen sulfite ions formed by hydrolysis:

$$HgO + 2HSO_3^- = [Hg(SO_3)_2]^= + H_2O.$$

Crystalline complex salts, $M^I_2[Hg(SO_3)_2]$ (sulfitomercurates(II), formerly called 'alkali mercury sulfonates'), are obtained from the solution on evaporation. These compounds are very strongly complexed, as is evident from the fact that no mercury is precipitated from fresh solutions by most reagents—e.g., caustic potash, sodium carbonate, sodium phosphate, etc. It is possible that these complex salts may involve covalent bonding of mercury to sulfur:

$$\begin{matrix} O \\ O \leftarrow S - Hg - S \rightarrow O \\ HO \qquad\qquad OH \end{matrix} \qquad \text{(the parent acid).}$$

Solutions of the salts gradually decompose, especially when warmed, according to the equation $[Hg(SO_3)_2]^= = Hg + SO_2 + SO_4^=$.

(iv) *Carbonates*. Yellow mercury(I) carbonate is precipitated from solutions of mercury(I) salts on the addition of alkali carbonate. It decomposes readily into Hg, HgO and CO_2. Neutral mercury(II) carbonate has never been prepared, but basic carbonates of variable composition are precipitated from mercury(II) nitrate solution by alkali carbonate or alkali hydrogen carbonate.

(*v*) *Mercury(II) oxalate*, HgC_2O_4, a sparingly soluble white powder, has some importance because of its photo-decomposition. This property is utilized in the Eder photometer. Eder's solution contains 80 g of $(NH_4)_2C_2O_4$ and 50 g $HgCl_2$ in 3 liters of water*. The oxalate is present in this solution as a soluble double compound with mercuric chloride. The reaction which occurs on exposure to light is:

$$2HgCl_2 + (NH_4)_2C_2O_4 = Hg_2Cl_2 + 2CO_2 + 2NH_4Cl.$$

As long as the composition of the reactants may be considered as constant, the quantity of calomel deposited is proportional to the quantity of incident light. Eder's photometer thus permits a quantity of light to be determined by means of the analytical balance.

(j) Mercury-Nitrogen Compounds

As has already been mentioned, mercury can not only add ammonia on to certain of its compounds, but is also able to replace the hydrogen of ammonia. The most important example of an ammonia addition compound is the 'fusible white precipitate', and important representatives of mercury-substituted ammonia or ammonium derivatives are the 'infusible white precipitate' and 'Millon's base'.

(*i*) *Fusible White Precipitate*. If a mercuric chloride solution, containing much ammonium chloride, is treated with ammonia, a white crystalline precipitate is obtained. Since this can be melted (with decomposition) when it is heated, it has received the name of 'fusible white precipitate'. It is a well defined compound, with the composition $HgN_2H_6Cl_2$. It may also be prepared by the action of liquid ammonia on mercuric chloride. The compounds dissolves readily in dilute nitric, sulfuric, and acetic acids, forming a mixture of mercury(II) chloride and the ammonium salt of the acid concerned. This substance is almost certainly to be regarded as *diamminemercury(II) chloride*:

$$\begin{array}{ccc} NH_3 & & Cl \\ & \diagdown Hg \diagdown & \\ NH_3 & \diagup & \diagup Cl \end{array}$$

Analogous compounds are formed by organic amines and also by hydrazine. The hydrazine compound, prepared from ether-alcohol solution, has the composition $Cl_2Hg \cdot N_2H_4$. Thus hydrazine replaces 2 molecules of ammonia, as it ordinarily does in coordination compounds. Other negative radicals may be combined with mercury in place of the chlorine. However, the ammines of mercury salts which are ordinarily highly ionized (e.g., mercury(II) nitrate) are formed only with pyridine or hydrazine as ligands, in place of ammonia. Tetrammine salts, $[Hg(NH_3)_4]X_2$ have also been prepared by the action of ammonia on mercury(II) salts suspended in saturated ammonium salt solutions.

(*ii*) *Infusible White Precipitate*. If ammonia is added to a mercuric chloride solution which contains little or no ammonium chloride, a white precipitate having the composition $HgNH_2Cl$ is formed:

$$HgCl_2 + 2NH_3 = Hg(NH_2)Cl + NH_4Cl.$$

This decomposes into mercury(I) chloride, ammonia, and nitrogen, without melting, when it is heated, and was therefore given the name 'infusible white

* Roloff has shown that the sensitivity towards light can be considerably increased by the addition of $Hg(NO_3)_2$.

precipitate'. It is insoluble in water, but soluble in dilute strong acids in the presence of a little sodium chloride. All the nitrogen is evolved as ammonia when it is boiled with potassium hydroxide and all the chlorine can be precipitated with silver nitrate. This substance is to be regarded as a *mercury(II) amido-chloride*, NH_2—Hg—Cl.

Analogous bromine compounds are known, and the constitutions assigned to these substances on chemical grounds have been confirmed by determinations of the crystal structures of $(NH_3)_2HgBr_2$ and NH_2—HgBr (Rüdorff and Brodersen, 1952). These substances have structures which are very closely related to one another and to the structure of ammonium bromide. The apparently, amorphous character, insolubility, and infusibility of the amido-compounds suggest that they are polymerized substances. This is, indeed, the case. In NH_2HgBr, NH_2 groups take the place of NH_4^+ ions of ammonium bromide, and mercury atoms are so disposed between them that there are infinite strings NH_2—Hg←NH_2—Hg←NH_2—, etc., running through the crystal. In the diammine, there are only one half as many mercury atoms, and the nitrogen groups are neutral NH_3 groups instead of the univalent (covalently bound) NH_2 radical. There are thus discrete linear complex ions, $[NH_3{\rightarrow}Hg{\leftarrow}NH_3]^{2+}$. The Hg↔N distances are very similar: 2.17 Å in NH_2HgBr, 2.11 Å in $[(NH_3)_2Hg]Br_2$. These bond lengths agree with covalent bond lengths calculated from the covalent radii. Because of the similarity in the structure of the solid substances, it is possible to obtain mixed crystals of $Hg(NH_3)_2Br_2$ and ammonium bromide, or of $Hg(NH_3)_2Br_2$ with $Hg(NH_2)Br$—e.g., by co-precipitation or by heating the substances in a sealed tube.

(iii) Millon's Base. The action of ammonia solution on yellow mercury oxide yields a bright yellow microcrystalline powder ($d = 4.08$), practically insoluble in water and other solvents, and having the composition $Hg_2NH_5O_3$. This is also a well defined compound, which has the property of reacting with acids to form salts. It is known as Millon's base, and according to Hofmann (1899) it is probably to be formulated as *dihydroxomercury(II)-ammonium hydroxide*,

$$\left[\begin{matrix} HO—Hg \\ HO—Hg \end{matrix}{>}NH_2\right] OH.$$

Its salts—which contain one molecule less water—would be oxodimercury(II)-ammonium salts,

$$\left[O{<}\begin{matrix} Hg \\ Hg \end{matrix}{>}NH_2\right] X.$$

The corresponding base, $[OHg_2NH_2]OH$, is formed from Millon's base by dehydration over caustic potash in an atmosphere of ammonia. It is a dark yellow powder ($d = 7.42$), which loses a further molecule of water when it is heated to $125°$ in a current of ammonia. The product is a dark brown powder ($d = 8.52$) which explodes if it is touched, and which no longer has the property of forming salts with acids. Hofmann gave it the formula $\begin{matrix} Hg \\ Hg \end{matrix}{>}NOH$; it may well be a substance of high molecular complexity.

All these compounds are rather unstable. Millon's base decomposes when exposed to light, and crackles when it is powdered in a mortar. The second base is still more photosensitive; the third is so unstable that in the dry condition it explodes with a sharp report if it is touched or heated to $130°$.

Salts derived from Millon's base (X = F, Cl, Br, I, CN, NO_2, NO_3, $\frac{1}{2}SO_4$, $\frac{1}{2}CO_3$, etc.) can be made not only by the action of acids on the base, (whereby double decomposition occurs, but not dissolution), but also directly. Those compounds of this type which are derived from highly ionized mercury(II) salts are invariably formed when a solution of the salt concerned is treated with ammonia, e.g.,

$$2Hg^{++} + NO_3^- + H_2O + 4NH_3 = [OHg_2NH_2]NO_3 + 3NH_4^+ \qquad (1)$$

Corresponding treatment of the weakly ionized mercury(II) salts, however, leads first to compounds analogous to the 'white precipitates'. If these are boiled with water or caustic alkali (eqns. 2 and 3), they are converted into salts of Millon's base. The action of ammonia on basic mercury salts leads to the same products (eqn. 4), as also does the action of ammonia and alkali hydroxide on the complex halides of mercury (eqn. 5).

$$2[Hg(NH_3)_2]Cl_2 + H_2O \quad = [OHg_2NH_2]Cl + 3NH_4Cl \qquad (2)$$
fusible white precipitate

$$2[NH_2HgCl] + H_2O \quad = [OHg_2NH_2]Cl + NH_4Cl \qquad (3)$$
infusible white precipitate

$$HgCl_2 \cdot HgO + 2NH_3 \quad = [OHg_2NH_2]Cl + NH_4Cl \qquad (4)$$

$$2[HgI_4]^= + NH_3 + 3OH^- \quad = [OHg_2NH_2]I + 7I^- + 2H_2O \qquad (5)$$

The reaction represented by the last equation constitutes the basis of the detection of ammonia by Nessler's reagent.

According to Egidius (1936), the black solids formed by the action of ammonia on mercury(I) salts contain compounds corresponding to the fusible and infusible white precipitates, and to the chloride of Millon's base; the compounds are stated to be $[Hg_2(NH_3)_2]Cl_2$, NH_2—Hg_2—Cl, and $[OHg_4NH_2]Cl$, respectively. Evidence for the formation of such derivatives of unipositive mercury (which has been much disputed) was found in analytical data and X-ray diffraction patterns, although no complete structure determinations have been carried out. It is not possible to isolate the compounds in the pure state; NH_2Hg_2Cl, the analogue of infusible white precipitate, is the initial black product of the action of ammonia on Hg_2Cl_2, but slowly decomposes into $NH_2HgCl + Hg$ when kept in the moist state.

(k) Mercury Alkyls

Mercury reacts more readily than any other metal with organic compounds, to form Hg—C bonds, and mercury alkyls—e.g., $Hg(CH_3)_2$ *mercury dimethyl* ($d = 3.07$, b.p. 95°)— are very readily formed, e.g. by the action of magnesium alkyl halides on mercuric chloride. Mercury alkyl halides are thereby formed as intermediates. Mercury also reacts directly with some alkyl iodides (e.g., methyl iodide) in sunlight, to give the same products, and can replace hydrogen in a wide variety of organic compounds (direct 'mercuration'). The mercury alkyls are colorless liquids, with a weak and characteristic smell. Their vapors are *very toxic*. Unlike the zinc alkyls, they are not attacked by air or water at ordinary temperature. This is primarily because of the very low affinity of mercury for oxygen. The mercury alkyls are actually less stable than the zinc alkyls, as shown by the fact that they react with metallic zinc to form zinc alkyls: $Hg(CH_3)_2 + Zn = Zn(CH_3)_2 + Hg$.

The alkyl mercury halides are reactive solids. Thus *methyl mercury iodide*, CH_3HgI (colorless leaflets, m.p. 152°, volatile in steam) reacts with moist silver oxide, forming *methyl mercury hydroxide*, CH_3HgOH (m.p. 137°). This substance is soluble in water and alcohol and is a moderately strong base. It reacts with acids to form salts which resemble the mercury(II) salts in that the derivatives of strong oxyacids are highly ionized (e.g., CH_3HgNO_3, molar conductivity of 0.1-normal solution $= 101.0$ at 25°), whereas the

$$\overset{O}{\underset{\parallel}{}}$$

halogen salts and salts of weak oxyacids are but little ionized (CH_3HgOC—CH_3, molar conductivity of 0.1-normal solution $= 4.9$).

10. Analytical (Mercury)

Mercury is very readily detected in the course of analysis. Mercury compounds, when heated with soda in an ignition tube, yield a grey distillate consisting of fine droplets of mercury, which can readily be recognized under a microscope if necessary. Mercury can be deposited from dilute solutions, by dipping in a strip of thin copper foil, which can subsequently be rolled up and heated in an ignition tube.

Air may be tested for the presence of mercury by passing it through a plug of pure gold leaf, which is then introduced into a gas discharge tube. After evacuation, the gold leaf plug is heated and the spectrum of the residual gases is excited. Even minimal traces of mercury give rise to a green line at $\lambda = 546$ mμ, and also an indigo blue line at $\lambda = 456$ mμ if the metal is present in larger amounts. Stock has worked out a procedure for the detection and exact determination of very minute amounts of mercury (down to 0.01γ) [*Angew. Chem*; 44 (1931) 200]. This involves the electrolytic deposition of the metal on a copper wire and distillation in a glass tube of special form. The mercury is collected in one end of the tube, which is drawn out into a capillary and cooled, and the size of the resulting droplet is measured with the eyepiece micrometer of a microscope. In such a determination of minute quantities of any substance, it is particularly necessary to test all the reagents used, to ensure that they are free from mercury ('blank' determinations), and also to avoid any possibility of subsequent contamination of reagents through the air of the laboratory.

In the systematic analysis for cations, unipositive mercury is precipitated as calomel on the addition of dilute hydrochloric acid, and this may be recognized by the black coloration produced with aqueous ammonia. Dipositive mercury is precipitated by hydrogen sulfide from dilute hydrochloric acid solution as the black sulfide, HgS. This may be distinguished from all the other sulfides precipitated, in that it is insoluble both in ammonium polysulfide and in dilute nitric acid.

If mercury or the sulfide is dissolved in aqua regia, evaporated down and treated with cobalt nitrate and potassium thiocyanate (without diluting too much), beautiful deep blue needles of cobalt mercury(II) thiocyanate, $CoHg(SCN)_4$, are formed. This reaction is so sensitive that (carried out as a micro-reaction) it permits the detection of 0.04γ of mercury. The blue to violet color given with diphenylcarbazone is still more sensitive. This test, best carried out as a drop reaction, can also be adapted for use as a colorimetric or microcolorimetric method for the quantitative determination of amounts of mercury down to 0.05γ (Stock, 1928–9).

In gravimetric analysis, mercury is precipitated as mercury(I) chloride, Hg_2Cl_2, or as the sulfide, HgS, depending on whether it is present in the $+1$ or $+2$ state. Both compounds are weighed as such after appropriate drying. It should be noted that solutions containing mercury must not be evaporated to dryness with hydrochloric acid, since very appreciable quantities of the metal may be lost by volatilization in the form of chloride. It is often most convenient, therefore, to deposit mercury electrolytically, as metal. This can be done in a flask with a platinum wire, to serve as the cathode, fused through the base. This is covered with mercury at the beginning of electrolysis, and the increase in weight as a result of electrolysis gives the amount of mercury deposited.

References

1 R. VANDERSCHUEREN, *Le zinc; Les minerais de zinc, la technologie de zinc, le zinc métal*, Paris 1934, 397 pp.
2 O. BARTH, *Die Metallverflüchtigungsverfahren, mit besonderer Berücksichtigung der Herstellung von Zinkoxyd*, Halle 1935, 261 pp.
3 O. C. RALSTON, *Zinkelektrolyse und nassmetallurgische Zinkverfahren*, Halle 1938, 282 pp.
4 A. BURKHARDT, *Technologie der Zinklegierungen*, 2nd. Ed., Berlin 1940, 324 pp.
5 H. BABLIK, *Das Feuerverzinken*, Vienna 1941, 271 pp.
6 A. GÜNTERSCHULZE, *Galvanische Elemente*, Halle 1927, 184 pp.
7 E. W. BAADER and E. HOLSTEIN, *Das Quecksilber; seine Gewinnung, Verwendung und Giftwirkung*, Berlin 1933, 239 pp.

CHAPTER 10

THE LANTHANIDE SERIES

Atomic numbers	Elements	Symbols	Atomic weights	Valence states	Atomic numbers	Elements	Symbols	Atomic weights	Valence states
58	Cerium	Ce	140.13	III, IV	65	Terbium	Tb	158.93	III, IV
59	Praseodymium	Pr	140.92	III, IV	66	Dysprosium	Dy	162.51	III
60	Neodymium	Nd	144.27	III	67	Holmium	Ho	164.94	III
61	Promethium (unstable)	Pm	145†	III	68	Erbium	Er	167.27	III
62	Samarium	Sm	150.35	II, III	69	Thulium	Tm	168.94	III
63	Europium	Eu	152.0	II, III	70	Ytterbium	Yb	173.04	II, III
64	Gadolinium	Gd	157.26	III	71	Lutetium	Lu	174.99	III

† Atomic weight of longest lived isotope (half-life ca. 30 years).

1. General

(a) Introduction

The *lanthanide* (or 'rare earth') series comprises the elements of atomic number 58–71 inclusive, as listed in the above table. In their properties, these elements closely resemble the Group IIIA element lanthanum (atomic number 57), which immediately precedes them in the sequence of atomic numbers. In general, their properties are intermediate between those of lanthanum and its lower homologues, yttrium and scandium, and these three elements, together with the fourteen elements from cerium to lutetium are included in the series of the *rare earth metals**.

Among the rare earths, the fourteen elements listed above are marked out by their position in the Periodic Table, in which they form a special series. This, in turn, is the result of a very characteristic feature of their atomic structure, which was discussed in the introduction to Vol. II of this book:—the lanthanide elements are characterized by the *progressive filling of the 4f levels* of their electronic configuration. [1]. As has already been seen, this necessarily means that the atomic number increases by 14 units from the first to the last element of the group. If each of these values of the nuclear charge corresponded to a stable nucleus, there would be 14 elements in the naturally occurring lanthanide series. However, the rules of nuclear stability (cf. p. 592) make it probable that there is no stable nucleus with the atomic number 61, and element 61 has, in fact, never been found

* The higher homologue of lanthanum, *actinium*, might also be included in the rare earth series, on the basis of its chemical behavior. It is not usual to do so, however, since this excessively rare element is important in a quite different context—namely that of radioactivity. It was, however, formerly customary to include *thorium* among the rare earths, as being usually associated with them in Nature. Although there are some relationships, there is no justification for regarding thorium as a raré earth element.

in Nature*. A 'gap' therefore remained within the lanthanide series (just as in the case of the next homologue of manganese), until those elements which were unstable, and therefore not found in Nature, could be prepared by nuclear transmutation.

(b) Valence

With the exception of cerium, all the lanthanides are normally tripositive in their compounds. Cerium can be not only tripositive, but also tetrapositive in many of its compounds. A few of the other lanthanides constitute exceptions to this rule, in that they form certain compounds in which they exercise other valences than three. Praseodymium and terbium can also function as tetrapositive elements, whereas samarium, europium, and ytterbium can be dipositive**.

The trivalence common to all the lanthanides can be explained in terms of their atomic structure in the following way. Spectroscopic evidence shows that, in the neutral lanthanum atom, two electrons occupy $6s$ orbits, and one electron a $5d$ orbit. It is also proved by magnetic susceptibility measurements that the occupation of the $4f$ intermediate level, which begins with cerium, is complete in lutetium, i.e., with the last element of the lanthanide series. It follows therefore that—unlike the processes which give rise to the Sub-groups of the Periodic System—on the average one electron is added to the $4f$ level as the atomic number increases, throughout the lanthanide group.*** The ground states of all the lanthanide elements are not known with certainty, but even if it is not strictly accurate to say that the outer configuration of all the lanthanide elements is the same as that of the lanthanum atom, the ionization energies are such that three, and only three, electrons are readily abstracted to form a positive ion. The lanthanides thus have essentially the same valence properties as lanthanum. The electrons of the $4f$ shell are more firmly bound, and only in exceptional cases can they function as valence electrons.**** This is exemplified by the properties of cerium. Although this is the first element in which the nuclear charge has reached a value great enough to stabilize the $4f$ levels, the binding energy of a $4f$ electron is already such that the reduction process $Ce^{++++} + e \rightarrow Ce^{+++}$ is strongly exothermic whereas the addition of a further electron to the 'outer shell' by the process $Ce^{+++} + e \rightarrow Ce^{++}$ is attended with the expenditure of a considerable amount of energy*****.

The ability of certain of the lanthanides to function as tetra- or dipositive is intimately connected with their position in the sequence, as is clearly shown by Fig. 55. As in Fig. 28, Vol. I (p. 127), the electron number of the neutral atoms and in the ions are plotted as a function of atomic number. Applying the considerations set out on pp. 127–128, Vol. I, it would appear that in the range covered by the lanthanides, there are *three* electron systems of especial stability. These are, first, the electron systems found in the La^{3+} and Lu^{3+} ions,

* The reported discovery of element 61 by Harris, Yntema and Hopkins ('illinium'), and by Rolla and Fernandes ('florentium') in 1926 has not been confirmed. The optical absorption spectra and X-ray emission spectra ascribed to these elements were really due to other known elements (Prandtl, 1926–37; Noddack, 1934). Since fission of uranium does occur in Nature, a certain amount of fission-product promethium (cf. p. 593) must be present in uranium minerals, but the quantity is far below the limit of ordinary chemical detection.

** This statement refers to the valence state in solid, salt-like compounds. As regards the existence of +2 lanthanide ions in solution, see p. 481.

*** In the Sub-groups (as was explained on p. VI, Introduction), a re-arrangement of levels leads to the completion of the d level in the eighth or ninth element of each transition series, although a total of 10 electrons is added to the d level.

**** In some lanthanides the $5d$ shell is unoccupied (see Table 77, p. 611). In these cases *one* of the $4f$ electrons is only loosely bound and behaves almost as does a $5d$ electron in the other lanthanides.

***** These considerations apply to the energy of creation of ions in solution, or in ionic crystals, where the total energy change is the sum of the change in ionization energy and the change in the solvation energy or lattice energy that accompanies an alteration in the valence of the ion.

corresponding to the xenon configuration (with completely empty $4f$ level) and the xenon configuration + filled $4f$ level, respectively. Whereas scandium and yttrium are followed by a series of elements which can lose elec-trons until the resulting atomic cores have the configuration of the preceding inert gas, this does not happen after lanthanum, be-cause of the relatively high binding energy of the $4f$ electrons. There are, however, two elements which can lose $4f$ electrons— cerium (cf. p. 485) and praseodymium. In the case of the latter, however, the removal of *one* electron is much more difficult than with cerium, and it is very uncertain whether both $4f$ electrons are ever lost, giving +5 praseodymium (cf. Prandtl and Rieder, 1938). In none of the next elements is the lattice energy sufficient to compensate for the energy expended on ionizing a fourth electron. Corresponding considerations govern the behavior at the end of the lanth-anide group. Here, the extra stability of a completed $4f$ shell leads to a relatively firm retention of one of the valence electrons in ytterbium. In appropriate circumstances, the free energy of a compound in which ytterbium is dipositive may be less than that of a compound in which it is tripositive and it is possible to isolate compounds of +2 ytterbium. It is also very evident that in the center of the lanthanide series there is also a configuration which recurs in several ions—namely that of the gadolinium ion, with just seven electrons in the $4f$ level. Terbium, the element following gadolinium, attains the same configuration in the Tb^{4+} ion, whereas europium, in the Eu^{2+} ion, also acquires the Gd^{3+} core by the loss of two electrons. The influence of this stable electron number shows itself also in the behavior of the next-preceding element, samarium, which shows a tendency at least to approxiamte to the Gd^{3+} structure by functioning as a dipositive element.

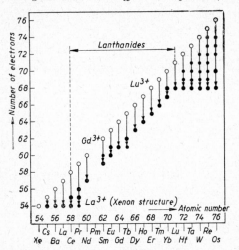

Fig. 55. Numbers of electrons in the atoms and ions of the lanthanides and adjacent elements.

Open circles represent the number of electrons present in the *neutral atoms.* Filled circles give the numbers of electrons in the *charged atoms* (ions) formed from the various valence states. Valence states of low stability are represented by smaller circles.

The gadolinium ion thus occupies a special place in the series of lanthanides, correspond-ing (although to a much lesser degree) to the place of the inert gases in the Periodic System as a whole. Klemm (1929) was the first to draw attention to this phenomenon, which shows itself not only in the regularities between the valences, but in other properties also:—e.g., in the absorption spectra (p. 484) and in the magnetic susceptibilities of the tri-valent ions. It can be justified on theoretical grounds. The magnetic moments of the lanth-anide ions indicate that, as the $4f$ level is filled up, the seven sub-levels of which it is com-posed, characterized by different values of the magnetic quantum number m (Vol. I, p. 97), become at first each occupied by *one* electron. Only when all the $4f$ levels are 'singly' occu-pied does a second electron (with opposed spin) enter an already tenanted level. Gado-linium, as the seventh element of the lanthanide series, is the element in which all the $4f$ levels are singly occupied. From what has already been seen of the sequence of ionization potentials in the series B-Ne, Al-A, Ga-Kr, (Vol. I, p. 17, Fig. 4), it is clear that whenever all the sub-levels of a shell are singly occupied, the binding energy of the electrons is in-creased (as with the atoms of N, P and As, in the instances cited). The same effect can there-fore be expected for the Gd^{3+} ion*.

* The sequence of ionization energies $M^{3+} \rightarrow M^{4+}$ among the lanthanides is not yet known from direct measurements or from spectral data. However, it is possible to predict with fair confidence that the curve of 4th-stage ionization energy versus atomic number will at least show a break at Gd^{3+} (as in the case of Sb on the curve for In to Xe, Fig. 4 of Vol. I), even if it does not actually have a peak.

The compounds of the $+2$ lanthanides are not stable in solution, in the strict sense, since all the bivalent lanthanide ions decompose water, with the evolution of hydrogen (i.e., $M^{++} + H_2O \rightarrow M^{+++} + OH^- + \frac{1}{2}H_2$). However, the decomposition takes place more or less sluggishly. For europium(II) compounds, the reaction is so retarded that their solutions are practically stable at ordinary temperature. It is likely that all the lanthanide elements form bivalent ions as in intermediate stage in the discharge of their trivalent ions (in aqueous solution) to the metals. This conclusion was drawn by W. Noddack (1937) from current-voltage curves obtained in the discharge of the ions at a dropping mercury cathode. Fig. 56 shows the reduction potentials $(M^{+++} \rightarrow M^{++})$ and the deposition potentials $(M^{++} \rightarrow M_{amalgam})$ as determined by Noddack. The difference between the two quantities (i.e., the distance between corresponding points on curves I and II) give a measure of

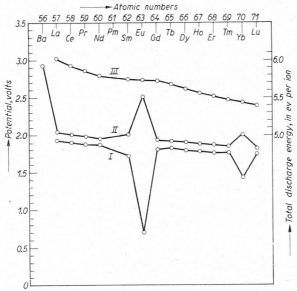

Fig. 56. Reduction potentials (curve I) of trivalent lanthanide ions, deposition potentials of the bivalent ions (curve II), and total discharge potentials of the trivalent ions (curve III). Ordinates for curves I and II are marked on the left hand side, ordinates for curve III on the right hand side of the diagram.

the relative stabilities of the various lanthanide ions in the bivalent state. It is clear that those elements which are known to form well defined compounds from the bivalent state (Eu, Yb and Sm) are also considerably more stable in solution as bivalent ions than the other lanthanides. For the latter, the reduction and discharge potentials lie very close together (0.065 to 0.145 volts apart). Except for Eu, Yb and Sm, the biggest difference between reduction and discharge potentials was found for gadolinium. As an electrolytic ion, in aqueous solution, gadolinium is thus more easily reduced than most of the other lanthanides. It does not follow that this is necessarily in contradiction with the special stability ascribed to the free Gd^{3+} ion. The effects of hydration energy on the free energy of formation of electrolytic ions are such that reduction potentials and discharge potentials do not always run parallel with the ionization potentials of the free ions. The special place of gadolinium shows itself in the series of electrolytic ions if the total energy to be expended in discharging the ion is compared. This is given (in electron-volts per g. ion) by $E_{3 \rightarrow 2} + 2E_{2 \rightarrow 0}$ (cf. Vol. I, p. 142, eqn. (3a)). This quantity, plotted as curve III (Fig. 56) shows a distinct break at gadolinium.

(c) Basicity of the Hydroxides

The hydroxides of the lanthanides are all definitely *basic* in character. They resemble the alkaline earth hydroxides more closely than aluminum hydroxide in this respect, but they are much more insoluble than the alkaline earth hydroxides, and are quantitatively precipitated by alkali hydroxides, even in the presence of ammonium salts. If very dilute alkali hydroxide or ammonia is added drop by drop, precipitation takes place in the order Sc^{+++}, Lu^{+++}, Yb^{+++}, Tm^{+++}, Er^{+++}, Ho^{+++}, Dy^{+++}, Tb^{+++}, Sm^{+++}, Gd^{+++}, Eu^{+++}, Y^{+++}, Nd^{+++}, Pr^{+++}, Ce^{+++}, La^{+++}. It is not possible to bring about a complete separation of the various elements by this means, but it is possible to observe a partial separation, following the sequence just given, especially if the fractional precipitation is repeated several times. This sequence corresponds to the order of increasing magnitude of the solubility products, and therefore to increase in basic strength. It may be seen that the hydroxides of the lanthanide elements are intermediate in basic strength between those of scandium and lanthanum. Cerium(III) hydroxide, the strongest base of the lanthanide series proper, differs very little from lanthanum hydroxide. Lutetium hydroxide, the weakest base, precedes scandium hydroxide, but is a sufficiently stronger base to permit the separation of scandium relatively easily from the other rare earth metals, by fractional precipitation with ammonia.

The order of basic strength among the oxides of the lanthanides may also be decided by other criteria—e.g., from the order of thermal decomposition of the nitrates when they are heated. In general, the greater the basic strength of an oxide, the more stable is the nitrate derived from it. It is possible to effect a separation between rare earths which differ sufficiently in basicity, by repeated application of this and similar methods. The precipitation processes based on differences in basicity were therefore very important historically, for the discovery of the rare earths. If it is required, not to separate them, but to determine the order of basic strength as accurately as possible, there are other, more suitable methods, which enable measurements to be carried out on preparations of the pure, already separated rare earths. Determinations of this kind were carried out by James (1914, 1921) and Endres (1932). The former determined the amount of iodine liberated from an iodide-iodate mixture by solutions of the various sulfates, and also measured the rate of evolution of carbon dioxide from sodium carbonate, by boiling solutions of the rare earth sulfates. He thus obtained the following sequence, in order of decreasing basicity: La, Ce, Pr, Nd, Sm, Eu, Gd, Tb, Dy, Y, Er, Tm, Yb, Sc. The basic strength thus decreases regularly within the rare earth group, in the order of increasing atomic number.

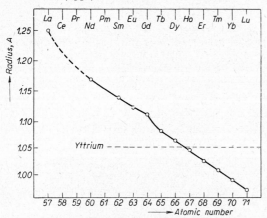

Fig. 57. Variation of ionic radius of the lanthanide elements with atomic number.

This order is closely connected with the decrease in ionic radius, shown in Fig. 57. The smaller the radius of a positive ion, the more firmly, *ceteris paribus*, will a hydroxyl ion be bound to it, according to Coulomb's law. The decrease in ionic radius, called by Goldschmidt the 'lanthanide contraction', is in its turn the result of the increasing attraction of the nucleus for the outer electrons as the nuclear charge increases. Whereas the beginning of new electron shells—i.e., the entry of electrons into orbits of higher principal quantum number—is attended with an increase in atomic radius, the progressive completion of an inner shell, along the rare earth series, is accompanied by a continuous decrease in atomic radius. One consequence of this is that the last elements of the series actually have smaller atomic and ionic radii than yttrium, which has a much smaller nuclear charge, and therefore contains far fewer electrons. This is why, in all properties which depend on ionic radius (or ionic volume), yttrium falls in the middle of the lanthanide series.

Endres measured directly the relative magnitudes of the solubility products of the hydroxides. His measurements confirm the parallel course of basicities and ionic radii, except for the case of yttrium. He found the solubility product of yttrium hydroxide to be somewhat greater than that of dysprosium hydroxide, whereas the ionic radius of yttrium is slightly smaller than that of dysprosium (Fig. 57).

(d) Oxides and Salts

The oxides of the lanthanides resemble those of lanthanum, yttrium, and scandium in being practically insoluble in water. They are strongly basic, and as just stated, the basicity diminishes regularly with increasing atomic number, so that the oxide of the last element of the lanthanide group, Lu_2O_3, falls about midway between Y_2O_3 and Sc_2O_3.

The sesquioxides of lanthanum and the elements immediately following have hexagonal structures (La_2O_3 or 'Type A' structure; see Fig. 15, p. 38). Cell dimensions are:

	La_2O_3	Ce_2O_3	Pr_2O_3	Nd_2O_3
$a =$	3.93	3.88	3.85	3.84 Å
$c =$	6.12	6.06	6.00	6.01 Å

The other sesquioxides, and also Sc_2O_3 and Y_2O_3, form cubic oxides ('Type C' structure; Sc_2O_3, Mn_2O_3, or Tl_2O_3 structure; see p. 37–8), with the cell dimensions shown in Table 49 (Zachariasen, 1928; Klemm, 1935). Some workers (e.g., Goldschmidt) have reported the formation of a third modification, said to be the form stable at low temperatures for the elements from Ce to Nd, and at high temperatures for Sm to Gd. The structure of this 'Type B' modification is unknown. Under suitable conditions (e.g., when their nitrates are ignited at low temperatures) La_2O_3 and Nd_2O_3 can also be obtained in the cubic modification (Lohberg, 1935). Ce_2O_3 and Pr_2O_3 can also be obtained in the cubic form (Klemm, 1939), but only under special conditions, because of the ease with which they undergo oxidation.

TABLE 49

CELL DIMENSIONS OF THE CUBIC MODIFICATION OF THE RARE EARTH
SESQUIOXIDES, AND APPARENT IONIC RADII OF THE TRIVALENT IONS
OF THE RARE EARTH ELEMENTS (IN Å)

	Sc_2O_3	Y_2O_3	La_2O_3	Ce_2O_3	Pr_2O_3	Nd_2O_3	Sm_2O_3	Eu_2O_3	Gd_2O_3
$a =$	9.79	10.60	11.4	?	?	11.050	10.893	10.842	10.797
$r =$	0.83	1.06	1.25	—	—	1.17	1.14	1.12	1.11

	Tb_2O_3	Dy_2O_3	Ho_2O_3	Er_2O_3	Tm_2O_3	Yb_2O_3	Lu_2O_3
$a =$	10.69	10.629	10.58	10.505	10.455	10.408	10.375
$r =$	1.08	1.07	1.06	1.04	1.02	1.00	0.99

Whereas La_2O_3, Y_2O_3, and Sc_2O_3 are colorless, the oxides of the lanthanides are mostly colored, as also are the ions of these elements in solution. Their solutions have characteristic absorption spectra, with relatively sharp and intense absorption bands (Table 50).

The ions Ce^{+++}, Gd^{+++}, Yb^{+++} and Lu^{+++} have no visible color but the other ions are more or less intensely colored. Cerium is only colorless in the trivalent state; compounds of

+4 cerium are orange to intense red in color in solution. Cerium dioxide, however, is practically colorless. A few other lanthanides which form colored ions also give rise to white oxides. If the light reflected from the colored oxides is examined spectroscopically, it is found to show absorption bands very similar to those found for light transmitted through the solutions.

TABLE 50

ABSORPTION SPECTRA OF THE LANTHANIDES IN AQUEOUS SOLUTION

Wave lengths in mμ. Where the limits of the bands are not given, the wave length is that of the band center or position of maximum extinction

Cerium	No visible absorption. Strong bands in ultraviolet
Praseodymium	597　589 **481.9** (sharp) 469 (sharp) **444** (broad)
Neodymium	730　687.7　678.6　637　629　623　**594–562　534–498** 480.2　475.5　469.1 461 (diffuse) 433　427.2　418.2
Samarium	559　499.5　489.2　487–472 **464** 451　443 (diffuse) 418　416　**402** 390.5　374.6 362
Europium	536　525　465.6　465.1　464.7　394.3　385.3　380.9　376.6　374.9　361.7
Gadolinium	No absorption bands in visible region
Terbium	573　523　487.5　379.7–375.2　369.4
Dysprosium	753　479–468　453.4　450.1　427.4　398　388 (broad) 365
Holmium	641　550–533　485.2　480　473　468　**458–443** 422　417　390　386　361
Erbium	667　653　648　549　541.3　**523　521** 491.5　487.1　453.4　449.7　442.2　407　405　**379.4** 364　359　356
Thulium	699 **682** 658.3　464.2
Ytterbium	No absorption bands in the visible region
Lutetium	No absorption bands in the visible region

It has been shown, by extending absorption spectrum measurements into the ultraviolet region, that the absorption bands are displaced in regular fashion as the atomic number increases (Prandtl, 1934). From praseodymium to gadolinium, the bands shift progressively from the red to the ultraviolet, and from gadolinium to thulium they move back towards the red. In addition to these, there are some bands which do not undergo any considerable and regular shift, but which are peculiar in that they diminish in intensity from praseodymium to europium, are completely absent from the spectrum of gadolinium, and then reappear in increasing strength in the succeeding elements. Thus the lanthanides group themselves symmetrically about the gadolinium ion, in their optical absorption. This shows itself in the color of the ions in the following way:

$_{57}$La^{3+} colorless
$_{58}$Ce^{3+} (u.v. absorption only)
$_{59}$Pr^{3+} (yellow green)
$_{60}$Nd^{3+} (red violet)
$_{61}$Pm^{3+} (color unknown)
$_{62}$Sm^{3+} (yellow)
$_{63}$Eu^{3+} (nearly colorless)
$_{64}$Gd^{3+} (colorless)
$_{65}$Tb^{3+} (nearly colorless)
$_{66}$Dy^{3+} (pale yellow green)
$_{67}$Ho^{3+} (brownish yellow)
$_{68}$Er^{3+} (pink)
$_{69}$Tm^{3+} (pale green)
$_{70}$Yb^{3+} (u.v. absorption only)
$_{71}$Lu^{3+} (colorless).

Whereas the salts of scandium, yttrium, and lanthanum are all *diamagnetic*, the lanthanides (except for compounds of Lu^{3+}) are *paramagnetic* in all salts in which they are trivalent (cf. Vol. I, Table 60, p. 304). The elements from gadolinium to thulium inclusive actually have a higher magnetic susceptibility than the most strongly paramagnetic salts of iron, cobalt, nickel, or manganese.

Salts of the lanthanides (including Sc, Y, and La) generally crystallize as hydrates. The chlorides, sulfates, and nitrates are freely soluble in water. The oxalates, fluorides, carbonates, and phosphates, especially, are very sparingly soluble. The highly colored sulfides, which can be prepared by reduction of the sulfates, are decomposed by boiling them with water.

Most salts of the rare earth elements form double salts with the corresponding salts of the alkali metals and ammonium, and also in many cases with the bivalent metals. Many of these double salts possess excellent crystallizing powers.

A number of these double salts have been found very suitable for effecting a separation of the rare earths from one another by *fractional crystallization*: examples are the *double nitrates* (e.g., the ammonium and magnesium double nitrates), *double sulfates*, *double oxalates*, and *double carbonates* (with the alkali metals). Fractional crystallization of the ammonium double nitrates had already been used by Mendeléeff for the separation of lanthanum from 'didymium' (cf. p. 490), and was again employed by Auer von Welsbach when he succeeded, for the first time, in resolving 'didymium' into neodymium and praseodymium.

Cerium, the first element of the lanthanide group, occupies a rather special position, since the compounds of $+4$ *cerium* differ completely from the typical compounds of the lanthanides. They are more closely analogous to the compounds of thorium, and cerium was therefore formerly often considered to be a homologue of thorium, occupying a position in the Periodic System roughly corresponding to that which is today filled by hafnium.

If the properties of their compounds are compared, $+4$ cerium is, in fact, more closely analogous to thorium than is hafnium. This is quite comprehensible, in terms of the theory of atomic structure. It is, in the first place, not surprising that the first element of the lanthanide series should function as tetrapositive, as well as tripositive. With one electron in the $4f$ shell, it is to be expected, by analogy with the entry of electrons to the $3d$ shell with scandium, that this electron should be not very firmly bound. As soon as this electron has been ionized off, giving the quadrivalent ion Ce^{4+}*, the cerium atom has *lost the essential characteristic of the lanthanide atoms*. As Hevesy has commented, the Ce^{4+} ion is in some respects a foreigner among the lanthanide group. If it were not for the occupation of the $4f$ levels between lanthanum and lutetium, Sub-group IV would contain a homologue of thorium, following lanthanum, not with the properties of hafnium, but with exactly the behavior of tetrapositive cerium. This would undoubtedly have a closer resemblance to thorium than hafnium has; for the effect of the lanthanide contraction is that in atomic and ionic radii, and all the properties connected with them, hafnium is remarkably close to zirconium, and correspondingly different from thorium. Hafnium also differs from thorium in that its positive ion does not have the inert gas configuration, whereas the Ce^{4+} ion resembles thorium in this respect.

(e) Properties of the Metals

The lanthanide elements are prepared in metallic form by the same methods as lanthanum. 'Mischmetall', which is commonly prepared technically, is a mixture consisting chiefly of the cerite earths. An average composition is: 50% Ce, 40% La, 3% other rare earth metals, and 7% Fe. It is 'pyrophoric', i.e., small particles filed off it ignite in air, and burn with a bright light and much heat. It is therefore

---

* It is, of course, assumed that this electron is more deeply seated than the valence electrons present in the lanthanum configuration, and will therefore be lost only in the conversion of the Ce^{3+} ion into the Ce^{4+} ion.

suitable for use in friction igniters (as in pocket lighters, etc.). For this purpose it is alloyed with more iron, since it is otherwise too soft. The sparking of mischmetall and its iron alloys (cerium-iron) when struck with roughened steel, etc., is due to the low igniton temperature and high heat of combustion of the rare earth metals.

Cerium and cerium mischmetall ignite in pure oxygen at about 150°; lanthanum does so at about 450°. Heats of formation per mol of oxide are

La_2O_3	CeO_2	Pr_2O_3	PrO_2	Nd_2O_3	Sm_2O_3	
539	233	439	231	435	430	kcal

—i.e., are considerably higher in certain cases than the heat of formation of aluminum oxide (402.9 kcal). These high heats of formation correspond to the strongly electropositive character of the rare earth metals, which shows itself also in their high deposition potentials.

The metals are relatively fusible. The melting point of cerium is 815°.

Until a few years ago, only lanthanum, cerium, praseodymium, and neodymium had been prepared in the metallic state. Urbain prepared gadolinium in 1935. Klemm and Bommer (1937) prepared small amounts of all the lanthanide elements, and determined their crystal structure and magnetic susceptibility. The method employed by Klemm was the reduction of the chlorides with metallic potassium (in a few cases with rubidium or cesium) at a moderately low temperature in an evacuated or argon-filled sealed glass tube. The excess of alkali metal was distilled off after the reduction, but the alkali chloride formed by the reaction was not separated off, since it did not interfere with subsequent measurements. Crystal structures, cell dimensions, and calculated densities of the rare earth metals are listed in Table 51.

More recently, the development of ion exchange methods for separating the rare earths (see below) has made it possible for Spedding to prepare kilogram quantites of very pure rare earth metals. The anhydrous chlorides and anhydrous fluorides were reduced by means of pure metallic calcium at 1300–1500°. By this method, lanthanum, cerium, praseodymium, neodymium, gadolinium, yttrium, erbium, holmium, dysprosium, terbium, and thulium have been prepared in the form of ingots [Spedding and Daane, *J. Am. Chem. Soc.*, 74 (1952) 2783; *J. Electrochem. Soc.*, 100 (1953) 442]. Approximate melting points of those elements which had not hitherto been obtained as the free metals were found to be: Y 1450°, Er 1400–1500°, Ho 1400–1525°, Dy 1400–1525°, Tm 1500–1600°. The three lanthanide elements samarium, europium, and ytterbium, which display considerable stability in the bivalent state, cannot be isolated by the calcium reduction of the chlorides or fluorides; reduction does not go beyond $SmCl_2$, $EuCl_2$ and YbF_2 (or $YbCl_2$) at any temperatures. It was found that when a preparation containing 98% $GdCl_3$ + 2% $SmCl_3$ was reduced to metal, the resulting specimen of gadolinium contained no detectable samarium.

Klemm's magnetic measurements show that all the lanthanides are paramagnetic in the metallic state. For the metals from Eu to Tm, paramagnetism passes over into ferromagnetism below a certain temperature (the Curie temperature—cf. Vol. I, p. 303), but the property is strongly developed only in Gd. The Curie temperatures are:

Eu	Gd	Tb	Dy	Ho	Er	Tm	
15°	302°	205°	150°	—	40°	10°	K

They thus vary regularly with atomic number, and the special position of gadolinium in the series shows itself in the magnetic properties also.

TABLE 51

CRYSTAL STRUCTURES OF LANTHANIDE ELEMENTS

Element	Structure	Cell dimensions		Density	Element	Structure	Cell dimensions		Density
α-Ce	hex.c.p.	$a = 3.65$	$c = 5.96$	6.78	Tb	hex.c.p.	$a = 3.585$	$c = 5.664$	8.332
β-Ce	f.c.c.	$a = 5.140$		6.810	Dy	hex.c.p.	$a = 3.578$	$c = 5.648$	8.562
α-Pr	hex.c.p.	$a = 3.657$	$c = 5.924$	6.776	Ho	hex.c.p.	$a = 3.557$	$c = 5.620$	8.764
β-Pr	f.c.c.	$a = 5.151$		6.805	Er	hex.c.p.	$a = 3.532$	$c = 5.589$	9.164
Nd	hex.c.p.	$a = 3.655$	$c = 5.880$	7.004	Tm	hex.c.p.	$a = 3.523$	$c = 5.564$	9.346
Sm	?	—		6.93	Yb	f.c.c.	$a = 5.468$		7.010
Eu	b.c.c.	$a = 4.573$		5.244	Lu	hex.c.p.	$a = 3.509$	$c = 5.559$	9.740
Gd	hex.c.p.	$a = 3.622$	$c = 5.748$	7.948					

[hex.c.p. = hexagonal closest packing,   f.c.c. = face centered cubic structure, b.c.c. = body centered cubic structure]

In Fig. 58, the *atomic volumes* of the lanthanides are plotted against atomic number. Europium, ytterbium, and samarium, which differ from the other elements in their ability to function as di-positive (cf. Fig. 56), also stand out from the other lanthanides in the metallic state by their much greater atomic volumes. The two first mentioned elements are the analogues of barium, not of lanthanum in this respect, as is clear from the diagram. It may, in fact, be deduced from the magnetic properties (cf. Vol. I p. 306) that europium and ytterbium are present in the metallic crystal lattice as doubly charged positive ions, as is barium. Samarium, which is far less readily converted to the bivalent state than europium or ytterbium, deviates far less than these elements from the regular trend of atomic volumes in the lanthanide series.

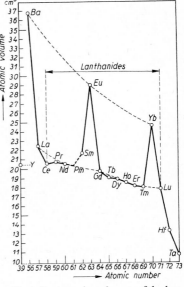

Fig. 58. Atomic volumes of the lanthanides and adjacent elements.

If all the lanthanide metals were present as triply charged positive ions in their metallic crystals, like lanthanum, the atomic volume curve would follow approximately the dotted curve in Fig. 58, starting from lanthanum. Ce deviates from this curve in the opposite sense from Eu and Yb. In agreement with this, magnetic measurements indicate the presence of Ce^{4+} ions in its metallic lattice, although Ce^{3+} ions predominate considerably at ordinary temperature. The atomic volume of cerium therefore lies much closer to the dotted curve than do the values for Eu and Yb, and the deviations are much smaller still for Pr and Tb. It is legitimate to assume, however, that these deviations arise from the same cause as in Ce. Pr and Tb share with Ce the ability, although less strongly developed, to become tetrapositive. The chemical properties of the lanthanide elements are reflected to a remarkable extent in the atomic volume curve.

Except for those differences which result from the capacity of a few of the elements to exercise valences other than 3, the known chemical properties of the lanthanides are completely analogous to those of lanthanum. They form hydrides with hydrogen, resembling lanthanum hydride, and combine with nitrogen to form nitrides.

TABLE 52

HEATS OF FORMATION OF LANTHANIDE(III) COMPOUNDS

(kcal per g-equivalent)

	Ce	Pr	Nd	Sm	Gd	Dy	Ho	Er	Tm	Lu
Chlorides	86.6	85.9	84.8	80.3?	81.7	79.2*	77.6	77.3	76.5	75.9
Iodides	54.5	53.8	52.8	51.8	49.2	48.1	47.3	46.6	45.9	44.4
Sulfides			43.8							
Nitrides	26.0									
Hydrides	21.1									

* Modification stable at room temperature.

Heats of formation of a few simple lanthanide compounds are given in Table 52. Data for the oxides have already been given (p. 486).

(f) Alloys

The behavior of the lanthanide elements towards other metals, as known at present, is summarized in Table 53.

Lanthanum is included in Table 53, as well as the first two elements of the lanthanide series. It may be seen that there is a very close analogy between the intermetallic compounds of the lanthanide group elements and those formed by lanthanum itself. Not only are most of the compounds similar in composition, but the melting points are very close in every case. In addition to those listed in the table, a few other intermetallic compounds of lanthanum are known—LaGa, LaGa$_2$, LaPb$_3$, La$_3$Ni (m.p. 515°), La$_2$Ni, LaNi (m.p. 685°) and LaNi$_5$ (m.p. 1325°). It is very probable that similar compounds are formed by Ce, Pr, and other lanthanides, but these systems have not yet been investigated. The compositions assigned in the literature to the Tl compounds of cerium and praseodymium are only based on analogy, and require confirmation. As is shown by the data in Table 53, there are certain points of difference between the various individual systems, in spite of the far-reaching similarity between them.

The metals also form compounds of quasi-metallic character with boron. These have the composition MB$_6$ (LaB$_6$, etc.), and are isotypic with CaB$_6$; ThB$_6$ has the same structure. They are steel-blue in color, whereas the alkaline earth borides are lustrous black to dark brown. They have a considerable metallic conductivity, like all compounds of this crystal structure.

The lanthanide elements combine with silicon to give compounds of the type MSi$_2$; those formed by the first elements of the series are iso-structural with LaSi$_2$. The lanthanide contraction shows up very clearly in the regular increase in density (calculated from X-ray data) in the series LaSi$_2$ – CeSi$_2$ – PrSi$_2$ – NdSi$_2$ – SmSi$_2$, from 5.41 to 6.26 (Brauer, 1952).

With the elements of Group V, the lanthanides (La, Ce, Pr, and Nd, have been investigated) form compounds of the type MX (and other types also), with the NaCl structure. They also form compounds of the type MX with the metals Zn and Cd, but these have the CsI structure (Iandelli, 1936–37).

Four of the lanthanides are feebly radioactive. These are *lanthanum* (which changes by electron capture), *neodymium* (a β-emitter), *samarium* (an α-emitter), and *lutetium* (a β-emitter). The radioactivity of these elements depends upon the presence in each of them of an extremely small amount of an unstable isotope (see p. 538). The nuclide ^{147}Sm, upon the disintegration of which the radioactivity of samarium depends, is the only moderately heavy natural nuclide which is an α-emitter*.

* Several of the artificially prepared lanthanides are α-emitters.

TABLE 53

MISCIBILITY AND COMPOUND FORMATION BETWEEN LANTHANUM AND LANTHANIDE ELEMENTS AND OTHER METALS

(For meaning of symbols see Table 9, p. 49)

	Mg	Al	Tl	Si	Sn	Bi	Fe	Cu	Ag	Au	Zn	Cd	Hg
La	$liq < \infty$ $s\ o$ $La_4Mg[]$ 503° $LaMg$ 743° $LaMg_3$* 766° $LaMg_9$* 662°	$s\ o$ $LaAl$* 859° $LaAl_2$ 1424° $LaAl_4$ 1222°	$s\ o$ La_2Tl 1200° $LaTl$ 1182° $LaTl_3$ 1096°	—	$s\ o$ La_2Sn 1420° La_2Sn_3 1188° $LaSn_2$ 1150° $LaSn_3$?	$LaBi$	—	$s\ o$ $LaCu$* 551° $LaCu_2$ 834° $LaCu_3$* 793° $LaCu_4$ 902°	$s\ o$ $LaAg$ 885° $LaAg_2$* 864° $LaAg_3$ 955°	$s\ o$ La_2Au* 665° $LaAu$ 1360° $LaAu_2$ 1214° $LaAu_3$ 1204°	$LaZn$ 815° $LaZn_2$ 855° $LaZn_3$ 960° $LaZn_4$ 871° $LaZn_5$ $LaZn_8$	$LaCd$	$LaHg_4$
Ce	$s\ o$ $CeMg$ 740° $CeMg_3$ 780° $CeMg_9$* 622°	$s\ o$ Ce_3Al 614° Ce_2Al* 593° $CeAl$* 780° $CeAl_2$ 1465° $CeAl_4$* 1250°	— Ce_2Tl? $CeTl$? $CeTl_3$?	$s\ o$? $CeSi$ 1530°	$s\ o$ Ce_2Sn 1400° Ce_2Sn_3 1165° $CeSn_2$ 1140° $CeSn_3$	$s\ o$ Ce_3Bi* 1400° Ce_4Bi_3 1630° $CeBi$* 1525° $CeBi_2$ 883°	$s < \infty$ $CeFe_2$* 773° Ce_2Fe_5 1090°	$s\ o$ $CeCu$* 515° $CeCu_2$ 820° $CeCu_4$* 780° $CeCu_6$ 940°	— —	— —	Ce_4Zn? Ce_2Zn? $CeZn$	$CeCd$	$s\ o$? $CeHg_4$
Pr	$s\ o$ $PrMg$ 767° $PrMg_3$ 798°	$s\ o$ $PrAl$* 906° $PrAl_2$ 1442° $PrAl_4$ 1244°	Pr_2Tl? $PrTl$? $PrTl_3$?	— —	$PrSn_3$	$PrBi$	— —	$s\ o$ $PrCu$* 563° $PrCu_2$ 841° $PrCu_4$* 824° $PrCu_6$ 962°	$s\ o$ $PrAg$ 928° $PrAg_2$* 878° $PrAg_3$ 956°	$s\ o$ Pr_2Au* 710° $PrAu$ 1350° $PrAu_2$ 1210° $PrAu_4$ 1200°	$PrZn$	$PrCd$	$PrHg_4$

(g) History

The first two stages in the history of the discovery of the rare earths have been described in Chap. 2. These were the discovery of the yttrium earths and the cerite earths, within a short time, and then, 40 years later, the resolution of each of these two earths into three new earths, by Mosander. The latter history of the rare earths is characterized by the same features as the early stages. Every improvement in the techniques of separation and characterization led to a fresh resolution of what had hitherto been considered to be homogeneous substances until, ultimately, as shown schematically in Table 54, the number of rare earths had been raised to that at present recognized as occuring naturally—namely 16. Three of these 16 rare earths—the three elements belonging properly to Group IIIB—were known and prepared in a state of purity at a relatively early stage, whereas some of the lanthanide elements were discovered within recent times.

TABLE 54

SCHEMATIC SUMMARY OF THE HISTORY OF THE DISCOVERY
OF THE RARE EARTH ELEMENTS

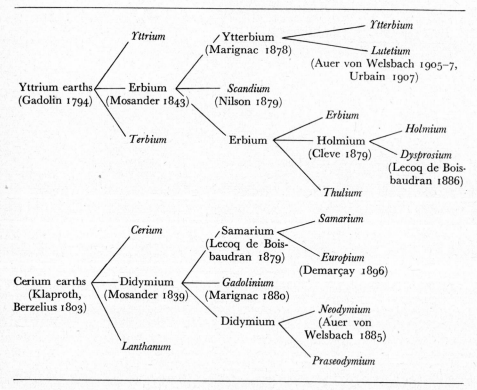

About the same interval of time as elapsed between the discovery of the yttrium and cerite earths, and their resolution into the oxides of 6 metals in all (yttrium, erbium, and terbium; cerium, didymium, and lanthanum) had to pass before a fresh stage in the history of their discovery began about 1880. In the meanwhile, the introduction of spectral analysis by Bunsen had furnished a powerful accessory for the identifying the elements and testing their purity. Moreover, new and more readily accessible minerals had been discovered, which enabled larger quantities of the rare earths to be extracted—e.g., *samarskite,*

from which Lecoq de Boisbaudran isolated the oxide of samarium in 1879. In the previous year, Delafontaine had concluded on the basis of spectral analysis that 'didymium oxide' prepared from this mineral was not homogeneous (by comparing its absorption spectrum with that of didymium preparations from other sources). In 1880 Marignac separated yet another oxide from samarskite—that of *gadolinium*. Two years earlier Marignac had succeeded in isolating a new oxide from erbia, up till then regarded as a pure substance. Unlike erbium oxide, which was pink, the new oxide was colorless, and as its properties were intermediate between those of erbium and yttrium oxides, it was called *ytterbium*, oxide. Nilson, in 1879, separated yet another oxide—that of *scandium*—from the old erbia and in the same year Cleve obtained two more new oxides from the same source—the oxides of *thulium* and *holmium*. The latter, in its turn, proved to be inhomogeneous. It still contained *dysprosium*, the characteristic absorption spectrum of which was recognized by Lecoq de Boisbaudran (1886) as belonging to a new element.

In 1885, Auer von Welsbach succeeded in resolving didymium into two components—*praseodymium* and *neodymium*—by the fractional crystallization of the ammonium double nitrates. Until this time, the method of fractional crystallization had received but little attention, but it has subsequently become of very great importance, for both the analytical and the preparative separation of the rare earths. Technical utilization of minerals containing the rare earths (especially monazite sand) in the gas mantle industry began at about the same time as Auer von Welsbach developed the technique of fractional crystallization.

Since that time, rare earth chemistry has been concerned with the preparation of the elements in a state of genuine purity, and the closer study of their properties, rather than with the discovery of new members of the group. There were, in fact, few left to be discovered, since 14 of the 16 stable elements now known to comprise the group (including Sc, Y, and La) had been discovered by 1890. Demarçay, in 1896, was able to separate a new element, with a characteristic absorption spectrum, from samarium. This subsequently received the name of *europium*. In 1905, Auer von Welsbach inferred from spectroscopic evidence that Marignac's ytterbium was not a pure element, and he isolated a new element from it by fractional crystallization; the oxide of this element was obtained practically pure in 1907, and the element was called *cassiopeium*. Urbain effected the same resolution practically simultaneously, and named the element *lutetium**.

When Moseley, in 1913, discovered the phenomena and principles of X-ray spectroscopy, it became possible for the first time to determine how many rare earth elements still awaited discovery, from the atomic numbers found for the known elements. It emerged that between barium (atomic number 56) and tantalum (atomic number 73), there must be a total of 16 elements. All of these were already known, except the elements of atomic numbers 61 and 72. Bohr's atomic theory led to the conclusion that element 72 should not be a metal of the rare earth series. Recognition of this fact quickly led to the discovery of element 72 (hafnium) in zirconium minerals, by von Hevesy and Coster. No reliable evidence has ever been advanced for the discovery of the rare earth, element 61, in Nature. With our increasing knowledge of the stability of those nuclear species with mass numbers which might correspond to nuclei of element 61, it has become apparent on systematic grounds that all nuclei of nuclear charge 61 are almost certainly unstable. It is probable, therefore, that element 61 does not occur in significant amount in the earth's crust, although radioactive species of this element (*promethium*) have been isolated in weighable quantities from the products of uranium fission.

(h) Subdivision of the Rare Earths

The first few elements of the lanthanide series bear a particularly close resemblance to lanthanum, especially in the solubilities of their salts, whereas the others (roughly from europium onwards) are more closely related to yttrium. The rare earths as a whole thus fall into two subdivisions. One of these is known as the *cerite earths*, after cerium, its technically most important member, whereas the

---

* Auer von Welsbach's name of cassiopeium has been retained in German usage, but *lutetium* has now been accepted as the internationally agreed name for element 71.

other is called the *yttrium earths*. The yttrium earths, in turn, can be yet further subdivided on the basis of the more or less close relationship between the individual members. The eventual classification is shown in Table 55.

TABLE 55

CLASSIFICATION OF THE RARE EARTHS

Cerite earths	Yttrium earths				
Oxides of Lanthanum Cerium Praseodymium Neodymium (Promethium) Samarium	1. *Yttrium oxide*	2. *Terbium earths* Oxides of Europium Gadolinium Terbium	3. *Erbium earths* Oxides of Dysprosium Holmium Erbium Thulium	4. *Ytterbium earths* Oxides of Ytterbium Lutetium	5. *Scandium oxide*

This older form of subdivision, based essentially on the analytical and preparative behavior of the rare earth compounds, may be compared with a classification introduced by Klemm in 1929, on the basis of the ideas of atomic structure.

From the standpoint of atomic structure, it is evident, in the first place, that scandium, yttrium, and lanthanum constitute a special subdivision (or part of a special group, namely Group IIIB of the Periodic System) as compared with the rest of the rare earths. The latter also constitute a subdivision on their own, that which has been given the name of the *lanthanide* series in this chapter. This has long been customary, and the name expresses the close relationship between the lanthanide elements and those that fit into the normal sequence of the Periodic Table. Klemm divides the lanthanides further into two sub-groups, on the basis of the special place in the series occupied by gadolinium. These two groups comprise the elements in the order of their atomic numbers, from cerium to gadolinium and from terbium to lutetium, respectively.

When the variation in properties of the lanthanides or their ions along the series is considered from this standpoint, it is found that three types of variation can be discerned.

[1] The properties may show a clear *periodicity*, in such a way that the trend followed in the first sub-group is repeated in the second.

This sort of variation is seen, for example, in the valence properties. As shown by the following summary, the valences exhibited in isolable compounds of the second sub-group are substantially the same as in the first, except that the tendency to exhibit other valences than three is generally more weakly developed in the second sub-group.

Sub-group I	Valence	III, IV	III, IV	III	III	II, III	II, III	III
	Element	$_{58}$Ce	$_{59}$Pr	$_{60}$Nd	$_{61}$Pm	$_{62}$Sm	$_{63}$Eu	$_{64}$Gd
Sub-group II	Valence	III, IV	III	III	III	III	II, III	III
	Element	$_{65}$Tb	$_{66}$Dy	$_{67}$Ho	$_{68}$Er	$_{69}$Tm	$_{70}$Yb	$_{71}$Lu

The curves I and II in Fig. 56 also show a clear periodicity, as also does the atomic volume curve of Fig. 58 when it is compared with the atomic volume curve for the elements as a whole (Fig. 2, Vol. I). The magnetic susceptibilities, of both ionic compounds of the lanthanides and of the metals, are also properties which follow a clearly periodic trend.

[2] It is also possible for the properties of the lanthanides to vary in a symmetrical manner, so as to change with atomic number in the opposite sense, for elements standing to the left or to the right of gadolinium, respectively.

For properties which vary in this manner with atomic number, a particularly close similarity is found between elements which are equidistant from the center of the lanthanide group, and not between elements which lie one above the other in the sub-groups as written above. The change in color of the lanthanide ions—i.e., the displacement of optical absorption bands—with increasing atomic number is a typical example of a property which varies symmetrically within the lanthanide series.

[3] There are also properties which alter monotonically within the lanthanide series. Where this is the case, however, the curves which represent the variation in any property with atomic number frequently have a distinct break at gadolinium.

Examples are found in the curves of ionic radii (Fig. 57) and of total ionization energy for the trivalent ions (curve III of Fig. 56). The special position of gadolinium shows itself in these curves, as it does also in the curves for the heats of solution of the rare earth metals in hydrochloric acid, and for the heats of formation of the anhydrous chlorides (Klemm and Bommer, 1941).

The analytical-preparative behavior of the rare earths—i.e., their separability, based on differences of solubility of their compounds—is based essentially on a single property of their ions, namely the ionic radii*, whereas the structure of the atom, in addition to ionic radius, determines all the chemical and physical properties. A classification based on atomic structure would be expected to furnish a better correlation of the properties of the elements, and an understanding of the connection between them, than is possible from the limited viewpoint of the older classification. The relation between the older classification and Klemm's systematics is roughly the same as that between the analytical separation of the metals and the classification of the metals as a whole by means of the Periodic System. For this very reason, the older form of subdivision has its uses for preparative and practical purposes.

(i) Occurrence

The mode of occurrence of the rare earths in Nature is determined primarily by the same properties as the analytical and preparative subdivision, and by the resulting gradations in the relations of the individual rare earths to one another. (For this reason, the older classification is also said to be 'mineralogical'). It is true that all the rare earths are invariably associated with one another to a greater or less extent, but there are minerals in which the elements of one or other of the analytical-preparative groups greatly predominate. This is the origin of the old division into cerite earths, which are the main constituents of cerite, and yttrium earths, which occur chiefly in *ytterbite* (gadolinite), although Berzelius (1814) proved that ytterbite contained subordinate amounts of cerium also.

The most important minerals in which cerite earths predominate are *cerite* itself (a hydrated silicate of cerite earths), *orthite* (a double silicate of cerite earths, aluminum and calcium oxides), and *monazite* (a phosphate of cerite earths, containing on the average 5% thorium oxide).

Minerals in which yttrium earths predominate are *gadolinite* or *ytterbite* (a basic silicate, which is always black because of its iron content), *xenotime* (a phosphate of the yttrium earths, analogous to monazite), *yttrotantalite*, *samarskite*, and *fergusonite* (isomorphous mixtures of niobates and tantalates, containing yttrium earths), and *euxenite* (containing titanic acid, as well as niobic and tantalic acids).

In addition to the rare earth minerals proper, many minerals contain minor amounts of rare earths. These include, in particular, the phosphates, arsenates, vanadates, molybdates,

* The varying magnitude of their polarizing power, as well as the ionic radii, is of some influence. The ease of conversion of cerium to the $+4$ state is utilized in the separation of cerium. This important property of cerium finds no expression in the older from of subdivision of the group.

and tungstates of calcium, strontium, iron, and lead. In some cases, the relative amounts of the various rare earth elements differ from what is found in the minerals listed previously. In many cases, for example, it has been found that the oxides of Sm, Eu, and Gd preponderate, although these elements are present in very minor amounts in the rare earth minerals proper.

(j) Preparation and Uses

Monazite sand is the most usual starting material for the technical preparation of the rare earths [2]. After mineral dressing (gravity separation, magnetic separation), it contains, on the average, about 5% thorium oxide, 60% of cerite earths, and 3–4% of yttrium earths. It was originally worked up chiefly for its thorium content, and the rare earths were obtained only as by-products of the manufacture of thorium oxide or nitrate (cf. p. 81 et seq.). At the present time, the rare earths (especially cerium and its compounds) have such extensive industrial applications that they now constitute the main products extracted from monazite.

The rare earths, together with thorium, are usually first precipitated as oxalates from acid solution. The thorium can be dissolved out by treating the precipitate with a warm saturated solution of ammonium oxalate. The remaining insoluble oxalates of the rare earths are generally then ignited to the oxides, which are dissolved in acid and converted (by addition of appropriate salts) into suitable double salts (e.g., ammonium double nitrates, magnesium double nitrates). Until recently, in so far as it was necessary to effect any separation of the rare earths from one another, this was carried out almost exclusively by fractional crystallization.

Cerium can be relatively easily separated from the other rare earths, by taking advantage of the very different properties of cerium(IV) compounds*. The preparation of other rare earth elements in a state of purity, by the methods of fractional precipitation or fractional crystallization is very difficult and tedious. Other methods of separation have been proposed in recent years. Solvent extraction methods, making use of the differences in partition coefficient of certain compounds between two immiscible solvents, was proposed by Fischer (1937). Europium (and less satisfactorily, ytterbium) can be relatively rapidly separated from other rare earths by cathodic reduction in the presence of sulfate ions (Brukl, 1936). Europium(II) sulfate is thereby precipitated as an insoluble compound. If europium is present in the solution only in low concentration, the solubility product of $EuSO_4$ may not be attained; precipitation can then be achieved by the addition of $SrCl_2$ during the electrolytic reduction. The $SrSO_4$ which is precipitated then incorporates $EuSO_4$ by mixed crystal formation. When the precipitate is ignited, Eu_2O_3 is formed, and the europium can readily be extracted as chloride. It is necessary to repeat the electrolytic reduction without the addition of $SrCl_2$, since $SrSO_4$ appears able to take up small amounts of other rare earth(III) sulfates in solid solution. If europium is present in fairly high concentration (e.g., in suitable tail fractions from a fractional crystallization of the magnesium double nitrates), reduction of europium(III) to europium(II) can be carried out with amalgamated zinc in a Jones reductor, and the hydrated chloride, $EuCl_2 \cdot 2H_2O$, can be precipitated by adding an excess of concentrated hydrochloric acid (McCoy, 1939).

The more readily reducible lanthanide elements can be concentrated from mixtures by forming their *amalgams*, according to a method first introduced by Marsh (1942). This is based on the close analogy between the lanthanides, in their bivalent state, and the alkaline earth metals. When a Ba or Sr salt solution is treated with sodium amalgam, the alkaline earth is partially displaced by sodium. Barium or strontium amalgam is formed, and a displacement equilibrium is established—e.g., $2Na(Hg) + Sr^{++} \rightleftharpoons Sr(Hg) + 2Na^+$. In the

* Conditions of formation of cerium(IV) compounds must be carefully controlled, however, if this separation is to be almost quantitative.

same way, when a solution of mixed rare earth acetates, acidified with acetic acid, is shaken with 0.3% sodium amalgam, the reducible rare earths (samarium, europium and—in mixtures of the heavier lanthanides—ytterbium) are rapidly collected in the amalgam phase, practically free from the other lanthanide elements.

Another recently introduced method for achieving a fairly rapid and complete separation of the lanthanide elements makes use of the formation of complex ions in solution, with such powerful complex forming reagents as ethylenediaminetetracetic acid, $(HO \cdot CO \cdot CH_2)_2$ $N \cdot CH_2 \cdot CH_2 \cdot N(CH_2COOH)_2$ ('enta-acid' or 'versene'). This tetrabasic acid is a powerful complexing agent. It acts as a 'sexadentate' group—i.e., it occupies six coordination positions around a metallic ion—and forms anionic complexes with almost all bivalent and trivalent metals—e.g., $H_2[CuY]$, $H_2[CaY]$, where Y represents the anion of fully ionized 'enta-acid', H_4Y. The equilibrium constants for the formation of these complexes vary from one element to another. The trivalent lanthanide elements behave similarly, and in a solution there will be two completing equilibria:

$$H_4Y \rightleftharpoons 4H^+ + Y^{4-}; \qquad \frac{[Y^{4-}][H^+]^4}{[H_4Y]} = K_a$$

$$\text{and} \quad M^{3+} + Y^{4-} \rightleftharpoons MY^-; \qquad \frac{[MY^-]}{[M^{3+}][Y^{4-}]} = K_{MY}$$

The values of the equilibrium constant K_{MY} spread over a much wider range than is usual for the properties of the lanthanide elements:

	La	Ce	Nd	Sm	Gd	Dy	Er	Yb	(Y)
$\log_{10}K_{MY}$	14.7	15.4	16.1	16.6	16.7	17.6	18.0	18.7	17.4

The extent to which the various lanthanide elements, in a mixture, are complexed by 'enta-acid' will therefore differ markedly even for successive lanthanides. In addition, the equilibria will be shifted markedly by changes in $[H^+]$ concentration. It is therefore possible so to control conditions that the lanthanides are successively liberated from their complexes, and made accessible to other reagents. The 'enta-acid' complexes can also be separated by crystallization (Marsh, 1950, Vickery, 1951, Schwarzenbach and Spedding, 1953).

Probably the most powerful method now available for the separation and purification of the lanthanide elements is the chromatographic process, using columns of cation-exchange resins such as Amberlite IR-100 or Dowex 50, or their equivalents. As has been mentioned earlier (Vol. I, p. 59), these cation-exchange resins consists of polymerized organic compounds bearing phenolic, sulfonic, or carboxylix acid groups. The hydrogen atoms of these groups are replaceable by other cations, which are mutually replaceable, and when a solution is brought in contact with such a resin, an equilibrium is established between cations in the solution and cations adsorbed on the resin. E.g., $n HR + M^{n+} \rightleftharpoons nH^+ + MR_n$ (where HR represents the resin in its 'acid form', and M^{n+} is an n-valent cation). In any system, the position of equilibrium depends upon the activities of the competing cations, H^+ and M^{n+}, in both the solution and resin phases, and on the affinity of the ions for the resin. In general, the latter increases with the charge on the cation. It also depends on the ionic radius, and alters systematically along the series of trivalent lanthanide cations. Hence a mixture of lanthanide ions is subject to a selective adsorption on a cation exchange column. The selective effect is greatly magnified by *eluting* the adsorbed cations from the resin again, by a suitable reagent, so that they pass progressively down the resin column. A fresh equilibrium is set up between resin phase and solution as the 'front' passes down the column, so that the column is equivalent to a great number of stages of fractional adsorption. The least firmly bound cation is therefore eluted first, and then the others in systematic order. Under appropriate conditions a substantial proportion of each lanthanide may be obtained in the pure state, before the next element to be eluted appears in the solution in appreciable concentration.

The selectivity exhibited by the adsorption of lanthanide ions on a cation exchange resin may be combined with the variations in the stability of complex ions formed in the solution. This enables the activity of the lanthanide cations in the solution to be controlled at such differing values that the differentiation between successive members of the lanthanide series is very greatly enhanced. Spedding (1947 and later) has shown that it is possible by this means to prepare the lighter lanthanide elements (cerite earths) in a pure state, and on a kilogram scale, in a single operation from monazite residues. The resin is first brought into the hydrogen form or sodium form, and the lanthanide cations are completely adsorbed on the top of the column, from chloride solutions. They are then eluted with a citric-acid-ammonium citrate solution (0.1 to 5%, according to conditions), and at pH 3 to 7, depending on the dilution of the citrate. The heavier and more strongly complexed lanthanides are eluted first. In citrate solutions, the lanthanide ions form complexes, e.g.:

$$M^{3+} + 3\ Citr^{3-} + 6H^+ \rightleftharpoons [M(H_2Citr)_3].$$

Hence the activity of the M^{3+} ion is controlled by the citrate concentration, hydrogen ion concentration, and the equilibrium constant for complex formation, which varies along the series.

By taking advantage of the much wider range of variation of the 'enta-acid' complex equilibrium constants, it is possible to effect a more or less selective adsorption of the less strongly complexed cations. This brings about a very rapid, but not quite complete resolution of a complex mixture [Wheelwright and Spedding, *J. Amer. Chem. Soc.*, 75 (1953) 2529, see Spedding, *Disc. Faraday Soc.*, 7 (1949) 214].

The ion exchange method is particularly suitable for the final purification of lanthanide preparations. It has been found that material prepared by the traditional methods of fractionation, and believed to be pure, generally contains small amounts of the neighboring lanthanide elements which cannot readily be removed by any other method than a final treatment on a cation exchange column.

The gas mantle industry requires pure cerium oxide, or ceria which is free, at least, from colored oxides. For many other technical purposes, however, it suffices to obtain mixtures of the rare earths, in the proportions in which they occur in monazite. Preparations made from such mixtures are known as 'cerite compounds'. A mixture of the oxides is used for the decolorization of glass. The nitrates, mixed with magnesium powder, are used in flash powders, and the fluorides are used for spot light and searchlight arc carbons. Carbon rods impregnated with rare earth fluorides give a particularly brilliant and pure white light. Motion picture arc lamp carbons may have up to 60% of cerite fluorides in their cores. The oxalates are used for pharmaceutical purposes (peremesin). A mixture of the metals themselves, 'cerium mischmetall' (cf. made by electrolysis of the molten chlorides, p. 485) is used as an alloying component or as a deoxidant for light metal alloys. Cerium-iron is used as a pyrophoric alloy in gas lighters, cigarette lighters, tracer ammunition, etc.

Pure compounds of cerium, praseodymium, and neodymium are used in coloring glass and enamels, and as underglaze colors for porcelain. Cerium compounds are used as carotting agents for hair in the felt hat industry, and as analytical reagents (cf. p. 499). The visual colors of praseodymium and neodymium are closely complementary, and mixtures of praseodymium and neodymium oxides are used ('didymium oxide') in the manufacture of Crookes' glass.

2. Most Important Compounds of the Rare Earths [4]

This brief survey of the most important compounds of the rare earths follows the older, 'mineralogical' classification, as being most convenient for a first study of the group. The relationships underlying Klemm's systematic classification have already been dealt with in the general section of the chapter.

(a) Cerite Earths

(i) *Lanthanum*, the first element of the cerite group, has already been discussed. *Cerium* is the most important member of the group technically. The remaining elements (praseodymium, neodymium, and samarium) form colored trivalent ions, with characteristic absorption spectra.

(ii) *Praseodymium* received its name from the green color of its salts (Gk. πρασαῖος, leek green). *Neodymium* salts are violet-red, the oxide is light blue. Salts of *samarium* are topaz yellow. Samarium imparts a rose-pink color to the electric arc. Praseodymium and samarium are among the elements which can form compounds from other than the tri-valent state.

When praseodymium salts of volatile acids are ignited in air, a black-brown oxide with the composition Pr_6O_{11} is formed. This can be converted to the black dioxide, PrO_2, by fusing it with sodium chlorate, or by heating it in nitrogen dioxide, or in oxygen at high pressure; if it is heated in hydrogen, the yellow sesquioxide, Pr_2O_3, is obtained. The dioxide crystallizes with the fluorite structure ($a = 5.36$ Å). The oxide Pr_6O_{11} has a very similar cubic crystal structure ($a = 5.53$ Å). These higher oxides undergo reduction when they are treated with acids, a salt of trivalent praseodymium being formed, and oxygen being evolved.

Prandtl (1938) reported that when a mixture of Pr_2O_3 and Y_2O_3 was heated in oxygen under pressure, the uptake of oxygen was considerably greater than corresponded to the formation of PrO_2. This would imply that praseodymium can be oxidized up to a valence state still higher than the quadrivalent state. In some experiments, Prandtl observed an oxygen uptake of up to 90% of that needed to oxidize Pr_2O_3 to Pr_2O_5, and he attributed the function of yttria in promoting the uptake of oxygen as probably being due to salt for-mation, to give $YPrO_4(= Y_2O_3 \cdot Pr_2O_5)$. Prandtl's observations have not been confirmed by other workers, however (e.g., Marsh, 1945).

Anhydrous samarium trichloride, $SmCl_3$, is reduced to the red-brown dichloride, $SmCl_2$, when it is heated in hydrogen. The dichloride reacts immediately with water; hydrogen is liberated, and samarium reverts to the trivalent state. The diiodide is also known.

(b) Cerium Compounds

Compounds of trivalent cerium are exactly similar to those of the other elements of the rare earth group, especially those of lanthanum. They resemble the latter in being colorless. The salts are stable in air, but cerium(III) oxide hydrate has a powerful tendency to undergo oxidation. It rapidly turns deep violet in air, and is ultimately converted to yellow cerium(IV) oxide hydrate.

Salts of $+4$ cerium are mostly rather unstable, unlike those of cerium(III). They are also rather extensively hydrolyzed in solution, so that basic salts are often precipitated when the solutions are diluted. Most cerium(IV) salts are also easily reduced to cerium(III) salts. Complex salts are often more stable than the simple cerium(IV) salts—especially the *nitrato* salts, $M^I_2[Ce(NO_3)_6]$, which have excellent crystallizing properties. The most stable simple cerium(IV) salt in solution is the *sulfate*. The chloride is known only in the form of the deep red solution of the com-plex acid, $H_2[CeCl_6]$, and as the salts of this acid. It readily reverts to the colorless cerium(III) chloride, with evolution of chlorine.

The oxidation potential $Ce^{+++} \rightarrow Ce^{++++}$ is -1.61 volts (relative to the normal hydrogen electrode) under the standard conditions given in Table 103, p. 765, Vol. I. Cerium(IV) salts are thus even stronger oxidants than gold(III) salts (cf. p. 412) and are thus reduced even by weak reducing agents—e.g., by iron(II) salts in acid solution, or by hydrogen peroxide.

(i) *Oxides and Oxide Hydrates. Cerium dioxide*, CeO_2, is formed by igniting the cerium salts of volatile acids in air, or by burning the metal. It is an almost white powder, with a weak yellowish tint, and turns lemon yellow when it is heated. It crystallizes with the fluorite structure ($a = 5.40$ Å); its density is 7.2. Ignited cerium dioxide is completely insoluble in hydrochloric acid and nitric acid. Very concentrated sulfuric acid converts it to cerium(IV) sulfate, whereas sulfuric acid of lower concentration reacts to bring about partial reduction and evolution of oxygen. In the presence of reducing agents, cerium dioxide also dissolves in the other acids mentioned. Cerium dioxide is converted into a dark blue intermediate oxide, of the composition Ce_4O_7, when it is not too strongly ignited in hydrogen (e.g., at 750°)*. Ce_4O_7 is reconverted to the dioxide when it is heated in air. Like the intermediate oxide of praseodymium, Ce_4O_7 is cubic, with a pseudo-fluorite 'defective' structure. Cerium(IV) oxide hydrate is precipitated from solutions of cerium(IV) salts, by ammonia or caustic alkalis, as a yellowish slimy precipitate. The same reagents precipitate cerium(III) hydroxide (or oxide hydrate) from cerium(III) salt solutions, but this is rapidly oxidized in air, whereby deep violet oxides are formed as intermediate products. If the solution of a cerium(III) salt is treated with hydrogen peroxide and ammonia, an intensely red-brown *cerium peroxide hydrate* is formed as a slimy precipitate. This reaction makes possible the detection of very small quantities of cerium, from the yellow coloration of the solution obtained on addition of hydrogen peroxide.

CeO_2 combines with Na_2O, forming a double oxide Na_2CeO_3 with the rock salt structure, having the cations statistically distributed (cf. p. 275). This double oxide has been obtained by heating CeO_2 directly with Na_2O (Zintl and Morawietz, 1940), or with NaOH (D'Ans, 1930). Zintl obtained the corresponding oxide of praseodymium, Na_2PrO_3, by heating Pr_2O_3 with Na_2O in oxygen.

(ii) *Chlorides. Cerium(III) chloride*, $CeCl_3$, is obtained in the anhydrous state as a crystalline deliquescent mass by heating the oxide in a stream of chlorine or hydrogen chloride. It is very soluble in water and alcohol. Hydrates crystallize from the aqueous solution, and these furnish *cerium oxychloride* when they are heated in air. Cerium(III) chloride can form complex salts with the chlorides of organic bases, of the type $M^I[CeCl_4]$. Double or complex chloro salts of tetrapositive cerium are also known—e.g., $[C_5H_5NH]_2[CeCl_6]$ (pyridinium hexachlorocerate(IV)), a yellow crystalline substance.

The action of elementary fluorine on anhydrous $CeCl_3$ or Ce_2S_3 yields *cerium tetrafluoride*, CeF_4. This is a colorless substance ($d = 4.77$), insoluble in water. It loses no fluorine when it is heated to 400° in a vacuum, but is reduced to CeF_3 (m.p. 1460°) by hydrogen at 300°.

(iii) *Nitrates. Cerium(III) nitrate*, $Ce(NO_3)_3 \cdot 6H_2O$, crystallizes as a colorless, deliquescent mass from solutions—e.g., of cerium carbonate in nitric acid. It is freely soluble in water and alcohol. It forms very well crystallized double salts with ammonium nitrate, magnesium nitrate, and the nitrates of other bivalent metals. *Cerium(IV) nitrate*, apart from its complex salts, is known only as the basic nitrate, $Ce(OH)(NO_3)_3 \cdot 3H_2O$. This is obtained in red crystals by dissolving cerium(IV) oxide hydrate in concentrated nitric acid, and evaporating the solution. Its aqueous solution is yellow, and acid as a result of hydrolysis. A freshly prepared solution turns red when nitric acid is added; addition of alkali nitrate has the same effect. Bright red anhydrous *nitrato salts*, of the type $M^I_2[Ce(NO_3)_6]$ crystallize from solutions containing alkali nitrates. Nitrato salts of similar type, but crystallizing with $8H_2O$, are formed with bivalent metals. *Ammonium nitratocerate(IV)*, $(NH_4)_2[Ce(NO_3)_6]$, which is very soluble in water but relatively sparingly soluble in nitric acid, is used to effect a relatively simple separation of cerium from admixed rare earths, the ammonium double nitrates of which are considerably more soluble in nitric acid.

(iv) *Sulfates. Cerium(III) sulfate*, $Ce_2(SO_4)_3$, obtained by fuming down the nitrate or oxalate with sulfuric acid, forms a white hygroscopic powder when it is anhydrous. Like other anhydrous rare earth sulfates, it dissolves freely in water at 0°, to give a solution which is highly supersaturated with respect to the dodecahydrate, the stable solid phase at this temperature. There are several other hydrates, as well as the 12-hydrate. These furnish the anhydrous salt when they are heated to 400–450°, but decomposition begins if they are heated above 500°, and the pure dioxide is ultimately obtained by heating to a white heat. Cerium(III) sulfate forms complex salts with the alkali sulfates—e.g., $K[Ce(SO_4)_2] \cdot H_2O$,

---

* Cerium(III) oxide, Ce_2O_3, is obtained if CeO_2 is heated to 1250° in hydrogen.

a colorless, sparingly soluble crystalline powder (other complex salts, such as $K_6[Ce_4(SO_4)_9]\cdot$ $8H_2O$ and $K_5[Ce(SO_4)_4]$ are formed at the same time). The ammonium complex sulfates, $(NH_4)[Ce(SO_4)_2]$ and $(NH_4)_5[Ce(SO_4)_4]$ (or $(NH_4)_8[Ce_2(SO_4)_7]$?) are more soluble than the potassium compounds. The ammonium disulfato salt usually separates as the unstable tetrahydrate, but the monohydrate and anhydrous salt are the stable phases in contact with the solution (Schröder, 1938).

Cerium(IV) sulfate, $Ce(SO_4)_2$, is obtained anhydrous, as a deep yellow crystalline powder, insoluble in concentrated sulfuric acid, by heating cerium dioxide with concentrated sulfuric acid. It dissolves readily in water, giving a brownish yellow solution, from which the 12-hydrate crystallizes below 3°, the 8-hydrate between 3° and 42.5°, and the 4-hydrate above 42.5°. Hydrolysis occurs if the solution is greatly diluted, and leads to the precipitation of sparingly soluble basic salts. Fairly stable sols of these can be prepared under suitable conditions (Janek, 1933). A freshly prepared solution of cerium(IV) sulfate turns deep red on the addition of sulfuric acid. Jones (1935) showed by transport number measurements that this color change was due to the formation of complex acids. Complex sulfates (*sulfato-cerate(IV) salts*) can be crystallized from solutions containing alkali sulfates—e.g., $K_4[Ce(SO_4)_4]\cdot 2H_2O$, very sparingly soluble, orange yellow monoclinic crystals, $(NH_4)_6[Ce(SO_4)_5]\cdot 2H_2O$, orange monoclinic crystals.

Cerium(IV) sulfate is a valuable reagent for oximetric titrations (Furman, 1928 and later, Willard, 1928 and later), and is now widely used in analytical work. It is also often used as an oxidant in preparative organic chemistry.

(*v*) *Cerium carbonate.* Only the carbonate of $+3$ cerium is known. It is formed as a micro-crystalline precipitate, with the composition $Ce_2(CO_3)_3\cdot 5H_2O$, when ammonium carbonate is added to a cerium(III) salt solution. With excess of ammonium carbonate a complex salt, $NH_4[Ce(CO_3)_2]\cdot 3H_2O$, is obtained. Similar salts are formed with sodium and potassium carbonates.

(*vi*) *Cerium oxalate.* Cerium forms an oxalate from the trivalent state only. Oxalic acid at once reduces $+4$ cerium. Cerium(III) oxalate, $Ce_2(C_2O_4)_3\cdot 10H_2O$ forms as a white crystalline precipitate when oxalic acid is added to a solution of a cerium salt. As is typical of rare earth oxalates, it is not merely practically insoluble in water, but dissolves only to a very slight extent in dilute strong acids. 0.041 mg of the anhydrous oxalate dissolves in 100 cc of water at 25°, or 164 mg in 100 cc of 1-normal sulfuric acid at 20°. In general, the solubility of the rare earth oxalates in water increases as the basic strength of the oxides diminishes. Scandium oxalate is thus the most soluble in water. The solubility in acids, however, increases in the opposite direction, and is greatest for lanthanum oxalate.

(*vii*) *Cerium sulfide*, Ce_2S_3, is obtained by strongly heating the anhydrous chloride or sulfate in hydrogen sulfide. It is a black-violet powder ($d = 5.10$). If the temperature during the preparation is lower, a compound richer in sulfur is obtained, Ce_2S_4 (dark brown powder). As its chemical properties and magnetic susceptibility show*, this is to be regarded as resulting from a polysulfide-like addition of sulfur to Ce_2S_3, and is derived from trivalent cerium. It is therefore not the sulfide analogue of CeO_2. At about 720°, Ce_2S_4, loses sulfur and forms Ce_2S_3. A similar polysulfide, La_2S_4 (golden-brown powder), is formed also by lanthanum, which cannot undergo any change of valence.

(c) Yttria Earths

Yttrium itself is by far the most abundant of the elements of the yttria earth group—usually present in rare earth minerals.

The amount of yttrium oxide, in atom per cent, is about three times as great as the combined total for terbium, erbium, and ytterbium earths together. Scandium, which forms the most weakly basic of the rare earth oxides, provides a transition from the rare earths to alumina. It has already been considered (Chapter 2).

---

* Since the Ce^{4+} ion has the xenon configuration, an ionic cerium(IV) compound ought to be diamagnetic, as is the case for CeO_2. Ce_2S_4 is almost as strongly paramagnetic as Ce_2S_3 (Klemm, 1930), and has practically the same molecular susceptibility as typical Ce(III) salts, such as $Ce_2(SO_4)_3$.

The ionic radius of yttrium is such that its properties bring it into the middle of the yttria earth group. All the compounds of the terbium, erbium and ytterbium earths are so closely related to those of yttrium that only a brief survey is necessary to make their characteristics clear.

(i) *Terbium Earths*. This group comprises europium, gadolinium, and terbium. None of these elements give very characteristic absorption spectra, but their arc spectra are highly characteristic. Gadolinium imparts a carmine red, and terbium a yellowish white color to the carbon arc. Europium gives a fine red coloration even to the Bunsen flame.

Salts of europium are pale pink. Their solutions show only a few, rather weak absorption bands. Gadolinium and terbium salts are colorless*. Terbium forms a dark brown higher oxide, which corresponds in composition approximately to the formula Tb_4O_7. This has a similar crystal structure to Pr_6O_{11}, and Ce_4O_7 (defective fluorite structure, with $a = 5.28$ Å). It is obtained when the terbium salts of volatile acids are ignited in air. The intense color of this oxide makes it possible to detect the presence of small amounts of terbium in the other oxides of the group. Tb_2O_3 is white, as is Gd_2O_3, whereas Eu_2O_3 is a very pale pink.

Tb_4O_7 can be oxidized further, approximately to Tb_6O_{11}, by heating it in oxygen under pressure. Prandtl (1938) found that more oxygen was taken up if Y_2O_3 were also present, and the maximum uptake corresponded approximately with the formation of TbO_2. Prandtl concluded that terbium differed from praseodymium in that its maximum valence state could not exceed four (but see above, under praseodymium). The higher oxide of terbium is, in any case, much less stable than the higher oxide of praseodymium as is evident from the much greater ease with which it is reduced by hydrogen to the sesquioxide.

Europium is the rarest of the rare earth elements, except for the unstable element 61 (Promethium). It is about 500 times rarer than yttrium, and 150 times rarer than cerium.

The chloride $EuCl_3$ can be reduced to the colorless dichloride $EuCl_2$ by heating it in hydrogen (cf. samarium dichloride). It is more stable than $SmCl_2$, and is oxidized by water only when it is heated. It is readily soluble in water, and crystallizes as the dihydrate, $EuCl_2 \cdot 2H_2O$. It can be obtained in solution by direct reduction of the trichloride with zinc and hydrochloric acid.

$EuCl_2$ is isotypic with $PbCl_2$, whereas EuF_2 (obtained like $EuCl_2$, or by heating $EuSO_4$ with concentrated NaF solution) has the fluorite structure ($a = 5.796$ Å). $EuBr_2$ and EuI_2 have also been prepared (Klemm, 1938), as has EuS (Nowacki, 1938). The latter, which is formed when $Eu_2(SO_4)_3$ is heated in a stream of H_2S, has the NaCl structure ($a = 5.957$ Å). EuSe and EuTe have the same structure. $EuSO_4$, which is readily prepared by electrolytic reduction of europium(III) sulfate solutions, is sparingly soluble, and is isomorphous with $SrSO_4$ and $BaSO_4$.

Anhydrous $EuCl_2$ adds on ammonia, forming the octammine $EuCl_2 \cdot 8NH_3$, as does $SrCl_2$, and europium(II) compounds in general show a close resemblance to those of strontium. This is the consequence of the small difference in ionic radius between the Eu^{2+} and the Sr^{2+} ions.

(ii) *Erbium Earths*. Dysprosium, holmium, erbium, and thulium make up this sub-group. All these elements form colored ions, with characteristic absorption spectra (Table 50). Solutions of dysprosium salts are yellow to greenish yellow, holmium salts brownish yellow, erbium salts an intense pink, and thulium salts green. The ultraviolet absorption spectrum of dysprosium is more characteristic than the visible spectrum, and there are also two strong bands in the near infrared (at 810 mμ and 912 mμ). Dysprosium and holmium, in the form of their compounds, are the most strongly paramagnetic of the rare earths. Holmium received its name from Stockholm, and thulium from Thule, the old name for Scandinavia.

(iii) *Ytterbium Earths*. Ytterbium and lutetium oxides, which comprise this sub-group, are colorless, as are the salts. Their arc spectra are characteristic however, and ytterbium salt solutions show absorption in the near infrared. Ytterbium gives a green color, and lutetium a blue-green color in the electric arc. The ytterbium earths were separated from the other rare earths by Marignac in 1878, by taking advantage of their very weakly basic character as displayed in the fractional decomposition of the nitrates on heating. Auer von Welsbach in 1905 deduced from observations on the arc spectra that Marignac's

* Gadolinium salt solutions have an ultraviolet absorption band at 272.9 mμ.

old 'ytterbium' should be capable of resolution, and in 1907 he succeeded in separating ytterbium and lutetium by a very long series of fractional crystallizations of the ammonium double oxalates. Urbain effected the resolution at about the same time, by fractional crystallization of the bromates.

Klemm (1929) showed that $YbCl_3$ could be easily reduced to $YbCl_2$ by heating it in hydrogen to 550°. $YbBr_2$ and YbI_2 were later prepared by Jantsch (1931). YbI_2 has the cadmium iodide structure ($a = 4.48$, $c = 6.96$ Å). The chalcogenides are far less readily reduced to ytterbium(II) compounds than are the halides. Yb_2O_3 cannot be reduced at all, and Yb_2S_3 only incompletely. Yb_2Se_3 and Yb_2Te_3, however, are completely reduced by heating them in hydrogen, giving YbSe and YbTe, respectively (Klemm, 1939). These compounds have sodium chloride-type crystal structures, almost identical in cell dimensions with CaSe and CaTe, respectively (YbSe, $a = 5.867$ Å; YbTe, $a = 6.340$ Å). The Yb^{2+} ion is thus almost identical in radius with the Ca^{2+} ion (1.06 Å, cf. Yb^{3+}, radius 1.00 Å).

3. Analytical (Rare Earths) [4]

The characteristic reaction of the whole group is the relative insolubility of their oxalates in acids. The presence of rare earths can therefore be detected in the course of the systematic separation of the cations, by adding oxalic acid to the hydrochloric acid solution (which should not be too strongly acidic), after precipitation of the hydrogen sulfide group.

Those rare earths which form colored salts can often be identified by observation of the absorption spectra of solutions. It has to be borne in mind that the position of the absorption bands may be modified by interaction with other substances present in the solution, even when these latter are themselves colorless. The absorption spectra of praseodymium, and erbium are particularly characteristic, with sharp, intense bands. With suitable precautions, it is possible to determine Pr, Nd, Sm, Eu, Gd, Dy, Ho, Er, Tm, and Yb quantitatively, by measuring the extinction coefficient of the strongest bands (cf. Rodden, *J. Res. Nat. Bur. Standards*, 28 (1942) 265). The emission spectra of the rare earths (arc and spark spectra) are extraordinarily rich in lines. In such cases, it is best to identify the elements present by means of their 'raies ultimes'—that is, the last lines to disappear as the concentration of the element in question is continuously diminished. Emission spectroscopy is therefore useful in the last stages of purification. The simplest and least uncertain means of identification, proof of purity, and quantitative analysis of rare earth preparations is often afforded by X-ray spectroscopy. However, the lower limit of detection of any rare earth element by X-ray spectroscopy is usually about 0.1%, although a greater sensitivity can sometimes be reached.

As mentioned on p. 498, cerium may be detected by means of its reaction with hydrogen peroxide.

References

1 G. VON HEVESY, *Die seltenen Erden vom Standpunkt des Atombaus*, Berlin 1927, 140 pp.
2 C. R. BÖHM, *Die Darstellung der seltenen Erden*, 2 vols., Berlin 1905, 492 and 484 pp.
3 R. J. MEYER and O. HAUSER, *Die Analyse der seltenen Erden und der Erdsäuren* [Vol. 14–15 of *Die chemische Analyse*], Stuttgart 1912, 320 pp.
4 D. M. YOST, H. RUSSELL and C. S. GARNER, *The Rare Earth Elements and Their Compounds*, New York 1947, 92 pp.

CHAPTER 11

RADIOACTIVITY AND ISOTOPY

1. General

(a) Definition of Radioactivity

Radioactive elements are those which possess the property of continuously and spontaneously emitting energy (i.e., without excitation by any external means). The rays emitted by radioactive preparations commonly have the property of penetrating materials which are opaque to ordinary light (such as black opaque paper or metal foils), of ionizing air and thereby discharging electrified objects, and of blackening photographic plates. In many cases (as, for example, the rays from radium) they can excite the emission of light from certain luminescent materials. Radioactivity [1–7] is an *atomic* property—i.e., the kind and the intensity of the radiation emitted by any radioactive element is absolutely the same whether the element is present as such, or as any one of its compounds.

(b) Discovery of Radioactivity

Shortly after Röntgen's discovery of X-rays in 1895, it occurred to Becquerel that since an X-ray tube glows with ordinary fluorescent light at the same time as it emits Röntgen rays, it might be worth determining whether fluorescent substances inherently possessed the property of emitting radiation which, like the X-rays, could penetrate opaque materials. He tested various fluorescent substances to this end, including, among others, salts of *uranium*. He found that these last did, indeed, have the property of fogging a photographic plate through a wrapping of black opaque paper. The other fluorescent substances examined by Becquerel did not show this property, but in the case of uranium the effect was exhibited not only by the fluorescent uranyl salts, but also by the non-fluorescent uranates. It was ultimately established not only that all uranium compounds, and the metal itself, emitted a peculiar penetrating radiation, but also that (unlike the phenomena of fluorescence and phosphorescence) the effect was completely independent of any previous illumination, or previous treatment of the preparations. Becquerel also observed that the radiation rendered the air electrically conducting, and discharged electrified objects—a property which has since proved of great importance in the measurement of radioactive phenomena. It was particularly striking that Becquerel's *radioactive radiations*, as they were later termed, appeared to be emitted with unchanging intensity over long periods of time. This was at first difficult to reconcile with the law of the conservation of energy, but later found its explanation (and, at the same time, its restrictions) in the disintegration hypothesis put forward by Rutherford and Soddy in 1903.

The search for other elements with the same radioactive properties of as uranium soon led to the discovery of the radioactivity of *thorium* by Schmidt and, almost simultaneously, by Mme. Curie (1898). In the same year, Pierre and Mme. Curie discovered two *new elements, polonium* and *radium*, which were far more intensely radioactive than uranium.

As has been mentioned previously (Vol. I, p. 244), Mme Curie investigated uranium minerals for their radioactive properties (using the ionization of air by the radiations as a measure of activity), and observed that whereas the activity of artificially prepared uranium compounds was proportional to their uranium content, the uranium minerals generally displayed a considerably higher activity than would be expected on this basis. Pitchblende from Joachimsthal, for example, was three times as radioactive as metallic uranium. From the proportionality between radioactivity and uranium content of pure uranium compounds, irrespective of their chemical nature, Mme. Curie inferred that radioactivity was an atomic property. If this be so, the greater activity of naturally occurring uranium compounds could be explained only by the hypothesis that they contained at least one element which was more strongly radioactive than uranium. Mme. Curie and her husband, with the assistance of Bémont, at once undertook a chemical analysis of pitchblende, in which they traced the course followed by the radioactive constituents by measuring their ionizing power. Uranium is precipitated in the ammonium sulfide group, but they found that a strongly radioactive substance revealed its presence in the precipitate obtained with hydrogen sulfide in acid medium. The radioactivity accompanied bismuth in the subsequent course of the analysis. Bismuth itself is not radioactive, however; hence the activity must have been due to a new element resembling bismuth in properties. This element was named *polonium* by Mme. Curie, after her native land. In the further course of the separation, a second active element was found to follow barium closely, and this received the name of *radium* from its very intense emission of radioactive rays.

Attempts to isolate radium were soon begun, but were very difficult because radium is present in uranium ores only in very minute amounts. Joachimsthal pitchblende is relatively rich in radium, but even so contains only 1 mg radium in 5 kg of ore. However, the Austrian government presented Mme. Curie with two truck loads of pitchblende residues, practically without charge. From this, as a result of laborious work, she succeeded in preparing 100 mg of a pure radium salt (radium bromide).

Very shortly after the discovery of polonium and radium, Debierne (1899) found yet another radioactive element in pitchblende residues—*actinium* (ἀκτινόεις giving out rays). In the analytical separation, this was precipitated along with iron and the rare earths. The actinium content of pitchblende is much smaller even than its radium content, and it did not prove possible until very recently to isolate weighable amounts of actinium compounds.

(c) General Properties of Radioactive Radiations

The penetrating radiation emitted by radium and other radioactive substances reveals itself principally by three effects—[i] its *luminescent* effects, in exciting luminescence in certain materials (e.g., Sidot's blende) at ordinary temperature; [ii] its *ionizing properties*; and [iii] its *action on the photographic plate*.

After development, a photographic plate is found to be blackened at the places exposed to the radioactive rays. A few milligrams of a radium salt can produce blackening in their immediate neighborhood within a minute. The ionizing property of radium is far more intense, and $\frac{1}{1000}$ mg of radium will discharge an electroscope almost instantaneously.

Luminescent effects of the radiations can generally be seen only in the dark, with a rested eye. The more concentrated the activity, the stronger are the effects.

Crystalline radium compounds themselves show a pronounced blue luminescence, under the action of their own radiations. Pure fused radium chloride or bromide is so strongly luminescent that the effect is noticeable in full daylight. Radium preparations containing barium luminesce even more brightly than pure radium preparations.

More detailed investigation showed that radium preparations emitted three kinds of rays, though it may be remarked that these are not all directly emitted by radium itself.

[i] The first type, the *α-rays*, has only a low penetrating power (e.g., 3 to 7 cm of air), and can be deflected to a small extent by a magnetic or electric field. The direction of deflection indicates that they are positively charged. They thus show some analogy to the canal rays or positive rays which can be produced in a discharge tube, which were discovered by Goldstein (1886) and extensively investigated by J. J. Thomson. Like the canal rays, they consist of massive particles moving with high velocities and bearing a positive charge. The velocities of α-rays are much higher than for canal rays (half or more of the velocity of light), and their ratio of charge to mass (in fundamental units) is always constant, $= \frac{1}{2}$. As has since been proved, the α-particles are doubly charged helium atoms—i.e., helium atoms which have lost both their electrons, or *bare helium nuclei*.

[ii] The second type of rays, called *β-rays*, is much more penetrating than the α-rays, and is relatively strongly deflected by magnetic or electric fields in a direction which proves the rays to be negatively charged*. They correspond in every way to the cathode rays, but have much higher velocities (i.e., higher energies). They are electrons moving with high velocities (up to 0.998 of the velocity of light, in the case of β-rays from radium-C).

[iii] The third type of radiation, the *γ-rays*, has very great penetrating power, and sustain no deflection in electrical or magnetic fields. They are thus not charged, and consist of electromagnetic radiation, similar to the hard X-rays, but of shorter wave length.

2. Characteristic Properties of Radioactive Materials

Before giving any more detailed account of the three types of radiation, the effects produced by the radiations from radium and other radioactive materials may be considered more fully [6].

(a) Ionization of Gases

Preparations of radium (and also of thorium and uranium) owe their ability to ionize air and other gases chiefly to the α-rays. The effects due to α-rays are observed only if the preparations are not shielded (i.e., are not surrounded by some material which completely absorbs the α-radiation). The β-rays have a weaker ionizing action, and the smallest ionizing effect is exhibited by γ-rays.

---

* The term β-rays is now sometimes used to include not only the electron rays but also the emission of *positrons* (i.e., particles with the same mass as an electron, but bearing a unit positive charge). If this terminology is used, the radiations are distinguished as β^+ and β^- rays, respectively. In discussing the natural radioelements, the term 'β-rays' will always be used in the sense of β^- rays.

This statement is true with regard to the number of ions produced along equal lengths of path.* If a very thin layer of a radium compound is spread over the lower of a pair of plates, 5 cm apart, then the ionization of the air by the α, β and γ particles is roughly as 10,000 : 100 : 1, as measured by the electrical conductivity.

(b) Observation of Gaseous Ions

Ions in gases act like dust particles, in that they serve as nuclei for the condensation of water vapor. This is the basis of the method developed by C.T. R. Wilson, in 1911, for making the ions visible, and hence for observing the tracks of the ionizing particles (cf. Fig. 59).

When a compressed gas is allowed to expand against the atmospheric pressure, it is familiar that it cools down through performing external work. If the gas were initially saturated with moisture, it becomes supersaturated in expanding, and deposits the excess moisture as a fog. If the degree of adiabatic cooling is not excessive, condensation occurs only if the gas contains nuclei for the condensation of the droplets of water which make up the fog. These nuclei are ordinarily provided by particles of dust, but it has been found that gaseous ions can act in the same way. If all dust particles are removed (as can readily be done by allowing water droplets to condense on them and sink to the bottom of the chamber), a constant degree of supersaturation (readily controlled by the amount of adiabatic expansion) will subsequently bring about condensation only if the gas is ionized—e.g., by X-rays, or by a radioactive preparation. It is possible, by means of the Wilson cloud chamber, to distinguish between positive and negative ions, or determine the relative numbers of each, since it is found that a greater degree of supersaturation is needed for condensation on positive ions than on negative. It has been shown, for example, that positive and negative ions are always formed in equal amounts. The 'cloud tracks' formed by the passage of ionizing particles are usually registered photographically (Fig. 59).

Fig. 59. Trajectories of α-particles, shown by cloud tracks.

(c) Penetrating Properties of Radioactive Radiations

α-Rays are completely absorbed by 0.1 mm of aluminum foil, or by a flake of mica or a sheet of paper. β-rays are absorbed only by 5 mm of aluminium or 1 mm of lead foil, whereas γ-rays are very little absorbed by these materials. The absorption of radiation by various materials is roughly proportional to their density (i.e., to the mass of absorber per unit length of path), but substances of very high density absorb rather better than would be predicted from this rule.

* Whereas the β-rays produce far fewer ions per cm of path than do the α-rays, they also execute much longer tracks in one and the same medium. The *total ionization* is practically the same for α and β-rays of equal energy—namely one ion pair per 30 ev (approximately). The same is true of other types of high energy particles (e.g., accelerated protons, see p. 517 and 565).

(d) Luminescence

Sidot's blende (crystalline zinc sulfide) gives a vivid blue luminescence under the impact of radioactive radiations. Barium tetracyanoplatinate(II) gives a green fluorescence, and natural or artificial willemite (zinc silicate, Zn_2SiO_4) also a green fluorescence. Scheelite, diamond, and most fluorites give a blue fluorescence. Luminosity is also excited in a wide variety of other substances, including minerals, glasses, and a great number of organic compounds. Fluorescence is also excited in the retina of the eye*.

In most cases the principal luminescent effects are due to the α-rays. There are certain substances, however, which can be excited by β-rays, but do not luminesce with α-rays—e.g., the mineral *kunzite* (a variety of triphane discovered by the American, Kunz) and antipyrine salicylate. It was formerly stated that γ-rays had practically no luminescent effects. The excitation of luminescence depends on absorption of the rays in the phosphor, and γ-rays therefore excite little luminescence at the surface of a phosphor. However, if a γ-ray source is immersed in a luminescent substance (e.g., in a cavity in a sodium iodide crystal, or in a solution of terphenyl in benzene), luminosity is excited throughout the medium, and can be used for scintillation counting of γ-rays (see below).

The luminescence of many substances ceases immediately the radioactive source is removed. Others show an after-glow (phosphorescence). Under the action of β-rays, Sidot's blende gives a diffuse phosphorescence with a slow decay. The luminescence produced in this material by α-rays, however, is so sharply defined, both spatially and temporally, that the flashes produced can be used to count the α-particles colliding with the phosphor.

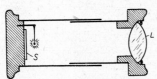

Fig. 60. Spinthariscope.
S is a zinc sulfide screen, in front of which is placed the radium preparation on the tip of a needle. α-Rays from this preparation strike the screen and cause scintillations which are observed by means of the lens L.

(e) Scintillations

Viewed under a lens or low-power microscope, a screen coated with Sidot's blende may be seen not to luminesce over its whole surface when exposed to a source of α-rays. Flashes of light appear and disappear at random at different places. These scintillations (latin, *scintilla*, a spark) are apparently due to the impact of individual α-particles, and it is possible by direct counting to determine the number of α-particles reaching the luminescent surface per unit time, and hence the number emitted by the radioactive preparation (*spinthariscope* of Crookes, Fig. 60). The phosphor may be observed, not with the naked eye but with a photomultiplier tube, whereby each scintillation is converted to an electrical pulse which is registered by a suitable circuit. In this form, the *scintillation counter* has become an important instrument for the measurement of both α-ray and β-ray activities (see below).

(f) Photographic Effects

Where no absorber is interposed between the plate and the radioactive preparation, the blackening of photographic plates is due largely to α-radiation. With an absorber interposed—e.g., a thick layer of air, or a thin layer of solid material—the effects are due chiefly to β-particles. Since the latter have considerable penetrating power, but are absorbed to different degrees in substances of varying density, sources of β-radiation can be used for radiography, to reveal the inner contours of opaque objects. The radiographs lack definition, however, owing to the considerable scattering sustained by β-rays in their

* The approach of a radioactive preparation can be detected by the closed eye. This is, however, a hazardous proceeding, and in practice all unnecessary and uncontrolled exposure to radiation should be rigorously avoided.

passage through matter. Hence more satisfactory radiographs can be made using strong γ-ray sources. The natural and artificial radioelements offer a number of convenient γ-ray sources of different penetrating powers, equivalent to those of the X-rays from high tension X-ray sets. In recent years, γ-ray sources have come into wide use for industrial radiography —e.g., for the inspection of welds in pressure vessels.

(g) Coloration Produced by Radioactive Radiations

Glass and other substances gradually become colored under the prolonged action of radioactive radiations. The blue color occasionally displayed by rock salt (Vol. I, p. 188) is probably attributable to the action of radioactivity. Under the action of the radiations, a very small proportion of the positively charged sodium atoms in the crystal become neutralized. Under these conditions, the metal separates out in the matrix in a state of extreme subdivision, and shows the intense color characteristic of colloidal systems. The same effect can be brought about artificially by exposure to radioactive radiation. Elster and Geitel observed that potassium sulfate, colored green by exposure to the rays from radium, displayed a strong photoelectric effect, like that of the free alkali metal (i.e., it emitted electrons when irradiated with ultraviolet light).

It is often observed that the coloration produced by radioactive rays is reversed by heating. The ability of substances to luminesce under the action of radioactive rays is considerably diminished when they become colored, but is restored when the coloration is reversed by heating.

If a substance contains inclusions of small particles of radioactive material (as is often the case with minerals), it is only the immediate environment of the radioactive inclusion that acquires a coloration, since the α-particles are the most efficient in producing this effect also, and have only a very small range in solids. A halo, which displays pleochroism in polarized light, and is therefore known as a *pleochroic halo*, is ultimately formed around each inclusion. These pleochroic haloes enable extraordinarily small quantities of radioactive elements to be detected. Thus a little as $5 \cdot 10^{-10}$ g of uranium oxide, which is in radioactive equilibrium (see below) with only 10^{-16} g (or about $3 \cdot 10^5$ atoms) of radium, in mica of the Devonian age, is surrounded by a fully developed, pleochroic halo even though on the average only one α-particle is emitted every 10 hours. Even as little as 10^{-17} g of radium, emitting a single α-particle every 4 or 5 days, has been in some cases detected by a rudimentary pleochroic halo. These observations show that other elements, which are frequently present in mica in much greater quantities, than these, cannot be undergoing radioactive disintegration with emission of α-particles, since no effects due to their radiations can be detected even after geological epochs.

(h) Other Chemical Effects of Radioactive Radiations

The effects of radioactive radiations on gases resemble those of the silent electric discharge. In this case also, the α-rays produce the greatest effect. Pierre and Marie Curie noticed the ozonizing action of the rays from radium in 1899. With sufficiently active preparations, the formation of ozone is evident from its smell. Carbon monoxide, carbon dioxide, ammonia, hydrogen chloride, and other compounds are decomposed by the radiations. Conversely, these compounds are synthesised from their elements. The detailed study of the chemical effects of α-particles in gases has been largely due to Lind. Water is decomposed, the primary products probably being hydrogen atoms and hydroxyl radicals, and the end products chiefly oxygen and hydrogen (with some hydrogen peroxide). Moist radium preparations, or solutions of radium salts, should therefore not be stored for prolonged periods in sealed tubes, since the pressure may build up until the tubes burst. In working with the emanations, it is found that carbon dioxide is liberated by the action of their α-rays on tap greases. Vaseline and paraffin are also affected chemically. Intense α-emitting preparations gradually make quartz glass brittle, and affect other materials also. Radioactive preparations can also undergo secondary chemical changes as a result of their own radiations. Thus radium bromide exposed to air loses bromine, either as a direct result of its radiations or through the ozone formed, and is gradually converted to carbonate.

(i) Physiological Effects

Radioactive radiations have serious physiological effects, which may be cumulative over a period of time. *It is therefore essential to use the greatest care in handling any radioactive preparations.* Even short exposure to intense sources of radiation is sufficient to cause painful inflammation. In general, abnormal cells are more susceptible to destruction and damage than healthy cells, and the penetrating radiation from γ-ray emitting sources has been widely used in radiotherapy for the treatment of cancer. At the same time, exposure to radiation can have carcinogenic effects and, by affecting the cells of the bone marrow, can bring about leukemia.

Work with quantities of radioactive substances equivalent to more than a few milligrams of radium ('millicurie level') requires suitable techniques, with the experimenter adequately shielded from penetrating radiations. Work with pure β-emitters can be carried out behind transparent 'perspex' shields; preparations emitting γ-rays must be manipulated behind lead shielding, using remote handling methods if necessary. For work with intense sources, equivalent to grams of radium, properly designed laboratories are essential.

3. Properties and Nature of the Three Kinds of Radiation

(a) Alpha Rays

(i) *Range, Energy, and Ionizing Power.* α-rays move through gases in almost rectilinear paths (cf. Fig. 59). At some point they vanish—i.e., beyond a certain point it is not possible to detect any ionization or other effects of the α-rays. The distance from the source to the point at which their ionizing action ceases is known as their *range*, and since all the α-rays emitted from a radioactively homogeneous source have practically the same range, the *range of the α-particles is an important characteristic of each α-emitter.*

All the α-particles are ejected with the same velocity from a homogeneous preparation. This velocity can be determined, as with cathode rays and canal rays, by observing their deflection in electrical and magnetic fields. The velocity diminishes progressively as the particle traverses its path through the medium, and near the extremity of the range, the decrease in velocity is rapid. The particle is thus suddenly slowed down. The number of α-particles causing ionization remains constant almost to the extremity of the range, and then suddenly drops to zero. The ionizing effect of the α-particles is inversely proportional to their velocity—i.e., the longer the time taken to traverse an atom, the greater is the probability of ionization.

Fig. 61 shows how the ionization of a gas by α-particles varies with distance from the source, and hence with the velocity of the particles. The data (taken from measurements by Geiger) relate to the ionization of air at 12° C and 760 mm pressure by the α-particles from radium C′, one of the disintegration products of radium. The number of ions formed per millimeter of path is plotted as ordinate. The ionization clearly increases progressively with distance from the source— i.e., as the velocity drops off—and reaches a sharp maximum immediately before the point at which the slowing down takes place. It then falls abruptly to zero. An α-particle from radium C′, with a range of about 7 cm, produces about 237,000 ions before it is absorbed. The number of α-particles emitted by the radium C′ which is in equilibrium with 1 g of radium was found by Rutherford and Geiger to be $3.4 \cdot 10^{10}$ per second. 1 g of radium, together with all its decay products including radium C′, emits four times as many α-particles as this—i.e., about $8 \cdot 10^{12}$ particles per minute. The particles were counted by detecting the ionization of the air produced by each particle which passed through a diaphragm defining a known solid angle with the source as vertex. Every particle passing through the opening set up a pulse in the electrometer. The number found by Rutherford and Geiger in this way agreed with that determined by counting scintillations.

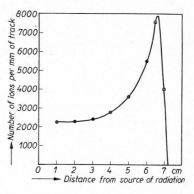

Fig. 61. Dependence of ionization due to α-particles on particle velocity.

The range of an α-particle is four times as long in hydrogen as in oxygen at the same temperature and pressure. In general, Bragg and Kleeman found that the atomic stopping power of any element was proportional to the square root of the atomic weight. For mixtures or compounds, the molecular stopping power was almost equal to the sum of the atomic stopping powers of the components.

The range of α-particles in any gas is proportional to the absolute temperature and inversely proportional to the pressure. The range at 15 °C is 1.055 times as long as that at 0 °C. The variation of α-particle range with temperature has been used as a means of determining the temperature of flame gases.

Geiger deduced that the range of α-particles was roughly proportional to the third power of their initial velocity. If ϱ is the range at 0 °C (in cm), and v the initial velocity (in cm per sec), $\varrho \sim 9.25 \cdot 10^{-28} v^3$ or $v \sim 1.026 \cdot 10^9 \sqrt[3]{\varrho}$.

The *relative stopping power* of any substance for α-rays is defined as the ratio $S = \varrho_a/\varrho_s$, where ϱ_a is the range of the α-rays in air, and ϱ_s their range in the substance concerned. Thus for aluminium $S = 1660$, for mica $S = 2000$, and for gold $S = 4800$. As a measure of the range of α-rays in a solid substance, it is usual to specify the weight (in mg per cm²) of a foil of the substance which is just thick enough to absorb the α-rays. If this 'range' expressed in mg per cm² is divided by $1000 d_s$ (d_s = density of solid), the quotient gives ϱ_s in cm, since the mass of a sheet of q cm² cross section, ϱ_s cm thick is $m = \varrho_s \cdot q \cdot d_s$ g. The range of α-particles in air is given by the product $\varrho_s \cdot S$, where ϱ_s is their range in a solid of relative stopping power S. For example, the α-rays from radioactively pure thorium are just absorbed in an aluminium foil weighing 4.43 mg per cm². The density of aluminium is 2.70, so that the range of thorium α-rays is $\dfrac{4.43}{1000 \cdot 2.70} = 1.64 \cdot 10^{-3}$ cm in aluminium, and $1.64 \cdot 10^{-3} \cdot 1660 = 2.72$ cm in air. Instead of the 'range in mg per cm²', the stopping power of materials is sometimes expressed in terms of the mass (in mg per cm²) which has stopping power for α-rays equivalent to that of 1 cm of air. Thus, for aluminium this is 1.63 mg per cm². The ratio of the 'range in mg per cm²' to this quantity is equal to the range in air

(e.g. $\dfrac{4.43}{1.63} = 2.72$).

The air equivalent of a substance is not quite independent of the velocity of the α-rays; the same is true of the relative stopping power. For accurate measurements, the range is determined directly in dry air, using reduced pressures to obtain the longest possible range. The well defined range possessed by an α-particle of given initial velocity is due to the fact that it loses a definite amount of kinetic energy every time it collides with an atom and produces ionisation. This loss of energy amounts to 35 e.v. per ion pair produced in air.

Occasionally collisions occur in which much greater amounts of energy (e.g. up to 1000 e.v.) may be transferred to the atom struck. These collisions involving large energy losses are statistically distributed over all the trajectories of the particles, with the result that their lengths fluctuate about a mean value. Fig. 62

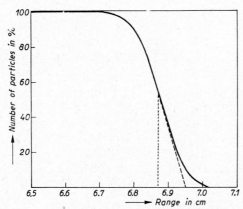

Fig. 62. Range variation of the α-particles from radium C′ (according to I. Curie and Mercier)

shows this for the α-particles from radium C′ (according to measurements of I. Curie and Mercier). The percentage of all α-particles having a given range is plotted (as ordinate) against the range. All the α-particles originally emitted are still present at the end of a track 6.70 cm long in air; all have disappeared after an air path of 7.02 cm. The decrease in their number is greatest at the point where the curve has its inflexion—e.g. at the range 6.87 cm. It is usual to give for the range of α-rays from radium C′ not this most common ('modal') value, but the value 6.95 cm, obtained by extrapolating the tangent at the point of inflexion to intersect the axis of abscissae (dotted line). A value for the range which usually differs only slightly from the foregoing is obtained by plotting the number of ions produced per unit length of path, as a function of distance from the source (Fig. 61), the steeply falling limb of the resultant curve is then extrapolated linearly to intersect the axis.

(ii) Charge on the α-Particle. Rutherford and Geiger determined directly the quantity of electricity transported from a radioactive source by the α-radiation. Since they had already determined the number of α-particles emitted, their experiments at once gave the charge on a single particle, whereas from the electrical and magnetic deflections it is possible to determine only the ratio of mass to charge. The charge on the α-particle proved to be twice the (positive) elementary quantum of electricity—i.e., twice the charge on the hydrogen ion or hydrogen nucleus. Similar experiments were carried out by Regener, with the same result.

(*iii*) *Nature of the α-Particle*. From its deflection in the magnetic and electric fields, it was shown that the ratio of charge to mass for the α-particle was about half as great as the ratio between the charge of the hydrogen ion and the mass of hydrogen atom. Since each α-particle bears 2 unit positive charges, it must have about 4 times the mass of the hydrogen atom. The only known element with mass number 4 is *helium*. Rutherford proved that helium was, in fact, formed during the emission of α-rays, by enclosing radium emanation, an α-ray emitter, in a glass capillary with such thin walls that the α-particles passed through them. It was shown spectroscopically that helium gradually collected on the outer side of the capillary. It was established that the thin glass capillary was not permeable to gas. The experiment proved conclusively that α-particles were helium atoms, with a twofold positive charge. From the spectroscopic discussion of Chapter 4, Vol. I, it follows that the helium atom loses all its extranuclear electrons in acquiring a double positive charge; only the nucleus remains. Hence the α-particles must be regarded as *helium nuclei*, ejected with great velocity.

As stated above, 1 g radium emits $13.6 \cdot 10^{10}$ helium nuclei per second. Hence it produces $13.6 \cdot 10^{10}/27 \cdot 10^{18} = 5.03 \cdot 10^{-9}$ cc of helium per second, or 159 mm³ of helium per year.

(b) Beta Rays

Unlike the α-rays, the β-rays are not homogeneous even in the case of a radioactively pure substance; they consist of electrons with velocities (i.e., kinetic energies) varying continuously over a certain range. The most energetic have velocities closely approaching the velocity of light ($0.998\ c$ for the fastest β-rays of RaC). The direction in which β-rays are deflected in magnetic or electrical fields shows them to bear a negative charge.

Kaufman (1901–1906) determined the ratio of charge to mass for cathode ray particles and for β-particles, and showed that the mass varied with the velocity as predicted by the theory of relativity.

(*i*) *Maximum Energy of β-Particles*. Although the β-particles emitted in the course of some particular radioactive disintegration may have any kinetic energy within certain limits, the different values for kinetic energy are distributed in a definite manner. If the number of particles possessing any given energy is plotted as a function of the energy, the resulting curve has the same form for all β-ray emitting substances. The maximum energy found among the β-particles emitted in the particular radioactive disintegration process, denoted by E_{max}, is a characteristic constant for each β-ray emitting substance, in the same way as the range of the α-particles is characteristic of each α-emitter. See further p. 521 *et seq.*

(*ii*) *Ionizing Power*. Over equal lengths of path, the ionizing action of β-rays is much less than that of α-rays. As for α-rays, the ionization decreases as the velocity of the β-particle increases*.

The α-rays from radium C′, which are exceeded in velocity only by those of thorium C′, produce nine times as many ions in a given length of track as do β-particles of the same velocity, and more than three hundred times as many as the fastest β-particles.

---

* The *total ionization* increases with increasing particle velocity, as for α-rays. It is proportional to the kinetic energy of the particle. Cf. footnote p. 505.

About 50 to 100 ion pairs per centimeter of track are produced in air at atmospheric pressure by the fastest β-particles, as compared with 20,000 to 40,000 by the fastest α-particles.

As the ionization increases with falling velocity, it attains a maximum at about $3 \cdot 10^8$ cm per sec. If the velocity falls below this, the ionization becomes weaker and disappears at rather lower speeds.

(*iii*) *Absorption of Beta Rays.* The absorption of β-rays follows a different law from that of α-rays. They do not suffer an abrupt stop, but are gradually attenuated, partly by absorption and partly by 'scattering', i.e., by the deflection of the β-particles from their rectilinear path.

If β-rays are traversing an absorbing medium in the direction x (see accompanying sketch), the decrease in the intensity I (as measured by the ionization produced in passing through a given layer of air) is represented by the equation

$$-\frac{\mathrm{d}I}{\mathrm{d}x} = \mu I \tag{1}$$

where the proportionality factor μ is known as the *absorption coefficient* of the given medium for β-rays. μ increases with the density of the material; it is also dependent upon the energy of the β-particles*, and is therefore different for the rays emitted by different radioelements. Thus the absorption coefficient of aluminum is 14 for the β-rays emitted from UX_2, 150 for those from UX_1, and 5500 for those from RaD. Since the velocity of the β-particles falls off as they traverse the absorbing material, the ionizing effect of each particle per unit length of track increases. At the same time, the number of particles moving in the x-direction diminishes as a result of scattering. These two effects are superposed, and the net result is that the coefficient μ for a beam of β-rays of given initial velocity remains roughly constant. If μ remains constant, integration of (1) gives

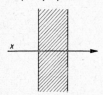

$$I = I_0 \, e^{-\mu x} \tag{2}$$

where I_0 is the intensity of the rays before entering the absorbing medium, x is the length of path within the medium, and e the base of natural logarithms.

(c) Gamma Rays

The penetrating power of γ-rays is not merely much greater than that of the α- and β-rays; in general, it is much greater than that of X-rays. The existence of γ-radiation was established by Villard, in 1900, for radium. γ-radiation from radium is readily detected by its photographic effect, and by the fluorescence produced on a screen coated with barium cyanoplatinate(II) or willemite. β-rays must be screened off, by interposing a 2 mm thick lead foil. Detection of γ-radiation from weakly radioactive preparations, such as those of uranium or thorium, is less easy**.

Emission of γ-radiation is not characteristic of all the radioelements, but is observed with certain of them only. γ-radiation may be produced in the course of either α-particle or β-particle decay, and is explained in terms of the theory of atomic disintegration in the

* The greater the kinetic energy, the smaller is μ. Use is made of this relation for approximately determining the energy of β-radiation. (See p. 521.)
** γ-rays are not emitted directly either by thorium or uranium, but originate from disintegration products of these elements.

following way. After the emission of an α- or β-particle, the new nucleus may, in many instances, be left in an excited state. It may then dispose of its surplus energy in the form of a quantum of electromagnetic radiation of very high frequency—i.e., in the form of γ-rays. In accordance with this mode of origin, primary γ-rays* always furnish a *line spectrum*, which enables inferences to be drawn concerning energy states in the *atomic nucleus*, in the same way as optical line spectra furnish information as to energy states of the extranuclear structure.

Absorption of γ-rays follows an exponential law, corresponding to that of β-rays, except that the coefficient μ is much smaller:

$$I = I_0 \cdot e^{-\mu x} \tag{3}$$

μ depends both on the nature of the absorbing material and on the wave length (i.e., the energy) of the γ-radiation. The smaller the wave length, the smaller is μ in any medium— i.e., the more penetrating is the radiation.

4. The Disintegration Theory of Radioactivity

In 1903, Rutherford and Soddy, in England, put forward an interpretation of the phenomena of radioactivity, in terms of their *disintegration theory*. The essential nature of radioactivity is considered to be that the atoms of radioactive elements are unstable, and undergo spontaneous disintegration. In breaking up, they eject radiations (α- or β-particles, or quanta of γ-radiation), and the number of particles corresponds to the number of atoms disintegrating. In the case of the natural radioelements, species of the atoms formed in the process are themselves generally still unstable, and are therefore radioactive. By their disintegration, they yield yet other nuclei until finally a stable, and therefore inactive, nuclear species results.

This conception explains why radioactive substances continuously radiate energy, without acquiring any energy from external sources: an experimentally established fact that at first appeared to contravene the principle of conservation of energy; the phenomena are now attributed to processes which lower the internal potential energy of the atomic nucleus. It implies, further, that the energy emission, e.g., from radium, cannot remain strictly constant; if, during observable periods, only a vanishingly small fraction of the atoms has decomposed, the activity will apparently remain constant. If, however, an appreciable proportion of the atoms has decayed within a short time, the decrease in the intensity of the emitted radiation can readily be observed. Measurement of the rate at which it declines is, indeed, one of the most important ways of characterizing many radioactive substances.

Transformations of the atomic nuclei give products which differ from the original not only in their radioactive properties, but also in chemical properties, as was proved at an early stage. Thus radioactive processes involve the *transmutation of the elements*.

(a) The Disintegration Series

One element may be transformed by disintegration into another, also radioactive, and this in turn into a third radioelement, and so on. A succession of the products of radioactive processes therefore results, and the whole is known as the *disintegration series* derived from the original radioelement. Three** disintegration

---

* See p. 517 on the subject of secondary γ-rays.
** Strictly speaking, a fourth disintegration series is represented in Nature in minute amounts. Consideration of this series is deferred to Chapter 13.

series are derived from the radioelements that exist in Nature—the *thorium* (or '4n') *disintegration series* (Table 57, p. 528, the *uranium-radium* (or '4n + 2') *disintegration series* (Table 59, p. 533), and the '4n + 3' or *actinium disintegration series* (Table 60, p. 537). Each of these series terminates with an inactive product. The element from which some other is directly formed is frequently spoken of as its *parent element.* The immediate disintegration product of a radioelement is known as its *daughter.*

(b) Half-lives and Decay Constants

As is required by the disintegration theory, the intensity of the radiation emitted by some element decreases continuously with time. The time elapsing until the intensity of the radiation has fallen to one-half of its initial value is known as the *half-life* of the radioelement in question.

Suppose that the number of atoms of the radioelement present at the time $t = 0$ (i.e., at the commencement of observations) is N_0, and at some later time t the number of atoms remaining unchanged is N_t. Then according to the disintegration theory (cf. p. 525),

$$N_t = N_0 \cdot e^{-\lambda t} \tag{4}$$

In (4), λ is a constant, known as the *decay constant* or *disintegration constant*, which depends upon the nature of the radioelement. The reciprocal of λ is the *mean lifetime* of the atoms of the element in question. The *half-life*, $t_{1/2}$,—i.e., the time at which just one-half of the original number of atoms remains, so that the intensity of the radiations has fallen to half its initial value,—is related to the value of the decay constant. Substituting in (4),

$$N_{t_{1/2}} = \frac{N_0}{2} = N_0 \cdot e^{-\lambda t_{1/2}}$$

$$\text{Hence } t_{1/2} = \frac{1}{\lambda} \log_e 2 = 0.69315 \cdot \frac{1}{\lambda} \tag{5}$$

At a time $t = 10 t_{1/2}$, $N_t : N_0 = 1 : 2^{10} = \dfrac{1}{1024}$. Thus after a time equivalent to ten half-lives, the number of atoms of the radioelement present will have fallen to about one-thousandth of the number originally present. In many cases, therefore (except where very highly active preparations are involved), it may be taken that the radiations due to the element in question will have practically decayed away to zero after ten half lives.

(i) Determination of Disintegration Constants. Equation (4) may be written in the form

$$\log N_t = \log N_0 - 0.4343 \, \lambda t \tag{6}$$

This is the equation of a straight line. Hence if the values of $\log N_t$ are plotted, as ordinates, against the corresponding times t, as abscissas, the experimental observations should fall on a straight line. The disintegration constant λ is given by the slope of this line:

$$0.4343 \, \lambda = \frac{\log N_0 - \log N_t}{t} \tag{7}$$

In determining disintegration constants by this method, it is not necessary to know the actual number of radioactive atoms in the preparation. It is sufficient to know the number of particles emitted in a sufficiently short interval of time. This is proportional to the total quantity of the radioactive species, and the factor of proportionality disappears when the logarithms are subtracted from one another. The point of time taken as zero time, $t = 0$, is arbitrary.

Equation (6) applies to a homogeneous radioactive substance. If a mixture of several radioactive species is present, having different decay constants, a curved plot is obtained, instead of a straight line. In such cases, the individual λ values for the several radioactive species can often be derived by mathematical analysis. Provided that the decay constants are sufficiently different, and the relative proportions of the several species not too disparate, it is usually found that the log N_t versus t curve resolves itself into a set of linear segments. Except when it is possible to effect a complete resolution of a mixture of radioelements by chemical means, graphical or mathematical methods of resolution are usually employed for determination of half-lives.

For relatively long lived species, disintegration constants are determined in a different manner, since it would take years to obtain the necessary experimental data. If the half-life is of the order of magnitude of years, the quantity of radioactive material is practically unaltered during a period of a few hours. Hence the differential quotient $\dfrac{dN}{dt}$ (cf. eqn. (11), p. 525) may be replaced by the difference quotient, $\dfrac{\Delta N}{\Delta t}$. Then

$$\lambda = -\frac{\Delta N}{\Delta t} \cdot \frac{1}{N} \tag{8}$$

N is known if the weight and composition of the preparation are known, together with the atomic weight of the radioactive species. The decrease in N $(= -\Delta N)$ during the time Δt is equal to the number of particles emitted during this period*.

It is not practicable to determine the half-life or decay constant directly for all radioactive substances. The very long lived elements, such as thorium and uranium (uranium I) show absolutely no changes is the intensity of their radiations over any measurable period. It is possible, in such cases, to deduce the decay constants for the long lived species from the relative amounts of daughter products in radioactive equilibrium with them (see below).

It is often possible to estimate the approximate decay constants of α-emitting substances with an enormously high disintegration rate, by using a relationship discovered empirically by Geiger and Nuttall (1911):

$$\log \lambda = A + B \log \varrho \tag{9}$$

where A and B are constants, and ϱ is the range of the α-particles. The value of A is somewhat different for each of the three natural disintegration series. The Geiger-Nuttall relation expresses the inner connection between two fundamental magnitudes which enter into radioactive changes, namely the disintegration constant and the velocity with which α-particles are ejected from the nucleus (i.e., with the energy liberated in radioactive change). Gamow (1928) has given an explanation of the Geiger-Nuttall relation.

* This method therefore is applicable only to weighable quantities of the pure radioelement, or to sources of known isotopic composition; it also requires a knowledge of absolute counting rates. The method applicable for short half-lives can be used for unweighably small quantities of material, in the presence of a relatively much larger quantity of carrier if necessary. It also does not require a knowledge of the efficiency of counting, and the absolute rate of emission of particles, since the same source may be counted successively in the same counting device.

(*ii*) *Sargent Curves.* The Canadian physicist Sargent (1933) discovered a relationship between the maximum energy E_{max} of β-rays, and the decay constants of β-emitters, corresponding to that holding between the range of α-particles and the decay constants of α-emitters. If the values of $\log \lambda$ for the various natural β-emitting radioelements are plotted against the corresponding E_{max}, it is found that most of the points lie on one or other of two straight lines. It has been shown that these Sargent curves are a consequence of the theory of β-particle disintegration developed by Fermi (1934) and Pauli (1931). On this theory, $\lambda = k(E_{max})^5$ (approximately). Hence

$$\log \lambda = \log k + 5 \log E_{max} \tag{10}$$

In this equation, k is a parameter that may assume several values, depending on whether the nuclear transition from a higher to a lower energy state, with emission of a β-particle, is a process of high probability or of low probability. If two different values are taken for k, eqn. (10) leads to two straight lines, the course of which corresponds roughly to the two Sargent curves derived from the E_{max} and λ values for the β-emitters of the natural disintegration series. The theory leaves open the possibility that k might assume more than two values. When the β-emitting species obtained by artificial nuclear transformations are included, it appears that they do indeed lie on additional Sargent curves.

(c) Radioactive Equilibrium

If the various members of a disintegration series are not separated from each other, but each radioactive daughter product is left in contact with its parent element, a *state of equilibrium* between the individual members of the series is established in course of time. This is reached when the amount of each species decaying per unit of time is equal to the amount of that species newly formed from its parent. If the first member of the series is a long-lived element, such as radium or thorium, the total activity of the series remains practically constant over long periods of time. The subject of radioactive equilibrium is further treated below.

5. Secondary Radiations

It might be expected from the disintegration theory that emission of an α-particle would leave the residual daughter atom bearing a twofold negative charge. It is found, however, that the products are quite definitely attracted by a negative electrode, and so are positively charged. This property can be utilized to effect a concentration of the products at a required place—e.g., on the surface of a metal wire, by bringing the negatively charged wire close to a radioactive preparation.

(a) Recoil Rays

It can be shown that the daughter atoms are indeed originally negatively charged, but subsequently acquire a positive charge. According to the laws of mechanics, the massive atoms must recoil when they emit α- or β-particles, in the same way as a gun recoils when it fires a shell. This recoil ejects some of the daughter atoms into free space, with a very high velocity, especially in the case of α-particle emission. In the latter case, the velocity of the residual atom is about 2% of that of the α-particle, or of the order of magnitude of $1/_{1000}$ of that of light. This high velocity makes the recoil atoms fairly efficient in ionizing the gas molecules with which they may collide. In particular, these collisions strip them of their two excess electrons, and usually of another electron also, so that they are left with a net positive charge. If collisions with gas molecules are avoided, by working in high vacuum, the reversal of charge does not take place.

Recoil atoms lose their very high velocity and their ionizing power after traversing a path a fraction of a millimeter long through a gas at atmospheric pressure. Once their velocity has dropped to that of normal gas molecules, they behave like ordinary gaseous ions. The recoil atoms from β-ray transformations do not at any time possess substantially higher velocities than gas molecules.

(b) Negative Secondary Radiation

Recoil rays are one example of secondary radiations arising from radioactive processes. Both negative and positive secondary radiations of other kinds also occur. *Negative secondary rays* are produced when α-, β- or γ-rays collide with atoms of other elements. They consist chiefly of electrons, of lower velocity than the β-particles. J. J. Thomson, who first observed them in 1904, termed them δ-rays. These δ-rays (or slow β-rays) are produced chiefly by α-radiation, which can also give rise to a very weak secondary γ-radiation.

Stronger γ-radiation is generated in the absorption of β-rays. Its origin is in every way analogous to the generation of X-rays when cathode rays impinge on an anticathode. The 'secondary β-radiation' which accompanies β-rays emission consists chiefly of reflected and scattered β-rays. Genuine secondary β-rays may be emitted simultaneously; these are essentially the same as the δ-rays.

γ-rays can give rise both to secondary β-rays and to secondary γ-rays. (X-rays do the same). The secondary β-rays liberated by the γ-rays have penetrating power similar to that of primary β-rays. The ionizing action of γ-rays is largely due to these secondary β-rays and their effects.

(c) Positive Secondary Radiation

The positive secondary radiation produced when α-particles collide with other atoms is of particular interest. If the mass of the bombarded atoms is similar to, or smaller than, that of the α-particle, these secondary particles acquire a considerable range. It may be calculated from the dynamics of collisions that, for the most favored case of central collision* with a hydrogen atom, the 'knocked on' proton has a velocity 1.6 times as great as that of the colliding α-particle. Since the range is proportional to the cube of the velocity, these protons should have a range about four times as long as that of the α-particles producing them. This is, in fact, borne out by observation, as was first shown by Marsden (1914) for the passage of α-particles through hydrogen, and later for their passage through any compound containing hydrogen.

6. Determination of Disintegration Energy

Disintegration energies were originally measured by calorimetric methods—i.e. the heat liberated in radioactive decay was measured directly. Calorimetric methods are not applicable in all cases, however; moreover, they do not usually permit of any exact assignment of the energy to the various types of radiation (α-rays, β-rays, γ-radiation, secondary radiation). A knowledge of how the energy is distributed amongst the different types of radiation is fundamentally important, however, for any deeper understanding of the nature of radioactive processes. Methods which give this information directly are now invariably used. The sum

---

* For a central elastic collision,

$$m_1c_1 + m_2c_2 = m_1v_1 + m_2v_2$$

and

$$\tfrac{1}{2}m_1c_1{}^2 + \tfrac{1}{2}m_2c_2{}^2 = \tfrac{1}{2}m_1v_1{}^2 + \tfrac{1}{2}m_2v_2{}^2$$

where m_1, m_2 are the masses of the two bodies, c_1, c_2 their velocities before collision, and v_1, v_2 their velocities after collision. If $m_1 = 4$, $m_2 = 1$ and $c_2 = 0$, $v_2 = \tfrac{16}{10}c_1$.

of the several contributions to the energy of disintegration is the same as the energy determined by calorimetric methods.

For the relation between the disintegration energy and the loss of mass associated with the disintegration, see p. 567.

The energy liberated in α-decay appears as the kinetic energy of the disintegration products—i.e. is equal to the sum of the kinetic energy of the α-particle and that of the product atom. As follows from the law of conservation of momentum, the ratio between the kinetic energy of the product atom and that of the α-particle is equal to the ratio of the mass of the α-particle to that of the product atom. Hence, to determine the disintegration energy it is necessary to determine only the kinetic energy of the ejected α-particles; this is then multiplied by $\dfrac{M + 4}{M}$, where M is the mass number of the disintegration product.

The matter is rather less simple in the case of β-decay, since the β-particles do not all have the same energy. It follows from considerations set out later that the energy liberated in disintegration and emission of β-rays is given by the *maximum energy* of the ejected β-particles.

Emission of γ-radiation frequently accompanies both α- and β-decay; it may originate either in the *external shell* of the atom, and then has the character of a secondary radiation, or in the *nucleus*. In the latter case, the γ-radiation arises from a transition of the emitting nucleus from a state of high energy to a state of lower energy. The phenomenon is exactly analogous to the transitions of electrons in the external shell of the atom from high energy states to lower energy states, with emission of light or X-rays. It is thus possible to recognise the existence of a number of discrete energy levels in the nucleus of the atom, just as in the external shell—a fact of importance in the theory of the nuclear forces which hold the structural units of the nucleus together. The existence of such energy levels can also be deduced from certain phenomena observed for α-rays.

(a) Energies of α-Rays

The energy of α-particles can be determined exactly from their deflection in a magnetic field. It can be found more conveniently, and with sufficient accuracy, from their range. Since the relation between the energy and the range of α-particles is rather complicated, it is best represented graphically. Such curves are given by Livingstone and Bethe [*Rev. Mod. Physics*, 9 (1937) 266], Holloway and Livingstone [*Phys. Rev.*, 54 (1938) 31] and Riezler (ref. on p. 573). Geiger's formula (p. 515) is only approximate, and even if only low accuracy is desired its application is limited to α-particles with a range between 3 cm and 7 cm in air. For smaller ranges the initial velocity of the particles is considerably higher, and for greater ranges it is smaller than would be deduced from the Geiger formula.

In general, the α-particles emitted by a given radioactive species are all *identical* in initial velocity (initial energy). In certain cases, however, several groups of α-particles are emitted, having markedly different energies. Thus six groups of α-particles are emitted from thorium C, each characterised by a different initial energy. Their values are listed in the first column of Table 56, whilst the second column shows the fraction of the α-particles in each group. The substance is said to have an *α-ray spectrum*. In the example quoted, the differences in the energies of the α-particles are explained by the hypothesis that the nucleus of thorium C″ (the product of α-decay of the thorium C nucleus) is frequently formed in an 'excited' state—i.e. in a state of higher energy than the ground state of the nucleus. The highest value for the disintegration energy of thorium C—and hence for the energy of the ejected α-particles—will be obtained if the disintegration leads directly to the formation of a thorium C″ nucleus in its *ground state*. If an *excited* nucleus of thorium C″ is formed, the energy of disintegration is diminished by the energy difference between the excited and the unexcited nucleus (Fig. 63). The differences between the energy levels of the excited nuclei and the ground state

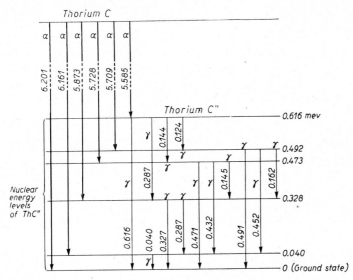

Fig. 63. Origin of the α-particle spectrum of ThC. Relation between the nuclear energy levels in ThC″ deduced from this evidence, and the γ-ray quanta emitted in process of formation of ThC″. The numbers beside the arrows give the quantities of energy (in mev) liberated. Beside the energy levels of the excited state are shown the differences in energy between excited and ground states, as given by the energies of the ThC particles.

are shown in column 4 (Table 56). Disintegration energies listed in column 3 are obtained by multiplying the values of column 1 by the factor $(208 + 4)/(208)$. The excited nucleus usually gives up its excess of energy after an immeasurably short time in the form of γ-radiation, and thus passes over either directly or by intermediate steps into the ground state. Hence the energies of the γ-rays must be equal to the differences between energy levels. Measurement of the energies of the γ-rays thus provides a second and quite independent method of determining the energy levels of the excited thorium C″ nucleus. The γ-ray energies thereby obtained are shown on the right hand side of Fig. 63; they agree well with the observed values. As shown by Fig. 63 not all the possible transitions between the energy levels of the thorium C″ nucleus are, in fact, observed, but only certain of them. It may be inferred from this that, as for transitions in the shell of the atom, certain 'selection rules' apply to the nucleus, whereby particular transitions may be 'forbidden'—i.e. are highly improbable.

TABLE 56

SPECTRUM OF α-PARTICLES FROM THORIUM C

Energy of α-particles in mev	Frequency, as percentage of total number	Disintegration energy in mev	Difference in energy, relative to maximum
6.084	27.15	6.201	—
6.044	69.78	6.161	0.040
5.762	1.81	5.873	0.328
5.620	0.16	5.728	0.473
5.601	1.09	5.709	0.492
5.480	0.01	5.585	0.616

An α-particle spectrum can also arise if some of the nuclei which emit the α-particles are in an excited state at the instant of disintegration. In addition to the normal α-particles, there will then be some of higher energy. These will be very few in number, as compared with the normal particles, since in general the transition of an excited nucleus to the ground state, with emission of γ-radiation, is much more rapid than disintegration and ejection of an α-particle. For this reason, the α-decay of excited nuclei is only detectable in the case of extremely short lived α-emitters.

(b) Energy of γ-Rays

The energy of a γ-quantum is given by the product $h\nu$, like that of a light quantum. It is only in exceptional cases, however, that the frequency ν (or corresponding wave length λ) can be measured by diffraction from the lattice planes of a crystal, as in X-ray spectroscopy. This method cannot be used for γ-rays with frequencies very much greater than the frequencies of X-rays, as is generally the case for primary γ-rays. An approximate measure of the γ-ray energy is afforded by their penetrating power. The harder the radiation—i.e. the higher its energy—the less is it absorbed by matter. For accurate measurements, however, the method is to determine the energy of the electrons that are ejected from the external shells of the atoms with which the γ-ray quantum collides. These 'secondary β-rays' have discrete, homogeneous energies, unlike the primary β-rays which originate in the nucleus of the atom. An electron liberated by a γ-quantum possesses kinetic energy equivalent to the difference between the energy of the γ-quantum and the original binding energy of the electron*. The latter is found from the characteristic X-radiation which is emitted as a result of the electron being ejected from its original energy level (cf. Vol. I, Chap. 7). The kinetic energy of the liberated electron is determined by measuring its deflection in a magnetic field. Since the electrons liberated by the primary γ-rays possess definite and discrete energies, it follows that the same is true of the γ-rays emitted from the

---

* This is true for electrons liberated by the so-called *photoeffect*. In certain cases, the phenomenon known as the *Compton effect* may occur instead, in which the γ-quantum gives up only a portion of its energy to the electron and is itself reflected according to the laws of elastic collision. In this case the frequency of the γ-quantum is diminished, by an amount corresponding to the transfer of energy in the collision. The original energy of the γ-quantum can also be determined from the Compton effect.

nucleus. This, in turn, necessarily implies that the nucleus of the atom can assume definite and discrete energy states, and that when it undergoes a transition from a state of high energy to a state of lower energy, the difference in energy is emitted as a quantum of γ-radiation. One example of such a process has already been considered (Fig. 63). In this the γ-radiation was associated with an α-decay, but γ-radiation may also appear as a result of a β-decay. Hence the β-decay process can also lead to the formation either of normal or of excited nuclei.

Internal Conversion of γ-Rays. It is often found that a γ-quantum causes the ejection of an electron from the electronic shell of the nucleus from which the γ-ray was emitted. The characteristic X-ray spectrum of the daughter nucleus is thereby excited. This process is known as the *internal conversion* of the γ-ray. As a rule, only a small fraction of the γ-quanta emitted in radioactive decay undergo such internal conversion, and this fraction usually becomes smaller as the energy of the γ-rays increases. The β-rays produced by the internal conversion of γ-rays should be regarded as secondary rays, by reason of their origin. Like the γ-rays, they differ from primary β-rays in possessing discrete energies.

(c) Energy of β-Particles

Primary β-rays—i.e. electrons ejected directly from nuclei in the act of radioactive disintegration—do not possess discrete energies. They possess every possible velocity, from zero up to a maximum which is different for each emitter. If the numbers of β-particles possessing a given energy are plotted as a function of energy, a curve such as that of Fig. 64 is obtained. This is known as the *spectrum of the β-particles*. The general form of the spectrum is the same, irrespective of the nature of the source emitting the β-particles; it is only the end point of the curve, corresponding to the *maximum energy* found for any β-particles from the given source (indicated in Fig. 64 as E_{max}), that differs for β-particles from different sources. Very careful measurements of the distribution of numbers of β-particles over the different energies are

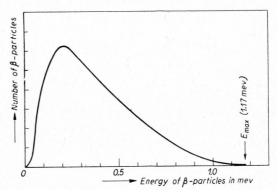

Fig. 64. Beta spectrum of radium E.

needed to determine this maximum value, since the curve meets the energy axis at a very acute angle. In most cases, the maximum velocity of the β-particles emitted from different β-active sources lies between 25% and 99% of the velocity of light. These correspond to maximum energies of 0.025 to 3.15 mev; the most usual values are about 1 mev.

Several methods are available for determining the velocity or kinetic energy of β-particles, and the distribution of β-particles over the energy spectrum. One of the most usual is to employ a so-called magnetic lens—i.e. a coil through which an electric current flows—and a Geiger counter. As they traverse the coil, the β-particles are deviated from their original trajectory and describe spiral paths, the curvature of which depends upon the velocity of the particle and the strength of the magnetic field. For a given current through the lens, only particles of a well defined uniform velocity can enter the counter tube. By altering the current, it is possible to bring the whole range of particles, from those with the lowest velocity to those with the highest, on to the counter, and thus to determine the fraction of the total number possessing each velocity.

The energy liberated in β-decay is given by the *maximum* energy of the β-particles, and not by their average energy. This follows from the mass-energy equivalence law (p. 567) and the loss of mass in β-decay processess, and from other evidence (cf. p. 529). Most of the β-particles thus possess much less energy than is liberated by the β-disintegration of the nucleus. To explain this disappearance of energy*, the hypothesis has been put forward that in the act of β-decay, the nucleus emits not only positrons or electrons, but also another type of particle which is uncharged and which has not yet been directly detected. The mass of this particle is certainly much smaller than that of the neutron, and probably considerably smaller than that of the electron; it is possible that, like a light quantum, its rest-mass is zero. This hypothetical particle is known as the *neutrino*.

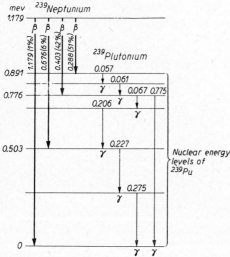

Fig. 65. β-Decay of neptunium (^{239}Np), and γ-radiation of the excited plutonium nucleus which is thereby formed.
Numbers beside the arrows show the maximum energies of the various groups of β-particles, or the energies of the γ-rays (both in mev). To the left of the energy levels of the excited state are shown the differences between excited and ground states, as inferred from the β-particle energies. Shown in brackets are the percentages of the total number of β-particles to be assigned to each group.

It has already been mentioned that, as with α-decay, the nuclei of the products of β-disintegration processes are frequently formed in an excited state. When this is the case, it is usually found that several different groups of β-particles are emitted, each of which possesses a continuous β-ray spectrum of the type shown in Fig. 64, but characterised by different values for the maximum energy E_{max}. Since the various β-particle spectra are superposed on one another, it is often difficult or quite impossible to determine the maximum energies of the various groups of particles directly. However, in order to determine the disintegration energy from the energy of the β-particles, it is usually sufficient to know the maximum energy for a single group of particles, provided that there is trustworthy evidence as to which of the γ-rays that are always observed in such cases is to be associated with the emission of the particular selected group of β-particles. Fig. 65 represents a case in which it has been possible to evaluate the maximum energy for every group of β-particles emitted. The nuclear energy levels deduced from these measurements are in excellent accord with those inferred from the observed

* If there were no emission of neutrinos, the β-decay process would infringe the principle of conservation of angular momentum, as well as the law of conservation of energy.

γ-radiation*. The particular decay scheme represented in Fig. 65 is that of the transuranic element neptunium; it is relatively simple, and corresponds rather closely to the scheme shown in Fig. 63 for disintegration and emission of several groups of α-particles. A rather more complex decay scheme is shown in Fig. 73 (p. 580). This relates to an example in which the nucleus emitting the β-particle and the nucleus formed as the product of β-decay may both exist in several different energy states.

7. Emanations and Active Deposits; the Thorium Disintegration Series

Evidence that radioactive processes were associated with transmutation of the elements was first provided by the study of the *emanations* and the *active deposits*.

The term *emanations* is applied to gaseous substances, themselves radioactive, which are continuously evolved from the elements radium, actinium, and thorium. As has previously been stated, these emanations have the chemical character of inert gases (Vol. I, p. 122).

The first radioactive gas to be discovered was *thorium emanation*, detected by Owens and Rutherford (1899–1900) shortly after the observation that thorium was radioactive. *Radium emanation* and *actinium emanation* were discovered shortly afterwards.

The processes associated with the formation and decay of the emanations proved to be of fundamental importance in formulating the disintegration theory. It is convenient to consider them in more detail with reference to the thorium series (Table 57, p. 528), since the systematics were first worked out for the substances of this disintegration series.

The discovery of thorium emanation was made in the following way. A number of observers had noticed that the radioactivity of thorium compounds was often very variable, when the substances were handled in open vessels, whereas when they were investigated in closed containers it was found that the activity grew until it reached a constant final value. If, however, a stream of air was passed through the vessel, the intensity of the radiation diminished again. Owens showed that the radiations could pass through a porous paper which would completely absorb the ordinary α-radiation. Rutherford explained all the observations by the hypothesis that thorium preparations imparted to the air some kind of particles which were themselves the sources of radioactive radiations. He was able to show that these particles behaved just like a gas: they were transported in a stream of air, could pass through a cotton wool filter and were not removed from it by bubbling through water; they could diffuse through paper, but were completely stopped by a thin sheet of mica.

Rutherford found that any object coming into contact with the emanation itself became radioactive. This *'induced radioactivity'*, as Rutherford showed, was attributable to the deposition of radioactive solid particles, the *'active deposit'*, from the emanation on the surface of the object. He proved, further, that this active deposit must be composite in character. The form of the curve relating the induced activity with time was found to depend upon the duration of the exposure of the object to the emanation (cf. Fig. 66). Rutherford inferred from the form of the activity-time curve that the active deposit of thorium must consist of at least two

---

* Note that the γ-radiation leads not directly to the energy levels themselves, but to the differences between them.

products, of different half-lives. These decay products are now known as *thorium B* and *thorium C*.

It was soon found to be possible to separate these two substances through their different volatilities; thorium B volatilizes at about 1000°. The remaining thorium C then gives a half life of about 1 hour. The curve obtained after long times of exposure, showing the decay of the induced activity (Curve II, Fig. 66), yields a half-life of about 12 hours, which must be attributed to thorium B. The two curves of Fig. 66 can be simply explained by the assumption that thorium C is formed from thorium B, which has a half-life of about 12 hours (10.6 hours according to more recent measurements) but decays without itself emitting α-rays. After short exposures, the deposit consists almost entirely of thorium B. The α-ray activity is at first zero; but grows in the same measure as thorium C accumulated through the decay of thorium B. However, since thorium C decays rapidly, the amount of it begins to diminish again as soon as the quantity of thorium B has become so small that the rate at which thorium C is formed from it can no longer keep pace with the rate of decay of thorium C. After a long exposure to emanation, however, the quantity of thorium C corresponding to the thorium B is already present in the deposit from the start of the observations. All that is then observed is a straightforward decay of the activity, as shown in curve II. This is determined by the

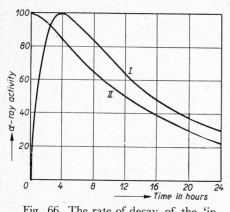

Fig. 66. The rate of decay of the 'induced' activity is dependent upon the length of exposure. Curve I is obtained after short exposure times, curve II after long exposure times.

decreasing rate of production of thorium C from thorium B. It thus represents the decrease in the quantity of thorium B, even though the α-particles counted are not those of thorium B but of its daughter product.

In 1906, Hahn concluded that thorium C must also be complex in nature, from the fact that it emits α-particles of two different ranges. During the following ten years, the theory of the *dual* or *branching* disintegration of thorium C was worked out. According to this, thorium C may disintegrate either by ejecting a α-particle and then an β-particle, or alternatively first by β-emission and then by α-emission. In either case, the end product is an inactive isotope of lead, thorium D (thorium lead). The second manner of disintegration furnishes a species of extraordinarily short half-life, thorium C'; it is impossible to measure the half-life directly, but by application of the Geiger-Nuttall relation (p. 515) is has been estimated at about $3 \cdot 10^{-7}$ seconds. Another short-lived element, although far longer lived than thorium C', is thorium A. Its existence was first inferred by Geiger and Rutherford in 1911, from Geiger and Marsden's observation that thorium emanation emits α-particles of two distinctly different ranges.

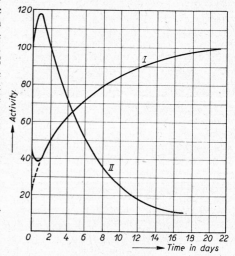

Fig. 67. The activity of thorium, after separation of thorium X, increases with time (curve I) in the same measure as the activity of the separated thorium X decays (curve II). [The departure of the curves from their normal course during the first two days is associated with the formation and decay of other disintegration products of thorium—the active deposits.]

In the meanwhile, the species in the thorium disintegration series which precede thorium emanation had also been investigated. Sir William Crookes, in 1900, showed that it was possible by chemical means to isolate from uranium a substance which had a far stronger action on the photographic plate than had uranium itself. This substance was called uranium X (now known as uranium X_1). In 1902, Rutherford and Soddy found that they could isolate some substance chemically from thorium, which was several thousand times as active as the thorium from which it had been obtained. By analogy with uranium X, this was called thorium X. When the radioactivity of the thorium which had been freed from thorium X was compared with the activity of the thorium X removed from it, the remarkable observation was made that the activity of the thorium X gradually decayed by an amount exactly equivalent to the progressive recovery of the activity of the thorium (Fig. 67). It was the closer study of this phenomenon, in relation to thorium X and uranium X, that led Rutherford to the formulation of the *atomic disintegration theory of radioactivity*.

Rutherford showed that the peculiar relation between the changes of activity of the thorium and the thorium X could be explained if the thorium continuously produced thorium X, at a constant rate, while the activity of the latter decayed according to an exponential law.

Let there be N_1 atoms of thorium, and N_2 atoms of thorium X in the preparation at any instant, and let the fraction λ_1 of the thorium atoms change into thorium X atoms each second. The decay constant of thorium X is λ_2, i.e., the rate of disintegration of the thorium X atoms at any instant is given by $\lambda_2 \cdot N_2$. Then the rate of change of the thorium X present is

$$\frac{dN_2}{dt} = \text{(rate of formation)} - \text{(rate of disintegration)} = \lambda_1 \cdot N_1 - \lambda_2 \cdot N_2.$$

For a pure preparation of separated thorium X, $N_1 = 0$.

Hence

$$\frac{dN_2}{dt} = -\lambda_2 \cdot N_2 \qquad \text{or} \qquad \frac{dN_2}{N_2} = \lambda_2 \cdot dt \qquad (11)$$

Integrating this equation over the time interval $t = 0$ to $t = t$, we obtain the general law for the decay of a radioelement, already referred to on p. 514:

$$\log_e \frac{N_{2t}}{N_{2_0}} = \lambda_2 \cdot t \qquad \text{or} \qquad N_{2t} = N_{2_0} \cdot e^{-\lambda_2 t} \qquad (12)$$

where N_{2_0} is the number of atoms of thorium X present at zero time, and N_{2_t} is the number remaining after the time t.

For the more general case, we may write (11) in the form

$$\frac{dN_2}{dt} = -\lambda_2 \left(N_2 - \frac{\lambda_1 N_1}{\lambda_2} \right) \qquad \text{or} \qquad \frac{dN_2}{N_2 - \frac{\lambda_1 N_1}{\lambda_2}} = -\lambda_2 dt.$$

Integrating this equation as before, and taking the initial value of N_2 as zero, to represent the growth of thorium X activity in the separated thorium,

$$N_{2t} = \frac{\lambda_1}{\lambda_2} N_1 \left(1 - e^{-\lambda_2 t} \right) \qquad (13)$$

It is evident that after a long time, as $t \to \infty$, $e^{-\lambda_2 t} \to 0$, so that

$$N_2 = \frac{\lambda_1}{\lambda_2} N_1, \text{ i.e., } \lambda_2 N_2 = \lambda_1 N_1 \tag{14}$$

The rate at which thorium X is being formed is then equal to the rate at which it is disintegrating; (14) is accordingly obtained by writing $\dfrac{dN_2}{dt} = 0$ in eqn. (11).

Equation (14) brings out an important relationship between the quantities of the various members of a radioactive disintegration series. The decay of each member of any disintegration series is governed by the first-order kinetic law embodied in eqns. (11) and (12), with the appropriate values for the decay constant λ. For each successive member, the rate of disintegration is proportional to the quantity of that species present, and this in turn, for all except the first member of the series, depends upon the rate at which it is being formed from its parent element. Ultimately a state of *radioactive equilibrium* is attained, in which the quantity of each radioactive element formed in unit time exactly balances the amount decaying. If the rate of formation is less than the rate of decay, the quantity of the element present decreases until the decay rate is correspondingly reduced. Conversely, if the rate at which it is produced is greater than the rate of decay, the quantity present increases, and with it the rate of disintegration, until equilibrium is reached. Equation (14) gives the general relation between the number of atoms of each species, N_1, N_2, N_3 . . . , etc., and their decay constants, λ_1, λ_2, λ_3, etc.:

$$\lambda_1 N_1 = \lambda_2 N_2 = \lambda_3 N_3, \text{ etc.} \tag{14a}$$

or, in terms of the half-lives t'_1, t'_2, t'_3, etc.,

$$\frac{N_1}{t'_1} = \frac{N_2}{t'_2} = \frac{N_3}{t'_3}, \text{ etc.} \tag{14b}$$

The condition for attainment of equilibrium is that the first member of the series should be an element of very long half-life, so that its activity is sensibly constant. If elements with a long half-life (i.e., a small disintegration constant) occur in the middle of the disintegration series, long periods of time must elapse before radioactive equilibrium is attained.

In a series of investigations (1905 and later) by Hahn and others, it was shown that thorium X is not produced directly from thorium, but is actually formed by the decay of another strongly radioactive substance, *radiothorium*. Radiothorium was discovered almost simultaneously by Hahn, in the residues obtained in the extraction of thorium from thorianite, and by Blanc in sediments from radioactive mineral water. Subsequent study revealed that radiothorium was so closely identical with thorium, in its chemical properties, that it was impossible to separate thorium and radiothorium by any chemical means. It is possible, however, to isolate radiothorium free from thorium, because the former is produced by the decay of yet another chemically quite distinct element, *mesothorium*.

Even before it had been successfully separated from thorium, the existence of the last-mentioned intermediate product had been inferred by Hahn, from the fact that commercial thorium preparations (which were generally older preparations) often differed very appreciably in α-particle activity from those which investigators had themselves prepared from minerals; other workers had previously made the same observation. Radiothorium is initially present in freshly prepared thorium in amounts corresponding to radiochemical equilibrium. If it were produced directly from thorium, the radiothorium would remain constant in amount and activity, whereas it is found to decrease at first, and then to build up again to the amount corresponding to radioactive equilibrium. This behavior can only be explained if, between thorium and radiothorium, some other substance is interposed, with a relatively long half-life as compared with radiothorium. This cannot be an α-emitter, and must be absent from fresh thorium preparations; hence it is chemically distinct from thorium. Shortly afterwards Boltwood confirmed this hypothesis by showing that mesothorium could readily be separated chemically from thorium. The half-life of mesothorium is 6.7 years; that of radiothorium is 1.9 years.

In 1908, Hahn found that mesothorium could be chemically resolved into two distinct elements, mesothorium 1 and mesothorium 2. The latter is a decay product of the former, and rapidly changes into radiothorium.

The decay of thorium to its ultimate inactive end-product (thorium lead) thus involves a great number of intermediate steps. In thorium preparations of sufficient age—and especially in thorium minerals—all of these have come into radioactive equilibrium.

Table 57 shows the species formed successively by the radioactive disintegration of thorium; the genetic relations between them, and the direction of the changes, are indicated by the arrows. Against each arrow is shown the type of radiation emitted during the disintegration. The other columns give the ranges of the α-particles emitted, half-lives, and atomic numbers and atomic weights of the disintegration products, as deduced from the displacement law (p. 529). The atomic weight of the stable end-product, *thorium lead*, has been verified experimentally with lead isolated from thorium minerals, and has been shown to agree with expectations based on the disintegration theory. As may be seen from Table 57, alternative modes of disintegration may be followed in the case of thorium C and also of thorium A. The dual decay of thorium C has long been known, but that of thorium A was discovered relatively recently. Just as thorium C' and thorium C'' both change to the same product, thorium D, so both the products of the dual decay of thorium A are transformed to the same species, thorium C. In column 7 of Table 57 are shown the quantities of the individual disintegration products present at equilibrium in a metric ton of thorium*. The last column is discussed on p. 530.

One of the disintegration products of thorium, mesothorium (the mixture of $MsTh_1$ and $MsTh_2$ in radioactive equilibrium) has practical applications. It is prepared from monazite sand, as is thorium. Since α-emitting disintegration products (RdTh, etc) are rapidly formed from it, it can be used for many purposes in place of radium. It has the disadvantage, however, that its activity diminishes to an appreciable extent after only a few years. Because it is short-lived, mesothorium is relatively much more active than radium, and for a given level of activity, fresh mesothorium preparations are much cheaper than radium preparations. Mesothorium is therefore used in radioactive self-luminous paints, and also to some extent for medical irradiation purposes.

It may be seen from the Table that dual disintegration processes occur at several places in the thorium series. The disintegration is said to undergo *branching*. The side branches shown in Table 57 all later reunite in the main series. If the same decay product is formed by way of alternative intermediates, it follows that the summ of the several disintegration energies must be the same for both paths. Thus the net energy liberated must be the same, whether thorium C changes into thorium D by way of thorium C' (i.e., as a result of a β-decay followed by an α-decay), or whether the thorium D is formed by way of thorium C'' (i.e. by an α-decay process followed by a β-decay). It can be seen from the data listed in

* These values are obtained from the half-lives listed in column 6, together with the ratios of the atomic weights to that of thorium. Due allowance has to be made for branching ratios in the case of the decay products of thorium A and thorium C.

TABLE 57

THORIUM DISINTEGRATION SERIES

Element	Symbol	Atomic number	Atomic weight	Range of α-particle*	Half-life	Amt. in radioactive equilibrium with 1000 kg of thorium	Isotopes
Thorium	Th	90	**232.05**	2.72 cm	$1.39 \cdot 10^{10}$ yrs	1000 kg	UX$_1$, UY, Io, RdAc, RdTh
α↓ Mesothorium 1	MsTh$_1$	88	228	—	6.7 yrs	$4.7 \cdot 10^{-4}$ g	Ra, AcX, ThX
β↓ Mesothorium 2	MsTh$_2$	89	228	—	6.13 hrs	$4.94 \cdot 10^{-8}$ g	Ac
β↓γ Radiothorium	RdTh	90	228	3.95 cm	1.90 yrs	$1.34 \cdot 10^{-4}$ g	UX$_1$, UY, Io, RdAc, Th
α↓γ Thorium X	ThX	88	224	4.30 cm	3.64 days	$6.92 \cdot 10^{-7}$ g	Ra, AcX, MsTh$_1$
α↓ Thorium emanation	ThEm	86	220	5.02 cm	54.5 sec	$1.18 \cdot 10^{-10}$ g	RaEm, AcEm
α↓ Thorium A ⌐ 0.014% Thorium- → β astatine	ThA	84	216	5.66 cm	0.158 sec	$3.35 \cdot 10^{-13}$ g	RaA, AcA, RaC', AcC', ThC', RaF
99.986% α↓	ThAt	85	216	6.84 cm	ca. 10^{-3} sec	ca. 10^{-19} g	RaAt, AcAt
Thorium B β↓γ	ThB	82	212	—	10.6 hrs	$7.94 \cdot 10^{-8}$ g	RaB, AcB, RaD, AcD, ThD, RaG, Pb
Thorium C β↓γ	ThC	83	212	4.79 cm	60.5 min	$7.56 \cdot 10^{-9}$ g	RaC, AcC, RaE, Bi
65% α↓ 35% α↓ Thorium C'	ThC'	84	212	8.61 cm	$3 \cdot 10^{-7}$ sec	$4 \cdot 10^{-19}$ g	RaC' AcC', RaA, AcA, ThA, RaF(Po)
α↓ Thorium C" β, γ	ThC"	81	208	—	3.1 min	$1.3 \cdot 10^{-10}$ g	RaC", AcC", Tl
β, γ Thorium D (thorium lead)	ThD	82	**208.0**	—	stable	∞	RaD, AcD, RaB, AcB, ThB, RaG

* In air at 15° and 760 mm.

Table 58 that this is, in fact, the case. The thorium D nucleus formed by β-decay of thorium C″ is initially formed in an excited state, from which it passes to the ground state with the emission of γ-radiation*. This γ-radiation must, of course, be allowed for in drawing up the total energy balance. It is only possible to obtain agreement for the total energy of disintegration by the two paths if, in each case, the energy of the β-decay step is taken as being the *maximum* kinetic energy of the β-particles, not the mean energy. This is one of the experimental grounds for identifying the maximum energy, not the mean, with the energy of β-decay processes.

TABLE 58

ENERGY BALANCE FOR THE FORMATION OF THORIUM D FROM THORIUM C
VIA THORIUM C′ AND VIA THORIUM C″ RESPECTIVELY

	Energy in mev		Energy in mev
β-Particles from ThC	2.256	a-Particles from ThC	6.084
		Energy of recoil atom	0.117
a-Particles from ThC′	8.776	β-Particles from ThC″	1.792
Energy of recoil atoms	0.166	γ-Rays from excited ThD nucleus	3.204
Total	11.198	Total	11.197

8. Chemical Nature of the Disintegration Products

(a) The Displacement Laws

The chemical nature of the disintegration products is governed by the two displacement laws first framed by Soddy and Fajans.

(1) After any radioactive transformation involving emission of an a-particle, the resulting element belongs to the group standing two places to the left of that containing the parent element.

(2) A transformation involving emission of a β-ray displaces the resulting element one place to the right in the Periodic System.

Thus radium, atomic number 88, belonging to Group II of the Periodic System, ejects an a-particle to form radon (radium emanation), atomic number 86, which belongs to Group 0 (or Main Group VIII). This, by a further a-particle emission, yields an element (atomic number 84) having properties which allocate it to the VIth Group of the Table. This is radium A, which decays by ejecting yet another a-particle, to give the element of atomic number 82, belonging to Group IV (i.e., radium B). This emits a β-particle, so that the resulting element, radium C, must belong to Group V, and must have the atomic number 83 (cf. Table 59). The atomic numbers of the elements of the thorium series, shown in Table 57, follow from the displacement laws in an analogous manner.

These displacement laws are a direct consequence of the disintegration theory,

---

* There are actually at least six excited states in which the ThD nucleus may be produced by the β-decay of thorium C″. In the great majority of disintegrations, however, the product nucleus is left in the state in which it has an energy 3.204 mev above the ground state.

if it is assumed that the α- and β-particles originate in the nucleus of the disintegrating atoms. They have been amply confirmed by many chemical investigations of the elements formed in radioactive change.

(b) Atomic Weights of the Disintegration Products

The disintegration theory predicts not only the place of the disintegration products in the Periodic Table, but also their atomic weights. The emission of an α-particle must decrease the mass of the nucleus, and therefore the atomic weight, by about 4 units, whereas emission of a β-particle results in a negligible change in mass. Thus α-emission produces an element with atomic weight about 4 units lower than that of the parent element, whereas β-emission produces an element with practically the same atomic weight as the parent element (*isobaric* with it).

In brief: when an atom loses an α-particle, it is transformed into a new atomic species with an atomic number (nuclear charge) 2 units lower than the original, and an atomic weight 4 units less; loss of a β-particle raises the atomic number (nuclear charge) by one unit, but does not sensibly affect the atomic weight.

For the loss of mass which, according to the mass-energy equivalence principle, must be associated with the liberation of energy in radioactive disintegration, see p. 567.

(c) Isotopy

All the products of radioactive change are chemical elements, but in assigning them to positions in the Periodic System it is in many cases necessary to place them in positions already allocated to other atomic species, differing in the stability of their atoms, and also, as a rule, in atomic weight. This phenomenon is known as *isotopy*. *Isotopes* are thus elementary substances in the chemical sense, which must be assigned *to the same position* of the Periodic Table (ἐν τῷ ἴσῳ τόπῳ) since their atoms bear the same nuclear charge. The last column of Table 57 shows the nuclear species which are isotopic with the substances listed in column 1. The nuclear charge determines the number and arrangement of the electrons in the extranuclear structure of the atom, and these in turn define the chemical properties. It follows that *isotopic substances are practically identical in chemical properties*. Once they have been mixed together, it is therefore impossible to separate isotopic substances by chemical means.

Hence, from the chemical standpoint, a *mixture of isotopes* must be regarded as a *single chemical element*. However, as will be shown in the next chapter, the statement that mixtures of isotopes are inseparable by chemical processes is true only with reservations. For this reason a chemical element is now no longer to be defined as a substance which *cannot be resolved by any chemical means*, but rather as one of which all the atoms possess the *same nuclear charge* (or atomic number).

(d) Fajans' Precipitation Rule and Paneth's Adsorption Rule

In many cases, the elements produced by radioactive change can be obtained only in the most minute amounts, because of their instability. It is therefore impossible to carry out one of the most important types of chemical reaction—namely precipitation from solutions—because it is impossible to attain solutions of sufficient concentration for the solubility product to be attained, even with a great excess of reagents. Fajans found, however, that the behavior of radioelements conformed with the following rule:

Even when present in unweighably small quantities, a radioelement will be co-precipitated with some other substance if the conditions of its precipitation are such that the radioelement would form a sparingly soluble compound if it were present in weighable quantities.

This phenomenon is to be explained on the basis of Paneth's adsorption rule:

A solid substance strongly adsorbs those radioelements which give rise to insoluble or sparingly soluble compounds with the electronegative component (acid radical) of the adsorbent.

Since they were first stated, many exceptions to these rules have been discovered. According to Hahn, the following modified rules are strictly true.

(*i*) *Precipitation Rule*. However great the dilution, an element will be carried down by a crystalline precipitate, provided that it can be built into the crystal lattice of the precipitate. If it cannot be so incorporated, it will remain in solution, even though its compound with the oppositely charged component of the crystal lattice is insoluble in the solvent concerned.

In considering the applicability of this rule, it must be borne in mind that foreign substances can be incorporated in minimal concentrations in a crystal lattice through the formation of 'anomalous mixed crystals', and by 'inner adsorption' (p. 541).

(*ii*) *Adsorption Rule*. An element is strongly adsorbed on a precipitate, even from very highly dilute solution, if the precipitate bears a surface charge of opposite sign to that borne by the element in question, and provided that the adsorbed compound is very sparingly soluble in the solvent.

A precipitate bears a positive or negative surface charge, according to whether the cation or the anion of the compound is present in slight excess in the solution with which it is in contact. Thus calcium sulfate is obtained in a surface-active form when it is rapidly precipitated by adding alcohol to a solution containing an excess of $SO_4^=$ ions, which confer on it a negative surface charge. Under these conditions, it carries ThB almost quantitatively down from solution by adsorption. If, however, it is precipitated in the same way from a solution containing excess Ca^{++} ions, it carries down but very little ThB. If the calcium sulfate is allowed to crystallize slowly, so that adsorption becomes negligible because of the small surface area, no ThB is carried down at all, even though ThB, as an isotope of lead, would form an insoluble compound under just the same conditions, if it were present in weighable amounts. Lead sulfate cannot be incorporated in the crystal lattice of calcium sulfate, hence ThB is not built in to the structure. Therefore ThB is not precipitated, in accordance with Hahn's precipitation rule, whereas on the basis of Fajans' rule it would be expected that ThB should be carried down even when precipitation and crystallization occur slowly. ThX, which is isotopic with radium, behaves similarly to ThB under the same conditions.

9. The Uranium-Radium Disintegration Series

Table 59 shows the substances resulting successively from the disintegration of uranium, ending with the stable product, lead. As in Table 57, the genetic relationships are denoted by arrows. It may be seen that uranium gives rise to radium, by way of a number of intermediate steps. For this reason, radium always accompanies uranium in Nature.

The rate of transformation of uranium into radium is extremely slow, since the half-life of uranium is about $4.5 \cdot 10^9$ years. Hence the intensity of the radiations emitted by uranium is very small as compared with that from radium, with its much more rapid decay rate (half-life 1590 years). It follows from the ratio of their half-lives that radium cannot accumulate in any quantity in uranium ores. From eqn. (14), p. 526, it may be calculated that only 0.332 g of radium will be present at equilibrium with the uranium I contained in 1000 kg (= 1 metric ton) of uranium.

One of the intermediate products in the transformation of uranium into radium is isotopic with the starting material, and like the first member of the series it has a very long half-life. Ordinary uranium is therefore invariably a mixture of these two isotopes. In radiochemistry, these isotopes are distinguished as uranium I and uranium II, respectively. Natural uranium also contains a third isotope, of mass 235, which is also known as actinouranium since it is the first member of the actinium decay series.

Evidence that naturally occurring uranium consists of three isotopes was first obtained from the study of its radiations. Thus the existence of the two isotopes U I and U II was inferred from the observation that uranium which was quite pure in the chemical sense emitted two distinctly different groups of α-rays, having ranges of 2.6 cm and 3.2 cm, respectively*. By application of the Geiger-Nuttall relation, the ratio of U I to U II can be worked out as approximately 1000 : 1.

Recent determinations of the disintegration constants of U I and U II lead to a more precise value for the atomic ratio $^{238}U : ^{234}U$ in natural uranium, $1 : 5.14 \cdot 10^{-5}$. The composition of uranium in atoms per cent then becomes 99.280% ^{238}U, 0.715% ^{235}U, 0.005% ^{234}U, or in weight per cent, 99.289% ^{238}U, 0.706% ^{235}U and 0.005% ^{234}U. Using the masses of the uranium isotopes listed in Table 59, the isotope ratios lead to a physical atomic weight of 238.108 for natural uranium, and a chemical atomic weight of 238.04. The atomic weight determined by chemical methods is 238.07.

Uranium II is derived from uranium I by an α-particle change followed by two β-particle changes. The two β-emitters formed as intermediates are known as *uranium X_1* and *uranium X_2*. Uranium X_1, an isotope of thorium, can readily be separated chemically from uranium, and it was discovered by Crookes as early as 1900. The existence of uranium X_2 was originally inferred from the displacement law, since the isotopy of uranium II with uranium I necessarily implied that the α-particle transformation of U I must be followed by two, and not one, β-particle changes. The chemical separation of uranium X_2 from its parent uranium X_1 was first achieved by Fajans, in 1913. The chemical behavior of uranium X_2 and the displacement law both indicated that it must be assigned to a position in the Periodic System, as that time unoccupied, belonging to a homologue of tantalum. As a new chemical element, it was given a special name—'brevium', in reference to its short half-life. Since the discovery of the long-lived protactinium, it has been usual to consider the latter as the typical chemical representative of the species with atomic number 91. There is therefore no need for a special name for uranium X_2.

Radioactive disintegration of uranium X_1, by emission of a β-particle, can give rise to another nuclear species, known as uranium Z, instead of to uranium X_2**. Uranium Z was

---

* The range of α-particles from ^{235}U differs very little from that of the particles from U I. The existence of the third isotope was first deduced from that fact that, like radium, actinium is always present in uranium minerals, and the actinium : uranium ratio is constant. At first it appeared possible that U I or U II might undergo a dual or branching decay, and thereby give rise to the actinium disintegration series. This possibility had to be abandoned when the atomic weight of protactinium was found to be an odd number (231), since that implied that the parent member of the whole series must also have an odd atomic weight, whereas U I and U II have even mass numbers.

**This phenomenon, known as *nuclear isomerism* has subsequently been met with in many instances among the artifically produced radioactive elements.

discovered by Hahn in 1921, by applying to uranium salts (which were practically free from protactinium) the same methods of separation as had been used for the extraction of protactinium from pitchblende residues (cf. p. 117). He added a small amount of a tantalum salt to the uranium salt solution, and precipitated the tantalum 'carrier'. Uranium Z differs from uranium X_2 in its much longer half-life (6.7 hrs), and from protactinium both by its much shorter half-life and in that it decays with emission of a β-particle. The activity of uranium Z is very small as compared with that of uranium X_1 with which it is in radio-

TABLE 59
URANIUM-RADIUM DISINTEGRATION SERIES

Element	Symbol	Atomic number	Atomic weight	Range of α-particles in air	Half-life	Amount in radioactive equilibrium with 1000 kg of uranium
Uranium I $\alpha\downarrow$	U I	92	**238.07**	2.63 cm	$4.515 \cdot 10^9$ yrs	992.9 kg
Uranium X_1 $\beta\downarrow\gamma$	UX_1	90	234	—	24.10 days	$1.426 \cdot 10^{-5}$ g
Uranium X_2 ⌐ 0.15% 99.85% | $\downarrow\gamma$ (i.t.) β | Uranium Z	UX_2	91	234	—	1.14 min	$4.68 \cdot 10^{-10}$ g
$\downarrow$ $\beta\downarrow$	UZ	91	234	—	6.7 hrs	$2.5 \cdot 10^{-10}$ g
Uranium II $\alpha\downarrow$	U II	92	234	3.22 cm	$2.32 \cdot 10^5$ yrs	50.1 g
Ionium $\alpha\downarrow\gamma$	Io	90	230	3.13 cm	$8.3 \cdot 10^4$ yrs	17.6 g
Radium $\alpha\downarrow\gamma$	Ra	88	**226.05**	3.30 cm	1590 yrs	0.332 g
Radium emanation $\alpha\downarrow$ (radon)	RaEm	86	222	4.06 cm	3.825 days	$2.147 \cdot 10^{-6}$ g
Radium A ⌐ 99.97% $\alpha\downarrow$ β | 0.03%	RaA	84	218	4.67 cm	3.05 min	$1.16 \cdot 10^{-9}$ g
Radium B |	RaB	82	214	—	26.8 min	$1.01 \cdot 10^{-8}$ g
β|γ ⌐ Radium- $\downarrow$ $\alpha\downarrow$ astatine	RaAt	85	218	5.53 cm	a few sec	ca. $1 \cdot 10^{-14}$ g
Radium C ⌐ 0.04% | 99.96% | $\beta\downarrow$ 0.1% α | β | Emanation	RaC	83	214	4.1 cm	19.7 min	$7.4 \cdot 10^{-9}$ g
$\downarrow\alpha\downarrow$ 218	^{218}Em	86	218	6.1 cm	0.019 sec	ca. $1 \cdot 10^{-18}$ g
Radium C′ $\downarrow$ α	RaC′	84	214	6.95 cm	$1.5 \cdot 10^{-4}$ sec	$9.1 \cdot 10^{-16}$ g
Radium C″ $\beta\downarrow\gamma$	RaC″	81	210	—	1.32 min	$1.95 \cdot 10^{-13}$ g
Radium D ←| $\beta\downarrow\gamma$	RaD	82	210	—	22.3 yrs	$4.32 \cdot 10^{-3}$ g
Radium E ⌐ $\beta\downarrow\gamma$ α | $5 \cdot 10^{-5}$%	RaE	83	210	—	4.95 days	$2.63 \cdot 10^{-6}$ g
Radium F (Polonium) $\downarrow$	RaF (Po)	84	210	3.85 cm	138.8 days	$7.37 \cdot 10^{-5}$ g
α|γ Thallium 206 $\downarrow$ $\beta\downarrow$	^{206}Tl	81	206	—	4.23 min	$7.6 \cdot 10^{-16}$ g
Radium G (Uranium lead)	RaG (^{206}Pb)	82	**206.0**	—	stable	∞

active equilibrium (only about $^1/_{1000}$ as strong), and it must therefore be very rigorously purified from uranium X_1 if it is to be identified by its decay curve.

Uranium Z is formed from uranium X_1 by a β-decay process, as also is uranium X_2. The β-decay of uranium X_2 leads to uranium II, and the β-decay of uranium Z must also give rise to uranium II, since uranium Z (like uranium X_2) comes from uranium X_1 itself, and not from one of its isotopes. Uranium X_2 and uranium Z thus present us with a peculiar situation, in that two atomic species are not only chemically identical and have the same atomic weight (i.e., are both isotopic and isobaric), but they emit the same qualitative kind of radiation (β-rays, but of different energies) and furnish the same product.

This phenomenon is known as *nuclear isomerism*. It has since been found to occur frequently among the artificially produced radioactive species.

It was formerly considered that the formation of uranium X_2 and uranium Z from uranium X_1 represented a dual decay scheme for the latter. It has been found, however, that the actual course is rather different. Uranium Z is not formed directly from uranium X_1, but from a portion of the uranium X_2, which is the primary decay product. The greater part (99.85%) of the uranium X_2 disintegrates directly by a β-decay process to uranium II. A small fraction (0.15%) of the uranium X_2, however, undergoes a different change. This involves emission of γ-radiation, and is known as an *isomeric transition* (i.t.) (cf. p. 578). Uranium Z is formed by an isomeric transition of uranium X_2–i.e., by a process of pure γ-ray emission, without α- or β-radiation. Formation of nuclear isomers by similar processes has been observed in numerous instances among artificially produced radioelements. Nuclear isomers can also be formed by other processes, besides isomeric transitions. These will be considered in Chap. 13.

'Branching', or dual decay schemes, are found at several places in the uranium-radium disintegration series. The dual decay of radium A and radium C corresponds to the decay of thorium A and thorium C, respectively. There is also a third branching which has no counterpart in the thorium series. Radium-astatine, one of the products of the dual decay of radium A, also has two alternative modes of disintegration: to form either radium C or a gaseous disintegration product, an *emanation*. The latter has the same atomic number as radium emanation, but a different atomic weight and a very different half-life. Thus there are two emanations within the radium decay series. The emanation isotope formed from radium-astatine is produced in such extraordinarily small amounts, however, that it can not be detected without the use of special techniques.

A fourth branching occurs at radium E. This disintegrates principally by β-decay forming radium F (polonium). A small fraction, however, (0.00005%) disintegrates by α-decay, giving rise to a short-lived thallium isotope (^{206}Tl). This disintegrates by β-decay, radium F by α-decay. In both cases there is formed radium G, the end product of the uranium-radium disintegration series.

Ionium, the decay product of uranium II and the parent of radium, was first discovered by Boltwood, in 1907, in uranium minerals. Since ionium is isotopic with thorium and cannot be separated from it, and since thorium minerals almost always contain a little uranium, together with its decay products, thorium preparations almost invariably contain radioactively significant amounts of ionium. The half-life of ionium is so long that its activity does not diminish detectably in any times available for observation, whereas all the other thorium isotopes are very short-lived.

The most important member of the uranium disintegration series is *radium*, the chemical properties of which have been discussed in Vol. I. The decay of radium follows a scheme very similar to that of thorium. The first disintegration product is a gas, *radium emanation* or *radon*, which is transformed by way of radium A, radium B, and radium C (which has a dual decay) into radium D, in the same way as thorium emanation is transformed into thorium D. Whereas the latter is the stable end-product of the decay series, radium D undergoes further decay. It is converted via radium E and radium F (identical with the polonium which can be extracted from uranium ores) into radium G, the stable end-product of the uranium-radium series.

From the Soddy-Fajans displacement rules, radium G should have atomic number 82 and atomic weight 206. Hence it should be chemically identical with lead, although it has a different atomic weight. The same applies to thorium D, which should have the atomic number 82 and atomic weight 208 according to the displacement laws. Convincing evidence in support of the displacement rules—and hence for the theory of atomic disintegration upon which they were based—was forthcoming when it was proved experimentally that the lead present in uranium ores (uranium lead) had the atomic weight 206, and that in thorium ores (thorium lead) had the atomic weight 208, as compared with ordinary lead, with the atomic weight 207.2.

Uranium lead is formed from uranium (atomic weight 238.07) by way of 8 α-particle transformations in all. According to the displacement rules, the mass lost by each atom should be eight times the mass of a helium atom, giving a value of $238.07 - (8 \times 4.003) = 206.05$ for the atomic weight of the end-product*. In the same way, thorium lead should have the atomic weight $232.05 - (6 \times 4.003) = 208.03*$. Richards and Hönigschmid (1914 and later) quite independently found from chemical determinations of the atomic weight that lead from very pure uranium minerals and from the purest possible thorium minerals had atomic weights close to these values. Thus Hönigschmid found the value 206.05 for lead from the purest East African pitchblende, and 206.06 for lead from Norwegian pitchblende, as compared with 207.90 for lead from a Norwegian thorite, containing 30.1% thorium but only 0.45% uranium, and 207.97 for a lead from a completely uranium-free thorium mineral.

As has already been indicated, isotopic species of atoms are symbolized by writing the corresponding mass numbers against the chemical symbol**. Thus uranium lead (a lead isotope of mass 206) is represented by the symbol ^{206}Pb, and thorium lead by ^{208}Pb. As was first found by Aston (1927), using the mass spectrograph, ordinary lead of atomic weight 207.21 is a mixture of isotopes, with the mass numbers 206, 207 and 208.

Aston found the ratio $^{206}Pb : ^{207}Pb : ^{208}Pb = 100 : 11 : 4$ in a uranium lead. The mineral from which the lead had been extracted contained a little thorium, and therefore the ^{208}Pb isotope was not completely absent. A significant feature, however, is that the ^{207}Pb isotope is considerably more abundant than the ^{208}Pb isotope, whereas the reverse is true of ordinary lead. It follows that the ^{207}Pb content cannot be attributed to contamination with ordinary lead, and it must be concluded that ^{207}Pb is formed from uranium by radioactive transformation, as is ^{206}Pb. It was an obvious assumption that ^{207}Pb must have been formed from a uranium isotope ^{235}U, by way of the actinium series. The existence of this isotope, and its genetic relation to the actinium series have subsequently been established by other means. In Tables 57, 59 and 60, those atomic weights which have been determined by chemical means are printed in heavy type. Atomic weights given in ordinary type are those deduced from the displacement law.

It has been found that a fourth isotope, ^{204}Pb, is present in small amounts in ordinary lead, in addition to the three isotopes mentioned above. According to

---

* Allowance should also be made for a small change in mass as a result of the enormous energy loss associated with the radiation (cf. p. 567). This has not been included in the above considerations, and amounts to 0.05 and 0.04 mass units for the two cases.

** At the 17th Conference of the International Union of Pure and Applied Chemistry, Stockholm 1953, the Committee for Nomenclature established the following rules for inorganic chemical nomenclature:

left upper index	–	mass number
left lower index	–	atomic number
right lower index	–	number of atoms
right upper index	–	ionic charge (indicated by n+ rather than by +n).

It has been the convention in British literature to write the mass number *before* the chemical symbol (e.g., ^{206}Pb), whereas in American literature the mass number appears *after* the symbol (e.g., Pb^{206}). In this book, the Stockholm rules have been followed throughout.

Collins [*Phys. Rev.*, 88 (1952) 1275], the ratio of the four isotopes in ordinary lead, by atoms, is $^{204}Pb : ^{206}Pb : ^{207}Pb : ^{208}Pb = 100 : 1845 : 1561 : 3840$.

The isotopic composition of lead in atoms per cent, given in Table 62, is based on these figures. The relative proportions of the lead isotopes are found to display certain variations, however, in lead ores which are free from thorium and uranium, depending on their source. Nier [*Phys. Rev.*, 60 (1941) 112] has shown that the probable explanation of this is that the primeval rock, from which the lead minerals have been differentiated in the course of the earth's evolution, contained thorium and uranium. The later the stage at which differentiation of the primeval mass occurred, the more 'radiogenic' lead was admixed with the original lead. This explanation is supported by the observation that in the ores investigated by Nier, the isotopes ^{206}Pb and ^{208}Pb were always enriched in practically the same ratio, as compared with the non-radiogenic ^{204}Pb. Furthermore, this ratio was about what would be expected if the Th : U ratio in the primeval rock corresponded to the average value now found for this ratio in the earth's crust.

10. The Actinium Disintegration Series

The members of the actinium disintegration series are shown in Table 60. The first member of the series is ^{235}U (actinouranium), and although this is present in natural uranium only to the extent of 0.71%, the rates of conversion of natural uranium to actinium and to radium, respectively, are in the ratio 4.6 : 95.4, since ^{235}U has an appreciably shorter half-life than uranium I.

^{235}U emits α-rays, and is converted into *uranium Y*, a fairly short-lived isotope of thorium. Uranium Y emits β-rays, to form *protactinium*, which was discussed in Chap. 4. As this is a long-lived element, it is relatively highly enriched in uranium minerals, as is radium. *Actinium*, on the other hand, has a relatively short half-life, and its concentration in uranium ores is accordingly very minute. Actinium undergoes a dual decay. *Radioactinium*, formed by β-decay, occupies the same place in the actinium series as does radiothorium in the thorium decay series, or ionium in the uranium-radium disintegration series. About 1% of the actinium atoms undergo the alternative α-particle decay, to form *francium*, the homologue of cesium (cf. p. 593).

Hyde and Ghiorso (1953) have shown that francium also undergoes dual decay. The greater proportion changes into actinium X, by emission of β-radiation. At the same time, some emits α-rays and is thereby converted into astatine. There are thus two astatine isotopes in the actinium decay series, distinguished from one another in Table 60 as AcAtI and AcAtII. Actinium astatine II is formed by the β-decay of actinium A, together with a very much larger amount of actinium B produced by α-emission. It occupies the same place in the actinium decay series as do the astatine isotopes in the thorium and uranium-radium decay series, and it was discovered roughly at the same time as the latter (by Karlik and Bernert, 1943). Except for the differences occasioned by the existence of francium, the decay processes in the actinium series, from radioactinium onwards, correspond exactly with those in the thorium series. They are also essentially similar to those of the uranium-radium series. Like the thorium and uranium-radium series, the actinium series ends with a lead isotope, actinium D.

TABLE 60

ACTINIUM DISINTEGRATION SERIES

Element	Symbol	Atomic number	Atomic weight	Range of α-particle in air	Half-life	Amount in radioactive equilibrium with Ac in 1000 kg of uranium
Actinouranium	^{235}U	92	235	2.7 cm	$7.07 \cdot 10^8$ yrs	7.06 kg
α ↓						
Uranium Y	UY	90	231	—	24.6 hrs	$2.75 \cdot 10^{-8}$ g
β ↓						
Protactinium	Pa	91	**231**	3.62 cm	$3.2 \cdot 10^4$ yrs	0.314 g
α ↓						
Actinium	Ac	89	227	—	22 yrs	$2.1 \cdot 10^{-4}$ g
98.8% β↓ α 1.2%						
Radioactinium	RdAc	90	227	4.67 cm	18.6 days	$4.85 \cdot 10^{-7}$ g
α |γ						
Francium	Fr	87	223	—	21 min	$4.5 \cdot 10^{-12}$ g
β ↓ 99.996%						
Actinium X	AcX	88	223	4.33 cm	11.3 days	$2.93 \cdot 10^{-7}$ g
α α ↓ 0.004%						
Actinium astatine I	AcAtI	85	219	4.97 cm	0.9 min	$8 \cdot 10^{-18}$ g
β ↓ 3%						
Actinium emanation α | 97%	AcEm	86	219	5.75 cm	3.92 sec	$1.16 \cdot 10^{-12}$ g
α						
Bismuth 215	^{215}Bi	83	215	—	8 min	$6 \cdot 10^{-17}$ g
β ↓						
Actinium A	AcA	84	215	6.49 cm	0.00183 sec	$5.3 \cdot 10^{-16}$ g
α↓ 99.9995% β 0.0005%						
Actinium B	AcB	82	211	—	36.1 min	$6.15 \cdot 10^{-10}$ g
β |γ Actinium astatine II	AcAtII	85	215	8.0 cm	ca. 10^{-4} sec	ca. 10^{-22} g
α ↓						
Actinium C	AcC	83	211	5.46 cm	2.16 min	$3.68 \cdot 10^{-11}$ g
β↓ 0.32% α 99.68%						
Actinium C′	AcC′	84	211	6.59 cm	ca. 0.005 sec	ca. $5 \cdot 10^{-18}$ g
α | Actinium C″	AcC″	81	207	—	4.76 min	$7.93 \cdot 10^{-11}$ g
β ↓γ						
Actinium D (actinium lead)	AcD (^{207}Pb)	82	207	—	stable	∞

Although uranium lead always contains the isotope ^{207}Pb formed from actinouranium, as well as the isotope ^{206}Pb formed from uranium I, the former is present in such relatively small amount that it exerts but little influence on the atomic weight of the uranium lead. With the use of isotopic ratios given on p. 536 for the ^{206}Pb and ^{207}Pb in uranium lead, the final atomic weight works out at 0.09 mass units higher than was calculated without allowing for the actinium lead content. If allowance is made at the same time for the loss of mass in the form of the energy radiated (0.05 mass units), the atomic weight of uranium lead calculated from the disintegration theory is 206.09. This is in good agreement with Hönigschmid's experimentally determined atomic weight of 206.05 or 206.06.

A fourth decay series is found in Nature, in addition to the thorium, uranium-radium and actinium series. This is the *neptunium* series. The occurrence of neptunium and its decay products in Nature—namely in uranium minerals—was detected by Peppard (1952). They are only present in infinitesimal quantities, the maximum neptunium content of pitchblende being $1.8 \cdot 10^{-10}$ % of its uranium content. For this reason, the natural occurrence of this element and its decay products is not of practical significance. The radioactive properties of the members of the neptunium series have therefore not been studied from the naturally occurring material, but exclusively using radioactive materials of artificial origin. The neptunium series differs from the other three decay series in that the stable end member is not an isotope of lead, but of *bismuth*. The neptunium decay series is further discussed at the end of Chap. 14. It will there be found that the other three decay series have also been extended by a considerable number of artificially prepared nuclear species, which are genetically related to the natural radio-elements.

11. Place of the Radioactive Elements in the Periodic System

Although radioactive transformations lead to the formation of a large number of different substances, all of which are distinctly different in their radioactive properties (e.g., in their half-lives and in the nature of their decay products), these substances are actually distributed among only a few places in the Periodic System, as a result of the phenomenon of isotopy. All the radioactive elements of the three disintegration series discussed above, including the first and last members of each series, have atomic numbers ranging from 81 to 92. At least one radioactive species, and in most cases several (up to as many as seven) has to be assigned to each of the positions 81 to 92. Positions 81 and 82 are occupied by stable elements (thallium and lead), as well as the radioactive species. *All the nuclear species corresponding to atomic numbers of 83 and above are unstable*.

Only seven of the elements of atomic number lower than 81 have been found to be radioactive. These are lutetium (at. no. 71), samarium (at. no. 62), neodymium (at. no. 60), lanthanum (at. no. 57), indium (at. no. 49), rubidium (at. no. 37) and potassium (at. no. 19). These seven elements are all very weakly radioactive as compared with the elements of high atomic number. Thorium, which has the longest half-life of any of the elements heading the disintegration series, emits about 100 times as many rays per g atom in unit time as does samarium, and about 1000 times as many as potassium.

The radioactivity of lutetium, samarium, neodymium, lanthanum, indium, rubidium, and potassium arises from the presence of unstable isotopes in the naturally occurring elements. These unstable isotopes (with the exception of ^{40}K) all have smaller decay rates than that of thorium. *Lutetium* contains the unstable isotope ^{176}Lu, a β-emitter, with a half-life of $2.4 \cdot 10^{10}$ years, present to the extent of 2.5%. *Samarium* consists, to the extent of 14.62%, of the unstable ^{147}Sm. This emits α-rays, with a half-life of $1.4 \cdot 10^{11}$ years. *Neodymium* contains 5.69% of the isotope ^{150}Nd, a β-emitter with half-life $5 \cdot 10^{10}$ years. *Lanthanum* and *indium* are very weakly radioactive indeed. The former contains 0.089% of ^{138}La, which decays by an electron capture process with the half-life $2 \cdot 10^{11}$ years. Indium consists to the extent of 95.77% of ^{115}In, which emits β-rays and has a half-life of $6 \cdot 10^{14}$ years. *Rubidium* contains 27.2% of the β-emitting ^{87}Rb (half-life $6.3 \cdot 10^{10}$ years). ^{40}K, present to the extent

---

* Bismuth (at. no. 83), which occurs in Nature as a pure element, is actually radioactive, although extraordinarily weakly so (half-life $2.5 \cdot 10^{17}$ years). The statement that there is a stable isotope of polonium (Holubei and Cauchois, 1940) must be regarded as doubtful in the extreme.

of 0.011% in *potassium*, has a half-life of $1.18 \cdot 10^9$ years. It is chiefly a β-emitter, but undergoes some K-electron capture also (cf. Vol. I, p. 159 *et seq.*).

12. Chemistry of the Radioelements [7]

The whole group of atomic species which have the same atomic number, and must therefore be assigned to the same position in the Periodic System, although they differ in stability and atomic weight, is sometimes referred to as a 'pleiad'. It is named after the member with the longest half-life. Thus the natural radioelements contain the pleiads of thallium, lead, bismuth, polonium, radon, radium, actinium, thorium, protactinium, and uranium. The last column of Table 57 indicates the elements which belong to the individual pleiads.

The elements after which the various pleiads are named have already been discussed. The chemical properties of the other members of each pleiad are thereby specified, since all the members of each pleiad are chemically identical. The differences which are manifested in *radioactive* properties have no effect on the *chemical* properties. In chemical investigations of the products of radioactive change it is, however, necessary to allow for the fact that they are usually present in immeasurably low concentrations. In order to precipitate them from solution, it is essential to add some of the inactive isotopes as carriers, in order to obtain visible and filtrable precipitates. If no inactive isotopes are available, it is possible to use as carrier some other inactive element which conforms to the criteria of the Fajans or Hahn precipitation rules.

Radioactive substances are *detected* by means of their radiation, with the use of some form of detector which counts the number of α- or β-particles or γ-ray quanta emitted from a source—e.g., a Geiger-Müller counter for β-particles, or an ionization chamber for γ-particles. Substances are commonly *identified* by determination of their decay constants, or half-lives; with β-emitting elements the absorption curve and maximum energy of the β-rays can also be determined. α-emitting elements of long life are characterized by determining the energy of the radiations in a counter which determines the total ionization along the track of each particle. The *quantity* of a radioelement present is found from the counting rate in a counter of standardized geometry, whereby the rate of occurrence of disintegrations is measured. [10] The detection and determination of elements by their radioactive radiations is by far the most sensitive method available. With elements of moderately short half-life, 10^4 to 10^9 atoms can be identified, determined and (with suitable carriers) investigated chemically, whereas the ordinary methods of microchemistry (e.g., microgram scale operation) require a sample of 10^{15} to 10^{16} atoms*.

---

* The statistical nature of the process of radioactive change sets a lower limit to the usefully measurable rate of disintegration. In a large assembly of atoms the rate of disintegration is sensibly uniform, but with a small assembly of atoms – i.e., a relatively low disintegration rate – the number of events in unit time shows random fluctuations. The statistical error attached to a count of $\mathcal{N}$ events in t seconds is $\mathcal{N}^{-1/2}$. Hence to obtain 1% accuracy in determining the amount of any radioelement from the counting rate, it is necessary to continue counting until a minimum of 10,000 disintegrations has been recorded. Furthermore, every counting apparatus has a 'natural background' of counts, due to cosmic rays, radioactive contamination of the materials of construction, and other effects. If the rate of disintegration of a sample is very low it may therefore become difficult to obtain a counting rate which is significantly different from the background, when due allowance is made for statistical errors.

When once mixed together, it is impossible to effect a chemical separation between the members of a pleiad, since they are chemically identical. The individual isotopes can, however, be isolated by preparing them by the radioactive decay of a pure starting material. Thus pure radium emanation, free from thorium or actinium emanations, is obtained from pure radium compounds. Similarly, radium emanation furnishes the subsequent decay products of the radium decay series, free from the isotopes of the thorium and actinium series.

Decay products which are not isotopic with one another can be separated either by chemical reactions or by using other properties which lend themselves to separation processes—e.g., their differences in volatility when heated. It is often possible to obtain a decay product in the pure state by taking advantage of differences in rates of formation and decay.

Thus radium A is obtained almost pure when a negatively charged wire is exposed for a few seconds to the action of radium emanation. If, however, the wire is exposed to the emanation for several hours, the radium A deposited on the wire decays away almost completely within 20–30 minutes of removal of the wire from the gas, and the residual activity is due to radium B and radium C. The latter is left in a state of purity if the wire is heated to a dull red heat for a short time. Above 600° radium B volatilizes rapidly, whereas radium C begins to vaporize only at 1100°. Radium B can most conveniently be obtained pure by placing a negatively charged plate close to a second plate, on which is a deposit of radium A. In its disintegration, radium A produces recoil rays, consisting of the atoms of radium B, and these are caught by the negatively charged plate. These examples afford some indication of the sort of methods used for isolating radioactive decay products in the state of purity which is essential if they are to be used in certain types of chemical investigation (see below). For more detailed consideration of the methods of radiochemical separation, reference should be made to the monographs cited at the end of this chapter.

13. Applications of Radioactive Methods in Chemistry

The intense radiations emitted by short-lived radioactive substances make it possible to detect very much smaller quantities than is possible by any chemical reactions. It is therefore possible by using radioactive substances to carry out investigations which would not be practicable by any other methods, on account of their inadequate sensitivity [8, 9].

(a) Radioactive Substances as Indicators

Radioactive indicator or *tracer* methods are based on the isotopy of radioactive atoms with stable elements. The radioactive isotopes may be employed in place of the stable species, in studying the behavior of the latter. Alternatively, some of the radioactive isotope is added to the inactive element. It is then inseparable in all subsequent processes, and its radiations enable even unweighably small quantities of the inactive element to be determined.

As an example of this, Paneth and Hevesy determined the solubility of lead chromate and lead sulfide in water. They precipitated the chromate (or sulfide) from a solution containing a known weight of lead chloride, together with an electroscopically measured quantity of radium D. The precipitates were shaken with water until saturation was reached, and the activity of the residue obtained by evaporating the saturated solution was determined. From this was deduced the quantity of lead which had gone into solution as the chromate or sulfide, since the ratio of total lead to radium D (referred to the same time*) must be the same as in the initial solution.

* In such determinations as this, with relatively short-lived radioelements, it is obviously necessary to allow for the progressive decrease in the quantity of the radioelement, as calculated from its disintegration constant.

In other instances, radioactive tracers have been used to adduce evidence for the existence of compounds which could not be prepared in sufficient amounts for identification by other methods, until the optimum conditions for their preparation were known. By the use of radioactive isotopes it was not only possible to prove the existence of such compounds, but also to work out the most favorable conditions for their preparation. With this knowledge, it became possible to obtain the inactive compounds in detectable quantities. This was the procedure by which Paneth succeeded in isolating the hydrides of bismuth and lead (cf. Vol. I, pp. 682, 557).

Another interesting application of radioelements as indicators is the study of *exchange reactions* of substances which are homogeneous in the chemical sense. Thus Hevesy (1920) showed that the lead atoms in solid lead, or the lead ions in solid lead salts are not immutably fixed in their positions, but change places with one another more or less frequently, depending upon the temperature. This conclusion follows from the fact that radium D, an isotope of lead, diffuses from the surface of metallic lead into the interior. In the same way, radium D diffuses in the form of an ion into the interior of a crystal of lead chloride.

The possibility of using radioactive substances as indicators was originally very limited, since there were only three stable elements which were isotopic with radioactive elements of the natural decay series. With the discovery of *artificial radioactivity* (p. 574 *et seq.*), the number of radioactive atomic species has been very greatly increased. Radioactive isotopes of suitable half-life are now known for a great number of the stable elements, and the scope of radioactive tracer methods has been greatly extended accordingly.

(b) Behavior of Very Minute Quantities of Matter

The ease with which radioactive substances can be detected makes it possible to use them in studies of the general laws governing the *behavior of substances at extreme dilutions*. For example, such investigations lead to an understanding of the co-precipitation of foreign substances when a precipitate is thrown out of solution. Thus it has been shown that this co-precipitation may be due to the inclusion of the liquid, to the formation of mixed crystals, to adsorption on the surface of the precipitate after its formation, and also to several other phenomena. According to Hahn (1934), in cases where the formation of mixed crystals could not be expected from the laws governing the behavior of matter in larger quantity, it may also be due to the formation of so-called '*anomalous mixed crystals*'. It is also possible for '*inner adsorption*' to occur—i.e., the inclusion of foreign substances within the crystal, in that they are deposited from solution under the action of forces closely related to those which bring about adsorption on surfaces.

(c) Emanation Methods

The extent of the surface area of substances, the changes which the surface area undergoes on heating, and the 'loosening' of the crystal lattice at higher temperatures can all be studied by so-called *emanation methods*. This is based on the homogeneous incorporation, within the substance concerned, of some radioelement which decays to an emanation (e.g., radium, radiothorium which forms thorium X and then emanation, or radioactinium). The ratio of the quantity of emanation which escapes from the solid in a given time, to the quantity of emanation generated within it in the same time is known as the *emanating power*. The more finely divided the solid, the greater is its emanating power. Practically no emanation can escape at ordinary temperature from a compact ionic crystal, whereas a substance of spongy structure, or a highly defective crystal (cf. p. 743 *et seq.*) possesses a high emanating power. It has been possible to show by the emanation method, for example, that that the temperature at which two solids will react with one another is the temperature at which the concentration of lattice defects in one or other of the crystals begins to increase rapidly (cf. Chap. 19).

This 'loosening up' temperature, for a great many substances, is about one-half of the absolute melting point, as Tammann had already earlier concluded from observations on the reactivity of solids and the sintering of powders. Hahn (1934) used the emanation method to study aging phenomena in preparations of iron(III) oxide. He was able to show that iron(III) oxide precipitated from aqueous solution undergoes a rapid spontaneous crystallization at about 400°. Above 800° there is a 'loosening' of the crystal lattice, which is accompanied by the expulsion of the last traces of water from the preparation. Development of the perfectly ordered crystal lattice of hematite is completed by about 1000° (cf. p. 273 and Vol. I, p. 306).

The phenomena observed in the emanation method can be understood as follows. In the expulsion of an α-particle from a disintegrating atom of radium or its isotopes, the newly formed atom of emanation recoils with considerable velocity, and it can be shown that in a typical ionic solid—e.g., a metallic oxide—the recoil path is 200–300 Å in length. The emanation atom may escape from the solid by one of two processes: It will be ejected directly, if its recoil path crosses the surface of the solid particle; if the recoil path terminates within the solid particle, the emanation atom may still escape if it diffuses to the surface of the particle before itself undergoing disintegration. The total emanating power is thus made up of two portions—a recoil fraction, and a fraction due to diffusion. The former depends only on the average depth of the parent radium (or ThX or AcX) atoms beneath the surface of the particle. It is thus dependent upon particle size, but not upon the temperature. The second factor depends upon the average length of the diffusion path to the surface (and therefore upon particle size) and also upon the diffusion coefficient of emanation atoms within the solid. The latter, like all diffusion processes in solids is an activated process, and strongly temperature dependent. Hence the diffusion emanating power increases rapidly in comparison with the recoil emanating power at higher temperatures. If the diffusion coefficient is great enough, every emanation atom may be able to diffuse out before it decays; Gregory (1951) has shown that this is the case for the very open-structured crystals of barium stearate. Any physical or chemical change that alters either particle size or the diffusion coefficient is revealed by the consequent change in emanating power. Thus grain growth and recrystallization increase the necessary recoil path and diffusion path, and greatly reduce the emanating power. The solid product of a chemical reaction undergone by a solid—e.g., the calcium oxide formed by decomposition of calcium carbonate—is frequently formed initially in a state of extreme subdivision. The occurrence of reaction is therefore attended with an abrupt increase in emanating power; as soon as recrystallization of the reaction product begins, the emanating power diminishes. If no grain growth, recrystallization, or chemical reaction occurs, the emanating power of a solid is found to increase exponentially with temperature, as is typical of the process of activated diffusion*.

* See Cook, *Z. physik. Chem.*, B 42 (1939) 221; Gregory, *Trans. Faraday Soc.*, 48 (1952) 643.

References

1 St. Meyer and E. R. von Schweidler, *Radioaktivität*, 2nd Ed., Leipzig 1927, 722 pp.
2 G. von Hevesy and F. Paneth, *Lehrbuch der Radioaktivität*, 2nd Ed., Leipzig 1931, 287 pp.
3 K. W. F. Kohlrausch, *Radioaktivität* [Vol. 15 of the *Handbuch der Experimentalphysik*, edited by W. Wien and F. Harms], Leipzig 1928, 885 pp.
4 M. Curie, *Radioactivité*, Paris 1935, 564 pp.
5 J. Chadwick, *Radioactivity and Radioactive Substances*, 3rd Ed., New York 1948, 128 pp.
6 E. Rutherford, J. Chadwick and A. C. D. Ellis, *Radiations from Radioactive Substances*, Cambridge 1951, 599 pp.
7 F. Henrich, *Chemie und chemische Technologie radioaktiver Stoffe*, Berlin 1918, 351 pp.
8 O. Hahn, *Applied Radiochemistry*, London 1936, 278 pp.
9 F. Paneth, *Radio-Elements as Indicators, and other selected topics in inorganic chemistry*, New York 1928, 164 pp.
10 H. Geiger and W. Makower, *Messmethoden auf dem Gebiete der Radioaktivität*. Braunschweig 1920, 156 pp.

CHAPTER 12

ISOTOPY OF THE STABLE ELEMENTS

1. Detection of Isotopy

The phenomenon of isotopy is observed among the stable elements, as well as among the radioactive elements. Whereas radioactive elements which are isotopic with one another differ in their radioactive properties, as well as in their atomic weights, the isotopes of stable elements differ only in their atomic weights. The atomic weights are proportional to the masses of the isolated atoms. Various methods can be used to investigate whether an element consists of atoms of various different masses, or whether it is composed of atoms all having the same mass. The most important methods are the so-called *mass spectroscopy* and the analysis of *band spectra*.

(a) Mass Spectrography

Mass spectrography[1] is based on the fact that the canal rays are deflected by electrical and magnetic fields. In both cases, the deflection is proportional to the ratio e/m—i.e., to the ratio between the charge and the mass of the particles composing the canal rays or positive rays—but depends in a different manner upon the velocity v. Thus the deflection in a magnetic field is inversely proportional to the velocity of the particles, whereas the deflection by an electrostatic field is inversely proportional to the square of the velocity. The deflections produced by the magnetic and electrostatic fields also take place in different directions; the electrical deflection is in the direction of the field, and the magnetic deflection is perpendicular to the fields (just as a wire carrying an electric current is displaced sideways in a magnetic field). Consider a beam of canal rays, consisting solely of particles with the same charge to mass ratio e/m, but of various velocities. We allow the beam to traverse a homogeneous electrostatic field, at right angles to the trajectory of the rays—e.g., the field between a pair of parallel, oppositely charged plates. The rays are thereby deviated and separated from one another, in the direction of the lines of force. A photographic plate, set up perpendicular to the direction of the rays, will be blackened by the incident beam; instead of a single black spot, it will show a continuous black line if the canal rays contain particles with velocities distributed over a continuous range. Experiment shows that this is usually the case. If the electric field is in a horizontal direction, a horizontal line is obtained on the photographic plate. If the electrical field is now switched off, and replaced by a magnetic field in the same direction, a continuous black trace is once more obtained upon the photographic plate, but now at right angles to the original trace. If both fields be applied together, so that every particle strikes the plate after traversing both the electrical and the magnetic field, a transverse and a vertical deviation are imposed simultaneously. The trace upon the photographic plate is then a curve, such as is shown in Fig. 68. This curve is one limb of a parabola, since the coordinates of every point are given by

$$x = a \cdot \frac{e}{m} \cdot \frac{1}{v^2} \qquad y = b \cdot \frac{e}{m} \cdot \frac{1}{v}$$

The constants a and b are here determined by the experimental conditions. Combining the two expressions,

$$y^2 = \frac{b^2 e}{a\ m}\ x \quad \text{or} \quad y^2 = 2px \tag{1}$$

where the constant quantity $\dfrac{b^2 e}{2a\ m} = p.$

Equation (1) is familiar from coordinate geometry as the equation of a parabola. If a beam of positive rays contains particles with various values of e/m, the constant p assumes various values, and several parabolas are obtained, each corresponding to a single value for the ratio e/m of the particles in the positive ray beam. Appropriate values for the constants a and b may be computed from the experimental conditions. Hence the value of e/m can be calculated for each of the parabolas. This method was first worked out for *cathode rays*,

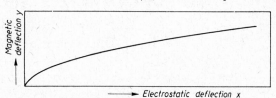

Fig. 68. Deflection of canal (positive) rays in electrostatic and magnetic fields.

which are deflected in the same way as the canal rays but in the opposite direction, since they bear a charge of opposite sign. The ratio between the charge and mass of the electron was first determined by such measurements on cathode rays. The ratio e/m was first determined for canal rays or positive rays for the purpose of ascertaining the charges borne by the atoms in such rays. It was found that the atoms in the canal rays generally carried only a small charge, usually of only 1 or 2 units. By using sufficiently strong fields, the parabolas corresponding to a doubling of the charge could be widely separated. In 1912, Thomson investigated the canal rays produced in a discharge tube which contained neon. Such canal rays gave parabolas corresponding chiefly to singly and doubly charged neon, i.e., to the very different charge to mass ratios $e/m = 1/20$ and $e/m = 2/20$. Close to the parabola for the singly charged neon, however, Thomson observed another blackening curve which would correspond to particles with a charge to mass ratio $e/m = 1/22$. He at first believed that this result must be attributed to impurities, but it was proved that this was not the case. Thomson therefore finally concluded that ordinary neon, with atomic weight 20 might be mixed with a small amount of a second inert gas with the atomic weight 22. His conclusion has subsequently been fully confirmed. This was the first instance in which differences in deflection sustained by the canal rays revealed that an element was composite in nature, and composed of several isotopes.

The canal ray method has subsequently been further developed, especially by Aston (from 1919 onwards). Aston achieved greater sensitivity by the device of having the magnetic field perpendicular to the electrostatic field, instead of parallel to it. The experimental conditions are so selected that for rays of the same e/m, but of all velocities, the electrostatic deflection is exactly compensated by a magnetic deflection in the opposite direction. The result of this is to contract each parabolic curve into a single point, so that the whole blackening effect of the rays is concentrated into a number of discrete points or (when slits are used as collimating apertures) into discrete lines. These are accordingly sharper and better defined than the extended parabolas. The photographs obtained by this means bear a certain resemblence to photographs of spectra, and a mixture of isotopes is resolved into its individual components in much the same way as inhomogeneous light resolved by a spectrograph. Aston's description of the method as '*mass spectrography*' is therefore appropriate. The technique of mass spectrography has now been so perfected that determinations of the mass of individual atomic species made by this means are considerably more precise than atomic weight determinations carried out by the classical methods of chemistry. The results of the different procedures can only be compared, in the case of mixed elements, if the relative proportions of the several isotopic species of atoms are known, as well as their masses. The proportions cannot be determined as accurately as the atomic masses, and only in exceptional cases are the average mass numbers determined by physical methods more accurate than chemical atomic weight determinations.

Dempster (1918) worked out a method of canal ray analysis in which a beam of positive ions of uniform velocity is deflected by a magnetic field alone. Ions of uniform velocity are produced by causing slow ions, such as are emitted from heated salts, etc., to be accelerated in a strong electrical field. The ratio e/m is then given by the magnetic deflection alone. At a later stage (1935), Dempster combined the 'prism' action of a homogeneous magnetic field with the 'lens' action of a radial electrical field. This 'double focussing' principle has since been incorporated in nearly all forms of precision mass spectrograph now used*. Of these, the mass spectrometer designed by Nier (1949–51) is of particular interest. The particles of differing mass are not recorded at different points on a photographic plate, but their masses are measured in terms of the potential which must be applied to a radial electrical field, in order that the particles should pass through a slit at the end of the apparatus. The Nier instrument has achieved great importance both for the analysis of isotopic mixtures and for direct chemical analysis (e.g., of hydrocarbon mixtures).

(b) Isotopy and Band Spectra

The total energy content of a polyatomic gas molecule is made up of the sum of its *translational energy* (i.e., the kinetic energy of its random translational motion), its *rotational energy*, the *vibrational energy* of its atoms, and the *binding energy* of its electrons. Quantized changes in the three last-mentioned contributions to the molecular energy give rise to all the spectral lines observed in the band spectrum of the molecule. In the long wave length range (infrared), where the light quanta have relatively small energies, no changes in electronic states can be involved, and the 'rotation-vibration' spectra are observed. In these, the position of the bands is determined by quantized changes in the vibrational energy, whereas the frequency interval between the lines which together make up each band is determined by changes in the rotational energy. Band systems corresponding to the rotation-vibration spectrum also appear in the shorter wave length region (visible and ultraviolet); here they are superposed upon the lines which would arise from changes in the electronic energy alone. The vibrational energy of a molecule depends upon the masses of the vibrating atoms; the rotational energy depends upon the moment of inertia, and thus again upon the mass of the atoms rotating about a common center. It follows that two molecules which are chemically alike, but are not identical in the masses of their component atoms, must give rise to different band spectra. Thus the position of the bands in the spectrum of $^{14}N^{15}N$ is somewhat different from that in the spectrum of $^{14}N_2$, whereas the displacement of the bands is still greater for $^{15}N_2$. In the same way, a molecule $^{13}C^{16}O$ has a slightly different band spectrum from that of $^{12}C^{16}O$, and this in turn from that of $^{12}C^{18}O$. Whereas the interval between successive lines within any band is only slightly affected by differences in the masses of the isotopic atoms, the shift in the positions of the bands, due to alterations in vibrational energy, can be quite considerable. It is found to be greatest for the elements of low atomic weight. Thus, for the molecules $^{14}N_2$ and $^{15}N_2$ it amounts to about 120 Å in the band system around 3600 Å. The study of band spectra has thus made it possible to detect the isotopically composite nature of the light elements, even when the minor isotopes are present in extremely low concentration. Thus, in 1929 it was inferred from observations on band spectra that carbon and oxygen did not consist entirely of atoms with the same masses, as had been concluded up till that date from the evidence of mass spectrography. The sensitivity of the mass spectrographic method has subsequently been greatly increased, and it has the advantage of being far more widely applicable than the optical method. In both methods, the relative proportions of the isotopes in a mixture can be found by measuring the degree of blackening of a photographic plate.

It follows from eqn. (13a), p. 87 Vol. I, that the difference in mass of the isotopes must show up in the line spectrum of the atoms also. With the heavier elements the effect is so minute that it has no general significance for the detection of isotopes. However, the existence of the heavy isotope of hydrogen (deuterium) was first proved through this displacement of the Balmer lines (p. 554).

---

* A survey of the newer types of mass spectrograph, and the principles upon which they have been designed, is given in a 'Symposium on Mass Spectroscopy in Physics Research' (National Bureau of Standards, Washington, 1951).

2. **Separation of Isotopes**

(a) **General**

The only processes which can be used to effect a separation of isotopes are those in which the *masses* of the atoms exert a strong influence. As a rule, therefore, chemical reactions do not come into question, since the masses of the atoms have practically no effect.

The valence forces which determine the course of chemical reactions depend primarily upon the number and the mode of binding (i.e., energy states) of the electrons in the 'outer shell' of the atom. The number of electrons is fixed by the nuclear charge; their binding energies and ground states are practically completely dependent on the nuclear charge also, as follows from the theory of atomic structure. Although, as already mentioned, the binding energy of an electron is not absolutely independent of the mass of the atomic nucleus, the effect on the binding energy is extraordinarily small, even in the case of hydrogen, for which it is greater than with any other element. In general, it can be neglected as of no significant effect whatever upon the chemical properties of hydrogen and, *a fortiori*, negligible for all other elements.

It may be foreseen, however, that the masses of the atoms will exert some influence on chemical behavior by virtue of the dependence of the vibrational and rotational energy of a molecule upon the masses of its atoms. These factors must affect the free energy changes in chemical reactions, and therefore exert an influence on the position of chemical equilibrium (cf. Vol. I, Chap. 5). This influence is very small, and is detectable in practice only when the ratio of the masses of the isotopic atoms is appreciably different from unity. This mass ratio is greatest for the isotopes of hydrogen. With increasing atomic weight, it rapidly falls to a number very close to 1, since the isotopes of any element can differ in mass from one another by only a few atomic weight units. Whereas the mass ratio of ^2H and $^1H =$ 2 : 1, it has already dropped to 1.17 : 1 for the lithium isotopes ^7Li and ^6Li. Hence hydrogen is the only element for which it may be said that the isotopes display different chemical properties (in a quantitative sense only, not in a qualitative sense).

For all the other light elements, the effect of isotopic masses is very small. However Urey [*J. Am. Chem. Soc.*, 57, (1935) 321] calculated the equilibrium constants for the distribution of certain isotopic molecules in reversible chemical reactions, and showed that the very small differences might be made a basis for the separation of isotopes. Equilibrium constants calculated for a few reactions are listed below:

$$S^{16}O_2 \text{ (g)} + 2H_2^{18}O \text{ (l)} \rightleftharpoons S^{18}O_2 \text{ (g)} + 2H_2^{16}O \text{ (l)} \qquad K = 1.040 \text{ at } 273.1° \text{ K}$$

$$C^{16}O_2 \text{ (g)} + 2H_2^{18}O \text{ (l)} \rightleftharpoons C^{18}O_2 \text{ (g)} + 2H_2^{16}O \text{ (l)} \qquad 1.097 \text{ at } 273.1° \text{ K}$$

$$C^{16}O_2 \text{ (g)} + 2H_2^{18}O \text{ (g)} \rightleftharpoons C^{18}O_2 \text{ (g)} + 2H_2^{16}O \text{ (g)} \qquad 1.128 \text{ at } 273.1° \text{ K}$$

$$^{13}CO_2 \text{ (g)} + {}^{12}CO_3^= \text{ (sol)} \rightleftharpoons {}^{12}CO_2 \text{ (g)} + {}^{13}CO_3^= \text{ (sol.)} \qquad 1.015 \text{ at } 273.1° \text{ K}$$

$$^{15}NH_3 \text{ (g)} + {}^{14}NH_4^+ \text{ (sol)} \rightleftharpoons {}^{14}NH_3 \text{ (g)} + {}^{15}NH_4^+ \text{ (sol.)} \qquad 1.023 \text{ at } 273.1° \text{ K}$$

Differences in mass exert their most direct effect upon the behavior of charged particles in an electrical or magnetic field. Hence the mass spectrograph resolves a mixture of isotopes directly and completely into its separate components. The amounts of the separated isotopes which can be prepared by this means are so minute however that, except for certain special cases, it has not been practicable to use the method as a means of preparing separated isotopes*. Pure isotopes, or

---

* Mass spectrographs of very high intensity have been built by a number of workers (e.g., Smyth and Rumbough, 1934; Walcher, 1938) for the purpose of separating pure isotopes for experiments (such as nuclear transmutations) which can be carried out with minimal quantities of material. During the 1939–45 war, this method of direct electromagnetic separation was developed on a larger scale, for the separation of ^{235}U in quantity from natural uranium, and can now be applied in the United States and in Britain to the isolation of separated isotopes of elements which could not be resolved by any other means.

highly enriched isotopic mixtures, can be applied to the investigation of a wide range of chemical problems (cf. pp. 553, 558–9). Considerable interest therefore attaches to methods which can be used for the preparation or enrichment of isotopes on a considerable scale.

The following are the principle methods available for the separation of isotopes.

(b) Diffusion Process

According to Graham's law, light molecules diffuse more rapidly through a porous wall than do heavy molecules. If the two species of molecules making up the gaseous mixture differ but little from one another in mass, a considerable change in the proportions of the mixture can be brought about only by continuing the diffusion process for a long period, and by carrying it out in such a way that a large number of diffusion cells are connected one after the other, in series (making use of the same principle as in the 'fractionating columns' used industrially for distillation). An apparatus of this kind was built by Hertz (1932). Some disadvantages attach to the use of a porous wall, so that in a later apparatus (1934) Hertz replaced the assembly of diffusion cells, each provided with a porous porcelain tube, by a group of mercury vapor diffusion pumps. Each of these pumped away the lighter component of the gas mixture undergoing separation, more readily than the heavier component. Using an apparatus of this kind with 48 diffusion pumps, Barwich (1936) was able to make a quantitative separation of the neon isotopes ^{20}Ne and ^{22}Ne within 5 hours. De Hemptinne and Capron, with a Hertz apparatus containing 51 diffusion pumps, enriched the isotope ^{13}C, which is present in normal carbon to the extent of 1.1%, up to a concentration of 75% in 30 hours. The amounts of the separated isotopes which are obtained in practice by this method are, however, quite small, since it is necessary to operate at very low pressures, i.e., with very dilute gases.

(c) Distillation Process

A partial separation of isotopes can be brought about by 'ideal distillation'. This process is based upon the fact that more atoms of the lighter isotope than of the heavier isotope pass from the liquid to the gas phase per unit time. If it is so arranged that all atoms entering the gas phase are condensed out before they can undergo molecular collisions and be returned to the liquid (as may be achieved, for example, by placing a chilled surface above the surface of the liquid, at a distance not greatly exceeding the mean free path of the molecules), the lighter isotope is enriched in the condensate and the heavier in the residue. This process is simple to carry out, but does not lead to a high degree of enrichment, and yields only small quantities of the enriched product. It has, however, been used successfully for certain purposes (cf. Vol. I, p. 159).

Large quantities of enriched isotopes can be obtained by fractional distillation, repeated many times in an efficient rectification column (fractionating column). The process operates most efficiently in the neighborhood of the triple point (cf. Vol. I, p. 51), since the vapor pressure differences between the isotopes are greatest in the solid state. Even under these conditions the differences in vapor pressure are mostly very small, so that fractionating columns with a large number of stages must be used. Starting from 420 liters of ordinary neon, containing 90.5% ^{20}Ne and 9.2% ^{22}Ne (with 0.3% ^{21}Ne), and using a fractionating column with 85 stages, Keesom (1934) obtained 4 liters of neon in which the lighter isotope had been enriched to 95.5%, and 5 liters of neon in which the heavier isotope, ^{22}Ne, had been enriched to 83.5%. By use of fractionating columns of very high efficiency, corresponding to a very large number of 'theoretical plates' and operating under almost total reflux, it is possible to prepare material enriched almost to 100% in the heavier carbon isotope, ^{13}C, by the triple point distillation of CO. This process is carried out on a 100 g to kilogram scale.

(d) Separation by Thermal Diffusion

The process worked out by Clusius and Dickel (1938) is notable for the simplicity of the apparatus and the efficiency of separation achieved. It is based on the phenomenon known as *thermal diffusion*, namely that if a temperature gradient is established in a gas, a concentration gradient is also set up at the same time*. Under certain conditions this may lead to a diffusion process in the gas, wherein the lighter molecules move faster than the heavier. The separating tube devised by Clusius consists of a vertical glass (or metal) tube 6—10 meters long, containing the gas mixture to be separated at atmospheric pressure; it is surrounded by a water-cooled jacket, and a wire is stretched along its axis. When this wire is heated electrically, the temperature gradient from wiere to wall sets up a concentration gradient in the opposite direction, which results in diffusion from the wall towards the wire. The lighter molecules thereby diffuse more rapidly, and are enriched in the neigh-borhood of the wire. The heating of the gas by the hot wire causes a convective streaming of the gas, upwards along the wire and downwards along the wall, whereby the equilibrium of thermal diffusion is continuously disturbed. The result is that although the un-mixing brought about by thermal diffusion is extremely minute in extent, it goes on continuously until nearly all the lighter molecules have collected in the upper part of the separation tube, and the heavier molecules in the lower part. Using a battery of several such tubes, Clusius enriched the chlorine isotope ^{37}Cl from 24.6% to 99.4% at the lower end, while the isotope ^{35}Cl was enriched from 75.4% to 99.6% at the other end. In continuous operation it was possible to withdraw between 10 and 30 cc of enriched products daily from the two ends of the tube. Clusius obtained 2.5 liters of each of the isotopes ^{20}Ne and ^{22}Ne by the same method, as also pure ^{84}Kr and ^{86}Kr, and also (1943) $^{18}O_2$ and $^{14}N^{15}N$. The thermal diffusion process, although efficient, is costly in operation through its high consumption of electric current, and the triple point distillation process is now preferred for the relatively large-scale separation of isotopes.

(e) Partition Processes

When any substance is partitioned between two solvents, molecules which have different masses but are chemically identical do not display identical behavior. If the process is repeated many times (in some form of column), it is possible to bring about a more or less extensive change in the isotope ratio. Similar considerations apply to partition between a solvent and the gas phase. Thus Urey (1937), by treating ammonia with a solution of ammonium sulfate** in a rectifying column of 621 stages, was able to raise the concentration of ^{15}N in nitrogen from 0.38% to 2.35%. Later (1939) by connecting several columns in tandem, he was able to produce 2.2 g of nitrogen per day, containing 70.6% of ^{15}N. G. N. Lewis succeeded in raising the concentration of the ^6Li isotope in lithium from 7.9% to 16.3%, by exchange of lithium between lithium amalgam and a solution of lithium chloride in ethyl alcohol.

(f) Electrolytic Separation

A separation, or at least an appreciable enrichment, of light isotopes can be brought about by prolonged electrolysis. In the case of the *hydrogen isotopes*, the electrolytic separa-tion is made still more favorable by the superposition of various other effects, and is one of the most efficient separation processes at present available (see p. 554). For other elements, the degree of separation achieved by electrolysis is much smaller. However, it has been shown that it is possible appreciably to alter the proportions of the isotopes ^6Li and ^7Li by electrolysis of aqueous solutions of lithium salts. Even with the rather heavier potassium, a marked change in the proportions of the isotopes ^{39}K and ^{40}K has been produced electro-lytically using several cells in series (Harteck, 1953). There is also some enrichment of the oxygen isotopes ^{18}O and ^{17}O during the electrolysis of water (Johnston, 1935).

3. Pure Elements and Mixed Elements

Those elements which consist entirely of atoms of the same mass are known as *pure elements*; those in which atoms of different masses are present are called *mixed*

---

* It is also possible to observe the converse effect—the establishment of a temperature gradient, as a result of the concentration gradient when two gases diffuse into one another [Clusius, *Naturwissenschaften* 30, (1942) 711.]

** The effect of the ammonium sulfate is to improve the separation, since in the equili-brium $^{15}NH_3 + [^{14}NH_4]^+ \rightleftharpoons {}^{14}NH_3 + [^{15}NH_4]^+$ the concentration of $[^{15}NH_4]^+$ ions is rather greater than that of $[^{14}NH_4]^+$ ions. The result is that $^{15}NH_3$, which is rather more soluble in water than $^{14}NH_3$, is retained to an enhanced extent by the aqueous solution.

elements. Table 61 shows the elements which are considered to be pure elements, according to the present state of our knowledge. In most cases it has been proved that they do not contain more than 0.2 atom-% of any other isotopes.

TABLE 61

PURE ELEMENTS

Atomic number Z	Element	Mass value	Mass defect (m.m.u.)	Chemical atomic weights (on basis natural oxygen = 16.000)	
				calc. from mass values	from international atomic wt. table
4	Be	9.01505	62.48	9.01257	9.013
9	F	19.00443	158.79	18.99920	19.00
11	Na	22.9964	201.1	22.9901	22.991
13	Al	26.99011	241.65	26.9827	26.98
15	P	30.98358	282.45	30.9751	30.975
21	Sc	44.97010	416.72	44.9577	44.96
25	Mn	54.95581	517.53	54.9407	54.94
27	Co	58.9510	556.6	58.9348	58.94
33	As	74.943	703	74.922	74.91
39	Y	(88.940)	—	88.916	88.92
41	Nb	(92.942)	—	92.916	92.91
45	Rh	102.949	938	102.921	102.91
53	I	126.944	1153	126.909	126.91
55	Cs	(132.947)	—	132.910	132.91
59	Pr	140.951	1266	140.912	140.92
65	Tb	(158.978)	—	158.934	158.93
67	Ho	(164.987)	—	164.942	164.94
69	Tm	(168.993)	—	168.947	168.94
73	Ta	181.003	1562	180.953	180.95
79	Au	197.039	1664	196.985	197.0
83	Bi	209.047	1762	208.990	209.00
89	Ac	227.098	1867	227.036	227
91	Pa	231.108	1892	231.044	231

Column 3 of Table 61 gives the *mass values* of the elements listed in columns 1 and 2*. The *mass value* of an atom (also termed the physical atomic weight) is the ratio of the mass of the atom to $^1/_{16}$ of the mass of the oxygen isotope ^{16}O. The last two columns show the (chemical) atomic weights as calculated from the atomic masses determined by mass spectrographic or nuclear physical methods, and as determined by chemical or physico-chemical methods (i.e., the atomic weights as listed in the international atomic weight table), both referred to ordinary oxygen = 16**. The significance of column 4 (mass defects) will be considered later (p. 567). It may be seen from the table that there is excellent agreement between the atomic weights determined by chemical methods and those based on mass spectrographic or nuclear physical measurement. It may also be noted that the mass values of the atoms differ only very slightly from whole numbers. Nevertheless,

* The values given in brackets have been calculated from the approximated masses and the packing fractions, as discussed on p. 568 *et seq.*
** Mass values referred to ^{16}O = 16.000 are converted to chemical atomic weights, on the scale ordinary oxygen = 16.000, by dividing them by 1.000275 (cf. p. 550).

the probable error of the data listed in Table 61 is in most cases only a few units in the last decimal place, and the deviations from whole number values lie outside the experimental error. The mass values of the atoms, rounded off to whole numbers, have an important physical significance, as will be seen later. By analogy with the atomic numbers, these rounded-off values are called the *mass numbers* of the atoms.

It is striking that, with the exception of beryllium (which certainly contains less than 0.001% of any isotope, as shown by the most recent work), all the pure elements are of *odd atomic number*. This is connected with the fact that the elements of odd atomic number are, in general, less stable than the elements of even atomic number which are adjacent to them in the Periodic Table. This rule shows itself in the distribution of the elements in Nature (cf. Fig. 84 p. 650). Unstable (i.e., radioactive) isotopes of the elements Bi, Ac, and Pa are known, but no stable isotopes. The isotopes of the radioactive elements Ac and Pa have decay constants which are so much greater than those of the principal representatives of these atomic numbers that actinium and protactinium exist in Nature as pure elements.

Table 62 gives the isotopic composition of the stable (i.e., not radioactive) mixed elements.

The five weakly radioactive elements K, Rb, Nd, Sm and Lu, which consist almost entirely of stable nuclear species, are included in Table 62. So also are In and Re. For both these elements, the predominant isotope is actually unstable. However, they are both so weakly radioactive (the half lives are $6 \cdot 10^{14}$ and $4 \cdot 10^{12}$ years respectively) that they behave practically as stable elements. The remaining radioactive elements are omitted since their isotopy has already been discussed in dealing with the disintegration series. The third column of Table 62 gives the proportions of the individual isotopes in the mixed elements, the isotopes being denoted by superscript figures written before the chemical symbol. The 'average mass values' listed in column 4 of the table are calculated from the mass values of the separate isotopes, according to the usual mixture rule; the deviation of the mass values from whole numbers has been taken into account in so doing. The remaining columns correspond to those of Table 61.

The masses of the isotopes of mixed elements, like those of the pure elements, are all very close to whole numbers. The chemical atomic weights of most elements differ considerably from whole numbers for the reason that the elements are composed of isotopes of different masses.

The two heavier isotopes of oxygen have the exact masses $^{17}O = 17.004520$ and $^{18}O = 18.004875$, when the mass of the lighter isotope ^{16}O is taken as exactly 16.00000. From the isotopic composition of the mixed element oxygen, as given in Table 62, the average mass is found to be 16.00447. *The unit of measurement for chemical atomic weights is one sixteenth of this value.* This unit is 1.000279 times as great as the unit used for measurement of the mass values, or physical atomic weights. Hence the latter must be divided by the factor 1.000279 in order to convert them to atomic weights on the chemical scale;[*] i.e., *chemical atomic weight* (relative to

ordinary oxygen $= 16.0000) = \dfrac{1}{1.000279} \times$ *physical atomic weight* (mass relative to $^{16}O = 16.00000$).

* It is intended to fix the conversion factor by convention, on account of the variations observed in the isotopic composition of natural oxygen (cf. Vol. I, p. 689). The factor 1.000275, which is based on older measurements, will probably be used when the convention is adopted. In fact this factor does not correspond exactly to any known occurrence of natural oxygen, but since 1940 it has been used continuously in converting to the chemical scale data recorded in terms of the physical scale (cf. *J. Amer. Chem. Soc.* 76 (1954) 2034).

TABLE 62

COMPOSITION OF THE STABLE MIXED ELEMENTS

z	Element	Composition in atoms per cent	Mean mass value	Chemical atomic weight on basis O = 16	
				Calc. from mean mass value	Value from atomic wt. table
1	H	99.984 % ^{1}H, 0.016 % ^{2}H (D)	1.00831	1.00802	1.0080
2	He	$1.3 \cdot 10^{-4}$ % ^{3}He, 99.99987 % ^{4}He	4.00387	4.00276	4.003
3	Li	7.30 % ^{6}Li, 92.70 % ^{7}Li	6.9451	6.9431	6.940
5	B	18.83 % ^{10}B, 81.17 % ^{11}B	10.8251	10.8221	10.82
6	C	98.9 % ^{12}C, 1.1 % ^{13}C	12.0149	12.0115	12.011
7	N	99.62 % ^{14}N, 0.38 % ^{15}N	14.0113	14.0074	14.008
8	O	99.7577 % ^{16}O, 0.0392 % ^{17}O, 0.2031 % ^{18}O	16.00447	16.00000	16.0000
10	Ne	90.51 % ^{20}Ne, 0.28 % ^{21}Ne, 9.21 % ^{22}Ne	20.1767	20.171	20.183
12	Mg	78.60 % ^{24}Mg, 10.11 % ^{25}Mg, 11.29 % ^{26}Mg	24.3195	24.313	24.32
14	Si	92.18 % ^{28}Si, 4.71 % ^{29}Si, 3.11 % ^{30}Si	28.095	28.087	28.09
16	S	95.060 % ^{32}S, 0.742 % ^{33}S, 4.182 % ^{34}S, 0.016 % ^{36}S	32.0738	32.0649	32.066
17	Cl	75.4 % ^{35}Cl, 24.6 % ^{37}Cl	35.471	35.462	35.457
18	A	0.35 % ^{36}A, 0.08 % ^{38}A, 99.57 % ^{40}A	39.960	39.948	39.944
19	K	93.260% ^{39}K, 0.011% ^{40}K (β- and K-active), 6.729% ^{41}K	39.111	39.100	39.100
20	Ca	96.92 % ^{40}Ca, 0.64 % ^{42}Ca, 0.129 % ^{43}Ca, 2.13 % ^{44}Ca, 0.003 % ^{46}Ca, 0.178 % ^{48}Ca	40.091	40.080	40.08
22	Ti	7.95 % ^{46}Ti, 7.75 % ^{47}Ti, 73.45 % ^{48}Ti, 5.51 % ^{49}Ti, 5.34 % ^{50}Ti	47.889	47.876	47.90
23	V	0.25 % ^{50}V, 99.75 % ^{51}V	50.958	50.944	50.95
24	Cr	4.31 % ^{50}Cr, 83.76 % ^{52}Cr, 9.55 % ^{53}Cr, 2.38 % ^{54}Cr	52.014	52.000	52.01
26	Fe	5.81 % ^{54}Fe, 91.64 % ^{56}Fe, 2.21 % ^{57}Fe, 0.34 % ^{58}Fe	55.866	55.850	55.85
28	Ni	67.77 % ^{58}Ni, 26.16 % ^{60}Ni, 1.25 % ^{61}Ni, 3.66 % ^{62}Ni, 1.16 % ^{64}Ni	58.729	58.712	58.71
29	Cu	68.94 % ^{63}Cu, 31.06 % ^{65}Cu	63.570	63.552	63.54
30	Zn	48.89 % ^{64}Zn, 27.81 % ^{66}Zn, 4.07 % ^{67}Zn, 18.61 % ^{68}Zn, 0.62 % ^{70}Zn	65.408	65.390	65.38
31	Ga	60.16 % ^{69}Ga, 39.84 % ^{71}Ga	69.750	69.730	69.72
32	Ge	21.2 % ^{70}Ge, 27.3 % ^{72}Ge, 7.9 % ^{73}Ge, 37.1 % ^{74}Ge, 6.5 % ^{76}Ge	72.654	72.633	72.60
34	Se	0.87 % ^{74}Se, 9.02 % ^{76}Se, 7.58 % ^{77}Se, 23.52 % ^{78}Se, 49.82 % ^{80}Se, 9.19 % ^{82}Se	79.015	78.99	78.96
35	Br	50.53 % ^{79}Br, 49.47 % ^{81}Br	79.933	79.911	79.916
36	Kr	0.342 % ^{78}Kr, 2.228 % ^{80}Kr, 11.50 % ^{82}Kr, 11.48 % ^{83}Kr, 57.02 % ^{84}Kr, 17.43 % ^{86}Kr	83.833	83.810	83.80
37	Rb	72.8 % ^{85}Rb, 27.2 % ^{87}Rb (β-active)	85.474	85.451	85.48
38	Sr	0.55 % ^{84}Sr, 9.75 % ^{86}Sr, 6.96 % ^{87}Sr, 82.74 % ^{88}Sr	87.648	87.623	87.63
40	Zr	51.46 % ^{90}Zr, 11.23 % ^{91}Zr, 17.11 % ^{92}Zr, 17.40 % ^{94}Zr, 2.80 % ^{96}Zr	91.257	91.231	91.22
42	Mo	15.84 % ^{92}Mo, 9.04 % ^{94}Mo, 15.72 % ^{95}Mo, 16.53 % ^{96}Mo, 9.46 % ^{97}Mo, 23.78 % ^{98}Mo, 9.63 % ^{100}Mo	95.92	95.89	95.95
44	Ru	5.68 % ^{96}Ru, 2.22 % ^{98}Ru, 12.81 % ^{99}Ru, 12.70 % ^{100}Ru, 16.98 % ^{101}Ru, 31.34 % ^{102}Ru, 18.27 % ^{104}Ru	101.06	101.03	101.1
46	Pd	0.8 % ^{102}Pd, 9.3 % ^{104}Pd, 22.6 % ^{105}Pd, 27.1 % ^{106}Pd, 26.7 % ^{108}Pd, 13.5 % ^{110}Pd	106.569	106.54	106.4
47	Ag	51.92 % ^{107}Ag, 48.08 % ^{109}Ag	107.901	107.871	107.880
48	Cd	1.215 % ^{106}Cd, 0.875 % ^{108}Cd, 12.39 % ^{110}Cd, 12.75 % ^{111}Cd, 24.07 % ^{112}Cd, 12.26 % ^{113}Cd, 28.86 % ^{114}Cd, 7.58 % ^{116}Cd	112.460	112.428	112.41
49	In	4.23 % ^{113}In, 95.77 % ^{115}In (β-active)	114.856	114.824	114.82

Continued

TABLE 62

COMPOSITION OF THE STABLE MIXED ELEMENTS

z	Element	Composition in atoms per cent	Mean mass value	Chemical atomic weight on basis O = 16	
				Calc. from mean mass value	Value from atomic table
50	Sn	0.94 % ^{112}Sn, 0.65 % ^{114}Sn, 0.33 % ^{115}Sn, 14.36 % ^{116}Sn, 7.51 % ^{117}Sn, 24.21 % ^{118}Sn, 8.45 % ^{119}Sn, 33.11 % ^{120}Sn, 4.61 % ^{122}Sn, 5.83 % ^{124}Sn	118.767	118.734	118.70
51	Sb	57.25 % ^{121}Sb, 42.75 % ^{123}Sb	121.798	121.764	121.76
52	Te	0.09 % ^{120}Te, 2.43 % ^{122}Te, 0.85 % ^{123}Te, 4.59 % ^{124}Te, 6.98 % ^{125}Te, 18.70 % ^{126}Te, 31.85 % ^{128}Te, 34.51 % ^{130}Te	127.674	127.638	127.61
54	Xe	0.095 % ^{124}Xe, 0.088 % ^{126}Xe, 1.916 % ^{128}Xe, 26.235 % ^{129}Xe, 4.051 % ^{130}Xe, 21.24 % ^{131}Xe, 26.925 % ^{132}Xe, 10.52 % ^{134}Xe, 8.93 % ^{136}Xe	131.344	131.308	131.30
56	Ba	0.102 % ^{130}Ba, 0.098 % ^{132}Ba, 2.42 % ^{134}Ba, 6.59 % ^{135}Ba, 7.81 % ^{136}Ba, 11.32 % ^{137}Ba, 71.66 % ^{138}Ba	137.372	137.33	137.36
57	La	0.089 % ^{138}La (K-active), 99.911 % ^{139}La	138.96	138.92	138.92
58	Ce	0.19 % ^{136}Ce, 0.25 % ^{138}Ce, 88.49 % ^{140}Ce, 11.07 % ^{142}Ce	140.158	140.12	140.13
60	Nd	27.10 % ^{142}Nd, 12.14 % ^{143}Nd, 23.84 % ^{144}Nd, 8.29 % ^{145}Nd, 17.26 % ^{146}Nd, 5.74 % ^{148}Nd, 5.63 % ^{150}Nd (β-active)	144.295	144.25	144.27
62	Sm	2.95% ^{144}Sm, 14.62% ^{147}Sm (α-active), 10.97% ^{148}Sm, 13.56 % ^{149}Sm, 7.27 % ^{150}Sm, 27.34 % ^{152}Sm, 23.29 % ^{154}Sm,	150.473	150.43	150.35
63	Eu	47.77 % ^{151}Eu, 52.23 % ^{153}Eu	152.014	151.97	152.0
64	Gd	0.20 % ^{152}Gd, 2.15 % ^{154}Gd, 14.78 % ^{155}Gd, 20.59 % ^{156}Gd, 15.71 % ^{157}Gd, 24.78 % ^{158}Gd, 21.79 % ^{160}Gd	156.98	156.94	157.2
66	Dy	0.05 % ^{156}Dy, 0.05 % ^{158}Dy, 0.1 % ^{160}Dy, 21.1 % ^{161}Dy, 26.6 % ^{162}Dy, 24.8 % ^{163}Dy, 27.3 % ^{164}Dy	162.53	162.49	162.51
68	Er	0.1 % ^{162}Er, 1.5 % ^{164}Er, 32.9 % ^{166}Er, 24.4 % ^{167}Er, 26.9 % ^{168}Er, 14.2 % ^{170}Er	167.23	167.18	167.27
70	Yb	0.140 % ^{168}Yb, 3.04 % ^{170}Yb, 14.34 % ^{171}Yb, 21.88 % ^{172}Yb, 16.18 % ^{173}Yb, 31.77 % ^{174}Yb, 12.65 % ^{176}Yb	173.09	173.04	173.04
71	Lu	97.5 % ^{175}Lu, 2.5 % ^{176}Lu (β-active)	175.028	174.98	174.99
72	Hf	0.19 % ^{174}Hf, 5.26 % ^{176}Hf, 18.51 % ^{177}Hf, 27.16 % ^{178}Hf, 13.79 % ^{179}Hf, 35.09 % ^{180}Hf	178.54	178.49	178.50
74	W	0.16 % ^{180}W, 26.35 % ^{182}W, 14.32 % ^{183}W, 30.68 % ^{184}W, 28.49 % ^{186}W	183.981	183.93	183.86
75	Re	37.07 % ^{185}Re, 62.93 % ^{187}Re (β-active)	186.275	186.22	186.22
76	Os	0.018 % ^{184}Os, 1.582 % ^{186}Os, 1.64 % ^{187}Os, 13.27 % ^{188}Os, 16.14 % ^{189}Os, 26.38 % ^{190}Os, 40.97 % ^{192}Os	190.312	190.26	190.2
77	Ir	38.5 % ^{191}Ir, 61.5 % ^{193}Ir	192.27	192.22	192.2
78	Pt	0.006 % ^{190}Pt, 0.784 % ^{192}Pt, 30.2 % ^{194}Pt, 35.2 % ^{195}Pt, 26.6 % ^{196}Pt, 7.21 % ^{198}Pt	195.185	195.13	195.09
80	Hg	0.15 % ^{196}Hg, 10.12 % ^{198}Hg, 17.04 % ^{199}Hg, 23.25 % ^{200}Hg, 13.18 % ^{201}Hg, 29.54 % ^{202}Hg, 6.72 % ^{204}Hg	200.65	200.60	200.6
81	Tl	29.46 % ^{203}Tl, 70.54 % ^{205}Tl	204.468	204.41	204.39
82	Pb	1.36 % ^{204}Pb, 25.12 % ^{206}Pb, 21.25 % ^{207}Pb, 52.27 % ^{208}Pb	207.272	207.214	207.2

4. Applications of Isotopes [2]

Pure isotopes, or enriched isotopic mixtures, can often be usefully employed in the study of chemical reactions.

Thus Polanyi (1934), by using $H_2{}^{18}O$, was able to show that the hydrolysis of esters took place by the reaction mechanism (A), and not according to (B).

(A)
$$CH_3-C\overset{O}{\underset{O-C_5H_{11}+H-{}^{18}O-H}{}} = CH_3-C\overset{O}{\underset{{}^{18}OH}{}} + C_5H_{11}OH$$

(B)
$$CH_3-C\overset{O}{\underset{O-C_5H_{11}+H-{}^{18}O-H}{}} = CH_3-C\overset{O}{\underset{OH}{}} + C_5H_{11}{}^{18}OH$$

The amyl alcohol formed in the hydrolysis contained ordinary oxygen, whereas the acetic acid which was the other product of reaction contained oxygen with the same ^{18}O content as the water used for the hydrolysis.

Compounds with an abnormal isotopic composition—especially with respect to ^{18}O, ^{13}C and ^{15}N—can be used in a similar manner for the elucidation of a variety of biological and physiological problems also.

The isotope which has hitherto proved of the greatest use in work of this kind is the heavy isotope of hydrogen, deuterium.

5. Deuterium and Deuterium Oxide [3-6]

The heavy isotope of hydrogen, ^2H, is given the special name of 'heavy hydrogen', or deuterium (Gk. τὸ δεύτερον, the second one). It is therefore usual to employ the symbol D rather than ^2H. The hydrogen isotope ^1H is known as 'light hydrogen' or protium (Gk. τὸ πρῶτον, the first one). The atomic nucleus of protium, H^+, is the proton; that of deuterium, D^+, is the deuteron (some authors use the term deuton). The name 'heavy water' is commonly employed for deuterium oxide, D_2O $(={}^2H_2O)$.

A third isotope of hydrogen, ^3H, known as tritium (Gk. τὸ τρίτον, the third one), can be obtained by artificial atomic transmutation. Tritium (denoted by the chemical symbol T) is unstable, and decays with the emission of β-radiation to give ^3He. The most recent measurements give a value of 12.41 years for its half-life (W. M. Jones, 1951). Although its half-life is rather short, tritium is found in Nature in very minute amounts. It is produced in the uppermost regions of the atmosphere, through nuclear reactions which are brought about by the cosmic rays; it gradually reaches the earth's surface by diffusion and convection (in the form of HT molecules). According to Harteck and Faltings (1950), 10 cc of air contains, on the average, about 1 atom of tritium; about 1 mole of tritium is present in the whole of the atmosphere. Tritium may be enriched about a millionfold in highly concentrated heavy water (A. V. Grosse, 1951). Ice with a high tritium content is luminescent. Tritium can be prepared artificially by the nuclear reactions shown in Table 73 (p. 588).

(a) Historical

Giauque discovered the isotopy of oxygen in 1929. It became apparent that if Aston's value for the mass of hydrogen, determined mass spectrographically (1927), were recalculated on the new basis for the chemical atomic weight scale, it must lead to an atomic weight 0.02% lower than that which had been found from chemical determinations. Although this discrepancy was small, it lay outside the limits of error of the very careful determinations concerned, and Birge (1931) therefore concluded that a heavier isotope must be contained in very small amount in hydrogen. Towards the end of 1931 Urey subjected a quantity of liquid hydrogen to the distillation process, in order to enrich it in the supposed heavier isotope. In the atomic spectrum of the residue he detected two very weak lines, which would be obtained if, in eqn. (13a) (Vol. I, p. 87), the value of M were twice that which leads to the lines of the ordinary Balmer spectrum. Washburn and Urey shortly afterwards found that the heavy hydrogen isotope was relatively rapidly enriched during the electrolysis of water, and in 1933 G. N. Lewis and MacDonald were able to obtain a few cubic centimeters of practically pure 'heavy water', D_2O, by long-continued electrolysis.

(b) Preparation

One of the most useful processes for the preparation of heavy water has proved to be the exhaustive electrolysis of water (containing NaOH as electrolyte). As the volume of water diminishes, the solution becomes proportionately enriched in D_2O. Practically pure deuterium oxide is now obtainable commercially.

The hydrogen evolved in the electrolysis of water is enriched in light hydrogen, ^1H. The ratio $^1H : ^2H$ in the gas evolved may be between 3 and 20 times as great as the ratio in the solution. The reasons for this strikingly high separation effect are not yet fully understood. The ratio $^1H : ^2H$ in hydrogen gas which has been brought into exchange equilibrium with water (cf. p. 558) is only 3 times as high as in water. It is generally assumed that the preferential liberation of the lighter isotope ^1H during electrolysis can be attributed to different rates of discharge of the ions H^+ and D^+. These rates would differ if the potential barrier to be overcome in the act of discharging the ions differed in height for the two isotopes (cf. p. 700 et seq.). It is also probable that the rate at which the atoms combine to form molecules is greater for H than for D. In comparison with all other isotope separation processes, the separation achieved in the electrolysis of water is very high. Nevertheless, assuming the separation factor* to have the value 5, only 10 cc of 98% D_2O can be obtained from 1000 liters of ordinary water. The electrical energy necessary to decompose 1000 liters of water, with a cell voltage of 3.6 volts, is approximately 1000 kWh. Preparation of 10 cc of 99.99% D_2O would necessitate starting with five times as much ordinary water, and would therefore require five times as much electrical energy. In practice, the usual starting material is not ordinary water, but water from technical electrolysis apparatus—e.g., from accumulators which have been in use for a prolonged period—which is already somewhat enriched in D_2O**. The hydrogen evolved from solutions of high D_2O content is also burned, and the resulting water, which is enriched in D_2O, is again subjected to electrolysis. It is possible by this means to cut down the consumption of electrical energy to about 100 kWh per gram of D_2O.

Gaseous deuterium may be prepared by decomposing deuterium oxide with sodium, red hot iron or tungsten, or by the electrolysis of a solution of anhydrous sodium carbonate in deuterium oxide. It is also possible to obtain deuterium directly from hydrogen gas by the processes of diffusion or distillation considered on p. 547. This method of preparation, however, is far more troublesome than the extraction of deuterium from water, by way of deuterium oxide, since the latter is relatively easily obtained pure, by electrolysis.

(c) Properties

Deuterium differs very appreciably in its properties from 'light hydrogen', and therefore from ordinary natural hydrogen also. The deuterium content of ordinary

---

* Separation factor $= \dfrac{\text{Ratio } ^1H : ^2H \text{ in evolved gas}}{\text{Ratio } ^1H : ^2H \text{ in solution}}$

** If fresh (ordinary) water is continuously added to the electrolysis vessel to replace that which is decomposed, the water becomes enriched continuously in D_2O until the ratio of ^1H to D in the evolved gases becomes equal to the ratio of ^1H to D in the ordinary water.

hydrogen is so small that the properties of the latter are not significantly different from those of pure protium.

The accurate value for the chemical atomic weight of D is 2.01419, and that of ^{1}H is 1.00787*.

Ordinary hydrogen contains the molecules HD, as well as those of ^{1}H$_2$ and D$_2$. Since the reaction H$_2$ + D$_2$ = 2HD does not take place with measurable velocity at ordinary temperature, it is also possible to obtain HD ('deuterium hydride') in the pure state. A few properties of these three substances are collected together in Table 63.

Deuterium (but *not* HD) resembles light hydrogen in that it exists in two different modifications—*ortho*- and *para-deuterium*. At ordinary temperature, and in a state of equilibrium, deuterium consists of 2 parts of o-D$_2$ and 1 part of p-D$_2$. At absolute zero pure ortho-deuterium is the stable form. The relation between the two forms is thus the converse of that found for light hydrogen (cf. Vol. I, p. 44).

The values found for the moments of inertia** of the molecules D$_2$, HD and H$_2$ (cf. Table 63) indicate that the interatomic distances in these molecules cannot differ by more than a few hundredths of an Ångstrom unit, at the most.

Table 64 gives the wave lengths of the atomic spectrum of deuterium, calculated from eqn. (13a), Vol. I, p. 87, together with the wave lengths measured by Urey (1932), Rank (1932), and Ballard (1933). The difference between the lines of the deuterium and hydrogen spectra implies that the ionization potential of the D atom is 0.00365 volt higher than that of the H atom. The electrolytic solution potential of deuterium (in contact with deuterium oxide) is also about 0.003 volt higher*** than that of hydrogen in contact with water. Deuterium is thus slightly nobler than hydrogen.

TABLE 63

PROPERTIES OF DEUTERIUM, D$_2$, DEUTERIUM HYDRIDE, HD, AND PROTIUM, H$_2$

	D$_2$	HD	^{1}H$_2$
Molecular weight	4.02838	3.02206	2.01574
Boiling point (°K)	23.6	—	20.38
Melting point (°K)	18.65	16.60	13.95
Vapor pressure at the m.p. (mm)	121	—	54
Vapor press. at triple point (mm)	128.5	95	53.8
Heat of sublimation at 0 °K (cal per g)	274.0	228	183.4
Normal potential (volts)	+0.003	—	0.000
Overvoltage on Hg	0.93	—	0.80
Heat of formation of molecule from atoms (kcal per mol)	104.5	103.5	102.7
Moment of inertia (g.cm^2)	$9.25 \cdot 10^{-41}$	$6.19 \cdot 10^{-41}$	$4.67 \cdot 10^{-41}$
Thermal conductivity λ at 0 °K	$2.92 \cdot 10^{-4}$	—	$4.12 \cdot 10^{-4}$

* These values are obtained from the physical atomic weights (mass values) of Table 67, by conversion to the chemical atomic weight scale.

** The moment of inertia I of a diatomic molecule is given by $I = r^2\left(\dfrac{m_1 \cdot m_2}{m_1 + m_2}\right)$

where m_1, m_2 are the masses of the atoms.

*** If it is assumed that the solution potentials of D$_2$ and H$_2$ have approximately the same temperature coefficient, the fact that the ionization potentials and the solution potentials differ by about the same amount would imply that the energy of hydration of the D$^+$ ion must be about 1 kcal greater than that of the H$^+$ ion. The data indicate that the dissociation of D$_2$ into 2D (at 0° K) requires the expenditure of 0.90 kcal per g-atom more than for the dissociation of H$_2$. This extra energy must be roughly compensated, if the foregoing assumption is valid, by the greater heat of hydration of the D$^+$ ion.

The activation energy (cf. p. 700 *et seq.*) is about 0.85 kcal per mol greater for D_2 than for H_2. This is one reason why reactions with D_2 always take place more slowly than reactions with H_2 (cf. p. 557).

TABLE 64
LINE SPECTRUM OF THE DEUTERIUM ATOM

Lyman series $(n_2 = 1)$

	$n_1 = 2$	$n_1 = 3$	$n_1 = 4$	$n_1 = 5$	$n_1 = 6$	$n_1 = 7$
λ calculated, Å	1215.33	1025.44	972.27	949.48	937.55	930.49
λ observed, Å	1215.33	1025.44	972.27	949.48	937.53	930.49

Balmer series $(n_2 = 2)$

	$n_1 = 3$	$n_1 = 4$	$n_1 = 5$	$n_1 = 6$
λ calculated, Å	6561.00	4860.00	4339.28	4100.62
λ observed, Å	6560.99	4860.00	4339.27	4100.63

The most important properties of 'heavy water' are collected in Table 65. Comparison with the data given in Table 11, Vol. I, p. 49 , will show how the properties differ very significantly in some cases from those of ordinary water. Thus the melting point is 3.8° higher, and the maximum density occurs at 11.6 °C instead of at 4 °C.

D_2O crystallizes with the same structure as H_2O (cf. Vol. I, p. 50). Megaw (1934) gives for the dimensions of the unit cell at 0°: $a = 4.517$, $c = 7.354$ Å in the case of D_2O, and $a = 4.514$, $c = 7.352$ Å for H_2O.

The properties of H_2O-D_2O mixtures are derived practically additively from those of the pure components. In particular, there is only a very minute change in volume on mixing D_2O and H_2O. The density of water is thus very nearly a linear function of its D_2O content.

An equimolar mixture of H_2O and D_2O consists chiefly of HOD molecules, which are formed, slightly exothermically, by the reaction $H_2O + D_2O \rightleftharpoons 2HOD + 0.036$ kcal. The ion product $[D_3O^+] \cdot [OD^-]$ is about 8 times smaller than the ion product $[H_3O^+] \cdot [OH^-]$, at 25°. The electrolytic mobilities of ions are rather smaller in D_2O than in H_2O. The ions D_3O^+ and H_3O^+ display a considerable difference in mobility, the latter being about $^3/_2$ times as mobile as the D_3O^+ ion. The solubility of salts in D_2O is usually somewhat lower than in H_2O, when referred to equimolar quantities of the solvent. However, there are some salts which are more soluble in D_2O than in H_2O.

The D_2O (or DOH) content of water is usually determined either by measurement of the density or interferometrically (by taking advantage of the change of refractive index). The former method is generally the more convenient, and is extremely precise. The concentration of deuterium in gaseous hydrogen is usually found by measurement of the thermal conductivity.

In ordinary water the atomic ratio D : H is 1 : 6000. In general, it exhibits no significant variations, irrespective of the origin of the water. However, the D-content of certain highly saline lakes and inland seas is perceptibly enhanced (e.g., by about 0.003 atom-per cent in the Dead Sea). Deuterium also seems to be slightly enriched at great depths in the oceans (by about 0.0023 atom-per cent at 3000 meters). Certain animal and vegetable fluids (milk, blood, fruit juices) have also been found to show enrichment to the extent of a few thousandths atom per cent, as have certain minerals also.

TABLE 65
PROPERTIES OF DEUTERIUM OXIDE

Freezing point (at 760 mm press.)	$3.82°$
Boiling point (at 760 mm press.)	$101.42°$
Critical temperature	$371.5°$
Temp. of maximum density	$11.6°$
Density at 20 °C	1.1050
Specific heat at 15° (cal per g)	1.02
Latent heat of evaporation at 0° (cal per g)	556.6
Surface tension at 20° (dyne/cm)	67.8
Viscosity at 20° (centipoise)	1.26
Dielectric constant at 25°	77.5
Refractive index, n_D, at 20°	1.3284
Ion product at 25°	$1.6 \cdot 10^{-15}$
Molecular freezing point depression	$2.05°$

Since ordinary oxygen is a mixture of three isotopes, ordinary water contains not only $H_2{}^{16}O$, $HD^{16}O$ and $D_2{}^{16}O$, but also the corresponding molecules with ^{17}O and ^{18}O. There are thus 9 different species of distinct molecules in all. Since water has so often been used as a reference substance for physical constants, it is most important to prove that the proportions of these various molecular species in ordinary water is always the same. During the electrolysis of water, the heavier isotopes of oxygen may undergo enrichment, as well as the deuterium. The excess of the heavier isotopes of oxygen can be removed by isotopic exchange with some anhydrous substance which contains the oxygen isotopes in their normal proportions (e.g., with ordinary sulfur dioxide). The physical constants listed for D_2O in Table 65 refer to deuterium oxide in which the oxygen isotopes have been 'normalized' to their ordinary, natural proportions.

(d) Chemical Properties

Deuterium and deuterium compounds are qualitatively identical with hydrogen and its compounds in their chemical behavior. There are, however, significant differences in the magnitudes of the equilibrium constants and especially in reaction velocities.

The differences in reaction rates usually become more pronounced, the lower the temperature at which reaction occurs. Thus H_2 reacts with Cl_2, in the presence of CO as catalyst, roughly 13 times as fast as deuterium at 0°, and about 10 times as fast at 35° (Rollefson, 1934). In the presence of nickel, H_2 reacts with O_2 at 180° 2.4 times as rapidly as D_2, and with N_2O at 160° 2.5 times as fast as does D_2 (Melville 1934). The conversion of p-H_2 to o-H_2 at room temperature, in the presence of oxygen as catalyst, is 16 times as fast as the corresponding conversion of o-D_2 into p-D_2.

A consequence of the lower reactivity of deuterium compounds, as compared with hydrogen compounds, is that reactions may take place considerably more slowly in deuterium oxide solution than in ordinary water, if the reaction investigated must be preceded by an exchange of H for D. Thus the decomposition of nitramide, catalysed by OH^- ions ($H_2N \cdot NO_2 = H_2O + N_2O$) is much slower in heavy water than in ordinary water, since in the former case *deuteronitramide*, $D_2N \cdot NO_2$, is first formed. Thus in heavy water the reaction observed is the decomposition of deuteronitramide, which takes place about 5 times slower than that of ordinary nitramide. There are also differences in reaction rates in ordinary water and heavy water—though of rather smaller magnitude—in cases where the water only influences the velocity of the reaction through its physical properties (viscosity, dielectric constant, solvating properties, etc.). In such instances also, the rates in heavy water solution are always lower than in ordinary water. There are some reactions, however, which are faster in heavy water than in ordinary water. These are mostly reactions which are catalyzed by the H_3O^+ and D_3O^+ ions, or by OH^- and OD^- ions. It may be deduced from the theory of homogeneous catalysis that the stronger binding of the deuteron, as compared with the proton, can result in an enhanced rate of reaction in such cases (Reitz, 1938).

H_2 and D_2 do not react with each other directly at ordinary temperature. They do so at high temperatures, however, and also in presence of a heated nickel wire.

This reaction does not go to completion, but leads to an equilibrium:

$$H_2 + D_2 \ \rightleftharpoons \ 2HD; \quad \frac{[HD]^2}{[H_2] \cdot [D_2]} = K_1.$$

K_1 has the value 3.3 at room temperature, and 3.8 at 400°. Thus the equilibrium shifts in favor of formation of HD as the temperature is raised.

The equilibria in the reactions $H_2 + DCl \rightleftharpoons HD + HCl$, and $H_2 + DI \rightleftharpoons HD + HI$ are given by

$$\frac{[HD] \cdot [HCl]}{[H_2] \cdot [DCl]} = K_2, \quad \text{and} \quad \frac{[HD] \cdot [HI]}{[H_2] \cdot [DI]} = K_3.$$

At 125°, $K_2 = 1.5$ and $K_3 = 2.0$; with rise of temperature these equilibria are also displaced in such a way as to favor formation of HD.

The equilibrium constant K_4 of the reaction

$$H_2O + HD \rightleftharpoons HOD + H_2; \quad \frac{[HOD] \cdot [H_2]}{[H_2O] \cdot [HD]} = K_4$$

has the value 3.1 at 20°. Thus hydrogen gas has a deuterium content only one third of that of water with which it is in equilibrium. This equilibrium can be attained at ordinary temperature in the presence of suitable catalysts. In certain cases conditions may be such that this exchange equilibrium can be reached during the evolution of hydrogen from solutions—for example, in the decomposition of sodium formate by palladium black or by enzymes ($HCO_2Na + H_2O = H_2 + CO_2 + NaOH$), or in the electrolytic generation of hydrogen at an electrode covered with platinum black. Farkas (1934) proved that under these conditions the ratio H : D in the gas evolved corresponded to the above equilibrium. As a rule, however, the proportion of light hydrogen liberated in the electrolysis of water is considerably higher than the equilibrium amount.

(e) Exchange Reactions with Deuterium Oxide

If compounds containing ionically bound hydrogen are introduced into heavy water, the hydrogen undergoes instantaneous exchange for deuterium at ordinary temperature. Thus HCl reacts immeasurably fast with D_2O:

$$HCl + D_2O = DCl + HOD.$$

It has been possible to show by means of such exchange reactions that all the H atoms in ammonia are exchangeable. It does not necessarily follow that in NH_3 itself the hydrogen is ionically bound. Exchange may proceed by way of the NH_4^+ ion, and it may be inferred that all the atoms of the NH_4^+ ion are equivalent:

$$NH_3 + D_2O = NH_3D^+ + OD^- = NH_2D + HOD, \text{ etc.}$$

Derivatives of ammonia such as methylamine, CH_3NH_2, aniline, $C_6H_5NH_2$ and acetanilide, $CH_3CONHC_6H_5$, behave similarly. In these and other compounds, the H atoms linked to carbon do not undergo exchange by D at ordinary temperature. Hydrogen combined with oxygen or sulfur can be exchanged for deuterium in the same way as that bound to nitrogen or halogen. Thus both H atoms of hydrogen peroxide, and the H atoms of organic hydroxyl groups (e.g., in acids, phenols, and alcohols) are exchangeable.

Where H atoms are joined to C atoms which are adjacent to CO- or NO_2-groups, it is found that although they do not undergo instantaneous exchange on treatment with D_2O, there is a gradual exchange of H for D. The rate at which exchange takes place depends upon the rate at which the keto form of the compound concerned undergoes conversion to the enol form (e.g., CH_3—CO—CH_3 into CH_3—$C(OH)$=CH_2), or the nitro-compound changes to the aci-form (e.g., CH_3—NO_2 into CH_2=$NO(OH)$).

At higher temperatures, or through the action of catalysts, it is possible to effect exchange of H atoms which are not ionically bound. Examples of such exchange processes include those considered above—the reaction of H_2 with D_2, of H_2 with DCl and DI, and the reaction between H_2 and HOD. In the case of organic compounds it is often possible to bring about a slow exchange of H atoms bound to carbon, by treatment with concentrated 'heavy sulfuric acid', D_2SO_4 or $HDSO_4$.

Equilibria are established in the exchange of ionically bound hydrogen by deuterium, just as in the reactions of H_2 discussed previously. Account must be taken of the position of equilibrium when exchange reactions are carried out with 'fairly heavy' water—i.e., water containing considerable quantities of H_2O as well as D_2O.

Since it has been observed that only ionically bound hydrogen atoms undergo immediate exchange with heavy water, it is possible, in ambiguous cases, to use exchange reactions, to distinguish between ionic and non-ionic hydrogen. Thus it has been established that only *one* H atom of the acid H_3PO_2 (cf. Vol. I, p. 631) can be exchanged for deuterium. It had already been inferred from the other properties of this compound that it contained only one ionically bound hydrogen atom. The result of the exchange experiment with heavy water confirms this conclusion.

Exchange reactions with heavy water can frequently be employed in the study of reaction mechanisms. They are also of value in biological and physiological chemistry. It has been found that deuterium is built into the organism in just the same way as hydrogen (though relatively less D than H is taken up). Deuterium oxide has a toxic effect at high concentrations, possibly because it disturbs the normal reaction equilibria in the organism; it is fatal to higher organisms at high concentrations. It is quite harmless if highly dilute, and it is therefore possible to use deuterium oxide as an indicator or 'tracer' in metabolic studies. The course followed in the organism by any compound containing firmly bound (i.e., not exchangeable) deuterium can be accurately followed.

(f) Other Deuterium Compounds

Numerous other deuterium compounds are known, in addition to the oxide[6]. These may be prepared by essentially the same reactions as are used for the ordinary hydrogen compounds. The deuterated analogues of compounds containing ionically bound hydrogen are most readily prepared from the latter by exchange reactions with D_2O.

Like deuterium oxide, other deuterium compounds often differ quite considerably in properties from the ordinary hydrogen compounds. They often have lower melting points than the latter (although higher in some instances, as with D_2O). The vapor pressure of the deuterium compounds is generally lower than that of the hydrogen compounds at low temperatures, but greater at high temperatures. The latent heats of fusion and evaporation also generally differ perceptibly from those of the hydrogen compounds. Differences are especially marked in the temperatures at which phase transformations occur in the solid state, and there are corresponding differences in the heats of transformation. Thus CH_4 undergoes a transformation in the solid state, at $20.4°$ K, with a heat of transformation of 15.7 cal per mol. The corresponding transition of CD_4 occurs at $26.3°$ K, with a heat effect of 58.7 cal per mol. ND_4Br is notable in that it undergoes a transition at $-104°$ C, which has no counterpart in the observed behavior of NH_4Br. Corresponding to the transition points of NH_4Br at $+141°$ C and $-38.8°$ C, ND_4Br undergoes transitions at $+132°$ and $-58.5°$ C (Clusius, 1938).

6. Artificial Atomic Transmutations

(a) Atomic Transformations Due to Alpha Particles

In 1919 Rutherford proved that it was not only when α-particles passed through hydrogen compounds that hydrogen nuclei (protons) were obtained as positively charged secondary rays. They were also obtained when the rays passed through air or nitrogen, but not through oxygen. Furthermore, nitrogen furnished 25% more protons than air. From these results Rutherford inferred that the protons had been ejected from the nuclei of the nitrogen atoms by the action of the α-rays. Proof that the particles observed were protons was afforded by measurement of their deflection in a magnetic field. This result was of outstanding importance, since it represented the *first observation of an artificially caused atomic transmutation*. In radioactive processes, the atoms are undergoing spontaneous disintegration. This was the first instance in which an atomic nucleus had been transmuted—i.e., in which one chemical element had been changed into another—at will, by subjecting it to bombardment with α-rays. Corresponding transformations were subsequently observed in the case of many other elements on bombardment with α-rays.

Later investigations showed that the transformations undergone by the atomic nuclei in these processes did not consist merely of the expulsion of a hydrogen nucleus from the atomic nucleus struck. As a general rule, the α-particle responsible for expulsion of the proton was built into the atomic nucleus in its place. The nucleus resulting from the transformation was thus, in most cases, not the species with mass and charge each one unit smaller than the original, but had a mass 3 units greater, and a charge 1 unit greater (corresponding to the difference in mass numbers and nuclear charges between $^4_2He^{2+}$ and $^1_1H^+$). Since the nuclear charge is equal to the atomic number, it follows that the product obtained by α-particle bombardment of $^{14}_7N$ would be the oxygen isotope $^{17}_8O$, and not the carbon isotope $^{13}_6C$, as might at first have been expected.

(b) Neutrons

In 1930, Bothe found that in many of the transformations brought about by α-rays, and especially in the α-ray irradiation of beryllium, protons were not ejected from the nuclei, but a new type of rays appeared. Shortly afterwards, Chadwick proved that these rays consisted of particles, with a mass almost equal to the mass of the proton, but *uncharged*. These received the name of *neutrons* (symbol n). In the α-irradiation of beryllium, neutrons are formed by the process $^9_4Be + ^4_2He = ^{12}_6C + ^1_0n$. Since the neutron is virtually an atomic nucleus without any charge, it might also be regarded as the chemical element of atomic number zero.

Together with protons, neutrons constitute the elementary structural units out of which all atomic nuclei are built up (cf. p. 566). The neutron is unstable in the free state, however, and after a mean life of about 20 min. it 'disintegrates' spontaneously into a proton and an electron. As can be seen from the data of Table 66, this 'disintegration' involves a not inconsiderable loss of mass.

Since neutrons bear no charge, they can pass without interaction through the electron clouds of the atoms. Thus they do not strip any electrons out of the extranuclear structure— i.e., have *no ionizing action*. In consequence of this, they are very penetrating. If their initial velocity is high enough, they will pass even through a layer of lead half a meter thick. They are stopped only by the rare event of collision with the *nucleus* of an atom. As follows from the laws of elastic collisions*, the maximum loss of velocity is sustained by collision of the neutron with a nucleus of similar mass, and especially with a *hydrogen nucleus*. This accords with the results of experiment, that compounds containing hydrogen are particularly efficient in stopping neutrons. Thus a layer of paraffin wax 20 cm thick completely stops even neutrons of high initial velocity.

Neutrons which are in thermal equilibrium with their surroundings—i.e., neutrons having a kinetic energy equal to that of gas molecules at the given temperature—are known as *thermal neutrons*. According to eqn. (14), Chap. 17, the mean kinetic energy of gas molecules—and hence of thermal neutrons—is $^3/_2 kT$ (where $k(= R/N_A)$ is Boltzmann's constant: $k = 1.38 \cdot 10^{-16}$ ergs or $8.61 \cdot 10^{-5}$ electron volts). At $T = 293°$ K, the mean kinetic energy of gas molecules is thus 0.038 ev. In the case of neutrons, it is usual to cite not the mean kinetic energy, but the 'probable kinetic energy'—i.e., the value for the kinetic energy corresponding to the most probable velocity at some given temperature. This is equal to kT, and has the value 0.025 ev at ordinary temperature. Even at 3000° K, the kinetic energy of thermal neutrons (0.258 ev) is still very small compared with the energy of neutrons emitted in nuclear reactions, which often have energies of a few million electron volts. Neutrons emitted in nuclear processes are therefore called '*fast neutrons*', whereas those which have kinetic energies in the range covered by thermal neutrons are called '*slow neutrons*'.

(c) Positrons

Yet another type of radiation is emitted in certain atomic transformations— namely *positron* rays. These consist of high velocity particles, with a mass equal to that of the electron, but bearing an opposite—i.e., a *positive* charge. Positrons are therefore often referred to as *positive electrons*, as distinguished from *negative electrons* (i.e., ordinary electrons).

The positron was discovered in 1932 by Anderson, in the course of investigations on cosmic rays**. Shortly afterwards it was found that positrons were also emitted in the nuclear reactions brought about by *a*-rays. The positron has only a very short life. It is, indeed, not unstable in the strict sense, but disappears if it collides with an electron. The two particles undergo mutual annihilation, their masses being converted into a quantum of *γ*-radiation (cf. p. 567 further upon this point).

(d) Mesons

Massive particles of still another kind are also formed through the action of cosmic rays— the so-called *mesons* (that is, 'intermediate' particles, because they have a mass between that

---

* If two spheres of equal mass collide, they exchange velocities in the case of a central elastic collision. If the sphere struck was initially at rest, then after the collision the impinging sphere has zero velocity. If, however, the colliding sphere strikes a solid wall, it rebounds with unchanged velocity. Quite generally, for a central elastic collision,

$$v_1 = 2\frac{m_1 c_1 + m_2 c_2}{m_1 + m_2} - c_1,$$ where c_1, c_2 are the velocities of the spheres of masses m_1, m_2 before the collision, and v_1 is the velocity of the body of mass m_1 after the collision.

** The cosmic rays [7]—discovered by Hess in 1912—are the radiations of extraordinarily great penetrating power, which reach the earth from outer space.

of the electron and that of the proton). There are several kinds of mesons. The mesons most frequently observed have a rest-mass about 210 times as great as that of the electron (m_e). These mesons—μ-mesons—have a mean life of about $2.15 \cdot 10^{-6}$ sec. They may be either positively or negatively charged; the charge on the meson is equal to that on the electron. *Neutral mesons* have also been detected—e.g., the π^0-mesons, of mass $265 m_e$, and mean life 10^{-14} secs. Both positively and negatively charged mesons of mass $276 m_e$ are also known (π^+ and π^--mesons, mean life $2.5 \cdot 10^{-8}$ sec). It is probable that there are still heavier mesons (with masses up to 2200 times that of the electron). They decay extremely rapidly, however, with lives of 10^{-10} to 10^{-9} sec., forming lighter mesons. The mesons of mass $210 m_e$ decay to form electrons—the negatively charged μ^--mesons giving an ordinary electron, and the positively charged μ^+-meson a positron. It is assumed that two neutrinos are formed simultaneously in each case (cf. pp. 522, 590).

(e) The Elementary Particles [7]

Electrons, positrons, mesons, neutrons, and protons are termed the *elementary* or *fundamental particles*, since they represent the smallest particles that have been observed in nuclear phenomena. Table 66 sets out the accurate mass values of these particles. Since deuterons and helium nuclei, as well as the fundamental particles, are frequently ejected from the atomic nucleus during transmutations, the masses of these species are included.

TABLE 66

MASSES AND CHARGES OF THE FUNDAMENTAL PARTICLES, AND
OF THE DEUTERON AND HELIUM NUCLEUS

Name	Symbol	Charge*	Mass value
Helium nucleus or α-particle	α	$+2$	4.002779
Deuteron	D$^+$ or d	$+1$	2.014192
Proton	H$^+$ or p	$+1$	1.007596
Neutron	n	0	1.008979
Electron	e$^-$	-1	0.0005486
Positron	e$^+$	$+1$	0.00055

* In multiples of the elementary quantum of electricity.

Photons (i.e., quanta of electromagnetic radiation, $h\nu$) and *neutrinos* are also now included among the elementary particles. The neutrino is an uncharged particle, with a rest-mass certainly less than one-thousandth of the rest mass of the electron; it is possible that it may have zero rest-mass, like the photon. The real nature of the elementary particles is perplexing. None of them can be regarded as immutable, ultimate units; on the other hand, it appears impossible to suppose that they are built up from some still more fundamental particles. The current tendency is to regard the different kinds of fundamental particle as different 'modes of appearance' or 'states' of some entity which cannot be more closely specified, but which can exist both in the form of matter and as energy.

(f) Different Sorts of Atomic Transformations; Nuclear Chemistry

It has been shown that nuclear reactions can be brought about not only by α-rays, but by protons, deuterons and more massive charged particles*, by neutrons, and γ-rays.

* Nuclear transmutations have also been produced by electrons of very high energy. ^{3}He nuclei have occasionally been employed in place of α-particles (^{4}He nuclei). At Berkeley (United States), Birmingham (England) and Stockholm (Sweden), 'stripped' atoms of carbon, nitrogen and oxygen—e.g., $^{12}_6$C^{6+}, $^{14}_7$N^{6+} and $^{16}_8$O^{6+}—have also been successfully accelerated to high energies and used for nuclear reactions.

A proton or deuteron can indeed penetrate the atomic nucleus with considerably greater ease than can a helium nucleus (α-particle), since the smaller charge borne by the proton or deuteron is subject to correspondingly smaller forces of electrostatic repulsion. The neutron can enter a nucleus with much greater ease still. Advantage is taken of this especially in the artificial production of radioactive species of atoms.

Nuclear reactions [9–11] may be represented by equations which correspond exactly to those employed for chemical reactions. That section of nuclear physics which deals particularly with transmutations of the elements, occurring in nuclear reactions, has appropriately been termed 'nuclear chemistry'*.

The fundamental law regulating nuclear reactions is that in all nuclear transmutations, *the sum of the mass numbers of the reacting particles and the sum of their charges remain unchanged as a result of reaction.*

The following equations give examples of some nuclear reactions.

$$(1) \quad {}^{10}_{5}B + {}^{4}_{2}\alpha = {}^{12}_{6}C + {}^{2}_{1}d$$
$$(2) \quad {}^{10}_{5}B + {}^{4}_{2}\alpha = {}^{13}_{6}C + {}^{1}_{1}p$$
$$(3) \quad {}^{10}_{5}B + {}^{4}_{2}\alpha = {}^{13}_{7}N + {}^{1}_{0}n$$
$$(4) \quad {}^{10}_{5}B + {}^{2}_{1}d = {}^{11}_{5}B + {}^{1}_{1}p$$
$$(5) \quad {}^{10}_{5}B + {}^{2}_{1}d = {}^{11}_{6}C + {}^{1}_{0}n$$

$$(6) \quad {}^{27}_{13}Al + {}^{2}_{1}d = {}^{25}_{12}Mg + {}^{4}_{2}\alpha$$
$$(7) \quad {}^{27}_{13}Al + {}^{2}_{1}d = {}^{28}_{13}Al + {}^{1}_{1}p$$
$$(8) \quad {}^{27}_{13}Al + {}^{2}_{1}d = {}^{28}_{14}Si + {}^{1}_{0}n$$
$$(9) \quad {}^{27}_{13}Al + {}^{1}_{1}p = {}^{24}_{12}Mg + {}^{4}_{2}\alpha$$
$$(10) \quad {}^{27}_{13}Al + {}^{1}_{0}n = {}^{27}_{12}Mg + {}^{1}_{1}p$$
$$(11) \quad {}^{27}_{13}Al + {}^{1}_{0}n = {}^{28}_{13}Al + \gamma$$

$$(12) \quad {}^{7}_{3}Li + {}^{1}_{1}p = {}^{8}_{4}Be + \gamma$$
$$(13) \quad {}^{79}_{35}Br + {}^{1}_{0}n = {}^{78}_{35}Br + 2{}^{1}_{0}n$$

$$(14) \quad {}^{2}_{1}D + {}^{4}_{2}\alpha = {}^{1}_{1}H + {}^{1}_{0}n + {}^{4}_{2}\alpha$$
$$(15) \quad {}^{2}_{1}D + \gamma = {}^{1}_{1}p + {}^{1}_{0}n$$

It is customary to denote the particles used for bombardment of the nucleus, and the particles ejected in the reaction, by small letter symbols—i.e., protons by p, deuterons by d, helium nuclei by α, and neutrons by n.

Mass numbers and nuclear charges have been written against all the symbols in the foregoing equations, for the sake of clarity. It is usual to specify only the mass numbers of the atoms involved (e.g., ${}^{27}Al + d = {}^{25}Mg + \alpha$), since the other quantities are implicit in the symbols used. An abbreviated symbolism is very often used to denote nuclear reactions. In this, the bombarding particle and the ejected particle (in that order) are written in a bracket, and this is set between the symbol for the atom subjected to bombardment and the symbol for the product atom. Thus the reaction (6) referred to above would be represented as: ${}^{27}Al \, (d, \alpha) \, {}^{25}Mg$. Transformations brought about by bombardment with deuterons, and resulting in the ejection of α-particles are referred to briefly as (d, α) changes; those brought about by deuteron bombardment with ejection of protons as (d, p) reactions, etc.

The examples given above serve to show that one and the same atom may undergo a considerable number of different nuclear reactions. These may give rise to nuclei of either higher or lower atomic number, or to nuclei of the same atomic number but different mass (i.e., *isotopes*); the products can also be atoms of different atomic number, but the same mass (*isobars*). Fig. 69 shows how the mass number A and the atomic number Z of a given nucleus may be changed in the most important types of transformations brought about by reaction with p-, d-, n-, α- and γ-radiations. The arrows show the amount and the direction of the changes in A and Z for the types of reaction given. Thus, the mass number increases by 3 units, and the atomic number by 1 unit in an (α, p) process; a (d, p) or (n, γ) reaction increases the mass number by 1 unit without changing the atomic number; in an (n, p)

* The essential goal of *nuclear physics* [12–21] is to interpret the nature of nuclear forces. One important approach to this objective is the study of transmutations occurring in nuclear reactions, and the magnitude of the energy liberated in such processes. Originally this subject could be investigated only by purely physical methods, and as such could be considered a section of pure physics. With the discovery of artificial radioactivity, the subject was opened up for chemical studies, and at the same time became of great importance to the chemist, since it provided a powerful new technique which could be applied to the investigation of chemical problems.

reaction the atomic number is decreased by 1 unit without change in the mass number.

The particles ejected in nuclear reactions are usually counted by means of a *Geiger counter*. This consists essentially of a wire, stretched axially in a glass or metal tube, and charged to a potential of a few thousand volts relative to the tube. The ions produced by the high energy particles are so accelerated by this potential gradient between wire and tube wall that, in their turn, they ionize further gas molecules. The effect of the ions primarily produced is thereby greatly intensified, so that the entry of each individual ionizing particle into the counter tube gives rise to a distinct pulse of current, which can readily be detected, amplified and recorded by electronic methods. The observation of particle tracks in the Wilson cloud chamber has also proved a very valuable aid to the accurate investigation of nuclear transformations [22].

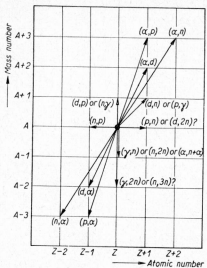

Fig. 69. Changes undergone by nuclear charge and mass number during artificial nuclear transformations.

Since it is possible to count very accurately the number of particles emitted during radioactive decay, and hence to determine the number of atoms disintegrating per unit time, this measurement is used as a measure of the intensity of radioactivity. The activity of a natural or artificial radioactive preparation is therefore expressed in *disintegrations per second* (dps). Intense activities are usually expressed in *curies* (smaller units being the millicurie, 10^{-3} curies, and the microcurie, 10^{-6} curies). The curie is defined by 1 *curie* = 3.7000 · 10^{10} *dps*; it was formerly defined as measuring that quantity of a radioactive substance which underwent the same number of disintegrations per second as 1 g of radium. Although approximately correct, this definition has been given up, for the reason that the factor for converting dps into curies would be dependent upon the value accepted at any time as the most reliable measurement of the half-life of radium.

Many artificially produced nuclear transformations, especially those brought about by charged particles, are of the type known as *exchange reactions*—i.e., reactions in which the colliding particle becomes incorporated in the nucleus, while some other particle is ejected. These nuclear reactions correspond to the substitution reactions of ordinary chemistry. There are also reactions analogous to the addition reactions of ordinary chemistry and to thermal and photochemical dissociations.

In the equations given above for typical nuclear reactions, processes (1) to (10) are exchange reactions. Processes (11) and (12) are addition reactions; in both cases the colliding particle is captured by the nucleus, without any other particle being ejected. The energy liberated in the reaction is liberated in each case as a quantum of γ-radiation. Equations (13) and (14) represent nuclear dissociations, attributable to the kinetic energy of the bombarding particle, in the same way as thermal dissociation is caused in gas molecules. Reaction (15) is a photochemical dissociation of the nucleus, caused by γ-rays.

Several examples of reversible nuclear reactions are already known—e.g.,

$$^{1}H + n \rightleftharpoons D + \gamma; \qquad ^{7}Li + a \rightleftharpoons ^{10}B + n; \qquad ^{11}B + a \rightleftharpoons ^{14}N + n.$$

(g) Energies of Nuclear Transformations

The quantities of energy liberated in nuclear reactions are enormous in comparison with those set free in ordinary chemical reactions. Thus, in the bombardment of beryllium with protons, giving the lithium isotope ^{6}Li and helium, the energy liberated is 2.13 million electron volts (2.13 mev), which is equivalent to 49.1 million kcal per g atom of Be (cf. Vol. I, p. 113). Values for the energy liberated in some other reactions are:

$$^{18}\text{O}\,(\text{p},\,a)^{15}\text{N} \qquad 3.97 \text{ mev} = \quad 91.5 \cdot 10^6 \text{ kcal per g-atom}$$
$$^{15}\text{N}\,(\text{p},\,a)^{12}\text{C} \qquad 5.02 \text{ mev} = \quad 115.7 \cdot 10^6 \text{ kcal per g-atom}$$
$$^{7}\text{Li}(\text{p},\,a)^{4}\text{He} \qquad 17.33 \text{ mev} = \quad 399.5 \cdot 10^6 \text{ kcal per g-atom}$$
$$^{6}\text{Li}(\text{d},\,a)^{4}\text{He} \qquad 22.28 \text{ mev} = \quad 513.6 \cdot 10^6 \text{ kcal per g-atom.}$$

Thus the energies liberated in nuclear reactions are a million times as great as those involved in ordinary chemical reactions. There are also nuclear reactions which are endothermic, and the energy change in these is of the same order as in the exothermic processes. Thus in the process $^{14}\text{N}(a,\,\text{p})^{17}\text{O}$ the energy absorbed is 1.20 mev ($= 27.7 \cdot 10^6$ kcal per g-atom), and the process $\text{D}(\gamma,\,\text{n})\text{H}$ is endothermic to the extent of 2.19 mev ($= 50.5 \cdot 10^6$ kcal per g-atom).

Although the quantities of energy liberated in nuclear reactions are so great, the *extremely small yield* of the reactions makes it impossible in general to utilize this source of energy. The diameter of the nucleus is very small, as compared with that of the atom (cf. Vol. I, p. 83); only a minute proportion of the particles which traverse an atom in the course of their track actually collide with the nucleus. Furthermore, only a fraction of such collisions results in the bombarding particle penetrating the nucleus, thereby resulting in a nuclear transformation. For example, an a-particle traverses about 10,000 molecules per cm of its path; the yield of proton rays is only *one* proton per 100,000 a-particles (i.e., collision between the a-particle and the nucleus takes place in only one case for every 10^9 molecules traversed). Under comparable conditions, the yield of protons from the bombardment of N_2 molecules with a-rays is ten times smaller still. This implies that only $^1/_{10}$ of the collisions between a-particles and nitrogen nuclei results in a transformation, giving a total yield of one collision leading to reaction in 10^{10} collisions between a-particles and N_2 molecules. Quite generally, the yields (i.e., the number of atoms transmuted per incident particle) found hitherto in nuclear reactions produced by a-, d- or p-rays are between 10^{-5} and 10^{-11}. The yields of reactions involving neutrons may be substantially greater. However, the neutrons themselves must be produced by reactions which have a much smaller yield (cf. p. 575).

The foregoing statements concerning yields of nuclear reactions apply to bombarding particles which have a sufficiently great kinetic energy to overcome the strong repulsive forces which the nucleus exerts upon an approaching positively charged particle, according to Coulomb's law. The a-particles emitted from radioactive substances, by means of which the first nuclear reactions were carried out (and which are still widely used), have kinetic energies of a few million electron volts. It has been shown, indeed, that energies of a few hundred thousand electron volts may suffice to bring about reaction with protons and deuterons. Even although collisions involving such energies are sufficient, it is not a simple matter to produce a beam of uniformly accelerated particles. Bombarding particles with energies of 1 mev and above can be obtained only by using costly high-voltage apparatus.

(h) Production of Bombarding Particles

The oldest method for producing highly accelerated bombarding particles is the secondary acceleration method introduced by Wien (1925). This was applied to nulcear reactions by Cockroft and Walton (1932). The underlying principle is that canal (positive) rays, from a discharge in helium, hydrogen, or deuterium are subjected to acceleration in a strong electric field, after they have emerged from the discharge vessel proper. By this means, it is possible to obtain the particles in relatively high concentration (in beams carrying a current up to several hundred microamperes). It is difficult to increase the energy of the individual particles much above a few hundred thousand electron volts, if orthodox methods are used to generate the requisite high tension accelerating voltage. Special types of high voltage generators have accordingly been developed to give accelerating voltages of 1 mev and above. One of the most important of these is the high voltage machine first constructed by Van de Graaff (1933), which works on the same principle as the old static electric machines. Other generators involve combinations of rectifiers and condensers in voltage-doubler circuits. Particles of the highest energies are produced by the *cyclotron*, devised by Lawrence (1932). This is an apparatus in which positive ions (H^+, D^+, He^{2+}, and more recently C^{6+}) are accelerated by making them traverse a spiral orbit in a strong magnetic field. In the course of this, they pass a few hundred or a few thousand times through an electrical field of

10–100 kilovolts. The period of orbital rotation of the ions in the magnetic field (which is independent of the radius of the orbit) is arranged to synchronizing with the alternating polarity of the electric field, with the result that the ions always traverse the electric field in the same direction. By this means, the accelerations received by the particles at each passage across the electric field are additive, and the ions can be accelerated to velocities which would correspond to a voltage drop of several million volts. Development of the principle of the cyclotron has made it possible to observe drastic nuclear dissociations brought about by He^{2+} ions of 300–400 mev energy. The maximum energy as yet obtainable in accelerating machines is 2000–3000 mev; the 'nuclear chemical' effects of such high energy particles have been but little explored.

(i) The Structure of the Nucleus [16–21]

Disintegration of the natural radioelements is attended by the emission of either helium nuclei or electrons; it was at first inferred, accordingly, that the nuclei of the heavier elements were built up from helium nuclei and electrons. However, smaller particles than helium nuclei are ejected in the course of artificial nuclear reactions—namely protons and neutrons. The current view, therefore, is that *protons* and *neutrons* are the basic structural units of all nuclei. It follows from nuclear physical investigations that, within the nucleus, the Coulomb repulsion between protons must be outweighed by very powerful forces of attraction, which are operative only at very short range. It is therefore not necessary to postulate that negatively charged particles (i.e., electrons) must be present within the nucleus to confer stability. On the contrary, there are many objections to such an assumption [16], and protons and neutrons are therefore now regarded as the sole structural units in the nucleus; as such, they are described by the common designation of *nucleons*.

The fact that electrons are very commonly ejected from the nucleus in the act of radio-active disintegration finds its explanation in the transformation of a neutron into a proton and an electron: $n = H^+ + e^-$. It must not be assumed that the neutron is *built up from* a proton and an electron, for the proton can, conversely, be transformed into a neutron and a positron: $H^+ = n + e^+$. The present tendency is rather to consider that the proton and the neutron are two different states of a single entity, the nucleon, differing in their charge and mass. The emission of helium nuclei from the nuclei of the heavy elements during their radioactive disintegration is to be explained by the particularly great stability of the helium nucleus (cf. p. 568). It is quite probable that the helium nuclei are not merely formed in the act of disintegration, but are already present within the nuclei of the heavy elements.

Thus the results of nuclear physics support the hypothesis advanced in 1815 by the physician Prout—although in a very much altered form; namely, that in the ultimate analysis all the elements were built up from hydrogen, as the primary form of matter. This hypothesis was originally based upon the assumption that the atomic weights of all the elements were whole-number multiples of the atomic weight of hydrogen. However, accurate determinations of the atomic weights soon showed that this assumption was incorrect. We now know why the atomic weights deviate from integral values. This is in part due to most elements being mixtures of isotopes. However, even the individual isotopes and the pure elements have atomic weights which are definitely not whole numbers when referred to hydrogen = 1. This is because of the loss of mass which occurs, according to the Einstein mass-energy equivalence, as a result of the enormous amount of energy evolved in the formation of atomic nuclei from protons and neutrons.

According to the principle of *equivalence between mass and energy* (Hasenöhrl, 1904, Einstein, 1905), every body which loses energy simultaneously suffers a decrease in mass, given by

$$M = \frac{E}{c^2} \qquad (2)$$

where M is the change in mass (in g), E the change in energy (in ergs), and c the velocity of light (cm-sec^{-1}). An energy evolution of 1 mev ($= 1.6015 \cdot 10^{-6}$ erg—cf. Vol. I, p. 112) therefore corresponds to a loss of mass of

$$\frac{1.6015 \cdot 10^{-6}}{(2.9978 \cdot 10^{10})^2} = 1.7820 \cdot 10^{-27} \text{ g}.$$

This is the loss of mass of a single atom; for 1 g-atom (multiplying by N_A) it corresponds to a mass loss of $1.0736 \cdot 10^{-3}$ g per 1 mev energy per atom evolved. The loss of mass (or *mass defect*) in nuclear reactions is usually expressed in m.m.u. ($=$ milli-mass units, 10^{-3} units on the physical atomic weight scale). Since 1 m.m.u. $= {}^1/1000.275$ of the chemical atomic weight unit used as a basis for the foregoing calculation, the numbers given above should be multiplied by 1000.275. The evolution of 1 mev energy therefore corresponds to a mass defect of 1.0739 m.m.u. Conversely, a mass defect of 1 m.m.u. represents the evolution of 0.9312 mev energy in a nuclear reaction.

It may now be stated that the principle of mass-energy equivalence has been thoroughly confirmed by experiment. It has been possible in whole series of nuclear reactions to prove directly that the energy of the colliding particle and the energy of the ejected particle (as obtained from the range of the rays) differ by an amount which is quantitatively related to the loss of mass in the reaction by the relation of eqn (2). The loss of mass accompanying the liberation of energy is particularly striking in the reaction of a positron with an electron. The charges are thereby neutralized, and the mass of both particles disappears simultaneously, while two quanta of γ-radiation are emitted; the energy of each of these is given by the relation $h\nu = mc^2$. The converse of this process, 'pair production', the creation of a pair of massive particles (electron + positron) from γ-radiation of sufficiently high frequency, has also been observed*.

(j) Mass Defect and Packing Fraction

The Einstein mass-energy equivalence law implies that mass is necessarily lost in the process of forming a stable nucleus from protons and neutrons, since energy is set free in the process. The difference between the mass of the nucleus, and the sum of the masses which the constituent protons and neutrons would possess in the free state, is known as the *mass defect* of the nucleus concerned.

The deuteron, which is built up from 1 proton and 1 neutron, as follows from the nuclear reaction given on p. 564, has the mass defect $1.007596 + 1.008979 - 2.014192 = 0.002383$ (cf. Table 66, p. 562) mass units, or 2.383 m.m.u. Since the helium nucleus, He^{2+}, has about four times the mass of the proton, and twice the charge, it may be assumed that it is composed of 2 protons and 2 neutrons. It follows that the mass defect is $2(1.007596 + 1.008979) - 4.00278 = 0.03037$ mass units, or 30.37 m.m.u. It follows immediately from the formation of the ^{3}He isotope, by means of the nuclear reaction D(d,n)^{3}He, that the nucleus of ^{3}He is composed of 2 protons and 1 neutron. Its mass is 3.01588, and the corresponding mass defect is 8.29 m.m.u. Mass defects can be calculated exactly similarly for the nuclei of the other elements, provided that the nuclear masses of the individual isotopes have been determined mass spectrographically with sufficient accuracy. Conversely, it is possible to determine the

---

* The frequency of 'annihilation radiation' can be calculated by substituting the appropriate values of h, m and c in the above equation: $\nu = 1.23 \cdot 10^{-20}$ sec^{-1}, or wave length $\lambda = 0.0244$ Å. For 'pair production', the frequency must be at least twice as great as this, since the two massive particles are created from a single γ-quantum.

mass defects by determination of the energies liberated or absorbed in nuclear reactions, and thence to find the accurate masses*.

It has been possible to deduce the number of protons and neutrons involved in the structure of the nuclei of a considerable number of elements, directly from observed nuclear reactions. From this knowledge there emerges a very simple principle, which enables the number of protons and of neutrons in the nucleus to be stated in every case, even when not directly derivable from observation. This principle is that *the total number of nucleons is the same as the mass number of the atom.* Since electrons have no part in the constitution of the nucleus, *the number of protons in the nucleus is equal to the nuclear charge* (i.e., *the atomic number*). *Hence the number of neutrons in any nucleus is equal to the difference between the mass number and the atomic number.*

This simple relationship between the mass number of any atom—i.e., its rounded-off atomic weight—and the total number of its nucleons is due to the fortunate circumstance that, following Berzelius, the atomic weights of the elements have been referred to oxygen = 16. If atomic weights had been measured relative to hydrogen = 1 instead, the result of the mass defects would have been to do away with any simple relationship between the rounded off atomic weight and the number of nucleons. For example, bismuth, with 209 nucleons, would have had the atomic weight 207.31 instead of 209.00. The atomic weights of the pure elements, and of the individual isotopes would not have approximated to whole numbers of this scale.

The ratio $(M - A) : A$, where M is the accurate mass value and A the mass number of any atom, is known as the *packing fraction f* of the atom. When these packing fractions, obtained from the difference between the accurately determined mass values and the mass numbers, are plotted as a function of the mass numbers, the resulting curve enables the packing fractions to be approximately estimated for those atomic species for which accurate mass values have not been experimentally determined (Aston). The required mass values can then be obtained with fairly high accuracy from the equation

$$M = A \ (1 + f) \tag{3}$$

Table 67 gives the packing fractions and the mass defects of the individual isotopes of the lightest atoms. For these elements, the mass values are known with very great accuracy, in part from mass spectographic measurements and in part from measurement of the energy liberated in nuclear reactions. The *energies of formation of the nuclei* from their nucleons, as calculated from the mass defects, are given in the Table, in units of millions of kcal per g-atom to indicate the enormous amounts of energy liberated in process of building up the nuclei of the atoms from protons and electrons. It is usual in nuclear physics to express the mass defect in mev.; the corresponding values are obtained by multiplying the mass defect in m.m.u. by the factor 0.9312 (cf. p. 567). In the last column of Table 67 are also given the ratios between the mass defects (in m.m.u.) and the total number A of nucleons. These figures represent the average mass defect—i.e., the average energy of formation of the nucleus—per individual nucleon. It is apparent that this quantity is especially large for the helium nucleus. It may be inferred that the nucleus of helium is possessed of particular stability. It is true that in the heavier atoms the mean energy of formation per nucleon is rather greater still. However, if this were not so, the nuclei of these atoms would not be stable at all; thus the nuclei of ^{12}C, ^{16}O, and ^{20}Ne, for which the proton numbers and neutron numbers are exact multiples of those in the helium nucleus, would break up spontaneously into

---

* It is permissible to use in the calculations the masses of the *neutral atoms* (i.e., nucleus + outer electrons) instead of the masses of the *nuclei*. In all nuclear transformations (except radioactive decay with β-emission, and K-electron capture) the sum of the nuclear charges does not change during reaction; hence the masses of the outer electrons cancel out when the change in mass is evaluated. See p. 575 for mass changes during radioactive decay.

TABLE 67

MASS VALUES, PACKING FRACTIONS AND MASS DEFECTS OF ATOMIC SPECIES
OF THE ELEMENTS HYDROGEN TO ZINC

[In the case of mixed elements, the predominant isotope is marked by heavy type.]

Atomic Number	Element	Species	Neutron Number	Mass value	Packing fraction $f \cdot 10^4$	Mass defect in m.m.u.	Mass defect in 10^6 kilocal per g-atom	Ratio Mass defect: Mass number
1	Hydrogen	^{1}**H**	0	1.008 145	81.45	—	—	—
	Deuterium	^{2}H(D)	1	2.014 741	73.705	2.383	51.2	1.191
	Tritium	^{3}H(T)	2	3.017 00	56.67	9.10	195.5	3.034
		unstable						
2	Helium	^{3}He	1	3.016 98	56.60	8.29	178.0	2.763
		^{4}**He**	2	4.003 876	9.69	30.372	652.22	7.593
3	Lithium	^{6}Li	3	6.016 95	28.25	34.42	739.2	5.737
		^{7}**Li**	4	7.018 17	25.96	42.18	905.8	6.026
4	Beryllium	^{9}**Be**	5	9.015 05	16.72	62.42	1340.5	6.936
5	Boron	^{10}B	5	10.016 11	16.11	69.51	1492.7	6.951
		^{11}**B**	6	11.012 81	11.65	81.79	1756.4	7.435
6	Carbon	^{12}**C**	6	12.003 842	3.202	98.90	2123.9	8.242
		^{13}C	7	13.007 51	5.78	104.21	2237.9	8.016
7	Nitrogen	^{14}**N**	7	14.007 538	5.384	112.33	2412.2	8.024
		^{15}N	8	15.004 902	3.268	123.94	2661.6	8.263
8	Oxygen	^{16}**O**	8	16.000 000	0	136.99	2941.8	8.562
		^{17}O	9	17.004 520	2.659	141.45	3037.6	8.321
		^{18}O	10	18.004 875	2.708	150.07	3222.8	8.337
9	Fluorine	^{19}**F**	10	19.004 427	2.330	158.67	3407.3	8.351
10	Neon	^{20}**Ne**	10	19.998 775	—0.6125	172.46	3704	8.623
		^{21}Ne	11	21.000 45	+0.214	179.77	3860	8.560
		^{22}Ne	12	21.998 36	—0.745	190.84	4098	8.674
11	Sodium	^{23}**Na**	12	22.996 4	—1.57	200.9	4315	8.74
12	Magnesium	^{24}**Mg**	12	23.992 8	—3.00	212.7	4567	8.86
		^{25}Mg	13	24.994 3	—2.28	220.2	4728	8.81
		^{26}Mg	14	25.989 9	—3.88	233.5	5015	8.98
13	Aluminum	^{27}**Al**	14	26.990 11	—3.663	241.48	5186	8.944
14	Silicon	^{28}**Si**	14	27.985 80	—5.071	253.94	5453	9.069
		^{29}Si	15	28.985 71	—4.928	263.00	5648	9.069
		^{30}Si	16	29.983 31	—5.563	274.38	5957	9.146
15	Phosphorus	^{31}**P**	16	30.983 58	—5.297	282.26	6061	9.105
16	Sulfur	^{32}**S**	16	31.982 25	—5.547	291.73	6265	9.117
		^{33}S	17	32.982 03	—5.445	300.93	6462	9.119
		^{34}S	18	33.978 73	—6.256	313.21	6726	9.212
17	Chlorine	^{35}**Cl**	18	34.980 05	—5.700	320.04	6873	9.144
		^{37}Cl	20	36.977 67	—6.035	340.37	7309	9.199
18	Argon	^{36}A	18	35.979 00	—5.833	329.23	7070	9.145
		^{38}A	20	37.974 91	—6.603	351.28	7544	9.244
		^{40}**A**	22	39.975 14	—6.215	369.01	7924	9.225
19	Potassium	^{39}**K**	20	38.976 06	—6.138	358.27	7694	9.187
		^{40}K	21	39.976 54	—5.865	366.77	7876	9.169
		unstable						
		^{41}K	22	40.974 90	—6.122	377.39	8104	9.205
20	Calcium	^{40}**Ca**	20	39.975 29	—6.177	367.19	7885	9.180
		^{42}Ca	22	41.972 16	—6.629	388.28	8338	9.245
		^{43}Ca	23	42.972 51	—6.393	396.91	8523	9.230

continued

TABLE 67

MASS VALUES, PACKING FRACTIONS AND MASS DEFECTS OF ATOMIC SPECIES
OF THE ELEMENTS HYDROGEN TO ZINC

[In the case of mixed elements, the predominant isotope is marked by heavy type.]

Atomic Number	Element	Species	Neutron Number	Mass value	Packing fraction $f \cdot 10^4$	Mass defect in m.m.u.	Mass defect in 10^6 kilocal per g-atom	Ratio Mass defect: Mass number
20	Calcium	^{44}Ca	24	43.969 24	—6.991	409.16	8786	9.299
		^{46}Ca	26	(45.968 5)	(—6.85)	(427.9)	(9188)	(9.301)
		^{48}Ca	28	47.967 78	—6.712	446.53	9589	9.303
21	Scandium	^{45}**Sc**	24	44.970 10	—6.644	416.44	8943	9.254
22	Titanium	^{46}Ti	24	45.966 97	—7.180	427.72	9185	9.298
		^{47}Ti	25	46.966 68	—7.089	436.98	9384	9.298
		^{48}**Ti**	26	47.963 17	—7.673	449.47	9652	9.364
		^{49}Ti	27	48.963 58	—7.433	458.04	9836	9.348
		^{50}Ti	28	49.960 77	—7.846	469.83	10089	9.397
23	Vanadium	^{50}V	27	49.963 3	—7.34	466.5	10017	9.329
		^{51}**V**	28	50.960 52	—7.741	478.23	10270	9.377
24	Chromium	^{50}Cr	26	49.962 10	—7.580	466.83	10025	9.337
		^{52}**Cr**	28	51.957 07	—8.256	489.82	10519	9.420
		^{53}Cr	29	52.957 72	—7.977	498.15	10698	9.399
		^{54}Cr	30	53.956 3	—8.09	508.5	10921	9.418
25	Manganese	^{55}**Mn**	30	54.955 81	—8.035	517.18	11106	9.403
26	Iron	^{54}Fe	28	53.957 04	—7.956	506.14	10869	9.373
		^{56}**Fe**	30	55.952 72	—8.443	528.42	11348	9.436
		^{57}Fe	31	56.953 59	—8.142	536.53	11522	9.413
		^{58}Fe	32	57.952 0	—8.28	547.1	11749	9.433
27	Cobalt	^{59}**Co**	32	58.951 0	—8.31	556.2	11945	9.428
28	Nickel	^{58}**Ni**	30	57.953 45	—8.026	543.98	11682	9.379
		^{60}Ni	32	59.949 01	—8.498	566.38	12163	9.440
		^{61}Ni	33	60.949 07	—8.349	575.30	12354	9.431
		^{62}Ni	34	61.946 81	—8.579	586.54	12596	9.460
		^{64}Ni	36	63.947 55	—8.195	603.75	12965	9.434
29	Copper	^{63}**Cu**	34	62.949 26	—8.054	592.23	12718	9.400
		^{65}Cu	36	64.948 35	—7.946	611.10	13123	9.402
30	Zinc	^{64}**Zn**	34	63.949 55	—7.883	600.09	12887	9.376
		^{66}Zn	36	65.947 22	—7.997	620.37	13322	9.400
		^{67}Zn	37	66.948 15	—7.739	628.42	13495	9.379
		^{68}Zn	38	67.946 86	—7.815	638.69	13716	9.393
		^{70}Zn	40	69.947 79	—7.459	655.72	14081	9.367

helium nuclei, if it were possible to form them by any process. Fig. 70 shows the ratio of mass defect to mass number, as a function of mass number. It may be seen that the curve has a strong peak at ^{4}He, and a second, but much less prominent peak, at ^{16}O; the curve rises slightly to chromium, and then falls gradually. The fact that elements with higher atomic weight than chromium are stable, although the average mass defect per nucleon becomes progressively smaller, is attributable to the presence in their nuclei of an increasing preponderance of neutrons as compared with protons. Thus the decrease in binding energy per nucleon is smaller than the gain in energy per extra neutron built into the nucleus. Calcium is the last element to possess an isotope containing equal numbers of protons and neutrons ($^{40}_{20}$Ca). For more than 20 protons in the nucleus, stability is attainable only with an excess of neutrons, and this excess of neutrons must become progressively greater as the number of protons rises. This is why (cf. Vol. I. p. 232) from calcium onwards, the atomic numbers of the elements become progressively smaller than one half the atomic weights.

Fig. 70 shows that for elements heavier than carbon, the average mass defect per nucleon varies only between 8.00 and 9.44 m.m.u. In ^{16}O, the mass defect per nucleon is 8.56 m.m.u. In the elements following carbon, therefore, each nucleon is at most 0.056% of a mass unit (i.e., $^1/_{16} \times$ ^{16}O) lighter, or at most 0.088% of a mass unit heavier, than in the oxygen isotope ^{16}O. This is why the mass values of the various atomic species differ so surprisingly little from whole numbers—i.e., from their mass numbers; if protons and neutrons had the same masses in the nuclei of all other elements as they have in ^{16}O, all the mass values would be identical with the integral mass numbers. As compared with the masses of the proton and neutron in the free state, however, these nucleons suffer a loss of mass of between 0.79% and 0.94% in the nuclei of the elements considered.

By application of the mass-energy equivalence principle, it is possible to calculate the masses of every member of a radioactive disintegration series with great accuracy, from the measured disintegration energies, provided that the mass of some one member of the series is known. Table 68 shows masses computed in this way, chiefly for the species of atoms making up the natural disintegration series. The differences between the mass defects given in the Table for the various species* are subject to errors of, at most, a few hundredths of an mev or m.m.u. Individual values of the masses may, however involve a rather larger uncertainty, although this will be the same for all; the magnitude of this uncertainty depends upon the accuracy of the masses as determined with the mass spectrograph. In the case of members of the thorium and uranium-radium decay series, this uncertainty is probably not more than a few tenths of an mev, so that the values given for the masses can be regarded

* Values for differences between the mass defects (in mev) are taken from the compilation of Pflügge and Mattauch (*Phys. Z., 44 (1943) 181). Individual values differ from those given in that reference, however, by an amount of 9.31 mev, which has been added to bring the figures into harmony with more recent determinations of the masses of $^1_0$n, $^1_1$H, $^{208}_{82}$Pb, $^{232}_{90}$Th, $^{234}_{92}$U and $^{238}_{92}$U (Bainbridge, 1951; Fowler, 1951; Nier, 1951; Stanford, 1951; Duckworth, 1952; Geiger, 1952; Richards, 1952; Ogata, 1953).

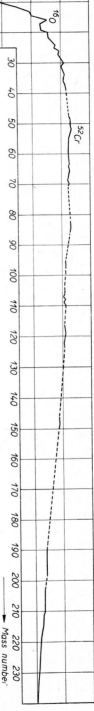

Fig. 70. Variation of the average mass defect per nucleon with the total mass number.

TABLE 68

MASS VALUES, PACKING FRACTIONS AND MASS DEFECTS OF THE NATURALLY
OCCURRING RADIOACTIVE SPECIES

Atomic Number	Species	Mass Number	Neutron Number	Mass Value	Packing fraction $f \cdot 10^4$	Mass defect			Ratio Mass defect (in m.m.u.): Mass number
						in m.m.u.	in mev	in 10^6 kcal per g-atom	
81	AcC″	207	126	207.0424	2.049	1748.68	1628.14	37 552	8.4478
	ThC″	208	127	208.0481	2.314	1751.94	1631.17	37 622	8.4228
	RaC″	210	129	210.0578	2.752	1760.25	1638.91	37 800	8.3822
82	RaG	206	124	206.0384	1.865	1742.87	1622.73	37 427	8.4606
	AcD	207	125	207.0400	1.933	1750.26	1629.61	37 586	8.4554
	ThD	208	126	208.0420	2.020	1757.23	1636.10	37 736	8.4482
	RaD	210	128	210.0512	2.437	1766.03	1644.29	37 925	8.4097
	AcB	211	129	211.0560	2.653	1770.21	1648.18	38 014	8.3896
	ThB	212	130	212.0603	2.843	1774.89	1652.54	38 115	8.3721
	RaB	214	132	214.0697	3.256	1783.45	1660.51	38 299	8.3339
83	RaE	210	127	210.0503	2.393	1766.11	1644.36	37 926	8.4100
	AcC	211	128	211.0536	2.542	1771.72	1649.59	38 047	8.3968
	ThC	212	129	212.0588	2.772	1775.56	1653.16	38 129	8.3753
	RaC	214	131	214.0778	3.634	1774.52	1652.19	38 107	8.2921
84	RaF(Po)	210	126	210.0482	2.294	1767.36	1645.53	37 953	8.4160
	AcC′	211	127	211.0521	2.470	1772.39	1650.21	38 061	8.3999
	ThC′	212	128	212.0556	2.622	1777.91	1655.35	38 180	8.3864
	RaC′	214	130	214.0635	2.969	1787.91	1664.66	38 394	8.3547
	AcA	215	131	215.0680	3.162	1792.44	1668.88	38 492	8.3369
	ThA	216	132	216.0716	3.317	1797.77	1673.84	38 606	8.3230
	RaA	218	134	218.0802	3.679	1807.16	1682.58	38 808	8.2897
86	An	219	133	219.0794	3.626	1815.27	1690.14	38 982	8.2889
	Tn	220	134	220.0825	3.748	1821.19	1695.65	39 109	8.2781
	Rn	222	136	222.0902	4.061	1831.45	1705.20	39 329	8.2498
87	AcK(Fr)	223	136	223.0918	4.118	1837.92	1711.22	39 468	8.2418
88	AcX	223	135	223.0907	4.067	1838.23	1711.51	39 475	8.2432
	ThX	224	136	224.0926	4.136	1845.26	1718.06	39 626	8.2378
	Ra	226	138	226.0990	4.382	1856.82	1728.82	39 874	8.2160
	MsTh$_1$	228	140	228.1059	4.646	1867.90	1739.14	40 112	8.1926
89	Ac	227	138	227.1013	4.462	1862.73	1734.32	40 001	8.2058
	MsTh$_2$	228	139	228.1050	4.607	1867.95	1739.18	40 113	8.1927
90	RdAc	227	137	227.1002	4.414	1862.97	1734.55	40 006	8.2069
	RdTh	228	138	228.1025	4.497	1869.62	1740.74	40 149	8.2001
	Io	230	140	230.1081	4.701	1881.98	1752.25	40 415	8.1825
	UY	231	141	231.1118	4.838	1887.32	1757.22	40 529	8.1702
	Th	232	142	232.1145	4.935	1893.58	1763.05	40 664	8.1620
	UX$_1$	234	144	234.1216	5.198	1904.40	1773.12	40 896	8.1385
91	Pa	231	140	231.1108	4.795	1887.49	1757.38	40 533	8.1710
	UX$_2$	234	143	234.1206	5.153	1904.61	1773.32	40 901	8.1394
92	U II	234	142	234.1172	5.010	1907.13	1775.66	40 954	8.1501
	AcU	235	143	235.1206	5.134	1912.69	1780.84	41 074	8.1391
	U I	238	146	238.1301	5.467	1930.16	1797.11	41 449	8.1099

as accurate to a few thousandth per cent. [Cf. the data recently given by Glass, Thompson and Seaborg, *J. Inorg. Nucl. Chem.*, 1 (1955) 3.] The mass M is calculated from the mass defect ΔM (in m.m.u.) by the formula

$$M = Z \cdot 1.008145 + (A - Z) \cdot 1.008979 - \frac{\Delta M}{1000},$$

where Z is the nuclear charge and A the mass number, 1.008145 the mass of the hydrogen atom ^{1}H, and 1.008979 the mass of the neutron. The equation can be employed, conversely, to calculate the mass defect of atoms for which accurate masses have been found directly with the mass spectrograph. Such values provide the basis for calculation of the remaining mass defects, using the mass defect differences found from disintegration energies.

References

1 F. W. Aston, *Mass Spectra and Isotopes*, London 1933, 248 pp.
2 J. Mattauch and A. Hammersfeld, *Isotopenbericht*, Tübingen 1949, 243 pp.
3 A. Farkas, *Orthohydrogen, Parahydrogen and Heavy Hydrogen*, Cambridge 1935, 215 pp.
4 H. Mark, *Das schwere Wasser*, Leipzig 1934, 32 pp.
5 I. Kirshenbaum, *Physical Properties and Analysis of Heavy Water* [edited by H. C. Urey and G. M. Murphy], New York 1951, 438 pp.
6 G. Champetier, *Deutérium et Composées de Deutérium*, Paris 1938, 80 pp.
7 E. Miehlnickel, *Höhenstrahlung (Ultrastrahlung)*, Dresden 1938, 316 pp.
8 R. A. Millikan, *Electrons (+ and —), Protons, Photons, Neutrons and Cosmic Rays*, Chicago 1935, 492 pp.
9 A. Haas, *Die Umwandlungen der chemischen Elemente*, Berlin 1935, 118 pp.
10 L. Meitner and M. Delbrück, *Der Aufbau der Atomkerne; Natürliche und künstliche Kernumwandlungen*, Berlin 1935, 62 pp.
11 E. N. da C. Andrade, *The New Chemistry*, London 1936, 58 pp.
12 W. Riezler, *Einführung in die Kernphysik*, 3rd Ed., Leipzig 1944, 248 pp.
13 H. Kallmann, *Einführung in die Kernphysik*, Leipzig 1938, 216 pp.
14 H. A. Bethe, *Elementary Nuclear Theory*, New York 1947, 147 pp.
15 H. Dänzer, *Einführung in die theoretische Kernphysik*, Karlsruhe 1948, 187 pp.
16 C. F. von Weizsäcker, *Die Atomkerne, Grundlagen und Anwendungen ihrer Theorie*, Leipzig 1937, 214 pp.
17 W. Heisenberg, *Die Physik der Atomkerne*, 3rd Ed., Berlin 1949, 192 pp.
18 G. Gamow, *Structure of Atomic Nuclei and Nuclear Transformations*, New York 1937, 284 pp.
19 G. Gamow and C. L. Critchfield, *Theory of the Atomic Nucleus and Nuclear Energy Sources*, 3rd Ed., Oxford 1949, 344 pp.
20 F. Rasetti, *Il Nucleo Atomico*, Bologna 1936, 232 pp.
21 F. Rasetti, *Elements of Nuclear Physics*, New York 1936, 327 pp.
22 W. Gentner, H. Maier-Leibnitz and W. Bothe, *Atlas typischer Nebelkammerbilder*, Berlin 1940, 125 pp.

CHAPTER 13

ARTIFICIAL RADIOACTIVITY AND NUCLEAR CHEMISTRY

1. Introduction

In 1934, during an investigation of the positron rays emitted when aluminum is bombarded with α-particles, Irene Curie and M. Joliot discovered that the emission of positrons did not cease at once when the α-emitting preparation (polonium) was removed. It was found to decay gradually with time, following the law characteristic of radioactive decay (eqn. (4), p. 514). Their supposition that the observations were caused by an *artificially induced radioactivity* [1-4] was immediately confirmed by other experiments. It was found that the activation of aluminum under the action of α-rays was due to the initial production of an unstable isotope of phosphorus; this was then spontaneously converted into a stable isotope of silicon, with the emission of a positron, the half-life of the active phosphorus being 3.2 minutes:

$$^{27}_{13}\text{Al} + ^4_2\alpha = ^{30}_{15}\text{P} + ^1_0\text{n}; \qquad ^{30}_{15}\text{P} \xrightarrow{\text{3.2 min.}} ^{30}_{14}\text{Si} + \text{e}^+.$$

A very large number of artificially produced radioactive species of atoms is now known. There are, indeed, very few elements which have not yet been 'activated'— i.e., converted to unstable, spontaneously disintegrating species.

2. Types of Radioactive Disintegration

(a) β-Decay

Most of the artificially produced radioactive nuclei undergo transformation into stable species by emission of either electrons or positrons. Both of these types of transformation are grouped together under the general name of β-decay.

Nuclei which owe their instability to too high a neutron content, in proportion to the number of protons in the nucleus, decay by ejection of an electron (β^--decay). Those which are unstable because they contain an excess of protons, relative to neutrons, eject a positron (β^+-decay). The former are stabilized by conversion of a neutron into an electron (which leaves the nucleus) and a proton. The latter undergo conversion of a proton into a neutron plus a positron, which is emitted from the nucleus.

Positrons ejected in the β^+-disintegration of nuclei possess all possible velocities, in the same way as the electrons emitted in β^--decay processes. The energy of the β^+-disintegration process is likewise given by the maximum kinetic energy of the positrons. As with β-emission, it must be assumed that the difference between this maximum value and the smaller kinetic energy possessed by most of the positrons is emitted in the form of neutrinos.

It is possible for stability to be achieved in another manner, by the entry of an electron into the nucleus from the extranuclear structure, and the reaction of this electron with a proton, to form a neutron. The energy liberated in this process is emitted in the form of a neutrino* (cf. pp. 522, 590). Decay processes of this kind are generally known as electron capture, or K-capture processes, since the electron captured by the nucleus generally originates in the innermost electron orbits, i.e., the K-level.

Radioactive transformations involving positron emission or K-electron capture have been observed only in the case of the artificially produced radioelements. Of the two processes occurring in the natural radioelements, the emission of electrons is also very commonly exhibited by the artificial radioelements. On the other hand, radioactive change by emission of α-particles has not been observed for any of the artificially produced radioactive species, except those that are identical with members of the natural disintegration series (including the transuranic elements).

In radioactive transformations where an electron is emitted, the atomic number of the atom increases by unity; conversely, the atomic number decreases by one unit as a result of the emission of a positron or of K-electron capture. In all these cases the mass of the atom remains unchanged.

Examples:

$$\ce{^6_2He} \xrightarrow{\text{0.8 sec.}} \ce{^6_3Li} + e^-, \quad \ce{^{17}_9F} \xrightarrow{\text{72 sec.}} \ce{^{17}_8O} + e^+$$

Electron capture:

$$\ce{^7_4Be} + e^- \xrightarrow{\text{53 days}} \ce{^7_3Li} + \nu \quad (\nu = \text{neutrino})$$

Written above the arrows of the reactions are the half-lives of the unstable species. As with the natural radioelements, these half-lives are highly characteristic properties of each nuclear species. They are thus independent of the external conditions, and also independent of the process by which the unstable nucleus was produced.

In each radioactive process, the atomic mass is diminished by an amount which is proportional to the magnitude of the energy radiated. Apart from this, the mass of the neutral atom does not change at all as a result of the emission of an electron, since the nuclear charge is thereby increased by one unit, and an additional electron must enter the extranuclear structure. The mass of the atom also remains unaltered in K-electron capture processes, except for the loss of mass by radiation of energy. Positron emission, on the other hand, entails a diminution of the mass of the neutral atom by twice the rest-mass of the electron—i.e., by the mass of the positron and that of one electron, lost from the extranuclear structure as a result of the diminution of the nuclear charge.

Radioactive species are produced by the same methods as are used for other nuclear reactions. In particular, irradiation with neutrons is very frequently employed. In most laboratories, a source of neutrons is provided either by the action of α-rays from radium or radon on beryllium ($\ce{^9_4Be} + \ce{^4_2\alpha} = \ce{^{12}_6C} + \ce{^1_0n}$), or by the nuclear reaction between highly accelerated deuterons (from cyclotron, Van de Graaff generator, or Cockcroft-Walton accelerator) and deuterium, the latter usually being used in the form of D_2O-ice ($\ce{^2_1D} + \ce{^2_1d} = \ce{^3_2He} + \ce{^1_0n}$). A far more intense neutron source is provided by atomic reactors (see below). For many purposes, slow neutrons are more efficient in bringing about nuclear reactions

---

* γ-radiation is also often emitted as a secondary result of the change, as it also frequently is in α- and β-decay processes. A further consequence of nuclear transformation by electron capture is the emission of the characteristic X-ray spectrum of the element formed by the transformation: the K-level is filled up again by an electron transition from some higher level.

TABLE 69

SUMMARY OF THE MOST IMPORTANT UNSTABLE ATOMIC SPECIES

Z	Element	Unstable species	Decay	Half-life	Decay products	Mode of formation of unstable nuclei
1	Tritium	^3_1H	e^-	12.41 yrs	^3_2He	^6_3Li (n, α); ^3_2He (n, p); ^2_1H (d, p)
4	Beryllium	^7_4Be	K	43 days	^7_3Li	$^{10}_5B$ (p, α); ^7_3Li (p, n); ^6_3Li (d, n)
6	Carbon	$^{11}_6C$	e^+	20.4 min	$^{11}_5B$	$^{14}_7N$ (p, α); $^{10}_5B$ (d, n); $^{12}_6C$ (n, 2n)
	,,	$^{14}_6C$	e^-	$5.6 \cdot 10^3$ yrs	$^{14}_7N$	$^{11}_5B$ (α, p): $^{13}_6C$ (n, γ); $^{14}_7N$ (n, p); $^{17}_8O$ (n, α)
7	Nitrogen	$^{13}_7N$	e^+	10.1 min	$^{13}_6C$	$^{10}_5B$ (α, n); $^{13}_6C$ (p, n); $^{12}_6C$ (d, n); $^{14}_7N$ (n,2n)
8	Oxygen	$^{15}_8O$	e^+	129 sec	$^{15}_7N$	$^{14}_7N$ (p, γ); $^{14}_7N$ (d, n); $^{16}_8O$ (n, 2n); $^{16}_8O$ (γ, n)
9	Fluorine	$^{18}_9F$	e^+	112 min	$^{18}_8O$	$^{18}_8O$ (p, n); $^{16}_8O$ (t, n); $^{20}_{10}Ne$ (d, α); $^{19}_9F$ (n, 2n)
11	Sodium	$^{24}_{11}Na$	e^-	14.8 hrs	$^{24}_{12}Mg$	$^{26}_{12}Mg$(d, α); $^{23}_{11}Na$ (d, p); $^{27}_{13}Al$ (n, α); $^{24}_{12}Mg$(n, p); $^{23}_{11}Na$ (n, γ)
12	Magnesium	$^{27}_{12}Mg$	e^-	9.58 min	$^{27}_{13}Al$	$^{25}_{12}Mg$(d, p); $^{27}_{13}Al$ (n, p); $^{26}_{12}Mg$(n, γ)
13	Aluminum	$^{28}_{13}Al$	e^-	2.3 min	$^{28}_{14}Si$	$^{25}_{12}Mg$(α, p); $^{27}_{13}Al$ (d, p); $^{31}_{15}P$ (n, α); $^{28}_{14}Si$ (n, p); $^{27}_{13}Al$ (n, γ)
14	Silicon	$^{31}_{14}Si$	e^-	157 min	$^{31}_{15}P$	$^{30}_{14}Si$ (d, p); $^{34}_{16}S$ (n, α); $^{31}_{15}P$ (n, p); $^{30}_{14}Si$ (n, γ)
15	Phosphorus	$^{32}_{15}P$	e^-	14.07 days	$^{32}_{16}S$	$^{29}_{14}Si$ (α, p); $^{34}_{16}S$ (d, α); $^{31}_{15}P$ (d, p); $^{35}_{17}Cl$ (n, α); $^{32}_{16}S$ (n, p); $^{31}_{15}P$ (n, γ)
16	Sulfur	$^{35}_{16}S$	e^-	87.1 days	$^{35}_{17}Cl$	$^{34}_{16}S$ (d, p); $^{35}_{17}Cl$ (n, p); $^{37}_{17}Cl$ (d, α)
17	Chlorine	$^{38}_{17}Cl$	e^-	38.5 min	$^{38}_{18}A$	$^{37}_{17}Cl$ (d, p); $^{41}_{19}K$ (n, α); $^{37}_{17}Cl$ (n, γ)
19	Potassium	$^{42}_{19}K$	e^-	12.4 hrs	$^{42}_{20}Ca$	$^{41}_{19}K$ (n, γ); $^{41}_{19}K$ (d, p); $^{45}_{21}Sc$ (n, α); $^{42}_{20}Ca$ (n, p)
20	Calcium	$^{45}_{20}Ca$	e^-	152 days	$^{45}_{21}Sc$	$^{44}_{20}Ca$ (d, p); $^{45}_{21}Sc$ (n, p); $^{44}_{20}Ca$ (n, γ)
	,,	$^{49}_{20}Ca$	e^- / e^-	30 min / 150 min	$^{49}_{21}Sc$ / $^{49}_{21}Sc$	$^{48}_{20}Ca$ (n, γ); $^{48}_{20}Ca$ (d, p):
21	Scandium	$^{46}_{21}Sc$	e^-	85 days	$^{46}_{22}Ti$	$^{43}_{20}Ca$ (α, p); $^{45}_{21}Sc$ (d, p); $^{45}_{21}Sc$ (n, γ)
23	Vanadium	$^{52}_{23}V$	e^-	3.74 min	$^{52}_{24}Cr$	$^{51}_{23}V$ (d, p); $^{55}_{25}Mn$(n, α); $^{52}_{24}Cr$ (n, p); $^{51}_{23}V$ (n, γ)
25	Manganese	$^{56}_{25}Mn$	e^-	2.59 hrs	$^{56}_{26}Fe$	$^{53}_{24}Cr$ (α, p); $^{58}_{26}Fe$ (d, α); $^{59}_{27}Co$ (n, α); $^{56}_{26}Fe$ (n, p); $^{55}_{25}Mn$(n, γ)
26	Iron	$^{59}_{26}Fe$	e^-	46.3 days	$^{59}_{27}Co$	$^{58}_{26}Fe$ (d, p); $^{58}_{26}Fe$ (n, γ); $^{59}_{27}Co$ (n, p)
27	Cobalt	$^{57}_{27}Co$	e^+	240 days	$^{57}_{26}Fe$	$^{56}_{26}Fe$ (p, γ); $^{56}_{26}Fe$ (d, n)
	,,	$^{60}_{27}Co$	e^-	5.24 yrs	$^{60}_{28}Ni$	$^{59}_{27}Co$ (n, γ); $^{59}_{27}Co$ (d, p)
28	Nickel	$^{63}_{28}Ni$	e^-	300 yrs	$^{63}_{29}Cu$	$^{62}_{28}Ni$ (d, p); $^{66}_{30}Zn$ (n, α); $^{63}_{29}Cu$ (n, p); $^{62}_{28}Ni$ (n, γ)

TABLE 69   (*Continued*)

SUMMARY OF THE MOST IMPORTANT UNSTABLE ATOMIC SPECIES

z	Element	Unstable species	Decay	Half-life	Decay products	Mode of formation of unstable nuclei
29	Copper	$^{62}_{29}Cu$	e^+	9.9 min	$^{62}_{28}Ni$	$^{59}_{27}Co$ (α, n); $^{62}_{28}Ni$ (p, n); $^{63}_{29}Cu$ $(n, 2n)$
	,,	$^{64}_{29}Cu$	$\nearrow K$ $\rightarrow e^+$ $\searrow e^-$	12.8 hrs 12.8 hrs 12.8 hrs	$^{64}_{28}Ni$ $^{64}_{28}Ni$ $^{64}_{30}Zn$	$^{64}_{28}Ni$ (p, n); $^{63}_{29}Cu$ (n, γ); $^{65}_{29}Cu$ (γ, n); $^{64}_{30}Zn$ (n, p)
30	Zinc	$^{63}_{30}Zn$	e^+	36 min	$^{63}_{29}Cu$	$^{60}_{28}Ni$ (α, n); $^{63}_{29}Cu$ (p, n); $^{63}_{29}Cu(d, 2n)$; $^{64}_{30}Zn$ $(n, 2n)$
33	Arsenic	$^{74}_{33}As$	$\nearrow e^-$ $\searrow e^+$	17.5 days 17.5 days	$^{74}_{34}Se$ $^{74}_{32}Ge$	$^{75}_{33}As$ $(n, 2n)$; $^{75}_{33}As$ (γ, n); $^{76}_{34}Se$ (d, α)
34	Selenium	$^{81}_{34}Se$	e^-	57 min	$^{79}_{35}Br$	$^{80}_{34}Se$ (d, p); $^{81}_{35}Br$ (n, p); $^{80}_{34}Se$ (n, γ); $^{82}_{34}Se$ $(n, 2n)$
35	Bromine	$^{80}_{35}Br$	$\nearrow e^-$ $\searrow e^+$	18.5 min 18.5 min	$^{80}_{36}Kr$ $^{80}_{34}Se$	$^{80}_{34}Se$ (p, n); $^{79}_{35}Br$ (d, p); $^{79}_{35}Br$ (n, γ); $^{81}_{35}Br$ $(n, 2n)$; $^{81}_{35}Br$ (γ, n)
47	Silver	$^{106}_{47}Ag$	$\nearrow K$ $\searrow e^+$	8.2 days 24.3 min	$^{106}_{46}Pd$ $^{106}_{46}Pd$	$^{103}_{45}Rh$ (α, n); $^{106}_{46}Pd$ (p, n); $^{105}_{46}Pd$ (d, n); $^{106}_{48}Cd$ (n, p); $^{107}_{47}Ag(n, 2n)$; $^{107}_{47}Ag$ (γ, n)
48	Cadmium	$^{115}_{48}Cd$	$\nearrow e^-$ $\searrow e^-$	2.4 days 43 days	$^{115}_{49}In$ $^{115}_{49}In$	$^{114}_{48}Cd$ (n, γ); $^{114}_{48}Cd$ (d, p); $^{115}_{49}In$ (n, p)
50	Tin	$^{125}_{50}Sn$	e^-	11.8 min	$^{125}_{51}Sb$	$^{124}_{50}Sn$ (d, p); $^{124}_{50}Sn$ (n, γ)
51	Antimony	$^{124}_{51}Sb$	e^-	60 days	$^{124}_{52}Te$	$^{126}_{52}Te$ (d, α); $^{123}_{51}Sb$ (d, p); $^{123}_{51}Sb$ (n, γ); $^{127}_{53}I$ (n, α)
52	Tellurium	$^{127}_{52}Te$	$\nearrow e^-$ $\searrow e^-$	9.3 hrs 90 days	$^{127}_{53}I$ $^{127}_{53}I$	$^{126}_{52}Te$ (d, p); $^{127}_{53}I$ (n, p); $^{128}_{52}Te$ $(n, 2n)$; $^{126}_{52}Te$ (n, γ)
53	Iodine	$^{128}_{53}I$	e^-	25 min	$^{128}_{54}Xe$	$^{128}_{52}Te$ (p, n); $^{127}_{53}I$ (n, γ)
56	Barium	$^{139}_{56}Ba$	e^-	84 min	$^{139}_{57}La$	$^{138}_{56}Ba$ (d, p); $^{138}_{56}Ba$ (n, γ); $^{139}_{57}La$ (n, p)
57	Lanthanum	$^{140}_{57}La$	e^-	41.4 hrs	$^{140}_{58}Ce$	$^{139}_{57}La$ (d, p); $^{139}_{57}La$ (n, γ)
78	Platinum	$^{197}_{78}Pt$	$\nearrow e^-$ $\searrow e^-$	17.4 hrs 82 days	$^{197}_{79}Au$ $^{197}_{79}Au$	$^{196}_{78}Pt$ (d, p); $^{200}_{80}Hg$ (n, α); $^{196}_{78}Pt$ (n, γ)
79	Gold	$^{198}_{79}Au$	e^-	2.7 days	$^{198}_{80}Hg$	$^{198}_{80}Hg$ (n, p); $^{197}_{79}Au$ (n, γ)
80	Mercury	$^{197}_{80}Hg$	K	23 hrs	$^{197}_{79}Au$	$^{196}_{80}Hg$ (n, γ); $^{196}_{80}Hg$ (d, p); $^{197}_{79}Au$ (p, n); $^{194}_{78}Pt$ (α, n)

than fast neutrons (their capture cross section—see below—is greater, since they remain for a longer time in the neighborhood of the nucleus); they are generally slowed down by paraffin wax or water. For certain types of nuclear reaction [e.g., for (n, 2n) and (n, 3n)], it is necessary to employ neutrons of very high energy, such as are obtained when a lithium target is bombarded with highly accelerated deuterons—^7Li (d, n)^8Be.

Table 69 shows a selection of the unstable species of atoms that have been produced artificially. The last column of the table gives the stable nuclear species from which they may be prepared, by means of the nuclear reactions shown. This table contains only a small number of the numerous unstable species that have already been identified [20]. Thus at least six different radioactive isotopes of copper have been obtained (with half-lives ranging from 4.34 min to 2.44 days), and at least seven unstable isotopes of zinc (with half-lives between 36 minutes and 250 days). In compiling Table 69, preference has been given to those unstable nuclei which, because of their mode of formation and half-lives, appear to be of special interest as 'tracers', or indicators in chemical research.

Many examples of *dual decay processes* have been observed among the artificially produced unstable nuclei, just as for the natural radioactive decay series. Several examples are listed in Table 69 (^{64}Cu, ^{74}As and ^{80}Br)*. In addition to the dual decay schemes, where *two different disintegration products* are formed with the same half-life**, a range of examples has been found in which one and the *same decay product* is formed by processes of *different half-life* (e.g. from ^{49}Ca, ^{106}Ag, ^{115}Cd, ^{127}Te, $^{197}_{78}$Pt—cf. Table 69). Nuclei having the same mass number and atomic number, but decaying with different half-lives to the same decay product, are said to be *nuclear isomers*, e.g.

$$\underset{20}{^{49}}\text{Ca} \xrightarrow{\text{30 min}} \underset{21}{^{49}}\text{Sc} + e^- \quad \text{and} \quad \underset{20}{^{49}}\text{Ca} \xrightarrow{\text{2.5 hrs}} \underset{21}{^{49}}\text{Sc} + e^-.$$

(b) Nuclear Isomerism

The first example of nuclear isomerism was recognised by Hahn, in 1921, namely the decay products uranium X_2 and uranium Z appearing in the uranium-radium series. For a long time this remained the only example, but numerous other instances have been found since it became possible to prepare unstable nuclear species by means of nuclear reactions.

The phenomenon of nuclear isomerism arises from the fact that the nucleus can exist in several different energy states, so that excited states are detectable as well as the ground state. In general, a nucleus reverts immediately (i.e., probably within 10^{-12} sec) from an excited state to its ground state, the energy difference between the two states being emitted as γ-radiation. However, just as for electronic transitions in the outer shell of the atom, there are certain 'forbidden transitions'—i.e. transitions that are highly improbable, and take place only rarely or not at all. If the transition from some excited state to the ground state is very improbable, the state is said to be *metastable. Isomeric nuclei* are nuclei with the same mass and charge,

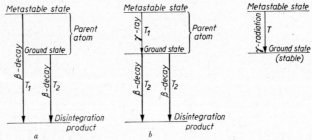

Fig. 71. Three types of radioactive change of isomeric nuclei.
(a) Independent decay
(b) Disintegration following isomeric transition
(c) Isomeric transition from a metastable state of a stable nucleus to the ground state. T = half-life.

* ^{64}Cu is an instance of dual decay, since two different disintegration products are formed (^{64}Ni and ^{64}Zn). The fact that one of them is produced by two different mechanisms (by K electron capture and by positron emission) is immaterial. A different situation is presented e.g. by ^{49}Ca, where the two transformations, both leading to the same end product, have different half lives. ^{49}Ca is thus an instance of nuclear isomerism.

** It follows directly from the definition of the disintegration constant λ that two different disintegration constants (i.e. different half-lives) cannot be involved in dual decay schemes. According to equation (11), p. 525, λ is determined by the decrease in amount of the active species during the time interval dt. This definition contains no assumptions as to whether this decrease is the consequence of a *single* radioactive process, or of several concurrent processes.

one of which is in the normal or ground state, and the other in a *metastable excited state* §. The disintegration of a nucleus which is in a metastable excited state (e.g., by ejection of a β-particle) results in just the same disintegration product as would be formed by a similar decay process from the nucleus in its ground state. The decay takes place, however, with a different half-life. This mode of disintegration is represented in Fig. 71a, and is known as *nuclear isomerism with independent decay*, as distinct from the type shown in Fig. 71b, in which the decay product is not formed directly from the metastable excited parent nucleus. The latter first undergoes a γ-ray transition from the excited state to the ground state, with the half-life T_1, and subsequently forms the disintegration product (e.g. by ejection of a β-particle) with the half-life T_2. The more improbable the transition of the parent nucleus from its excited state to its ground state, the longer is the half-life T_1. This type of transition is known as an *isomeric transition*, and in such a case the isomer which is present in the excited state is symbolized by appending an asterisk to its chemical symbol. Thus

$$^{44}Sc^* \xrightarrow[\gamma]{2.44 \text{ days}} {}^{44}Sc$$

symbolises the fact that the metastable excited nucleus of the scandium isotope of mass 44 changes into its ground state with a half-life of 2.44 days. The scandium isotope ^{44}Sc is unstable in its ground state also; it decays with a half-life of 3.92 hours, emitting a positron and forming the calcium isotope ^{44}Ca:

$$^{44}_{21}Sc \xrightarrow[\beta^+]{3.92 \text{ hrs}} {}^{44}_{20}Ca.$$

Whereas, in the decay of the normal ^{44}Sc nucleus, the observed half-life is 3.92 hours, the half-life observed for the decay of the isomeric $^{44}Sc^*$ nucleus is 2.44 days, since in this case the rate-determining process is the formation of normal ^{44}Sc nuclei from the isomeric $^{44}Sc^*$ nuclei. The considerations are the same as in consecutive chemical reactions, in which the reaction that proceeds most slowly determines the over-all velocity. If the isomeric transition takes place more rapidly than the disintegration and ejection of a massive particle, the latter is the rate-determining step. An example of this is provided by the radioactive cobalt isotope ^{60}Co, of which the metastable isomer $^{60}Co^*$ fairly rapidly passes over to the ground state with emission of γ-radiation, whilst the resulting ground state ^{60}Co undergoes β-decay to form ^{60}Ni over the course of years:

$$^{60}_{27}Co^* \xrightarrow[\gamma]{10.7 \text{ min}} {}^{60}_{27}Co \xrightarrow[\beta^-]{5.24 \text{ years}} {}^{60}_{28}Ni.$$

At the same time, a certain proportion of the $^{60}Co^*$ (about 10%) decays directly with the emission of β-rays†:

$$^{60}_{27}Co^* \xrightarrow[\beta^-]{10.7 \text{ min}} {}^{60}_{28}Ni.$$

On the whole, the nuclear isomerism of cobalt accords with the scheme of Fig. 71a, except that the ground state of ^{60}Co is also linked with the excited state through an isomeric transition. The details of this decay scheme of the two cobalt isomers is

§ There are also cases where two metastable states exist, as well as the ground state. There will then be three isomeric nuclei.

† The half-life of an excited nucleus for decay by emission of a massive particle is necessarily the same as for its isomeric transition with γ-ray emission, since both processes result in a decrease in the number of atoms of the atomic species. The argument is the same as for dual decay processes.

shown in Fig. 72. It may be seen that neither of the two β-transformations yields the
^{60}Ni nucleus directly in its ground state. In both cases, the ^{60}Ni nucleus is initially
formed in an excited state, from which it immediately passes into the ground state
by the emission of γ-radiation.

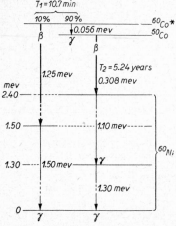

Fig. 72. Decay scheme for the
two isomers of ^{60}Co.

Relations between the isomeric nuclei UX$_2$
and UZ (Fig. 73) exactly parallel those between
^{60}Co* and ^{60}Co. UX$_2$ represents a metastable
excited state of the ^{234}Pa nucleus, whereas UZ is
the ground state of the ^{234}Pa nucleus. 99.85% of
the UX$_2$ nuclei transform directly into U II:

$$^{234}_{91}\text{Pa*} \xrightarrow[\beta^-]{\text{1.14 min}} {}^{234}_{92}\text{U}$$

whereas 0.15% first undergo an isomeric transition into normal ^{234}Pa (= UZ), which then
decays further to form ^{234}U (= U II):

$$^{234}_{91}\text{Pa} \xrightarrow[\beta^-]{\text{6.7 hrs}} {}^{234}_{92}\text{U}.$$

The only difference from the example discussed
previously is that, as shown in Fig. 73, UX$_2$ and UZ emit several groups of
β-rays, whereas ^{60}Co* and ^{60}Co
each emit only a single group of
β-rays.

A third type of radioactive transformation that results from nuclear
isomerism is observed when a nucleus which can be produced in a
metastable excited state is *stable* in
its ground state—i.e., does not undergo radioactive disintegration.
In this case, only a single half-
life is observed, corresponding to
the γ-ray isomeric transition of
the nucleus from its metastable
state to the ground state (Fig. 71c).
Table 70 lists some examples of the
three different kinds of radioactive
transformation arising from the
occurrence of nuclear isomerism.

(c) Mirror Image Nuclei

An interesting group of positron
emitters consists of those nuclides
in which the proton number is one
unit greater than the neutron
number. By emission of a posi-
tron, these are transformed into nuclei with the neutron number one unit

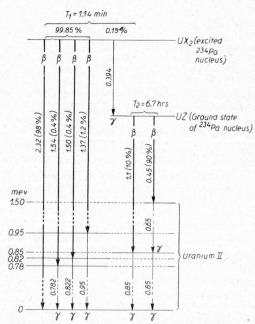

Fig. 73. Decay scheme for the nuclear isomers
UX$_2$ and UZ. Numbers beside the arrows show
the energies of disintegration or transition, in
mev. Relative amounts of the various groups
of β-rays are shown in brackets.

TABLE 70

RADIOACTIVE CHANGES OF ISOMERIC NUCLEI

Independent decay		Decay scheme involving isomeric transition followed by disintegration	
$\xrightarrow[2.5 \text{ hrs}]{\beta^-} {}^{49}_{21}\text{Sc}$	${}^{49}_{20}\text{Ca} \xrightarrow[30 \text{ min}]{\beta^-} {}^{49}_{21}\text{Sc}$	${}^{44}_{21}\text{Sc}^* \xrightarrow[2.44 \text{ days}]{\gamma} {}^{44}_{21}\text{Sc} \xrightarrow{\beta^+} {}^{44}_{20}\text{Ca}$	${}^{44}_{21}\text{Sc} \xrightarrow[3.92 \text{ hrs}]{\beta^+} {}^{44}_{20}\text{Ca}$
$\xrightarrow[6.5 \text{ days}]{K} \Big) \quad \xrightarrow[6.5 \text{ days}]{\beta^+} {}^{52}_{24}\text{Cr}$	${}^{52}_{25}\text{Mn} \xrightarrow[21 \text{ min}]{\beta^+} {}^{52}_{24}\text{Cr}$	${}^{69}_{30}\text{Zn}^* \xrightarrow[13.8 \text{ hrs}]{\gamma} {}^{69}_{30}\text{Zn} \xrightarrow{\beta^-} {}^{69}_{31}\text{Ga}$	${}^{69}_{30}\text{Zn} \xrightarrow[57 \text{ min}]{\beta^-} {}^{69}_{31}\text{Ga}$
$\xrightarrow[6.70 \text{ hrs}]{\beta^+} {}^{93}_{41}\text{Nb}$	${}^{93}_{42}\text{Mo} \xrightarrow[17 \text{ min}]{\beta^+} {}^{93}_{41}\text{Nb}$	${}^{81}_{34}\text{Se}^* \xrightarrow[57 \text{ min}]{\gamma} {}^{81}_{34}\text{Se} \xrightarrow{\beta^-} {}^{81}_{35}\text{Br}$	${}^{81}_{34}\text{Se} \xrightarrow[13.6 \text{ min}]{\beta^-} {}^{81}_{35}\text{Br}$
$\xrightarrow[62 \text{ days}]{K} {}^{95}_{42}\text{Mo}$	${}^{95}_{43}\text{Tc} \xrightarrow[20 \text{ hrs}]{K} {}^{95}_{42}\text{Mo}$	${}^{80}_{35}\text{Br}^* \xrightarrow[4.54 \text{ hrs}]{\gamma} {}^{80}_{35}\text{Br} \begin{smallmatrix} \xrightarrow{\beta^-} {}^{80}_{36}\text{Kr} \\ \xrightarrow{\beta^+} {}^{80}_{34}\text{Se} \end{smallmatrix}$	${}^{80}_{35}\text{Br} \begin{smallmatrix} \xrightarrow{\beta^- \;\; 18.5 \text{ min}} {}^{80}_{36}\text{Kr} \\ \xrightarrow{\beta^+ \;\; 18.5 \text{ min}} {}^{80}_{34}\text{Se} \end{smallmatrix}$
$\xrightarrow[8.2 \text{ days}]{K} {}^{106}_{46}\text{Pd}$	${}^{106}_{47}\text{Ag} \xrightarrow[24.3 \text{ min}]{\beta^+} {}^{106}_{46}\text{Pd}$	${}^{104}_{45}\text{Rh}^* \xrightarrow[4.34 \text{ min}]{\gamma} {}^{104}_{45}\text{Rh} \xrightarrow{\beta^-} {}^{104}_{46}\text{Pd}$	${}^{104}_{45}\text{Rh} \xrightarrow[41.8 \text{ sec}]{\beta^-} {}^{104}_{46}\text{Pd}$
$\xrightarrow[43 \text{ days}]{\beta^-} {}^{115}_{49}\text{In}$	${}^{115}_{48}\text{Cd} \xrightarrow[2.4 \text{ days}]{\beta^-} {}^{115}_{49}\text{In}$	${}^{114}_{49}\text{In}^* \xrightarrow[48.5 \text{ days}]{\gamma} {}^{114}_{49}\text{In} \xrightarrow{\beta^-} {}^{114}_{50}\text{Sn}$	${}^{114}_{49}\text{In} \xrightarrow[1.2 \text{ min}]{\beta^-} {}^{114}_{50}\text{Sn}$
$\xrightarrow[54.1 \text{ min}]{\beta^-} {}^{116}_{50}\text{Sn}$	${}^{116}_{49}\text{In} \xrightarrow[13 \text{ sec}]{\beta^-} {}^{116}_{50}\text{Sn}$	${}^{127}_{52}\text{Te}^* \xrightarrow[90 \text{ days}]{\gamma} {}^{127}_{52}\text{Te} \xrightarrow{\beta^-} {}^{127}_{53}\text{I}$	${}^{127}_{52}\text{Te} \xrightarrow[9.3 \text{ hrs}]{\beta^-} {}^{127}_{53}\text{I}$
$\xrightarrow[6 \text{ days}]{K} {}^{120}_{50}\text{Sn}$	${}^{120}_{51}\text{Sb} \xrightarrow[15 \text{ min}]{\beta^+} {}^{120}_{50}\text{Sn}$	${}^{129}_{52}\text{Te}^* \xrightarrow[35.5 \text{ days}]{\gamma} {}^{129}_{52}\text{Te} \xrightarrow{\beta^-} {}^{129}_{53}\text{I}$	${}^{129}_{52}\text{Te} \xrightarrow[1.12 \text{ hrs}]{\beta^-} {}^{129}_{53}\text{I}$
$\xrightarrow[2.66 \text{ days}]{K} {}^{197}_{79}\text{Au}$	${}^{197}_{80}\text{Hg} \xrightarrow[23 \text{ hrs}]{K} {}^{197}_{79}\text{Au}$	${}^{131}_{52}\text{Te}^* \xrightarrow[1.2 \text{ days}]{\gamma} {}^{131}_{52}\text{Te} \xrightarrow{\beta^-} {}^{131}_{53}\text{I}$	${}^{131}_{52}\text{Te} \xrightarrow[25 \text{ min}]{\beta^-} {}^{131}_{53}\text{I}$

Isomeric transition of stable nuclei from metastable excited state to stable ground state

$${}^{72}\text{Ge}^* \xrightarrow[5 \cdot 10^{-7} \text{ sec}]{\gamma} {}^{72}\text{Ge}; \qquad {}^{83}\text{Kr}^* \xrightarrow[1.88 \text{ hrs}]{\gamma} {}^{83}\text{Kr}; \qquad {}^{87}\text{Sr}^* \xrightarrow[2.75 \text{ hrs}]{\gamma} {}^{87}\text{Sr};$$

$${}^{107}\text{Ag}^* \xrightarrow[44.3 \text{ sec}]{\gamma} {}^{107}\text{Ag}; \qquad {}^{109}\text{Ag}^* \xrightarrow[39.2 \text{ sec}]{\gamma} {}^{109}\text{Ag}; \qquad {}^{113}\text{Cd}^* \xrightarrow[2.3 \text{ min}]{\gamma} {}^{113}\text{Cd};$$

$${}^{113}\text{In}^* \xrightarrow[1.74 \text{ hrs}]{\gamma} {}^{113}\text{In}; \qquad {}^{115}\text{In}^* \xrightarrow[4.5 \text{ hrs}]{\gamma} {}^{115}\text{In}; \qquad {}^{124}\text{Te}^* \xrightarrow[1.2 \cdot 10^{-3} \text{ sec}]{\gamma} {}^{124}\text{Te};$$

$${}^{137}\text{Ba}^* \xrightarrow[2.64 \text{ min}]{\gamma} {}^{137}\text{Ba}; \qquad {}^{197}\text{Au}^* \xrightarrow[7.4 \text{ sec}]{\gamma} {}^{197}\text{Au}; \qquad {}^{204}\text{Pb}^* \xrightarrow[1.10 \text{ hrs}]{\gamma} {}^{204}\text{Pb}.$$

greater than the proton number. If a nucleus containing p protons and n neutrons is termed a $p + n$ nucleus, it transforms into a $n + p$ nucleus (for $p = n + 1$); thus a $6 + 5$ nucleus (${}^{11}_{6}\text{C}$) decays into a $5 + 6$ nucleus (${}^{11}_{5}\text{B}$). This group of positron emitters is known as 'mirror nuclei', and an unbroken sequence is known for mirror nuclei with charges from 6 to 22. The half-lives for disintegration diminish regularly along this series; the disintegration energies increase in the same sequence, in such a way that there is a roughly linear proportionality between the logarithm of the decay constant and the logarithm of the disintegration energy. The mirror nuclei thus lie on a Sargent curve (with a few exceptions), and furthermore this curve forms the approximately linear continuation of one of the two curves on to which the natural radioelements fit (cf. p. 516).

(d) Successive Disintegrations

If the disintegration product of an unstable nucleus is itself also unstable, so that it decays radioactively in its turn, successive disintegrations occur, making up a decay series or decay chain. E.g.,

$$\ce{^{83}_{34}Se} \xrightarrow{25\ min} \ce{^{83}_{35}Br} + e^-, \qquad \ce{^{83}_{35}Br} \xrightarrow{2.5\ hrs} \ce{^{83}_{36}Kr} + e^-.$$

The occurrence of decay series of more than two steps among the artificially produced radioelements of low mass is exceptional. Disintegration series comprising many members are, however, observed especially among the transuranic elements (cf. Chap. 14). This is associated with the occurrence of α-particle decay (as well as β-decay) among these elements, just as it is among the natural radioelements ^{238}U, ^{235}U and Th. It is also found that disintegration series of considerable length occur in the decay of the nuclei of moderately high mass, which are formed when the heaviest nuclei undergo fission by neutrons.

(e) 'Spallation'

In general, only one or two (or in exceptional cases three) corpuscles—i.e., neutrons, protons, deuterons, or α-particles—are ejected from a nucleus by a colliding particle. It has been shown, however, that colliding particles of extremely high energy can bring about processes in which far greater numbers of massive particles are torn from the nucleus. Such an event is spoken of as a multiple disintegration or 'spallation' of the nucleus (Seaborg). Numerous examples of these multiple disintegration processes are now known. Thus $^{75}_{33}As$ is converted directly into $^{55}_{25}Mn$ by bombardment with deuterons of 200 mev energy, and $^{238}_{92}U$ is converted directly into $^{187}_{74}W$ (among other products) by α-particles of 380 mev energy:

$$^{75}_{33}As + ^{2}_{1}d = ^{56}_{25}Mn + 9^{1}_{1}p + 12^{1}_{0}n;$$
$$^{238}_{92}U + ^{4}_{2}\alpha = ^{187}_{74}W + 20^{1}_{1}p + 35^{1}_{0}n.$$

Thus an α-particle of extremely high energy, in a single act of collision with the uranium nucleus, can split off 55 nucleons—20 protons and 35 neutrons. In the usual abbreviated symbolism, the foregoing equations would be represented as ^{75}As (d, 9p 12n) ^{56}Mn, and ^{238}U (α, 20p 35n) ^{187}W.

Other examples of this type of multiple disintegration are ^{31}P (d, 9p 7n) ^{17}N, ^{32}S (d, 10p 7n) ^{17}N, ^{75}As (d, 10p 16n) ^{51}Cr, ^{70}Ge (d, p 5n) ^{66}Ge, ^{121}Sb (d, 6p 16n) ^{101}Pd, ^{238}U (α, 6p 12n) ^{224}Ra (ThX). A large number of different multiple disintegration processes are often induced concurrently. Thus, when $^{75}_{33}As$ is bombarded with deuterons of 190 mev energy, it has been observed that no fewer than 38 different atomic species are produced. Except for quite a small number, formed by nuclear reactions of the ordinary kind, all of these were the product of 'spallation' reactions.

3. Uses of Artificially Activated Elements [5-14]

The artificially produced radioactive atomic species can be used as *tracers* or indicators, in the same way as the natural radioelements. Indeed, they enable the radioactive tracer method to be extended—in principle, at least—to practically any required element. This possibility has introduced new and powerful techniques, not only to chemical research (especially in the field of reaction kinetics), but also to biology and medicine. Increasing use is being made of radioactive tracers in industrial work also. In addition to their value as tracers, the artificial radioelements find wide application as γ-ray irradiation sources, in place of X-ray tubes or radium preparations, both in radiotherapy and for industrial radiology (e.g., inspection of welds). These γ-ray sources are far less expensive than radium, and have advantages over high-voltage X-ray equipment in being more compact and much more mobile, and capable of being designed to meet the needs of special applications. It is possible that artificial radioelements may largely displace X-ray apparatus and radium preparations, as irradiation sources for therapy.

The wide application of artificially activated elements has been made possible by the fact that many of them—in particular those formed in the fission of uranium, and those which are readily obtained by irradiation with neutrons—can be continuously produced in nuclear reactors. These radioelements—or compounds containing them—are therefore obtainable commercially, like ordinary chemicals. Tables 71 and 72 show a number of the artificial radioelements; only those species are cited which emit either a pure β-radiation or a pure γ-radiation. The latter property is important if the elements are to be used as irradiation sources. Radioactive isotopes of a number of other elements are commercially obtainable, in addition to those mentioned in the two Tables. The most widely used radio-isotopes are iodine-131, phosphorus-32, carbon-14, sodium-24, gold-198, sulfur-35, cobalt-60, potassium-42, calcium-45, iron-55 and -59, strontium-89 and -90, bromine-82, and iridium-182.

^{60}Co has found especially widespread use in industry as an irradiation source, e.g., for materials testing; it has an extremely penetrating γ-radiation, and a relatively long half-life.

Radioactive isotopes find uses in therapy not only for external irradiations, but are also, in many cases, introduced into the body. Certain elements are secreted preferentially in particular organs of the body—e.g., iodine in the thyroid gland, and phosphorus in the bones. Thus if medicaments containing radio-iodine or radio-phosphorus (which are obtainable, already prepared for use) are administered to the body, these elements are accumulated at those sites where it is intended that their radiations should be applied. It is considered possible that radioactive radiations might be produced by nuclear reactions carried out directly within the body of the patient. Thus the attempt has been made to enrich cancerous tissue in boron, and then to convert the boron to lithium by irradiation with neutrons. The α-rays produced by the nuclear reaction $^{10}_{5}$B (n, α) $^{7}_{3}$Li then exert a concentrated biological effect on the tissues in the immediate neighborhood of the boron atoms. Radioactive sources are used in biology, e.g., to bring about mutations, through the action of the radiations on the chromosomes. Experiments of this type have proved of value in the study of genetics.

TABLE 71

COMMERCIALLY AVAILABLE RADIOISOTOPES WITH PURE β-RADIATION

Species	Half-life	Maximum energy of β-rays	Species	Half-life	Maximum energy of β-rays
$^{3}_{1}$H	12.41 yrs	0.0186 mev	$^{89}_{38}$Sr	54.5 days	1.463 mev
$^{14}_{6}$C	5589 yrs	0.155 mev	$^{90}_{39}$Y	2.54 days	2.18 mev
$^{31}_{14}$Si	157 min	1.48 mev	$^{111}_{47}$Ag	7.5 days	1.04 mev
$^{32}_{15}$P	14.07 days	1.69 mev	$^{143}_{59}$Pr	13.7 days	0.922 mev
$^{35}_{16}$S	87.1 days	0.169 mev	$^{147}_{61}$Pm	2.26 yrs	0.223 mev
$^{45}_{20}$Ca	152 days	0.254 mev	$^{204}_{81}$Tl	2.7 yrs	0.783 mev

TABLE 72

COMMERCIALLY AVAILABLE RADIOISOTOPES WITH ESPECIALLY PENETRATING γ-RADIATION

Species	Half-life	Decay	Maximum energy of β-rays (mev)	Energies of γ-rays (mev)
$^{24}_{11}$Na	15.10 hrs	e^-	1.39	2.755, 1.380
$^{42}_{19}$K	12.44 hrs	e^-	3.58, 1.99	1.51
$^{56}_{25}$Mn	2.59 hrs	e^-	2.86, 1.05, 0.75	1.77, 0.822
$^{59}_{26}$Fe	46.3 days	e^-	0.46, 0.257	1.30, 1.10
$^{60}_{27}$Co	5.26 yrs	e^-	0.318	1.33, 1.17
$^{64}_{29}$Cu	12.88 hrs	e^- e^+ K	0.57 (e$^+$ 0.66)	1.34
$^{76}_{33}$As	1.187 days	e^-	3.15, 2.56, 1.5, 0.4	1.70, 1.20, 0.55
$^{86}_{37}$Rb	19.5 days	e^-	1.80, 0.724	1.08
$^{124}_{51}$Sb	60 days	e^-	2.29, 1.69, 0.95, 0.68, 0.50	2.04, 1.71, 0.730
$^{140}_{57}$La	1.67 days	e^-	2.26, 1.67, 1.32	2.26, 1.67, 0.87
$^{142}_{59}$Pr	19.1 hrs	e^-	2.23, 0.66	1.74, 0.49, 0.424
$^{182}_{73}$Ta	117 days	e^-	0.53	1.22, 1.13, 0.22, 0.15

The chemical use of artificial radioactive elements has chiefly been as tracers. For example it has been shown (Erbacher, 1935) by using ^{198}Au as tracer that gold and platinum are not quantitatively separated by the action of H_2O_2, although from experiments on the individual elements it is found that only gold is reduced from its compounds by H_2O_2, and not platinum. Use of the radioactive sulfur isotope ^{35}S made it possible to prove the presence in thiosulfuric acid of two sulfur atoms with completely different chemical properties, in accordance with the constitutional formula ascribed to the compound (cf. Vol. I, p. 720); these sulfur atoms cannot undergo exchange with each other. It is also noteworthy that manganese in the form of MnO_4^- ions does not undergo exchange with manganese which is present as Mn^{++} or Mn^{+++} ions, whereas an immediate exchange of charges takes place between Mn^{++} and Mn^{+++} ions (Polissar, 1936). It has also been shown that iodine is not ionically bound in organic compounds such as C_2H_5I. If C_2H_5I is irradiated with slow neutrons, ^{128}I is formed, and this is ejected from the molecule (because of the recoil effect of emitting a quantum of γ-radiation). On treatment with water containing a reducing agent, the active iodine is converted into I$^-$ ions, and can readily be separated in this form from the C_2H_5I (Szilard, 1934)*. If the iodine in C_2H_5I were bound in a form which could be ionized, such a separation would be impossible. Bromine and iodine behave similarly, and in this manner it is possible to isolate the radioactive halogens in unweighably small ('carrier-free') quantities.

During the last few years, the radioactive indicator method has proved especially useful in the fields of organic chemistry and biochemistry. It has been possible by this means to trace for the first time the mechanism of certain reactions. Radioactive tracers have found similar applications in biology. Thus the use of the radioactive phosphorus isotope ^{32}P has made it possible to trace how phosphorus, taken up (in the form of its compounds) by the animal organism, is distributed in the course of time between the various organs, and how it is eventually excreted.

Artificially activated elements have also found applications as tracers in industry. For example, they have been used to control and indicate the flow of liquids, such as petroleum, in pipe lines. They also make it possible to follow the course taken by some selected element in large scale operations. Thus, by means of the appropriate tracer it has been possible to

---

* This so-called *Szilard-Chalmers process* is a valuable method of obtaining certain artificial radioelements in a state of high specific activity, from relatively weak neutron sources. The process is applicable only to elements which are present in covalently-bound form, and which undergo (n, γ) reactions with a sufficiently high recoil energy.

follow the rate of germanium, which is often present in zinc ores, through every step of
the extraction of zinc, even though the germanium content may be extremely small.
Radioactive tracers have also been used in studies of the friction of rubbing surfaces, the
wear of tires and of road surfaces.

4. Reactions of 'Hot' Atoms

The splitting-off of neutron activated halogen atoms from organic compounds,
cited above, is an example of the so-called 'hot atom' reactions. This name is given
to chemical reactions which are brought about directly by nuclear processes, in
that the energy liberated causes the rupture of chemical bonds. In certain cases,
hot atom reactions make it possible to separate nuclear isomers. Generally speaking
it is impossible to separate nuclear isomers, since these have not only the same
chemical properties but the same masses. However, in the case of a pair of nuclear
isomers related by an isomeric transition (cf. pp. 534, 578), the γ-rays emitted in
the isomeric transition can frequently be used to effect a separation. Consider, for
example, a tellurate solution containing the nuclear isomers ^{131}Te* and ^{131}Te, with
half-lives of 30 hours and 25 minutes, both present as tellurate ions. If a small
quantity of inactive alkali tellurite is added, and then separated again, from the
tellurate, the resulting preparation is found to contain only that nuclear isomer of
half-life 25 minutes. The explanation is that a portion of the γ-rays emitted in the
transformation of the ^{131}Te* nucleus into the normal ^{131}Te nucleus undergoes
internal conversion (p. 521). An electron is thereby ejected from the K or L shell
of the resulting ^{131}Te atom (of half-life 25 min). Through a process known as the
Auger effect, the energy set free in filling up this inner shell again is transferred to
the valence electrons, and leads to the breaking of a chemical bond between
tellurium and oxygen. The tellurate ion is thereby converted to a tellurite ion,
with the result that part of the 25-minute activity can be separated in the form of
tellurite. An analogous separation has been found to take place with other genetically related pairs of nuclear isomers.

5. Cross Sections in Nuclear Reactions

The number of neutrons crossing unit area (1 cm^2) each second, in a particular direction,
is known as the *neutron flux*, F. If the neutrons traverse an absorbing medium, the number
absorbed increases with the number of neutrons entering the medium. The absorption is
also dependent upon the velocity of the neutrons, v. If the thickness of the absorbing layer is
small, so that v is not appreciably diminished in traversing the medium, the decrease in
neutron flux along the x-direction can be represented as

$$-\frac{\mathrm{d}F}{\mathrm{d}x} = \sigma \cdot N_{\mathrm{L}} \cdot F \tag{1}$$

where N_{L} is the number of nuclei of the absorbing element per cubic centimeter (sometimes
termed Loschmidt's number), and σ is a constant, known as the *cross section* of the absorbing
nuclear species. This constant is of the dimensions cm^2, and can be schematically thought of
as the target area presented by the nucleus concerned to the neutron beam. If every neutron
were to be absorbed if its mid-point intersected a surface defined by the projected area of
an atomic nucleus, but were not absorbed if it did not fall within this area, then the total
cross section would be the same as the actual target area of the nucleus. This is of the order
of magnitude 10^{-24} cm^2. The cross section σ may, however, be very much smaller than this,

or may in some cases be much larger. The quantity 10^{-24} cm² is taken as the unit for measurement of σ, and is known as the *barn* (1 barn = 10^{-24} cm²).

The cross section defined by eqn. (1) is strictly referred to as the *total cross section*, σ_t. The attenuation of a beam of neutrons, by which σ_t is defined, is compounded of the diminution of neutron flux due to reflection and scattering of neutrons by the nuclei, and of the diminution due to the absorption of neutrons by the nuclei. A variety of nuclear reactions may be brought about by this absorption of neutrons, and for each mode of reaction it is possible to define, and to determine experimentally, a corresponding capture cross section.

Interaction cross sections can be determined for the absorption of other colliding particles (protons, deuterons, helium nuclei, and also photons of γ-radiation), and for the nuclear reactions that they induce, in the same way as for the absorption of neutrons and the various nuclear reactions caused by neutrons. It is of some importance to determine the cross sections for the individual nuclear reactions, since (as will be shown below) a knowledge of the cross sections makes it possible to predict the yield of artificially produced atomic species that can be obtained from a given flux of radiation.

The interaction cross section depends upon the energy of the colliding particles ($\frac{1}{2}mv^2$ or $h\nu$, respectively), as well as upon the type of particle and the nature of the nucleus bombarded. In the case of neutrons, the probability of capture by a nucleus, generally increases in proportion to the time spent by the neutron in the neighborhood of the nucleus. It may therefore usually be assumed that the capture cross section for neutrons decreases with increase of neutron energy. For positively charged particles, however. the cross section for nuclear reaction at first increases with increasing kinetic energy of the particles, and then in most cases begins to decrease again gradually at rather high values of the kinetic energy. The reason for this is that in order to penetrate into the nucleus, a colliding positively charged particle must possess sufficient translational energy to overcome the forces of repulsion initially exerted upon the particle by the similarly charged nucleus. Below a certain minimum 'threshold' energy, the cross section for endothermic nuclear reactions is zero. Unless the bombarding particle possesses sufficient energy to be used for such a reaction, it cannot enter the nucleus at all.

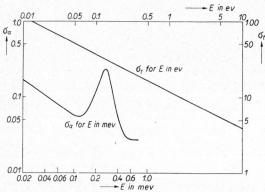

Fig. 74. The cross section of the lithium nucleus as a function of the kinetic energy of the neutron. E = kinetic energy of colliding neutron, in ev or mev

σ_a = cross section (in barns) for the nuclear reaction $^6_3\text{Li} + ^1_0\text{n} = ^3_1\text{H} + ^4_2\text{He}$

σ_t = total capture cross section for the absorption of neutrons by lithium

E and σ are plotted on a logarithmic scale along the coordinate axes.

If the energy liberated by the penetration of a colliding particle into the nucleus just suffices to raise the new nucleus so formed to an 'excited' state—i.e., into a state of higher energy from which it may undergo a transition to the ground state, with emission of γ-radiation, the probability of capturing the colliding particle is particularly great. This shows up in that the curve relating the cross section with the energy of the colliding particles displays steep maxima. These maxima are termed *resonance peaks*, since the phenomenon can be interpreted on a quantum mechanical basis as a resonance effect. For the capture of neutrons, it is usual to find very high resonance peaks only for neutron energies below about 10 ev. In the range up to 10 ev, however, the cross section for neutron capture may rise, at the resonance peaks, to a value 100,000 times as great as the actual target area of the atomic nucleus.

For the scattering of very slow neutrons—especially those of thermal energies—the cross section also depends upon the crystal structure of the substance bombarded. According to the fundamental equation of wave mechanics, $\lambda = \dfrac{h}{m \cdot v}$, the wave length λ to be ascribed to a thermal neutron is of the same order of magnitude as the interatomic distances in crystals. When a beam of such neutrons passes through a crystal, diffraction phenomena therefore take place. The attenuation of the neutron beam along different crystallographic directions therefore depends upon the crystal structure. The study of neutron diffraction is

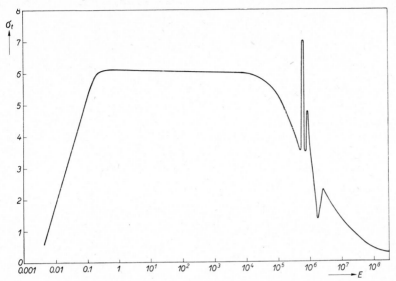

Fig. 75. Variation of the total cross section σ_t of the Be nucleus for neutrons of different energies. σ_t in barns; E in ev, on logarithmic scale.

analogous to, and supplementary to, the study of X-ray diffraction, as a means of investigating the structure of solids.

Figs. 74–77 exemplify the variation of the nuclear cross section with the velocity (or kinetic energy) of the particles. As the upper curve of Fig. 74 shows, the total cross section σ_t of lithium for neutrons decreases regularly with increase in the neutron energy. From the slope of the curve, it is seen that the cross section σ_t is a linear function of $\dfrac{1}{v}$ —i.e., of the time spent by the neutron close to the nucleus. This is true of many other elements also—e.g.,

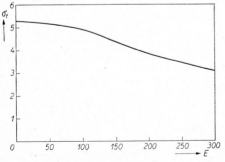

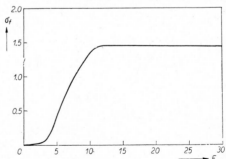

Fig. 76. Variation of the total cross section σ_t of the uranium nucleus for neutrons of different energies. σ_t in barns; E in mev

Fig. 77. Cross section of the ^{237}Np nucleus for fission by neutrons of different energies. σ_f in barns; E in mev

for boron. It may happen, however, that σ is practically independent of the neutron velocity, over a wide range. An example of this is given in Fig. 75. Resonance peaks may be seen at several places on the curve of Fig. 75—in this instance, rather exceptionally, at quite high E values. The lower curve of Fig. 74 shows how the cross section for the reaction $^6Li + n = ^3H + \alpha$ varies with neutron energy; here again there is a distinct resonance peak at quite high neutron energies. Figs. 76 and 77 give other examples of the variation of cross section with neutron energy. The curve of Fig. 77 refers to the cross section of ^{237}Np. As indicated by the curve, ^{237}Np is one of those nuclei for which fission can be brought about by fast neutrons, but not by slow neutrons. From the figures given in Table 73, it may be seen that the cross sections vary over an extremely wide range, according to the reactions to which they relate.

TABLE 73

CROSS SECTIONS FOR VARIOUS NUCLEAR REACTIONS

Reaction	Energy of bombarding particle		Cross-section, barns	Energy of bombarding particle		Cross-section, barns
$^2H + n = {}^3H + \gamma$	0.025	ev	0.0003	—		—
$^2H + d = {}^3He + p$	0.1	mev	0.023	1	mev	0.09
$^2H + \gamma = {}^1H + n$	2.6	mev	0.00066	6.2	mev	0.0012
$^6Li + n = {}^3H + \alpha$	0.025	ev	900	1	mev	0.33
$^7Li + p = 2\,\alpha$	0.1	mev	0.00008	1	mev	0.00009
$^7Li + p = {}^7Be + n$	2	mev	0.23	—		—
$^{10}B + n = {}^7Li + \alpha$	0.025	ev	2500	1	mev	0.45
$^{14}N + n = {}^{14}C + p$	0.025	ev	1.2	2.8	mev	0.04
$^{23}Na + n = {}^{24}Na + \gamma$	0.025	ev	0.41	1000	ev	3.33
$^{27}Al + n = {}^{28}Al + \gamma$	0.025	ev	0.1	0.5	mev	0.0018
$^{107}Ag + n = {}^{108}Ag + \gamma$	0.025	ev	3	1	mev	0.09
$^{157}Gd + n = {}^{158}Gd + \gamma$	0.025	ev	270,000	0.2	ev	19,200

6. Yield in Nuclear Reactions

The number N of bombarding particles absorbed per second by a thin layer of substance, of area q, is given by eqn (1) as

$$N = -\frac{dF}{dx} q \cdot dx = \sigma_t \cdot N_L \cdot F \cdot q \cdot dx.$$

The number of particles absorbed per second by one cubic centimeter of the sample is then

$$\frac{\sigma_t \cdot N_L \cdot F \cdot q \cdot dx}{q \cdot dx} = \sigma_t \cdot N_L \cdot F.$$

This also represents the number of nuclei N_L which would be produced per second per unit volume, if the bombarding particles disappeared by this reaction alone. However, only the fraction $\frac{\sigma_R}{\sigma_t}$ is used up in the nuclear reaction, where σ_R, σ_t represent the cross section for the reaction and the total cross section, respectively. Hence the yield of nuclei from the nuclear reaction is

$$N_R = \sigma_R \cdot N_L \cdot F \tag{2}$$

Therefore the capture cross section σ_R provides a direct measure of the yield of a nuclear reaction from the given element, in the particle flux F.

If eqn (2) is multiplied by the time of irradiation (in seconds), the result is the total number of nuclear transmutations effected in that time, on the assumption that the number of nuclei available for reaction has not changed significantly during the irradiation period;

if this assumption is not valid, the variation of the number of reactant nuclei with time must be allowed for.

A case of considerable practical importance is the calculation of the yield in the production of artificial radioactive nuclei. For this calculation, it is necessary to subtract from the number of nuclei produced by the bombarding particles, during the time interval dt, the number which have disappeared through radioactive disintegration. The increase in the number of radioactive nuclei, $\dfrac{dN}{dt}$, per cm³ of material irradiated, is then represented by

$$\frac{dN}{dt} = \sigma_R \cdot N_L \cdot F - N\lambda \tag{3}$$

(where λ is the disintegration constant of the radioactive species which is formed in the reaction, with cross section σ_R). Integration of eqn. (3) yields

$$N_t = \frac{1}{\lambda} \cdot \sigma_R \cdot N_L \cdot F \left(1 - e^{-\lambda t}\right) \tag{4}$$

where N_t is the number of radioactive nuclei per cm³ which have accumulated during the irradiation time t. Equation (4) has the same form as eqn. (13), p. 525, which represents the increase in the number of atoms of a radioactive decay product from a long-lived mother substance. The maximum amount N_∞ of the radioactive species which can be formed per cm³ of substance is found by substituting $t = \infty$ in eqn. (4):

$$N_\infty = \frac{1}{\lambda} \sigma_R \cdot N_L \cdot F \tag{5}$$

This yield is achieved by making the time or irradation long in comparison with the half-life of the radioelement formed. It may be seen from eqn. (5) that the maximum concentration of the artificial radioelement which can be obtained from a nuclear reaction is determined essentially by the particle flux F. The neutron flux which can be obtained in nuclear reactors, and used for the technical production of artificial radioelements is in most cases of the order of 10^{11} to 10^{12} neutrons per cm² per second. The quantities of material which may be subjected to neutron irradiation under these conditions are ordinarily 1 to 10 g or more— e.g., for the production of phosphorus (which is required in considerable quantity for medical work) by the $S(n, p)$ reaction, sulfur is irradiated in kilogram quantities.

7. Nuclear Stability

It is now generally assumed that the forces of interaction between the constituents of the nucleus are resonance forces in the wave mechanical sense, and that these are strong enough to hold the nucleus together against the Coulomb repulsive forces between the protons. However, the resonance forces of attraction only outweigh the Coulomb repulsion if the ratio of number of protons to number of neutrons lies within certain limits. Nuclei with a small number of protons are only stable, as a rule, if the number of neutrons is equal to, or one greater than, the number of protons. With increasing numbers of protons, the maximum stability progressively shifts in favor of an increasing preponderance of neutrons over protons. Fig. 78 shows the relation between the number of neutrons and the number of protons in the nuclei of stable and artificially produced unstable elements; from the Figure it may be seen that a nucleus is unstable both when the ratio of neutrons to protons is too small, and when it is too great.

It does not follow that an unstable nucleus disintegrates immediately; provided that the energy of the nucleus is not too greatly in excess of the energy of its disintegration products, there is only a certain *probability of disintegration*. This probability is quantitatively expressed by the *disintegration constant* [eqn. (4), p. 514]. The facts can be interpreted in the following manner, as was first proposed by Gamow (1928) for the emission of α-particles in radioactive decay. If the potential energy of a positively charged particle is plotted as a function of its distance from the center of the nucleus, a curve such as Fig. 79 is obtained. Outside the

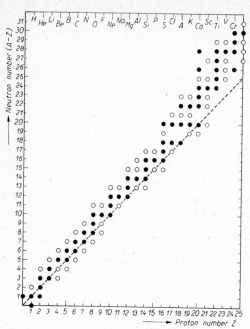

Fig. 78. Proton numbers and neutron numbers in the nuclei of the elements of atomic numbers 0–25.

Stable nuclides are represented by black circles, unstable species by open circles. Nuclei for which the neutron numbers and proton numbers are equal fall on the dotted line.

nucleus, the variation of the potential follows Coulomb's law. Inside a *stable* nucleus, the potential energy of an electropositive particle is *lower* than it would be outside the nucleus, even at an infinite distance. In an *unstable* nucleus, however, there are positively charged particles with a higher potential energy than they would possess at a great distance from the nucleus. Nevertheless, if the potential of such a particle is lower than the summit of the potential wall which surrounds the nucleus, it is unable to escape from the nucleus, according to classical mechanics, unless some external forces act upon the particle and lift it over the potential barrier. According, to wave mechanics however, there is a certain probability that the particle may 'leak' through the barrier*. The number of atoms to which this happens, in unit time, is directly proportional to the probability of the event. Hence the radioactive disintegration constant is numerically equal to the probability of escape of the corresponding particle from the nucleus. The smaller the breadth of the potential barrier through which the particle must tunnel, the greater is the probability of escape; correspondingly, the kinetic energy which the particle acquires after leaving the nucleus is also greater, since— graphically expressed—it slides down

the potential hill from a greater height. This conclusion leads at once to the Geiger-Nuttall relationship [eqn. (9), p. 515].

In α-emission from identical nuclei, all the particles are emitted with the same velocity. In β-decay, however, both positrons and electrons leave the nucleus with velocities which are distributed continuously over a wide range. The maximum kinetic energy of the β-particles in this process corresponds to the energy loss of the nucleus, as calculated from the loss of mass and the mass-energy equivalence law. The majority of the β-particles have a smaller energy than this, however, and in order to explain this disappearance of energy, the hypothesis has been advanced that in β-decay, other particles (in this case uncharged) are emitted from the nucleus, in addition to positrons or electrons. These hypothetical particles are termed *neutrinos* as already described (pp. 522 and 562).

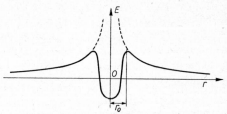

Fig. 79. Gamow potential barrier. *E* is the energy which would be liberated by removing a positively charged particle to the distance $r = \infty$. The nuclear radius r_0 is defined as the distance from the mid-point of the nucleus to that at which the potential function begins to deviate from that based on Couomb's law (shown by the dotted curve).

* There is a corresponding possibility of a particle leaking into the nucleus from outside. This is why collision with a positively charged particle may lead to penetration into the nucleus, even although the kinetic energy of the particles is too low to overcome the calculated Coulomb repulsion.

8. Energy of Formation of Nuclei

The mass defect of a nucleus provides an exact measure of the energy liberated in forming a nucleus from its constituent nucleons. However, the mass values are not known for all atomic species with sufficient accuracy for the calculation of energies of formation of the nuclei. It is therefore of some interest that it is possible to calculate the energy of formation, to a satisfactory approximation, from the theory of nuclear forces. This theory is, indeed, not sufficiently developed to permit of an exact calculation of nuclear energies without the introduction of empirical data; however, the empirically determined constants that appear in the calculations are the same for all the elements. The formula given below therefore makes it possible to compute energies of nuclear formation even for species which disintegrate at once, under the conditions in which they are formed, so that any experimental determination is completely impossible. Use will be made of this possibility in a later section, in the discussion of nuclear fission.

The formula for approximate calculation of the energy of formation of the nucleus is derived as follows. As already mentioned, it is assumed that the intra-nuclear forces are wave mechanical exchange forces. Under the conditions obtaining within the nucleus, these have only an extremely minute range of action. This leads to the result that the energy liberated from the association of the nucleons under the action of these forces is approximately proportional to the mass number A. The constant of proportionality will be denoted by a_1.

The idea that strong exchange forces operate between the nucleons is based on the assumption that a proton can transfer its charge to a neutron, so that if the two nucleons are in the closest proximity, the states p-n and n-p continually alternate with each other. Just as the exchange of an electron between two protons gives rise to a chemical bond, so the alternation of the states p-n and n-p gives rise to binding forces of very great magnitude. Binding forces may also operate between a pair of neutrons, but these are very much weaker than between a proton and a neutron. Hence the binding energy of a nucleus which contains an excess of neutrons is smaller than $a_1 A$. It can be shown that, when account is taken of the weaker neutron-neutron binding forces, the energy is less than $a_1 A$ by an amount $\dfrac{a_2(A - 2Z)^2}{A}$. In this expression, a_2 is another constant, and $A - 2Z$ is the excess of neutrons over protons.

Because of their like charge, the protons in the nucleus repel one another. The work to be expended, in overcoming this repulsion and building the protons into the nucleus amounts to $a_3 \dfrac{Z^2}{r}$, where a_3 is another constant proportionality factor, and r is the radius of the nucleus. Since r^3 is proportional to the mass of the nucleus, r is proportional to $A^{1/3}$. Therefore the binding energy given in the preceding paragraph must be further diminished by an amount $a_3 \dfrac{Z^2}{A^{1/3}}$.

Allowance must also be made for the looser binding of nucleons on the surface of the nucleus than in the interior of the nucleus. This surface effect is proportional to r^2, or to $A^{2/3}$, so that a further amount $a_4 A^{2/3}$ is to be deducted from the binding energy.

Finally the spin effect must be taken into account. Like the electrons in the core of an atom, a quantized spin must be assigned to the nucleons in the nucleus. Protons with opposite spin can pair up, just as electrons achieve 'spin coupling'. The same is true of neutrons, but there is no spin coupling between a proton and a neutron. In a nucleus which contains both an even number of protons and an even number of neutrons (an 'even-even' nucleus), all the nucleons are paired. The especial stability of even-even nuclei is attributed to this coupling of all spins. If a nucleus contains an even number of protons and an odd number of neutrons (an 'even-odd' nucleus), one of the neutrons remains unpaired. In general, such a nucleus is less stable than a nucleus in which all spins are coupled. If there is an odd number of protons and an odd number of neutrons in the nucleus ('odd-odd' nucleus), it contains two unpaired nucleons. It is exceptional for nuclei of this type to be stable. In calculating the energy of formation of the nucleus, the spin effect can be included by inserting in the expression a term $\dfrac{a_5}{A}$, in which a_5 may be either positive or negative in sign, according as all nucleons are paired, or as two are left unpaired. If the nucleus contains a single unpaired nucleon, a_5 takes the value $a_5 = 0$.

Thus the formula for calculation of the energy of formation of the nucleus E_k (in mev) takes the form

$$E_k = a_1 A - a_2 \frac{(A - 2Z)^2}{A} - a_3 \frac{Z^2}{A^{1/3}} - a_4 A^{2/3} + \frac{a_5}{A} \tag{6}$$

Values for the proportionality constants are

$$a_1 = 14.0, \qquad a_2 = 19.3, \qquad a_3 = 0.585, \qquad a_4 = 13.05,$$
$$a_5 = +130 \text{ for even-even nuclei}, \qquad a_5 = -130 \text{ for odd-odd nuclei},$$
$$a_5 = 0 \text{ for even-odd nuclei and odd-even nuclei}.$$

By means of this formula, for example, the energies of formation of the nuclei of $^{40}_{20}\text{Ca}$, $^{52}_{24}\text{Cr}$, $^{63}_{29}\text{Cu}$, $^{232}_{90}\text{Th}$, and $^{238}_{92}\text{U}$ may be calculated as 342, 452, 544, 1760, and 1796 mev, respectively. These agree satisfactorily with the values 342, 456, 552, 1763 and 1797 mev found from the mass defects of these nuclei.

9. Rules of Nuclear Stability

The theory of nuclear forces [16-21, Chap. 12] has not yet arrived at that stage of quantitative development at which it is possible to make a general prediction of the mass numbers associated with any particular nuclear charge. However, it is now possible, from the experimental data, to formulate certain rules concerning the composition of stable nuclei. The most important rules are as follows.

[1] There are no pairs of stable isobars differing in nuclear charge by one unit (Mattauch's rule).

This rule, which was deduced empirically by Mattauch in 1934, can be regarded as a strict law, according to current nuclear physical knowledge (Jensen, 1939). A few apparent exceptions are simulated by the existence of long-lived isotopes, the nuclei of which probably undergo transformation by electron capture.

[2] Among nuclei of even mass number, the stable isobars invariably have an even proton number and even neutron number (Harkins' rule). It is a corollary that an odd proton number is never associated with an odd neutron number.

A few light nuclei (^2_1H, ^6_3Li, $^{10}_5\text{B}$ and $^{14}_7\text{N}$) form exceptions to this rule.

[3] Every odd mass number is represented by only a single stable nucleus.

Nuclei with even proton numbers may possess either even or odd mass numbers. Nuclei with odd proton numbers, however, can only have odd mass numbers, according to rule 2; furthermore, according to rule 3, certain of these must be excluded, namely those that also exist as nuclei with other proton numbers. This is why the elements of odd atomic number are, for the most part, represented only by a single stable nucleus (cf. Table 61, p. 549). Such mixed elements as exist among the elements of odd atomic number never possess more than two stable nuclear species (cf. Table 62, p. 551–2). This accords with the next rule.

[4] Every chemical element is represented by at least one nucleus of odd mass, but never by more than two nuclei of odd mass (Aston's rule).

The elements of atomic numbers 18, 58, 43, and 61 contravene this rule. The first two (A and Ce) possess only isotopes of even mass number. The other two elements are unstable.

10. Artificial New Elements

The first atomic species to be produced artificially were only stable or unstable isotopes of already known chemical elements. It later proved possible to produce, by means of artificial nuclear transformations, elements which had not, up to that time, been discovered in Nature. It has, in fact, been found that certain of these do not exist in Nature at all. The artificial production of these elements enabled all the remaining gaps in the Periodic Table to be filled up, and elements have also been prepared with element numbers higher than that of uranium, until then the last element in the Periodic Table. These are the *transuranic elements*, which will be discussed in the next chapter.

The first element to be prepared by artificially induced nuclear reactions, but not found in Nature, was the element of atomic number 43, *technetium*. When the formation of this element was first observed, through the bombardment of molybdenum with deuterons (1937), there were altogether four gaps in the Periodic System; the places for the elements of atomic numbers 61, 85 and 87 were also vacant. Since that date, many *unstable* nuclear species of these elements have been discovered. Studies of nuclear systematics make it very probable that there are, in fact, no stable nuclei with these atomic numbers.

The existence of stable nuclei with these atomic numbers would be contrary to Mattauch's rule. From the position of element 43 in the Periodic Table, it may be expected that the most stable isotopes must have mass numbers of about 98. However, every mass number between 94 and 102 is already represented among the stable isotopes of the two neighboring elements, 42 and 44 (Mo and Ru) (cf. Table 62, p. 551). Hence, according to Mattauch's rule no stable isotope of atomic number 43 can exist. Similar considerations apply to element 61. Its most stable isotopes must have mass numbers about 147, but every mass number between 142 and 150 is already found among the stable isotopes of elements 60 and 62. These two examples show once again that the nuclei with even proton number are more stable than the nuclei of odd proton number. Isotopes of element 61 were first definitely characterized in the fission products of uranium (Marinsky, Glendenin, and Coryell (1947)). The element has been given the name of *promethium*. For the artificial preparation of element 43, see p. 231–2.

Elements 85 and 87 occur in that region of the Periodic Table where every element is unstable. It may be assumed, therefore, that the only nuclear species of elements 85 and 87 which can exist will also be unstable, and will be rather short-lived, since the elements of odd atomic number are usually shorter-lived than the adjacent elements of even atomic number. It is only possible for elements of fairly short half-life to exist in Nature in detectable quantities if they are formed continuously by the decay of elements with long half-lives. The only possible parents are uranium (^{238}U and ^{235}U) and thorium. Hence it was to be expected that if elements 85 and 87 exist in Nature, they must be members of the natural disintegration series. In 1939, Perey proved that actinium underwent a dual decay scheme, about 1% of the atoms emitting an α-particle, while the remainder underwent β-decay. According to the displacement rule, actinium must be converted by α-emission into an element with atomic number 87 (and mass number 223)—i.e., into *eka-cesium*, which has since been given the name *francium* (Fr). Francium is itself a β-emitter, of 21 minute half-life, which decays to form AcX. It was found that francium could readily be co-precipitated with rubidium or cesium perchlorate or chloroplatinate as carrier—i.e., that its properties correspond to those of an alkali metal. Element 85 (*eka-iodine*, now called *astatine*) was first obtained artificially by Segrè (1940), by the bombardment of bismuth with high energy α-particles— $^{209}_{83}$Bi $(\alpha, 2n)$ $^{211}_{85}$At. It has a half-life of 7.5 hrs, and undergoes dual decay. 60% of the atoms are converted into ^{207}Pb, by way of ^{211}Po (AcC′), and 40% change via ^{207}Bi, probably with ^{207}Pb as final end product. In 1943, Karlik and Bernert proved that astatine also occurred as a member of the natural disintegration series. They showed that RaA, ThA and AcA all underwent a dual disintegration, to form $^{218}_{85}$At (branching ratio 0.03%), $^{216}_{85}$At (branching

TABLE 74

NUCLEAR SPECIES OF THE ELEMENTS 43, 61, 85 AND 87

	Nuclear species	Half-life	Decay	Decay Products	Formation processes	
Isotopes of *technetium*, atomic number = 43	$^{92}_{43}$Tc?	4.5 min	e^+	$^{92}_{42}$Mo	$^{92}_{42}$Mo(d, 2n)	
	$^{93}_{43}$Tc?	2.7 hrs	e^+	$^{93}_{42}$Mo	$^{92}_{42}$Mo(d, n)	
	$^{94}_{43}$Tc*	50 min	γ	$^{94}_{43}$Tc	$^{94}_{42}$Mo(p, n)	$^{94}_{42}$Mo(d, 2n)
	$^{94}_{43}$Tc	< 50 min	e^+, K	$^{94}_{42}$Mo	$^{94}_{43}$Tc* $\longrightarrow \gamma$	
	$^{95}_{43}$Tc	20 hrs	K	$^{95}_{42}$Mo	$^{92}_{42}$Mo(α, p) $^{95}_{42}$Mo(d, 2n)	$^{95}_{42}$Mo(p, n) $^{95}_{44}$Ru $\longrightarrow e^+$
	$^{95}_{43}$Tc	62 days	99.2 % $\longrightarrow K$ 0.8 % $\longrightarrow e^+$	$^{95}_{42}$Mo $^{95}_{42}$Mo	$^{92}_{42}$Mo(α, p) $^{95}_{42}$Mo(d, 2n)	$^{94}_{42}$Mo(d, n) $^{95}_{42}$Mo(p, n)
	$^{96}_{43}$Tc	4.2 days	K	$^{95}_{42}$Mo	$^{95}_{42}$Mo(d, n) $^{96}_{42}$Mo(p, n)	$^{96}_{42}$Mo(d, 2n) ^{93}Nb (α, n)
	$^{97}_{43}$Tc*	91 days	γ	$^{97}_{43}$Tc	$^{96}_{42}$Mo(d, n) $^{97}_{44}$Ru $\longrightarrow e^+$	$^{97}_{42}$Mo(p, n)
	$^{98}_{43}$Tc?	40 min	e^-	$^{98}_{44}$Ru	$^{97}_{42}$Mo(d, n)	$^{98}_{42}$Mo(d, 2n)
	$^{99}_{43}$Tc*	6.6 hrs	γ	$^{99}_{43}$Tc	U(n, f) Th(n, f)	$^{99}_{42}$Mo $\longrightarrow e^-$
	$^{99}_{43}$Tc	$2.12 \cdot 10^5$ yrs	e^-	$^{99}_{44}$Ru	$^{99}_{43}$Tc* $\longrightarrow \gamma$	
	$^{100}_{43}$Tc?	1.33 min	e^-?	$^{100}_{44}$Ru?	$^{100}_{42}$Mo(d, 2n)	
	$^{101}_{43}$Tc	16.5 min	e^-	$^{101}_{44}$Ru	U(n, f) $^{102}_{44}$Ru (γ, p)	$^{100}_{42}$Mo(d, n) $^{101}_{42}$Mo $\longrightarrow e^-$
	$^{105}_{43}$Tc	15 min	e^-	$^{105}_{44}$Ru	U(n, f)	$^{105}_{42}$Mo $\longrightarrow e^-$
Promethium isotopes at. number = 61	$^{143}_{61}$Pm	~ 200 days	K	$^{143}_{60}$Nd	$^{141}_{59}$Pr(α, 2n)	$^{142}_{60}$Nd (d, n)
	$^{145}_{61}$Pm	~ 30 yrs	K	$^{145}_{60}$Nd	$^{144}_{62}$Sm(n, γ)	$^{145}_{62}$Sm $\longrightarrow K$
	$^{147}_{61}$Pm	3.7 yrs	e^-	$^{147}_{62}$Sm	U(n, f)	$^{147}_{60}$Nd $\longrightarrow e^-$
	$^{148}_{61}$Pm	5.3 days	e^-	$^{148}_{62}$Sm	$^{145}_{60}$Nd (α, p) $^{148}_{60}$Nd (p, n)	$^{148}_{60}$Nd (d, 2n) $^{147}_{61}$Pm(n, γ)
	$^{149}_{61}$Pm	2.00 days	e^-	$^{149}_{62}$Sm	U(n, f)	$^{149}_{60}$Nd $\longrightarrow e^-$
Astatine isotopes atomic number = 85	$^{210}_{85}$At	8.3 hrs	K	$^{210}_{84}$Po	$^{209}_{83}$Bi (α, 3n)	
	$^{211}_{85}$At	7.5 hrs	α	$^{207}_{83}$Bi	$^{209}_{83}$Bi (α, 2n)	$^{238}_{92}$U (α, 9p 22n)
	$^{212}_{85}$At	0.25 sec	α	$^{208}_{83}$Bi	$^{209}_{83}$Bi (α, n)	
	$^{214}_{85}$At	v. short	α	$^{210}_{83}$Bi	$^{218}_{87}$Fr $\longrightarrow \alpha$	
	$^{215}_{85}$At	~ 10^{-4} sec	α	$^{211}_{83}$AcC	$^{215}_{84}$AcA $\longrightarrow \beta$	$^{219}_{87}$Fr $\longrightarrow \alpha$
	$^{216}_{85}$At	~ 10^{-3} sec	α	$^{212}_{83}$ThC	$^{216}_{84}$ThA $\longrightarrow \beta$	$^{220}_{87}$Fr $\longrightarrow \alpha$
	$^{217}_{85}$At	0.021 sec	α	$^{213}_{83}$Bi	$^{221}_{87}$Fr $\longrightarrow \alpha$	
	$^{218}_{85}$At	few sec	99.9 % $\longrightarrow \alpha$ 0.1 % $\longrightarrow \beta$	$^{214}_{83}$RaC $^{218}_{86}$Em	$^{218}_{84}$RaA $\longrightarrow \beta$	
	$^{219}_{85}$At	0.9 min	97% $\longrightarrow \alpha$ 3% $\longrightarrow \beta$	$^{215}_{83}$Bi $^{219}_{86}$AcEm	$^{223}_{87}$Fr $\longrightarrow \alpha$	
Francium isotopes, at. number = 87	$^{218}_{87}$Fr	v. short	α	$^{214}_{85}$At	^{222}Ac $\longrightarrow \alpha$	
	$^{219}_{87}$Fr	~ 10^{-4} sec	α	$^{215}_{85}$At	^{223}Ac $\longrightarrow \alpha$	
	$^{220}_{87}$Fr	~ 30 sec	α	$^{216}_{85}$At	^{224}Ac $\longrightarrow \alpha$	
	$^{221}_{87}$Fr	4.8 min	α	$^{217}_{85}$At	^{225}Ac $\longrightarrow \alpha$	
	$^{223}_{87}$Fr	21 min	0.004% $\longrightarrow \alpha$ 99.996% $\longrightarrow \beta$	$^{219}_{85}$At $^{223}_{88}$AcX	^{227}Ac $\longrightarrow \alpha$	

ratio 0.014%) and $^{215}_{85}$At (branching ratio 0.0005%), respectively. These three isotopes of astatine are even shorter lived than $^{211}_{85}$At, which was first obtained artificially. They

rapidly undergo conversion into RaC, ThC and AcC, respectively, with emission of α-particles (ranges 5.53 cm, 6.84 cm, and 8.0 cm). At least 10 isotopes of astatine are now known; $^{210}_{85}At$, the longest-lived of these, has a half-life of 8.3 hrs, and changes into polonium by electron capture.

Table 74 gives a summary of the various nuclear species now known for the elements 43, 61, 85 and 87. It includes only those for which the mass numbers are known with certainty, or at least (in the cases marked with ?) with a high degree of probability. In fact, a considerably larger number of isotopes is known for most of these elements—e.g., at least 17 isotopes of technetium and 7 isotopes of promethium. These can be identified by their differing half-lives, but not all the mass numbers are known. The species ^{215}At, ^{216}At, ^{218}At, ^{219}At and ^{223}Fr form part of the natural disintegration series. Where the symbol in column 1 is marked with an asterisk, it implies that the nuclear species concerned is formed in a metastable state by the reactions of column 5; it then changes into a state of lower energy, with the half-life shown in column 2, without the emission or absorption of any corpuscle—i.e., by a γ-ray change. This kind of nuclear transformation is marked γ in column 3. The other spontaneous nuclear reactions are shown in column 3 by the same symbols as were used in Table 69, p. 576, except that that for the nuclei occurring in the natural decay series, the emission of an electron is termed β-decay. Some of the nuclear species listed in Table 74 have been detected among the products of the fission of uranium. This mode of formation is indicated as U (n, f). It may be seen from the Table that as a rule the isotopes of technetium and of promethium all undergo transformations of the same type as found for other artificially prepared unstable isotopes of the light and moderately heavy elements, but that the isotopes of the heavy elements astatine and francium undergo either α-decay or β-decay, even in the case of the artificially prepared isotopes which are not members of the natural disintegration series.

11. Nuclear Fission [15-18]

(a) General

It was first recognized by Hahn (1939) that when uranium is irradiated with neutrons, two alternative nuclear reactions may occur. The neutron may be incorporated in the uranium nucleus, to form an unstable isotope with higher mass number or, alternatively, there may be a *fission* of the nucleus into two nuclei, both of moderately large mass; these are unstable, and undergo further radioactive decay in due course. It has been found that the incorporation of a neutron takes place only in the case of the uranium isotope ^{238}U, and only when uranium is exposed to *slow neutrons*. ^{238}U reacts with slow neutrons to give the unstable ^{239}U, which subsequently changes into element 93 (neptunium) by emission of a β-ray. (This element will be more fully discussed in the next chapter). The isotope ^{238}U only undergoes fission of the nucleus when it is irradiated with *fast neutrons*. The nucleus of ^{235}U, on the other hand, undergoes fission by reaction with both *slow and fast neutrons*. This is of great importance, since the isotope ^{235}U is present in natural uranium, although only in small concentration (0.72 atom-% or 0.71 weight %).

The *fission products*, or fragments formed in the fission of the uranium nucleus, have been proved to comprise one or more isotopes of every chemical element between $_{30}Zn$ and $_{63}Eu$. For most of these elements, several different unstable, isotopic nuclear species are formed, but it is probable that only a few of these unstable nuclei should be regarded as the primary fission fragments, formed directly from the uranium nucleus—e.g., by processes such as

$$_{92}U + _0n = _{56}Ba + _{36}Kr, \quad \text{or} \quad _{92}U + _0n = _{54}Xe + _{38}Sr.$$

Most of the observed fission products are formed subsequently, by the secondary radioactive disintegration of the primary fission fragments e.g.,

$$^{140}_{54}Xe \xrightarrow[\beta]{15 \text{ min}} ^{140}_{55}Cs \xrightarrow[\beta]{33 \text{ min}} ^{140}_{56}Ba \xrightarrow[\beta]{300 \text{ hrs}} ^{140}_{57}La \xrightarrow[\beta]{36 \text{ hrs}} ^{140}_{58}Ce \text{ (stable)}.$$

In these successive β-ray transformations, the mass numbers of the fission products remain unaltered, although their chemical nature changes. In all, not less than 34 different elements have been identified among the products of fission of uranium by slow neutrons. As shown in Fig. 80, the mass numbers of the various atomic species extend over the range from 72 to 162. Fig. 80 shows how the experimentally determined yields of the various fission products varies with mass number. It is evident that the fission of a ^{235}U nucleus into two roughly equal fragments is a very rare occurrence. By far the largest proportions of the fission fragments have mass numbers close to the values 95 and 139. As a rule, then, the fission of the nucleus takes place unsymmetrically, to give one fragment with relatively smaller mass, and one of relatively larger mass.

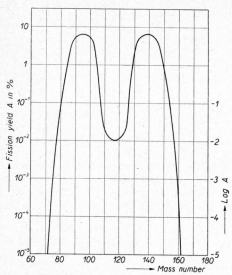

Fig. 80. Yield of fragments of various masses from the fission of the ^{235}U nucleus by means of slow neutrons.
(Data from Siegel, *J. Amer. Chem. Soc.*, 68 (1946) 2411).

It had originally been assumed that all the products of the nuclear reactions of uranium with neutrons had larger nuclear charges than the uranium nucleus. They were therefore at first considered to be 'transuranic elements' (Fermi, 1934; Hahn and Meitner, 1935). A closer chemical examination proved, however, that many of the substances concerned must have much lower atomic numbers than uranium. Thus the element originally taken to be 'eka-platinum' proved to be chemically identical with iodine. It was thus proved by the chemical evidence that it was possible for the nuclei of the heaviest elements to break up into quite large fragments—an inference of the greatest importance for nuclear physics. It was at once recognized that this discovery brought with it the possibility of utilizing atomic energy,—a possibility that was very soon translated into reality.

The fission of the heaviest nuclei—especially that of ^{283}U—is of fundamental importance for the utilization of atomic energy because *excess neutrons* are liberated in the process. The nuclei of the heaviest elements contain a higher ratio of neutrons to protons than do the nuclei of medium mass, which are formed as fission products. Under appropriate conditions, every neutron set free in the fission of uranium can bring about the fission of another nucleus of ^{235}U, and thereby liberate yet more neutrons. If care is taken to eliminate all impurities that would absorb neutrons, and if the quantity of uranium is sufficient to avoid the loss of the majority of the neutrons by leakage, so that a sufficient proportion of the neutrons collide with further ^{235}U nuclei and cause fissions, the number of neutrons increases continuously. If the process is not controlled, it accelerates with great speed (e.g., in an atomic bomb); if the process is controlled, it can be used to liberate useful energy (in a nuclear reactor). The nuclear reaction takes place explosively, for example, in a sufficiently large quantity of pure ^{235}U. If it takes place in natural uranium, containing only 0.71% of ^{235}U, some of the neutrons are used up in forming plutonium from ^{238}U (by way of ^{239}U and ^{239}Np) (cf. next chapter). In this case, the process can be so controlled that the number of neutrons does not exceed a certain maximum value. The amount of energy liberated in fission of ^{235}U is about 180 mev, or about $4 \cdot 10^9$ kcal, per g-atom. Further very large quantities of energy are set free during the radioactive decay of the fission products, and also in the conversion of ^{238}U into plutonium in nuclear reactors. The practical output of energy in a nuclear reactor is about 10^9 kcal per kg of uranium, equivalent to the heat output from about 100 tons of coal.

The fission of plutonium follows a very similar course to the fission of uranium. Thorium and protactinium have also proved to be *fissile*. Like the uranium isotope ^{238}U, however, these nuclei undergo fission only by *fast neutrons*, with energies in excess of 1 mev. For various reasons to be considered in a later section, fast neutrons are generally not suitable for reactions such as those involved in the production of plutonium and the utilization of atomic energy. Fission can also be brought about by protons, deuterons, or α-particles, if these are accelerated to very high energies. Very energetic γ-radiation can also produce fission. By means of highly accelerated particles it has been possible to bring about fission of the nuclei, of bismuth, lead, thallium, mercury, gold, platinum, and tantalum. However, the kinetic energy of the particles needed to cause fission increases progressively as the atomic numbers of the elements diminish. It is necessary to employ α-particles with a kinetic energy of 400 mev to cause fission in tantalum, and even particles with such enormous kinetic energy do not suffice to bring about the fission of nuclei of lower atomic number than tantalum. The reason for this will be considered below. It is, however, apparent why there should be only a very limited number of nuclear species with can undergo fission with slow neutrons, as well as with fast neutrons.

(b) Calculation of the Energy of Nuclear Fission

Nuclear fission is attended with the disappearance of a certain amount of mass. It therefore follows from the principle of mass-energy equivalence (p. 567) that energy must be liberated when the heavy nucleus breaks up into two fragments of medium mass. The amount of energy liberated naturally depends upon the particular pair of nuclei formed in fission. As has already been stated, the most abundant species of atoms formed by the slow neutron fission of ^{235}U are those with mass numbers 95 and 139. These mass numbers add up to 234, instead of 236, because for *each neutron* captured by the ^{235}U nucleus, *two neutrons* are emitted in the formation of these fission fragments. The stable species of atoms, formed as the ultimate products of β-ray decay of fission fragments of masses 95 and 139, are ^{95}Mo and ^{139}La. Substitution of accurate mass values in the equation

$$^{235}U + {}^1n = {}^{95}Mo + {}^{139}La + 2\ {}^1n$$

gives for the loss of mass:

$$\Delta m = 235.124 + 1.009 - 94.945 - 138.955 - 2.018 = 0.215 \tag{7}$$

or 215 m.m.u. From p. 567, this is equivalent to $215 \cdot 0.931 = 200$ mev.

Almost the same value has been found experimentally, from measurements of the kinetic energy of the particles formed in fission, and the energy of the γ-radiation. Values so obtained lie between 174 and 182 mev. These measurements relate to the energy released *directly* by fission; the quantity of energy evolved subsequently, by the radioactive disintegration of the fission products, amounts to an additional 21 mev approximately. The total of 195 to 203 mev thus obtained from experiment is in good agreement with the calculated value.

In this calculation of the fission energy, attention has been confined to the formation of ^{95}Mo and ^{139}La, i.e., to the formation of only one pair of fission products. The experimentally determined energy release related to the formation of the wide range of fission products that is actually produced. However, this does not make any significant difference to the energy yield in the fission process, as can be shown from the following more general method of calculation. The particular fission products formed in any event may be ignored, the treatment being based only on the experimental fact that the mass numbers of the fission fragments lie between 72 and 162 (cf. Fig. 80), and that the masses of the lighter and heavier fission fragments add up, on the average, to 234 (or a number close to 234). From Fig. 70 (p. 571), it can be seen that in a nucleus of mass 236 (i.e., the mass of the ^{235}U nucleus after capturing a neutron, but before undergoing fission) there is an average mass defect of 8.05 m.m.u. per nucleon. Over the range of masses which comprises the fission products, the mass defect averages 9.05 m.m.u. per nucleon. Suppose, therefore, that two nuclei within this range of mass numbers are to be built up from their separate nucleons. If the mass numbers of the two nuclei together total 234, a total of 234 nucleons will be required to construct them. Each of these nucleons sustains a loss of mass, of 9.05 m.m.u., in the process of forming the two nuclei, so that the process involves a total loss of $234 \cdot 9.05$ m.m.u. $= 2118$ m.m.u. Creation of the ^{236}U nucleus from its individual nucleons would involve a mass loss of $236 \cdot 8.05 = 1900$ m.m.u. The difference between these quantities 218 m.m.u., gives the average loss of mass in the fission of the ^{235}U nucleus by slow neutrons. Hence the average energy liberated, from this calculation, is $218 \cdot 0.931 = 203$ mev per atom of ^{235}U. Thus the amount of energy evolved in fission is about 200 mev per atom, irrespective of the particular fission products formed. It has been assumed that the mass numbers of the fission fragments, taken in pairs, add up to 234—i.e., that 2 neutrons, on the average, are emitted in each fission process. For every additional neutron emitted, the loss of mass is diminished by 9.05 mmu. It cannot, however, be concluded that the total energy per fission is correspondingly decreased by about 4% for each additional neutron emitted; if—as is usually the case—the neutrons have an opportunity to combine with other nuclei, the energy thereby liberated compensates, more or less, for the smaller energy of formation of the fission fragments. The fission energy of 200 mev per atom is equivalent to $4.6 \cdot 10^9$ kcal of heat per g-atom, or electrical energy of about 23,000 kilowatt hours per gram of ^{235}U.

(c) Theory of Nuclear Fission

The fact that there is a loss of mass in the fission of the nuclei of the heavy elements—i.e., that energy is evolved—implies that these heavy nuclei are unstable or metastable. Since fission does not take place immediately and spontaneously, it follows further that some potential barrier must be surmounted in the process, just as in radioactive disintegration (cf. p. 590), although the barrier opposing fission is often far higher. If the requisite energy of activation is imparted to the nucleus (cf. p. 599), it may surmount the potential barrier; for certain nuclei, the energy liberated when a neutron is incorporated in the atom is sufficient for this. Fission is induced in such nuclei even by slow neutrons. In general, however, this increment of energy is insufficient; additional energy is needed, and can be imparted to the nucleus by fast neutrons, in the form of kinetic energy.

The forces by which the nucleons are held together within the nucleus are formally comparable with the forces operating between the molecules of a liquid. Just as a drop of liquid normally assumes a spherical shape, and the surface tension opposes any change of shape, so the atomic nucleus will normally display spherical symmetry under the action of the nuclear forces, and will resist any deformation. If any force is applied to a fluid droplet, so that it is stretched in one direction, it first assumes an ellipsoidal form. It tends, however, to return to the spherical shape, and in some circumstances deformation-oscillations may be set up about this most stable form. The stronger the force inducing the oscillations, the greater will be the departure of the drop from the spherical shape—i.e., the more elongated will be the ellipsoidal form. The elongation of the ellipsoid cannot proceed beyond a certain limit, however. When a certain degree of elongation is reached, the ellipsoid begins to constrict itself in the middle, and as soon as such a constriction appears, it intensifies spontaneously until the drop breaks at the 'waist'. Two drops are thereby formed, each of which proceeds to take up a spherical form. The nucleus of an atom, in which deformation-oscillations are excited by imparting sufficient energy, behaves in the same way. The fission of the nucleus was interpreted on the basis of this physical analogue by Meitner and Fricke, as soon as the facts had been established by Hahn and Strassmann. The theory was very shortly afterwards put on a quantitative footing by Bohr and Wheeler.

One difference between the rupture of a liquid drop and the fission of an atomic nucleus is that energy must be expended to bring about the first process, whereas the second is generally attended with the liberation of energy*. However, the two processes are quite analogous up to the stage where the potential barrier is crossed—i.e., until the drop begins to constrict. Energy must be expended to deform the droplet in both cases until this state is attained. The only difference is that in the rupture of the liquid drop, the energy falls from that of the transition state to a value which is higher than the initial energy, before deformation commenced; in the fission of the nucleus the final energy is generally much lower than the initial energy.

The activation energy required to surmount the potential barrier (i.e., the work performed in deforming the drop so much that a constriction begins to form) can be calculated, in the case of the liquid drop, from the radius and the surface tension of the liquid; for the atomic nucleus, it can be calculated from the number of nucleons and the nuclear forces. It is evident that the weaker the total cohesive forces between the nucleons, the more readily can any nucleus be deformed. Forces of attraction are exerted between the neutrons and protons of a nucleus, and the gross magnitude of these forces increases with the number of nucleons—i.e., with the mass number A. The protons, however, exert a mutual repulsion, because of their like charges, and this repulsion is roughly proportional to Z^2, where Z is the total number of protons—i.e., the nuclear charge, or atomic number. It follows from the theory based on the drop model that the activation energy for nuclear fission decreases as the ratio Z^2/A rises. According to the calculation of Bohr and Wheeler, it would become zero for $Z^2/A = 45$. This would be the value for an element with nuclear charge number about 120. Hence any element with a nuclear charge of 120 units, or higher, would be not merely metastable, like the other heavy elements, but absolutely unstable. If it were possible to prepare such an element, it would immediately (or more strictly, within 10^{-12} sec.) undergo fission into elements of medium atomic weight.

For ^{235}U the value of Z^2/A is 36, and the activation energy for fission is 5.3 mev. For ^{238}U, $Z^2/A = 35.6$, and the activation energy 5.9 mev. Z^2/A for ^{239}Pu is 37.0, for ^{233}U is 36.4, and for Ta is about 29.5. The activation energy for fission of ^{239}Pu and ^{233}U is thus less than that for ^{235}U; the significance of these substances in processes employing nuclear fission will be discussed later. On the other hand, the activation energy for the fission of the Ta nucleus must be very much higher; this accounts for the fact that fission can be brought about only by particles of exceptionally high kinetic energy (e.g., 400 mev α-particles).

The difference of 0.6 mev between the activation energies of ^{238}U and ^{235}U is not in itself sufficient to account for the fission of ^{235}U by slow neutrons (e.g., with 0.025 ev energy), whereas neutrons with an energy of at least 1 mev are required for the fission of ^{238}U. It is also necessary to take into account the differences in the amounts of energy released when an additional neutron is built into the different nuclei. By means of eqn. 6, p. 592, it is possible to calculate the binding energies of the nuclei ^{235}U, ^{236}U, ^{238}U, and ^{239}U. The difference in energy of formation between ^{236}U and ^{235}U is thereby found to be 6.6 mev, and that between the energies of formation of ^{238}U and ^{239}U is 5.3 mev. If a neutron is added to the nucleus of ^{235}U, energy amounting to 6.6 mev is released, and this is far greater than the energy of activation for fission of the ^{235}U nucleus. In this case, therefore, it is not necessary for the incoming neutron to bring with it any additional energy, and fission can therefore be brought about by slow neutrons. If a neutron is incorporated in the nucleus of ^{238}U, however, the energy released is 5.3 mev, whereas the activation energy for fission is 5.9 mev in this case. Thus additional energy must be imparted to the nucleus from some source—i.e., fission is induced only by fast neutrons, which can supply the deficit of energy from their own kinetic energy.

The difference in the amounts of energy released in the two reactions

$$^{235}_{92}\text{U} + \text{n} \longrightarrow {}^{236}_{92}\text{U} \qquad \text{and} \qquad {}^{238}_{92}\text{U} + \text{n} \longrightarrow {}^{239}_{92}\text{U}$$

can be attributed chiefly to the last term in eqn. (6). For the conversion of an even-odd nucleus into an even-even nucleus, this term is positive, whereas for the conversion of an even-even nucleus into an even-odd nucleus, this term is negative in sign. This term is also negative for the addition of a neutron to an odd-odd nucleus (which is thereby transformed

---

* It has recently been found that fission can be induced in relatively light nuclei, by means of protons of extremely high energy (e.g., ^{65}Cu + ^{1}p = ^{38}Cl + ^{27}Al + ^{1}n). These reactions are endothermic.

into an odd-even nucleus). This is why the two odd-even nuclei ^{233}U and ^{239}Pu are fissile by slow neutrons, whereas the nuclei ^{231}Pa and ^{237}Np, which both belong to the odd-odd type, are fissile only by fast neutrons. Nevertheless, the activation energy for ^{231}Pa is but little larger than that for ^{235}U, whereas the activation energy of ^{237}Np is actually lower ($Z^2/A =$ 35.9 for ^{231}Pa, 36.6 for ^{237}Np). The nucleus of $^{232}_{90}$Th, which is of the even-even type, is also fissile by fast neutrons only. The energy liberated by addition of a neutron is about 5.3 mev in this case, but the activation energy for fission is even higher than for ^{238}U ($Z^2/A = 35.0$ for ^{232}Th).

(d) Spontaneous Fission

It was mentioned on p. 590 that, according to wave mechanics, there is a finite probability that an α-particle may 'leak through' the potential barrier surrounding the nucleus. The same applies to the fragments which result from fission, except that the probability that such heavy particles can 'tunnel' through the potential barrier is very much smaller than for α-particles. The probability is not zero, however, so that nuclear fissions must occur spontaneously, even if extremely rarely. It has already been stated that spontaneous fission of uranium was observed very shortly after the discovery of fission by slow neutrons. Numerous other examples of spontaneous fission have been discovered subsequently. They follow the same law as ordinary radioactive decay, so that it is possible, in an analogous manner, to determine the *half-lives* of the nuclei for spontaneous fission—i.e., the time which must elapse before half the atomic species concerned has undergone fission. Table 75 gives the fission half-lives for several atomic species.

TABLE 75

HALF-LIVES FOR SPONTANEOUS NUCLEAR FISSION

Nucleus	^{230}Th	^{232}Th	^{231}Pa	^{232}U	^{233}U
Half-life (years)	$1.5 \cdot 10^{17}$	$1.4 \cdot 10^{18}$	$1.2 \cdot 10^{16}$	$3.6 \cdot 10^{12}$	$> 3 \cdot 10^{17}$
Nucleus	^{234}U	^{235}U	^{238}U	^{237}Np	^{239}Np
Half-life (years)	$2 \cdot 10^{16}$	$1.9 \cdot 10^{17}$	$8.0 \cdot 10^{15}$	$> 4 \cdot 10^{16}$	$> 5 \cdot 10^{12}$
Nucleus	^{236}Pu	^{238}Pu	^{239}Pu	^{241}Am	^{240}Cm
Half-life (years)	$3.5 \cdot 10^{9}$	$2.45 \cdot 10^{10}$	$5.6 \cdot 10^{15}$	$1.2 \cdot 10^{12}$	$1.9 \cdot 10^{4}$
Nucleus	^{242}Cm	^{244}Cm	^{250}Cf	^{252}Cf	^{254}Fm
Half-life (years)	$7.2 \cdot 10^{6}$	$1.4 \cdot 10^{7}$	$5 \cdot 10^{3}$	$1 \cdot 10^{2}$	0.55

Seaborg (1952) has shown that, in the case of the even-even nuclei, there is substantially a linear relationship between the half-life for spontaneous fission and the value of Z^2/A. If this is extrapolated to $Z^2/A = 45$, the value found for the half-life of such a species corresponds roughly with that predicted theoretically by Bohr and Wheeler, on the basis of the drop model.

The nuclei with an odd number of nucleons have a smaller tendency to undergo spontaneous fission than those with an even number of nucleons. Thus ^{235}U has an appreciably greater half-life than ^{238}U. The conditions for spontaneous fission are therefore different from those involved in fission by neutrons. This would be expected from the outset, since spontaneous and induced fission are essentially different processes. In order to bring about the artificial fission of ^{235}U, the nucleus ^{236}U must first be formed (in a highly excited state) by incorporation of an additional neutron, and it is this species which finally undergoes fission. In the spontaneous fission of ^{235}U, however, it is this nucleus which is itself concerned in the whole event.

(e) Nuclear Chain Reactions

If two neutrons are set free from every nucleus undergoing fission, and if the conditions are such that each of these neutrons can in turn bring about a fission, it follows that the number of neutrons in the system concerned must increase at a rapidly accelerating rate. Under the conditions specified, the number of neutrons would be doubled in each generation of fissions, so that—starting with a single neutron—the number of neutrons after the n-th series of fissions would be 2^n. Thus after the 85th generation of fissions, the number of neutrons would have grown to about $4 \cdot 10^{25}$. This number would suffice to bring about the fission of about 15 kg of ^{235}U, with the liberation of about $3 \cdot 10^{10}$ kcal of heat. Since the time elapsing for the completion of 85 generations of nuclear fission would be less than 1 microsecond, it follows that such a reaction would take place explosively, and with enormous violence.

Reactions of the type considered, in which each act of reaction brings about one or more subsequent acts of reaction are known generally as *chain reactions*, and are very frequently involved in chemical processes. The nuclear transformation considered is an example of a nuclear chain reaction; such processes may be either controlled or uncontrolled. A nuclear explosion, such as was considered above, represents an uncontrolled chain reaction. A controlled chain reaction may be brought about by so choosing the conditions that, of the two or more fresh neutrons produced in each fission, only *one* is able to bring about another fission. Under these conditions, the process takes place at a constant rate, and with a constant population of neutrons in the system. The means of securing this result will be discussed in the following chapter, in connection with plutonium, which is prepared by utilizing a controlled chain reaction.

The number of neutrons set free in the fission of a ^{235}U nucleus is often greater than 2, the average neutron yield being about 2.5 per fission caused by slow neutrons. The average number of neutrons from the fission of ^{239}Pu is still higher—approximately 3. On the average, only about 2 of these can bring about further fissions, since the interaction of neutrons with ^{239}Pu nuclei leads, in about one case in three, to the reaction $^{239}Pu + n = {}^{240}Pu + \gamma$, instead of fission. Similarly, the capture of a neutron by the ^{235}U nucleus results in the formation of ^{236}U by a (n, γ) process in about $1/_6$ of the cases. These proportions are valid for slow neutrons; such differences as arise for fast neutrons may be neglected here. Thus the effective yield of neutrons from each fission of ^{235}U or ^{239}Pu is about 2 in each case, and each of these could initiate another fission. Two conditions must be fulfilled in order that this may occur, however. First, no other substance which would absorb neutrons must be present, mixed with the ^{235}U or ^{239}Pu; i.e., the process must take place in pure ^{235}U or ^{239}Pu. In the second place, no neutrons must be lost from the system, since the number of neutrons leaking out of the system would diminish the number of fissions correspondingly. Thus, if the ^{235}U or plutonium is in the form of a sphere, every neutron liberated at the center of the sphere which is not used up in forming ^{236}U or ^{240}Pu, must be prevented from passing through the surface of the sphere without being brought into collision with a nucleus, and causing a fission. If the mid-point of each nucleus is joined to the center of the sphere, and if the nuclear cross section is projected on to the surface of the sphere along this line*, then in a

* Strictly speaking, the required projection is the intersection of the surface of the sphere with the prolongation of a cone described with the center of the sphere as apex, and the cross section of each nucleus as base. The simpler procedure considered here gives a somewhat high value of r for a given value of σ.

sphere of sufficient size, the whole surface of the sphere will be covered by the projected nuclear cross sections. This condition is satisfied for a sphere of radius r (in cm), consisting of material with the fission cross section σ_f (in barns), according to the relation

$$r = \frac{3 \cdot 10^{24}}{\sigma_f \cdot N_L} \tag{8}$$

(N_L = number of fissile nuclei per cm³). The number of fissile nuclei in a sphere of radius r is $N = \frac{4}{3}\pi r^3 \cdot N_L$. Equating the total cross section of these nuclei with the area of the spherical surface, $N \cdot \sigma_f \cdot 10^{-24} = \frac{4}{3}\pi\, r^3 \cdot N_L \cdot \sigma_f \cdot 10^{-24} = 4\pi\, r^2$. Equation (8) follows directly from this. For a material consisting of a pure fissile isotope,

$$N_L = \frac{N_A \cdot d}{A} \tag{9}$$

where N_A is Avogadro's number, d the density, and A the atomic weight of the fissile species. For ²³⁵U, with $d = 19$, r is approximately $\dfrac{60}{\sigma_f}$ cm. If the aggregate of the nuclear cross sections were equal to only half the area of the sphere, half of the neutrons which might bring about fissions would escape from the system. The chain reaction would then not take place explosively, but the number of neutrons would remain constant. The radius of sphere for which this would hold can be found by equating the aggregate cross section to $2\pi r^2$ in the above equation, thereby obtaining a value for r half that found from eqn. (8). If the value of r is greater than this value, the number of neutrons must increase in each generation. The value of r for which the chain reaction becomes divergent is known as the *critical radius*; more generally, since analogous considerations apply to systems which are not spherical, a *critical size* may be defined. If a system is below the critical size, it is impossible for a nuclear chain reaction to take place explosively; on the contrary, if fissions are induced by irradiation with neutrons, the process rapidly dies away completely, as soon as the source of neutrons is removed. From the foregoing crude calculation, the critical size for a sphere of ²³⁵U is about $\dfrac{30}{\sigma_f}$ cm. If we take an arbitrary value of 5 barns for σ_f, the corresponding value for $r = 6$ cm. This would mean that a quantity of 15 kg of pure ²³⁵U would not be explosive, since it would form a sphere of less than 6 cm radius. Twice that quantity would be explosive, however, since it would form a sphere more than 7 cm in radius—i.e., above the critical size. A single neutron, liberated by a spontaneous fission or by cosmic rays, would initiate a divergent chain reaction. With the quantity of ²³⁵U considered, this would liberate about $6 \cdot 10^{11}$ kcal (equivalent to the explosion of 600,000 tons of TNT) within a very short interval of time. Such an explosion, taking place nearly adiabatically, would attain a temperature of about 10^7 °, comparable with that in the interior of the sun.

The extremely high temperatures attainable in nuclear explosions are theoretically capable of bringing about *thermonuclear* reactions, such as the processes

$$^3T + {}^3T = {}^4He + 2n + 11.4 \text{ mev},$$

$$^3T + {}^2D = {}^4He + n + 17.6 \text{ mev}.$$

The energy liberated per unit weight of material by these nuclear reactions is considerably greater than that from the fission of uranium or plutonium. It is likely that the 'hydrogen bomb' makes use of thermonuclear reactions of this type, initiated by the high temperature attained in the nuclear explosion of plutonium.

The neutrons liberated in nuclear fission have kinetic energies of more than 1 mev. They are therefore capable of bringing about the fission of ²³⁸U nuclei, as well as of ²³⁵U and ²³⁹Pu. Nevertheless, a chain reaction of the type just described can not be set up in ordinary uranium, for the reason that the majority of the *fast* neutrons do not cause fission when they collide with ²³⁸U nuclei, but are merely reflected or scattered. In the act of scattering, the neutrons lose energy to the uranium nuclei, which are excited to emit γ-rays. After being slowed down in this way, the neutrons can no longer cause fission of ²³⁸U nuclei, and are effective in causing fission of ²³⁵U only under certain specified conditions. These will be discussed in the next chapter.

(f) Nuclear Reactors

In order to utilize atomic energy for the generation of power, the production of plutonium and for other purposes, it is necessary to employ *controlled chain reactions*, in place of the uncontrolled chain reaction of a nuclear explosion. A *nuclear reactor* or *atomic pile* is a device for carrying out self-sustaining nuclear chain reactions at a constant velocity—i.e., with a constant flux of neutrons.

Every nuclear reactor requires for its operation some so-called 'nuclear fuel'—i.e., a substance in which the nuclear chain reaction can take place, and which serves as the source of energy. Most reactors built hitherto also contain some 'fertile material'—i.e., some substance which cannot itself sustain a nuclear chain reaction, but which is converted by reaction with neutrons into a fissile material. The fissile material (or 'fuel') and the fertile material together constitute the working charge of the reactor. Natural uranium contains 0.7% of the fissile material (^{235}U), the remainder being essentially a fertile material. As the ^{235}U in a natural uranium reactor is consumed, a portion of the fertile ^{238}U is converted into plutonium (^{239}Pu), which is another nuclear fuel.

One nuclear species that is likely to become important as a nuclear fuel is the uranium isotope ^{233}U. Like ^{235}U and ^{239}Pu, this is fissile by slow neutrons. It can be produced from thorium, which is much more abundant in Nature than uranium, by the nuclear reactions

$$^{232}_{90}Th + ^{1}_{0}n = ^{233}_{90}Th \quad \xrightarrow[\text{23 min}]{\beta} \quad ^{233}_{91}Pa \quad \xrightarrow[\text{27.4 days}]{\beta} \quad ^{233}_{92}U.$$

Uranium-233 is an α-emitter, with half-life of 163,000 years. It is now produced technically, by surrounding a uranium reactor with a mantle of thorium. The slow neutrons escaping from the core of the reactor are thereby utilized to form ^{233}Th according to the reaction given above.

(g) Technical Production of Uranium-235

During the last World War, two methods were worked out for the production of pure uranium-235 on a full manufacturing scale. These were the *electromagnetic separation* and the *diffusion process*. The former utilized a type of mass spectrograph (the 'Calutron'), operating on the same principle as the cyclotron. With the first experimental device of Lawrence (December 1941) the output was 1 microgram per hour of nearly pure ^{235}U. The full-scale calutron process was later operated at Oak Ridge, near Clinton, Tennessee, where a diffusion plant was also constructed, for the separation of gaseous UF_6 (cf. pp. 192, 547). Harteck and Groth in Germany achieved success in the separation of uranium isotopes by the intense gravitational field of a high speed centrifuge. The processes worked out for the complete separation of uranium-235 are of great importance since, in simplified form, they may be employed for the enrichment of natural uranium in the fissile isotope. It is far less costly to bring about a considerable enrichment of the uranium-235, than to separate the fissile isotope completely. The use of enriched uranium as 'fuel' in nuclear reactors offers considerable advantages, since it decreases the size and weight of the reactor and, by decreasing the number of neutrons captured in ^{238}U, imposes less stringent limitations on the 'competitive absorption' of neutrons in structural materials of the reactor (cf. Chap. 14).

12. Radiation Protection

In view of the increasing applications of radioactive materials in the laboratory and in industry, it is necessary to draw attention to the very serious *health hazard* which can arise from working with these substances unless certain necessary precautions are rigidly observed. The dangers are all the more serious in that the body receives no sensory stimulus to give a warning of damage from radioactive radiations. It is characteristic of the damage caused by radiations that there is a long delay before detectable effects are produced. The full consequences may not be evident for months, or even years, and it is possible for incurable damage to ensue before harmful effects are observed.

The extent of damage increases with the duration of exposure to radiation. As long as the effects are slight, the body can fully recover in the course of time; within limits, the total permissible exposure to radiation does not vary greatly with the dose rate. However, radiation which would be completely harmless for a single short exposure results in serious damage if exposure is continued over a long period of time.

Slight damage shows itself after a short time by a reddening of the skin (erythema) of the exposed parts. If irradiation is more intense, or more prolonged, blisters may be formed, and often give rise to wounds that heal badly. More serious cases can ultimately lead to malignant tumours, although these may only be formed after a period of years. Serious blood diseases (e.g. leukemia) can also result if the permissible dose of radiation is exceeded, especially in cases of chronic over exposure.

In order to avoid the harmful effects of radiation, it is essential in all work with radioactive materials to know the permitted *tolerance doses* of radiation, and to take the necessary precautions not to exceed these tolerances. No difficulties arise when the activity of the preparations used is about 1 microcurie or less, as is usually the case when using radioactive materials as tracers or indicators. It then suffices to wear rubber gloves, and to handle vessels containing active preparations by means of long tongs. With weakly radioactive materials (i.e. those with a low γ-ray activity), the simplest possible radiation protection is provided by keeping the materials always sufficiently far from the bodily organs. Even though there may be no appreciable absorption of the rays in a comparatively short air path, the intensity of the radiation falls off with the square of the distance, provided that the source is small compared with the distance. Thus if a tube containing a small amount of radioactive substance is held directly in the hand, its distance from the flesh is perhaps 3 mm. Use of ordinary crucible tongs would increase this to about 170 mm. Hence the intensity of radiation is diminished in the ratio $170^2 : 3^2 = 3200 : 1$, and the time taken to incur a tolerance dose is increased in the same proportion.

The *tolerance dose* is the integrated dosage of radiation which must not be exceeded if all damage to health is to be avoided. The radiation dose for X-rays is usually measured in Röntgens per hour (r/h), where 1 *Röntgen* (1 r) is defined as the quantity of X-rays or γ-rays which will produce ions bearing a total charge of 1 e.s.u., of each sign, in 1 cc of dry air at 0 °C and 1 atmosphere pressure. Experiments have proved that the production of 1 ion pair in air involves the average expenditure of 32.5 e.v. Since a singly charged ion bears a charge of $4.80 \cdot 10^{-10}$ e.s.u. (Vol. I, p. 86), and 1 e.v. $= 1.602 \cdot 10^{-12}$ erg, the energy expended in 1 g of air through the absorption of 1 röntgen of radiation is

$$\frac{32.5 \cdot 1.602 \cdot 10^{-12}}{0.001293 \cdot 4.80 \cdot 10^{-10}} = 83.8 \text{ ergs.}$$

For the biological action of radiation, the significant factor is the energy converted within the body tissues. The quantity of energy per röntgen absorbed in body tissues is greater than 84 ergs per gram—e.g. X-radiation imparts about 93 ergs per g to water, and about 150 ergs per g to bone substance. Furthermore, the action of the corpuscular radiations (α-rays, β-rays, neutrons) on the body tissues differs significantly from that of X-rays or γ-rays, being considerably greater. Following the suggestion of H. M. Parker (1948), two other units of measurement—the *r.e.p.* and the *r.e.m.*—have been introduced to make allowance for these differences.

1 r.e.p. (= *roentgen equivalent physical*) is the dose of radiation which imparts the same amount of energy to the substance considered as would be imparted by a dose of 1 r to 1 g of air at 0° and 760 mm—i.e., 83.8 ergs.

1 r.e.m. (= *roentgen equivalent man*) is the dose of radiation which produces the same biological effect as 1 r of X-rays or γ-rays.

In general, for β-rays, 1 rep = 1 rem; for α-rays, however, 1 rep = 10 to 20 rem*, for slow neutrons (thermal neutrons) 1 rep = 5 rem, and for fast neutrons, 1 rep = 10 rem.

Figures for the permissible dosage of radiation are: for X-rays, γ-rays, β-rays or slow neutrons, 0.3 rep per week of about 40 working hours; for fast neutrons 0.03 rep per week, and for α-rays 0.015 rep per week. These tolerance doses are for whole-body irradiation; rather higher dose rates may be tolerated if only small portions of the body are exposed. Thus if exposure is restricted to the hand or wrist, the permitted tolerance is taken as five times that for whole body exposure. Very much higher doses of radiation are used in radiotherapy—e.g. in treating skin cancers. The radiation is then limited, however, to a quite small region of the body, and is moreover used for the very reason that it exerts a biological effect. Why intensive irradiation should both give rise to malignant growths and destroy existing ones is a question that is not yet satisfactorily answered.

Special experimental methods must be employed in working with radioactive materials at levels much above 1 microcurie. These are necessary not only as health-protective measures, but also to prevent the radioactive contamination of laboratories and equipment. This is particularly necessary when the level of activity exceeds 1 millicurie. At high levels of activity, work must be carried out in a laboratory (a 'hot' laboratory) specially adapted for the purpose. This must be provided with ventilating system that prevents the contamination of the laboratory and the remainder of its building with active dust; all operations with open sources must then be carried out under draught hoods, whereas short-lived α-active materials must normally be handled in dust-boxes. The laboratory must also have provision for heavy shielding of γ-active materials, and for the performance of all chemical operations by remote handling methods (e.g. by the use of long tongs, or the transfer of liquids by air pressure) behind such shielding.

---

* Because of their low penetrating power, α-rays constitute a very slight hazard from *external* sources, provided the simplest health measures are observed. On the other hand, the *inhalation* of α-active materials, as dust, can cause serious internal damage. So also can the administration of α-active preparations for medicinal purposes unless the tolerance dose is observed.

References

1 W. HANLE, *Künstliche Radioaktivität, Kernphysikalische Grundlagen und Anwendungen*, 2nd Ed., Stuttgart 1952, 239 pp.
2 P. B. MOON, *Artificial Radioactivity*, Cambridge 1949, 102 pp.
3 J. M. CORK, *Radioactivity and Nuclear Physics*, New York 1950, 415 pp.
4 K. DIEBNER and E. GRASSMANN, *Künstliche Radioaktivität; experimentelle Ergebnisse*, Leipzig 1939, 87 pp.
5 G. VON HEVESY, *Radioactive Indicators*, New York 1948, 574 pp.
6 G. K. SCHWEITZER and J. B. WHITNEY, *Radioactive Tracer Techniques*, New York 1949, 241 pp.
7 W. E. SIRI, *Isotopic Tracers and Nuclear Radiation*, New York 1949, 653 pp.
8 K. E. ZIMEN, *Angewandte Radioaktivität*, Berlin 1952, 124 pp.
9 G. H. GUEST, *Radioisotopes; Industrial Applications*, London 1951, 185 pp.
10 E. BRODA, *Advances in Radiochemistry*, Cambridge 1950, 152 pp.
11 A. C. WAHL and N. A. BONNER, *Radioactivity applied to Chemistry*, New York 1951, 604 pp.
12 G. FRIEDLANDER and J. W. KENNEDY, *Nuclear and Radiochemistry* [revised version of *Introduction to Radiochemistry*], New York 1955, 468 pp.
13 M. CALVIN, C. HEIDELBERGER, J. C. REID, B. M. TOLBERT and P. F. YANKWICH, *Isotopic Carbon, Techniques in Its Measurement and Chemical Manipulation*, New York 1949, 376 pp.
14 W. H. WHITEHOUSE, and J. L. PUTMAN, *Radioactive Isotopes*, Oxford 1953, 424 pp.
15 S. GLASSTONE, *Sourcebook on Atomic Energy*, London 1950, 546 pp.
16 G. GAMOW, *Atomic Energy in Cosmic and Human Life*, Cambridge 1947, 161 pp.
17 W. E. STEPHENS, *Nuclear Fission and Atomic Energy*, Lancaster, Pa., 1948, 275 pp.
18 J. R. M. PANZER, *Atomic Energy Regulation*, New York 1950, 900 pp.

CHAPTER 14

THE TRANSURANIC ELEMENTS

Atomic Numbers	Elements	Symbols	Atomic Weights	Valence states
93	Neptunium	Np	237	II, III, IV, V, VI
94	Plutonium	Pu	242	II, III, IV, V, VI
95	Americium	Am	243	II, III, IV, V, VI
96	Curium	Cm	245	III
97	Berkelium	Bk	249	III, IV (?)
98	Californium	Cf	249	III
99	Einsteinium	E	255	III
100	Fermium	Fm	255	III
101	Mendelevium	Mv	256	III

[Atomic weights given in this table refer to the most stable isotope of the elements, as at present known. This is not necessarily the most accessible isotope.]

1. Introduction [1]

(a) General

All the transuranic elements—i.e., the elements with atomic numbers higher than that of uranium—have been obtained artificially, as the products of nuclear transmutations. It has been shown subsequently that two of them—neptunium and plutonium—do occur in Nature, in uranium ores. The amounts present are so minute, however, that they are without significance for the extraction of the elements, and these elements, like the other transuranics, can be obtained in practice only by means of nuclear reactions.

All the transuranic elements are *unstable*. For most of them a considerable number of isotopes is known and Table 76 sets out the different nuclear species of the transuranic elements which are best known at the present time.* The Table also gives the most important nuclear reactions by which the various isotopes are produced, and their most important radioactive properties. Every one of these nuclear species undergoes radioactive disintegration, and none of them has a half-life which, in order of magnitude, approaches that of uranium, the last element of the whole series to exist in Nature in considerable amounts.

The most important element of the transuranic series is *plutonium*. The longest-lived plutonium isotope has a half-live of $5 \cdot 10^5$ years. Since the age of the earth is more than a thousand times as long as this, it follows that no more plutonium would remain even if the earth originally contained appreciable amounts. Such

---

* Other isotopes which may be mentioned are ^{244}Pu, ^{245}Pu, ^{246}Pu, ^{245}Am, ^{246}Am, ^{241}Cm, ^{249}Cm, ^{250}Cf, ^{252}Cf, ^{246}E.

TABLE 76

SOME NUCLEAR SPECIES OF TRANSURANIC ELEMENTS

Element	Species	Half-life	Decay	Energy of radiation (mev)	Mode of formation
Neptunium $Z = 93$	^{231}Np	53 min	$<^K_\alpha$	α 6.2	^{238}U (d, 9n)
	^{234}Np	4.4 days	K	γ 1.9	^{235}U (p, 2n); ^{235}U (d, 3n), etc.
	^{235}Np	240 days	$<^K_\alpha$	α 5.06	^{235}U (d, 2n); ^{235}U (α, p 3n), etc.
	^{236}Np	20 hrs	β	0.5	^{235}U (d, n); ^{237}Np(n, 2n)
	^{237}Np	$2.25 \cdot 10^6$ yrs	α	4.75	^{237}U $\to \beta$; ^{241}Am $\to \alpha$
	^{238}Np	2.0 days	β	β 1.0; γ 1.1	^{238}U (d, 2n); ^{235}U (α, p), etc.
	^{239}Np	2.31 days	β	1.179; 0.676 0.403; 0.288	^{239}U $\to \beta$; ^{238}U (d, n)
Plutonium $Z = 94$	^{232}Pu	22 min	α	6.6	^{235}U (α, 7n)
	^{234}Pu	8.5 hrs	$<^K_\alpha$	α 6.2	^{233}U (α, 3n)
	^{236}Pu	2.7 yrs	α	5.7	^{236}Np $\to \beta$; ^{235}U (α, 3n), etc.
	^{237}Pu	40 days	K	—	^{235}U (α, 2n); ^{238}U (α, 5n), etc.
	^{238}Pu	ca. 50 yrs	α	5.493	^{238}Np $\to \beta$; ^{238}U (α, 4n), etc.
	^{239}Pu	$2.42 \cdot 10^4$ yrs	α	5.140	^{239}Np $\to \beta$; ^{238}U (α, 3n)
	^{240}Pu	$6.6 \cdot 10^3$ yrs	α	5.1	^{238}U (α, 2n)
	^{241}Pu	13.0 yrs	$<^\alpha_\beta$	β 0.02	^{238}U (α, n)
	^{242}Pu	$5 \cdot 10^5$ yrs	α	4.88	^{241}Pu (n, γ)
	^{243}Pu	5.0 hrs	β	0.5	^{242}Pu (n, γ)
Americium $Z = 95$	^{238}Am	1.3 hrs	K	—	^{237}Pu (d, n)
	^{239}Am	12 hrs	$<^K_\alpha$	γ 0.285 α 5.77	^{239}Pu (p, n); ^{237}Np(α, 2n)
	^{240}Am	2.1 days	K	γ 1.3	^{239}Pu (d, n); ^{237}Np(α, n)
	^{241}Am	470 yrs	α	5.45	^{241}Pu $\to \beta$; ^{245}Bk $\to \alpha$
	^{242}Am*	18 hrs	β	1.0	^{241}Am(n, γ)
	^{242}Am	ca. 400 yrs	$<^\beta_\alpha$	β 0.63	^{241}Am(n, γ)
	^{243}Am	10^4 yrs	α	5.21	^{242}Am(n, γ); ^{243}Pu $\to \beta$
	^{244}Am	26 min	β	1.5	^{243}Am(n, γ)
Curium $Z = 96$	^{240}Cm	27 days	α	6.25	^{239}Pu (α, 3n); ^{244}Cf $\to \alpha$
	^{242}Cm	162.5 days	α	6.11; 6.06	^{239}Pu (α, n); ^{242}Am $\to \beta$; ^{246}Cf $\to \alpha$
	^{243}Cm	ca. 100 yrs	$<^\alpha_{K?}$	α 5.99; 5.78; 5.73	^{242}Cm(n, γ); ^{243}Bk $\to K$
	^{244}Cm	ca. 10 yrs	α	5.78	^{241}Am(α, p); ^{244}Am $\to \beta$
	^{245}Cm	ca. $2 \cdot 10^4$ yrs	α	5.36	^{245}Bk $\to K$

TABLE 76 (continued)

SOME NUCLEAR SPECIES OF TRANSURANIC ELEMENTS

Element	Species	Half-life	Decay	Energy of radiation (mev)	Mode of formation
Berkelium $Z = 97$	^{243}Bk	4.6 hrs	$\begin{cases} \alpha \\ K \end{cases}$	α 6.72; 6.55; 6.20	^{241}Am$(\alpha, 2n)$
	^{245}Bk	5.0 days	α	6.33; 6.15; 5.90	^{244}Cm(d, n)
	^{249}Bk	ca. 1 yr	$\begin{cases} \alpha \\ \beta \end{cases}$	α 5.4 β 0.10	^{249}Cm $\rightarrow \beta$
	^{250}Bk	3.13 hrs	β	β 1.9 γ 0.9	^{249}Bk (n, γ)
Californium $Z = 98$	^{244}Cf	45 min	α	7.15	^{242}Cm$(\alpha, 2n)$; ^{238}U$(^{12}$C$, 6n)$; ^{238}U$(^{14}$N, p 7n$)$
	^{246}Cf	35.7 hrs	α	6.75	^{244}Cm$(\alpha, 2n)$; ^{238}U$(^{12}$C$, 4n)$
	^{248}Cf	225 days	α	6.26	^{238}U$(^{14}$N, p 3n$)$
	^{249}Cf	ca. 400 yrs	α	5.82; 6.0	^{249}Bk $\rightarrow \beta$
	^{253}Cf	ca. 20 days	β	—	^{252}Cf (n, γ)
Einsteinium $Z = 99$	^{247}E	7.3 min	α	7.47	^{238}U$(^{14}$N, 5n$)$
	^{253}E	20 days	α	6.63	^{253}Cf $\rightarrow \beta$
	^{254}E	36 hrs	β	1.1	^{253}E (n, γ)
	^{255}E	ca. 30 days	β	—	^{254}E (n, γ)
Fermium $Z = 100$	^{250}Fm	ca. 30 min	α	7.7	^{238}U$(^{16}$O, 4n$)$
	^{254}Fm	3.2 hrs	α	7.22	^{254}E $\rightarrow \beta$
	^{255}Fm	15 hrs	α	7.1	^{255}E $\rightarrow \beta$
Mendelevium $Z = 101$	^{256}Mv	few hrs	α	—	^{253}E (α, n)

plutonium as is now contained in uranium minerals must have been formed subsequently, and it has presumably been formed by the same processes as are involved in the technical production of plutonium.

The neutrons required for the nuclear reaction may be furnished by the spontaneous fission of uranium. Although spontaneous fission is an extremely slow process—i.e., is a very rare atomic event—its occurrence was proved by Flarow and Petriak, 1940, very shortly after the discovery of the fission of uranium by neutrons. These workers found the half-life of uranium for spontaneous fission to be about 10^{14} years; later investigations have given the more accurate value of $0.80 \cdot 10^{14}$ years. This means that for every million atoms of uranium which disintegrate with the ejection of an α-particle, *one* atom disintegrates by breaking up into two fragments of roughly equal size. Thus, in a gram of ordinary uranium, there is on the average one spontaneous fission per minute, so that there are about two neutrons liberated per minute per gram of uranium.

Another possible source of the neutrons required to produce Pu from ^{238}U is the reaction of the α-particles emitted by uranium and its disintegration products with the nuclei of the light elements contained in the minerals.

The transuranic elements form a series on their own, not only because they are all unstable, and can only be prepared by artificial means, but also because their chemical properties mark them out as a definite series. The transuranic elements *are not homologues of rhenium and the platinum metals*. Unlike the latter, they cannot be precipitated by means of H_2S from acid solution, neither do neptunium and plutonium form volatile oxides as do rhenium and osmium. The special place of the transuranic elements in the Periodic System is strictly parallel to the special place occupied by the lanthanides, and it can be similarly interpreted in terms of atomic structure. Just as the filling up of the $4f$ level takes place along the lanthanide series, so the filling of the $5f$ level leads to the transuranic elements. It can readily be understood that with increase in atomic number the transuranic elements show an increasing resemblance in chemical properties to the lanthanide elements.

The first three members of the transuranic series, neptunium, plutonium, and americium, can all exist in a higher oxidation state (the hexapositive state) than is observed for the lanthanide elements. This can be understood, since the binding of the $5f$ electrons in the atoms of the transuranic elements is less strong than that of the $4f$ electrons in the lanthanide elements. The position is much the same as that of the $3d$, $4d$ and $5d$ electrons in the transition metals. With increase of principal quantum number, the energy of binding of the d electrons diminishes, with the result that the ability of the transition elements to exist in higher oxidation states increases regularly in every group from top to bottom. The question as to whether the filling up of the $5f$ levels begins with the element immediately following actinium (i.e., with thorium—Seaborg 1949), or commences with neptunium (Dawson 1952) has not been finally settled. It does, however, seem probable (e.g., on the evidence of magnetic susceptibilities) that even if the filling of the $5f$ level begins with neptunium, the next element (plutonium) must have at least five electrons in the $5f$ level. From plutonium onwards, the number of $5f$ electrons increases regularly with the atomic number. The divergence of views between Dawson and Seaborg concerns only the electronic configuration of the elements preceding the transuranics; the configurations proposed for the transuranic elements themselves are essentially the same. In particular, both Dawson and Seaborg assign to curium the configuration $5f^76d7s^2$, which corresponds exactly to the structure of gadolinium, $4f^75d6s^2$. This leads to the conclusion that curium should occupy a special position among the transuranic or *actinide* elements, corresponding to that of gadolinium among the lanthanides. (cf. p. 480). This accords with the properties of the elements which precede curium, and especially those which follow it. It is therefore possible to trace the homology between the actinide elements and the corresponding members of the lanthanides, from their positions with respect to curium:

$_{61}$Pm	$_{62}$Sm	$_{63}$Eu	$_{64}$**Gd**	$_{65}$Tb	$_{66}$Dy	$_{67}$Ho	$_{68}$Er	$_{69}$Tm
$_{93}$Np	$_{94}$Pu	$_{95}$Am	$_{96}$**Cm**	$_{97}$Bk	$_{98}$Cf	$_{99}$E	$_{100}$Fm	$_{101}$Mv

The relationships with the corresponding lanthanides are especially clearly shown by curium, berkelium, and californium, in so far as the chemistry of these elements is at present known (pp. 612, 635). Resemblances are also to be found with neptunium, plutonium, and americium, but appear only in those compounds formed from the same valence states as are assumed by the corresponding lanthanides. The resemblance of neptunium, plutonium, and americium to the lanthanides is obscured by the ability of these elements to exist in higher oxidation states. In particular, all these elements can be hexapositive, and the properties of the compounds from this state of oxidation show a close resemblance to those of uranium, their left-hand neighbor in the Periodic System. The tendency to assume the $+6$ state is not so strong as with uranium, however, and falls off progressively from uranium to americium.

Table 77 shows how far the similarities between transuranic and lanthanide elements appear not only in electronic configuration, but also in valence properties.

TABLE 77

Element	Electronic confign.	Valence states	Element	Electronic confign. (Seaborg)*	(Dawson)	Valence states		
$_{58}$Ce	$4f\,5d\,6s^2$	**III, IV**	$_{90}$Th	$5f\,6d\,7s^2$	$6d^2 7s^2$	II,	III,	**IV**
$_{59}$Pr	$4f^3\,6s^2$	**III**, IV	$_{91}$Pa	$5f^2 6d\,7s^2$	$6d^3 7s^2$		IV,	**V**
$_{60}$Nd	$4f^4\,6s^2$	**III**	$_{92}$U	$5f^3 6d\,7s^2$	$6d^4 7s^2$	II,	III,	**IV, V, VI**
$_{61}$Pm	$4f^5\,6s^2$	**III**	$_{93}$Np	$5f^4 6d\,7s^2$	$6d^5 7s^2$**	II (?),	III,	**IV, V, VI**
$_{62}$Sm	$4f^6\,6s^2$	II, **III**	$_{94}$Pu	$5f^5 6d\,7s^2$	$5f^5 6d\,7s^2$**	II (?),	**III,**	**IV,** V, VI
$_{63}$Eu	$4f^7\,6s^2$	II, **III**	$_{95}$Am	$5f^7 7s^2$	$5f^6 6d\,7s^2$	II,	**III,**	IV, V, VI
$_{64}$Gd	$4f^7 5d\,6s^2$	**III**	$_{96}$Cm	$5f^7 6d\,7s^2$	$5f^7 6d\,7s^2$		**III**	
$_{65}$Tb	$4f^8 5d\,6s^2$	**III**, IV	$_{97}$Bk	$5f^8 6d\,7s^2$	$5f^8 6d\,7s^2$		III	
$_{66}$Dy	$4f^9 5d\,6s^2$	**III**	$_{98}$Cf	$5f^9 6d\,7s^2$	$5f^9 6d\,7s^2$		III	
	Lanthanides				Actinides			

* In addition to the configurations given, Seaborg considers the possibility of others—e.g., $5f6d^2 7s^2$ for Pa.

** From magnetic evidence, Dawson considers the possibility that there may be two d electrons in Pu—i.e., $5f^4 6d^2 7s^2$—and correspondingly for Np. Direct evidence for the configuration of Np is lacking and it is alternatively possible that Np might have four or five $5f$ electrons as supposed by Seaborg.

The Table includes the actinide elements preceding neptunium, together with the lanthanide elements which, according to Seaborg, are homologous with them. It may be seen that neptunium, plutonium and americium bear a very close resemblance to uranium. It must also be borne in mind that knowledge of the chemistry of curium, berkelium, and californium is very fragmentary indeed.

Since the distribution of electrons between the $5f$ and $6d$ levels of the actinides is not yet established, Table 77 sets out the configurations alternatively proposed by Seaborg [*Nucleonics*, 5, (1949) 16] and by Dawson [*Nucleonics*, 16 (1952) 39]. It is not possible to draw any unambiguous conclusions on this matter from the chemical properties of the elements. Even in the lanthanide group, the distribution of electrons between the $4f$ and $5d$ levels is uncertain in a number of cases. Chemical properties are determined chiefly by the *ionization energies* of the valence electrons. It is of little significance for the chemical properties whether the energy of binding of the electrons is greater in the $5f$ or the $6d$ state; in any case, in the ground state of the neutral atom the electrons will all be in the lowest levels—i.e., in those states in which their energy of binding is a maximum. The proper assignment of the electrons to the $5f$ or $6d$ levels is important, however, in determining many of the properties of the compounds of the transuranic elements, and especially their magnetic susceptibility.

All the transuranic elements can function in the tripositive state. Uranium can also be tripositive, although uranium(III) compounds are not very stable in aqueous solution. The neptunium(III) compounds are appreciably more stable, and with further increase in atomic number there is a very marked increase in the stability of compounds formed from the $+3$ oxidation state. Plutonium forms many compounds in which it is tripositive, and with americium the $+3$ state is the favored state. As far as is known at present, curium and californium exist only in the trivalent state, and it is interesting to compare this behavior with that of their formal 'homologues' of the lanthanide group, gadolinium and dysprosium*. The trihalides, MX_3 (X = F, Cl, Br, I) of the actinide elements are all isomorphous with each other and with the corresponding trihalides of actinium (q.v.). They are iso-structural with the corresponding trihalides of the rare earth metals.

Neptunium, plutonium and americium, like uranium, all form compounds from the +4 oxidation state. The tetrapositive state is the most stable valence state of neptunium and plutonium. Uranium(IV) compounds in solution readily undergo oxidation to uranium(VI) compounds, and americium(IV) compounds are readily converted to americium(III) compounds. The tetrachlorides, where they are known, are isomorphous with each other and with $ThCl_4$ and UCl_4.

The elements from uranium to americium also share the ability to function in the +5 state. The compounds derived from the +5 state are, however, generally less stable than the compounds derived from the other valence states, and tend to undergo disproportionation in solution. The least stable compounds of this type are the americium(V) compounds.

Like uranium, neptunium, plutonium, and americium can all function as hexapositive elements. Whereas the +6 oxidation state is the most stable valence state of uranium in aqueous solution, it becomes progressively less stable than the +4 state, and ultimately than the +3 state, as one passes through neptunium and plutonium to americium. The neptunyl- and plutonyl-compounds, $[NpO_2]X_2$ and $[PuO_2]X_2$, corresponding to the uranyl compounds $[UO_2]X_2$, have similar crystal structures. The double acetates $Na[UO_2]Ac_3$, $Na[NpO_2]Ac_3$, $Na[PuO_2]Ac_3$ and $Na[AmO_2]Ac_3$ (Ac = acetate group) are all isostructural.

In general, the chemical properties of americium are as closely related to those of the lanthanide elements as to uranium, and the elements following americium appear to resemble the rare earths very closely (but see footnote *).

Table 78 gives the atomic volumes and the apparent atomic radii of some of the actinides, as deduced from the densities of the metals. The ionic radii given are derived largely from the measurements of Zachariasen**. The radii of the hexapositive ions have been obtained by subtracting the radius of the oxygen ion (1.33 Å) from the M—O distance as determi-

* In making such a comparison, it must be borne in mind that knowledge of the chemistry of curium, berkelium, and californium is very scanty. Owing to the short half-lives of the accessible isotopes of these elements, investigation is limited at present to the microgram scale (Cm) or the extreme tracer scale (Bk and Cf). Moreover, the intense α-particle activity of concentrated preparations of these elements brings about a vigorous radiochemical decomposition of the solvent, water, so that it is not possible to study their reactions in solution except in the presence of a large excess of hydrogen peroxide. This reducing environment over-emphasizes the relative stability of the +3 state. It is noteworthy that increasing knowledge of the chemistry of americium has shown that the +4 and +6 states are far more important than was at first suspected—i.e., have emphasized the relationship to uranium. While the general trend in oxidation-reduction potentials along the series undoubtedly favors the lower valence states to an increasing extent, the abrupt difference in chemical behavior which appears to set in between americium and curium may be a reflection of incomplete knowledge, rather than of the properties of half-filled $5f$ levels.

** Ionic radii given here will be found not to agree with those listed by Zachariasen, but are 'Goldschmidt' radii, derived directly from measured M—O or M—F distances, and the standard ionic radii of oxygen (1.33 Å) and fluorine (1.36 Å). They are therefore directly comparable with the ionic radii of other elements tabulated elsewhere in this book.

TABLE 78

DENSITIES, ATOMIC VOLUMES, ATOMIC RADII, AND IONIC RADII OF ACTINIDE ELEMENTS

Element	$_{90}$Th	$_{91}$Pa	$_{92}$U	$_{93}$Np	$_{94}$Pu	$_{95}$Am
Density	11.71	—	19.0	19.5	—	11.7
Atomic volume	19.8	—	12.5	12.15	—	20.6
Atomic radius	1.82	—	1.48	1.32	—	—
Radius of 3-valent ion	—	—	1.04	1.02	1.01	1.00 Å
Radius of 4-valent ion	1.10	—	1.03	1.02	1.00	0.99 Å
Radius of 6-valent ion	—	—	0.58	0.57	0.56	— Å

ned by X-rays*. The data of Table 78 show that there is a steady decrease in the ionic radii of the actinide elements with increase of atomic number, as there is, e.g., along the lanthanide series. The variation of atomic volume with atomic number also resembles that found in the lanthanide series, americium having a particularly high atomic volume, as has its supposed homologue, europium.

The relationship between the actinide and lanthanide series of elements shows up also in the *magnetic properties*. Table 79 gives the magnetic susceptibilities of the ions of the two groups. The numbers of f electrons shown in the Table are based on the assumption that in every case the f electrons are the most firmly bound electrons in the outer shell. Thus it is assumed that, in the plutonium atom, with the electronic configuration $5f^5 6d 7s^2$, the two s-electrons and the d-electron are lost more readily than any of the f-electrons, so that in the formation of the Pu^{3+} ion the configuration becomes $5f^5$. There is no doubt that in all the actinide atoms, the $7s$ electrons are the most loosely bound. The hypothesis that the electronic configurations $5f^5$, $5f^4$, and $5f^3$ are present in the ions Pu^{3+}, Pu^{4+} and Pu^{5+}, respectively, is broadly supported by the magnetic evidence. In the ions with either 1 or 2 outer electrons, the measured values of the magnetic susceptibility tend to lie between those calculated for d- and f-electrons. It has been shown by Dawson (1950–52), however, that in magnetically dilute solids (e.g., dilute solid solutions of UO$_2$ in ThO$_2$, UF$_4$ in ThF$_4$, Na[PuO$_2$]Ac$_3$ in Na[UO$_2$]Ac$_3$, etc.), which can be investigated over a much wider range of temperature than is possible for aqueous solutions, the susceptibilities agree very closely with the 'spin only' susceptibility for a pair of d-electrons (cf. Table 79). It is probably not possible to

TABLE 79

MOLAR MAGNETIC SUSCEPTIBILITIES, χ_{MOL}, OF LANTHANIDE AND ACTINIDE IONS IN AQUEOUS SOLUTION AT ROOM TEMPERATURE

Ion	Ce^{4+}	Ce^{3+}		Pr^{3+}	Nd^{3+}	Pm^{3+}	Sm^{3+}	Eu^{3+}	Gd^{3+}
Number of 4f-electrons	0	1		2	3	4	5	6	7
$\chi_{mol} \cdot 10^3$	diamag.	+2.35		+5.05	+5.05	—	+1.0	+3.60	+24.6

Ion	Th^{4+}	U^{6+}	Np^{6+}	U^{4+}	Np^{5+}	Pu^{6+}	Np^{4+}	Pu^{4+}	Pu^{3+}	Am^{3+}	Cm^{3+}
Number of 5f or 6d electrons	0	0	1	2	2	2	3	4	5	6	7
$\chi_{mol} \cdot 10^3$	diamag.	diamag.	+2.05	+3.75	+4.10	+3.50	+3.95	+1.65	+0.25	+0.70	+26.5

* There is some reason to consider that the M—O bonds in the [MO$_2$]$^{2+}$ groups are not purely ionic, and the M^{6+} radii given in Table 78 are therefore somewhat arbitrary.

decide unambiguously from the available magnetic measurements which is the configuration of these ions. In many of the ions of the actinides, it is likely that the $5f$- and $6d$-levels lie very close together, so that both sets of levels may be occupied. In general, the magnetic susceptibility of the actinide ions varies with the number of outer electrons in much the same way as is observed in the lanthanide series. The very high susceptibility of the gadolinium ion is paralleled by that of the curium ion—i.e., the ion considered to be the homologue of gadolinium—whereas the very low paramagnetism of the Sm^{3+} ion corresponds to the low value found for the magnetic susceptibility of Pu^{3+}. The similarity goes so far that, just as the magnetic suseptibility of samarium compounds passes through a minimum at a certain temperature, there is also a flat minimum in the χ-T curve of plutonium(III) compounds, as found by Dawson (1951).

The compounds of the actinide elements, like those of the lanthanides, have highly characteristic *absorption spectra*, with sharp and intense absorption bands. The structures of the absorption spectra, in a number of instances, show a resemblance to those of the homologous lanthanides. It is interesting that in the transuranic series, as in the lanthanides, one ionic species is found in the middle of the series which is completely colorless in aqueous solution. This is the Cm^{+++} ion which, like the corresponding Gd^{+++} ion, has no optical absorption bands in the visible region, whereas all the other transuranic-(III) ions are colored, as also are the other ions in the lanthanide series.

(b) History [2]

The first of the transuranic elements to be discovered was *neptunium*. In 1940, McMillan and Abelson, in the United States, found that when uranium was irradiated with slow neutrons, a β-emitting species, having a half-life of 2.3 days, was formed; this could be separated chemically from uranium. The immediate product of irradiation of uranium with slow neutrons is a short-lived β-emitter (half-life 23 minutes), which is chemically inseparable from uranium; formation of this substance had been observed by Hahn and Meitner, in 1936. As was proved later, this is the uranium isotope ^{239}U, formed by neutron capture [(n,γ) reaction] in the ^{239}U nucleus: $^{238}_{92}U + ^{1}_{0}n = ^{239}_{92}U + \gamma$. The element produced by the β-particle disintegration of $^{239}_{92}U$ must necessarily have the atomic number 93, but it was not at first possible to identify this product. McMillan and Abelson were able to show that the new radioactive species which they had discovered was the element of atomic number 93, formed by the β-decay of ^{239}U. They called it neptunium, since it followed uranium, in the same way as the planet Neptune is the next one beyond Uranus in the solar system.

It was initially possible to study the chemical properties of neptunium only by the methods of radiochemistry (tracer techniques). This situation was changed when (in 1942) a long-lived isotope of neptunium was discovered (^{237}Np, with half life $2.25 \cdot 10^{6}$ years). This isotope is formed when uranium is irradiated with fast neutrons, and is produced by the

$$\text{reaction } ^{238}_{92}U \text{ (n,2n) } ^{237}_{92}U \xrightarrow[\text{7 days}]{\beta} {}^{237}_{93}Np.$$

The discovery of this long-lived isotope made it possible to work with pure, carrier-free neptunium compounds, even though quantities of only a few micrograms were available. It was subsequently found that ^{237}Np was formed as a by-product of the nuclear reactions taking place in the atomic pile. A few milligrams of neptunium were first isolated from this source in 1944, and several tenths of a gram later became available for chemical study.

Plutonium, the second element of the transuranic series, was discovered a few months after the identification of neptunium. Seaborg, McMillan, Wahl and Kennedy, at the end of 1940, irradiated uranium (in the form of U_3O_8) in a cyclotron, with high-energy deuterons, and noticed that a very short-lived neptunium isotope, ^{238}Np, was formed. This underwent β-particle decay, and produced a new element, emitting α-particles with a half-life of about 50 years. This new element, with the atomic number 94, was given the name of plutonium, after the planet Pluto which follows Neptune.

The plutonium isotope ^{239}Pu, resulting from the β-decay of ^{239}Np, was identified by Kennedy, Seaborg, Segrè and Wahl in 1941, as an α-emitter with a half-life of about 24,000 years. It was found that this plutonium isotope, like actino-uranium (^{235}U), undergoes fission by slow neutrons, and can be used for the same purposes. Very shortly after the discovery of ^{239}Pu, therefore, its large scale preparation was projected, and ^{239}Pu is now produced in considerable quantities in several countries.

The elements of atomic numbers 95 and 96 were discovered in 1944 by Seaborg, James, Morgan and Ghiorso. The first of these was named *americium*, after its supposed chemical similarity to europium in the lanthanide group. The second was given the name of *curium*, after Marie and Pierre Curie, with the intention also of emphasizing its relationship with the homologous lanthanide element gadolinium, which was named after the Finnish chemist Gadolin. The first of this pair of transuranic elements to be discovered was curium, which was first prepared by irradiating ^{239}Pu with high energy α-particles in a cyclotron:

$$^{239}_{94}\text{Pu} + {}^{4}_{2}\text{He} = {}^{242}_{96}\text{Cm} + {}^{1}_{0}\text{n}.$$

When it was attempted to produce ^{241}Pu from ^{238}U by a similar process, it was found that the resulting plutonium isotope, a weak β-emitter, decayed to form an α-emitter with a half-life of about 500 years:

$$^{238}_{92}\text{U} + {}^{4}_{2}\text{He} = {}^{241}_{94}\text{Pu} + {}^{1}_{0}\text{n} ; \quad {}^{241}_{94}\text{Pu} \xrightarrow[13 \text{ y.}]{\beta} {}^{241}_{95}\text{Am} \xrightarrow[470 \text{ y.}]{\alpha}$$

A second americium isotope was discovered shortly afterwards during the neutron irradiation of ^{241}Am; this undergoes β-decay, to yield ^{242}Cm:

$$^{241}_{95}\text{Am} + {}^{1}_{0}\text{n} = {}^{242}_{95}\text{Am} + \gamma; \quad {}^{242}_{95}\text{Am} \xrightarrow[18 \text{ hrs}]{\beta} {}^{242}_{96}\text{Cm} \xrightarrow[162 \text{ days.}]{\alpha}$$

Curium prepared by this means provided the first sample to be obtained free from any carrier, and was used by Werner and Perlman (1947) for ultramicrochemical investigations. Prior to this, the chemistry of curium had been studied only by tracer techniques.

In the latter part of 1949, Thompson, Ghiorso and Seaborg, using the cyclotron of the University of California, at Berkeley, prepared two additional transuranic elements, to which they gave the names of californium and berkelium. These elements were formed by irradiation of americium and curium with 35 mev α-particles:

$$^{241}_{95}\text{Am} + {}^{4}_{2}\text{He} = {}^{243}_{97}\text{Bk} + 2{}^{1}_{0}\text{n} ; \quad {}^{242}_{96}\text{Cm} + {}^{4}_{2}\text{He} = {}^{244}_{98}\text{Cf} + 2{}^{1}_{0}\text{n}. \text{ *}$$

These two elements, like the other transuranic elements, are both unstable. Berkelium undergoes transformation chiefly by K-electron capture, forming curium; a certain proportion disintegrates by α-particle emission, to form americium. The half-life of berkelium(^{243}Bk) is 4.6 hrs. Californium(^{244}Cf) is an α-emitter, with a half-life of 45 minutes. In the original discovery of californium, the amounts prepared and identified amounted to only about 5000 atoms, obtained by the irradiation of only a few micrograms of curium. A californium isotope with a rather longer half-life (^{246}Cf, half-life 35.7 hrs.) was later prepared (1951) by the irradiation of uranium with very high energy carbon nuclei, accelerated in the cyclotron:

$$^{238}_{92}\text{U} + {}^{12}_{6}\text{C} = {}^{246}_{98}\text{Cf} + 4{}^{1}_{0}\text{n}.$$

This isotope is also an α-emitter, and disintegrates to form ^{242}Cm. It can, however, also be produced from curium (in this case ^{244}Cm) by α-particle bombardment. It should be noted that the quantities of berkelium and californium available for chemical investigation have been (and are likely to remain) so excessively minute that their reactions have been studied only on the extreme tracer scale; their similarity to the lanthanide elements consists in following closely the reactions of a rare earth carrier, and is not based on any observations on the isolated elements at normal concentrations.

* Although formulated as $(\alpha, 2n)$ reactions, it is possible that these nuclear transformations take place with the ejection of only one neutron. The mass numbers are therefore uncertain to the extent of one unit.

The same is true of elements 99, 100 and 101. The first two were discovered almost simultaneously about the end of 1952 by various research groups (University of California Radiation Laboratory, Argonne National Laboratory and Los Alamos Scientific Laboratory) during work on the transformation products of uranium which had been subjected to a very high instantaneous neutron flux in a thermonuclear explosion. In honour of Albert Einstein and Enrico Fermi elements 99 and 100 were named *einsteinium* (symbol E) and *fermium* (symbol Fm). Their separation from the chemically similar transuranic elements Cf, Bk, Cm and Am was achieved by the use of a suitable cation exchange resin and elution with ammonium citrate solution. In this way their cations are more easily eluted than those of the elements mentioned above, and the whole series is eluted in the following order: $_{100}Fm^{3+}$ $_{99}E^{3+}$ $_{98}Cf^{3+}$ $_{97}Bk^{3+}$ $_{96}Cm^{3+}$ $_{95}Am^{3+}$. Under the same conditions the lanthanides give the series: $_{68}Er^{3+}$ $_{67}Ho^{3+}$ $_{66}Dy^{3+}$ $_{65}Tb^{3+}$ $_{64}Gd^{3+}$ $_{63}Eu^{3+}$. From the elution behaviour it was possible to infer the atomic numbers of the new atomic species on the basis of their characteristic half lives and disintegration energies. Both were α-emitters with half-lives of 20 days and 15 hours, and mass numbers 253 and 255 respectively. Later, further isotopes of both elements were discovered. Thus by irradiation of uranium with highly accelerated $^{16}O^{6+}$ particles Melander and his coworkers in Stockholm discovered in 1954 a fermium isotope with mass number 250 and half-life 30 minutes. By irradiation of ^{238}U with highly accelerated $^{14}N^{6+}$ particles, Ghiorso and his coworkers (1953) obtained two short-lived einsteinium isotopes, ^{246}E and ^{247}E. The longest-lived einsteinium isotope known at present is ^{255}E, a β-emitter with a half-life of about 30 days.

In 1955, Ghiorso prepared element 101 in the form of an atomic species with mass number 256 by irradiation of ^{253}E with α-particles accelerated to very high energies in the cyclotron. It is an α-emitter with a half-life of only a few hours. In honour of Mendeléeff it has been named *mendelevium* (symbol Mv).

(c) Occurrence

It has been shown that the transuranic elements do, in fact, occur in Nature, although only in minimal quantities. In 1941–42 Seaborg and Perlman established that pitchblende and carnotite contained minute amounts of plutonium in the form of ^{239}Pu, and it was later shown that all uranium minerals contain some plutonium, the amount bearing a roughly constant ratio to the uranium content. On the average, the amount of plutonium is about $10^{-12}\%$ of the uranium content.

It is generally assumed that the naturally occurring plutonium in uranium minerals arises from the same processes by which plutonium is produced technically—i.e., by the nuclear reactions:

$$_{92}^{238}U + _{0}^{1}n = _{92}^{239}U \xrightarrow[23 \text{ min.}]{\beta} {}_{93}^{239}Np \xrightarrow[2.3 \text{ days}]{\beta} {}_{94}^{239}Pu.$$

If this is so, it follows that ^{239}Np must also be present in the uranium minerals, although the quantity present at radiochemical equilibrium will be only $2.6 \cdot 10^{-5}\%$ of the plutonium content. 1 g of pitchblende will therefore contain fewer than 6 atoms of ^{239}Np. Such minute quantities cannot be detected even by the very sensitive methods employed for the measurement of radioactive radiations, as is evident from the fact that the quantity of ^{239}Np present in 1 kg of pitchblende will emit less than one β-particle per minute. On the other hand, it has been possible to establish the presence of another neptunium isotope in uranium minerals—namely the long-lived ^{237}Np. This is also present only in minimal quantities—namely to the extent of about $2 \cdot 10^{-10}\%$ of the uranium content. It has probably been formed by the reactions $_{92}^{238}U$ (n, 2n) $\longrightarrow$ $_{92}^{237}U \xrightarrow[6.63 \text{ d}]{\beta} {}_{93}^{237}Np$. This neptunium isotope is the parent member of the neptunium disintegration series shown on p. 639. By loss of an α-particle it is converted to a protactinium isotope, $_{91}^{233}Pa$, which furnishes ^{233}U by β-decay. The amount of ^{233}U found in natural uranium is estimated as about 0.003% of the ^{238}U content. It is considered that ^{233}U has been formed in Nature not only by way of ^{237}Np, but also by another process—namely by the nuclear reactions:

$$_{90}^{232}Th \text{ (n, } \gamma) \longrightarrow {}_{90}^{233}Th \xrightarrow[23.0 \text{ min}]{\beta} {}_{91}^{233}Pa \xrightarrow[27.4 \text{ d}]{\beta} {}_{92}^{233}U.$$

It cannot be excluded that other transuranic elements, as well as plutonium and neptunium, are formed in Nature as products of nuclear reactions. However, their half-lives are so short that they can hardly exist in Nature in sufficient amounts to make their direct detection possible.

2. Neptunium (Np)

(a) Metallic Neptunium

Of all the transuranic elements, neptunium displays the closest resemblance to uranium. It gives rise to the same oxidation states, but differs from uranium in that the tetrapositive state is the preferred oxidation state.

The relative stabilities of the individual oxidation states can be summarized by the following potential scheme, based on the measurements of Cohen and Hindman (1951).

$$
\begin{array}{c}
\overline{\hspace{4cm}-0.679\hspace{1.5cm}} \\
\overline{\hspace{2.5cm}-0.452\hspace{0.5cm}\downarrow\hspace{1cm}}\hspace{1cm}\downarrow \\
\mathrm{Np^{III}}\xrightarrow{-0.150}\mathrm{Np^{IV}}\xrightarrow{-0.754}\mathrm{Np^{V}}\xrightarrow{-1.132}\mathrm{Np^{VI}} \\
\underline{\hspace{4.5cm}-0.943\hspace{1.5cm}}
\end{array}
$$

These values represent oxidation-reduction potentials in volts at $25°$, referred to the standard hydrogen electrode, for the ions involved in the following electrode processes:

$$\mathrm{Np^{+++}} \rightleftharpoons \mathrm{Np^{++++}} + e; \quad \mathrm{Np^{++++}} + 2\mathrm{H_2O} \rightleftharpoons [\mathrm{NpO_2}]^+ + 4\mathrm{H^+} + e;$$

$$[\mathrm{NpO_2}]^+ \rightleftharpoons [\mathrm{NpO_2}]^{++} + e.$$

For the equilibria into which $\mathrm{H^+}$ ions enter, the potentials are those which hold for pH $=0$. Although $\mathrm{Np^{++++}}$ ions are more stable than $\mathrm{U^{++++}}$ ions, the oxidizing power of the $\mathrm{Fe^{+++}}$ ion is sufficiently great to convert them to $[\mathrm{Np^VO_2}]^+$ ions (Huizenga and Magnusson, 1951). Conversely, Np(V) compounds in concentrated perchloric acid solution undergo disproportionation: $2\mathrm{Np^V} \rightleftharpoons \mathrm{Np^{IV}} + \mathrm{Np^{VI}}$, as was shown by Hindman (1951).

The first investigations of carrier-free neptunium compounds were carried out in 1944 by Magnusson and LaChapelle, who employed a few micrograms of the long-lived isotope [237]Np, obtained by irradiation of uranium with fast neutrons. To assist in detection of the neptunium, a small amount of the strongly radioactive [239]Np was added as an indicator. As has already been mentioned, it later became possible to work with much greater quantities of material, when it was found that neptunium-237 was a by product of the technical preparation of plutonium. The yield of neptunium is said to be about 0.1% of the quantity of plutonium isolated.

Neptunium is a silvery, lustrous metal, fairly stable in air ($d_{\mathrm{pyc}} = 19.5$, m.p. $640°$). It exists in several distinct modifications. The form stable at ordinary temperature (α-Np) has an orthorhombic structure, which can be regarded as a deformed face-centered cubic structure. The unit cell, containing 8 atoms of Np, has the dimensions $a = 4.723$, $b = 4.887$, $c = 6.663$ Å ($d_{\mathrm{x-ray}} = 20.45$). At $278\ °\mathrm{C}$ this changes into a tetragonal modification (β-Np), and above $550°$ there is a third modification (γ-Np), which is stable up to the melting point (Zachariasen, 1952). Neptunium is a fairly strongly electropositive metal; it stands close to aluminum in the electrochemical potential series, being a little nobler than aluminum and also more noble than uranium and plutonium (cf. p. 621). Metallic neptunium can be prepared by heating NpF_3 to 1200 °C with Ba.

(b) Neptunium Compounds

Neptunium trifluoride, NpF_3, is obtained by heating NpO_2 to 500° in a mixture of HF and H_2:

$$NpO_2 + 3HF + \tfrac{1}{2}H_2 = NpF_3 + 2H_2O.$$

The *tetrafluoride*, NpF_4, is formed when either NpO_2 or NpF_3 is heated in a mixture of HF and O_2:

$$NpO_2 + 4HF = NpF_4 + 2H_2O; \quad NpF_3 + HF + \tfrac{1}{4}O_2 = NpF_4 + \tfrac{1}{2}H_2O.$$

Neptunium tetrachloride, $NpCl_4$, can be prepared by heating NpO_2 in CCl_4 vapor. When heated to 450° in hydrogen, the tetrachloride undergoes reduction to the trichloride:

$$NpO_2 + 2CCl_4 = NpCl_4 + 2COCl_2; \quad NpCl_4 + \tfrac{1}{2}H_2 = NpCl_3 + HCl.$$

These reactions correspond exactly to those used for the preparation of UCl_4 and UCl_3. The neptunium compounds also correspond very closely in properties to the corresponding halides of uranium. Similarly, *neptunium hexafluoride*, NpF_6, corresponds exactly in its properties to UF_6; it may be obtained by heating NpF_3 in a stream of fluorine.

Neptunium tribomide, $NpBr_3$, is formed by the action of bromine vapor on a mixture of NpO_2 and an excess of metallic Al. It can be obtained pure by first volatilizing away the $AlBr_3$ which is formed simultaneously at 250°, and then heating more strongly (to about 800°) to sublime the $NpBr_3$. The *triiodide*, NpI_3, is obtained in a similar manner. If a mixture of NpO_2 and Al is heated in bromine vapor, without having any excess of aluminum present, the *tetrabromide*, $NpBr_4$, is obtained. This sublimes at 500°. X-ray structure determinations have shown that all these neptunium halides are identical in structure with the corresponding compounds of uranium.

Neptunium combines with nitrogen and carbon, forming the compounds NpN ($a = 4.887$ Å, Zachariasen, 1949) and NpC ($a = 5.004$ Å, Templeton, 1952), both of which crystallize with the NaCl structure. The compounds Np_3P_4 and $NpSi_2$ are formed with phosphorus and silicon. Shaft and Fried (1949) obtained the sulfide Np_2S_3, isomorphous with U_2S_3, by heating NpO_2 in a mixture of H_2S and CS_2.

The most stable *oxide* of neptunium is the dioxide NpO_2. This can be heated in air without undergoing any change in composition. It can be converted to the higher oxide Np_3O_8 (e.g., by heating it in NO_2—Katz, 1949), but oxidation takes place far less readily than with UO_2. Np_3O_8 is isomorphous with U_3O_8. The monoxide NpO, corresponding to UO, appears not to be very stable.

Neptunium compounds derived from the $+3$, $+4$, $+5$, and $+6$ oxidation states can also be obtained in aqueous solution. Neptunium(III) compounds can be obtained in solution, for example, by electrolytic reduction of Np(IV). Air must be excluded during the preparation since atmospheric oxygen reoxidizes neptunium(III) compounds in aqueous solution to neptunium(IV) compounds.

Neptunium(IV) fluoride, iodate, and hydroxide (or oxide hydrate) are practically insoluble in water. They can be dissolved in sulfuric acid, however, and oxidized to neptunium(VI) compounds by the addition of bromate. Neptunium(V) compounds can be obtained either by electrolytic oxidation or by oxidation with iron(III) salts. Sulfurous acid reduces both neptunium(VI) and neptunium(V) compounds, to neptunium(IV) compounds.

Hexapositive neptunium gives rise both to *neptunates* and to *neptunyl* compounds. The former correspond to the uranates, and the latter to the uranyl compounds. Like the uranyl salts, the neptunyl salts, $[NpO_2]X_2$, have a tendency to form double and complex salts—e.g., sodium neptunyl acetate, $Na[NpO_2](C_2H_3O_2)_3$, which is isomorphous with sodium uranyl acetate.

Double and complex salts are also formed from the neptunium(IV) and the neptunium(V) salts; most of the latter contain the cationic radical $[NpO_2]^+$.

Addition of OH^- ions to neptunium(V) salt solutions precipitates neptunium(V) hydroxide, from which Gibson, Gruen and Katz (1952) prepared pure neptunium(V) oxalate, $[NpO_2]HC_2O_4 \cdot 2H_2O$, by dissolving the hydroxide in 1-N hydrochloric acid and adding a solution of oxalic acid in t-butyl alcohol. The aqueous solution of the oxalate gave the absorption spectrum characteristic of neptunium(V) compounds, but also afforded evidence of the formation of oxalato-complex ions in solution, as well as the ions of the simple salt.

The intermediate oxide Np_3O_8 is formed either by the action of NO_2 on neptunium (IV) hydroxide, neptunium(V) hydroxide or neptunium(IV) nitrate at 300°, or by heating ammonium dineptunate, $(NH_4)_2Np_2O_7 \cdot H_2O$, to 275–400° in air. If the temperature is raised to 770°, NpO_2 is formed. Whereas the action of NO_2 on U_3O_8 above 250° C leads to UO_3, it has not proved possible to obtain the corresponding oxide of neptunium, NpO_3. When Np_3O_8 is heated in air, it at first loses oxygen continuously, without change in crystal structure. At 600° there is a discontinuity, and increased loss of oxygen, and NpO_2 is formed. This latter undergoes no further decomposition when it is heated in a vacuum to 1100°. Np_3O_8 dissolves in $HClO_4$ to form a mixture of neptunium(V) and neptunium(VI) perchlorates, whereas when U_3O_8 is dissolved in acid it produces a mixture of uranium(IV)- uranium(V)- and uranium(VI) salts (Miller and Dean, 1952).

(c) Separation from Other Elements

Neptunium can be separated from plutonium and the other transuranic elements, and also from lanthanum and its congeners, by adding sodium bromate and then hydrofluoric acid to a sulfuric acid solution of neptunium and the accompanying elements. At ordinary temperature, only the neptunium is oxidized by the sodium bromate, forming neptunyl ions, $[NpO_2]^{++}$. These are not precipitated by F^- ions, whereas all the tri- and tetrapositive transuranic elements, and also lanthanum and the other lanthanides, are precipitated under these conditions as their insoluble fluorides. When it is required to purify neptunium from small traces of plutonium, etc., a small quantity of lanthanum salt may first be added, so that the precipitated lanthanum fluoride acts as a carrier for the very small quantities of transuranic element fluorides.

3. Plutonium (Pu)

(a) General

Plutonium, like neptunium, can exist in all the oxidation states displayed by uranium, but is converted to the +6 state even less readily than is neptunium. In general, plutonium strongly favors the tetrapositive state, although in combination with certain elements the tripositive state is still more stable. Thus the trihalides of plutonium are obtained under exactly the same conditions as would lead, with uranium and neptunium, to the formation of the tetrachloride and tetrabromide. The compositions of the oxysulfides—$U^{VI}O_2S$, $Np^{IV}OS$, $Pu^{III}_2O_2S$—also shows the progressive tendency, going from uranium to plutonium, to favor the lower valence states.

On passing from uranium to plutonium, and then to the higher transuranic elements americium and curium, the diminishing tendency to exist in higher oxidation states is well exemplified by the compositions of the oxides, as can be seen from the following summary table.

UO	NpO	PuO	AmO	
		Pu_2O_3		Cm_2O_3
		Pu_4O_7		
UO_2	NpO_2	PuO_2	AmO_2	
U_4O_9				
U_2O_5				
U_3O_8	Np_3O_8			
UO_3				

Plutonium forms a hydride, PuH_3, resembling uranium hydride. It combines with nitrogen and carbon, to form the compounds PuN and PuC, which are isostructural with the corresponding uranium compounds.

Plutonium resembles neptunium in that it can exist as ions in aqueous solution, in all its oxidation states except the bivalent state. Except where complex ions are formed, the species of ions present in acid solution are Pu^{+++}, Pu^{++++}, PuO_2^+ and PuO_2^{++}.

Their relative stabilities, as compared with the corresponding ions of uranium, can be inferred from the following potential schemes.

$$\text{Pu}^{III} \xrightarrow{-0.982} \text{Pu}^{IV} \xrightarrow{-1.134} \text{Pu}^{V} \xrightarrow{-0.912} \text{Pu}^{VI}$$

with bracketed potentials -1.058, -1.023, -1.009.

$$\text{U}^{III} \xrightarrow{+0.63} \text{U}^{IV} \xrightarrow{-0.55} \text{U}^{V} \xrightarrow{-0.06} \text{U}^{VI}$$

with bracketed potentials $+0.04$, -0.31, $+0.01$.

The oxidation-reduction potentials for the plutonium ions (in volts at 25° and referred to the standard hydrogen electrode) are based on the measurements of Connick (1944–51) and Hindman (1944). It may be seen for example, that the reaction (in solution at pH = 0)

$$PuO_2^{++} + H^+ + \tfrac{3}{2}H_2 = Pu^{+++} + 2H_2O$$

involves a decrease in free energy (of 69.8 kcal per g-ion of PuO_2^{++}). The corresponding free energy decrease for the reduction of the NpO_2^{++} ion to Np^{+++} ions, by H_2, is 47.0 kcal per g-ion. Oxidation of Pu^{+++} ions to Pu^{++++} ions requires the expenditure of 22.65 kcal of energy per g-ion, whereas oxidation of Np^{+++} ions to Np^{++++} ions is endothermic only to the extent of 3.5 kcal per g-ion. Except when stabilized by complex formation, the U^{+++} ion is thermodynamically unstable in aqueous solution, since it can be oxidized directly by H^+

ions to U^{++++} ions. At unit hydrogen ion activity, this process is exothermic to the extent of 14.5 kcal per g-ion.

Heats of reaction have also been determined for the reduction of the $[PuO_2]^{++}$ and Pu^{++++} ions to Pu^{+++} ions (Evans, 1945; Connick, 1951). These are

$$[PuO_2]^{++} + H^+ + {}^3H_2 = Pu^{+++} + 2H_2O + 77.8 \text{ kcal,}$$

$$Pu^{++++} + \tfrac{1}{2}H_2 = Pu^{+++} + H^+ \quad + 13.5 \text{ kcal.}$$

In the one case the heat of reaction is greater, and in the other case less than the free energy change. Hence the first of these reactions must be associated with a decrease in the entropy, and the second with an increase in entropy ($\triangle S = -26.6$ and $+30.6$ cal/°, respectively).

Equilibrium can be established between the Pu^{+++} ions and $[PuO_2]^{++}$ ions, the position depending on the pH of the solution:

$$2Pu^{+++} + [PuO_2]^{++} + 4H^+ \rightleftharpoons 3Pu^{++++} + 2H_2O.$$

For the equilibrium constant, $K_1 = \dfrac{[Pu^{+++}]^2 \cdot [(PuO_2)^{++}]}{[Pu^{++++}]^3}$, Kasha (1945) gives the

value $K = 0.040$ in a solution 1-normal in $HClO_4$, and $K_1 = 40.2$ in a solution 0.1 normal in $HClO_4$, both at 25° and unit ionic strength. The observed figures make K_1 proportional to $[H^+]^{-3}$, whereas the reaction equation should make K_1 proportional to $[H^+]^{-4}$. The reason for this discrepancy is not known.

There is an analogous equilibrium for the reaction:

$$Pu^{+++} + 2[PuO_2]^{++} + 2H_2O \rightleftharpoons 3[PuO_2]^+ + 4H^+.$$

The constant $K_2 = \dfrac{[Pu^{+++}] \cdot [(PuO_2)^{++}]^2}{[(PuO_2)^+]^3}$ was found by Connick (1944) to have the

value $1.2 \cdot 10^4$ at 25°, in a solution 0.5-normal in HCl. It follows that $[PuO_2]^+$ ions can be present only in extremely low concentration in strongly acid solutions. However, if the H^+ ion concentration is lowered, the equilibrium shifts strongly in favor of $[PuO_2]^+$ ions. According to Kasha, in a 1-normal $HClO_4$ solution about 0.5%, and in a 0.01-normal $HClO_4$ solution about 60% of the plutonium will be present in the $+5$ state, in equilibrium with the ions of the other oxidation states. (Measurements apply to solutions of unit ionic strength). It is, however, possible to achieve a fairly high concentration of $[PuO_2]^+$ ions in quite strongly acid solution—e.g., through the electrolytic reduction of $[PuO_2]^{++}$ ions— since the disproportionation whereby equilibrium is set up, as just considered, is a relatively slow process.

Plutonium can form anionic complexes from all the oxidation states in which it can exist in aqueous solution. Oxalate, acetate, and sulfate ions have an especially strong tendency to form such complexes, whereas chloride and nitrate ions have a weak tendency to do so. It has been stated that plutonium ions form no complexes with perchlorate ions.

Metallic plutonium can be prepared by heating plutonium fluoride in barium vapor. Plutonium is a fairly strongly electropositive metal. Its place in the electrochemical series follows from its standard potential, referred to the standard hydrogen electrode:

Al/Al^{+++}	U/U^{+++}	Pu/Pu^{+++}	Np/Np^{+++}	Mn/Mn^{++}
$+1.69$ volts	$+1.6$ volts	$+1.5$ volts	$+1.4$ volts	$+1.1$ volts

The *technical importance* of plutonium is based on the fact that the ^{239}Pu isotope is fissile, both by slow and by fast neutrons. This property is common to the nuclei of ^{239}Pu, ^{235}U, and ^{233}U. Although actinouranium (^{235}U) occurs naturally in large amounts, as a constant constituent of natural uranium (0.7%), whereas plutonium must be made artificially by nuclear reactions from ^{238}U, it is nevertheless easier to prepare plutonium than pure ^{235}U on an industrial scale. This is because plutonium can be separated chemically from uranium, whereas ^{235}U can only be obtained pure by using the methods involved in the separation of isotopes. The translation of such processes from the laboratory scale to the industrial scale is a matter of extreme technical difficulty. Furthermore, other artificial radioactive elements are obtained as by-products in the large-scale production of plutonium, while the enormous amount of atomic energy liberated during the production of plutonium can be applied to the generation of electrical energy (*nuclear power reactors*).

(b) Production of Plutonium

(*i*) *Nuclear Reactions.* The technical preparation of plutonium hinges on the occurrence of the following nuclear reactions, through the interaction of slow neutrons on the more abundant isotope of natural uranium, ^{238}U:

$$^{238}_{92}U + {}^{1}_{0}n \rightarrow {}^{239}_{92}U \xrightarrow[\text{23 min.}]{\beta} {}^{239}_{93}Np \xrightarrow[\text{2.3 d}]{\beta} {}^{239}_{94}Pu.$$

The neutrons needed for this process arise from the nuclear fission of ^{235}U, which is present to the extent of 0.7% in natural uranium.

In the nuclear fission of ^{235}U, at least *two* fresh neutrons are generated for every neutron which is used to bring about fission. If the conditions were such that each one of these neutrons could bring about the fission of another nucleus, the number of free neutrons would increase with extraordinary rapidity, as was discussed in Chap. 12. The chain reaction would be self-accelerating, and would take place explosively, with very great violence. This is the principle of the atomic bomb. Whereas the uncontrolled chain reaction leads to an explosion, a nuclear reactor such as is used for generation of power, or for the production of plutonium, makes use of a *controlled chain reaction*.

When nuclear fission takes place, not in pure ^{235}U, but in natural uranium, which contains far more ^{238}U than ^{235}U, the liberated neutrons collide far more frequently with ^{238}U nuclei than with ^{235}U nuclei. Even if only a small fraction of the collisions with ^{238}U lead to the capture of neutrons by the nuclei, with resulting formation of ^{239}U, it necessarily follows that only a fraction of the liberated neutrons can collide with ^{235}U nuclei, to bring about fresh nuclear fissions. Suppose, for simplicity, that *two* neutrons are liberated for every nucleus of ^{235}U undergoing fission, and that on each occasion, *one* of these two neutrons reacts with a ^{238}U nucleus to give ^{239}Pu, (by way of ^{239}U and ^{239}Np), whereas the other neutron brings about the fission of another ^{235}U nucleus. Then a fresh pair of neutrons is regenerated, and the number of free neutrons remains constant. The nuclear reaction will then be propagated with *constant velocity*. This is approximately the state of affairs achieved in the technical production of plutonium.

Only half the neutrons are absorbed by the ^{238}U nuclei, although these latter are about 139 times as abundant in natural uranium as are the ^{235}U nuclei. This is because the *capture cross section* (see p. 585 *et seq.*) of ^{238}U is very much smaller than that of ^{235}U. This is true for so-called *thermal neutrons*, having a kinetic energy of about 0.025 ev at ordinary temperature, and since in general the cross section varies inversely as the square of the neutron energy, the difference in capture efficiency is still greater for neutrons of high kinetic energy.

Between the range of thermal energies and of very high energies, however, the capture cross section of ^{238}U passes through a so-called *resonance peak* (at about 5 ev)—i.e., a narrow range within which the capture cross section rises to very high values indeed (several thousand barns). The neutrons ejected in the act of nuclear fission initially have very high kinetic energies (1 to 2 mev). When they collide with ^{238}U nuclei, they undergo elastic collisions in almost every case—i.e., they are reflected without being captured*. At each collision, whether with ^{238}U or with any other nuclei, the velocity of the neutrons is diminished, since they share their kinetic energy with the collision partner (cf. p. 561). Their kinetic energy ultimately drops to a value around 5 ev, and if at this stage they collide with ^{238}U nuclei, they will be almost quantitatively absorbed. In order to conserve sufficient nuclei to carry on the fission of ^{235}U it is therefore necessary to ensure that the decrease in kinetic energy of the fission neutrons takes place within some material that has the smallest possible neutron capture cross section. A material with properties suitable for this purpose is called a *moderator*, and the slowing-down of the neutrons is referred to as *thermalization*. The essential property of a good moderator is not only that it shall have as small a capture cross section as possible, but also that the number of collisions needed to slow the neutrons down to thermal energies shall be as small as possible, since even with a very small capture cross section a certain proportion of the collisions must lead to capture. It follows from the laws of dynamics that the best decelerating properties will be exhibited by the nuclei with the smallest mass numbers (p. 561). From Table 80 it is apparent that out of the various substances coming into practical consideration as moderators for the production of plutonium, graphite and heavy water are the most suitable. At the present time, graphite is more readily prepared than heavy water, in the very large quantities that are required. The graphite must, however, be of extremely high purity, since even minute traces of impurities with high cross sections very seriously increase the total capture cross section of the material.

TABLE 80

SLOWING DOWN EFFECT OF VARIOUS NUCLEI FOR NEUTRONS

Nucleus	H	D	He	Be	C	O
Mass number	1	2	4	9	12	16
Percentage energy loss at each collision	63%	52%	35%	18%	14%	11%
Number of collisions needed for thermalization	18	25	42	90	114	150
Capture cross section, barns	0.3	0.001	0	0.01	0.005	0.002

Fig. 81 represents diagrammatically the processes which, according to the foregoing discussion, are involved in the production of plutonium. According to this scheme, 1 atom of plutonium should be produced for each atom of ^{235}U that undergoes fission. This is approximately true in practice.

The condition that the process shall be self-sustaining is that out of the neutrons liberated in each fission, exactly *one* neutron, on the average, brings about another fission. This condition is not fulfilled simply by introducing a moderator into the system; other factors must also be taken into account, as is briefly considered below.

(ii) The Multiplication Factor. Assume that at a given instant, the number of fissions occurring in a system $= z$. Of the neutrons which are thereby liberated, let the number $k \cdot z$ bring about further fissions. Then the factor k is said to be the *multiplication factor* for the given system. If, for example, 100 neutrons bring about fissions at a given instant, and 105 of the neutrons so liberated ultimately causes a

* A very small fraction of the fast neutrons is captured by the ^{238}U nuclei, and brings about fission of ^{238}U.

second generation of fissions. The number of neutrons causing a third generation of fissions will then be 100 · 1.05 · 1.05, and the number of fissions in the nth generation will be $100 \cdot 1.05^{n-1}$, the multiplication factor being 1.05. If a single neutron were introduced into a system with the multiplication factor $k = 1.05$, the number of neutrons would be over 16,000 by the two-hundredth generation, and about $3.8 \cdot 10^{10}$ at the beginning of the five-hundredth generation. If k is appreciably greater than unity, the number of neutrons thus grows with extreme rapidity. Conversely, if k is less than 1, the number of neutrons rapidly falls; if $k = 0.95$, the number of neutrons would fall from 10^9 to 1 in four hundred generations. To maintain the nuclear reaction in a steady, self-sustaining state for the production of plutonium, it is therefore necessary to have a means of controlling the multiplication factor continuously, to have a value very close to 1.000...; the reaction may otherwise die out or become uncontrollable.

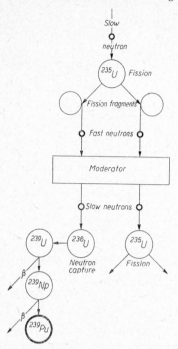

Fig. 81. Scheme of events in the production of plutonium.

The magnitude of the multiplication factor is determined by a number of processes. How far it is possible to modify the multiplication factor by changes in these, can be seen from the following.

[1] It has been assumed that 2 fast neutrons are liberated for each slow neutron absorbed in bringing about fission. In fact, the yield of neutrons per fission is appreciably greater than 2, and varies from one fissile material to another. This factor may be denoted by f_1. If z slow neutrons bring about a corresponding number of fissions, the number of free neutrons is thereby raised from z to $f_1 \cdot z$.

[2] The neutrons liberated in fission initially have a kinetic energy of 1–2 mev. They are therefore capable of inducing fission of ^{238}U nuclei, and although the corresponding fast fission cross section is very small, such fissions occur. The fast neutrons may also collide with ^{235}U nuclei, and can again bring about fission. The effect of these processes is to increase the number of neutrons further, by a *fast fission factor*, f_2, from $f_1 \cdot z$ to $f_1 \cdot f_2 \cdot z$. f_2 usually has a value about 1.03.

[3] In the course of being slowed down, and before they have been completely thermalized by the moderator, a certain proportion of the neutrons will inevitably collide with ^{238}U nuclei, while possessing energies in the resonance region. Such neutrons are absorbed, and the total number of neutrons is thereby reduced from $f_1 \cdot f_2 \cdot z$ to $f_1 \cdot f_2 \cdot f_3 \cdot z$. The factor f_3 (which is always < 1) measures the probability that a neutron will escape absorption by ^{238}U before attaining thermal energies. Since the absorption depends largely on the resonance effect for electrons of about 5 ev, the factor f_3 is termed the *resonance escape factor*. The magnitude of f_3 can be controlled by modifying the spatial distribution of fissile material and moderator. The neutrons are slowed down in course of diffusion through the moderator, and it is necessary to minimize the probability that they shall collide with uranium again before executing a sufficiently long diffusion path to reduce their energy below the resonance value. This factor therefore determines the manner in which the uranium rods are disposed in a regular array in channels through the moderator of a nuclear reactor.

[4] A small, but not insignificant, fraction of the neutrons is absorbed by the moderator in the process of thermalization. A further, very appreciable, proportion is used up to convert ^{238}U into plutonium, after thermalization. In addition, a very important proportion (which must be minimized as far as possible) is captured by impurities in the uranium and moderator, by structural materials of the reactor, and by the air or liquid coolant used to carry off the energy liberated by fission. The effect of this competitive absorption is to reduce the number of neutrons from $f_1 \cdot f_2 \cdot f_3 \cdot z$ to $f_1 \cdot f_2 \cdot f_3 \cdot f_4 \cdot z$. The larger the value of f_4, the greater is the number of thermal neutrons available to generate fresh neutrons by fission of ^{235}U. This factor is therefore known as the *thermal utilization factor*.

In a system of infinite size, the ratio $\dfrac{f_1 \cdot f_2 \cdot f_3 \cdot f_4 \cdot z}{z}$ would be the overall fraction of the neutrons coming from the original z fissions, which themselves bring about a fresh generation of fissions. This by definition, is the multiplication factor k. Hence

$$k = f_1 \cdot f_2 \cdot f_3 \cdot f_4 \tag{1}$$

Of these four factors which enter into the multiplication factor, f_3, f_4 and to some extent f_2 depend on the experimental conditions. It is evident that f_4 depends on the nature of the moderator, and also that the condition $k = 1.000\,000\ldots$ can be achieved by altering at will the amount of competitive absorption in the pile, by a suitable control device. It is also apparent that a high value for the resonance escape factor f_3 is favored by having the uranium present in compact pieces, embedded in the graphite, rather than as a more or less homogeneous mixture of finely divided uranium and graphite.

(*iii*) *Critical Size of a Nuclear Reactor*. Equation (1) above is valid for a system of infinite size, whereas an actual nuclear reactor involves a system of finite size. In a finite system there is also a loss of neutrons which diffuse out of the boundaries of the system, and the multiplication factor may be very considerably diminished from this cause. The number of neutrons leaking out of a reactor is proportional to the surface area, whereas the number produced is proportional to the volume of the core. In the case of a spherical core, the ratio of neutrons leaking out to neutrons produced is inversely proportional to the radius. If the multiplication factor k of eqn (1) is greater than unity, it is possible to select such dimensions for any reactor that the loss of neutrons by leakage exactly compensates their multiplication by fission. The over-all multiplication factor, which may be represented by k', then has a value of exactly 1.000. Suppose, for example, that 100 fissions produce such a number of neutrons that 105 neutrons should be available for the second generation of fissions if there were no loss from leakage. If 5 of these 105 neutrons are lost from the system by leakage from the core, there remain 100 neutrons actually available for the second generation of fissions. The effect of neutron leakage has then been to reduce the multiplication factor from $k = 1.05$ to $k' = 1.00$. The nuclear chain reaction will then not follow an auto-accelerating course, as it would in an infinite system, but proceeds at a constant rate. The dimensions of a nuclear reactor for which the loss by leakage is exactly compensated by the multiplication which would obtain in an infinite system, is known as the *critical size*. The critical size depends upon several factors— e.g., the isotopic composition of the fissile material, the nature and amount of the moderator, and the competitive absorption by impurities. The isotopic composition of the fissile material markedly affects the resonance escape factor f_3, and it is apparent that a relatively small degree of enrichment of ^{235}U, as compared with the proportion in natural uranium, should appreciably diminish the critical size. The critical size of a reactor using heavy water as moderator is appreciably smaller than for a graphite-moderated reactor.

(*iv*) *Control Rods*. A self-sustaining nuclear reaction can only take place if the reactor is at least of critical size. In practice, reactors are built so as to be rather above the critical size, but a fraction of the neutrons is absorbed in 'control rods' of cadmium or boron steel. Cadmium and boron have extremely large capture cross sections for slow neutrons, as compared with, e.g., aluminum which has an extremely small cross section. The neutron flux in a reactor is then kept constant by inserting the control rods more or less deeply into the core, thereby changing the thermal utilization factor f_4. If the flux rises, the control rods are pushed into a greater depth. If the flux falls, the control rods are withdrawn, to reduce the competitive absorption.

(v) *Delayed Neutrons.* The possibility of regulating the rate of the chain reaction in a nuclear reactor by means of control rods hinges upon the fact that a fraction of the neutrons released by fission is not emitted instantaneously. The enormous speed with which any adventitious change in the multiplication factor would reflect itself in a growth or diminution in the rate of fission would make control of the system impossible if this were not the case.

The neutrons emitted in the process of fission are not ejected from the nucleus undergoing fission, but from the fission fragments. However, these fragments emit most of the neutrons instantaneously, during the act of fission (probably within 10^{-12} sec). A small proportion, however (about 0.75%), is liberated after some delay (of the order of seconds). This means that if a nuclear reactor is so adjusted that its multiplication factor at a given instant is $k' = 1.0055$, the number of fissions in each generation increases in the ratio 10,055 : 10,000; but the number of neutrons instantaneously liberated to induce the next generation of fissions is only 9980. Thus the original 10,000 fissions will be followed within an extremely short interval of time by 9980 fissions, whereas the remaining 75 fissions of the second generation take place over an interval of time, falling off exponentially over several seconds. However, the increase in the number of fissions at each generation is due solely to the delayed neutrons, and takes place relatively slowly. The position of the control rods may be changed so that the multiplication factor is diminished, and the rate of increase in the neutron flux (and therefore in the energy released) is prevented from getting out of control.

(iv) *Atomic Piles.* The nuclear reactors used for the technical production of plutonium are commonly spoken of as *piles*. The name originated with the first self-sustaining prototype reactor, which was built up from alternate layers of graphite and uranium lumps (also some uranium oxide). Later reactors have been built from large graphite blocks, through which run channels. Rods of metallic uranium, sheathed in aluminum, are placed in these channels, through which also flows air or some other coolant to carry off the great amount of energy released in fission. The uranium is discharged from the reactor for processing from time to time, when it has a content of about 0.1% of plutonium (including such neptunium as is in course of decay).

It is not possible to continue the irradiation of the charge so long that all the ^{235}U is 'burned up' and converted to the corresponding amount of plutonium. As plutonium accumulated, it would itself undergo fission by slow neutrons to an increasing extent, until it would ultimately be consumed by fission at the same rate as it was being produced from ^{238}U. More important is the fact that as the fission products of ^{235}U accumulate, they act increasingly as impurities with very high neutron absorption. They thereby reduce the multiplication factor, until eventually k would drop below 1, and the reactor would cease to function.

The first reactors to be constructed used graphite as moderator; examples of experimental graphite-moderated piles are those at Harwell, in England, and Brookhaven, in the United States. Heavy water-moderated piles may be exemplified by those at Chalk River (Canada) Chatillon (France) and Argonne (United States) (See *Nucleonics*, 11, No. 5 (1953) 21). Such piles are used as neutron sources for physical experiments and for the study of radiation-induced reactions, and as means of preparing artificial radioelements for therapy, tracer studies, radioactivation analysis, etc.

A diagrammatic cross section of such a reactor is given in Fig. 82, which shows the lattice structure of graphite blocks, penetrated by channels bearing the uranium rods sheathed in aluminum. The release of energy due to fission in such a reactor may be of the order of 6,000 kilowatts, and the maximum neutron flux 10^{11} to 10^{12} neutrons per cm^2 per sec. Fig. 82 shows schematically the adjustable control rods, and also the provision for the production of artificial radioelements such as ^{32}P and ^{14}C. For this purpose, an experimental channel penetrating the reactor is furnished with a rectangular graphite block into which are fitted aluminum capsules. The materials to be irradiated (sulfur for ^{32}P, tellurium for ^{131}I, etc.) are loaded into these capsules, subjected to irradiation with neutrons in the pile, and subsequently removed for direct use or for chemical processing, as may be required.

Around the 'core' of the reactor is a thick layer of graphite blocks, which serves as a neutron reflector; the multiplication factor of the reactor is thereby raised. The pile is an

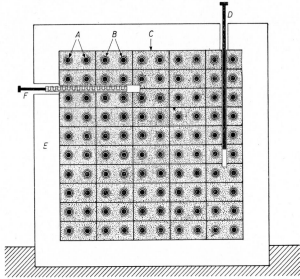

Fig. 82. Schematic cross section of a graphite-moderated atomic pile.

A   Uranium bars
B   Aluminum cans enclosing uranium
C   Graphite moderator
D   Control rod
E   Concrete shielding ('biological shield')
F   Graphite holder containing aluminum capsules, filled with
       materials undergoing neutron activation in the pile.

intense source not only of neutrons but also of γ-radiation, and the whole is surrounded by a thick 'biological shield' of dense concrete.

Whereas piles designed as experimental instruments, or solely for plutonium production, may operate at low temperatures with air cooling (as at Harwell and Brookhaven) or water cooling, reactors for power production must operate at higher temperatures, and may be cooled by carbon dioxide, helium, or liquid sodium. In such cases the coolant is passed through a heat exchanger in which steam is generated, as the motive power for turbines.

The first experimental installation for carrying out a self-sustaining fission reaction was erected at Columbia University, and operated in July 1941. It consisted of lumps of 'technical pure' uranium oxide, totalling 7,000 kg weight, embedded in graphite. It was not capable of a self-sustaining reaction, since too many neutrons were captured by impurities. A self-sustaining chain reaction (without any external source of neutrons) was first achieved in the purely experimental reactor built at the University of Chicago, and put into operation in December 1942. The production of important quantities of plutonium became possible with the pile built at Oak Ridge, Tenn., which started up in November 1943 at a power level of 800 kilowatts, subsequently raised, after considerable modifications to over 2000 kilowatts by the middle of 1944. The first production piles came into operation at Hanford (Washington, United States) in the autumn of 1944. This plant comprised three water-cooled production piles, with a total heat output of about 1,000 megawatts, which enabled plutonium to be produced on the kilogram scale. Nuclear reactors have since then been constructed in a number of countries other than the United States, some for plutonium production, some purely as experimental reactors, and others principally for the generation of power.

(*vii*) *Extraction of Pure Plutonium*. The irradiated uranium rods, containing neptunium and plutonium, are dropped into deep water tanks when they are removed from the reactor. They are allowed to 'cool' in these for a long period, until the conversion of neptunium into plutonium has become complete. At the same time, all the shorter-lived members of the complex mixture of fission products decay away. This is most desirable, since the level of γ-ray activity is at first excessively high, but decreases greatly during the cooling period. Even so, the intensity of the radiations remains so great that all chemical processing must be carried out in a plant heavily shielded with concrete, and provided with remote control for all operations.

In principle, the separation depends upon the similarity of plutonium and uranium in the $+6$ state, and the far greater stability of plutonium(IV) compounds as compared with uranium(IV) compounds. Thus the irradiated uranium can be dissolved in acid, and oxidized to uranium(VI) salts under such conditions that the plutonium remains unoxidized. The plutonium(IV) can then be precipitated from solution, in the presence of some suitable carrier. The precipitate can be centrifuged off, redissolved, and oxidized so that Pu(IV) is converted to Pu(VI). The carrier is then precipitated, and carries down with it a certain proportion of the fission products. Those fission products which are not taken down by the carrier should not be precipitated if, after removal of the carrier, the plutonium is reduced to Pu(IV) once more, and re-precipitated under the same conditions. In order to obtain absolutely pure plutonium, decontaminated from fission products, these oxidation-, reduction- and precipitation processes must be repeated several times.

Although these chemical operations appear simple in principle, the preparation of pure plutonium, free from β- and γ-ray activity, involves about thirty steps. In view of the health hazard presented by the radioactive radiations, and by the plutonium, all operations must be performed by remote control—behind heavy shielding until all fission products have been completely removed, and within totally enclosed 'dry boxes' even after the β- and γ-ray emitting fission products have been eliminated.

The design for the original Hanford plant had to be considered before plutonium or any of its compounds had been seen. Plutonium compounds, free from carrier, were first obtained in visible amounts in August 1942, and the chemical separation process originally used at Hanford was based upon experiments for which the total available quantity of plutonium was about 0.5 mg; this was prepared by means of a cyclotron, and not in a nuclear reactor. Most of the experimental work was carried out on the microgram scale. Translation of these laboratory experiments to the production basis involved an increase in scale of 1,000 million : 1.

(c) Toxicity of Plutonium

Plutonium and the other α-emitting elements of moderately long half-life are dangerously toxic substances, since they form insoluble phosphates and became incorporated in the growing bone cells if they are ingested, or enter the blood stream through an injury. Plutonium(IV) phosphate is so insoluble that, once precipitated, the element is excreted from the system only very slowly indeed. The effect is therefore cumulative. The intense chemical effects produced along the tracks of the α-particles, emitted from a speck of plutonium compound, can produce serious lesions in the bone marrow cells which generate the red blood corpuscles. Although the radiations from plutonium compounds are 'soft', and require no shielding, it is necessary to carry out all chemical work, or fabrication of plutonium under conditions such that exposure to air-borne dust, etc. (such as can arise from the drying of droplets of liquid when experimenting in the open laboratory), is absolutely prevented, since the cumulative life-time tolerance dose of plutonium is of the order of only 1 microgram. *Similar precautions are essential for the safe manipulation of radium, protactinium, and other α-emitters of similar half-life.*

4. Important Compounds of Plutonium

(a) General

Salts of $+3$, $+4$, and $+6$ plutonium are readily prepared, but the salts of $+5$ plutonium are obtained only with difficulty, because of the tendency of the pentapositive ion to undergo disproportionation. It has not proved possible to obtain plutonium(II) salts.

Plutonium(III) salts are partially hydrolyzed by water, unless the hydrolysis is repressed by the addition of an excess of acid. Plutonium(IV) salts have a much stronger tendency to undergo hydrolysis. Except for some strongly complexed compounds, the plutonium(VI) salts contain the plutonyl radical, $[PuO_2]^{2+}$, corresponding to the uranyl radical. Plutonyl compounds, $[PuO_2]X_2$, are hydrolyzed only to a very slight extent, and usually crystallize well.

The salts of plutonium, and their solutions, have characteristic colors, and display absorption spectra similar to those of the rare earths. Solutions of plutonium(III) salts are bright blue, those of plutonium(IV) salts are pale red (except when complex formation occurs), and those of plutonium(VI) are pink at high dilutions. Complex formation produces a change of color, and plutonium(IV) salts in nitric acid solution, for example, are deep green.

In accordance with the redox potentials (p. 620, and Table 103, Vol. I), plutonium(III) in aqueous solution is oxidized by atmospheric oxygen to plutonium(IV). At ordinary temperature, permanganate also brings about oxidation as far as plutonium(IV), but above 60° it oxidizes plutonium(III) or (IV) salts to plutonium(VI) salts (plutonyl salts). Uranium(IV) reduces plutonium(IV) ions to plutonium(III).

Plutonium(III) ions have a much smaller tendency to form complexes than plutonium(IV) ions, which very readily form a wide range of anionic complexes. For this reason, the oxidation potential of the Pu(IV) ion in (e.g.,) hydrochloric acid is smaller (by about 0.025 volt) than in perchloric acid. Plutonium ions, in any of their oxidation states, do not form complexes with perchlorate ions, at the concentrations ordinarily employed. Plutonium complexes with chloride ions only very weakly in the $+3$ state, but appreciably more strongly in the tetrapositive state. The free energy change associated with the conversion of Pu^{IV} to Pu^{III} is thereby diminished. Conversely, the free energy change for the conversion of Pu^{VI} to Pu^{IV} is greater in hydrochloric acid solution than in perchloric acid solution, since the greater tendency of Pu^{IV} to undergo complex formation favors the reaction. The difference between the oxidation potentials in hydrochloric acid and in perchloric acid is somewhat smaller (0.019 volt) in this case, however, since there is a smaller, but appreciable tendency for complex formation with chloride ions to occur in plutonium(VI) solutions also. The formation of complex ions in plutonium solutions is very clearly shown by changes in the absorption spectra, which can lead to very pronounced alterations in the color. Thus plutonium(IV) nitrate solutions are green, as a result of the formation of nitrato-complexes, instead of displaying the characteristic pale red color of the hydrated plutonium(IV) ion.

From solutions of plutonium(III) salts, *plutonium(III) hydroxide* is precipitated by ammonia as a dull bluish precipitate which readily undergoes oxidation in the air. From plutonium(IV) salt solutions, ammonia precipitates the pale green, slimy *plutonium(IV) hydroxide*, which is converted by ignition into dark brown, insoluble *plutonium dioxide*, PuO_2. The dioxide is also formed by ignition of the nitrate or iodate. Plutonium is only precipitated by alkali hydroxides from plutonium(VI) salt solutions (plutonyl salt solutions), as plutonate or polyplutonate, when it is present in relatively high concentration. The alkali plutonates differ from the corresponding uranium compounds in being much more soluble. Barium plutonate is much more sparingly soluble than the alkali plutonates. It has not been prepared in the pure state, but appears to be a polyplutonate—e.g., $BaPu_3O_{10}$.

(b) Oxides and Sulfides

(*i*) *Oxides*. The known oxides of plutonium are PuO, Pu_2O_3, Pu_4O_7, and PuO_2. The oxides PuO and PuO_2 are isostructural with the analogous oxides of uranium, neptunium, and americium. The monoxides of these metals have the NaCl structure, and the dioxides (as also thorium dioxide) have the fluorite structure. Pu_2O_3 is not isostructural with samarium sesquioxide (as might be expected from the supposed homology between plutonium and samarium). It has the hexagonal structure of the Type A rare earth oxides —e.g., La_2O_3—with $a = 3.840$, $c = 5.957$ Å. Pu_2O_3 forms mixed crystals with all the rare earth oxides crystallizing with this structure (Templeton, 1952).

Plutonium peroxide. Plutonium(IV) salt solutions turn blood red in color on the addition of hydrogen peroxide, by reason of the formation of peroxidic complex compounds. If the hydrogen peroxide concentration is sufficiently high, an intensely green precipitate is thrown down; neglecting its water content, this has a composition corresponding to the formula PuO_4. It contains plutonium in the tetrapositive state, and is isomorphous with the thorium compound of analogous composition (Koshland, 1945). It would appear that the same compound is precipitated from plutonium(VI) salt solutions, although much more slowly (Harvey, 1947).

(*ii*) *Sulfides*. Plutonium forms a *sesquisulfide*, Pu_2S_3, and an *oxysulfide*, Pu_2O_2S. The sesquisulfide is formed by heating $PuCl_3$ in a current of hydrogen sulfide. It crystallizes cubic, and is isostructural with La_2S_3, Ce_2S_3, Ac_2S_3, and Am_2S_3. The oxysulfide, obtained by heating PuO_2 to about $1250°$ in hydrogen sulfide, is isostructural with La_2O_2S and Ce_2O_2S. All these compounds have hexagonal structures which are related to the structures of the sesquisulfide (Zachariasen, 1949).

(c) Halides

(*i*) *Fluorides*. *Plutonium trifluoride*, PuF_3, is obtained as a violet precipitate when alkali fluorides are added to solutions of plutonium(III) salts. It undergoes oxidation, to form PuO_2 and PuF_4, when it is heated in oxygen to about $600°$. With sodium fluoride, the trifluoride forms the complex $NaPuF_4$. *Plutonium tetrafluoride* is formed as a flesh-colored precipitate having the composition $PuF_4 \cdot 2\frac{1}{2}H_2O$, by the action of fluoride ions on plutonium(IV) salt solutions. It is converted into yellow, anhydrous PuF_4 by gentle heating. PuF_4 undergoes decomposition when heated above $900°$ in a vacuum, forming the trifluoride according to the equation $PuF_4 \rightarrow PuF_3 + \frac{1}{2}F_2$. Some observations suggest that PuF_5 is formed as an intermediate stage in this reaction (Fried). No plutonium compound corresponding to UF_6 has yet been described.

Plutonium tetrafluoride was the first plutonium compound to be obtained in the carrier-free state. It was separated by Cunningham (1942) from uranyl nitrate which had been subjected to prolonged irradiation with neutrons from a beryllium cyclotron target, the neutrons being produced by the reaction:

$$\ce{^{9}_{4}Be + ^{2}_{1}d = ^{10}_{5}B + ^{1}_{0}n.}$$

Most of the unchanged uranium was first extracted, as uranyl nitrate, by shaking the solution with ether. A weak reducing agent was added, and neptunium and plutonium were then precipitated as fluorides, together with cerium and lanthanum fluorides as carriers. The sparingly soluble fluorides were converted to the soluble sulfates by fuming them down with concentrated sulfuric acid, and the aqueous solution of the sulfates was treated with bromate at ordinary temperature. Neptunium was thereby oxidized to $[NpO_2]^{++}$ ion, and remained in solution when the fluoride precipitation was repeated. The plutonium fluoride, with its rare earth fluoride carrier, was once more converted to the soluble sulfate, and the solution was treated with ammonia. The precipitated hydroxides were dissolved in nitric acid, and the plutonium was oxidized to $[PuO_2]^{++}$ by means of peroxysulfate and silver nitrate. Repetition of the fluoride precipitation now left the plutonium in solution, while the carriers were precipitated as fluorides. The solution was again fumed down with sulfuric acid, whereupon the addition of hydrofluoric acid gave a precipitate of pure plutonium(IV) fluoride. This procedure is worthy of note, as showing how a separation can be most simply effected between plutonium and the adjacent elements, by taking advantage of specific differences in their behavior towards oxidizing and reducing agents.

Plutonium tetrafluoride has a strong tendency to form double- or complex fluorides (fluoro salts), whereas the trifluoride has little tendency. Examples are $Na[PuF_5]$, $K[PuF_5]$ and $Rb[PuF_5]$ (all isotypic with $K[UF_5]$). Plutonium does not form a fluoro salt corresponding to $Cu[CeF_5]_2$ (Anderson, 1949). In addition to the pentafluoro salts, there are other compounds of the type $M^IPu_2F_9$.

PuF_4 forms anomalous mixed crystals with LaF_3. Hence Pu^{++++} ions can be co-precipitated with LaF_3 as well as with CeF_4. It has been shown (Schlyter and Sillen, 1950) by X-ray methods that CeF_4 and UF_4 (containing cations very similar in radius to the Pu^{4+} ion) form anomalous mixed crystals with LaF_3 also.

(ii) *Chlorides. Plutonium tetrachloride*, plutonium(IV) chloride, can only be obtained in solution. It is unstable in the solid state, although Pu^{III} in hydrochloric acid solution can very readily be oxidized to Pu^{IV}. When solutions of Pu^{IV} in hydrochloric acid are evaporated, decomposition occurs, resulting either in hydrolysis or—if hydrolysis is prevented—in the evolution of chlorine. Direct combination of plutonium with chlorine yields $PuCl_3$, which remains unchanged even when heated with liquid chlorine under pressure. Nevertheless, $PuCl_3$ volatilizes more readily in an atmosphere of chlorine than in a vacuum, and it may be inferred that although $PuCl_4$ is unstable in the solid state, it may exist at high temperatures (600–800°) in the gaseous state, in equilibrium with $PuCl_3$ and Cl_2.

Plutonium trichloride, $PuCl_3$, is formed not only by the direct combination of the elements, but also, e.g., by heating PuO_2 in a mixture of H_2 and HCl, or in a stream of CCl_4 or S_2Cl_2 vapor. The anhydrous compound is most simply prepared by evaporating a solution of plutonium(III) chloride to dryness, and dehydrating the resulting hydrate by heating it in HCl gas. Solutions of the trichloride are purple; the anhydrous compound is greenish blue. $PuCl_3$ combines with water vapor to form hydrates with 1, 3, and $6H_2O$. It reacts with NH_3 and with H_2S at 800–1000°, forming PuN and Pu_2S_3, respectively.

Plutonium oxychloride, PuOCl, is formed by heating $PuCl_3 \cdot 6H_2O$ in a sealed tube. It is greenish blue in color, like $PuCl_3$, from which it differs in being insoluble in water. It is dissolved by acids.

(iii) *Bromides. Plutonium tribromide*, $PuBr_3$, is obtained by methods similar to those used for $PuCl_3$, which it resembles in physical and chemical properties. It is hygroscopic, first forming the green hexahydrate $PuBr_3 \cdot 6H_2O$, and ultimately deliquesces to yield a purple solution. $PuBr_3 \cdot 6H_2O$ is isomorphous with $PuCl_3 \cdot 6H_2O$ and with $NdCl_3 \cdot 6H_2O$.

An *oxybromide*, PuOBr, corresponding to the oxychloride PuOCl, has also been obtained. As with $PuCl_4$, it is impossible to isolate $PuBr_4$ in the solid state, although plutonium(IV) bromide can be obtained in aqueous solution. Plutonium(IV) bromide is reduced by HBr, even in aqueous solution, to give plutonium(III) bromide.

$PuBr_3$ crystallizes in the orthorhombic system, and is isostructural with $NdCl_3$, $SmBr_3$, LaI_3, UI_3, and PuI_3. $PuCl_3$, however, forms hexagonal crystals ($a = 7.380$, $c = 4.238$ Å), isostructural with the following halides:

	UCl_3	$PrCl_3$	UBr_3	$PrBr_3$
a	7.428	7.41	7.926	7.92 Å
c	4.312	4.25	4.432	4.38 Å

Additional data on the plutonium trihalides are given in Table 81.

TABLE 81

MELTING POINTS, HEATS OF FUSION, EVAPORATION AND SUBLIMATION, AND
ENTROPIES OF FUSION OF PLUTONIUM TRIHALIDES

Compound	Melting point °C	Heat of fusion kcal/mol	Heat of evaporation kcal/mol	Heat of sublimation kcal/mol	Entropy of fusion cal/mol/°
PuF_3	1170	7.9	88.7	96.6	5.5
$PuCl_3$	760	15.2	57.6	72.8	14.7
$PuBr_3$	680	13.4	56.5	69.9	14.0

(d) Other Plutonium Salts

(i) *Plutonium(III) salts* include both the salts of oxyacids and also the trihalides already considered. Properties of a few salts of oxyacids are described below. The bright brown iodate, $Pu(IO_3)_3$, is sparingly soluble. Plutonium(III) trichloroacetate, $Pu(CCl_3 \cdot CO_2)_3$ (pale red), is also slightly soluble in water, but soluble in acetone. Plutonium(III) perchlorate is very soluble, and the violet plutonium(III) oxalate and the sulfate are fairly soluble. The latter gives rise to complex salts, many of which are only slightly soluble, as indicated by the following examples (figures are solubilities in g of Pu per liter of solution):

$KPu(SO_4)_2 \cdot 5H_2O$ $K_5Pu(SO_4)_4$ $RbPu(SO_4)_2 \cdot 4H_2O$ $CsPu(SO_4)_2 \cdot 4H_2O$

(lavender)   (almost colorless)   (bright lavender)   (bright lavender)

4.8   0.54   0.50   0.34

$NH_4Pu(SO_4)_2 \cdot 4H_2O$

(light blue)

1.30

(ii) *Plutonium(IV) Salts.* The *nitrate* is noteworthy among the plutonium(IV) salts, because of its marked tendency to form nitrato complexes, since such complex formation is, in general, rather uncommon. According to Hindman (1944), the ions present in aqueous solutions of plutonium(IV) nitrate include the complex anion $[Pu(NO_3)_6]^=$ and the cation $[Pu(NO_3)(H_2O)_x]^{+++}$. The hexanitrato salt $(NH_4)_2[Pu(NO_3)_6]$ is isomorphous with $(NH_4)_2[Ce(NO_3)_6]$ and $(NH_4)_2[Th(NO_3)_6]$. *Plutonium(IV) chloride* is present in the form of the complex ion $[PuCl_6]^=$ in plutonium(IV) solutions containing a sufficiently high concentration of Cl^- ions. According to Hindman, the ion $[PuCl_3]^+$ is also present in solution. In very dilute solutions, there is a weaker tendency to form complexes with the Cl^- ion than with the NO_3^- ion. Bright yellow hexachloro salts, such as $Cs_2[PuCl_6]$ and $[(CH_3)_4N]_2[PuCl_6]$, can be crystallized from not too dilute solutions.

Plutonium(IV) sulfate crystallizes from aqueous solution as the coral red to red brown tetrahydrate, $Pu(SO_4)_2 \cdot 4H_2O$. The sulfates of uranium(IV), cerium(IV), thorium, and zirconium form similar tetrahydrates. Plutonium(IV) sulfate forms fairly sparingly soluble double salts with the alkali sulfates—$M^I_4[Pu(SO_4)_4] \cdot nH_2O$. These are green, and correspond in composition to the alkali double sulfates formed by uranium(IV), cerium(IV), thorium, and zirconium. Hydrolysis of the neutral sulfate yields the light grey-green $Pu_2O(SO_4)_3 \cdot 8H_2O$—again paralleling the behavior of uranium, thorium, and zirconium, which form corresponding basic salts. Gelatinous, almost colorless precipitates are obtained when phosphate ions are added to solutions of plutonium(IV) salts; these phosphates vary in composition, depending on the conditions of precipitation. Smith (1944) was able to isolate the light brown compound $Pu_3(PO_4)_4 \cdot xH_2O$ in the crystalline state; this was found to be isotypic with the phosphates of cerium(IV) and thorium. Like almost all plutonium(IV) salts, the phosphate can form complex ions whereby its solubility may be very considerably increased.

(iii) *Plutonyl salts,* i.e., compounds of the general formula $[PuO_2]X_2$, are far less stable than the uranyl salts, but in general resemble the latter in their properties. Soluble plutonyl salts are obtained from solutions of plutonium(IV) salts by the action of strong oxidants such as cerium(IV) salts, dichromate, warm permanganate, or electrolytic oxidation. They can be isolated, and crystallize well, by the evaporation of the solutions. The sparingly soluble plutonyl salts are formed from the soluble salts by double decompositions. Examples of plutonyl salts are the acetate, $[PuO_2](C_2H_3O_2)_2 \cdot 2H_2O$ and its double salt with sodium acetate, already referred to, $Na[PuO_2](C_2H_3O_2)_3$, the readily soluble nitrate, $[PuO_2](NO_3)_2 \cdot 6H_2O$, and the colorless, gelatinous fluoride, $[PuO_2]F_2 \cdot xH_2O$. The last is slightly soluble in water, but dissolves in hydrofluoric acid, with the formation of complex ions. Addition of alkali fluoride to the pink solution so obtained yields pink double- or complex fluorides. Plutonyl hexacyanoferrate(III), $[PuO_2]_3[Fe(CN)_6]_2$, is obtained as a red-brown precipitate when $[Fe(CN)_6]^{3-}$ ions are added to a solution of plutonyl nitrate. The $[PuO_2]^{++}$ ion is reduced by $[Fe(CN)_6]^{4-}$ ions, with the deposition of a black precipitate.

5. Americium (Am)

In terms of Seaborg's actinide theory, the place occupied by americium should make it the homologue of europium. Like europium, it is most stable in the tri-positive state, but can also function as dipositive. Unlike europium, however, americium also forms compounds in which it exhibits valence states of +4, +5, and +6 like the preceding transuranic elements. However, the +4 state has not been detected in solution.

Americium differs from uranium, neptunium, and plutonium in that it can exist in aqueous solution as a bivalent positive ion. Dipositive americium is formed by the action of strong reducing agents, such as sodium amalgam, on ameri-cium(III) compounds. $AmSO_4$ can be coprecipitated from solution with $EuSO_4$.

The special stability of the +3 state is well shown by the halides of americium. Americium forms the trihalides under just the same conditions as furnish the tetrahalides of neptunium. For example, americium trifluoride, AmF_3, is formed when AmO_2 is heated in a mixture of HF and O_2, and the action of CCl_4 vapor on AmO_2 gives americium trichloride, $AmCl_3$. The tribromide, $AmBr_3$, is formed by heating a mixture of AmO_2 and Al in bromine vapor (cf. $NpBr_4$), and the triiodide, AmI_3, is formed similarly. It has not proved possible to prepare the tetrahalides of americium by other methods. Thus Fried (1951) found that no tetrafluoride could be made by heating AmF_3 in elementary fluoride to 500–700°.

The americium halides are isostructural with the trihalides of neptunium and plutonium. AmF_3 is light red, $AmCl_3$ and $AmBr_3$ are colorless, and AmI_3 is yellow. $AmCl_3$ sublimes at 850°, $AmBr_3$ between 850 and 900°, and AmI_3 at about 900°.

Americium is usually tripositive in other salts, as well as in the halides. In solution, the americium(III) salts are red. Westrum and Eyring (1951) found the heat of formation of the Am^{+++} ion from the reaction:

$$Am + 3HCl + aq. = Am^{+++} + 3Cl^- + \tfrac{3}{2}H_2, + aq.$$

to be 160 kcal per mol (for 1.5 normal hydrochloric acid, at 25 °C); corresponding figures for the heats of formation of some lanthanide ions are: La^{+++} 167 kcal; for Pr^{+++} 166 kcal per mol. The heat of formation of the Am^{++++} ion works out at about 112 kcal per g-ion.

Americium does not favor the +3 state in combination with oxygen. Although ammonia or sodium hydroxide precipitates the red, gelatinous hydroxide, $Am(OH)_3$, from americium(III) salt solutions, the only anhydrous oxides are the monoxide AmO and the dioxide AmO_2. The monoxide has the sodium chloride, structure ($a = 4.95$ Å), and is thus isostructural with UO, NpO, and PuO. The dioxide, which has the fluorite structure (like UO_2, NpO_2, and PuO_2) is formed when americium(III) hydroxide or nitrate is heated in air. It loses no oxygen when it is heated to $1000°$.

If the black dioxide is treated with dilute hydrochloric acid, gas is evolved, the solid swells, and is converted to a light red product which has not been identified (possibly AmOCl). This gradually dissolves, to give a solution with the red color characteristic of americium(III) compounds.

$Am(OH)_3$ is soluble in concentrated K_2CO_3. If NaOCl is added to the solution, a dark precipitate is formed on warming; this is probably americium(IV)- or americium(V) hydroxide. It is converted into AmO_2 when it is ignited.

Fried obtained the sulfide Am_2S_3 (isostructural with La_2S_3) by heating AmO_2 to $1400-1500°$ in a mixture of H_2S and CS_2.

Like uranium, neptunium and plutonium, americium can also form compounds from the $+5$ and $+6$ valence states. However, the americium(V)- and americium(VI) compounds are less stable than the corresponding compounds of plutonium. The relationship between americium and uranium is therefore less pronounced than is the case with the two preceding transuranic elements. Except for their lower stability, the compounds of $+5$ and $+6$ americium are very similar in properties to the corresponding compounds of uranium.

Americium(V) compounds can be obtained from americium(III) compounds by electrolytic oxidation. They are very unstable, but do not show such a strong tendency to undergo disproportionation as do the plutonium(V) compounds, since the americium(VI) compounds are also rather unstable. However, Asprey, Stephanou and Penneman (1951) found that when a solution of americium(V) in 0.3-normal $HClO_4$ was made 4-normal with respect to H_2SO_4, disproportionation occurred ($3Am^V \rightarrow Am^{III} + 2Am^{VI}$), as shown by the change in absorption spectrum. Americium(III) salt solutions are characterized by absorption bands at 504 mμ and 811 mμ; americium(V) salts absorb at 514 mμ and 714 mμ; americium(VI) salt solutions have an intense narrow absorption band at $991-992$ mμ, and also absorb strongly in the ultraviolet.

Asprey obtained pure americium(VI) salt solutions by oxidation of Am^{III}, in hydrochloric acid or perchloric acid solution, with ammonium peroxydisulfate. The resulting solutions are pure yellow, and give no precipitate with F^- ions. Addition of sodium acetate yields tetrahedral crystals of sodium americyl acetate, $Na[AmO_2](C_2H_3O_2)_3$; this is cubic ($a = 10.6$ Å), and is isomorphous with the corresponding uranyl, neptunyl, and plutonyl double acetates. Americyl nitrate can be extracted from aqueous solution by means of ether, as can uranyl acetate.

Metallic americium was first isolated by Westrum and Eyring (1951), by reduction of AmF_3 with metallic barium at $1100°$. It was prepared in quantities of 0.04 to 0.2 mg, in small beryllia crucibles in a vacuum microfurnace. Americium is a silver-white ductile metal of density 11.7. It combines with hydrogen when gently warmed, forming *americium hydride*.

It is possible to separate americium from the rare earths which are used as carriers during its preparation, by fractional precipitation of the fluorides, since AmF_3 is rather more soluble than the fluorides of the rare earth metals. A more convenient method of separation is based on the use of ion exchange resins. A hydrochloric acid solution containing americium and the rare earths is passed slowly down a column packed with a suitable cation-exchange resin, whereby the Am^{+++} and rare earth ions are taken up at the top of the column. The column is then eluted with a weakly acid ammonium citrate solution (cf. p. 495); the first ions to be eluted are those which form the most stable citrate complexes. The first fractions leaving the column contain lanthanum and cerium. Americium is eluted next, followed by the other rare earths.

6. Curium (Cm), Berkelium (Bk), and Californium (Cf)

The transuranic elements $_{96}Cm$, $_{97}Bk$, and $_{98}Cf$ are formally the homologues of $_{64}Gd$, $_{65}Tb$, and $_{66}Dy$, and on present evidence show a chemical resemblance to these typical lanthanide elements. The separation of these elements from one another, and from the rare earth elements, is a matter of some difficulty. It must be effected by methods similar to those used for the separation of the rare earth elements, and in particular, by use of ion exchange columns. Work with curium, berkelium, and californium is further complicated by their very strong radio-activity. This leads to a very vigorous decomposition of their aqueous solutions, with the evolution of hydrogen and oxygen, and formation of considerable quantities of hydrogen peroxide. It also limits severely the quantities of the elements that can conveniently be studied.

Curium is tripositive in all the compounds as yet prepared. In this it resembles the homologous gadolinium (but see p. 612). Solutions of curium salts are colorless and absorb only in the ultraviolet, as do gadolinium salts. From this, and from the magnetic behavior, it has been concluded that the Cm^{3+} ion has a configuration corresponding to that of the Gd^{3+} ion, i.e., $5f^7$, with the $5f$ levels just half-filled.

Addition of ammonia to solutions of curium salts precipitates the hydroxide $Cm(OH)_3$. Hydrofluoric acid similarly precipitates the insoluble fluoride, CmF_3. The sesquioxide, Cm_2O_3, is formed when the hydroxide is ignited in air. Unlike americium, no dioxide is formed, under these conditions at least. Metallic curium has been prepared by Crane.

It has been reported, on the evidence of tracer experiments only, that *berkelium*, like its formal homologue terbium, is normally tripositive but can be converted to the $+4$ state.

Californium, on the evidence of tracer experiments, exists in the $+3$ state only. It closely follows rare earth carriers, and therefore gives the reactions typical of the rare earths.

7. Einsteinum (E), Fermium (Fm), and Mendelevium (Mv)

(a) Preparation

From the discussion of nuclear stability and nuclear binding energies given on p. 591 it follows that the increasing probability of spontaneous fission sets a natural limit to the list of elements, but that all the elements of the $5f$ series may be capable of existence. Their more stable isotopes are likely to be those of high mass number*. Extension of the Periodic Table beyond californium thus involves nuclear reactions whereby a large number of neutrons can be introduced into the nuclei successively formed. Two methods of doing so are possible. (i) By a succession of (n, γ) reactions and β-decay processes. The occurrence of x consecutive neutron

* Thus Seaborg has predicted that the most stable isotope of element 102, with respect to a-, β- or K-decay processes should be $^{266}102$ the lighter isotopes having much higher disintegration rates. In this case, the maximum stability towards spontaneous fission should be found for $^{260}102$, with a-half-life and spontaneous fission half life both about 1 hr. See G. T. Seaborg *et al.*, *J. Inorg. and Nuclear Chem.*, 1 (1955) 3.

capture processes depends upon the cross sections at each stage, and upon the xth power of the neutron concentration; it can therefore be realised only in the most intense neutron flux. (ii) Corresponding to (a, n) or (a, p) reactions, the highly accelerated nuclei of carbon, nitrogen etc. (i.e. $^{12}_{6}C^{6+}$, $^{14}_{7}N^{7+}$ etc.) can be used to bombard heavy elements in the cyclotron. Complex nuclear reactions can be thereby brought about, whereby large changes in nuclear mass and charge are produced. E.g.

$$^{238}_{92}U + {}^{14}_{7}N \quad \diagup\diagdown \quad \begin{array}{l} ^{243}_{97}Bk + a + 5n \\ ^{247}_{99}E + 5n \end{array} \qquad \text{etc.}$$

Isotopes of the elements $_{99}E$, $_{100}Fm$ and $_{101}Mv$ have now been prepared by both these methods. Thus in the very intense neutron flux of the USAEC Materials Testing Reactor, the californium isotope ^{252}Cf is built up from ^{238}U or ^{239}Pu, and undergoes the further sequence of reactions:

$$^{252}Cf\ (n, \gamma) \to {}^{253}Cf\ \beta\ (20\ d) \to {}^{253}E\ (n, \gamma) \to {}^{254}E\ \beta\ (36\ hr) \to {}^{254}Fm\ a\ (3.2\ hr) \to$$
$$^{50}Cf\ a\ (12\ y) \to {}^{246}Cm\ \text{etc. (cf. Table 82)}.$$

(b) Chemical Properties

Chemical identification of elements 99, 100 and 101 turns upon their conformity with the actinide hypothesis, and the analogy that can therefore be drawn between their reactions and those of the corresponding lanthanide elements. Owing to their short half lives and their mode of formation, it has hitherto been possible to obtain only very few atoms at any one time. Their chemical properties have therefore been investigated on 'weightless' samples, at extreme dilution.

Coprecipitation reactions indicate that all three elements exist in solution as trivalent ions. In accordance with the trend of oxidation potentials in the actinide series, there is no evidence that Cf, E, Fm or (presumably) Mv can be raised to any higher valence state in aqueous solution. The fluorides EF_3, FmF_3 and MvF_3 are coprecipitated with LaF_3 under conditions that afford quantitative precipitation of the heavier rare earth elements.

On cation exchange columns, using ammonium citrate or lactate as eluant, the elements are sharply separated from each other and from the preceding actinides, in the sequence of decreasing atomic number. Differences in the stability of coordination complexes of adjacent elements diminishes with rise in the atomic number (just as in the lanthanide series), so that separations are less effective using anion exchange resins, or using concentrated hydrochloric acid as eluant on cation exchange resins.

8. Radioactive Disintegration Series of the Transuranic Elements

(a) Characteristics of the Decay Series

As has been discussed, the heaviest of the naturally occurring unstable elements each give rise to long radioactive disintegration series. The same is true of the transuranic elements, and all the disintegration series starting from the transuranic elements merge into those of the natural radioelements. After a certain number of transformations, it is found in every case that the decay products of the transuranic elements consist of nuclear species which are found also in the natural radioactive disintegration series. The subsequent course of their decay therefore follows the same course as that of the natural radioelements.

Radioactive disintegration of the heavy elements is always attended with either a decrease in mass number of 4 units (a-decay), or with no change in mass number (β-decay, K electron capture). Hence the mass numbers of all the members of any one series can be represented by a common formula. The isotope of thorium which heads the thorium decay series has the mass number 232. This is a multiple of 4, so that all the subsequent members of this decay series must have mass numbers which are whole number multiples of 4—i.e., all members of the thorium series have mass numbers of the form 4n, where n is a whole number. Similarly, the elements of the uranium-radium series have mass numbers of the form 4n + 2, and those of the actinium series have the form 4n + 3.

(b) Neptunium Decay Series

There is also a disintegration series characterized by mass numbers of the form 4n + 1. This is also known as the *neptunium series*, after its longest-lived member, the neptunium isotope ^{237}Np. Since this does occur in Nature (although in very minute amounts), the neptunium decay series is also one of the natural decay series. It consists entirely of nuclear species which do not occur in the other, more abundant, decay series which have been identified from the study of nuclear reactions. The stable end product of the neptunium disintegration series is *bismuth*. In this, the neptunium series differs notably from the three other natural decay series, in each of which the stable end product is an isotope of *lead*.

(c) Decay Series of the Other Transuranic Elements

Table 82 shows how the various nuclear species of the transuranic elements fit into the natural disintegration series. Disintegration of transuranic nuclides with mass numbers of the form $M = 4n$ invariably leads, eventually, to nuclear species belonging to the thorium decay series (4n series). Similarly, transuranic nuclides with $M = 4n + 1$ finally yield members of the neptunium decay series (4n + 1 series); those with $M = 4n + 2$ terminate in the uranium decay series (4n + 2 series), and those with $M = 4n + 3$ terminate in the actinium series (4n + 3 series).

Among the decay products of ^{232}Pu, a member of the 4n series, there is a radioactive inert gas ^{216}Em, isotopic with radon and thorium emanation. No emanation is found in the direct line of the neptunium series, although such a daughter product is formed by *branching*, as a decay product of ^{229}U. This uranium isotope does not occur in Nature, and can be obtained only by nuclear transformations. Among the decay products of ^{234}Pu there is also an emanation (^{218}Em); with this nuclide the disintegration series commencing with ^{234}Pu merges into the natural uranium-radium series. There is thus a second emanation in the uranium-radium series. Merging into the uranium-radium decay series, in addition to the disintegration series beginning with the transuranic elements, is yet another decay chain consisting of the daughter products of the decay of an artificially produced isotope of protactinium

$$^{226}_{91}\text{Pa} \xrightarrow[\text{1.5 min}]{a} {}^{222}_{89}\text{Ac} \xrightarrow[\text{short}]{a} {}^{218}_{87}\text{Fr} \xrightarrow[\text{v. short}]{a} {}^{214}_{85}\text{At} \xrightarrow[\text{v. short}]{a} {}^{210}_{83}\text{RaE}$$

RaF, formed in the natural decay series from RaE, is also produced by the decay of an artificially formed isotope of astatine,

$$^{210}_{85}\text{At} \xrightarrow[\text{8.3 hrs}]{K \text{ capture}} {}^{210}_{84}\text{RaF.}$$

TABLE 82

RADIOACTIVE DISINTEGRATION SERIES OF THE TRANSURANIC ELEMENTS

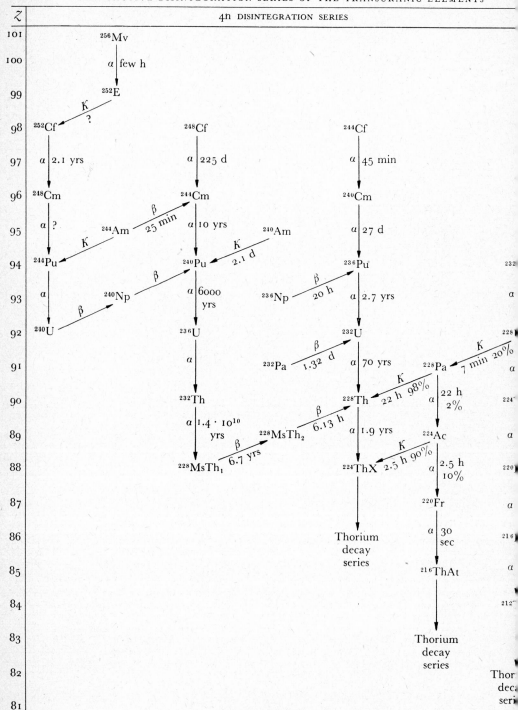

| z | 4n DISINTEGRATION SERIES |

TABLE 82 (continued)

RADIOACTIVE DISINTEGRATION SERIES OF THE TRANSURANIC ELEMENTS

4n + 1 OR NEPTUNIUM DISINTEGRATION SERIES	Z

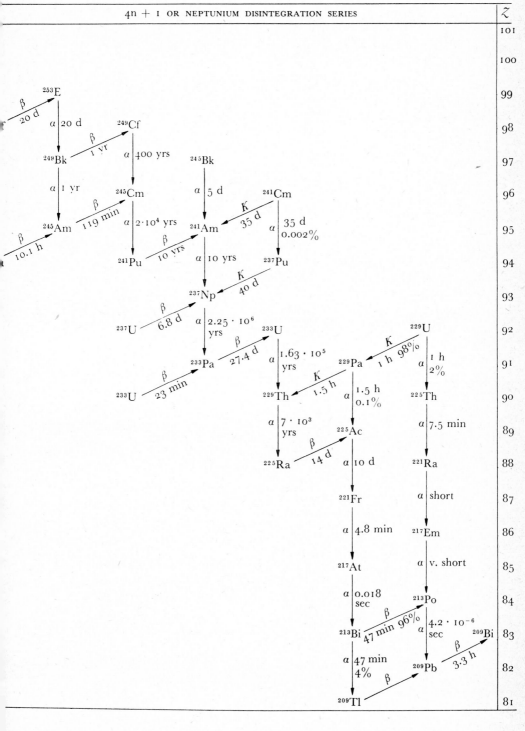

	101
	100
	99
	98
	97
	96
	95
	94
	93
	92
	91
	90
	89
	88
	87
	86
	85
	84
	83
	82
	81

TABLE 82 (continued)

RADIOACTIVE DISINTEGRATION SERIES OF THE TRANSURANIC ELEMENTS

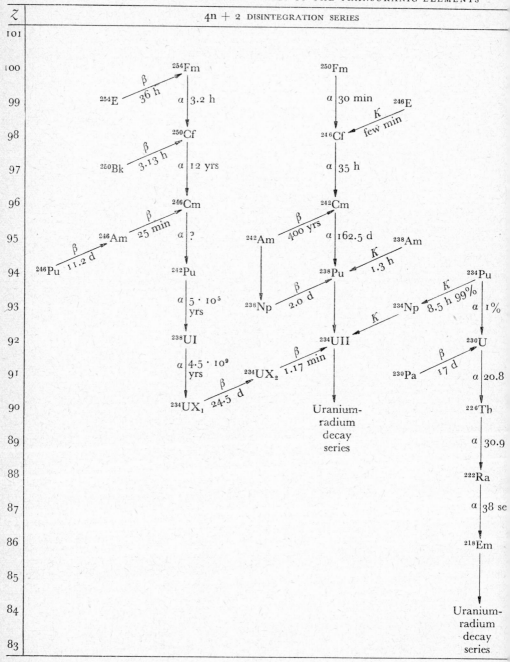

TABLE 82 (continued)

RADIOACTIVE DISINTEGRATION SERIES OF THE TRANSURANIC ELEMENTS

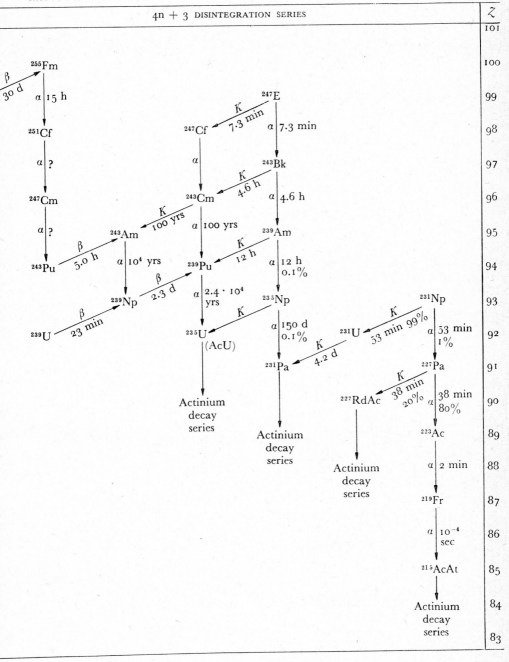

In this connection it may be noticed that RaE does not disintegrate to give RaF (polonium) as exclusive product, but that a very small proportion of disintegrations (0.00005%) is transformed by way of a short-lived thallium isotope into RaG (uranium lead).

$$^{210}_{83}\text{RaE} \quad \xrightarrow[\text{5 d.}]{\alpha} \quad ^{206}_{81}\text{Tl} \quad \xrightarrow[\text{4.2 min}]{\beta} \quad ^{206}_{82}\text{RaG(Pb)}.$$

References

1 G. L. SEABORG, J. J. KATZ and W. M. MANNING (Editors), *The Transuranium Elements*, Research Part 1 and 2 (National Nuclear Energy Series, Manhattan Project Technical Section, Division IV, Plutonium Project Record, Vol. 14B), New York 1949, 859 pp.
2 O. HAHN, *New Atoms; Progress and Some Memories* [edited by W. GAADE], Amsterdam 1950, 184 pp.

CHAPTER 15

DISTRIBUTION OF THE ELEMENTS; GEOCHEMISTRY

1. Composition of the Earth's Crust

There are very wide differences between the amounts of the various chemical elements present in the earth's crust, [1–5] as will already have become apparent from the discussions of the elements in the foregoing chapters. Table 84 shows the proportions in which the elements are present (by weight) in the earth's crust. The 'earth's crust' is taken as including the rocky crust of the earth down to a depth of 16 km below the surface, together with the oceanic and inland masses of water, and the enveloping air. The latter is the *atmosphere*, the sum total of the water masses make up the *hydrosphere*, and the rocky crust is the *lithosphere*.

The lithosphere comprises 93.06% by weight of the earth's crust (down to 16 km), the hydrosphere 6.91%, and the atmosphere 0.03%.

The rocks of the lithosphere are of varied composition, but according to F. W. Clarke, a representative petrological analysis would be 95% igneous rocks, 4.0% schists, 0.75% sandstones, and 0.25% limestones.*

The composition of the hydrosphere is determined chiefly by the composition of the oceans. [6] V. M. Goldschmidt (1924) gives the figures, in weight per cent, collected in Table 83.

TABLE 83
COMPOSITION OF SEA WATER

Oxygen	85.89%	Nitrogen	$1.7 \cdot 10^{-5}\%$	Selenium	$1 \cdot 10^{-7}\%$
Hydrogen	10.82%	Rubidium	$2 \cdot 10^{-5}\%$	Thorium	$5 \cdot 10^{-8}\%$
Chlorine	1.898%	Lithium	$1.2 \cdot 10^{-5}\%$	Molybdenum	$5 \cdot 10^{-8}\%$
Sodium	1.056%	Aluminum	$1 \cdot 10^{-5}\%$	Cerium	$4 \cdot 10^{-8}\%$
Magnesium	0.127%	Phosphorus	$5 \cdot 10^{-6}\%$	Silver	$3 \cdot 10^{-8}\%$
Sulfur	0.088%	Iodine	$5 \cdot 10^{-6}\%$	Vanadium	$3 \cdot 10^{-8}\%$
Calcium	0.040%	Arsenic	$1.5 \cdot 10^{-6}\%$	Lanthanum	$3 \cdot 10^{-8}\%$
Potassium	0.038%	Barium	$1 \cdot 10^{-6}\%$	Yttrium	$3 \cdot 10^{-8}\%$
Bromine	$6.5 \cdot 10^{-3}\%$	Zinc	$5 \cdot 10^{-7}\%$	Copper	$2 \cdot 10^{-8}\%$
Carbon	$2.8 \cdot 10^{-3}\%$	Manganese	$5 \cdot 10^{-7}\%$	Nickel	$1 \cdot 10^{-8}\%$
Strontium	$1.3 \cdot 10^{-3}\%$	Lead	$4 \cdot 10^{-7}\%$	Scandium	$4 \cdot 10^{-9}\%$
Boron	$4.8 \cdot 10^{-4}\%$	Iron	$2 \cdot 10^{-7}\%$	Mercury	$3 \cdot 10^{-9}\%$
Silicon	$2.0 \cdot 10^{-4}\%$	Cesium	$2 \cdot 10^{-7}\%$	Gold	$4 \cdot 10^{-10}\%$
Fluorine	$1.4 \cdot 10^{-4}\%$	Uranium	$1.5 \cdot 10^{-7}\%$	Radium	$7 \cdot 10^{-15}\%$

* There is inevitably some arbitrariness in deciding what constitutes a representative rock sample of the earth's crust. Furthermore, analyses for many of the minor elements in silicate rocks are relatively few in number, and—because of analytical difficulties—not all equally reliable. Differences in assessing such factors lead to discrepancies between tables of abundance as drawn up by different authorities.

TABLE 84

RELATIVE ABUNDANCE OF THE CHEMICAL ELEMENTS

[Composition of the earth's crust, including hydrosphere and lithosphere, in weight per cent.]

1.	Oxygen	48.6%	47.	Arsenic	$5 \cdot 10^{-4}\%$
2.	Silicon	26.3%	48.	Hafnium	$4.5 \cdot 10^{-4}\%$
3.	Aluminum	7.73%	49.	Dysprosium	$4.5 \cdot 10^{-4}\%$
4.	Iron	4.75%	50.	Uranium	$4 \cdot 10^{-4}\%$
5.	Calcium	3.45%	51.	Argon	$3.6 \cdot 10^{-4}\%$
6.	Sodium	2.74%	52.	Cesium	$3.2 \cdot 10^{-4}\%$
7.	Potassium	2.47%	53.	Ytterbium	$2.7 \cdot 10^{-4}\%$
8.	Magnesium	2.00%	54.	Erbium	$2.5 \cdot 10^{-4}\%$
9.	Hydrogen	0.76%	55.	Bromine	$2.5 \cdot 10^{-4}\%$
10.	Titanium	0.42%	56.	Tantalum	$2 \cdot 10^{-4}\%$
11.	Chlorine	0.14%	57.	Holmium	$1.1 \cdot 10^{-4}\%$
12.	Phosphorus	0.11%	58.	Europium	$1 \cdot 10^{-4}\%$
13.	Carbon	0.087%	59.	Antimony	$1 \cdot 10^{-4}\%$
14.	Manganese	0.085%	60.	Terbium	$9 \cdot 10^{-5}\%$
15.	Fluorine	0.072%	61.	Lutetium	$7.5 \cdot 10^{-5}\%$
16.	Sulfur	0.048%	62.	Mercury	$5 \cdot 10^{-5}\%$
17.	Barium	0.040%	63.	Thallium	$3 \cdot 10^{-5}\%$
18.	Nitrogen	0.030%	64.	Iodine	$3 \cdot 10^{-5}\%$
19.	Rubidium	0.028%	65.	Thulium	$2 \cdot 10^{-5}\%$
20.	Zirconium	0.020%	66.	Bismuth	$2 \cdot 10^{-5}\%$
21.	Chromium	0.018%	67.	Cadmium	$1.8 \cdot 10^{-5}\%$
22.	Strontium	0.015%	68.	Indium	$1 \cdot 10^{-5}\%$
23.	Vanadium	0.015%	69.	Silver	$1 \cdot 10^{-5}\%$
24.	Nickel	0.010%	70.	Selenium	$9 \cdot 10^{-6}\%$
25.	Zinc	0.008%	71.	Palladium	$1 \cdot 10^{-6}\%$
26.	Copper	0.007%	72.	Platinum	$5 \cdot 10^{-7}\%$
27.	Lithium	0.0065%	73.	Neon	$5 \cdot 10^{-7}\%$
28.	Cerium	0.004%	74.	Gold	$5 \cdot 10^{-7}\%$
29.	Tin	0.004%	75.	Helium	$3 \cdot 10^{-7}\%$
30.	Cobalt	0.004%	76.	Tellurium	$2 \cdot 10^{-7}\%$
31.	Yttrium	0.0028%	77.	Iridium	$1 \cdot 10^{-7}\%$
32.	Neodymium	0.0024%	78.	Rhodium	$1 \cdot 10^{-7}\%$
33.	Niobium	0.002%	79.	Ruthenium	$1 \cdot 10^{-7}\%$
34.	Lanthanum	0.0018%	80.	Osmium	$1 \cdot 10^{-7}\%$?
35.	Lead	0.0016%	81.	Rhenium	$1 \cdot 10^{-7}\%$?
36.	Thorium	0.0015%	82.	Krypton	$2 \cdot 10^{-8}\%$
37.	Gallium	0.0015%	83.	Xenon	$2.4 \cdot 10^{-9}\%$
38.	Boron	0.001%	84.	Radium	$1.3 \cdot 10^{-100}\%$
39.	Tungsten	0.001% ?	85.	Protactinium	$8 \cdot 10^{-110}\%$
40.	Molybdenum	$7.5 \cdot 10^{-4}\%$	86.	Actinium	$3 \cdot 10^{-140}\%$
41.	Germanium	$7 \cdot 10^{-4}\%$	87.	Polonium	$3 \cdot 10^{-140}\%$
42.	Samarium	$6.5 \cdot 10^{-4}\%$	88.	Radon	$6 \cdot 10^{-160}\%$
43.	Gadolinium	$6.5 \cdot 10^{-4}\%$	89.	Neptunium	$4 \cdot 10^{-170}\%$
44.	Beryllium	$6 \cdot 10^{-4}\%$	90.	Plutonium	$2 \cdot 10^{-190}\%$
45.	Praseodymium	$5.5 \cdot 10^{-4}\%$	91.	Francium	$7 \cdot 10^{-230}\%$
46.	Scandium	$5 \cdot 10^{-4}\%$	92.	Astatine	$4 \cdot 10^{-230}\%$

Elements 1–24 bracketed: 99.47%
Elements 13–39 bracketed: 0.47%
Elements 25–39 bracketed: 0.05%

At the earth's surface, the atmosphere has an average water vapor content of 0.27%*, and contains 75.31% nitrogen, 22.95% oxygen, 1.43% inert gases and 0.03 to 0.04% carbon

* All percentages are by weight.

dioxide. The water vapor content can vary over wide limits, according to the temperature and other conditions (at 0° the atmosphere would be saturated with 0.48% by weight of water vapor, whereas at 40° it could take up 5.82%), but the composition of the atmosphere is remarkably uniform with respect to the main constituents. This is primarily due to the effects of atmospheric turbulence. The composition should, in theory, change with altitude. The carbon dioxide content and water vapor content do, indeed, fall off with height, and above 12 km the water content has fallen practically to zero. It has been found that the temperature of the atmosphere falls off with altitude, up to a height of about 12 km, at which it is —54°C, and then remains uniform at greater heights. Whereas the temperature gradients in the lower atmosphere or *troposphere* cause efficient mixing, it was thought that the uniform temperature of the *stratosphere* indicated the absence of turbulence. Under such conditions, the partial pressures of the atmospheric gases should fall off inversely as their densities, and at 100 km it would be expected that helium would become a major constituent. This expectation has not been realized. Analyses of air samples from altitudes up to 70 km do not differ much from those of air at sea level, although the relative proportions of the inert gases begin to change in the expected direction in samples of air collected above 60 km.*

The density of the air diminishes rapidly with increasing altitude, apart from any effects in changes of composition.** The weight of the atmosphere above 100 km is therefore vanishingly small compared with the total weight. The presence of a tenuous atmosphere at altitudes greater than 200 km is evidenced by the luminosity of meteorites, heated by friction as they enter the atmosphere at such altitudes, and also by the occurrence of ionized oxygen and nitrogen (ionized by absorption of very short wavelength ultraviolet radiation from the sun), which are of importance for the transmission of radio waves.

It is evident from Table 84 that by far the greatest proportion by weight of the earth's crust is made up of a small number of elements. Almost half the earth's crust, by weight, is oxygen, and over a quarter is silicon. The 12 most abundant elements together make up 99.5% by weight of the whole, i.e., the remaining 80 elements amount to only 5 parts per thousand by weight. When the next fourteen elements are allowed for, the rest constitute only 5 parts in ten thousand. The 53 elements following tungsten in Table 84—i.e., 58% of all the elements occurring in Nature—amount to only one ten thousandth of the weight of the earth's crust.

It is also noteworthy that certain elements which are commonly considered to be rare, are actually relatively abundant. This is especially true of titanium, and also of zirconium and vanadium. Even the elements of the rare earths actually make up a larger fraction of the earth's crust than do many of the heavy metals which are generally regarded as 'common'—e.g., bismuth, mercury, cadmium and antimony. Titanium is, indeed, one of the most widely distributed of all the elements, roughly equalling (by weight) the total of the 80 elements following phosphorus in the table.

It is instructive to examine the relative abundance in terms of relative numbers of *atoms* of each element. (Table 85). Oxygen and silicon are still the most abundant, but hydrogen then occupies the third place. Carbon and nitrogen also move up in the sequence, while iron drops back to seventh place, below sodium and calcium. It is noteworthy that titanium retains its importance in this mode of presenting the data.

Table 85 gives the 17 most abundant elements, with their proportions in the earth's crust expressed in atom-per cent. Not only are these the most abundant elements, but (except for titanium) they are also the most important for human life. The table lists all the

* See Paneth, *J. Chem. Soc.* (1952) 3651, Glueckauf, *Compendium of Meteorology*, American Meteorological Society (1951).

** The total atmospheric pressure should be approx. 0.001 mm at 100 km, if no change in composition takes place.

TABLE 85

RELATIVE ABUNDANCE OF THE MOST WIDE-SPREAD ELEMENTS

Element	Atoms of element per 100 atoms of silicon	Proportion in earth's crust, in atom-%
1. Oxygen	296	53.8 atom-%
2. Silicon	100	18.2 atom-%
3. Hydrogen	74	13.5 atom-%
4. Aluminum	30.5	5.55 atom-%
5. Sodium	12.4	2.26 atom-%
6. Calcium	9.2	1.67 atom-%
7. Iron	9.1	1.64 atom-%
8. Magnesium	8.8	1.60 atom-%
9. Potassium	4.4	0.80 atom-%
10. Titanium	0.9	0.16 atom-%
11. Phosphorus	0.38	0.07 atom-%
12. Chlorine	0.30	0.054 atom-%
13. Carbon	0.27	0.048 atom-%
14. Manganese	0.18	0.032 atom-%
15. Nitrogen	0.16	0.027 atom-%
16. Sulfur	0.16	0.027 atom-%
17. Fluorine	0.14	0.025 atom-%

elements which are essential for living organisms—C, O, H, N, K, Ca, Mg, Fe, P and S—without which vegetable or animal life could not exist.*

The 17 elements listed in Table 85 make up 99.9 atom percent of the earth's crust.

2. Distribution and Accessibility of the Elements

Many elements such as titanium, zirconium, and vanadium are commonly thought of as rare, although they make up a relatively considerable proportion of the earth's crust. Such metals as lead, tin, bismuth, antimony or mercury, on the other hand, are not looked upon as rare, even though these five metals together represent (in terms of our present knowledge) a smaller amount by weight than vanadium alone. This is because the metals of common use are highly concentrated in certain ore deposits, and are therefore accessible, whereas it is relatively un-common to find titanium, zirconium or vanadium in concentrated form, and in large quantities. These latter are present in small amounts in almost all rocks, with the result that the total quantity present in dispersed form greatly exceeds the amount of metals which are largely concentrated in ore deposits.

The fact that many metals are found largely in ore deposits, in highly concentrated, but very localized form, causes some difficulty in comparing their abundance with that of ele-ments which are widely dispersed in small concentrations as constituents of the rock-forming minerals. The distribution of the rock-forming minerals, and the abundance of the

* These are the elements which are concerned in the structures of living tissues and in the main reactive systems of living organisms (e.g., magnesium in the chlorophyll of the photosynthetic system in plants). It is true that other elements which are present in very small amounts, such as cobalt, may be quite essential for life. Their role in enzyme systems or in other complex reacting systems in the living cell may be out of all proportion to the total amount present.

elements contained in them (including the rare elements), are fairly well known.* On the other hand, the number and the extent of the ore deposits can only roughly be guessed. For this reason the relative abundance of those elements which are not ordinarily present in detectable amounts in the igneous rocks can only be estimated with some uncertainty. The figures given in Table 84 for these 'ore forming' elements are therefore not as reliable as those for the 'rock forming' elements. This is particularly true for the rarest metals, such as silver, mercury, bismuth, and gold. The discovery of a single major ore deposit could, in some of these cases, change the figure to be assigned for the abundance. By contrast, the values assigned to the abundances of the principal rock-forming elements today differ very little from those which were calculated by F. W. Clarke in 1916.

Whereas many metals are found in quantity only in a few ore deposits, most (if not all) the elements are also present in minute concentrations in the rock-forming minerals. When this is so, this disperse occurrence, which was overlooked until relatively recently, is actually the most important consideration in assessing the average concentration of such elements in the earth's crust. Thus it was formerly assumed, on the basis of the occurrence of actual zinc ores and minerals, that the zinc content of the earth's crust amounted to about 0.005%. As was first pointed out by Noddack, zinc is present in traces in igneous rocks; the best figure for the zinc content of igneous rocks was given by Noddack as about 0.02%, but is now considered to be rather lower. Since the igneous rocks make up about 88.4% of the mass of the crust (including the hydrosphere and the atmosphere), the total amount of zinc present at this low concentration considerably outweighs the quantity present as zinc minerals in ore deposits. The abundance of the platinum metals is still more strongly influenced by taking account of their occurrence in minimal amounts in ultrabasic rocks** (e.g., olivine rocks) and in fairly widely distributed ores, such as chromite and molybdenite. Noddack has estimated that this mode of occurrence increases the abundance of ruthenium by a factor of about 50,000, of rhodium by about 6000, of osmium by about 1000, and of palladium and platinum by more than 100, as compared with the abundance calculated from the occurrence of these elements in ore deposits.

For elements such as these, which are rare in the absolute sense, but which are found in very low concentrations in many rocks, the abundance data depend upon the accuracy with which these very low concentrations can be determined. It follows that the accepted values for the abundance of these elements have changed considerably during recent years, whereas the data for the abundant elements have remained substantially unchanged. In many cases, the *relative* abundances within a certain group of elements is known with considerably greater precision than their *absolute* proportion in the composition of the earth's crust. This is true of the lanthanide elements, for example (see below). According to V. M. Goldschmidt (1937), the relative amounts of the following noble metals can be taken as established: $Rh : Ir : Pd : Pt : Ag : Au = 1 : 1 : 10 : 5 : 100 : 5$. Their absolute concentrations in the earth's crust are, however, probably right only in order of magnitude.

From recent work, the proportions $Pb : Th : U$ in the earth's crust are as $1 : 0.584 : 0.156$.

Even for the industrially useful metals, which are found in enormous quantities in ore deposits, the amounts so segregated are vanishingly small as compared with the total content of the thin layer of the earth's crust which could readily be mined. This is shown in Table 86, which sets out the world's ore reserves of certain metals, as recognized in 1930, with their annual consumption. It is obvious that new

* Knowledge of the minor constituents of silicate minerals is still fragmentary. Analyses of silicates have not often been extended to include the rarer of the elements that commonly form sulfide ores, for example, and such determinations are very difficult analytically.

** For example, G. Berg (1929) estimated the abundance of the platinum metals (in g per ton) as Ru $3.6 \cdot 10^{-7}$, Rh 10^{-6}, Pd $6 \cdot 10^{-5}$, Os $9 \cdot 10^{-6}$, Ir $5 \cdot 10^{-6}$, Pt $1.2 \cdot 10^{-4}$.

deposits must continually be discovered, since the reserves of certain metals would otherwise already have been exhausted. No doubt many extensive ore deposits will also be discovered in the future. However, even if it be supposed that the figures for the world's resources should be multiplied tenfold or a hundredfold, the contents of the ore deposits remain very small as compared with the total quantities of these metals in the earth's crust. The ratio is shown in the last column of the Table. That proportion of the metal which is distributed throughout the rocks of the earth's crust, and not segregated in ore deposits, is known as the *dispersed* amount. As the rich ore deposits are exhausted, so the dispersed portion of our resources will become increasingly important for the technical extraction of the metals. Even at the present time, methods (e.g., flotation) for the enrichment of low-grade ores play a decisively important part in the extraction of many metals (e.g., copper). The last column of Table 86 shows no very wide variations between the elements as to the proportion found concentrated in ore deposits. In general, the greater the abundance of any metal in the earth's crust, the larger will be the quantity of that metal found in the form of workable ores.

TABLE 86

THE WORLD'S RESOURCES OF SOME METALS IN ORE DEPOSITS
(after I. and W. Noddack, 1930)

Metal	World resources in ore deposits	World consumption, 1930	Total quantity in 1 km layer of earth's crust	Ratio Amount of ore deposits: Amount in earth's crust
Fe	$73 \cdot 10^9$ t	$0.08 \cdot 10^9$ t	$1.3 \cdot 10^{16}$ t	$5.6 \cdot 10^{-6}$
Mn	$460 \cdot 10^6$ t	$2.0 \cdot 10^6$ t	$2.2 \cdot 10^{14}$ t	$2.1 \cdot 10^{-6}$
Cr	$1.5 \cdot 10^6$ t	$0.17 \cdot 10^6$ t	$9.2 \cdot 10^{13}$ t	$1.6 \cdot 10^{-8}$
Ni	$51 \cdot 10^6$ t	$0.055 \cdot 10^6$ t	$5.0 \cdot 10^{13}$ t	$1.0 \cdot 10^{-6}$
Cu	$93 \cdot 10^6$ t	$1.35 \cdot 10^6$ t	$2.8 \cdot 10^{13}$ t	$3.3 \cdot 10^{-6}$
Zn	$26 \cdot 10^6$ t	$1.40 \cdot 10^6$ t	$5.6 \cdot 10^{13}$ t	$4.6 \cdot 10^{-7}$
Pb	$13 \cdot 10^6$ t	$1.63 \cdot 10^6$ t	$2.3 \cdot 10^{12}$ t	$5.7 \cdot 10^{-6}$
Sn	$5.6 \cdot 10^6$ t	$0.18 \cdot 10^6$ t	$1.7 \cdot 10^{12}$ t	$3.3 \cdot 10^{-6}$

Closely related to the average content of the various elements in the earth's crust is their occurrence in particular minerals, of which they form essential constituents. In general, those elements that are most abundant also form the largest number of minerals. The number of minerals characterized by different chemical composition can be only arbitrarily defined. W. and I. Noddack recognize 1800 mineral species, of which 1097 (61%) contain oxygen as an essential constituent, followed by silicon (21%), iron (20%), sulfur (19.8%) and aluminum (17%) as the elements forming the largest numbers of distinctive mineral species. Rarer elements, such as beryllium, chromium, zirconium, tin, niobium, tantalum, molybdenum, tungsten, form far fewer minerals (less than 20 in each case). Rare elements, such as scandium, germanium, ruthenium and palladium each form only one or two compounds that exist as minerals, whereas no minerals at all are known of gallium, indium, rhodium or rhenium. However, the parallelism between abundance and proneness to form minerals is not complete, and is to some extent illusory. Thus silver is no more abundant than some of the elements that form no minerals of their own, yet at least 71 different silver minerals are known. In many cases the number of known minerals is small in proportion to the abundance of an element because, in the process of mineral formation, it has for the most part been incorporated isomorphously in the minerals of a more abundant element. This may result in the existence of no distinctive minerals of elements which are not really rare. For example, there are no known minerals of rubidium or hafnium, although the former is much more abundant, and the latter as abundant, as beryllium, which forms 20 minerals. There are, however, many minerals containing rubidium or hafnium in isomorphous substitution.

3. Regularities in the Abundance of the Elements; Harkins' Rule

It is evident from Table 84 that the most abundant elements are principally those of the first series in the Periodic Table, while the rarest elements are chiefly from the last period of the Periodic System—i.e., the most abundant elements on earth are those of lowest atomic weight. Furthermore, the non-metals and light metals are more important in the earth's crust than the heavy metals. Thus the 12 most abundant elements in Table 84 comprise 5 non-metals, 6 light metals, and only 1 heavy metal (iron). This generalization is still more marked when the *relative numbers of atoms* are compared, instead of weight percentages. 9 of the first 17 elements, on this basis, are non-metals.

In 1917, W. D. Harkins drew attention to the rule that, in general, *the elements of odd atomic number are rarer than the adjacent elements of even atomic number*. In Fig. 83

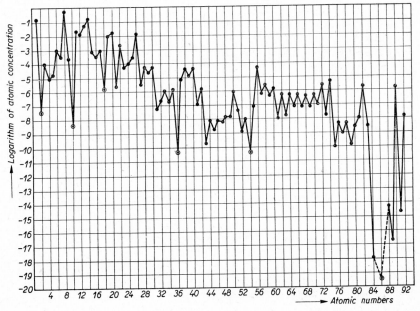

Fig. 83. Relation between the abundance of the elements and their atomic numbers.

the atomic percentages of the elements are plotted (on a logarithmic scale) against the atomic numbers, showing the applicability of Harkins' rule to the composition of the earth's crust. It is clear that (with a few exceptions, such as the inert gases where special factors apply), the troughs correspond to the elements of odd atomic number, and the peaks to elements of even atomic number.

V. M. Goldschmidt has shown that the relative abundances of the rare earth metals provide a particularly striking illustration of Harkins' rule. The particular significance of this lies in that the very close chemical relationship between the rare earths ensures that they have behaved similarly in geochemical and mineral forming processes. The relative abundances as found should thus provide a true measure of the relative amounts of the rare earth elements in the earth's substance. Fig. 84 shows the relative abundance of lanthanum and the succeeding elements, expressed as number of atoms per 100 atoms of yttrium. The line joining up the points is a well defined zig-zag, with the elements of even atomic number at the peaks in every case.

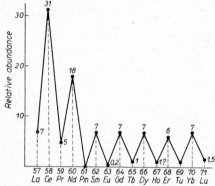

Fig. 84. Relative abundance of rare earth elements (according to Goldschmidt)

There are several obvious exceptions to Harkins' rule—e.g., the inert gases. The points corresponding to these are ringed around in Fig. 83. Although of even atomic number, they are decidely rarer than the adjacent elements*.

The low concentration of inert gases in the atmosphere is to be explained, according to P. Harteck [*Angew. Chem.* 63 (1951) 1], by the action of highly ionized atoms shot out by solar prominences. These continuously 'tap off' a portion of the earth's atmosphere. Certain gases—e.g., oxygen, nitrogen, hydrogen—are continuously supplied to the atmosphere in amounts that balance the loss, but the effect has been to deplete the atmosphere steadily of the other gases throughout the earth's history.

The particles emitted from the sun are chiefly protons, and also helium nuclei. Some of these reach the vicinity of the earth, and enter the earth's atmosphere with velocities of 1000 to 1500 km per sec., the density of such particles being about 10^4 to 10^5 particles per cm³. These velocities are so high that a considerable number of molecules (or atoms) in the earth's atmosphere, which suffer collision with these fast protons, may receive sufficient kinetic energy to escape from the earth's gravitational field. Close to the earth's surface, the kinetic energy needed for any massive particle to escape from the earth's gravitational field corresponds to a velocity of 11.5 km per second. In the upper levels of the atmosphere, the velocity required for escape into space is correspondingly less. As follows from the Maxwell distribution law (Chap. 17), only an infinitesimal fraction of the molecules in the earth's atmosphere will have the requisite velocity at ordinary temperature. Hydrogen atoms provide the only exception to this statement, and since all

---

* The other exceptions are probably due to the fact that Fig. 83 represents the proportions of the elements in the *crust* of the earth, and not in the whole earth or in a representative sample of cosmic matter. As determined from the average composition of meteorites, the abundances of Be (at. no. = 4), Mg (at. no. = 12) and S (at. no. = 16) no longer present exceptions to Harkins' rule.

gases exist predominantly in the atomic state in the upper regions of the atmosphere, it follows that hydrogen must continuously escape from the earth's atmosphere, even without any transport of energy from the solar prominences. The effect of this transport of energy is to increase the fraction of hydrogen escaping from the atmosphere and also to make possible the escape of heavier atoms from the upper levels of the atmosphere, out of the earth's gravitational field. Hydrogen atoms and oxygen atoms are continuously supplied to the upper atmosphere by the photochemical decomposition of water vapor (cf. p. 645). Nitrogen must also be supplied to the atmosphere by the decomposition of nitrogen compounds at the earth's surface, in at least the same measure as it escapes from the atmosphere. The situation of the inert gases is different, however. There is practically no source of supply of neon, krypton, and xenon. It is true that small quantities of krypton and xenon must enter the atmosphere from the products of spontaneous fission of uranium, but spontaneous fission is such a slow process that the amounts cannot balance the losses of these gases by escape into space; at the most, it may have led to some change in the isotopic composition of these elements. Far more argon must also have escaped from the atmosphere during the earth's history than has been formed by radioactive decay of ^{40}K, although the production of ^{40}A from potassium explains the predominance of this isotope over the other argon isotopes at the present time. Next to hydrogen, helium is the element which is most strongly tapped off from the atmosphere through the action of protons from the sun. For this reason helium is present only in very small amount in the earth's atmosphere, although it is so abundant in the sun and the stars. V. M. Goldschmidt (1938) has estimated that the amount of helium in the atmosphere is only 1% of the quantity produced by radioactive disintegration (in the form of a-rays) since the formation of the earth. Hence the effect of the corpuscular radiation emitted from the sun has been the loss not only of all the helium present in the earth's original atmosphere, but also of most of the helium produced subsequently by radioactive decay, in so far as the latter entered the atmosphere and was not retained within the earth's crust.

4. Composition of the Earth's Interior

Whereas the average density of the crustal rocks of the earth is no higher than 2.7, the mean density of the whole earth is much greater—5.5*. It necessarily follows that the interior of the earth must consist of specifically denser materials than the crust, and it becomes almost certain that it must differ considerably in composition from the crust. The difference in density cannot be explained solely as a result of the enormous pressure exerted by the outer shell; silicate rocks (the main constituents of the crust) or other oxidic minerals such as magnetite or chromite (which might become of increasing importance at great depths) are too incompressible to account for the observed density. Hence it may be inferred that the interior of the earth consists of elements of considerably higher density than the rocks of the crust, and it is assumed that the core of the earth consists largely of *iron*. Evidence in support of this is afforded by the predominance of the

---

* The earth has been weighed by comparing the gravitational constant with the measured force of attraction between known masses.

lines of iron in the sun's spectrum, and especially by the composition of meteorites*. Meteorites are of two types—the stony meteorites, and the iron meteorites, which consist essentially of an iron-nickel alloy with about 91% iron, 8–9% nickel. It is assumed that the earth's core is similar in composition, consisting essentially of an iron-nickel alloy, with relatively small amounts of other elements (see below).

In addition to iron and nickel, the iron meteorites contain small amounts of other elements, especially Co, Cu, Mn, P, S, C, and distinctly detectable amounts of the platinum metals. These minor constituents amount to only about 0.5 to 1% by weight of the iron meteorites. The stony meteorites are more common than the iron meteorites; the relative amounts of silicate and iron phases in meteoritic matter are not easily estimated but were assumed by V. M. Goldschmidt to be about as 6 : 1. The average composition of stony meteorites is similar to that of the igneous rocks. H. C. Urey [*Phys. Rev.*, 88, (1952) 248] accepts the values listed in Table 87 as representative of the average composition of meteorites. The mean uranium content of meteorites is probably less than 10^{-5}%. More important than this is the ratio He : U, since this provides a means of determining the age of meteorites (see p. 658).

TABLE 87

AVERAGE COMPOSITION OF METEORITES

Elements arranged in order of abundance. The figures represent proportions in weight per cent.

1. O	36.3	15. Ti	0.058	29. Se	$6.7 \cdot 10^{-4}$	43. Yb	$1.7 \cdot 10^{-4}$	57. Ho	$6.0 \cdot 10^{-5}$
2. Fe	24.1	16. Cl	0.047	30. Y	$5.5 \cdot 10^{-4}$	44. Hf	$1.6 \cdot 10^{-4}$	58. Lu	$5.4 \cdot 10^{-5}$
3. Si	17.96	17. Cu	0.018	31. Sc	$4.9 \cdot 10^{-4}$	45. Cd	$1.6 \cdot 10^{-4}$	59. Tb	$5.2 \cdot 10^{-5}$
4. Mg	13.55	18. Zr	0.008	32. Ga	$4.6 \cdot 10^{-4}$	46. B	$1.5 \cdot 10^{-4}$	60. Rh	$4.7 \cdot 10^{-5}$
5. S	2.01	19. Zn	0.0076	33. Li	$4.4 \cdot 10^{-4}$	47. Ru	$1.4 \cdot 10^{-4}$	61. Nb	$4.1 \cdot 10^{-5}$
6. Ni	1.45	20. Ge	0.0053	34. Mo	$3.6 \cdot 10^{-4}$	48. Ag	$1.35 \cdot 10^{-4}$	62. Ir	$3.8 \cdot 10^{-5}$
7. Ca	1.43	21. V	0.0048	35. Nd	$3.0 \cdot 10^{-4}$	49. I	$1.25 \cdot 10^{-4}$	63. Tm	$3.1 \cdot 10^{-5}$
8. Al	1.42	22. F	0.004	36. Ba	$2.9 \cdot 10^{-4}$	50. Os	$1.2 \cdot 10^{-4}$	64. Ta	$2.8 \cdot 10^{-5}$
9. Na	0.75	23. Br	0.0025	37. Ce	$2.1 \cdot 10^{-4}$	51. Cs	$1.1 \cdot 10^{-4}$	65. Eu	$2.7 \cdot 10^{-5}$
10. Cr	0.27	24. Sr	0.0023	38. Dy	$2.1 \cdot 10^{-4}$	52. Sm	$1.1 \cdot 10^{-4}$	66. Au	$2.5 \cdot 10^{-5}$
11. Mn	0.24	25. As	0.0018	39. La	$1.9 \cdot 10^{-4}$	53. Be	$9.3 \cdot 10^{-5}$	67. In	$2 \cdot 10^{-5}$
12. P	0.15	26. W	0.0016	40. Pt	$1.9 \cdot 10^{-4}$	54. Pd	$9.2 \cdot 10^{-5}$	68. Bi	$2 \cdot 10^{-5}$
13. Co	0.11	27. Sn	0.0014	41. Gd	$1.7 \cdot 10^{-4}$	55. Pr	$8.8 \cdot 10^{-5}$	69. Tl	$1.5 \cdot 10^{-5}$
14. K	0.09	28. Rb	0.0008	42. Er	$1.7 \cdot 10^{-4}$	56. Sb	$6.4 \cdot 10^{-5}$	70. Te	$1.3 \cdot 10^{-5}$
								71. Re	$0.8 \cdot 10^{-5}$
1–14 together	99.83%	15–28 together	0.1631%	29–42 together	0.0047%	43–56 together	0.0017%		

Observations on the transmission of earthquake waves through the earth have revealed the existence of several discontinuities in the density, compressibility and rigidity of the earth's matter at certain depths. The first region of discontinuity is fairly close to the earth's surface, at a depth of about 100 km. It is generally considered that at this depth, and down at least to the second discontinuity at about

* Other evidence frequently advanced is not unambiguous. The earth's magnetism is probably not a property of the core, but of the crust. The density of the core cannot be estimated entirely independently, without some assumptions as to the elastic properties of the core material. In fact, calculations by Jensen (1938) suggest that the density of iron would increase rather steeply with pressure at pressures greater than 10^6 atmospheres, and would range from about 9 to 12 (as compared with the ordinary value of 8) under the conditions obtaining in the center of the earth.

1000 km, the composition of the earth does not differ drastically from that at the earth's surface, as represented by the ultra-basic igneous rocks. However, under the very great pressures obtaining at these depths the stable mineral species may differ from those which are stable at the surface, and minerals of higher density will be favored. Certain minerals which are only stable at high pressures (metastable under ordinary conditions) are occasionally found at the earth's surface, where they may have been transported by some orogenic action. The diamond is an example. This group of minerals, as a whole, is sometimes known as the *eclogites*, and the shell of dense oxidic minerals—possibly denser, high pressure modifications of silicates—is known as the eclogite shell (see Fig. 85).

The core of the earth, consisting essentially of nickel-iron alloy, begins below the third surface of discontinuity at a depth estimated by Bullen (1938–1950) as 2900 km below the surface. This metallic core is rigid, although the temperature of the interior of the earth is high, because of the enormous pressure to which it is subjected. Its mass amounts to about 30–33% of the total mass of the earth, and its density is believed to increase from about 9.4 at the boundary to about 12 at the center of the earth, where the pressure is approximately $3 \cdot 10^6$ atmospheres. If the core or any other region of the earth were not solid, it would influence the period of the earth's rotation.

The nature of the shell enclosed between the second and third surfaces of discontinuity is still unsettled. It was originally considered that the 1000 km discontinuity corresponded to an abrupt change in density from 3.5–4 to about 5–6, but according to Bullen the density may increase fairly regularly from about 3.5 to 5.5 throughout the outer and inner mantles. There is, correspondingly, some uncertainty as to the chemical nature of the inner mantle. Goldschmidt (1922) made the plausible assumption that it consisted of the oxides and sulfides of the heavy metals, principally of iron. On the usual assumption that the material of the earth at one time constituted a molten mass, he compared the system with the melt in a blast furnace, which undergoes separation into three liquid phases—a metallic layer at the bottom, an intermediate zone of sulfides and oxides, and a covering layer of silicates. Very similar views, based on metallurgical experience, were developed by G. Tammann at about the same time. Goldschmidt's views also found support from his interpretation of analyses of meteorites. Iron meteorites often contain smaller amounts of a material known as *troilite*, consisting essentially of FeS. It is, however, now generally considered that this view is not correct, and that it over-estimates the proportion of the sulfide phase. It is believed rather that oxidic minerals persist right down to the boundary of the iron-nickel core, and that the progressive change in density and elasticity in the inner mantle corresponds to a region in which, through incompleteness of phase separation, there are increasing amounts of the iron-nickel material intermingled with silicates and oxides. These matters are very uncertain, however, and involve alternative conceptions as to the origin and history of the earth*.

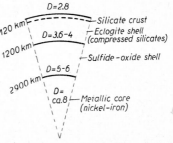

Fig. 85. Cross section through the earth (according to Goldschmidt)

* See, for example, H. C. Urey, *The Planets*, University of Chicago Press, 1952.

5. The Laws of Geochemical Distribution of the Elements

The study of the composition and chemical transformations of the earth as a whole constitutes the science of *geochemistry*. The objective of this study is to determine the distribution of the elements in the earth, and to decipher the processes whereby this distribution was achieved.

It may be assumed that the distribution of the elements as it is now observed is the result of the action, on the whole, of three influences. The first is *gravity*, which (in so far as there has been a transport of matter) must have resulted in the gradual accumulation of the heaviest elements in the interior of the earth while the lighter elements have been displaced to the surface. Secondly, there is the *chemical specificity* of the elements. As the mass of the earth has cooled off, this has led to the formation of compounds between the elements, in accordance with the thermodynamic properties revealed by laboratory experiments. This has provided a mechanism whereby certain metals of high atomic weight and high density, but with a great affinity for oxygen, have been preferentially concentrated in the crust of the earth (e.g., barium). Thirdly, there are effects resulting from differences in ease of *crystallization* and in differences of *crystal structure*. As a result of these, the elements which have been enriched in the outer regions of the earth during solidification have been distributed in very different ways among the minerals and rocks. In consequence, we find that certain elements are almost always associated with each other, whereas other elements are sharply separated.

One decisive factor in determining the distribution of the individual elements between the heavy metallic core of the earth, and the lighter silicate layer, at the time when these made up a liquid system, was the variation in electrochemical character from one element to another. Because of their great affinity for oxygen, the most strongly electropositive metals must have been concentrated in the silicate layer, while the most weakly electropositive metals must have been enriched in the iron of the core. Since the silicate layer is rich in iron, metals which are baser than iron can be present only in traces in the iron core. Conversely, the metals that are nobler than iron may be present in the core in much higher concentration than in the silicate layer. The composition of the iron meteorites supports this contention, e.g., for nickel. If it can be assumed that Goldschmidt's hypothesis of a sulfide layer is valid, similar considerations would determine the partition of the metals between the silicate and the sulfide layers. (Table 88). Tammann (1923) used evidence of this type, derived from the composition of the crustal materials, as a basis for his conclusions concerning the composition of the interior of the earth.

The composition of the earth's crust, and the differences between the crust and the interior arising from the different density of the elements, are thus assumed to be due to the operation jointly of gravity and chemical specificity. If so, the differing distributions of the elements has arisen in the same manner as the segregation of products in laboratory experiments, during the solidification of melts. During the solidification of the earth's crust, this segregation was all the more effective in that the rate of cooling was extremely slow in comparison with any laboratory experiment. Here again, the force of gravity played an important role. As the first minerals crystallized, the heaviest of them, such as magnetite, chromite, pyroxene, olivine, sank to the bottom. Only relatively rarely have they later been transported to the earth's surface again—e.g., by plutonic action. The lighter silicates, however, would normally have been transported to the top, as would products separating in the gaseous state, such as volatile compounds of the heavy metals (e.g., their halides). This is one reason why the heavy metals have been transported upwards to some extent, instead of sinking. Heavy metal compounds dissolved in light melts, or in aqueous solutions will also have ascended. It can probably be assumed that most of the ore deposits of the heavy metals within our reach arose in this manner. This is especially true of deposits of lead, zinc, tin, molybdenum, tungsten, silver, and gold ores. Further processes of separa-

TABLE 88

Siderophile elements	Chalcophile elements	Lithophile elements	Atmophile elements	Biophile elements
Au	Cu Ag	Li Na K Rb Cs	H	H (Na) (K)
Ge Sn (Pb)	Zn Cd Hg	Be Mg Ca Sr Ba	Hg	(Mg) (Ca)
C		Ra (Zn) (Cd)		
P (As)	Ga In Tl	B Al Sc Y Rare	C	(B)
		Earth Metals Ac		
		Ga (In) (Tl)		
Mo (W)	(Ge) (Sn) Pb	C Si Ti Zr Hf Th	N	C (Si)
Re	As Sb Bi	(Ge) (Sn) (Pb)	O	N P
Fe Co Ni	(Mo) (Re)	V Nb Ta Pa P (As)	He Ne A Kr Xe	O (S)
Ru Rh Pd	S Se Te	O Cr Mo W U Pu		(F) (Cl) (I)
Os Ir Pt	Fe (Co) (Ni)	Mn (Fe) (Co) (Ni)		(Mn) (Fe) (Cu)
	(Ru) (Pd) (Pt)	H F Cl Br I		

tion have taken place—and are still occurring—in and on the solid crust of the earth, through the action of the atmosphere and hydrosphere, as a result of the processes of weathering and sedimentation (see Goldschmidt, *J. Chem. Soc.* (1937) 655).*

The possibility of forming mixed crystals is always important in the crystallization of melts. It has been especially important in determining the composition of the crystalline minerals of the earth's crust, and the resulting association of the elements with each other. It is familiar that chemically similar elements can mutually replace each other in crystals. Goldschmidt showed, however, that this possibility dependend not only on chemical similarity, but also on the atomic and ionic radii. In general, elements can only replace each other in crystalline compounds to any considerable extent when they are not merely chemically similar (i.e., are similar in the configuration of their outer electron shells), but also agree closely in atomic or ionic radius. This similarity in atomic size is often quite as important as chemical similarity. Thus beryllium, the homologue of magnesium, never occurs in isomorphous admixture in magnesium minerals, whereas nickel does, although it is far removed from magnesium in the Periodic System. As shown by Fig. 3, Vol. I, p. 16, the ionic radius of beryllium is very different from that of magnesium, whereas that of nickel is very similar. The invariable association of the rare earths with one another is also due to the fact that their ions not only bear the same charge, but also differ very little in ionic radius. Aluminum, however, has a much smaller ionic radius than the rare earth metals, and is therefore never

* The considerations of this section, based on the distribution coefficients of the elements between high temperature liquid phases, and the subsequent crystallization processes, are quite general. If, in accordance with the current trend in ideas, the earth was built up by accretion of interstellar matter, rather than by the condensation of some uniform high temperature fluid, the distribution between phases may not be so directly applicable to the whole mass of the earth. The processes of differentiation will still apply to the cooling of a magmatic melt in the crust of the earth. The materials accessible to investigation are all derived from the outermost crust—this being true even of the ultrabasic rocks, and the materials from which orebodies have been derived.

associated with them although it is also a trivalent metal. Gallium, on the other hand, differs little from aluminum in ionic radius, and is always associated with aluminum in Nature.

It is possible for isomorphism between compounds to lead to the invariable association of certain pairs (or groups) of elements in the crystals deposited during solidification of the earth's crust; if the elements were originally present in very different concentrations in the solidifying magma, their concentrations will also differ widely in the resulting minerals. If, furthermore, the difference, in chemical properties between such a pair of elements is very slight, it is possible for the more abundant element to mask the presence of the relatively less abundant so completely that the latter is hard to detect and to isolate. The rarer element is said to be 'camouflaged'. The typical example is the camouflage of hafnium by zirconium, which is so complete that hafnium was not discovered until 1924, although it is present in all zirconium minerals and is not at all rare (cf. Table 84). In this example, the discovery was rendered difficult in that hafnium and zirconium are remarkably similar chemically, as well as forming isomorphous compounds. More often, where the chemical similarity is not so close, the camouflage is favored by the overwhelming preponderance of one of the elements. This is the case in the masking of gallium by aluminum. Gallium is usually considered to be one of the rarest elements, although it is, in fact, more abundant than silver, bismuth, or mercury. It accompanies aluminum in Nature, however, and is present in all aluminum minerals, although always in quite minimal amounts. It is therefore very widely distributed, but is among the elements that are relatively very inaccessible. In this sense it can be called a rare element.

In the uppermost layers of the earth's crust, the distribution of the elements is also modified by the intervention of living organisms [4]. Selective uptake of certain elements ('biophile elements') by organisms, and the ability of organisms to store up certain elements, have resulted in processes of un-mixing on a very large scale through the operation of vital processes. Examples are provided by the enormous deposits of limestones, largely due to the deposition of calcium carbonate by plants and animals (algae, mussels, snails, corals, etc.). Enrichment of very rare elements has also taken place in the same way, as shown by the accumulation of germanium in vegetation of the Carboniferous period in certain places, from which it has passed to the coal deposits.

6. Distribution of the Radioelements and their Effect on the Heat Balance of the Earth

One aspect of the data given in Table 84 that needs further comment is the distribution of the radioelements in the earth's crust. This enables us to judge how far the evolution of heat from radioactive disintegration affects the heat balance of the earth. 1 g of uranium, in equilibrium with its disintegration products, evolves 0.79 cal of heat per year. 1 g of thorium similarly generates 0.23 cal per year, 1 g of potassium $1.24 \cdot 10^{-4}$ cal, and 1 g rubidium $2.38 \cdot 10^{-4}$ cal. Since 1 g of the earth's crust contains, on the average, $2 \cdot 10^{-2}$ g potassium, $4.5 \cdot 10^{-6}$ g rubidium, $2 \cdot 10^{-6}$ g thorium and $4 \cdot 10^{-7}$ g uranium, the heat produced in the crust from radioactive processes amounts to about $3.76 \cdot 10^{-6}$ cal per g per year,* of which $3.0 \cdot 10^{-6}$ cal per g per year is due to potassium, $0.3 \cdot 10^{-6}$ cal per g. per year to uranium, $0.46 \cdot 10^{-6}$ cal per g per year to thorium, and less than $0.01 \cdot 10^{-6}$ cal per g per year to the rubidium. Although its radioactivity is very feeble, the potassium is the

* This figure is based on the abundance of the elements in the silicate layer, as given by H. Brown, *Rev. Modern Physics*, 21 (1949) 625. Older compilations of geochemical data assigned a higher abundance to thorium, but it is now generally accepted that thorium and uranium contribute about the same amount to the radioactive heat production.

principal contributor to the heat evolved, because of its abundance. However, the radio-elements are not uniformly distributed through the earth's crust, but are enriched in the upper layers. The production of heat from radioactivity is then correspondingly greater in the upper layers. According to A. Holmes and R. W. Lawson (1926), the average for granitic rocks is $15.9 \cdot 10^{-6}$ cal per g per year, of which $7.1 \cdot 10^{-6}$ cal is due to uranium, $4.6 \cdot 10^{-6}$ cal to thorium, and $4.2 \cdot 10^{-6}$ cal to potassium.

This heat production from radioactive elements in the earth's crust is of great significance for the earth's heat balance. More than three quarters of the heat lost by the earth through radiation into space is of radioactive origin. If it were not for this internal source of heat, the age of the earth deduced from geological and palaeontological data, and confirmed by radioactive measurements of the age of minerals (i.e., about $3 \cdot 10^9$ years) would be quite inexplicable.

7. Radiochemical Determination of the Age of Minerals

It is possible to make rather accurate determinations of the age of certain minerals—and therefore of the rocks in which they occur—from the end products of radioactive decay. The 'age' of the mineral here means the time which has elapsed since the mineral was laid down in the solid state. From the disintegration rate of uranium (Chap. 11) it follows that $1.35 \cdot 10^{-10}$ g of lead (^{206}Pb) is formed per year per gram of uranium. Hence the age of the mineral is given by the weight of uranium lead (^{206}Pb) per gram of uranium, divided by $1.35 \cdot 10^{-10}$. The age can similarly be found from the thorium lead (^{208}Pb) : thorium ratio, or from the acti-nium lead (^{207}Pb) : ^{235}U ratio. It is, of course, necessary to know the isotopic composition of the lead in the mineral concerned. The amount of lead that is not of radiogenic origin can be found from the ^{204}Pb content. This, multiplied by the factors given on p.535, gives the corresponding quantities of ^{206}Pb, ^{207}Pb, and ^{208}Pb which are not radiogenic. The quantity of lead formed by radioactive decay is then obtained by subtracting the 'normal' lead content from the total lead. In minerals of very great age, account must also be taken of the decrease in the quantity of uranium and thorium in course of time, as a result of their decay. This is simply found by using equation (4), p. 514. If the present uranium content (more strictly, uranium I content) of the mineral is $\mathcal{N}_{UI}$, and the content of uranium lead is $\mathcal{N}_{RaG}$, the original uranium content was $\mathcal{N}_{UI} + \mathcal{N}_{RaG}$, since one atom of uranium I was destroyed for each atom of uranium lead formed. Then equation (4) takes the form

$$\mathcal{N}_{UI} = (\mathcal{N}_{UI} + \mathcal{N}_{RaG})\, e^{-\lambda t}$$

(λ = disintegration constant of uranium I, t = age of mineral). By rearrangement of this expression,

$$t = \frac{1}{\lambda} \cdot 2.303\, \log_{10} \left(\frac{\mathcal{N}_{RaG}}{\mathcal{N}_{UI}} + 1 \right) \tag{1}$$

Corresponding expressions are obtained by substituting the experimentally determined ratios

$$\frac{\mathcal{N}_{ThD}}{\mathcal{N}_{Th}} \quad \text{or} \quad \frac{\mathcal{N}_{AcD}}{\mathcal{N}_{235}{}_U}$$

together with the disintegration constants of thorium and ^{235}U, respectively. Hence three independent expressions may be obtained for the age of the mineral from the isotopic composition of the lead, together with the uranium and thorium contents.

Thus, in a samarskite of known origin, Nier (1941) found the isotopic ratio ^{204}Pb : ^{206}Pb : ^{207}Pb : ^{208}Pb = 0.167 : 100 : 7.60 : 21.3. Thus, of every 100 g-atoms of ^{206}Pb, $0.167 \cdot 18.45 = 3.08$ g-atoms were non-radiogenic in origin—i.e., there were 96.92 g-atoms of radiogenic ^{206}Pb. Corresponding figures for ^{207}Pb and ^{208}Pb are $0.167 \cdot 15.61 = 2.61$ g-atoms, and $0.167 \cdot 38.40 = 6.41$ g-atoms of non-radiogenic origin, respectively, and hence $(7.60 - 2.61) = 4.99$, and $21.3 - 6.4 = 14.9$ g-atoms of AcD and ThD. Chemical analysis of the mineral revealed 6.91% U, 3.05% Th and 0.314% Pb (by weight). From its isotopic composition, the mean atomic weight of lead in the mineral was 206.39. From the data given, 96.92 g-atoms of RaG corresponded to 100 g-atoms of ^{206}Pb, i.e., to 129.07 g-atoms of total lead. Therefore the atomic ratio RaG : UI is given by

$$\frac{\text{RaG}}{\text{UI}} = \frac{0.314 \cdot 238.12 \cdot 96.92}{206.39 \cdot 6.91 \cdot 129.07} = 0.0394.$$

From eqn. (1), with $\lambda = 1.535 \cdot 10^{-10}$ year^{-1}, $t = 2.52 \cdot 10^{8}$ years. Similarly,

$$\frac{\text{ThD}}{\text{Th}} = \frac{0.314 \cdot 232.11 \cdot 14.89}{206.39 \cdot 3.05 \cdot 129.07} = 0.01336,$$

whence ($\lambda_{\text{Th}} = 4.99 \cdot 10^{-11}$ year^{-1}) $t = 2.66 \cdot 10^{8}$ years. A third value, $t = 2.80 \cdot 10^{8}$ years, was found from the ratio AcD : RaG. Nier used this ratio in preference to the ratio AcD : ^{235}U, in order to eliminate the disintegration constant of ^{235}U from the calculation, since it is not known with great accuracy. For a Canadian pitchblende Nier found an age of 1000 million years (from the values $1.00 \cdot 10^{9}$, $0.95 \cdot 10^{9}$, $1.03 \cdot 10^{9}$ years). A much greater age—2500 million years—was found for a Canadian monazite.

This direct determination of the age of minerals provides a check on estimates derived from geological reasoning. It has been shown that the age of the oldest sedimentary (pre-Cambrian) rocks is about 3000 million years. If these rocks are correctly regarded as metamorphosed sedimentaries, they were deposited from the oceans, and their age represents a lower limit for the time elapsed since the solidification of the earth's surface and the formation of the oceans.

It is also possible to determine the age of minerals from measurements of the ratio helium : uranium (or helium : radium) in the minerals. The calculation is based on the fact that 1 g of uranium, with its disintegration products, generates $1.1 \cdot 10^{-7}$ cc of helium per year. Ages based on helium measurements are minimum ages, since it cannot always be guaranteed that no helium has been lost by diffusion from the minerals during geological ages and processes. The method has been used by Paneth (1928 and later) for determinations of the age of meteorites. The oldest of these gave an apparent age of about 3000 million years. This is about the magnitude of the age given by astronomers for the solar system.

Other methods for radiochemical age determinations are based on the formation of ^{87}Sr by β-decay of ^{87}Rb (*strontium method*), on the formation of ^{40}Ca by β-decay of ^{40}K, or the formation of ^{40}A from ^{40}K by electron capture (*potassium-calcium method* and *potassium-argon method*). The *rhenium-osmium method* can be applied to minerals containing rhenium and is based on the decay of the 62.9 % of ^{187}Re in rhenium to give ^{187}Os. In contrast to radiogenic osmium, ordinary osmium can be activated by neutrons, and the two can thus be readily distinguished. The rhenium content can also be determined by neutron activation.

The ^{14}C-method can be applied to relatively recent limestones (e.g., stalactites and stalagmites). A method of determining when rocks arrived at the surface of the earth or were formed there is based on the determination of the content of ^{36}Cl, which is formed from ^{35}Cl by the action of cosmic rays through neutron capture—^{35}Cl (n, γ)^{36}Cl.

8. Occurrence of the Elements Outside the Earth

Evidence for the extra-terrestrial abundance of the elements is derived from two sources—the meteorites reaching us from space, and spectroscopic analysis of the luminous stars. As has already been discussed, the analysis of meteorites shows them to consist of the same elements as the earth's crust in very similar relative amounts to that found, e.g., in the average of igneous silicate rocks. Many of the elements have also been identified spectroscopically in the luminous stars, although the relative proportions of the elements may differ very markedly from that found on earth. Thus, in the sun's atmosphere the most abundant elements (according to H. N. Russell (1929), A. Unsöld (1948) and K. O. Wright (1948)) are hydrogen, helium, oxygen, magnesium, and iron, followed approximately by silicon, sodium and calcium. It would appear, however, that in the hottest stars the lightest elements—hydrogen and helium—greatly preponderate. With decreasing temperature of the stars, the heavier elements, such as potassium, calcium, iron, and titanium show themselves in increasing number and quantity in the spectrum.

Any quantitative statement about the cosmic abundance of the various chemical elements is attended with great uncertainty. Recent results of a number of investigators lead, however, to values that are in reasonably good agreement. This may be seen from Table 89, which sets out the relative abundance of the elements as given by H. Brown [*Rev. Mod. Physics* 21, (1949) 625] and G. P. Kuiper, based on astronomical observations and on the composition of meteorites. The elements are listed in order of atomic number, z. Lithium, beryllium, and boron are omitted, as being completely unstable at the temperatures obtaining in the interior of stars. In the case of other elements not listed, it may be assumed that their abundance, relative to that of silicon, does not differ greatly from that found in meteorites.

TABLE 89

RELATIVE ABUNDANCE OF THE ELEMENTS IN THE COSMOS

The numbers represent the number of gram atoms of the element concerned per gram atom of silicon.

z	Element	Abundance after Brown	Abundance after Kuiper	z	Element	Abundance after Brown	Abundance after Kuiper
1	H	35,000	37,000	19	K	0.0069	0.0057
2	He	3,500	5,100	20	Ca	0.067	0.072
6	C	8	4.1	21	Sc	0.000 018	—
7	N	16	8.9	22	Ti	0.0026	0.0028
8	O	22	15	23	V	0.00025	—
9	F	0.009	0.12	24	Cr	0.0095	0.0081
10	Ne	0.9–24	17	25	Mn	0.0077	0.0082
11	Na	0.046	0.062	26	Fe	1.83	0.62
12	Mg	0.89	1.8	27	Co	0.0099	0.0032
13	Al	0.088	0.083	28	Ni	0.134	0.038
14	Si	1.000	1.000	29	Cu	0.00046	—
15	P	0.013	0.020	30	Zn	0.00016	0.0029
16	S	0.35	0.46	36	Kr	0.000 087	—
17	Cl	0.017	0.40	54	Xe	0.0000015	—
18	A	0.013–0.22	2.5				

Table 89 does not take account of such material as may be distributed through-out interstellar space in the form of cosmic dust or in the gaseous state. The mass of this interstellar matter is estimated at 30–50% of the total mass of the universe, but very little is known of its composition. Table 90 lists the relative proportions of a few elements, based on the data of Dunham (1939) and Struve, who have assumed that the ratio Ca : Si in interstellar matter is the same as in stellar matter.

There are certain relations between the nuclear properties of the elements and their relative abundance in the cosmos since, according to certain cosmological

TABLE 90

ATOMIC ABUNDANCE OF SOME ELEMENTS IN INTERSTELLAR MATTER
(g-atoms per g-atom of Si)

Element	H	C	O	Na	K	Ca	Ti
g-atoms per 1 g-atom Si.	200,000 to 1,000,000	1.5	1000.	1	0.2	0.07	0.003

theories, the nuclei were formed in thermonuclear reactions from the primary 'stuff' of the universe. Their relative abundance thus represents the proportions which were in thermodynamic equilibrium with each other at the lowest temper-ature permitting of rapid thermonuclear processes. It is therefore of great interest for nulcear physics to have as accurate as possible a knowledge of these relative abundances. Moreover, from a comparison of the terrestrial and the cosmic abundances it is possible to derive evidence as to the origin of the earth, the distribution of the elements between the core and the surface of the earth, and the general systematics of the geochemical distribution of the elements*.

* See, especially, H. C. Urey, *The Planets*, Chicago (1952), and G. P. Kuiper, *The Atmosphere of the Earth and the Planets*, Chicago (1952, 2nd Edition).

References

1 G. BERG, *Das Vorkommen der chemischen Elemente auf der Erde*, Leipzig 1932, 204 pp.
2 F. W. CLARKE, *The Data of Geochemistry*, 5th Ed., Washington 1924, 841 pp.
3 G. BERG, *Vorkommen und Geochemie der mineralogischen Rohstoffe*, Leipzig 1929, 414 pp.
4 W. J. VERNADSKY, *Geochemie in ausgewählten Kapiteln*, Leipzig 1930, 370 pp.
5 V. M. GOLDSCHMIDT, *Geochemistry* (edited by A. MUIR), Oxford 1954, 744 pp.
6 H. W. HARVEY, *Recent Advances in the Chemistry and Biology of Sea Water*, Cambridge 1945, 168 pp.

CHAPTER 16

COLLOIDS AND SURFACE CHEMISTRY

1. Introduction

(a) General

In his studies of the diffusion of dissolved substances Thomas Graham, in 1860, discovered that membranes such as parchment paper would hold back certain substances such as gelatin or gum arabic, which ordinarily are obtained in the amorphous state. On the other hand, substances which crystallized well were found to pass freely through these membranes, when the same solvent was present on both sides. Graham perceived that these classes of materials differed in a range of other properties also. He called the substances which behaved like gelatin 'colloids' (greek κόλλα = glue), and the others *crystalloids*. To solutions of colloids he gave the name of *sols*, to distinguish them from true solutions (i.e., from the solutions of crystalloids).

It has since been found that the capacity to exhibit the properties characteristic of colloids—i.e., to form sols—is not a peculiarity of certain substances. It is generally held that, in principle, all substances can form colloidal dispersions under appropriate conditions. It is true that certain substances such as gelatin, silica, aluminum oxide, etc., do so particularly readily, but it has been found possible to prepare colloidal dispersions of substances which ordinarily exhibit the properties of crystalloids in their most typical form—e.g., sodium chloride (cf. Vol. I, p. 188). We therefore no longer classify *substances* as being colloidal or crystalloidal, but recognize the colloidal and crystalloidal *states* of subdivision.

Colloid science embraces a wide field of study [1–15]. Phenomena associated with the behavior of matter in the colloidal state are encountered in all phases of chemistry. In particular, colloidal chemical phenomena are often of great importance in technical processes [16]. The interpretation of the colloidal state and the discovery of the laws governing colloids has been indispensable for the development of biological chemistry.

Properties especially characteristic of matter in the colloidal state include, in addition to their very small diffusion coefficients, the very low osmotic pressure of their solutions, their property (usually very marked) of scattering incident light (Tyndall effect, cf. p. 667), and the ease with they may be thrown out of solution ('flocculated') by relatively slight changes in conditions (e.g., by addition of electrolytes or rise of temperature). The properties first mentioned are the direct consequence of the fact that the *particle size* of substances in the colloidal state is far greater than the sizes of atoms, and therefore larger than the sizes of ordinary molecules. It is also possible for colloidal dispersions to be as transparent as true solutions. No discontinuties can be detected when such dispersions are examined microscopically by transmitted light. They will also pass through ordinary filters without any particles being retained. The maximum resolving power of a microscope is a half the wave length of the light used—i.e., down to about 200 mμ, whereas the dimensions of molecules do not ordinarily exceed 1 mμ (10 Å). Hence 200 mμ and 1 mμ may be taken as the upper and lower limits respectively for the dimensions of colloid particles. More precise measurement of particle size has shown that colloidal particles may have any sizes in this range. All solutions containing particles with sizes between these upper and lower limits display the properties characteristic of colloidal dispersions. Thus the characteristic of the colloidal state of matter is that, in this state, the particle size is defined between the given limits*.

There are certain molecules whose diameter exceeds 1 mμ in all directions ('giant molecules'). Thus the diameter of the molecule of egg albumen is 4.34 mμ, and that of snail hemocyanine is as much as 24 mμ (Svedberg). Inorganic giant molecules are formed by the heteropolyacids and their salts, by the phosphonitrile fluorides, etc. Substances of this kind (called *eucolloids* by Ostwald) have the properties of typical colloids when they are present in solution in molecular dispersion.

The stability of colloidal dispersions falls off rapidly in the range 100 to 500 mμ, and their properties pass into those of the suspensions. A suspension contains coarser particles, which may be resolved under the microscope, and which are retained by filter paper. They differ from colloidal dispersions in that they are not stable over long periods. The suspended particles gradually settle out under the influence of gravity (sedimentation).

Microscopically visible suspended particles are sometimes referred to as *microns*. Particles which can not be seen by the ordinary microscope, but which can be detected by the ultramicroscope are called *submicrons* (size range 1 to 200 mμ), and those which are not microscopically detectable are *amicrons*.

(b) Classification of Colloidal Systems

According to the state of subdivision (*degree of dispersion*)**, matter is defined as *coarsely disperse*, *colloidally disperse*, and *molecular disperse*. The substance which is present in a more or less fine state of subdivision is the *disperse phase*, the homogeneous medium in which it is present is the *dispersion medium*, and the whole constitutes a *disperse system*. Colloidal systems can be grouped into eight categories, depending on the state of aggregation of the disperse phase and the dispersion medium (Table 91, Ostwald's classification).

In the following sections, the most important properties of colloidal systems will be discussed with reference to colloidal solutions, since these are the best known and most fully studied colloidal systems. The most distinctive properties of other colloidal systems are given on p. 686 *et seq*.

---

* The figures given (size < 200 mμ> 1 mμ) must not be taken as *sharply* defining the colloidal state. Near the lower and upper limits, the properties of the systems pass continuously into those of true solutions and particulate suspensions, respectively. The shape of the particles, their other properties, and the nature of the dispersion medium are also important, but these factors do not sensibly affect the size range of the colloidal state.

** Wo. Ostwald defined the degree of dispersion D as the ratio of the total surface A of the disperse phase to its volume V—i.e., $D = A/V$.

TABLE 91

CLASSIFICATION OF COLLOIDAL SYSTEMS

Dispersion medium	Disperse phase	Description	Examples
Solid	Solid	Solid sols (vitreosols, crystallosols, etc.)	Ruby glass (p. 414). Blue rock salt (Vol. I, p. 188
	Liquid	—	—
	Gaseous	Solid colloidal foams	Pumice (partly coarsely disperse)
Liquid	Solid	Colloidal solutions or sols (hydrosols, organo-sols, etc.)	Colloidal gold dispersion (p. 413). Aquadag, oildag, (Vol. I, p. 422).
	Liquid	Colloid emulsions (emulsoids)	Milk (partly coarsely disperse). Gelatin solution.
	Gaseous	Colloidal foams	Soap lather (partly coarsely disperse)
Gaseous	Solid	Colloidal dusts (aerosols)	Chemical warfare smokes
	Liquid	Fogs	Atmospheric mists and fogs; sulfuric acid fog (Vol. I, p. 707)

c) Historical

Graham may be regarded as the founder of colloid science; the fundamental concepts are due to him, even though there were a few isolated investigations in the field prior to his work. Since the beginning of the present century, colloid chemistry has been far more widely studied. The invention of the ultramicroscope by Siedentopf and Zsigmondy, in 1903, made it possible to observe colloidal particles directly for the first time, and at about the same time the first real laws governing the behavior of colloid particles were formulated theoretically by Sutherland (1905) and von Smoluchowski (1906). The systematic organization of the field, and its development into a distinct branch of scientific activity was due largely to Wo. Ostwald, whereas Bredig, von Weimarn, Perrin, Duclaux, Svedberg, Langmuir, Freundlich and others brought about great advances both in experimental methods and in the theory. More recently, the correlation of colloid science with the study of surface phenomena (capillary phenomena and adsorption) has greatly deepened our understanding of the colloidal state.

2. Colloidal Solutions

Following Graham's terminology, a colloidal solution (*sol*) is known as a *hydrosol*, *alcosol*, *etherosol*, or generally as an *organosol*, depending upon whether the dispersion medium is water, alcohol, ether, or some other kind of organic liquid. In the same way, solid colloidal-disperse systems are termed crystallo-sols or vitreosols (latin *vitreus*, glassy), etc. The voluminous precipitates formed from sols by flocculation (*gels*) are similarly distinguished as *hydrogels*, *alcogels*, etc. A gel which can be reconverted into a sol is said to be *reversible*; otherwise, it is *irreversible*.

Systems in which all the dispersed particles are of the same size are called *isodisperse*; if particles of various sizes are present simultaneously, the systems are *polydisperse*. In general, colloidal solutions tend to be more or less polydisperse, and it is necessary to control conditions very closely if essentially isodisperse sols are to be obtained.

(a) **Preparation** [*17*]

Colloidal solutions may be prepared either by processes of *dispersion* in which coarser particles are broken down, or by *condensation*, starting from substances in molecular dispersion—i.e., true solution.

(*i*) Among the *dispersion methods*, one of the most important is that of electrical sputtering, first used by Bredig (1898). It is particularly suitable for the preparation of metal sols, and consists in striking an electric arc between two metal wires, beneath cooled water. Svedberg has found that for the preparation of organosols it is better to disperse the metal by high frequency sparks, such as are obtained from the discharge of a Leyden jar condenser.

It is often found that a substance passes into a colloidally disperse solution merely on contact with a liquid. This effect can show itself in a troublesome form during the washing of analytical precipitates. In particular, some sulfides (CuS, HgS, FeS, ZnS) and oxide hydrates have a tendency to form hydrosols when they are washed with pure water, and thereby to pass through the filter. This can generally be avoided by washing with water containing some electrolyte. Conversely, electrolytes can bring about the dissolution of some substances, to form hydrosols*. Thus b-stannic acid can be brought into colloidal solution by the action of dilute potassium hydroxide; it also goes into colloidal solution when it is treated first with concentrated hydrochloric acid and then with water (Vol. I, p. 532). The action is said to be one of *peptization*, by analogy with the dissolution of albumen under the action of the enzyme pepsin present in the gastric juices.

Peptization by electrolytes occurs especially readily with substances which are in the form of gels, but it is also possible in some instances to peptize crystalline materials. Thus hydrosols of elementary Ti, Zr, Mo, and W can be made by treating the finely powdered metals alternately with acids and alkalis, washing each time with water (cf. p. 162).

It is also possible to disperse materials colloidally by purely mechanical means—i.e., by very fine grinding. The process is assisted by adding as a diluent during the grinding some inert solid substance which is easily soluble in the dispersion medium. Thus von Weimarn (1923) prepared hydrosols of sulfur, selenium, tellurium, and various metals by this means, adding glucose as a neutral diluent.

(*ii*) *Condensation methods* can be used quite generally for the preparation of sols. Most of them depend on the production of a sparingly soluble precipitate by some chemical reaction, under conditions which do not favor the rapid growth of the particles as they first separate, and excluding as far as possible those factors which bring about coagulation of the colloidal dispersion (cf. p. 672).

Sols may be produced by condensation by other processes than chemical reactions. Thus sols of sulfur or phosphorus can be made by pouring alcoholic solutions of these substances into water (von Weimarn, 1911). Schalnikoff (1927) prepared metal hydrosols by condensing the metals and the dispersion medium together, from their supersaturated vapors, on a surface cooled with liquid air.

The following substances form aqueous colloidal solutions with especial ease:

* See p. 669 *et seq.* for an explanation of such phenomena.

[i]   the sulfides of arsenic, antimony, tin, bismuth, copper, mercury, nickel, zinc, and cadmium:

[ii] the oxides or oxide hydrates of iron and aluminum:

[iii] the dioxides of silicon, germanium, tin, titanium, and zirconium, the pentoxides of vanadium and antimony, and the trioxides of molybdenum and tungsten:

[iv] noble and semi-noble metals, such as gold, silver, platinum, palladium, mercury, and copper:

[v]   many organic substances such as albumen and starch.

The substances included under [i] to [iii] have such a strong tendency to form sols that they must be precipitated under certain specified conditions (temperature, addition of electrolytes) in order to ensure that they do not remain partly or completely in colloidal solution. This is particularly necessary if precipitation must be effected from very dilute or very concentrated solutions. In both these extreme cases, as shown by von Weimarn (1908 and later), circumstances are favorable for the formation of colloidal particles in precipitation processes.

The metals listed under [iv] can readily be obtained as hydrosols either by electrical sputtering or by reduction of their compounds in solution. In some cases this property is utilized for analytical purposes, and in other cases metallic sols are used as catalysts or for medicinal purposes.

When prepared by chemical reactions (or, in some cases, by other methods), the desired sols are not obtained directly in a pure state, but contain as impurities greater or less amounts of electrolytes or other substances dissolved in molecular dispersion. These impurities often have an adverse influence on stability, and they are commonly removed by dialysis.

(b) Separation Methods

(i) *Dialysis* (διάλυσις, separation) is the process of separating substances from one another, by utilizing their different rates of diffusion through membranes. Graham's dialyzer consists of a glass vessel, closed at the bottom by a sheet of parchment paper; the solution to be dialyzed is placed in the vessel, which is dipped into a rather wider vessel containing the pure solvent—e.g., water. In modern laboratory practice, bags made of parchment paper, collodion, etc., are generally used, dipping into a beaker of the pure solvent.

Small bags of this kind are readily prepared by Zsigmondy's method [6]. The inside of a very clean, dry test tube is wetted all over with a collodion solution. When this has dried (but not too completely), water is poured in, whereupon the small collodion bag detaches itself after a short time from the glass. For the dialysis of larger quantities of material, efficient and rapid equipment is obtainable commercially.

Dialysis is often employed not only to purify colloidal solutions but also to prove the colloidal character of some dissolved substance. Dialysis is now widely used industrially, for the purification of both naturally occurring and artificial colloids.

The velocity of dialysis depends upon the ratio between the area of the membrane and the quantity of liquid to be dialyzed, and on the difference in concentration on either side of the membrane with respect to the substances for which the membrane is permeable. The solvent on the outer side must therefore be frequently renewed or, better, a flow of solvent may be used. Dialysis is often speeded up by raising the temperature (Schmidt 1930).

Dialysis can be considerably expedited by application of an electric potential. The process of *electrodialysis* [23] based on this enables electrolytes to be far more completely removed than is possible by ordinary dialysis. It is therefore finding increasing application for the purification of the biocolloids (proteins, polysaccharides, pectic substances, enzymes, hormones, etc.). A simple laboratory technique for electrodialysis has been described by Reiner (1926) [cf. Dhéré, *Kolloid Z.*, 41 (1927) 243, 315].

(c) Ultrafiltration

It is possible to make use of the impermeability of collodion membranes towards colloidal substances to effect a separation not only from dissolved materials present in molecular dispersion, but also from the dispersion medium. Such membranes are permeable with respect to the solvent, although a certain pressure must be applied in order to force the solvent through. Experience shows that the atmospheric pressure usually suffices, and may be applied by exposing the other side of the membrane to a rough vacuum (filter pump). This the principle of *ultrafiltration*.

Ostwald (1918) has described a method for preparing ultrafilters by impregnating filter paper with a collodion layer [5]. Ultrafilters prepared in this way enable colloidal solutions to be filtered without suction ('spontaneous ultrafilters'). The 'membrane filter' of Zsigmondy and Bachmann (1918) is obtainable commercially in various pore sizes. Unlike the ordinary collodion filters, these can be dried without change in permeability. Ultrafilters of the Bechold-König type (1924) are also often used.

Ultrafiltration can be utilized in many ways. Apart from colloid chemical work, it may be employed in analytical and preparative chemistry for the rapid separation of gelatinous or exceptionally finely crystalline precipitates (G. Jander, 1919 and later). The technique is widely used in biochemistry, physiological chemistry, bacteriology, and medicine.

(d) Optical Properties of Sols

(*i*) *General*. When light is incident on a sol, it is partly reflected from the colloidal particles, partly scattered sideways, partly absorbed (being converted thereby into other forms of energy, chiefly heat), and partly transmitted.

As in the case of true solutions, the decrease in intensity of light as a result of absorption, in passing through a sol is governed by Lambert's law, which states that the extinction is proportional to the thickness of the layer. 'Extinction' signifies the expression $\log (I_0/I)$, where I_0 is the intensity of the incident light, I is that of the transmitted light. Thus

$$\log \frac{I_0}{I} = a \cdot x \tag{1}$$

where x is the thickness of the layer, and a a proportionality factor known as the *extinction coefficient*. Beer's law also holds; this states that the optical absorption of any substance is independent of the length of path (of constant cross section) over which it is distributed. It follows that the extinction is proportional to the concentration:

$$\log \frac{I_0}{I} = \varepsilon \cdot c \cdot x \tag{2}$$

If the concentration x is expressed in mols per liter, ε is the molar extinction coefficient. Beer's law is obeyed only if changes in concentration bring about no changes in the absorbing particles (e.g., changes in the degree of dispersion). Hence, by measuring the light absorption of sols at various concentrations it is possible (as for true solutions) to investigate the action of the solvent on the disperse phase. For any substance, starting from the molecular disperse state, the light absorption at first increases with increasing particle size, and then diminishes again. It usually reaches its maximum value in the colloidal range. Selectively absorbing (colored) substances therefore usually display the greatest intensity of color in the colloidal state. This is important, for example, in preparing artists' pigments.

The tone of the color generally shifts from red to blue with increasing particle size, as Faraday first observed in the case of gold solds. If the gold particles have a diameter smaller than 80 mμ, gold sols usually have an intense red color. Addition of substances which bring about an increase in particle size by partial coagulation change the color through violet to deep blue. Other factors (e.g., the shape of the particles) as well as the particle size, affect the color, and it is therefore not possible to deduce the size of the particles from the color alone.

One property which is particularly characteristic of particles in colloidal solution is their strong *light-scattering* power. The scattered light is always found to be partially polarized. The path of a beam of light through a colloidal solution is therefore sharply delineated by the scattered light. The phenomenon is known as the *Tyndall effect*, since it was first investigated by Tyndall in 1869. Substances present in a state of molecular dispersion do not exhibit the Tyndall effect*.

(*ii*) *Ultramicroscopy.* In consequence of the diffraction of light, colloida particles appear as self-luminous points when viewed transversely. They are therefore microscopically visible, although they are so small, provided that observations are made, not along the direction of the rays of light, but perpendicular thereto. This is the principle of the *ultramicroscope* inverted by Siedentopf and Zsigmondy**, which enables particles of only a few mμ diameter to be observed if the illumination is sufficiently intense.

(e) Number, Size and Shape of Particles in Sols

The *number of colloidal particles* in a given volume can be counted directly under the ultramicroscope. If the quantity of substance present in a sol, in colloidal form, is also known, the size of the individual particles can be computed from their number. It may readily be seen that, for cubical particles, the diameter (or edge length) l is given by

$$l = \sqrt[3]{\frac{c}{n \cdot d}} \tag{3}$$

n = number of colloid particles per cc.
d = density of substance present in colloidal form
c = concentration of substance, g per cc.

The diameter of spherical particles may be obtained by multiplying the expression under the root sign by $6/\pi$. The application of (3) involves the assumption that the sol is isodisperse. It is often possible to resolve polydisperse sols into roughly isodisperse fractions, by filtering them through Zsigmondy membrane filters of graduated porosity. More exact determinations of the particle size can, however, be obtained from the ultracentrifuge in such cases. The determination of particle size in sols, suspensions, and other disperse systems is known generally as 'dispersoid analysis' [*19*]. Various other methods may be used, in addition to those mentioned.

Particles which have roughly the same dimensions in all three directions are known generally as 'spheroidal', as distinct from disc-shaped or rod-shaped particles. Except in systems where the dispersion medium is a gas, deviations from the spherical shape, unless they are very considerable, do not usually exert an appreciable influence on the properties. Sols which contain disc-shaped or rod-shaped particles become double-refracting when

* Light is, indeed, scattered by the molecules of gases and liquids (so-called 'Rayleigh scattering'), but the intensity of the light scattered in this way is negligible as compared with the scattering produced by colloidal particles.
** The so-called 'ultracondenser', made for use with an ordinary microscope, depends on essentially the same principle.

they flow (streaming anisotropy). When such a sol is observed between crossed Nicol prisms, the field of view becomes illuminated as soon as streaming is induced. The character of the streaming double refraction is different for disc sols and rod sols (negative for platelets, positive for rods). With sols of very decidedly aspherical particles, it is often possible to recognize the streaming anisotropy even with the naked eye, from a shimmering, or formation of striae on stirring. Sols made of very elongated particles—e.g., an aged V_2O_5 sol— have a beautiful silky luster when they are stirred. Particles which deviate greatly from the spherical shape, and have a lively Brownian movement (see below) often twinkle in a characteristic manner in the ultramicroscope.

As with true solutions, the number of particles in sols can naturally vary between wide limits. It is generally between 10^8 and 10^{14} per cc in sols of the usual concentration. The smaller the particle size, the greater must the number of particles be for a given analytical concentration. Except, however, in the sols of the eucolloids and the lyophilic substances related to them, the number of particles can rarely be raised much above 10^{16} per cc, even in sols of especially fine particle size, without flocculation taking place. This upper limit corresponds roughly to the ionic concentration per cc in a 0.00001 normal KCl solution. An average figure for the number of particles in sols of lyophobic colloids would be less than one millionth of that in solutions of crystalloids of ordinary concentration.

(f) Osmotic Properties of Colloidal Solutions

Substances in colloidal solution exert only a very small osmotic pressure, corresponding to the low concentration of particles in their solutions. The resulting lowering of the vapor pressure, lowering of the freezing point, and elevation of the boiling point are therefore hardly measurable. Freezing point depressions of a few thousandths of a degree, such as have been recorded for sols of hydrophobic colloids, must probably be attributed largely to the traces of electrolytes always present in such sols. These also affect the direct measurement of osmotic pressures. In itself, measurement of osmotic pressure of colloidal solutions presents little difficulty, since it is not difficult to prepare a membrane which is quite impermeable to the colloidal substance, but completely permeable to other substances present. The osmotic pressures of typical colloids, although much smaller than those of crystalloids, are of conveniently measurable magnitude as may be calculated from the number of particles. Thus, a sol containing 50 g per liter of As_2S_3, with a particle size of 2.5 mμ, would have an osmotic pressure of about 15 cm of water. This assumes that electrolytes present in the sol exert no influence on the osmotic pressure although, as was first deduced theoretically by Donnan in 1911, there must be some effect even when the membrane is completely permeable to electrolytes [22]. Particle numbers or particle sizes of the colloid in a sol can therefore be derived from osmotic pressure measurements only when allowance is made for this 'Donnan membrane equilibrium'.*

(g) Brownian Movement

Ultramicroscopic observation shows that colloidal particles suspended in a liquid or a gas are in a state of continual, lively, and quite irregular motion. The phenomenon can be observed not only with colloidal particles, but also with particles visible under the ordinary microscope, but becomes more vigorous as the particles get smaller. It was discovered for pollen grains by the botanist Robert Brown in 1827. It has been shown that the Brownian movement is produced by the collisions of molecules with the small particles. The amplitude of the Brownian movement becomes larger as the dispersion medium is made more tenuous, and the phenomenon is therefore particularly marked for particles suspended in gases.

The Brownian movement explains why suspended substances, provided that they are sufficiently finely divided, do not settle out under the influence of gravity, but remain dispersed in the medium. Just as the density of a gas (such as air) decreases in a regular

* The difficulty is not so acute in dealing with eucolloids, and measurements of osmotic pressure are widely employed in studying high-molecular lyophilic systems.

fashion with increasing altitude, so does the concentration of suspended particles, as was proved experimentally by Perrin [20]. The kinetic theory of gases requires that the height of the layer in which the gas density falls in some definite ratio (e.g., to half its value) should be inversely proportional to the molecular weight of the gas. Similarly, the height of the layer in which the concentration of (uncharged) suspended particles falls off by a definite ratio is inversely proportional to the particle weight.* For particles of mastic, of 1.2 μ diameter, in water, Perrin found that the half-value layer was 6 μ high. For similar particles, of 0.12 μ diameter, this layer was 6 mm high, and for particles of 0.012 μ diameter it was 6 meters.

(h) Ultracentrifuge

The force of gravity, to which the fall off of particle concentration with height is due, may be replaced by centrifugal force. The thickness of the half-value layer in the sedimentation equilibrium is thereby diminished in the same proportion as the sedimenting force is increased. This is the principle underlying Svedberg's ultracentrifuge [18], whereby it is possible to produce centrifugal fields of up to 10^6 g. Under these high forces, sedimentation equilibria can be established not only with colloids, but even with molecular-disperse substances (with molecular weights down to about 200), in the same way as those observed by Perrin for microscopically visible particles. The weights and sizes of particles in highly disperse sols, and the molecular weights of dissolved substances—especially of those with high molecular weights—can thereby be determined. If substances of different molecular weights are present together in solution, they undergo sedimentation at different rates in the centrifugal force field.

Particle size determinations can be carried out on sols of less highly disperse particles, by means of ordinary laboratory centrifuges [19]

(i) Electrical Properties of Colloids. Coagulation and Peptization

(i) *General.* Colloidal particles suspended in a liquid migrate in an electric field; hence they must be electrically charged. Colloidal *metals* and *metal sulfides* generally bear a negative charge, colloidal *metal oxides* and *hydroxides* are usually positively charged.

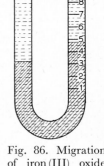

Fig. 86. Migration of iron (III) oxide hydrate sol in an electric field.

The migration of colloidal or more coarsely disperse particles in an electric field is known as *cataphoresis* or *electrophoresis* [23]. It may be demonstrated in a simple manner by means of the apparatus shown in Fig. 86. A colloidal solution of iron(III) oxide hydrate is placed in the U-tube and covered with pure water in each limb, and a sufficiently high voltage is applied to the electrodes dipping into each limb (e.g., 220 v). It will be found that both boundaries of the brown colloidal solution are gradually displaced in the direction of the negative electrode. Hence the iron oxide particles are positively charged. If the experiment is continued until the particles come in contact with the negative electrode, they are there discharged and flocculated. The velocity of cataphoretic migration of colloidal particles is

* If the concentrations of particles at heights h_1, h_2 are c_1, c_2, respectively, the mass and density of the particles being m and d, and the density of the dispersion medium being d_0, then

$$2.3 \log \frac{c_1}{c_2} = \frac{\mathcal{N}_A}{RT} m \left(1 - \frac{d_0}{d} \right) g(h_2 - h_1)$$

where $\mathcal{N}_A$ is Avogadro's number, R the gas constant and T the absolute temperature; g = gravitational constant.

usually of the same order of magnitude as that of electrolytic ions—generally $3 \cdot 10^{-4}$ to $5 \cdot 10^{-4}$ cm $\cdot$ sec^{-1} per volt $\cdot$ cm^{-1}. Corresponding values for the Na$^+$ and K$^+$ ions are $4.4 \cdot 10^{-4}$ and $6.7 \cdot 10^{-4}$ cm $\cdot$ sec^{-1} per volt $\cdot$ cm^{-1} at $18°$.

The electrical charge on colloidal particles is due to ions adsorbed on their surface. Thus H$^+$ ions are usually adsorbed on the surface of iron(III) oxide in colloidal solution. If very dilute potassium hydroxide is added to the colloidal solution, the charges on the colloidal particles are partially or completely neutralized, through the combination of OH$^-$ ions with adsorbed H$^+$ ions. This can be detected by a decrease in the cataphoretic velocity of the particles. As this decreases the stability of the sol diminishes, and when the migration rate becomes zero, the sol suddenly flocculates out*. Similar phenomena can be observed for the sols of a great many substances, and especially for most inorganic colloids. It may be concluded that the sols of these substances are stable only when the particles are electrically charged.[21]

The electric charge provides an explanation of why the colloid particles do not unite with one another when they collide (or approach one another very closely) as a result of their Brownian movement, but rebound. In the absence of an electric charge, the surface forces tend to reduce the total surface area as far as possible. This they do through the association of the particles into coarsely disperse aggregates, which sink to the bottom—i.e., flocculation occurs.

It follows from this theory of coagulation that the *cations* of the added electrolyte are responsible for the flocculation of *negatively charged* particles, whereas the *anions* are concerned with *positively charged* particles. The flocculating power of ions increases rapidly with their charge (Hardy-Schulz rule). Uni-, bi- and tri-valent inorganic ions have about the same effect at molar concentrations in the ratio of $500 : 10 : 1$, respectively. Inorganic ions of the same charge differ little in their effects, except that the OH$^-$ and especially the H$^+$ ion have a far greater flocculating power than other univalent ions.

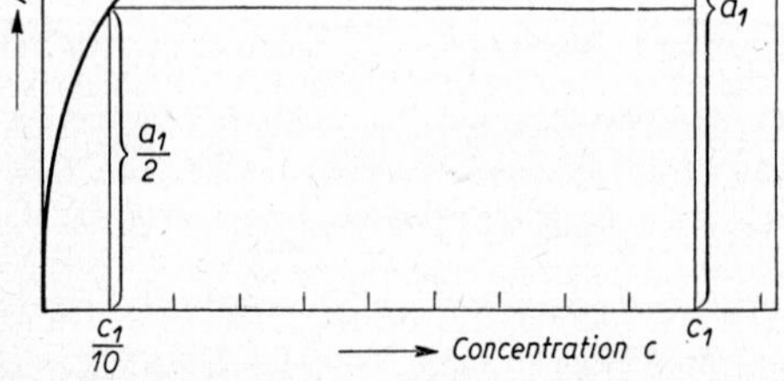

Fig. 87. Relation between concentration and adsorption.

It is quite understandable that the flocculating power of ions should increase with their valence, but some explanation is required for the extraordinary rapidity of the increase. As was shown by Freundlich (1907), this is a consequence of the way in which the adsorbability of ions depends on their concentration. This can be seen from Fig. 87, in which the bold curve represents the variation of amount adsorbed with concentration (cf. p. 679). If an amount a_1 of a univalent ion is absorbed by some colloid from a solution in which it is present at a concentration c_1, the same neutralization of charge would be effected by the adsorption of just one half of this quantity of a bivalent ion. However, the ordinate $\dfrac{a_1}{2}$ corresponds to a concentration represented by the absicssa c_2

* Flocculation generally takes place rather before the charge has fallen quite to zero.

which, as the curve is drawn, is only $\frac{1}{10} c_1$. It is thereby assumed that the ions would be adsorbed to the same extent, irrespective of their valence—as is approximately true for most inorganic ions. The H^+ and OH^- ions are much more strongly adsorbed than other inorganic ions, and their flocculating power is correspondingly greater than that of other univalent (inorganic) ions. The adsorbability of organic ions varies from one example to another, and there is therefore no simple relation between their charge and their coagulating power. In general, they are more strongly adsorbed than inorganic ions, and are therefore much more effective flocculents. If the relative flocculating powers of ions are compared on the basis of the electrolyte adsorbed by the colloid, and not on that present in the solution, then (expressing this amount in equivalents), the same figures, within experimental error, are obtained for all electrolytes with any one colloid. Thus the amount of any electrolyte which is effective in flocculating a colloid is the amount which is necessary to lower its electrical charge to a value just below what is necessary to prevent coalescence of the particles, by electrostatic repulsion. The minimum value of the potential difference between particle and solution, necessary to inhibit flocculation, is known as the *critical potential* in the sol concerned.

TABLE 92

FLOCCULATING POWER OF ELECTROLYTES

The numbers in the Table give the concentrations of electrolyte in millimols per liter, which coagulate a standard sol (1.85 g of As_2S_3 per liter) in the same time. (From measurements by Freundlich).

Electrolytes with univalent cations		Electrolytes with bivalent cations		Electrolytes with trivalent cations		Electrolytes with organic cations	
LiCl	58	$MgCl_2$	0.72	$AlCl_3$	0.093	Anilinium chloride	2.5
NaCl	51	$MgSO_4$	0.81	$Al(NO_3)_3$	0.095	Morphine chloride	0.42
KCl	50	$CaCl_2$	0.65	$\frac{1}{2}Al_2(SO_4)_3$	0.096	Crystal violet	0.16
KNO_3	50	$SrCl_2$	0.64	$Ce(NO_3)_3$	0.080	New magenta	0.11
$\frac{1}{2}K_2SO_4$	65	$BaCl_2$	0.69			Benzidinium dinitrate	0.087
		$ZnCl_2$	0.69			Quinine sulfate	0.24
HCl	31	$UO_2(NO_3)_2$	0.64				
$\frac{1}{2}H_2SO_4$	30						

Table 92 shows the flocculating power of some electrolytes towards an As_2S_3 sol. Since the colloidal particles in this sol are negatively charged (through adsorption of S^{2-} ions), it is the cation, in each case, which determines the flocculating power of the electrolytes. The quantities of electrolyte needed to bring about flocculation are quite small, as the figures show. The data show that the flocculating power depends also to some extent on the valence of the anion. This is because the adsorbability of the ions depends not on the analytical concentrations, but on the activities of the ions, and these are influenced by the anions also.

The addition of an electrolyte can often bring about a *reversal of charge* on the colloidal particles. Thus the particles of a gold sol, ordinarily negatively charged, become positively charged if a solution of an aluminum salt is added. A very small amount of the latter discharges and flocculates the gold particles. A larger quantity does not produce flocculation, however, and it is subsequently found that the direction of migration of the particles in an electric field is reversed.

The charge on the colloidal particles may often be increased by the addition of very small amounts of ions with like charge, and the sol is thereby stabilized. Thus metal sols can be stabilized by the addition of very dilute alkali hydroxide solution, since OH^- ions are strongly adsorbed even from very weakly alkaline solution. If the alkali hydroxide concentration is raised, the cations are adsorbed as well, and flocculation occurs. The effect of the cation predominates from the outset in the case of polyacidic bases, such as $Ba(OH)_2$.

Peptization can be explained in terms of the conferment of a charge to the colloidal particles, through the adsorption of ions. Ions which are strongly adsorbed, by their action on a gel which has been flocculated by discharge of its particles, or on some other substance which is porous enough to be penetrated by electrolytes, establish a charge on the particles, which then repel one another so that the substance goes into colloidal solution. Since H^+ and OH^- ions are especially strongly adsorbed, they are particularly effective as peptizing agents.

Peptization can lead to an equilibrium between sol and peptisable material, similar to that between dissolved substance and solid phase in a saturated true solution. There is, however, a striking difference from the case of the true solution, in that for given amounts of liquid and of peptizing agent, *the concentration of the sol depends upon the quantity of the solid phase present*. The amount peptized is a maximum for a moderate amount of the solid phase (Ostwald's rule). The explanation of this is as follows. The charge on the particles produced by adsorption of ions reaches a maximum for a certain concentration of electrolyte (p. 686). Hence there is an optimum peptizing action at some particular electrolyte concentration. The larger the quantity of the solid phase, the greater is the amount of peptizing agent adsorbed, so that the amount of peptizing agent remaining in solution depends on the quantity of the solid, as also, therefore, does the difference between the final concentration and that corresponding to the optimum peptizing effect.

The number of collisions between particles, and therefore the *rate of coagulation*, increases with rising temperature and with increase in the concentration of the sol. The principal means of flocculating colloids (other than by an applied electric potential) are thus (a) the addition of electrolytes, (b) boiling the sols, and (c) evaporation.

Addition of a sol with oppositely charged particles may also bring about coagulation. Unrestricted mixing of colloidal solutions is thus not possible.

In order to precipitate substances which remain in colloidal solution in the course of analytical and preparative work, it is common to boil the solutions for some time with the addition of a suitable electrolyte (generally HCl, $CH_3 \cdot CO_2H$, or $NH_3 \cdot H_2O$). If this is not effective, coagulation can be achieved by evaporation to dryness.

A substance which goes up into colloidal solution again when the gel is treated with water (or, more generally, with the dispersion medium from which the gel was flocculated) is said to be a *reversible* colloid, as distinct from an *irreversible* colloid, which can not be brought into solution again without a peptizing agent. Most inorganic substances form irreversible colloids; if they are reversible when in the finely divided and highly hydrated state, they can usually be converted to the irreversible state by particle growth. With silicon dioxide, for example, this growth can be brought about by repeated evaporation with concentrated hydrochloric acid. This procedure is adopted in analytical chemistry to secure quantitative separation of silica.

(*ii*) *Micelles*. Since the colloidal particles in a sol are electrically charged, whereas the sol as a whole is electrically neutral, the sol must necessarily contain crystalloidal particles (ions) with a charge opposite to the colloidal particles. These ions will cluster as closely as possible round the colloidal particles, because of electrostatic attraction, without combining directly with the oppositely char ionsged on the particles (as takes place in adsorption)*.

* The same forces can be considered responsible for this lack of combination as hold the ions apart in ordinary electrolyte solutions.

The whole complex of colloidal particle plus its accompanying, neutralizing ions (known as *'gegenions'*) is called a *micelle*, as suggested by Malfitano (1904) and Duclaux (1907). The liquid in which the micelles are suspended is the *inter-micellar liquid*. Fig. 88 is a diagrammatic representation of a micelle in an As_2S_3 sol, formed by passing H_2S into an arsenic(III) salt solution.

The intermicellar liquid can be separated from the micelles by ultrafiltration. If coagulation does not take place in this process, the micelles remaining intact, the colloid can re-form the sol directly, on the addition of a neutral dispersion medium such as pure water. If the micelles are destroyed by coagulation, however, it is only possible to bring the colloidal particles back into solution (in the case of a lyophobic colloid) by giving them a fresh electric charge and thereby reforming the micelles.

Fig. 88.
Schematic diagram of micelle.

(*iii*) *Lyophilic and Lyophobic Colloids.* As already indicated, a charge on the particles is not a necessary condition for the stability of *all* sols. There are substances which are maintained in colloidal solution by essentially the same forces as are the crystalloids—i.e., by the forces responsible for solvation. These are said to be lyophilic colloids, as distinct from the lyophobic colloids (greek, φιλεῖν, to love, and φοβεῖν, to flee) [24]. With particular reference to their behavior towards water, the terms *hydrophilic* and *hydrophobic* are used.

Typical hydrophilic colloids include glue, gelatin, agar, starch, dextrin, and many other organic substances (in particular most proteins). Among inorganic substances, the hydrates of silica, tin dioxide, and antimony pentoxide, the finely disperse forms of molybdic and tungstic acids, and molybdenum and tungsten blues are considered to be hydrophilic colloids. These inorganic substances are very similar to the hydrophobic colloids, however, and it is not possible to draw a sharp distinction between hydrophilic and hydrophobic colloids.

The typical lyophilic and hydrophilic colloids can be discharged by H^+ and OH^- ions without being flocculated. Their particles are strongly solvated, and therefore differ very little optically from the dispersion medium. They can therefore be discerned only with difficulty, and often not at all, under the ultramicroscope. For the same reason, the sols often display only a very weak Tyndall effect. The viscosity of the dispersion medium is very greatly increased by lyophilic colloids, whereas it is practically unaffected by lyophobic colloids. Lyophilic colloids also differ from lyophobic systems in that the surface tension of the dispersion medium is strongly modified.

The high viscosity of lyophilic sols is probably due, in large measure, to the high degree of solvation of the colloidal particles. Under otherwise equal circumstances, the viscosity increases with increasing particle size, in both the molecular disperse and the colloidally disperse range. If the radius of the particles is increased by solvation, there will be a resulting increase in the viscosity. A lyophilic substance generally forms a sol of higher viscosity in a dispersion medium which does not dissolve it well, than in one which is a good solvent. This does not necessarily contradict the assumption made as to the effect of solvation, since it can often be shown that the poorer dispersion medium is less effective in producing a subdivision of the dispersoid substance than is the better solvent. The following figures are cited as examples of the high viscosities found for lyophilic sols.

Pure water (20°)	As_2S_3 in water (3% at 20°)	gelatin in water (1%)	rubber in benzene (1%)	nitrocellulose in butyl alcohol (1%)	cellulose in Schweitzer's reagent (1%)
0.01005	0.01043	3	35	up to 65	180 poises

Thus gelatin in 1% solution raises the viscosity of water almost 10,000 times as much as does arsenic trisulfide in 3% solution. However, the viscosities of sols of lyophilic colloids can vary between fairly wide limits for any one substance at the same concentration and temperature. It depends to a considerable extent on the mode of preparation of the sol, and often on its age. This is because the micelles in such sols often undergo very protracted changes—often in the nature of an aggregation process*, but sometimes of the nature of an increase in the degree of dispersion. These and other changes in the micelles (e.g., changes in the degree of solvation) can show themselves very markedly in changes of the viscosity of the sols. Viscosity measurements are therefore of great value in studying the sols of lyophilic colloids. They are also used industrially as a method of testing substances of this type. [6, 22].

Lyophilic colloids, like lyophobic colloids, are usually electrically charged**. This charge can often be neutralized by the addition of H^+ or OH^- ions (according to its sign), so that the colloid no longer migrates in an electric field, but without destroying the stability of the sol. The pH of the solution in which this occurs is known as the *isoelectric point* of the colloid in question, as suggested by Michaelis***. If the pH is further changed, the direction of migration of the colloid (and there-fore its charge) becomes reversed. The ease with which such a reversal of charge can be brought about by adding H^+ or OH^- ions is very characteristic of lyophilic colloids.

Sols of lyophilic colloids are generally somewhat insensitive towards electrolytes, in accordance with the relatively unimportant role played by their charge in determining stability. Coagulation occurs only after adding rather high concentrations of electrolytes. It is then not so much the charge of the flocculating ions that is important (in conformity with the Hardy-Schulz rule), but other properties which show themselves also in their effect on the solubility of crystalloidal dissolved substances. The most important factor is probably the *hydration* (or, generally, the solvation) of the flocculating ions.

3. Gels and Jellies

(a) Gelation and Swelling

It is highly characteristic of lyophilic colloids that their sols (when not too dilute) will solidify homogeneously into compact, soft, fairly elastic masses known as *gels* or *jellies*. The process is known as *gelation* or gelatinization. A gel may contain very considerable quantities of the dispersion medium (over 99% in some cases). If the concentration of colloidal substance in the gel is not too high, the gel may behave almost exactly the same as the liquid sol, apart from its mechanical properties. Thus the rate of diffusion of crystalloids in the gel is practically the same as in the sol, the electrical conductivity is unchanged, and chemical reactions take place in the gel just as they do in the ordinary solution†. As the concentration of the

* Aggregation here implies combination between particles, to lower the degree of dispersion, but without producing flocculation.

** The charge on lyophilic colloids is often due, not to the adsorption of ions, but to electrolytic dissociation, as in the case of molecular-disperse substances. Discharge of the colloids then simply involves the repression of the dissociation equilibrium.

*** An isoelectric point can be similarly defined for amphoteric crystalloids.

† The only difference is that all interfering effects of convection are absent in the gel. When one substance diffuses into a gel containing a second compound, with which it can react to form a precipitate, so-called Liesegang rings are obtained—e.g., when silver nitrate diffuses into gelatin containing dichromate. For explanations of Liesegang rings see W. Ostwald, *Z. physik. Chem.*, 23 (1897), 365, Wo. Ostwald, *Kolloid Z.*, 36 (1925), Suppl. Vol. p. 380; 40 (1926) 144. The same layered precipitate is obtained if convection is eliminated by other means—e.g., by carrying out precipitation in a capillary tube.

colloidal constituent of the gel is increased, however, the rate of diffusion of dissolved crystalloids diminishes progressively.

Jellies formed by the homogeneous solidification of a sol differ from the products obtained by flocculation of colloids, not only in the manner of their formation, but also in properties. The latter are known as gels in a narrower sense. The term is, however more widely used, to embrace both flocculation gels and jellies. It has been suggested that the gels formed by flocculation (or coagulation) should be called 'coagels', but the term has not come into common usage.

Jellies can be dried, to form quite solid masses, and thereby undergo extensive shrinkage. In many cases (e.g., glue or gelatin) horny products are obtained, which can be reconverted into jellies by the absorption of water. The re-formation of jellies in this way is known as *swelling**.

Other jellies, such as hydrated silica gels, shrink only to a certain degree when they are dried out. On further drying, the jelly suddenly becomes opaque, as air enters the capillary spaces vacated by the water (cf. p. 684 and Vol. I, p. 494). Shrinkage does not proceed further, to any appreciable degree. The void spaces can be filled up again by the absorption of water. However, the dried-out silica gel does not thereby undergo any increase in volume, and thus cannot be reconverted into a real jelly. It has thus acquired essentially the same properties as a silica gel obtained by flocculation. Jellies which lose their swelling power (and therefore their jelly character) when they are dried are called *non-swelling* jellies, as opposed to the *swelling jellies* first mentioned.

Many jellies can be liquefied by warming them, and solidify again after some time when they are cooled to the original temperature. This behavior is familiar in the case of gelatin jellies. There are also jellies and gels which can be liquefied at constant temperature by shearing stresses (e.g., by shaking, or better, by ultrasonic vibrations). This phenomenon is known as thixotropy (θίξις, touch and τρόπος, change).

(b) Syneresis

Jellies may shrink to a greater or less extent, not only when they are dried but, in course of time, when they are allowed to stand in moist air (or air saturated with the vapor of the dispersion medium). Under these conditions, a dilute solution of the colloid, and of crystalloids which may be present in solution, is spontaneously squeezed out. This phenomenon, which was first noticed by Graham, is called *syneresis* (συναιρεῖν, to squeeze).

(c) Structure of Jellies

Investigation of jellies by microscopy or ultramicroscopy is considerably impeded by the fact that, in their swollen state, the substances which form jellies differ very little in refractive index from the dispersion medium itself. Bütschli (1892) found a method of circumventing this difficulty in that he subsequently displaced the water from hydrogels by other dispersion media such as alcohol or benzene. From microscopic observations made in this manner he reached the conclusion that jellies have a network or honeycomb structure. He interpreted the honeycomb structure as signifying that the colloidal material forms a continuous foam-like mass, with its interstices filled with the dispersion medium. Ultramicroscopic observation of jelly formation (Zsigmondy and Bachmann, 1911) subsequently showed that the honeycomb was formed by lamellae of liquid—i.e., by the intermicellar liquid which holds the micelles apart, as they approach very closely together during jelly

---

* The term swelling is also applied to the absorption of water by strongly surface-active substances, which occurs with considerable increase in volume, but without the formation of an actual jelly.

formation. This would support the converse picture—a foam-like structure of the dispersion medium or intermicellar liquid, with the interstices filled with the micelles of the colloid. It is probable however, that at the points of closest contact the micelles also grow together to a greater or less extent. If it is assumed that this occurs on an extensive scale, one gets back to the conception of a net-work or sponge-like structure, as was originally suggested by Bütschli for many jellies. Von Weimarn (1908) and Poole (1925), on different grounds, arrived at essentially the same view of the structure of jellies. The essential feature of the modern view of jellies is the assumption that the dispersion medium forms a continuous phase throughout the whole jelly, without being subdivided into droplets surrounded by membranes. Herein is the explanation of the unhindered diffusion through dilute jellies. Other phenomena observed with jellies—such as those associated with the drying of silica jellies—are also comprehensible in terms of this structure.

(d) Plastic Masses

Jellies are closely related in structure to the so-called plastic masses—e.g., to the clay slip sused in the production of ceramic products, to resins and synthetic plastics, to cellulose products such as celluloid, lacquers, and to chemically homogeneous materials such as rubber and gutta percha. Colloid science has become of great importance in arriving at an understanding of the properties of such substances. [*25*, *26*]

(e) Oxide Hydrates [*27*, *15*]

Colloid chemical and X-ray investigations have done much to clear up the nature of the so-called *oxide hydrates*—the gelatinous precipitates obtained when many elements are thrown out of aqueous solution in the form of their oxides or hydroxides. Typical instances of such oxide hydrates are the gels of SiO_2, GeO_2, SnO_2, TiO_2, ZrO_2, Al_2O_3, Cr_2O_3, and Fe_2O_3. In the form of their freshly prepared gels, these substances are in such a fine state of subdivision that they appear amorphous towards X-rays. They accordingly have an enormous surface area, on which water is so firmly bound by adsorption (and, to some extent, by capillary action—cf. p. 684) that it has been possible only in a few isolated cases to determine whether water is also present in the precipitates in a chemically bound state (cf. Vol. I, pp. 353 *et seq.*, 495).

If the gels are left in contact with the liquid from which they were precipitated, most of them are converted in course of time into solids of crystalline texture, as detected by X-rays. This 'aging' of precipitates has been demonstrated by Fricke, Hüttig, Biltz, Böhm, Simon, Kohlschütter, Weiser, and others. In many cases (SiO_2, GeO_2, SnO_2 TiO_2, ZrO_2, etc.) the atoms thereby take up the arrangement found in the crystal structures of the oxides, whereas in other cases (Al_2O_3, ZnO, CdO, etc.) the hydroxides are formed. It is impossible to decide whether the gels contain the oxides in the one case, and the hydroxide in the other, in amorphous form. The name 'oxide hydrate' applied to these amorphous precipitates (to distinguish them from the hydroxides) expresses this indeterminacy.

As the examples show, the development of the diffraction pattern of the oxide, and not of the hydroxide, during aging of the oxide hydrates, is particularly characteristic of the elements of Group IV of the Periodic System. The ability to form well defined hydroxides increases rapidly in the main groups on either side of Group IV. In the sub-groups, however, the formation of oxides by the aging of precipitates has been observed for a number of elements outside Group IV.

From X-ray investigations of such gels, it is possible not only to ascertain whether the substance is amorphous or crystalline (in the latter case, comparison with the diffraction patterns of known compounds also indicates which compound is present), but also to determine the *particle size*. As was stated in Vol. I, Chap. 7, the smaller the size of the diffracting crystallites, the more do the lines of a Debye-Scherrer pattern become broadened. Provided that certain conditions are fulfilled, it is possible to calculate the particle size from the breadth of the diffraction lines. In comparing particle sizes so determined with those measured in other ways (cf. pp. 667, 669) it must be borne in mind that the 'X-ray particle size' signifies the volume of space throughout which the crystal lattice is regular and continuous. Colloidal particles or micelles are generally not single crystals, but aggregates, so that it is not usually possible to infer the size of the micelles from the X-ray particle size.

(f) Protective Colloids

Lyophilic colloids confer their own insensitivity towards electrolytes on any lyophobic colloid with whose particles they unite. This can be utilized for the stabilization of lyophobic colloids. Lyophilic colloids used for this purpose are called *protective colloids*.

If, for example, 1cc of 10% NaCl solution is added to 10 cc of a 0.005% red gold hydrosol, the color rapidly changes to violet, and complete flocculation of the gold particles soon ensues. However, if a hydrophilic colloid such as gelatin is previously added to the gold solution, the color change due to particle growth does not take place, and there is consequently no flocculation. The amount of the protective colloid, in mg, which just suffices to protect a gold hydrosol, under specified conditions, from the color change which heralds flocculation, was called by Zsigmondy its *gold number*. Corresponding use may be made of the silver number, sulfur number, prussian blue number, etc. (or quite generally the protective number), depending on the hydrophobic colloid used for test of protective action. Under the conditions given above, gelatin and glue have a gold number of 0.005, albumen of 0.2, potato starch of 25. The quantities necessary to achieve a certain protective effect thus differ widely for different protective colloids, and the protective action also depends upon the state of the protective colloid. Thus freshly precipitated tin dioxide hydrate has a protective effect towards gold (purple of Cassius, cf. p. 421), but the aged dioxide hydrate has no effect at all. The gel obtained by evaporating a sol which contains a protective colloid can be brought directly into solution again, provided that the protective colloid has suffered no change. This property is used in preparing 'soluble' gels of lyophobic substances, especially the metals (e.g., protargol, p. 396). The necessary quantities of protective colloid for stabilization of such a sol are often extremely small. As the above figures show, 0.005 mg of gelatin suffice to protect 100 times that weight of gold against flocculation by NaCl in 0.2 normal solution.

If lyophilic colloids are used in amounts which are considerably less than the protective figures, many sols display an enhanced sensitivity towards electrolytes ('sensitization').

(g) Surface Activity of Colloidal Substances

Colloidal particles have an enormous surface area, as compared with coarsely disperse substances. 10.5 g of silver, which would form a cube of 1 cm side and 6 cm² area in the compact form, could be subdivided into 10^{15} cubes of 100 mμ side, with a total area of 60 sq. meters. If subdivision were taken further, 10^{21} cubes of 1 mμ side could be formed, with a total surface of 6000 sq. meters. In passing from the coarsely disperse to the colloidally disperse state there is thus an extraordinary increase in surface area, and this results in a corresponding increase the so-called surface forces. Substances in the colloidal state are therefore always 'surface-active', as is shown, for example, by their strong adsorptive properties. This is the basis of the wide use of colloidal substances as adsorbents. It is possible to draw certain conclusions as to the surface area from measurements of adsorptive power, and this provides another means of determining the particle size of colloidal materials. Surface activity is also associated with the strongly enhanced catalytic properties exhibited by many substances when present in the colloidal state (Chap. 17).

4. Surface Phenomena

The existence of surface forces [*8, 33, 34*] is to be explained by the incomlpete saturation of the valence forces of atoms or ions in the surface layers. In sodium chloride, for example, every ion in the interior of the crystal is surrounded by 6 oppositely charged ions, whereas an ion lying in a cube face has only 5 neighbors (cf. Vol. I, p. 209, Fig. 44). The ions in the surface are thus coordinatively un-saturated. In substances which form body-centered cubic structures, such as Mo, W, or α-Fe, every atom in the surface has 4 unoccupied coordination positions, and the same applies to elements crystallizing in the face-centered cubic system, such as Cu, Ag, Au, and γ-Fe. For atoms or ions situated on edges or corners, the number of unoccupied coordination positions is still greater—e.g., in the NaCl structure it is 2 for ions on cube edges, or 3 for ions on corners. The energy liberated by the saturation of the free coordinative valences at the surface (including edges and corners) is the *surface energy*. It follows from the foregoing that the surface energy of any substance depends upon the physical form of the substance, and rises rapidly with the degree of subdivision: not only does the surface (in the narrower sense) increase, but the total length of the edges and the number of corners thereby in-crease even more*.

Surface forces are also exerted by crystals which are built up from molecules, and arise in this case from the unsaturated Van der Waals force fields of the molecules in the surface. The same is true of liquids.

The surface energy can be decreased (i.e., the surface forces may be saturated) either by the addition of foreign molecules to the surface (adsorption) or by a decrease in surface area. The latter can be achieved by a change in shape or by a decrease in the degree of subdivision. Liquids take up a spherical shape, except in so far as they are hindered by external forces, since the sphere has the smallest sur-face: volume ratio. Small droplets unite to form bigger drops (agglomeration). Crystallites of solids attempt to grow similarly, either by growing together directly (cf. recrystallization, Chap. 19), or by the growth of large crystals at the expense of smaller ones, which dissolve or evaporate.

It follows from the increase in energy with the degree of subdivision that the vapor pressure and the solubility of any substance must increase as the particle size is reduced. However, such changes in vapor pressure or solubility make themselves felt only as the state of subdivision approaches that of the colloidally disperse state. With moderately sparingly soluble substances such as gypsum, the increase of solubility for very small particle sizes is quite measurable. With very insoluble substances it is usually impossible to measure the solubility accurately enough to determine the effect of particle size on solu-bility. However, the decrease in surface area that accompanies grain growth can generally be detected by the decrease in adsorptive properties as precipitates age**. Particle growth takes place so rapidly for freely soluble substances that these are, in practice, never obtained in a very fine state of subdivision.

* If a cube of 1 cm side is subdivided into cubes of 100 mμ, the surface area is increased in the ratio $10^5 : 1$, the total edge length by $10^{10} : 1$, and the total number of corners by $10^{15} : 1$. It is to be noted, however, that a real solid does not have a completely smooth surface, so that a real cube always has a larger surface, and a greater length of edge and number of angles than would be assigned to its idealized form.
** Aging of precipitates is, however, often accompanied by other effects than grain growth—e.g., by transformation to other modifications.

(a) Adsorption and Absorption

(*i*) *Adsorption* [35–37] is the concentration of a gaseous or dissolved substance at the *surface* of a solid or liquid. If the substance penetrates into the *interior* (as in the pick-up of hydrogen by palladium, or the dissolution of a gas in a liquid), the process is known as *absorption*. It is also possible for gases or vapors to be condensed by the action of capillary forces in the void spaces of porous substances (p. 684), and this is known as *capillary condensation*. It is not always known which of these phenomena is responsible for the observed effects, and their actions may be superposed. In such cases the general term *sorption* is used.

The body on which adsorption takes place is called the *adsorbent*, and the substance which is adsorbed is the *adsorbate*. For the case of sorption, a similar distinction is drawn between the *sorbent* and the *sorbate*. The process of removal of a sorbed substance from or out of the sorbent is called *desorption*.

Adsorption is accompanied by the evolution of heat (the *heat of adsorption*), since the surface energy of the solid is lowered. It follows from Le Chatelier's principle that adsorption diminishes with rise of temperature. The same is generally true for sorption (cf. Fig. 90, p. 681).

At constant temperature, the adsorption is proportional to the quantity (more strictly, to the surface area) of the adsorbent. It also depends on the concentration of the adsorbate in the gas phase or in the solution from which adsorption occurs. The mode of dependence upon concentration differs in a characteristic way from that which governs the equilibrium of a gaseous or dissolved substance with a solvent. The latter is usually directly proportional to the density (or partial pressure) of the gas, or to the concentration of the dissolved substance. Adsorption generally rises very rapidly at low concentrations of the adsorbate, and then increases relatively little at higher concentrations. This is illustrated by Fig. 89,

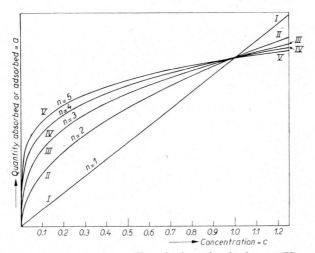

Fig. 89. Absorption isotherm (I) and adsorption isotherms (II) to (V).

where the straight line I represents the relation between the pressure of a gas and its absorption by a liquid as governed by Henry's law, while curves II–V represent the adsorption of gases by solids, also as a function of the pressure.

Henry's law states that the solubility of a gas in a liquid or solid is proportional to its partial pressure—i.e.,

$$c = p \times \text{const.} \tag{4}$$

where c is the concentration in the liquid, and p is the partial pressure in the gas phase. Henry's law is valid when the dissolved substance neither combines with the solute, nor undergoes dissociation in the process of dissolution. It also holds for the absorption of gases by solids when these assumptions are valid. If, as in the absorption of hydrogen by palladium and other metals, the molecule dissociates into two atoms, the solubility is proportional to the square root of the pressure of the undissociated gas: $c = p^{1/2} \times \text{const.}$

The partition of a substance between two solvents is governed by a law similar to that defining distribution between a solvent and the gas space:

$$\frac{c_1}{c_2} = \text{const.} \tag{5}$$

(c_1, c_2 are concentrations in solvents 1, 2) i.e., a substance distributes itself in a constant ratio between two solvents, provided it has the same molecular complexity in each. If it is present in monomolecular form A in one solvent, and as polymeric molecules A_n in the other, so that the distribution involves the equilibrium $nA \rightleftharpoons A_n$, then

$$\frac{c_1}{c_2^{1/n}} = \text{const.} \tag{6}$$

(Nernst's distribution law).

A formally similar relation has been found for the variation of adsorption with gas pressure or concentration:

$$a = k \cdot c^{1/n} \tag{7}$$

(Ostwald's adsorption isotherm).

Here a is the quantity adsorbed per unit weight of the adsorbent*, when the concentration of adsorbate in the gas space or the solution is c. k and n are constants, depending on the nature of adsorbent and adsorbate, and on the temperature. The similarity between eqns (6) and (7) is purely formal, however, and does not imply any underlying relationship between the two cases. Eqn. (7) cannot be deduced theoretically, as (6) can be, and it can indeed be shown that (7) only holds over a limited range of concentrations [8, 35]. This restriction was pointed out by Ostwald (1905) when he first put forward the empirical formula. The isotherm is often used in practice, and n generally has values between 1.5 and 5. In logarithmic form, eqn (7) becomes $\log a = \text{const.} + \frac{1}{n} \log a$—i.e., the equation of a straight line.

Langmuir (1918) and others have framed adsorption isotherms with some theoretical foundation. On the assumption that adsorption takes place chiefly in such a way as to form monomolecular layers, Langmuir arrived at the formula

$$a = \frac{\alpha \cdot \beta \cdot p}{1 + \beta \cdot p} \tag{8}$$

where a, as before, is the quantity adsorbed per unit weight of adsorbent, p is the pressure of adsorbate in the gas phase, and α and β are constants depending on the other experimental conditions. Under the conditions where the basic assumptions are valid, the Langmuir isotherm agrees well with the experimental facts. In practice, however, the underlying assumptions are often not valid, and eqn (8) therefore often does not reproduce the observations any better than Ostwald's isotherm, especially for strongly adsorbed substances. The same limitation applies to other isotherms with a theoretical basis [35]. Krohn (1937)

---

* Since the true surface area of solids is usually unknown (cf. footnote 1 to p. 678), the quantity adsorbed is generally referred to unit weight of solid, on the assumption that for any one adsorbent the surface area is proportional to the weight of the sample.

proposed an empirical isotherm which holds over a wider range than Ostwald's, and is more suitable for strongly adsorbed substances than Langmuir's. This is

$$(a + a)^2 = K(p + \beta), \qquad \text{or} \qquad 2 \log(a + a) = \log K + \log(p + \beta). \qquad (9)$$

(*ii*) *Sorption of Gases.* Extensive use is now made of the sorption of gases by surface-active substances—e.g., for the recovery of the vapors of volatile solvents. For this purpose, porous sorption agents are used, with strongly developed internal surfaces—e.g., specially prepared wood charcoals (active charcoals) and silica gel. In such materials the phenomenon of capillary condensation is often superposed on the adsorption of gases. Fig. 90 (cf. also Vol. I, p. 425) shows how sorption by active carbon varies with the nature of gases or vapors, and with temperature.

The rate of adsorption of gases is generally very high, and for materials with an exposed surface, equilibrium is usually established within a few seconds. With substances such as charcoal and silica gel, in which adsorption on inner surfaces can take place only after the occurrence of diffusion through narrow and relatively long capillaries, adsorption equilibrium is established correspondingly slowly.

(*iii*) *Activated Adsorption.* In many cases it is found that the quantity of gas adsorbed falls off with rising temperature only until some definite temperature is reached, and that adsorption then *increases* strongly with further rise of temperature. Since the heat of adsorption is positive, this can only mean that some other process involving absorption of heat, is super-imposed on the adsorption. In certain instances it has been possible to show that the process concerned is the dissociation of the adsorbed molecules into atoms or free radicals. In other cases, the phenomenon is due to an endothermic dissolution of the adsorbed substance in the adsorbent. The two processes (dissociation and dissolution in the solid) may be associated with each other. Thus H_2 is only adsorbed on Ni or Cu at temperatures up to about 100° K. Above this temperature, the adsorption involves dissociation into H atoms to an increasing extent, and at the same time the H atoms dissolve in the metal (Benton, 1931). The same phenomena may be observed when the adsorbed substance combines chemically with the adsorbent (*chemisorption*), since such combination must in general be preceded by the dissociation of the reacting molecule. In either case, the increase of adsorption with rising temperature is due to the fact that some process occurs which requires an *energy of activation* (cf. p. 698). For this reason, H. S. Taylor (1931) called the phenomenon *activated adsorption*.

(*iv*) *Adsorption from Solutions.* There is no simple and general relation between adsorption of gases and adsorption of substances from solution. It does not follow that an adsorbent which is very efficient for gases will necessarily be particularly good for dissolved substances, or vice versa. The nature of the *solvent* is very important in adsorption from solutions. A solvent which is itself very strongly adsorbed decreases the adsorbability of substances dissolved in it.

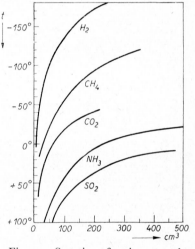

Fig. 90. Sorption of various gases by active charcoal at various temperatures (according to Remy and Hene, 1932).
Ordinates: temperatures in °C.
Abscissas: quantity of gas absorbed, in cm³, referred to 15 °C and 1 atm pressure.

(*v*) *Adsorption of Electrolytes.* One *species* of ion is generally adsorbed preferentially from electrolyte solutions—either the anion or the cation, depending on the nature of the adsorbent. It is also common for the adsorption of electrolytes to result in a change in the pH of the solution. Thus if manganese dioxide hydrate is introduced into potassium chloride solution, the potassium ion is preferentially adsorbed. At the same time, the solution becomes acidic. This is because the manganese dioxide hydrate exchanges the K^+ ions which it adsorbes for H^+ ions which it gives up to the solution. Electrolytes are generally adsorbed only in relatively small amounts. Simple ions of the same sign do not differ very widely among themselves in adsorbability. The lower the degree of hydration of ions, and the larger their charge, the more strongly are they usually adsorbed.

(vi) *Selective Adsorption.* Adsorption of a strongly adsorbed material greatly diminishes the capacity of an adsorbent to take up a less strongly adsorbed substance. For this reason, it is often possible in practice to take out only the most strongly adsorbed component from a mixture of several substances, which are adsorbed to different extents. (Selective adsorption).

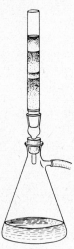

Fig. 91. Chromatographic adsorption.

The process of *chromatographic separation* is based on this selective adsorption of dissolved substances [*38*, *39*]*. A solution containing substances A, B, C... etc., which differ in their adsorbability, is sucked through a glass tube which is packed with a suitable adsorbent. The uppermost layer retains practially only the most strongly adsorbed substance *A*, and only when this has been removed from solution does adsorption of the next substance *B* take place on a lower layer of adsorbent, the surface of which has not yet been covered. Adsorption of *C*, in turn, takes place still lower down the column. If the adsorbed substances are colored, their selective adsorption results in the formation of a series of colored rings, which render the separation very evident. (Hence the name 'chromatography'). The extent of separation can be greatly enhanced by a subsequent elution with a suitable pure solvent (selective desorption).

The value of chromatography as a method for separating inorganic ions was first demonstrated by some instructive experiments of Schwab and Jockers (1937). The technique has since been considerably developed, especially through the use of cellulose (in the form of filter paper strips, or columns of cellulose pulp) as the adsorption medium (Burstall 1949 and later). [*40*–*42*] In this form it can be used to effect microchemical separations in analytical work, and for the extreme purification of many metals. [See Burstall, Davies, Linstead and Wells, *J. Chem. Soc.*, (1950) 516].

Selective adsorption from aqueous solutions provides an explanation for the observation frequently made in geology and soil science, that certain compounds are often highly enriched in strata containing large amounts of colloidal material.

The industrial recovery of volatile solvents, by adsorbing their very dilute vapors from air, by means of active carbon or silica gel, is another example of selective adsorption. It is also possible to apply the chromatographic principle, of successive adsorption on a column from a flowing system, to the separation of adsorbable gases or vapors.

(b) Monomolecular Surface Layers [*34*]

The spreading of oil films, etc., on water is related to the phenomena of adsorption. If the surface of the water is large enough, in comparison with the quantity of oil used, spreading results in the formation of a monomolecular layer of oil on the surface. If polar molecules, such as those of the fatty acids, are spread on water, the long-chain molecules take up a definite orientation within the monolayer, such that they are almost perpendicular to the water surface, with their polar ends (i.e., the —CO_2H groups, in the case of fatty acids) directed towards the aqueous layer. The nature of such films was established by Langmuir (1917), and has been extensively investigated by Adam and others. There is no doubt that

* The method of chromatographic adsorption was devised in 1906 by the Russian botanist Tswett, who was able to separate the pigments of leaves, by using powdered $CaCO_3$ as an adsorbent. The process was later highly developed by Kuhn, Zechmeister, Karrer, and others, and has been used with great success for the separation of organic natural products.

similar phenomena are associated with adsorption at solid surfaces—i.e., there is the same tendency to form monomolecular layers, in which polar molecules are adsorbed in such a way as to be oriented.

(c) Surface Tension and Capillary Forces

(*i*) *Surface Tension.* The free surface energy per unit area is known as the *surface tension*. The surface tension of liquids can be measured directly, since it is possible for their surfaces to be enlarged or decreased by changes of shape.

The increase $\triangle F$ in the free energy of a liquid, as a result of an increase $\triangle A$ in the area of its surface, is given by the quantity of work which must be expended to enlarge the surface by this amount, i.e.,

$$\triangle F = \sigma \cdot \triangle A \tag{10}$$

where σ is the surface tension. Surface tension can therefore be defined as work per unit area, or as force per unit length, and may be measured in ergs per cm² or, more usually, dynes per cm.

The surface tension depends on the nature of both substances which meet at an interface. Thus the surface tension of water is 72.7 dynes per cm against air, or 413 dynes per cm against mercury (at 20°). Surface tension diminishes with rising temperature, indicating that heat is absorbed in the process of increasing the surface. The total surface energy is given by the sum of the heat added to the system and the work done. The latter measures only the free surface energy.

(*ii*) *Gibbs Adsorption Law.* Gibbs deduced from thermodynamic reasoning a direct relation between the surface tension of liquids and adsorption. This can briefly be stated as follows. A gas or dissolved substance becomes concentrated at the surface of a liquid (is 'positively' adsorbed) if it lowers the surface tension of the liquid. It is 'negatively' adsorbed (i.e., its concentration at the surface is lower than in the interior of the liquid) if it raises the surface tension. Dissolved inorganic salts raise the surface tension of water towards air. Hence the concentration of a salt at the surface of a solution is smaller than in the interior. Some observations suggest that the surface layer in many salt solutions consists essentially of pure water.

(*iii*) *Capillary Tubes.* The rise of liquids in narrow tubes (provided that the liquids wet the walls) is explained by surface tension. If a capillary tube is dipped into a liquid which wets it, the whole wall becomes covered by a film of liquid. The surface tension decreases the area of this liquid film by sucking the liquid up into the capillary. The work done in raising a column of liquid of density d and height h in a tube of radius r by an amount dh is: $\pi r^2 \cdot h \cdot g \cdot d \cdot dh$ (where g is the gravitational constant). In rising by the amount dh within the capillary, the area of the liquid is decreased by $2\pi r \cdot dh$, and the free energy is thereby diminished by $2\pi r \cdot \sigma \cdot dh$. Hence

$$\pi r^2 \cdot h \cdot g \cdot d \cdot dh = 2\pi r \cdot \sigma \cdot dh, \qquad \text{or} \qquad \sigma = \tfrac{1}{2} r \cdot h \cdot g \cdot d \tag{11}$$

The surface tension of liquids can therefore be conveniently determined by measuring the capillary rise h in a tube of known radius.

If the liquid does not wet the wall of the tube (as, for example, mercury in contact with glass), the liquid meniscus within the tube is depressed as compared with a free surface. The displacement is a capillary depression. In this case also the non-wetting substance attempts to decrease its surface. It can be shown that the surface tension is given by (11), except that h is now the capillary depression.

(*iv*) *Surface Curvature and Vapor Pressure.* It is familiar that the liquid meniscus in a tube is concave upwards if the liquid wets the walls, and convex if it is non-wetting. In the former case, since the liquid in the tube stands at a higher level than that outside, the concave meniscus surface supports a column of vapor of smaller height than that bearing on the flat liquid surface outside the tube. In the equilibrium state, the height of vapor bearing on the liquid surface at each point must be the vapor pressure of the liquid. It therefore follows that the vapor pressure of any one liquid, must be smaller when the surface is concave than when it is flat. Conversely, a convex surface has a higher vapor pressure than a flat surface. This is quite generally true—e.g., for the surface of droplets. The vapor pressure over a convex surface increases as the curvature gets greater and Lord Kelvin (1871) deduced the thermodynamic relation

$$\ln \frac{p_2}{p_1} = \frac{M}{RT} \cdot \frac{2\sigma}{d} \left(\frac{1}{r_2} - \frac{1}{r_1} \right) \tag{12}$$

p_1, p_2 are here the vapor pressures of drops of radius r_1, r_2, respectively, σ is the surface tension, d the density of the liquid and M the molecular weight of its vapor. The increase of vapor pressure due to convexity of the surface becomes at all appreciable only in the size range of colloidal subdivision. Water drops with a radius of 10^{-4} cm have a vapor pressure only about 0.1% higher than that of a flat water surface. If the radius is reduced to 10^{-6} cm, the vapor pressure rises by 10%, and if the radius of the droplet is reduced to 10^{-7} cm (10 Å), it is about 100% higher than that of a flat surface.

(v) *Capillary Condensation.* One result that may follow from the lowering of vapor pressure over concave surfaces is that unsaturated vapors may undergo condensation in very fine capillaries such as are found in dried out gels and other fine pored materials (e.g., wood charcoal). This phenomenon is known as imbibition or capillary condensation. It may be observed, for example, when water vapor is taken up by a dried out silica gel at constant temperature and progressively higher vapor pressures (Fig. 92).

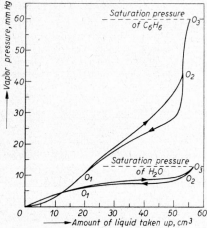

Fig. 92. Adsorption and desorption of vapor by silica gel (according to Anderson).

Water is at first adsorbed on the surface of the capillaries which penetrate the gel. During this stage the vapor pressure rises very markedly, but the gel remains clear and transparent. From the point O_1 onwards the gel becomes turbid, but becomes transparent once more at the point O_2. Between these two 'transformation points' the vapor pressure rises but little for a considerable increase in water uptake. It is also found that over this range, and with a gel containing air in its pores, the uptake and lose of water vapor are never strictly reversible. As shown by the arrows in Fig. 92, different vapor pressure curves are traced out during the uptake and the loss of water ('hysteresis'). Beyond the point O_2, the uptake and loss of water are again quite reversible, and the vapor pressure again rises much more strongly with increases in water content until at O_3 the saturation vapor pressure of water at the experimental temperature is reached. From this point onwards, of course, any amount of water vapor can be condensed on to the gel at constant pressure.

Except for the hysteresis, the form of the curves resembles those obtained in the isothermal formation or degradation of a chemical compound (cf. Vol. I, p. 74, Fig. 15). However, the almost horizontal course of the curve between the points O_1 and O_2 in Fig. 92 is not due to the formation of a chemical compound, but to the fact that water vapor is taken up by capillary condensation within this range. If all the capillaries in the gel were of the same radius, condensation of water vapor would take place at constant pressure. They actually vary in diameter, so that the finest capillaries are filled first, and then at progressively higher pressures the larger ones in which the liquid meniscus is less strongly curved, and in which the vapor pressure is lowered to a smaller extent. Between the transformation points the gel is opaque, as a result of the existence of air bubbles in the incompletely filled capillaries. The observed hysteresis is due to the work required for the expulsion of air from the capillaries. If the air is completely removed before the experiment, no hysteresis is observed (Patrik, 1920). If other vapors, such as benzene vapor, are admitted to the gel in place of water vapor, the point O_2 is reached when the *same volume of liquid* has been taken up—i.e., when the void spaces have been filled up (cf. Fig. 92).

It is possible to calculate the diameter of the capillaries from the vapor pressures at which condensation occurs. For the gel to which the curves of Fig. 92 relate, a capillary diameter of 2.6 mμ is obtained from the point O_1, and point O_2 corresponds to a diameter of 5.6 mμ. The values deduced are almost independent of whether the calculations were based on the vapor pressures of water, alcohol, or benzene at the points O_1, O_2. These values indicate that capillary condensation plays an important part only for extremely fine-pored substances, and even then only at vapor pressures close to the saturation pressure of the absorbed vapor.

(vi) Electrocapillary Phenomena [23]. If an electrical potential is applied between the ends of a capillary filled with liquid, the liquid is displaced along the tube— evidence that the liquid is charged relative to the wall of the tube. The direction of migration (i.e., whether the charge is positive or negative with respect to the wall) depends upon the substances making up the liquid and wall, respectively. The phenomenon can be particularly clearly demonstrated if a system of capillaries, such as a porous pot, is used instead of a single capillary. If this is filled with water or other liquid, and closed with a stopper bearing a vertical tube, the liquid will rise in the tube when the direction of the applied potential is appropriate. The phenomenon is known as *electroosmosis*.

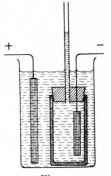

Fig. 93.
Electro-osmosis.

The electrical charge on the liquid relative to the wall—or relative to an immobile layer of liquid immediately adjacent to the wall—is due to a transference of ions to this layer from the interior of the liquid, or vice versa. It thus involves an electrical charge effect arising from adsorption, and studies of the effect may throw light on the nature of the adsorption process. The phenomenon of *cataphoresis* discussed earlier has the same basis as electro-osmosis; in one case the solid particles move with respect to the liquid, and in the other the liquid is displaced with respect to a solid wall.

From the principle of action and reaction, it follows that if an applied potential brings about the displacement of a liquid in a capillary tube, then conversely the movement of a liquid through a capillary must give rise to a potential difference between the two ends of the tube. This can, in fact, be observed when a liquid flows through a capillary, and is known as a *streaming potential*.

From measurements of the streaming potential it is possible to derive directly the difference in charge between a liquid and a solid in contact with it, or the potential difference which corresponds to this difference in charge. According to Helmholtz (1879), this potential difference ζ is given by

$$\zeta = \frac{4\pi \cdot \eta \varkappa E}{PD} \tag{13}$$

where η is the viscosity, $\varkappa$ the specific electrical conductivity, D the dielectric constant of the streaming liquid, P the pressure under which the liquid is forced through the capillary, while E is the measured streaming potential.

An exactly similar formula holds for electroosmosis. The quantity of liquid (measured by its volume v) transported per unit of time through a capillary of diaphragm of cross section q, under the action of a potential gradient ε, is given by

$$v = \frac{q\,\zeta\,\varepsilon\,D}{4\pi\,\eta} \qquad (14)$$

The velocity of movement of the liquid within the capillary, u, is

$$u = \frac{\zeta\,\varepsilon\,D}{4\pi\,\eta} \qquad (15)$$

Equation (15) also holds for the velocity of cataphoretic migration of particles suspended in a liquid, except that for spherical particles the factor 6 is substituted for the factor 4 (which is valid for cylindrical particles). D, ζ and η have the same meanings in eqns (14) and (15) as in eqn (13).

The potential difference ζ (the 'zeta-potential') gives a direct measure of the charge conferred on a solid by a liquid, as a result of adsorption of ions. Measurements of this afford a direct insight into the processes of adsorption and particle charging which are involved in the flocculation and peptization of colloids. It can be proved, for example, that when electrolytes are added to a liquid, ions of one or other sign are always preferentially adsorbed (either the cations or the anions), but that at the same time adsorption of the other ions also takes place, to a smaller extent. This is shown in that the zeta-potential always decreases again with increasing concentration of electrolyte:—i.e., the charge on the solid phase, with respect to the liquid, diminishes. This phenomenon can be understood from the typical form of the adsorption curves (Fig. 89, p. 679). When the preferentially adsorbed ion has reached a certain concentration in the solution, the additional amount adsorbed increases but little when the concentration is raised further, whereas for the more weakly adsorbed ion the part of the adsorption curve with a small slope has not yet been reached. When the concentration of the electrolyte is raised further, the adsorption of the less strongly adsorbed ion is therefore augmented, whereas that of the more strongly adsorbed ion remains practically unchanged. The ratio between the amounts of the ions in the surface layer is thus displaced in favor of the more weakly adsorbed ions, and the resultant charge due to the more strongly adsorbed ion is therefore diminished.

5. Emulsions and Emulsoids

A liquid containing suspended droplets of another liquid is called an *emulsion*. Following Wo. Ostwald's usage, colloidal emulsions are called *emulsoids**, as distinct from *suspensoids*, which are colloidal solutions of solid substances. Milk, in which droplets of fat are dispersed, is an example of an emulsion.

Formation of emulsions is especially favored between liquids which do not differ too much in density, and which have only a small interfacial tension. The interfacial tension** results in a strong tendency of the droplets to agglomerate. Emulsions are therefore generally stable only when they contain added substances, known as 'emulsifiers', which are adsorbed at the surface of the droplets, and thereby alter the conditions. [8, 28]

Emulsions are often of great technical importance [29], and play an especially great role in the oil and fat, rubber, paint and varnish, adhesives, and plastics industries. They

---

* The terms 'emulsoid' and 'suspensoid' are often applied also to the sols of lyophilic and lyophobic colloids, respectively, without regard to the state of aggregation of the disperse phase. However, it is better to avoid the usage of these expressions in this altered sense, in order to prevent misunderstanding.

** I.e., the surface free energy of a liquid-liquid interface.

also find uses in the cellulose, paper, and fiber industries and in leather manufacture. Aqueous emulsions of bitumen and similar materials are used in road surfacing.

The destruction ('breaking') of emulsions is an important problem in sewage purification. For the removal of oil from the condenser water of steam engines, in which it forms an extremely stable emulsion, use is being increasingly made of electrolytic methods. These are based on the cataphoretic migration of the negatively charged oil droplets towards the anode, where they are adsorbed by a precipitate (e.g., of basic iron carbonate, when iron electrodes are used) which is deposited simultaneously.

It is probable that a large number of the sols of lyophilic organic colloids belong to the class of colloidal emulsions or emulsoids. Thus it is generally assumed that gelatin is present in the liquid state in its hydrosols. Mercury sols can be cited as an example of emulsoids of a hydrophobic colloid.

6. Aerodisperse Systems

(a) Foams

Foams are disperse systems in which a gaseous disperse phase is enclosed by thin lamellae of the liquid dispersion medium. The gas bubbles of the disperse phase may be visible to the naked eye, or may be microscopic or ultramicroscopic in size. In the last case, a *colloidal foam* is formed.

Surface tension is even more important for the stability of foams than for that of emulsions. In accordance with this, stability of foams is favored by a low surface tension at the liquid-gas interface. The viscosity of the liquid is also of some importance.

The dispersoid is charged relative to the dispersion medium in foams also, and the greater the charge, the greater is the stability of a foam. Substances dissolved in the dispersion medium alter the surface tension and the charge, and thereby exercise a strong influence on the stability of foams. Because of its high surface tension, pure water is unable to form a stable foam with air. Addition of soaps and similar substances can transform water into a good foaming medium.

Solid substances present in a fine state of subdivision can have an extraordinarily great effect on the stability of foams, and it is the *lyophobic* solids which stabilize them. Lyophilic solids have little or no effect on foam stability. These effects of solid particles and other additions on the formation of foams and froths are of great importance in the flotation of ores (cf. p. 161).

Certain glucosides, the *saponins*, have the property of depressing the surface tension of water especially strongly, and therefore act as highly efficient foaming agents. Addition of 10 mg of saponin per liter makes water foam strongly. The action of these substances at such low concentrations is explained by the fact that, in accordance with Gibbs' adsorption law, they are concentrated almost exclusively in the surface layer.

(b) Electrostatic Charging of Gas Bubbles in Liquids

When gas bubbles are allowed to percolate through a column of liquid, they acquire an electric charge, as was first proved by Coehn (1914). The magnitude and the sign of the charge are dependent on the concentration of the solution through which the bubbles pass (cf. Remy and Koch, 1924). In general, however, the bubbles pick up a positive charge in acid solutions, and a negative charge in alkaline solutions. If they are passed into an alkaline solution through a porous wall—e.g., through a ceramic filter, or fritted glass disc, the bubbles formed are so fine that the liquid acquires a milky turbidity. If they are passed into an acid solution, they form big bubbles which stick to the wall. This can be explained by the fact that most porous media become negatively charged with respect to both acid and alkaline solutions. If the gas bubbles also become negatively charged, they are at once repelled from the walls as they emerge from the capillaries, whereas with the converse charge they are attracted to the wall, and can therefore grow to a considerable size (Kautsky, 1926). This phenomenon has to be borne in mind when it is desired to bring gases or aerosols into intimate contact with some absorbent liquid.

(c) Aerosols [30–32]

Aerosols are systems formed by suspensions of solid or liquid particles in air or other gases. Systems with solid suspended particles are *dusts*, those with liquid suspended particles are called *fogs*. Very finely disperse aerosols are often formed by the combustion of solid or liquid materials, and systems containing the solid products of combustion are known as *smokes**.

Atmospheric fogs and clouds are examples of aerosols. Finely disperse aerosols are only slightly turbid, except in very thick layers. Aerosols are encountered extremely frequently. Thus ordinary air is, strictly speaking, not a pure gas but an aerosol. A strong beam of light shining through it shows up the suspended particulate matter (motes of dust in sunlight— cf. p. 667). Denser aerosols are often formed in chemical and physical processes involving gases and liquids (or their vapors), or by the reaction of gases with solids at high temperatures.

The amplitude of the Brownian movement of the suspended particles is often an important factor in determining the stability of aerosols. If this is small, the aerosols are quite stable, provided that the gravitational settling of the suspended particles is opposed by convection of the gas which carries them. Particles suspended in aerosols of this kind are not readily influenced by chemical means. Thus a fog containing suspended droplets of sulfuric acid will pass almost unchanged through water, and is still less absorbed by concentrated potassium hydroxide (Remy, 1924, 1927). This is because by far the greatest number of the particles are never brought into contact with the absorbent, by reason of their slight Brownian movement. This is of importance in the analytical investigation of gases, when tests have to be made for impurities present as particulate suspensions.

It was formerly considered that the vapor pressure equilibrium at the surface of the droplets was an important factor in the stability of fogs. As shown on p. 683 *et seq.*, the vapor pressure over a convex surface is greater than that over a plane surface. It is therefore possible for a solution in the form of sufficiently fine droplets to have the same vapor pressure as a plane, or slightly curved, surface of the pure solvent. However, this has no significance for the stability of fogs. If the droplets coalesce with each other or with the solvent, the vapor pressure falls in any case, as a result of the decrease in curvature. Thus energy is necessarily liberated by coalescence of droplets.

Very finely disperse suspended particles in gases clump together fairly rapidly if they are uncharged and if the aerosol is not too highly diluted. In this process, fog particles tend towards a fairly definite particle size ($2.5 \cdot 10^{-4}$ cm), and when this has been reached the process of agglomeration practically ceases, even in relatively concentrated aerosols (Remy, 1924).

Suspended particles formed by chemical processes usually bear no electric charge (Remy, 1924), unlike the atmospheric fogs. The particles of artificial aerosols acquire a charge only

* The term 'smokes' is now often extended to include colloidal dusts. However, the aerosols formed by combustion processes are of such great practical importance that a special name is needed for them, and it is better to reserve the name 'smoke' for this usage. Both solid and liquid suspended particles are often present together in the smokes formed by combustion.

after some time, through adsorption of gaseous ions from the air, and roughly equal numbers of positively and negatively charged particles are formed thereby (G. Jander, 1933). It is likely that the charge is important for stability in the case of very finely disperse aerosols, with a lively Brownian movement, but not for the less finely disperse systems. No effect of surface tension on stability has as yet been observed.

Two principal methods are used industrially for the removal of particulate matter suspended in gases. Relatively coarse dusts are removed by sedimentation (in dust chambers). Finely dispersed suspensions are most efficiently treated by *electrostatic precipitators* (Cottrell dust separators), in which the aerosol passes over a series of 'spray electrodes' supplied with rectified (direct current) high voltage current. The suspended particles are thereby charged electrically, and settle out on an oppositely charged (or earthed) electrode.

Cloth filters are often also used to remove dust from flue gases or roaster gases. 'Cyclone' separators are also used. In the latter, a rotary motion is imparted to the gas by a fan, and the suspended particles are flung by centrifugal force to the walls, where they are deposited. It is probable that the action of cloth filters on finely dispersed particles depends on the same effect, brought into play by passage through a labyrinth of narrow channels with many changes of direction. Coarser particles are, naturally, retained by the cloths by a pure sieve action.

This impact, or centrifugal, effect on passage through a labyrinth of fine passages is also used for the removal of particulate materials in certain respirator filters, such as are used industrially to give protection against certain health hazards, and for military purposes, against particulate chemical warfare agents. Ordinary respirators only hold back poisons present in gaseous form, since their active fillings operate by adsorption, and by chemical absorbents. Certain types of respirators are provided, in addition, with a labyrinth-type filter, to remove solid or liquid particulate matter from the air breathed.

7. Difform Systems

The surface area of a body may be considerably increased not merely by sub-division, but by *change of shape*. If a cube is rolled out into a thin sheet, its surface is increased to one third of that obtained by cutting it up into cubes with an edge length equal to the thickness of the sheet. If drawn into a wire, its area is two thirds of that of the total surface of the cubes with an edge the same as the diameter of the wire. Thus, by dividing a centimeter cube into cubes of 10 mμ side, the surface area is increased from 6 cm² to 600 m²; by drawing it into wire of 10 mμ diameter the surface becomes about 400 m²; or by rolling it into 10 mμ foil the surface becomes 200 m². In systems such as these (called by Wo. Ostwald *difform* systems), surface forces are of great importance, as in colloidal systems. The monomolecular surface layers discussed on p. 682 represent an extreme case of a difform system.

References

1 W. OSTWALD, *Die Welt der vernachlässigten Dimensionen*, 11th Ed., Dresden 1937, 325 pp.;
       W. OSTWALD, *Grundriss der Kolloidchemie*, 7th Ed., Dresden 1922, 525 pp.
2 H. BECHOLD, *Einführung in die Lehre von den Kolloiden*, Dresden 1934, 160 pp.
3 R. E. LIESEGANG, *Kolloidchemie*, Dresden 1928, 176 pp.
4 A. VON BUZAGH, *Kolloidik*, Dresden 1936, 323 pp.
5 W. OSTWALD, P. WOLSKI and A. KUHN, *Kleines Praktikum der Kolloidchemie*, 8th Ed.,
       Dresden 1935, 174 pp.

6 R. Zsigmondy, *Kolloidchemie*, 5th ed., Vol. I, Leipzig 1925, 246 pp.; Vol. II, Leipzig 1927, 256 pp.

7 T. Svedberg, *Colloid Chemistry*, 2nd Ed., New York 1928, 302 pp.

8 H. Freundlich, *Kapillarchemie*, 4th Ed., Vol. I, Leipzig 1930, 566 pp.; Vol. II, Leipzig 1932, 955 pp.

9 A. W. Thomas, *Colloid Chemistry*, New York 1934, 512 pp.

10 J. C. Ware, *The Chemistry of the Colloidal State*, 2nd Ed., New York 1936, 334 pp.

11 J. Alexander, *Colloid Chemistry, Principles and Practice*, 4th Ed., New York 1927, 528 pp.

12 P. Bary, *Les Colloides*, 2nd Ed., Paris 1933, 586 pp.

13 J. Alexander, *Colloid Chemistry, Theoretical and Applied*, 6 vols., New York 1926–46.

14 R. Zsigmondy and P. A. Thiessen, *Kolloidforschung in Einzeldarstellungen*, 11 vols., Leipzig 1925 and later.

15 H. B. Weiser, *Inorganic Colloid Chemistry*, Vol. I, *The Colloidal Elements*, London 1933, 389 pp.; Vol. II, *The Hydrous Oxides and Hydroxides*, London 1935, 429 pp.

16 R. E. Liesegang, *Kolloidchemische Technologie*, 2nd Ed., Dresden 1932, 1085 pp.

17 T. Svedberg, *Die Methoden zur Herstellung kolloider Lösungen anorganischer Stoffe*, 3rd Ed., Dresden 1922, 507 pp.

18 T. Svedberg and K. O. Pedersen, *Die Ultrazentrifuge*, Dresden 1940, 433 pp.

19 F. V. von Hahn, *Dispersoidanalyse*, Dresden 1928, 553 pp.

20 J. Perrin, *Atoms*, 2nd Ed., London 1923, 231 pp.

21 W. Pauli and E. Valko, *Elektrochemie der Kolloide*, Vienna 1929, 647 pp.

22 T. R. Bolam, *The Donnan Equilibrium*, London 1932, 161 pp.

23 P. H. Prausnitz and J. Reitstötter, *Elektrophorese, Elektroosmose, Elektrodialyse*, Dresden 1931, 307 pp.

24 M. H. Fischer and M. O. Hooker, *Die lyophilen Kolloide*, Dresden 1935, 233 pp.

25 E. C. Bingham, *Fluidity and Plasticity*, New York 1922, 440 pp.

26 R. Houwink, *Elasticity, Plasticity and Structure of Matter*, 2nd Ed., Cambridge 1954, 368 pp.

27 R. Fricke and G. F. Hüttig, *Hydroxyde und Oxyd-Hydrate*, Leipzig 1937, 641 pp.

28 W. Clayton, *The Theory of Emulsions*, 3rd Ed., Philadelphia 1935, 458 pp.

29 C. Philipp, *Technisch verwendbare Emulsionen*, 2nd Ed., 2 vols., Berlin 1940, 587 and 895 pp.

30 A. Winkel and G. Jander, *Schwebestoffe in Gasen (Aerosole)*, Stuttgart 1934, 116 pp.

31 W. E. Gibbs, *Clouds and Smokes*, London 1924, 240 pp.

32 R. Whytlaw-Gray and H. S. Patterson, *Smoke*, London 1932, 192 pp.

33 N. K. Adam, *The Physics and Chemistry of Surfaces*, 3rd Ed., Oxford 1941, 448 pp.

34 A. Marcellin and R. Köhler, *Oberflächenlösungen (Zweidimensionale Flüssigkeiten und monomolekulare Schichten)*, Dresden 1933, 160 pp.

35 E. Hückel, *Adsorption und Kapillarkondensation*, Leipzig 1928, 308 pp.

36 J. W. Macbain, *The Sorption of Gases and Vapours by Solids*, London 1932, 577 pp.

37 F. Krczil, *Adsorptionstechnik*, Dresden 1935, 132 pp.

38 L. Zechmeister and L. Cholnoky, *Die chromatographische Adsorptionsmethode*, Vienna 1937, 231 pp.

39 E. and M. Lederer, *Chromatography, A Review of Principles and Applications*, 2nd Ed., Amsterdam 1956, approx. 600 pp.

40 F. Cramer, *Papierchromatographie*, Weinheim 1951, 81 pp.

41 J. N. Balston and B. E. Talbot, *A Guide to Filter Paper and Cellulose Powder Chromatography*, London 1952, 145 pp.

42 R. J. Block, R. LeStrange and G. Zweig, *Paper Chromatography, A Laboratory Manual*, New York 1952, 195 pp.

CHAPTER 17

CATALYSIS AND REACTION KINETICS

1. Introduction

(a) Catalysis

The term *catalysis* [1–5] is applied to the acceleration of the rate of chemical reactions by the addition of substances which are not themselves consumed by the reactions—i.e., which appear to act by virtue of their mere presence. Such substances are called *catalysts*. The action of catalysts is shown in a particularly striking manner by certain reactions which do not take place at all at a measurable rate if catalysts are not present. Thus, in the absence of other substances, hydrogen peroxide appears to be absolutely stable, i.e., no perceptible decomposition takes place over the period of time involved in an observation. When certain substances are added, however, (e.g., finely divided platinum) vigorous decomposition at once ensues (cf. Vol. I, p. 55). A similar situation applies to the initiation of the combustion of electrolytic gas (detonating gas) at ordinary temperature by platinum black. In such a case it is said that the reaction is initiated by the added substance, and from this usage the added substance receives the name of a 'catalyst' (Greek ϰαταλύειν = to initiate). If, as in the instances quoted, the catalyst is not present in the same phase as the system in which reaction takes place, the process is described as *heterogeneous catalysis* and the catalyst is sometimes referred to as a *contact material*. If, on the other hand, the catalyst and the reacting substances form a single phase (homogeneous) system—as, for example, in the acceleration of the decomposition of hydrogen peroxide by OH^- ions—the process is referred to as *homogeneous catalysis* [2]. The substance, or mixture of substances whose reaction is speeded up by the catalyst, is termed the *substrate* of the catalysis.

It is frequently possible for several different reactions to occur in one system. Thus, a mixture of CO and H_2 could react to form CH_3OH (cf. Vol. I, p. 429), or $CH_4 + H_2O$ (cf. Vol. I, p. 444), or liquid hydrocarbons (cf. Vol. I, p. 431). According to the catalyst which is chosen, one or other of these reactions can be so accelerated that it takes place almost exclusively. In such cases it is said that the reaction is directed ('steered') by the catalysis.

Since the catalyst is not used up in the reaction which it brings about or accelerates, it should in principle be possible to bring about reactions between unlimited quantities of reactants by means of an indefinitely small quantity of catalyst. In practice, however, limits are set to the catalytic power by the fact that the catalyst may be gradually consumed by reactions other than that which is being catalyzed, but taking place simultaneously (cf. Vol. I, p. 708). Moreover, the catalyst may have its characteristics altered, and be thereby made ineffective, by impurities which are present accidentally in the reaction mixture. This is referred to as 'poisoning' the catalyst, and the impurities responsible are described as 'catalytic poisons'. The poisoning of catalysts is particularly important as an interfering factor which accompanies heterogeneous catalysis, as the consumption of the catalyst in side reactions is important in homogeneous catalysis.

In order to have any effect, the catalyst must necessarily come into contact with the molecules whose reaction is to be accelerated. The catalytic acceleration of the reaction is therefore naturally proportional to the amount of catalyst used, since the larger the quantity of catalyst, the greater is the number of molecules of substrate which can reach it by diffusion in unit time. The number of molecules of substrate which can come into contact with the catalyst per unit of time depends upon the concentration of catalyst, in the case of homogeneous catalysis, and upon its surface area in heterogeneous catalysis. The efficiency of a given contact material therefore increases with its degree of subdivision (surface per unit weight, *specific surface*). This fact accounts for the especially great catalytic activity of *colloidal* materials [*10*]. The catalytic efficiency of substances of appropriate chemical nature may be still greater if they are dissolved in a state of *molecular dispersion*. According to Bredig, colloidal platinum can catalytically decompose many million times its own amount of hydrogen peroxide. However, the catalytic acceleration produced by copper ions upon the autoxidation of sulfurous acid can still be observed at a dilution of 1 in 10^9, and the catalytic acceleration of silver ions upon hemolysis (dissolution of the red blood corpuscles) can be detected even at a dilution of 1 in 10^{15} (the so-called 'oligodynamic' action of silver).

Catalytic processes have become of enormous importance in chemical industry. In addition to the large scale synthetic processes discussed in Vol. I, there are innumerable other branches of inorganic and organic chemical industry which now make use of catalytic reactions [*8, 9, 11, 12*].

(b) Biological Importance of Catalysis [*6, 7*]

Catalysis is of very great importance in the vital processes of the living organism. The organic substances which function as catalysts in the organism are known as *ferments* or *enzymes*. It was formerly held that there was a fundamental difference between ferments and inorganic catalysts in that the former were *specific*—i.e., acted only on certain specified substances, in a particular manner. However, it has been found that even ferments are not invariably specific in their action, while by suitable modifications in their composition it has proved possible to confer on inorganic catalysts an increasing degree of specificity.

(c) Historical [*4*]

Isolated instances of catalytic processes were already observed during the latter half of the eighteenth century. Thus Scheele (1782) discovered the catalytic effect of mineral acids (i.e., of H⁺ ions) upon esterification and saponification. The observation made by Döbereiner in 1823, that hydrogen would ignite in air at ordinary temperature in contact with platinum sponge ('Döbereiner's lamp') created particular interest. A concept of catalysts was first formulated in 1835 by Berzelius, who defined catalysts as substances which 'by their mere presence would bring about chemical effects that would not take place if they were absent'. It was due to W. Ostwald (1888) that the factor of *time* was introduced into the definition of catalysts, whereby these are regarded as substances with the property of *accelerating* chemical reactions without themselves being consumed. This definition was the first to indicate how the nature of catalytic action might be explained, but only more recently, as a result of the detailed study of the kinetics and the mechanism of chemical reactions, has it become possible to gain some insight into the nature of the actual processes involved. The investigation of composite catalysts ('promoted' catalysts), which has been of such immense importance in the technical development of catalysis, received its stimulus largely from the great role which such catalysts have played in the ammonia synthesis. Work in this field was initiated especially by Mittasch, since 1909 [cf. *Ber.* 59, (1926) 13 and *Z. Electrochem.* 36, (1930) 569].

(d) Theories of Catalysis [*1, 2*]

Even in the time of Berzelius, many investigators expressed the view that the essential role of the catalyst lay in its combination with the substances between which it induced reaction, whereby highly reactive *intermediate compounds* were formed, from which the catalyst was once more regenerated by their subsequent decomposition. In the lead chamber process the occurrence of such intermediate compounds, especially nitrosyl sulfuric acid,* had already been demonstrated by Humphrey Davy in 1812.

As opposed to this, the occurrence of catalytically accelerated reactions on platinum and similar materials was cited. It appeared highly improbable that the formation of intermediate compounds with hydrogen, oxygen, or hydrogen peroxide could take place with such a noble metal as platinum. Many workers—as, for example, Liebig—went so far as to exclude from the category of genuine catalytic reactions those processes in which it was possible to detect the formation of intermediate compounds between the substrate and the substance inducing the reaction. However, the intermediate reaction theory justified itself for the interpretation of *homogeneous* catalysis when the formation of intermediate compounds between catalyst and substrate had been proved in a large number of cases. The action of contact substances in *heterogeneous catalysis* was nevertheless regarded until very recently as residing principally in their ability to increase the concentration of the substrate at their surfaces, as a result of adsorption.

The formation of 'intermediate compounds' in a broad sense is today assumed not only in homogeneous catalysis, but also in heterogenoeus catalysis: these intermediates are, however, not chemical compounds in the ordinary sense, but products of various types arising from the action of 'secondary valence' forces. They frequently have no well defined stoichiometric composition; even when such is the case, their life time is often so brief that it would be quite impossible to isolate them in weighable amounts. The methods which have been evolved for studying the mechanism of chemical reactions have rendered it possible to detect the formation of excessively short-lived intermediate products of this kind.

From such studies it has emerged that the course of chemical reactions—quite irrespective of whether or not they involve a catalyst—is considerably more complicated than is expressed by the ordinary chemical equations which, in general, only show the initial substances and the final products of reaction. Even when the chemical equations are broken down to represent intermediate reactions in the ordinary sense, they still take account only of the intermediates which can be obtained in weighable quantities. In reality, as the study of reaction mechanisms has demonstrated, chemical reactions very frequently proceed by way of *free radicals* or other intermediate species which are far too unstable to be isolated. It has been shown that one essential function of a catalyst is to promote the formation of such substances, or to favor their subsequent reactions in some well defined direction.

The study of *reaction rates*, and the elucidation of *reaction mechanisms* have thus become of fundamental importance for an interpretation of catalysis. We shall therefore precede a discussion of the present theory of catalytic processes by a brief review of the most important findings of *reaction kinetics* [*14–18*]—i.e., the study of reaction rates and reaction mechanisms.

* cf. Vol. I, p. 708.

2. **Chemical Reaction Kinetics** [*19*]

(a) **Reaction Velocity**

If a reacting substance is present at a concentration c, the rate of a reaction which it undergoes is measured by the change in concentration with time, i.e., by the expression dc/dt, where dc is the change of concentration during an infinitesimal interval of time dt*. The decrease in concentration of the substance consumed in the reaction ($-dc$) is equivalent to the increase in concentration ($+dc$) of the products of reaction. The rate of a reaction depends upon the temperature, and on the concentrations of the reacting substances. The manner in which it depends on the concentrations determines the so-called *order* of the reaction.

(b) **The Order of Reactions**

If the concentration of only one of the starting materials alters appreciably during the course of reaction, it is frequently found that

$$\frac{-dc}{dt} = kc \tag{1}$$

where k represents a constant (the *rate constant*) depending upon the nature of the reacting substance and on the experimental conditions. The velocity of the reaction is thus proportional to the concentration of the substance which is undergoing transformation. Such a reaction is said to be *of the first order*.

If two substances react according to the general equation $A + B = C$**, reactions in a homogeneous system (homogeneous reactions) commonly follow the expression:

$$\frac{-dc}{dt} = k \cdot c_A \cdot c_B \tag{2}$$

where c_A, c_B stand for the concentrations of A and B, respectively. Such reactions are called *reactions of the second order*.

If *three* substances react according to $A + B + C = D$, the rate of the reaction may be proportional to each of the concentrations c_A, c_B, c_C (i.e., to their product) so that

$$\frac{-dc}{dt} = k \cdot c_A \cdot c_B \cdot c_C \tag{3}$$

The reaction is then *of the third order*.

Reactions of higher order than the third are but rarely met with. However, it is not infrequently found, especially with reactions occurring at phase boundaries (heterogeneous

---

* In defining the rate of the reaction, the concentration change must refer to an infinitesimal interval of time, since c is itself changing with time. The change in concentration which may be determined experimentally is given by integrating the differential equation over a finite interval of time. If the substances consumed in the reaction are present at equal concentrations, the results of integrating equations (1) to (4) are:

(1a) $\ln(c_0/c_t) = kt$; (2a) $1/c_t - 1/c_0 = kt$; (3a) $\frac{1}{2}(1/c_t^2 - 1/c_0^2) = kt$; (4a) $c_0 - c_t = kt$.

In these expressions, c_0 is the initial concentration of reactants, and c_t the concentration after a time t.

** It is immaterial whether one reaction product is formed or several products, provided that the starting material is not being re-formed to any important extent, by a reaction proceeding in the reverse direction. The more nearly a position of equilibrium is reached, the more the effect of the back reaction makes itself noticeable. In that case, the measured reaction velocity is equal to the difference between the ratio of the opposed forward and back reactions.

reactions) that the rate of reaction is practically independent of the concentration of the substances reacting:

$$\frac{-dc}{dt} = k \tag{4}$$

In this case, the reaction is said to be *of zero order*.

If the reactants are present in equal concentration, the expressions on the right hand side of equations (2) and (3) can be written in the form kc^2 and kc^3. The corresponding expressions in equations (1) and (4) could be written in the form kc^1 and kc^0. The order of the reaction is thus given by the exponent of c. Reactions are also known for which non-integral values are found for the exponent of c (cf. p. 697).

(c) Molecular Kinetic Interpretation of Reaction Order

In a reaction which follows the general equation $A + B = C$—e.g., for the reaction

$$H_2 + I_2 = 2HI \tag{5}$$

the velocity of the reaction is apparently determined by the frequency with which the molecules of the reactants collide with each other. The number of collisions between two molecules of the same kind, according to the kinetic theory of gases, is proportional to the square of their number within a given volume—i.e., to the square of their molecular concentration—and the number of collisions between two molecules of different kinds is proportional to the product of their respective molecular concentrations. If, then, it is found that eqn (2) describes the velocity of the reaction, this indicates that the transformation is brought about by collisions between two molecules of different kinds. This is so, for example, in the reaction between hydrogen and iodine in the gas phase. Equation (5) thus represents not only the quantitative relation between reactants and reaction products, but also the *mechanism of the reaction*. As has already been remarked, this is by no means true of the ordinary equations symbolizing most of the reactions of chemistry.

The decomposition of HI into H_2 and I_2 must also come about by the collision of two molecules (in this case 2HI). Equation (2) thus applies to this reaction also, with $c_A = c_B = c_{HI}$. The rate of decomposition is therefore proportional to the square of the hydrogen iodide concentration. As further examples of second-order homogeneous reactions, there may be quoted the thermal decomposition of nitrous oxide, according to the equation $2N_2O = 2N_2 + O_2$, and of chlorine monoxide: $2Cl_2O = 2Cl_2 + O_2$.

If, in a reaction following the general equation $A + B = C$, the substance B is present at a much higher concentration than substance A, so that its concentration does not sensibly alter, the concentration c_B in eqn (2) can be incorporated in the constant k. Equation (2) then passes over into equation (1). This case is frequently found for homogeneous reactions taking place in solution, when the solvent participates in the reaction.

The inversion of cane sugar may be cited as an example; it may be represented by

$$C_{12}H_{22}O_{11} + H_2O = 2C_6H_{12}O_6.$$

At the concentrations ordinarily used for its investigation, the amount of water used up in the reaction with sucrose is vanishingly small in comparison with the total amount of water, and the reaction therefore appears to be a first order process.

It is frequently found that reactions take place by processes of lower order than would be expected from the equations representing the over-all chemical reaction. Thus it has been established that the decomposition of nitrogen pentoxide, by the process $2N_2O_5 = 4NO_2 + O_2$ takes place both in the gas phase and in indifferent solvents by a *first order* reaction, and not by a *second order* reaction, as would be expected if the foregoing equation represented the reaction mechanism. In such cases, it has been possible to show that the reaction proceeds by way of *intermediate stages*, and if it is desired to express the mechanism of the reaction, it is necessary to break the over-all chemical equation down into several partial equations. It can readily be perceived that the rate of the reaction, as measured, must be determined by that member of the chain of consecutive reactions which takes place most slowly.

It is also possible for the reaction velocity, as measured, to be of a *higher* order than would be indicated by the partial reactions. Thus, according to Trautz (1924), the reaction

$$2NO + Cl_2 = 2NOCl \tag{6}$$

probably takes place in two stages:

$$NO + Cl_2 \rightleftharpoons NOCl_2 \tag{6a}$$

$$NOCl_2 + NO = 2NOCl \tag{6b}$$

even though, from the experimental measurements, it appears to be a third order reaction. If it is assumed that reaction (6b) is slow as compared with (6a), the latter can lead to an equilibrium governed by the Mass Action Law:

$$[NO] \cdot [Cl_2] = k_1[NOCl_2].$$

The rate of reaction (6b) is given by:

$$-\frac{d[NO]}{dt} = k_2[NOCl_2][NO].$$

Substituting the concentration of $NOCl_2$ deduced from the Mass Action law, the velocity of the reaction is found to be

$$-\frac{d[NO]}{dt} = \frac{k_2}{k_1}[NO]^2[Cl_2];$$

this depends upon the concentration in exactly the same manner as would be deduced for a reaction proceeding directly, according to equation (6), by ternary collisions.

Ternary collisions—i.e. simultaneous collisions of three molecules are very much less frequent than ordinary (binary) collisions. In a gas at atmospheric pressure, the ratio of ternary collisions to binary collisions is about 1 : 1000*. In general, therefore, it is much more probable that a reaction involving three molecules should take place by way of two binary collisions than through a single ternary collision.

* This holds for molecules built up from small number of atoms. The ratio of the probabilities of a molecule A colliding with a molecule B exactly during its collision with a second molecule of B, and of its colliding with an isolated molecule of B is the same as the ratio of the diameter of the molecule B to its mean free path.

(d) **Molecularity of Reactions**

Reactions in which only *one* molecule undergoes transformation in each individual act are said to be *unimolecular*; those involving a collision between two molecules are called *bimolecular*, and those involving ternary collisions *trimolecular*. Reactions higher than trimolecular would appear to be extremely improbable, because four-body collisions are excessively rare events. Even trimolecular reactions are very uncommon.

Among simple reactions, those which take place by way of intermediate stages are distinguished as *composite reactions*, and the individual stages are termed *partial reactions* or *consecutive reactions*. E.g.,

$$\text{Composite reaction} \qquad A + B + C = D$$
$$\text{Consecutive reactions} \quad \begin{cases} A + B = E \\ E + C = D \end{cases}$$

Reactions which really represent the processes between the molecules, and thus express not merely the result of the chemical change, but also its mechanism, are spoken of as *elementary* or *primary* reactions.

If one of a set of consecutive reactions is very slow as compared with the rest, then, as has already been stated, this reaction alone practically determines the rate of the total reaction. In this case, the reaction as a whole is found to be of an order which is given by the molecularity of the slowest partial reaction. If, on the other hand, the rates of the consecutive reactions differ but little, the order of the reaction is determined by the partial reactions as a whole. For this reason it is not uncommon to find orders of reaction which are not whole numbers.

Many reactions which, as ordinarily represented, would appear to constitute simple examples of homogeneous gas reactions have proved on closer study either not to proceed at all in the gas phase under ordinary conditions, but on the surfaces of the containing vessel ('wall reactions'), or else to take place by quite a different mechanism from that expressed by the ordinary chemical equation. Thus the reaction $2CO + O_2 = 2CO_2$ ordinarily takes place as a wall reaction. The *explosive* reaction does, indeed, take place in the gas phase, but involves the presence of water vapor. The latter makes possible the following bimolecular consecutive reactions:

$$CO + H_2O = CO_2 + H_2; \qquad H_2 + O_2 = H_2O_2; \qquad 2H_2O_2 = 2H_2O + O_2 \text{*}.$$

This reaction also exemplifies the important role of water vapor in catalytically accelerating gas reactions, as has frequently been observed**.

If the number of collisions between gas molecules, as calculated from the kinetic theory of gases, is compared with the number of collisions necessary to produce the

---

* There is now no doubt that fast gas-phase reactions proceed by free-radical reactions and chain mechanisms. It is now generally agreed that the second reaction represented above involves the following steps: initiating reaction forming free radicals, e.g., $H_2 + O_2 = 2OH$; followed by $OH + H_2 = H_2O + H$, $H + O_2 = OH + O$ (propagation); $O + H_2 = OH + H$ (branching); $H + O_2 + X = HO_2 + X$ (X = third body); $HO_2 + H_2 = H_2O + OH$; $HO_2 + H_2 = H_2O_2 + H$. There may also be other radical recombination mechanisms on the wall. The recurring steps in the chain may be $H + O_2 = HO_2$, followed by $HO_2 + H_2 = H_2O_2 + H$ [23].

** It has occasionally been assumed that reactions are catalyzed by very small traces of water vapor when, in reality, the inhibition of reaction brought about by intensive drying has been due to the poisoning of the system by impurities. See, for example, Bodenstein, *Z. Physik. Chem.* B. 20, (1933) 451 and Rodebush, *J. Am. Chem. Soc.*, 55, (1933) 1742.

amount of chemical change, as determined from the observed velocity of reaction, it is found that in gas reactions which proceed at a conveniently measurable speed at atmospheric pressure, only about 1 collision in 10^{10} to 10^{12} leads to reaction. Thus it is often only a very small fraction of the colliding molecules which are capable of reacting with one another. It is an obvious assumption that these molecules which are capable of reaction differ in some way from the rest. For the present they may be termed 'activated' molecules; the nature of the activation will become clear from the subsequent discussion.

(e) Reaction Velocity and Temperature

Arrhenius (1889) found that the relation between the velocity of a reaction and the temperature could be expressed, over a wide range of temperature, by the expression:

$$\frac{d \ln k}{dt} = \frac{A}{RT^2} \tag{7}$$

In this expression, k is the velocity constant of the reaction, T the absolute temperature, R the gas constant, and A is a quantity which has the dimensions of energy*. Arrhenius called this energy the 'energy of activation'—i.e., the amount of heat which must be imparted to the molecules in order to raise them from their ordinary state, in which they are not capable of reacting, into the 'activated' state. The Arrhenius equation holds both for homogeneous and for heterogeneous reactions. Just as energies of activation can be calculated from the velocity of reaction, it is conversely possible, if the energy of activation is known, to predict the rate of reaction.

The need for 'activating' the gas molecules explains why it is very often necessary to impart energy in order to initiate a gas reaction, even when the reaction itself is highly exothermic. If this energy is added in the form of heat, the initiation of reaction is spoken of as *ignition*.

(f) Photochemical Reactions [13]

Reactions which are initiated by *irradiation with light* are called *photochemical reactions*. Since the increment of energy, $\varepsilon_2 - \varepsilon_1$, which a molecule gains by absorption of light is given (just as for the atom, Vol. I, p. 85), by the frequency of the absorbed light:

$$\varepsilon_2 - \varepsilon_1 = h\nu \tag{8}$$

it is possible to determine the energy to be expended *for the activation of a single molecule* from the study of photochemical reactions. Further, the total amount of light energy of frequency ν with which the system is irradiated determines the *number of molecules activated*. The quotient $E/h\nu$, where E is the quantity of energy supplied, represents the number of light quanta, and since each molecule absorbs a single quantum $h\nu$ for its activation, the quotient is also the number of molecules activated. If, from other studies it is known how much energy must be expended in

---

* Thus equation is not only formally analogous to Van't Hoff's reaction isochore (eqn. (5) on p. 37 of Vol. I), but can be derived from it. It follows from the derivation that (7) is strictly valid only when it contains an extra term; in practice, however, this term can generally be neglected (cf. footnote, p. 707).

order to produce some definite change in the molecule (e.g., to excite an electron to a higher energy level, or to dissociate the molecule into atoms), then the frequency ν of the light which initiates some particular reaction reveals also *the nature of the transformation* which the molecule must undergo in order to react with the other substance concerned.

This may be illustrated with reference to the photochemical hydrogen-chlorine reaction. Combination between hydrogen and chlorine $H_2 + Cl_2 = 2HCl$ is induced by irradiation with light of wave length 4785 Å or less. From this, it follows in the first place that only the *chlorine* need be activated, and not the hydrogen, since hydrogen does not absorb light of this wave length. The light quantum corresponding to the wave length given has a magnitude of $hc/\lambda = 4.10 \cdot 10^{-12}$ erg (cf. Vol. I, p. 84 *et seq.*). The energy necessary to break up the Cl_2 molecule into 2 Cl atoms is $3.93 \cdot 10^{-12}$ ergs, as follows from the data listed in Table 104, Vol. I, p. 773, dividing the figure given by Avogadro's number and the thermal equivalent of the erg. Thus the light quantum supplies the energy necessary to dissociate the Cl_2 molecule into atoms*. Cl atoms, unlike Cl_2 molecules, are capable of reacting with H_2 molecules at ordinary temperature. Irradiation with light therefore initiates the reaction between hydrogen and chlorine at room temperature.

(g) Chain Reactions [20, 21]

When the quantity of hydrogen chloride formed in the photochemical combination of hydrogen and chlorine is compared with the total amount of light energy absorbed, it is found that the yield of hydrogen chloride is greater, by an enormous factor, than the number of Cl atoms produced by irradiation of the Cl_2 molecules. Each quantum of light can dissociate only a *single* Cl_2 molecule. Nevertheless, Bodenstein (1913 and later) found that often more than 100,000, and in some circumstances as many as 3,000,000 Cl_2 molecules react with hydrogen for every quantum of light. As was first shown by Nernst (1918), this can be explained in the following way. The primary formation of chlorine atoms (9)

$$Cl_2 + h\nu = 2Cl \quad (\text{'primary process'}) \tag{9}$$

is followed by a sequence of self-reproducing reactions of the following kind:

$$\left. \begin{array}{l} Cl + H_2 = HCl + H \\ H + Cl_2 = HCl + Cl \\ Cl + H_2 = HCl + H \\ H + Cl_2 = \ldots \ldots \ldots \text{ etc.} \end{array} \right\} \text{('Reaction Chain')} \quad \begin{array}{l} (10a) \\ (10b) \\ (10c) \end{array}$$

The reactions (10a) and (10b) alternate with one another indefinitely until the H and Cl atoms are removed from the system by some accidental side reaction, such as

$$H + Cl = HCl \quad (\text{'Chain breaking process'}) \tag{11}$$

It has subsequently been found that *chain reactions* of this kind are of extremely frequent occurrence; free radicals may enter into them as intermediate products of reaction, in place of free atoms. [22]

The longer, on the average, that the chain breaking process is deferred, the greater is the quantity of material entering into reaction from a given number of primary acts. Since, in

---

* That the individual Cl_2 molecules absorb rather more light energy than is necessary to split them into atoms is due to the fact that the photochemical dissociation furnishes one ordinary and one 'excited' Cl atom—i.e., one atom in which an electron occupies a higher energy level than in the ground state of the atom.

the example taken, the H or Cl atoms liberated in one of the component partial reactions are always close to the Cl_2 or H_2 molecules with which they can react, the 'chain breaking' reaction (11) can only very rarely take place. The prospects for propagation of the chain are far less favorable, however, if the gas mixture contains impurities or added substances which are capable of capturing the free radicals or atoms liberated as chain-propagating intermediates. Even minimal concentrations of such impurities (e.g., of oxygen in the case of the H_2-Cl_2 reaction) can diminish the average chain length by a considerable factor. Bodenstein states that, in the hydrogen-chlorine reaction, chain breaking as a result of reaction (11) practically never occurs*; it is invariably brought about by impurities which are present in traces in the reacting mixture. Quite generally, the great sensitivity displayed towards impurities in the reacting substances is a characteristic feature of chain reactions.

Chain reactions are not by any means involved in *all* photochemical processes. Processes are known in which each quantum of light absorbed leads to the reaction of exactly 1 molecule. On the other hand, chain reactions are not in any way limited to photochemical reactions, they may be initiated by other forms of energy, as well as by light energy—e.g., by the energy of molecular collisions (in the form of thermal energy, or of α-rays). The reaction chain which was quoted as an example can itself be initiated by introducing atomic hydrogen into the reaction system. The reaction

$$Na + Cl_2 = NaCl + Cl$$

can also serve as the primary process.

Chain reactions play an especially important role in the processes of combustion in flames. Their multiplication ('branching') can be responsible for the transformation of combustion into an *explosion* [24–26]. This can often be inhibited by adding other substances which lead either to the early termination of the reaction chains, or to the suppression of those primary reactions which initiate particularly efficient chain processes. The action of 'antiknocks' in fuels for internal combustion engines (cf. Vol. I, p. 557) depends on this effect.

(h) Activation Energy and Potential Barriers

(*i*) *General*. Fig 94 shows in a diagrammatic form how the chemical potential (i.e., the ability to perform work by means of chemical reaction) changes with time during the photochemically initiated hydrogen-chlorine reaction. For the sake of simplicity we consider only the course of the reaction between 2 molecules of Cl_2 and 2 molecules of H_2, and assume that the chain is terminated by the process $H + Cl = HCl$, after the third member of the reaction chain.

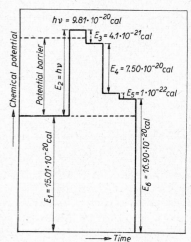

Fig. 94. Schematic diagram showing how potential barrier of chemical reaction can be surmounted as result of photochemical activation.
Example: hydrogen-chlorine reaction.

In this case, the entire course is represented by equations (9), (10a), (10b), (10c), and (11) (p. 699). The chemical energy of the system

* The probability of recombination in the gas phase is very much smaller for free atoms than for free radicals, since, unlike the latter, free atoms can recombine only as the result of ternary collisions. It is essential for a third body (which may be of any species) to be involved in the collision, in order to carry away a portion of the energy of the collision. A pure elastic collision must otherwise take place, the two atoms flying apart again, with the same relative velocities with which they collided.

is greater at the inception of the reaction than in the final state, by the amount of the energy of formation of 4HCl:

$$\frac{4 \cdot 22750}{6.064 \cdot 10^{23}} = 15.01 \cdot 10^{-20} \text{ cal.}$$

This difference in energy between initial and final states is shown as E_1 in Fig. 94.

In spite of the rather large gain of energy associated with the formation of HCl (22.75 kcal for 1 mol of HCl), the system cannot pass from the initial state ($2H_2 + 2Cl_2$) into the final state (4HCl). Experience shows that, provided 'wall reactions' are prevented, a mixture of chlorine and hydrogen can be storted in-definitely in the dark without reaction taking place. If we leave out of account the possibility of initiating reaction catalytically, it is necessary to raise the system to a *state of higher energy* in order to start the reaction. This is graphically expressed in the statement that the initial and final states are separated by a *'potential barrier'*. The system must first be 'lifted' over this before the exothermic processes, which lead to the final state, can set in.

The conditions are similar to those obtaining for radioactive atoms; as we have seen, the spontaneous disintegration of their nuclei is likewise hindered by a 'potential wall'. There is a certain difference, in that the potential barrier shown in Fig. 79 (p. 590) represents the potential of the nucleus as a function of *distance* from the center of the nucleus, whereas Fig. 94 represents the chemical potential of the system as a function of *time*.

In the case under consideration, the potential barrier is due to the fact that Cl_2 molecules as such cannot react with H_2 molecules; only Cl *atoms* can do so. Thus energy must first be expended in dissociating a Cl_2 molecule into Cl atoms, before reaction can be initiated. The energy expended in bringing about dissociation is the *activation energy* of the system considered; in this particular case, as already mentioned, it is $3.93 \cdot 10^{-12}$ erg or $9.41 \cdot 10^{-20}$ cal.* This quantity then determines the height of the potential barrier which prevents the spontaneous reaction of chlorine with hydrogen.

In order to start the reaction, it is sufficient to supply $9.41 \cdot 10^{-20}$ cal of energy to a single Cl_2 molecule. If, however, the molecule is dissociated photochemically, one of the two atoms, for atom-mechanistic reasons, is liberated in an 'excited' state; the energy of dissociation is thereby increased, in this case, to $9.81 \cdot 10^{-20}$ cal (E_2 in Fig. 94). We will (quite arbitrarily) assume that the 'excited' Cl atom is the first to react with a hydrogen molecule, according to equation (10a). Energy equal to $E_3 = 0.41 \cdot 10^{-20}$ cal is thereby liberated**. The succeeding reaction (10b) takes place with liberation of $E_4 = 7.50 \cdot 10^{-20}$ cal. In the meanwhile, the second Cl atom originating from the primary process has also entered into reaction with H_2. The energy set free ($E_5 = 0.01 \cdot 10^{-20}$ cal) is smaller than E_3, by the amount equivalent to the 'excitation energy'***. In accordance with our assumptions, the process now terminates with the reaction $H + Cl = HCl$. Although this is strongly exothermic ($E_6 = 16.90 \cdot 10^{-20}$ cal), it actually takes place only very rarely, because it is a process of low probability (cf. p. 700). The total energy set free in the reaction, $E_3 + E_4 + E_5 + E_6$ is equal to the initial energy of the system E_1 augmented by the radiant energy

---

* Since activation energies are generally measured in heat units (cal or kcal) they are often called 'heats of activation'.
** The quantities of energy E_3, E_4, E_5 and E_6 can readily be obtained from Tables 7, 104 and 105 of Vol. I, in the same way as E_1.
*** In the example considered, the excited state of the atom does not appear to be of any importance for its reactivity.

absorbed, $E_2 = h\nu$. The light energy is thus *used up* in the reaction. *In this respect the photo-chemical processes differ fundamentally from catalytic processes.*

If the photochemical primary process is followed by a chain reaction of 200,000 steps, the number of H_2 and Cl_2 molecules reacted is 100,000 times as many as in the example just considered. In that case, E_2 is infinitesimally small compared with E_1, which represents the total chemical energy of all the molecules entering into the reaction; nevertheless, the statement still remains true that the energy liberated in the reaction is greater by the amount $h\nu$ than the energy which would be liberated if the reaction proceeded without irradiation.

(ii) Thermal Activation. As already stated, the energy of activation can be imparted to the molecules which are to be activated not only in the form of light energy, but in other forms—e.g., as *thermal energy.* It is possible, from the kinetic theory of gases, to calculate the fraction of all the molecules of a gas which, at any given temperature possess a thermal energy in excess of the energy of activation.

According to the kinetic theory of gases [27], the heat content Q of an ideal monatomic gas is given by the sum of the kinetic energies of its molecules. If the number of molecules is N,

$$Q = N \cdot \tfrac{1}{2} m u^2 \qquad (12)$$

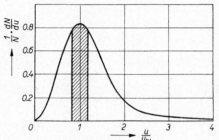

Fig. 95. Distribution of velocities among the molecules of a gas, according to the Maxwell distribution law.

where m is the mass of a single molecule, and $\overline{u^2}$ is the so-called 'mean square velocity' of their translational movement. This is because, the molecules do not all have the same velocity u, but the velocities of the individual molecules are distributed as shown in the curve Fig. 95 (Maxwell's distribution law). In Fig. 95, the velocities of the individual molecules are represented along the abscissa, the unit of measurement being the velocity most frequently represented, the so-called 'most probable (or *modal*) velocity' u_w. The ordinates give the values of $\dfrac{dN}{N} \cdot \dfrac{1}{du}$, i.e., the fraction of all the molecules having velocities between u and $u + du$, divided by the breadth du of this range of velocities. For example, the number of molecules with velocities between $0.8\ u_w$ and $1.2\ u_w$, as represented by the shaded area in Fig. 95, is 0.8×0.4, or about one third of the total number. In the form represented in Fig. 95, the shape of the distribution curve is independent of the temperature. The mean value of all the velocities, u_m, is rather greater than the most probable velocity u_w. The value entering into formula (12), however, is not the square of the mean velocity, u_m but the average of the squares of the individual velocities*. This is represented by the symbol $\overline{u^2}$, and the square root ('root mean square' velocity) by $\overline{u}$.

According to the kinetic theory of gases, the pressure p of a gas is given by:

$$p = \tfrac{1}{3} \cdot n\, m\, \overline{u^2} = \tfrac{1}{3}\, \frac{N}{v}\, m\, \overline{u^2} \qquad (13)$$

where $n = \dfrac{N}{v}$ represents the number of molecules per unit volume. For 1 gram molecule of gas (13) becomes

$$p = \tfrac{1}{3}\, \frac{N_A}{v} \cdot m\, \overline{u^2} \quad (N_A = \text{Avogadro's number}) \qquad (13a)$$

* The values u_w, u_m or $\overline{u}$ and $\overline{u^2}$ stand in the ratios $u_m = \dfrac{2}{\sqrt{\pi}}\, u_w = 1.128\, u_w$; $\overline{u^2} = \tfrac{3}{2}\, u_w^2$; $\overline{u} = \sqrt{\overline{u^2}} = 1.225\, u_w = 1.085\, u_m$.

Combining (13a) with the gas equation $pv = RT$, again referred to 1 gram molecule, it follows that

$$\tfrac{1}{3} N_A m \overline{u^2} = RT \qquad \text{or} \qquad \tfrac{1}{2} m \overline{u^2} = \tfrac{3}{2} \cdot \frac{R}{N_A} \cdot T \tag{14}$$

The mean kinetic energy (so-called 'translational energy') of the molecules is directly proportional to the absolute temperature. The most probable velocity u_w (as also u) thus increases, proportional to $T^{\frac{1}{2}}$. If, then, the probabilities of the velocities are plotted directly against the velocity u, and not against the ratio u/u_w as was done in Fig. 95, the maximum probability shifts progressively towards higher velocities with rise in temperature.

Whereas equations (13) and (14) are applicable to ideal gases generally, equation (12) holds only for *monatomic* ideal gases. Poly-atomic gases can contain heat not only in the form of *translational energy*, but also in the form of *rotational energy* of their molecules and *vibrational energy* of the atoms of which the molecules are composed. As is well known, this manifests itself in the differing ratios of the specific heats at constant pressure and at constant volume (c_p/c_v) found for monatomic and poly-atomic gases, respectively.* The following discussion, in which only the translational energy is considered, apply essentially to these other forms of energy also.

It may be deduced from the Maxwell distribution law that the fraction of molecules a with kinetic energy such that the component along the line of centers during a molecular collision is greater than A is given by

$$a = e^{-A/RT} \tag{15}$$

This expression also gives the *probability* of collision between two molecules such that $(A_1 + A_2)$, the sum of the components of their translational energy, along the line of centers, is greater than A. Thus (15) indicates the ratio of the number of collisions in which $A_1 + A_2 \geqq A$ to the total number of molecular collisions. If, for A we substitute the activation energy necessary for some chemical reaction, a then signifies the fraction of molecules which, in the most favorable case, can react with one another on collision. The requisite amount of energy to bring about chemical reaction is available at the instant of collision only to those molecules which bring to the collision a total energy equal to or greater than the activation energy. The reaction velocity is thus proportional to the value of a. Table 93 gives values of a, calculated from equation (15) at different temperatures and for several activation energies. It may be seen how very strongly dependent upon both temperature and activation energy these values of a are.

The velocities of, e.g., homogeneous bimolecular gas reactions can be calculated from the values of a, if it be assumed that *every collision* in which the collisional energy exceeds the activation energy leads to reaction**. For the sake of simplicity, consider a reaction between two identical molecules (as exemplified, for example, by the decomposition of hydrogen iodide, $2HI = H_2 + I_2$). The velocity of reaction is then twice as great as the number of collisions leading to reaction, since

---

* See, for example, Eggert, *Lehrbuch der physikal. Chemie* 7th. Edition, 1949, Partington, *Treatise on Physical Chemistry* Vol. I (Macmillan, London, 1950), Fowler, *Statistical Mechanics*, Cambridge University Press, 1936, Taylor and Glasstone, *Treatise on Physical Chemistry*, Vol. I, 1942, for the variation of rotational and vibrational energy with temperature.

** A comparison of calculated and observed reaction velocities indicates that the error implicit in this assumption is compensated, in practice, for molecules of simple structure, by the error which is introduced by neglecting their rotational and vibrational energy; cf. Hinshelwood [15].

TABLE 93

VARIATION OF THE QUANTITY a $(= e^{-A/RT})$ WITH THE ACTIVATION ENERGY A (in cal per mol) AND WITH THE TEMPERATURE (in °K)

Only those values of a inside the framed section of the table lead to reactions which take place at measurable speeds at ordinary pressure.

$A =$	1,000	10,000	20,000	40,000	50,000	100,000
$T = 300°$	$1.87 \cdot 10^{-1}$	$5.09 \cdot 10^{-8}$	$2.59 \cdot 10^{-15}$	$6.73 \cdot 10^{-30}$	$3.43 \cdot 10^{-37}$	$1.18 \cdot 10^{-73}$
$600°$	$4.32 \cdot 10^{-1}$	$2.26 \cdot 10^{-4}$	$5.09 \cdot 10^{-8}$	$2.59 \cdot 10^{-15}$	$5.86 \cdot 10^{-19}$	$3.43 \cdot 10^{-37}$
$1000°$	$6.04 \cdot 10^{-1}$	$6.49 \cdot 10^{-3}$	$4.21 \cdot 10^{-5}$	$1.77 \cdot 10^{-9}$	$1.15 \cdot 10^{-11}$	$1.32 \cdot 10^{-22}$
$1500°$	$7.15 \cdot 10^{-1}$	$3.48 \cdot 10^{-2}$	$1.21 \cdot 10^{-3}$	$1.53 \cdot 10^{-6}$	$5.09 \cdot 10^{-8}$	$2.59 \cdot 10^{-15}$
$2000°$	$7.77 \cdot 10^{-1}$	$8.05 \cdot 10^{-2}$	$6.49 \cdot 10^{-3}$	$4.21 \cdot 10^{-5}$	$3.18 \cdot 10^{-6}$	$1.15 \cdot 10^{-11}$
$3000°$	$8.45 \cdot 10^{-1}$	$1.87 \cdot 10^{-1}$	$3.48 \cdot 10^{-2}$	$1.21 \cdot 10^{-3}$	$2.26 \cdot 10^{-4}$	$5.09 \cdot 10^{-8}$

two molecules are transformed in each such collision. It is thus represented by

$$\frac{-dn}{dt} = 2a\,\frac{n \cdot Z}{2} \tag{16}$$

where n is the total number of reactant molecules per cm³, and Z is the 'collision frequency' (see below) of a single molecule*. It turns out that if a is greater than 10^{-7}, reactions take place instantaneously, whereas if a is less than 10^{-18}, no reaction is detectable. These numbers apply to reactions at ordinary pressure; their orders of magnitude are not affected if reaction between dissimilar molecules is considered, instead of reactions between molecules of the same kind. The limits of measurable reaction are shifted somewhat by a change in pressure. They are also different in the case of chain reactions, by a factor corresponding to the mean length of chain which follows the primary reaction: thus they are changed by 5 powers of ten for a chain of 100,000 members.

The limits just quoted for the range of reaction velocities which is accessible to investigation** are based on the arbitrary, but convenient definitions that a reaction is said to be 'instantaneous' when its rate is so fast that, after 1 second, at most 0.1% of the initial quantities have not yet entered into reaction:

$$n_t \leqq 0.001 n_0 \quad \text{(for } t = 1 \text{ sec)} \tag{17a}$$

On the other hand, a reaction is taken as 'no longer observable' when less than 0.1% of the reactants enter into reaction in a day:

$$n_0 - n_t \leqq 0.001 n_0 \quad \text{(for } t = 8.64 \cdot 10^4 \text{ sec)} \tag{17b}$$

In (17a) and (17b), n_0 is the number of molecules per cm³ of initial reactant at the time 0, and n_t the number remaining at a time t. The a values which fulfil these conditions can be

---

* If one molecule collides with Z other molecules each second, n molecules will make nZ collisions per second. If n is the total number of molecules per cm³, the number of separate collisions is $\frac{1}{2} nZ$, since *two* molecules participate in each collision.

** These limits do not imply that reaction velocities which fall within the range can necessarily be measured accurately. The range over which this is possible depends on the nature of the reaction, and the experimental technique employed, but in general is likely to be much narrower than the range defined by equations (17a), (17b) above.

calculated from (16). The collision number Z entering into this calculation is the number of collisions sustained by a single molecule in each second; according to the kinetic theory of gases, it is given by

$$Z = \sqrt{2} \cdot \pi \cdot \sigma^2 \cdot \overline{u} \cdot n \tag{18}$$

where $n = $ the number of molecules per cm³, $\overline{u}$ is the root mean square velocity, and σ the molecular diameter. A few collision numbers are set down in Table 94, in order to indicate the order of magnitude of this factor Z.*

TABLE 94

COLLISION FREQUENCIES Z IN SEVERAL GASES AT 0° C AND 1 ATM. PRESSURE

	O_2	Cl_2	NO	N_2O	H_2O
$Z = $	$4.14 \cdot 10^9$	$6.21 \cdot 10^9$	$4.86 \cdot 10^9$	$5.91 \cdot 10^9$	$8.85 \cdot 10^9$ sec^{-1}

	CO_2	CS_2	NH_3	CH_4
$Z = $	$5.76 \cdot 10^9$	$8.67 \cdot 10^9$	$8.36 \cdot 10^9$	$7.69 \cdot 10^9$ sec^{-1}

Substituting eqn (18) in (16), we obtain

$$\frac{-dn}{dt} = a \cdot \sqrt{2} \cdot \pi \cdot \sigma^2 \, \overline{u} \, n^2 \tag{19}$$

and from this, by integration

$$\frac{1}{n_t} - \frac{1}{n_0} = a \sqrt{2} \cdot \pi \cdot \sigma^2 \, \overline{u} \, t \tag{19a}$$

In (19a), n_t may be expressed in terms of n_0, using relation (17a). It is then found that

$$a_1 \gtrless \frac{1000 - 1}{\sqrt{2} \cdot \pi \cdot \sigma^2 \cdot \overline{u} \cdot n_0}.$$

Similarly, using relation (17b),

$$a_2 \lesseqgtr \frac{0.001}{8.64 \cdot 10^4 \cdot \sqrt{2} \cdot \pi \, \sigma^2 \, \overline{u} \, n_t}.$$

In these expressions, $\sqrt{2} \cdot \pi \cdot \sigma^2 \cdot \overline{u} \cdot n_0$ stands for the collision number at $t = 0$, and $\sqrt{2} \cdot \pi \cdot \sigma^2 \cdot \overline{u} \cdot n_t$ for the collision number at time t. In an order of magnitude calculation we can substitute, for both cases, the rounded off average from Table 94, i.e., $7 \cdot 10^9$. It is then found that

$$a_1 \gtrless \frac{999}{7 \cdot 10^9} = 1.43 \cdot 10^{-7}$$

and

$$a_2 \lesseqgtr \frac{10^{-3}}{8.64 \cdot 10^4 \cdot 7 \cdot 10^9} = 1.65 \cdot 10^{-18}.$$

* The collision numbers given in Table 94 are calculated from the viscosities η of the gases by means of the formula

$$Z = \frac{\overline{u}}{l} = 0.30967 \cdot d \cdot \overline{u^2}/\eta$$

which also follows from the kinetic theory. l is here the mean free path of the gas molecules and d the gas density.

The values of a which lead to reactions of measurable velocity thus lie roughly between the limits 10^{-7} and 10^{-18}. These limits, like the average collision number assumed, apply to $0°$ and 760 mm pressure. Within the range of temperature covered by Table 93, however, their order of magnitude does not change with rising temperature. This is because $\bar{u}$ is proportional to $T^{\frac{1}{2}}$, so that Z increases only about 3.3 fold if T is raised from $273°$ to $3000°$ K*. As follows from eqn (18) however, the value of Z is proportional to the pressure, since u is proportional to pressure.

If the activation energy is known, it follows from the foregoing section that the reaction velocity can be calculated directly from gas kinetic data. Combining it with eqn (15), eqn (19) can be written in the form

$$\frac{-dc}{dt} = \frac{\sqrt{2}}{1000} N_A \cdot \sigma^2 \, \bar{u} \, e^{-A/RT} \cdot c^2 \tag{20}$$

where c is the concentration in mols per liter, and N_A is Avogadro's number. Since $\bar{u}$ varies but little with temperature, the expression

$$\frac{\sqrt{2}}{1000} N_A \, \pi \cdot \sigma^2 \cdot \bar{u}$$

can be replaced by a constant C, so that (20) is transformed into

$$\frac{-dc}{dt} = C \, e^{-A/RT} \cdot c^2 \tag{21}$$

The product $C \cdot e^{-A/RT}$ corresponds to the coefficient k in eqn (2), p. 694, when referred to concentrations expressed in mols per liter.

Equation (20) is true of bimolecular reactions between like molecules (e.g., $2HI = I_2 + H_2$). For bimolecular reactions between unlike molecules (e.g., $H_2 + I_2 = 2HI$), $\bar{u}$ in eqn (20) must be replaced by $\sqrt{u_1^2 + \overline{u_2^2}}$, σ by the mean value $\dfrac{\sigma_1 + \sigma_2}{2}$, and c^2 by the product $c_1 \cdot c_2$.

It should be noticed that the derivation of eqn (20) presupposed that, in a collision, the energy effective in producing activation could be equated, without serious error, with the sum of the translational energies of the two molecules. For molecules of complicated structure, in which the average vibrational energy is considerably greater than the translational energy, the vibrational energy must also be taken into account. A considerably more complex expression must then be inserted in eqn (20), in place of $e^{-A/RT}$.

From the equation

$$k = C \, e^{-A/RT} \tag{22}$$

it follows that

$$\ln k = \ln c - \frac{A}{RT} \tag{23}$$

Differentiating (23),

$$\frac{d \ln k}{dt} = \frac{A}{RT^2} \tag{24}$$

Equation (24) is identical with the empirically established eqn (7), p. 698, which was the basis upon which Arrhenius interpreted the quantity A as a thermal energy of activation.

* It is necessary, however, to take into account the temperature variation of σ, as well as that of u, for the purpose of accurate calculations.

(iii) Activation Energy Dependent on Mechanism of Reaction. The activation energy of a reaction can be calculated from eqn (23), if the velocity constant k has been determined at several temperatures*. If $\ln k$ is plotted graphically against $1/T$, a straight line is obtained, and the slope of this is equal to A/R. If the activation energy for the decomposition of hydrogen iodide is calculated in this way from the values quoted in Table 95, the result obtained is $A = 44$ kcal per mol of HI. The energy necessary to dissociate hydrogen iodide molecules into free atoms amounts to 70.8 kcal/mol. Thus, unlike the photochemically initiated hydrogen-chlorine reaction, the thermal decomposition of hydrogen iodide does not take place by a dissociation into free atoms. A sufficiently energetic collision between two HI molecules is apparently adequate to bring about a rearrangement of the atoms in accordance with the scheme

$$\begin{pmatrix} H \\ I \end{pmatrix} + \begin{pmatrix} H \\ I \end{pmatrix} \rightleftharpoons \quad \begin{matrix} \boxed{H\ H} \\ + \\ \boxed{I\ I} \end{matrix}$$

If a dissociation of HI molecules into atoms were necessary, in order to initiate reaction, the energy of activation would be 71 kcal/mol, and not 44 kcal/mol. This example shows that, unlike the *heat of reaction*, which is independent of the route by which the reaction takes place, the *energy of activation* depends upon the actual course taken by the reaction. This fact is of fundamental importance in understanding the nature of catalysis.

TABLE 95

VELOCITY COEFFICIENTS FOR THE FORMATION OF HI $(= k_1)$ AND
DECOMPOSITION OF HI $(= k_2)$, AFTER BODENSTEIN

k_1 and $k_2 = \dfrac{c_0 - c_t}{c_0 \cdot c_t \cdot t}$, referred to the same initial concentrations of starting materials

(c_0, c_t in mols per liter; t in sec; $T =$ temp. in °K)

	556°	575°	629°	647°	666°	683°	700°	716°	781°
1	$3.52 \cdot 10^{-7}$	$1.22 \cdot 10^{-6}$	$3.02 \cdot 10^{-5}$	$8.59 \cdot 10^{-5}$	$2.20 \cdot 10^{-4}$	$5.12 \cdot 10^{-4}$	$1.16 \cdot 10^{-3}$	$2.50 \cdot 10^{-3}$	$3.95 \cdot 10^{-2}$
2	$4.44 \cdot 10^{-5}$	$1.32 \cdot 10^{-4}$	$2.52 \cdot 10^{-3}$	$5.23 \cdot 10^{-3}$	$1.41 \cdot 10^{-2}$	$2.46 \cdot 10^{-2}$	$6.42 \cdot 10^{-2}$	$1.40 \cdot 10^{-1}$	$1.34 \cdot 10^{0}$

The combination of iodine with hydrogen to form HI also takes place by a direct rearrangement of the atoms in molecular collision. In this case, indeed, the energy of activation (40 kcal/mol of I_2) would be great enough to dissociate I_2 into 2I atoms (heat of dissociation 35.4 kcal/mol). However, the process

$$I + H_2 = HI + H - 32 \text{ kcal} \tag{25}$$

is strongly endothermic. It could thus take place only if the I atom collided with a H_2 molecule bearing at least the indicated amount of thermal energy. From eqn (15), the probability of such a collision is only about 10^{-7} at 1000° K. The probability of collision between two I atoms, from the degree of dissociation at 1000° as given in Table 104, p. 773 (Vol. I), is $5.4 \cdot 10^{-2}$ of the total number of collisions. If it be assumed that one collision in

---

* If the variation of A with temperature is taken into account (although this is not usually necessary), eqn (23) is to be replaced by $\ln k = \ln c - \dfrac{A}{RT} + \frac{1}{2} \ln T$.

1000 between I atoms is a three-body collision, so that recombination could occur*, it would follow that for every 500 I atoms which recombined to form I_2 molecules, *one*, at the most, would react with H_2 to form HI (eqn 25). However, a direct molecular-kinetic calculation of the rate of reaction from eqn (20), shows that almost *every* collision, in which the collisional energy exceeds 40 kcal/mol, leads to reaction. Thus, in spite of the relatively considerable thermal dissociation of iodine, the reaction proceeds *directly* between I_2 and H_2 molecules, and not by a mechanism involving free I-atoms. For this reason it is also understandable that the activation energies for the reactions in the two opposed directions should be very little different in magnitude. If both reactions took place by way of free atoms, the energy of activation for the decomposition of HI would necessarily be much greater than that for the formation of HI.

(*iv*) *Wave Mechanical Calculation of Activation Energies.* In simple cases, it is possible, in terms of the wave mechanical or quantum mechanical theory of the chemical bond (Vol. I, Chap. 5), to represent the total chemical energy of a system as a function of the internuclear distances between the atoms, and the mode of binding of the electrons. On this basis, it should often then be possible to calculate the activation energy necessary to transform the system from one given state to another. Hitherto such calculations have been carried out successfully only for especially simple cases. Nevertheless, it has been shown, for example, that it is possible for the system $I_2 + H_2$ to pass continuously into the system 2HI, without the necessity for any dissociation into atoms to occur as an intermediate stage. It is not possible to discuss these calculations more fully at this place; short treatments of the basis of the wave mechanical calculation of activation energies are given in the books of Schumacher [*14*] and Frankenburger [*2*], listed on p. 718.

3. Reaction Kinetic Interpretation of Catalysis

(a) General

The foregoing considerations apply directly to gaseous systems. They can, however, be applied in their essential features to liquid and to multi-phase systems**. It follows that the rate of reaction in a system at any given temperature is determined by the *energy of activation*, as long as energy is supplied in no other form than as heat.

Since, by definition, a catalyst can subsequently be separated unchanged from the system in which it has brought about reaction, it follows that the catalyst cannot have given up any energy to the system***. It can act, therefore, only in as far as it alters the factors which determine the *rate* of the reaction—i.e., the factors a and Z in eqn (16), together with an additional factor which takes account of the probability of a fruitful transformation following from a collision to which the molecules bring the necessary amount of energy. In eqn (16), this 'reaction probability' has been put equal to 1; for molecules of more complex structure it may be much smaller. This factor may be increased by a catalyst (for an example,

---

* See footnote p. 700. If the third partner is a H_2 molecule, the process $2I + H_2 = 2HI$ also appears possible, but it has never been demonstrated experimentally that this reaction plays a significant role.

** In multiphase (heterogeneous) systems, the over-all reaction velocity depends also upon the rate of diffusion of reactants and products to and away from the phase boundary. If the velocity of diffusion is small in comparison with the rate of the chemical change, then the total rate of reaction may be determined by the processes of material transport alone (cf. p. 748 *et seq.*).

*** It is possible for energy to be conferred *transiently* by the catalyst to the reacting system, but it must subsequently be recovered completely in the course of further reaction (that is, neglecting fortuitous incidental effects, which have nothing to do with the essential nature of catalysis).

see p. 712), but the resulting acceleration of reaction is not very large. The collision number Z can be increased in magnitude by an added substance, in particular by reason of adsorption of the reactants on its surface, or condensation in capillaries. Even with highly surface-active catalysts, however, this effect is usually of subordinate importance as compared with the effect of the catalyst on the factor a—that is (from eqn (15)) on the activation energy A (cf. p. 711–2). *In the majority of cases the action of a catalyst arises exclusively, or at least predominantly, from the fact that it lowers the activation energy.* Since the activation energy depends on the path followed in the reaction, this can only mean that, in the presence of the catalyst, the reaction proceeds by a different route from that taken if no catalyst is present. Depending on the nature of the system and the kind of catalyst, there are various possible ways in which the modified reaction path may come about. It is essential, however, that the catalyst itself *must be transiently involved in the reaction.* As a rule, the lowering of the activation energy is brought about through the formation of some intermediate product which is not formed in the absence of the catalyst, the activation energy involved in the formation of this product being less than that necessary for the rate-determining step of the reaction path followed when no catalyst is present. It is of minor importance whether this 'intermediate product' is a chemical compound in the ordinary sense, or whether it is a solid solution formed on the surface of the catalyst, or some other system originating from the interaction of catalyst or reactants; indeed, its nature will differ from one example to another. As far as the principles of catalysis are concerned, it is also unimportant whether the entire reaction path is different in the catalyzed reaction from that of the normal reaction, or whether only a part is modified.

In general, every catalytically accelerated reaction of the form

$$A + B = AB \tag{26}$$

is composed of two consecutive reactions, which (denoting the catalyst by K) can be schematically represented as

$$A + K = AK \tag{26a}$$

$$AK + B = AB + K \tag{26b}$$

AK is the product of some interaction with the catalyst, and need not be a chemical compound in the strict sense. It is, of course, implicit that the catalyst is regenerated in its original from by the reaction symbolized in (26b).

In order that the reaction may be accelerated by the formation of 'intermediate compounds' between catalyst and substrate (or some constituent of the latter), an essential point is that the intermediates should be *unstable.* If a relatively stable compound is formed, its formation generally brings the reaction to a standstill, whereas a relatively unstable compound reacts further. Substances such as (e.g.) the alkali metals, which in general form very stable compounds, therefore show but little catalytic power. The suppression of catalytic action through the formation of stable compounds is also illustrated by the behavior of cobalt towards CH_4 and CO (p. 292): only above the range of temperature at which a carbide of cobalt is stable can the metal catalyze the decomposition of methane or carbon monoxide.

The reaction in which the catalyst participates need not necessarily be the *initial step* in the reaction path. It may be preceded by partial reactions which can themselves take place sufficiently fast in the absence of a catalyst, although they generally lead to an equilibrium state before reaction has proceeded to any considerable extent, because at some stage the reaction path involves a high potential barrier.

In many cases, the rate of a reaction is accelerated because the catalytically induced reaction, unlike the normal mechanism, is a *chain reaction*. Even in this case, however, the essence of the action of the catalyst lies in that it lowers the activation energy of some reaction—namely, the one which leads to the establishment of a reaction chain*. It cannot be inferred directly from the extent of the acceleration brought about by a catalyst that the latter initiates a chain reaction. As shown in Table 93 (p. 704), a considerable lowering of the energy of activation may result in an enormous increase in reaction velocity.

The mechanism of action of catalysts will be exemplified in the following section, by reference to a few examples. It should be noted, however, that whereas the *nature* of catalytic action may be considered to have been explained in its essentials from the study of reaction kinetics, it is still possible to make any definite pronouncements on the *detailed mechanism* of catalysis in only a limited number of cases.

(b) Homogeneous Catalysis

Genuine catalytic homogeneous gas reactions (that is, reactions which take place, under the influence of a gaseous catalyst, exclusively in the gas phase, and not at surfaces—e.g., the walls of the vessel), are rather rare. As an example, there may be cited the reaction between nitric oxide and chlorine, catalyzed by bromine (von Kiss, 1923 *et seq.*). In the absence of bromine, the reaction follows the course

$$2NO + Cl_2 = 2NOCl \tag{27}$$

with $NOCl_2$ as an intermediate product (cf. p. 696). In the presence of bromine, the course of the reaction is shown by (28) and (29):

$$2NO + Br_2 = 2NOBr \tag{28}$$

$$2NOBr + Cl_2 = 2NOCl + Br_2 \tag{29}$$

The energies of activation for reactions (28) and (29) are smaller than for reaction (27). Hence the formation of NOCl is accelerated by the formation of NOBr as an intermediate, the bromine being subsequently regenerated from this by reaction with Cl_2.

The thermal decomposition of dimethyl ether, which is catalytically accelerated by hydrogen (Hinshelwood, 1927), is an example of a homogeneous gas catalysis in which there is no real intermediate product formed, involving the catalyst. In this instance the catalyst acts simply by transferring *energy* in the act of collision, to the molecules requiring activation; this energy is recovered in later collisions with other molecules. The decomposition follows the equation

$$CH_3 \cdot O \cdot CH_3 = CH_4 + CO + H_2$$

and is a reaction of the first order. The rate of decomposition depends upon the number of $(CH_3)_2O$ molecules possessing sufficient rotational and vibrational energy to break the

---

* In this case, we may encounter the phenomenon of a catalyzed reaction (the chain reaction) which, even in the presence of the catalyst, has a higher activation energy than the normal reaction (not involving a chain mechanism). The catalyzed reaction is faster than the normal reaction, in this case, by virtue of being a chain reaction. However, the fact that a chain reaction occurs at all is still dependent on the catalyst making possible some reaction of lower activation energy—whether it be the primary reaction or the chain-propagating reactions.

C—O bond. As experiment has shown, the requisite energy can be acquired not only from collisions with other $(CH_3)_2O$ molecules, but from collisions between $(CH_3)_2O$ and H_2 molecules, although not from collisions with various other molecules—e.g., He, N_2, or CO_2. Since the hydrogen recovers in subsequent collisions the energy which it transfers, the reaction can be considered as a genuine catalysis.

Homogeneous catalytic reactions in *liquids* are very much commoner than homogeneous gas catalysis. Of particular importance among these are the acceleration of reactions (especially oxidation and reduction processes) by *heavy metal salts* and catalysis by *acids and bases* and by *dysprotide and emprotide substances* (cf. Chap. 18).

The catalytic action of heavy metal salts frequently hinges on the ability of their cations, by giving up or accepting electrons, to convert the molecules of the substrate into free radicals or unstable ions. In this way, *chain reactions* are often initiated (cf. p. 768 et seq., Vol. I). In addition, this process may be associated with a considerable diminution of the energy of activation, so that a very considerable increase in reaction velocity would be produced even in the absence of chain reactions.

In catalysis by acids, the mechanism usually involves the intermediate formation of addition compounds of the substrate molecules with protons, these having a high reactivity (low activation energy). Since (with a few exceptions), protons, and not hydronium ions, combine with the substrate, it is not uncommonly found that the catalytic action of acids is even stronger in weakly dissociating or indifferent solvents than in aqueous solution (cf. p. 720 ff). For the same reason, a catalytic action is often exercised by substances such as NH_4^+ ions, which can give up protons, as well as by acids in the ordinary sense. The catalytic effect of bases frequently arises from the ability of their hydroxyl ions to *abstract protons* from the molecules of the substrate. Where this is the case, other substances which have a tendency to bind protons—e.g., the ions of weak acids—display the same catalytic action as bases.

(c) Heterogeneous Catalysis

The only essential difference between heterogeneous catalysis and homogeneous catalysis lies in that the reaction takes place in the former, at the boundary between two phases (catalyst and substrate, respectively). The reason that the reaction is accelerated in heterogeneous catalysis is, again, usually a lowering of the energy by the action of the catalyst. Substrate molecules are converted into more reactive (that is, more easily activated) forms under the action of the surface forces or the free valences (cf. p. 677) acting at the phase boundary. These forces may bring about the dissociation of adsorbed molecules into atoms, but the more usual effect is a loosening of the binding between atoms, as compared with the free molecules. Such a 'deformation' can, however, lead to a considerable increase in reactivity, as can be shown by wave mechanics.

Since, when solids are involved, the number of free valences is especially high at the *edges* and *corners*, these parts of the surface are usually particularly active catalytically ('active centers' of Langmuir and H. S. Taylor). For this reason, the fine subdivision of the catalyst (e.g., use in colloidal form), or a roughening of the surface, may lead to an increased catalytic activity, not merely because of the increase in total surface, but because of the increased number of active centers created.

As well as a lowering of the activation energy by the deformation effect, there are other ways in which the reaction velocity can possibly be altered at phase boundaries. These are operative, however, only in exceptional cases. Thus, a solid wall may act as the third partner necessary to carry away energy released in the recombination of free atoms (cf. p. 700 footnote). Further, an *orientation* of the molecules, as a result of adsorption, may increase (or diminish) the probability that the reactive centers of the molecules may be brought together in the act of collision (acceleration of reaction by modification of the 'probability of reaction', cf. p. 708). The mere increase in concentration of the substrate at the surface of the catalyst, as a result of adsorption, may lead to enhanced reaction rate (by increasing the collision frequency). In practice, however, the effect is rarely significant, and as a rule is quite inadequate to explain the increases in reaction rate that are observed. Thus, it may be deduced from their heats of adsorption that the gases H_2 and O_2 are adsorbed in a monomolecular layer on porcelain at $900°$ K, giving about a 16-fold increase in concentration. If this condensation alone were responsible for the catalytic acceleration of their combination, the rate of the electrolytic gas reaction would be accelerated in proportion to the increase of collision frequency—i.e., for the termolecular reaction involved, in proportion to the 3rd power of the degree of condensation, or by about 4000 : 1. In fact, as found by Bodenstein (1899), the reaction of H_2 with O_2 is accelerated by porcelain under the conditions cited by about 20,000,000 to 1. In accordance with the fact that the condensation of the substrate by adsorption on the catalyst surface is usually unimportant for catalysis, as compared with the *deformation* effect, there is, as a rule, no relation between the adsorptive properties of different substances and their efficiency as catalysts.

The most important processes of heterogeneous catalysis are those in which the catalyst is a solid, and the substrate is in the gaseous or liquid (or dissolved) state. Use is often made, also, of the catalysis of gas reactions by liquid phase or dissolved catalysts. Examples in which the catalyst is in a less highly condensed state of aggregation than the substrate are rare, however. The acceleration of many reactions between solid substances by water vapor is an example of such catalysis, however.

It is frequently found that chemical reactions or physical transformations (change of modification, allotropic changes) of *solids* can be accelerated by added solid substances. There may be various reasons for this. It may be that the added substances increase the number of 'defects' ('Lockerstellen') in the crystal lattice of the reacting substance (cf. p. 743). The added substance may also accelerate the reaction by lowering the melting point of one or more of the initial reactants, as a result of forming solid solutions. It may also facilitate the crystallization of the reaction product, as a result of nucleation. The last mentioned, in particular, is the usual reason for the frequently observed *autocatalytic* acceleration of those reactions of solids in which solid reaction products are formed.

(d) Autocatalysis

The term autocatalysis is applied to the catalytic acceleration of a reaction by one of the reaction products. The phenomenon shows itself in that reaction rate *speeds up* during the initial stages (corresponding to the increase in the amount of catalyst as reaction proceeds), and begins to diminish again only after some time, when the reactants have decreased so much in concentration that this last factor predominates. When a reaction is autocatalytically accelerated, it is often found that a considerable time elapses before reaction begins at an appreciable speed, but when once started the rate often increases very rapidly, until the reaction is proceeding very vigorously. If some of the reaction product which is responsible for the autocatalysis is added to the reaction mixture at the outset, the reaction at once commences at a correspondingly rapid rate.

The phenomenon of autocatalysis is fairly widespread. In addition to the group of reactions already mentioned, it may be exemplified by the acceleration of the oxidizing action of nitric acid by the addition of nitrogen oxides, and the acceleration of potassium per-manganate oxidations by the addition of manganese(II) salts.

(e) Catalytic 'Direction' of Reactions

Whenever a substance, or a reaction mixture, is capable of undergoing several possible reactions, the reaction which is initiated when the temperature is gradually raised is that which has the lowest energy of activation. This follows directly from the discussion of reaction kinetics. The same consideration applies to reactions which take place in the presence of a catalyst. However, it need not necessarily follow that the reaction for which the energy of activation is lowerd by the catalyst (or the one for which the greatest diminution of activation energy is produced) will be the reaction having the lowest activation energy in the absence of the cata-lyst. The catalyst may so strongly modify the activation energy for some reaction which does not lead to the same end product as the normal non-catalytic reaction, that this reaction now has a lower activation energy than the normal reaction. In such a case, the reaction occurring in the presence of the catalyst will be different from the uncatalyzed reaction. Furthermore, different catalysts may selectively diminish the energy of activation for different reactions. The particular reaction to occur will then depend upon the nature of the catalyst. In this way the reaction is 'steered' or 'directed' by catalysis.

For example, methyl ethyl ether, $CH_3 \cdot O \cdot C_2H_5$, undergoes thermal decomposition in the absence of a catalyst according to:

$$H_3C-\underset{\underset{H}{|}}{\overset{\overset{H}{|}}{C}}-O-CH_3 \quad \longrightarrow \quad CH_4 + CO + CH_4.$$

In the presence of certain catalysts, however, (e.g., iodine) the primary decomposition* is:

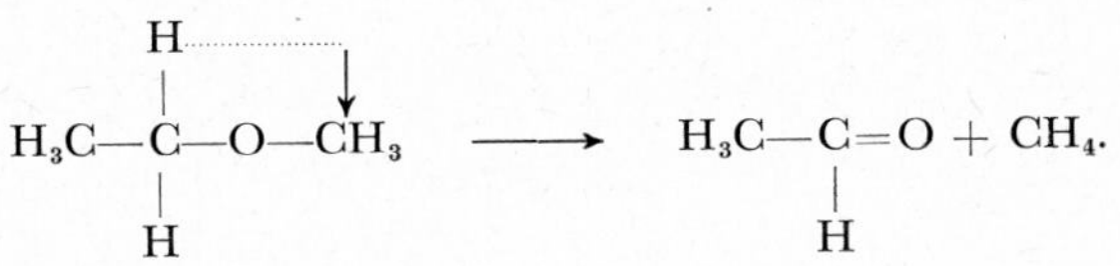

$$H_3C-\underset{\underset{H}{|}}{\overset{\overset{H}{|}}{C}}-O-CH_3 \quad \longrightarrow \quad H_3C-\underset{\underset{H}{|}}{C}=O + CH_4.$$

Others ethers behave in a similar way (Clusius and Hinshelwood, 1930); e.g., diethyl ether:

Without catalyst

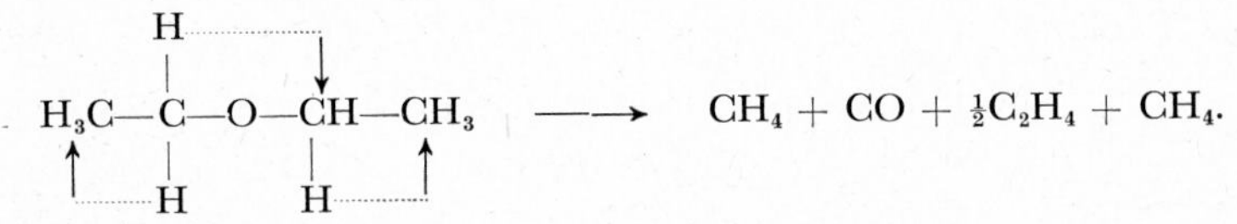

$$H_3C-\underset{\underset{H}{|}}{\overset{\overset{H}{|}}{C}}-O-\underset{\underset{H}{|}}{CH}-CH_3 \quad \longrightarrow \quad CH_4 + CO + \tfrac{1}{2}C_2H_4 + CH_4.$$

In the presence of catalyst

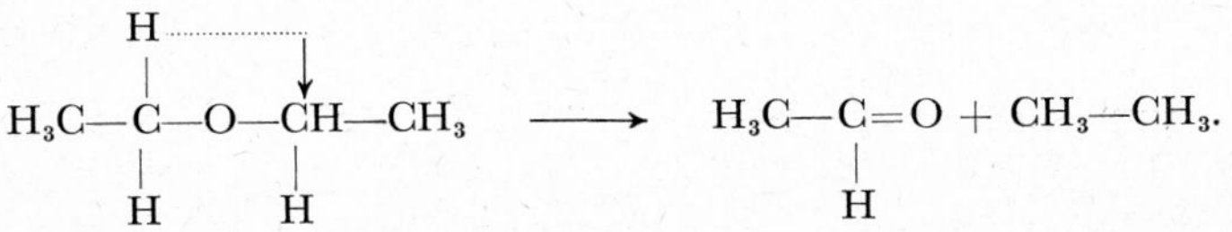

$$H_3C-\underset{\underset{H}{|}}{\overset{\overset{H}{|}}{C}}-O-\underset{\underset{H}{|}}{CH}-CH_3 \quad \longrightarrow \quad H_3C-\underset{\underset{H}{|}}{C}=O + CH_3-CH_3.$$

* The acetaldehyde formed enters into a secondary decomposition reaction $CH_3CHO = CH_4 + CO$.

In the absence of a catalyst, the molecule breaks up simultaneously at several points, when it acquires the necessary energy of activation (47 kcal for $CH_3 \cdot O \cdot C_2H_5$; 53 kcal for $C_2H_5 \cdot O \cdot C_2H_5$). The effect of the catalyst is to loosen *one* of the bonds, so that a smaller energy of activation (38 kcal and 34 kcal, respectively) is sufficient to cause rupture. In accordance with this, reaction takes place in the presence of the catalyst not only at a considerably lower temperature, but in a different manner. The decomposition of dimethyl ether referred to on p. 710 (energy of activation 65 kcal) cannot be catalytically accelerated by iodine. In this case, therefore, the addition of iodine does not modify the *mode* of decomposition.

(f) Catalytic Poisons

Substances which diminish or inhibit the activity of catalysts are termed *catalytic poisons*. Their action may be due to their ability to combine with the catalyst, forming less active, or completely inactive substances; in addition, where chain reactions are involved, they may also enter into reaction with the species formed as intermediate products, in such a way that the chains are prematurely terminated. In neither case is it essential that true chemical compounds should be formed with the catalytic poison; in the case of heterogeneous catalysis, it is sufficient that the poison should be strongly adsorbed on the surface of the catalyst. In general, solid catalysts are particularly sensitive to poisons, since the adsorption of quite small amounts of foreign substances may considerably diminish the extent of their available surface. Furthermore, adsorption takes place preferentially on the 'active centers', i.e., on just those surface sites which have the highest catalytic power.

Poisoning is described as 'temporary' or 'permanent', depending on whether the effect is reversed or persists, when the poison is removed from the reaction system. One reason for the occurrence of temporary poisoning, for example, is the excessive accumulation of the reaction product at the surface of the catalyst. Such poisoning can usually be overcome without difficulty by increase in the rate of passage of reactants, or the rate of stirring.

Use is often made of catalytic poisons, to suppress unwanted reactions ('protective poisoning').

(g) Negative Catalysis

According to the definition of a catalyst, it may be said that negative catalysis takes place when a reaction which takes place without a catalyst is *retarded* by some added substance. This may arise, for example, if the added substance acts as a 'chain breaker' for a chain reaction which normally occurs. As already mentioned, the action of 'anti-knocks' in fuels depends upon this effect.

Although the concept of a 'negative catalyst' is quite different from that of a catalytic poison, it is often difficult, in practice, to decide whether any given instance is a case of negative catalysis or of poisoning. This is because the possibility cannot always be definitely excluded that a reaction which apparently takes place as an uncatalyzed process is not, in reality, brought about by traces of some (positive) catalyst; the action of this catalyst may then be inhibited by the added foreign substance.

(h) Multiple Catalysis

The catalytic action of a mixture of two or more substances is not, as a rule, equivalent to the sum of their several effects. It may be less, but may also be considerably greater.

For example, Remy (1924) found for the catalytic efficiencies of Rh and Pt, for the hydrogen-oxygen reaction, the relative values 22 and 34, but the value 100 for the efficiency of an alloy of the two metals. The efficiency of Rh can be raised from 22 to 34, or under certain conditions from 56 to 137, by alloying it with Ru, which is itself quite without catalytic action on the hydrogen-oxygen reaction at ordinary temperature. This example demonstrates that the efficiency of a catalyst may be augmented by mixing it with a substance which is itself quite inactive as a catalyst for the reaction concerned.

Quite small amounts of an added substance often suffice to bring about a considerable increase in catalytic efficiency. Such added substances are known as 'activators' or 'promoters'. The efficiency of substances which are used in the solid state as catalysts can further be raised to a considerable extent by depositing them upon materials with a large surface area—e.g., platinum on asbestos. Such materials, which operate essentially by increasing the surface area of the catalyst, are termed 'carriers'. A carrier may also act, at the same time, as a promoter, as is shown by the fact that different carriers, with almost the same specific surface, may influence the efficiency of catalysts deposited on them to quite different degrees.

Multiple catalysts (mixed catalysts, promoted catalysts) have become of very great technical importance in connection with heterogeneous catalysis. Indeed, it is now exceptional for a technical contact process to operate with a one-component catalyst. Multiple catalysts are now almost invariably employed, since it has been shown that the activity of versatile catalysts, such as iron, can be enhanced for various specified reactions by means of suitable carriers and promoters. Thus iron is made especially effective in the ammonia synthesis by addition of aluminum oxide; by adding bismuth oxide it is adapted for the catalytic oxidation of ammonia to nitric oxide; and by adding chromium oxide its activity is enhanced for the conversion of carbon monoxide-steam mixtures into carbon dioxide and hydrogen. The sensitivity of catalysts towards catalytic poisons can also be considerably reduced, and often almost eliminated by suitable additions.

As an example of multiple catalysis in homogeneous systems, reference may be made to the observation made by Price, in 1898, that the oxidation of hydrogen iodide by peroxysulfate in aqueous solution was speeded up when copper and iron salts were added simultaneously, by twice as much as if the effects of the two substances were merely additive. Although there were already several recorded observations of the enhancement of catalytic activity by the addition of other substances to the catalyst, Price was the first to state clearly that the efficiency of multiple catalysts was not merely additive. Brode (1901), from a continuation of Price's experiments, recognized for the first time that the action of a catalyst could be influenced by substances which, by themselves, have very little or no catalytic effect.

More recently multiple catalysis has been intensively studied, especially by Mittasch. Promoter action may arise from various causes. In reactions between two substances, it may happen that catalyst A activates only (or practically exclusively) the molecules of the one reactant, whereas catalyst B activates the other. If *both sorts of molecules* are simultaneously activated by a mixture of A and B, the consequent increase in the rate of reaction may be especially great. With solid contact materials, the action of activators, is often due to their influence on the particle size of the catalyst, or to the fact that they increase the number of active centers by causing lattice imperfections*. Furthermore, an alloy between two elements may

---

* The particularly high catalytic efficiency of substances in the 'active intermediate states' often encountered in reactions between solids (cf. p. 746 *et seq.*) is the consequence of such lattice imperfections and similar effects.

exert quite specifically different surface forces from those of its components, in that the spheres of action of its atoms mutually overlap and modify one another.

(i) Raney Catalysts

Raney (1925–27) has devised a process by which certain metals can be obtained in an especially active catalytic form. In this process, the catalytically active metal is alloyed with a catalytically inactive one, the alloy pulverized, and then the alloying metal dissolved out. The most widely used Raney metal is nickel. The preparation of Raney nickel usually involves the melting together of nickel and aluminum, treatment of the coarse or finely ground alloy with alkali solution, and careful washing of the residue; this treatment gives a black or grey-black powder. The product is sensitive to air, often even pyrophoric, and must, therefore, be stored under water, alcohol, hydrocarbons, or the like.

Raney nickel does not differ structurally from ordinary nickel, but it exists in an especially fine crystalline form (diameters of the crystallites, 40–80 Å). It still contains a small amount of aluminum (up to several per cent), which seems to be necessary for its activity. Most important, however, Raney nickel contains a considerable quantity of hydrogen (up to 33 atom %) and its electrochemical behavior corresponds to platinum black saturated with hydrogen. In contrast to the behavior of hydrogen-free nickel, it amalgamates with mercury. Nitrite and nitrate are reduced to ammonia by Raney nickel, and permanganate to manganese dioxide. Chlorate, bromate, iodate, molybdate, and tungstate may also be reduced by Raney nickel. It reacts slowly with water with the evolution of hydrogen and the formation of nickel(II) hydroxide. In the presence of readily oxidizable substances, which are capable of acting as oxygen acceptors, Raney nickel remains unchanged and behaves merely as an oxidation catalyst. For example, hypophosphite in the presence of Raney nickel is oxidized by water to phosphite:

$$H_2PO_2^- + H_2O = HPO_3^= + H^+ + H_2.$$

This reaction proceeds quantitatively and may be employed for the acidimetric determination of hypophosphite. Oxidations by means of atmospheric oxygen may also be catalytically accelerated by Raney nickel. In industry, Raney nickel is used primarily as a hydrogenation catalyst. Because of its especially high activity it acts rapidly. In addition, it permits the hydrogenation at lower temperatures of such substances which at higher temperatures give troublesome side reactions.

Raney metals also find use as dehydrogenation catalysts. Raney cobalt is an excellent catalyst for the synthesis of gasoline from CO and H_2. It acts therein not only as a hydrogenation catalyst but also at the same time as a polymerization catalyst, especially when it is prepared by way of a Co-Si alloy rather than through a Co-Al alloy. According to Herglotz and Lissner (1949), Raney cobalt is also extremely useful for the removal of organically bound sulfur by catalytic conversion to hydrogen sulfide. In Raney cobalt the metal is present in the hexagonal modification, whereas cobalt contact catalysts prepared in the usual manner (e.g., by heating CoO in a current of H_2) contain the metal in the cubic form. Both modifications of cobalt are catalytically active, whereas in nickel only the cubic form is active.

(j) Measurement of Catalytic Power

The efficiency of a catalyst is given by the ratio of the velocity constants of the catalytically induced and the uncatalyzed reaction, respectively. The velocity constants must, of course, be compared at the same temperature. Furthermore, the efficiency of the catalyst must be related, in the case of homogeneous catalysis, to some definite concentration of catalyst, or, in the case of heterogeneous catalysis, to a definite surface area of the catalyst. However, in the case of solid substances, which are of such importance as catalysts, the extent of the true surface is, at best, only approximately known. Furthermore, for a given total surface there may be wide variations in the number of 'active centers', which strongly influence the catalytic properties. In such cases, the catalytic efficiency is generally represented much more exactly by specifying the decrease in the energy of activation brought about by the catalyst. This comparison is, of course, only valid on the assumption that the increased reaction velocity is not due, in any appreciable degree, to other causes than a change in the energy of activation.

Since the temperature at which reaction takes place at an appreciable rate is also lowered by a decrease in activation energy, the lowering of the reaction temperature can also be used as a rough measure of catalyst efficiency. In the case of reactions which follow a different course in the presence and in the absence of a catalyst, and in other cases where the activation energy of the uncatalyzed reaction is not known, it is possible to compare only the *relative efficiencies* of different catalysts. In practice, however, this is the information usually required.

(k) Catalysis and Chemical Equilibrium

As has already been stated, it follows from the definition of a catalyst that it neither imparts energy to nor takes energy from the system in which it accelerates a reaction. It therefore follows, further, that a catalyst *does not displace the equilibrium* in any system, since any disturbance of equilibrium alters the energy content of a system (cf. Vol. I, p. 143). According to the kinetic concept of chemical equilibrium, in the equilibrium state, the forward and back reactions are proceeding at equal velocities, e.g., the formation of a compound takes place at the same rate as it decomposes. It therefore follows, further, that a catalyst must speed up the reaction in both directions to the same extent. Thus a catalyst which promotes the formation of a compound must also accelerate its decomposition by the same amount. This principle is often utilized in investigating the efficiency of catalysts. Thus, in studying the suitability of various contact materials for the ammonia synthesis, it is convenient to measure the extent to which the *decomposition* of ammonia is accelerated, rather than the *synthesis* of ammonia. However, the rule that the efficiency of a catalyst is independent of the direction in which reaction is taking place is strictly valid only for systems which are not too far removed from equilibrium.

4. Induced Reactions

Reactions which are initiated or accelerated by some added substance are said to be *induced* reactions if the added substance is *consumed* by the reaction upon which its efficiency depends (i.e., not merely used up in side reactions). Whereas only reactions in which energy is liberated can be brought about by the action of a catalyst (since the latter imparts no energy to the system in which it accelerates reaction), processes which *absorb* energy may be involved in induced reactions. The energy which must be acquired by the system before reaction can occur is then made available by chemical change of the substance which induces reaction; this chemical change must therefore always be a process which furnishes energy.

The oxidation of arsenious acid by bromic acid in aqueous solution may be cited as an example of an induced reaction:

$$HBrO_3 + 3H_3AsO_3 = 3H_3AsO_4 + HBr.$$

Although this reaction is exothermic, it takes place spontaneously only with extreme slowness. It is rapid, however, if sulfurous acid is added to the reaction mixture. The sulfurous acid is thereby also oxidized:

$$3H_2SO_3 + HBrO_3 = 3H_2SO_4 + HBr.$$

In this respect, the process differs fundamentally from a catalytic reaction.

If the process initiated by the added substances cannot take place spontaneously, the maximum quantity of the reactants that can be induced to react must stand in some simple stoichiometric ratio to the quantity of reaction-inducing substance consumed. In this case, the two processes are said to be 'coupled reactions'. Examples of coupled reactions are the autoxidation processes accompanied by the formation of hydrogen peroxide, such as the autoxidation of titanium(III) compounds in alkaline solution, discussed on p. 55 in which at most $\frac{1}{2}$ molecule of H_2O_2 is formed for every Ti^{III} atom oxidized. In this instance, the induced reaction is the oxidation of water to H_2O_2. However, if the induced reaction is one which can also proceed spontaneously, it is possible for the 'induction factor' (i.e., the ratio between the quantity of substance brought to reaction and the quantity of reaction-inducing substance consumed) to vary between wide limits, and to attain relatively high values. Thus, in the oxidation of iodide ions in solution induced by vanadic acid, the induction factor may vary between $1/_{160}$ and 12, according to the experimental conditions (Bray, 1933).

References

1 G. M. SCHWAB, *Katalyse vom Standpunkt der chemischen Kinetik*, Berlin 1931, 249 pp.
2 W. FRANKENBURGER, *Katalytische Umsetzungen in homogenen und enzymatischen Systemen*, Leipzig 1937, 444 pp.
3 R. H. GRIFFITH, *The Mechanism of Contact Catalysis*, 2nd Ed., Oxford 1946, 262 pp.
4 A. MITTASCH, *Kurze Geschichte der Katalyse*, Berlin 1939, 139 pp.
5 G. M. SCHWAB (Editor), *Handbuch der Katalyse*, 7 vols, Vienna 1940–52.
6 A. MITTASCH, *Über Katalyse und Katalysatoren in Chemie und Biologie*, Berlin 1936, 65 pp.
7 W. LANGENBECK, *Die organischen Katalysatoren und ihre Beziehungen zu den Fermenten*, Berlin 1935, 112 pp.
8 T. P. HILDITCH, *Die Katalyse in der angewandten Chemie*, Leipzig 1932, 355 pp.
9 H. BRÜCKNER, *Katalytische Reaktionen in der organisch-chemischen Industrie*, Dresden 1930, 168 pp.
10 W. HÜCKEL, *Katalyse mit kolloiden Metallen*, Leipzig 1927, 86 pp.
11 F. KRCZIL, *Technische Adsorptionsstoffe in der Kontaktkatalyse*, Leipzig 1938, 726 pp.
12 V. N. IPATIEFF, *Catalytic Reactions at High Pressures and Temperatures*, London 1936, 786 pp.
13 K. F. BONHOEFFER and P. HARTECK, *Grundlagen der Photochemie*, Dresden 1933, 295 pp.
14 H. J. SCHUMACHER, *Chemische Gasreaktionen*, Dresden 1938, 487 pp.
15 C. N. HINSHELWOOD, *The Kinetics of Chemical Change*, Oxford 1940, 282 pp.
16 L. S. KASSEL, *Kinetics of Homogeneous Gas Reactions*, New York 1932, 330 pp.
17 R. N. PEASE, *Equilibrium and Kinetics of Gas Reactions*, Princeton 1942, 236 pp.
18 A. SKRABAL, *Homogenkinetik*, Dresden 1941, 231 pp.
19 K. J. LAIDLER, S. GLASSTONE and H. EYRING, *The Theory of Rate Processes*, New York 1941, 611 pp.
20 K. CLUSIUS, *Kettenreaktionen* [*Fortschritte der Chemie, Physik, Physikalische Chemie*, edited by A. EUCKEN], Berlin 1932, 73 pp.
21 N. SEMENOFF, *Chemical Kinetics and Chain Reactions*, Oxford 1935, 492 pp.
22 W. A. WATERS, *Chemistry of Free Radicals*, Oxford 1946, 295 pp.
23 C. N. HINSHELWOOD and A. T. WILLIAMSON, *The Reaction Between Hydrogen and Oxygen*, Oxford 1934, 107 pp.
24 W. JOST, *Explosions- und Verbrennungsvorgänge in Gasen*, Berlin 1939, 608 pp.
25 B. LEWIS and G. VON ELBE, *Combustion, Flame and Explosions of Gases*, Cambridge 1938, 415 pp.
26 G. JAHN, *Der Zündvorgang in Gasgemischen*, Munich 1934, 69 pp.
27 L. BOLTZMANN, *Vorlesungen über Gastheorie*, 2nd Ed., Vol. I, Leipzig 1910, 204 pp.; Vol. II, Leipzig 1912, 265 pp.

CHAPTER 18

REACTIONS IN NON-AQUEOUS SOLUTIONS

1. Introduction

(a) General

Although water plays a dominating role as a solvent in inorganic chemistry, there are other solvents which are also important. Non-aqueous solvents are employed for preparative purposes and for molecular weight determinations, especially for substances which are insoluble in, or decomposed by, water. In addition to this, however, the general study of the properties of substances in non-aqueous solutions, and a comparison with their behavior in aqueous systems, have contributed considerably to our knowledge of the nature of solutions, and of the reactions which take place in solution. It has thereby been found that solutions in water are in no sense the prototype of all solutions, but merely a special case (albeit a very important case practically).

Neglecting molten metals, solvents can be classified into two groups:—those which can bring about the electrolytic dissociation of substances dissolved in them, and those in which no electrolytic dissociation (or extremely slight dissociation) occurs. The latter are known as indifferent solvents. It is not possible, however, to draw a sharp distinction between the two groups.

In the process of dissolution, it is possible for a decomposition of the dissolved substance to be coupled with an addition reaction with the constituents of the solvent, after the style of a double decomposition:

$$AB + CD \rightleftharpoons AC + BD.$$

Such a case is known as a *solvolysis*. The process often leads to establishment of an equilibrium, as with hydrolysis processes (Vol. I, p. 822), which constitute a special case of solvolysis.

(b) Indifferent Solvents [1, 2]

It is familiar that non-ionizing solvents, or solvents in which ionization is very slight, play the most important part in organic chemistry. Their use in inorganic chemistry is limited, however. This is because the indifferent solvents generally dissolve significant amounts only of those substances which, in the crystalline state, form molecular lattices or structures allied to them. Only a rather small proportion of inorganic substances are included under this heading, however, and by far the greater proportion of inorganic compounds is therefore insoluble in indifferent solvents—especially those used at or slightly above ordinary temperature. The indifferent solvents must commonly used for inorganic compounds are carbon tetrachloride, chloroform, and carbon disulfide. Ether, which is practically a non-ionizing solvent, can also be included. Dioxane (the cyclic 1,4-diethylene dioxide) resembles ether in this respect, and has been frequently used as a solvent for inorganic compounds in recent years.

As has been mentioned earlier, (Vol. I, p. 76), the dissolution of any substance is accompanied by its solvation. Indeed, as a rule it is probably conditional upon solvation, since it has been shown that the solubility of substances in indifferent solvents generally runs parallel to their heats of solvation. The fact that dissolved substances obeyed quantitative laws which were formally similar to those governing substances in the gaseous state (and especially the formal analogy between osmotic pressure and gas pressure) led for a long time to the relative neglect of the interaction between the solute and the solvent. Dissolved substances were regarded as being in a state that corresponded almost completely to the gaseous state. More recently, various authors (e.g., Fredenhagen, 1932) have drawn attention to the untenability of this view.

In Vol. I it was shown that solvation had a very marked influence on the deposition potentials of the metals from (aqueous) solutions. The considerations advanced there apply in principle to non-aqueous solutions also. This effect is, of course, the smaller as the heat of solvation (of the ions, in this case) becomes smaller.

A marked influence of the solvent can be discerned on the course of chemical reactions. It has long been known that the rate of chemical reactions is markedly dependent on the nature of the solvent, even when comparison is restricted to solvents which have no detectable ionizing properties.

It is of considerable practical importance that the *direction* in which a chemical change proceeds is also often dependent on the solvent. For example, the reaction

$$2HI + S \rightleftharpoons H_2S + I_2$$

goes from left to right in the gas phase, but from right to left in aqueous solution. The latter is determined by the heat of solution of HI in water. H_2S and I_2 do not react at all with each other in dry carbon tetrachloride, nor in ether if water is completely absent, except in so far as HI is gradually consumed in the formation of ethyl iodide.

The effect of the solvent on the direction of a chemical reaction often arises from the fact that some substance is formed, when the reaction proceeds in one or other direction, which is insoluble in that solvent. In accordance with Le Chatelier's principle, this results in displacing the equilibrium so as to form more of the substance which is thrown out of solution. This principle is widely used for preparative purposes in non-aqueous solutions, as well as in aqueous solutions.

There is a great deal of quantitative information concerning the solubility of inorganic and organic compounds in indifferent and other non-aqueous solvents. Relatively few systematic investigations have been carried out, however, on the interaction between indifferent solvents and the substances dissolved in them, apart from determinations of the composition of crystalline solvates isolated from such solutions. Extensive studies of the *electrolytic conductivity* of non-aqueous solutions have been made, especially by Walden [3, 4]. The behavior of acids (and of other substances which correspond to the acids of aqueous systems) in non-aqueous solvents is now fairly well understood as a result of recent work.

2. Behavior of Acids in Non-Aqueous Solvents [5]

It has long been known that in non-aqueous solutions (and especially those in which they are dissociated only to an extremely minute degree), acids often have stronger catalytic properties than in aqueous solution. Since the catalytic action of acids in aqueous solution (e.g., their catalysis of the decomposition of diazoacetic ester) is proportional to the hydrogen ion concentration of the aqueous solution (Vol. I, p. 77 et seq.), this observation remained puzzling as long as the interaction between solvent and dissolved substance was overlooked. In aqueous solution, free H^+ nuclei (protons) can exist only in extraordinarily small concentration, because of the strong affinity of H_2O molecules for protons (i.e., formation of $[H_3O]^+$ or $[H_9O_4]^+$ ions), whereas in non-aqueous solutions the protons may be more loosely bound. When this is taken into account, it is possible to understand why, for reactions catalyzed by protons*—such as the diazoacetic ester decomposition—solutions of acids in non-aqueous solvents should be more efficient than aqueous solutions**.

The fact that catalytic efficiency of acids in aqueous solutions is, none the less, generally proportional to the $[H_3O]^+$ ion concentration, arises from the proportionality between the proton concentration and the $[H_3O]^+$ ion concentration in aqueous solutions. Since the water traps the protons, forming H_3O^+ (or $[H_9O_4]^+$) ions, it *lowers* the catalytic efficiency of protons. Quite generally, the more weakly the solvent combines with protons, the greater is the catalytic efficiency of a hydrogen compound in that solvent, if the separation of protons is involved. Hence those properties of hydrogen compounds which are the consequence of their ability to give up protons are decisively modified by *proton enchange reactions* with the solvent.

(a) Proton Exchange. Brönsted's Theory of Acids and Bases

Substances such as acids (in the familiar sense), which give up protons when they are dissolved in water are called *proton donors*, or Brönsted acids (see below) while those, such as the H_2O molecule, which add on protons, are said to be *proton acceptors*, or Brönsted bases. Proton donors are also said to be *dysprotic* and proton acceptors *emprotic* substances. When a proton donor gives up a proton, $D \rightleftharpoons A + H^+$, it is converted into the *conjugate base A*. Combination of *A* with a proton turns it into the *conjugate acid D*.

Thus the F^- ion is the conjugate base of the proton donor HF, and the $[NH_4]^+$ ion is the conjugate acid of the proton acceptor NH_3. In water, the OH^- ion is the conjugate base of the proton donor H_2O, but this in turn is the conjugate base of the proton donor $[H_3O]^+$. As is shown by the last example, one and the same species may function as a proton donor or a proton acceptor, according to the reaction considered. Substances which can do so are said to be *amphiprotic*. By definition, conjugate acid-base pairs must always differ in their ionic charge.

Since free protons are never formed in appreciable amounts in chemical reactions, conversion of a weighable amount of a proton donor D_1 into its conjugate proton acceptor A_1 necessarily implies the presence of a second proton acceptor A_2; this takes up the protons which are split off, and is itself converted into its conjugate proton donor D_2:

$$D_1 + A_2 \rightleftharpoons A_1 + D_2 \tag{1}$$

* Almost all reactions catalyzed by 'hydrogen ions' are, in fact, catalyzed by protons. Only in exceptional cases do the $[H_3O]^+$ or $[H_9O_4]^+$ ions have a specific catalytic effect.

** Similar considerations apply to determinations of the strength of acids which depend directly on the measurement of the proton activity, and give only an indirect measure of the $[H_3O]^+$ or $[H_9O_4]^+$ ion activity—e.g., the potentiometric determination of the 'hydrogen ion concentration' in acid solutions.

Proton transfer reactions of this kind may be exemplified by the following:

Donor 1	Acceptor 2		Donor 2	Acceptor 1
HCl +	NH$_3$	=	[NH$_4$]$^+$ +	Cl$^-$
HCl +	H$_2$O	=	[H$_3$O]$^+$ +	Cl$^-$
H$_2$O +	NH$_3$	=	[NH$_4$]$^+$ +	OH$^-$

When, following Brönsted (1923), reactions of these kinds are considered as proton transfer processes, it emerges that not merely related, but fundamentally identical processes are involved in a series of reactions which formerly appeared to be quite distinct—the neutralization of an acid by an anhydrobase (1st example), the electrolytic dissociation of an acid (2nd example), and the formation of a base from an anhydrobase and water (3rd example). A second advantage is that the behavior of acids in different solvents can be considered from a single standpoint. If acidic character is defined as the ability to give up protons, there is no fundamental difference at all between the behavior of acids towards water and towards ammonia or any other proton acceptor*. It is also important that the ability to act as a proton donor is in no way restricted to those compounds which function as acids in aqueous solution. Not only do ions such as [NH$_4$]$^+$ have the character of proton donors, but—depending on the nature of the solvent and of the other partner in the reaction —other substances can act as proton donors, although quite unable to give up a proton directly to H$_2$O molecules. Thus Bonhoeffer (1936) showed that elementary hydrogen could act as a proton donor towards the hydroxyl ion. He found that proton transfer occurred to a considerable extent when normal hydrogen, H$_2$, was allowed to react at 100° with heavy water, D$_2$O, especially in presence of alkali:

$$DO^- + H — H + D —OD = DOH + HD + OD^-.$$

From the concept of proton transfer, the following important consequence results. I we retain the traditional definition of acids and bases (although making them more precise) —i.e., if we define acids as substances which give up protons to form hydronium ions, [H$_3$O]$^+$ (or [H$_9$O$_4$]$^+$), in aqueous solution, and bases as substances which split off hydroxyl ions, OH$^-$, in aqueous solution—we can only assign a meaning to the exercise of acidic or basic character in aqueous solutions. No hydronium ions are formed by giving up protons in non-aqueous solutions, and the substances which function as bases in aqueous solutions do not differ in any respect from the salts when they are dissolved in non-aqueous solvents. In aqueous solution, however, both acids and bases (in the traditional sense) differ fundamentally from salts in that the electrolytic dissociation of acids and bases furnishes *ions belonging to the solvent itself*—[H$_3$O]$^+$ and OH$^-$, respectively. The strength of acids and bases is then to be measured by the extent to which [H$_3$O]$^+$ or OH$^-$ ions are formed (i.e., the degree of dissociation of the aqueous solutions), whereas *neutralization* consists essentially in the combination of [H$_3$O]$^+$ and OH$^-$ ions with each other.

On the basis of this definition, for non-aqueous solutions we can (following G. Jander) speak only of properties 'analogous to acidic' or 'analogous to basic', but not of 'acidic' or 'basic' properties. If acids are defined as substances capable of giving up protons (without reference to the solvent), the essential nature of neutralization must be viewed in terms of a *complete proton transfer reaction*. 'Bases' are then substances which can neutralize 'acids'—i.e., proton acceptors. From this point of view, substances which furnish hydroxyl ions do not differ essentially from salts even in aqueous solution. It is also necessary, in terms of the second definition, to include among the acids the ions formed from salts, if they are able to lose protons—e.g., the [NH$_4$]$^+$ ion. There are only two courses open. Either the concept of

---

* In addition to the substances which form salt-like addition compounds with acids— e.g., ethers, aldehydes, ketones, etc.—(cf. Vol. I, p. 696), many other compounds in which acids are soluble behave as proton acceptors (although in widely varying degree). It must not be overlooked, admittedly, that some of the practically important properties of acids are closely connected with the very special nature of water—namely those properties, such as the high conductivity of aqueous solutions of acids, which are due to the formation of [H$_3$O]$^+$ or [H$_9$O$_4$]$^+$ ions. Cf. footnote on previous page.

'acid' and 'base' must be restricted to aqueous solutions, or else the traditional definition must be given up, and proton donors generally must be called 'acids', proton acceptors be called 'bases', and compounds which furnish hydroxyl ions in aqueous solution must be considered to be salts. This second course of action was followed by Brönsted, who developed the theory of proton transfer processes in 1923, and called it the theory of the *acid-base function*. Water is of such overwhelming practical importance as a solvent that special names are desirable for the two classes of substances which occupy a special position, as furnishing the ions proper to the solvent itself. It would hardly be possible to replace the traditional names 'acid' and 'base' for this purpose. Care must therefore be taken to avoid confusion with the new concepts of 'Brönsted acids' and 'Brönsted bases', for proton donors and proton acceptors, respectively*. Some authors have accordingly preferred to introduce the names 'dysprotides' (= proton donors) and 'emprotides' (= proton acceptors), respectively**.

Indifferent solvents, according to Brönsted, are those which are neither proton donors nor proton acceptors. They may be called 'aprotic'. When any substance which can accept or donate protons is dissolved in an aprotic solvent, it is evident that no electrolytic ions will be formed, irrespective of whether the solvent has a high or a low dielectric constant.

When pure picric acid is dissolved in an indifferent solvent, such as benzene, the solution remains colorless, since undissociated picric acid is colorless. If a proton acceptor, such as aniline or methyl alcohol, is added, picrate ions are formed in accordance with the general scheme for proton transfer (eqn. (1), p. 721), and the solution turns yellow:

$$C_6H_2(NO_2)_3OH + C_6H_5NH_2 = C_6H_5NH_3^+ + C_6H_2(NO_2)_3O^- \quad (2)$$

$$\quad D_1 \qquad\qquad\quad A_2 \qquad\qquad\qquad D_2 \qquad\qquad\quad A_1$$

$$\text{picric acid} \qquad \text{aniline} \qquad\qquad \text{anilinium} \qquad \text{picrate ion}$$
$$\text{ion}$$

$$C_6H_2(NO_2)_3OH + CH_3OH = CH_3OH_2^+ + C_6H_2(NO_2)_3O^- \quad (3)$$

$$\quad D_1 \qquad\qquad\quad A_2 \qquad\qquad\quad D_2 \qquad\qquad\quad A_1$$

$$\text{picric acid} \qquad \text{methanol} \qquad\quad \text{methyl-} \qquad \text{picrate ion}$$
$$\text{hydronium ion}$$

The dissociation equilibria of color indicators can be treated similarly. Picric acid is, indeed, a member of this class. The shift in dissociation equilibrium (and change of color) of these indicators when the $[H_3O]^+$ ion concentration in aqueous solution is changed, is just a special case of the general proton transfer scheme:

$$C_6H_2(NO_2)_2OH + H_2O \rightleftharpoons H_3O^+ + C_6H_2(NO_2)_3O^- \quad (4)$$

$$\text{donor } D_1 \quad \text{acceptor } A_2 \quad \text{donor } D_2 \quad \text{acceptor } A_1$$

Whether the proton donor D_2 is the H_3O^+ ion or some other Brönsted acid merely influences the *position* of the equilibrium, but has no significance otherwise. Hence acids can be titrated in indifferent solvents, using color indicators, just as well as in aqueous solution (i.e., using anhydrobases or other proton acceptors). Very weak acids, or proton donors without acidic character, can, in fact only be titrated in non-aqueous solutions. Just as color indicators can be used for the colorimetric determination of H_3O^+ ion concentration in aqueous solution, they can also be used in non-aqueous (in particular, in indifferent) solvents to determine colorimetrically the loss of protons from proton donors. Picric acid is a proton donor. Similar considerations apply to proton-acceptor indicators, except that in

---

* In Brönsted's terminology, the $[NH_4]^+$ and $[H_3O]^+$ ions are acids, the $[SO_4]^{2-}$ and Cl$^-$ ions are bases, while NaOH and Ca(OH)$_2$ are salts.
** Brönsted has occasionally used the words 'protogenic' and 'protophilic' substances.

this case the proton donor D_1 is not the undissociated molecule, but the Brönsted acid formed from it by union with a proton.

(b) Proton Affinity

The proton affinity of a molecule or ion is the maximum work which can be performed by adding on a proton to it. Direct measurement of proton affinities is difficult, but the proton affinities of different substances can readily be compared—e.g., by finding the position of equilibrium in the proton transfer process of eqn. (1). Comparing two proton donors D_1 and D_2 in the same solvent, then the greater the proton affinity of the acceptor A_2 compared with A_1 (i.e., the more readily the donor D_1 gives up protons and the less readily D_2 does so), the more is the equilibrium displaced from left to right*. From the ease with which they split off protons, proton donors may be arranged in a sequence—e.g., (according to Brönsted):

1. Hydrogen chloride	6. Salicylic acid	11. Phenylacetic acid
2. Trichloracetic acid	7. o-Chlorobenzoic acid	12. Benzoic acid
3. Dichloracetic acid	8. m-Chlorobenzoic acid	13. Acetic acid
4. Picric acid	9. Benzylammonium ion	14. iso-Amylammonium ion
5. Monochloracetic acid	10. Formic acid	15. Piperidinium ion

In this series, which applies to solutions in an indifferent solvent, benzene, the ease with which protons are lost decreases from hydrogen chloride to the piperidinium ion. The order is not very different from that found in aqueous solutions, in which the strength is usually measured by the extent to which the H_3O^+ ion is formed. There is a significant difference in the case of strong acids, in that measurement in indifferent solvents shows up considerable differences in the ease with which protons are lost, which do not appear (or show up rather indefinitely) in aqueous solution, where the proton transfer reaction with H_2O takes place almost quantitatively. Hantzsch had earlier pointed out that water exerts a 'levelling up' effect on the strength of acids.

It was shown by Schwarzenbach (1930) and Wiberg (1934) that proton donor properties can be determined potentiometrically, since the potential difference between a hydrogen electrode and a solution (aqueous or non-aqueous) is determined by the proton activity (i.e., roughly by the proton concentration) in the solution. It is possible to set up a potential series of proton donors, in which they are arranged in the order of their ease of giving up protons, just as the metals and other reducing agents are arranged in a series, according to their potentiometrically measured ease of giving up electrons.

3. 'Water-like' Solvents [6]

(a) General

Considerable light on the nature of reactions occurring in aqueous solution has been shed by the study of reactions in solvents which resemble water in certain properties—strong solvent power, a low intrinsic conductivity, and high conductivity of solutions, arising from the ability to bring about electrolytic dissociation of dissolved substances. Jander [*Naturwissenschaften*, 26 (1938) 779] has suggested

---

* The position of equilibrium depends upon the solvent. Assume that a proton acceptor X^- has the same proton affinity as NH_3. Then equal amounts of HX, NH_3, NH_4^+ and X^- would coexist in the equilibrium $HX + NH_3 \rightleftharpoons NH_4^+ + X^-$ if the solvent were without influence. If water is used as solvent, however, part of the acceptor NH_3 is used up by the reaction $NH_3 + H_2O \rightleftharpoons NH_3 \cdot H_2O$, to form hydrated ammonia molecules. According to the mass action law, this displaces the other equilibrium from right to left. Quite generally, solvation must have the effect of shifting the equilibrium. In the same way, the association of the dissolved molecules which is often observed in weakly solvating solvents must also displace the proton transfer equilibrium.

that such should be called 'water-like' solvents. They include liquid ammonia, liquid hydrogen fluoride, and liquid sulfur dioxide. Liquid hydrogen sulfide can, in a sense, be included also, although its properties in many ways closely resemble those of the indifferent solvents. Thus the characteristics of water-like solvents are found to a greater or less degree in the hydrides of the three elements adjacent to oxygen in the Periodic System. They are not restricted, however, to the hydrides— as the inclusion of sulfur dioxide shows—but may be found among compounds derived from water by replacement of either the hydrogen or the oxygen by other elements. Liquid sulfur dioxide is, in fact, far more 'water-like' than liquid hydrogen sulfide. The condition that a solvent shall act as a water-like solvent is that auto-dissociation shall occur as in water (although only to a minute extent in the pure solvent):

$$2H_2O \rightleftharpoons H_3O^+ + OH^- \tag{5}$$

$$2NH_3 \rightleftharpoons NH_4^+ + NH_2^- \qquad\qquad 2H_2S \rightleftharpoons H_3S^+ + SH^-$$
$$2HF \rightleftharpoons H_2F^+ + F^- \qquad\qquad 2SO_2 \rightleftharpoons SO^{2+} + SO_3^{2-} \tag{6}$$

The common feature of these ionization processes is that one molecule of the solvent transfers an elementary ion to another molecule, so as to form an oppositely charged ion pair.

It is evident that these dissociation processes go beyond the idea of the proton transfer reactions, and the analogies between the reactions occurring in the different 'water-like' solvents therefore comprise a wider range of phenomena than do the proton transfer reactions. They include total neutralization processes* (including processes analogous to neutralization in water-like solvents), the phenomena of solvolysis, reactions analogous to the behavior of amphoteric hydroxides, reactions analogous to the formation of bases from anhydrobases, and of acids from anhydroacids, and indeed the whole range of formation of addition and substitution products by reaction between solute and solvent.

In general, the higher the dielectric constant of a solvent, the more readily can it bring about electrolytic dissociation of dissolved substances. This is only true when dissociation takes place in a similar way in the solvents under comparison, and this is not always the case (cf. p. 731-2). Table 96 shows the dielectric constants and some other properties of the water-like solvents listed above. Of the substances given in the table, only hydrogen fluoride has a very high dielectric constant, like that of water. Hydrogen sulfide and sulfur dioxide differ from water not only in their much smaller dielectric constants, but also in their considerably greater molecular volumes. This last is probably one reason why they generally have a far less profound action on solutes than do water, hydrogen fluoride, and liquid ammonia. Except at extreme dilution, they apparently never bring about complete dissociation, but only partial dissociation. Furthermore, substances which are completely hydrolyzed in water only partially undergo the processes analogous to hydrolysis in these solvents. Moreover, parallel to the dissociation, it is frequently found that the molecules of dissolved substances are associated in these solvents. Conversely, formation of addition compounds with molecules of the solvent takes place to a far smaller extent with hydrogen sulfide and sulfur dioxide than with water or ammonia. However, the addition of solvent molecules to the dissolved substances (*solvation*) in liquid hydrogen sulfide and liquid sulfur dioxide also occurs, as can be inferred from the fact that such addition compounds can often be isolated from solution, even though most of them decompose at or a little above ordinary temperature.

---

* The only processes of neutralization (in the usual sense) covered by proton transfer reactions are reactions between acids and anhydrobases.

TABLE 96

COMPARISON OF SOME PROPERTIES OF 'WATER LIKE' SOLVENTS WITH THOSE OF WATER

	Water	Ammonia	Hydrogen fluoride	Hydrogen sulfide	Sulfur dioxide
Molecular weight	18.016	17.032	20.01	34.08	64.06
Melting point	0.00°	—77.7°	—83.1°	—85.60°	—72.5°
Boiling point	100.00°	—33.35°	+19.54°	—60.75°	—10.0°
Density at the boiling point	0.958	0.683	0.991	0.993	1.46
Molecular volume at the boiling point	18.81	24.94	20.19	34.32	43.9
Dielectric constant	87.8 (at 0°)	22 (at —34°)	83.6 (at 0°)	6 (at 0°, in the liquid state)	13.8 (at +14.5° in the liquid state)
Electrical conductivity in $\text{ohm}^{-1} \cdot \text{cm}^{-1}$	$4 \cdot 10^{-8}$ (at 18°)	$4 \cdot 10^{-10}$ (at —15°)	$5 \cdot 10^{-4}$ (at —15°)	$4 \cdot 10^{-11}$ (at —78°)	$1 \cdot 10^{-7}$ (at 0°)
Viscosity in $\text{cm}^{-1} \cdot \text{g} \cdot \text{sec}^{-1}$	0.0029 (at 100°)	0.0026 (at —34°)	—	—	0.0039 (at 0°)
Molecular boiling point elevation, in 1000 g of solvent	0.513°	0.34°	1.90°	0.63°	1.45°
Product of dissociation	$[H_3O]^+ + [OH]^-$	$[NH_4]^+ + [NH_2]^-$	$[H_2F]^+ + F^-*$	$[H_3S]^+ + [SH]^-$	$[SO]^{2+} + [SO_3]^{2-}$

* Cf. footnote on p. 731.

Analogies between processes in aqueous solutions and in water-like systems are particularly clearly developed in the 'ammono' system—i.e., in liquid ammonia— and in the 'sulfito' system—i.e., in liquid sulfur dioxide. The principal kinds of reactions which are involved will be briefly discussed.

(b) Reactions Analogous to Neutralization

(*i*) *Characteristics.* The characteristic feature of neutralization in aqueous solutions is that the process represented by eqn. (5), p. 725, runs practically quantitatively from right to left. In the same way, in the 'neutralization-like' reactions, the processes shown by eqn. (6) also take place from right to left. Since substances which produce H_3O^+ ions in aqueous solution are called acids, and those which produce OH^- ions are called bases, it is convenient to adopt Jander's suggestion, and to give the names 'acid analogues' and 'base analogues' to substances which, in water-like solvents, form the ions typical of the auto-ionization of the solvent itself.

Thus in liquid ammonia, acid analogues are compounds which form NH_4^+ ions by proton transfer to the solvent; in liquid hydrogen fluoride they form H_2F^+ ions; in liquid hydrogen sulfide they form H_3S^+ ions; and in liquid sulfur dioxide they form SO^{2+} ions. Base analogues in these systems are compounds which form NH_2^-, F^- (or HF_2^-), SH^-, and SO_3^{2-} ions, respectively. Hence acid analogues and base analogues react to form undissociated solvent, together with a salt. E.g.,

In water
$$HCl + K[OH] = KCl + H_2O$$
$$[H_3O][ClO_4] + K[OH] = KClO_4 + 2H_2O$$

In ammonia
$$[NH_4][Cl] + K[NH_2] = KCl + 2NH_3$$

In sulfur dioxide
$$[SO]Cl_2 + K_2[SO_3] = 2KCl + 2SO_2$$
$$[SO][SCN]_2 + K_2[SO_3] = 2K[SCN] + 2SO_2$$
$$[SO]Br_2 + [(CH_3)_4N]_2[SO_3] = 2[(CH_3)_4N]Br + 2SO_2$$

(*ii*) *Solvolysis.* Corresponding to hydrolysis in aqueous solutions ('aquo system') is *ammonolysis* in the ammono system, or quite generally, solvolysis.

Hydrolysis $SiCl_4 + 4H_2O = 4HCl + Si(OH)_4$ $(SiO_2 + 2H_2O)$
Ammonolysis $SiCl_4 + 8NH_3 = 4[NH_4]Cl + Si(NH_2)_4$ $(Si(NH)_2 + 2NH_3)$
Hydrolysis $GeCl_4 + 2H_2O = 4HCl + GeO_2$
Ammonolysis $GeCl_4 + 6NH_3 = 4[NH_4]Cl + Ge[NH]_2$

(*iii*) *Ansolvoacids and Ansolvobases.* By analogy with anhydroacids and anhydrobases, these terms are applied to compounds from which acid analogues or base analogues are formed, by combination with molecules of the solvent:

Anhydrobase → base $NH_3 + H[OH] = [NH_4]^+ + [OH]^-$
(in aqueous system)

Ansolvobase → base analogue $H_2O + HF = [H_3O]^+ + F^-$
(in liquid hydrogen fluoride)

The number of compounds which function as ansolvobases towards hydrogen fluoride is extremely large.

(*iv*) *Behavior of Amphoteric Hydroxides.* There is an exact analogy between the behavior of amphoteric hydroxides towards strong bases in water and that of the corresponding substances in water-like solvents. Addition of a little potassium hydroxide to aqueous aluminum chloride first precipitates aluminum oxide hydrate, which then dissolves in excess potash to form the hydroxoaluminate. In the same way, aluminum chloride dissolved in liquid sulfur dioxide reacts with tetramethylammonium sulfite, a base analogue in this system; sparingly soluble aluminum sulfite (also a base analogue) is first precipitated, but this goes into solution again as sulfito-aluminate when an excess of the tetramethylammonium sulfite is added:

$$AlCl_3 + 3KOH = Al(OH)_3 + 3KCl, \quad Al(OH)_3 + 3KOH = K_3[Al(OH)_6].$$

$$AlCl_3 + \tfrac{3}{2}[(CH_3)_4N]_2[SO_3] = \tfrac{1}{2}Al_2[SO_3]_3 + 3[(CH_3)_4N]Cl$$

$$\tfrac{1}{2}Al_2[SO_3]_2 + \tfrac{3}{2}[(CH_3)_4N]_2[SO_3] = [(CH_3)_4N]_3[Al(SO_3)_3].$$

In the example taken, the analogy is very far-reaching. The precipitates in both cases are gelatinous, with strong adsorptive properties. In contact with the solution, the precipitated aluminum sulfite gradually changes into a more insoluble form, just as the oxide hydrate does in contact with the aqueous solution.

The formation of potassium tetramidozincate is an example of an analogous reaction in the liquid ammonia system:

$$ZnI_2 + 4KNH_2 = K_2[Zn(NH_2)_4] + 2KI.$$

(v) *Double Decomposition Reactions.* Salts, in water-like solvents, are electrolytes which have no ions in common with the solvent. Thus the ammonium compounds are not salts in the ammono system, but are the analogues of acid hydrates in the aquo system. Similarly, when reactions in liquid hydrogen fluoride are considered, fluorides (e.g., NaF) are not salts, but base analogues.

As in aqueous solutions, the occurrence of double decomposition reactions between salts in water-like solvents depends upon differences in solubility. Solubility relations often differ considerably from those in aqueous solutions, so that reactions in non-aqueous solvents are frequently useful for preparative purposes. The fact that such reactions between salts are less interfered with by solvolysis is also often of some value.

(c) Discussion of Some Water-like Solvents

(i) *Liquid ammonia* is a good solvent for many inorganic and organic compounds. It occupies a special place among water-like solvents, since it is readily accessible, cheap, and can be used for many purposes. No great experimental difficulties are involved in carrying out studies in liquid ammonia. [7, 8]

Researches on chemical reactions in liquid ammonia solution started with the observation of Weyl (1864) that the alkali metals are soluble therein. More extensive studies of the solvent properties of liquid ammonia were carried out by Gore (1872–73) and Divers (1873). In his experiments, Divers used not pure liquid ammonia, but a solution of ammonium nitrate in the liquid, which is formed directly by the action of ammonia gas on ammonium nitrate at ordinary temperature (or better, at 0°); the process is analogous to the formation of a calcium chloride solution by the action of steam or moist air on calcium chloride. A systematic investigation of the properties of liquid ammonia as a solvent, and of the reactions taking place in it, has been carried out since 1898 by Franklin, Kraus, and others (chiefly in the United States). From a comparison with processes occurring in aqueous solutions, Franklin (1912) suggested the idea of an ammono system of compounds which behave towards one another in liquid ammonia solution in the same way as acids, bases, and salts do in water (the aquo system). [8]

Franklin gave the names *ammonobases* and *ammonoacids* to the base analogues and acid analogues respectively in the ammono system. NH_2-compounds which dissociate to form NH_2^- ions—i.e., primarily *metal amides*—behave as ammono-bases. Compounds which form NH_4^+ ions by proton transfer in liquid ammonia act as ammonoacids. As already noted, ammonium compounds, $[NH_4]X$, which are salts in the aquo system, correspond in liquid ammonia to the acid hydrates of the aquo system. Hence ammonium salts in liquid ammonia react in the same way as acids in aqueous solution.

A typical example of the 'acidic' behavior of ammonium salts in liquid ammonia is provided by their use in reactions of the type:

$$Mg_2Si + 4NH_4Br = 2MgBr_2 + 4NH_3 + SiH_4$$

(Johnson, 1935).

Organic acid amides and most of the amides of non-metals also behave as ammonoacids, just as the corresponding hydroxides in aqueous solution are acids. Organic amines in the ammono system correspond to the alcohols in the aquo system. Imides and imines bear the same relation to amides and amines as, in the aquo system, the oxides do to the hydroxides; they differ in so far as they can lose yet more ammonia to form nitrides, which can also be considered as the analogues of the oxides.

Like water, ammonia has a strongly marked capacity to form addition compounds (*ammoniates*) with other substances. Our knowledge of this class of compound has been greatly extended by the study of liquid ammonia solutions, in that higher ammoniates are often formed under these conditions than can be obtained with gaseous ammonia.

Liquid ammonia has such a great affinity for water that it is often possible to extract water, by treatment with liquid ammonia, from substances which cannot otherwise be directly dehydrated without extensive decomposition (cf. Vol. I, pp. 354, 495).

Liquid ammonia can also bring about the *ammonolysis* of many compounds. Amides and imides can be made in this way, and from them, by loss of ammonia, the nitrides. Formation of nitrides in this way corresponds exactly to the formation of oxides from hydroxides.

The following solvolysis corresponds to the formation of basic salts by hydrolysis:

$$2PbI_2 + 6NH_3 = Pb(NH_2)_2 \cdot Pb(NH_2)I + 3NH_4I.$$

Electrolytic dissociation of typical salts in liquid ammonia is usually less than in aqueous solution. There are, however, some compounds which dissociate electrolytically when dissolved in liquid ammonia, although they are not perceptibly dissociated in aqueous solution. Thus acetylene has a distinct electrolytic conductivity in liquid ammonia solution. Benzaldehyde, C_6H_5CHO, and nitromethane, CH_3NO_2, also conduct, as do other nitro compounds.

The high conductivity of solutions of the alkali and alkaline earth metals in liquid ammonia is noteworthy. The dilute solutions are deep blue in color and the concentrated solutions golden. The alkaline earth metals are present in them as ammoniates—e.g., $Ca(NH_3)_6$; the state of dissolved alkali metals is not certain. Solutions of the alkali metals change more or less rapidly into colorless solutions of the metal amides, M^INH_2 (a cesium solution within a few minutes, a sodium solution in the course of months).

The solutions of the metals in liquid ammonia are highly reactive. Thus, with the alkali metals (except lithium) dissolved in liquid ammonia, oxygen forms the alkali higher oxides, NO produces the hyponitrites, $M^I_2N_2O_2$, PH_3 forms phosphinides, M^IPH_2.

A variety of reactions also takes place with organic compounds—e.g., with weak proton donors:

$$(C_6H_5)_3CH + Na = (C_6H_5)_3CNa + \tfrac{1}{2}H,$$

$$3C_2H_2 + 2Na = 2NaHC_2 + C_2H_4.$$

Sodium-ammonia solutions react with organic halogen compounds in the manner $RX + Na = NaX + R$. The free radical R thus liberated may either react with the solvent to form an amine RNH_2, an imine R_2NH, or a nitride R_3N, and the hydrocarbon RH, or two radicals may unite to form the higher hydrocarbon R—R. In some cases* the free radical remains in solution as such, and the possibility of obtaining free organic or inorganic radicals by reactions of this type has been widely exploited. It has also been possible by analogous reactions to prepare inorganic compounds with relatively long chains—e.g.,

$$Sn(CH_3)_3—Sn(CH_3)_2—Sn(CH_3)_2—Sn(CH_3)_2—Sn(CH_3)_3.$$

Organic compounds with proton donor properties—acid amides, imides, phenols, thiophenols, alcohols, hydrazines, etc.,—undergo direct replacement of H-atoms by alkali metals dissolved in liquid ammonia, and the metal-substituted compounds** readily undergo reaction with alkyl halides in the same solvent. Liquid ammonia is therefore a good solvent for alkylation reactions. Liquid ammonia is a particularly good solvent for nitrogen compounds and for reactions carried out with them (Audrieth, 1930). This can be understood in that ammonia stands to these compounds in the same relation as does water to the oxygen compounds, for which water is known to have a specific solvent power.

It is often possible by studying liquid ammonia solutions, to prove that compounds undergo electrolytic dissociation even when the substances cannot be investigated in aqueous solution, because they undergo decomposition—e.g., sodium amide, $NaNH_2$, and sodium acetylide, $NaHC_2$.

Many intermetallic compounds also dissolve in liquid ammonia, and undergo electrolytic dissociation. Thus Zintl (1931) found that the compounds $\overline{NaPb_3}$ was dissolved by ammonia, to form $[Na(NH_3)_9]_4[Pb_9]$. The existence of the anion $[Pb_9]^{4-}$ of this salt-like compound is noteworthy, both because it shows that lead can acquire a negative charge, as might be expected from its position in the Periodic Table, and because of its analogy to the polysulfide and polyiodide anions. The $[Pb_9]^{4-}$ ions are aggregated in solution, into submicrons detectable by the ultramicroscope. Zintl was able to prepare the compounds $[Na(NH_3)_8]_4[Sn_9]$ and $[Na(NH_3)_6]_3[Sb_7]$ in a similar manner. If the ammonia is driven off, the alloy phases with no salt-like character are regenerated.

It is possible to obtain intermetallic compounds directly from solution, by reaction between metals and metallic salts in liquid ammonia. E.g.,

$$9Na + 4Zn(CN)_2 = NaZn_4 + 8NaCN.$$

For double decompositions in liquid ammonia, the relative solubilities of the substances involved are an essential determining factor. As in aqueous systems, there are also numerous reactions occurring in homogeneous solution, which are the result of other factors—e.g., the formation of salts from base analogues and acid analogues (or by reaction between ammonium compounds and metal amides), due to the tendency for recombination of NH_4^+ and NH_2^- ions to form NH_3.

Examples of double decomposition reactions in liquid ammonia are given below. In all the cases shown, the completion of reaction is due to the insolubility of one of the reaction products. The compounds insoluble in liquid ammonia are therefore underlined.

* With triarylmethyl halides, for example, where the free radical is stabilized by 'resonance'.

** These can be formed by the action of the alkali metal amides in liquid ammonia, as well as by the alkali metals. The reaction of the amides with proton donors corresponds to that of bases with acids in water.

$$Ba(NO_3)_2 + 2AgCl = \underline{BaCl_2} + 2AgNO_3$$
$$Ba(NO_3)_2 + 2KOC_2H_5 = \underline{Ba(OC_2H_5)_2} + 2KNO_3$$
$$AgNH_2 + HN{=}C(NH_2)_2 = \underline{AgN{=}C(NH_2)_2} + NH_3$$
$$AgNO_3 + KNH_2 = \underline{AgNH_2} + KNO_3$$
$$Pb(NO_3)_2 + 2KNH_2 = \underline{PbNH} + NH_3 + 2KNO_3$$
$$3Hg(NO_3)_2 + 6KNH_2 = \underline{Hg_3N_2} + 4NH_3 + 6KNO_3$$

Insoluble compounds may be converted to soluble compounds by complex formation, as in aqueous systems:

$$\underline{Zn(NH_2)_2} + 2KNH_2 = K_2[Zn(NH_2)_4].$$

The solubility of inorganic compounds in liquid ammonia depends on their *heat of solvation*, in the same way as do solubilities in water. The solubilities of some compounds in liquid ammonia and in water are compared in Table 97.

TABLE 97

SOLUBILITY OF ALKALI SALTS IN LIQUID AMMONIA AND IN WATER

(in mols of salt per 1000 g of solvent, after Linhard and Stephan, 1932)

Solvent	NaCl	NaBr	NaI	NaNO$_3$	KCl	KBr	KI	KNO$_3$
Liquid ammonia	2.20	6.21	8.80	15.00	0.0177	2.26	11.09	1.04
Water	6.10	7.71	10.72	8.62	3.76	4.49	7.72	1.30

Liquid ammonia has been successfully used as a solvent for electrolysis. Many metals can be electrolytically deposited from solutions of their salts in liquid ammonia, although it is impossible to deposit them from aqueous solution—e.g., beryllium (Booth, 1930–31). Free radicals such as $HgCH_3$ (Kraus, 1913) and $Cr(C_6H_5)_4$ (Hein, 1926) have also been liberated electrolytically from liquid ammonia solution. When a solution of a quaternary ammonium salt $[NR_4]X$ in liquid ammonia is electrolyzed, a deep blue coloration is obtained at the cathode, similar to that which is characteristic of the solutions of the alkali metals in liquid ammonia. It has been inferred from this that the free NR_4 radicals must have a transient free existence in such solutions (Schlubach, 1920).

(ii) *Liquid Hydrogen Fluoride.* Our knowledge of liquid hydrogen fluoride as a solvent is due largely to K. Fredenhagen (1928 and later). Although the dielectric constant of liquid hydrogen fluoride is not quite so great as that of water (Table 96), the solvent properties and dissociating power of hydrogen fluoride are far greater than those of water. Not only does it dissolve many inorganic compounds, but it is an excellent solvent for the majority of organic compounds. Its behavior towards compounds which can undergo electrolytic dissociation is often rather different from that of water towards the same substances. Many compounds which in aqueous solution have a definitely acidic character react with hydrofluoric acid in a manner resembling the reaction of anhydrobases with water. The behavior of liquid hydrogen fluoride towards salts is largely determined by this peculiarity.

On the assumption that the auto-ionization of liquid hydrogen fluoride follows the equation:

$$2HF \rightleftharpoons [H_2F]^+ + F^-, \quad *$$

* The properties of hydrogen fluoride in aqueous solution make it more probable that auto-ionization takes place chiefly by the process $3HF \rightleftharpoons [H_2F]^+ + [HF_2]^-$. If this is so, the hydrogen fluorides $M^I[HF_2]$, and not the normal fluorides M^IF, are the true base

fluorides play the part of base analogues in liquid hydrogen fluoride, since they furnish F^- ions. There are only a few substances which can function as acid analogues in liquid hydrogen fluoride.

The difference in behavior of hydrogen fluoride and water towards many substances arises from the fact that hydrogen fluoride is a much more powerful proton donor than water. It donates protons to many compounds which can abstract a proton from water. Thus it reacts with alcohols and phenols according to the scheme:

$$ROH + HF \rightleftharpoons [ROH_2]^+ + F^-.*$$

Thus phenols, which act as acids in aqueous solution, behave in liquid hydrogen fluoride in the manner of anhydrobases in water, being converted into base analogues, $[ROH_2][F]$ by reaction with the solvent. Liquid hydrogen fluoride reacts similarly with water, formic acid, acetic acid, etc., and also with acetates, and even with nitrates:

$$
\begin{aligned}
H_2O + HF &= [H_3O]^+ + F^- \\
HCOOH + HF &= [HC(OH)_2]^+ + F^- \\
CH_3COOH + HF &= [CH_3C(OH)_2]^+ + F^- \\
CH_3COOK + 2HF &= [CH_3C(OH)_2]^+ + K^+ + 2F^- \\
KNO_3 + 2HF &= [NO(OH)_2]^+ + K^+ + 2F^-.
\end{aligned}
$$

The reaction shown with the salts corresponds to the reaction of water with metal amides:

$$KNH_2 + 2HOH = [NH_4]^+ + K^+ + 2OH^-,$$

except that in the latter case the weak base NH_4OH is formed together with the strong base KOH: in the reactions in liquid hydrogen fluoride both base analogues, KF and $[CH_3C(OH)_2]F$ or $[NO(OH)_2]F$, respectively, are strongly dissociated.

Work with liquid hydrogen fluoride as solvent is attended by far greater experimental difficulties than work in liquid ammonia. However, the strong proton donor properties of hydrogen fluoride lend a special character to its reaction with dissolved substances, and an extension of our knowledge (especially of solutions of organic compounds) is likely to yield important results.

(iii) *Liquid Hydrogen Sulfide.* As befits its relatively large molecular volume and low dielectric constant, liquid hydrogen sulfide does not possess nearly such great solvent power or ionizing properties as the solvents just considered. Only a limited number of inorganic compounds will dissolve in liquid hydrogen sulfide, but it is a rather better solvent for organic compounds. Among inorganic compounds, a few halides such as PCl_5, PCl_3, PBr_3, PI_3, and $BiCl_3$ are soluble; their solutions have only a very low electrolytic conductivity. Substances with marked proton acceptor properties—e.g., organic amines—form better conducting solutions. It is curious that the molar conductivity of these solutions decreases with dilution. In many cases, substances dissolved in liquid hydrogen sulfide decompose, by reaction with the solvent. These decompositions often take place extremely slowly—often only in the course of weeks at room temperature (in a pressure tube).

---

analogues in the hydrogen fluoride system. The normal fluorides would correspond to the base anhydrides (oxides) of the aqueous system. However, it has not yet been proved that $[HF_2]^-$ ions are formed by the dissociation of *liquid* hydrogen fluoride, and the reactions are usually formulated in terms of the assumption that F^- ions are present.

* Parallel with this, the reactions

$$[ROH_2]^+ + F^- \rightleftharpoons RF + H_2O; \quad H_2O + HF \rightleftharpoons [H_3O]^+ + F^-$$

may also take place to some extent, and lead to a further increase in the electrical conductivity of such solutions.

It is not likely that liquid hydrogen sulfide will be of practical importance in any field, since it does not have good solvent properties, and its low boiling point occasions a good deal of experimental difficulty. For this reason it has not been widely studied, and little is known of reactions between dissolved substances in hydrogen sulfide. Wilkinson (1925 and later) had carried out much of the work on hydrogen sulfide as a solvent.

(iv) Liquid Sulfur Dioxide. As was found by Walden (1902), liquid sulfur dioxide is a good solvent for many organic and inorganic compounds. Among inorganic compounds, the iodides, bromides, chlorides, thiocyanates, and acetates of certain light and heavy metals are fairly soluble, as are most salts of substituted ammonium radicals. Organic compounds containing oxygen and nitrogen are often very soluble, but only cyclic or unsaturated hydrocarbons. The differing solubilities of different classes of hydrocarbons in liquid sulfur dioxide are utilized in the Edeleanu process for refining petroleum.

A systematic investigation of reactions between substances dissolved in liquid sulfur dioxide has been made during recent years by G. Jander (1936 and later). This has shown that in all respects the processes are analogous to those taking place in aqueous solution, when allowance is made for the fact that in this solvent thionyl compounds, SOX_2, act as acid analogues and the sulfites, $M^I_2[SO_3]$, as base analogues. Jander was able to show by conductometric titrations that these substances mutually neutralize each other, like acids and bases in aqueous solutions. Acid analogues and base analogues in liquid sulfur dioxide are dissociated to about the same extent as are the weak acid and bases in water.

Water behaves in liquid sulfur dioxide as ammonia does in aqueous solution—i.e., like an anhydrobase:

$$2H_2O + 2SO_2 \rightleftharpoons [(H_2O)_2SO]^{2+} + [SO_3]^{2-}.$$

The base analogue compound, $[(H_2O)_2SO][SO_3]$, thereby produced, can form salts with acid analogues.

$$[(H_2O)_2SO][SO_3] + SOBr_2 = [(H_2O)_2SO]Br_2 + 2SO_2.$$

This reaction can be followed conductometrically. Just as salts which are insoluble in water can often be converted into soluble complex salts, by substitution of ammonia, so the same occurs in liquid sulfur dioxide by substitution of water:

$$Co(SCN)_2 + 2H_2O = [Co(H_2O)_2](SCN)_2.$$

The reaction of zinc ethyl with liquid sulfur dioxide provides an illustration of solvolysis:

$$Zn(C_2H_5)_2 + SO_2 = ZnO + OS(C_2H_5)_2.$$

Liquid sulfur dioxide has a much weaker solvolytic power than water, however, so that substances such as $SiCl_4$, $SnCl_4$, and $SbCl_3$, which have a great tendency to undergo hydrolysis, form clear solutions in liquid sulfur dioxide, with no sign of solvolysis.

(iv) Other Water-like Solvents. Table 98 summarizes some properties of a number of other solvents which are more or less water-like in behavior. The solvents are set out in the order of their dielectric constants. Liquid hydrogen cyanide is notable for its very high dielectric constant. Anhydrous ('glacial') acetic acid is a widely used non-aqueous solvent, and its behavior towards dissolved substances has been extensively studied.

TABLE 98

PROPERTIES OF SOME IONICALLY DISSOCIATING SOLVENTS

Name	Formula	Dielectric Constant at 25°	Auto-conductivity, in ohm^{-1}·cm^{-1} at 25°	Viscosity in cm^{-1}·g·sec^{-1} at 25°	Melting point	Boiling point
Hydrogen cyanide	HCN	152 (at 0°)	$4.5 \cdot 10^{-7}$	0.00242 (at 0°)	$- 15°$	$+ 26.5°$
Formamide	$HCONH_2$	> 84	$2 \cdot 10^{-7}$	0.0077 (at 105°)	$+ 2°$	$+ 105°$ (at 11 mm press.)
Nitric acid	HNO_3	> 81 (at 0°)	$1.36 \cdot 10^{-2}$ (at 0°)	0.0227 (at 0°)	$- 41.1°$	$+ 84°$
Formic acid	$HCOOH$	58	$6.2 \cdot 10^{-5}$	0.0163	$+ 8.43°$	$+ 100.6°$
Hydrazine	N_2H_4	51.7	$2.5 \cdot 10^{-6}$	0.00905	$+ 1.4°$	$+ 113.5°$
Methanol	$CH_3 \cdot OH$	30.3	$6 \cdot 10^{-8}$	0.00546	$- 97.8°$	$+ 64.7°$
Ethyl alcohol	$C_2H_5 \cdot OH$	23	$9 \cdot 10^{-9}$	0.01084	$- 114.1°$	$+ 78.3°$
Acetone	$CH_3 \cdot CO \cdot CH_3$	21	$5.5 \cdot 10^{-8}$	0.00316	$- 94.3°$	$+ 56.3°$
Hydrogen chloride	HCl	10.2 (at −108°)	$2 \cdot 10^{-7}$ (at −100°)	0.00504 (at −100°)	$- 114.8°$	$- 84.9°$
Acetic acid	$CH_3 \cdot COOH$	7	$6 \cdot 10^{-9}$	0.0016	$+ 16.7°$	$+ 118.1°$
Phosgene	$COCl_2$	4.3	—	—	$-126°$	$+ 8.2°$

4. Reactions in Melts

Molten substances can be included among solvents in the most general sense. Melts of indifferent substances, such as are sometimes used in molecular weight determinations (e.g., camphor), do not differ in any essential respect from ordinary indifferent solvents, except that they melt at higher temperatures. *Molten salts* and similar substances are very different, however, and their melts possess a considerable electrolytic conductivity. They occupy a special position in relation not only to the indifferent solvents, but also to the water-like solvents, since the intrinsic conductivity of the latter is generally very small as compared with that of dissolved electrolytes. Of substances which are liquid at the ordinary temperature, it is possible that anhydrous nitric acid ('nitronium nitrate', cf. Vol. I, p. 600) should be included among the 'molten salts'.

Molten salts are of great technical importance in the electrolytic preparation of the metals, and especially the light metals. The conditions for the electrodeposition of metals from such melts have therefore been extensively studied [9, 10]. The decomposition potential of a pure compound in the molten state is generally not identical with the decomposition potential of the compound when present in solution in the melt of some other substance. This accords with the general rule that the decomposition potential depends on the solvent. The composition of the melt does not usually exert a very large effect on the decomposition potential of a dissolved compound, but it cannot be neglected in precise measurements.

The study of reactions occurring between molten metals and molten salts or oxides with which they are in contact (displacement equilibria) is also of considerable technical importance, since it can be applied to the interaction between metals and slags in metallurgical processes. It is usually possible to apply the chemical law of mass action, and the Nernst distribution law, directly to equilibria of this kind, without making allowance for deviations from the laws of ideal solutions, such as might arise from the association of molecules in the melt. In many cases, however, the deviations from the ideal laws are very considerable, and association between ions frequently occurs in melts. Thus Jellinek (1933) concluded from substitution equilibria that ions such as Li_2^{2+}, Cl_2^{2-}, Cl_4^{4-}, Br_2^{2-}, and Br_4^{4-} were present in molten salts. The exact nature of the associated ions, and of the structure of the liquids, is not yet fully explained; it is most likely that the association involves a loose coordination between positive and negative ions, so that the 'short range' structure around each ion is not very different from that existing in the solid salts. [See, for example [9], and Heymann, *Proc. Roy. Soc.*, A 188, (1947) 393; A 194 (1948) 210].

5. The Lewis Theory of Acids and Bases

For many purposes it is advantageous to consider acid-base reactions and related processes independently of the solvent. This is made possible by the acid-base theory of G. N. Lewis. According to this theory, the *acid character* of a substance arises from the ability of one of its atoms to *accept a share in a pair of electrons* from another substance, with the resultant formation of a coordinate covalent bond; the substance *donating* a share of an electron pair is said to possess *basic character*. *Amphoteric substances* are therefore those which can both accept and also supply an electron pair for the formation of a coordinate covalent bond.

The reaction of an acid with a base in aqueous solution, i.e., reaction between the hydronium and hydroxyl ions, depends then, according to Lewis, on the following process:

$$\left[\begin{array}{c} H : \overset{..}{O} : H \\ H \end{array} \right]^{+} + \left[: \overset{..}{\underset{..}{O}} : H \right]^{-} = H : \overset{..}{\underset{..}{O}} : H : \overset{..}{\underset{..}{O}} : H^* = 2H_2O \qquad (7)$$

* This is merely a hypothetical intermediate product; it is not to be regarded that hydrogen can be bound to two oxygen atoms by covalent bonds.

In the consideration of the neutralization process, Lewis therefore emphasizes not the formation of the end product (H_2O), but rather of the intermediate (and in many cases, hypothetical) product of the reaction, i.e., that substance arising from the initial formation of the coordinate covalent bond. From the Lewis viewpoint there is a complete analogy between neutralization reactions in aqueous solution and reactions involving salt formation without participation of solvent. For example, the formation of NH_4Cl from HCl and NH_3. (combination of an acid with an anhydrobase), and the production of $CaCO_3$ from CO_2 and CaO (reaction be_ tween an acid anhydride and a base anhydride), may be regarded as neutraliza_ tion reactions:

$$
\begin{array}{c}
\text{H} \\
\text{H:N:} \\
\text{H}
\end{array}
+ \quad \text{H:Cl:} \quad = \quad
\begin{array}{c}
\text{H} \\
\text{H:N:H:Cl:} \\
\text{H}
\end{array}
\quad = \quad [\text{NH}_4]^+ \ \text{Cl}^- \qquad (8)
$$

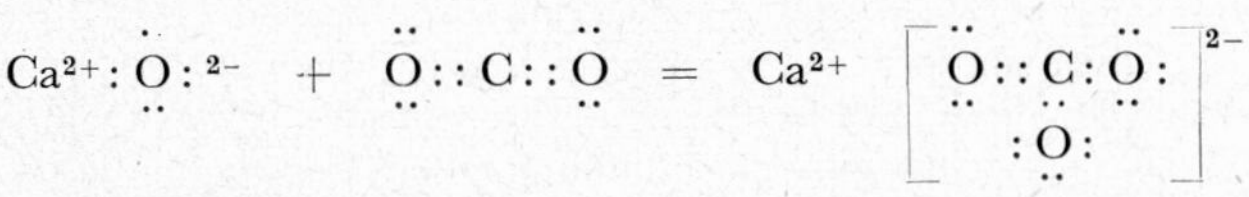

$$
\text{Ca}^{2+}\text{:O:}^{2-} \ + \ \text{O::C::O} \ = \ \text{Ca}^{2+}
\left[
\begin{array}{c}
\text{O::C:O:} \\
\text{:O:}
\end{array}
\right]^{2-} \qquad (9)
$$

A corresponding formulation results for the reaction of acid analogues with base analogues; for example,

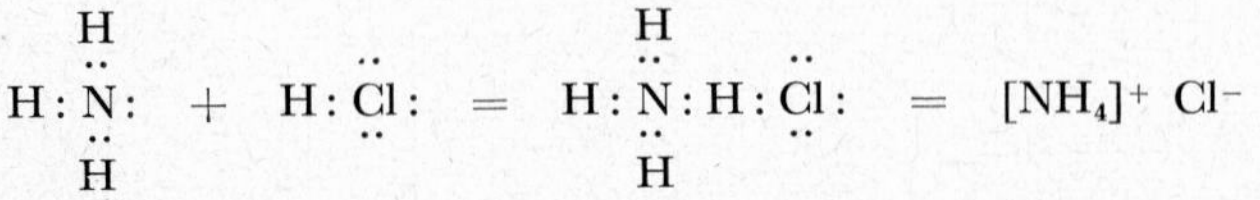

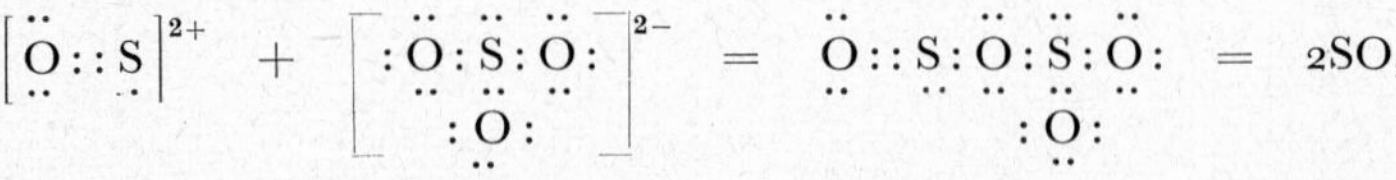

$$
\left[\text{O::S} \right]^{2+} +
\left[
\begin{array}{c}
\text{:O:S:O:} \\
\text{:O:}
\end{array}
\right]^{2-} = \
\begin{array}{c}
\text{O::S:O:S:O:} \\
\text{:O:}
\end{array}
\ = \ 2\text{SO}_2 \qquad (10)
$$

Many species of complex compounds fit into the same pattern and can therefore be regarded as neutralization products of acid-base reactions. As illustrations, the formation of the BF_4^- ion, as well as of the addition compound of BF_3 with NH_3, may be given:

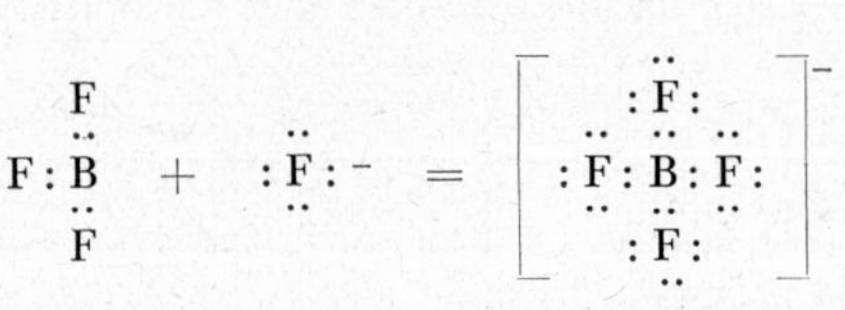

$$
\begin{array}{c}
\text{F} \\
\text{F:B} \\
\text{F}
\end{array}
+ \quad \text{:F:}^- \quad = \quad
\left[
\begin{array}{c}
\text{:F:} \\
\text{:F:B:F:} \\
\text{:F:}
\end{array}
\right]^-
\qquad (11)
$$

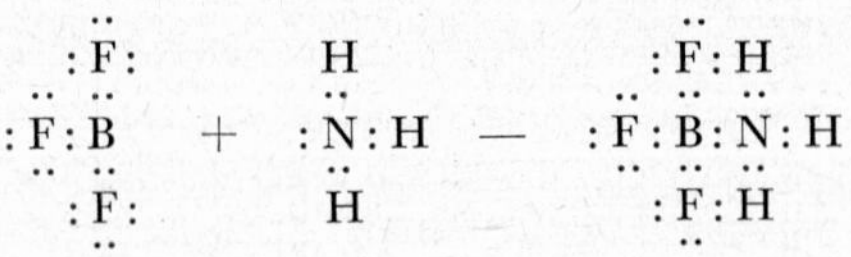

$$
\begin{array}{c}
\text{:F:} \\
\text{:F:B} \\
\text{:F:}
\end{array}
+ \quad
\begin{array}{c}
\text{H} \\
\text{:N:H} \\
\text{H}
\end{array}
\quad - \quad
\begin{array}{c}
\text{:F:H} \\
\text{:F:B:N:H} \\
\text{:F:H}
\end{array}
\qquad (12)
$$

The apparent analogies between reactions of the types represented in equations (7)–(12) shed light on many reactions which play an important role especially in organic chemistry. Thus the action of aluminum chloride as a catalyst in certain reactions (e.g., in Friedel-Crafts syntheses) becomes understandable when one considers that this compound has, in common with the excellent catalyst the $[H_3O]^+$ ion, the ability to accept an electron pair with the formation of a covalent bond.

In general, acids react instantaneously with bases, without an apparent dependence on temperature. But there are cases in which an activation energy is necessary for reaction to occur and consequently the speed of reaction is tempera-

ture dependent. Substances which apparently must be activated before they mani-
fest their acid (or base) character have been termed *secondary* acids (or bases) by
Lewis. He has assumed that where activation is necessary for reaction, electromeric
forms of the substances exist, and it is in these forms only that the acid or base
function may be manifested.

References

1 T. H. DURRANS, *Solvents*, 4th Ed., London 1938, 238 pp.

2 J. H. HILDEBRAND, *Solubility of Non-electrolytes*, 2nd Ed., New York 1936, 203 pp.

3 P. WALDEN, *Elektrochemie nichtwässeriger Lösungen*, Leipzig 1924, 516 pp.

4 P. WALDEN, *Das Leitvermögen der Lösungen* [OSTWALD-DRUCKER, *Handbuch der allgemeinen
Chemie*, Vol. IV], 3 vols., Leipzig 1924, 383, 346, and 397 pp.

5 L. F. AUDRIETH and J. KLEINBERG, *Non-Aqueous Solvents*, New York 1953, 284 pp.

6 G. JANDER, *Die Chemie in wasserähnlichen Lösungsmitteln*, Berlin 1949, 367 pp.

7 J. BRONN, *Verflüssigtes Ammoniak als Lösungsmittel*, Berlin 1905, 252 pp.

8 E. C. FRANKLIN, *The Nitrogen System of Compounds*, New York 1935, 339 pp.

9 P. DROSSBACH, *Elektrochemie geschmolzener Salze*, Berlin 1938, 144 pp.

10 *Die technische Elektrolyse im Schmelzfluss* [*Handbuch der technischen Elektrochemie*, edited by
V. ENGELHARDT, Vol. III], Leipzig 1934, 565 pp.

CHAPTER 19

REACTIONS OF SOLID SUBSTANCES

1. Introduction

(a) General

The reactions of solids [1-3]—that is to say, reactions in which one or more of the reactants is a solid substance—can be divided into two classes, depending on whether reaction takes place *at the surface* of the solid reactant or *in the interior* of the solid. The reactions of the second kind are commonly referred to as reactions *in* solids or 'solid-phase' reactions*. When all the reactants are solids (i.e., when no liquids or gases are involved) the reactions are referred to as reactions *between* solids.

Examples of reactions at the surface of solids, in which the solid is consumed are the combustion of carbon in oxygen to form CO_2, the dissolution of zinc in hydrochloric acid, and the rusting of iron. These reactions show that gaseous or liquid (or dissolved) substances can take part in the reaction, and the products may or may not be solid. However, if a solid reaction product is formed, the chemical change can continue at the surface of the reacting solid only if the reaction product is sufficiently porous to allow the access of fresh gaseous or liquid reactant to the surface.

It is only possible for two solid substances to react directly, without the intervention of gaseous, liquid or dissolved substances, if chemical change can take place in the *interior* of the solid. The same applies to reactions of solids with gases or liquids and solutions, if the reaction product forms a compact coherent layer on the surface.

Reactions taking place *at the surface* of solids have long been familiar to chemists. Little attention was formerly given to the fact that reactions can also take place *in* solids, although thermal decomposition reactions of solids, attended with the evolution of gas, are of this kind, and many of them—such as the decomposition of $CaCO_3$ into CaO and CO_2—have been applied to practical ends since the earliest times. Reactions involving *only* solids—i.e., in which the reactants, the product and any intermediate stages are all solids—were formerly considered to be impossible. Although certain reactions of this kind have long had practical application—e.g., the cementation of iron—it was held that solids were, by their nature, incapable of reaction, in accordance with the old rule 'corpora non agunt nisi fluida'. It is, indeed, true that gases and liquids are generally much more reactive than solids. Nevertheless, there are many examples of reactions

---

* There is no general descriptive name for reactions of the first class. The name 'reactions *on* solids' or 'surface reactions' is used for reactions which take place at the surface of solids without the solid being consumed in the reaction—i.e., for catalytic reactions.

between solids which occur rapidly and strongly exothermically. This generally occurs, indeed, only at elevated temperatures, but far below the melting point. The systematic study of such reactions and the associated phenomena has produced results of considerable technical importance, as well as adding much to our knowledge of the nature of the crystalline state, and of the mechanism of chemical reactions in general.

Among reactions proceeding in the solid state there must be included the electrolytic decomposition of solid salts and similar compounds. As was shown by Bruni (1913) and Tubandt (1920), many salts have a considerable electrolytic conductivity far below their melting point—e.g., AgI at temperatures as low as 146°. The mobility of the silver ion in solid silver iodide at 146° is about the same as that of the silver ion in aqueous solution at 25°. It is noteworthy that (at lower temperatures, especially) as a rule only one of the two species of ion making up the crystal of a compound is concerned in the transport of current, whereas the other species of ion remains immobile in the crystal lattice.

(b) Historical

The processes which play a part in reactions taking place at the surface of solids were largely explained in the classical epoch of physical chemistry. Reactions and other transformations occurring in the interior of solids, and especially the reactions involving only solid substances, have only received systematic study during recent years. Processes characteristic of such reactions were first investigated for *metals* by Tammann (1905 and later). The diffusion of solid metals into one antoher was first recognized by Roberts-Austen (1895) for the interdiffusion of lead and gold; Bruni (1911) proved the occurrence of diffusion in metals, from the decrease in electrical conductivity that accompanies mixed crystal formation. The wider study of reactions between solids was first taken up by Hedvall (1910), for the case of *non-metallic solids* (oxides, salts, etc.). Following Hedvall, the course and mechanism of reactions of this type has been investigated by Tammann (since 1925), W. Jander (since 1927), Fischbeck (since 1927), and others. From the knowledge of the structure of solids derived from X-ray crystallography and an understanding of the forces acting between the constituents of the crystal lattice, von Hevesy (1922), Smekal (1925), Jost (1926), Wagner (1930), and others were able to find a rigorous physical basis for the interpretation of the phenomena observed in the reactions of solids. It also became clear why reactions and transformations occuring in the solid state should pass through intermediate states characterized by high reactivity, such that many substances prepared by solid-state reactions can be obtained in forms with high energy content and high reactivity (Hedvall, 1922 and later; Hüttig, 1931 and later). 'Tarnish reactions' and other processes can also be given an exact treatment. This theoretical advance is due largely to Wagner (1933).

(c) Diffusion

Processes of diffusion are of fundamental importance for reactions taking place both at the surface and in the interior of solids. *Diffusion* signifies the progressive mixing of different substances in contact with one another, as a result of molecular motion, without the operation of any external forces.

Diffusion in solutions is governed by the following law. The quantity dn of a dissolved substance which diffuses in the time dt through the cross section q, from the direction of high concentration to that of low concentration, is proportional to the concentration gradient in the direction of diffusion, $-dc/dx$, and the cross section q:

$$\frac{dn}{dt} = -D \cdot q \cdot \frac{dc}{dx} \tag{1}$$

D is known as the *diffusion coefficient* of the substance, and defines the quantity (in grams or gram-molecules, according to the units chosen) diffusing in unit time through a cross section of 1 sq. cm, when the concentration gradient is 1. The concentration must be expressed in the same units.

The diffusion coefficient depends upon the temperature, the nature of the dissolved substance, and the solvent. Since dissolved substances alter the properties of the solvent, the diffusion coefficient of any substance is not independent of other substances contained in the same solution. For the same reason, it is not independent of the concentration of the diffusing substance. At constant temperature, and in some one solvent, the diffusion coefficients of different substances are inversely proportional to the square root of their molecular weights, provided that the substances are chemically similar, of similar structure, and not too highly solvated in solution*.

$$\frac{D_1}{D_2} = \frac{\sqrt{M_2}}{\sqrt{M_1}} \tag{2}$$

where D_1, D_2 are the diffusion coefficients, and M_1, M_2 the molecular weights of substances 1, 2. In electrolyte solutions containing different quantities of other electrolytes,

$$\frac{D_1 \cdot \eta_1}{D_2 \cdot \eta_2} = \frac{\sqrt{M_2}}{\sqrt{M_1}} \tag{2a}$$

where η_1, η_2 are the different viscosities of the solutions resulting from the differing contents of foreign electrolyte.

The fundamental law of diffusion represented by eqn (1) was stated by Fick in 1855. It applies also to the diffusion of miscible liquids into one another, and for the mutual diffusion of gases.

When Fick's law is applied to the diffusion of gases into one another, the quantity substituted for c in eqn (1) is the density of the gas diffusing through unit cross section. The same diffusion coefficient applies to the diffusion of the second gas, in the opposite direction, but the density-gradients are different for the two gases. Thus equal volumes of the two gases diffuse across any cross section, but different quantities by weight. The diffusion coefficients of gases depend upon the temperature and on the pressure, but vary only very slightly with the composition of gaseous mixtures.

Diffusion of gases into one another should not be confused with the permeation of gases through a fine-pored wall (e.g., of unglazed porcelain), which is also commonly called 'diffusion'. The diffusion of gases through porous membranes follows the law of Graham and Bunsen, which states that the diffusion coefficient of a gas is inversely proportional to the square root of its molecular weight. The same law applies to *effusion*—i.e., the flow of a gas out of a very fine aperture in a thin wall. Since the density of a gas is proportional to its molecular weight (Avogadro's law), Graham's law can be put in the form:—the rates of

---

* The limitations to the validity of this law are not yet well understood. Since the molecules in a liquid do not move in free space, conditions are very different from those governing diffusion of gases, in which the diffusion coefficient is proportional to the molecular velocity (i.e., to $M^{-\frac{1}{2}}$). It might, at first sight, be supposed that molecular *size* would be more important than molecular *weight* in solutions, where the solute molecules must escape at each step from a 'cage' of solvent molecules of comparable size. For colloidal particles, however, where there is an order of magnitude difference in size between diffusing particles and solvent molecules, the diffusion coefficient does accord with eqn (2). The theoretical justification for eqn (2) is thus weak, and it has not been very rigorously tested, although it has been widely used for determinations of molecular and ionic weights. See Souchay, *Bull. Soc. chim. France*, 14 (1947), 914.

effusion of gases, and of their diffusion through porous membranes, are inversely proportional to the square root of their molecular weights:

$$\frac{v_1}{v_2} = \sqrt{\frac{M_2}{M_1}} \qquad (3)$$

This simple relation does not hold for the rate of diffusion of two gases into one another in the absence of a membrane. However, the kinetic theory of gases leads to a fairly simple connection between the rate of diffusion of gases into each other and the mean free paths and velocities of the individual gas molecules. This last is, indeed, inversely proportional to the square root of the molecular weight, so that, in general, the smaller their molecular weights, the faster do gases diffuse into one another.

Diffusion in solids [4, 5] is also formally described by Fick's diffusion law, although the mechanism of the diffusion process is quite different from that applying in gases, liquids, or solutions.

Table 99 gives some values for diffusion coefficients in gases, solutions, and solids. It may be seen that diffusion in gases occurs much more rapidly than in liquids. In solids, even at high temperatures, diffusion rates are generally far smaller than in liquids, but there are certain cases in which the diffusion coefficients in solids at only slightly elevated temperatures actually exceed the values found for liquids or solutions at ordinary temperature. Diffusion coefficients in solid substances thus vary over a very wide range of magnitudes. Whereas the diffusion coefficient of Au in Pb reaches a value of $3.5 \cdot 10^{-7}$ cm²sec⁻¹ at only 250°, that of Mo in W is only $1.0 \cdot 10^{-10}$ cm²sec⁻¹ at 2000°. The lower limit of measurability is about 10^{-17} cm²sec⁻¹*, but measurement of diffusion coefficients much smaller than 10^{-12} cm²sec⁻¹ is attended with some difficulty.

Diffusion coefficients in solids vary with temperature according to the formula

$$D = D_0 \cdot e^{-\frac{Q}{RT}} \qquad (4)$$

where D_0 and Q are constants depending on the nature of the material**. Q has the dimensions of energy, and can be regarded as the activation energy of the (activated) diffusion process (cf. Chap. 17). Fig. 96 shows the logarithm of the diffusion coefficient of Mo in W, plotted against the absolute temperature; if $\log D$ is plotted against $\frac{1}{T}$, a straight line is obtained. Whereas, between 2600° K and 2300° K, the diffusion coefficient of molybdenum decreases by only about one power of 10, it falls by more than seven powers of 10 between 1000° K and 700° K. This extraordinarily rapid diminution in diffusion coefficient below a certain temperature range is a consequence of the form of eqn (4), and is therefore observed for most solids. For this reason, even for solids in which diffusion occurs fairly readily at only a few hundred degrees centigrade, the diffusion coefficient has generally fallen to immeasurably small values at ordinary temperature.

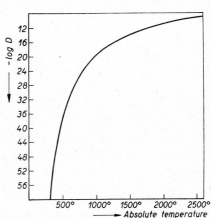

Fig. 96. Temperature variation of the diffusion coefficient of molybdenum in tungsten.

* In a few special cases, where diffusion can be measured by the α-particle recoil method (see ref. [4]), this can be extended to 10^{-20} cm²sec⁻¹.

** The empirically determined constants D_0 and Q can also be calculated on certain assumptions—cf. Van Liempt, *Rec. trav. chim.*, 51 (1932) 114.

TABLE 99

DIFFUSION COEFFICIENTS OF VARIOUS SUBSTANCES, in $cm^2\ sec^{-1}$

Gases in air at 0° and 760 mm pressure	Dissolved substances in water at 18°	Ions of salts, in crystals		Metals in solid metals	Non-metals in metals
H_2　0.665	KCl　$1.67 \cdot 10^{-5}$	Cu^+	in AgI at 480°　$3.9 \cdot 10^{-5}$	Au　in Pb　$6.5 \cdot 10^{-5}$ at 320°	H in Pd　$0.5 \cdot 10^{-5}$ at 25°
H_2O　0.198	LiCl　$1.07 \cdot 10^{-5}$	Ag^+	in Cu_2S at 223°　$3.1 \cdot 10^{-6}$	Ag　in Pb　$9.1 \cdot 10^{-8}$ at 285°	N in Fe　$1.4 \cdot 10^{-7}$ at 1000°
NH_3　0.198	$MgSO_4$　$0.56 \cdot 10^{-5}$	Ag^+	in Cu_2Te at 330°　$3.6 \cdot 10^{-7}$	Zn　in Pb　$1.6 \cdot 10^{-10}$ at 285°	C in Fe　$1.1 \cdot 10^{-6}$ at 1000°
CO_2　0.139	C_2H_5OH　$1.10 \cdot 10^{-5}$	Cd^{2+}	in $ZnWO_4$ at 950°　$1.6 \cdot 10^{-10}$	Cu　in Au　$9.4 \cdot 10^{-11}$ at 560°	
CS_2　0.088	$CO(NH_2)_2$　$1.01 \cdot 10^{-5}$	Sr^{2+}	in $CaMoO_4$ at 950°　$1.7 \cdot 10^{-11}$	Au　in Pt　$1.4 \cdot 10^{-11}$ at 900°	
C_6H_6　0.075 Benzene	$C_6H_{12}O_6$　$0.57 \cdot 10^{-5}$ Glucose	MoO_4^{2-}	in $BaWO_4$ at 950°　$9.6 \cdot 10^{-11}$	Au　in Ag　$5 \cdot 10^{-17}$ at 490°	

2. **Reactions Between Solids** [2]

(a) **General**

When a mixture of two powdered solids is heated, it is often found that a strongly exothermic process sets in at some temperature. It shows itself in a sudden, and often very considerable, rise on the *heating curve*, obtained by plotting the temperature of the mixture against the time, while the mixture is heated steadily (cf. Fig. 97). This evolution of heat, which may take place at a temperature far below the melting points or decomposition temperatures of the constituents of the mixture, represents the heat of reaction of some chemical reaction taking place between the solids—i.e., in the example of Fig. 97, in the reaction

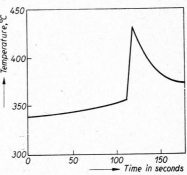

Fig. 97. Heating curve for BaO + CuSO₄ (according to Hedvall).

$$BaO + CuSO_4 = BaSO_4 + CuO.$$

Analysis of the materials after heating proves that this reaction has, in fact, taken place.

Even when the heat of reaction is too small, or the rate of reaction too sluggish, to show itself by a rise in the heating curve, it is possible to determine the temperature at which reaction sets in from analyses of mixtures heated to different temperatures.

It has been proved (Hedvall, 1924; Tammann, 1927) that reaction in cases such as that cited is not dependent upon the presence of lower-melting impurities, nor need it proceed by way of gaseous decomposition products, formed in traces by thermal dissociation. The processes involved are true solid-phase reactions. Reactions occurring in the same way as that cited include those between basic and acidic oxides, to form salts or double oxides; reactions between non-volatile acidic oxides and carbonates; the substitution of metals in salts, oxides, or sulfides by baser metals; double decompositions between salts (e.g., $BaCO_3 + CaSO_4 = BaSO_4 + CaCO_3$); and double decompositions between metal oxides and metal sulfides, phosphides, carbides, or silicides.

To explain these reactions, it is assumed that, with rise in temperature the atoms or ions in crystals execute vibrations of continually increasing amplitude about their mean positions. When the amplitude of these vibrations becomes great enough, they result in a more or less frequent exchange of positions in the crystal lattice. [4]

This place-exchange occurs especially readily in 'loosened' or 'defective' crystals—i.e., those in which there are sub-microscopic channels, surfaces of discontinuity, etc.,—since in these the free surfaces (in the interior of the crystal) offer possibilities of freer displacement of the atoms. Even in completely homogeneous crystals, however, the lattice defects (interstitial atoms and vacant lattice positions), which are always present to some extent in a state of thermodynamic equilibrium, make it possible for place-exchange to occur. At points where different crystals are in contact, atoms or ions can pass from one crystal to another by place-exchange. If the crystal powder consists of a single substance this process can lead to cohesion or *sintering*, such as is utilized in powder metallurgy (cf. p. 2). Associated with this is a certain measure of realignment of the crystalline structures and a recrystallization of the grains (cf. Vol. I, p. 571). In a powder mixture, place-exchange at the points of contact of chemically dissimilar particles results in a chemical change, which is then propagated into the interior of the crystal by a continuation of the process from the interface. This propagation of the chemical change is necessarily dependent upon the rate at which atoms or ions can migrate into the crystal by place-exchange processes—i.e., on their *diffusion rate* in the crystal.

It has been shown that the rate of propagation of reactions in solids, and the temperature dependence of the reaction rates, are fairly well described by the laws of diffusion. In particular, they usually follow the parabolic ($t^{1/2}$) law quite closely (p. 754). The variation of reaction rate with temperature is given by eqn (4) above. According to the theoretical model, the occurrence of recrystallization and of chemical reaction in solids are conditioned by the same factor. Hence the idea that the progress of chemical reactions is determined by *place-exchange processes* is supported by Tammann's observation that the temperature of inception of reaction between solids is almost always the temperature at which one or other reactant begins to recrystallize.

Sintering, which was formerly regarded as a consequence of the partial fusion of crystal powders, can (as was proved by Hedvall) take place at temperatures below those at which any fusion is possible. In such cases, the caking together of powder grains is the result of place-exchange and diffusion processes at the points of contact of crystals, and the consequent recrystallization. The latter very often leads to a considerable decrease in the volume occupied by the powder—a process known in ceramic technology as shrinkage. These processes often take place with enhanced ease in mixtures of powders in which reaction can occur, since the place-exchange at points of contact is favored by the chemical affinity of the system. The interpretation of these phenomena is of considerable technical importance— e.g., in understanding the phenomena observed in the firing of Portland cement, and in ceramics.

When they are taken to completion, reactions between solids always result in the complete disappearance of one of the components, except when a complete series of mixed crystals can be formed.

It can be shown from the Phase Rule that this must be so. Consider, for example, a mixture of BaO and $CuSO_4$, in which the reaction $BaO + CuSO_4 = BaSO_4 + CuO$ can occur. The system contains 3 components (BaO, CuO and SO_3). The proportions of the starting materials are arbitrary, but are not subject to change when a mixture has been made up, so that there remain two degrees of freedom (temperature and pressure). The number of solid phases that can coexist in a state of equilibrium is thus *three*. Hence *only one* other solid phase can be present in equilibrium with CuO and $BaSO_4$, and this may be either BaO or $CuSO_4$, depending on which was present in excess in the original mixture.

If it is possible for several compounds to be formed between two substances, the product of the reaction in the solid state depends not only on the composition of the mixture, but also on the rates of diffusion through the various phases of the ions undergoing exchange. For example, from CaO and SiO_2 four compounds, with the compositions $3CaO \cdot SiO_2$, $2CaO \cdot SiO_2$, $3CaO \cdot 2SiO_2$, and $CaO \cdot SiO_2$, can be formed by crystallization from melts. It was shown by W. Jander that by reaction in the solid state the primary product is always the orthosilicate, $2CaO \cdot SiO_2$, whatever the composition of the mixture. If SiO_2 is present in excess, the orthosilicate is subsequently converted into the silicate corresponding to the composition of the mixture. In a mixture with a 1 : 1 molar ratio, the orthosilicate is first converted to the silicate $3CaO \cdot 2SiO_2$, and from this the metasilicate, $CaO \cdot SiO_2$, is formed (cf. Fig. 98).

According to Jander, the course of the reaction is as follows. Where a CaO crystal is in contact with a grain of SiO_2, a thin crystalline layer of orthosilicate is first formed, by a place-exchange process (cf. Fig. 99a). In order that further reaction shall occur, it is neces-

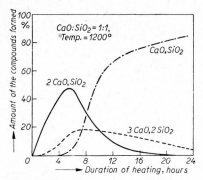

Fig. 98. Reaction between CaO and SiO_2 in the solid state. Variation in yield of products with the duration of heating (after Jander).

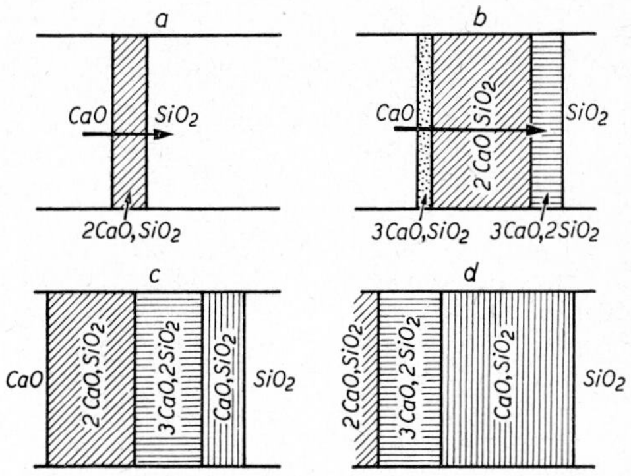

Fig. 99. Schematic representation of the course of the reaction between CaO and SiO_2 in the solid state.

sary for CaO (which is more mobile than SiO_2, as a variety of observations indicate) to diffuse through the $2CaO \cdot SiO_2$, as shown by an arrow in the sketch. Provided that the orthosilicate layer is not too thick, this diffusion occurs so rapidly that continued formation of $2CaO \cdot SiO_2$ occurs at the interface between this phase and the SiO_2. $3CaO \cdot SiO_2$ can be formed at the interface between $2CaO \cdot SiO_2$ and CaO, but this compound is formed so slowly at $1200°$ that it could not be detected in the experiments forming the basis of Fig. 98*. As the orthosilicate layer grows in thickness, the arrival of CaO at the interface between $2CaO \cdot SiO_2$ and SiO_2 eventually fails to keep pace with the production of $3CaO \cdot 2SiO_2$ at that point. A thin layer of this last compound therefore begins to form, and its further growth is subject to the same considerations as the formation of orthosilicate. The latter goes on forming at its interface with the $3CaO \cdot 2SiO_2$ layer, depending upon the transport of CaO by diffusion. This stage of reaction is illustrated by Fig. 99b. When the arrival of CaO at the interface between the SiO_2 and the $3CaO \cdot 2SiO_2$ layer no longer suffices to continue formation of the latter, the metasilicate, $CaO \cdot SiO_2$, begins to form there (Fig. 99c). When, finally, the CaO has completely disappeared, diffusion of CaO out of the orthosilicate, and through the $3CaO \cdot 2SiO_2$ and $CaO \cdot SiO_2$ layers, towards the SiO_2, diminishes. If SiO_2 is present in excess, the same applies to the $3CaO \cdot 2SiO_2$ phase (Fig. 99d). If the original mixture had a 1 : 1 molar ratio, only the metasilicate eventually remains.

If the rates of formation of $3CaO \cdot 2SiO_2$ and of $CaO \cdot SiO_2$ were vanishingly small compared with the rate of diffusion of CaO through the rapidly formed orthosilicate, the latter alone would be obtained (together, of course, with excess of one or other component), irrespective of the composition of the starting materials. With many reactant systems, this appears to be the case. Thus Dykerhoff (1925) found that only the compound $CaO \cdot Al_2O_3$ was formed when solid mixtures of CaO and Al_2O_3 were heated, whatever the proportions, although these substances can form four compounds ($3CaO \cdot Al_2O_3$, $5CaO \cdot 3Al_2O_3$, $CaO \cdot Al_2O_3$ and $3CaO \cdot 5Al_2O_3$).

Rates of reactions in solids are strongly dependent on the particle size of the powders, since they are dependent upon the speed of diffusion of atoms or ions through the crystalline particles. The more finely divided a powder mixture, the more rapidly do reactions proceed to completion. The presence of lattice disturbances also promotes reactions, since place-exchange processes are thereby facilitated.

* $3CaO \cdot SiO_2$ decomposes into CaO and $2CaO \cdot SiO_2$ rather below $1200°$.

It can be both deduced theoretically and experimentally proved that lattice disturbances favor place-exchange. Thus, in ionic compounds the electrolytic conductivity, which gives a direct measure of the mobility of ions, is greatly increased when the concentration of lattice disturbances is raised by incorporating foreign ions. The reactivity towards other solids is thereby increased simultaneously.

So called 'amorphous' substances, in which there is no ordered arrangement of atoms (except within very minute regions), and which are often very much 'loosened' by the presence of channels, internal surfaces, and other effects of 'microstructure', are generally especially reactive.

Substances can often be obtained in such a 'loosened' state by drying out gels, or by the thermal decomposition of suitable starting materials at low temperatures*. Thus 'amorphous' silica, obtained by gentle heating of silica gel, is enormously more reactive than the most finely divided quartz. In the same way, iron(III) oxide, prepared by thermal decomposition of iron(III) sulfate, has a particularly high reactivity, as long as it is not too strongly heated. This material is crystalline, as its X-ray diffraction pattern shows, but the crystals are very much 'loosened'. This is shown, for example, in the very considerable decrease in volume which occurs when it is fairly strongly heated. The void spaces thereby diappear by recrystallization, and at the same time the reactivity diminishes.

(b) Active Intermediate States

Associated with the last mentioned phenomena is the occurrence of 'active intermediate states', in the course of reactions between solids. If two solids (e.g., ZnO and Fe_2O_3) which can combine with each other (to form $ZnFe_2O_4$ in this instance)

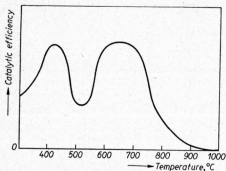

are gradually heated, it is often possible to observe the occurrence of intermediate states in which the properties differ both from those of the initial mixture and from those of the ultimate product of reaction. For example, after a mixture of ZnO and Fe_2O_3 has been heated for a long time at 400°, its catalytic effect on the reaction $2CO + O_2 = 2CO_2$ is considerably greater than that of a mixture not subjected to the same pre-treatment (cf. Fig. 100). If, however, the mixture is heated at 500° instead of at 400°, its catalytic activity is not greater than that of an unheated

Fig. 100. Catalytic activity of a ZnO-Fe_2O_3 mixture at 250°, as a function of the temperature at which the mixture was previously heated (according to Hüttig).

mixture, but is actually less. On heating at 650°, its catalytic activity is again considerably augmented. At this temperature, the formation of the spinel $ZnFe_2O_4$ begins. When formation of the compound is completed, the catalytic activity is completely lost. The hygroscopicity of this particular mixture also depends upon its pretreatment, and similar changes in the intermediate states can also be observed for other properties. The augmented catalytic efficiency in certain inter-

* The properties of solids obtained by these methods depend to some extent simply on their very large surface: volume ratio, and to some extent on lattice disturbances and intimate microstructure. The relative importance of the two factors varies from one case to another, and is often difficult to assess.

mediate states is of considerable practical importance in connection with the use of such oxide mixtures as catalysts.

Hüttig believes that the occurrence of active intermediate states can be attributed to two distinct causes, depending upon whether the active state is formed at a temperature at which there is no evidence at all for the formation of a compound, or at some higher temperature where inter-diffusion of the components (and therefore compound formation) has already begun. In the former case, at points where the different constituents are in close contact, the atoms or ions may be brought into a reactive state by the interaction of their fields of force; alternatively, there may be some surface diffusion of ions from one crystal to another. Either mechanism could lead to a modification of those properties which depend upon the nature of the *surface*. These properties include the catalytic activity, adsorptive properties towards gases and vapors, and in some instances the color. When the temperature is raised further, the atoms or ions held on the surface by a sort of activated adsorption penetrate into the surface, and the catalytic activity is lost. This accounts for the rapid falling off in catalytic activity in the foregoing example, when the mixture is heated above 400°. On further increase in temperature, diffusion of the atoms or ions of one substance into the other takes place to an increasing extent. However, the growth of the crystal lattice of the new compound may lag to a greater or less extent behind this diffusion process. As long as this process is going on close to the surface, it can modify the surface properties again. Since the new compound may be formed in a largely disordered state, this leads to a renewed increase of catalytic properties, but for reasons different from those involved in the first case. Hence the catalytic properties run parallel to a fresh set of other properties. Thus, in ZnO-Fe_2O_3 mixtures, the second increase in activity is coupled with an increase in paramagnetism, and the paramagnetism changes to ferromagnetism close to the maximum catalytic activity. The diffraction lines characteristic of $ZnFe_2O_4$ begin to appear in the diffraction pattern simultaneously. As these lines grow in intensity, and as the ions in the surface and adjacent layers therefore take on increasingly perfect order, the catalytic properties diminish again.

The occurrence of two active intermediate states, separated by a minimum of activity, has been found for other systems—e.g., for $CuO + Fe_2O_3$, $MgO + Cr_2O_3$, $CuO + Cr_2O_3$ —as well as for $ZnO + Fe_2O_3$. A single maximum in activity is often found; one cause of activity may be lacking, or the two activity maxima may merge into one, in that the intermediate state of surface diffusion passes directly into that due to inner diffusion, without any range of deactivation intervening. This happens, for example, with the system $MgO + Fe_2O_3$.

Within certain groups of substances, especially for metals, the temperatures at which 'loosening' of the surface and 'loosening' of the crystal lattice occur can be very roughly expressed as a uniform fraction of the (absolute) melting point. Hüttig [*Kolloid-Z.*, 98 (1942) 263] gives the 'loosening' temperature of the surface as 0.29 T_m, and that of the crystal lattice as 0.42 T_m (T_m = melting point, in °K). [See also L. Lichtenecker, *Z. Elektrochem.*, 48 (1942) 669].

Thermal decomposition of solids also provides the conditions in which a largely disordered arrangement of atoms in and near the surface can give rise to active intermediate states. A state of enhanced chemical reactivity is also traversed in the course of changes from one modification of a substance to another.

Thus, as has been shown by Hedvall, $AgNO_3$ reacts with the oxides CaO, SrO, and BaO, not at the temperatures at which these oxides react with other nitrates—which are different for the three oxides concerned—but at roughly the same temperature in each case, just above the transition temperature of $AgNO_3$ (160°).

It may be stated generally that, whenever solid substances undergo a transformation involving the rearrangement of their atoms, they pass through states in which they are more reactive than in the initial or final states (the 'Hedvall effect').*

(c) Special Reactivity of Co-precipitated Solids

It is often possible to prepare particularly reactive mixtures of solid substances by precipitating them simultaneously, in stoichiometric proportions, from solution. For example, Fricke and Dürr (1939) found that, whereas the formation of a spinel from a simple mixture of Fe_2O_3 and ZnO could de detected only above 600°, it could be detected directly in a mixture of the two oxides formed by co-precipitation at 60°. Fricke and Schröder (1948) were able to synthesize *chrysoberyl* by heating a co-precipitated mixture of $Be(OH)_2$ and γ-AlO(OH) to 1200°. By combining X-ray investigations with measurements by the emanation method (cf. p. 541–2), they were able to show that the AlO(OH) acted as a protective colloid towards the $Be(OH)_2$ with which it was precipitated, and hindered the crystallization of the latter on heating. This protective action was still exercised at temperatures high enough for these substances to be converted to the oxides by loss of water. Up to the temperature (950°) at which chrysoberyl began to form, only the diffraction pattern of γ-Al_2O_3 could be observed, and not that of BeO. At 1200°, the diffraction pattern of γ-Al_2O_3 had disappeared, and that of well crystallized chrysoberyl remained.

3. Reactions of Solids with Liquids

If the product of reaction goes into solution, the reactions of solid substances with liquids follow essentially the same law as governs the dissolution of a solid in a liquid, in the ordinary sense.

Provided that a uniform distribution of the dissolved substance throughout the solvent is maintained by stirring, the rate of dissolution is proportional to the area of contact between solid and solution, and proportional to the difference between the concentration of the solution at any moment and the saturation concentration:

$$\frac{dc}{dt} = k \cdot A \ (c_s - c) \tag{5}$$

where A = area of the solid, c_s = concentration of the saturated solution, c = concentration of solution at the time t, and k is the velocity constant (A. A. Noyes, 1897).

This equation, when compared with eqn (1), p. 739, implies that the rate of diffusion of the dissolved substance in the solvent determines the rate at which dissolution occurs, provided that equilibrium is established at the surface of the solid so rapidly that the layer of solution immediately adjacent to the solid surface is always saturated. This condition is generally fulfilled. It was shown by Brunner (1904) that $k = \dfrac{D}{\delta \cdot v}$, where D is diffusion

---

* The effects described in this section are established as *phenomena*, but the interpretations advanced by Hüttig and Hedvall are not intended to be more than tentative. As noted earlier, the reactivity and properties of solids depend enormously on the length of the diffusion path and on the specific surface. Chemical reaction or allotropic change at low temperatures is likely to lead to a product in a state of extreme subdivision, with a high surface energy. Effects due to this factor will be superimposed on any effects due to differences of *fine structure*, in surface or crystal lattice, such as may be involved in the fundamental processes by which reaction occurs. The effects are thus not at all well understood, and much work remains to be done.

coefficient of the dissolving substance, δ is the thickness of the liquid layer adhering to the surface of the crystals (i.e., the layer which is not affected by stirring), and v is the volume of the solution.

If the product of reaction does not dissolve, and reaction is propagated into the interior of the solid, its velocity is determined by the rate at which either the molecules of the liquid, or the atoms or ions of the solid reactant, can diffuse through the solid product of reaction. At ordinary temperature, rates of diffusion in solids are usually so small that reactions with liquids, following this mechanism, are rarely of any practical significance. Conditions are very different, however, if the product of reaction remains undissolved, but is porous or non-adherent. The liquid can then continually penetrate through the channels and cracks to freshly exposed portions of the surface. In this way, a reaction can often take place fairly rapidly, until the starting material is fully consumed. This happens, for example, in the action of water on metallic magnesium.

4. Reaction Between Solids and Dissolved Substances

(a) General

If the product of reaction goes into solution, the reaction, in this case also is confined to the surface of the solid reactant. The equation of Nernst (1904) then described the process:

$$\frac{dn}{dt} = D \cdot A \cdot \frac{c}{\delta} \tag{6}$$

where D is the diffusion coefficient of the dissolved reactant, A the area of the solid reactant, δ the thickness of the adherent layer of liquid at the surface of the solid, and c the concentration of the reactant in the solution (outside the adherent layer). dn is then the increase in amount of reaction product (e.g., in g-mols) during the time dt. The equation holds provided that the concentration of reaction product in the solution is kept uniform, by stirring. The thickness of the adherent layer, δ, depends upon the rate of stirring, but is usually of the order 10^{-3} cm. In the derivation of the Nernst equation, it is assumed that the velocity of the reaction is infinitely great as compared with the diffusion velocity of the reagent, so that the concentration of reagent at the immediate surface of the solid can be taken as practically zero. Experience shows that this assumption is usually fulfilled.

Equation (6) implies that, under the conditions assumed, the rate at which a solid is consumed by reaction with a dissolved substance (e.g., MgO with hydrochloric acid) is determined entirely by the rate at which the molecules or ions which react with the solid can reach the surface by diffusion. It follows from Fick's law of diffusion, by substitution of A, the area of the solid surface, for q, and the concentration gradient of the dissolved reactant in the adherent layer, $(c - o)/\delta = \frac{c}{\delta}$, in place of $\frac{dc}{dx}$. By integration of eqn (1) on the assumption that the area of the solid surface remains constant during dissolution, and writing $c_t = \frac{n_0 - n_t}{v}$, where $n_0 =$ quantity of reagent in the total volume v of solution at the beginning of the experiment, and n_t = quantity of reagent consumed during the time t, we obtain:

$$\ln\left(\frac{n_0}{n_0 - n_t}\right) = \frac{D \cdot A \cdot t}{\delta \cdot v} \tag{7}$$

From this equation it is possible to predict the rate at which a solid will be dissolved away by any dissolved reagent, if the thickness δ has previously been determined for the selected rate of stirring.

Since $k = \dfrac{D}{\delta \cdot v}$ and $c = \dfrac{n}{v}$, it follows from a comparison of eqns (5) and (6) that the rate at which a substance dissolves in pure water, and the rate at which a saturated solution of the same substance will react with any other solid must be equal (Brunner, 1904). If the reaction at the surface does not take place with essentially infinite velocity, neither the Nernst equation nor the foregoing rule will be valid.

Reactions which proceed in the solid phase between solid substances and dissolved reagents accord with what has already been stated for the corresponding reactions of solids with pure liquids.

(b) Dissolution of Metals in Acids

The dissolution of metals in acids occupies a special place among the reactions of solids with dissolved reagents, because of the role played by local couple formation. The importance of this has already been considered for the case of zinc (p. 432), where the effects are strikingly manifested.

Palmaer (1929) has shown that the efficiency of local couples quite generally are the determining factor in the rate at which metals dissolve in acids. Since the number and kind of local couples changes continually during dissolution, according to the impurities that may be laid bare, it is not usually possible to find a kinetic expression which reproduces the rate of dissolution over any long interval of time. For a given state of the surface, Palmaer found that the rate at which metal goes into solution is given by

$$\frac{dx}{dt} = \frac{E \cdot \varkappa \cdot A}{C} \tag{8}$$

where E is the e.m.f. set up by the local couple (into which the hydrogen over-voltage enters as an important factor), $\varkappa$ is the specific conductivity of the electrolyte solution, A the area of the surface, and C a coefficient depending on the other experimental conditions. Among other factors, this takes account of the average spacing between electrodes in the local couples. Thus the effect of a given quantity of impurity, embedded between the crystallites of a metal, increases as the size of the individual crystallites diminishes.

Acceleration of the dissolution of a metal by the action of added metallic impurites may be bound up with the question of the hydrogen over-potential. Thus the dissolution of zinc in acids is *not* facilitated by alloying with small amounts of lead or cadmium, or by amalgamating with mercury. The deposition potential of hydrogen on these metals is so great that it is not attained by the solution potential of zinc. Again, alloying with those more noble metals with which zinc forms mixed crystals or compounds does not promote the dissolution in acid, since local elements can only be set up between substances which are present as different phases.

(c) Passivity

In the *passive state* exhibited by certain metals, they resemble the noble metals in being stable towards reagents which would attack them in the normal state. They also display a considerably lower solution potential than in the normal or *active* state. The conversion of a metal from the active to the passive state is known as *passivation*, and may be brought about by anodic polarization or by chemical means—namely, by the action of certain oxidants such as concentrated nitric acid.

Electrolytic (or galvanic) polarization is the phenomenon whereby in electrolysis, or in a galvanic cell, an electromotive force is set up in the opposite direction to that which is responsible for the flow of current. The opposed electromotive force may be due to changes either of the electrode surface, or of the immediately adjacent layer of electrolyte solution. Changes of the anode brought about by the passage of current are known as anodic polarization, changes of the cathode as cathodic polarization. A simple example of electrolytic polarization is the establishment of a potential difference between two platinum wires dipping into dilute sulfuric acid, after an electric current has been passed through the solution. The electrodes take up different potentials with respect to the solution, as a result of the deposition of hydrogen on the cathode and of oxygen on the anode. If the metallic circuit is completed after switching off the primary current, a current flows in the opposite direction until the gases covering the electrodes are used up. Electrical accumulators represent a practical application of the use of electrolytic polarization as a source of current.

Only the metal used as anode can be passivated by polarization. When made the cathode, not only can a metal not be passivated, but it may be reconverted from the passive to the active condition. Passivation by dissolved chemical reagents can be attributed to polarization by local currents, in terms of the theory of local couples. However, the origin of the polarization when passivation occurs is not known with certainty. Many metals, especially chromium and iron, become passivated to a certain extent simply on exposure to air (spontaneous passivation). It is possible to prepare alloys which are very strongly passivated spontaneously (e.g., V2A and 18–8 stainless steels).

Passivated metals are very readily activated by the action of protons. Hence the electrolytic evolution of hydrogen at the metal (cathodic polarization) has a very strong activating effect. Activation is also produced by heating in hydrogen, by the action of acids, by treatment with reducing agents, and especially by the action of 'nascent' hydrogen. Solutions of salts, and especially of chlorides, may also destroy the passive state, as also can mechanical treatment (in contact with solutions)—e.g., scratching, the effect of blows, and especially of vibration, and ultrasonic excitation (Schmid, 1937).

The phenomenon of passivity was discovered, for the case of iron, towards the end of the 18th century. Faraday (1836) attempted to explain it on the assumption of a protective layer of oxide or similar compound. Schönbein opposed this, with the view that passivity was due to a physical alteration of the metal (e.g., conversion into an allotropic modification). Belck (1887) later suggested that the passive behavior was due to a thin film of gaseous oxygen, covering the surface of the metal. These three different conceptions are still advocated, although in more or less altered forms, although a number of other explanations have also been put forward. Most of these, however, have been based on inadequate experimental evidence. Thus some workers have contended that, for the metals in question, the passive state is the normal state, and the active state is due to hydrogen dissolved in the metal. Although many have concerned themselves with the study of passivity, especially since the beginning of the present century, its origin is not yet fully explained in all points.

The hypothesis advanced by Smits (1923) and, independently by Russell (1926), was in some respects a translation of Schönbein's ideas into the terms of the modern outlook. It attributed the difference between the active and the passive state to differences in electron configuration of the atoms in the surface of the metal. Smits showed that passivity could be at once explained if it may be assumed that in the surface of a metal (e.g., iron) atoms or ions are present, in equilibrium, in states of different potential energy (due to differences in the quantum states of the valence electrons), but the equilibrium may be sluggishly attained. A metal in which such equilibrium was slowly established would behave like a solid solution of two different metals. When the metal is dissolved anodically, one kind of atom will go more rapidly into solution, and the solution potential will thereby be shifted in the direction of the more noble 'pseudo-component', until this species alone is present in the surface. The metal has then reached a state of maximum passivity. Any influence which reestablishes equilibrium between the two states of the metal atoms—e.g., the catalytic effect of protons— restores the active state. As a purely formal theory, this explains, among other things, why metals in the passive state often dissolve to give ions of different valence from those formed by the active metal. It is, however, not readily put into the terms of the later theory of metals; electronic processes in general proceed rapidly.

The idea that the passivating action of oxidizing agents is due to the formation of an *oxide skin* is supported by many, but is strongly disputed by others. On the basis of his own observations, Müller [7] has adapted and generalized it, attributing the passivity in many cases to the formation of a coating in the metal, which is penetrated by a not very large number of fine pores. This coating need not necessarily consist of the metal oxide, but can be a salt. Müller attributes the considerable decrease of anodic potential that accompanies passivation to the fact that, once the coating has formed over the anode, the conduction of current can take place only through the pores. This 'film' theory of passivity is in harmony with a variety of phenomena observed in the passivation of metals by anodic polarization. It does not explain why a passive metal has a lower solution potential, even when no current is flowing, than the same metal in the active state. It may be deduced thermo- dynamically, as experiment also shows, that the open-circuit potential ('resting' potential) of a metal is unaffected by covering it with a porous coating (e.g., of a more noble metal), yet this lower 'resting potential' is characteristic of the passive condition. Müller meets the difficulty by assuming that the layer of oxide or salt modifies the electronic configuration of the atoms in the metallic surface. In this, his views resemble those of Smits and Russell: the essential difference is that Müller regards the formation of the layer of oxide or salt not as more or less incidentally associated with the passivation but (in most cases, at least) as the necessary condition for passivation to occur. One of the principal objections to the oxide film theory of passivity is that, on most metals, the formation of an oxide film does not result in passivation. Even a metal such as chromium, which has a strong tendency to become passive, is found not to be passive when it has been coated, by heating it in air, with an oxide film thick enough to show interference colors. In fact, chromium which has received an oxide coating, by heating it in air, retains its activity better than the polished metal. These facts were demonstrated by Hittorf as long ago as 1899.

The objections to the 'oxide skin' theory are met by the assumption that passivity is due to a surface covering of oxygen on the surface of the metal (or on the very thin film of oxide which is always present after exposure to air, cf. p. 754). A variety of experimental facts, and theoretical considerations favor the existence of an adsorbed film. The passivating effect of such a film is now usually attributed to the saturation of the free valence forces of surface atoms. For this, a unimolecular layer would suffice. It is both possible and probable that the adsorbed layer consists of *oxygen ions*, which do not, however, displace the atoms of the metal from the lattice structure, as is the case when the oxide is formed. A surface covered with such an adsorbed layer will have different properties from one which is merely covered with a layer of the oxide. It is probable that all metals at once become covered with a thin oxide layer, as soon as they come into contact with air. If passivity is regarded as being due to an oxide layer, it must therefore be assumed that the oxide film produced by passivation differs in nature from that ordinarily formed when the metal is oxidized in air. On the other hand, the adsorbed gas theory relies only on the known phenomena of adsorption in order to explain the difference in properties between passivated metals and metals which have been oxidized in air in the ordinary way. It also agrees with the fact that mere contact with air often produces more or less complete passivity, whereas when oxidation is carried further the passivation is not intensified, but is actually lost.

Another peculiar phenomenon can be correlated with the ability of many metals to exist in states of different activity—namely the periodic changes in the rate of dissolution of such metals. This phenomenon (first observed by W. Ostwald for metallic chromium, prepared by aluminothermic reduction) is illustrated by Fig. 101, in which time is plotted as abscissa, while the ordinate represents the rate of dissolution of chromium in hydrochloric acid at each moment, as measured by the rate of evolution of hydrogen. This is based on the work

of Brauer (1901). It was found that variations in the dissolving rate were paralleled by variations in the solution potential, although the latter were very much smaller than the difference in potential between completely active and completely passive chromium. Ostwald suggested that the effect might be associated with the presence of impurities in the chromium, and this has been confirmed subsequently. The phenomenon can also be observed with certain

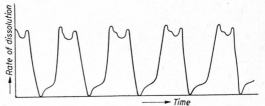

Fig. 101. Periodic fluctuations in the rate of dissolution of chromium in 2-normal hydrochloric acid.

varieties of iron, and may be due to the alternate passivation of the metal by anodic polarization due to local currents (arising from impurities embedded between the crystallites), and subsequent reactivation by the evolved hydrogen (probably as a consequence of the diffusion of protons in the metal). Rathert (1914) was able to produce periodic activation of an iron electrode, which was kept passive by immersion in dilute chromic acid, by allowing hydrogen to diffuse through from the back.

5. Reactions of Solids with Gases

(a) General [6]

Since diffusion occurs much more rapidly in gases than in solutions, it is found that the reaction rates of solid substances with gases, to give gaseous products, are much more frequently determined by the speed of the actual chemical process than are reactions of solids with dissolved substances. This is especially true of solid-gas reactions at fairly low temperatures.

Even at high temperatures, it is often found that the mass flow of the gas is far more important than diffusion proper. Thus, in the combustion of coal, it is familiar that the rate of burning is much increased by increasing the flow of air. The air supply cannot be raised indefinitely, however, since too much heat would be carried away by the air flow. It is therefore better to increase the rate of combustion by increasing the surface area of the coal (use of pulverized fuel). In general, when the products of reaction are gaseous, the reaction of a solid with a flowing gas follows a course very similar to that of the dissolution of a solid in a stirred liquid. The velocity of the actual chemical process at the surface of the solid then plays a role similar to that of diffusion through the adherent layer in the case of dissolution. As the latter depend on the flow rate of the liquid outside the adherent layer, so the former depends on the rate of flow of the gas at the surface of the solid. Almost identical pictures are obtained, when the flow lines are made visible by the 'schlieren' method, around a cube of carbon burning in air, and around a cube of salt dissolving in water (Rosin, 1931). In both cases, it is the end of the cube directed away from the direction of gas- or liquid flow which is the more attacked, since the turbulence is greater there. Researches on the course of reactions of this kind are of technical importance in relation to combustion, the generation of producer gas, the roasting of ores, and the operation of the blast furnace.

When one product of reaction is *liquid*, propagation of the reaction depends on the rate at which the gas diffuses through the liquid. This case is of very minor importance, since evaporation as a result of the heat of reaction generally prevents the formation of a liquid layer.

Reactions of solids with gases, with formation of *solid* reaction products, are of great practical importance. If the reaction product forms a coherent coating, reaction can only continue as a result of diffusion through the layer of reaction product. Processes of this type are involved in the tarnishing of metals, when they are heated in air or other gases with which they combine to form compact, insoluble coatings. At ordinary temperature, when the rate of diffusion is practically zero, such tarnish layers protect the metal from further attack by the air. If this were not the case, most of the common metals would be rapidly be converted into their oxides, for oxygen reacts extraordinarily rapidly with fresh surfaces of metals at ordinary temperature.

The time taken to form a unimolecular oxide layer on the surface of metals is generally only a small fraction of a second. By contrast, it may be calculated from the temperature coefficient of the tarnish reaction that it would take about 10^9 years for copper, and no less than $5 \cdot 10^{19}$ years for nickel, before the oxide film had grown thick enough, in dry air at ordinary temperature, to show the first clearly visible (1st order yellow*) interference color (Tammann, 1922).

As has already been mentioned, diffusion in solids—and hence the growth of tarnish layers—follows the same law as diffusion in solutions. If eqn (1), p. 739, is applied to a reaction which depends upon the diffusion of one reactant through a compact layer, we must notice that the thickness x of the layer through which diffusion occurs is now not a constant, but is proportional to the quantity n of substance which has reacted: $x = a \cdot n$. On the other hand, there is a constant difference between the concentrations c_1, c_2 of the diffusing substance at the two boundaries of the layer. It may also be assumed that the concentration gradient within the layer is constant. Then at each point within the layer, $-\dfrac{dc}{dx} = \dfrac{(c_2 - c_1)}{x}$. The cross section of the diffusion layer can be taken as being the surface area A of the solid, and hence a constant. Under these conditions, and taking the quantity of material reacted $= 0$ at $t = 0$,

$$n^2 = \frac{2D \cdot A \cdot (c_2 - c_1) \cdot t}{a} \qquad \text{or} \qquad n = K \cdot t^{\frac{1}{2}} \qquad (9)$$

$$\text{where } K = \sqrt{\frac{2D \cdot A \cdot (c_2 - c_1)}{a}}.$$

This 'square root law', as has already been stated, holds both for reactions between solids and gases, and for reactions between solids and dissolved substances, in cases where the product forms a compact layer. It also holds for reactions between solids, in which case we may put $c_1 = 1$, $c_2 = 0$.

No change is involved in the general form of eqn (9) if it is assumed that both reactants diffuse through the layer. Two diffusion coefficients must then be substituted. The case of a single diffusing species appears to be the more general, for ionic compounds at least.

Compact oxide layers are formed by the action of dry air on metals if the oxide has a greater volume than the equivalent amount of metal (Pilling, 1923). If this is not so, a porous oxide layer is formed, through which the oxygen can attack the surface of the metal directly. Oxidation can then continue *at constant velocity*, and is generally perceptible even at ordinary temperature. This is the case for the metals Li, Na, K, Rb, Cs, Mg, Ca, Sr, and Ba.

Pilling's rule does not hold for the behavior of the metals in moist air. Porous layers, which do not impede further oxidation, are very commonly formed on

---

* In a medium of refractive index 1, this interference color corresponds to a layer 0.15 μ thick, or $\dfrac{1}{n}$ times this thickness in a medium of refractive index n.

metals in moist air. It is even possible for the rate of oxidation to increase with time, if the products of reaction catalyze the oxidation. This happens in the rusting of iron.

The behavior of solids (especially metals) towards gases, to form solid reaction products, thus falls into three different types, as is demonstrated clearly by the curves in Fig. 102. This shows the increase in weight of Cu, Zn, and Fe as a result of oxidation in moist air, as a function of time. Oxidation of copper follows the 'square root' law, as shown by the parabolic form of the curve— i.e., with formation of a compact layer on the metal. Zinc is oxidized at constant velocity—i.e., with the formation of a porous coating. Iron not merely gives a porous product, but the reaction is autocatalytic, so that the rate increases with time.

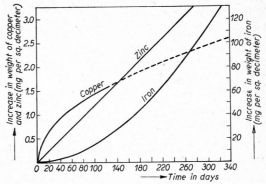

Autocatalysis in the rusting of iron is generally attributed to the formation of local couples. With non-metals, autocatalytic acceleration of a reaction in which a solid product is formed may be due either to the elimination of supersaturation at the phase boundary, through the action of crystal nuclei of the product (catalysis of nucleation), or to an increase in the chemical reactivity of a solid as a result of contact with the nuclei of the reaction product.

Fig. 102. Increase in weight of copper, zinc, and iron during oxidation in moist air (according to Vernon)

Hüttig (1942) has found that in the *de-gassing* of solids (i.e., the removal of firmly held volatile impurities), the loss of gas takes place most rapidly at the 'loosening' temperature referred to on p. 747.

(b) Corrosion and its Prevention

One field in which an understanding of the mechanism of the reactions of solids is of immense importance is that of *corrosion research*. The economic value of this field of research can be judged from the enormous assets destroyed annually by corrosion. It has been estimated that the total world production of 1766 million tons of iron between 1890 and 1923, has to be compared with the loss of no less than 718 million tons by rusting, during the same period.

Corrosion [8–21] is the destruction of a solid body by unwanted chemical or electrochemical attack on its surface. Metals, in particular, are subject to corrosion, and among the metals it is the corrosion of iron which is of the greatest importance.

The *rusting of iron* is attributed to the action of local couples. Since the iron oxide so produced itself creates new local couples in which it functions as the cathode, rusting is an autocatalytically accelerated process. The current set up by a local couple depends on the specific conductivity of the solution in contact with the metal (cf. eqn (8), p. 750), so that salt solutions—e.g., sea water—are more corrosive than ordinary water. Coatings of more noble metals, such as tin or copper, afford protection against rusting as long as they are undamaged, but accelerate rusting if they are broken since they then set up local couples in which the iron is anodic. A coating of zinc, however, affords protection even when it is imperfect, since in this case it is the less noble metal, zinc, which is attacked. The zinc is only slowly consumed, however, since there is no autocatalysis of the corrosion as there is in rust formation. However, the protective action may be impaired by the formation of zinc oxide.

It is not, at first sight, easy to reconcile the observation that compact iron is attacked neither by dry oxygen (because of the formation of an impervious oxide layer) nor by pure air-free water, although it is attacked by the simultaneous action of air and water, with the fact that when rusting occurs through the action of moisture, the iron is most strongly corroded at those points to which air has the least ready access. Piled up iron plates are most heavily rusted where they touch one another. Wire ropes may be corroded internally while appearing quite bright on the outside. It can readily be shown experimentally that iron charged with oxygen takes up a lower potential than in the uncharged state. Thus the iron must go into solution (by anodic attack) chiefly at those points which are not charged with oxygen, and the observed phenomena are fully accounted for by the theory.

Corrosion research is not concerned only with the electro-chemical theory of corrosion and the general principles based on metallography. Studies of the surface coatings formed on metals under various conditions—e.g., in tarnish reactions—, an extension of our knowledge of passivity and related phenomena, and an understanding of surface reactions such as is derived from researches on adsorption and catalysis, all contribute to it.

The most important measures taken to combat corrosion of metals [22, 23] include

(i) the use of self-passivating alloys such as the stainless steels;
(ii) the use of alloys which spontaneously form protective coatings (e.g., hydro-nalium and other corrosion resistant aluminum alloys);
(iii) the artificial production or intensification of protective coatings (e.g.,temper colors on steel, anodizing of aluminum);
(iv) the application of corrosion resistant metal coatings. This is generally done either electrolytically [24–26] (e.g., chromium and nickel plating, zinc galva-nizing) or by metal spraying [27]. Other methods may also be used for certain metals (cf. galvanizing of iron, p. 434).
(v) Enamelling (Vol. I, p. 507).
(vi) Coating with lacquers or paints [28].

A considerable measure of protection against corrosion by atmospheric oxygen at high temperatures (up to 1200°) can be conferred on iron by allowing aluminum to diffuse into the surface ('alitizing'). The thin film of aluminum oxide then formed by heating in air greatly retards the propagation of the oxidation. Thus the useful life may be extended by 6 to 8 fold at 1000°. This process is applied, for example, to grates.

Iron water pipes are usually protected from corrosion through the formation of a thin layer of calcium carbonate. This layer forms, however, only when the water contains sufficient calcium hydrogen carbonate and not too great an excess of carbonic acid, and when dissolved oxygen is also present. The role played by oxygen is not fully understood.

6. Topochemical Reactions

(a) General

Topochemical reactions [29] are reactions in which the properties of the solid product of reaction are largely determined by the fact that the reaction took place on a solid—i.e., was localized [Gk, τόπος = place]. Such reactions often produce substances which are impossible to obtain from reactions in gases or solutions. In other cases, the products may have the same composition as those obtainable from reactions in gases or solutions, but they may differ strikingly in their properties when obtained from reactions on the surface or in the interior of a solid. This may be because the chemical nature or the physical texture (e.g., the dispersity) of the solid starting material determines the course of the reaction, or that its structure and texture affect the corresponding properties of the solid reaction product.

Whereas most reactions are classified according to their kinetics, topochemical reactions are marked by a different principle of classification—the relation between the properties and texture of the starting materials and the product—which follows from the site at which reaction takes place. For this reason they have a considerable preparative interest, but their study has contributed considerably to an understanding of the mechanism of reactions, and of the relation between structure and chemical properties.

The concept of topochemical reactions was introduced by V. Kohlschütter in 1919; relatively few processes of the kind were then recognized. The connections between conditions of formation and texture have subsequently been demonstrated by V. Kohlschütter and a number of others (especially Feitknecht, Fricke, U. Hofmann and H. W. Kohlschütter). Many examples of such processes have already been discussed; a few typical topochemical reactions, which bring out their essential features, will be considered in this section.

Three types of topochemical reaction may be considered; these accord to the manner in which the properties of the solid product are influenced by those of the starting material.

[i] Reactions in which it is only the *degree of dispersion, specific surface area*, and the kind and extent of *lattice imperfections* in the reaction product that are dependent on the nature of the starting material.

[ii] Reactions in which the initial material exerts an influence on the *crystal structure* of the product.

[iii] Reactions in which the nature of the starting material also determines the *chemical composition* of the product. These reactions give rise to compounds which cannot be obtained at all by other methods, or at least not by analogous reactions in the gas phase or in solution.

It is often found that the starting material has an influence on the degree of dispersion and the specific surface of the products of thermal decomposition reactions.

Thus Feitknecht found that when various calcium compounds were heated to the same temperature (900°), they gave rise to powdered calcium oxide of completely different texture. The products are structurally identical, but quite different in their degree of dispersion. Thus the bulk volumes occupied by the same weights of CaO

$$\text{from the sources:} \quad CaC_2O_4 \quad Ca(OH)_2 \quad CaCO_3 \quad \text{and} \quad Ca(NO_3)_2$$
$$\text{vary in the ratio:} \quad 100 \quad : \quad 78 \quad : \quad 59 \quad : \quad 34.5$$

The rate at which the various powders absorbed water vapor at ordinary temperature diminished with their voluminousness. The volumes of the hydroxide formed from the above oxides were (relative to the above and to each other) $164 : 127 : 102 : 74.5$. When slaked with liquid water, the CaO powders obtained from CaC_2O_4, $Ca(OH)_2$, and $CaCO_3$ formed plastic $Ca(OH)_2$ gels ('milk of lime'). That from $Ca(NO_3)_2$ did not, and it was not possible to obtain milk of lime by subsequent addition of liquid water to any of the $Ca(OH)_2$ powders formed by absorption of water vapor.

It has already been mentioned that the adsorptive and catalytic properties of compounds prepared by thermal decomposition may be very different from those of the same compounds prepared by reactions in solution. The example just given shows, further, that these properties may be developed in very different degree for the same compound prepared by thermal decomposition, but from different starting materials. The fact that very different temperatures may be needed for the decomposition of different starting materials may be of importance in this connection, since the specific surface* and the reactivity are considerably influenced by this factor. The increasing homogenization and recrystallization of the particles that result from heating to higher temperatures are only relevant to topochemistry if the original microstructure of the particles was markedly dependent on the nature of the starting material.

An application of topochemistry which is of some technical importance is the relation between the color of certain substances and their mode of preparation. This has long been used in the manufacture of mineral pigments. Thus it has long been known that an iron(III) oxide of a beautiful red shade could be prepared by careful heating of iron(III) sulfate or similar compounds: such a material was used as a paint in ancient times (Pompeian red). This iron oxide consists of fine scales, and not of minute rhombohedral crystals like the usual material, although according to Hedvall (1922) there is no difference in crystal structure. The difference lies only in its particular habit.

It is reasonable to assume that, in this group of topochemical reactions, the properties of the product are affected by the starting material chiefly in so far as the structure (and possibly the degree of subdivision) of the latter modify the processes of *nucleation, growth of nuclei,* and *diffusion.* There may also be some influence on the type and the concentration of lattice imperfections and disorder in the newly formed solid phase. As long as the atomic arrangement corresponding to the newly formed phase is not everywhere perfected, it may be assumed that there will be some atoms or ions, even if few in number, which retain more or less their original mutual dispositions or something like it. This would explain why it is often easier to reconvert the products of solid reactions into their starting materials than into other substances. Thus strontium chloride prepared by the careful dehydration of $SrCl_2 \cdot 6H_2O$ absorbs water vapor more rapidly than ammonia, whereas that prepared from $SrCl_2 \cdot 8NH_3$ takes up ammonia gas faster than it does water vapor. Furthermore, aluminum oxide prepared by thermal decomposition of the nitrate at not too high temperatures dissolves in dilute nitric acid faster than in acetic acid, whereas aluminum oxide obtained by thermal decomposition of the basic acetate dissolves more rapidly in acetic acid. Related to such effects is the common observation that a solid often has the optimum catalytic activity towards some gaseous system when it has been prepared in an atmosphere of this same system.

An example of the influence of the starting material on the crystal structure of the product is provided by Feitknecht's observation that the metastable α-zinc hydroxide (p. 438) is always formed when certain basic zinc salts (e.g., basic zinc iodide, $ZnI_2 \cdot 4Zn(OH)_2$) are leached with water. On the other hand, hydrolysis of zinc salts in aqueous solution usually results in completely amorphous precipitates, which may change either into zinc oxide or into zinc hydroxide (usually ε-$Zn(OH)_2$) as a result of aging.

Basic zinc salts have layer lattice structures, as also has the α-$Zn(OH)_2$. It is therefore understandable that the formation of the α-$Zn(OH)_2$ structure should be favored by the pre-existing arrangement of the ions in the basic salts. However, a prerequisite condition for the occurrence of the reaction without great disturbance of the original arrangement of the ions, is that there should be sufficient distance between the layers for the hydroxyl ions to enter. This condition is fulfilled in the case of zinc iodide by the incorporation of the bulky iodide ions between layers.

(b) Graphite Compounds

Topochemical reactions which lead to compounds not obtainable by other means are of special interest, both preparatively and from the standpoint of valence and structural chemistry. The oldest example of such reactions is the for-

* The *specific surface* of a solid is the area of total surface of 1 g of powder, as measured, e.g., by gas adsorption or similar methods. A powder consisting of particles 1 μ in size has a specific surface of 15 sq.meters per g, if the density of the solid = 4.

mation of the so-called 'graphitic acid' by oxidation of graphite with potassium chlorate and nitric acid. (Vol. I, p. 421).

The structure of graphitic acid has been explained through the X-ray work of U. Hofmann (1930 and later). It is derived from that of graphite, in that oxygen atoms are inserted between the sheets of carbon atoms—probably in such a way that an O atom is bound to a pair of adjacent C atoms, alternately above and below the sheet. Thus every layer of O atoms is followed by a sheet of C atoms (at a distance of 1.4 Å), and this in turn by another layer of O atoms at the same distance. Next to this, but at a considerably greater distance, is another layer of oxygen atoms, belonging to the next sheet of C atoms (cf. Fig. 49, Vol. I, p. 216, which represents an essentially similar layer structure). The interatomic distance C ↔ O is 1.57 Å. Within each sheet, the arrangement of C atoms is the same as in graphite, but the distance between sheets of C atoms is enlarged to practically double that found in graphite, as a result of the insertion of the layers of O atoms. The distance between sheets can readily undergo yet further expansion, if H_2O molecules are allowed to diffuse in ('swelling' of graphitic acid, which must not be confused with the 'tumescence' produced by gas evolution, when it is heated dry). It is apparently not possible to obtain completely anhydrous samples of graphitic acid. This is probably because —OH groups, as well as O atoms, are built in between the C sheets. This assumption appears to be substantiated both by the acid reaction of moist graphitic acid towards litmus, and by its ability to exchange protons for other cations (Thiele, 1937). These acidic properties may be due to some extent to the presence of carbonyl groups at the edges of the C sheets (i.e., at the crystal boundaries). For this reason, Hofmann regards 'graphitic acid' as being *graphitic oxide*, contaminated to a greater or less degree by acidic 'edge compounds'. However, the existence of —OH groups between the sheets is supported by the observation (U. Hofmann, 1939) that methylation of the —OH groups leads to a not inconsiderable enlargement of the distance between sheets.

By treating graphitic oxide with hydrogen sulfide, Hofmann was able to effect an exchange of O for S, so as to produce a corresponding graphitic sulfide.

It is also possible for HSO_4^- ions to be incorporated between the graphitic sheets. *Graphite hydrogen sulfate* is thereby produced (Hofmann, 1934). It is formed when graphite is treated with concentrated sulfuric acid in the presence of a small amount of nitric acid or other oxidizing agent:

$$nH_2SO_4 + graphite + \frac{n}{2} = O \frac{n}{2} H_2O + [graphite \cdot HSO_4]_n.$$

Graphite hydrogen sulfate can only be formed if some oxidizing agent is present, since in order to combine with HSO_4^- ions, it is necessary for the sheets of C atoms to lose electrons and acquire a positive charge. Graphite hydrogen sulfate can be degraded to graphite again. This takes place in stages, all the HSO_4^- ions between certain pairs of sheets being lost. Graphite hydrogen sulfate has a deep blue color by reflected light. Like the graphitic oxide, it is always contaminated by 'edge compounds'. Insertion of the HSO_4^- groups causes an expansion of the graphite structure in the direction of the *c*-axis. This, together with the oxidation of the edges of the sheets to form carboxyl groups which takes place simultaneously, leads to a far-reaching dispersion, to form *colloidal graphite*, when graphite is subjected to prolonged treatment with sulfuric acid and oxidizing agents. The strong attack undergone by graphite anodes in sulfuric acid is due to the same cause; they are not destroyed in solutions of halogen acids.

Graphite hydrogen sulfate is only stable under concentrated sulfuric acid. Water decomposes it, with evolution of CO_2. When HSO_4^- ions have been introduced between all the sheets of C atoms, graphite hydrogen sulfate has the composition $C_{24}^+HSO_4^- \cdot 2H_2SO_4$. By treating it with other strong acids, it is possible to exchange the molecules and anions of sulfuric acid reversibly, thereby forming graphite nitrate, graphite perchlorate, graphite selenate, etc.

The structure and composition of *graphite fluoride* (carbon monofluoride), which was prepared by Ruff by the action of fluorine on graphite (Vol. I, p. 423), are also determined by the greater strength of the bonds within each sheet than between sheets.

In $[CF]_n$, according to Ruff (1934) and Rüdorff (1947), fluorine atoms enter between the carbon sheets, so that every C atom is joined to a F atom, alternately above and below the carbon layer. Since every C atom then forms four bonds (three to adjacent C atoms, one to an F atom), the valence of the carbon atoms resembles that in paraffinic compounds or the diamond structure; the sheets of C atoms are no longer planar, but puckered, and the semi-metallic optical and electrical properties of graphite have disappeared.

Compounds, which cannot be obtained in any other way, are also formed by the action of alkali metal vapor on graphite, and have their structure determined by that of the starting material. The compounds KC_8, RbC_8, and CsC_8 have been prepared (Fredenhagen, 1926).

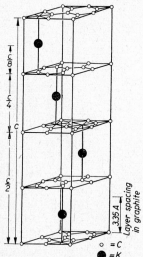

Fig. 103. Unit cell of KC_8. $a = 4.94$, $c = 21.34$ Å. The scale of the diagram is half as great as that of the graphite structure in Fig. 80, Vol. I, p. 423.

As Schleede showed by X-ray investigations (1932), these compounds can be derived from graphite by the introduction of a single layer of alkali metal atoms between every pair of carbon sheets of the graphite structure (cf. Fig. 103). The distance between sheets is thereby increased from 3.35 Å to 5.34, 5.68 and 5.94 Å, respectively. When they are degraded by heating, the compounds M^IC_8 change discontinuously into compounds M^IC_{16}, which differ from the former in that a sheet of metal atoms is introduced between alternate pairs of carbon sheets.

Graphite fluoride and the alkali graphites can also be obtained from carbon black (active charcoal or lamp black). This is in agreement with the conclusion that in carbon black there are the same hexagonal sheets of C atoms as in graphite (cf. Vol. I, p. 425).

These examples show what determines the possibility of preparing, by topochemical processes, compounds which are not obtainable in other ways. The starting material must involve some atomic arrangement, peculiar to the solid state, which is so stable that it persists unchanged through reactions occurring under mild conditions. The structure of the starting materal must contain not only these particularly strong bonds, but other weaker bonds which may be broken by other reagents, with the formation of addition compounds. These conditions are most readily fulfilled by layer lattice structures, since in these the binding forces in one direction differ from those in the other two directions. The necessary conditions may, however, often be fulfilled also by substances with chain, ribbon, or channelled structures.

The topochemical formation of a compound not obtainable by other means, from a starting material with a channelled structure, is exemplified by the preparation of the sulfide and polysulfide zeolites related to ultramarine. Cation exchange reactions of zeolitic substances can also be regarded as topochemical processes. Montmorillonite, which can also undergo cation exchange, has a well developed layer lattice structure (U. Hofmann, 1933–37). The reactions of siloxene (Vol. I, p. 477) are other topochemical processes of the same type.

References

1 C. H. DESCH, *The Chemistry of Solids*, Oxford 1934, 213 pp.

2 J. A. HEDVALL, *Reaktionsfähigkeit fester Stoffe*, Leipzig 1938, 234 pp.

3 K. HAUFFE, *Reaktionen in und an festen Stoffen*, Berlin 1955, 696 pp.

4 W. JOST, *Diffusion und chemische Reaktion in festen Stoffen*, Dresden 1937, 231 pp.

5 W. SEITH, *Diffusion in Metallen (Platzwechselreaktionen)*, Berlin 1939, 151 pp.

6 C. J. SMITHELLS, *Gases and Metals*, London 1937, 218 pp.

7 W. J. MÜLLER, *Die Bedeckungstheorie der Passivität der Metalle, und ihre experimentelle Begründung*, Berlin 1933, 102 pp.

8 O. KRÖHNKE, E. MAASS and W. BECK, *Die Korrosion, unter Berücksichtigung des Materialschutzes*, Leipzig 1929, 208 pp.

9 F. TÖDT, *Messung und Verhütung der Metallkorrosion*, Berlin 1941, 164 pp.

10 F. N. SPELLER, *Corrosion, Causes and Prevention*, 2nd Ed., London 1935, 694 pp.

11 U. R. EVANS, *An Introduction to Metallic Corrosion*, London 1948, 206 pp.; *Metallic Corrosion. Passivity and Protection*, 2nd Ed., London 1948, 863 pp.

12 W. PALMAER, *The Corrosion of Metals*, Vol. I, Stockholm 1929, 347 pp.; Vol. II, Stockholm 1931, 198 pp.

13 O. BAUER, O. KRÖHNKE and G. MASING (Editors), *Die Korrosion metallischer Werkstoffe*, 4 Vols., Leipzig 1936–40.

14 G. GUIDI and G. GUZZONI, *La Corrosione dei Metalli*, Milan 1937, 374 pp.

15 J. C. HUDSON, *The Corrosion of Iron and Steel*, New York 1940, 319 pp.

16 F. RITTER, *Korrosionstabellen metallischer Werkstoffe*, Berlin 1937, 193 pp.

17 E. RABALD, *Corrosion Guide*, Amsterdam 1951, 629 pp.

18 E. HERZOG, *Les Méthodes d'Essai de Corrosion des Métaux et Alliages*, Paris 1937, 78 pp.

19 R. J. McKAY and R. WORTHINGTON, *Corrosion Resistance of Metals and Alloys*, New York 1936, 492 pp.

20 G. L. GRARD, *La Corrosion en Métallurgie*, Paris 1936, 348 pp.

21 A. SIEGEL, *Korrosionen an Eisen und Nichteisenmetallen*, Berlin 1938, 86 pp.

22 A. GUÉRILLOT, *Méthodes modernes de Protection des Métaux contre la Corrosion*, Paris 1934, 200 pp.

23 W. MACHU, *Metallische Überzüge*, 2nd Ed., Leipzig 1943, 644 pp.

24 W. PFANHAUSER, *Galvanotechnik*, 8th Ed., 2 Vols., Leipzig 1941, 983 and 607 pp.

25 J. BILLITER, *Prinzipien der Galvanotechnik*, Vienna 1934, 326 pp.

26 H. KRAUSE, *Galvanotechnik (Galvanostegie und Galvanoplastik)*, 10th Ed., Leipzig 1941, 269 pp.

27 M. U. SCHOOP and C. H. DAESCHLE, *Handbuch der Metallspritztechnik*, Zürich 1935, 170 pp.

28 M. RAGG, *Die Schiffsboden- und Rostschutz-Farben*, Berlin 1925, 256 pp.

29 W. FEITKNECHT, *Topochemische Umsetzungen fester Stoffe in Flüssigkeiten* [A. EUCKEN (Editor), *Fortschritte der Chemie, Physik und physikalischen Chemie*, Vol. 21, No. 2], Berlin 1930, 56 pp.

Rules for the Nomenclature of Inorganic Compounds

In 1938, a committee of the International Union for Pure and Applied Chemistry issued a report, *Rules for the Nomenclature of Inorganic Compounds*, with the object of securing uniformity of nomenclature and doing away with incorrect and outmoded names. The rules were published in German, English, French, Italian and Spanish by the General Secretariat of the IUPAC, in 1940, and were also reproduced in the *Journal of the American Chemical Society* [63 (1941) 889] and the *Journal of the Chemical Society* [1940, p. 1404]. A new version of the rules of the International Commission is in course of preparation, but is not yet completed. The rules of 1940 will be supplemented rather than basically altered by the new rules. A summary of the most important rules is given below. The amendments to the 1940 Rules have been incorporated as far as possible.

A. General

Names and Formulae.

Inorganic compounds may be designated either by *formulae* or by *names*.

Use of formulae is especially to be recommended in the case of complicated compounds. It is also particularly important in giving preparative details, since it obviates any possibility of confusion.

In naming compounds, either (a) *rational names* or (b) *trivial names* may be assigned.

Where *rational names* are used, it is not always necessary that the name should express the stoichiometric composition of the substance. Numeral prefixes, indications of valency and so on may frequently be omitted, as being superfluous in a given context. *E.g.*, aluminum sulphate instead of aluminum(III) sulphate; potassium chloroplatinate instead of potassium hexachloroplatinate(IV), potassium cyanoferrate(II) instead of potassium hexacyanoferrate(II).

Some *trivial names* do not involve any incorrect chemical conceptions, but are in most cases drawn in expressive fashion from the common applications and properties of the substances—for example *nitre, quick lime, caustic soda, chamber crystals*. It is permissible to use trivial names of this kind. On the other hand, it is *no longer permissible* to employ *trivial names which are derived from false conceptions*. Names must be regarded as incorrect in derivation when they were originally framed in order to make definite statements about the composition or constitution of the compounds concerned, which are contradicted by our present knowledge and theoretical concepts. Names of this sort are sulfate of magnesia, carbonate of lime, nitrate of potash, acetate of alumina. Although such names are in common use (*e.g.* in industry), they are scientifically incorrect.

Use of hyphens.

Names should not be separated into their individual component parts by means of hyphens (*e.g.*, *chloroplatinic acid*, not chloro-platinic acid). In English, no hyphen is used where a Roman numeral is interposed to denote the valency of an element (*e.g.*, *iron(III) sulfate, ammonium iron(III) sulfate*). Where the word *hydrate* is accompanied by an arabic numeral, a hyphen is interposed—*e.g.*, *copper sulfate 5-hydrate*.

B. *Nomenclature of binary compounds*

1. *Order of constituents in names and formulae.*

As far as it can be inferred from the character of the compound, the *electropositive constituent* of salt-like and other electrovalent binary compounds *should be written first both in the formula and in the name* of the compound. This should be followed by the name of the more electro-negative constituent, with the suffix -*ide*. In the case of covalent compounds, or where it is uncertain which of the constituents should be regarded as the more electropositive, that element should be named first which appears earlier in the sequence: B, Si, C, Sb, As, P, N, H, Te, Se, S, I, Br, Cl, O, F. Thus, *sodium chloride; silver sulfide; lithium hydride; boron carbide; oxygen difluoride.*

Certain long-established trivial names for some hydrogen compounds (water, ammonia, hydrazine) are still acceptable. For some other hydrogen compounds the following names are accepted:

B_2H_6 diborane, B_2H_4 diborine, BH_3 (as type): borine
SiH_4 silane, Si_2H_6 disilane, etc.
PH_3 phosphine, P_2H_4 diphosphine,
AsH_3 arsine, As_2H_4 diarsine, SbH_3 stibine.

2. *Indication of stoichiometric proportions of constituents.*

The relative atomic proportions of the component elements in a compound may be denoted in three ways: by indicating the *valency*, by denoting the *stoichiometric composition*, or by a *functional* nomenclature.

The use of endings -*ous* and -*ic* to represent lower and higher oxidation states of electropositive elements is ambiguous and unsatisfactory. Hence Stock's system of denoting oxidation states by the use of Roman numerals should now be used exclusively—thus $CuCl$, *copper(I) chloride;* $CuCl_2$, *copper(II) chloride;* Fe_3O_4, *iron(II,III) oxide.*

To indicate stoichiometric composition, Greek numbers are used as prefixes, joined to the name of the relevant element without a hyphen. In most instances it is possible to omit the prefix 'mono'. The number 8 is denoted by 'octa', and 9 by 'ennea'. For numbers higher than 12 it is preferable to use the appropriate arabic numerals in place of a written prefix. Arabic figures are also to be used where some fractional proportion of a molecule is to be expressed, although '$1/2$' can be denoted by 'hemi-'. Examples of these rules are: N_2O, *dinitrogen (mon)oxide;* NO, *nitrogen oxide;* NO_2, *nitrogen dioxide;* N_2O_4, *dinitrogen tetroxide;* FeS_2, *iron disulfide;* $Co_2(CO)_8$, *dicobalt octacarbonyl.*

Since the degree of molecular complexity of many substances may vary with temperature, state of aggregation etc. the name to be used shall be based upon the simplest possible formula of the substance.

3. *Nomenclature of intermetallic compounds.*

It is recommended that *the formulae only* should be used, and as far as possible the exact numbers of atoms should be shown. If it is not possible to specify the numbers of atoms exactly, or for intermetallic compounds which exist over a wide range of homogeneity, simplified or idealized formulae may be used to represent compounds. Compounds of variable composition are termed 'Berthollide' or 'non-Daltonide' compounds, to distinguish them from 'Daltonide' compounds (of constant composition). If there is any risk of confusion with a compound of constant composition, it is recommended that variability of composition should be indicated by putting the sign $\sim$ before or above the idealized formula (*e.g.*, $\sim$ ZnAu for the β-phase of the Zn-Au system, which may vary in composition between $Zn_{41}Au_{59}$ and $Zn_{58}Au_{42}$).

4. *Designation of mass, atomic number and ionization state on atomic symbols.*

Since a subscript index on the right of the atomic symbol has long been used to show the *number of atoms* in a formula, and a superscript figure to the right of the symbol has been used to give the *charge* on the atom or molecule, the corresponding positions to the left of the symbol are logically available for showing the atomic number and mass number. Where it is required to show the isotopic mass, it is recommended that this should be written as a superscript on the left side of the symbol, with the atomic number as a subscript. Thus the symbol $^{35}_{17}Cl_2^{1+}$ would represent an ionized chlorine molecule, bearing a single positive charge, consisting of two chlorine atoms of mass 35, and atomic number 17. While this recommendation of the International Union has been followed to some extent (*e.g.*, by British chemists), the practice of writing the atomic mass above and to the right of the symbol (e.g. $_{17}Cl^{35}$) has become so prevalent (*e.g.*, in the American literature) that uniformity can hardly be attained. It may be noted that the inclusion of both atomic numbers and mass numbers is rarely necessary, except in formulating nuclear reactions.

5. *Group names.*

It is recommended that binary compounds of the halogens should be termed *halogenides*, not halides. The elements oxygen, sulfur, selenium and tellurium are the *chalcogens*, and their binary compounds should be *chalcogenides*. The *alkali metals* and *alkaline earth metals* should not be referred to as 'alkalis' and 'alkaline earths' respectively, since these are the outmoded names for their oxides, which should be avoided in scientific writing. On the other hand, it is permissible and usual to employ 'alkali-' or 'alkaline earth-' as an abbreviated (adjectival) form in compound words (*e.g.*, *alkali halogenides, alkaline earth carbonates*).

The name "*rare earth metals*" may be used for the elements Sc, Y and La to Lu inclusive; the name "*lanthanum series*" for the elements 57–71 (La to Lu, inclusive), and the name "*lanthanides*" for the elements 58–71 (Ce to Lu, inclusive) are recommended. The elements 89 (actinium) to 103 form the "*actinium series*" and the name "*actinides*" is reserved for the elements in which the 5*f* shell is filled up. The name "*transuranium elements*" is also sanctioned for the elements following uranium.

C. Compounds of more than two elements

Rules for the nomenclature of binary compounds apply in principle to the naming of compounds with more than two elements. Radicals known by special names are treated as elementary constituents in building up the names of compounds. Thus NH_4Cl, *ammonium chloride*; KCN, *potassium cyanide*; $Pb(N_3)_2$, *lead azide*.

If several electropositive constituents are bound to one electronegative element or radical, the most strongly electropositive is named first, in accordance with the rule for binary compounds. Conversely, if a compound contains several electronegative constituents, these should be named in the sequence given in *B*,1. For compounds consisting of discrete molecules, however, or for radicals the sequence must be in accordance with the order in which the atoms are actually bound, e.g. NCS^-, not CNS^-.

Sulfur replacing oxygen in acid radicals is generally denoted by '*thio-*'. Thus compounds of the SCN radical are thiocyanates, not sulfocyanates (nor sulfocyanides).

Names such as *lead chlorofluoride* (instead of lead chloride fluoride), *lead sulfochloride* (for lead sulfide chloride) may be used for mixed salts.

In naming silicates, names such as alumino-, boro-, beryllo-silicates should be used *only* for compounds in which Al, B, Be etc. replace Si in the crystallographic sense. Thus *orthoclase*, $K[AlSi_3O_8]$, is an aluminosilicate (potassium aluminotrisilicate); *spodumene*, $LiAl[Si_2O_6]$, is an aluminum silicate (lithium aluminum disilicate); *muscovite*, $KAl_2[AlSi_3O_{10}](OH)_2$, is an aluminum aluminosilicate.

D. Names for Ions and Radicals

1. Cations.

a. Monatomic cations should be named as the corresponding element, without change or suffix.

b. This principle should also apply to those polyatomic cations for which radical names are preserved, e.g. NO^+ and NO_2^+, the nitrosyl and nitryl cations respectively.

c. Polyatomic cations formed from *monatomic cations* by addition of *other ions or neutral atoms or molecules* (ligands) will be regarded as complex and will be named according to the rules given in section *G*, (e.g. $[Al(H_2O)_6]^{3+}$, hexaquo-aluminium ion and $[CoCl(NH_3)_5]^{2+}$, chloro-pentamminecobalt ion). For some important polyatomic cations special radical names may be used, e.g. for UO_2^{2+} the name "uranyl ion" instead of "dioxouranium(VI)-ion".

d. Names for polyatomic cations derived *by addition of protons to monatomic anions* are formed by adding -onium to the root name of the characteristic element (e.g. phosphonium, arsonium, stibonium, oxonium, sulfonium, selenonium, telluronium, and iodonium). Organic compounds derived by substitution in these primitive compounds should be named as such, whether the primitive itself be known or not, e.g. $(CH_3)_4Sb^+$, the tetramethylstibonium ion.

The ion H_3O^+, which is in fact the monohydrated proton, is to be known as the *oxonium ion* when it is believed to have this constitution, as for example in $[H_3O]^+ [ClO_4]^-$, *oxonium perchlorate*. The widely used term '*hydronium*' should be kept for cases where it is wished to denote an indefinite degree of hydration, i.e. as in aqueous solution. If however the hydration is of no particular importance to the matter under consideration, the simpler term 'hydrogen ion' may be used. The latter name may also be used for the indefinitely solvated proton in non-aqueous solvents; definite ions, such as $CH_3OH_2^+$ and $(CH_3)_2 OH^+$ may be treated as derivatives of the oxonium ion, i.e. as methyloxonium and dimethyloxonium ions respectively.

e. The name '*ammonium*' for the ion NH_4^+ does not conform to subsection *d* but should be retained. Substituted ammonium ions, derivable from nitrogen bases with names ending in -amine will receive names formed by changing -amine into -ammonium, e.g. $HONH_3^+$ the hydroxylammonium ion. Where however the nitrogen base is known by a name ending in a termination other than -amine, the ending -ium should be added (if necessary a final *e* or other vowel may be omitted), e.g. hydrazinium, anilinium, glycinium, pyridinium, guanidinium, imidazolium.

f. Where more than one ion is derived from one base, e.g. $N_2H_5^+$ and $N_2H_6^{2+}$, their ionic charges should be indicated on their names, as the hydrazinium $(1+)$ and hydrazinium $(2+)$ ion respectively.

g. Cations formed by adding protons to other molecules may also be given names formed by adding -ium to the name of the compound. In the case of cations formed by adding protons to acids, the word acidium is to be added after the name of the anion; e.g. $H_2NO_3^+$ is the nitrate acidium ion.

2. Anions.

a. The names for monatomic anions should consist of the name with the termination -ide. Thus:

H^-	hydride ion	O^{2-}	oxide ion	N^{3-}	nitride ion
F^-	fluoride ion	S^{2-}	sulfide ion	P^{3-}	phosphide ion
Cl^-	chloride ion	Se^{2-}	selenide ion	As^{3-}	arsenide ion
Br^-	bromide ion	Te^{2-}	telluride ion	Sb^{3-}	antimonide ion
I^-	iodide ion				

Expressions such as e.g. 'chlorine ion' are used, particularly in connection with crystal structure work and spectroscopy. The Nomenclature Commission recommends that whenever the charge corresponds with that indicated above, the termination -ide should be used.

b. Certain polyatomic ions also have special names ending in -ide. These are:

OH^-	hydroxide ion	N_3^-	azide ion
O_2^{2-}	peroxide ion	NH^{2-}	imide ion
O_2^-	hyperoxide ion	NH_2^-	amide ion
O_3^-	ozonide ion	$NHOH^-$	hydroxylamide ion
S_2^{2-}	disulfide ion	$N_2H_3^-$	hydrazide ion
I_3^-	triiodide ion	CN^-	cyanide ion
ICl_2^-	dichloroiodide ion		

Names for other polysulfides and polyhalides may be formed analogously.

Ions such as HS^- and HO_2^- will be called the hydrogen sulfide ion and the hydrogen peroxide ion respectively in accordance with section *F*,2 and do not require special names.

c. The names for polyatomic anions will consist of the name of the central atom with the termination -*ate*. Atoms (or groups) other than oxygen attached to this central atom should be treated as ligands in a complex ion (section *G*). It is quite practicable to treat oxygen also in this way, although custom hitherto has been to ignore it altogether in the name, and to indicate its proportion by prefixes (per-, hypo- etc.) and sometimes the suffix -ite in place of -ate. These old-established names are often convenient, and may be preserved as trivial names. For all new ions, however, it is preferable to use the system given in section G.

3. Radicals.

a. Generally, radicals or atomic groups bear the same names as the corresponding ions, e.g. the ammonium radical; the sulfate group. There are a few exceptions to this rule:

> hydroxyl group corresponding with the hydroxide ion
> mercapto or thiol group corresponding with the hydrogen sulfide ion
> amino group corresponding with the amide ion

b. Certain radicals containing oxygen or other chalcogens have special names ending in -yl:

HO	hydroxyl	S_2O_5	pyrosulfuryl
CO	carbonyl	SeO	seleninyl
NO	nitrosyl	SeO_2	selenonyl
NO_2	nitryl	CrO_2	chromyl
PO	phosphoryl	UO_2	uranyl
VO	vanadyl	NpO_2	neptunyl
SO	thionyl	PuO_2	plutonyl
SO_2	sulfuryl		

When oxygen is substituted by other chalcogens this is indicated by prefixing thio-, seleno- etc., e.g. thiocarbonyl, CS.

In the cases where such radicals may have different valencies, the valency of the characteristic element should be indicated by means of the Stock-number, e.g. UO_2 may be either uranyl(VI) or uranyl(V) corresponding with the ions UO_2^{2+} and UO_2^+ respectively. Similarly VO may be either vanadyl(V), vanadyl(IV) or vanadyl(III).

E. Oxoacids

For most of the important oxoacids, there are names established by long usage, which it is not necessary to alter. In the following six cases, however, there is some confusion, due to the existence of alternative designations, some of which are based upon false conceptions. The recommended nomenclature is as follows.

1. $H_2S_2O_4$ *Dithionous acid* (not hydrosulfurous or hyposulfurous acid). Salts—*dithionites*.
2. $H_2S_2O_3$ *Thiosulfuric acid* (not hyposulfurous acid). Salts—*thiosulfates*.
3. H_2SO_5 *Peroxomonosulfuric acid* (not persulfuric acid). The acid $H_2S_2O_8$ should also be termed *peroxosulfuric acid* (or better, *peroxodisulfuric acid*), and not persulfuric acid.
4. H_2NO_2 *Nitroxylic acid* (not hydronitrous acid). Salts—*nitroxylates*.
5. $H_2B_4O_7$ *Tetraboric acid* (not pyroboric acid). Salts—*tetraborates*.
6. $H_4P_2O_6$ *Hypophosphoric acid* (not subphosphoric acid). Salts—*hypophosphates*.

Acids which are derived from oxygen acids by the replacement of oxygen by sulfur are to be known as *thioacids* (not sulfoacids). Their salts are *thiosalts*. Examples—H_2CS_3 *trithiocarbonic acid.* Na_3SbS_4 *trisodium tetrathioantimonate* (or sodium thioantimonate).

F. Salts

1. General.

Salts are named by appending to the name of the metal or electropositive radical that of the acid, with the suffix -*ate*, -*ite* or -*ide*. Examples—$AgNO_3$, *silver nitrate;* $NaNO_2$, *sodium nitrite*, $MgSO_4$, *magnesium sulfate;* MnS, *manganese sulfide;* Na_2SO_2, *sodium sulfoxylate*, KCN, *potassium cyanide*. Mixed salts may be named by applying the principles set out in *C*—for example, $KNaCO_3$, potassium sodium carbonate; $KCaPO_4$, potassium calcium phosphate; NH_4MgPO_4, ammonium magnesium phosphate.

Salts of nitrogen compounds can, in many cases, be regarded as coordination compounds, analogous to the ammonium salts. They are then designated -*onium* or -*inium* compounds. Examples—$[N(CH_3)_4]Cl$, *tetramethylammonium chloride;* $[N_2H_5]Cl$, *hydrazinium (mono)chloride;*

[N_2H_6]Cl_2, *hydrazinium dichloride;* [C_5H_5NH]Cl, *pyridinium chloride.* If nitrogen compounds are regarded as addition compounds, however, they must be named according to the rule given in *G*, 5.

2. *Acid salts (hydrogen salts).*

The rational names for acid salts are formed by insertion of 'hydrogen' for the hydrogen atoms they contain, the hydrogen always being the last of the electropositive constituents specified (*e.g.*, $KHSO_4$, *potassium hydrogen sulfate*, Na_2HPO_4, *disodium hydrogen phosphate*, NaH_2PO_4, *sodium dihydrogen phosphate*, $NaNH_4HPO_4$, *sodium ammonium hydrogen phosphate*). If it is desired to emphasize the general class of compound rather than the composition, the terms *acid salt*, *primary*, *secondary* or *tertiary* salt etc. may be used.

Specification of the acid: base ratio by the prefix *bi-* is not in harmony with the principles of rational nomenclature, and it is therefore *incorrect* to use names such as bicarbonate, bisulfate etc.

3. *Basic salts.*

Basic salts which are proved to contain hydroxyl groups (or ions) are termed *hydroxy salts.* In the light of available structural evidence, this nomenclature cannot be taken as implying that hydroxysalts are constitutionally addition compounds of hydroxides with neutral salts. Example Cd(OH)Cl, *cadmium hydroxychloride.* Where hydroxyl groups are bound within a coordination complex, Werner's terminology is used, and the hydroxyl groups are designated *hydroxo* groups.

Basic salts containing both acid radicals and oxygen atoms linked to the metal are termed *oxysalts.* In a few cases (e.g. the BiO^+ and UO_2^{++} groups) it is generally accepted that definite radicals, or their polymers in solid compounds, exist as structural units. In such cases, names based on the radicals, with the suffix *-yl*, are used. Examples—BiOCl, *bismuth oxychloride* or *bismuthyl chloride*; $UO_2(NO_3)_2$, *uranium dioxydinitrate* or *uranyl nitrate.* Werner's system is used for oxygen atoms bound in the coordination complex, which are termed *oxo-atoms.*

G. Compounds of higher order

1. *Complex compounds or coordination compounds.*

Werner's system of nomenclature is used, except insofar as the oxidation state of the central atom is now to be indicated by a roman numeral, as is now general throughout inorganic nomenclature. The cation is to be named first, followed by the anion. The names of negative coordinated groups are given the ending *-o*, neutral ligands being given no characteristic endings. Water and ammonia as neutral ligands in complexes are referred to as *aquo* and *ammine* respectively. The order of listing ligands is (i) negative groups, (ii) neutral groups (Example—[$CoCl_2(NH_3)_4$]$^+$, *dichlorotetrammine cobalt(III) ion*). The oxidation state of the central atom of complex cations or neutral complexes is shown as in the example above; for complex anions, the oxidation state is indicated as in $H_2[PtCl_6]$, *hydrogen hexachloroplatinate(IV)*, $K_3[Co(NO_2)_6]$, *potassium hexanitrocobaltate(III)*.

2. *Isopolyacids and their salts.*

Where the composition and constitution of the structural units of isopolyacids and their salts are known, nomenclature follows the rules already given. In many cases, however, the constitution is not yet known, and there is need for uniformity in representing the stoichiometric composition of the compounds concerned. Three methods may be used: (a) the composition, referred to the simplest empirical formula, is indicated by Greek numerical prefixes, as for other compounds; (b) the compound may be named by adding the numbers of the several atoms in brackets after the name; (c) the ratio basic anhydride: acid anhydride is shown by means of arabic figures, in parentheses. This form of representation has been widely used. In all cases, the basic constituent is named before the acidic. Acid hydrogen atoms are indicated by 'hydrogen', and should always be specified if present. The number of hydrogen atoms should also be given in naming the free acids, if possible.

Examples:

Empirical formula	Use of Greek prefixes	Use of numbers in brackets
Na_2MoO_4	disodium molybdate	sodium molybdate (2.1.4)
$Na_2Mo_3O_{10}$	disodium trimolybdate	sodium molybdate (2.3.10)
$Na_2W_4O_{13}$	disodium tetrawolframate	sodium wolframate (2.4.13)
$Na_2V_4O_{11}$	disodium tetravanadate	sodium vanadate (2.4.11)
NaV_3O_8	sodium trivanadate	sodium vanadate (1.3.8)
$Na_5P_3O_{10}$	pentasodium triphosphate	sodium phosphate (5.3.10)
$Na_2B_4O_7$	disodium tetraborate	sodium borate (2.4.7)
$Na_2B_8O_{13}$	disodium octaborate	sodium borate (2.8.13)
NaB_5O_8	sodium pentaborate	sodium borate (1.5.8)

Resolved formula	Use of anhydride ratios	Resolved formula	Use of anhydride ratios
Na_2O,MoO_3	sodium(1 : 1)molybdate	$5Na_2O,3P_2O_5$	sodium(5 : 3)phosphate
$Na_2O,3MoO_3$	sodium(1 : 3)molybdate	$Na_2O,2B_2O_3$	sodium(1 : 2)borate
$Na_2O,4WO_3$	sodium(1 : 4)wolframate	$Na_2O,4B_2O_3$	sodium(1 : 4)borate
$Na_2O,2V_2O_5$	sodium(1 : 2)vanadate	$Na_2O,5B_2O_3$	sodium(1 : 5)borate
$Na_2O,3V_2O_5$	sodium(1 : 3)vanadate		

3. Heteropolyacids and their salts.

As with the isopolyacids, the constitution of a considerable number of heteropolyacids is now known; the compounds can be named according to the rules already given, or may in many cases be conveniently referred to by formula wherever possible. Nomenclature based on stoichiometric composition follows, as for isopolyacids, from the simplest formula expressing the analytical composition. The empirical formula of the *acids* may be resolved into nonmetallic acid, oxide of the acid forming metal, water. That of salts may be broken up into salt of nonmetallic acid (further resolved into the ratio of basic anhydride to acid anhydride, if necessary), oxide of acid forming metal, water. The numbers of atoms of the two acid-forming elements in this simplest formula are then expressed either by greek numeral prefixes, or by arabic figures. The suffix -o is added to the the stem of the name of the acid forming metal; the nonmetallic acid is named last, in conformity with the fact that this is now known to form the central atom or group of a particular type of complex.

$R_3PO_4,12MoO_3$	Dodecamolybdophosphate	$3R_2O·P_2O_5·24MoO_3$	24-molybdo-2-phosphate or 12-molybdophosphate*
$2R_5PO_5,17WO_3$	17-wolframodiphosphate	$5R_2O·P_2O_5·17WO_3$	17-wolframo-2-phosphate
$R_5BO_4,12WO_3$	Dodecawolframoborate	$5R_2O·B_2O_3·24WO_3$	24-wolframo-2-borate or 12-wolframoborate*
$R_8SiO_6,12WO_3$	Dodecawolframosilicate	$4R_2O·SiO_2·12WO_3$	12-wolframo-silicate*

4. Double salts.

Names of double salts are formed by citing the cationic constituents in order of decreasing electropositivity. Constituents common to both components need be named only once. Examples—$KCl·MgCl_2$, *potassium magnesium chloride;* $Na_2SO_4·CaSO_4$; *sodium calcium sulfate,* $KCl·MgSO_4$, *potassium chloride-magnesium sulfate.*

5. Hydrates, and other addition compounds.

Compounds containing molecules H_2O, as constituents are given the generic name of *hydrates.* The number of molecules present may be denoted either by Greek numerical prefixes

---

* Note that in these cases the name derived directly from the resolved formula can obscure the constitutional relationship known to exist between them, as members of the 12-wolframo- (or 12-molybdo-) type, $[XM_{12}O_{40}]^{x-}$, with X = P, B, Si etc., M = W or Mo.

or by arabic figures. Example—CaCl$_2$·6H$_2$O, *calcium chloride hexahydrate* or *calcium chloride 6-hydrate*. If the name is intended, however, to express the fact that the relevant molecules are bound *in a coordination complex*, the compounds must be termed *aquo*-compounds.

The ending *-ate* should not be used for addition compounds of H$_2$O$_2$ or NH$_3$. For example CaCl$_2$·8NH$_3$ should be named *calcium chloride 8-ammonia*, and not calcium chloride 8-ammoniate.

Other addition compounds may be named in similar manner, e.g.—AlCl$_3$·4C$_2$H$_5$OH aluminium chloride tetraethanol or aluminium chloride 4-ethanol. In general types of addition compounds, such as are formed by PCl$_3$, NOCl, H$_2$S, C$_2$H$_5$OH etc., are best referred to by formulae rather than by special names. Alternatively, such a compound as AlCl$_3$·NOCl may be referred to as 'the compound of aluminium chloride with nitrosyl chloride'.

H. State of aggregation—polymorphism

1. Polymorphism.

In order to distinguish clearly between different polymorphic modifications of a single substance, it often will suffice to indicate the crystal system after the name or formula; e.g. zinc sulfide (cub) = zinc blende or sphalerite, and ZnS (hex) = wurtzite. The abbreviations should be made so that confusion is avoided, and the terms trigonal (trig) and orthorhombic (orh) are to be preferred to rhombohedral and rhombic. It is also possible to indicate simple structures as being face-centred, body-centred, close-packed, etc.

Examples: cubic face-centred (cub. f.c.); tetragonal body-centred (tetr. b.c.); hexagonal close-packed (hex. c.p.). A slight distortion of a simpler lattice could be indicated by "dist." —e.g. "dist. cub. f.c.".

2. Allotropes.

For the nomenclature of solid allotropic forms the foregoing rule concerning polymorphism can be used.

If rational names for gaseous and liquid modifications are required, these should be based on the size of the molecule which can be indicated by Greek numerical prefixes. If the number of atoms be great and unknown, 'poly' can be used.

Symbol	Old name	New name
H	atomic hydrogen	monohydrogen
O$_2$	oxygen (common)	dioxygen
O$_3$	ozone	trioxygen
P$_4$	white (or yellow) phosphorus	tetraphosphorus
S$_8$	λ-sulfur	octasulfur
S$_n$	μ-sulfur	polysulfur

The new names need only be used when it is wished to emphasise that a definite allotropic modification is meant.

Distribution of Electrons among the Energy Levels in the Ground States of Free Atoms

The numbers of electrons in the orbitals of the ground states of the elements are given. The quantum numbers of the orbitals are given at the head of the Table. The rare gas configurations are printed in heavy type. Data as yet uncertain are given in brackets.

The *term symbol*, quoted in the last column, denotes the energy state of the whole atom, resulting from vectorial addition of the subsidiary quantum numbers for the individual electrons. (Closed shells and sub-shells always add up to zero and therefore do not contribute to the resultant.) The individual quantum numbers l_1, l_2 etc. give a resultant quantum number L, and this combines with the resultant spin quantum number Sp for the atom as a whole to give a resultant J. Atomic energy states with $L = 0$, 1, 2, 3, 4 etc. are denoted as S, P, D, F, G etc. states respectively. The value of J is written as a subscript following the letter which denotes the L value. The superscript before the letter denotes the "multiplicity", i.e. the number of sub-levels which is equal to $2\,Sp + 1$.

Atomic Numbers	X-ray levels	K	L		M			N				O				P			Q	Term symbols
	Princ. quant. number $n =$ Subs. quant. number $l =$ Symbols for the energy levels	1 0 s	2 0 s	2 1 p	3 0 s	3 1 p	3 2 d	4 0 s	4 1 p	4 2 d	4 3 f	5 0 s	5 1 p	5 2 d	5 3 f	6 0 s	6 1 p	6 2 d	7 0 s	
1	Hydrogen ...	1																		$^2S_{1/2}$
2	**Helium**	**2**																		$^1S_0$
3	Lithium	2	1																	$^2S_{1/2}$
4	Beryllium	2	2																	$^1S_0$
5	Boron	2	2	1																$^2P_{1/2}$
6	Carbon	2	2	2																$^3P_0$
7	Nitrogen	2	2	3																$^4S_{3/2}$
8	Oxygen......	2	2	4																$^3P_2$
9	Fluorine	2	2	5																$^2P_{3/2}$
10	**Neon**	2	**2**	**6**																$^1S_0$
11	Sodium	2	**2**	**6**	1															$^2S_{1/2}$
12	Magnesium ..	2	**2**	**6**	2															$^1S_0$
13	Aluminum ...	2	**2**	**6**	2	1														$^2P_{1/2}$
14	Silicon	2	**2**	**6**	2	2														$^3P_0$

Atomic Numbers	X-ray levels	K	L		M			N				O				P			Q	Term symbols
	Princ. quant. number $n =$	1	2	2	3	3	3	4	4	4	4	5	5	5	5	6	6	6	7	
	Subs. quant. number $l =$	0	0	1	0	1	2	0	1	2	3	0	1	2	3	0	1	2	0	
	Symbols for the energy levels	s	s	p	s	p	d	s	p	d	f	s	p	d	f	s	p	d	s	
15	Phosphorus ..	2	2	6	2	3														$^4S_{3/2}$
16	Sulfur	2	2	6	2	4														$^3P_2$
17	Chlorine	2	2	6	2	5														$^2P_{3/2}$
18	**Argon**	2	2	6	2	6														$^1S_0$
19	Potassium ...	2	2	6	2	6		1												$^2S_{1/2}$
20	Calcium	2	2	6	2	6		2												$^1S_0$
21	Scandium ...	2	2	6	2	6	1	2												$^2D_{3/2}$
22	Titanium	2	2	6	2	6	2	2												$^3F_2$
23	Vanadium ...	2	2	6	2	6	3	2												$^4F_{3/2}$
24	Chromium ..	2	2	6	2	6	5	1												$^7S_3$
25	Manganese...	2	2	6	2	6	5	2												$^6S_{5/2}$
26	Iron	2	2	6	2	6	6	2												$^5D_4$
27	Cobalt	2	2	6	2	6	7	2												$^4F_{9/2}$
28	Nickel	2	2	6	2	6	8	2												$^3F_4$
29	Copper	2	2	6	2	6	10	1												$^2S_{1/2}$
30	Zinc	2	2	6	2	6	10	2												$^1S_0$
31	Gallium	2	2	6	2	6	10	2	1											$^2P_{1/2}$
32	Germanium .	2	2	6	2	6	10	2	2											$^3P_0$
33	Arsenic	2	2	6	2	6	10	2	3											$^4S_{3/2}$
34	Selenium	2	2	6	2	6	10	2	4											$^3P_2$
35	Bromine	2	2	6	2	6	10	2	5											$^2P_{3/2}$
36	**Krypton**	2	2	6	2	6	10	2	6											$^1S_0$
37	Rubidium ...	2	2	6	2	6	10	2	6			1								$^2S_{1/2}$
38	Strontium....	2	2	6	2	6	10	2	6			2								$^1S_0$
39	Yttrium	2	2	6	2	6	10	2	6	1		2								$^2D_{3/2}$
40	Zirconium ...	2	2	6	2	6	10	2	6	2		2								$^3F_2$
41	Niobium	2	2	6	2	6	10	2	6	4		1								$^6D_{1/2}$
42	Molybdenum.	2	2	6	2	6	10	2	6	5		1								$^7S_3$
43	Technetium .	2	2	6	2	6	10	2	6	(5)		(2)								$(^6S_{5/2})$
44	Ruthenium ..	2	2	6	2	6	10	2	6	7		1								$^5F_5$
45	Rhodium	2	2	6	2	6	10	2	6	8		1								$^4F_{9/2}$
46	Palladium ...	2	2	6	2	6	10	2	6	10										$^1S_0$
47	Silver	2	2	6	2	6	10	2	6	10		1								$^2S_{1/2}$
48	Cadmium ...	2	2	6	2	6	10	2	6	10		2								$^1S_0$
49	Indium	2	2	6	2	6	10	2	6	10		2	1							$^2P_{1/2}$
50	Tin	2	2	6	2	6	10	2	6	10		2	2							$^3P_0$
51	Antimony ...	2	2	6	2	6	10	2	6	10		2	3							$^4S_{3/2}$
52	Tellurium....	2	2	6	2	6	10	2	6	10		2	4							$^3P_2$
53	Iodine	2	2	6	2	6	10	2	6	10		2	5							$^2P_{3/2}$
54	**Xenon**	2	2	6	2	6	10	2	6	10		2	6							$^1S_0$
55	Cesium	2	2	6	2	6	10	2	6	10		2	6			1				$^2S_{1/2}$
56	Barium	2	2	6	2	6	10	2	6	10		2	6			2				$^1S_0$

Atomic Numbers	X-ray levels	K	L	M	N	O	P	Q	Term symbols
	Princ. quant. number n =	1	2 2	3 3 3	4 4 4 4	5 5 5 5	6 6 6	7	
	Subs. quant. number l =	0	0 1	0 1 2	0 1 2 3	0 1 2 3	0 1 2	0	
	Symbols for the energy levels	s	s p	s p d	s p d f	s p d f	s p d	s	
57	Lanthanum ..	2	2 6	2 6 10	2 6 10	2 6 1	2		$^2D_{3/2}$
58	Cerium	2	2 6	2 6 10	2 6 10 (1)	2 6 (1)	(2)		$(^3H_4)$
59	Praseodymium	2	2 6	2 6 10	2 6 10 (3)	2 6	(2)		—
60	Neodymium .	2	2 6	2 6 10	2 6 10 4	2 6	2		$^5I_4$
61	Promethium..	2	2 6	2 6 10	2 6 10 (5)	2 6	(2)		—
62	Samarium ...	2	2 6	2 6 10	2 6 10 6	2 6	2		$^7F_0$
63	Europium....	2	2 6	2 6 10	2 6 10 7	2 6	2		$^8S_{7/2}$
64	Gadolinium ..	2	2 6	2 6 10	2 6 10 7	2 6 1	2		^9D
65	Terbium	2	2 6	2 6 10	2 6 10 8	2 6 1	2		—
66	Dysprosium ..	2	2 6	2 6 10	2 6 10 (9)	2 6 (1)	(2)		—
67	Holmium	2	2 6	2 6 10	2 6 10 (10)	2 6 (1)	(2)		—
68	Erbium	2	2 6	2 6 10	2 6 10 (11)	2 6 (1)	(2)		—
69	Thulium	2	2 6	2 6 10	2 6 10 13	2 6	2		$^2F_{7/2}$
70	Ytterbium ...	2	2 6	2 6 10	2 6 10 14	2 6	2		$^1S_0$
71	Lutetium	2	2 6	2 6 10	2 6 10 14	2 6 1	2		$^2D_{3/2}$
72	Hafnium.....	2	2 6	2 6 10	2 6 10 14	2 6 2	2		$^3F_2$
73	Tantalum ...	2	2 6	2 6 10	2 6 10 14	2 6 3	2		$^4F_{3/2}$
74	Tungsten	2	2 6	2 6 10	2 6 10 14	2 6 4	2		$^5D_0$
75	Rhenium	2	2 6	2 6 10	2 6 10 14	2 6 5	2		$^6S_{5/2}$
76	Osmium	2	2 6	2 6 10	2 6 10 14	2 6 6	2		$^5D_4$
77	Iridium......	2	2 6	2 6 10	2 6 10 14	2 6 7	2		$^4F_{9/2}$
78	Platinum	2	2 6	2 6 10	2 6 10 14	2 6 9	1		$^3D_3$
79	Gold	2	2 6	2 6 10	2 6 10 14	2 6 10	1		$^2S_{1/2}$
80	Mercury	2	2 6	2 6 10	2 6 10 14	2 6 10	2		$^1S_0$
81	Thallium	2	2 6	2 6 10	2 6 10 14	2 6 10	2 1		$^2P_{1/2}$
82	Lead	2	2 6	2 6 10	2 6 10 14	2 6 10	2 2		$^3P_0$
83	Bismuth	2	2 6	2 6 10	2 6 10 14	2 6 10	2 3		$^4S_{3/2}$
84	Polonium	2	2 6	2 6 10	2 6 10 14	2 6 10	2 4		$^3P_2$
85	Astatine	2	2 6	2 6 10	2 6 10 14	2 6 10	2 5		$^2P_{3/2}$
86	**Radon**	2	2 6	2 6 10	2 6 10 14	2 6 10	2 6		$^1S_0$
87	Francium ...	2	2 6	2 6 10	2 6 10 14	2 6 10	2 6	1	$^2S_{1/2}$
88	Radium	2	2 6	2 6 10	2 6 10 14	2 6 10	2 6	2	$^1S_0$
89	Actinium	2	2 6	2 6 10	2 6 10 14	2 6 10	2 6 1	2	$(^2D_{3/2})$
90	Thorium.....	2	2 6	2 6 10	2 6 10 14	2 6 10	2 6 (2)	(2)	$(^3F_2)$
91	Protactinium .	2	2 6	2 6 10	2 6 10 14	2 6 10	2 6 (3)	(2)	$(^4F_{3/2})$
92	Uranium	2	2 6	2 6 10	2 6 10 14	2 6 10	2 6 (4)	(2)	$(^5D_0)$
93	Neptunium ..	2	2 6	2 6 10	2 6 10 14	2 6 10 (5)	2 6	(2)	—
94	Plutonium ...	2	2 6	2 6 10	2 6 10 14	2 6 10 (5)	2 6 (1)	(2)	—
95	Americium ..	2	2 6	2 6 10	2 6 10 14	2 6 10 (6)	2 6 (1)	(2)	—
96	Curium......	2	2 6	2 6 10	2 6 10 14	2 6 10 (7)	2 6 (1)	(2)	—
97	Berkelium....	2	2 6	2 6 10	2 6 10 14	2 6 10 (8)	2 6 (1)	(2)	—
98	Californium .	2	2 6	2 6 10	2 6 10 14	2 6 10 (9)	2 6 (1)	(2)	—
99	Einsteinium ..	2	2 6	2 6 10	2 6 10 14	2 6 10 (10)	2 6 (1)	(2)	—
100	Fermium	2	2 6	2 6 10	2 6 10 14	2 6 10 (11)	2 6 (1)	(2)	—
101	Mendelevium.	2	2 6	2 6 10	2 6 10 14	2 6 10 (12)	2 6 (1)	(2)	—

Suggestions for Further Reading

General

See Volume I, p. 840.

Chapter 1 (see also Vol. I, p. 843, *Chapter 13*).

Battelle Memorial Institute, Columbus (Ohio), *Metals of Tomorrow* [Ti, Mo, Si, Se, V, Zr, Ce, Ge, Li], *The Iron Age*, 170 (1952) 259–90.

W. BAUKLOH, *Die physikalisch-chemischen Grundlagen der Metallurgie*, Berlin 1949.

A. BONDI, The Spreading of Liquid Metals on Solid Surfaces. Surface Chemistry of High-Energy Substances, *Chem. Revs.*, 52 (1953) 417–58.

C. A. HAMPEL (Editor), *Rare Metals Handbook*, New York 1954.

C. R. HAYWARD, *An Outline of Metallurgical Practice*, New York 1946.

Institution of Mining and Metallurgy, *The Refining of Nonferrous Metals*, London 1950.

E. P. POLUSHKIN, *Defects and Failures of Metals*, Amsterdam 1956.

Chapter 3

J. A. DEMENT and H. C. DAKE, *Rarer Metals*, London 1950.

B. LUSTMAN and F. KERZE, Jr., *Metallurgy of Zirconium*, New York 1955.

T. MOELLER, G. K. SCHWEITZER and D. D. STARR, The Analytical Aspects of Thorium Chemistry, *Chem. Revs.*, 42 (1948) 63–105.

Chapter 5

J. S. ANDERSON, Chemistry of the Metal Carbonyls, *Quart.Revs.*, 1 (1947) 331–57.

R. S. ARCHER et al., *Molybdenum: Steels, Irons, Alloys*, New York 1948.

K. COHEN, *The Theory of Isotope Separation as Applied to the Large-Scale Production of* ^{235}U, New York 1951.

G. H. DIEKE and A. B. F. DUNCAN, *Spectroscopic Properties of Uranium Compounds*, New York 1949.

K. C. LI and C. U. WANG, *Tungsten*, 3rd Ed., London 1955.

A. H. SULLY, *Chromium* [H. M. FINNISTON (Editor), *Metallurgy of the Rarer Metals*, Vol. I], New York 1954.

A. TANNENBAUM (Editor), *Toxicology of Uranium; Survey and Collected Papers*, New York 1951.

C. VOEGTLIN and H. C. HODGE, *Pharmacology and Toxicology of Uranium Compounds*, New York 1949.

F. H. WESTHEIMER, The Mechanisms of Chromic Acid Oxidations, *Chem.Revs.*, 45 (1949) 419–51.

Chapter 6

R. S. DEAN, *Electrolytic Manganese and Its Alloys*, New York 1952.

A. H. SULLY, *Manganese* [H. M. FINNISTON (Editor), *Metallurgy of the Rarer Metals*, Vol. III], New York 1955.

R. C. YOUNG and J. W. IRVINE, Reduction of Perrhenate, *Chem. Revs.*, 23 (1938) 187–91.

Chapter 7

J. S. ANDERSON, see *Chapter 5*.

R. S. NYHOLM, Recent Stereochemistry of the Group VIII Elements, *Quart. Revs.*, 3 (1949) 321–44.

R. S. NYHOLM, The Stereochemistry and Valence States of Nickel, *Chem. Revs.*, 53 (1953) 263–308.

H. L. RILEY, Carbides, Nitrides, and Carbonitrides of Iron, *Quart. Revs.*, 3 (1949) 160–72.
J. L. SNOEK, *New Developments in Ferromagnetic Materials,* 2nd Ed., Amsterdam 1949.
R. S. YOUNG, *Cobalt*, New York 1948.

Chapter 8

A. BUTTS (Editor), *Copper; The Science and Technology of the Metal, Its Alloys and Compounds.*
  New York 1954.
J. V. N. DORR and F. L. BOSQUI, *Cyanidation and Concentration of Gold and Silver Ores,* 2nd
  Ed., New York 1950.

Chapter 10

J. K. MARSH, The Separation of the Lanthanons (Rare Earth Elements), *Quart. Revs.*,
  (1947) 126–43.
R. C. VICKERY, *Chemistry of the Lanthanons*, London 1953.

Chapter 11

K. COHEN, see *Chapter 5.*
S. C. CURRAN, The Determination of Geological Age by Means of Radioactivity, *Quart.
  Rev.*, 7 (1953) 1–18.
E. FEENBERG, *Shell Theory of the Nucleus*, Princeton (N.J.) 1955.
W. M. GARRISON and J. G. HAMILTON, Production and Isolation of Carrier-Free Radio-
  isotopes, *Chem. Revs.*, 49 (1951) 237–72.
J. B. HOAG, *Electron and Nuclear Physics*, 3rd Ed., New York 1948.
A. G. MADDOCK, Radioactivity of the Heavy Elements, *Quart. Revs.*, 5 (1951) 270–314.
J. SACKS, Radioactive Isotopes as Indicators in Biology, *Chem. Revs.*, 42 (1948) 411–56.
H. SEMAT, *Introduction to Atomic and Nuclear Physics*, 3rd Ed., New York 1954.
R. S. SHANKLAND, *Atomic and Nuclear Physics*, New York 1955.

Chapter 12

M. DOLE, The Chemistry of the Isotopes of Oxygen, *Chem. Revs.*, 51 (1952) 263–300.

Chapter 13

See *Chapter 11.*

Chapter 14

M. W. LISTER, Chemistry of the Transuranic Elements, Quart Revs., 4 (1950) 20–44.
G. T. SEABORG and J. J. KATZ (Editors), Actinide Elements, New York 1954.

Chapter 15

S. C. CURRAN, see *Chapter 11.*
D. T. GIBSON, The Terrestrial Distribution of the Elements, *Quart. Revs.*, 3 (1949) 263–91.
B. MASON, *Principles of Geochemistry*, New York 1952.

Chapter 16

S. BERKMAN and G. EGLOFF, The Physical Chemistry of Foams, *Chem. Revs.*, 15 (1934)
  377–424.
A. BIONDI, see *Chapter 1.*
J. H. DE BOER, *Dynamical Character of Adsorption*, Oxford 1953.
B. JIRGENSONS and M. E. STRAUMANIS, *Short Textbook of Colloid Chemistry*, New York 1954,
H. B. KLEVENS, Solubilization, *Chem. Revs.*, 47 (1950) 1–74.
H. R. KRUYT (Editor), *Colloid Science*, 2 vols., Amsterdam 1949, 1952.
C. L. MANTELL, *Adsorption*, 2nd Ed., New York 1951.
F. H. POLLARD and J. F. W. McOMIE, *Chromatographic Methods of Inorganic Analysis*, London
  1953.

E. K. Rideal et al., *Colloid Science*, New York 1947.
O. C. Smith, *Inorganic Chromatography*, New York 1953.
K. H. Stern, The Liesegang Phenomenon, *Chem. Revs.*, 54 (1954 79–99.

Chapter 17

Advances in Catalysis and Related Subjects, 6 vols., New York 1948–54.
E. S. Amis, *Kinetics of Chemical Change in Solution*, New York 1949.
J. O. Edwards, Rate Laws and Mechanisms of Oxyanion Reactions with Bases, *Chem. Revs.*, 50 (1952) 455–82.
P. H. Emmett (Editor), *Catalysis*, 2 vols., New York 1954–55.
J. A. Hedvall, Changes in Crystal Structure and Their Influence on the Reactivity and Catalytic Effect of Solids, *Chem. Revs.*, 15 (1934) 139–68.
J. A. A. Ketelaar, *Chemical Constitution; An Introduction to the Theory of the Chemical Bond*, Amsterdam 1953.
K. J. Laidler, *Chemical Kinetics*, New York 1950.
E. A. Moelwyn-Hughes, *Kinetics of Reactions in Solution*, 2nd Ed., Oxford 1948.
L. C. Pauling, *The Nature of the Chemical Bond and the Structure of Molecules and Crystals, An Introduction to Modern Structural Chemistry*, Ithaca (N.Y.) 1940.
G. K. Rollefson and M. Burton, *Photochemistry and the Mechanisms of Chemical Reactions*, New York 1955.
R. W. Stott, *Electronic Theory and Chemical Reactions; An Elementary Treatment*, 3rd Ed., London 1953.
B. M. W. Trapnell, *Chemisorption*, New York 1955.

Chapter 19

S. Brunauer, *The Physical Adsorption of Gases and Vapours*, Oxford 1944.
G. Cohn, Reactions in the Solid State, *Chem. Revs.*, 42 (1948) 527–79.
W. E. Garner, *Chemistry of the Solid State*, New York, 1955.
J. A. Hedvall, see *Chapter 17*.
P. W. M. Jacobs, Ionic Conductance in Solid Salts, *Quart. Revs.*, 6 (1952) 238–61.
A. R. Miller, *The Adsorption of Gases on Solids*, Cambridge 1949.
A. L. G. Rees, *The Chemistry of the Defect Solid State*, London 1954.
R. Smoluchowski (Editor), *Phase Transformations in Solids* (Symposium held at Cornell University, 1948), New York 1951.

NAME INDEX

SUBJECT INDEX

A

Absorption, 679
— isotherm, 679
Acanthite, 407
Acetates, 732
Acetatoaurates, 420
Acetatotrichromium (III) salts, 141
Acetatotriiron (III) complex, 286
— salts, 285
Acetic acid, 732
Acetylides, 32
Acidoaquotetramminechromium (III) compounds, 147
Acidodiaquotriamminechromium (III) compounds, 147
Acidomanganate (IV) salts, 225, 226
Acidoniobates, 108
Acidopentamminechromium (III) compounds, 147
Acidopentaquochromium (III) compounds, 147
Acidotriaquodiamminechromium (III) compounds, 147
Acidovanadates, 93
Actinide elements, iv, 610
— — atomic structure and valence states, 611
— — electronic configuration, 610
— — history, 614
— — magnetic properties, 613
— series, i, v
Actinium, 30, 42, 503
— compounds, 43
— disintegration series, 536, 537
— emanation, 42
— lead, 537
Actinon, 42
Actinouranium, 189, 536, 537
Activated adsorption, 681
Activation energy, 700, 701
— — wave mechanical calculation, 708
Activators, 715
Active deposits, 523

Active intermediate states, 746, 747
— states, 273
Adsorption, 679
— from solutions, 681
— isotherms, 679
— of electrolytes, 681
— rule, 531
— selective, 682
Aerodisperse systems, 687
Aerosols, 688
Affination, 394
Age hardening, 27, 28
Alabandite, 221
Alitizing, 756
Alkali mercury sulfonates, 473
— permanganates, solubility, 228
Alloys, 2
— improvement, 27
Alluvial or placer gold, 409
Alpha-particle, 563
— charge on, 511
— energy, 518, 519
— nature of the, 511
— range, 509, 510
— spectrum, 520
Alpha-rays, 504, 508
Aluminothermic reduction, 2
Aluminum-bronzes, 375
Amalgams, 459, 460
Americium, 611, 612, 615, 633, 634
— (III) compounds, 634
— (V) compounds, 634
— hydride, 634
— metallic, 634
— oxides, 634
— trichloride, 633
— trifluoride, 633
Ammonia, liquid, 728, 729, 730, 731
Ammoniates, 729
Ammonium chromate, 156
— dichromate, 156
— molybdate, 164, 168
— molybdophosphate, 168, 182
— nitratocerate (IV), 498
— pertechnetate, 232, 233

Ammonium uranate, 201
Ammonoacids, 728, 729
Ammonobases, 728
Ammonolysis, 727, 729
Ammono system, 727
Amphiprotic, 721
Amphoteric hydroxides, behavior, 727
Anatase, 59
Anhydrobase, 722
Annihilation radiation, 567
Anomalous mixed crystals, 531, 541
Ansolvoacids, 727
Ansolvobases, 727
Anti-knock, 289, 714
Antimony, 645, 646
Apatite, 102
Aprotic, 723
Aquoamminocobalt(III) ions, 305
Aquopentamminechromium (III) compounds, 146
Aquopentamminecobalt(III) chloride, 304
Arbor Dianae, 394
Argentite, 392
Arite, 313
Arsenic trioxide, 274
Artificial atomic transmutations, 560
— new elements, 593
— radioactivity, 541, 574
Artificially activated elements, uses, 582, 583
Assaying, 4
Astatine, 593
Aston's rule, 592
Atacamite, 390
Atmosphere, 643
Atomic piles, 603, 626
— graphite-moderated, 627
Auger effect, 585
Aurates, 418
Auric, see Gold (III)
Austenite, 263, 265, 269
Auto-catalysis, 374, 712
Auto-dissociation, 725
Auto-ionization, 727
Autoxidation, 718
Azurite, 367, 389